Pearson MyLab

Math for College, Career, and Life

We all use math in our day-to-day lives. The goal of this book is to improve students' mathematical literacy so that they can use math more effectively in everyday life. Mathematics can help students better understand a variety of topics and issues, making them more aware of the uses and abuses of numbers. The ultimate goal is to help them become better educated citizens who are successful in their college experiences, their careers, and their lives.

In Your World boxes focus on topics that students are likely to encounter in the world around them—in the news, in consumer decisions, or in political discussions. The connection to the real world is further enhanced by **In Your World** exercises, designed to spur additional research or discussion that will help students relate the unit's topics to the themes of college, career, and life.

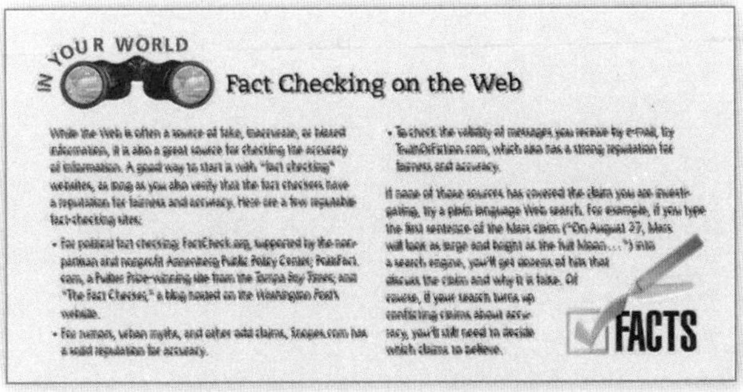

IN YOUR WORLD

45. **Political Action.** This unit outlined numerous budgetary problems facing the U.S. government, as they stood at the time the text was written in 2017. Has there been any significant political action to deal with any of these problems? Learn what, if anything, has changed since 2017; then write a one-page position paper outlining your own recommendations for the future.

46. **Debt Problem.** How serious a problem is the gross debt? Find arguments on both sides of this question. Summarize the arguments, and state your own opinion.

Concepts are brought to life in five **In Your World** videos, assignable within MyLab Math. These videos provide further connections to students' everyday lives in an entertaining and engaging way.

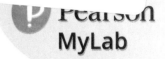

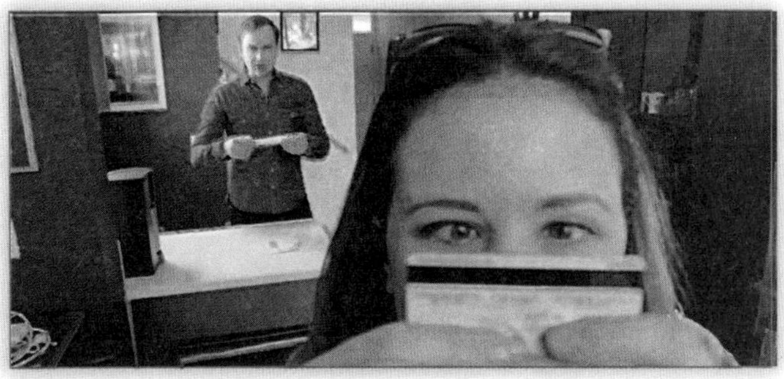

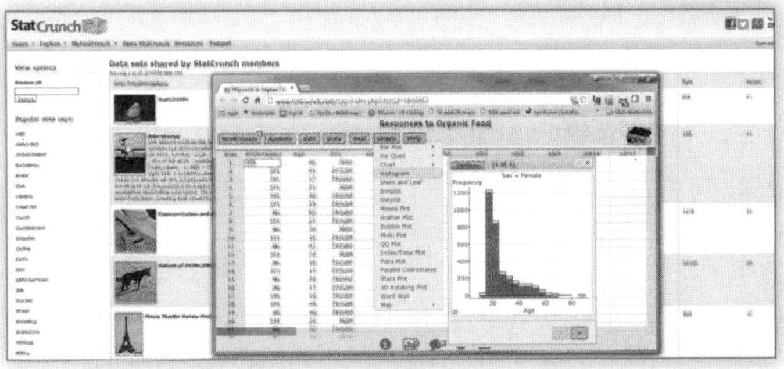

NEW! Integrated StatCrunch

StatCrunch, powerful web-based statistical software, allows students to harness technology to perform data analysis. This edition includes StatCrunch exercises in relevant exercise sets and some questions assignable in MyLab Math.

pearson.com/mylab/math

Why Should You Care About Quantitative Reasoning?

Quantitative reasoning is the ability to interpret and reason with information that involves numbers or mathematical ideas. It is a crucial aspect of literacy, and it is essential in making important decisions and understanding contemporary issues.

The topics covered in this text will help you work with quantitative information and make critical decisions. For example:

- You should possess strong skills in critical and logical thinking so that you can make wise personal decisions, navigate the media, and be an informed citizen. For example, do you know why you'd end up behind if you accepted a temporary 10% pay cut now and then received a 10% pay raise later? This particular question is covered in Unit 3A, but throughout the book you'll learn how to evaluate quantitative questions on topics ranging from personal decisions to major global issues.
- You should have a strong number sense and be proficient at estimation so that you can put numbers from the news into a context that makes them understandable. For example, do you know how to make sense of the more than $20 trillion federal debt? Unit 3B discusses how you can put such huge numbers in perspective, and Unit 4F discusses how the federal debt grew so large.
- You should possess the mathematical tools needed to make basic financial decisions. For example, do you enjoy a latte every morning before class? Sometimes two? Unit 4A explores how such a seemingly harmless habit can drain more than $2400 from your wallet every year.
- You should be able to read news reports of statistical studies in a way that will allow you to evaluate them critically and decide whether and how they should affect your personal beliefs. For example, how should you decide whether a new opinion poll accurately reflects the views of Americans? Chapter 5 covers the basic concepts that lie behind the statistical studies and graphics you'll see in the news, and discusses how you can decide for yourself whether you should believe a statistical study.
- You should be familiar with basic ideas of probability and risk and be aware of how they affect your life. For example, would you pay $30,000 for a product that, over 20 years, will kill nearly as many people as live in San Francisco? In Unit 7D, you'll see that the answer is very likely yes—just one of many surprises that you'll encounter as you study probability in Chapter 7.
- You should understand how mathematics helps us study important social issues, such as global warming, the growth of populations, the depletion of resources, apportionment of congressional representatives, and methods of voting. For example, Unit 12D discusses the nature of redistricting and how gerrymandering has made congressional elections less competitive than they might otherwise be.

In sum, this text will focus on understanding and interpreting mathematical topics to help you develop the quantitative reasoning skills you will need for college, career, and life.

7th EDITION

Using & Understanding MATHEMATICS

A Quantitative Reasoning Approach

Jeffrey Bennett
University of Colorado at Boulder

William Briggs
University of Colorado at Denver

Pearson

330 Hudson Street, NY NY 10013

Director, Portfolio Management: Anne Kelly
Courseware Portfolio Specialist: Marnie Greenhut
Courseware Portfolio Management Assistant: Stacey Miller
Content Producer: Patty Bergin
Managing Producer: Karen Wernholm
TestGen Content Manager: Mary Durnwald
Manager, Content Development (Math XL): Robert Carroll
Media Producer: Nick Sweeny
Product Marketing Manager: Kyle DiGiannantonio
Field Marketing Manager: Andrew Noble
Marketing Assistant: Hanna Lafferty
Senior Author Support/Technology Specialist: Joe Vetere
Manager, Rights and Permissions: Gina Cheselka
Manufacturing Buyer: Carol Melville, LSC Communications
Composition, Illustrations, and Text Design: Integra
Production Coordination: Jane Hoover/Lifland et al., Bookmakers
Cover Design: Barbara T. Atkinson
Cover Image: Arata Photography/Moment/Getty Images

Library of Congress Cataloging-in-Publication Data

Names: Bennett, Jeffrey O., author. | Briggs, William L., author.
Title: Using and understanding mathematics: a quantitative reasoning approach / Jeffrey Bennett and William Briggs.
Description: 7th edition. | Boston: Pearson, [2019]
Identifiers: LCCN 2017048514| ISBN 9780134705187 (alk. paper) | ISBN 0134705181 (alk. paper)
Subjects: LCSH: Mathematics–Textbooks.
Classification: LCC QA39.3 .B46 2019 | DDC 510–dc23
LC record available at https://lccn.loc.gov/2017048514

2 18

Pearson

ISBN-13: 978-0-13-470518-7
ISBN-10: 0-13-470518-1

This book is dedicated to everyone who wants a better understanding of our world, and especially to those who have struggled with mathematics in the past. We hope this book will help you achieve your goals.

And it is dedicated to those who make our own lives brighter, especially Lisa, Julie, Katie, Grant, and Brooke.

CONTENTS

ABOUT THE AUTHORS

Jeffrey Bennett served as the first director of the program "Quantitative Reasoning and Mathematical Skills" at the University of Colorado at Boulder, where he developed the groundbreaking curriculum that became the basis of this textbook. He holds a BA in biophysics (University of California, San Diego) and an MS and a PhD in astrophysics (University of Colorado), and has focused his career on math and science education. In addition to co-authoring this textbook, he is also the lead author of best-selling college textbooks on statistical reasoning, astronomy, and astrobiology, and of more than a dozen books for children and adults. All six of his children's books have been selected for NASA's "Story Time From Space" (storytimefromspace.com), a project in which astronauts on the International Space Station read books aloud and videos are posted that anyone in the world can watch for free. His most recent books include *I, Humanity* for children and *Math for Life* and *A Global Warming Primer* for the general public. Among his many other endeavors, Dr. Bennett proposed and co-led the development of the Voyage Scale Model Solar System, which is located outside the National Air and Space Museum on the National Mall in Washington, DC. Learn more about Dr. Bennett and his work at www.jeffreybennett.com.

William Briggs was on the mathematics faculty at Clarkson University for 6 years and at the University of Colorado at Denver for 23 years, where he taught both undergraduate and graduate courses, with a special interest in applied mathematics. During much of that time, he designed and taught courses in quantitative reasoning. In addition to this book, he has co-authored textbooks on statistical reasoning and calculus, as well as monographs in computational mathematics. He recently completed the book *How America Got Its Guns* (University of New Mexico Press). Dr. Briggs is a University of Colorado President's Teaching Scholar and the recipient of a Fulbright Fellowship to Ireland; he holds a BA degree from the University of Colorado and an MS and a PhD from Harvard University.

PREFACE

Human history becomes more and more a race between education and catastrophe.

—H. G. Wells
The Outline of History, 1920

To the Student

There is no escaping the importance of mathematics in the modern world. However, for most people, the importance of mathematics lies not in its abstract ideas, but in its application to personal and social issues. This textbook is designed with such practical considerations in mind. In particular, this book has three specific purposes:

- to prepare you for the mathematics you will encounter in other **college** courses, particularly core courses in social and natural sciences;
- to develop your ability to reason with quantitative information in a way that will help you achieve success in your **career**; and
- to provide you with the critical thinking and quantitative reasoning skills you need to understand major issues in your **life**.

We hope this book will be useful to everyone, but it is designed primarily for those who are *not* planning to major in a field that requires advanced mathematical skills. In particular, if you've ever felt any fear or anxiety about mathematics, we've written this book with you in mind. We hope that, through this book, you will discover that mathematics is much more important and relevant to your life than you thought and not as difficult as you previously imagined.

Whatever your interests—social sciences, environmental issues, politics, business and economics, art and music, or any of many other topics—you will find many relevant and up-to-date examples in this book. But the most important idea to take away from this book is that mathematics can help you understand a variety of topics and issues, making you a more aware and better educated citizen. Once you have completed your study of this book, you should be prepared to understand most quantitative issues that you will encounter.

To the Instructor

Whether you've taught this course many times or are teaching it for the first time, you are undoubtedly aware that mathematics courses for non-majors present challenges that differ from those presented by more traditional courses. First and foremost, there isn't even a clear consensus on what exactly should be taught in these courses. While there's little debate about what mathematical content is necessary for science, technology, engineering, and mathematics (STEM) students—for example, these students all need to learn algebra and calculus—there's great debate about what we should teach non-STEM students, especially the large majority who will *not* make use of formal mathematics in their careers or daily lives.

As a result of this debate, core mathematics courses for non-STEM students represent a broad and diverse range. Some schools require these students to take a traditional, calculus-track course, such as college algebra. Others have instituted courses focused on some of the hidden ways in which contemporary mathematics contributes to society, and still others have developed courses devoted almost exclusively to financial literacy. Each of the different course types has its merits, but we believe there is a better option, largely because of the following fact: The vast majority (typically 95%) of non-STEM students will *never* take another college mathematics course after completing their core requirement.

Given this fact, we believe it is essential to teach these students the mathematical ideas that they will *need* for their remaining college course work, their careers, and their daily lives. In other words, we must emphasize those topics that are truly important to the future success of these students, and we must cover a broad range of such topics. The focus of this approach is less on formal calculation—though some is certainly required—and more on teaching students how to think critically with numerical or mathematical information. In the terminology adopted by MAA, AMATYC, and other mathematical organizations, students need to learn *quantitative reasoning* and to become *quantitatively literate*. There's been a recent rise in the popularity of quantitative reasoning courses for non-STEM students. This book has been integral to the quantitative reasoning movement for years and continues to be at the forefront as an established resource designed to help you succeed in teaching quantitative reasoning to your students.

The Key to Success: A Context-Driven Approach

Broadly speaking, approaches to teaching mathematics can be divided into two categories:

- A *content-driven approach* is organized by mathematical ideas. After each mathematical topic is presented, examples of its applications are shown.

- A *context-driven approach* is organized by practical contexts. Applications drive the course, and mathematical ideas are presented as needed to support the applications.

The same content can be covered through either approach, but the context-driven approach has an enormous advantage: It motivates students by showing them directly how relevant mathematics is to their lives. In contrast, the content-driven approach tends to come across as "learn this content because it's good for you," causing many students to tune out before reaching the practical applications. For more details, see our article "General Education Mathematics: New Approaches for a New Millennium" (*AMATYC Review*, Fall 1999) or the discussion in the Epilogue of the book *Math for Life* by Jeffrey Bennett (Big Kid Science, 2014).

The Challenge: Winning Over Your Students

Perhaps the greatest challenge in teaching mathematics lies in winning students over—that is, convincing them that you have something useful to teach them. This challenge arises because by the time they reach college, many students dislike or fear mathematics. Indeed, the vast majority of students in general education mathematics courses are there not by choice, but because such courses are required for graduation. Reaching your students therefore requires that you teach with enthusiasm and convince them that mathematics is useful and enjoyable.

We've built this book around two important strategies that are designed to help you win students over:

- Confront negative attitudes about mathematics head on, showing students that their fear or loathing is ungrounded and that mathematics is relevant to their lives. This strategy is embodied in the Prologue of this book (pages P1–P13), which we urge you to emphasize in class. It continues implicitly throughout the rest of the text.
- Focus on goals that are meaningful to students—namely, on the goals of learning mathematics for *college, career*, and *life*. Your students will then learn mathematics because they will see how it affects their lives. This strategy forms the backbone of this book, as we have tried to build every unit around topics relevant to college, career, and life.

Modular Structure of the Book

Although we have written this book so that it can be read as a narrative from beginning to end, we recognize that many instructors might wish to teach material in a different order than we have chosen or to cover only selected portions of the text, as time allows, for classes of different length or for students at different levels. We have therefore organized the book with a modular structure that allows instructors to create a customized course. The 12 chapters are organized broadly by contextual areas. Each chapter, in turn, is divided into a set of self-contained *units* that focus on particular concepts or applications. In most cases, you can cover chapters in any order or skip units that are lower priority for your particular course. The following outline describes the flow of each chapter:

Chapter Overview Each chapter begins with a two-page overview consisting of an introductory paragraph and a multiple-choice question designed to illustrate an important way in which the chapter content connects with the book themes of *college, career*, and *life*. The overview also includes a motivational quote and a unit-by-unit listing of key content; the latter is designed to show students how the chapter is organized and to help instructors decide which units to cover in class.

Chapter Activity Each chapter next offers an activity designed to spur student discussion of some interesting facet of the topics covered in the chapter. The activities may be done either individually or in small groups. A new Activity Manual containing additional activities is available with this seventh edition in print form and also in MyLab Math.

Numbered Units Each chapter consists of numbered units (e.g., Unit 1A, Unit 1B, ...). Each unit begins with a short introduction and includes the following key features:

- **Headings to Identify Key Topics.** In keeping with the modularity, each subtopic within a unit is clearly identified so that students understand what they will be learning.
- **Summary Boxes.** Key definitions and concepts are highlighted in summary boxes for easy reference.
- **Examples and Case Studies.** Numbered examples are designed to build understanding and to offer practice with the types of questions that appear in the exercises. Each example is accompanied by a "Now try ..." suggestion that relates the example to specific similar exercises. Occasional case studies go into more depth than the numbered examples.
- **Exercises.** Each unit concludes with a set of exercises, subdivided into the following categories:
  - **Quick Quiz.** This ten-question quiz appears at the end of each unit and allows students to check whether they understand key concepts before starting the exercise set. Note that students are asked not only to choose the correct multiple-choice answer but also to write a brief explanation of the reasoning behind their choice. Answers are included in the back of the text.

- **Review Questions.** Designed primarily for self-study, these questions ask students to summarize the important ideas covered in the unit and generally can be answered simply by reviewing the text.
- **Does It Make Sense?** These questions ask students to determine whether a short statement makes sense, and explain why or why not. These exercises are generally easy once students understand a particular concept, but difficult otherwise; they are therefore an excellent probe of comprehension.
- **Basic Skills & Concepts.** These questions offer practice with the concepts covered in the unit. They can be used for homework assignments or for self-study (answers to most odd-numbered exercises appear in the back of the book). These questions are referenced by the "Now try ..." suggestions in the unit.
- **Further Applications.** Through additional applications, these exercises extend the ideas and techniques covered in the unit.
- **In Your World.** These questions are designed to spur additional research or discussion that will help students relate the unit content to the book themes of college, career, and life.
- **Technology Exercises.** For units that include one or more Using Technology features, these exercises give students an opportunity to practice calculator or software skills that have been introduced. Some of these exercises are designed to be completed with StatCrunch (StatCrunch), which comes with the MyLab Math course. Applications using StatCrunch, powerful Web-based statistical software that allows users to collect data, perform analyses, and generate compelling results, are included in this edition for the first time.

Chapter Summary Appearing at the end of each chapter, the Chapter Summary offers a brief outline of the chapter's content, including page numbers, that students can use as a study guide.

Additional Pedagogical Features In addition to the standard features of all chapters listed above, several other pedagogical features occur throughout the text:

- **Think About It.** These features pose short conceptual questions designed to help students reflect on important new ideas. They also serve as excellent starting points for classroom discussions and, in some cases, can be used as a basis for clicker questions.
- **Brief Review.** This feature appears when a key mathematical skill is first needed; topics include fractions, powers and roots, basic algebraic operations, and more. The word "review" indicates that most students will have learned these skills previously, but many will need review and practice. Practice is available in the exercise sets, with relevant exercises identified by a "Now try ..." suggestion at the end of the Brief Review.
- **In Your World.** These features focus on topics that students are likely to encounter in the world around them, whether in the news, in consumer decisions, or in political discussions. Examples include how to understand jewelry purchases, how to invest money in a sensible way, and how to evaluate the reliability of pre-election polls. (*Note*: These features are not necessarily connected directly to the In Your World exercises, but both have direct relevance to students' world.)
- **Using Technology.** These features give students clear instructions in the use of various technologies for computation, including scientific calculators, Microsoft Excel, and online technologies such as those built into Google. Book-specific TI Tech Tips containing instructions for performing computations with a graphing calculator, such as the TI-83 or TI-84, are available in the Tools for Success section of MyLab Math.
- **Caution!** New to the seventh edition, these short notes, integrated into examples or text, highlight common errors that students should be careful to avoid.
- **Mathematical Insight.** This feature, which occurs less frequently than the others, builds on mathematical ideas in the main narrative but goes somewhat beyond the level of other material in the book. Examples of the topics covered are proof of the Pythagorean theorem, Zeno's paradox, and derivations of the financial formulas used for savings plans and mortgage loans.
- **Margin Features.** The margins contain several types of short features: **By the Way**, which offers interesting notes and asides relevant to the topic at hand; **Historical Note**, which gives historical context to the topic at hand; and **Technical Note**, which offers details that are important mathematically, but generally do not affect students' understanding of the material. The margins also contain occasional quotations.

Prerequisite Mathematical Background

Because of its modular structure and the inclusion of the Brief Review features, this book can be used by students with a wide range of mathematical backgrounds. Many of the units require nothing more than arithmetic and a willingness to think about quantitative issues in new ways. Only a few units use techniques of algebra or geometry, and those skills are reviewed as they arise. This book should therefore be accessible to any student who has completed two or more years of high school mathematics. However, *this book is not remedial*: Although much of the book relies on mathematical techniques from secondary school, the techniques arise in applications that students generally are not taught in high school and that require students to demonstrate their critical thinking skills.

For courses in which students do require more extensive prerequisite review, we have created a version of the *Using & Understanding Mathematics* MyLab Math course called *Using & Understanding Mathematics with Integrated Review* that includes just-in-time review of selected prerequisite topics.

Note on "Developmental Math" We are often asked whether this text can be used by students for whom placement tests suggest that they belong in developmental mathematics courses. In most cases, we believe the answer to be a resounding "yes." Our experience suggests that many students who do poorly on mathematics placement tests are not really as weak as these tests may suggest. Most students *did* learn basic mathematical skills at one time, and if the skills arise with context (as they do in this book), we've found that students can quickly relearn them. This is especially true if you provide the students with a little bit of extra practice as offered in our Brief Review features or by the resources in MyLab Math or MyLab Math with Integrated Review. Indeed, we believe that most students in this situation will learn basic mathematical skills *better* by taking a quantitative reasoning course based on this textbook than they will by taking a developmental course.

Changes in the Seventh Edition

We've been pleased by the positive responses from so many users of previous editions of this text. Nevertheless, a book that relies heavily on facts and data always requires a major updating effort to keep it current, and we are always looking for ways to improve clarity and pedagogy. As a result, users of prior editions will find many sections of this book to have been substantially revised or rewritten. The changes are too many to list here, but some of the more significant changes are the following.

Chapter 1 We significantly revised Units 1A and 1E with the particular goal of helping students evaluate media information and recognize "fake news."

Chapter 2 We reorganized and significantly rewrote this entire chapter to introduce a basic problem-solving strategy in Unit 2A. Moreover, we modified the four-step strategy presented in previous editions to create a simpler three-step strategy called "Understand-Solve-Explain." We have found that this strategy is easier for students to remember and therefore easier for them to put into practice.

Chapters 3 and 4 These two chapters contain several units that revolve around economic data such as demographic data, the Consumer Price Index, interest rates, taxes, and the federal budget. These data obviously required major updates given the changes that have occurred in the U.S. economy in the four years since the last edition. In addition, we've added basic ideas about health insurance to our discussion of personal finances in Unit 4A.

Chapters 5 and 6 These chapters focus on statistical data, which means we updated or replaced large sections of the chapter content to include more current data.

Chapter 7 We significantly revised Section 7D on risk, both for greater clarity and to update data.

Chapters 8 and 9 Units 8B, 8C, and 9C all rely heavily on population data, which means we revised significant portions of these units to reflect the latest global demographic data.

Chapter 12 The 2016 election provided numerous new examples for our discussion of the electoral college in Unit 12A. Other recent examples of the intersection of mathematics and politics also provide interesting new examples and exercises throughout this chapter.

In Your World We've added seven new In Your World features, so every chapter now has at least one, further showcasing math for college, career, and life.

Caution! These short notes highlighting common errors are new to this edition.

Exercise Sets We've thoroughly revised the exercise sets: Over 30% of the exercises are changed or new.

StatCrunch StatCrunch has been newly integrated into the MyLab Math course and relevant Technology Exercises.

Video Program The seventh edition is accompanied by an all-new video program consisting of both familiar lecture-style videos for every example and innovative concept videos.

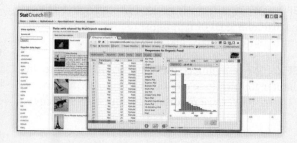

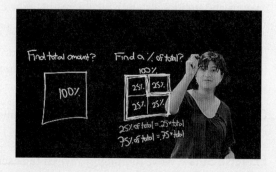

Pearson MyLab

Resources for Success

Instructor Resources

MyLab Math with Integrated Review

This MyLab Math course option can be used in co-requisite courses, or simply to help students who enter the quantitative reasoning course lacking prerequisite skills or a full understanding of prerequisite concepts.

- For relevant chapters, students begin with a Skills Check assignment to pinpoint which prerequisite developmental topics, if any, they need to review.
- Those who require additional review proceed to a personalized homework assignment that focuses on the specific prerequisite topics on which they need remediation.
- Students can also review the relevant prerequisite concepts using videos and Integrated Review Worksheets in MyLab Math. The Integrated Review Worksheets are also available in printed form as part of the Activity Manual with Integrated Review Worksheets.

Specific to the *Using & Understanding Mathematics* MyLab Math course:

- NEW! Completely new lecture video program with corresponding assessment
- NEW! Dynamic concept videos
- NEW! Interactive concept videos with corresponding assessment
- NEW! Animations with corresponding assessment
- NEW! Integration of StatCrunch in the left-hand navigation of the MyLab Math course makes it easy to access the software for completion of the Technology Exercises that use StatCrunch.
- Bonus unit on mathematics and business, including assessment

Instructor's Edition

(ISBNs: 0-13-470522-X /978-0-13-470522-4)

The Instructor's Edition of the text includes answers to all of the exercises and Quick Quizzes in the back of the book.

The following resources are ONLINE ONLY and are available for download from the Pearson Higher Education catalog at www.pearson.com/us/sign-in.html or within your MyLab Math course.

Instructor's Solution Manual

James Lapp

This manual includes answers to all of the text's Think About It features, Quick Quizzes, Review Questions, and Does It Make Sense? questions and detailed, worked-out solutions to all of the Basic Skills & Concepts, Further Applications, and Technology Exercises (including StatCrunch exercises).

Instructor's Testing Manual

Dawn Dabney

The Testing Manual provides four alternative tests per chapter, including answer keys.

TestGen

TestGen® (www.pearsoned.com/testgen) enables instructors to build, edit, print, and administer tests using a computerized bank of questions developed to cover all the objectives of the text. TestGen is algorithmically based, allowing instructors to create multiple but equivalent versions of the same question or test with the click of a button. Instructors can also modify test bank questions or add new questions. The software and test bank can be downloaded from Pearson's Instructor Resource Center.

PowerPoint Lecture Presentation

These editable slides present key concepts and definitions from the text. Instructors can add art from the text located in the Image Resource Library in MyLab Math or add slides they have created. PowerPoint slides are fully accessible.

Image Resource Library

This resource in the MyLab Math course contains all the art from the text for instructors to use in their own presentations and handouts.

Pearson MyLab

Student Resources

Student's Study Guide and Solutions Manual

(ISBNs: 0-13-470524-6 /978-0-13-470524-8)

James Lapp

This manual contains answers to all Quick Quiz questions and to odd-numbered Review Questions and Does It Make Sense? questions, as well as worked-out solutions to odd-numbered Basic Skills & Concepts, Further Applications, and Technology Exercises (including StatCrunch exercises).

NEW! Activity Manual with Integrated Review Worksheets

(ISBNs: 0-13-477664-X /978-0-13-477664-4)

Compiled by Donna Kirk, The College of St. Scholastica

More than 30 activities correlated to the textbook give students hands-on experiences that reinforce the course content. Activities can be completed individually or in a group. Each activity includes an overview, estimated time of completion, objectives, guidelines for group size, and list of materials needed. Additionally, the manual provides the worksheets for the Integrated Review version of the MyLab Math course.

pearson.com/mylab/math

Acknowledgments

A textbook may carry its authors' names, but it is the result of hard work by hundreds of committed individuals. This book has been under development for more than 30 years, and even its beginnings were a group effort, as one of the authors was a member of a committee at the University of Colorado that worked to establish one of the nation's first courses in quantitative reasoning. Since that beginning, the book has benefited from input and feedback from many faculty members and students.

First and foremost, we extend our thanks to Bill Poole and Elka Block, whose faith in this project from the beginning allowed it to grow from class notes into a true textbook. We'd also like to thank other past and present members of our outstanding publishing team at Pearson Education, including Greg Tobin, Anne Kelly, Marnie Greenhut, Patty Bergin, Barbara Atkinson, Kyle DiGiannantonio, Stacey Miller, Hannah Lafferty, and Nick Sweeny. We thank Rhea Meyerholtz and Paul Lorczak for an excellent job on accuracy checking, and Shane Goodwin of BYU–Idaho for his help in preparing the Using Technology boxes (and for many other suggestions he has made as well).

We'd like to thank the following people for their help with one or more editions of this book. Those who assisted with this seventh edition have an asterisk before their names.

*Merri Jill Ayers, *Georgia Gwinnett College*
Lou Barnes, *Premier Mortgage Group*
Carol Bellisio, *Monmouth University*
Bob Bernhardt, *East Carolina University*
Terence R. Blows, *Northern Arizona University*
*Loi Booher, *University of Central Arkansas*
W. Wayne Bosché, Jr., *Dalton College*
Kristina Bowers, *University of South Florida*
Michael Bradshaw, *Caldwell Community College and Technical Institute*
Shane Brewer, *Utah State University–Blanding Campus*
W. E. Briggs, *University of Colorado, Boulder*
Annette Burden, *Youngstown State University*
Ovidiu Calin, *Eastern Michigan University*
Susan Carr, *Oral Roberts University*
*Henry Chango, *Community College of Rhode Island*
Margaret Cibes, *Trinity College*
Walter Czarnec, *Framingham State College*
Adrian Daigle, *University of Colorado, Boulder*
Andrew J. Dane, *Angelo State University*
*Dr. Amit Dave, *Georgia Gwinnett College*
Jill DeWitt, *Baker College of Muskegon*
Greg Dietrich, *Florida Community College at Jacksonville*
Marsha J. Driskill, *Aims Community College*
John Emert, *Ball State University*
Kathy Eppler, *Salt Lake Community College*
Kellie Evans, *York College of Pennsylvania*
Fred Feldon, *Coastline Community College*
Anne Fine, *East Central University*
David E. Flesner, *Gettysburg College*
Pat Foard, *South Plains College*
Brian Gaines, *University of Illinois*
*Jose Gimenez, *Temple University*
Shane Goodwin, *Brigham Young University–Idaho*
Barbara Grover, *Salt Lake Community College*
Louise Hainline, *Brooklyn College*
Ward Heilman, *Bridgewater State University*
Peg Hovde, *Grossmont College*
Andrew Hugine, *South Carolina State University*
Lynn R. Hun, *Dixie College*
Hal Huntsman, *University of Colorado, Boulder*
Joel Irish, *University of Southern Maine*
David Jabon, *DePaul University*
Melvin F. Janowitz, *University of Massachusetts, Amherst*
Craig Johnson, *Brigham Young University–Idaho*
Vijay S. Joshi, *Virginia Intermont College*
Anton Kaul, *University of South Florida*
Bonnie Kelly, *University of South Carolina*
William Kiley, *George Mason University*
*Donna Kirk, *The College of Saint Scholastica*
Jim Koehler, *University of Colorado, Denver*
*Charlotte Koleti, *Georgia Gwinnett College*
Robert Kuenzi, *University of Wisconsin, Oshkosh*
Erin Lee, *Central Washington University*
R. Warren Lemerich, *Laramie County Community College*
Deann Leoni, *Edmonds Community College*
Linda Lester, *Wright State University*
Paul Lorczak, *MathSoft, Inc.*
Jay Malmstrom, *Oklahoma City Community College*
*Howard Mandelbaum, *John Jay College*
Erich McAlister, *University of Colorado, Boulder*
*Meghan McIntyre, *Wake Technical Community College*
Judith McKnew, *Clemson University*
Lisa McMillen, *Baker College*

Patricia McNicholas, *Robert Morris College*
Phyllis Mellinger, *Hollins University*
Elaine Spendlove Merrill, *Brigham Young University–Hawaii*
*Dillon Miller, *San Jacinto College*
*Mehdi Mirfattah, *Long Beach City College*
Carrie Muir, *University of Colorado, Boulder*
Colm Mulcahy, *Spelman College*
*Bette Nelson, *Alvin Community College*
Stephen Nicoloff, *Paradise Valley Community College*
Paul O'Heron, *Broome Community College*
L. Taylor Ollmann, *Austin Community College*
*Diane Overturf, M.S., *Viterbo University*
A. Dean Palmer, *Pima Community College*
Mary K. Patton, *University of Illinois at Springfield*
Frank Pecchioni, *Jefferson Community College*
*Michael Polley, *Southeastern Community College*
Jonathan Prewett, *University of Wyoming*
Evelyn Pupplo-Cody, *Marshall University*
Scott Reed, *College of Lake County*
Frederick A. Reese, *Borough of Manhattan Community College*
Nancy Rivers, *Wake Technical Community College*
Anne Roberts, *University of Utah*
*Michelle Robinson, *Fayetteville Technical Community College*
Sylvester Roebuck, Jr., *Olive Harvey College*
*Sheri Rogers, *Linn-Benton Community College*
Lori Rosenthal, *Austin Community College*
Hugo Rossi, *University of Utah*
*Robin Rufatto, *Ball State University*
*Ioana Sancira, *Olive Harvey College*
Doris Schraeder, *McLennan Community College*
Dee Dee Shaulis, *University of Colorado, Boulder*
Judith Silver, *Marshall University*
Laura Smallwood, *Chandler-Gilbert Community College*
Sybil Smith-Darlington, *Middlesex County College*
Alu Srinivasan, *Temple University*
John Supra, *University of Colorado, Boulder*
Scott Surgent, *Arizona State University*
Timothy C. Swyter, *Frederick Community College*
Louis A. Talman, *Metropolitan State College of Denver*
David Theobald, *University of Colorado, Boulder*
Robert Thompson, *Hunter College (CUNY)*
Terry Tolle, *Southwestern Community College*
Kathy Turrisi, *Centenary College*
*Claudio Valenzuela, *Southwest Texas Junior College*
Christina Vertullo, *Marist College*
Pam Wahl, *Middlesex Community College*
Ian C. Walters, Jr., *D'Youville College*
Thomas Wangler, *Benedictine University*
Richard Watkins, *Tidewater Community College*
Charles D. Watson, *University of Central Kansas*
*Dr. Gale Watson, *East Georgia State College*
*Dr. Beverly Watts, *McDowell Technical Community College*
Emily Whaley, *DeKalb College*
*John Williamson, *Sandhills Community College*
David Wilson, *University of Colorado, Boulder*
Robert Woods, *Broome Community College*
Fred Worth, *Henderson State University*
Margaret Yoder, *Eastern Kentucky University*
Marwan Zabdawi, *Gordon College*
Fredric Zerla, *University of South Florida*
Donald J. Zielke, Concordia Lutheran College

Prologue

LITERACY FOR THE MODERN WORLD

Equations are just the boring part of mathematics.
—Stephen Hawking, physicist

If you're like most students enrolled in a course using this text, you may think that your interests have relatively little to do with mathematics. But as you will see, nearly every career today requires the use and understanding of some mathematics. Furthermore, the ability to reason quantitatively is crucial for the decisions that we face daily as citizens in a modern technological society. In this Prologue, we'll discuss why mathematics is so important, why you may be better at it than you think, and how this course can provide you with the quantitative skills needed for your college courses, your career, and your life.

Q

Imagine that you're at a party and you've just struck up a conversation with a dynamic, successful lawyer. Which of the following are you most likely to hear her say during your conversation?

- **A** "I really don't know how to read very well."
- **B** "I can't write a grammatically correct sentence."
- **C** "I'm awful at dealing with people."
- **D** "I've never been able to think logically."
- **E** "I'm bad at math."

A

We all know that the answer is E, because we've heard it so many times. Not just from lawyers, but from businesspeople, actors and athletes, construction workers and sales clerks, and sometimes even teachers and CEOs. It would be difficult to imagine these same people admitting to any of choices A through D, but many people consider it socially acceptable to say that they are "bad at math." Unfortunately, this social acceptability comes with some very negative social consequences. (See the discussion about Misconception Seven on page P-7.)

ACTIVITY

Job Satisfaction

Each chapter in this textbook begins with an activity, which you may do individually or in groups. For this Prologue, the opening activity will help you examine the role of mathematics in careers. Additional activities are available online in MyLab Math.

Top 20 Jobs for Job Satisfaction
1. Mathematician
2. Actuary (works with insurance statistics)
3. Statistician
4. Biologist
5. Software engineer
6. Computer systems analyst
7. Historian
8. Sociologist
9. Industrial designer
10. Accountant
11. Economist
12. Philosopher
13. Physicist
14. Parole officer
15. Meteorologist
16. Medical laboratory technician
17. Paralegal assistant
18. Computer programmer
19. Motion picture editor
20. Astronomer

Source: JobsRated.com.

Everyone wants to find a career path that will bring lifelong job satisfaction, but what careers are most likely to do that? A recent survey evaluated 200 different jobs according to five criteria: salary, long-term employment outlook, work environment, physical demands, and stress. The table to the left shows the top 20 jobs according to this survey. Notice that most of the top 20 jobs require mathematical skills, and all of them require an ability to reason with quantitative information.

You and your classmates can conduct your own smaller study of job satisfaction. There are many ways to do this, but here is one procedure you might try:

1. Each of you should identify at least three people with full-time jobs to interview briefly. You may choose parents, friends, acquaintances, or just someone whose job interests you.
2. Identify an appropriate job category for each interviewee (similar to the categories in the table to the left). Ask each interviewee to rate his or her job on a scale of 1 (worst) to 5 (best) on each of the five criteria: salary, long-term employment outlook, work environment, physical demands, and stress. You can then add the ratings for the five criteria to come up with a total job satisfaction rating for each job.
3. Working together as a class, compile the data to rank all the jobs. Show the final results in a table that ranks the jobs in order of job satisfaction.
4. Discuss the results. Are they consistent with the survey results shown in the table? Do they surprise you in any way? Will they have any effect on your own career plans?

What Is Quantitative Reasoning?

Literacy is the ability to read and write, and it comes in varying degrees. Some people can recognize only a few words and write only their names; others read and write in many languages. A primary goal of our educational system is to provide citizens with a level of literacy sufficient to read, write, and reason about the important issues of our time.

Today, the abilities to interpret and reason with **quantitative information**—information that involves mathematical ideas or numbers—are crucial aspects of literacy. These abilities, often called **quantitative reasoning** or **quantitative literacy**, are essential to understanding issues that appear in the news every day. The purpose of this textbook is to help you gain skills in quantitative reasoning as it applies to issues you will encounter in

- your subsequent coursework,
- your career, and
- your daily life.

Quantitative Reasoning and Culture

Quantitative reasoning enriches the appreciation of both ancient and modern culture. The historical record shows that nearly all cultures devoted substantial energy to mathematics and to science (or to observational studies that predated modern science). Without a sense of how quantitative concepts are used in art, architecture, and science, you cannot fully appreciate the incredible achievements of the Mayans in Central America, the builders of the great city of Zimbabwe in Africa, the ancient Egyptians and Greeks, the early Polynesian sailors, and many others.

Similarly, quantitative concepts can help you understand and appreciate the works of the great artists. Mathematical concepts play a major role in everything from the work of Renaissance artists like Leonardo da Vinci and Michelangelo to the pop culture of television shows like *The Big Bang Theory*. Other ties between mathematics and the arts can be found in both modern and classical music, as well as in the digital production of music. Indeed, it is hard to find popular works of art, film, or literature that do not rely on mathematics in some way.

Mathematics knows no races or geographic boundaries; for mathematics, the cultural world is one country.
—David Hilbert (1862–1943), German mathematician

Quantitative Reasoning in the Work Force

Quantitative reasoning is important in the work force. A lack of quantitative skills puts many of the most challenging and highest-paying jobs out of reach. Table P.1 defines skill levels in language and mathematics on a scale of 1 to 6, and Table P.2 (on the next page) shows the typical levels needed in many jobs.

Note that the occupations requiring high skill levels are generally the most prestigious and highest paying. Note also that most of those occupations call for high skill levels in *both* language and math, refuting the myth that if you're good at language, you don't have to be good at mathematics, and vice versa.

TABLE P.1 Skill Levels

Level	Language Skills	Math Skills
1	Reads signs and basic news reports; writes and speaks simple sentences	Addition and subtraction; simple calculations with money, volume, length, and weight
2	Can read short stories and instruction manuals; writes compound sentences with proper grammar and punctuation	Arithmetic; can compute ratios, rates, and percentages; can draw and interpret bar graphs
3	Reads novels and magazines; writes reports with proper format; speaks well before an audience	Basic geometry and algebra; can calculate discounts, interest, profit and loss
4	Reads novels, poems, and newspapers; prepares business letters, summaries, and reports; participates in panel discussions and debates	Has true quantitative reasoning abilities: understands logic, problem solving, ideas of statistics and probability, and modeling
5	Reads literature, scientific and technical journals, financial reports, and legal documents; can write editorials, speeches, and critiques	Calculus and statistics
6	Same types of skills as level 5, but more advanced	Advanced calculus, modern algebra, and advanced statistics

Source: Adapted from levels described in the *Wall Street Journal*.

TABLE P.2 Skill-Level Requirements

Occupation	Language Level	Math Level	Occupation	Language Level	Math Level
Biochemist	6	6	Web page designer	5	4
Computer engineer	6	6	Corporate executive	5	5
Mathematician	6	6	Computer sales agent	4	4
Cardiologist	6	5	Athlete's agent	4	4
Social psychologist	6	5	Management trainee	4	4
Lawyer	6	4	Insurance sales agent	4	4
Tax attorney	6	4	Retail store manager	4	4
Newspaper editor	6	4	Cement mason	3	3
Accountant	5	5	Poultry farmer	3	3
Personnel manager	5	4	Tile setter	3	3
Corporate president	5	5	Travel agent	3	3
Weather forecaster	5	5	Janitor	3	2
Secondary teacher	5	5	Short-order cook	3	2
Elementary teacher	5	4	Assembly-line worker	2	2
Financial analyst	5	5	Toll collector	2	2
Journalist	5	4	Laundry worker	1	1

Source: Data from the *Wall Street Journal.*

Misconceptions About Mathematics

Do you consider yourself to have "math phobia" (fear of mathematics) or "math loathing" (dislike of mathematics)? We hope not—but if you do, you aren't alone. Many adults harbor fear or loathing of mathematics, and unfortunately, these attitudes are often reinforced by classes that present mathematics as an obscure and sterile subject.

In reality, mathematics is not nearly so dry as it sometimes seems in school. Indeed, attitudes toward mathematics often are directed not at what mathematics really is but at some common misconceptions about mathematics. Let's investigate a few of these misconceptions and the reality behind them.

Misconception One: Math Requires a Special Brain

We are all mathematicians … [your] forte lies in navigating the complexities of social networks, weighing passions against histories, calculating reactions, and generally managing a system of information that, when all laid out, would boggle a computer.
—A. K. Dewdney, *200% of Nothing*

One of the most pervasive misconceptions is that some people just aren't good at mathematics because learning mathematics requires special or rare abilities. The reality is that nearly everyone can do mathematics. All it takes is self-confidence and hard work—the same qualities needed to learn to read, to master a musical instrument, or to become skilled at a sport. Indeed, the belief that mathematics requires special talent found in a few elite people is peculiar to the United States. In most other countries, particularly in Europe and Asia, *all* students are expected to become proficient in mathematics.

Of course, different people learn mathematics at different rates and in different ways. For example, some people learn by concentrating on concrete problems, others by thinking visually, and still others by thinking abstractly. No matter what type of thinking style you prefer, you can succeed in mathematics.

Misconception Two: The Math in Modern Issues Is Too Complex

Some people claim that the advanced mathematical concepts underlying many modern issues are too complex for the average person to understand. It is true that only a few people receive the training needed to work with or discover advanced mathematical concepts. However, most people are capable of understanding enough about the mathematical basis of important issues to develop informed and reasoned opinions.

The situation is similar for other fields. For example, years of study and practice are required to become a proficient professional writer, but most people can read a book. It takes hard work and a law degree to become a lawyer, but most people can understand how the law affects them. And though few have the musical talent of Mozart, anyone can learn to appreciate his music. Mathematics is no different. If you've made it this far in school, you can understand enough mathematics to succeed as an individual and a concerned citizen.

Skills are to mathematics what scales are to music or spelling is to writing. The objective of learning is to write, to play music, or to solve problems—not just to master skills.

—from ***Everybody Counts***, a report of the National Research Council

Misconception Three: Math Makes You Less Sensitive

Some people believe that learning mathematics will somehow make them less sensitive to the romantic and aesthetic aspects of life. In fact, understanding the mathematics that explains the colors of a sunset or the geometric beauty in a work of art can only enhance our aesthetic appreciation of these things. Furthermore, many people find beauty and elegance in mathematics itself. It's no accident that people trained in mathematics have made important contributions to art, music, and many other fields.

It is impossible to be a mathematician without being a poet in the soul.

—Sophia Kovalevskaya (1850–1891), Russian mathematician

Misconception Four: Math Makes No Allowance for Creativity

The "turn the crank" nature of the problems in many textbooks may give the impression that mathematics stifles creativity. Some of the facts, formalisms, and skills required for mathematical proficiency are fairly cut and dried, but *using* these mathematical tools takes creativity. Consider designing and building a home. The task of construction requires specific skills to lay the foundation, frame in the structure, install plumbing and wiring, and paint walls. But the full process involves much more: Creativity is needed to develop the architectural design, respond to on-the-spot problems during construction, and factor in constraints based on budgets and building codes. The mathematical skills you've learned in school are like the skills of carpentry or plumbing. Applying mathematics is like the creative process of building a home.

Tell me, and I will forget. Show me, and I may remember. Involve me, and I will understand.

—Confucius (c. 551–479 B.C.E.)

Misconception Five: Math Provides Exact Answers

A mathematical formula will yield a specific result, and in school that result may be marked right or wrong. But when you use mathematics in real-life situations, answers are never so clear cut. For example:

> *A bank offers simple interest of 3%, paid at the end of 1 year (that is, after 1 year the bank pays you 3% of your account balance). If you deposit \$1000 today and make no further deposits or withdrawals, how much will you have in your account after 1 year?*

A straight mathematical calculation seems simple enough: 3% of \$1000 is \$30; so you should have \$1030 at the end of a year. But will you? How will your balance be affected by service charges or taxes on interest earned? What if the bank fails? What if the bank is located in a country in which the currency collapses during the year? Choosing a bank in which to invest your money is a *real* mathematics problem that doesn't necessarily have a simple or definitive solution.

Probably the most harmful misconception is that mathematics is essentially a matter of computation. Believing this is roughly equivalent to believing that writing essays is the same as typing them.

—John Allen Paulos, mathematician

IN YOUR WORLD

People Who Studied Mathematics

The critical thinking skills developed through the study of mathematics are valuable in many careers. The following list represents a small sample of people who studied mathematics but are more famous for work in other fields. (Many of the names come from "Famous Nonmathematicians," a list compiled by Steven G. Buyske, Rutgers University.)

Ralph Abernathy, civil rights leader, BS in mathematics, Alabama State University

Tammy Baldwin, U.S. Senator (Wisconsin), BA in mathematics, Smith College

Sergey Brin, co-founder of Google, BA in mathematics, University of Maryland

Mayim Bialik, actress on *The Big Bang Theory*, studied mathematics in working toward her PhD in neuroscience

Harry Blackmun, former Supreme Court justice, summa cum laude in mathematics, Harvard University

James Cameron, film director, studied physics before leaving college, works in oceanic and space research

Lewis Carroll (Charles Dodgson), mathematician and author of *Alice in Wonderland*

Felicia Day, actress, BA in mathematics, University of Texas

David Dinkins, former mayor of New York City, BA in mathematics, Howard University

Alberto Fujimori, former president of Peru, MS in mathematics, University of Wisconsin

Art Garfunkel, musician, MA in mathematics, Columbia University

Reed Hastings, founder and CEO of Netflix, BA in mathematics, Bowdoin College

Grace Hopper, computer pioneer and first woman rear admiral in the U.S. Navy, PhD in mathematics, Yale University

Mae Jemison, first African-American woman in space, studied mathematics in working toward her BS in chemical engineering, Stanford University

John Maynard Keynes, economist, MA in mathematics, Cambridge University

Hedy Lamarr, actress, invented and patented the mathematical technique called "frequency hopping"

Lee Hsien Loong, Prime Minister of Singapore, BA in mathematics, Cambridge University

Brian May, lead guitarist for the band Queen, completed his PhD in astrophysics in 2007, Imperial College

Danica McKellar, actress, BA with highest honors in mathematics, UCLA, and co-discoverer of the Chayes-McKellar-Winn theorem

Andrea Merkel, Chancellor of Germany, studied mathematics in working toward her PhD in physics, University of Leipzig

Harvey Milk, politician and activist for gay rights, BA in mathematics, State University of New York

Edwin Moses, three-time Olympic champion in the 400-meter hurdles, studied mathematics in working toward his BS in physics, Morehouse College

Florence Nightingale, pioneer in nursing, studied mathematics and applied it to her work

Natalie Portman, Oscar-winning actress, semifinalist in Intel Science Talent Search and co-author of two published scientific papers

Sally Ride, first American woman in space, studied mathematics in working toward her PhD in physics, Stanford University

David Robinson, basketball star, BA in mathematics, U.S. Naval Academy

Alexander Solzhenitsyn, Nobel prize–winning Russian author, degrees in mathematics and physics, University of Rostov

Bram Stoker, author of *Dracula,* BA in mathematics, Trinity University, Dublin

Laurence Tribe, Harvard law professor, summa cum laude in mathematics, Harvard University

John Urschel, NFL offensive lineman (Baltimore Ravens) who retired at age 26 to pursue a PhD in mathematics, MIT

Virginia Wade, Wimbledon champion, BA in mathematics, Sussex University

Misconception Six: Math Is Irrelevant to My Life

No matter what your path in college, career, and life, you will find mathematics involved in many ways. A major goal of this text is to show you hundreds of examples in which mathematics applies to everyone's life. We hope you will find that mathematics is not only relevant but also interesting and enjoyable.

Neglect of mathematics works injury to all knowledge ... .
—Roger Bacon (1214–1294), English philosopher

Misconception Seven: It's OK to Be "Bad at Math"

For our final misconception, let's return to the multiple-choice question on the opening page of this Prologue. You'll not only hear many otherwise intelligent people say "I'm bad at math," but it's sometimes said almost as a point of pride, with no hint of embarrassment. Yet the statement often isn't even true. A successful lawyer, for example, almost certainly did well in *all* subjects in school, including math, and so is more likely expressing an attitude than a reality.

You must be the change you wish to see in the world.
—Mahatma Gandhi (1869–1948)

Unfortunately, this type of attitude can cause a lot of damage. Mathematics underlies nearly everything in modern society, from the daily financial decisions that all of us must make to the way in which we understand and approach global issues involving the economy, politics, and science. We cannot possibly hope to act wisely if we approach mathematical ideas with a poor attitude. Moreover, it's an attitude that can easily spread to others. After all, if a child hears a respected adult saying that he or she is "bad at math," the child may be less inspired to do well.

So before you begin your coursework, think about your own attitudes toward mathematics. There's no reason why anyone should be "bad at math" and every reason to develop skills of mathematical thinking. With a good attitude and some hard work, by the end of this course you'll not only be better at math, but you'll be helping future generations by making it socially unacceptable for anyone to be "bad at math."

What *Is* Mathematics?

In discussing misconceptions, we identified what mathematics is *not*. Now let's look at what mathematics *is*. The word *mathematics* is derived from the Greek word *mathematikos,* which means "inclined to learn." Literally speaking, to be mathematical is to be curious, open-minded, and always interested in learning more! Today, we tend to look at mathematics in three different ways: as the sum of its branches, as a way to model the world, and as a language.

Mathematics as the Sum of Its Branches

As you progressed through school, you probably learned to associate mathematics with some of its branches. Among the better-known branches of mathematics are these:

- **logic**—the study of principles of reasoning;
- **arithmetic**—methods for operating on numbers;
- **algebra**—methods for working with unknown quantities;
- **geometry**—the study of size and shape;

- **trigonometry**—the study of triangles and their uses;
- **probability**—the study of chance;
- **statistics**—methods for analyzing data; and
- **calculus**—the study of quantities that change.

One can view mathematics as the sum of its branches, but in this text we'll focus on how different branches of mathematics support the more general goals of quantitative thinking and critical reasoning.

Mathematics as a Way to Model the World

Mathematics also may be viewed as a tool for creating *models*, or representations of real phenomena. Modeling is not unique to mathematics. For example, a road map is a model that represents the roads in some region.

Mathematical models can be as simple as a single equation that predicts how the money in your bank account will grow or as complex as a set of thousands of interrelated equations and parameters used to represent the global climate. By studying models, we gain insight into otherwise unmanageable problems. A global climate model, for example, can help us understand weather systems and ask "what if" questions about how human activity may affect the climate. When a model is used to make a prediction that does *not* come true, that failure points out areas where further research is needed. Today, mathematical modeling is used in nearly every field of study. Figure P.1 shows some of the many disciplines that use mathematical modeling to solve problems.

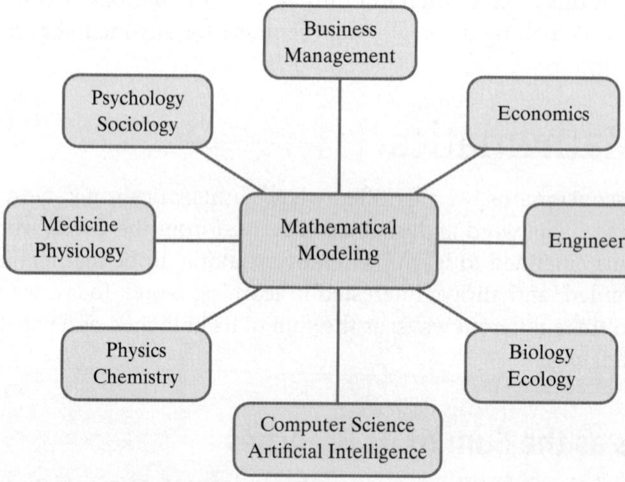

FIGURE P.1

Mathematics as a Language

The Book of Nature is written in the language of mathematics.
—Galileo

A third way to look at mathematics is as a language with its own vocabulary and grammar. Indeed, mathematics often is called "the language of nature" because it is so useful for modeling the natural world. As with any language, different degrees of fluency are possible. From this point of view, quantitative literacy is the level of fluency required for success in today's world.

The idea of mathematics as a language is also useful in thinking about how to *learn* mathematics. Table P.3 compares learning mathematics to learning a language and learning about art.

TABLE P.3 Learning Mathematics: An Analogy to Language and Art

Learning a Language	Learning the Language of Art	Learning the Language of Mathematics
Learn many styles of speaking and writing, such as essays, poetry, and drama	Learn to recognize many styles of art, such as classical, Renaissance, impressionist, and modern	Learn techniques from many branches of mathematics, such as arithmetic, algebra, and geometry
Place literature in a context based on history and the social conditions under which it was created	Place art in a context based on history and the social conditions under which it was created	Place mathematics in a context based on its history, purposes, and applications
Learn the elements of language—such as words, parts of speech (nouns, verbs, etc.), and rules of grammar—and practice their proper use	Learn the elements of visual form—such as lines, shapes, colors, and textures—and practice using them in your own artwork	Learn the elements of mathematics—such as numbers, variables, and operations—and practice using them to solve simple problems
Critically analyze language in forms such as novels, short stories, essays, poems, speeches, and debates	Critically analyze works of art including paintings, sculptures, architecture, and photographs	Critically analyze quantitative information in mathematical models, statistical studies, economic forecasts, investment strategies, and more
Use language creatively for your own purposes, such as writing a term paper or story or engaging in debate	Use your sense of art creatively, such as in designing your house, taking a photograph, or making a sculpture	Use mathematics creatively to solve problems you encounter and to help you understand issues in the modern world

How to Succeed in Mathematics

If you are reading this text, you probably are enrolled in a mathematics course. The keys to your success in the course include approaching the material with an open and optimistic frame of mind, paying close attention to how useful and enjoyable mathematics can be in *your* life, and studying effectively and efficiently. The following sections offer a few specific hints that may be of use as you study.

The Key to Success: Study Time

The single most important key to success in any college course is to spend enough time studying. A general rule of thumb for college classes is that you should expect to study about 2 to 3 hours per week *outside* class for each unit of credit. For example, a student taking 15 credit hours should spend 30 to 45 hours each week studying outside of class. Combined with time in class, this works out to a total of 45 to 60 hours per week—not much more than the typical time a job requires, and you get to choose your own hours. Of course, if you are working or taking care of a family while you attend school, you will need to budget your time carefully.

The following table gives some rough guidelines for how you might divide your studying time in your mathematics course:

If your course is	Time for reading the assigned text (per week)	Time for homework assignments (per week)	Time for review and test preparation (average per week)	Total study time (per week)
3 credits	1 to 2 hours	3 to 5 hours	2 hours	6 to 9 hours
4 credits	2 to 3 hours	3 to 6 hours	3 hours	8 to 12 hours
5 credits	2 to 4 hours	4 to 7 hours	4 hours	10 to 15 hours

If you are spending fewer hours than these guidelines suggest, you can probably improve your grade by studying more. If you are spending more hours than these guidelines suggest, you may be studying inefficiently; in that case, you should talk to your instructor about how to study more effectively.

Using This Textbook

The chapters in this textbook are structured to help you to study effectively and efficiently. To get the most out of each chapter, you might wish to use the following study plan.

- Begin by reading the assigned material *twice*:
  - On the first pass, read straight through to gain a "feel" for the material and concepts presented.
  - On the second pass, read more carefully while using the wide margins to take notes that will help later with homework and exams. Be sure to take notes by hand (or by typing if you have an e-text), rather than using a highlight pen (or highlighting tool), which makes it too easy to highlight mindlessly.
- Next go back and *work through* the examples. That is, don't just read them, but instead try to work them yourself, looking at the solutions only if you get stuck.
- You'll now be ready to try the end-of-chapter exercises. Start by making sure you can answer the Quick Quiz and Review Questions. Then do all assigned exercises, and ideally do even more; you can check answers for odd-numbered exercises in the back of the textbook.
- If you have access to MyLab Math, be sure to take advantage of the many additional study resources available on this website.

General Strategies for Studying

- Budget your time effectively. Studying 1 or 2 hours each day is more effective, and far less painful, than studying all night before homework is due or before exams. *Note*: Research shows that it can be helpful to create a "personal contract" for your study time (or for any other personal commitment), in which you specify rewards you'll give yourself for success and penalties you'll assess for failings.
- Engage your brain. Learning is an active process, not a passive experience. Whether you are reading, listening to a lecture, or working on homework assignments, always make sure that your mind is actively engaged. If you find your mind drifting or find yourself falling asleep, make a conscious effort to revive yourself or take a break if necessary.
- Don't miss class, and come prepared. Listening to lectures and participating in class activities and discussions is much more effective than reading someone else's notes or watching a video later. Active participation will help you retain what you are learning. Also, be sure to complete any assigned reading *before* the class in which it will be discussed. This is crucial, because class lectures and discussions are designed to reinforce key ideas from the reading.
- Start your homework early. The more time you allow yourself, the easier it is to get help if you need it. If a concept gives you trouble, first try additional reading or studying beyond what has been assigned. If you still have trouble, ask for help: You surely can find friends, peers, or teachers who will help you learn.

- Working together with friends can be valuable in helping you understand difficult concepts. However, be sure that you learn *with* your friends and do not become dependent on them.
- Don't try to multitask. Research shows that human beings simply are not good at multitasking: When we attempt it, we do more poorly at all of the individual tasks. And in case you think you are an exception, research has also shown that those people who believe they are best at multitasking are often the worst! So when it is time to study, turn off your electronic devices, find a quiet spot, and concentrate on your work. (If you must use a device to study, as is the case with an e-text or online homework, turn off e-mail, text, and other alerts so that they will not disrupt your concentration.)

Preparing for Exams

- Rework exercises and other assignments. Try additional exercises to be sure you understand the concepts. Study your assignments, quizzes, and exams from earlier in the semester.
- Study your notes from classes, and reread relevant sections in your textbook. Pay attention to what your instructor expects you to know for an exam.
- Study individually *before* joining a study group with friends. Study groups are effective only if *every* individual comes prepared to contribute.
- Don't stay up too late before an exam. Don't eat a big meal within an hour of the exam (thinking is more difficult when blood is going to the digestive system).
- Try to relax before and during the exam. If you have studied effectively, you are capable of doing well. Staying relaxed will help you think clearly.

Presenting Homework and Writing Assignments

All work that you turn in should be of *collegiate* quality: neat and easy to read, well organized, and demonstrating mastery of the subject matter. Future employers and teachers will expect this quality of work. Moreover, although submitting homework of collegiate quality requires "extra" effort, it serves two important purposes directly related to learning:

1. The effort you expend in clearly explaining your work solidifies your understanding. Writing triggers different areas of your brain than reading, listening, or speaking does. As a result, writing something down will reinforce your learning of a concept, even when you think you already understand it.
2. By making your work clear and self-contained (that is, making it a document that you can read without referring to the questions in the text), you will have a much more useful study guide when you review for a quiz or exam.

The following guidelines will help ensure that your assignments meet the standards of collegiate quality:

- Always use proper grammar, proper sentence and paragraph structure, and proper spelling. Do not use texting shorthand.
- All answers and other writing should be fully self-contained. A good test is to imagine that a friend is reading your work and to ask yourself whether the friend would understand exactly what you are trying to say. It is also helpful to read your work out loud to yourself, making sure that it sounds clear and coherent.

- In problems that require calculation:
  - Be sure to *show your work* clearly, so that both you and your instructor can follow the process you used to obtain an answer. Also, use standard mathematical symbols, rather than "calculator-ese." For example, show multiplication with the $\times$ symbol (not with an asterisk), and write 10^5, not 10^5 or 10E5.
  - *Word problems should have word answers.* That is, after you have completed any necessary calculations, any problem stated in words should be answered with one or more *complete sentences* that describe the point of the problem and the meaning of your solution.
  - Express your word answers in a way that would be *meaningful* to most people. For example, most people would find it more meaningful if you express a result of 720 hours as 1 month. Similarly, if a precise calculation yields an answer of 9,745,600 years, it may be more meaningful in words as "nearly 10 million years."
- Include illustrations whenever they help explain your answer, and make sure your illustrations are neat and clear. For example, if you graph by hand, use a ruler to make straight lines. If you use software to make illustrations, be careful not to clutter them up with unnecessary features.
- If you study with friends, be sure that you turn in your own work stated in your own words—you should avoid anything that might even give the *appearance* of possible academic dishonesty.

Prologue

DISCUSSION QUESTIONS

1. **Mathematics in Modern Issues.** Describe at least one way that mathematics is involved in each issue below.
   Example: An epidemic of a new virus. Mathematics is used to study the probability of being infected by the virus and to determine where the virus arose and how it spreads.
   a. The long-term viability of the Social Security system
   b. The appropriate level for the federal gasoline tax
   c. National health care policy
   d. Job discrimination against women or ethnic minorities
   e. Effects of population growth (or decline) on your community
   f. Possible bias in standardized tests (e.g., the SAT)
   g. The degree of risk posed by carbon dioxide emissions
   h. Immigration policy of the United States
   i. Violence in public schools
   j. Whether certain types of guns or ammunition should be banned
   k. An issue of your choice from today's news
2. **Quantitative Concepts in the News.** Identify a major unresolved issue discussed in today's news. List at least three areas in which quantitative concepts play a role in the policy considerations of this issue.
3. **Mathematics and the Arts.** Choose a well-known historical figure in a field of art in which you have a personal interest (e.g., a painter, sculptor, musician, or architect). Briefly describe how mathematics played a role in or influenced that person's work.
4. **Quantitative Literature.** Choose a favorite work of literature (poem, play, short story, or novel). Describe one or more instances in which quantitative reasoning is helpful in understanding the subtleties intended by the author.
5. **Your Quantitative Major.** Identify ways in which quantitative reasoning is important within your major field of study. (If you haven't yet chosen a major, pick a field that you are considering for your major.)
6. **Career Preparation.** Realizing that most Americans change careers several times during their lives, identify at least three occupations in Table P.2 that interest you. Do you have the necessary skills for them at this time? If not, how can you acquire the skills you lack?
7. **Attitudes Toward Mathematics.** What is your attitude toward mathematics? If you have a negative attitude, can you identify when that attitude developed? If you have a positive attitude, can you explain why? How might you encourage someone with a negative attitude to become more positive?
8. **"Bad at Math" as a Social Disease.** Discuss reasons why many people think being "bad at math" is socially acceptable and how we as a society can change those attitudes. If you were a teacher, what would you do to ensure that your students develop positive attitudes toward mathematics?

1 THINKING CRITICALLY

The primary goal of this text is to help you develop the quantitative reasoning skills you will need to succeed in other college courses, in your career, and in your life as a citizen in an increasingly complex world. Quantitative reasoning requires a combination of basic mathematical skills and an ability to approach problems in a critical and analytical way. For this reason, we devote this first chapter to ideas of logic that will help you think critically not only about quantitative issues but more generally about distinguishing fact from fiction or fake news in the media.

Q Perhaps you, like millions of others, have received this message: "On August 27, Mars will look as large and bright as the full Moon. Don't miss it, because no one alive today will ever see this again." This claim:

- **A** is true, because on this date Mars will be closer to Earth than any time in thousands of years.
- **B** is true, because on this date Mars will be closer to Earth than the Moon.
- **C** was true for the year 2012, but not for other years.
- **D** is false.
- **E** is partially true: Mars really will be this bright, but it happens every year on August 27, so you'll see it again.

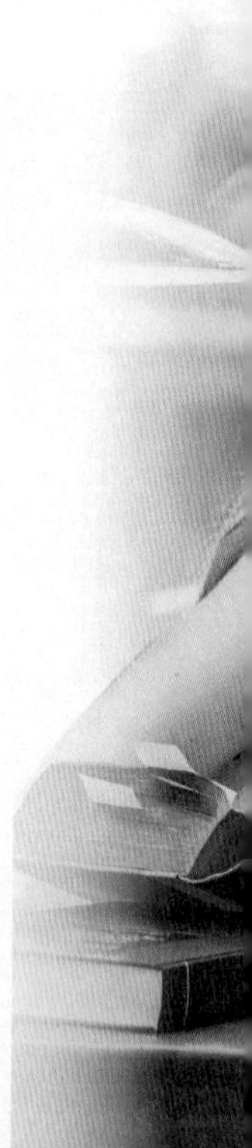

Mathematics is just logic with numbers attached.

—Marilyn vos Savant, American author

A If you're like most students, you may be wondering what this question has to do with math. The answer is "a lot." To begin with, logic is a branch of mathematics, and you can use logic to analyze the claim about Mars. Beyond that, the question also involves mathematics on several deeper levels. For example, the statement "Mars will look as large… as the full Moon" is a statement about angular size, which is a mathematical way of expressing how large an object appears to your eye. In addition, a full understanding of the claim requires understanding how the Moon orbits Earth and planets orbit the Sun, which means understanding that orbits have the mathematical shape called an *ellipse* and obey precise mathematical laws.

So what's the answer? Here's a hint: Think about the fact that Mars is a planet orbiting the Sun while the Moon orbits Earth. Given that fact, ask yourself when, if ever, Mars could appear as large and bright as the full Moon. To see the answer and discussion, go to Example 11 on page 10.

UNIT 1A

Living in the Media Age: Explore common fallacies, or deceptive arguments, and learn how to avoid them.

UNIT 1B

Propositions and Truth Values: Study basic components of logic, including propositions, truth values, truth tables, and the logical connectors *and, or,* and *if…then.*

UNIT 1C

Sets and Venn Diagrams: Understand sets, and use Venn diagrams to visualize relationships among sets.

UNIT 1D

Analyzing Arguments: Learn to distinguish and evaluate basic inductive and deductive arguments.

UNIT 1E

Critical Thinking in Everyday Life: Apply logic to common situations in everyday life.

ACTIVITY

Bursting Bubble

Use this activity to gain a sense of the kinds of problems this chapter will enable you to analyze. Additional activities are available online in MyLab Math.

You probably remember the "Great Recession," which officially lasted from 2007 to 2009 and led to massive bank bailouts, huge increases in unemployment, and many other severe economic consequences. While the recession had many causes, the clear trigger that set it off was a fairly sudden collapse in housing prices. This collapse led many homeowners to default on their home mortgages, which in turn created a crisis for banks and other institutions that bought, sold, or insured home mortgages. If we hope to avoid similar crises in the future, a key question is whether there were early warning signs that might have allowed both individuals and policy makers to make decisions that could have prevented the problems before they occurred.

Figure 1.A shows how average (median) home prices have compared to average income over the past several decades. A ratio of 3.0, for example, means that the average home price is three times the average annual household income of Americans; that is, if you had a household income of $50,000 per year and bought an average house, the price of your house would be $150,000. Notice that the ratio remained fairly stable and below about 3.5 until 2001, but it then shot upward, creating what in hindsight was an overpriced housing market. Economists refer to such unsustainably high housing prices as a *housing bubble*.

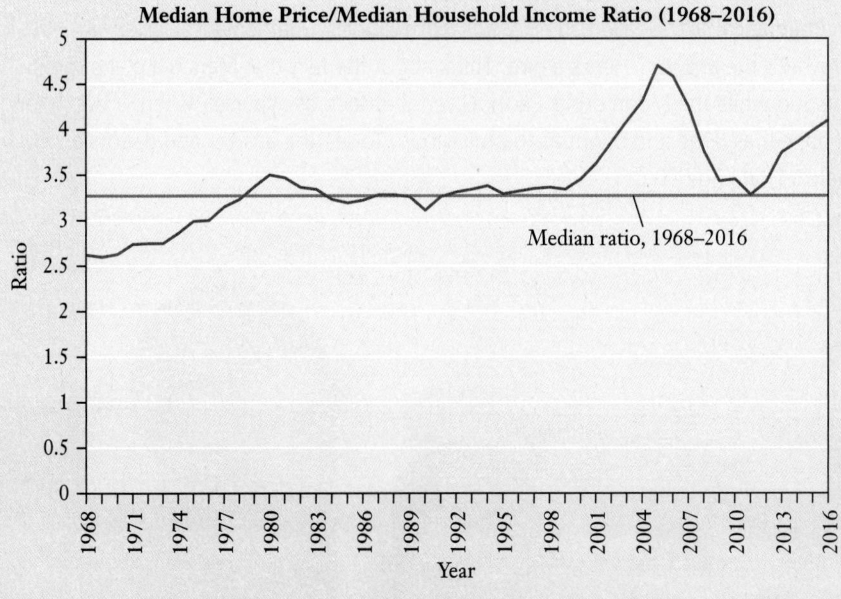

FIGURE 1.A *Source:* Data from *The State of the Nation's Housing 2017*, the Joint Center for Housing Studies of Harvard University.

Was the rise in the home price/income ratio a warning sign that should have been heeded? Use your powers of logic—the topic of this chapter—to discuss the following questions.

1. Consider a family with an annual income of $50,000. If they bought an average home, how much would they have spent in 2000, when the home price/income ratio was about 3.5? How much would they have spent in 2005, when the ratio was about 4.7?

2. In percentage terms, a rise in the ratio from 3.5 to 4.7 is an increase of nearly 35%. Because the ratio was below 3.5 for decades before 2001, we can conclude that the average home

was at least 35% more expensive relative to income in 2005 than it had been historically. What can you infer about how the percentage of income that a family spent on housing changed during the housing bubble?

3. In general, a family can increase the percentage of its income that it spends on housing only if some combination of the following three things happens: (1) its income increases, so it can afford to spend more of it on housing; (2) it cuts expenses in other areas; or (3) it borrows more money. Which of these three happened in most cases during the housing bubble? Explain how you know.
4. Overall, do you think it was inevitable that the bubble would burst? Defend your opinion.
5. Notice in Figure 1.A that the home price/income ratio began another sharp rise in 2013, but the figure only shows data through 2016 (the latest available when this book went to print). Briefly research whether home prices have continued to rise since that time. Based on the data shown and your research, do you think that we are in another housing bubble? Defend your opinion.
6. How could you use the data on the home price/income ratio to help you make a decision about how much to spend when you are looking to buy a home?
7. Additional Research: The data shown here reflect a nationwide average, but the home price/income ratio varies considerably in different cities and regions. Find data for a few different cities or regions, and discuss the differences.

UNIT 1A Living in the Media Age

We are living in what is sometimes called the "media age," because we are in almost constant contact with media of some sort. Some of the media content is printed in books, newspapers, and magazines and on billboards. Much more is delivered electronically through the Internet, tablets and smart phones, television, movies, and more. Most people rely on these media sources for information, which means they form opinions and beliefs based on these sources.

Unfortunately, much of the information in the media is either inaccurate or biased, designed less to inform us than to convince us of something that may or may not be true. As a result, making sense of modern media information requires a deep understanding of the many ways in which people may try to manipulate your views. In this first unit, we'll explore a few of the tools that can help you navigate media sources intelligently. These tools will also provide a foundation for the critical thinking and quantitative reasoning that we'll focus on in the rest of this book.

The Concept of Logical Argument

If you read the comments that follow many news articles on the Web, you'll often see heated discussions that might look much like this "argument" between two classmates.

Ethan: *The death penalty is immoral.*

Jessica: *No it isn't.*

Ethan: *Yes it is! Judges who give the death penalty should be impeached.*

People generally quarrel because they cannot argue.
—G. K. Chesterton (1874–1936), English author

Jessica: *You don't even know how the death penalty is decided.*

Ethan: *I know a lot more than you know!*

Jessica: *I can't talk to you; you're an idiot!*

This type of argument may be common, but it accomplishes little. It doesn't give either person insight into the other's thinking, and it is unlikely to change either person's opinion. Fortunately, there is a better way to argue. We can use skills of **logic**—the study of the methods and principles of reasoning. Arguing logically may still not change either person's position, but it can help them understand each other.

In logic, the term **argument** refers to a reasoned or thoughtful process. Specifically, an argument uses a set of facts or assumptions, called **premises**, to support a **conclusion**. Some arguments provide strong support for their conclusions, but others do not. An argument that fails to make a compelling case for its conclusion may contain some error in reasoning, or **fallacy** (from the Latin for "deceit" or "trick"). In other words, a fallacious argument tries to persuade in a way that doesn't really make sense when analyzed carefully.

Definitions

Logic is the study of the methods and principles of reasoning.

An **argument** uses a set of facts or assumptions, called **premises**, to support a **conclusion.**

A **fallacy** is a deceptive argument—an argument in which the conclusion is not well supported by the premises.

BY THE WAY

Advertisements are filled with fallacies, largely because there's usually no really good reason why you should buy some particular brand or product. Still, they must work, because U.S. businesses spend almost $200 billion per year—or nearly $700 per person in the United States—trying to get you to buy stuff.

Common Fallacies

Fallacies in the media are so common that it is nearly impossible to avoid them. Moreover, fallacies often sound persuasive, despite their logical errors, in part because public relations specialists have spent billions of dollars researching how to persuade us to buy products, vote for candidates, or support particular policies. We therefore begin our study of critical thinking by analyzing examples of a few of the most common fallacies. Each of these fallacies has a name, but learning the names is far less important than learning to recognize the faulty reasoning, because this skill will help you think critically about all types of arguments you may see in the media.

EXAMPLE 1 Appeal to Popularity

"Ford makes the best pickup trucks in the world. More people drive Ford pickups than any other light truck."

Analysis The first step in dealing with any argument is recognizing which statements are premises and which are conclusions. This argument tries to make the case that *Ford makes the best pickup trucks in the world*, so this statement is its conclusion. The only evidence it offers to support this conclusion is the statement *More people drive Ford pickups than any other light truck*. This is the argument's only premise. Overall, this argument has the form

Premise: More people drive Ford pickups than any other light truck.

Conclusion: Ford makes the best pickup trucks in the world.

Note that the argument as originally presented states the conclusion before the premise. Such "backward" structures are common in everyday speech and are perfectly legitimate as long as the argument is well reasoned. In this case, however, the reasoning is faulty.

The fact that more people drive Ford pickups does not necessarily mean that they are the best trucks.

This argument suffers from the fallacy of *appeal to popularity* (or *appeal to majority*), in which the fact that large numbers of people believe or act some way is used inappropriately as evidence that the belief or action is correct. We can represent the general form of this fallacy with a diagram in which the letter *p* stands for a particular statement (Figure 1.1). In this case, *p* stands for the statement *Ford makes the best pickup trucks in the world*. ▸ Now try Exercise 11.

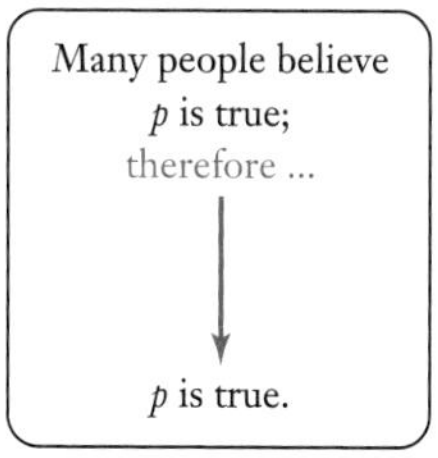

FIGURE 1.1 The fallacy of appeal to popularity. The letters *p* and *q* (used in later diagrams) represent statements.

EXAMPLE 2 False Cause

"I placed the quartz crystal on my forehead, and in five minutes my headache was gone. The crystal made my headache go away."

Analysis We identify the premises and conclusion of this argument as follows:

Premise: I placed the quartz crystal on my forehead.
Premise: Five minutes later my headache was gone.
Conclusion: The crystal made my headache go away.

The premises tell us that one thing (crystal on forehead) happened before another (headache went away), but they don't prove any connection between them. That is, we cannot conclude that the crystal caused the headache to go away.

This argument suffers from the fallacy of *false cause*, in which the fact that one event came before another is incorrectly taken as evidence that the first event *caused* the second event. We can represent this fallacy with a diagram in which *A* and *B* represent two different events (Figure 1.2). In this case, *A* is the event of putting the crystal on the forehead and *B* is the event of the headache going away. (We'll discuss how cause *can* be established in Chapter 5.) ▸ Now try Exercise 12.

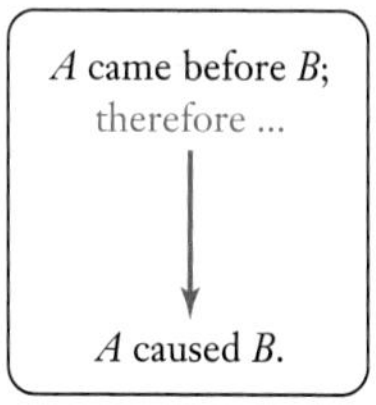

FIGURE 1.2 The fallacy of false cause. The letters *A* and *B* represent events.

EXAMPLE 3 Appeal to Ignorance

"Scientists have not found any concrete evidence of aliens visiting Earth. Therefore, anyone who claims to have seen a UFO must be hallucinating."

Analysis If we strip the argument to its core, it says this:

Premise: There's no proof that aliens have visited Earth.
Conclusion: Aliens have not visited Earth.

The fallacy should be clear: A lack of proof of alien visits does not mean that visits have not occurred. This fallacy is called *appeal to ignorance* because it uses ignorance (lack of knowledge) about the truth of a proposition to conclude the opposite (Figure 1.3). We sometimes sum up this fallacy with the statement "An absence of evidence is not evidence of absence." ▸ Now try Exercise 13.

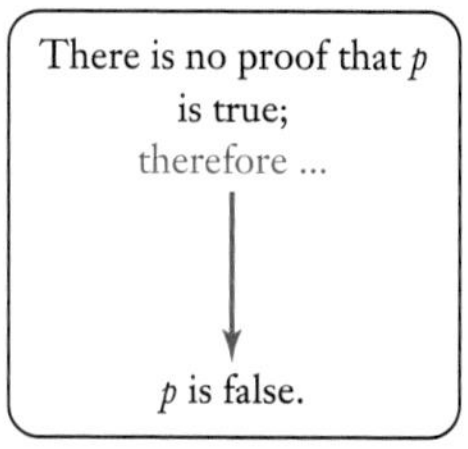

FIGURE 1.3 The fallacy of appeal to ignorance.

Think About It Suppose a person is tried for a crime and found not guilty. Can you conclude that the person is innocent? Why or why not? Why do you think our legal system demands that prosecutors prove guilt, rather than demanding that defendants (suspects) prove innocence? How is this idea related to the fallacy of appeal to ignorance?

EXAMPLE 4 Hasty Generalization

"Two cases of childhood leukemia have occurred along the street where the high-voltage power lines run. The power lines must be the cause of these illnesses."

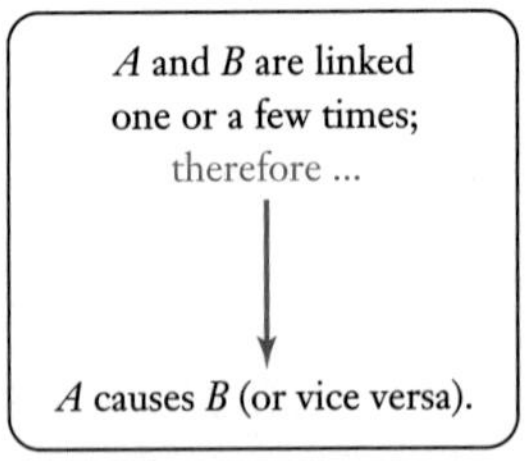

FIGURE 1.4 The fallacy of hasty generalization.

Analysis The premise of this argument cites two cases of leukemia, but two cases are not enough to establish a pattern, let alone to conclude that the power lines caused the illnesses.

The fallacy here is *hasty generalization*, in which a conclusion is drawn from an inadequate number of cases or from cases that have not been sufficiently analyzed. If any connection between power lines and leukemia exists, it would have to be established with far more evidence than is provided in this argument. (In fact, decades of research have found no connection between power lines and illness.) We can represent this fallacy with a diagram in which *A* and *B* represent two linked events (Figure 1.4).

▶ Now try Exercise 14.

EXAMPLE 5 Limited Choice

"You don't support the President, so you are not a patriotic American."

Analysis This argument has the form

Premise: You don't support the President.

Conclusion: You are not a patriotic American.

The argument suggests that there are only two types of Americans: patriotic ones who support the President and unpatriotic ones who don't. But there are many other possibilities, such as being patriotic while disliking a particular President.

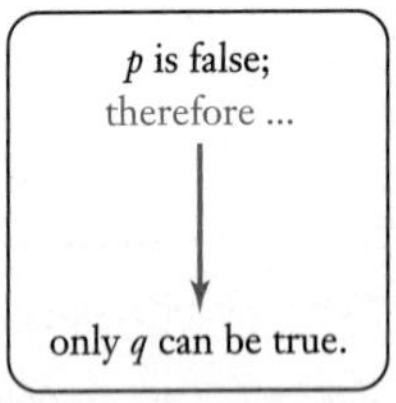

FIGURE 1.5 The fallacy of limited choice.

This fallacy is called *limited choice* (or *false dilemma*) because it artificially precludes choices that ought to be considered. Figure 1.5 shows one common form of this fallacy. Limited choice also arises with questions such as "Have you stopped smoking?" Because both yes and no answers imply that you smoked in the past, the question precludes the possibility that you never smoked. (In legal proceedings, questions of this type are disallowed because they attempt to "lead the witness.") Another simple and common form of this fallacy is "You're wrong, so I must be right."

▶ Now try Exercise 15.

EXAMPLE 6 Appeal to Emotion

In ads for Michelin tires, a picture of a baby is shown with the words "because so much is riding on your tires."

Analysis If we can consider this to be an argument at all, it has the form

Premise: You love your baby.

Conclusion: You should buy Michelin tires.

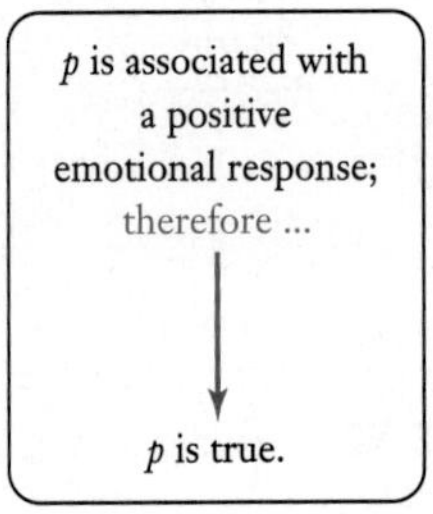

FIGURE 1.6 The fallacy of appeal to emotion.

The advertisers hope that the love you feel for a baby will make you want to buy their tires. This attempt to evoke an emotional response as a tool of persuasion represents the fallacy of *appeal to emotion*. Figure 1.6 shows its form when the emotional response is positive. Sometimes the appeal is to negative emotions. For example, the statement "If my opponent is elected, your taxes will rise" tries to convince you that electing the other candidate will lead to consequences you won't like. (In this negative form, the fallacy is sometimes called *appeal to force*.)

▶ Now try Exercise 16.

EXAMPLE 7 Personal Attack

Gwen: *You should stop drinking because it's hurting your grades, endangering people when you drink and drive, and destroying your relationship with your family.*

Merle: *I've seen you drink a few too many on occasion yourself!*

Analysis Gwen's argument is well reasoned, with premises offering strong support for her conclusion that Merle should stop drinking. Merle rejects this argument by noting that Gwen sometimes drinks too much herself. Even if Merle's claim is true, it is irrelevant to Gwen's point. Merle has resorted to attacking Gwen personally rather than arguing logically, so we call this fallacy *personal attack* (Figure 1.7). (It is also called *ad hominem*, Latin for "to the person.")

The fallacy of personal attack can also apply to groups. For example, someone might say, "This new bill will be an environmental disaster because its sponsors received large campaign contributions from oil companies." This argument is fallacious because it doesn't challenge the provisions of the bill, but only questions the motives of the sponsors. ▶ Now try Exercise 17.

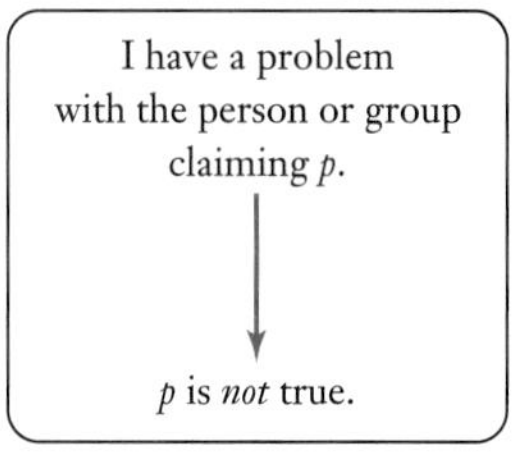

FIGURE 1.7 The fallacy of personal attack.

Think About It A person's (or group's) character, circumstances, or motives can sometimes be relevant to an argument. That is why, for example, witnesses in criminal cases often are asked questions about their personal lives. If you were a judge, how would you decide when to allow such questions?

EXAMPLE 8 Circular Reasoning

"Society has an obligation to provide health insurance because health care is a right of citizenship."

Analysis This argument states the conclusion (*Society has an obligation to provide health insurance*) before the premise (*health care is a citizen's right*), which is fairly common and is generally acceptable. But in this case, the premise and the conclusion both say essentially the same thing, because social obligations are generally based on definitions of accepted rights. This argument therefore suffers from *circular reasoning* (Figure 1.8). ▶ Now try Exercise 18.

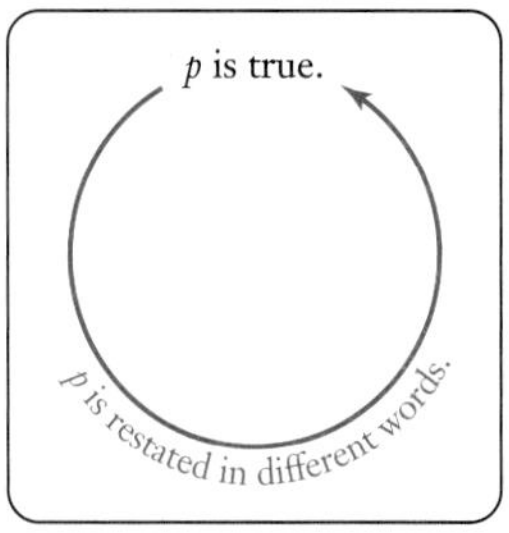

FIGURE 1.8 The fallacy of circular reasoning.

EXAMPLE 9 Diversion (Red Herring)

"We should not continue to fund cloning research because there are so many ethical issues involved. Strong ethical foundations are necessary to the proper functioning of society, so we cannot afford to have too many ethical loose ends."

Analysis The argument begins with its conclusion—we should not continue to fund cloning research. However, the discussion is all about ethics. This argument represents the fallacy of *diversion* (Figure 1.9) because it attempts to divert attention from the real issue (funding for cloning research) by focusing on another issue (ethics). The issue to which attention is diverted is sometimes called a *red herring*. (A herring is a fish that turns red when rotten. Use of the term *red herring* to mean a diversion can be traced back to the 19th century, when British fugitives discovered that they could divert bloodhounds that were pursuing them by rubbing a red herring across their trail.) Note that personal attacks (see Example 7) are often used as diversions. ▶ Now try Exercise 19.

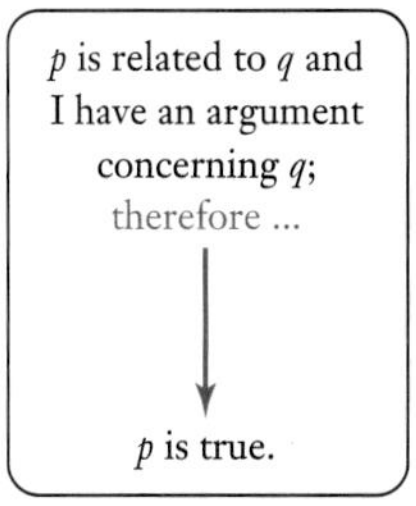

FIGURE 1.9 The fallacy of diversion.

EXAMPLE 10 Straw Man

Suppose that the mayor of a large city proposes decriminalizing drug possession in order to reduce overcrowding in jails and save money on enforcement. His challenger in the upcoming election says, "The mayor doesn't think there's anything wrong with drug use, but I do."

Analysis The mayor did not say that drug use is acceptable. His proposal for decriminalization is designed to solve another problem—overcrowding of jails—and tells us nothing about his general views on drug use. The speaker has distorted the mayor's views. Any argument based on a distortion of someone's words or beliefs is called a *straw man* (Figure 1.10). The term comes from the idea that the speaker has used a poor representation of a person's beliefs in the same way that a straw man is a poor representation

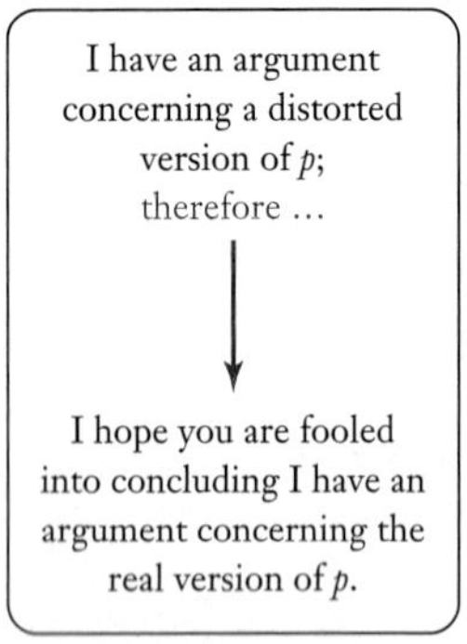
I have an argument concerning a distorted version of p; therefore ...
I hope you are fooled into concluding I have an argument concerning the real version of p.

FIGURE 1.10 The straw man fallacy.

of a real man. A straw man is similar to a diversion. The primary difference is that a diversion argues against an unrelated issue, while the straw man argues against a distorted version of the real issue. ▶ Now try Exercise 20.

Evaluating Media Information

The fallacies we've discussed represent only a small sample of the many tactics used by individuals, groups, and companies seeking to shape your opinions. There are no foolproof ways to be sure that a particular piece of media information is reliable. However, there are a few guidelines, summarized in the following box, that can be helpful. Keep these ideas in mind not only as you evaluate media information, but also as you continue your study in this course. As you will see, much of the rest of this book is devoted to learning to evaluate quantitative information using the same general criteria given in the box.

Five Steps to Evaluating Media Information

1. **Consider the source.** Are you seeing the information from its original source, and if not, can you track down the original source? Does this source have credibility on this issue? Be especially careful to make sure this source is what you think it is; many websites that spread fake news make themselves look like the legitimate sites of real news organizations.
2. **Check the date.** Can you determine when the information was written? Is it still relevant, or is it outdated?
3. **Validate accuracy.** Can you validate the information from other sources (such as major news websites)? Do you have good reason to believe it is accurate? Does it contain anything that makes you suspicious?
4. **Watch for hidden agendas.** Is the information presented fairly and objectively, or is it manipulated to serve some particular or hidden agenda?
5. **Don't miss the big picture.** Even if a piece of media information passes all the above tests, step back and consider whether it makes sense. For example, does it conflict with other things you think are true, and if so, how can you resolve the conflict?

BY THE WAY

The Mars claim has been circulated so much that it is known as the "Mars hoax." While it is untrue, Mars does become as bright or brighter than any star in our night sky for several weeks around the times when it comes closest to us in its orbit, which happens about every 26 months. Recent or upcoming dates when Mars reaches peak brightness are: July 27, 2018; October 13, 2020; and December 8, 2022.

EXAMPLE 11 Mars in the Night Sky

Evaluate the Mars claim from the chapter opener: "On August 27, Mars will look as large and bright as the full Moon. Don't miss it, because no one alive today will ever see this again."

Analysis Let's apply the five steps for evaluating media information:

1. **Consider the source.** The original source of the claim is not given, which means you have no way to know if the source is authoritative. This should make you immediately suspicious of the claim.
2. **Check the date.** Although the claim sounds specific in citing August 27 for the event, no year is given, so you have no way to know if the claim is intended to apply to this year, every year, or a particular past or future year. This should increase your concern about the claim.
3. **Validate accuracy.** The claim is easy to look up, and you'll find numerous websites stating that it is untrue and that the same false claim circulates annually. Of course, you should also check the validity of these websites before believing them, but you'll find some are reliable sources such as NASA or well-respected news sites. We therefore conclude that the correct answer to the chapter-opening question is D—the claim is false. But let's continue with the last two steps anyway.

IN YOUR WORLD

Fact Checking on the Web

While the Web is often a source of fake, inaccurate, or biased information, it is also a great source for checking the accuracy of information. A good way to start is with "fact checking" websites, as long as you also verify that the fact checkers have a reputation for fairness and accuracy. Here are a few reputable fact-checking sites:

- For political fact checking: FactCheck.org, supported by the nonpartisan and nonprofit Annenberg Public Policy Center; PolitiFact.com, a Pulitzer Prize–winning site from the *Tampa Bay Times*; and "The Fact Checker," a blog hosted on the *Washington Post*'s website.
- For rumors, urban myths, and other odd claims, Snopes.com has a solid reputation for accuracy.
- To check the validity of messages you receive by e-mail, try TruthOrFiction.com, which also has a strong reputation for fairness and accuracy.

If none of those sources has covered the claim you are investigating, try a plain language Web search. For example, if you type the first sentence of the Mars claim ("On August 27, Mars will look as large and bright as the full Moon . . .") into a search engine, you'll get dozens of hits that discuss the claim and why it is false. Of course, if your search turns up conflicting claims about accuracy, you'll still need to decide which claims to believe. We'll discuss strategies that can help you with this in Unit 1E.

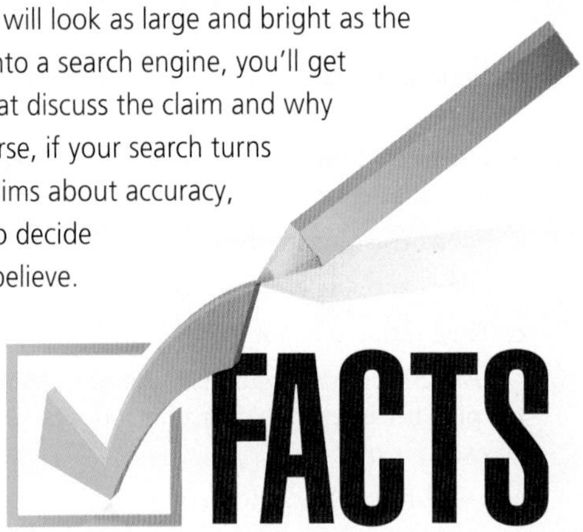

4. **Watch for hidden agendas.** In this case, there's no obvious hidden agenda. It seems more likely that the claim is just a misstatement of fact. A search on the claim will quickly reveal that it originally arose in 2003, when on August 27 Mars came slightly closer to Earth than it will come again for at least 200 years. However, Mars was still nowhere near as large and bright in our sky as the full Moon.
5. **Don't miss the big picture.** This step asks us to stand back and think about whether the claim make sense, which you can do by thinking about the hint in the chapter opener: Mars is a planet orbiting the Sun, while the Moon orbits Earth. This fact means that the Moon is always much closer to us than Mars; even at its closest, Mars is about 150 times as far from Earth as the Moon. You can then conclude that Mars could never appear as large and bright in our sky as the full Moon. (If you want to be more quantitative: At 150 times the distance of the Moon, Mars would have to be 150 times as large as the Moon in diameter in order to appear equally large in our sky. However, Mars is only about twice as large in diameter as the Moon.)

▸ Now try Exercises 21–24.

Quick Quiz 1A

Choose the best answer to each of the following questions. Explain your reasoning with one or more complete sentences.

1. A logical argument always includes
   a. at least one premise and one conclusion.
   b. at least one premise and one fallacy.
   c. at least one fallacy and one conclusion.
2. A fallacy is
   a. a statement that is untrue.
   b. a heated argument.
   c. a deceptive argument.
3. Which of the following could not qualify as a logical argument?
   a. a series of statements in which the conclusion comes before the premises
   b. a list of premises that do not lead to a conclusion
   c. a series of statements that generate heated debate
4. An argument in which the conclusion essentially restates the premise is an example of
   a. circular reasoning.
   b. limited choice.
   c. logic.

5. The fallacy of appeal to ignorance occurs when
   a. the fact that a statement p is true is taken to imply that the opposite of p must be false.
   b. the fact that we cannot prove a statement p to be true is taken to imply that p is false.
   c. a conclusion p is disregarded because the person who stated it is ignorant.
6. Consider the argument "I don't support the President's tax plan because I don't trust his motives." What is the conclusion of this argument?
   a. I don't trust his motives.
   b. I don't support the President's tax plan.
   c. The President is not trustworthy.
7. Consider again the argument "I don't support the President's tax plan because I don't trust his motives." This argument is an example of
   a. a well-reasoned, logical argument.
   b. the fallacy of personal attack.
   c. the fallacy of appeal to emotion.
8. The argument "You don't like soccer so you can't be a sports fan" is an example of
   a. a well-reasoned, logical argument.
   b. the fallacy of diversion.
   c. the fallacy of limited choice.
9. Suppose that the fact that event A occurred before event B is used to conclude that A caused B. This is an example of
   a. a well-reasoned, logical argument.
   b. the fallacy of false cause.
   c. the fallacy of hasty generalization.
10. When we speak of a straw man in an argument, we mean
    a. a misrepresentation of someone else's idea or belief.
    b. a person who has not used good logic.
    c. an argument so weak it seems to be made of straw.

Exercises 1A

REVIEW QUESTIONS

1. What is *logic*? Briefly explain how logic can be useful.
2. How do we define *argument*? What is the basic structure of an argument?
3. What is a *fallacy*? Choose three examples of fallacies from this unit, and, in your own words, describe how the given argument is deceptive.
4. Summarize the five steps given in this unit for evaluating media information, and explain how you can apply them.

DOES IT MAKE SENSE?

Decide whether each of the following statements makes sense (or is clearly true) or does not make sense (or is clearly false). Explain your reasoning.

5. In order to present a logical argument to Andrea, Julian had to shout at her.
6. I persuaded my father that I was right with a carefully constructed argument that contained no premises at all.
7. I didn't believe the premises on which he based his argument, so he clearly didn't convince me of his conclusion.
8. She convinced me she was right, even though she stated her conclusion before supporting it with any premises.
9. I disagree with your conclusion, so your argument must contain a fallacy.
10. Even though your argument contains a fallacy, your conclusion is believable.

BASIC SKILLS & CONCEPTS

11–20: Analyzing Fallacies. Consider the following examples of fallacies.

a. Identity the premise(s) and conclusion of the argument.
b. Briefly describe how the stated fallacy occurs in the argument.
c. Make up another argument that exhibits the same fallacy.

11. (Appeal to popularity) Apple's iPhones outsell all other smart phones, so they must be the best smart phones on the market.
12. (False cause) I became sick just hours after eating at Burger Hut, so its food must have made me sick.
13. (Appeal to ignorance) Decades of searching have not revealed life on other planets, so life in the universe must be confined to Earth.
14. (Hasty generalization) I saw three people use food stamps to buy expensive steaks, so abuse of food stamps must be widespread.
15. (Limited choice) He refused to testify by invoking his Fifth Amendment rights, so he must be guilty.
16. (Appeal to emotion) Thousands of unarmed people, many of them children, are killed by firearms every year. It's time we ban the sale of guns.
17. (Personal attack) Senator Smith's bill on agricultural policy is a sham, because he has accepted contributions from companies that sell genetically modified crop seeds.
18. (Circular reasoning) It's illegal to drive faster than the speed limit and breaking the law makes you a criminal, so drivers who exceed the speed limit are criminals.

19. (Diversion) Good grades are needed to get into college, and a college diploma is necessary for a successful career. Therefore, attendance should count in high school grades.
20. (Straw man) The mayor wants to raise taxes to fund social programs, so she must not believe in the value of hard work.

21–24: Media Claims. Each of the following claims can easily be checked on the Web. Do a check, state whether the claim is true or false, and briefly explain why.

21. Barack Obama's mother-in-law, Marian Robinson, receives a $160,000 pension for babysitting her granddaughters while Obama was president.
22. Two grandsons of the tenth U.S. President, John Tyler (1841–1845), were still alive in 2013.
23. On December 12, 2016, President Obama ordered a ban on sending Christmas cards to members of the military.
24. Actor Tom Hanks has a brother, Jim, who sounds so much like Tom that Jim does voice work for Tom.

FURTHER APPLICATIONS

25–40: Recognizing Fallacies. In the following arguments, identify the premise(s) and conclusion, explain why the argument is deceptive, and, if possible, identify the type of fallacy it represents.

25. I ate oysters for dinner and later that night I had a nightmare. Oysters caused my nightmare.
26. There are approximately 40,000 Chinese restaurants in America compared to about 14,000 McDonald's. Chinese food is preferable to hamburgers.
27. All the nurses in Belvedere Hospital are women, so women are better qualified for medical jobs.
28. The governor wants to sell public lands to an energy exploration company. But he's an untrustworthy opportunist, so I oppose this land sale.
29. My uncle never drank alcohol and lived to be 93 years old. That's how I know that avoiding alcohol leads to greater longevity.
30. The state has no right to take a life, so the death penalty should be abolished.
31. Miguel de Cervantes's book *Don Quixote* is popular because an estimated 500 million copies have been sold since it was published in 1612.
32. Claims that fracking causes earthquakes are ridiculous. I live near an oil well and have never felt an earthquake.
33. An audit of the last charity I gave to showed that most of the money was used to pay administrators in the front office. Charities cannot be trusted.
34. Prison overcrowding is a crisis that we must alleviate through increased use of capital punishment.
35. The senator is a member of the National Rifle Association, so I'm sure she opposes a ban on large-capacity magazines for ammunition.
36. Wider highways can relieve traffic congestion, so we should build wider highways to benefit the tourist industry.
37. Responding to Democrats who support doubling the federal minimum wage, a Republican says, "The Democrats think that everyone should have the same income."
38. The giant sea squid (*Architeuthis dux*) has never been observed in its deep-water habitat, so it must be extinct in the wild.
39. My little boy loves dolls and my little girl loves trucks, so there's no truth to the claim that boys are more interested in mechanical toys while girls prefer maternal toys.

40. Responding to Republicans who want to deregulate oil and gas exploration, a Democrat says, "Republicans don't think that government can improve society . . . ."

41–44: Additional Fallacies. Consider the following fallacies (which are not discussed in the text). Explain why the fallacy applies to the example and create your own argument that displays the same fallacy.

41. The fallacy of *division* has this form:

Premise:	X has some property.
Conclusion:	All things or people that belong to X must have the same property.
Example:	Americans use more gasoline than Europeans, so Jake, who is an American, must use more gasoline than Europeans.

42. The *gambler's fallacy* has this form:

Premise:	X has been happening more than it should.
Conclusion:	X will come to an end soon.
Example:	It has rained for 10 days, which is unusual around here. Tomorrow will be sunny.

43. The *slippery slope fallacy* has this form:

Premise:	X has occurred and is related to Y.
Conclusion:	Y will inevitably occur.
Example:	America has sent troops to three countries recently. Before you know it, we will have troops everywhere.

44. The *middle ground fallacy* has this form:

Premise:	X and Y are two extreme positions on a question.
Conclusion:	Z, which lies between X and Y, must be correct.
Example:	Senator Peters supports a large tax cut, and Senator Willis supports no tax cut. That means a small tax cut must be the best policy.

IN YOUR WORLD

45. **Evaluating Media Information.** Choose a current topic of policy discussion (examples might include gun control, health care, tax policy, or many other topics). Find a website that argues on one side or the other of the topic. Evaluate the arguments based on the five steps for evaluating media information given in this unit. Write a short report on the site you visited and your conclusions about the reliability of its information.

46. **Snopes.** Visit the Snopes.com website and choose one topic from its list of the "hottest urban legends." In one or two paragraphs, summarize the legend, whether it is true or false, and why.

47. **Checking Facts.** Visit the FactCheck.org or PolitiFact.com website, and choose one of the featured stories on a recent claim. In one or two paragraphs, summarize the story and its accuracy.

48. **Fallacies in Advertising.** Pick a single night and a single commercial television channel, and analyze the advertisements shown over a one-hour period. Describe how each advertisement tries to persuade the viewer, and discuss whether the argument is fallacious. What fraction of the advertisements involve fallacies? Are any types of fallacy more common than others?

49. **Fallacies in Politics.** Discuss the tactics used by both sides in a current or recent political campaign. How much of the campaign is/was based on fallacies? Describe some of the fallacies. Overall, do you believe that fallacies influenced (or will influence) the outcome of the vote?

50. **Personal Fallacies.** Describe an instance in which you were persuaded of something that you later decided was untrue. Explain how you were persuaded and why you later changed your mind. Did you fall victim to any fallacies? If so, how might you prevent the same thing from happening in the future?

51. **Comment Fallacies.** The "reader comments" that are posted with news stories are notorious for containing common fallacies. Choose one recent news story, view the comments section, and identify at least three comments that suffer from one or more fallacies. Explain the fallacies in each case.

52. **Fake News Sites.** Visit a fake news site that has a Web address that tries to make it look like a real news site (a search on "fake news web sites" will turn up numerous lists) and choose a current feature story. Discuss the ways in which the story is written to sound authentic. Discuss how you are able to recognize that the story is actually fake.

UNIT 1B Propositions and Truth Values

"Contrariwise," continued Tweedledee, "if it was so, it might be; and if it were so, it would be; but as it isn't, it ain't. That's logic."
—Lewis Carroll, *Through the Looking Glass*

Having discussed fallacies in Unit 1A, we now turn our attention to proper arguments. The building blocks of arguments are called **propositions**—statements that make (propose) a claim that may be either true or false. A proposition must have the structure of a complete sentence and must make a distinct assertion or denial. For example:

- *Joan is sitting in the chair* is a proposition because it is a complete sentence that makes an assertion.
- *I did not take the pen* is a proposition because it is a complete sentence that makes a denial.
- *Are you going to the store?* is not a proposition because it is a question. It does not assert or deny anything.
- *Three miles south of here* is not a proposition because it does not make any claim and is not a complete sentence.
- $7 + 9 = 2$ is a proposition, even though it is false. It can be read as a complete sentence, and it makes a distinct claim.

▶ Now try Exercises 13–18.

Definition

A **proposition** makes a claim (either an assertion or a denial) that may be either true or false. It must have the structure of a complete sentence.

Negation (Opposites)

The logical opposite of a proposition is called its **negation**. For example, the negation of *Joan is sitting in the chair* is *Joan is not sitting in the chair*, and the negation of $7 + 9 = 2$ is $7 + 9 \neq 2$. If we represent a proposition with a letter such as p, then its negation is *not* p (sometimes written $\sim p$). Negations are also propositions, because they have the structure of a complete sentence and may be either true or false.

A proposition has a **truth value** of either true (T) or false (F). If a proposition is true, its negation must be false, and vice versa. We can represent these facts with a simple **truth table**—a table that has a row for each possible set of truth values. The following truth table shows the possible truth values for a proposition p and its negation *not p*. It has two rows because there are only two possibilities.

p	*not p*	
T	F	← This row shows that if p is true (T), *not p* is false (F).
F	T	← This row shows that if p is false (F), *not p* is true (T).

HISTORICAL NOTE

The Greek philosopher Aristotle (384–322 B.C.E.) made the first known attempt to put logic on a rigorous foundation. He believed that truth could be established from three basic laws: (1) A thing is itself. (2) A statement is either true or false. (3) No statement is both true and false. Aristotle's laws were used by Euclid (c. 325–270 B.C.E.) to establish the foundations of geometry, and logic remains an important part of mathematics.

Definitions

Any proposition has two possible **truth values:** T = true or F = false.

The **negation** of a proposition p is another proposition that makes the opposite claim of p. It is written *not p* (or $\sim p$), and its truth value is opposite to that of p.

A **truth table** is a table with a row for each possible set of truth values for the propositions being considered.

EXAMPLE 1 Negation

Find the negation of the proposition *Amanda is the fastest runner on the team.* If the negation is false, is Amanda really the fastest runner on the team?

Solution The negation of the given proposition is *Amanda is* **not** *the fastest runner on the team.* If the negation is false, the original statement must be true, meaning that Amanda is the fastest runner on the team. ▶ Now try Exercises 19–22.

Double Negation

The Groucho Marx quotation in the margin may be an extreme example, but many everyday statements contain double (or multiple) negations. We've already seen that the negation *not p* has the opposite truth value of the original proposition p. The double negation *not not p* must therefore have the same truth value as the original proposition p. We can show this fact with a truth table. The first column contains the two possible truth values for p. Two additional columns show the corresponding truth values for *not p* and *not not p*.

I cannot say that I do not disagree with you.

—Groucho Marx

p	*not p*	*not not p*
T	F	T
F	T	F

In ordinary language, double negations rarely involve phrases like "not not," so we must analyze wording carefully to recognize them.

EXAMPLE 2 Radiation and Health

After reviewing data showing an association between low-level radiation and cancer among older workers at the Oak Ridge National Laboratory, a health scientist from the University of North Carolina (Chapel Hill) was asked about the possibility of a similar association among younger workers at another national laboratory. He was quoted as saying (*Boulder Daily Camera*):

My opinion is that it's unlikely that there is no association.

Does the scientist think there is an association between low-level radiation and cancer among younger workers?

Solution Because of the words "unlikely" and "no association," the scientist's statement contains a double negation. To see the effects of these words clearly, let's start with a simpler proposition:

p = *it's likely that there is an association (between low-level radiation and cancer)*

The word "unlikely" gives us the statement *it's unlikely that there is an association*, which we identify as *not p*. The words "no association" transform this last statement into the original statement, *it's unlikely that there is no association*, which we recognize as *not not p*. Because the double negation has the same truth value as the original proposition, we conclude that the scientist believes it is likely that there is an association between low-level radiation and cancer among younger workers.

▶ Now try Exercises 23–24.

BY THE WAY

Ernesto Miranda confessed to and was convicted of a 1963 rape and kidnapping. His lawyer argued that the confession should not have been admitted as evidence because Miranda had not been told of his right to remain silent. The Supreme Court agreed and overturned his conviction. He was then retried and again convicted (on the basis of evidence besides the confession). Miranda was stabbed to death during a bar fight after his release from prison. A suspect in his killing chose to remain silent upon arrest, and police never filed charges.

EXAMPLE 3 The Miranda Ruling

If you've ever watched a crime show, you are familiar with the Miranda rights that law enforcement officers recite to suspects ("You have the right to remain silent, ..."). These rights stem from a 1966 decision of the U.S. Supreme Court (*Miranda v. State of Arizona*), which the Court revisited in a case in 2000 (*Dickerson v. United States*). In his majority opinion, Chief Justice William Rehnquist wrote:

... [legal] principles weigh heavily against overruling [Miranda].

Based on this statement, did the Court support or oppose the original Miranda decision?

Solution The Chief Justice's statement contains a double negation. The first negation comes from the term "overruling," which alone would imply opposition to the original decision. But the statement argued *against* overruling, meaning that the original Miranda decision was upheld.

▶ Now try Exercises 25–28.

Logical Connectors

Propositions are often joined together with logical connectors—words such as *and*, *or*, and *if...then*. For example, consider the following two propositions:

p = *The test was hard.*

q = *I got an A.*

If we join the two propositions with *and*, we get a new proposition: *The test was hard and I got an A*. If we join them with *or*, we get this proposition: *The test was hard or I got an A*. Although such statements are familiar in everyday speech, we must analyze them carefully.

And Statements (Conjunctions)

An *and* statement is called a **conjunction**. According to the rules of logic, the conjunction *p and q* is true only if p and q are *both* true. For example, the statement *The test was hard and I got an A* is true only if it *was* a hard test and you *did* get an A.

The Logic of *And*

Given two propositions p and q, the statement *p and q* is called their **conjunction**. It is true only if p and q are *both* true.

To make a truth table for the conjunction *p and q*, we analyze all possible combinations of the truth values of the individual propositions *p* and *q*. Because *p* and *q* each have two possible truth values (true or false), there are $2 \times 2 = 4$ cases to consider. The four cases become the four rows of the truth table.

Truth Table for Conjunction *p and q*

p	*q*	*p and q*	
T	T	T	← case 1: *p*, *q* both true
T	F	F	← case 2: *p* true, *q* false
F	T	F	← case 3: *p* false, *q* true
F	F	F	← case 4: *p*, *q* both false

Note that the *and* statement is true only in the first case shown in the table, where both individual propositions are true. ▸ Now try Exercises 29–30.

EXAMPLE 4 *And* Statements

Evaluate the truth value of the following two statements.

a. The capital of France is Paris and Antarctica is cold.
b. The capital of France is Paris and the capital of America is Madrid.

Solution

a. The statement contains two distinct propositions: *The capital of France is Paris* and *Antarctica is cold*. Because both propositions are true, their conjunction is also true.
b. The statement contains two distinct propositions: *The capital of France is Paris* and *the capital of America is Madrid*. Although the first proposition is true, the second is false. Therefore, their conjunction is false. ▸ Now try Exercises 31–36.

EXAMPLE 5 Triple Conjunction

Suppose you are given three individual propositions *p*, *q* and *r*. Make a truth table for the conjunction *p and q and r*. Under what circumstances is the conjunction true?

Solution We already know that the statement *p and q* has four possible cases for truth values. For each of these four cases, proposition *r* can be either true or false. Therefore, the statement *p and q and r* has $4 \times 2 = 8$ possible cases for truth values. The following truth table contains a row for each of the eight cases. Note that the four cases for *p and q* each appear twice, once with *r* true and once with *r* false.

p	*q*	*r*	*p and q and r*
T	T	T	T
T	T	F	F
T	F	T	F
T	F	F	F
F	T	T	F
F	T	F	F
F	F	T	F
F	F	F	F

The conjunction *p and q and r* is true only if all three statements are true, as shown in the first row. ▸ Now try Exercises 37–38.

BY THE WAY

Logical rules lie at the heart of modern computer science. Computers generally represent the numbers 0 and 1 with electric current: No current in the circuit means 0, and current in the circuit means 1. Computer scientists then think of 1 as true and 0 as false and use logical connectors to design circuits. For example, an *and* circuit allows current to pass (its value becomes 1 = *true*) only if both incoming circuits carry current.

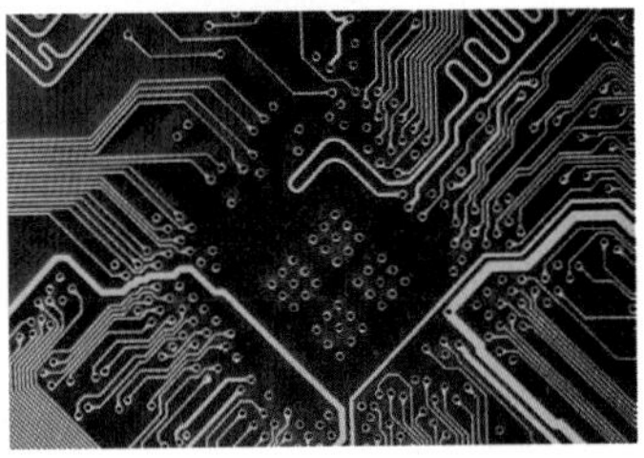

Think About It Given four propositions p, q, r, and s, how many rows are required for a truth table of the conjunction *p and q and r and s*? When is the conjunction true?

Understanding *Or*

The connector *or* can have two different meanings. If a health insurance policy says that it covers hospitalization in cases of illness *or* injury, it probably means that it covers either illness or injury or both. This is an example of the **inclusive *or*** that means "either or both." In contrast, when a restaurant offers a choice of soup *or* salad, you probably are not supposed to choose both. This is an example of the **exclusive *or*** that means "one or the other."

Two Types of *Or*

The word *or* can be interpreted in two distinct ways:

- An **inclusive *or*** means "either or both."
- An **exclusive *or*** means "one or the other, but not both."

In everyday life, we determine whether an *or* statement is inclusive or exclusive by its context. But in logic, we assume that *or* is inclusive unless told otherwise.

EXAMPLE 6 Inclusive or Exclusive?

Kevin's insurance policy states that his house is insured for earthquake, fire, or robbery. Imagine that a major earthquake levels much of his house, the rest burns in a fire, and his remaining valuables are looted in the aftermath. Would Kevin prefer that the *or* in his insurance policy be inclusive or exclusive? Why?

Solution He would prefer an inclusive *or* so that his losses from all three events (earthquake, fire, looting) would be covered. If the *or* were exclusive, the insurance would cover only one of the losses.

▶ Now try Exercises 39–44.

Or Statements (Disjunctions)

A compound statement made with *or* is called a **disjunction**. We assume that the *or* is inclusive, so the disjunction *p or q* is true if either or both propositions are true. A disjunction *p or q* is false only if both individual propositions are false.

The Logic of *Or*

Given two propositions, p and q, the statement *p or q* is called their **disjunction**. In logic, we assume that *or* is *inclusive*, so the disjunction is true if either or both propositions are true, and false only if *both* propositions are false.

These rules lead to the following truth table.

Truth Table for Disjunction *p or q*

p	q	*p or q*	
T	T	T	← case 1: *p*, *q* both true, so *p or q* is true
T	F	T	← case 2: *p* true, so *p or q* is true
F	T	T	← case 3: *q* true, so *p or q* is true
F	F	F	← case 4: *p*, *q* both false, so *p or q* is false

▶ Now try Exercises 45–50.

EXAMPLE 7 Smart Cows?

Consider the statement *Airplanes can fly or cows can read.* Is it true?

Solution The statement is a disjunction of two propositions: (1) *airplanes can fly*; (2) *cows can read*. The first proposition is clearly true, while the second is clearly false, which makes the disjunction *p or q* true. That is, the statement *Airplanes can fly or cows can read* is true.

▶ Now try Exercises 51–56.

BY THE WAY

Most search engines automatically connect words in the search box with the logical connector AND. For example, a search on *television entertainment* is really a search on "television AND entertainment," and it will return any Web page that has both words, regardless of whether the words come together. If you want to search for the exact phrase "television entertainment," you should include quotation marks when you enter it.

If . . . Then Statements (Conditionals)

Another common way to connect propositions is with the words *if...then*, as in the statement "If all politicians are liars, then Representative Smith is a liar." Statements of this type are called **conditional propositions** (or *implications*) because they propose something to be true (the *then* part of the statement) on the *condition* that something else is true (the *if* part of the statement).

We can represent a conditional proposition as *if p, then q*, where proposition *p* is called the **hypothesis** (or *antecedent*) and proposition *q* is called the **conclusion** (or *consequent*).

Let's use an example to discover the truth table for a conditional proposition. Suppose that while running for Congress, candidate Jones claimed:

If I am elected, then the minimum wage will increase.

This proposition has the standard form *if p, then q*, where p = *I am elected* and q = *the minimum wage will increase*. Because each individual proposition can be either true or false, we must consider four possible cases for the truth value of *if p, then q*:

1. ***p* and *q* both true.** In this case, Jones was elected (*p* true) and the minimum wage increased (*q* true). Jones kept her campaign promise, so her claim, *If I am elected, then the minimum wage will increase*, was true.
2. ***p* true and *q* false.** In this case, Jones was elected, but the minimum wage did not increase. Because things did not turn out as she promised, her claim, *If I am elected, then the minimum wage will increase*, was false.
3. ***p* false and *q* true.** This is the case in which Jones was not elected, yet the minimum wage still increased. The conditional statement makes a claim about what should happen in the event that Jones is elected. Because she was not elected, she surely did not break any campaign promise, regardless of whether or not the minimum wage increased. It is a rule of logic that we consider Jones's claim to be true in this case.
4. ***p* and *q* both false.** Now we have the case in which Jones was not elected and the minimum wage did not increase. Again, because she was not elected, she surely did not violate her campaign promise, even though the minimum wage did not increase. As in the previous case, Jones's claim is true.

In summary, the statement *if p, then q* is true in all cases except when the hypothesis *p* is true and the conclusion *q* is false. Here is the truth table:

Truth Table for Conditional *if p, then q*

p	*q*	*if p, then q*
T	T	T
T	F	F
F	T	T
F	F	T

▶ Now try Exercises 57–58.

The Logic of *if... then*

A statement of the form *if p, then q* is called a **conditional proposition** (or *implication*). Proposition p is called the **hypothesis** and proposition q is called the **conclusion**. The conditional *if p, then q* is true in all cases except the case in which p is true and q is false.

Think About It Suppose candidate Jones had made the following campaign promise: *If I am elected, I will personally eliminate all poverty on Earth.* According to the rules of logic for cases (3) and (4) above, we consider this statement to be true in the event that Jones is not elected. Is this logical definition of truth the same one that you would use if you heard her make this promise? Explain.

EXAMPLE 8 Conditional Truths

Evaluate the truth of the statement *if* $2 + 2 = 5$, *then* $3 + 3 = 4$.

Solution The statement has the form *if p, then q*, where p is $2 + 2 = 5$ and q is $3 + 3 = 4$. Both p and q are clearly false. However, according to the rules of logic, the conditional *if p, then q* is true any time p is false, regardless of what q says. Therefore, the statement *if* $2 + 2 = 5$, *then* $3 + 3 = 4$ is true. ▶ Now try Exercises 59–66.

Alternative Phrasings of Conditionals

In ordinary language, conditional statements don't always appear in the standard form *if p, then q*. In such cases, it can be useful to rephrase the statements in the standard form. For example, the statement *I'm not coming back if I leave* has the same meaning as *If I leave, then I'm not coming back*. Similarly, the statement *More rain will lead to a flood* can be recast as *If there is more rain, then there will be a flood.*

Two common ways of phrasing conditionals use the words *necessary* and *sufficient*. Consider the true statement *If you are living, then you are breathing*. We can rephrase this statement as *Breathing is necessary for living*. Note that this statement does not imply that breathing is the *only* necessity for living; in this case, it is one of many necessities (including eating, breathing, and having a heartbeat). More generally, any true statement of the form *if p, then q* is equivalent to the statement *q is necessary for p*.

Now consider the true statement *If you are in Denver, then you are in Colorado*. We can rephrase this statement as *Being in Denver is sufficient for being in Colorado*, because Denver is a city in Colorado. Note, however, that being in Denver is *not* necessary for being in Colorado, because you could also be in other places in the state. More generally, any true statement of the form *if p, then q* is equivalent to the statement *p is sufficient for q*.

HISTORICAL NOTE

The system of logic presented in this chapter was first developed by the ancient Greeks and is called *binary logic* because a proposition is either true or false, but not both and certainly nothing in between. Today, mathematicians also use systems of logic in which other truth values are possible. One form, called *fuzzy logic*, allows for a continuous range of values between absolutely true and absolutely false. Fuzzy logic is used in many new technologies.

Alternative Phrasings of Conditionals

The following are common alternative ways of stating *if p, then q*:

p is sufficient for q	*p will lead to q*	*p implies q*
q is necessary for p	*q if p*	*q whenever p*

EXAMPLE 9 Rephrasing Conditional Propositions

Recast each of the following statements in the form *if p, then q*.

a. A rise in sea level will devastate Florida.
b. A red tag on an item is sufficient to indicate it's on sale.
c. Eating vegetables is necessary for good health.

Solution
a. This statement is equivalent to *If sea level rises, then Florida will be devastated.*
b. This statement can be written as *If an item is marked with a red tag, then it is on sale.*
c. This statement can be expressed as *If a person is in good health, then the person eats vegetables.* ▶ Now try Exercises 67–72.

Converse, Inverse, and Contrapositive

The order of the propositions does not matter in a conjunction or disjunction. For example, *p and q* has the same meaning as *q and p*, and *p or q* has the same meaning as *q or p*. However, when we switch the order of the propositions in a conditional, we create a different proposition called the **converse**. The following box summarizes definitions for the converse and two other variations on a conditional proposition *if p, then q*.

Variations on the Conditional

Name	Form	Example
Conditional	*if p, then q*	If you are sleeping, then you are breathing.
Converse	*if q, then p*	If you are breathing, then you are sleeping.
Inverse	*if not p, then not q*	If you are not sleeping, then you are not breathing.
Contrapositive	*if not q, then not p*	If you are not breathing, then you are not sleeping.

We can determine the truth values for the converse, inverse, and contrapositive with a truth table. Because all the statements use the same two propositions (p, q), the table has four rows. The first two columns show the truth values for *p* and *q*, respectively. The next two columns show the truth values for the negations *not p* and *not q*, which are needed for the inverse and contrapositive. The fifth column shows the truth values found previously for the conditional proposition *if p, then q*. We then find the truth values for the converse, inverse, and contrapositive by applying the rule that a conditional is false only when the hypothesis is true and the conclusion is false.

p	q	not p	not q	if p, then q	if q, then p (converse)	if not p, then not q (inverse)	if not q, then not p (contrapositive)
T	T	F	F	T	T	T	T
T	F	F	T	F	T	T	F
F	T	T	F	T	F	F	T
F	F	T	T	T	T	T	T

Note that the truth values for the conditional *if p, then q* are the same as the truth values for its contrapositive. We therefore say that a conditional and its contrapositive are **logically equivalent**: If one is true, so is the other, and vice versa. The table also shows that the converse and inverse are logically equivalent.

Definition

Two statements are **logically equivalent** if they share the same truth values: If one is true, so is the other, and if one is false, so is the other.

EXAMPLE 10 Logical Equivalence

Consider the true statement *If a creature is a whale, then it is a mammal.* Write its converse, inverse, and contrapositive. Evaluate the truth of each statement. Which statements are logically equivalent?

Solution The statement has the form *if p, then q*, where *p* = *a creature is a whale* and *q* = *a creature is a mammal*. Therefore, we find

> **converse** (*if q, then p*): *If a creature is a mammal, then it is a whale.* This statement is false, because most mammals are not whales.
>
> **inverse** (*if not p, then not q*): *If a creature is not a whale, then it is not a mammal.* This statement is also false; for example, dogs are not whales, but they are mammals.
>
> **contrapositive** (*if not q, then not p*): *If a creature is not a mammal, then it is not a whale.* Like the original statement, this statement is true, because all whales are mammals.

Note that the original proposition and its contrapositive have the same truth value and are logically equivalent. Similarly, the converse and inverse have the same truth value and are logically equivalent.

▶ Now try Exercises 73–78.

Quick Quiz 1B

Choose the best answer to each of the following questions. Explain your reasoning with one or more complete sentences.

1. The statement *Mathematics is fun* is
   a. an argument.
   b. a fallacy.
   c. a proposition.
2. Suppose you know the truth value of a proposition *p*. Then you also know the truth value of this proposition's
   a. negation.
   b. truth table.
   c. conjunction.
3. Which of the following has the form of a *conditional* statement?
   a. *x or y*
   b. *x and y*
   c. *if x, then y*
4. Suppose you want to make a truth table for the statement *x or y or z*. How many rows will the table require?
   a. 2
   b. 4
   c. 8
5. Suppose the statement *p or q* is true. Then you can be certain that
   a. *p* is true.
   b. *q* is true.
   c. one or both of the statements are true.
6. Suppose statement *p* is false and statement *q* is true. Which of the following statements is false?
   a. *p and q*
   b. *p or q*
   c. *if p, then q*
7. The statement *If it's a dog, then it is a mammal* may be rephrased as
   a. Being a mammal is sufficient for being a dog.
   b. Being a mammal is necessary for being a dog.
   c. All mammals are dogs.
8. The statement *If the engine is running, then the car must have gas* is logically equivalent to the statement
   a. If the car has gas, then the engine is running.
   b. If the engine is not running, then the car does not have gas.
   c. If the car does not have gas, then the engine is not running.
9. Two statements are logically equivalent if
   a. they mean the same thing.
   b. they have the same truth values.
   c. they are both true.
10. Consider the statement *You've got to play if you want to win.* If you put this statement in the form *if p, then q*, the wording of *q* would be
    a. You've got to play.
    b. You want to win.
    c. You've got to play if you want to win.

Exercises 1B

REVIEW QUESTIONS

1. What is a *proposition*? Give a few examples, and explain why each is a proposition.
2. What do we mean by the *negation* of a proposition? Make up your own example of a proposition and its negation.
3. Define *conjunction*, *disjunction*, and *conditional*, and give an example of each in words.
4. What is the difference between an inclusive *or* and an exclusive *or*? Give an example of each.
5. Make a truth table for each of the following: *p and q*; *p or q*; *if p, then q*. Explain all the truth values in the tables.
6. Describe how to state the converse, inverse, and contrapositive of a conditional proposition. Make a truth table for each. Which statement is logically equivalent to the original conditional?

DOES IT MAKE SENSE?

Decide whether each of the following statements makes sense (or is clearly true) or does not make sense (or is clearly false). Explain your reasoning.

7. My logical proposition is a question that you must answer.
8. The mayor opposes repealing the ban on handguns, so he must support gun control.
9. We intend to catch him, dead or alive.
10. When Sally is depressed, she listens to music. I saw her today listening to music, so she must have been depressed.
11. Now that I've studied logic, I can always determine the truth of any statement by making a truth table for it.
12. If all novels are books, then all books are novels.

BASIC SKILLS & CONCEPTS

13–18: A Proposition? Determine whether the following statements are propositions, and give an explanation.

13. The sky is the limit.
14. To be or not to be.
15. Back to the future.
16. Some sycophants are soporific.
17. What were you thinking?
18. The deposed leader was corrupt.

19–22: Negation. Write the negation of the given proposition. Then state the truth value of the original proposition and its negation.

19. Asia is in the northern hemisphere.
20. Peru is in the northern hemisphere.
21. The Beatles were not a German band.
22. Earth is the center of the universe.

23–28: Multiple Negations. Explain the meaning of the given statement, which contains a multiple negation. Then answer the question that follows.

23. Sarah did not decline the offer to go to dinner. Did Sarah go to dinner?
24. The mayor opposes the ban on anti-fracking rallies. Does the mayor approve of allowing the rallies?
25. The President vetoed the bill to cut taxes. Based on this veto, will taxes be lowered?
26. The House failed to overturn the veto on a bill that would stop logging. Based on this vote, will logging continue?
27. Sue disavows association with any organization that opposes planting new trees in the park. Does Sue want new trees planted in the park?
28. The senator opposes overriding the governor's veto of the bill. Does the senator support the bill?

29–30: Truth Tables. Make a truth table for the given statement. The letters *p*, *q*, *r*, and *s* represent propositions.

29. *q and r*
30. *p and s*

31–36: *And* Statements. The following statements have the form *p and q*. State *p* and *q*, and give their truth values. Then determine whether the conjunction (the *and* statement) is true or false, and explain why.

31. Dogs are animals and oak trees are plants.
32. $12 + 6 = 18$ and $3 \times 5 = 8$.
33. Venus is a planet and the Sun is a star.
34. Emily Dickinson was a poet and Kanye West is a Major League pitcher.
35. All birds can fly and some fish live in trees.
36. Not all men are tall and not all women are short.

37–38: Truth Tables. Make a truth table for the given statement. Assume that *p*, *q*, *r*, and *s* represent propositions.

37. *q and r and s*
38. *p and q and r and s*

39–44: Interpreting *or*. State whether *or* is used in the inclusive or exclusive sense in the following propositions.

39. I will either walk or ride a bike to the park.
40. Before the entrée, you have a choice of soup or salad.
41. The next book I'll read will be the Bible or the Koran.
42. The oil change is good for 5000 miles or 3 months.
43. I would like to scuba dive or surf on my next vacation.
44. The insurance policy covers fire or theft.

45–50: Truth Tables. Make a truth table for the given statement. Assume that *p, q, r*, and *s* represent propositions.

45. *r or s*
46. *p or r*
47. *p and (not p)*
48. *q or (not q)*
49. *p or q or r*
50. *p or (not p) or q*

51–56: *Or* Statements. The following statements have the form *p or q*. State *p* and *q*, and give their truth values. Then determine whether the disjunction (*or* statement) is true or false, and explain why.

51. Elephants are animals or elephants are plants.
52. The Nile River is in Europe or the Ganges River is in Asia.
53. $3 \times 5 = 15$ or $3 + 5 = 8$.
54. $2 + 2 = 5$ or $3 + 3 = 7$.
55. Cars swim or dolphins fly.
56. Oranges or bananas are round.

57–58: Truth Tables. Make a truth table for the given statement. Assume that *p, q, r*, and *s* represent propositions.

57. *if p, then r*
58. *if q, then s*

59–66: *If... then* Statements. Identify the hypothesis and the conclusion in the following conditional propositions, and state their truth values. Then determine whether the entire proposition is true or false.

59. If trout can swim, then trout are fish.
60. If Paris is in France, then New York is in America.
61. If Paris is in France, then New York is in China.
62. If Paris is in Mongolia, then New York is in America.
63. If trees can walk, then birds wear wigs.
64. If $2 \times 3 = 6$, then $2 + 3 = 6$.
65. If dogs can swim, then dogs are fish.
66. If dogs are fish, then dogs can swim.

67–72: Rephrasing Conditional Statements. Express the following statements in the form *if p, then q*. Identify *p* and *q* clearly.

67. When it snows, I get cold.
68. A Bostonian lives in Massachusetts.
69. Breathing is a sufficient condition for being alive.
70. Breathing is necessary for being alive.
71. Being pregnant is sufficient for being a woman.
72. Being pregnant is necessary for being a woman.

73–78: Converse, Inverse, and Contrapositive. Write the converse, inverse, and contrapositive of each of the following conditional propositions. Of the four variations, state which pairs are logically equivalent.

73. If Tara owns a Cadillac, then she owns a car.
74. If the patient is alive, then the patient is breathing.
75. If Helen is the U.S. President, then she is a U.S. citizen.
76. If I am using electricity, then the lights are on.
77. If the engine is running, then there is gas in the tank.
78. If the polar ice caps melt, then the oceans will rise.

FURTHER APPLICATIONS

79–82: Famous Quotes. Rephrase the following quotations using one or more conditional statements (*if p, then q*).

79. "Without passion, you don't have energy. Without energy, you have nothing." —Donald Trump
80. "Change will not come if we wait for some other person or some other time." —Barack Obama.
81. "Even if you're flipping fries at McDonald's, if you are excellent, everybody wants to be in your line." —Oprah Winfrey
82. "Give me six hours to chop down a tree and I will spend the first four sharpening the axe." —Abraham Lincoln

83–87: Writing Conditional Propositions. Create your own example of a proposition that has the given property.

83. A true conditional proposition whose converse is false
84. A true conditional proposition whose converse is true
85. A true conditional proposition whose contrapositive is true
86. A false conditional proposition whose inverse is true
87. A false conditional proposition with a true converse
88. **Alimony Tax Laws.** The federal tax policy on alimony payments is as follows:
    (1) Alimony you receive after you remarry is taxable if the payer did not know you had remarried.
    (2) Alimony you receive after you remarry is not taxable if the payer did know you had remarried.
    (3) Alimony you pay is never deductible by the payer.

    Rephrase these three statements as conditional propositions.

89–92: Necessary and Sufficient. Write the following conditional statements in two forms: (a) *p* is sufficient for *q* and (b) *q* is necessary for *p*.

89. "If you believe, then you can achieve." —Tupac Shakur
90. "You can't climb the ladder of success with your hands in your pockets." —Arnold Schwarzenegger
91. "Once you have six children, you're committed." —Angelina Jolie
92. "If you need both of your hands for whatever it is you're doing, then your brain should probably be in on it too." —Ellen DeGeneres

93–98: Logical Equivalence. Consider the following pairs of statements in which *p, q, r* and *s* represent propositions. Make a truth table for each statement of the pair, and determine whether the two statements are logically equivalent.

93. *not (p and q); (not p) or (not q)*
94. *not (p or q); (not p) and (not q)*
95. *not (p and q); (not p) and (not q)*
96. *not (p or q); (not p) or (not q)*
97. *(p and q) or r; (p or r) and (p or q)*
98. *(p or q) and r; (p and r) or (q and r)*

99. **Logical Equivalence.** Explain why the contrapositive is called the inverse of the converse. Is the contrapositive also the converse of the inverse?

IN YOUR WORLD

100. **Logical *Or*.** Find a news article or advertisement in which the connector *or* is used. Is it used in an inclusive or exclusive sense? Explain its meaning in the given context.

101. **Multiple Negation.** Find a news article or advertisement in which a double (or multiple) negation is used. Explain the meaning of the sentence in which it occurs.

102. **Conditional News.** Find a news article or editorial in which a conditional statement is used. If necessary, rephrase the statement in standard *if p, then q* form. Comment on the truth of the individual propositions *p* and *q* and of the conditional proposition *if p, then q*.

UNIT 1C Sets and Venn Diagrams

We've seen that propositions come in many forms. The only general requirement is that a proposition must be a sentence that makes a clear claim (assertion or denial). In this unit, we will focus on propositions that claim a relationship between two categories of things. For example, the proposition *All whales are mammals* makes the claim that the category *whales* is entirely contained within the category *mammals*.

Propositions of this type are most easily studied with the aid of two key ideas. The first is the idea of a *set*, which is really just another word for a collection. The second is the idea of a *Venn diagram*, which is a simple way of visualizing relationships among sets. Both ideas are very useful for organizing information and hence are important tools of critical thinking.

Relationships Among Sets

A **set** is a collection of objects, living or nonliving. The **members** of a set are the specific objects within it. For example:

- Set: *days of the week*
  Members: Sunday, Monday, Tuesday, Wednesday, Thursday, Friday, Saturday
- Set: *American military services*
  Members: Army, Navy, Air Force, Marine Corps
- Set: *Academy Award–winning actresses*
  Members: all the individual actresses who have won at least one Academy Award

BY THE WAY

The four military services are organized under the U.S. Department of Defense. The Coast Guard is often grouped with these services, but it is administered by the Department of Homeland Security.

Set Notation

Sets are often described by listing their members within a pair of braces, { }, with each member separated from the next by a comma. For example, the set *American military services* can be written as

American military services: {Army, Navy, Air Force, Marine Corps}

Some sets have so many members that it would be difficult or impossible to list all of them. In that case, we can use three dots, . . . , to indicate that the list continues in the same basic manner. (The three dots are called an ellipsis, but most people just say "dot-dot-dot" when reading them.) If the dots come at the end of a list, they indicate that the list continues indefinitely. For example, we can write the set dog breeds as

dog breeds: {Rottweiler, German shepherd, poodle, . . .}

The three dots indicate that the list continues as you would expect—in this case, with the names of all other dog breeds. It does not matter how many members you list, as long as the pattern is clear. Listing three members, as in the dog breed list, is usually enough to make the point.

In other cases, the dots appear in the middle of the list, representing members that were not explicitly written. For example, we can use dots to allow us to write the set of all capital letters without having to write every single letter within the braces:

$$\textit{capital letters}: \{A, B, C, \ldots, Z\}$$

Although it is less common, the three dots can also be placed at the beginning of a list. For example, we can write the set of negative integers as

$$\textit{negative integers}: \{\ldots, -3, -2, -1\}$$

The whole is more than the sum of its parts.

—Aristotle, *Metaphysica*

Here, the dots indicate that the list continues to the left, to ever smaller (more negative) numbers.

Definitions

A **set** is a collection of objects; the individual objects are the **members** of the set. We write sets by listing their members within a pair of braces, { }. If there are too many members to list, we use three dots, ..., to indicate a continuing pattern.

Think About It How would you use braces to write the set of students in your mathematics class? How about the set of countries you have visited? Describe one more example of a set that affects you personally, and write it with braces notation.

EXAMPLE 1 Set Notation

Use braces to write the contents of each of the following sets:

a. the set of countries larger in land area than the United States
b. the set of years of the Cold War, generally taken to have started in 1945 and ended in 1991
c. the set of natural numbers greater than 5

Solution

a. *countries larger in land area than the United States*: {Russia, Canada}.
b. *years of the Cold War*: {1945, 1946, 1947, ..., 1991}. Here, the dots indicate that the list includes all the years between 1947 and 1991, even though they are not listed explicitly.
c. *natural numbers greater than 5*: {6, 7, 8, ...}. Here, the dots indicate that the list continues to ever-larger integers.

▶ Now try Exercises 29–36.

Illustrating Relationships with Venn Diagrams

HISTORICAL NOTE

John Venn was an ordained priest who wrote books on logic, statistics, and probability. He also wrote a history of Cambridge University, where he was both a student and a teacher. Though well respected for many of his contributions to logic, he is best known for the diagrams that bear his name.

The English logician John Venn (1834–1923) invented a simple visual way of describing relationships among sets. His diagrams, now called **Venn diagrams**, use circles to represent sets. Venn diagrams are fairly intuitive and best learned through examples.

Consider the sets *whales* and *mammals*. Because all whales are mammals, we say that the set *whales* is a **subset** of the set *mammals*. We represent this relationship in a Venn diagram by drawing a circle to represent the set *whales* inside a circle representing the set *mammals* (Figure 1.11). Notice that the diagram illustrates only the relationship between the sets. The sizes of the circles do not matter.

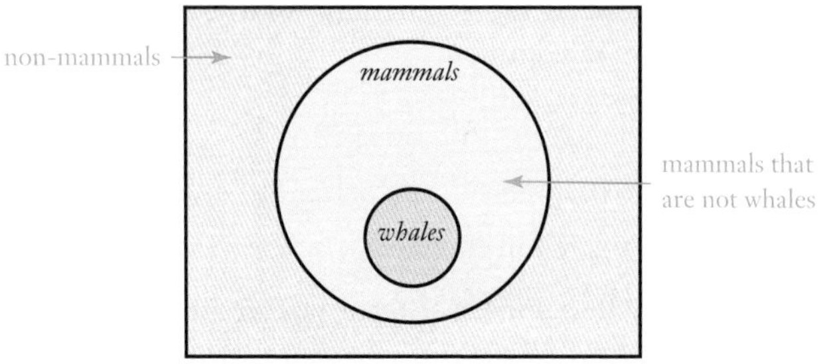

FIGURE 1.11 The set *whales* is a subset of the set *mammals*.

Technical Note
Some logicians distinguish between *Venn diagrams* and *Euler diagrams* (named for the mathematician Leonhard Euler, pronounced "oiler"), depending on the precise way the circles are used. In this book, we use the term Venn diagrams for all circle diagrams.

The circles are enclosed by a rectangle, so this diagram has three regions:

- The inside of the *whales* circle represents all whales.
- The region outside the *whales* circle but inside the *mammals* circle represents mammals that are not whales (such as cows, bears, and people).
- The region outside the *mammals* circle represents non-mammals; from the context, we interpret this region as representing animals (or living things) that are not mammals, such as birds, fish, and insects.

Next, consider the sets *dogs* and *cats*. A pet can be either a dog or a cat, but not both. We therefore draw the Venn diagram with circles that do not touch, and we say that *dogs* and *cats* are **disjoint sets** (Figure 1.12). Again, we enclose the circles in a rectangle. This time, the context suggests that the region outside both circles represents pets that are neither dogs nor cats, such as birds and hamsters.

For our last general case, consider the sets *nurses* and *women*. As shown in Figure 1.13, these are **overlapping sets** because it is possible for a person to be both a woman and a nurse. Because of the overlapping region, this diagram has four regions:

- The overlapping region represents people who are both women and nurses—that is, female nurses.
- The non-overlapping region of the *nurses* circle represents nurses who are not women—that is, male nurses.
- The non-overlapping region of the *women* circle represents women who are not nurses.
- From the context, we interpret the region outside both circles as representing people who are neither nurses nor women—that is, men who are not nurses.

Note that the sizes of the regions are not important. For example, the small size of the overlapping region in Figure 1.13 does *not* imply that female nurses are less common

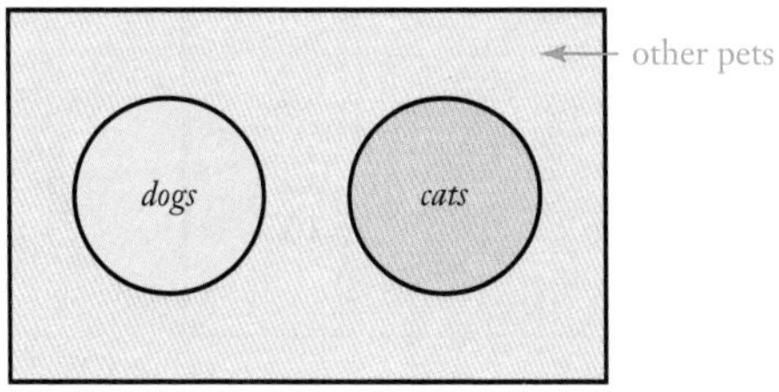

FIGURE 1.12 The set *dogs* is disjoint from the set *cats*.

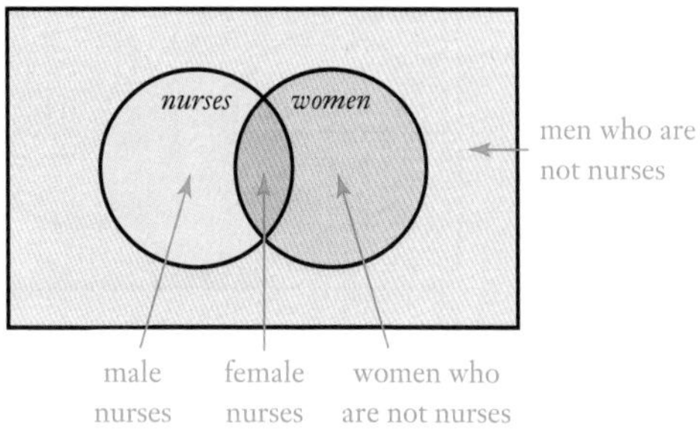

FIGURE 1.13 The sets *nurses* and *women* are overlapping.

than male nurses. In fact, it is possible that some of the regions might have no members at all. For example, imagine that Figure 1.13 represents a clinic with only four nurses, all of whom are male. In that case, the overlapping region, which represents female nurses, would have no members. Speaking more generally, we use overlapping circles whenever two sets might have members in common.

Technical Note

Mathematically, there is a fourth way that two sets can be related: They can be *equal,* meaning that they have precisely the same members.

Set Relationships and Venn Diagrams

Two sets *A* and *B* may be related in three basic ways:

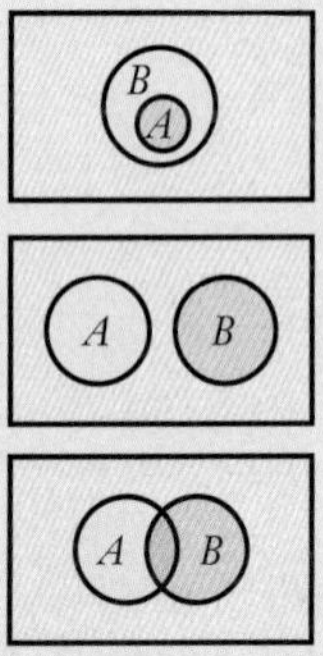

- *A* may be a **subset** of *B* (or vice versa), meaning that all members of *A* are also members of *B*. The Venn diagram for this case shows the circle for *A* inside the circle for *B*.
- *A* may be **disjoint** from *B*, meaning that the two sets have no members in common. The Venn diagram for this case consists of separated circles that do not touch.
- *A* and *B* may be **overlapping** sets, meaning that the two sets share some of the same members. The Venn diagram for this case consists of two overlapping circles. We also use overlapping circles for cases in which the two sets might share common members.

EXAMPLE 2 Venn Diagrams

Describe the relationship between the given pairs of sets, and draw a Venn diagram showing this relationship. Interpret all the regions of the Venn diagram.

a. *Democrats* and *Republicans* (party affiliations)
b. *Nobel Prize winners* and *Pulitzer Prize winners*

Solution

a. A person can be registered for only one political party, so the sets *Democrats* and *Republicans* are disjoint. Figure 1.14a shows the Venn diagram. The region outside both circles represents people who are neither Democrats nor Republicans—that is, people who are registered for other political parties, who are independent, or who are not registered.

b. It is possible for a person to win both a Nobel Prize and a Pulitzer Prize, so these are overlapping sets. Figure 1.14b shows the Venn diagram. The region outside both circles represents people who have not won either prize, which means most of humanity.

BY THE WAY

Among the people who have won both a Nobel Prize (in Literature) and a Pulitzer Prize are Toni Morrison, Pearl Buck, Ernest Hemingway, William Faulkner, John Steinbeck, and Saul Bellow.

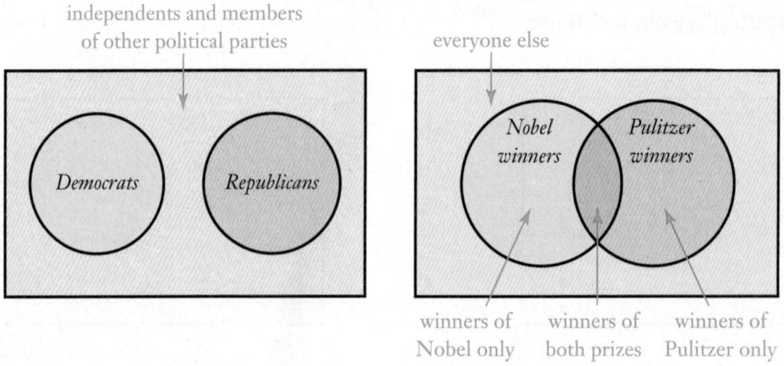

(a) (b)

FIGURE 1.14

▸ Now try Exercises 37–42.

EXAMPLE 3 Sets of Numbers

Draw a Venn diagram showing the relationships among the sets of natural numbers, whole numbers, integers, rational numbers, and real numbers. Where are irrational numbers found in this diagram? (If you've forgotten the meanings of these number sets, see the following Brief Review.)

Solution Natural numbers are also whole numbers, which means that the natural numbers are a subset of the whole numbers. Similarly, whole numbers are integers, so the whole numbers are a subset of the integers. The integers, in turn, are a subset of the

Brief Review Sets of Numbers

In mathematics, we commonly work with the following important sets of numbers.

The set of **natural numbers** (or *counting numbers*) is

$$\{1, 2, 3, \dots\}$$

We can represent the natural numbers on a *number line* with equally spaced dots beginning at 1 and continuing to the right forever.

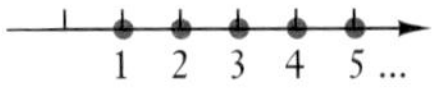

The set of **whole numbers** is the same as the set of natural numbers except it includes zero. Its members are

$$\{0, 1, 2, 3, \dots\}$$

We can represent the whole numbers on a number line with equally spaced dots beginning at 0 and continuing to the right forever.

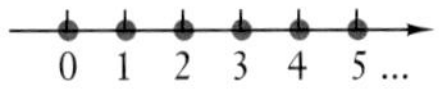

The set of **integers** includes the whole numbers and their negatives. Its members are

$$\{\dots, -3, -2, -1, 0, 1, 2, 3, \dots\}$$

On the number line, the integers extend forever both to the left and to the right.

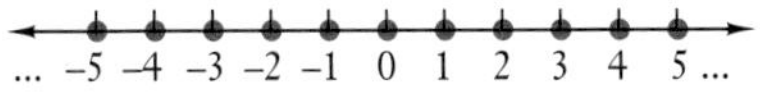

The set of **rational numbers** includes the integers and the fractions that can be made by dividing one integer by another, as long as we don't divide by zero. (The word *rational* refers to a *ratio* of integers.) In other words, rational numbers can be expressed in the form

$$\frac{x}{y}, \text{ where } x \text{ and } y \text{ are integers and } y \neq 0$$

(Recall that the symbol $\neq$ means "is not equal to.")

When expressed in decimal form, rational numbers are either terminating decimals with a finite number of digits (such as 0.25, which is $\frac{1}{4}$) or repeating decimals in which a pattern repeats over and over (such as $0.333\dots$, which is $\frac{1}{3}$).

Irrational numbers are numbers that *cannot* be expressed in the form x/y, where x and y are integers. When written as decimals, irrational numbers neither terminate nor have a repeating pattern. For example, the number $\sqrt{2}$ is irrational because it cannot be expressed exactly in a form x/y; as a decimal, we can write it as $1.414213562\dots$, where the dots mean that the digits continue forever with no pattern. The number π is also an irrational number, which as a decimal is written $3.14159265\dots$.

The set of **real numbers** consists of both rational and irrational numbers; hence, it is represented by the entire number line. Each point on the number line has a corresponding real number, and each real number has a corresponding point on the number line. In other words, the real numbers are the integers and "everything in between." A few selected real numbers are shown on the number line below.

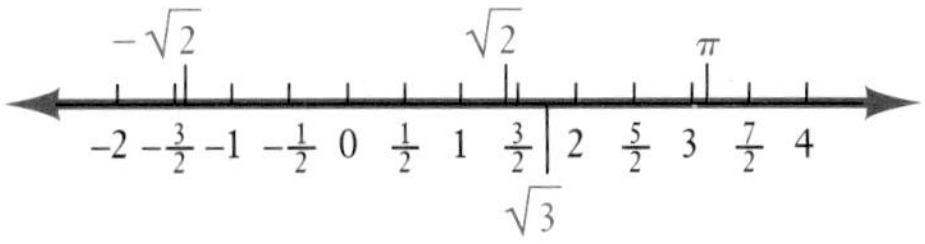

Examples:

- The number 25 is a natural number, which means it is also a whole number, an integer, a rational number, and a real number.
- The number -6 is an integer, which means it is also a rational number and a real number.
- The number $\frac{2}{3}$ is a rational number, which means it is also a real number.
- The number $7.98418\dots$ is an irrational number; the dots indicate that the digits continue forever with no particular pattern. It is also a real number.

▶ Now try Exercises 13–28.

rational numbers, and the rational numbers are a subset of the real numbers. Figure 1.15 shows these relationships with a set of nested circles. Because irrational numbers are real numbers that are not rational, they are represented by the region outside the rational numbers circle but inside the real numbers circle.

BY THE WAY

Are you wondering what goes outside the *real numbers* circle in Figure 1.15? The answer is numbers that are not real, called *complex numbers*. Complex numbers use the imaginary number $i = \sqrt{-1}$, which has no real number value. Complex and imaginary numbers are often used in science and engineering.

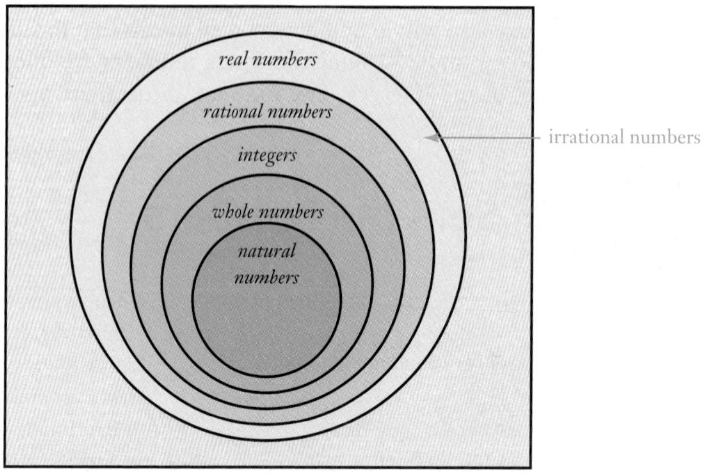

FIGURE 1.15

▸ Now try Exercises 43–44.

Categorical Propositions

Now that we have discussed general relationships between sets, we are ready to study propositions that make claims about sets. For example, the proposition *All whales are mammals* makes a very specific claim—namely, that the set *whales* is a subset of the set *mammals*. Propositions of this type are called **categorical propositions** because they claim a particular relationship between two categories or sets.

Like all propositions, categorical propositions must have the structure of a complete sentence. Categorical propositions also have another important general feature. Of the two sets in a categorical proposition, one set appears in the subject of the sentence and the other appears in the predicate. For example, in the proposition *All whales are mammals*, the set *whales* is the **subject set** and the set *mammals* is the **predicate set**. We usually use the letter S to represent the subject set and the letter P for the predicate set, so we can rewrite *All whales are mammals* as

All S are P, where S = *whales* and P = *mammals*.

Categorical propositions come in the following four standard forms.

The Four Standard Categorical Propositions

Form	Example	Subject Set (S)	Predicate Set (P)
All S are P.	All whales are mammals.	*whales*	*mammals*
No S are P.	No fish are mammals.	*fish*	*mammals*
Some S are P.	Some doctors are women.	*doctors*	*women*
Some S are not P.	Some teachers are not men.	*teachers*	*men*

Venn Diagrams for Categorical Propositions

We can use Venn diagrams to make visual representations of categorical propositions. Figures 1.16 to 1.19 show the diagrams for each of the four examples of propositions shown in the preceding box. Note the following key features of these diagrams:

- Figure 1.16 shows the Venn diagram for the proposition *All whales are mammals*. The circle for the set S = *whales* is drawn inside the circle for the set P = *mammals*, because S is a subset of P in this case. All propositions of the form *all S are P* have the same basic Venn diagram.
- Figure 1.17 shows the Venn diagram for the proposition *No fish are mammals*. In this case, the sets S = *fish* and P = *mammals* have no common members (they are disjoint). The Venn diagram therefore shows two separated circles.
- Figure 1.18 shows the Venn diagram for the proposition *Some doctors are women*, which requires overlapping circles for S = *doctors* and P = *women*. However, the overlapping circles alone do not represent the proposition, because they don't tell us which regions have members. In this case, the proposition asserts that there are some people (at least one) who are both doctors and women. We indicate this fact by putting an X in the overlapping region of the diagram. (Note that the proposition does not tell us whether the non-overlapping regions contain any members).
- Figure 1.19 shows the Venn diagram for the proposition *Some teachers are not men*, which also requires overlapping circles. This proposition asserts that there are some people (at least one) who are in the S = *teachers* circle but not in the P = *men* circle. Therefore, we put an X in the non-overlapping region of the *teachers* circle.

Mathematics is, in its way, the poetry of logical ideas.
—Albert Einstein

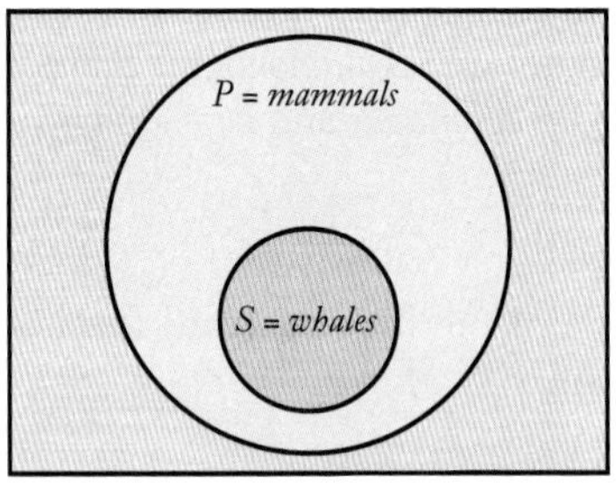

FIGURE 1.16 The Venn diagram for *all S are P*.

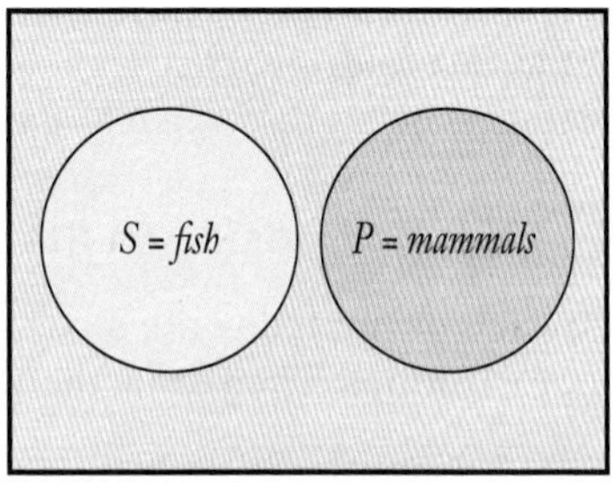

FIGURE 1.17 The Venn diagram for *no S are P*.

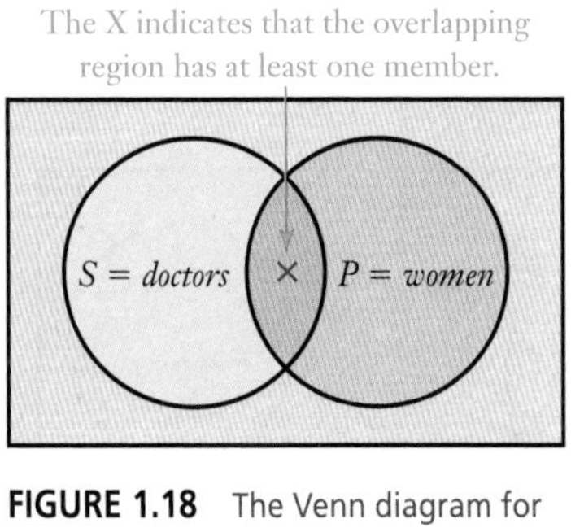

FIGURE 1.18 The Venn diagram for *some S are P*.

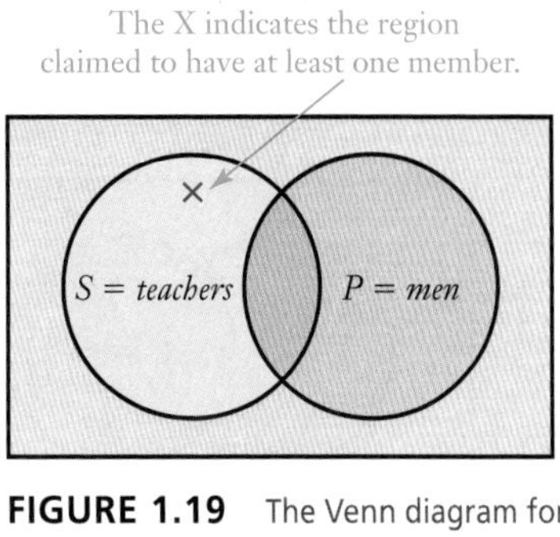

FIGURE 1.19 The Venn diagram for *some S are not P*.

EXAMPLE 4 Interpreting Venn Diagrams

Answer the following questions based only on the information provided in the Venn diagrams. That is, don't consider any prior knowledge you have about the sets.

a. Based on Figure 1.16, can you conclude that some mammals are not whales?
b. Based on Figure 1.17, is it possible that some mammals are fish?
c. Based on Figure 1.18, is it possible that all doctors are women?
d. Based on Figure 1.19, is it possible that no men are teachers?

Solution

a. No. The diagram shows only that all members of the set S = *whales* are also members of the set P = *mammals*, but it does not tell us whether there are any mammals that lie outside the *whales* circle.
b. No. The sets for S = *fish* and P = *mammals* are disjoint, meaning they cannot have common members.
c. Yes. The X in the overlap region tells us that some women are definitely doctors. But there are no Xs elsewhere, so it is possible that all other regions have no members, in which case all women would be doctors.
d. Yes. The X is outside the P = *men* circle, so we have no information on whether any men are teachers. ▶ Now try Exercises 45–46.

Think About It The principal of an elementary school states that at her school some teachers are not men. Can you conclude that some of the teachers are men? Why or why not? Can you conclude that none of the teachers is a man? Why or why not?

Putting Categorical Propositions in Standard Form

Many statements in everyday speech make claims about relationships between two categories, but don't look precisely like one of the four standard forms for categorical propositions. It's often useful to rephrase such statements in one of the standard forms. For example, the statement *All diamonds are valuable* can be rephrased to read *All diamonds are things of value*. Now the statement has the form *all S are P*, where S = *diamonds* and P = *things of value*.

EXAMPLE 5 Rephrasing in Standard Form

Rephrase each of the following statements in one of the four standard forms for categorical propositions. Then draw the Venn diagram.

a. Some birds can fly.
b. Elephants never forget.

Solution

a. *Some birds can fly* can be rephrased as *Some birds are animals that can fly*. It now has the form *some S are P* where S = *birds* and P = *animals that can fly*. Figure 1.20a shows the Venn diagram, with an X in the overlapping region to indicate the claim that there *are* some birds that can fly.

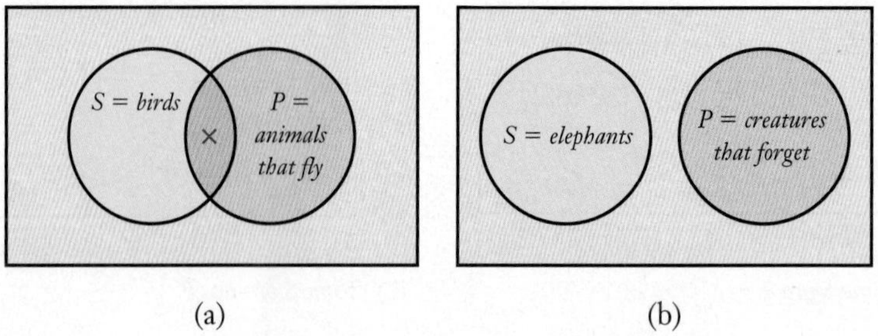

FIGURE 1.20

b. *Elephants never forget* can be rephrased as *No elephants are creatures that forget*. This proposition has the form *no S are P* shown in Figure 1.20b, where $S = elephants$ and $P = creatures\ that\ forget$. ▸ Now try Exercises 47–52.

Venn Diagrams with Three Sets

Venn diagrams are particularly useful for dealing with three sets that may overlap one another. For example, suppose you are conducting a study to learn how teenage employment rates differ between boys and girls and between honor students and others. For each teenager in your study, you need to record the answers to these three questions:

- Is the teenager a boy or a girl?
- Is the teenager an honor student or not?
- Is the teenager employed or not?

Figure 1.21 shows a Venn diagram that can help you organize this information. Notice that it has three circles, representing the sets *boys*, *honor students*, and *employed*. Because the circles overlap one another, they form a total of eight regions (including the region outside all three circles). Be sure that you understand the labels shown for each region.

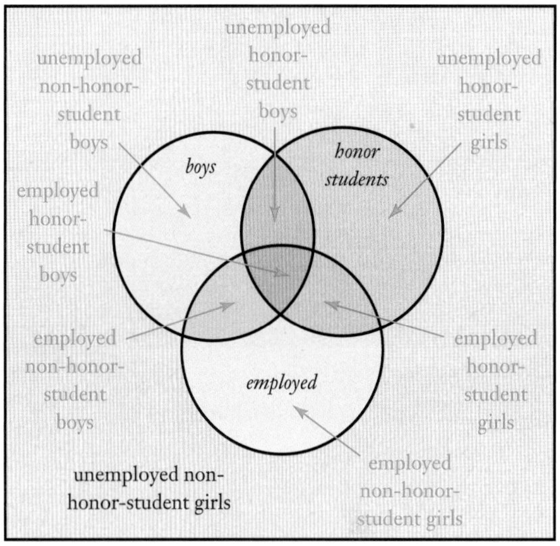

FIGURE 1.21 A Venn diagram for three overlapping sets has $2^3 = 8$ regions.

Think About It Could the information in Figure 1.21 still be presented if we replaced the *boys* circle in the Venn diagram with a *girls* circle (but the other two circles stayed the same)? If so, how, and if not, why not? Could the information be presented if the three circles were *girls*, *boys*, and *unemployed*? Explain.

EXAMPLE 6 Recording Data in a Venn Diagram

You hire an assistant to help you with the study of teenage employment described above. He focuses on a small group of teenagers who are all enrolled in the same school. He reports the following facts about this group:

- Some of the honor-student boys are unemployed.
- Some of the non-honor-student girls are employed.

Put Xs in the appropriate places in Figure 1.21 to indicate the regions that you can be sure have members. Based on this report, do you know whether any of the school's honor-student girls are unemployed? Why or why not?

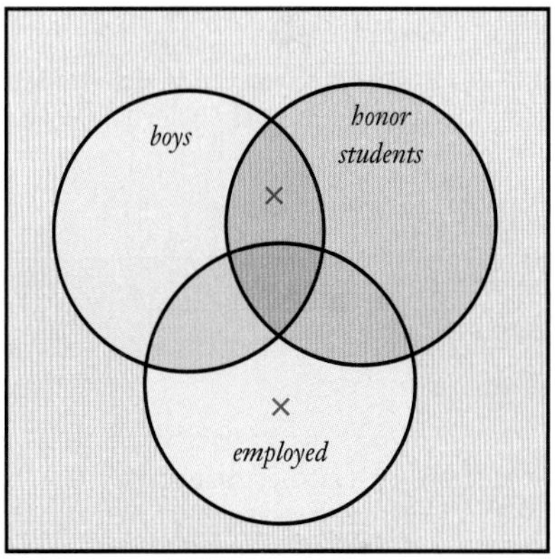

FIGURE 1.22 The Xs mark regions known to have members.

Solution Figure 1.22 shows the Xs in the correct regions of the diagram (notice how those regions are labeled in Figure 1.21). The region corresponding to unemployed honor-student girls is the pink region of the honor students circle. There is no X in this region because the given information does not tell us whether this region has members. Therefore, among the teenagers in the school, we do not know whether any honor-student girls are unemployed. ▸ Now try Exercises 53–55.

EXAMPLE 7 Color Monitors

Color television and computer monitors make all the colors you see by combining *pixels* (short for "picture elements") that display just three colors: red, green, and blue. In pairs, these combinations of colors give the following results (with the two colors at equal strength):

Combination	Result
Red-green	Yellow
Red-blue	Purple (or magenta)
Blue-green	Light blue (or cyan)

White is made by combining all three colors, and black is made by using none of the colors. Draw a Venn diagram to represent all this color information. (On older, low-resolution televisions, you can see the individual red, blue, and green dots with a magnifying glass.)

Solution The Venn diagram in Figure 1.23 represents the color combinations with three overlapping circles. The basic colors (red, green, and blue) appear in regions where there is no overlap. The red-green, red-blue, and blue-green combinations appear in regions where two circles overlap. White appears in the central region, where all three colors are present, while black appears outside all three circles, where no colors are present. A real monitor can make a much wider range of colors by varying the relative strengths of the colors.

▶ Now try Exercises 56–58.

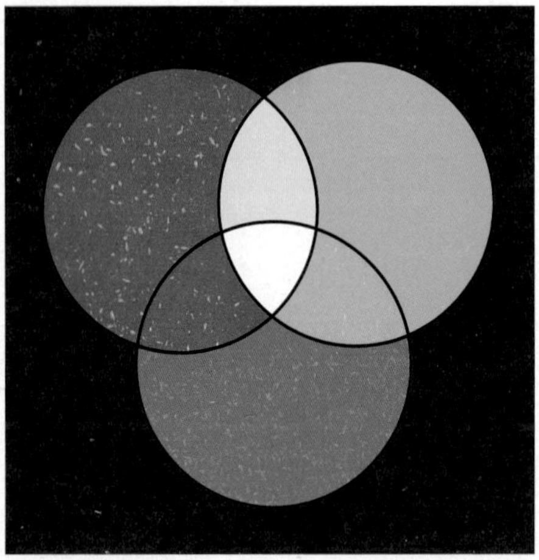

FIGURE 1.23 The color combinations possible from red, green, and blue (in equal strength).

Venn Diagrams with Numbers

So far, we have used Venn diagrams only to describe relationships, such as whether two sets overlap or whether overlapping sets share common members. Venn diagrams can be even more useful when we add specific information, such as the number of members in each set or overlapping region. The following examples illustrate some of the ways in which Venn diagrams can be used with numbers.

EXAMPLE 8 Smoking and Pregnancy

Consider the study summarized in Table 1.1, which was designed to learn whether a pregnant mother's status as a smoker or nonsmoker affects whether she delivers a baby with a low or a normal birth weight. This table is an example of what we call a **two-way table**, because it shows two variables: *mother's smoking status* and *baby's birth weight status*. The table caption tells us that the study involved 350 births, and the four data cells tell us the number of births with each of the four possible combinations of the two variables.

TABLE 1.1 Distribution of 350 Births by Birth Weight Status and Mother's Smoking Status

		Baby's Birth Weight Status	
		Low Birth Weight	Normal Birth Weight
Mother's Smoking Status	Smoker	18	132
	Nonsmoker	14	186

Source: Data from U.S. National Center for Health Statistics.

BY THE WAY

The reason that a mother's smoking can lead to low birth weight and other detrimental health effects in babies has been tied specifically to nicotine. This means that e-cigarettes are probably equally damaging when used during pregnancy, although research is still underway.

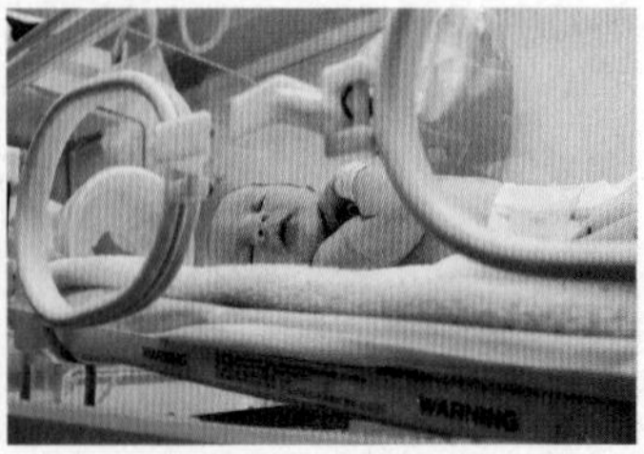

a. Make a list summarizing the four key facts shown in the table.
b. Draw a Venn diagram to represent the table data.
c. Based on the Venn diagram, briefly summarize the results of the study.

Solution

a. The four cells in the table tell us the following key facts:
- 18 babies with low birth weight were born to smoking mothers.
- 132 babies with normal birth weight were born to smoking mothers.
- 14 babies with low birth weight were born to nonsmoking mothers.
- 186 babies with normal birth weight were born to nonsmoking mothers.

b. Figure 1.24 shows one way of making the Venn diagram. The circles represent the sets *smoking mothers* and *low birth weight babies*. The numbers in the four regions correspond to the data in Table 1.1.

c. The Venn diagram makes it easy to see how smoking affected babies in the study. Notice that normal birth weight babies were much more common than low birth weight babies among both smokers (132 normal birth weight and 18 low birth weight) and nonsmokers (186 normal birth weight and 14 low birth weight). However, it is more useful to look at the *proportions*. The smoking mothers had a total of 132 + 18 = 150 babies, so the 18 low birth weight babies represent a proportion of 18/150, or 12%. The nonsmoking mothers had a total of 186 + 14 = 200 babies, so the 14 low birth weight babies represent a proportion of 14/200, or 7%. The significantly higher proportion of low birth weight babies among smoking mothers suggests that smoking increases the risk of having a low birth weight baby, a fact that has been borne out by careful statistical analysis of this and other studies.

▶ Now try Exercises 59–62.

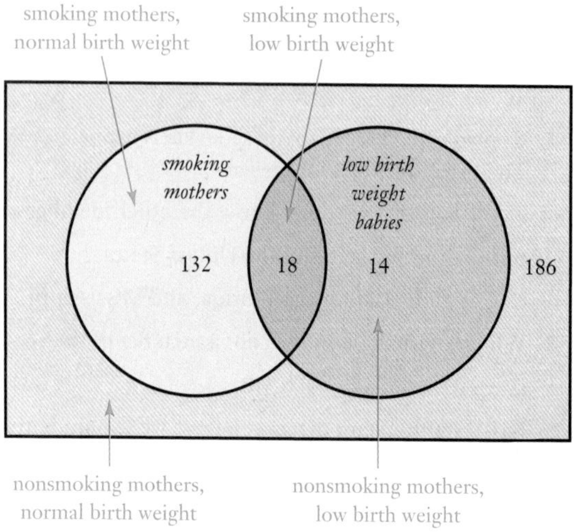

FIGURE 1.24 Venn diagram for the data in Table 1.1.

Think About It Explain why the Venn diagram in Example 8 also could be drawn with circles for the sets *nonsmoking mothers* and *normal birth weight babies*. Draw the diagram for this case, and put the four numbers from Table 1.1 in the right places.

EXAMPLE 9 Three Sets with Numbers—Blood Types

Human blood is often classified according to whether three antigens, A, B, and Rh, are present or absent. Blood type is stated first in terms of the antigens A and B: Blood containing only A is called type A, blood containing only B is called type B, blood containing both A and B is called type AB, and blood containing neither A nor B is called type O. The presence or absence of Rh is indicated by adding the word "positive" (present) or "negative" (absent) or the equivalent symbol (+ or −). Table 1.2 shows the eight blood types that result and the percentage of people with each type in the U.S. population. Draw a Venn diagram to illustrate these data.

TABLE 1.2 Blood Types in U.S. Population

Blood Type	Percentage of Population
A positive	34%
B positive	8%
AB positive	3%
O positive	35%
A negative	8%
B negative	2%
AB negative	1%
O negative	9%

Solution We can think of the three antigens as three sets *A*, *B*, and *Rh (positive)*. We therefore draw a Venn diagram with three overlapping circles. Figure 1.25 shows the eight regions, each labeled with its blood type and percentage of the population. For example, the central region corresponds to the presence of all three antigens (AB positive), so it is labeled with 3%. You should check that all eight regions are labeled according to the data from Table 1.2.

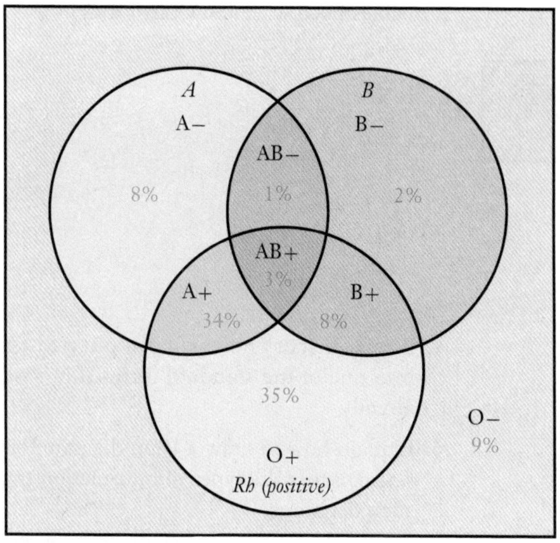

FIGURE 1.25 Venn diagram for blood types in the U.S. population.

▶ Now try Exercises 63–66.

HISTORICAL NOTE

Human blood groups were discovered in 1901 by Austrian biochemist Karl Landsteiner. In 1909, he classified the groups he had discovered as A, B, AB, and O. He also showed that blood transfusions could be done successfully if the donor and recipient were of the same blood type. For this work, he was awarded the 1930 Nobel Prize in Medicine.

Quick Quiz 1C

Choose the best answer to each of the following questions. Explain your reasoning with one or more complete sentences.

1. Consider the set {Alabama, Alaska, Arizona, . . . , Wyoming}. The three dots (. . .) represent
   a. the fact that we don't know the other members of the set.
   b. the other 46 states of the United States.
   c. Colorado, California, Florida, and Mississippi.
2. Which of the following is not a member of the set of integers?
   a. -107 b. 481 c. $3\frac{1}{2}$
3. Based on the Venn diagram below, we conclude that

   a. *C* is a subset of *D*. b. *D* is a subset of *C*.
   c. *C* is disjoint from *D*.
4. Suppose that *A* represents the set of all cats and *B* represents the set of all dogs. The correct Venn diagram for the relationship between these sets is

a.
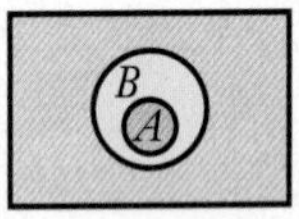

b.
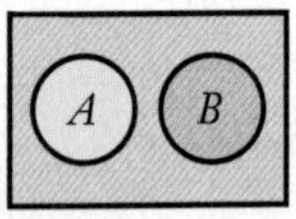

c.
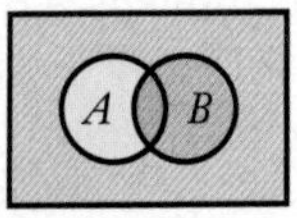

5. Suppose that *A* represents the set of all apples and *B* represents the set of all fruit. The correct Venn diagram for the relationship between these sets is

a.
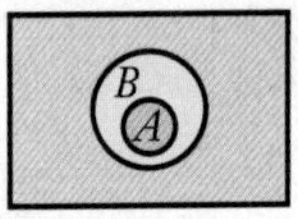

b.
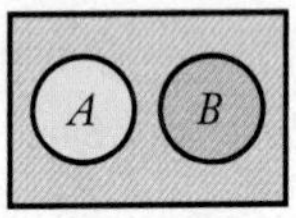

c.
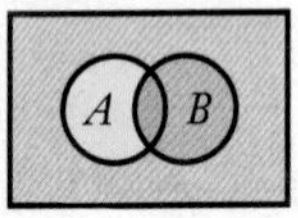

6. Suppose that *A* represents the set of all high school cross-country runners and *B* represents the set of all high school swimmers. The correct Venn diagram for the relationship between these sets is

a.
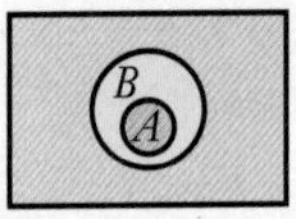

b.
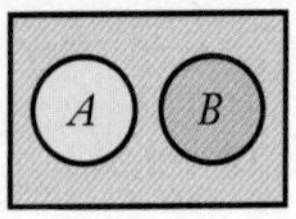

c.
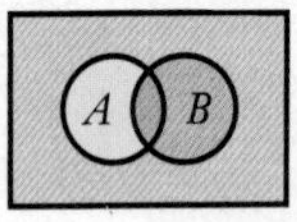

7. In the Venn diagram below, the X tells us that

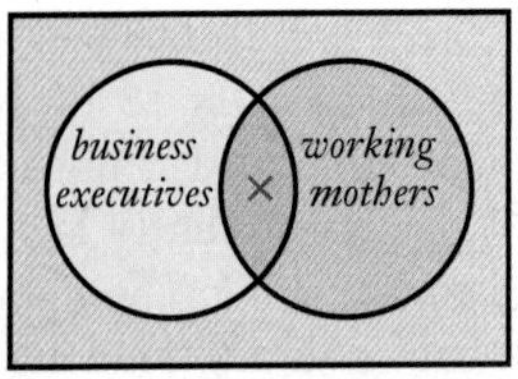

   a. some working mothers are business executives.
   b. no working mothers are business executives.
   c. some working mothers are not business executives.
8. The region with the X in the Venn diagram below represents

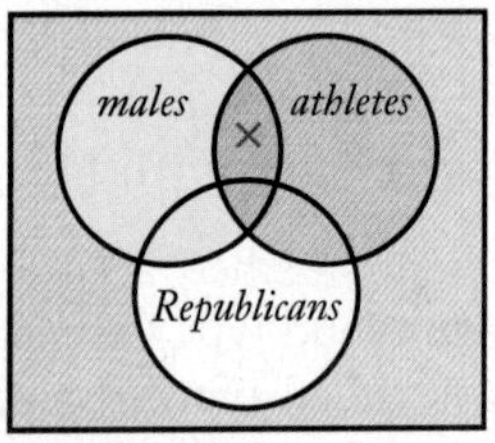

   a. male Republican athletes.
   b. male Republicans who are not athletes.
   c. male athletes who are not Republicans.
9. Consider again the Venn diagram from Question 8. The central region of the diagram represents people who are
   a. male and Republican and athletes.
   b. male or Republican or athletes.
   c. neither male nor Republican nor athletes.
10. Look at the data in Table 1.1 (p. 34). The total number of babies born with a low birth weight was
   a. 14.
   b. 18.
   c. 32.

Exercises 1C

REVIEW QUESTIONS

1. What is a *set*? Describe the use of braces for listing the members of a set.
2. What is a *Venn diagram*? How do we show that one set is a subset of another in a Venn diagram? How do we show disjoint sets? How do we show overlapping sets?
3. List the four standard categorical propositions. Give an example of each type, and draw a Venn diagram for each of your examples.
4. Briefly discuss how you can put a categorical proposition into one of the standard forms if it is not in such a form already.
5. Explain how to draw a Venn diagram for three overlapping sets. Discuss the types of information that can be shown in such diagrams.
6. Explain how to read a table such as Table 1.1 and how to show the information in a Venn diagram.

DOES IT MAKE SENSE?

Decide whether each of the following statements makes sense (or is clearly true) or does not make sense (or is clearly false). Explain your reasoning.

7. The people who live in Chicago form a subset of those who rent apartments in Chicago.
8. All jabbers are wocks, so there must be no wocks that are not jabbers.
9. I counted an irrational number of students in my statistics class.
10. I surveyed my class to find out whether students rode a bike on campus or not. Then I made a Venn diagram with one circle (inside a rectangle) to summarize the results.
11. My professor asked me to draw a Venn diagram for a categorical proposition, but I couldn't do it because the proposition was clearly false.
12. I used a Venn diagram with three circles to show how many students on campus are vegetarians, Republicans, and/or women.

BASIC SKILLS & CONCEPTS

13–28: Classifying Numbers. Choose the first set in the list *natural numbers, whole numbers, integers, rational numbers*, and *real numbers* that contains each of the following numbers.

13. 888	14. -23	15. $3/4$
16. $-6/5$	17. 3.414	18. 0
19. π	20. $\sqrt{8}$	21. -45.12
22. $\sqrt{98}$	23. $\pi/4$	24. $-34/19.2$
25. $-123/79$	26. -923.66	27. $\pi/129$
28. 93,145,095		

29–36: Set Notation. Use set notation (braces) to write the members of the following sets, or state that the set has no members. You may use three dots (...) to indicate patterns.

29. The dates of July
30. The odd numbers between and including 23 and 35
31. The states that share a border with the state of Mississippi
32. Every third number between 6 and 25, beginning with 6
33. The perfect squares between 10 and 40
34. The states that begin with the letter K
35. Even numbers between 2 and 35 that are multiples of 3
36. The vowels of the English alphabet

37–44: Venn Diagrams for Two Sets. Draw Venn diagrams with two circles showing the relationship between the following pairs of sets. Provide an explanation of the diagram you draw.

37. teachers and women
38. cage fighters and red-headed people
39. shirts and clothing
40. airliners and automobiles
41. poets and plumbers
42. women and American presidents
43. teenagers and octogenarians
44. novels and mysteries

45–52: Categorical Propositions. For the given categorical propositions, do the following.

a. If necessary, rephrase the statement in standard form.

b. State the subject and predicate sets.

c. Draw a Venn diagram for the proposition, and label all regions of the diagram.

d. Based only on the Venn diagram (not on any other knowledge you have), answer the question that follows each proposition.

45. All kings are men. Can you conclude that some men are not kings?
46. No carrots are fruit. Is it possible that some carrots are fruit?
47. Some surgeons are fishermen. Can you conclude that some fishermen are not surgeons?
48. Every fish can swim. Can you conclude that some swimmers are not fish?
49. Monks don't swear. Is it possible that some swearers are monks?
50. Some days are Tuesdays. Based on this proposition, can you conclude that some days are not Tuesdays?
51. Some sharpshooters are not men. Is it possible that at least one sharpshooter is a man?
52. Some shortstops are blonds. Can you conclude that there are red-headed shortstops?

53–58: Venn Diagrams for Three Sets. Draw Venn diagrams with three overlapping circles (eight regions) for the following groups of three sets. Describe the members of each region, or state that a region has no members.

53. women, Republicans, and chefs
54. hockey players, figure skaters, and men
55. poets, playwrights, painters
56. oceans, bodies of salt water, and bodies of fresh water
57. words that begin with *t*, nouns, and words with fewer than 5 letters
58. teachers, swimmers, and tall people
59. **Two-Way Table.** An average of many different studies of handedness indicate that in a random sample of adults, 12 percent of men are left-handed and 9 percent of women are left-handed. (Assume the rest are right-handed.) Suppose you have a sample of 200 women and 150 men, in which the numbers of left-handed and right-handed people reflect the average percentages. Make a two-way table that shows the distribution you would find.
60. **Two-Way Table.** According to exit polls from the 2016 Presidential election, Donald Trump received approximately 52% of the vote among white women, 4% of the vote among black women, 62% of the vote among white men, and 13% of the vote among black men. Show these data in a two-way table.

61–62: Two-Circle Venn Diagram with Numbers. Use the Venn diagram to answer the following questions.

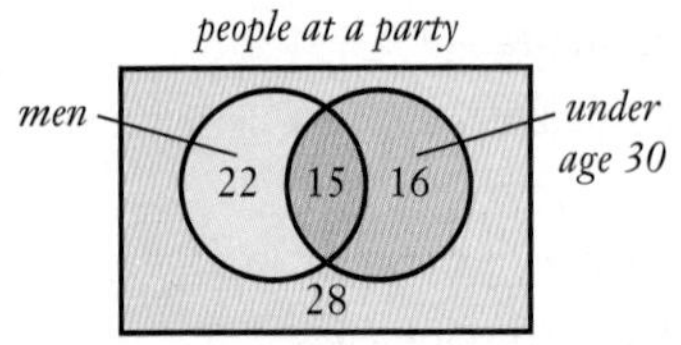

61. a. How many people at the party are under 30?
    b. How many women at the party are not under 30?
    c. How many men are at the party?
    d. How many people are at the party?

62. a. How many men at the party are not under 30?
    b. How many women are at the party?
    c. How many women at the party are under 30?
    d. How many people at the party are not under 30?

63–64: Three-Circle Venn Diagram with Numbers. Use the Venn diagram to answer the following questions.

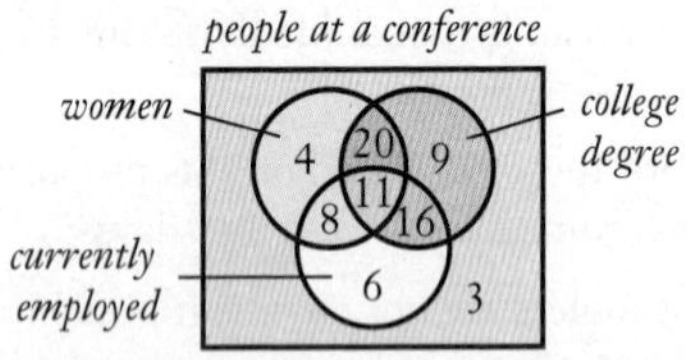

63. a. How many people at the conference are employed men with a college degree?
    b. How many people at the conference are unemployed women?
    c. How many people at the conference are employed women without a college degree?
    d. How many women are at the conference?

64. a. How many people at the conference are employed women with a college degree?
    b. How many people at the conference are employed men?
    c. How many people at the conference are unemployed men without a college degree?
    d. How many people are at the conference?

65. **Hospital Drug Use.** The following numbers of patients in a (hypothetical) hospital on a single day were taking antibiotics (A), blood pressure medication (BP), and/or pain medication (P):

A only	12	A and BP only	15
BP only	8	A and P only	24
P only	22	BP and P only	16
None	2	All three medications	20

    a. Draw a three-circle Venn diagram that summarizes the results in the table.
    b. How many patients took pain medication or blood pressure medication?
    c. How many patients took blood pressure medication but not antibiotics?
    d. How many patients took (at least) blood pressure medication?
    e. How many patients took only pain medication?
    f. How many patients took exactly two medications?

66. **Technology Survey.** A survey of 150 college students revealed the following data on the numbers of them who used smart phones, tablets, and/or laptops (based on data from the Pew Research Center).

Smart phone only	51	Smart phone and tablet only	35
Tablet only	8	Smart phone and laptop only	37
Laptop only	12	Tablet and laptop only	4
None	2	All three devices	1

    a. Draw a three-circle Venn diagram that summarizes the results of the survey.
    b. How many students used (at least) a smart phone?
    c. How many students used a tablet or a laptop?
    d. How many students used a smart phone and a laptop but not a tablet?
    e. How many students used exactly two devices?
    f. How many students used a laptop or a smart phone, but not a tablet?

FURTHER APPLICATIONS

67–70: Venn Diagram Analysis.

67. A movie critic reviewed 36 films; 12 were documentaries and 24 were feature films. She gave favorable reviews to 8 of the documentaries and unfavorable reviews to 6 of the feature films.
    a. Make a two-way table summarizing the reviews.
    b. Make a Venn diagram from the table in part (a).
    c. How many documentaries received unfavorable reviews?
    d. How many feature films received favorable reviews?

68. All runners who competed in a marathon were given a drug test after the race. Of the 24 who tested positive, 6 finished in the top 10. Thirty-five runners tested negative.
    a. Make a two-way table summarizing the test results.
    b. Make a Venn diagram from the table in part (a).
    c. How many runners who tested negative did not finish in the top 10?
    d. How many runners were tested?

69. One hundred people who grew up in either Nashville or San Francisco were surveyed to determine whether they preferred country music or blues ("both" and "neither" were not acceptable responses). Of those who grew up in San Francisco, 30 preferred blues and 19 preferred country. Of those who grew up in Nashville, 35 preferred country.
    a. Make a two-way table summarizing the survey.
    b. Make a Venn diagram from the table in part (a).
    c. How many Nashville respondents preferred blues?
    d. How many respondents preferred blues?

70. In a trial of a new allergy medicine, 75 people were given the medicine and 75 were given a placebo. Of those given the medicine, 55 showed improvement in their allergy symptoms. Of those given the placebo, 45 did not show improvement.

a. Make a two-way table summarizing the results.

b. Make a Venn diagram from the table in part (a).

c. How many people who received medicine did not improve?

d. How many people who received the placebo improved?

71–74: Two-Way Tables.

71–72: Election Results by Gender. Each table gives the approximate percentages of the national popular vote by gender for the two major-party candidates in recent presidential elections. In each case, draw a two-circle Venn diagram that represents the results. There is more than one correct way to choose your two circles. (The gender breakdown is based on exit polls, since gender is not known from individual ballots; totals don't add to 100% because of third-party candidates and people who chose not to answer the exit poll questions.)

71. 2012 Presidential Election

	Obama	Romney
Women voters	55%	44%
Men voters	45%	52%

72. 2016 Presidential Election

	Trump	Clinton
Women voters	42%	54%
Men voters	53%	41%

73. **Polygraph Accuracy.** Researchers Charles Honts (Boise State University) and Gordon Barland (Department of Defense Polygraph Institute) conducted a test of polygraph (lie detector) accuracy. The following table indicates the partial results when subjects were given a polygraph test and the researchers knew in advance whether they actually lied or actually did not lie.

	Subject did not lie.	Subject lied.	Total
Polygraph: lie		42	57
Polygraph: no lie	32		
Total		51	

a. Fill in the missing numbers in the table.

b. In what percentage of the cases did the polygraph correctly detect whether the subject actually lied or did not lie?

c. In what percentage of the cases did the polygraph fail to detect whether the subject actually lied or did not lie?

d. Based on these data, comment on the accuracy of polygraph testing.

74. **Driver Safety.** A study of texting and seat belt use when driving was conducted with high school students ("Texting while driving and other risky motor vehicle behaviors among U.S. high school students," O'Mally, Shults, and Eaton, *Pediatrics*, 131, 6). Partial results are shown in the table.

	Irregular seat belt use	Regular seat belt use	Total
Texted while driving	1737		3785
Did not text while driving		2775	
Total	3682		

a. Fill in the missing numbers in the table.

b. What percentage of students reported risky behavior in both categories (texting and irregular seat belt use)?

c. What percentage of students reported only texting while driving?

d. According to these data, comment on whether risky behavior in the two categories appears to be linked.

75–78: More Than Three Sets. Draw a Venn diagram that illustrates the relationships among the given sets. The diagram should have one circle for each set. A circle may lie entirely inside of other circles, may overlap other circles, or may be completely separate from other circles.

75. animals, house pets, dogs, cats, canaries

76. athletes, women, professional soccer players, amateur golfers, sedentary doctors

77. things that fly, birds, jets, hang gliders, eagles

78. painters, artists, musicians, pianists, violinists, abstract painters

IN YOUR WORLD

79. **Categorical Propositions.** Find at least three examples of categorical propositions in news articles or advertisements. State the sets involved in each proposition, and draw a Venn diagram for each proposition.

80. **Venn Diagrams in Your Life.** Describe a situation in your life that could be described or organized using a Venn diagram.

81. **Two-Way Table.** Find a news article or research report that can be summarized with a two-way table similar to Table 1.1. Draw a Venn diagram to represent the data in the table.

82. **State Politics.** Find out how many states have a Republican majority in the lower house of the state legislature and how many states have a Republican majority in the upper house (senate). Draw a Venn diagram to illustrate the situation.

83. **U.S. Presidents.** Collect the following facts about each past American President:

- Single or married (classify as married if married for part of the term)
- Inaugurated before or after age 50
- Served one term (or less) or more than one term

Make a three-circle Venn diagram to represent your results.

UNIT 1D Analyzing Arguments

Recall from Unit 1A that an argument uses a set of premises to support one or more conclusions. If an argument is well constructed, its conclusions follow from its premises in a compelling way. But how do we determine whether an argument is well constructed and compelling? We'll examine this question in this unit.

Two Types of Argument: Inductive and Deductive

Arguments come in two basic types, known as *inductive* and *deductive*. The following two arguments illustrate the two types. As you read each one, ask yourself whether its premises lead to its conclusion in a compelling way.

Argument 1 (Inductive)

Premise:	Birds fly into the air but eventually come back down.
Premise:	People who jump into the air fall back down.
Premise:	Rocks thrown into the air come back down.
Premise:	Balls thrown into the air come back down.
Conclusion:	What goes up must come down.

Argument 2 (Deductive)

Premise:	All politicians are married.
Premise:	Senator Harris is a politician.
Conclusion:	Senator Harris is married.

Note that Argument 1 begins with a set of fairly specific premises, each of which makes a claim about a specific type of object. The conclusion of Argument 1 is a more general statement about how any object might behave. This type of argument, in which the conclusion is formed by generalizing more specific premises, is called an **inductive argument**. (The term *inductive* has the root *induce*, which means "to lead by persuasion.")

In contrast, Argument 2 begins with a general statement about politicians and then draws a specific conclusion about a particular politician. Argument 2 is called a **deductive argument** because it allows us to *deduce* a specific conclusion from more general premises.

I think, therefore I am.
—René Descartes (1596–1650), French philosopher and mathematician

Definition

An **inductive argument** makes a case for a general conclusion from more specific premises.

A **deductive argument** makes a case for a specific conclusion from more general premises.

Evaluating Inductive Arguments

Let's start by examining Argument 1 in greater detail. Its premises are clearly true, and each premise lends support to the conclusion. The variety of specific examples cited by its premises makes the argument appear quite compelling—and, indeed, people long believed the conclusion *What goes up must come down* to be true. Nevertheless, we now know that the conclusion is false because a rocket launched with sufficient speed can leave Earth permanently. We thereby see a key fact about inductive arguments: No matter how strong an inductive argument may seem, it does not prove that its conclusion is true.

Speaking more formally, we evaluate an inductive argument in terms of its **strength**. A strong argument makes a compelling case for its conclusion, though it cannot prove its conclusion to be true. A weak argument is one in which the premises do not lend much support to the conclusion. Note that evaluating strength involves personal judgment. An argument that one person finds strong might appear weak to someone else.

Note also that the strength of an inductive argument is not necessarily related to the truth of its conclusion. Argument 1 was strong enough to convince people throughout most of history, but its conclusion turned out to be false. Conversely, a weak argument may have a true conclusion. For example, the simple argument *The sky is blue because I said so* is extremely weak, even though its conclusion is true.

BY THE WAY

The first artificial satellite to leave Earth was *Sputnik 1*, launched by the Soviet Union on October 4, 1957. *Sputnik 1* orbited Earth for several months before atmospheric drag caused it to fall back to Earth. By early 1959, both the United States and the Soviet Union had launched satellites that would never return to Earth.

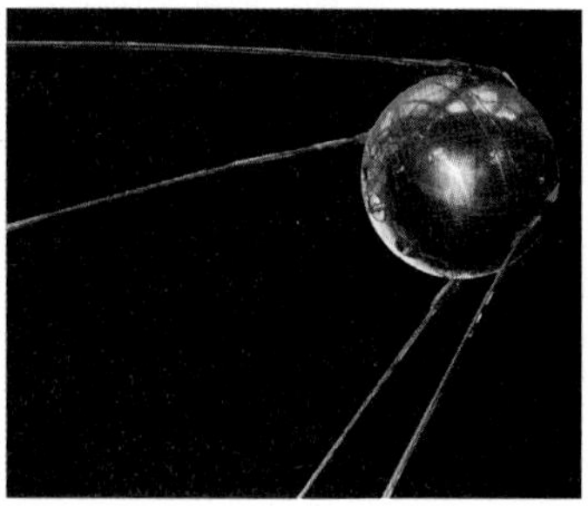

Evaluating an Inductive Argument

An inductive argument cannot prove its conclusion true, so we evaluate it only in terms of its **strength**. An argument is strong if it makes a compelling case for its conclusion. It is weak if its conclusion is not well supported by its premises.

EXAMPLE 1 Hit Movie

A movie director tells her producer (who pays for the movie to be made) not to worry—her film will be a hit. As evidence, she cites the following facts: She's hired big stars for the lead roles, she has a great advertising campaign planned, and it's a sequel to her last hit movie. Is this an inductive or deductive argument? Evaluate its strength.

Solution Each of the three pieces of evidence is a specific fact about the movie, and the director uses these facts to support the conclusion that her movie will be a hit. Because the conclusion is more general than the premises, the argument is inductive. However, this argument is relatively weak, because even the best-planned movies can flop.

▶ Now try Exercises 15–22.

Think About It Suppose you are the director in Example 1, and the producer says she needs more evidence that your movie will be a hit. How might you gather additional evidence to strengthen the case for your movie?

The long yellow line marks the San Andreas Fault; asterisks and dates identify locations of past large earthquakes.

EXAMPLE 2 Earthquake

Evaluate the following argument, and discuss the truth of its conclusion.

> *Geological evidence shows that, for thousands of years, the San Andreas Fault has suffered a major earthquake at least once every hundred years. Therefore, we should expect another earthquake on the fault during the next one hundred years.*

Solution This argument is inductive because it cites many specific past events as evidence that another earthquake will occur. The fact that the pattern has held for thousands of years suggests a strong likelihood that it will continue to hold. The argument does not prove that another earthquake will occur, but it makes another earthquake seem quite likely. The argument is strong.

▶ Now try Exercises 23–28.

Evaluating Deductive Arguments

Now let's turn to Argument 2 (p. 40). If you were to accept its two premises—*All politicians are married* and *Senator Harris is a politician*—then it would follow that Senator Harris is married. In this sense, the argument seems quite solid. However, the first premise is clearly untrue in this case (because there are many unmarried politicians), so we can't actually be sure that Senator Harris is married. This example tells us that evaluating a deductive argument requires answering *two* key questions:

- Does the conclusion follow necessarily from the premises?
- Are the premises true?

We can be sure that the conclusion is true only if the answer to both questions is yes.

In more formal terms, the first question above deals with the **validity** of the argument. A deductive argument is **valid** if its conclusion follows necessarily from its premises. Note that validity is concerned only with the logical structure of the argument. Validity involves no personal judgment and has nothing to do with the truth of the premises or conclusions.

If a deductive argument is valid *and* its premises are true, then we say that the argument is **sound**. Soundness represents the highest test of reliability of a deductive argument because, at least in principle, a sound argument proves its conclusion true. However, soundness may still involve personal judgment if the truth of the premises is debatable.

Argument 2 is valid because the conclusion follows necessarily from the premises. But it is not sound, because the first premise (*All politicians are married*) is false.

Evaluating a Deductive Argument

We apply two criteria when evaluating a deductive argument:

- The argument is **valid** if its conclusion follows necessarily from its premises, regardless of the truth of those premises or conclusions.
- The argument is **sound** if it is valid *and* its premises are all true.

Think About It Consider several decisions you made recently. For each case, decide whether the reasoning you used to reach the decision was inductive or deductive, and explain your reasoning process.

Tests of Validity

We used intuition to determine the validity of the argument about Senator Harris. But deductive arguments can be subtle, and it can be helpful to use Venn diagrams (see Unit 1C) to test their validity. This process requires two basic steps, both illustrated in Figure 1.26 for the Senator Harris argument:

Step 1: We begin by using a Venn diagram to represent the information contained in the premises. In this case:

- The first premise, *All politicians are married*, asserts that the set *politicians* is a subset of the set *married people*. We therefore represent this proposition by drawing a circle for *politicians* inside a circle for *married people*.
- The second premise tells us that Senator Harris is a politician. We indicate this fact by putting an X (to represent the senator) inside the *politicians* circle.

Step 2: We test validity by checking to see whether the conclusion is confirmed by the Venn diagram. In this case, the X is also inside the *married people* circle, meaning that Senator Harris is a married person—just as the conclusion claims. This concludes our Venn diagram test, because the diagram has shown that the argument is valid by demonstrating that the premises lead necessarily to the conclusion.

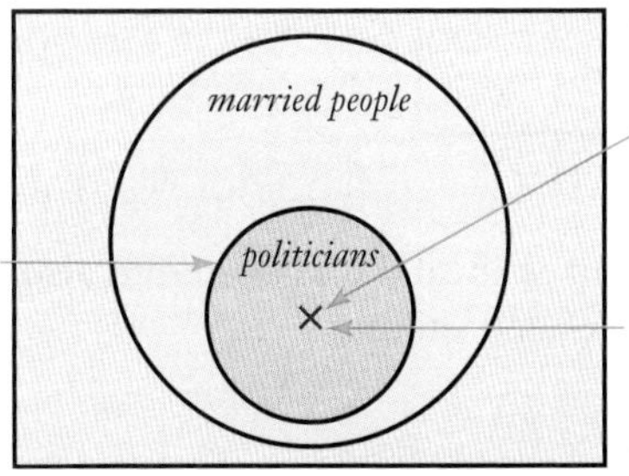
married people
politicians
×
Premise 1, *All politicians are married*, tells us to draw the *politicians* circle as a subset of the *married people* circle.
We represent Premise 2, *Senator Harris is a politician*, by placing an X (for Senator Harris) inside the *politicians* circle.
The conclusion, *Senator Harris is married*, requires the X to be inside the *married people* circle—and it is, so the argument is **valid**.

FIGURE 1.26 This Venn diagram shows why the argument about Senator Harris (Argument 2 on p. 40) is valid.

A Venn Diagram Test of Validity

To test the validity of a deductive argument with a Venn diagram:

1. Draw a Venn diagram that represents all the information contained in the premises.
2. Check to see whether the Venn diagram confirms the conclusion. If it does, then the argument is valid. Otherwise, the argument is not valid.

EXAMPLE 3 Invalid Argument

Evaluate the validity and soundness of the following argument.

Premise: All fish live in water.
Premise: Whales are not fish.
Conclusion: Whales do not live in water.

Solution It's fairly obvious that both premises are true, so the fact that the conclusion is false implies that the argument must have a logical flaw, making it invalid. We can see the flaw with a Venn diagram test (Figure 1.27):

Step 1: We represent the information contained in the premises as follows:

- The first premise, *All fish live in water*, tells us that the set *fish* is a subset of the set *things that live in water*, so we draw the *fish* circle inside the *things that live in water* circle.
- We can represent the second premise, *Whales are not fish*, by putting an X (representing whales) outside the *fish* circle. Notice, however, that the premises do not tell us whether or not whales belong in the circle for things that live in water. Therefore, to show that both possibilities remain open, we place the X on the border of this circle, indicating that it may be either inside or outside.

Step 2: We test validity by noting that the conclusion, *Whales do not live in water*, implies that the X representing whales should be *outside* the circle for *things that live in water*. But it isn't; we placed it on the border because the premises did not provide enough information to know whether it should be inside or outside the circle. Therefore, the conclusion does *not* follow necessarily from the premises, and the argument is invalid.

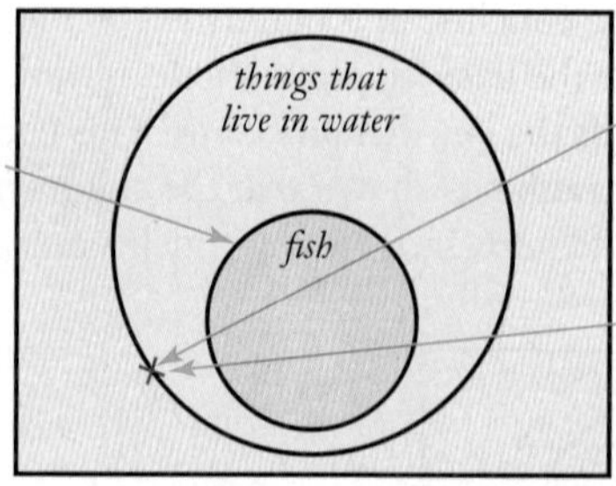
things that live in water
fish
Premise 1, *All fish live in the water*, tells us to draw the *fish* circle as a subset of the *things that live in water* circle.
We represent Premise 2, *Whales are not fish*, with an X (for whales) outside the *fish circle*. But the premises don't tell us whether the X belongs inside or outside the *things that live in water* circle, so we place it on the border of this circle.
The conclusion, *Whales do not live in the water*, requires the X to be *outside* the *things that live in water* circle—but it isn't, so the argument is **invalid**.

FIGURE 1.27

In summary, the argument is invalid and therefore also unsound, because soundness requires both true premises and validity. ▸ **Now try Exercises 29–32.**

EXAMPLE 4 Invalid but True Conclusion

Evaluate the validity and soundness of the following argument.

Premise: All 20th-century U.S. Presidents were men.
Premise: John Kennedy was a man.
Conclusion: John Kennedy was a 20th-century U.S. President.

Solution This argument might at first seem compelling, because both of its premises and its conclusion are all true. However, notice what happens in the Venn diagram test (Figure 1.28):

Step 1: We represent the information contained in the premises as follows:

- The first premise, *All 20th-century U.S. Presidents were men*, tells us that the set *20th-century presidents* is a subset of the set *men*, so we draw the *20th-century presidents* circle inside the *men* circle.
- We can represent the second premise, *John Kennedy was a man*, with an X (representing Kennedy) inside the *men* circle. However, because the premises do not tell us whether this X also belongs inside the *20th-century presidents* circle, we place it on the border of this circle.

Step 2: We test validity by noting that the conclusion implies that the X (representing Kennedy) should be *inside* the *20th-century presidents* circle. Because it's not (it's on the border), the argument is invalid.

In summary, the argument is invalid and therefore also unsound.

▸ **Now try Exercises 33–36.**

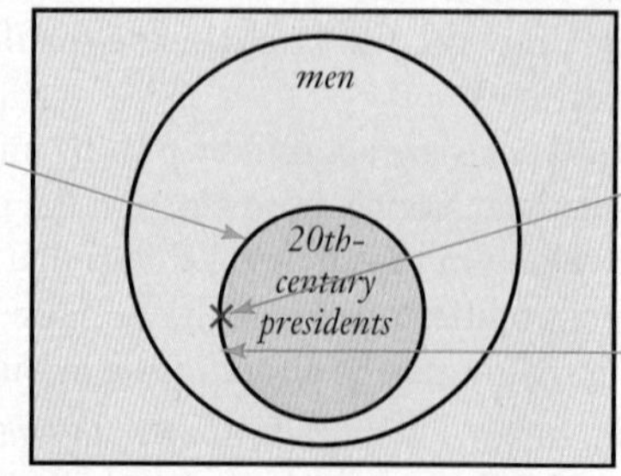
men
20th-century presidents
Premise 1, *All 20th-century U.S. Presidents were men*, tells us to draw the *presidents* circle as a subset of the *men* circle.
Premise 2, *John Kennedy was a man*, tells us to put an X (for Kennedy) inside the *men* circle but does not tell us whether this X also belongs inside the *presidents* circle; we therefore place the X on the border of this circle.
The conclusion, *John Kennedy was a 20th-century U.S. President*, requires the X to be *inside* the *presidents* circle — but it isn't, so the argument is **invalid**.

FIGURE 1.28

Think About It For the argument in Example 4, replace John Kennedy with the name of a man who not a U.S. President, such as Albert Einstein. Are the premises still true? How does such a replacement help you see the flaws in the argument?

Conditional Deductive Arguments

Consider the following argument:

Premise: If a person lives in Chicago, then the person likes windy days.

Premise: Carlos lives in Chicago.

Conclusion: Carlos likes windy days.

This type of deductive argument, in which the first premise is a conditional statement *if p, then q*, is among the most common and important types of argument. (In this case, p = *a person lives in Chicago* and q = *the person likes windy days*.) You can probably see that this argument is valid: If people who live in Chicago like windy days and Carlos lives in Chicago, then it follows that Carlos likes windy days.

Conditional arguments come in four basic forms. Each has a special name, which will make sense if you remember that p is the hypothesis and q is the conclusion in *if p, then q*. For example, the second premise of the above argument about Carlos asserts that the hypothesis is true for Carlos, so the argument is called *affirming the hypothesis*. The following box summarizes the general structure and validity of the four forms of conditional arguments.

Four Basic Conditional Arguments

	Affirming the Hypothesis*	Affirming the Conclusion	Denying the Hypothesis	Denying the Conclusion**
Structure	If p, then q.	If p, then q.	If p, then q	If p, then q.
	p is true.	q is true.	p is not true.	q is not true.
	q is true.	p is true.	q is not true.	p is not true.
Validity	Valid	Invalid	Invalid	Valid

*Also known by the Latin term *modus ponens*.
**Also known by the Latin term *modus tollens*.

We can use Venn diagrams to test the validity of conditional arguments much as we've already done. To see how, note that *if p, then q* means that if an object is a member of the set p, then that object is also a member of the set q. We therefore draw the Venn diagram for *if p, then q* by placing the p circle inside the q circle (Figure 1.29). The following four examples show how we test the validity of the four basic conditional arguments.

Technical Note
In order to use Venn diagrams to evaluate conditional arguments, we treat p and q as though they are sets rather than propositions in the conditional statement *if p, then q*. More specifically, the circle p in the Venn diagram represents the set of all cases in which p is true, and the circle q represents the set of all cases in which q is true.

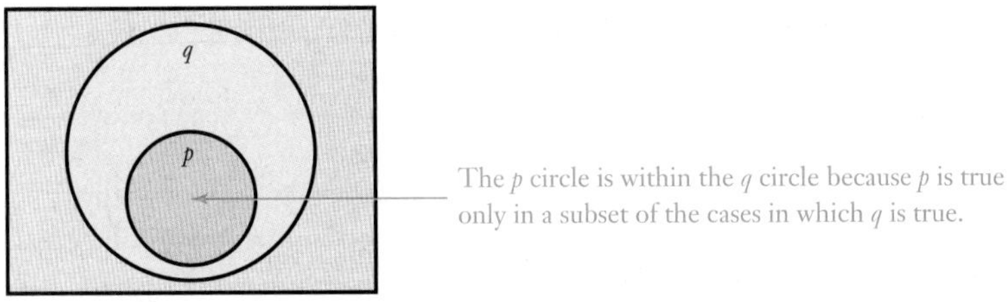
q
p
The p circle is within the q circle because p is true only in a subset of the cases in which q is true.

FIGURE 1.29 A Venn diagram for *if p, then q*.

EXAMPLE 5 Affirming the Hypothesis (Valid)

Use a Venn diagram test to show that the argument concerning Carlos and Chicago (which affirms the hypothesis) is valid.

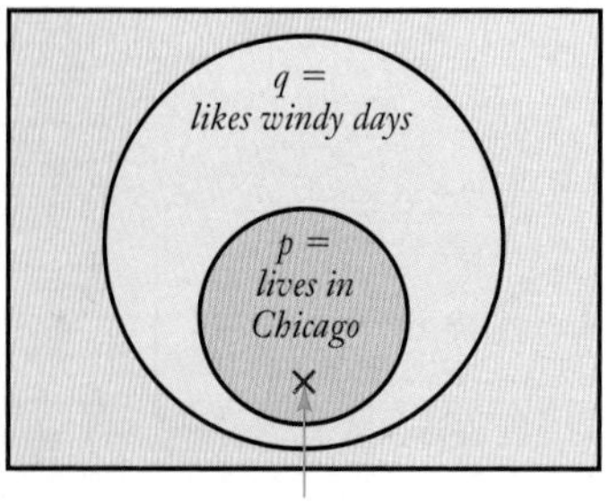

We put an X (for Carlos) in the *p* circle to show that Carlos lives in Chicago. The X is also inside the *q* circle, so we conclude that Carlos likes windy days.

FIGURE 1.30

Solution We conduct the test as shown in Figure 1.30:

Step 1: We represent the information contained in the premises as follows:

- The first premise is *if p, then q*, in which $p =$ *a person lives in Chicago* and $q =$ *the person likes windy days*. We therefore draw a Venn diagram with the p circle inside the q circle.
- We can represent the second premise, *Carlos lives in Chicago*, by placing an X (to represent Carlos) inside the p circle.

Step 2: We test validity by noting that the conclusion, *Carlos likes windy days*, implies that the X for Carlos should be inside the q circle. The X is indeed within the q circle, so the argument is valid. ▸ Now try Exercises 37–38.

EXAMPLE 6 Affirming the Conclusion (Invalid)

Use a Venn diagram to test the validity of the following argument, which affirms the conclusion.

Premise:	If an employee is regularly late, then the employee will be fired.
Premise:	Sharon was fired.
Conclusion:	Sharon was regularly late.

The premise *Sharon was fired* tells us that the X (for Sharon) must be inside the *q* circle. Because the premise does not tell us whether she was regularly late, we place the X on the border of the *p* circle.

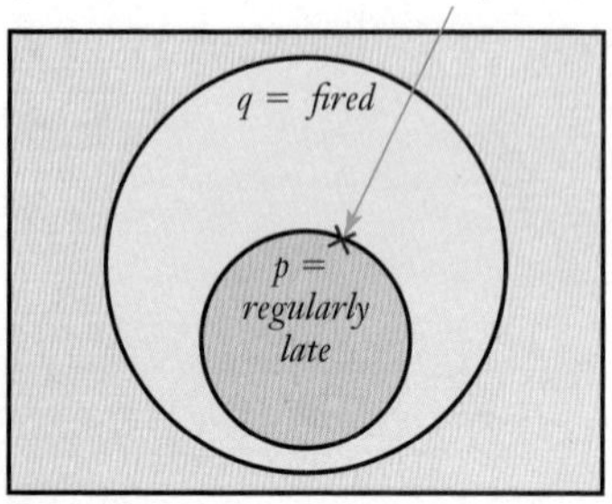

FIGURE 1.31

Solution Figure 1.31 shows the test:

Step 1: We again start with the Venn diagram for *if p, then q*, in which the p circle is inside the q circle. In this case, $p =$ *an employee is regularly late* and $q =$ *the employee will be fired*. We can represent the second premise, *Sharon was fired*, by placing an X for Sharon inside the q circle. However, because the premise does not tell us whether Sharon was regularly late, we place the X on the border of the p circle, indicating that we don't know whether the X belongs inside or outside this circle.

Step 2: The conclusion, *Sharon was regularly late*, claims that the X for Sharon should be inside the p circle. But it isn't (the X is on the border), so the argument is invalid. In other words, we cannot conclude that Sharon was fired because of lateness. She might have been fired for some other reason.

▸ Now try Exercises 39–40.

EXAMPLE 7 Denying the Hypothesis (Invalid)

Use a Venn diagram to test the validity of the following argument, which has the form called denying the hypothesis.

Premise:	If you liked the book, then you loved the movie.
Premise:	You did not like the book.
Conclusion:	You did not love the movie.

The premise *You did not like the book* tells us that the X (for you) must be outside the *p* circle. But we place the X on the border of the *q* circle because the premise does not tell us whether you loved the movie.

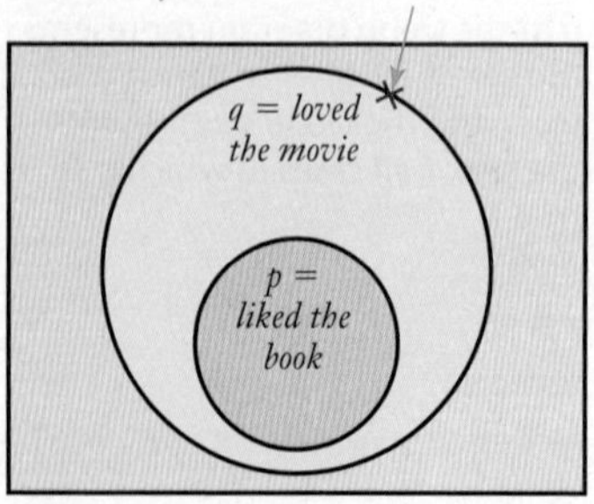

FIGURE 1.32 For the conclusion to follow from the premises, the X (for you) must be outside the *q* circle. However, the premises tell us only that the X is outside the *p* circle. The argument is invalid.

Solution Figure 1.32 shows the test:

Step 1: Once again, we start with the Venn diagram for *if p, then q*, with $p =$ *you liked the book* and $q =$ *you loved the movie*. The second premise tells us that you did not like the book, so we place an X (to represent you) outside the p circle. However, it does not tell us whether you loved the movie, so we place the X on the border of the q circle.

Step 2: The conclusion claims that you did *not* love the movie, in which case the X should be outside the q circle. It is not, so the argument is invalid. In other words, it's possible that you loved the movie even though you did not like the book. ▸ Now try Exercises 41–42.

EXAMPLE 8 Denying the Conclusion (Valid)

Use a Venn diagram to test the validity of the following argument, which has the form denying the conclusion.

Premise: A narcotic is habit-forming.

Premise: Aspirin is not habit-forming.

Conclusion: Aspirin is not a narcotic.

Solution Figure 1.33 shows the test:

Step 1: This time, we must start by rephrasing the first premise in standard conditional form: *If a substance is a narcotic, then it is habit-forming.* We identify p = *a substance is a narcotic* and q = *the substance is habit-forming* and draw the Venn diagram. The second premise asserts that aspirin is *not* habit-forming, so we place an X (representing aspirin) outside the q circle.

Step 2: Because the X is also outside the p circle (representing a narcotic), the conclusion follows and the argument is valid. ▸ Now try Exercises 43–44.

The premise *Aspirin is not habit-forming* tells us that the X (for aspirin) must be outside the q circle, which means it is also outside the p circle.

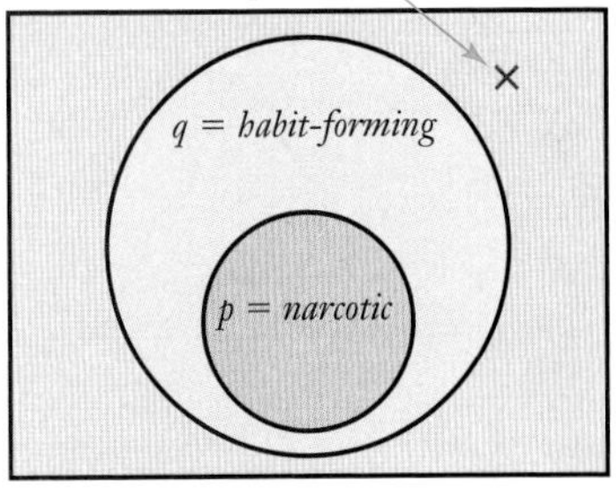

FIGURE 1.33

Think About It Suppose you replace aspirin with heroin in Example 8. Is the argument valid? Is it sound?

Deductive Arguments with a Chain of Conditionals

Another common type of deductive argument involves a chain of three or more conditionals. Such arguments have the following form:

Premise: If p, then q.

Premise: If q, then r.

Conclusion: If p, then r.

This particular chain of conditionals is valid: If p implies q and q implies r, it must be true that p implies r.

EXAMPLE 9 A Chain of Conditionals

Determine the validity of this argument: "If elected to the school board, Maria Lopez will force the school district to raise academic standards, which will benefit my children's education. Therefore, my children will benefit if Maria Lopez is elected."

Solution This argument can be rephrased as a chain of conditionals:

Premise: If Maria Lopez is elected to the school board, then the school district will raise academic standards.

Premise: If the school district raises academic standards, then my children will benefit.

Conclusion: If Maria Lopez is elected to the school board, then my children will benefit.

Cast in this form, the conditional propositions form a clear chain from p = *Maria Lopez is elected* to q = *the school district will raise academic standards* to r = *my children will benefit*. Therefore, the argument is valid. ▸ Now try Exercises 45–46.

EXAMPLE 10 Invalid Chain of Conditionals

Determine the validity of the following argument: "We agreed that if you shop, I make dinner. We also agreed that if you take out the trash, I make dinner. Therefore, if you shop, you should take out the trash."

Solution Let's assign p = *you shop*, q = *I make dinner*, and r = *you take out the trash*. Then this argument has the following form:

Premise: If p, then q.

Premise: If r, then q.

Conclusion: If p, then r.

The conclusion is invalid, because there is no chain from p to r.

▸ Now try Exercises 47–48.

Induction and Deduction in Mathematics

Perhaps more than any other subject, mathematics relies on the idea of *proof*. A mathematical proof is a deductive argument that demonstrates the truth of a certain claim, or *theorem*. A theorem is considered proven if it is supported by a valid and sound proof. Although mathematical proofs use deduction, theorems are often discovered by induction.

Consider the Pythagorean theorem, which applies to right triangles (those with one 90° angle). It says $a^2 + b^2 = c^2$, where c is the length of the longest side, or **hypotenuse**, and a and b are the lengths of the other two sides (Figure 1.34a). A geometric construction of the case where $a = 3$, $b = 4$, and $c = 5$ shows how the squares of the sides are related (Figure 1.34b). The same relationship can be found in any right triangle, so there is strong inductive evidence to suggest that the theorem is true for all right triangles.

The process of seeking inductive evidence can be very useful when you are having difficulty remembering whether a particular theorem or mathematical rule applies. It often helps to try a few test cases and see if the rule works. Although test cases can never constitute a proof, they often are enough to persuade you that the rule is true. However, the rule cannot be true if even one test case fails.

BY THE WAY

The Pythagorean theorem is named for the Greek philosopher Pythagoras (c. 580–500 B.C.E.) because he was the first person known to have proved it deductively. However, the theorem was known to many ancient cultures and had been used for at least a thousand years before the time of Pythagoras. These ancient cultures probably never bothered to prove it. Instead, they simply noticed that it was true every time they tested it and that it was very useful in art, architecture, and construction.

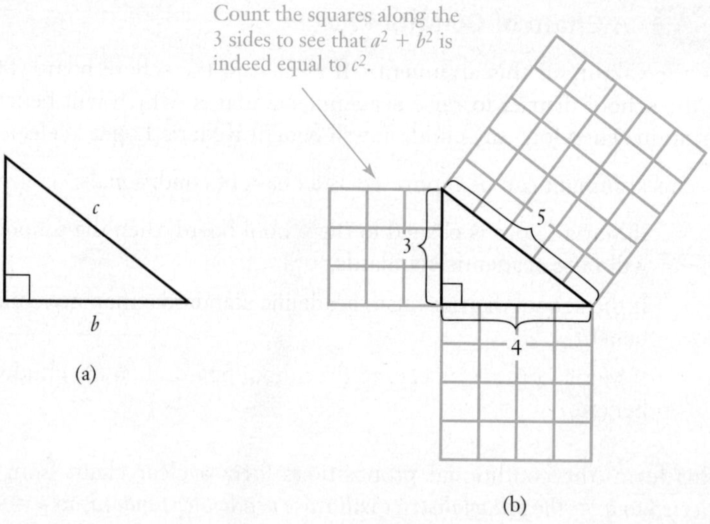
Count the squares along the 3 sides to see that $a^2 + b^2$ is indeed equal to c^2.
a
b
c
(a)
3
4
5
(b)

FIGURE 1.34

EXAMPLE 11 Inductively Testing a Mathematical Rule

Test the following rule: For all numbers a and b, $a \times b = b \times a$.

Solution We begin with some test cases, using a calculator as needed.

$$\text{Does } 7 \times 6 = 6 \times 7? \quad \Rightarrow \text{Yes!}$$
$$\text{Does } (-23.8) \times 9.2 = 9.2 \times (-23.8)? \quad \Rightarrow \text{Yes!}$$
$$\text{Does } 4.33 \times \left(\frac{1}{3}\right) = \left(\frac{1}{3}\right) \times 4.33? \quad \Rightarrow \text{Yes!}$$

The three test cases are each somewhat different (mixing fractions, decimals, and negative numbers), yet the rule works in all three cases. This outcome offers a strong inductive argument in favor of the rule. Although we have not proved the rule $a \times b = b \times a$, we have good reason to be confident that it is true. Our confidence would be strengthened by additional test cases that confirm the rule. ▶ Now try Exercises 49–50.

MATHEMATICAL INSIGHT

Deductive Proof of the Pythagorean Theorem

There are many ways to prove the Pythagorean theorem deductively, but one of the simplest is attributed to a 12th-century Hindu mathematician named Bhaskara. His proof begins with a large square, inside of which is a smaller square surrounded by four identical right triangles (Figure 1.35). Note that the diagram divides the large square into five separate regions (the small square and the four triangles) so that

$$\begin{matrix}\text{area of} \\ \textit{large}\text{ square}\end{matrix} = \begin{pmatrix}\text{area of} \\ \textit{small}\text{ square}\end{pmatrix} + 4 \times \begin{pmatrix}\text{area of} \\ \text{triangle}\end{pmatrix}$$

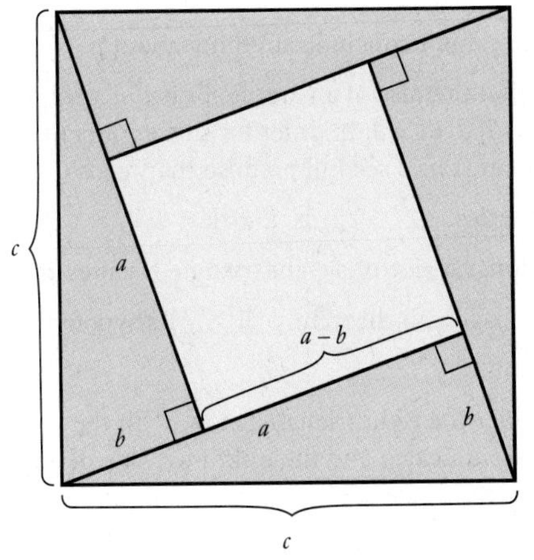

FIGURE 1.35

Now we write each of these areas as follows:

- The side length of the large square is c, so its area is c^2.
- The side length of the small square is $a - b$, so its area is $(a - b)^2$.
- The area of any triangle is given by the formula $\frac{1}{2} \times \text{base} \times \text{height}$ (see Unit 10A). Each of the four triangles has base length a and height b, so each has area $ab/2$.

Substituting these areas into the preceding equation gives

$$\underbrace{c^2}_{\text{area of large square}} = \underbrace{(a-b)^2}_{\text{area of small square}} + 4 \times \underbrace{\frac{ab}{2}}_{\text{area of triangle}} = (a-b)^2 + 2ab$$

We expand the first term on the right, $(a - b)^2 = a^2 - 2ab + b^2$, and substitute it in the previous equation:

$$\begin{aligned} c^2 &= (a-b)^2 + 2ab \\ &= a^2 - 2ab + b^2 + 2ab \\ &= a^2 + b^2 \end{aligned}$$

We have arrived at the Pythagorean theorem by following a deductive chain of logic from beginning to end. Legend has it that, when Bhaskara showed his proof to others, he accompanied it with just a single word: "Behold!"

EXAMPLE 12 Invalidating a Proposed Rule

Suppose you cannot recall whether adding the same amount to both the numerator and the denominator (top and bottom) of a fraction such as $\frac{2}{3}$ is legitimate. That is, you are wondering whether it is true that, for all numbers a,

$$\frac{2}{3} \stackrel{?}{=} \frac{2+a}{3+a}$$

Solution Again, we check the rule with test cases.

Suppose that $a = 0$. Is it true that $\frac{2}{3} = \frac{2+0}{3+0}$? $\Rightarrow$ Yes!

Suppose that $a = 1$. Is it true that $\frac{2}{3} = \frac{2+1}{3+1}$? $\Rightarrow$ No!

Although the rule worked in the first test case, it failed in the second. Therefore, it is not in general legitimate to add the same amount to the top and bottom of a fraction.

▶ Now try Exercises 51–52.

Quick Quiz 1D

Choose the best answer to each of the following questions. Explain your reasoning with one or more complete sentences.

1. To prove a statement true, you must use
   a. an inductive argument.
   b. a deductive argument.
   c. a conditional argument.
2. If a deductive argument is valid, its conclusion
   a. is true. b. is sound.
   c. follows from its premises.
3. A deductive argument cannot be
   a. both valid and sound. b. valid but not sound.
   c. sound but not valid.
4. Consider an argument in which Premise 1 is *All knights are heroes* and Premise 2 is *Paul is a hero.* If X represents Paul, which Venn diagram correctly represents the two premises?

a.
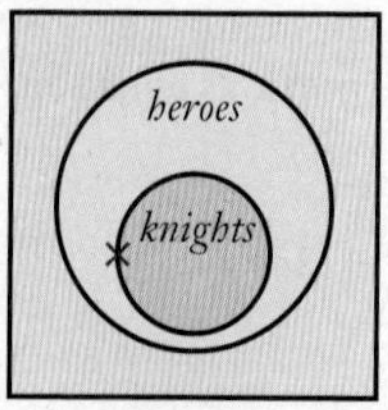

b.
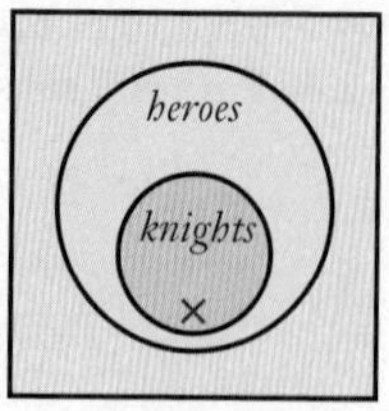

c.
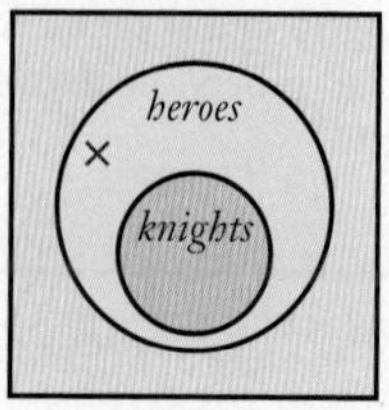

5. Consider again the argument from Question 4. Which of the following conclusions is true?
   a. Paul is a knight. b. Paul is not a knight.
   c. Paul may or may not be a knight.
6. Consider an argument in which Premise 1 is *If p, then q* and Premise 2 is *q is not true.* What can you conclude about p?
   a. p is true. b. p is not true.
   c. We cannot conclude anything about p.
7. Consider an argument in which Premise 1 is *If p, then q* and Premise 2 is *q is true.* What can you conclude about p?
   a. p is true. b. p is not true.
   c. We cannot conclude anything about p.
8. The first premise of an argument is *If a, then b.* The conclusion is *If a, then d.* In order for this argument to be valid, there must be a second premise that reads
   a. *If a, then c.* b. *If c, then d.* c. *If b, then d.*
9. The longest side of a right triangle is called the
   a. Pythagorean theorem. b. hypotenuse.
   c. slant.
10. Consider the right triangle below, with the two known side lengths indicated and the unknown side of length c. Which statement is true?
    a. $c = 6$
    b. $c^2 = 41$
    c. $c = 41$

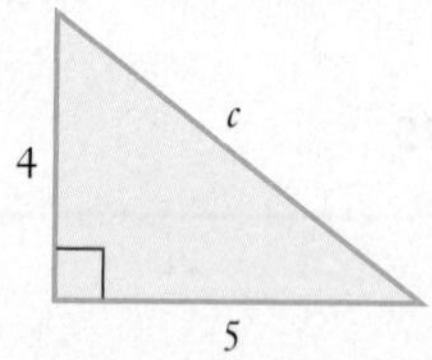

Exercises 1D

REVIEW QUESTIONS

1. Summarize the differences between deductive and inductive arguments. Give an example of each type.
2. Briefly explain the idea of strength and how it applies to inductive arguments. Can an inductive argument prove its conclusion true? Can an inductive argument be valid? Can it be sound?
3. Briefly explain the ideas of validity and soundness and how they apply to deductive arguments. Can a valid deductive argument be unsound? Can a sound deductive argument be invalid? Explain.
4. Describe the procedure used to test the validity of a deductive argument with a Venn diagram.
5. Create your own example of each of the four basic conditional arguments. Then explain why each of your arguments is valid or invalid.
6. What is a *chain* of conditionals? Give an example of a valid argument made with such a chain.
7. Can inductive logic be used to prove a mathematical theorem? Explain.
8. How can inductive testing of a mathematical rule be useful? Give an example.

DOES IT MAKE SENSE?

Decide whether each of the following statements makes sense (or is clearly true) or does not make sense (or is clearly false). Explain your reasoning.

9. My inductive argument provides absolute proof of your client's guilt.
10. The many examples of people whose cancer went away following chemotherapy make a strong case for the idea that chemotherapy can cure cancer.
11. My argument is deductively valid, so if you accept the truth of my premises, you must also accept the truth of my conclusion.
12. Because my argument is valid, you must accept the truth of my conclusion.
13. If you use logic, then your life will be organized. Therefore, if your life is organized, you must be using logic.
14. Fermat's Last Theorem was proposed in 1637, but mathematicians weren't sure if it was true until it was proved deductively in 1994.

BASIC SKILLS & CONCEPTS

15–22: Argument Type. Explain whether the following arguments are deductive or inductive.

15. I have never seen a bank open on the Fourth of July. All banks must be closed on the Fourth of July.
16. All banks are closed on national holidays. Today is Memorial Day, so my bank must be closed today.
17. All of John Grisham's books have been superb, so his next one is also bound to be superb.
18. Every day of my life, the sun has risen. Tomorrow the sun will rise.
19. If I don't sleep well, I wake up feeling absent-minded. If I am absent-minded, I forget to eat lunch. Therefore, if I don't sleep well, I miss lunch.
20. If the sum of the digits of a natural number is divisible by 9, then the number itself is divisible by 9. The sum of the digits of 279 is 18, which is divisible by 9, so we can conclude that 279 is divisible by 9.
21. It always rains on Sundays. Tomorrow is Sunday, so it will rain tomorrow.
22. Everyone in my family is right-handed, so my children will be right-handed.

23–28: Analyzing Inductive Arguments. Determine the truth of the premises of the following arguments. Then assess the strength of each argument and the truth of its conclusion.

23. Premise: Michelangelo, Manet, Monet, Modigliani, Munch, and Miró were great painters.
Conclusion: Painters with names that begin with M are great.

24. Premise: If I eat ice cream, I gain weight.
Premise: If I eat cake, I gain weight.
Conclusion: Eating dessert causes me to gain weight.

25. Premise: Robins and wrens are birds and they fly.
Premise: Eagles and falcons are birds and they fly.
Premise: Hawks and vultures are birds and they fly.
Conclusion: Penguins are birds and they fly.

26. Premise: Apes and baboons have hair and they are mammals.
Premise: Mice and rats have hair and they are mammals.
Premise: Tigers and lions have hair and they are mammals.
Conclusion: Animals with hair are mammals.

27. Premise: $2 + 3 = 5$
Premise: $5 + 4 = 9$
Premise: $7 + 6 = 13$
Conclusion: The sum of an even integer and an odd integer is an odd integer.

28. Premise: $(-6) \times (-4) = 24$
Premise: $(-2) \times (-1) = 2$
Premise: $(-27) \times (-3) = 81$
Conclusion: Whenever we multiply two negative numbers, the result is a positive number.

29–36: Analyzing Deductive Arguments. Consider the following arguments.

a. Draw a Venn diagram to determine whether the argument is valid.

b. Evaluate the truth of the premises, and state whether the argument is sound.

29. Premise: All European countries are countries in the European Union.
Premise: Great Britain is a European country.
Conclusion: Great Britain is a country in the European Union.

30. Premise: All fruits are foods with sugar.
Premise: Chocolate bars contain sugar.
Conclusion: Chocolate bars are fruit.

31. Premise: All states east of the Mississippi River are small states.
Premise: Iowa is a small state.
Conclusion: Iowa is east of the Mississippi River.

32. Premise: All queens are women.
Premise: Meryl Streep is a woman.
Conclusion: Meryl Streep is a queen.

33. Premise: All 8000-meter peaks are peaks in Asia.
Premise: Denali is not in Asia.
Conclusion: Denali is not an 8000-meter peak.

34. Premise: All vegetables are green.
Premise: Beans are green.
Conclusion: Beans are vegetables.

35. Premise: All movie stars are wealthy people.
Premise: Matt Damon is wealthy.
Conclusion: Matt Damon is a movie star.

36. Premise: All islands are countries.
Premise: Iceland is an island.
Conclusion: Iceland is a country.

37–44: Deductive Arguments with Conditional Propositions. Consider the following arguments.

a. Identify the form of each conditional argument, and determine its validity with a Venn diagram.

b. Evaluate the truth of the premises, and state whether the argument is sound.

37. Premise: If an animal is a cat, then it is a mammal.
Premise: Siamese cats are cats.
Conclusion: Siamese cats are mammals.

38. Premise: If it is a two-axle car, then it has four wheels.
Premise: Volkswagens are two-axle cars.
Conclusion: Volkswagens have four wheels.

39. Premise: If you live in Phoenix, you live in Arizona.
Premise: Bruno lives in Arizona.
Conclusion: Bruno lives in Phoenix.

40. Premise: If you make bread, then you must use an oven.
Premise: Jessica used the oven.
Conclusion: Jessica baked bread.

41. Premise: If you live in Phoenix, you live in Arizona.
Premise: Amanda does not live in Phoenix.
Conclusion: Amanda does not live in Arizona.

42. Premise: If a novel was written in the 19th century, then it was written by hand.
Premise: Jake finished his first novel last year.
Conclusion: Jake's first novel was not written by hand.

43. Premise: If a tree is deciduous, then it loses its leaves in the winter.
Premise: Fir trees do not lose their leaves in the winter.
Conclusion: Fir trees are not deciduous.

44. Premise: If you are literate, then you are a college graduate.
Premise: Tom is not a college graduate.
Conclusion: Tom is not literate.

45–46: Chains of Conditionals. For each of the following arguments, identify p, q, and r so that the argument has the structure of a chain of conditionals (*if p then q, if q then r, if p then r*). Then state whether the argument is valid or invalid.

45. Premise: If a natural number is divisible by 18, then it is divisible by 9.
Premise: If a natural number is divisible by 9, then it is divisible by 3.
Conclusion: If a natural number is divisible by 18, then it is divisible by 3.

46. Premise: If taxes are increased, then taxpayers will have less disposable income.
Premise: With less disposable income, spending will decrease and the economy will slow down.
Conclusion: A tax increase will slow down the economy.

47–48: Evaluating Chains of Conditionals. Determine the validity of each of the following arguments.

47. Premise: If everyone obeys the golden rule, then there will be fewer arguments.
Premise: If there are fewer arguments, then the world will be more peaceful.
Conclusion: If the world is more peaceful, then everyone is obeying the golden rule.

48. Premise: If taxes are cut, the U.S. government will have less revenue.
Premise: If there is less revenue, then the federal deficit will be larger.
Conclusion: Tax cuts will lead to a larger federal deficit.

49–52: Testing Mathematical Rules. Test the following rules with several different sets of numbers. If possible, try to find a counterexample (a set of numbers for which the rule is not true). State whether you think the rule is true.

49. Is it true for all real numbers a and b that $a + b = b + a$?

50. Is it true for all nonzero real numbers a, b, and c that

$$\frac{a}{b + c} = \frac{a}{b} + \frac{a}{c}?$$

51. Is it true for all positive real numbers a and b that $\sqrt{a + b} = \sqrt{a} + \sqrt{b}$?

52. Is it true for all positive integers n that

$$1 + 2 + 3 + \ldots + n = \frac{n \times (n + 1)}{2}?$$

FURTHER APPLICATIONS

53–58: Validity and Soundness. If possible, give a simple three-proposition deductive argument that has the following properties. If it is not possible, explain why not.

53. Valid and sound
54. Not valid and sound
55. Valid and not sound
56. Valid with false premises and a true conclusion
57. Not valid with true premises and a true conclusion
58. Valid with false premises and false conclusion

59–62: Make Your Own Argument. Create a simple three-proposition argument that has each of the following forms.

59. Affirming the hypothesis
60. Affirming the conclusion
61. Denying the hypothesis
62. Denying the conclusion

63–66: Conditionals in Books. Consider the following propositions, and answer the questions that follow.

63. "If we insist too adamantly on protecting our privacy, [then] we will sacrifice both free enterprise and security." —Ted Koppel, *Lights Out*

a. What is the logical conclusion if we insist too adamantly on protecting our privacy?

b. What is the logical conclusion if we don't sacrifice free enterprise and security?

c. What is the logical conclusion if we don't insist too adamantly on protecting our privacy?

64. "If [the Japanese] wanted to be as strong as the West[, then] they had to break from their current cultural norms." —Thomas Friedman, *Thank You for Being Late*

a. What is the logical conclusion if the Japanese wanted to be as strong as the West?

b. What is the logical conclusion if the Japanese did not break from their cultural norms?

c. What is the logical conclusion if the Japanese did not want to be as strong as the West?

65. "If such procedures are not followed[, then] apes … act as if they don't understand the problem at hand." —Frans de Waal, *Are We Smart Enough to Know How Smart Animals Are?*

a. What is the logical conclusion if the usual procedures are not followed?

b. What is the logical conclusion if apes act as if they understand the problem at hand?

c. What is the logical conclusion if the usual procedures are followed?

66. "If the Burgess Shale did not exist, [then] we would not be able to invent it." —Stephen Jay Gould, *Wonderful Life*

a. What is the logical conclusion if the Burgess Shale did not exist?

b. What is the logical conclusion if we were able to invent the Burgess Shale?

c. What is the logical conclusion if the Burgess Shale does exist?

67. **The Goldbach Conjecture.** Recall that a prime number is a natural number whose only factors are itself and 1 (examples of primes are 2, 3, 5, 7, and 11). The Goldbach conjecture, posed in 1742, claims that every even number greater than 2 can be expressed as the sum of two primes. For example, $4 = 2 + 2$, $6 = 3 + 3$, and $8 = 5 + 3$. A deductive proof of this conjecture has never been found. Test the conjecture for at least 10 even numbers, and present an inductive argument for its truth. Do you think the conjecture is true? Why or why not?

68. **Twin Primes Conjecture.** If you write out the first several prime numbers (2, 3, 5, 7, 11, 13, 17, 19, 23, …) you will see that occasionally the gap between two consecutive primes is 2 (for example, 5, 7 and 17, 19). These pairs of closely spaced primes are called *twin primes*. A famous conjecture states that the number of twin primes is infinite. While a deductive proof of this conjecture has never been found, twin primes with nearly 400,000 digits have been identified, which gives experimental support to the conjecture. Find the first ten pairs of twin primes and present an inductive proof for the truth of the conjecture. Do you think the conjecture is true? Why or why not?

IN YOUR WORLD

69. **The Pythagorean Theorem.** Learn more about the history of the Pythagorean theorem, and write a short report on one aspect of its history. For example, you might write about its use in cultures that predated Pythagoras or describe another proof of the theorem with its historical context.

70. **Deductive Reasoning in Your Life.** Give an example of a situation in which you used deductive reasoning in everyday life. Explain the situation, describe the steps in your thinking, and explain why it was deductive reasoning.

71. **Inductive Reasoning in Your Life.** Give an example of a situation in which you used inductive reasoning in everyday life. Explain the situation, describe the steps in your thinking, and explain why it was inductive reasoning.

72. **Editorial Arguments.** Find three simple arguments in editorials. State whether each is deductive or inductive, and evaluate it accordingly.

73. **Arguing Your Side.** Choose an issue that you feel strongly about, and create an argument in support of your position. Is your argument inductive or deductive? Evaluate your argument.

74. **Arguing the Other Side.** Choose an issue that you feel strongly about, and create an argument that tends to contradict your position. That is, try to create an argument for the other side. Can you make the argument convincing? Does the argument help you understand the other side of the issue?

UNIT 1E Critical Thinking in Everyday Life

The skills discussed in the preceding units are all useful in their own right. But **critical thinking** involves much more than isolated skills. It also involves careful reading (or listening), sharp thinking, logical analysis, good visualization, healthy skepticism, and more.

Because it is so wide ranging, critical thinking cannot be described by any simple step-by-step procedure. Instead, it is developed through experience and by questioning and analyzing every argument or decision you face. Nevertheless, a few general guidelines or hints can be useful, and we will discuss a few in this unit.

Hint 1: Read (or Listen) Carefully

Language can be used in complex ways that require careful effort to understand. Always read (or listen) carefully to make sure you've grasped precisely what was said. Also be sure that you've distinguished what was actually said, what was assumed, and what must be determined.

EXAMPLE 1 Confusing Ballot Wording

The following is the actual wording of a ballot question (Referendum 426) posed in 2016 to voters in Nebraska.

> *The purpose of Legislative Bill 268, passed by the First Session of the 104th Nebraska Legislature in 2015, is to eliminate the death penalty and change the maximum penalty for the crime of murder in the first degree to life imprisonment. Shall Legislative Bill 268 be repealed? [check one box below]*
>
> ☐ *Retain*
> ☐ *Repeal*

Consider a person who supports the death penalty. Should she vote to retain or repeal?

Solution Notice that the ballot question asks whether to retain or repeal a bill that had already eliminated the death penalty. Therefore:

- Voting to retain means the bill stays in force, and the death penalty remains eliminated.
- Voting to repeal means the bill that eliminated the death penalty is revoked, so the death penalty is restored.

In other words, a person who supports the death penalty should vote to *repeal* the bill. You can probably see that the wording of the question in essence involved a double negative, because someone in favor of the death penalty needed to vote to *repeal the elimination* of the death penalty. ▶ Now try Exercises 11–14.

BY THE WAY
Nebraska voters chose the "Repeal" option by a margin of about 61% to 39%, thereby restoring the death penalty in Nebraska.

Think About It If voters really did misunderstand the wording of the Nebraska ballot question, would the confusion have affected both sides equally, or would it have favored one side more than the other? Defend your opinion.

Hint 2: Look for Hidden Assumptions

The arguments we studied in Unit 1D consisted of clear premises leading to the conclusion. Many real arguments lack such clarity, relying instead on ambiguous terms or hidden assumptions. Often, the speaker (or writer) may think these premises are "obvious," but listeners (or readers) may not agree. Indeed, an argument that seems convincing to the speaker may actually be quite weak to a listener who is not aware of the hidden assumptions.

EXAMPLE 2 Building More Prisons

Analyze the following argument:

> *We should build more prisons because incarcerating more criminals will reduce the crime rate.*

Solution This argument looks deceptively simple because it is so short. Its conclusion is based on a single premise—that incarcerating more criminals will reduce the crime rate. If we recognize the premise as a conditional statement, the argument has the form

Premise:	If we incarcerate more criminals, then the crime rate will be reduced.
Conclusion:	We should build more prisons.

Viewed this way, the argument hardly makes any sense at all, because the premise says nothing about building prisons. Clearly, the speaker or writer must have intended some hidden assumptions to be "obvious" to the listeners or readers.

We can begin to make sense of the argument by identifying possible hidden assumptions. For example, a plausible argument might look like this:

Hidden Assumption 1:	If we build more prisons, then more criminals can be incarcerated.
Stated Premise:	If we incarcerate more criminals, then the crime rate will be reduced.
Hidden Assumption 2:	If the crime rate is reduced, then we will have a more desirable society.
Hidden Assumption 3:	If a policy will lead to a more desirable society, then it should be implemented.
Conclusion:	We should build more prisons.

With the three hidden assumptions, the argument is a long chain of conditionals that is deductively valid. But even if we assume the speaker intended these assumptions to be "obvious," the argument is sound only if you accept the truth of both the single stated premise and the three hidden assumptions. All are debatable.

For example, many people would dispute the first hidden assumption, because incarcerating more criminals requires not only more prison space, but also a more efficient court system. The stated premise is also open to debate: Studies do not agree on whether a higher incarceration rate reduces the crime rate. One could counter the second hidden assumption by arguing that crime reduction might not make a more desirable society if it comes at the price of less personal freedom. Even the last hidden assumption is debatable, because we might not choose to implement a beneficial policy if it has a high cost.

In summary, the original argument makes sense only if we add several hidden assumptions, and these assumptions are open to debate. As a result, the truth of the conclusion is also debatable. ▶ Now try Exercises 15–18.

Hint 3: Identify the Real Issue

It can be difficult to identify the real issue in a debate because people may be attempting to hide their true intentions. Fortunately, by analyzing arguments carefully, we can often determine whether the real issue is hidden, even if we don't know exactly what it is.

EXAMPLE 3 Banning Concerts

Analyze the following segment of an editorial from a local newspaper.

> *With last Saturday's sellout crowd at the Moonlight Amphitheater, it is clear that the parking problem has gotten worse. Concert goers parked along residential streets up to a mile away from the amphitheater, badly overcrowding sidewalks, blocking driveways, and disrupting traffic. In light of this parking problem, future concerts should be banned.*

Solution The argument makes several claims about parking problems, but stripped of the details the argument boils down to this:

Premise: There is a parking problem for concerts at the amphitheater.
Conclusion: Future concerts at the theater must be banned.

The rest of the argument simply lists reasons why the parking problem is serious. But is a parking problem a good enough reason to ban all future concerts? After all, a parking problem could have many solutions. For example, it might be possible to create new parking lots, to use shuttle buses, to step up enforcement of parking violations, or to encourage carpooling. The weakness of the argument should make us wonder whether the editorial writers are *really* concerned about parking or whether they are using this issue to oppose the concerts themselves. Or perhaps they are responding to a few vocal residents who complained about the parking problem. It's hard to know for sure, but it seems unlikely that parking is the only issue. ▶ Now try Exercises 19–20.

HISTORICAL NOTE

Can logic settle all arguments? The German mathematician Gottfried Wilhelm von Leibniz (1646–1716) thought so and took the first steps in a two-century search for a "calculus of reasoning." An outgrowth of this search called *symbolic logic* is now used extensively in mathematics and computing. But Leibniz's dream of resolving arguments through logic was not to be. In 1931, the Austrian mathematician Kurt Gödel discovered that no system of logic can solve all mathematical problems, let alone problems of ethics or morality.

Hint 4: Understand All the Options

We regularly make decisions in situations in which we have several options. For example, we face decisions about which insurance policy to choose, which auto loan to take, or which model of new computer to buy. The key to such decisions is making sure that you understand how each option would affect you.

EXAMPLE 4 Which Airline Ticket to Buy?

Airlines typically offer many different prices for the same trip. Suppose you are planning a trip six months in advance and discover that you have two choices in purchasing an airline ticket:

(A) The lowest fare is $400, but 25% of the fare is nonrefundable if you cancel the trip.
(B) A fully refundable ticket is available for $800.

Analyze the situation.

Solution We can think of each option as a pair of conditional propositions. Under Option A, you will lose 25% of $400, or $100, if you cancel your trip. That is, Option A represents the following pair of conditional propositions:

(1A) If you purchase ticket A and go on the trip, then you will pay $400.
(2A) If you purchase ticket A and cancel the trip, then you will pay $100.

Similarly, Option B represents the following pair of conditional propositions:

(1B) If you purchase ticket B and go on the trip, then you will pay $800.
(2B) If you purchase ticket B and cancel the trip, then you will pay $0.

Figure 1.36 represents the four possibilities. Clearly, Option A is the better buy if you go on the trip, and Option B is the better buy if you end up canceling your trip. However, because you are planning six months in advance, it's impossible to foresee all the circumstances that might lead you to cancel your trip. Therefore, you might want to analyze the difference between the two tickets under the two possibilities (going on the trip or canceling).

If you go on the trip: ticket B costs $400 more than ticket A.
If you cancel the trip: ticket A costs $100 more than ticket B.

In effect, you must decide which ticket to purchase by balancing the risk of spending an extra $400 if you go on the trip against spending an extra $100 if you cancel. How would you decide?

▶ Now try Exercises 21–22.

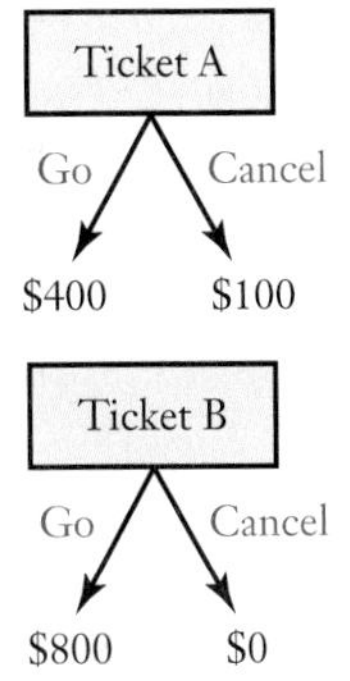

FIGURE 1.36

Think About It Consider the two ticket options in Example 4 from the point of view of the airline. How does offering the two options help the airline maximize its revenue?

Hint 5: Watch for Fine Print and Missing Information

There's an old saying that "the devil is in the details," and it is probably nowhere more applicable than with regard to the fine print of offers, deals, and contracts. What looks like a great deal without the fine print may be a poor one with it. Worse yet, sometimes even the fine print neglects to state important information. Use your powers of critical thinking to decide what information you need. If anything is missing, be sure to ask before you act.

EXAMPLE 5 A Safe Investment?

Marshall was nearing retirement age and was concerned about his retirement savings. He thought the stock market was too risky, so he put his money into a certificate of deposit (CD) at his local bank. Like most CDs, the one at his local bank was federally insured by the F.D.I.C. (which insures bank deposits in the United States), but its interest rate was quite low. Then he heard about this offer:

> *Our certificates of deposit have the highest interest rates* you'll find anywhere—more than double that of most other banks! For a combination of investment safety and high interest, there's no better choice.*
>
> **Our high yields come from our unique ability to invest in valuable offshore assets*

Marshall transferred all his savings into one of these high-yield CDs. Two years later, the bank offering these CDs failed, and Marshall learned that he would not be able to recover any of the money he had lost. What happened?

Solution Marshall lost his retirement savings because, unlike most CDs, this one was not insured by the F.D.I.C. Losses like this have affected millions of people in recent years, but these losses could have been avoided with additional thought. First, Marshall should have wondered how one bank could possibly have offered a rate so much higher

BY THE WAY

Example 5 is based on a real fraud case involving "certificates of deposit" sold by the Stanford Financial Group, which claimed to offer safe investments but actually invested in risky assets held in a lightly regulated bank outside the United States. The 2009 failure of the Stanford Financial Group cost investors more than $8 billion, with thousands of people losing their lifelong retirement savings.

than that of other banks. Second, the fine print stating that investments were "offshore" should have been a clue to the fact that this was not a normal CD. Third, because the offer did not say that the CD was insured, he should have inquired further before assuming that it was. ▶ Now try Exercises 23–24.

Hint 6: Are Other Conclusions Possible?

It is the peculiar and perpetual error of the human understanding to be more moved and excited by affirmatives than by negatives.
—Francis Bacon, 16th-century philosopher

One of the most common lapses of critical thinking involves what is called **confirmation bias**, in which a person tends to search only for evidence that supports some preexisting belief or opinion, and will ignore or try to explain away any evidence that is contrary to that belief or opinion. You may have seen confirmation bias in action even in your own thinking. It also frequently occurs in group deliberations, when people reinforce rather than challenge each others' shared beliefs or opinions; in these cases, it is often called *groupthink*.

The best way to counter confirmation bias is to always question conclusions. In particular, no matter how compelling the conclusion of some argument may seem, you should always ask whether there might be other conclusions possible with the same evidence, and look for other evidence that might reveal flaws in the original conclusion.

EXAMPLE 6 Phrenology

Beginning in the early 19th century, many psychiatrists and doctors in Europe and the United States came to believe in something called *phrenology*. The idea behind phrenology was that measurements of the human skull could reveal the size and importance of "organs" within the brain, which in turn were believed to control specific personality traits and intellectual faculties. Belief in phrenology was widespread both among professionals and the public, and it was a topic of numerous best-selling books. Although phrenology was discredited by the middle of the 19th century (and is now known to have had no basis in reality), it still cropped up on occasion long after that.

a. Rigorous methods for research in psychology did not yet exist in the early 19th century. Explain how this fact might have enabled confirmation bias to contribute to the popularity of phrenology.

b. Phrenology was associated with numerous social misconceptions of the day, including beliefs about the superiority of certain races that were used by some to justify slavery (and, later, eugenics). Explain how such associations might have increased confirmation bias.

c. Suggest an experiment that might have led phrenologists to suspect that there might be alternate explanations for differences in personality and intelligence.

Solution

a. Without careful experimental testing of the ideas of phrenology, it was easy for people to ignore competing ideas and there was no easy way to collect evidence that might have refuted phrenology. As a result, once the practice became popular, it was easy for belief in it to become self-reinforcing, because so many people were looking for confirmation of that belief. Without rigorous methods of investigation, it is easy to ignore competing theories and look only for supporting evidence.

b. People who supported institutions such as slavery could claim phrenology as a "scientific" validation of their beliefs, which in turn made them more likely to look for additional reason to support their belief in phrenology. This outlook would tend to suppress evidence that discredited phrenology.

c. Researchers could have compared the skull measurements of several people who had similar personality traits or intellectual abilities. They would then have discovered that the claimed correlations between skull measurements and traits of personality or intelligence do not actually exist. ▶ Now try Exercises 25–26.

Think About It Some people attribute the increased political polarization of Americans to what is sometimes called the "echo chamber," in which like-minded people only pay attention to Web and media sources that echo their own beliefs. How is the idea of the echo chamber similar to the idea of confirmation bias? What do you think can be done to counter the echo chamber and decrease political polarization?

Hint 7: Watch for Outright Fakery

Sadly, there is now a large industry that produces so-called **fake news**, which consists of stories designed to sound real but that are simply invented—either to maliciously deceive readers or as an exercise in satire. Satirical websites usually make their intentions clear, but many of those posting fake news maliciously go to great lengths to make their stories look and sound real. It's therefore important to know how to look for potential warning signs.

Fake news is made-up stuff, masterfully manipulated to look like credible journalistic reports that are easily spread online to large audiences willing to believe the fictions and spread the word.
—Angie Drobnic Holan, PolitiFact.com

EXAMPLE 7 Fake News Story

According to statistics on how widely various fake news stories spread, the top fake news story of 2016 came under the headline "Obama signs executive order banning the pledge of allegiance in schools nationwide." What signs might have told you that this story was fake?

Solution The first warning sign comes from the headline itself: No matter what you may have thought about President Obama, it's unclear why he would sign something so obviously controversial and with no clear benefit to him or others. This warning sign should cause you to look further. For example, the first step in the box "Five Steps to Evaluating Media Information" (on p. 10) tells you to consider the source. With a little digging, you would have found that the original source of the article was a website that calls itself "abc news" but that is not the real site for ABC News (though the site looks very much like the real ABC site, even using a similar logo and similar Web address). Moreover, even if you did not find the original source, you would have quickly noticed that this story was *not* being reported on any of the major news networks, which doesn't make sense for something that would be so controversial. These warning signs should have immediately made you suspect that the story might be fake, and you could have verified that at fact-checking websites (such as those listed in the In Your World box in Unit 1A, on p. 11).

▶ Now try Exercises 27–28.

Hint 8: Don't Miss the Big Picture

You've probably heard the expression "not seeing the forest for the trees," which means missing out on the big picture (the fact that there is a forest in front of you) because you are too focused on details (the individual trees). The details may be very important to a debate or argument, but you should always step back as you consider them to make sure you have not lost sight of the forest.

EXAMPLE 8 Burst of the Housing Bubble

Look back at the chapter-opening activity ("Bursting Bubble") about the collapse in the housing market that helped trigger a global recession. Economists and real estate professionals collected vast amounts of data as home prices rose, so everyone was well aware that prices were increasing at a rapid rate from 2000 to 2005. But people continued to invest in real estate at ever-higher prices, and lenders continued to loan money for those purchases, because so many people convinced themselves that real estate was a "safe" investment. Were these arguments reasonable, or did they miss the big picture?

IN YOUR WORLD Beware of "Up to" Deals

You've probably heard special cell phone offers such as "get up to 1 GB of free data per month." These offers can be good deals if you are able to take full advantage of them. But the words "up to" hold hidden danger. In this case, the data may be free only during certain hours of the day, or they may be subject to roaming charges. If you are not careful, this special deal may turn out to be quite expensive.

More generally, you should beware of any deal that contains the words "up to." For example, an offer of "up to five free oil changes if you pay for two now" may be structured to make it difficult for you to take advantage of all of them. "Up to" may also mean "no more than"; for example, when you hear that you can "earn up to $5000 a month while working from home," it probably means you're likely to earn far less. Until you've carefully thought through the full meaning of the offer, it's best to assume that "up to" deals are up to no good.

Solution The data on the home price/income ratio (see Figure 1.A on p. 4) suggest that the big picture was missed. Remember that the rise in this ratio that occurred during the housing bubble meant that families had to spend larger percentages of their income on housing, and the only way they could do that was through some combination of higher income, reduced expenses in other areas, and/or increased borrowing. Other data show that the average income (adjusted for inflation) of working families did not rise during the housing bubble, nor did families cut back significantly on other expenses. It's therefore reasonable to conclude that many families were buying homes that were more expensive than they could afford and that lenders were lending them money anyway. In light of this big picture, a significant drop in housing prices seems to have been inevitable.

Quick Quiz 1E

Choose the best answer to each of the following questions. Explain your reasoning with one or more complete sentences.

1. What does it mean to think critically about the issue of global warming?

   a. You should be critical of claims about the threat posed by global warming.

   b. You should carefully evaluate the evidence and arguments invoked in debates about global warming.

   c. You should assume that global warming is a critical threat to our survival.

2. "If you want to save the social services that would be lost with a property tax reduction, then vote against Proposition C." Based on this quote, a vote for Proposition C presumably means that

   a. you favor an increase in property taxes.

   b. you favor a decrease in property taxes.

   c. you support social services.

3. Suppose that an argument is deductively valid only if you assume that it contains one or more unstated, hidden assumptions. In that case, the original argument

   a. is also deductively valid. b. is inductively strong.

   c. is probably weak.

4. A teacher claims that, because spell checkers have weakened students' spelling skills, middle school students should not be allowed to use them. What hidden assumption is made in this argument?

   a. Traditional methods of teaching spelling are effective.

   b. Students should know how to spell as well as a spell checker.

   c. Spell checkers are not reliable.

5. You need to buy a car and are considering loans from two banks. Bank 1 offers a loan at 7% interest plus an application fee, and Bank 2 offers a loan at 7.5% interest plus an application fee. The loan terms are otherwise identical. Based on this information, you can conclude that

   a. Bank 1 offers the better deal.

   b. Bank 2 probably offers the better deal, since it probably has a lower application fee.

   c. there is no way to tell which deal is better without knowing the amounts of the application fees.

6. You get your hair cut at a shop that charges \$30. The owners offer you a deal: If you prepay \$150 for five haircuts, they'll give you a punch card that entitles you to a sixth haircut free. The punch card expires in one year. Is this a good deal?

   a. Yes, because you'll get a free haircut.

   b. It depends on whether you'll get six haircuts at this shop in the next year.

   c. No, because no one ever gives anything that's really free.

7. You buy a cell phone plan that gives you up to 1000 minutes of calling for \$20 per month. During a particular month, you use only 100 minutes. Your per-minute cost for that month is

   a. 2¢ b. 5¢ c. 20¢

8. You are planning a trip in six months. You can buy a fully refundable airline ticket for \$600 or a \$400 ticket of which 50% is refundable if you cancel your trip. If you purchase the \$400 ticket and end up canceling the trip, you will have spent

   a. \$200. b. \$400. c. \$600.

9. Auto insurance policy A has an annual premium of \$500 and a \$200 deductible for collision (meaning you pay the first \$200 for a collision claim). Auto insurance policy B has an annual premium of \$300 and a \$1000 deductible for collision. Which one of the following conclusions does not follow?

   a. You will spend less on premiums with policy B.

   b. Your expense for insurance and collision repairs over a year will be less with policy B.

   c. You will spend less for a \$900 collision repair with policy A.

10. The Smiths have a picnic every Saturday provided it does not rain. What can be concluded if today the Smiths did not have a picnic?

   a. Today is not Saturday.

   b. Today it rained.

   c. If it did not rain, then today is not Saturday.

Exercises 1E

REVIEW QUESTIONS

1. Describe critical thinking, and explain why it is important to everyone.
2. Summarize the hints given in this unit, and explain how each is important to critical thinking.
3. Give a few examples of situations in which you used critical thinking in your own life.
4. Give at least one case you know of in which a poor decision was made (by you or someone else) because of a lack of good critical thinking.

DOES IT MAKE SENSE?

Describe whether each of the following statements makes sense (or is clearly true) or does not make sense (or is clearly false). Explain your reasoning.

5. Reed was relieved because his insurance company chose not to deny his claim.
6. Although the plane crashed in Nevada, the survivors were buried in California.
7. Sue prefers the Red Shuttle because it gets her to the airport in an hour and a half, while the Blue Shuttle takes 80 minutes.
8. Alan decided to buy his ticket from Ticketmaster for \$33 plus a 10% surcharge rather than from the box office website, where it costs \$35 with no additional charges.
9. There was no price difference, so Michael chose the tires with a 5-year, 40,000-mile warranty over the tires with a 50-month, 35,000-mile warranty.
10. Auto policy A has \$30,000 worth of collision insurance with an annual premium of \$400. Auto policy B has \$25,000 worth of collision insurance with an annual premium of \$300. Clearly policy B is the better policy.

BASIC SKILLS & CONCEPTS

11. **Reading a Ballot Initiative.** The following is the actual wording of a ballot question posed in 1992 to Colorado voters.

    Shall there be an amendment to the Colorado constitution to prohibit the state of Colorado and any of its political subdivisions from adopting or enforcing any law or policy which provides that homosexual, lesbian, or bisexual orientation, conduct, or relationships constitutes or entitles a person to claim any minority or protected status, quota preferences, or discrimination?

Does a "yes" vote support or oppose gay rights? Explain.

12. **Interpreting Policies.** A city charter's sole policy on reelection reads as follows:

A person who has served three consecutive terms of four years each shall be eligible for appointment, nomination for or election to the office of councilmember no sooner than for a term beginning eight years after completion of that councilmember's third consecutive full term.

a. What is the maximum number of consecutive years that a councilmember could serve?

b. How many years must a councilmember who has served three consecutive full terms wait before running for office again?

c. Suppose a councilmember has served two consecutive full terms and is then defeated for reelection. According to this provision, is she or he required to wait eight years before running for office again?

d. Suppose a councilmember serves three consecutive full terms and is reelected 10 years later. According to this provision, how many consecutive terms can she or he serve at that time?

13. **Reading a Ballot Initiative.** Consider the following ballot initiative, which appeared in the 2010 statewide elections in Oklahoma and was passed.

This measure... requires that each person appearing to vote present a document proving their identity. The document must meet the following requirements. It must have the name and photograph of the voter. It must have been issued by the federal, state, or tribal government. It must have an expiration date that is after the date of the election. No expiration date would be required on certain identity cards issued to person 65 years of age or older. In lieu of such a document, voters could present voter identification cards issued by the County Election Board. A person who cannot or does not present the required identification may sign a sworn statement and cast a provisional ballot.

a. According to the initiative, would a state driver's license allow a person to vote?

b. According to the initiative, would a (federal) Social Security card allow a person to vote?

c. Without a "document proving their identity," what options for voting do citizens have?

d. What documents are required to obtain a voter identification card?

14. **Interpreting a Proposed Amendment.** The following is the actual wording of an amendment on which Florida voters were asked to vote in 2016:

Amendment 1—Rights of Electricity Consumers Regarding Solar Energy Choice. *This amendment establishes a right under Florida's constitution for consumers to own or lease solar equipment installed on their property to generate electricity for their own use. State and local governments shall retain their abilities to protect consumer rights and public health, safety and welfare, and to ensure that consumers who do not choose to install solar are not required to subsidize the costs of backup power and electric grid access to those who do.*

On initially reading this amendment, most people assumed that it was intended to expand solar energy in Florida. However, consider the following additional facts: (1) Although the first sentence says the amendment "establishes a right" for consumers to own or lease solar equipment, that right already existed in Florida. (2) Comments from amendment supporters suggested the intent of the second sentence was to allow utilities to start charging new fees that would have the effect of making it more expensive to install solar equipment.

a. What particular words in the amendment might have had the effect of allowing new fees for solar? Explain.

b. If the right in the first sentence already existed, what purpose might the sentence have served for those who wrote the proposed amendment? There may be more than one possible purpose.

c. Comment overall on whether you think the amendment was deliberately deceptive. Defend your opinion.

15–18: Hidden Assumptions. Identify at least two hidden assumptions in the following arguments.

15. Buying a house today makes good sense. The rent money you save can be put into a long-term investment.

16. I recommend giving to the United Way because it supports so many worthwhile causes.

17. Governor Reed has campaigned on tax cuts. He gets my vote.

18. I support increased military spending because we need a strong America.

19–20: Unstated Issues. The following arguments give several reasons for a particular political position. Identify at least one unstated issue that may, for at least some people, be the real issue of concern.

19. I oppose the President's spending proposal. Taxpayer money should not be used for programs that many taxpayers do not support. Excessive spending also risks increasing budget deficits. Greater deficits increase the federal debt, which in turn increases our reliance on foreign investors.

20. People who eliminate meat from their diet risk severe nutritional deficiencies. Eating meat is by far the easiest way to consume complete protein plus many other essential nutrients all in one food source. It makes sense: Our ancestors have been meat-eaters for thousands of years.

21. **Airline Options.** In planning a trip to New Zealand six months in advance, you find that an airline offers two options:

- Option A: You can buy a fully refundable ticket for $2200.
- Option B: You can buy a $1200 ticket, but you forfeit 25% of the price if the ticket is canceled.

Describe your options in the events that you do and do not make the trip. How would you decide which ticket to buy?

22. **Buy versus Lease.** You are deciding whether to buy a car for $18,000 or to accept a lease agreement. The lease entails a $1000 initial fee plus monthly payments of $240 for 36 months. Under the lease agreement, you are responsible for service on the car and insurance. At the end of the lease, you may purchase the car for $9000.

    a. Should the cost of service and insurance determine which option you choose?

    b. Does the total cost of purchasing the car at the end of the lease agreement exceed the cost of purchasing the car at the outset?

    c. What are some possible advantages of leasing the car?

23. **You've Won!** You receive the following e-mail notification: "Through a random selection from more than 20 million e-mail addresses, you've been selected as the winner of our grand prize—a two-week vacation in the Bahamas. To claim your prize, please call our toll-free number. Have your credit card ready for identification and a small processing fee." Does this sound like a deal worth taking? Explain.

24. **Reading a Lease.** Consider the following excerpt from the contract for the lease of an apartment:

    Landlord shall return the security deposit to resident within one month after termination of this lease or surrender and acceptance of the premises, whichever occurs first.

    Suppose your lease terminates on June 30, and you move out of the apartment on June 5. Explain whether or not the landlord has complied with the terms of the lease if you receive your security deposit back on

    a. June 28. b. July 2. c. July 7.

25. **Pyramidology.** For centuries, enthusiasts have looked for hidden codes and prophecies in the structure of the Egyptian pyramids. The measurements and orientations of the pyramids have been claimed to predict cosmic events, future events, or the end of the world. Explain how confirmation bias might have given these claims life and made them popular.

26. **Nuclear Deterrence.** Read the following historical argument, and discuss whether its conclusion is the only possible conclusion or whether others may also be possible.

    The development of nuclear weapons changed the way world leaders think about potential conflicts. A single nuclear weapon can kill millions of people, and the arsenals of the United States and the Soviet Union contained enough power to kill everyone on Earth many times over. This potential for catastrophic damage led to the idea of nuclear deterrence, which held that the United States and the Soviet Union would be deterred from direct warfare by the fear of nuclear war. For the more than 45 years of the Cold War, the United States and the Soviet Union never did fight directly. This was one of the longest periods in human history during which two major enemies avoided direct war. We can only conclude that nuclear deterrence prevented war between the United States and the Soviet Union.

27–28. Fake News. The following are fake news headlines that were widely circulated prior to the 2016 presidential election. For each case, suggest at least one reason you should have been immediately suspicious of the headline.

27. "Pope Francis Shocks World, Endorses Donald Trump for President, Releases Statement"

28. "Hillary Clinton Runs Secret Child Sex Ring at Comet Ping Pong Pizzeria in Washington, DC"

FURTHER APPLICATIONS

29–40: Read and Think Carefully. The following questions represent simple puzzles that can help you build your skills for reading critically. Give an answer and brief explanation for each.

29. José had 6 bagels and ate all but 4 of them. How many bagels were left?

30. Is it possible for a man to marry his widow's sister?

31. Paris Hilton's rooster laid an egg in Britney Spears' yard. Who owns the egg?

32. A large barrel is filled with 8 different kinds of fruit. How many individual fruits must you remove from the barrel (without looking) to be certain that you have two of the same fruit?

33. Suppose you go to a conference attended by 20 Canadians and 20 Norwegians. How many people must you meet to be certain that you have met two Norwegians?

34. Suppose you go to a conference attended by 20 Canadians and 20 Norwegians. How many people must you meet to be certain that you have met one Norwegian and one Canadian?

35. Suppose you go to a conference attended by 20 Canadians and 20 Norwegians. How many people must you meet to be certain that you have met two people of the same nationality?

36. A boy and his father were in a car crash and were taken to the emergency room. The surgeon looked at the boy and said, "I can't operate on this boy; he is my son!" How is this possible?

37. Suzanne goes bowling at least one day per week, but never on two consecutive days. List all the numbers of days per week that Suzanne could go bowling.

38. A race car driver completed the first lap in one minute and forty seconds. Despite a crosswind, driving at the same speed, he completed the second lap in 100 seconds. Give a possible explanation.

39. Half of the people at a party are women, and half of the people at the party love chocolate. Does it follow that one quarter of the people at the party are women who are chocolate lovers?

40. Half of a country's exports consist of corn, and half of the corn is from the state of Caldonia. Does it follow that one quarter of the exports consist of corn from Caldonia?

41–44: Decision Making. Analyze the following situations, and explain what decision you would make and why.

41. You and your spouse are expecting a baby. Your current health insurance costs $115 per month, but doesn't cover prenatal care or delivery. You can upgrade to a policy that will

cover prenatal care and delivery, but your new premium will be $275 per month. The cost of prenatal care and delivery is approximately $4000.

42. It's time to paint your living room. You and your nephew can do the job in four hours with no labor costs. However, you'll take those four hours off from your regular job that pays $40 per hour. Alternatively, you can hire a single painter who can paint the room in six hours at a rate of $30 per hour. Assume that the paint costs the same with either option.

43. Over a period of a year, you fly twice a month between two cities 1500 miles apart. Average round-trip cost on Airline A is $350. Airline B offers the same trip for only $325. However, Airline A has a frequent flyer program in which you earn a free round-trip ticket after you fly 15,000 miles. Airline B does not have a frequent flyer program.

44. One day, your auto insurance agent calls and says that your insurance rates will be increased. You have a choice of keeping your current deductible of $200 with a new premium of $450 per year or going to a higher deductible of $1000 with a premium of $200 per year. In the past 10 years, you have filed claims of $100, $200, and $600.

45. **IRS Guidelines on Who Must File a Federal Tax Return.** According to the IRS, a single person under age 65 (and not blind) must file a federal tax return if any of the following apply (numbers were for tax year 2012):
   (i) unearned income was more than $950.
   (ii) earned income was more than $5950.
   (iii) gross income was more than the larger of $950 or the person's earned income (up to$5650) plus $300.

   Determine whether the following single people (under age 65 and not blind) must file a return.

   a. Maria had unearned income of $750, earned income of $6200, and gross income of $6950.

   b. Van had unearned income of $200, earned income of $3000, and gross income of $3500.

   c. Walt had no unearned income and had earned and gross income of $5400.

   d. Helena had unearned income of $200, earned income of $5700, and gross income of $6000.

46. **IRS Guidelines on Dependent Children.** You may claim a child as a dependent on your federal tax return if that person is a qualifying child who meets the following requirements:
   (i) The child must be your son, daughter, stepchild, foster child, brother, sister, half-brother, half-sister, stepbrother, stepsister, or a descendant of any of them.
   (ii) The child must be under age 19 at the end of the year, under age 24 at the end of the year and a full-time student, or any age if permanently disabled.
   (iii) The child must have lived with you for more than half of the year, unless the absence is "temporary" for education, illness, vocation, business, or military service.
   (iv) The child must not have provided more than half of his/her support for the year.
   (v) The child cannot file a joint return.

   Determine whether you could claim a child as a dependent in the following situations.

   a. You have a 22-year-old stepdaughter who is a full-time student, lives year-round in another state, and receives full support from you.

   b. You have an 18-year-old son who works full-time writing software, lives with you, and supports himself.

   c. Your nephew (who cannot be claimed by anyone else) is a 20-year-old full-time student who lives with and is supported by you.

   d. Your half-brother is an 18-year-old part-time student who lived with you for eight months of the tax year and received two-thirds of his support from you.

47. **Credit Card Agreement.** The following rules are among the many provisions of a particular credit card agreement.

   For the regular plan, the minimum payment due is the greater of $10.00 or 5% of the new balance shown on your statement (rounded to the nearest $1.00) plus any unpaid late fees and returned check fees, and any amounts shown as past due on your statement.

   If you make a purchase under a regular plan, no finance charges will be imposed in any billing period in which (i) there is no previous balance or (ii) payments received and credits issued by the payment due date, which is 25 days after the statement closing date shown on your last statement, equal or exceed the previous balance. If the new balance is not satisfied in full by the payment due date shown on your last statement, there will be a finance charge on each purchase from the date of purchase.

   a. If the new balance in your account is $8 and you have $35 in unpaid late fees, what is your minimum payment due?

   b. Suppose you have a previous balance of $150 and you pay $200 one month after the statement closing date. Will you be assessed a finance charge?

   c. In part (b), if you make a purchase on the same day that you make the $200 payment, will a finance charge be assessed on that purchase?

48. **Apple® EULA.** An end-user license agreement (EULA) is a contract between a software manufacturer and a user that spells out the terms of use of the software (that most of us accept without reading!). Within the many pages of the EULA for Apple's iTunes Store are the following clauses.

   a. Apple reserves the right at any time to modify this Agreement and to impose new or additional terms or conditions on your use of the iTunes Service. Such modifications and additional terms and conditions will be effective immediately and incorporated into this Agreement. Your continued use of the iTunes Service will be deemed acceptance thereof.
   b. Apple is not responsible for typographic errors.

a. Do new conditions for the use of iTunes Store need to be approved by the user?

b. Are users notified of changes in the EULA?

c. What potential risks for the user do you see in clause (a)?

d. What potential risks for the user do you see in clause (b)?

49. **Texas Ethics.** In its "Guide to Ethics Laws," the Texas Ethics Commission states:

A state officer or employee should not accept or solicit any gift, favor, or service that might reasonably tend to influence the officer or employee in the discharge of official duties or that the officer or employee knows or should know is being offered with the intent to influence the officer's or employee's official conduct.

a. Imagine that you are a state representative. Do you believe it would be legal to accept a maximum campaign contribution from a person if you knew nothing about the person except her name?

b. Describe a situation in which you (as a state representative) would accept a contribution because it clearly conforms with this guideline. Then describe a situation in which you would not accept a contribution because it clearly violates this guideline.

50–54: Critical Thinking. Consider the following short arguments that support a conclusion (which may only be implied). Use critical thinking methods to analyze the arguments and determine whether the conclusion is convincing. Write out your analysis carefully.

50. Newspapers have been a mainstay of American life for two centuries. However, their very existence is being threatened by Internet news sources. There is no substitute for newspapers as a local news source. Furthermore, for national issues, citizens need the investigative reporting of trained and independent newspaper journalists, rather than the biased opinions of Internet bloggers.

51. The N.C.A.A. claims to represent the interests of athletes and universities, but it frequently acts to maximize revenue from athletic events while trying to hide scandals involving athletes and putting down any attempts by athletes to unionize. Clearly, its real purpose has become little more than protecting its lucrative brand while preventing its revenue-producing labor force—college football and basketball players—from making any money.

52. "A free people should be an armed people. It insures against the tyranny of the government. If they know the biggest army is the American people, then you don't have the tyranny that came from King George." —Representative Chris Gohmert, Texas

53. "We know that assault weapons and high-capacity ammunition magazines are weapons of choice in contemporary mass shootings. We know that law enforcement officers nationwide are increasingly finding themselves staring down the barrels of assault weapons in the hands of criminals … . The time has come for action. The enactment of Senator Feinstein's legislation would be a significant step toward ending the gruesome gun violence that is destroying our families and communities." —Coalition to Stop Gun Violence, "CSGV Statement on Introduction of Assault Weapons Ban Legislation," January 24, 2013

54. "The President is trying to avoid talking about the real subject that threatens our country—and that's the debt. … We have an illusion of wealth, but there is still great danger that we'll run into another crisis like we did in 2008. The wealth is built from fake money, it's built from manufacturing new money to paper over our debt." —Senator Rand Paul, Republican, Kentucky

55. **Poetry and Mathematics.** Consider the following poem by the English poet A.E. Housman (1859–1936).

LOVELIEST OF TREES

Loveliest of trees, the cherry now
Is hung with bloom along the bough
And stands about the woodland ride
Wearing white for Eastertide.

Now, of my threescore years and ten,
Twenty will not come again,
And take from seventy springs a score,
It only leaves me fifty more.

And since to look at things in bloom
Fifty springs are little room,
About the woodlands I will go
To see the cherry hung with snow.

How old was the poet at the time he wrote this poem? (*Hint:* A *score* is 20.) Based on your reading of this poem, how much longer does the poet expect to live? Explain.

IN YOUR WORLD

56. **Interpreting the Second Amendment.** Much of the debate over gun control revolves around interpretations of the Second Amendment to the U.S. Constitution. It reads:

A well-regulated militia being necessary to the security of a free state, the right of the people to keep and bear arms shall not be infringed.

Gun rights advocates tend to focus on the "right of the people to keep and bear arms." Gun control advocates tend to focus on "a well-regulated militia." Visit a few of the many websites on each side of this issue to find interpretations of the Second Amendment that both support and oppose gun control. Based on what you learn, do you believe that the Second Amendment allows for gun control laws? Defend your opinion.

57. **Ballot Initiatives.** Investigate a particular ballot initiative in your state, county, or city (or choose one in another area from a website). Explain the important arguments on each side of the issue. If possible, find a statement of the initiative as it appeared on the ballot. Is the statement of the initiative clear? Explain.

58. **Fine Print.** Several websites, including mouseprint.org, give recent examples of deceptive uses of fine print. Describe one case that you think is particularly deceptive or dangerous. Describe the case in detail, and explain how the fine print is important to understanding the situation.

59. **Argument Analysis.** Find a news article or editorial in which a clear position is taken on a controversial issue. Analyze the argument using the hints and strategies given in this unit. Whether or not you agree with the final conclusion, state whether the argument is effective.

60. **Personal Decisions.** Discuss a situation in your own life in which you needed to think carefully before making a critical decision. Did you use critical thinking strategies? In retrospect, would you have made a different decision using critical thinking strategies?

61. **Fake News.** Visit a fact-checking website (such as PolitiFact.com, FactCheck.org, or Snopes.com) and write a brief description of one recent fake news story that gained wide traction, explaining how we can know that the story is fake.

62. **Conspiracy Theories.** Choose some well-known conspiracy theory (such as the idea that President Kennedy was killed by the CIA) and read about how and why people continue to believe in it. Discuss with classmates how the conspiracy claim illustrates confirmation bias. Overall, does the claim you are investigating deserve any credence at all? Defend your opinion.

Chapter 1 Summary

UNIT	KEY TERMS	KEY IDEAS AND SKILLS
1A	logic (p. 6) argument (p. 6) premises, conclusions (p. 6) fallacy (p. 6)	**Identify** an argument's premises and conclusions. (p. 6) **Recognize** fallacious arguments, which contain logical flaws so that their conclusions are not well supported by their premises. (p. 6) **Evaluating media information** (p. 10) 1. Consider the source. 2. Check the date. 3. Validate accuracy. 4. Watch for hidden agendas. 5. Don't miss the big picture.
1B	proposition (p. 14) truth values, truth tables (p. 15) negation: *not p* (p. 15) conjunction: *p and q* (p. 16) disjunction: *p or q* (p. 18) conditional: *if p, then q* (p. 19) hypothesis, conclusion (p. 19) logical equivalence (p. 21)	**Understand** truth tables for negation, conjunction, disjunction, conditional. (p. 15) **Inclusive versus exclusive *or*** (p. 18) Inclusive *or* means *either or both.* Exclusive *or* means *one or the other, but not both.* **Variations on the conditional** *if p, then q* (p. 21) **Converse:** *if q, then p* **Inverse:** *if not p, then not q* **Contrapositive:** *if not q, then not p*
1C	members (of a set) (p. 25) set (p. 26) Venn diagrams (p. 26) categorical propositions (p. 30) subject set (p. 30) predicate set (p. 30)	**Set relationships** (p. 28) Subsets, disjoint sets, overlapping sets **Four categorical propositions** (p. 30) All *S* are *P*; No *S* are *P*; Some *S* are *P*; Some *S* are not *P* **Uses of Venn diagrams** (p. 31) Illustrating set relationships Organizing information
1D	inductive argument (p. 40) deductive argument (p. 40) strength (p. 41) validity (p. 42) soundness (p. 42)	**Evaluating arguments** (p. 40) Inductive arguments in terms of strength Deductive arguments in terms of validity and soundness Venn diagram tests of validity Chains of conditionals **Induction and deduction in mathematics** (p. 48) Propose a theorem inductively. Prove a theorem deductively.
1E	critical thinking (p. 54) confirmation bias (p. 58) fake news (p. 59)	**Guidelines for critical thinking** (p. 54) 1. Read (or listen) carefully. 2. Look for hidden assumptions. 3. Identify the real issue. 4. Understand all the options. 5. Watch for fine print and missing information. 6. Are other conclusions possible? 7. Watch for outright fakery. 8. Don't miss the big picture.

2 APPROACHES TO PROBLEM SOLVING

In your past mathematics classes, it might have seemed that mathematical problems involved only numbers and symbols. But the mathematical problems that you encounter in other classes, in jobs, and in daily life are almost always posed in words. That's why Chapter 1 focused on logic and critical thinking: These skills help you find the key ideas buried in problems that are posed in words. In this chapter, we begin a study of *quantitative* problem solving, in which the problems involve words *and* numbers.

Q

Consider an average-size man who is drinking beer. If all the alcohol in the beer were immediately absorbed into the man's bloodstream, how much beer could the man drink before he was legally intoxicated in the United States (blood alcohol content of 0.08)?

- **A** About three ounces
- **B** One 12-ounce bottle
- **C** Three 12-ounce bottles
- **D** One six-pack of 12-ounce bottles
- **E** Three six-packs of 12-ounce bottles

Nothing in life is to be feared. It is only to be understood.

—Marie Curie (1867–1934), only person to win Nobel Prizes in both physics and chemistry

A

Be sure to make your answer choice before you read on. Now, if you're ready to continue.... Most people are surprised to learn that the correct answer is A: There is enough alcohol in just three ounces of beer to make an average man legally intoxicated. (The amount is even less for an average woman.) In reality, it usually takes somewhat more alcohol to become intoxicated, because your body doesn't absorb it all immediately and begins to metabolize it once it is absorbed. Still, the answer shows that alcohol impairment can occur much more quickly and easily than most people guess.

If you're still surprised, it's easy to calculate the answer for yourself. We'll discuss the necessary techniques in this chapter and show you the calculation for this particular case in Unit 2B, Example 12 (p. 100).

UNIT 2A

Understand, Solve, and Explain: Learn a general problem-solving process, along with the basic principles of unit analysis that can be very useful to this approach.

UNIT 2B

Extending Unit Analysis: Review standardized units, and apply unit analysis to problems involving energy, density, and concentration.

UNIT 2C

Problem-Solving Hints: Explore a general set of hints for effective problem solving.

ACTIVITY

Global Melting

Use this activity to gain a sense of the kinds of problems this chapter will enable you to analyze. Additional activities are available online in MyLab Math.

The photo below shows a view of Antarctica from space. The continent is covered almost everywhere with thick white ice. What would happen to sea level if all this ice were to melt? This is an important question, because continued global warming will increase the rate of melting, though scientists suspect it would take thousands of years for the ice to melt entirely.

To answer a question like this one, you first need to gather some data. A quick Web search will tell you the following:

- The total land area of Antarctica is about 14 million square kilometers.
- The mean (average) thickness of Antarctic ice is about 2.15 kilometers.
- Earth's oceans cover a total surface area of about 340 million square kilometers.
- When ice melts into water, the resulting water volume is about 5/6 of the original ice volume.

Working individually or in small groups, try to answer the following questions. If you have difficulty, you might want to read ahead through parts of this chapter.

1. What is the total volume of the Antarctic ice sheet in cubic kilometers? (*Hint:* You need only two of the pieces of data listed above.)
2. If all this ice were to melt, what volume of liquid water would it represent?
3. Suppose all the water from the melted ice flowed into Earth's oceans. Assume that the total surface area of the oceans does not change—that is, the oceans stay in the same basin, rather than spreading out over the continents. How much would sea level rise? Give your answer in kilometers, meters, and feet. (Bonus: Do you think the assumption of the ocean surface area staying constant is valid? Why or why not?)
4. The Greenland ice sheet contains about 10% as much ice as the Antarctic ice sheet. How much would sea level rise if the Greenland ice sheet melted?
5. Discuss: While melting of the Antarctic or Greenland ice sheet would raise sea level, melting of the ice in the Arctic Ocean would not. Why? Are there other consequences of melting the ice in the Arctic Ocean?
6. Discuss: Global warming is expected to cause melting of the polar ice sheets, but scientists cannot yet predict how fast the ice will melt. Given this uncertainty, how should the danger of polar melting be dealt with in political discussions of global warming?

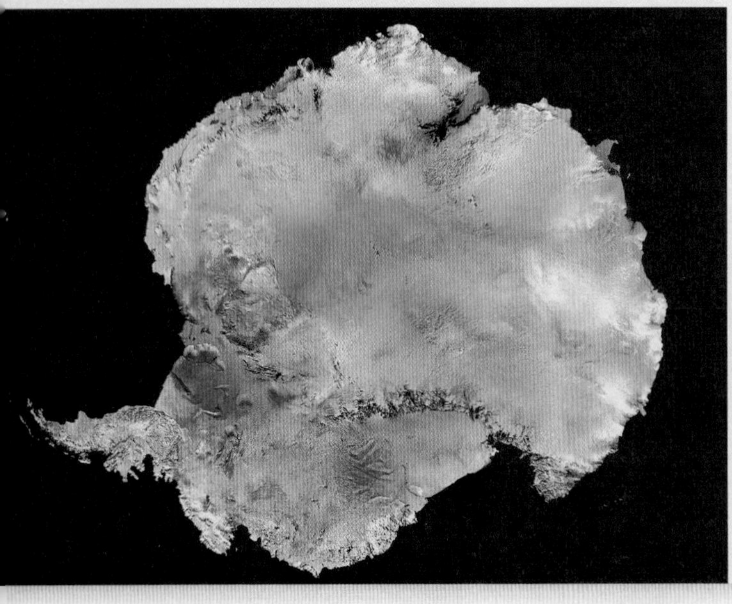

UNIT 2A Understand, Solve, and Explain

I think it's very important to have a feedback loop, where you're constantly thinking about what you've done and how you could be doing it better.

—Elon Musk

Problem solving is more of an art than a science, and there is no one method that will work for every problem. Instead, much as we did in Chapter 1 with critical thinking skills, we must develop quantitative problem-solving skills through experience and by learning to be careful, creative, questioning, and rigorous—all at the same time. Researchers and educators have developed numerous strategies that can help with this process, and you can find many books and articles if you want to delve into the art of problem solving in more detail. Here, we will focus on the three-step process described on the next page. This process is called Understand-Solve-Explain, which you can remember by its acronym: U-S-E.

Keep in mind the Understand-Solve-Explain process is designed as a framework for solving problems, and it is generally *not* necessary to write out the three steps explicitly. However, you'll probably find it useful to keep the three steps in mind whenever you are confronted by any problem, even if it is not mathematical. Moreover, if you get stuck, writing out the steps is often a good way to get unstuck.

The following example applies the Understand-Solve-Explain process to a relatively simple problem, designed primarily to show you how the process is intended to work. Note that this example uses only a few of the many ideas offered in the box that summarizes the process, and it also uses some ideas (such as working with units) that we'll discuss more formally after we have completed this example.

EXAMPLE 1 Light-Year

How far is a light-year?

Solution Let's follow the three-step Understand-Solve-Explain process. We'll include more detail than you will generally need, which should help later when you encounter more complex problems.

Understand: The question asks how *far,* so we know that we are looking for a *distance* as the answer, which means the answer should have units such as miles or kilometers. This should make sense if you know or look up the fact that a light-year is *defined* to be the distance that light can travel in 1 year, and that light travels at a speed of 300,000 kilometers per second (km/s).

The fact that we know the speed and want to find a distance tells us that we need an equation to calculate distance from speed. If you don't remember this equation, you can think of a simpler but analogous problem, such as: If you drive at 50 miles per hour, how far will you travel in 2 hours? The answer is fairly obviously 100 miles (50 miles in the first hour and 50 miles in the second hour), and you'll notice that you can get this answer by multiplying the speed by the time:

$$\text{speed} \times \text{time} = \text{distance}$$

Therefore, we can calculate the distance represented by a light-year simply by multiplying the speed of light by a time of 1 year:

$$1 \text{ light-year} = (\text{speed of light}) \times (1 \text{ yr})$$

This equation represents the path that will take us to the solution.

Solve: We have the required equation (above), but there's a problem with the consistency of our units: The speed of light is given in kilometers per *second*, but the time interval is 1 *year*. We need to make the time units consistent in order to carry out the

BY THE WAY

If you divide the speed of light (300,000 kilometers per second) by Earth's equatorial circumference of about 40,000 kilometers, you'll find that light travels so fast that it could circle Earth nearly 8 times in just 1 second.

Three-Step Problem Solving: Understand–Solve–Explain (U-S-E)

Step 1. UNDERSTAND the problem. You can't solve a problem until you first understand it, so always begin by thinking carefully about the nature of the problem. For example:

- Think about what the problem asks you to do. If it is a textbook-type problem, what exactly are you supposed to determine? If it is a more complex problem, can you distill it down to its essential details? Specifically, what information is given (the knowns or input) and what is asked for (the unknown or output)?
- Remember that drawing diagrams can often help you make sense of a problem. It also helps to consider the context of the problem: How does it relate to what you are learning (for textbook problems) or why has it been posed (for problems in the real world)?
- Ask yourself what the solution should look like. For example, if you are looking for a numerical answer, do you expect it to be large or small, and what units should it have? Should the answer be exact or an estimate? Be as specific as possible in visualizing the form of the solution.
- Try to map a path (either mentally or in writing) that will lead you from your understanding of the problem to its solution. List any information or data you need to solve the problem, and determine how you'll find that information or data if it's not given.
- As you move on to steps 2 and 3, continually revisit your understanding of the problem. If necessary, be prepared to rethink your understanding or your plan for reaching a solution.

Step 2. SOLVE the problem. Once you believe you understand the problem and have a path to its solution, carry out the required steps and calculations. For example:

- Obtain any needed information or data.
- For multi-step problems, be sure to keep an organized, neatly written record of your work, which will be helpful if you later need to review or revise your solution.
- Double-check each step as you work to avoid carrying errors through to the end of your solution.
- Constantly reevaluate your plan as you work. If you find a flaw in your approach or your understanding of the problem, return to step 1 and revise your plan.

Step 3. EXPLAIN your result. Although many people neglect this final step, it is arguably the most important. After all, a result is useless if it is wrong or misinterpreted, and the only way you can gain confidence in your results is by explaining them clearly:

- Be sure that your result makes sense. For example, be sure that it has the expected units, that its numerical value is sensible, and that it is a reasonable answer to the original problem.
- Once you are sure that your result is reasonable, it's a good idea to recheck your calculations once more or, even better, find an independent way to check your result.
- Identify and understand any potential sources of uncertainty in your result. If you made assumptions, were they reasonable?
- Write your solution clearly and concisely, using complete sentences to make sure the context and meaning are clear. If appropriate to the context, explain and discuss any relevant uncertainties or assumptions, as well as the implications of your result.

calculation. We'll discuss unit conversion in more detail shortly, but here is how it works in this case:

$$\begin{aligned}1\text{ light-year} &= (\text{speed of light}) \times (1\text{ yr})\\ &= \left(300{,}000\frac{\text{km}}{\cancel{\text{s}}}\right) \times \left(1\,\cancel{\text{yr}} \times \frac{365\,\cancel{\text{days}}}{1\,\cancel{\text{yr}}} \times \frac{24\,\cancel{\text{hr}}}{1\,\cancel{\text{day}}} \times \frac{60\,\cancel{\text{min}}}{1\,\cancel{\text{hr}}} \times \frac{60\,\cancel{\text{s}}}{1\,\cancel{\text{min}}}\right)\\ &= 9{,}460{,}000{,}000{,}000\text{ km}\end{aligned}$$

To be sure that you've carried out the numerical calculation correctly on your calculator, you should check it at least twice.

Explain: We begin by checking that the answer makes sense: It has the expected units of distance (kilometers) and it is a long distance, as we should expect given that light travels very fast. In order to state the answer clearly, we recognize that 9,460,000,000,000 can be expressed as 9.46 trillion, which is much easier to say and interpret. We also notice that we've used at least one approximation, because the length of a year is not exactly 365 days; it is closer to 365¼ days, which is why we have a leap year every 4 years. (The 300,000 kilometer per second value for the speed of light is also approximate; a more precise value is 299,792 km/s.) Finally, we can make the result easier to remember by rounding 9.46 up to 10 (that is, rounding to the nearest ten), leading to the following final answer in sentence form:

One light-year is approximately 9.46 trillion kilometers, or to make it easier to remember, about 10 trillion kilometers.

▶ Now try Exercises 19–20.

Principles of Unit Analysis

Example 1 and the Understand-Solve-Explain box both talk about "units," but what exactly does this word mean? In brief, it's the difference between abstract numbers—which is the familiar way in which numbers are used in many mathematics classes—and the real-world numbers, which usually represent a count or a measurement of *something*. For example, in an abstract sense, the number 5 is just a point on a number line, but in real life the number 5 might represent 5 apples, 5 dollars, or 5 hours.

The words that describe what we are measuring or counting—such as apples, dollars, or hours—are called the **units** associated with the number. Notice that units provide crucial context. For example, if you ask how long it will take to get your car fixed and the repairman tells you "5," it matters a lot whether he means 5 minutes, 5 hours, 5 days, or 5 weeks. The technique of working with units to help solve problems is called **unit analysis** (or *dimensional analysis*).

Definitions

The **units** of a quantity describe what that quantity measures or counts.

Unit analysis is the process of working with units to help solve problems.

Unit analysis is a powerful technique that can often help you map the path to a solution. To illustrate this idea, suppose you want to find a general formula for calculating speed. We can do so by considering again the simple analogy we used within Example 1, in which we recognized that you would drive a total of 100 miles if you

BY THE WAY

Another good way to recall the general formula speed = distance/time is to remember that speed limits are commonly stated in *miles per hour,* which means a distance (miles) divided by a time (hours). You can also rearrange the formula to find that distance = speed × time and time = distance ÷ speed.

drove 2 hours at a speed of 50 miles per hour. With these numbers, it's easy to see that you can get the speed if you divide the distance by the time:

$$100 \text{ mi} \div 2 \text{ hr} = \frac{100 \text{ mi}}{2 \text{ hr}} = 50 \frac{\text{mi}}{\text{hr}} \leftarrow \text{Read } \frac{\text{mi}}{\text{hr}} \text{ as "miles per hour."}$$

This tells us that the general formula for speed is

$$\text{speed} = \frac{\text{distance}}{\text{time}}$$

Brief Review Common Fractions

We can express a fraction in three basic ways: as a *common fraction* such as $\frac{1}{2}$, as a fraction in *decimal form* such as 0.5, and as a *percentage* such as 50%. Common fractions represent division and are written in the form a/b, where a and b can be any numbers as long as b is not zero. The number on top is the **numerator** and the number on the bottom is the **denominator**:

$$\begin{matrix}\text{numerator} \rightarrow \\ \text{denominator} \rightarrow\end{matrix} \frac{a}{b} \quad \text{means} \quad a \div b$$

Note that, when working with fractions, it's helpful to write integers as fractions with denominator 1. For example, we can write 3 as $\frac{3}{1}$ or -4 as $\frac{-4}{1}$.

Adding and Subtracting Fractions

If two fractions have the same denominator (a *common denominator*), we can add or subtract them by adding or subtracting their numerators. For example:

$$\frac{1}{5} + \frac{2}{5} = \frac{1+2}{5} = \frac{3}{5} \quad \text{or} \quad \frac{7}{9} - \frac{2}{9} = \frac{7-2}{9} = \frac{5}{9}$$

Otherwise, we must rewrite the fractions with the same denominator before adding or subtracting. For example, we can add $\frac{1}{2} + \frac{1}{3}$ by rewriting them as $\frac{3}{6}$ and $\frac{2}{6}$ (so both have a denominator of 6), respectively:

$$\frac{1}{2} + \frac{1}{3} = \frac{3}{6} + \frac{2}{6} = \frac{3+2}{6} = \frac{5}{6}$$

Multiplying Fractions

To multiply fractions, we multiply the numerators and denominators separately. For example:

$$\frac{1}{3} \times \frac{2}{5} = \frac{1 \times 2}{3 \times 5} = \frac{2}{15}$$

Sometimes we can simplify fractions at the same time we multiply them by *canceling* terms that occur in both the numerator and the denominator. For example:

$$\frac{3}{4} \times \frac{5}{3} = \frac{\cancel{3} \times 5}{4 \times \cancel{3}} = \frac{5}{4}$$

Reciprocals and Division

Two nonzero numbers are **reciprocals** if their product is 1. For example:

$$2 \text{ and } \frac{1}{2} \text{ are reciprocals because } 2 \times \frac{1}{2} = 1$$

$$\frac{4}{3} \text{ and } \frac{3}{4} \text{ are reciprocals because } \frac{4}{3} \times \frac{3}{4} = 1$$

In general, we find a reciprocal by switching the numerator and the denominator, remembering that integers have a denominator of 1:

- The reciprocal of a is $\frac{1}{a}$ $(a \neq 0)$.
- The reciprocal of $\frac{a}{b}$ is $\frac{b}{a}$ $(a \neq 0, b \neq 0)$.

We can replace division with multiplication by the reciprocal, which means we *invert and multiply*. For example:

$$10 \div \frac{1}{2} = \underbrace{10 \times \underbrace{\frac{2}{1}}_{\text{invert}}}_{\text{and multiply}} = 20$$

$$\frac{3}{4} \div \frac{2}{5} = \underbrace{\frac{3}{4} \times \underbrace{\frac{5}{2}}_{\text{invert}}}_{\text{and multiply}} = \frac{15}{8}$$

Rules Summary

Addition/subtraction (must have same denominator)	$\frac{a}{c} + \frac{b}{c} = \frac{a+b}{c}$ or $\frac{a}{c} - \frac{b}{c} = \frac{a-b}{c}$
Multiplication	$\frac{a}{b} \times \frac{c}{d} = \frac{a \times c}{b \times d}$
Division (invert and multiply)	$\frac{a}{b} \div \frac{c}{d} = \frac{a}{b} \times \frac{d}{c}$

▸ Now try Exercises 13–14.

Note that when working with units, it's easier to do the division by writing it in fraction form. Also note that when abbreviating units, we do not distinguish between singular and plural. For example, we use *mi* for both *mile* and *miles*.

Key Words: *Per* and *Of*

The speed example above shows that the word *per* (which means "for every") is a key word in mathematical problems, because it tells us to divide. A second important key word is *of*, which usually implies multiplication. For example, if you buy 10 apples at a price *of* \$2 per apple, the total price you pay is

$$10 \text{ apples} \times \frac{\$2}{\text{apple}} = \$20$$

Notice three important steps in this short calculation: First, we multiplied where we saw the word *of*. Second, we wrote the price of \$2 *per* apple as a division (in fraction form). And third, just as we may cancel a number that appears in the numerator and denominator of a fraction, we may do the same cancellation of units.

EXAMPLE 2 Using Key Words

Apply the three-step problem-solving process and use key words appropriately to answer the following questions.

a. You are buying 30 acres of farm land at a price of \$12,000 per acre. What is the total cost?
b. A car travels 25 miles every half-hour. How fast is it going?

Solution

a. Understand: The question asks about "total cost," so we expect an answer in *dollars*. The key word *of* suggests that we should multiply the acreage being purchased by the price.

Solve: We carry out the calculation; note that the price is given in dollars per acre, so we write the division by acres in fraction form. That allows "acres" to cancel, leaving the final answer in dollars:

$$30 \text{ acres} \times \frac{\$12{,}000}{1 \text{ acre}} = \$360{,}000$$

Explain: We have found that purchasing 30 acres of farmland at a price of \$12,000 per acre will cost a total of \$360,000.

b. Understand: The question asks "how fast," so we expect the answer to be a speed. We notice that "25 miles every half-hour" has the same meaning as "25 miles *per* half-hour." The key word *per* tells us that we should *divide* the distance of 25 miles by the time of 1 half-hour (or $\frac{1}{2}$ hour).

Solve: We carry out the calculation:

$$25 \text{ mi} \div \underbrace{\frac{1 \text{ hr}}{2}}_{\text{invert and multiply}} = 25 \text{ mi} \times \frac{2}{1 \text{ hr}} = 50 \frac{\text{mi}}{\text{hr}}$$

Notice that we interpreted $\frac{1}{2}$ hr as $\frac{1 \text{ hr}}{2}$. We then replaced the division with multiplication by the reciprocal ("invert and multiply"), because that makes it easier to see how the units work in the final answer.

Explain: We have found that a car that covers a distance of 25 miles in a half-hour travels at a speed of 50 miles per hour.

▶ Now try Exercises 21–24.

Squares, Cubes, and Hyphens

Other key words arise with units raised to powers. For example:

- To find the *area* of a room, we multiply its length by its width (Figure 2.1). If the room is 12 feet long and 10 feet wide, its area is

$$\begin{aligned} 12 \text{ ft} \times 10 \text{ ft} &= 120(\text{ft} \times \text{ft}) \\ &= 120 \text{ ft}^2 \end{aligned}$$

  We read this area as "120 square feet," in which the key word *square* implies the second power. Note that we multiply the numbers ($12 \times 10 = 120$) separately from the units ($\text{ft} \times \text{ft} = \text{ft}^2$) but keep track of both.

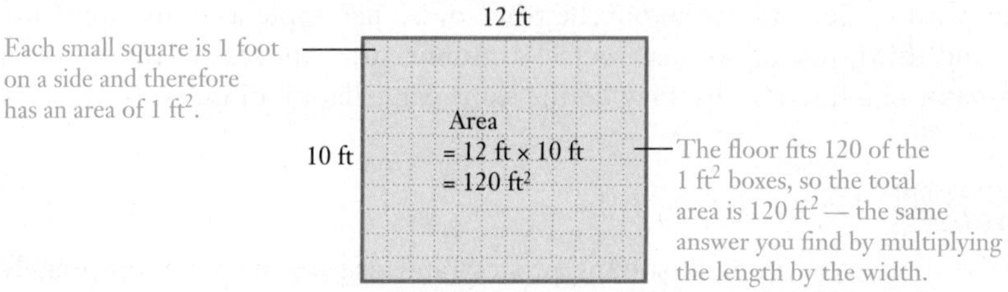

FIGURE 2.1 Multiplying a length in feet by a width in feet gives an area in *square feet* (ft^2).

- To find the *volume* of a box, we multiply its width, depth, and height (Figure 2.2). If the box is 6 inches wide, 4 inches deep, and 10 inches high, its volume is

$$6 \text{ in} \times 4 \text{ in} \times 10 \text{ in} = 240 \ (\text{in} \times \text{in} \times \text{in}) = 240 \text{ in}^3$$

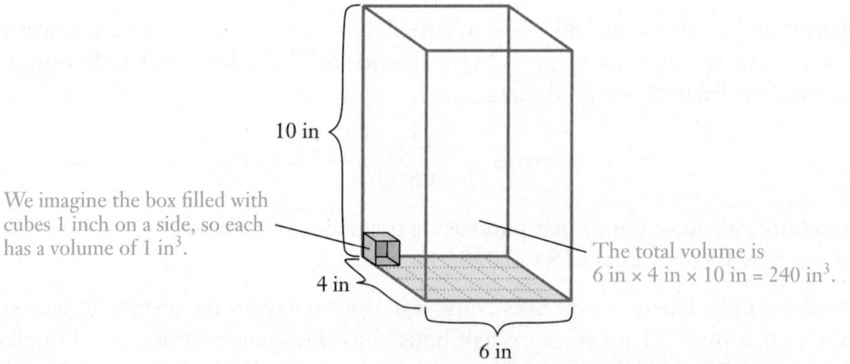

FIGURE 2.2 If we measure the side lengths of a box in inches, we calculate its volume in *cubic inches* (in^3).

  We read this volume as "240 cubic inches," noting that the key word *cubic* implies the third power. ▸ **Now try Exercises 25–26.**

So far, all the key words we've used with units should be familiar from everyday life. However, there is one more common key word, or rather key symbol, that often arises with units: a hyphen. For example, if you look at a utility bill, it will probably state electricity usage in units of "kilowatt-hours." The hyphen means multiplication. That is, if a movie set uses a 0.5-kilowatt light bulb for a time *of* 6 hours, its energy usage is

$$0.5 \text{ kilowatt} \times 6 \text{ hr} = 3 \text{ kilowatt} \times \text{hr} = 3 \text{ kilowatt-hr} \leftarrow \text{Read as "kilowatt-hours."}$$

EXAMPLE 3 Identifying Units

Identify the units of the answer in each of the following cases.

a. The price you paid for gasoline, found by dividing its total cost in dollars by the number of gallons of gas that you bought.
b. The area of a circle, found with the formula πr^2, where the radius r is measured in centimeters. (Note that π is a number without units.)
c. A volume found by multiplying an area measured in acres by a depth measured in feet.

Solution

a. The price of the gasoline has units of dollars divided by gallons, which we write as \$/gal and read as "dollars per gallon."
b. The area of the circle has units of centimeters to the second power, which we write as cm^2 and read as "square centimeters."
c. In this case, the volume has units of acres × feet, which we write and read as "acre-feet." This unit of volume is commonly used by hydrologists (water engineers) in the United States. ▶ Now try Exercises 27–32.

Think About It Find at least five numbers in news articles, and in each case identify the units of the number.

BY THE WAY

An *acre* was originally defined as the amount of land a pair of oxen could plow in a day. Today, it is defined as $\frac{1}{640}$ of a square mile, or 43,560 square feet (which is about 10% less than the area of an American football field without the end zones). An acre-foot is therefore equivalent to a volume of 43,560 ft^3.

Unit Analysis as a Problem-Solving Aid

You have probably already noticed that unit analysis can be very useful to our Understand-Solve-Explain process. In fact, unit analysis can be so powerful that scientists often start analyzing a problem by looking at its units, an approach that has led to important discoveries. Moreover, once we determine the expected units of an answer, we can use unit analysis to check our work and identify any errors. The following box summarizes some of the key ideas of unit analysis, and Examples 4 and 5 then illustrate how to use unit analysis to help find pathways to solutions.

Summary of Unit Analysis

- You generally cannot add or subtract numbers with different units, but you can combine different units through multiplication, division, or raising to powers. Always perform all operations on both the numbers and their associated units.
- The following key words will help you identify the appropriate operations:

Key Word or Symbol	Operation	Example
per	Division	Read miles ÷ hours as "miles per hour."
of or hyphen	Multiplication	Read kilowatts × hours as "kilowatt-hours."
square	Raising to second power	Read ft × ft, or ft^2, as "square feet" or "feet squared."
cube or *cubic*	Raising to third power	Read ft × ft × ft, or ft^3, as "cubic feet" or "feet cubed."

- It is easier to keep track of units if you replace division with multiplication by the reciprocal. For example, instead of dividing by 60 $\frac{\text{s}}{\text{min}}$, multiply by $\frac{1 \text{ min}}{60 \text{ s}}$.
- When you complete your calculations, make sure that your answer has the units you expected. If it doesn't, you've done something wrong.

EXAMPLE 4 Using Units to Find a Pathway

How many crates do you need to hold 2000 apples if each crate holds 40 apples?

Solution Although you might immediately know how to solve this problem, we'll apply unit analysis in the "understand" step to show how it leads us to a pathway to the solution.

Understand: The question asks "how many crates," so the answer should have units of *crates*. Notice that the statement "each crate holds 40 apples" implies that we can fit 40 apples *per* crate, which we write as 40 apples/crate. We now look for a way to combine 2000 apples and 40 apples/crate that gives us an answer with the correct units. If you try various operations, you'll find that the only one that makes sense and gets us to the correct final units (crates) is to divide 2000 apples by 40 apples/crate.

Solve: We carry out the calculation, dividing 2000 apples by the crate capacity of 40 apples/crate. As usual, we replace division with multiplication by the reciprocal ("invert and multiply"), which makes it much easier to see that the units of "apples" cancel and leave the final answer in units of *crates*.

$$2000 \text{ apples} \div \underbrace{\frac{40 \text{ apples}}{\text{crate}}}_{\substack{\text{invert} \\ \text{and multiply}}} = 2000 \cancel{\text{apples}} \times \frac{1 \text{ crate}}{40 \cancel{\text{apples}}} = 50 \text{ crates}$$

Explain: Using crates that hold 40 apples each, you need 50 crates to hold 2000 apples.

▸ **Now try Exercises 33–40.**

EXAMPLE 5 Exam Check

You are a grader for a math course. An exam question reads: "Eli purchased 5 pounds of apples at a price of 50 cents per pound. How much did he pay for the apples?" On the paper you are grading, a student has written: "50 ÷ 5 = 10. He paid 10 cents." Write a note to the student explaining what went wrong.

Solution Dear student: First, notice that your answer does not make sense. If 1 pound of apples costs 50¢, how could 5 pounds cost only 10¢? You could have prevented your error by keeping track of the units. In the exam question, the number 50 has units of *cents per pound* and the number 5 has units of *pounds*. Therefore, your calculation "50 ÷ 5 = 10" means

$$50\frac{¢}{\text{lb}} \div 5 \text{ lb} = 50\frac{¢}{\text{lb}} \times \frac{1}{5 \text{ lb}} = 10\frac{¢}{\text{lb}^2}$$

(As usual, we replaced the division with multiplication by the reciprocal.) Your calculation gives units of "cents per square pound," so it cannot be correct for a question that asks for a price. The correct calculation multiplies the price per pound by the weight in pounds:

$$50\frac{¢}{\cancel{\text{lb}}} \times 5 \cancel{\text{lb}} = 250¢ = \$2.50$$

The units now cancel as they should: The 5 pounds of apples cost $2.50.

▸ **Now try Exercises 41–44.**

Unit Conversions

Another important use of unit analysis comes in converting numbers from one unit to another unit, such as from miles to kilometers or quarts to cups. As a simple example, suppose we want to convert 2 feet to inches. Because 1 foot is the same as 12 inches, we do the conversion as follows:

$$2 \text{ ft} = 2 \cancel{\text{ft}} \times \underbrace{\frac{12 \text{ in}}{1 \cancel{\text{ft}}}}_{\substack{= 1, \text{ because} \\ 12 \text{ in} = 1 \text{ ft}}} = 24 \text{ in}$$

Notice that, in multiplying by $\frac{12 \text{ in}}{1 \text{ ft}}$, we've really just multiplied by 1, because 12 inches and 1 foot are equal. This idea extends to all unit conversions: We always multiply by an appropriate form of 1 so we don't change the meaning of the original expression. For example, the following are all different ways of writing 1:

$$1 = \frac{1}{1} = \frac{8}{8} = \frac{\frac{1}{4}}{\frac{1}{4}} = \frac{1 \text{ kilogram}}{1 \text{ kilogram}} = \frac{1 \text{ week}}{7 \text{ days}} = \frac{12 \text{ inches}}{1 \text{ foot}}$$

The last expression (which we used in our example) shows the necessity of stating units: 12 ÷ 1 is *not* 1, but 12 in ÷ 1 ft *is* 1.

Conversion Factors

The term $\frac{12 \text{ in}}{1 \text{ ft}}$, which is equal to 1, is often called a **conversion factor**. We can write this conversion factor in three equivalent ways:

$$12 \text{ in} = 1\text{ft} \qquad \text{or} \qquad \frac{12 \text{ in}}{1 \text{ ft}} = 1 \qquad \text{or} \qquad \frac{1 \text{ ft}}{12 \text{ in}} = 1$$

You can solve any unit conversion problem simply by identifying (or looking up) the conversion factor that gives you an appropriate form of 1.

EXAMPLE 6 Inches to Feet

Convert a length of 102 inches to feet.

Solution We start with the term 102 inches on the left, as shown below. Our goal is to multiply this term by 1 in a form that will change the units from inches to feet. We therefore use the conversion factor in the form that has *inches* in the denominator, so that *inches* cancel:

$$102 \text{ in} = 102 \cancel{\text{in}} \times \underbrace{\frac{1 \text{ ft}}{12 \cancel{\text{in}}}}_{1} = 8.5 \text{ ft}$$

A length of 102 inches is equal to 8.5 feet.

▶ Now try Exercises 45–46.

EXAMPLE 7 Seconds to Minutes

Convert a time of 3000 seconds into minutes.

Solution There are 60 seconds in 1 minute, so we can write the conversion factor in the following three forms:

$$1 \text{ min} = 60 \text{ s} \quad \text{or} \quad \frac{1 \text{ min}}{60 \text{ s}} = 1 \quad \text{or} \quad \frac{60 \text{ s}}{1 \text{ min}} = 1$$

We start with the term 3000 seconds on the left and note that the middle form of the conversion factor causes *seconds* to cancel, giving an answer in minutes:

$$3000 \text{ s} = 3000 \cancel{\text{s}} \times \frac{1 \text{ min}}{60 \cancel{\text{s}}} = \frac{3000}{60} \text{ min} = 50 \text{ min}$$

A time of 3000 seconds is equal to 50 minutes. ▶ Now try Exercises 47–48.

Think About It In Example 7, suppose you accidentally used the third form of the conversion factor ($\frac{60 \text{ s}}{1 \text{ min}}$). What units would your answer have? How would you know that you'd done the problem incorrectly?

EXAMPLE 8 Using a Chain of Conversions

How many seconds are there in one day?

Solution Most of us don't immediately know the answer to this question, but we do know that 1 day = 24 hr, 1 hr = 60 min, and 1 min = 60 s. We can answer the question by setting up a *chain* of unit conversions in which we start with *day* and end up with *seconds*:

$$1 \cancel{\text{day}} \times \frac{24 \cancel{\text{hr}}}{1 \cancel{\text{day}}} \times \frac{60 \cancel{\text{min}}}{1 \cancel{\text{hr}}} \times \frac{60 \text{ s}}{1 \cancel{\text{min}}} = 86{,}400 \text{ s}$$

Brief Review Decimal Fractions

For a fraction in decimal form, each digit corresponds to a certain place value, which is always a power of 10 (such as 10, 100, 1000, ...). The following example shows values for the decimal places in the number 3.141.

3	.	1	4	1
ones (1)	(decimal point)	tenths $(0.1 = \frac{1}{10})$	hundredths $(0.01 = \frac{1}{100})$	thousandths $(0.001 = \frac{1}{1000})$

Converting to Common Form

Converting a fraction from decimal to common form requires recognizing the value of the last digit in the decimal. For example:

$$0.4 = \frac{4}{10} \qquad 3.15 = \frac{315}{100}$$

$$0.097 = \frac{97}{1000}$$

Converting to Decimal Form

To convert a common fraction into decimal form, we carry out the division implied by the fraction. For example:

$$\frac{1}{4} = 1 \div 4 = 0.25$$

Many common fractions cannot be written *exactly* in decimal form. For example, the decimal form of $\frac{1}{3}$ contains an endless string of threes:

$$\frac{1}{3} = 0.3333333\ldots$$

In mathematics, we often represent such a pattern with a bar. For example, the bar over the 3 in $0.\overline{3}$ means that the 3 repeats indefinitely. In daily life, we usually round repeating decimals, and $\frac{1}{3}$ is often rounded to 0.33. ▶ Now try Exercises 15–18.

By using the conversion factors needed to cancel the appropriate units, we are left with the answer in *seconds*. (Notice the similarity to the conversion chain used in Example 1.) There are 86,400 seconds in one day. ▶ Now try Exercises 49–52.

Conversions with Units Raised to Powers

We must take special care when converting units raised to powers. For example, suppose we want to know the number of square feet in a square yard. We may not know the conversion factor between square yards (yd^2) and square feet (ft^2), but we know that 1 yd = 3 ft. Therefore, we can replace 1 yard by 3 feet when we write out 1 square yard:

$$1\ \text{yd}^2 = 1\ \text{yd} \times 1\ \text{yd} = 3\ \text{ft} \times 3\ \text{ft} = 9\ \text{ft}^2$$

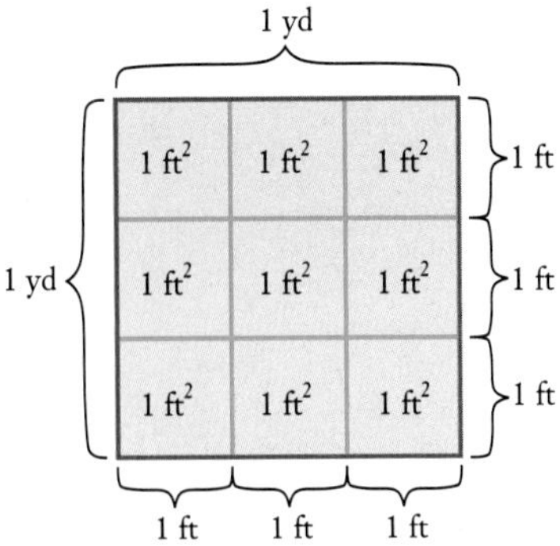

FIGURE 2.3 Notice that 1 square yard contains 9 square feet.

That is, 1 square yard is the same as 9 square feet. We can also find this conversion factor by squaring both sides of the yards-to-feet conversion:

$$1\ \text{yd} = 3\ \text{ft} \xrightarrow[\text{both sides}]{\text{square}} (1\ \text{yd})^2 = (3\ \text{ft})^2 \xrightarrow[\text{each term}]{\text{square}} 1\ \text{yd}^2 = 9\ \text{ft}^2$$

Figure 2.3 confirms that 9 square feet fit exactly into 1 square yard. As usual, we can write the conversion factor in three equivalent forms:

$$1\ \text{yd}^2 = 9\ \text{ft}^2 \quad \text{or} \quad \frac{1\ \text{yd}^2}{9\ \text{ft}^2} = 1 \quad \text{or} \quad \frac{9\ \text{ft}^2}{1\ \text{yd}^2} = 1$$

EXAMPLE 9 Carpeting a Room

You want to carpet a room that measures 10 feet by 12 feet, making an area of 120 square feet. But carpet is usually sold by the square yard. How many square yards of carpet do you need?

Solution We need to convert the room's area from units of *square feet* to *square yards*, so we use the conversion factor in the form that has square feet (ft^2) in the denominator:

$$120\ \text{ft}^2 \times \frac{1\ \text{yd}^2}{9\ \text{ft}^2} = \frac{120}{9}\ \text{yd}^2 \approx 13.3\ \text{yd}^2$$

The symbol ≈ means "approximately equal to."

Note that we rounded the answer to the nearest 0.1 square yard. Most stores will not sell fractions of a square yard, so you will need to buy 14 square yards of carpet for a 10-foot by 12-foot room. ▶ Now try Exercises 53–56.

Think About It Although we did not list the individual Understand-Solve-Explain steps in Example 9 (and won't in most other cases throughout this book), they are implicit in the solution. Describe how we could have written the three steps. (Note: If you need more problem-solving practice, you may find it useful to also identify the three steps on the examples that follow.)

EXAMPLE 10 Cubic Units: Purchasing Garden Soil

You are preparing a vegetable garden that is 40 feet long and 16 feet wide, and you need enough soil to fill it to a depth of 1 foot. The landscape supply store sells soil by the cubic yard. How much soil should you order?

Solution To find the volume of soil that should be ordered, we multiply the garden's length (40 feet), width (16 feet), and depth (1 foot):

$$40 \text{ ft} \times 16 \text{ ft} \times 1 \text{ ft} = 640 \text{ ft}^3$$

Because soil is sold by the cubic yard, we need to convert this volume from units of *cubic feet* to *cubic yards*. We know that 1 yd = 3 ft, so we find the required conversion factor by cubing both sides of this equation:

$$1 \text{ yd} = 3 \text{ ft} \xrightarrow[\text{both sides}]{\text{cube}} (1 \text{ yd})^3 = (3\text{ft})^3 \xrightarrow[\text{each term}]{\text{cube}} 1 \text{ yd}^3 = 27 \text{ ft}^3$$

In the last step, we recognized that $3^3 = 3 \times 3 \times 3 = 27$. As usual, we can write this conversion factor in three equivalent forms:

$$1 \text{ yd}^3 = 27 \text{ ft}^3 \quad \text{or} \quad \frac{1 \text{ yd}^3}{27 \text{ ft}^3} = 1 \quad \text{or} \quad \frac{27 \text{ ft}^3}{1 \text{ yd}^3} = 1$$

To convert the soil volume from cubic feet to cubic yards, we use the conversion factor that has cubic feet (ft^3) in the denominator:

$$640 \text{ ft}^3 \times \frac{1 \text{ yd}^3}{27 \text{ ft}^3} = \frac{640}{27} \text{ yd}^3 \approx 23.7 \text{ yd}^3$$

You will need to order about 24 cubic yards of soil for your garden.

▶ **Now try Exercises 57–60.**

Currency Conversions

A particularly important type of conversion in everyday life is from one country's money, or **currency**, to another's. Converting between currencies is a unit conversion problem in which the conversion factors are known as the *exchange rates*. Table 2.1 shows a typical table of currency exchange rates:

- Use the numbers in the column "Dollars per Foreign" to convert from foreign currency into U.S. dollars. For example, this column shows that 1 euro = \$1.058.
- Use the numbers in the column "Foreign per Dollar" to convert U.S. dollars into foreign currency. For example, this column shows that \$1 = 0.9449 euro.

USING TECHNOLOGY

Currency exchange rates are constantly changing, but you can get current rates by typing "exchange rates" into any search engine. Google will also do direct conversions; for example, typing "25 euros in dollars" will tell you today's dollar value of 25 euros.

Think About It Find the reciprocals of the numbers in the "Dollars per Foreign" column of Table 2.1. Are your results the numbers in the "Foreign per Dollar" column? Why or why not?

TABLE 2.1 Sample Currency Exchange Rates (January 2017)

Currency	Dollars per Foreign	Foreign per Dollar
British pound	1.221	0.8191
Canadian dollar	0.7586	1.318
European euro	1.058	0.9449
Japanese yen	0.008658	115.5
Mexican peso	0.04574	21.86

EXAMPLE 11 Price Conversion

At a French department store, the price for a pair of jeans is 85 euros. What is the price in U.S. dollars? Use the exchange rates in Table 2.1.

Solution In the "Dollars per Foreign" column in Table 2.1, we see that 1 euro = \$1.058. As usual, we can write this conversion factor in two other equivalent forms:

$$\frac{1 \text{ euro}}{\$1.058} = 1 \quad \text{or} \quad \frac{\$1.058}{1 \text{ euro}} = 1$$

We use the latter form to convert the price from euros to dollars:

$$85 \text{ euro} \times \frac{\$1.058}{\text{euro}} = \$89.93$$

The price of 85 euros is equivalent to \$89.93. ▶ Now try Exercises 61–62.

EXAMPLE 12 Buying Currency

You are on holiday in Mexico and need cash. How many pesos can you buy with \$100? Use the exchange rates in Table 2.1 and ignore any transaction fees.

Solution In the "Foreign per Dollar" column in Table 2.1, we see that \$1 = 21.86 pesos. We use this conversion in the form that puts dollars in the denominator so that the dollars cancel:

$$\$100 \times \frac{21.86 \text{ pesos}}{\$1} = 2186 \text{ pesos}$$

Your \$100 will buy 2186 pesos. ▶ Now try Exercises 63–64.

EXAMPLE 13 Gas Price per Liter

A gas station in Canada sells gasoline for CAD 1.34 per liter. (CAD is an abbreviation for Canadian dollars.) What is the price in dollars per gallon? Use the currency exchange rate in Table 2.1 and the fact that 1 gallon is equivalent to 3.785 liters.

Solution We use a chain of conversions to convert from CAD to dollars and then from liters to gallons. From Table 2.1, the currency conversion is \$0.7586 per CAD, and we use the fact that 1 gallon = 3.785 liters implies a conversion of 3.785 liters per gallon:

$$\frac{1.34 \text{ CAD}}{1 \text{ L}} \times \frac{\$0.7586}{1 \text{ CAD}} \times \frac{3.785 \text{ L}}{1 \text{ gal}} \approx \frac{\$3.85}{1 \text{ gal}}$$

The price of the gasoline is about \$3.85 per gallon. ▶ Now try Exercises 65–66.

BY THE WAY

The euro, which was first adopted in 1999 (though coins and notes did not circulate until 2002), is the official currency of the Eurozone, which includes most but not all of the nations in the European Union. The euro was created both to make financial transactions easier between nations (which previously all had their own currencies) and to help build a common market. However, critics believe its use has hurt countries with weaker economies, and as of 2017, several countries are threatening to withdraw from the Eurozone and return to using their own currencies.

IN YOUR WORLD

Changing Money in Foreign Countries

It costs money to exchange currency, so you should always look for the best deal. Two factors affect the cost of exchanging currency: (1) the exchange rate and (2) fees for the exchange.

Published exchange rates are usually "wholesale" rates available only to banks. Most money changers—including ATMs, airport exchange stations, hotels, and exchange booths on streets—make money by giving you a rate that is not as good as the wholesale rate. For example, if the wholesale rate is 21.9 pesos per dollar, a money changer might give you only 20.1 pesos per dollar.

Fees can also affect the cost of exchanging currency. Many money changers (especially in hotels, stores, and street booths) charge a fee every time you make an exchange. Here are a few general hints for exchanging currency:

- Before you travel, get a small amount of your destination's currency from your local bank. That way, if you need cash upon arrival, you won't be forced to accept a poor exchange rate or high fees.
- Once you get to your destination, banks usually offer better exchange rates than other money changers. The best deal may be your ATM card, but be sure to find out whether your bank charges a fee for foreign ATM transactions.
- Consider using a credit card for purchases and hotel and restaurant bills. Credit cards generally offer good exchange rates and, unlike cash, can be replaced if lost or stolen. However, most credit cards add fees for foreign purchases. Avoid using your credit card to get cash, unless it is an emergency, because fees and interest rates are usually especially high for cash advances.

Quick Quiz 2A

Choose the best answer to each of the following questions. Explain your reasoning with one or more complete sentences.

1. To end up with units of speed, you need to
   a. multiply a distance by a time.
   b. divide a distance by a time.
   c. divide a time by a distance.
2. What does the word *per* mean?
   a. divided by
   b. multiplied by
   c. in addition to
3. What does the word *of* mean?
   a. divided by
   b. multiplied by
   c. in addition to
4. You are given two pieces of information: (1) the price of gasoline in dollars per gallon and (2) the gas mileage of a car in miles per gallon. You are asked to find the cost of driving this car in dollars per mile. You should
   a. divide the price of gas by the car's gas mileage.
   b. multiply the price of gas by the car's gas mileage.
   c. divide the car's gas mileage by the price of gas.
5. Which of the following represents 9 square miles?
   a. a line of small squares that is 9 miles long
   b. a square 3 miles on a side
   c. a square 9 miles on a side
6. If you multiply an area in square feet by a height in feet, the result will have units of
   a. feet.
   b. $feet^2$.
   c. $feet^3$.
7. There are 1760 yards in a mile. Therefore, 1 cubic mile represents
   a. 1760 square yards.
   b. 1760^3 $yards^3$.
   c. 1760^3 yards.
8. One square foot is equivalent to
   a. 12 square inches.
   b. 120 square inches.
   c. 144 square inches.

9. You buy apples while traveling in Europe. The price is most likely to be quoted in
   a. euros per kilogram.
   b. euros per liter.
   c. euros per meter.

10. If the current exchange rate is \$1.058 per euro, then
   a. \$1 is worth more than 1 euro.
   b. 1 euro is worth more than \$1.
   c. \$1 is worth 0.8 euros.

Exercises 2A

REVIEW QUESTIONS

1. Briefly describe the Understand-Solve-Explain approach to problem solving, giving examples of things you should consider in each step.
2. What are *units*? Briefly describe how units can help you check your answers and solve problems. Give examples.
3. Describe the meaning of the key words *per*, *of*, *square*, and *cube*, and explain the meaning of a hyphen in units.
4. Explain why a unit conversion involves just multiplying by 1. Find the two forms of 1 that are equivalent to the conversion 1 lb = 16 oz.
5. Explain in words or with a picture why there are 9 square feet in 1 square yard and 27 cubic feet in 1 cubic yard. Then describe generally how to find conversion factors involving squares or cubes.
6. Describe how to read and use the currency data in Table 2.1.

DOES IT MAKE SENSE?

Decide whether each of the following statements makes sense (or is clearly true) or does not make sense (or is clearly false). Explain your reasoning.

7. I drove at a speed of 35 miles for the entire trip.
8. One dollar is worth about 115 Japanese yen, so 1 yen is worth more than \$1.
9. I figured out how long the airplane will take to reach Beijing by dividing the distance to Beijing by the airplane's speed.
10. I have a box with a volume of 2 square feet.
11. I bought the fabric at X-mart, because their price of \$3 per square foot was better than Y-mart's price of \$15 per square yard.
12. An actual commercial: "The Goodyear Aquatread tire can channel away 1 gallon of water per second. One gallon per second—that's 396 gallons per mile."

BASIC SKILLS & CONCEPTS

13–18: Math Review. The following exercises require the skills covered in the Brief Review boxes in this unit.

13. Evaluate the following expressions.
   a. $\frac{3}{4} \times \frac{1}{2}$ b. $\frac{2}{3} \times \frac{3}{5}$ c. $\frac{1}{2} + \frac{3}{2}$ d. $\frac{2}{3} + \frac{1}{6}$
   e. $\frac{2}{3} \times \frac{1}{4}$ f. $\frac{1}{4} + \frac{3}{8}$ g. $\frac{5}{8} - \frac{1}{4}$ h. $\frac{3}{2} \times \frac{2}{3}$

14. Evaluate the following expressions.
   a. $\frac{1}{3} + \frac{1}{5}$ b. $\frac{10}{3} \times \frac{3}{7}$ c. $\frac{3}{4} - \frac{1}{8}$ d. $\frac{1}{2} + \frac{2}{3} + \frac{3}{4}$
   e. $\frac{6}{5} + \frac{4}{15}$ f. $\frac{3}{5} \times \frac{2}{7}$ g. $\frac{1}{3} + \frac{13}{6}$ h. $\frac{3}{5} \times \frac{10}{3} \times \frac{3}{2}$

15. Write each of the following as a common fraction.
   a. 3.5 b. 0.3 c. 0.05 d. 4.1
   e. 2.15 f. 0.35 g. 0.98 h. 4.01

16. Write each of the following as a common fraction.
   a. 2.75 b. 0.45 c. 0.005 d. 1.16
   e. 6.5 f. 4.123 g. 0.0003 h. 0.034

17. Convert the following fractions to decimal form; round to the nearest thousandth if necessary.
   a. $\frac{1}{4}$ b. $\frac{3}{8}$ c. $\frac{2}{3}$ d. $\frac{3}{5}$
   e. $\frac{13}{2}$ f. $\frac{23}{6}$ g. $\frac{103}{50}$ h. $\frac{42}{26}$

18. Convert the following fractions to decimal form; round to the nearest thousandth if necessary.
   a. $\frac{1}{5}$ b. $\frac{4}{9}$ c. $\frac{4}{11}$ d. $\frac{12}{7}$
   e. $\frac{28}{9}$ f. $\frac{56}{11}$ g. $\frac{102}{49}$ h. $\frac{15}{4}$

19–20: Using Understand-Solve-Explain. Write out the three steps of the Understand-Solve-Explain process when answering the following questions.

19. How much does a typical water bed weigh? Useful data: 1 cubic foot of water weighs 64.2 pounds and a typical water bed holds 28 cubic feet of water.
20. How fast does Earth rotate? Useful data: In one full rotation of Earth, a point on the equator travels about 25,000 miles in 24 hours.

21–24: Using Key Words. Write out the three steps of the Understand-Solve-Explain process when answering the following questions. Show operations and units clearly.

21. How much will you pay for 2.5 pounds of apples at a price of \$1.25 per pound?
22. What is the total weight of 11 basketballs that weigh 22 ounces each?
23. If you are paid \$420 for 23 hours of work, what is your hourly wage?

24. How many buses are need to transport 495 people if each bus holds 45 people?

25. **Area and Volume Calculations.** Show clearly your use of units for the following calculations.

    a. A storage pod has a rectangular floor that measures 20 feet by 12 feet and a flat ceiling that is 8 feet above the floor. Find the area of the floor and the volume of the pod.

    b. A lap pool has a length of 25 yards, a width of 20 yards, and a depth of 2 yards. Find the pool's surface area (the water surface) and the total volume of water that the pool holds.

    c. A raised flower bed is 30 feet long, 6 feet wide, and 1.2 feet deep. Find the area of the bed and the volume of soil it holds.

26. **Area and Volume Calculations.** Show clearly your use of units for the following calculations.

    a. A warehouse is 60 yards long, 30 yards wide, and 6 yards high. What is the area of the warehouse floor? If the warehouse is filled to *half* its height with tightly packed boxes, what is the volume of the boxes?

    b. A room has a rectangular floor that measures 24 feet by 16 feet and a flat 8-foot ceiling. What is the area of the floor and how much air does the room hold?

    c. A grain silo has a circular base with an area of 260 square feet and is 22 feet tall. What is its total volume?

27–32: Identifying Units. Identify the units of the following quantities. State the units mathematically (for example, mi/hr) and in words (for example, miles per hour).

27. Your average speed on a bike ride, found by dividing distance traveled in miles by the elapsed time in hours

28. The per-mile price of an airline ticket, found by dividing the distance traveled in miles by the price of the ticket in dollars.

29. The flow rate of a shower head, found by dividing the volume of water that leaves the shower head in cubic inches during a period of time measured in seconds

30. The price of a bottle of French perfume, found by multiplying the unit price of the perfume in euros per milliliter by the volume of the bottle in milliliters

31. The cost of car trip, found by dividing the cost of gas in dollars per gallon by the gas mileage of the car in miles per gallon and multiplying that result by the length of the trip in miles

32. The number of bagels produced by a bakery, found by multiplying the production rate in units of bagels per baker per hour by the number of hours and by the number of bakers

33–40: Working with Units. Write out the three steps of the Understand-Solve-Explain process when answering the following questions.

33. What is the total cost of 1.2 cubic yards of soil if it sells for \$24 per cubic yard?

34. A hose fills a hot tub at a rate of 4.5 gallons per minute. How many minutes will it take to fill a 400-gallon hot tub?

35. How much would you pay for 2.5 ounces of gold at a price of \$1200 per ounce?

36. Suppose you earn \$10.30 per hour and work 24 eight-hour days in a month. How much do you earn in that month?

37. In 2016, 595,700 Americans died of (all forms of) cancer. Assuming a population of 321 million, what was the mortality rate in units of deaths per 100,000 people?

38. The city with the greatest population density is Manila in the Philippines. The city has a population of 1.8 million people and an area of 16.6 square miles. What is its population density in units of people per square mile?

39. If your car gets 32 miles per gallon, how much does it cost you to drive 30 miles when gasoline costs \$2.55 per gallon?

40. Los Angeles Dodgers pitcher Clayton Kershaw has a seven-year contract that gives him an average annual salary of \$30.7 million. Assuming he starts 30 games in a season (and his salary is solely for pitching), how much is he paid per game?

41–44: What Went Wrong? Consider the following exam questions and student solutions. Determine whether each solution is correct. If it is not correct, write a note to the student explaining why it is wrong and give a correct solution.

41. *Exam Question:* A candy store sells chocolate for \$7.70 per pound. The piece you want to buy weighs 0.11 pound. How much will it cost, to the nearest cent? (Neglect sales tax.)

    Student Solution: $0.11 \div 7.70 = 0.014$. It will cost \$0.014.

42. *Exam Question:* You ride your bike up a steep mountain road at 5 miles per hour. How far do you go in 3 hours?

    Student Solution: $5 \div 3 = 1.7$. I ride 1.7 miles.

43. *Exam Question:* You can buy a 50-pound bag of flour for \$11 or you can buy a 1-pound bag for \$0.39. Compare the per pound price for the large and small bags.

    Student Solution: The price of the large bag is $50 \div \$11 = \$4.55$ per pound, which is much more than the 39¢ per pound for the small bag.

44. *Exam Question:* The average person needs 1500 Calories per day. A can of Coke contains 140 Calories. How many Cokes would you have to drink to fill your daily caloric need? (*Note*: This diet would not meet other nutritional needs!)

    Student Solution: $1500 \times 140 = 210{,}000$. You would have to drink 210,000 Cokes to meet your daily caloric need.

45–52: Unit Conversions. Carry out the following unit conversions. Where necessary, round to the nearest hundredth.

45. Convert 32 feet to inches.

46. Convert 16 feet to yards.

47. Convert 35 minutes to seconds.

48. Convert 17 years to days (neglecting leap years).

49. Convert 4.2 hours to seconds.

50. Convert the Space Station's orbital speed of 17,200 miles per hour to miles per second.

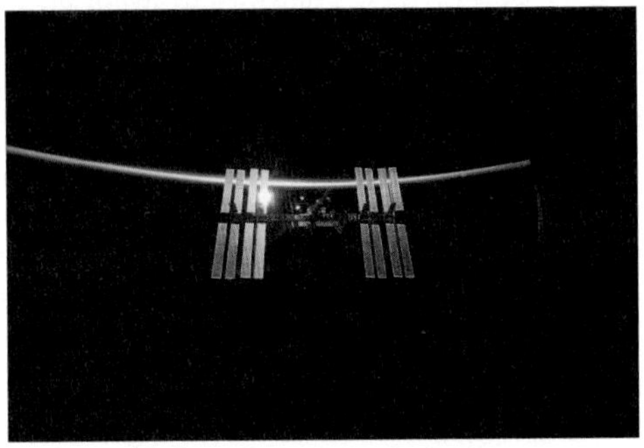

51. Convert 4 years to hours (neglecting leap years).

52. Convert 45,789 inches to miles, using these facts: 1 mi = 1760 yd, 1 yd = 3 ft, and 1 ft = 12 in.

53–60: Conversions with Square and Cubic Units.

53. Find a conversion factor between square feet and square inches. Write it in three forms.

54. Find the area in square feet of a rectangular yard that measures 20 yards by 12 yards.

55. How many square feet of land did you buy if you purchased a 3.5-acre lot? (1 acre = 43,560 ft^2)

56. What is the area in square feet of a 100 yard by 60 yard football field?

57. Find a conversion factor between cubic meters and cubic centimeters. Write it in three forms.

58. A new sidewalk will be 4 feet wide and 150 feet long and filled to a depth of 6 inches (0.5 foot) with concrete. How many cubic yards of concrete are needed?

59. An air conditioning system can circulate 350 cubic feet of air per minute. How many cubic yards of air can it circulate per minute?

60. A hot tub pump circulates 3.5 cubic feet of water per minute. How many cubic inches of water does it circulate each minute?

61–66: Currency Conversions. Use the currency exchange rates in Table 2.1 for the following questions.

61. Your dinner in London cost 82 British pounds. How much was it in U.S. dollars?

62. Your hotel rate in Tokyo is 45,000 yen per night. What is the nightly rate in U.S. dollars?

63. As you leave Paris, you convert 320 euros to dollars. How many dollars do you receive?

64. You return from Mexico with 2500 pesos. How much are they worth in U.S. dollars?

65. Gasoline sells for 1.5 euros/liter in Bonn. What is the price in U.S. dollars per gallon? (1 gal = 3.785 L)

66. You purchase fresh strawberries in Mexico for 28 pesos per kilogram. What is the price in U.S. dollars per pound? (1 kg = 2.205 lb)

FURTHER APPLICATIONS

67–71: More Unit Practice. Use unit analysis to answer the following questions.

67. A commuter train travels 45 miles in 34 minutes. What is its speed in miles per hour?

68. Competition speed skydivers have reached record speeds of 614 miles per hour. At this speed, how many feet would you fall every second?

69. In 2015, there were approximately 3,998,000 million births in the United States. Find the birth rate in units of births per minute. Assuming a U.S. population of 321 million, what was the annual birth rate in units of births per 1000 people?

70. If you sleep an average of 7.5 hours each night, how many hours do you sleep in a year?

71. Assume an average human heart beats 65 times per minute and an average human lifetime is 80 years. How many times does the average heart beat in an average lifetime?

72–75: Gas Mileage. Answer the following practical gas mileage questions.

72. You plan to take a 2000-mile trip in your car, which averages 32 miles per gallon. How many gallons of gasoline should you expect to use? Would a car whose gas mileage is half as much (16 miles per gallon) require twice as much gasoline for the same trip? Explain.

73. Two friends take a 3000-mile cross-country trip together, but they drive their own cars. Car A has a 12-gallon gas tank and averages 40 miles per gallon, while car B has a 20-gallon gas tank and averages 30 miles per gallon. Assume both drivers pay an average of \$2.55 per gallon of gas.

a. What is the cost of one full tank of gas for car A? For car B?

b. How many tanks of gas do cars A and B each use for the trip?

c. About how much do the drivers of cars A and B each pay for gas for the trip?

74. Suppose your car averages 38 miles per gallon on the highway if your average speed is 55 miles per hour, and it averages 32 miles per gallon on the highway if your average speed is 70 miles per hour.

a. What is the driving time for a 2000-mile trip if you drive at an average speed of 55 miles per hour? What is the driving time at 70 miles per hour?

b. Assume a gasoline price of \$2.55 per gallon. What is the gasoline cost for a 2000-mile trip if you drive at an average speed of 55 miles per hour? What is the gasoline cost at 70 miles per hour?

75. Suppose your car averages 32 miles per gallon on the highway if your average speed is 60 miles per hour, and it averages 25 miles per gallon on the highway if your average speed is 75 miles per hour.

a. What is the driving time for a 1500-mile trip if you drive at an average speed of 60 miles per hour? What is the driving time at 75 miles per hour?

b. Assume a gasoline price of $2.55 per gallon. What is the gasoline cost for a 1500-mile trip if you drive at an average speed of 60 miles per hour? What is the gasoline cost at 75 miles per hour?

76. **Professional Basketball Salaries.** In the 2016–2017 season, Lebron James earned $31 million for playing 80 games, each lasting 48 minutes. (Assume no overtime games.)

a. How much did James earn per game?

b. Assuming that James played every minute of every game, how much did he earn per minute?

c. Suppose, averaged over the course of a year, James spent a total of 40 hours training for and playing in each of the 80 games. Including the training time, what was his hourly salary?

77. **Full of Hot Air.** The average person breathes 6 times per minute (at rest), inhaling and exhaling half a liter of air each time. How much "hot air" (air warmed by the body), in liters, does the average person exhale each day?

78. **Busy Reading.** Suppose you have a tablet with a capacity of 16 gigabytes. For a book containing only plain text, 1 byte typically corresponds to one character and an average page contains 2000 characters. Assume all 16 gigabytes are used to hold plain-text books.

a. How many pages of text can the tablet hold?

b. How many 500-page books can the tablet hold?

79. **Landscaping Project.** Suppose that you are planning to landscape a portion of your yard that measures 60 feet by 35 feet. Determine the price of the needed items at a local store, and use those prices to answer the following questions.

a. How much would it cost to plant the region with grass seed (which is rated by the number of square feet that can be covered per pound of seed)?

b. How much would it cost to cover the region with sod?

c. How much would it cost to cover the region with high-quality topsoil and then plant two flowering bulbs per square foot?

IN YOUR WORLD

80. **Real-World Problem Solving.** Choose some major problem that has been in the news (for example, the cost of an infrastructure project or a tax policy change), and without actually solving it, describe how the Understand-Solve-Explain process might help in solving it.

81. **Teach Your Children.** Imagine that you are a teacher or parent dealing with a 7th grader who is stumped by "story problems" in a textbook. How would you go about helping this child? Be as specific as possible.

82. **Units on the Highway.** Next time you are on the highway, look for three signs that use numbers (such as speed limits or distances to nearest exits). Are the units of the numbers given? If not, how are you expected to know the units? In cases where the units are not given, do you think the units would be obvious to everyone? Why or why not?

83. **Are the Units Clear?** Find a news story that involves numerical data. Are all the numbers in the story given with meaningful units, or is the meaning of some of the units unclear? Briefly summarize how well (or how poorly) the news story uses units.

84. **South American Adventure.** Suppose you are planning an extended trip through many countries in South America. Use one of the many currency exchange sites on the Web to get all the exchange rates you'll need. Make a brief table showing each currency you'll need and the current value of each currency in dollars.

USING TECHNOLOGY

85. **Currency Conversions.** Use the Internet to convert $100 to the following currencies.

a. Brazilian reals
b. Israeli shekels
c. Moroccan dirhams
d. Russian rubles
e. Turkish lira
f. Chinese yuan
g. Colombian pesos
h. Indian rupees

UNIT 2B Extending Unit Analysis

We've seen how unit analysis can be useful in finding a path to the solution of a problem. We can do even more if we understand the units that commonly arise in a greater range of situations. For example, we can use units to help solve practical problems involving energy, density, and concentration—including the type of problem that opens the chapter on page 68. We'll now turn our attention to understanding the units for these problems, starting with an overview of the two commonly used systems of standardized units.

Standardized Unit Systems

Today, we take for granted that units like inches or feet have a clear and well-defined meaning. But most measurements were originally based on attributes that could vary from person to person. For example, the *foot* was once the length of the foot of whoever was doing the measuring, and our word *inch* comes from the Latin *uncia,* meaning "thumb-width." There are 12 inches in 1 foot because the Romans discovered that most adult feet are about 12 thumb-widths long. The Romans paced out longer distances. Our word *mile* comes from the Latin *milia passum*, meaning "one thousand paces."

Think About It How many of *your* thumb-widths fit along your bare foot? Were the Romans correct in believing there are 12 thumb-widths in a foot?

As you might imagine, lengths that vary from person to person can lead to difficulties. For example, if you are buying 10 feet of rope, should you measure the rope by the length of your foot or the seller's foot? For this reason, units eventually became standardized so that they would have the same meaning for everyone. Today, only two systems of standardized units enjoy substantial usage:

- In the United States, most of our daily measurements are made with what is often called the English system of units, but is more formally known as the *U.S. customary system* (USCS).
- The rest of the world uses the *international metric system*, known by the abbreviation SI (from the French *Système International d'Unités*), though customary units are sometimes still used for nonofficial purposes.

BY THE WAY
The United States is one of only three countries in the world that have not fully adopted the international metric system. The other two are Liberia and Myanmar, though both of them are in the process of conversion. However, use of the metric system has been legal in the United States since 1866, and the official definitions of U.S. customary units are based on their metric equivalents.

The U.S. Customary System

The U.S. customary system has roots dating back thousands of years, and its units became standardized in often surprising ways. For example, the modern length of 1 yard was defined by English King Henry I (1100–1135), who decreed it to be the distance from the tip of *his* nose to the tip of *his* thumb on *his* outstretched arm.

Table 2.2 summarizes the official U.S. customary system, showing standard units for length, weight, and volume. Note that the system is extremely complex, and

TABLE 2.2 The U.S. Customary System of Measurement (common abbreviations in parentheses)

LENGTHS	1 inch (in) = 2.54 centimeters	1 furlong = 40 rods = $\frac{1}{8}$ mile	
	1 foot (ft) = 12 inches	1 mile (mi) = 1760 yards = 5280 feet	
	1 yard (yd) = 3 feet	1 nautical mile = 1.852 km ≈ 6076.1 feet	
	1 rod = 5.5 yards	1 league (marine) = 3 nautical miles	
	1 fathom = 6 feet		
WEIGHTS	Avoirdupois	Troy	Apothecary
	1 grain = 0.0648 gram	1 grain = 0.0648 gram	1 grain = 0.0648 gram
	1 ounce (oz) = 437.5 grains	1 carat = 0.2 gram = 3.086 grains	1 scruple = 20 grains
	1 pound (lb) = 16 ounces	1 pennyweight = 24 grains	1 dram = 3 scruples
	1 ton = 2000 pounds	1 troy ounce = 480 grains	1 apoth. ounce = 8 drams
	1 long ton = 2240 pounds	1 troy pound = 12 troy ounces	1 apoth. pound = 12 ounces
VOLUMES	Liquid Measures		Dry Measures
	1 tablespoon (tbsp or T) = 3 teaspoons (tsp or t)		1 in^3 ≈ 16.387 cm^3
	1 fluid ounce (fl oz) = 2 tablespoons = 1.805 in^3		1 ft^3 = 1728 in^3 = 7.48 gallons
	1 cup (c) = 8 fluid ounces		1 yd^3 = 27 ft^3
	1 pint (pt) = 16 fluid ounces = 28.88 in^3		1 dry pint (pt) = 33.60 in^3
	1 quart (qt) = 2 pints = 57.75 in^3		1 dry quart (qt) = 2 dry pints = 67.2 in^3
	1 gallon (gal) = 4 quarts		1 peck = 8 dry quarts
	1 barrel of petroleum = 42 gallons		1 bushel = 4 pecks
	1 barrel of liquid = 31 gallons		1 cord = 128 ft^3

IN YOUR WORLD

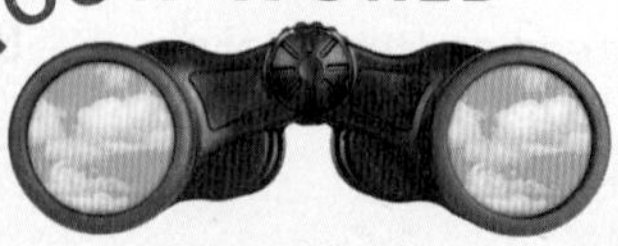

Gems and Gold Jewelry

If you've ever bought jewelry, you've probably seen labels stating karats or carats. But you may be surprised to know that these are not just alternative spellings of the same word. A *carat* is a unit of weight defined to be exactly 0.2 gram (see Table 2.2). A *karat* is a measure of the purity of gold: 24-karat gold is 100% pure, 18-karat gold is 75% pure, 14-karat gold is 58% pure (because $\frac{14}{24} \approx 0.58$), and so on.

If you are buying gold jewelry, look for a label stating both its purity in karats and its weight (usually given in grams); the amount of gold you actually buy depends on both. For example:

- Ten grams of 18-karat gold contains 7.5 grams of pure gold, because 18-karat gold is 75% pure.
- Ten grams of 14-karat gold contains 5.8 grams of pure gold, because 14-karat gold is 58% pure.

Although a higher karat value is purer, it comes with a practical tradeoff. Gold is a soft metal, and the metals (usually silver and copper) mixed into lower-karat gold give it added strength. That is why a 14-karat gold ring is generally stronger and more durable than an 18-karat gold ring.

You'll deal with carats (beginning with a c) if you are buying gems such as diamonds or emeralds. Because 1 carat is 0.2 gram, the number of carats tells you the precise weight of a gem. However, other factors besides weight will affect the gem's price, such as its shape and color. If you are gem shopping, you'll need to make tradeoffs among these factors to stay within your budget. For example, \$10,000 might buy either a 1-carat diamond with good color and clarity or a 2-carat diamond with poorer color and clarity. Given the cost of most gems, you should make careful price comparisons before buying.

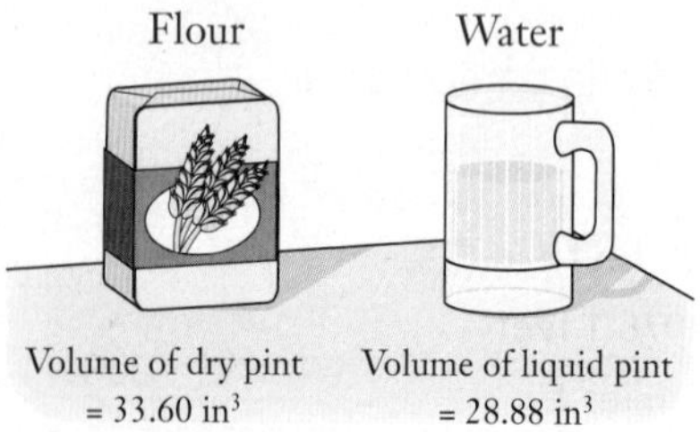

FIGURE 2.4 In the U.S. customary system, the volume of a *pint* depends on whether you are measuring dry materials or liquids.

the same words can have multiple meanings. For example, the volume of a *dry* pint is 33.60 cubic inches but a *liquid* pint is only 28.88 cubic inches, which means a container that holds 1 pint of water is too small for 1 pint of flour (Figure 2.4). (Also note that a "British pint" differs from both of these!) Even worse, the U.S. customary system has three official sets of units for weight: We most commonly use *avoirdupois* weights, but jewelers use *troy* weights and pharmacists traditionally used *apothecary* weights. (The basic unit of weight in all three sets is the *grain*, an ancient unit originally based on the weight of a typical grain of wheat.)

EXAMPLE 1 The Kentucky Derby

The length of the Kentucky Derby horse race is 10 furlongs. How long is the race in miles?

Solution From Table 2.2, 1 furlong $= \frac{1}{8}$ mile $= 0.125$ mile. As discussed in Unit 2A, we can write this conversion factor in two other equivalent forms:

$$\frac{1 \text{ furlong}}{0.125 \text{ mi}} = 1 \quad \text{or} \quad \frac{0.125 \text{ mi}}{1 \text{ furlong}} = 1$$

The second form allows us to convert furlongs to miles:

$$10 \cancel{\text{furlongs}} \times \frac{0.125 \text{ mi}}{1 \cancel{\text{furlong}}} = 1.25 \text{ mi}$$

The length of the Kentucky Derby race is 1.25 miles. ▶ Now try Exercises 15–20.

EXAMPLE 2 Price Comparison

You are planning to make pesto and need to buy basil. At the grocery store, you can buy small containers of basil priced at \$2.99 for each 2/3-ounce container. At the farmer's market, you can buy basil in bunches for \$12 per pound. Which is the better deal?

Solution To compare the prices, we need them both in the same units. Let's convert the small container price to a price per pound. We start with the fact that the container price is \$2.99 *per* 2/3 ounce, which means we need to divide. We then multiply by the conversion of 16 ounces per pound:

$$\text{container price} = \frac{\$2.99}{\frac{2}{3}\ \cancel{\text{oz}}} \times \frac{16\ \cancel{\text{oz}}}{1\ \text{lb}} = \frac{\$71.76}{\text{lb}}$$

The small containers are priced at almost \$72 per pound, which is six times as much as the farmer's market price. ▶ Now try Exercises 21–24.

The International Metric System

The international metric system was invented in France late in the 18th century for two primary reasons: (1) to replace many customary units with just a few basic units and (2) to simplify conversions through use of a decimal (base 10) system. The basic units of length, mass, time, and volume in the metric system are

- the **meter** for length, abbreviated m
- the **kilogram** for mass, abbreviated kg
- the **second** for time, abbreviated s
- the **liter** for volume, abbreviated L

Each of these basic units can be combined with a prefix that indicates multiplication by a power of 10. For example, *kilo* means 1000, so a kilometer is 1000 meters, and *micro* means *one-millionth*, so a microgram is a millionth of a gram. Table 2.3 lists common metric prefixes.

Think About It People often say that expensive things cost "megabucks." What does this statement mean literally? Can you think of other cases where metric prefixes have entered popular language?

TABLE 2.3 Common Metric Prefixes

Small Values			Large Values		
Prefix	Abbrev.	Value	Prefix	Abbrev.	Value
deci-	d	10^{-1} (one-tenth)	deca-	da	10^{1} (ten)
centi-	c	10^{-2} (one-hundredth)	hecto-	h	10^{2} (hundred)
milli-	m	10^{-3} (one-thousandth)	kilo-	k	10^{3} (thousand)
micro-	μ or mc*	10^{-6} (one-millionth)	mega-	M	10^{6} (million)
nano-	n	10^{-9} (one-billionth)	giga-	G	10^{9} (billion)
pico-	p	10^{-12} (one-trillionth)	tera-	T	10^{12} (trillion)

* Micro- is usually abbreviated with μ (the Greek letter mu), but in medical applications it is common to use "mc" instead.

BY THE WAY

The abbreviation for pound, *lb,* comes from the Latin *libra,* meaning "scales."

HISTORICAL NOTE

The French inventors of the metric system drew inspiration from the "founding fathers" of the United States. Thomas Jefferson and Benjamin Franklin both served as ambassadors to France and promoted decimal-based measurement; Jefferson devised our decimal-based currency, in which \$1 = 100¢ In 1790, with the support of President Washington, Jefferson (then Secretary of State) proposed adoption of the metric system to Congress. Had Congress agreed, the United States would have been the *first* country to adopt the metric system—ahead of France, which adopted it in 1795.

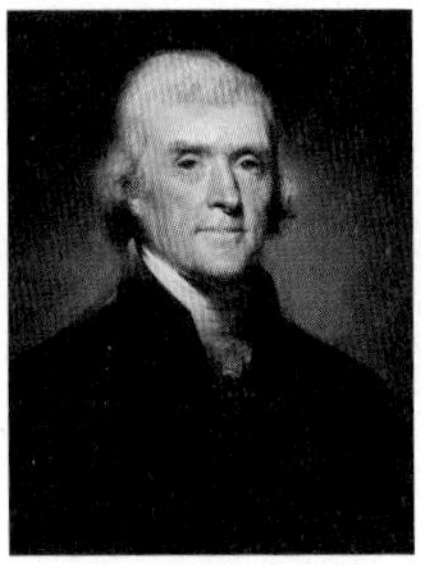

Brief Review Powers of 10

Powers of 10 indicate how many times to multiply 10 by itself:

$$10^2 = 10 \times 10 = 100$$
$$10^6 = 10 \times 10 \times 10 \times 10 \times 10 \times 10 = 1{,}000{,}000$$

Negative powers are reciprocals of positive powers:

$$10^{-2} = \frac{1}{10^2} = \frac{1}{100} = 0.01$$
$$10^{-6} = \frac{1}{10^6} = \frac{1}{1{,}000{,}000} = 0.000001$$

Notice that powers of 10 follow two basic rules:

1. *A positive exponent tells how many zeros follow the 1.* For example, 10^0 is a 1 followed by no zeros; 10^8 is a 1 followed by eight zeros.
2. *A negative exponent tells how many places are to the right of the decimal point, including the 1.* For example, $10^{-1} = 0.1$ has one place to the right of the decimal point; $10^{-6} = 0.000001$ has six places to the right of the decimal point.

Multiplying and Dividing Powers of 10

These examples show that we can multiply powers of 10 by adding exponents:

$$10^4 \times 10^7 = \underbrace{10{,}000}_{10^4} \times \underbrace{10{,}000{,}000}_{10^7} = \underbrace{100{,}000{,}000{,}000}_{10^{4+7} = 10^{11}}$$

$$10^5 \times 10^{-3} = \underbrace{100{,}000}_{10^5} \times \underbrace{0.001}_{10^{-3}} = \underbrace{100}_{= 10^{5+(-3)} = 10^2}$$

$$10^{-8} \times 10^{-5} = \underbrace{0.00000001}_{10^{-8}} \times \underbrace{0.00001}_{10^{-5}} = \underbrace{0.0000000000001}_{= 10^{-8+(-5)} = 10^{-13}}$$

We can divide powers of 10 by subtracting exponents:

$$\frac{10^5}{10^3} = \underbrace{100{,}000}_{10^5} \div \underbrace{1000}_{10^3} = \underbrace{100}_{= 10^{5-3} = 10^2}$$

$$\frac{10^3}{10^7} = \underbrace{1000}_{10^3} \div \underbrace{10{,}000{,}000}_{10^7} = \underbrace{0.0001}_{= 10^{3-7} = 10^{-4}}$$

$$\frac{10^{-4}}{10^{-6}} = \underbrace{0.0001}_{10^{-4}} \div \underbrace{0.000001}_{10^{-6}} = \underbrace{100}_{10^{-4-(-6)} = 10^2}$$

Powers of Powers of 10

We can use the multiplication and division rules to raise powers of 10 to other powers. For example:

$$(10^4)^3 = 10^4 \times 10^4 \times 10^4 = 10^{4+4+4} = 10^{12}$$

Or we can get the same result by multiplying the two powers:

$$(10^4)^3 = 10^{4\times 3} = 10^{12}$$

Adding and Subtracting Powers of 10

There is no shortcut for adding or subtracting powers of 10. The values must be written in longhand notation. For example:

$$10^6 + 10^2 = 1{,}000{,}000 + 100 = 1{,}000{,}100$$
$$10^8 + 10^{-3} = 100{,}000{,}000 + 0.001 = 100{,}000{,}000.001$$
$$10^7 - 10^3 = 10{,}000{,}000 - 1000 = 9{,}999{,}000$$

Summary

To multiply powers of 10, add exponents:	$10^n \times 10^m = 10^{n+m}$
To divide powers of 10, subtract exponents:	$\frac{10^n}{10^m} = 10^{n-m}$
To raise powers of 10 to other powers, multiply exponents:	$(10^n)^m = 10^{n\times m}$

▶ **Now try Exercises 13–14.**

EXAMPLE 3 Using Metric Prefixes

a. Convert 2759 centimeters to meters.
b. How many nanoseconds are in a microsecond?

Solution

a. Table 2.3 shows that *centi-* means 10^{-2}, so $1 \text{ cm} = 10^{-2} \text{ m}$ or, equivalently, $1 \text{ m} = 100 \text{ cm}$. Therefore, 2759 centimeters is the same as 27.59 meters:

$$2759 \text{ cm} \times \frac{1 \text{ m}}{100 \text{ cm}} = 27.59 \text{ m}$$

b. We compare the quantities by dividing the longer time (microsecond) by the shorter time (nanosecond):

$$\frac{1 \text{ μs}}{1 \text{ ns}} = \frac{10^{-6} \text{ s}}{10^{-9} \text{ s}} = 10^{-6-(-9)} = 10^{-6+9} = 10^3$$

A microsecond is 10^3, or 1000, times as long as a nanosecond, so there are 1000 nanoseconds in a microsecond.

▸ Now try Exercises 25–30.

HISTORICAL NOTE

People around the world use the same base 10 (*decimal*) counting system. This may seem natural because we have 10 fingers, but various ancient cultures used base 2, base 3, base 5, base 20, and other bases. Vestiges of other bases remain in our language. For example, we have 60 seconds in a minute and 60 minutes in an hour because the ancient Babylonians used base 60. Similarly, a dozen is 12 and a gross is $12 \times 12 = 144$, presumably because base 12 was once common in northern Europe.

Metric–USCS Conversions

We carry out conversions between metric and USCS units like any other unit conversions. Table 2.4 lists a few handy conversion factors. It's useful to memorize approximate conversions, particularly if you plan to travel internationally or if you work with metric units in sports or business. For example, if you remember that a kilometer is about 0.6 mile, you will know that a 10-kilometer road race is about 6 miles. Similarly, if you remember that a meter is about 10% longer than a yard, you'll know that a 100-meter race is about the same as a 110-yard race.

TABLE 2.4 USCS–Metric Conversions

USCS to Metric	Metric to USCS
1 in = 2.540 cm	1 cm = 0.3937 in
1 ft = 0.3048 m	1 m = 3.28 ft
1 yd = 0.9144 m	1 m = 1.094 yd
1 mi = 1.6093 km	1 km = 0.6214 mi
1 lb = 0.4536 kg	1 kg = 2.205 lb
1 fl oz = 29.574 mL	1 mL = 0.03381 fl oz
1 qt = 0.9464 L	1 L = 1.057 qt
1 gal = 3.785 L	1 L = 0.2642 gal

BY THE WAY

Technically, pounds are a unit of *weight* and kilograms are a unit of *mass*, so the given conversions between pounds and kilograms are valid only on Earth. For example, a 50-kilogram astronaut weighs about 110 pounds on Earth; in Earth orbit, the astronaut is weightless but still has a mass of 50 kilograms.

EXAMPLE 4 Marathon Distance

The marathon running race is about 26.2 miles. About how long is it in kilometers?

Solution Table 2.4 shows that 1 mi = 1.6093 km. We use the conversion in the form with miles in the denominator to find

$$26.2 \text{ mi} \times \frac{1.6093 \text{ km}}{1 \text{ mi}} = 42.2 \text{ km}$$

Rounded to one decimal place, the length of a marathon is 42.2 kilometers.

▸ Now try Exercises 31–34.

EXAMPLE 5 Square Kilometers to Square Miles

How many square kilometers are in one square mile?

Solution We square both sides of the conversion factor, 1 mi = 1.6093 km:

$$(1 \text{ mi})^2 = (1.6093 \text{ km})^2 \quad \Rightarrow \quad 1 \text{ mi}^2 = 2.5898 \text{ km}^2$$

One square mile is approximately 2.6 square kilometers.

▶ Now try Exercises 35–40.

EXAMPLE 6 Melting Antarctica

The chapter-opening activity asked about the impact of melting of ice in Antarctica and Greenland. If you completed its first two parts, you would have found that, if melted, the total volume of Antarctic ice would represent 25 million cubic kilometers of water. You were also told that this water would then spread out over Earth's 340 million square kilometers of ocean surface. Use these facts to determine how much sea level would rise if all the Antarctic ice melted into the oceans.

Solution Because this problem is more subtle, let's write out the steps in the Understand-Solve-Explain process.

Understand: We want to know how much sea level would rise, so we expect an answer with units of length, such as kilometers. At this point, we can use the power of unit analysis to find a path to the answer. Notice that the two pieces of information we are given are a volume, which has units of length^3, and an area, which has units of length^2. Because dividing length^3 by length^2 results in a length, we can find an answer with the correct units by dividing the volume of water by the surface area of the oceans. We now just need to ask whether this approach makes sense, and you can see that it does by conducting a similar but much simpler experiment: If you poured 150 cubic centimeters of water into a glass with a bottom area of 10 square centimeters, you'd find that the water would fill the glass to a height of 150 cm^3 ÷ 10 cm^2 = 15 centimeters (Figure 2.5).

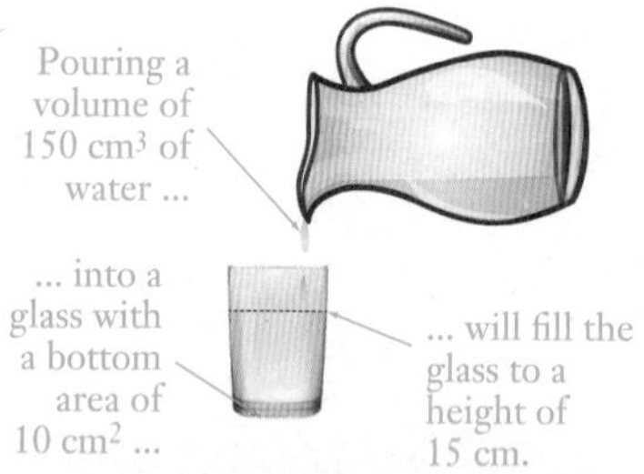
Pouring a volume of 150 cm³ of water ...
... into a glass with a bottom area of 10 cm² ...
... will fill the glass to a height of 15 cm.

FIGURE 2.5 A simple experiment, such as pouring a known volume of water into a glass, will demonstrate that you can calculate a water height by dividing the volume of water by the area over which it spreads.

Solve: We now carry out the calculation by dividing the given volume of water that would melt into the oceans by the given surface area of the oceans:

$$\frac{25 \text{ million km}^3}{340 \text{ million km}^2} \approx 0.074 \text{ km}$$

Explain: We have found an answer for the sea level rise, but because it is a small fraction of a kilometer, it will be easier to interpret if we convert it into meters and feet:

$$0.074 \text{ km} \times \frac{1000 \text{ m}}{1 \text{ km}} = 74 \text{ m} \qquad \text{and} \qquad 74 \text{ m} \times \frac{3.28 \text{ ft}}{1 \text{ m}} \approx 240 \text{ ft}$$

We have found that complete melting of Antarctica's ice would raise sea level by about 74 meters, or about 240 feet. In accord with the third step of the Understand-Solve-Explain process (p. 72), we should also discuss the implications of this answer, which are as follows: Global warming is expected to cause increased melting of ice around the world. Fortunately, even under the worst-case scenarios, scientists suspect that it would take at least a couple thousand years for the Antarctic ice to melt completely. Nevertheless, the fact that this complete melting would raise sea level by more than 200 feet forces us to confront the dismaying fact that unless we solve the problem of global warming, our future descendants will need deep-sea diving equipment to visit the ruins of our coastal cities.

▶ Now try Exercises 41–42.

BY THE WAY

Scientists do not fully understand the processes that determine the rate at which polar ice will melt as a result of global warming. Nevertheless, most scientists expect global warming to cause a sea level rise of at least 1 meter (3 feet) by the end of this century, and some models suggest the rise could be as much as about 6 meters (20 feet). Even small rises can be consequential. For example, sea level has already risen about 1 foot over the past century, a fact that is thought to have contributed to the damage that occurs along coastlines during storm surges.

Standardized Temperature Units

Another important set of standardized units are those we use to measure temperature. Three temperature scales are commonly used (Figure 2.6):

- The **Fahrenheit** scale, commonly used in the United States, is defined so water freezes at 32°F and boils at 212°F.
- The rest of the world uses the **Celsius** scale, which places the freezing point of water at 0°C and the boiling point at 100°C.
- In science, we use the **Kelvin** scale, which is the same as the Celsius scale except for its zero point, which corresponds to −273.15°C. A temperature of 0 K is known as **absolute zero**, because it is the coldest possible temperature. (The degree symbol [°] is not used with the Kelvin scale.)

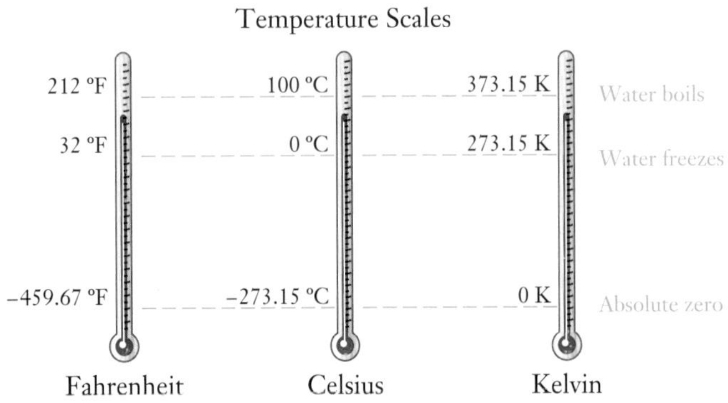

FIGURE 2.6 Three temperature scales.

The definitions of the three temperature scales lead to simple formulas for converting between them, summarized in the box below. Notice that conversions between Kelvin and Celsius simply require adding or subtracting the 273.15°C difference in their zero points. To understand the conversions between Celsius and Fahrenheit, notice that the Celsius scale has 100°C between the freezing and

USING TECHNOLOGY

Metric Conversions

You can always do unit conversions by using the appropriate conversion factors on your calculator. However, some technologies make it even easier.

Microsoft Excel Use the built-in function CONVERT by entering the number you want to convert along with the correct unit abbreviations in quotes. The screen shot below shows how to enter a conversion of 35 kilometers to miles. Clicking "Return" will put the numerical answer in the cell (in this case, 21.75). To find the necessary unit abbreviations, type "convert" into the search box on the Help menu of Excel.

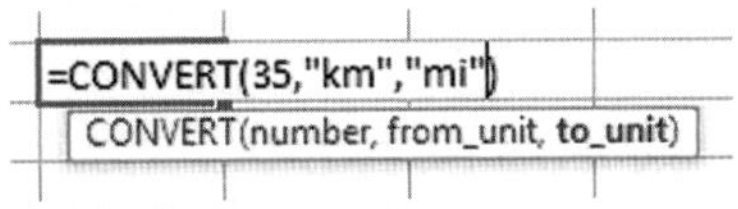

Microsoft Excel 2016, Windows 10, Microsoft Corporation.

Google You can do basic unit conversions simply by typing what you want to convert into the Google search box. The screen shot below shows how Google converts 50 liters into gallons. When you hit "Return," the result shows up below the search box.

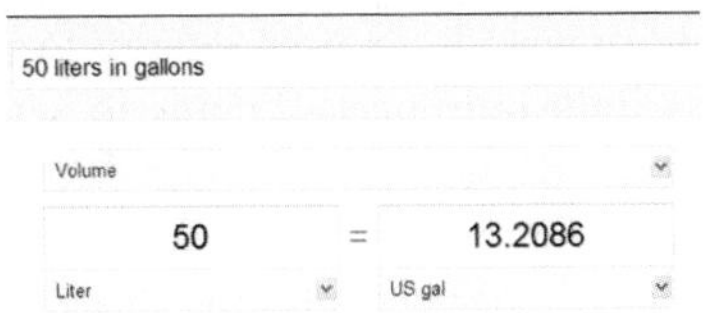

HISTORICAL NOTE
The Fahrenheit scale was invented by German-born scientist Gabriel Fahrenheit in 1714. He originally defined the scale with 0°F representing the coldest temperature he could create with a mixture of ice and salt, and 100°F representing a rough estimate of human body temperature. The Celsius scale was invented by Swedish astronomer Anders Celsius in 1742. The Kelvin scale was invented in 1848 by British physicist William Thomson, later known as Lord Kelvin.

boiling points of water, while the Fahrenheit scale has 212°F − 32°F = 180°F between those points. Each Celsius degree therefore represents $\frac{180}{100} = 1.8$ Fahrenheit degrees, which explains the factor of 1.8 (or $\frac{9}{5}$) in the conversions. The added or subtracted 32 accounts for the difference in the Celsius and Fahrenheit zero points.

Temperature Conversions

The conversions are given both in words and with formulas in which C, F, and K are Celsius, Fahrenheit, and Kelvin temperatures, respectively.

To Convert from	Conversion in Words	Conversion Formula
Celsius to Fahrenheit	Multiply by 1.8 (or $\frac{9}{5}$). Then add 32.	$F = 1.8C + 32$
Fahrenheit to Celsius	Subtract 32. Then divide by 1.8 (or $\frac{9}{5}$), or, equivalently, multiply by $\frac{5}{9}$.	$C = \frac{F - 32}{1.8}$
Celsius to Kelvin	Add 273.15.	$K = C + 273.15$
Kelvin to Celsius	Subtract 273.15.	$C = K - 273.15$

EXAMPLE 7 Human Body Temperature

Average human body temperature is 98.6°F. What is it on the Celsius and Kelvin scales?

Solution We convert from Fahrenheit to Celsius by subtracting 32 and then dividing by 1.8:

$$C = \frac{F - 32}{1.8} = \frac{98.6 - 32}{1.8} = \frac{66.6}{1.8} = 37.0°\text{C}$$

We find the Kelvin equivalent by adding 273.15 to the Celsius temperature:

$$K = C + 273.15 = 37 + 273.15 = 310.15\text{ K}$$

Human body temperature is 37°C or 310.15 K. ▸ **Now try Exercises 43–46.**

Think About It The local weather report says that tomorrow's temperature will be 59°, but does not specify whether it is being given in Celsius or Fahrenheit. Can you tell which it is? How?

Units of Energy and Power

We're now ready to extend unit analysis to problems involving energy, density, and concentration. We'll start with energy, which is familiar to all of us because we pay energy bills to power companies, we use energy from gasoline to run our cars, and we argue about the best alternative sources to replace the energy we currently get from fossil fuels. But what *is* energy?

Broadly speaking, energy is what makes matter move or heat up. We need energy from food to keep our hearts beating, to maintain our body temperatures, and to walk or run. A gasoline-powered car uses energy to move the pistons in its engine, which turn the wheels. A light bulb needs energy to generate light.

Technical Note
A food Calorie (uppercase C) is 1000 calories (lowercase c). The calorie was once used commonly in science, but scientists today almost always measure energy in joules.

For Americans, the most familiar energy unit is the food *Calorie* (note the uppercase C) used to measure the energy our bodies can draw from food. A typical adult uses about 2500 Calories of energy each day. The international metric unit of energy is the **joule**. One Calorie is equivalent to 4184 joules.

The Difference Between Energy and Power

The words *energy* and *power* are often used together, but they are not the same. Power is the *rate* at which energy is used, which means it has units of energy divided by time. The most common unit of power is the **watt**, defined as 1 joule per second.

Energy and Power

Energy is what makes matter move or heat up. The international metric unit of energy is the **joule.**

Power is the *rate* at which energy is used. The international metric unit of power is the **watt,** defined as

$$1 \text{ watt} = 1 \frac{\text{joule}}{\text{s}}$$

EXAMPLE 8 Pedal Power

As you ride an exercise bicycle, the display states that you are using 500 Calories per hour. Are you generating enough power to light a 100-watt bulb? (1 Calorie = 4184 joules)

Solution We use a chain of conversions to go from Calories per hour to joules per second:

$$\frac{500 \text{ Cal}}{1 \text{ hr}} \times \frac{4184 \text{ joule}}{1 \text{ Cal}} \times \frac{1 \text{ hr}}{60 \text{ min}} \times \frac{1 \text{ min}}{60 \text{ s}} \approx 581 \frac{\text{joule}}{\text{s}}$$

Your pedaling generates energy at a rate of 581 joules per second, which is a power of 581 watts—enough to light five (almost six) 100-watt bulbs.

▶ Now try Exercises 47–48.

Electric Utility Bills

On utility bills, electrical energy is usually measured in units of **kilowatt-hours**. Recall that the hyphen implies multiplication, and 1 kilowatt is 1000 watts, which is 1000 joule/s. We therefore find that:

$$1 \text{ kilowatt-hour} = \frac{1000 \text{ joule}}{1 \text{ s}} \times 1 \text{ hr} \times \frac{60 \text{ min}}{1 \text{ hr}} \times \frac{60 \text{ s}}{1 \text{ min}} = 3{,}600{,}000 \text{ joules}$$

BY THE WAY

If you purchase a gas appliance, such as a gas stove or a kerosene heater, its energy requirements may be labeled in *British thermal units,* or BTUs. One BTU is equivalent to 1055 joules.

Definition

A **kilowatt-hour** is a unit of energy:

$$1 \text{ kilowatt-hour} = 3.6 \text{ million joules}$$

EXAMPLE 9 Operating Cost of a Light Bulb

Your utility company charges 15¢ per kilowatt-hour of electricity. How much does it cost to keep a 100-watt light bulb on for a week? How much will you save in a year if you replace the bulb with an LED bulb that provides the same amount of light for only 25 watts of power?

Solution Let's start with the question about how much energy the bulb uses in a week. We are given the bulb's power usage (100 watts) and that it is left on for a week. Because power is the *rate* at which energy is used, we find the total energy usage by multiplying the power by the time. However, if we did that with the given units, we would end up with an answer in "watt-weeks." Energy usage is more commonly stated in kilowatt-hours, so we use a chain of conversions to convert 100 watts to kilowatts and 1 week to hours:

$$100 \text{ watt} \times \frac{1 \text{ kilowatt}}{1000 \text{ watt}} \times 1 \text{ week} \times \frac{7 \text{ day}}{1 \text{ week}} \times \frac{24 \text{ hr}}{1 \text{ day}} = 16.8 \text{ kilowatt-hours}$$

Knowing the total energy used, we can find the cost of this energy by multiplying it by the price of 15¢, or $0.15, per kilowatt-hour:

$$16.8 \text{ kilowatt-hr} \times 15 \frac{¢}{\text{kilowatt-hr}} = 252¢ = \$2.52$$

The electricity for the bulb costs $2.52 per week. If you replace the 100-watt bulb with a 25-watt LED, you'll use only $\frac{1}{4}$ as much energy, which means your weekly cost will be only $0.63. In other words, your savings will be $2.52 − $0.63 = $1.89 per week, so in a year you'll save about:

$$\text{Yearly savings with energy-efficient bulb: } \frac{\$1.89}{\text{wk}} \times 52 \frac{\text{wk}}{\text{yr}} \approx \$98/\text{yr}$$

If left on all the time, the more efficient bulb would save almost $100 per year. Of course, you probably use the light only a few hours each day.

▶ Now try Exercises 49–50.

Think About It Check a utility bill (yours or a friend's). Is the electricity usage metered in units of kilowatt-hours (often abbreviated kWh)? If not, what units are used? If so, what is the price per kilowatt-hour?

Units of Density and Concentration

The terms *density* and *concentration* arise in a variety of different contexts. **Density** is generally used to describe compactness or crowding, and the precise units depend on the context. For example:

- *Material density* is given in units of mass per unit volume, such as grams per cubic centimeter (g/cm^3). A useful reference is the density of water—about 1 g/cm^3. Objects with densities less than 1 g/cm^3 float in water, while higher-density objects sink.
- *Population density* is given by the number of people per unit area. For example, if 750 people live in a square region that is 1 mile on a side, the population density of the area is 750 people/mi^2.
- *Information density* is often used to describe how much memory can be stored by digital media. For example, each square inch on the surface of a dual-layer Blu-ray Disc holds about 1 gigabyte of information, so we say that the disc has an information density of 1 GB/in^2.

Think About It Use the concept of density to explain why you float better in a swimming pool when your lungs are filled with air than when you fully exhale.

Concentration is generally used to describe the amount of one substance that is mixed with another. For example:

- The *concentration of an air pollutant* is often measured by the number of molecules of the pollutant per million molecules of air. For example, if there are 12 molecules of carbon monoxide in each 1 million molecules of air, the carbon monoxide concentration is 12 parts per million (ppm). (The U.S. Environmental Protection Agency says that air is unhealthy if the carbon monoxide concentration is above 9 ppm.)

HISTORICAL NOTE

A king once asked the famed Greek scientist Archimedes (c. 287–212 B.C.E.) to test whether a crown was made of pure gold, as claimed by its goldsmith, or a mixture of silver and gold. Archimedes was unsure how to do it, but according to a story passed down by a later Roman writer (Vitruvius, b. 81 B.C.E.), he had a sudden insight while taking a bath one day. Knowing that silver is less dense than gold, he realized he could compare the rise in water level for the crown and for an equal weight of pure gold. Thrilled at this insight, he ran naked through the streets shouting "Eureka!" (meaning "I have found it"). Worth noting: Archimedes probably did not use the technique in the story, because it would have required more measurement accuracy than possible at the time. Instead, he may have used a more complex technique based on the principle of buoyancy, which he also discovered.

IN YOUR WORLD

Save Money and Save the Earth

You can save both money and energy by replacing standard (incandescent) light bulbs with more efficient bulbs, such as compact fluorescents or LEDs (light emitting diodes). These bulbs save energy because a much higher proportion of the energy they use goes into making light, compared to standard bulbs, which generate much more heat. In fact, compact fluorescents and LEDs typically need only about one-quarter as much energy as standard bulbs to generate the same amount of light. For example, a 25-watt compact fluorescent can produce as much light as a 100-watt standard bulb.

The energy savings can be quite substantial. In Example 9, we found that a single replacement bulb could save almost \$100 per year if the bulb were left on all the time. More realistically, using the bulb for an average of 3 hours each night—or 1/8 of each 24-hour day—would still save more than \$12 per year. Over the several years that compact fluorescents and LEDs typically last, this savings more than makes up for their higher initial cost.

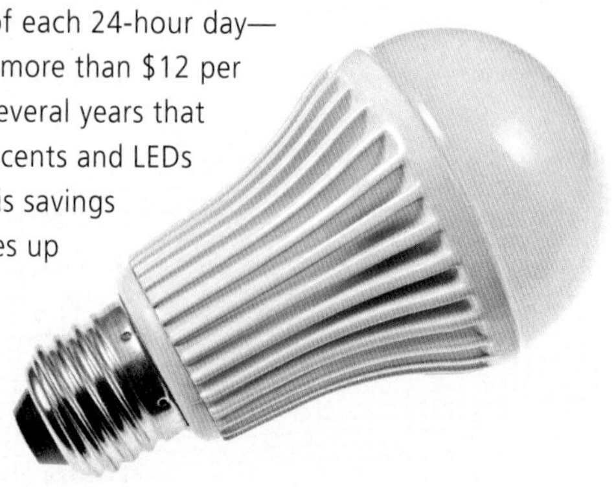

- Medicine dosages often require calculations based on a recommended concentration per kilogram of body weight or on the concentration of an active ingredient in a liquid suspension or IV drip. For example, a recommended dosage might be 2 milligrams per kilogram (2 mg/kg) of body weight, and the concentration of the medicine in a liquid suspension might be 10 milligrams per milliliter (10 mg/ml).
- **Blood alcohol content (BAC)** describes the concentration of alcohol in a person's body. It is usually measured in units of grams of alcohol per 100 milliliters of blood. For example, in most of the United States, a driver over age 21 is considered legally intoxicated if his or her blood alcohol content is at or above 0.08 gram of alcohol per 100 milliliters of blood (written as 0.08 g/100 mL).

BY THE WAY

Most countries have a lower legal limit than the United States for BAC while driving. For example, you are considered legally drunk at a BAC of 0.05 in much of Europe, Canada, Australia, and South Africa, and the legal driving limit is 0.02 in Norway and Sweden. The National Transportation Safety Board has predicted that lowering the U.S. limit to 0.05 would save about 7000 lives each year (by reducing fatalities caused by drunk drivers).

EXAMPLE 10 New York City

Manhattan Island has a population of about 1.6 million people living in an area of about 57 square kilometers. What is its population density? If there were no high-rise apartments, how much space would be available per person?

Solution We divide the population by the area to find:

$$\text{population density} = \frac{1{,}600{,}000 \text{ people}}{57 \text{ km}^2} \approx 28{,}000 \frac{\text{people}}{\text{km}^2}$$

Manhattan's population density is about 28,000 people per square kilometer. If there were no high rises, each resident would have 1/28,000 square kilometer of land. This number is easier to interpret if we convert from square kilometers to square meters:

$$\frac{1 \text{ km}^2}{28{,}000 \text{ people}} \times \left(\frac{1000 \text{ m}}{1 \text{ km}}\right)^2 = \frac{1 \cancel{\text{km}^2}}{28{,}000 \text{ people}} \times \frac{1{,}000{,}000 \text{ m}^2}{1 \cancel{\text{km}^2}} \approx 36 \frac{\text{m}^2}{\text{person}}$$

Without high rises, each person would have only 36 square meters, equivalent to a room 6 meters, or about 20 feet, on a side—and this does not include any space for roads, schools, or other common properties. Clearly, Manhattan Island could not fit so many residents without high rises.

▶ Now try Exercises 51–56.

EXAMPLE 11 Ear Infection

A child weighing 15 kilograms has a bacterial ear infection. A physician orders treatment with the antibiotic amoxicillin at a dosage of 30 milligrams per kilogram of body weight per day, divided into doses to be given every 12 hours.

a. How much amoxicillin should the child be given every 12 hours?

b. If the medicine is to be taken in a liquid suspension with concentration 25 mg/mL, how much should the child take every 12 hours?

Solution

a. The prescribed dosage is 30 mg/kg of body weight per day, but because it will be given in two doses (every 12 hours), each dose will be half of the total, or 15 mg/kg of body weight. Therefore, for a child weighing 15 kilograms, the dosage should be

$$\text{dose every 12 hours} = \underbrace{\frac{15\text{ mg}}{\cancel{\text{kg}}}}_{\substack{\text{doses per kg}\\ \text{of body weight}}} \times \underbrace{15\ \cancel{\text{kg}}}_{\substack{\text{child's body}\\ \text{weight}}} = 225\text{ mg}$$

b. The liquid suspension contains 25 milligrams of amoxicillin per milliliter (mL) of liquid, and from part (a) we know the total amount of amoxicillin in each dose should be 225 mg. We are looking for the total amount of liquid that the child should be given for each dose, so the answer should have units of milliliters. The only way to get the correct answer units is to divide, replacing division with multiplication by the reciprocal:

$$\text{liquid dose} = \underbrace{225\text{ mg}}_{\substack{\text{required}\\ \text{amoxicillin dose}}} \div \underbrace{\frac{25\text{ mg}}{\text{mL}}}_{\substack{\text{concentration of}\\ \text{liquid suspension}}} = 225\ \cancel{\text{mg}} \times \frac{1\text{ mL}}{25\ \cancel{\text{mg}}} = 9\text{ mL}$$

The child should be given 9 milliliters of the liquid every 12 hours.

▶ **Now try Exercises 57–58.**

EXAMPLE 12 Blood Alcohol Content (BAC)

We now consider the solution to our chapter-opening question (p. 68). An average-size man has about 5 liters (5000 milliliters) of blood, and an average 12-ounce can of beer contains about 15 grams of alcohol (assuming the beer is about 6% alcohol by volume). If all the alcohol were immediately absorbed into the bloodstream, what blood alcohol content would we find in an average-size man who quickly drank a single can of beer? How much beer would make him legally intoxicated (BAC of 0.08)?

Solution We are first asked to find the man's blood alcohol content (BAC) under the assumption that all the alcohol in one beer is absorbed immediately into his bloodstream. We are given both the amount of alcohol in the beer (15 grams) and the volume of the man's blood (5000 milliliters), so we can find the BAC by dividing:

$$\frac{15\text{ g alcohol}}{5000\text{ mL blood}}$$

We can convert our answer to standard units for blood alcohol in two steps. First, we carry out the division to find the concentration in grams per milliliter:

$$\frac{15\text{ g}}{5000\text{ mL}} = 0.003\,\frac{\text{g}}{\text{mL}}$$

Next, we multiply both the top and the bottom of the fraction by 100 to find the concentration in grams per 100 milliliters:

$$0.003\,\frac{\text{g}}{\text{mL}} \times \frac{100}{100} = 0.3\,\frac{\text{g}}{100\text{ ml}}$$

The man's blood alcohol concentration would be 0.3 gram per 100 mL of blood—almost four times the legal limit of 0.08 g/100 mL. Therefore, it would take only about one-quarter of the can, or 3 ounces of beer, to reach the legal limit. In reality, the man's blood alcohol content won't get this high from a single beer, because it takes some time for all the alcohol to be absorbed into the bloodstream (typically 30 minutes on an empty stomach and up to 2 hours on a full stomach) and because metabolic processes gradually eliminate the absorbed alcohol (at a rate of about 10 to 15 grams per hour). Nevertheless, a single beer is enough to cause impaired brain function—making it unsafe to drive—and this example points out how quickly and easily a person can become dangerously intoxicated. ▶ **Now try Exercises 59–60.**

Think About It Many college students have lost their lives by rapidly consuming several "shots" of strong alcoholic drinks. Explain why such rapid consumption of alcohol can lead to death, even when the total amount of alcohol consumed may not sound like a lot.

BY THE WAY

According to the American Medical Association, brain function is impaired when the blood alcohol content reaches 0.04 g/100 mL of blood—half the U.S. legal limit of 0.08 g/100 mL. A blood alcohol content at or above 0.3 g/100 mL can lead to coma and sometimes to death.

Quick Quiz 2B

Choose the best answer to each of the following questions. Explain your reasoning with one or more complete sentences.

1. The fact that 1 liter = 1.057 quarts can be written as the conversion factor
   a. 1.057 quart/liter.
   b. 1.057 liter/quart.
   c. 1 quart/1.057 liter.
2. Which is greater, 1200 meters or 3600 feet?
   a. 1200 m
   b. 3600 ft
   c. They are the same.
3. You are given two pieces of information: (1) the volume of a lake in cubic feet and (2) the average depth of the lake in feet. You are asked to find the surface area of the lake in square feet. You should
   a. multiply the volume by the depth.
   b. divide the volume by the depth.
   c. divide the depth by the volume.
4. Which of the following is *not* a unit of energy?
   a. joule
   b. watt
   c. kilowatt-hour
5. You want to know how much total energy is required to operate a 100-watt light bulb. Do you need any more information?
   a. No.
   b. Yes; you need to know the temperature of the light bulb when it is on.
   c. Yes; you need to know how long the light bulb is on.
6. New Mexico has a population density of about 12 people per square mile and an area of about 120,000 square miles. To estimate its total population, you should
   a. multiply the population density by the area.
   b. divide the population density by the area.
   c. divide the area by the population density.
7. A temperature of 110°C is
   a. typical of Phoenix in the summer.
   b. typical of Antarctica in the winter.
   c. hot enough to boil water.
8. The concentration of carbon dioxide in Earth's atmosphere might be stated in
   a. grams per meter.
   b. parts per million.
   c. joules per watt.

9. The guidelines for prescribing a particular drug specify a dose of 300 mg per kilogram of body weight per day. To find how much of the drug should be given to a 30-kg child every 8 hours, you should

a. multiply 300 mg/kg/day by 30 kg and multiply that result by 3.

b. divide 300 mg/kg/day by 30 kg and multiply that result by 3.

c. multiply 300 mg/kg/day by 30 kg and divide that result by 3.

10. A blood alcohol content (BAC) of 0.08 g/100 mL means

a. a person with 4 liters of blood has $0.08/4 = 0.02$ gram of alcohol in his or her blood.

b. a person with 4 liters of blood has $0.08 \times 40 = 3.2$ grams of alcohol in his or her blood.

c. a person with 4 liters of blood has $0.08/40 = 0.002$ gram of alcohol in his or her blood.

Exercises 2B

REVIEW QUESTIONS

1. What do we mean by *standardized units*, and why are they useful?
2. What are the basic metric units of length, mass, time, and volume? How are the metric prefixes used?
3. What is *energy*? List at least three common units of energy. Under what circumstances do the different units tend to be used?
4. Using examples, show how to convert among the Fahrenheit, Celsius, and Kelvin temperature scales.
5. What is the difference between *energy* and *power*? What are the standard units for power?
6. What do we mean by *density*? What do we mean by *concentration*? Describe common units of density, concentration, and blood alcohol content.

DOES IT MAKE SENSE?

Decide whether each of the following statements makes sense (or is clearly true) or does not make sense (or is clearly false). Explain your reasoning. *Hint:* Be sure to consider whether the units are appropriate to the statement, as well as whether the stated amount makes any sense. For example, a statement that someone is 15 feet tall uses the units (feet) appropriately, but does not make sense because no one is that tall.

7. I drank 2 liters of water today.
8. I know a professional bicyclist who weighs 300 kilograms.
9. My car's gas tank holds 12 meters of gasoline.
10. My daily food intake gives me about 10 million joules of energy.
11. The beach ball we played with had a density of 10 grams per cubic centimeter.
12. I live in a big city with a population density of 15 people per square kilometer.

BASIC SKILLS & CONCEPTS

13–14: Math Review. Carry out the following calculations.

13. a. $10^4 \times 10^7$ b. $10^5 \times 10^{-3}$ c. $\frac{10^6}{10^2}$ d. $\frac{10^8}{10^{-4}}$ e. $\frac{10^{12}}{10^{-4}}$ f. $10^{23} \times 10^{-23}$ g. $10^4 + 10^2$ h. $\frac{10^{15}}{10^{-5}}$

14. a. $10^{-2} \times 10^{-6}$ b. $\frac{10^{-6}}{10^{-8}}$ c. $10^{12} \times 10^{23}$ d. $\frac{10^{-4}}{10^5}$ e. $\frac{10^{25}}{10^{15}}$ f. $10^1 + 10^0$ g. $10^2 + 10^{-1}$ h. $10^2 - 10^1$

15–20: USCS Units. Answer the following questions involving conversions within the USCS system.

15. The Kentucky Derby distance is 10 furlongs. How far is the Kentucky Derby in (a) miles? (b) yards?
16. The depth of the Challenger Deep is 36,198 feet. How deep is it in (a) fathoms? (b) leagues (marine)? (Round to the nearest hundredth.)
17. One cubic foot holds 7.48 gallons of water, and 1 gallon of water weighs 8.33 pounds. How much does 6 cubic feet of water weigh in pounds? In tons?
18. A bag of garden soil weighs 40 pounds and holds 2 cubic feet. Find the weight of 15 bags in kilograms and the volume of 15 bags in cubic yards.
19. A Nimitz-class (nuclear-powered) aircraft carrier has a top speed of 30 knots and a displacement of approximately 102,000 tons. Express the top speed in miles per hour and the displacement in metric tonnes. (1 knot = 1 nautical mile per hour; 1 metric tonne = 1000 kilograms)
20. What is your average speed in miles per hour and in feet per second if you run a mile in 4 minutes?

21–24: Price Comparison. In each case, decide which of the two given prices is the better deal and explain why.

21. You can buy shampoo in a 6-ounce bottle for \$3.99 or in a 14-ounce bottle for \$9.49.
22. You can buy a box of 150 Sharpie pens for \$165 or a dozen for \$13.50.
23. You can fill a 15-gallon tank of gas for \$35.25 or buy gas for \$2.55/gal.
24. You can rent a storage locker for \$30/yd^2 per month or for \$1.90/ft^2 per week.

25–30: Metric Prefixes. Complete the following sentences with a power of ten greater than 1.

25. A kilometer is ________ times as large as a meter.
26. A kilogram is ________ times as large as a microgram.

27. A liter is _________ times as large as a microliter.

28. A kilometer is _________ times as large as a nanometer.

29. A square meter is _________ times as large as a square millimeter.

30. A cubic meter is _________ times as large as a cubic centimeter.

31–40: USCS–Metric Conversions. Convert the following quantities to the indicated units. Where necessary, round to the nearest tenth.

31. 13 liters to quarts
32. 3.5 meters to feet
33. 34 pounds to kilograms
34. 8.6 miles to kilometers
35. 3 square kilometers to square miles
36. 70 kilometers per hour to miles per hour
37. 47 miles per hour to meters per second
38. 200 cubic centimeters to cubic inches
39. 23 grams per cubic centimeter to pounds per cubic inch
40. 2500 acres to square kilometers

41. **Greenland Ice Sheet.** The Greenland ice sheet contains about 3 million cubic kilometers of ice. If completely melted, this ice would release about 2.5 million cubic kilometers of water, which would spread out over Earth's 340 million square kilometers of ocean surface. How much would sea level rise?

42. **Volcanic Eruption.** The greatest volcanic eruption in recorded history took place in 1815 on the Indonesian island of Sumbawa, when the volcano Tambora expelled an estimated 100 cubic kilometers of molten rock. Suppose all of the ejected material fell on a region with an area of 600 square kilometers. Find the average depth of the resulting layer of ash and rock.

43–44: Celsius-Fahrenheit Conversions. Convert the following temperatures from Fahrenheit to Celsius or vice versa.

43. a. 45°F b. 20°C c. −15°C d. −30°C e. 70°F

44. a. −8°C b. 15°F c. 15°C d. 75°F e. 20°F

45–46: Celsius-Kelvin Conversions. Convert the following temperatures from Kelvin to Celsius or vice versa.

45. a. 50 K b. 240 K c. 10°C

46. a. −40°C b. 400 K c. 125°C

47–48. Power Output. In each case, find your average power in watts.

47. Assume running consumes 150 Calories per mile. If you run 6-minute miles, what is your average power output, in watts, during a 1-hour run?

48. Assume that riding a bike burns 100 Calories per mile. If you ride at a speed of 20 miles per hour, what is your average power output, in watts?

49–50. Energy Savings. For these questions, assume that there are 365 days in a year.

49. Your utility company charges 13¢ per kilowatt-hour of electricity. What is the daily cost of keeping a 100-watt light bulb lit for 12 hours each day? How much will you save in a year if you replace the bulb with an LED bulb that provides the same amount of light using only 25 watts of power?

50. Suppose you have a clothes dryer that uses 4000 watts of power and you run it for an average of 1 hour each day. If you pay the utility company 14¢ per kilowatt-hour of electricity, what is the average daily cost to run your dryer? How much would you save in a year if you replaced it with a more efficient model that uses only 2000 watts?

51–56: Densities. Compute the following densities using the appropriate units.

51. A block of oak has a volume of 200 cubic centimeters and weighs 0.12 kilogram. What is its density? Will it float in water?

52. The density of Styrofoam is 0.03 ounce per cubic inch. What is its density in grams per cubic centimeter? Will it float in water?

53. Which has more people, County A with an area of 100 mi^2 and a population density of 25 people/km^2 or County B with an area of 25 km^2 and a population density of 100 people/mi^2?

54. The country with the highest population density is Monaco, where approximately 38,000 people live in an area of 1.95 square kilometers. What is the population density of Monaco in people per square kilometer? Compare this density to that of the United States, which is approximately 35 people per square kilometer.

55. New Jersey and Alaska have populations of about 9.0 million and 738,000, respectively (U.S. Census Bureau, 2008). Their areas are 7417 and 571,951 square miles, respectively. Compute the population densities of the two states.

56. A standard DVD has a surface area of 134 square centimeters. Depending on formatting, it holds either 4.7 (single sided) or 9.4 gigabytes (double sided). Find the data density in both cases.

57–58. Medication Doses.

57. The antihistamine Benadryl is often prescribed for allergies. A typical dose for a 100-pound person is 25 mg every 6 hours.

a. Following this dosage, how many 12.5 mg chewable tablets would be taken in a week?

b. Benadryl also comes in liquid form with a concentration of 12.5 mg/5 mL. Following the prescribed dosage, how much liquid Benadryl should a 100-pound person take in a week?

58. Suppose a penicillin dose of 9000 units per kilogram of body weight, taken every 6 hours, is prescribed for treatment of a bacterial infection. For penicillin, 400,000 units is equal to 250 mg.

a. Express the dose in milligrams per kilogram (mg/kg) of body weight.

b. How many milligrams of penicillin would a 20-kilogram child take in one day?

59. **Blood Alcohol Content: Wine.** A typical glass of wine contains about 20 grams of alcohol. Consider a 120-pound woman, with approximately 4 liters (4000 milliliters) of blood, who drinks two glasses of wine.

a. If all the alcohol were immediately absorbed into her bloodstream, what would her blood alcohol content be? Explain why it is fortunate that, in reality, the alcohol is not absorbed immediately.

b. Again assume that all the alcohol is absorbed immediately, but the woman's body then eliminates the alcohol (through metabolism) at a rate of 10 grams per hour. What is her blood alcohol content 3 hours after drinking the wine? Is it safe for her to drive at this time? Explain.

60. **Blood Alcohol Content: Hard Liquor.** Eight ounces of a hard liquor (such as whiskey) typically contain about 70 grams of alcohol. Consider a 180-pound man, with approximately 6 liters (6000 milliliters) of blood, who quickly drinks 8 ounces of hard liquor.

a. If all the alcohol were immediately absorbed into his bloodstream, what would his blood alcohol content be? Explain why it is fortunate that, in reality, the alcohol is not absorbed immediately.

b. Again assume that all the alcohol is absorbed immediately, but the man's body then eliminates the alcohol (through metabolism) at a rate of 15 grams per hour. What is his blood alcohol content 4 hours after drinking the liquor? Is it safe for him to drive at this time? Explain.

FURTHER APPLICATIONS

61. **The Metric Mile.** Consider the following world records (as of 2017) in track and field for the mile (1 USCS mile) and the 1500-meter (metric mile) races.

	Men	Women
Mile	3:43:13	4:12:56
1500 meters	3:26:00	3:50:07

a. Complete the sentence: The 1500-meter race is _______ % of the mile race in length.

b. Compute and compare the average speeds, in miles per hour, in the men's mile and 1500-meter races.

c. Compute and compare the average speeds, in miles per hour, in the women's mile and 1500-meter races.

d. If the average speed for the 1500-meter were run for the entire length of a mile race, would it result in a world record? Answer for both men and women.

62. **What Is a League?** In Jules Verne's novel *20,000 Leagues Under the Sea* (published in 1870), does the title refer to an ocean depth or a distance traveled? Explain your answer.

63–66: Currency Conversions. Answer the following question using the exchange rates given in Table 2.1.

63. A 0.8-liter bottle of Mexican wine costs 100 pesos. At that price, how much would a half-gallon jug of the same wine cost in dollars?

64. Carpet at a British home supply store sells for 16 pounds (currency) per square meter. What is the price in dollars per square yard?

65. The monthly rent on an 80-square-meter apartment in Monte Carlo, Monaco, is 1150 euros. The monthly rent on a 500-square-foot apartment in Santa Fe, New Mexico, is $800. In terms of price per area, which apartment is less expensive?

66. Gasoline costs 1.40 euro per liter in France. What is that price in dollars per gallon?

67–70: Gems and Gold. Use carats and karats, as discussed in In Your World on p. 90, to answer the following questions.

67. What is the weight of the 45.52-carat Hope diamond in grams and ounces?

68. What is the purity (as a percentage) of a 14-karat gold ring?

69. How many ounces of gold are in a 16-karat gold chain that weighs 2.2 ounces?

70. What is the weight in carats of a diamond that weighs 0.15 ounce?

71. **The Cullinan Diamond.** The Cullinan diamond is the largest single gem-quality rough diamond ever found and weighs 3106 carats. How much does the Cullinan diamond weigh in milligrams? in (avoirdupois) pounds?

72. **The Star of Africa.** The Star of Africa was cut from the Cullinan diamond and weighs 530.2 carats; it is part of the British crown jewels collection. How much does the Star of Africa weigh in milligrams? in (avoirdupois) pounds?

73. **Shower vs. Bath.** Assume that when you take a bath, you fill the tub to the halfway point and the tub measures 6 feet by 3 feet by 2.5 feet. When you take a shower, you use a showerhead with a flow rate of 1.75 gallons per minute and you typically spend 10 minutes in the shower. There are 7.5 gallons in one cubic foot.

a. Do you use more water taking a shower or taking a bath?

b. How long would you need to shower in order to use as much water as you use taking a bath?

c. Assuming your shower is in a bath tub, propose a non-mathematical way to compare, in one experiment, the amounts of water you use taking a shower and a bath.

74. **Supertankers.** An oil supertanker has a deadweight tonnage (the total amount that it can carry including crew, supplies, and cargo) of 300,000 long tons.

a. How many kilograms can the tanker carry?

b. Assume that the tonnage consists entirely of oil. If the density of oil is 850 kilograms per cubic meter, how many cubic meters of oil can the tanker carry?

c. Assume that 1000 liters of oil has a volume of 1 cubic meter. How many barrels of oil can the tanker carry? (Use data from Tables 2.2 and 2.4.)

d. Find the current price of oil in dollars per barrel. What is the value of the oil carried by a full tanker?

75. **Hurricane Katrina.** Experts estimate that when the levees around New Orleans broke in the aftermath of Hurricane Katrina in 2005, water flowed into the city at a peak rate of 9 billion gallons per day. There are 7.5 gallons in 1 cubic foot.

a. Find the flow rate in units of cubic feet per second (cfs). Compare this flow rate to the average flow rate of the Colorado River in the Grand Canyon, which is 30,000 cfs.

b. Assume that the flooded part of the city had an area of 6 square miles. Estimate how much (in feet) the water level rose in one day at the given flow rate.

76. **Glen Canyon Flood.** The Department of the Interior periodically releases a "spike flood" from the Glen Canyon Dam into the Colorado River. The purpose is to restore the river and the habitats along its banks, particularly in the Grand Canyon. The reservoir behind the dam (Lake Powell) holds approximately 1.2 trillion (1,200,000,000,000) cubic feet of water. During a recent week-long spike flood, water was released at a rate of 25,800 cubic feet per second. How much water was released during the 1-week flood? What percentage of the total water in the reservoir was released during the flood?

77. **Measuring Lumber.** The standard unit for measuring raw undried wood for lumber in the United States and Canada is the board-foot (abbreviated fbm, for foot board measure). One fbm is the volume of a 1-foot-by-1-foot-by-1-inch board.

a. Assume that a hemlock tree is approximated by a cylinder 15 inches in radius and 120 feet in height. Estimate the number of board-feet in the tree. The volume of a cylinder is $V = \pi r^2 h$.

b. Assuming no waste, how many whole 8-foot two-by-fours can be cut from 150 board-feet? The actual dimensions of a two-by-four are 1.5 inches by 3.5 inches.

c. A housing project requires 75 two-by-sixes, 12 feet in length. The actual dimensions of a two-by-six are 1.5 inches by 5.5 inches. How many board feet of lumber are needed?

78. **Cutting Timber.** A stand of fir trees occupies 60 acres, with an average density of 200 trees per 20 acres. A forester estimates that each tree will yield 400 board-feet of lumber. Estimate the yield of the stand in board-feet if one-tenth of the trees are cut.

79. **Fertilizing Winter Wheat.** Guidelines for the amount of supplementary nitrogen needed to grow winter wheat depend on the amount of nitrogen in the soil (as determined by a soil test), the price of fertilizer, and the price of wheat at harvest. Suppose the soil on a particular farm has a nitrogen content of 2 ppm (parts per million) and 50 acres will be planted in winter wheat. Consider two pricing scenarios.

- Case A: The price of fertilizer is \$0.25/lb, the price of wheat is \$3.50/bushel, and the expected yield is 60 bushels/acre.
- Case B: The price of fertilizer is \$0.50/lb, the price of wheat is \$4.50/bushel, and the expected yield is 50 bushels/acre.

In Case A, the guidelines recommend adding 100 pounds of nitrogen per acre, and in Case B, 70 pounds of nitrogen per acre.

Assuming all other factors are equal, compute and compare the net profits (income minus expenses) for the two scenarios.

80. **Metric Area.** When the metric system was first proposed, the unit of area was the *are* with 1 are $= 100 \text{ m}^2$. Today, the accepted metric unit of area is the hectare (ha), where 1 ha $=$ 100 are. This unit is used around the world in forestry and agriculture.

a. Using a power of 10, how many square meters are there in 1 hectare?

b. Express 1 km^2 in hectares.

c. Find the conversion factor between hectares and acres.

d. Using the currency conversions in Table 2.1, which is the more expensive price for land, \$5000/acre or 10,000 euro/ha?

81–82: Electric Bills. Consider the following electric bills.

a. Calculate the total electrical energy usage in joules.

b. Calculate your average power usage in watts.

c. Assume that your power supplier generates electricity by burning oil. Note that 1 liter of oil releases 12 million joules of energy. How much oil is needed to generate the electricity you use? Give your answer in both liters and gallons.

81. In May, you used 900 kilowatt-hours of energy for electricity.

82. In October, you used 1050 kilowatt-hours of energy for electricity.

83. **Human Wattage.** Suppose you require 2500 food Calories per day.

a. What is your average power, in watts? Compare your answer to the wattage of some familiar appliance.

b. How much energy, in joules, do you require from food in a year? Counting all forms of energy (such as gasoline, electricity, and energy for heating), the average U.S. citizen consumes about 400 billion joules of energy each year. Compare this value to the energy needed from food alone.

84. **Power Spa.** An outdoor spa (hot tub) draws 1500 watts to keep the water warm. If the utility company charges \$0.10 per kilowatt-hour, how much does it cost to operate the spa for 4 months during the winter (24 hours per day)?

85. **Nuclear Power Plant.** Operating at full capacity, the Columbia Generating Station, a nuclear power plant near Richland, Washington, can generate 1190 megawatts of power. Nuclear fission of 1 kilogram of uranium (in the form of uranium-235) releases 16 million kilowatt-hours of energy. How much energy, in kilowatt-hours, can the plant generate each month? How much uranium, in kilograms, is needed by this power plant each month? If a typical home uses 1000 kilowatt-hours per month, how many homes can this power plant supply with energy?

86. **Coal Power Plant.** A new coal-burning power plant can generate 1.5 gigawatts (billion watts) of power. Burning 1 kilogram of coal yields about 450 kilowatt-hours of energy. How much energy, in kilowatt-hours, can the plant generate each month? How much coal, in kilograms, is needed by this power plant each month? If a typical home uses 1000 kilowatt-hours per month, how many homes can this power plant supply with energy?

87–88: Solar Energy. Use these facts in the following exercises: Solar (photovoltaic) cells convert sunlight directly into electricity. If solar cells were 100% efficient, they would generate about 1000 watts of power per square meter of surface area when exposed to direct sunlight. With lower efficiency, they generate proportionally less power. For example, 10% efficient cells generate 100 watts of power in direct sunlight.

87. Suppose a 1-square-meter panel of solar cells has an efficiency of 20% and receives the equivalent of 6 hours of direct sunlight per day. How much energy, in joules, can it produce each day? What average power, in watts, does the panel produce?

88. Suppose you want to supply 1 kilowatt of power to a house (the average household power requirement) by putting solar panels on its roof. If you use the solar cells described in Exercise 87, how many square meters of solar panels would you need? Assume that you can make use of the average power from the solar cells (by, for example, storing the energy in batteries until it is needed).

89. **Wind Power: One Turbine.** Modern wind energy "farms" use large wind turbines to generate electricity from the wind. At a typical installation, a single modern turbine can produce an average power of about 2.5 megawatts. (This average takes wind variations into account.) How much energy, in kilowatt-hours, can such a turbine generate in a year? Given that the average household uses about 10,000 kilowatt-hours of energy each year, how many households can be powered by a single wind turbine?

90. **California Wind Power.** As of 2016, California has wind farms capable of generating a total of 6.1 gigawatts of power (roughly 5% of the state's total electricity).

    a. Assuming wind farms typically generate 30% of their capacity, how much energy, in kilowatt-hours, can the California wind farms generate in 1 year? Given that the average household uses about 10,000 kilowatt-hours of energy each year, how many households can be powered by these wind farms?

    b. One of the great advantages of wind power is that it does not produce the carbon dioxide emissions that contribute to global warming. On average, energy produced from fossil fuels generates about 1.5 pounds of carbon dioxide for every kilowatt-hour of energy. Suppose California did not have its wind farms and their energy output was instead produced from fossil fuels. How much more carbon dioxide would be entering the atmosphere each year?

91. **Solution Concentrations.**

    a. A 5% dextrose solution (D5W) contains 5 mg of dextrose per 100 mL of solution. How many milligrams of dextrose is in 750 mL of a 5% solution? How many milliliters of D5W solution should be given to a patient needing 50 mg of dextrose?

    b. Normal saline solution (NS) has a concentration of 0.9% sodium chloride, or 0.9 mg per 100 mL. How many milligrams of sodium chloride is in 1.2 L of NS? How many milliliters of NS should be given to a patient needing 15 mg of sodium chloride?

92. **Infusion Rates for Dextrose (D5W).** A 5% dextrose solution (5 mg per 100 mL of solution) is given intravenously. Suppose a total of 1.5 L of the solution is given over a 12-hour period.

    a. What is the flow rate in units of mL/hr? In units of mg (of dextrose)/hr?

    b. If each milliliter of the solution is equivalent to 15 drops (this drop factor is expressed as 15 gtt/mL, where the abbreviation *gtt* comes from the Latin *gutta*, for drop), what is the flow rate in units of gtt/hr?

    c. During the 12-hour period, how much dextrose is delivered?

93. **Infusion Rates for Normal Saline (NS).** A normal saline solution (0.9 mg of sodium chloride per 100 mL of solution) is given intravenously, and a total of 0.5 L of the solution is given over a 4-hour period.

    a. What is the flow rate in units of mL/hr? In units of mg (of sodium chloride)/hr?

    b. If each milliliter contains 20 drops (expressed as 20 gtt/mL), what is the flow rate in units of gtt/hr?

    c. During the 4-hour period, how much sodium chloride is delivered?

94. **Infusion Rates for Dopamine.** A solution consisting of 300 mg of dopamine in 200 mL of solution is administered at a rate of 10 mL/hr.

    a. What is the flow rate in units of mg (of dopamine)/hr?

    b. If a patient is prescribed a dosage of 60 mg of dopamine, how long should the infusion last?

95. **Administering an Antibiotic.** The antibiotic Keflex is available in a solution with a concentration of 250 mg per 5 mL of solution or in 250-mg capsules. The recommended dosage of Keflex is 25 mg/kg/day.

    a. How many capsules should a 40-kg person take every 6 hours?

    b. Suppose the antibiotic solution is given intravenously to a 40-kg person over a 6-hour period, using a microdrip system with a drop factor of 60 gtt/mL (60 drops/mL). What is the drop rate that should be used, in units of gtt/hr (drops/hr)?

96. **Administering Penicillin.** A doctor administers penicillin V to a 36-kg patient, using a dosage formula of 50 mg/kg/day. Assume that penicillin V is available in a 200 mg per 5 mL suspension or in 300-mg tablets.

    a. How many tablets should a 36-kg person take every 4 hours?

    b. A macrodrop system with a drop factor of 15 gtt/mL delivers the drug intravenously to the patient over an 8-hour period. What flow rate should be used, in units of gtt/hr?

97. **Measuring Drops.** Medication is often delivered through an intravenous (IV) drip line that allows a prescribed amount of a drug to be administered at a fixed rate. Each delivery system has a particular drop factor, which is the number of drops per milliliter of solution (abbreviated gtt/mL).

    a. A particular macrodrip system has a drop factor of 20 gtt/mL. How many drops are in a 0.5-L bag of normal saline?

    b. A particular microdrip system has a drop factor of 60 gtt/mL. How many drops are in a 1-L bag of D5W (dextrose) solution?

    c. Suppose an entire 1-L bag of normal saline is administered in 5 hours through a system with a drop factor of 15 gtt/mL. How many drops were delivered? What was the rate of infusion in gtt/min?

98. **Drug Dosage.** The label on a particular drug recommends 75–150 mg of the drug per kilogram of body weight per day. A doctor prescribes 200 mg of the drug every 8 hours for a 20-pound child. Is the prescription within the guidelines?

IN YOUR WORLD

99. **Should the United States Go Metric?** Research the history of attempts to convert the United States to the metric system. Do you think it will ever happen? Do you think it would be a good idea?

100. **Polar Ice Melting.** Starting with a search on "glaciers" or "glaciology" (the study of glaciers), use the Web to learn more about polar ice melting. Focus on one aspect of the issue, such as how much melting is caused by global warming, the environmental impacts of melting, or the geological history of ice ages. Write a one-page summary of what you learn.

101. **Utility Bill.** Analyze a utility bill. Explain all the units shown, and determine the relative costs of different energy uses. What changes would you recommend if the recipient of the bill wanted to lower energy costs?

102. **Electric Cars.** In terms of energy required to drive any particular distance, electric cars are far more efficient than gasoline-powered cars. Find out why this is the case, and what the tradeoffs are between buying an electric car or a gasoline-powered car. Write a one-page summary of your findings.

103. **Air Pollution.** Investigate the average concentrations of various pollutants in a major U.S. city of your choice. Find the EPA standards for each pollutant, and find some of the hazards associated with exposure to each pollutant. Track how the levels of pollution in this city have changed over the past 20 years. Based on your findings, do you think it is likely that pollution in this city will get better or worse over the next decade? Summarize your findings and your conclusions in a one- or two-page report.

104. **Alcohol Poisoning.** Research some aspect of the dangers of alcohol, such as drunk driving or alcohol poisoning. Find statistics related to this issue, especially data that relate blood alcohol content to specific dangers. Summarize your findings in a short report about how society might combat the danger.

UNIT 2C Problem-Solving Hints

The Understand-Solve-Explain process introduced in Unit 2A provides a procedural framework for problem solving, and the methods of unit analysis we have discussed can be very helpful. Still, problem solving is an art, and the only sure way to become more creative and improve your problem solving is through practice. In this unit, we'll discuss eight general hints that can be helpful in problem solving, each illustrated with an example.

Hint 1: There May Be More Than One Answer

How can society best reduce the total amount of greenhouse gases emitted into the atmosphere? We won't even attempt to answer this question, but it should make the point that no *single* best answer may be available. Indeed, many different political and economic strategies could yield similar reductions in greenhouse gas emissions.

Most people recognize that policy questions do not have unique answers, but the same is true of many mathematical problems. For example, both $x = 4$ and $x = -4$ are solutions to the equation $x^2 = 16$. Without further information and context, we have no way to determine whether both solutions are valid for a particular problem. Nonunique solutions often occur because not enough information is available to distinguish among a variety of possibilities.

EXAMPLE 1 Box Office Receipts

Tickets for a fundraising event were priced at \$10 for children and \$20 for adults. Shauna worked the first shift at the box office, selling a total of \$130 worth of tickets. However, she did not keep a careful count of how many tickets she sold for children and adults. How many tickets of each type (child and adult) did she sell?

Solution Let's proceed by trial and error. Suppose Shauna sold just one \$10 child ticket. In that case, she would have sold \$130 − \$10 = \$120 worth of adult tickets. Because the adult tickets cost \$20 apiece, this means she would have sold \$120 ÷ (\$20 per adult ticket) = 6 adult tickets. We have found an answer to the question: Shauna could have collected \$130 by selling 1 child and 6 adult tickets. But it is not the only answer, as we can see by testing other values. For example, suppose she sold three of the \$10 child tickets, for a total of \$30. Then she would have sold \$130 − \$30 = \$100 worth of adult tickets, which means 5 of the \$20 adult tickets. We have a second possible answer—3 child and 5 adult tickets—and have no way to know which answer is the actual number of tickets sold.

In fact, there are seven possible answers to the question. In addition to the two answers we've already found, other possible answers are 5 child tickets and 4 adult tickets; 7 child tickets and 3 adult tickets; 9 child tickets and 2 adult tickets; 11 child tickets and 1 adult ticket; and 13 child tickets with 0 adult tickets. Without further information, we do not know which combination represents the actual ticket sales.

▶ Now try Exercises 7–8.

Think About It Verify that each of the child/adult ticket combinations listed as a possible solution in Example 1 does indeed total to \$130. Then explain why there aren't any solutions that have an even number of child tickets.

Hint 2: There May Be More Than One Method

Just as there may be more than one right answer, there may be more than one method for finding an answer. But not all methods are equally efficient. As the following example illustrates, an efficient method can save a lot of time and work.

EXAMPLE 2 Jill and Jack's Race

Jill and Jack ran a 100-meter race. Jill won by 5 meters; that is, Jack had run only 95 meters when Jill crossed the finish line. They decide to race again, but this time Jill starts 5 meters behind the starting line. Assuming that both runners run at the same pace as before, who will win?

Solution Method 1 One approach to this problem is analytical—we analyze each race quantitatively. We were not told how fast either Jill or Jack ran, so we can choose some reasonable numbers. For example, we might assume that Jill ran the 100 meters in the first race in 20 seconds. In that case, her pace was 100 m ÷ 20 s = 5 meters per second (5 m/s). Because Jack ran only 95 meters in the same 20 seconds, his pace was 95 m ÷ 20 s = 4.75 m/s.

For the second race, Jill must run 105 meters (because she starts 5 meters behind the starting line) to Jack's 100 meters. We predict their times by dividing their respective race distances by their speeds from the first race:

$$\textit{Jill:}\quad 105\text{ m} \div 5\frac{\text{m}}{\text{s}} = 105\,\cancel{\text{m}} \times \frac{1\text{ s}}{5\,\cancel{\text{m}}} = 21\text{ s}$$

$$\textit{Jack:}\quad 100\text{ m} \div 4.75\frac{\text{m}}{\text{s}} = 100\,\cancel{\text{m}} \times \frac{1\text{ s}}{4.75\,\cancel{\text{m}}} \approx 21.05\text{ s}$$

Jill will win the second race by a slim margin.

Solution Method 2 Although the analytical method works, we can use a much more intuitive and direct solution. In the first race, Jill runs 100 meters in the same time that Jack runs 95 meters. Therefore, in the second race, Jill will pull even with Jack 95 meters from the starting line. In the remaining 5 meters, Jill's faster speed will allow her to pull away and win. Note how this insight avoids the calculations needed for the analytical method.

▶ Now try Exercise 9.

Hint 3: Use Appropriate Tools

You don't need a computer to check your tab in a restaurant, but you wouldn't want to use an abacus to do your income tax. For any given task, there is an appropriate level of power that is needed to accomplish it, and it is a matter of style and efficiency to neither underestimate nor overestimate that level. You usually will have a choice of tools to use in any problem. Choosing the tools most suited to the job will make your task much easier.

EXAMPLE 3 The Cars and the Canary

Two cars, 120 miles apart, begin driving toward each other on a long straight highway. One car travels at 20 miles per hour and the other at 40 miles per hour (Figure 2.7). At the same time, a canary, starting on one car, flies back and forth between the two cars as they approach each other. If the canary always flies at 150 miles per hour and turns around instantly at each car, how far has it flown when the cars meet?

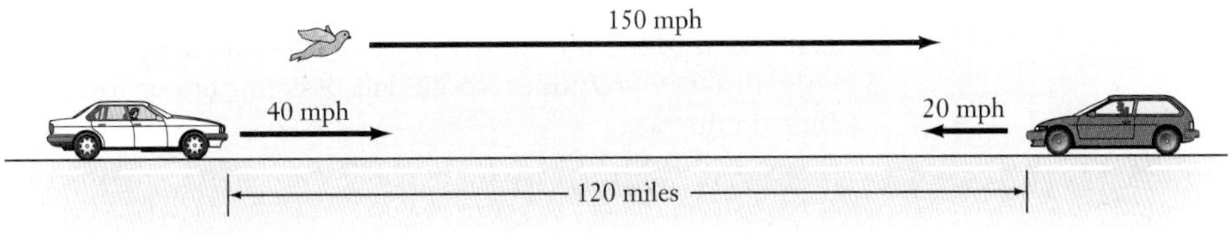

FIGURE 2.7 Set-up for the cars and canary problem.

Solution Because the problem asks "how far," we might be tempted to calculate the *distance* traveled by the canary on each back-and-forth trip between the cars. However, these trips get shorter as the cars approach each other and we would have to add up all the individual distances. In principle, we would need to add up an *infinite* number of ever-smaller distances—a task that involves the mathematics of calculus.

But note what happens if we focus on *time* rather than distance. The cars will approach each other at a relative speed of 60 mi/hr (because one car is traveling at 20 mi/hr and the other is traveling in the opposite direction at 40 mi/hr). Because they initially are 120 miles apart, they will meet in precisely 2 hours:

$$120 \text{ mi} \div 60\frac{\text{mi}}{\text{hr}} = 120 \cancel{\text{mi}} \times \frac{1 \text{ hr}}{60 \cancel{\text{mi}}} = 2 \text{ hr}$$

The canary is flying at a speed of 150 miles per hour, so in 2 hours it flies

$$2 \cancel{\text{hr}} \times 150 \frac{\text{mi}}{\cancel{\text{hr}}} = 300 \text{ mi}$$

The canary flies 300 miles before the cars meet. We could have found the answer with calculus, but why bother when we were able to do it with just multiplication and division?

▶ Now try Exercise 10.

MATHEMATICAL INSIGHT

Zeno's Paradox

The Greek philosopher Zeno of Elea (c. 460 B.C.E.) posed several paradoxes that defied solution for thousands of years. (A *paradox* is a situation or statement that seems to violate common sense or to contradict itself.) One paradox begins with an imaginary race between the warrior Achilles and a slow-moving tortoise. The tortoise is given a head start, but our common sense says that the swift Achilles will soon overtake the tortoise and win.

Zeno suggested a different way to look at the race. Suppose that, as shown in the figure, Achilles starts from point P0 and the tortoise starts from P1. During the time it takes Achilles to reach P1, the slow-moving tortoise will move ahead a little bit to P2. While Achilles continues on to P2, the tortoise will move ahead to P3. And so on. That is, Achilles must cover an infinite set of ever-smaller distances to catch the tortoise (that is, from P0 to P1, from P1 to P2, etc.). From this point of view, it seems that Achilles will never catch the tortoise.

This paradox puzzled philosophers and mathematicians for more than 2000 years. Its resolution depends on a key insight that became clear only with the invention of calculus in the 17th century: It does not necessarily require an infinite amount of time to cover an infinite set of distances. For example, imagine that the infinite set of distances covered by Achilles begins with 1 mile, then $\frac{1}{2}$ mile, then $\frac{1}{4}$ mile, and so on. Then the total distance he covers, in miles, is

$$1 + \frac{1}{2} + \frac{1}{4} + \frac{1}{8} + \frac{1}{16} + \frac{1}{32} + \frac{1}{64} + \frac{1}{128} + \frac{1}{256} + \frac{1}{512} + \frac{1}{1024} + \frac{1}{2048} + \cdots$$

This sum is called an **infinite series** because it is the sum of an infinite number of terms. You might guess that an infinite series would sum to infinity. But note what happens when we add just the first four terms, then the first eight terms, and then the first twelve terms. You can confirm the following results with a calculator.

$$1 + \frac{1}{2} + \frac{1}{4} + \frac{1}{8} = 1.875$$

$$1 + \frac{1}{2} + \frac{1}{4} + \frac{1}{8} + \frac{1}{16} + \frac{1}{32} + \frac{1}{64} + \frac{1}{128} = 1.9921875$$

$$1 + \frac{1}{2} + \frac{1}{4} + \frac{1}{8} + \frac{1}{16} + \frac{1}{32} + \frac{1}{64} + \frac{1}{128} + \frac{1}{256} + \frac{1}{512} + \frac{1}{1024} + \frac{1}{2048} = 1.99951171875$$

If you continue to add more terms in this infinite series, you will find that the sum gets closer and closer to 2, but never exceeds it. In fact, it can be proven deductively that the sum of this infinite series is 2. Therefore, even though the paradox has Achilles covering an infinite number of ever-shorter distances, the *total* distance he runs is finite—in this example, it is 2 miles. Clearly, it won't take him long to run a finite distance of 2 miles, so he will pass the slower tortoise and win the race.

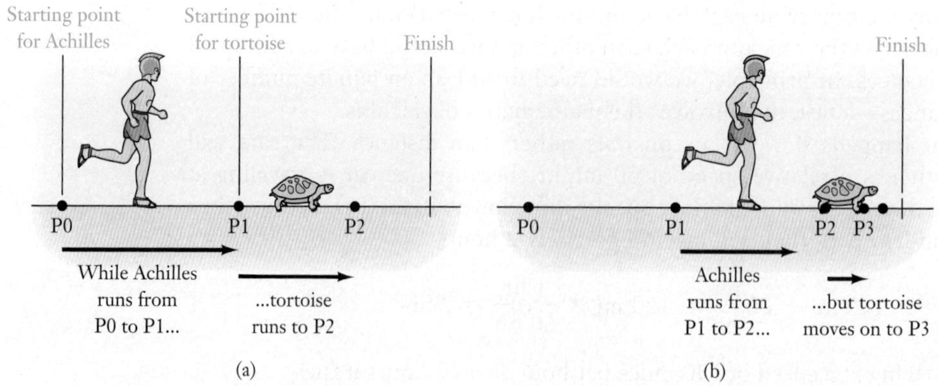

Hint 4: Consider Simpler, Similar Problems

Sometimes you are confronted with a problem that at first seems daunting. Our fourth hint is to consider a simpler, but similar, problem. The insight gained from solving the easier problem may help you understand the original problem.

EXAMPLE 4 Coffee and Milk

Suppose you have two cups in front of you: One holds coffee and one holds milk (Figure 2.8). You take a teaspoon of milk from the milk cup and stir it into the coffee cup. Next, you take a teaspoon of the mixture in the coffee cup and put it back into the milk cup. After the two transfers, there will be either (1) more coffee in the milk cup than milk in the coffee cup, (2) less coffee in the milk cup than milk in the coffee cup, or (3) equal amounts of coffee in the milk cup and milk in the coffee cup. Which of these three possibilities is correct?

FIGURE 2.8 The coffee and milk problem.

Solution A cup of either milk or coffee contains something like a *trillion trillion* molecules of liquid. Clearly, it would be difficult to visualize how such enormous numbers of molecules mix together, let alone to calculate the result. However, the essence of this problem is the *mixing* of two things. So one approach is to try a similar mixing problem that is much easier: mixing two piles of marbles.

Suppose that the black pile in Figure 2.9 has ten black marbles and represents the coffee. The white pile has ten white marbles and represents the milk. In this simpler problem, we can represent the first transfer—one teaspoon of milk into the coffee cup—by moving two white marbles to the black pile. This leaves the white pile with just eight white marbles, while the black pile now has ten black and two white marbles.

FIGURE 2.9 The white and black piles of marbles represent the milk and coffee, respectively. Moving two white marbles to the black pile represents putting a teaspoon of milk in the coffee.

Representing the second transfer—of one teaspoon from the coffee cup into the milk cup—involves taking any two marbles from the black pile and putting them in the white pile. We can then ask a question analogous to the original question: Are there more black marbles in the white pile or white marbles in the black pile?

Because the marbles represent molecules that mix thoroughly, the two marbles for the second transfer must be drawn at random, which presents three possible cases: The two marbles in the second transfer can be either both black, both white, or one of each.

But, as shown in Figure 2.10, in all three cases we end up with the same number of white marbles in the black pile as black marbles in the white pile. By analogy, we have the answer to our original question: After the two transfers, the amounts of coffee in the milk cup and milk in the coffee cup are equal.

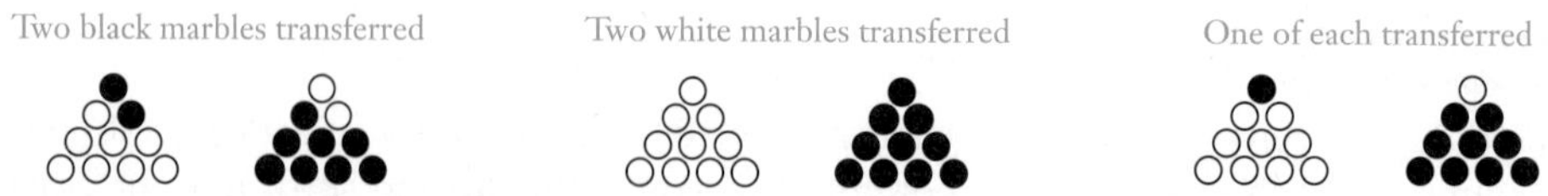

FIGURE 2.10 When we transfer two marbles back to the white pile, the two marbles may be both black (left), both white (center), or one black and one white (right). In all three cases, we end up with the same number of white marbles in the black pile as black marbles in the white pile.

The only remaining step is to confirm that the simpler problem is a reasonable representation of the real problem. Our choice of using two marbles to represent a teaspoon was arbitrary. If we redo this example with transfers of one, three, or any other number of marbles, we will find the same result: We always end up with the same number of black marbles in the white pile as white marbles in the black pile. Starting with ten marbles in each pile also was arbitrary; the conclusion remains the same if we start with twenty, fifty, or a trillion trillion marbles. Because molecules can be thought of as tiny marbles for the purpose of this problem, the real problem has no essential differences from the marble problem. ▶ **Now try Exercise 11.**

Think About It Most people are surprised by the result of the coffee and milk problem. Are you? Now that you know the solution, can you give a simple explanation of the real problem that would satisfy surprised friends?

Hint 5: Consider Equivalent Problems with Simpler Solutions

Replacing a problem with a similar, simpler problem can reveal essential insights about a problem, as we've just seen. However, "similar" is not good enough when we need a numerical answer. In that case, a useful approach to a difficult problem is to look for an *equivalent* problem. An equivalent problem will have the same numerical answer but may be easier to solve.

EXAMPLE 5 Party Decorations

Juan is decorating for a party in a room that has ten large cylindrical posts. The posts are 8 feet high and have a circumference of 6 feet. His plan is to wrap eight turns of ribbon around each post (from bottom to top, as shown on the left in Figure 2.11). How much ribbon does Juan need?

Solution The problem is difficult because it involves three-dimensional geometry. However, we can convert it to a simpler equivalent problem. Assume that each wrapped post is a hollow cylinder, and imagine cutting it down its length and unfolding it into a flat rectangle (as shown on the right in Figure 2.11). The width of the rectangle is the 6-foot circumference of the post, and its length is the 8-foot length of the post.

Now, instead of dealing with ribbon wrapped around a three-dimensional post, we have a simple rectangle with eight diagonal segments of ribbon. The total length of the eight segments will be the length of the ribbon required for one of the posts in the original problem. The height of each triangle is $\frac{1}{8}$ of the length of the rectangle, or 8 ft $\div$ 8 = 1 ft. The base of each triangle is the 6-foot width of the rectangle. The Pythagorean theorem tells us that

$$\text{base}^2 + \text{height}^2 = \text{hypotenuse}^2 \quad \text{or} \quad \text{hypotenuse} = \sqrt{\text{base}^2 + \text{height}^2}$$

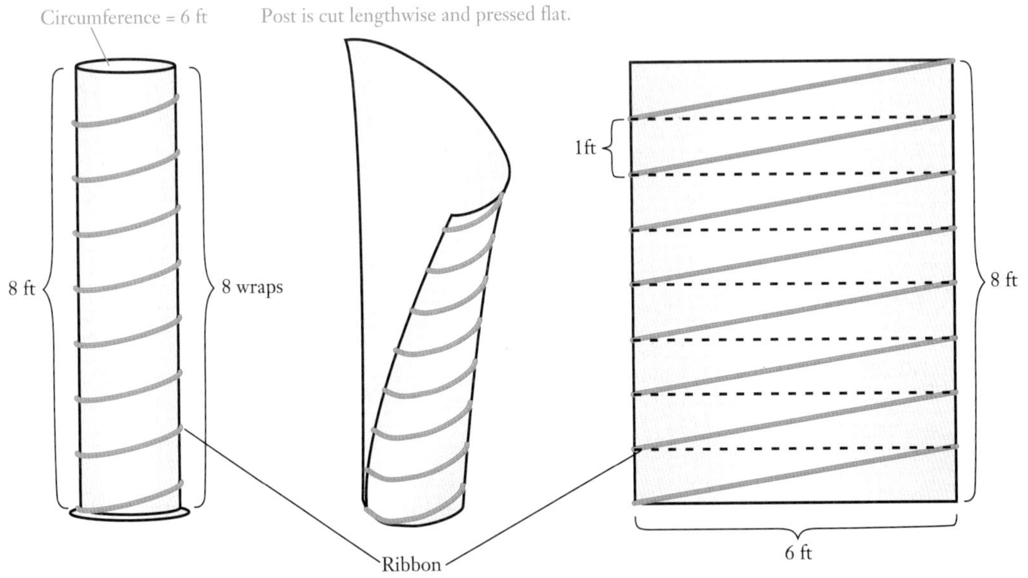

FIGURE 2.11 We want to know the length of ribbon needed to coil around a post (left). We can simplify the problem by imagining that we cut the wrapped post lengthwise (center). When the post is pressed flat, the ribbon lies in straight segments, each of which is the hypotenuse of a right triangle.

Substituting the 6-foot base and 1-foot height yields

$$\text{hypotenuse} = \sqrt{(6\text{ ft})^2 + (1\text{ ft})^2} \approx 6.1\text{ ft}$$

The length of each ribbon segment is 6.1 feet, and the total length of the ribbon is $8 \times 6.1\text{ ft} = 48.8\text{ ft}$. To decorate 10 posts, Juan needs $10 \times 48.8\text{ ft} = 488$ feet of ribbon. Note how much easier it was to solve this equivalent problem than the original one.

▸ Now try Exercise 12.

Hint 6: Approximations Can Be Useful

Another useful strategy is to make problems easier by using approximations. Most real problems involve approximate numbers to begin with, so an approximation often is good enough for a final answer. In other cases, an approximation will reveal the essential character of a problem, making it easier to reach an exact solution. Approximations also provide a useful check: If you come up with an "exact solution" that isn't close to the approximate one, something may have gone wrong.

EXAMPLE 6 A Bowed Rail

Imagine a mile-long bar of metal such as the rail along railroad tracks. Suppose that the rail is anchored on both ends (a mile apart) and that, on a hot day, its length expands by 1 foot. If the added length causes the rail to bow upward in a circular arc as shown in Figure 2.12a, about how high would the center of the rail rise above the ground?

Solution Because the added length is short compared to the original length, we can approximate the curved rail with two straight lines (Figure 2.12b). We now have two right triangles and the Pythagorean theorem applies. The bases of the two right triangles together give the original rail length of 1 mile, so each base is $\frac{1}{2}$ mile long. The two hypotenuses together represent the expanded length of 1 mile + 1 foot, so each hypotenuse is $\frac{1}{2}$ mile + $\frac{1}{2}$ foot long. Because there are 5280 feet in a mile, $\frac{1}{2}$ mile equals 2640 feet. The height of the rail off the ground is approximately the height of the triangles:

$$\text{height of triangle} = \sqrt{(2640.5\text{ ft})^2 - (2640\text{ ft})^2} \approx 51.4\text{ ft}$$

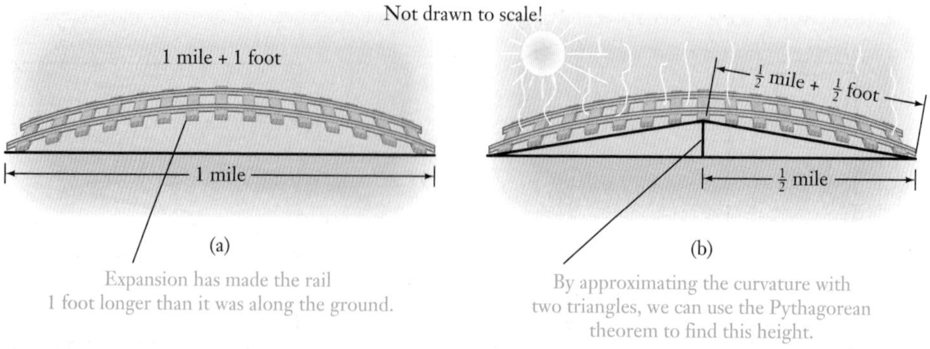
Not drawn to scale!
1 mile + 1 foot
1 mile
(a)
Expansion has made the rail 1 foot longer than it was along the ground.
½ mile + ½ foot
½ mile
(b)
By approximating the curvature with two triangles, we can use the Pythagorean theorem to find this height.

FIGURE 2.12 (a) If a rail lengthens while remaining anchored at its ends, it will bow upward as shown. (b) We can find the approximate height of the bow by ignoring its curvature and assuming that each side makes a triangle.

According to our approximation, the center of the rail would rise more than 50 feet off the ground! Because a triangle will stick up higher than a curve with the same base and length, the actual height is less than the approximate height. An exact solution shows that the top of the curved rail would be about 48 feet off the ground.

▶ **Now try Exercise 13.**

Think About It Are you surprised by the answer to Example 6? How does the use of the approximation tell you that the original question contained at least some unrealistic assumptions? Which assumptions do you think were unrealistic?

Hint 7: Try Alternative Patterns of Thought

Try to avoid rigid patterns of thought that tend to suggest the same ideas and methods over and over again. Instead, approach every problem with an openness that allows innovative ideas to percolate. In its most wondrous form, this approach is typified by what Martin Gardner (1914–2010), a well-known popularizer of mathematics, called an *aha! solution*. This is a solution that involves a penetrating insight that reduces the problem to its essential parts.

BY THE WAY

Hospital birth records show that boys and girls are born in nearly, but not precisely, equal numbers: Roughly 106 boys are born for every 100 girls. However, males have higher mortality rates than females at every age, so the numbers even out in adulthood and women outnumber men in old age.

EXAMPLE 7 China's Population Policy

In an effort to reduce population growth, from 1978 through 2015, China had a policy that allowed only one child per family. The policy succeeded in dramatically slowing China's population growth, but it also had a number of unintended consequences. One particularly troubling consequence was that, because of a cultural bias toward sons, many Chinese families found ways (primarily through selective abortions) to make sure that their one child was a boy, which caused China's population to end up with far more young men than young women. To address the unintended consequences, one popular proposal suggested replacing the one-child policy with a *one-son* policy, in which families would be allowed to continue having children until they had a son. That is, if a family's first child was a boy, the family would have met its limit of children, but if the first child was a girl, the family could have additional children until one was a boy.

a. Suppose that the one-son policy had been implemented and birth rates returned to their natural levels (about half boys and half girls). Compared to the one-child policy, how would the one-son policy change the population and proportions of boys and girls? Assume that most families would choose to have a son under the one-son policy.

b. China did not adopt a one-son policy; instead, as of 2016, China changed its one-child policy to a two-child policy. What is the population impact of a two-child policy compared to a one-son policy? Assume that most families would choose to have both children allowed under the two-child policy.

Solution

a. Most people initially approach this problem by noting that, under a one-son policy, half the families would have a boy as their first and only child. Of the remaining families, half would have a boy as their second child (reaching their one-son limit), while the other half would have a girl and go on to a third child. Half of these families would have a boy on the third try, while the other half would continue on after a third girl. And so on.

However, a moment of insight makes the problem much easier. Since we are assuming that the one-son policy would restore natural birth rates, roughly equal numbers of boys and girls would be born. And if we assume that most families would eventually have exactly one boy under this policy, then having equal numbers of boys and girls overall would lead to an average of one girl per family and hence an average of two children per family (one boy and one girl). Therefore, the one-son policy would double the average number of children per family from one to two while restoring the population of children to being about half boys and half girls.

b. If most families chose to have both children allowed under a two-child policy, then this policy obviously would double the total number of children compared to a one-child policy. So in terms of population growth, the impacts of a two-child policy and a one-son policy would be the same. The main difference is that under the two-child policy, many families would end up with both children being girls. (Probability tells us to expect one-fourth of the families to have two girls, because if we use B for boy and G for girl, GG is just one of four possible birth orders: GG, GB, BG, BB.) Therefore, a cultural bias toward sons might still lead families to look for ways to have a son, in which case China might still end up with more boys than girls being born.

▶ Now try Exercises 14–16.

Hint 8: Do Not Spin Your Wheels

Finally, everyone has had the experience of getting "bogged down" with a problem. When your wheels are spinning, let up on the gas! Often the best strategy in problem solving is to put a problem aside for a few hours or days. You may be amazed at what you see (and what you overlooked) when you return to it.

Quick Quiz 2C

Choose the best answer to each of the following questions. Explain your reasoning with one or more complete sentences.

1. A quantitative problem from daily life
   a. always has exactly one solution.
   b. may have more than one algebraic solution, but only one of these solutions can make sense.
   c. may have more than one correct answer.
2. Which of the following statements is *not* true?
   a. Mathematical problems with numerical answers always have a unique solution.
   b. There may be several methods that lead to the same solution of a mathematical problem.
   c. Sometimes it is possible to find a solution to a seemingly difficult problem by solving a simpler but similar problem.
3. You are asked to calculate how long a new flashlight battery (standard AA size) will last when the flashlight is on continuously. You should expect your answer to be
   a. a few minutes.
   b. between a few hours and a few days.
   c. at least several years.
4. You are asked to calculate the weight that a hotel elevator can safely carry. You should expect your answer to be
   a. less than 10 kilograms.
   b. several hundred kilograms.
   c. tens of thousands of kilograms.

5. If you add up an infinite number of ever-smaller fractions, the answer

   a. must be infinity. b. may be finite or infinite.

   c. must be zero.

6. The exterior surface area of a cylinder with a circumference of 10 inches and a length of 20 inches (and no end caps) is the same as the area of

   a. a rectangle measuring 10 inches by 20 inches.

   b. a circle with a circumference of 10 inches.

   c. a right triangle with its two shorter sides measuring 10 inches and 20 inches.

7. When confronted with a complex problem, why can it be useful to consider a simpler, similar problem?

   a. The answer to the simpler problem will always be the same as the answer to the complex problem.

   b. You'll be able to find the answer to the complex problem by multiplying the answer to the simpler problem by a scale factor.

   c. The simpler problem may provide you with insight that will help you approach the more complex problem.

8. Forty balls numbered 1 through 40 are mixed up in a barrel. How many balls must you draw from the barrel (without looking) to be sure that you have two even-numbered balls?

   a. 3 b. 40 c. 22

9. Karen arrives at the subway station every day at a random time and takes the first train that arrives. If she takes the A train, which arrives regularly every hour, she goes to the museum. If she takes the B train, which also arrives regularly every hour, she goes to the beach. A month later, she has gone to the beach 25 times and to the museum 5 times. Of the following options, which is the most likely explanation?

   a. The A train arrives on the hour and the B train always arrives 30 minutes later.

   b. The A train always arrives 10 minutes after the B train.

   c. The A train always arrives 10 minutes before the B train.

10. A small grill can hold two hamburgers at a time. If it takes 5 minutes to cook one side of a hamburger, what is the *shortest* time needed to cook both sides of three hamburgers?

    a. 20 minutes b. 15 minutes

    c. 10 minutes

Exercises 2C

REVIEW QUESTIONS

1. Summarize the hints for problem solving given in this unit, and give an example that applies each one.
2. Give an example of a problem that has more than one correct numerical answer.

DOES IT MAKE SENSE?

Decide whether each of the following statements makes sense (or is clearly true) or does not make sense (or is clearly false). Explain your reasoning.

3. My simple problem-solving recipe will work for any and all mathematical problems.
4. Whether it's a problem in mathematics or something else, it's best to start by taking time to make sure you understand the nature of the problem.
5. Mathematics requires precision, so approximations should never be used.
6. Once I decide my initial plan for solving a problem, I never change it.

BASIC SKILLS & CONCEPTS

7. **A Toll Booth.** A toll collector on a highway receives \$2 for cars and \$3 for buses. At the end of a 1-hour period, she has collected \$32. How many cars and buses passed through the toll booth during that period? List all possible solutions.
8. **Donations.** A public radio station offers two levels of support during a fundraiser. Level 1 membership requires a \$25 donation, and Level 2 membership requires a \$50 donation. At the end of the day, a fundraiser had pledges for \$350. How many Level 1 and Level 2 membership pledges were received during this fundraiser? List all possible solutions.
9. **A Second Race.** Jordan and Amari run a 200-meter race, and Jordan wins by 10 meters. They decide to run the 200-meter race again with Jordan starting 10 meters behind the starting line.

   a. Assuming both runners run at the same pace as they did in the first race, who wins the second race?

   b. Suppose Jordan starts 5 meters behind the starting line in the second race. Who wins the race?

   c. Suppose Jordan starts 15 meters behind the starting line in the second race. Who wins the race?

   d. How far behind the starting line must Jordan start so the second race is a tie? First estimate the distance as closely as possible. Then try to find it exactly.

10. **Catching Up.** A roadrunner is running at 5 m/s in pursuit of a coyote that is walking at 1 m/s along a straight road. The pursuit begins when the coyote is at Point A, 100 meters ahead of the roadrunner.

    a. How long does it take the roadrunner to reach Point A? During that time, how far does the coyote walk? The coyote's new position is Point B.

    b. Starting from Point A, how long does it take the roadrunner to reach Point B? During that time, how far does the coyote walk? The coyote's new position is Point C.

    c. Starting from Point B, how long does it take the roadrunner to reach Point C? During that time, how far does the coyote walk?

d. Based on the roadrunner's times to reach Points A, B, and C, estimate how long it takes the roadrunner to catch the coyote.

e. Suppose it takes the roadrunner 25 seconds to catch the coyote. How far has the roadrunner run? How far has the coyote walked?

11. **Mixing Marbles.** Consider the case in which each of two piles initially contains fifteen marbles (see Example 4 in this unit). Suppose that on the first transfer, three black marbles are moved to the white pile. On the second transfer, any three marbles are taken from the white pile and put into the black pile. Demonstrate, with diagrams and words, that you will always end up with as many white marbles in the black pile as black marbles in the white pile.

12. **Coiling Problem.** Eight turns of a wire are wrapped from bottom to top around a pipe with a length of 20 centimeters and a circumference of 6 centimeters. What is the length of the wire?

13. **Bowed Rail.** Suppose a railroad rail is 1 kilometer long and it expands on a hot day by 10 centimeters in length. Approximately how high would the center of the rail rise above the ground?

14. **China's Population.** To convince yourself that a one-son policy would lead to an average of two children per family, with equal numbers of boys and girls, do the following. Suppose that 10,000 families are having children according to the one-son policy. Describe the general makeup of all the families (that is, start with the fact that 5000 families have a boy as their first and therefore only child, and continue on). Use this process to show that the average number of children is two and that boys and girls are equal in number.

15. **Alternative Thinking: The Monk and the Mountain.** A monk set out from a monastery in the valley at dawn. He walked all day up a winding path, stopping for lunch and taking a nap along the way. At dusk he arrived at a temple on the mountaintop. The next day the monk made the return walk to the valley, leaving the temple at dawn, walking the same path for the entire day, and arriving at the monastery in the evening. Must there be at least one point along the path that the monk occupied at the same time of day on both the ascent and descent? (*Source: The Act of Creation*, Arthur Koestler)

16. **Crossing a Moat.** A castle is surrounded by a deep 10-feet-wide moat (see Figure 2.13). A knight on a rescue mission must cross the moat using only two $9\frac{1}{2}$- foot planks. Without using glue, nail, or supernatural means, how does he do it?

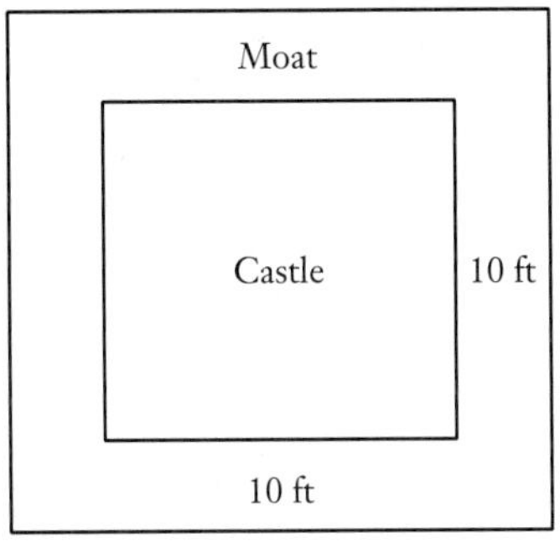

FIGURE 2.13

FURTHER APPLICATIONS

17. **Fencing a Yard.** Suppose you are designing a rectangular garden and you have 20 meters of fencing with which to enclose the garden.

a. Can you enclose a garden that is 7 meters long and 3 meters wide with the available fencing? What is the area of the garden?

b. Can you enclose a garden that is 8 meters long and 2 meters wide with the available fencing? What is the area of the garden?

c. By computing the area of other possible gardens that can be enclosed with 20 meters of fencing, find or estimate the dimensions of the garden that has the most area.

18. **Traffic Counter.** A thin tube stretched across a street counts the number of pairs of wheels that pass over it. A car with two axles registers two counts. A light truck with three axles registers three counts. During a 1-hour period, a traffic counter registered 35 counts. How many cars and light trucks passed over the traffic counter? List all possible solutions.

19. **An Ant's Journey.** Imagine a box-shaped room with an 8-foot ceiling and a rectangular floor that measures 12 feet by 10 feet. An ant sits high up on one of the 12-foot-wide walls, at a point that is 1 foot from the ceiling and 1 foot from a side wall. Her goal is to walk to a point on the opposite 12-foot-wide wall, 1 foot from the floor and 1 foot from the opposite side wall. What is the length of the shortest path? (*Hint:* Draw a diagram in which you unfold the room so the walls lay flat around the floor. Then identify the ant's start and end positions, draw a line between these points, and use the Pythagorean Theorem.)

20. **A Common Error: Averaging Speeds.** Suppose you run 2 miles from your house to a friend's house at a speed of 4 miles per hour. When the time comes to return home, you are tired and walk the same 2 miles home at 2 miles per hour.

a. How long did you spend running to your friend's house?

b. How long did you spend walking home?

c. Is it true that the average speed for the round trip is the average of 4 miles per hour and 2 miles per hour (which is 3 miles per hour)? (*Hint:* Did you spend more time traveling 4 miles per hour or 2 miles per hour?)

d. What is your average speed for the round trip?

21–32: Puzzle Problems. The following puzzle problems require careful reading and thinking. *Aha!* solutions are possible.

21. It takes you 30 seconds to walk from the first (ground) floor of a building to the third floor. How long will it take to walk from the first floor to the sixth floor (at the same pace, assuming that all floors have the same height)?

22. Reuben says, "Two days ago I was 20 years old. Later next year I will be 23 years old." Is this possible? If so, how, and if not, why not?

23. There are three kinds of apples all mixed up in a basket. How many apples must you draw (without looking) from the basket to be sure of getting at least two of one kind?

24. A woman bought a horse for \$500 and then sold it for \$600. She bought it back for \$700 and then sold it again for \$800. How much did she gain or lose on these transactions?
25. Three boxes of fruit are labeled *Apples, Oranges,* and *Apples and Oranges.* Each label is wrong. By selecting just one fruit from just one box, how can you determine the correct labeling of the boxes?
26. Each of ten large barrels is filled with golf balls that all look alike. The balls in nine of the barrels weigh 1 ounce and the balls in one of the barrels weigh 2 ounces. With only *one* weighing on a scale, how can you determine which barrel contains the heavy golf balls?
27. A woman is traveling with a wolf, a goose, and a mouse. She must cross a river in a boat that will hold only herself and one other animal. If left to themselves, the wolf will eat the goose and the goose will eat the mouse. How many crossings are required to get all four creatures across the river alive?
28. You are considering buying 12 gold coins that look alike, but you have been told that one of them is a heavy counterfeit. How can you find the heavy coin in three weighings on a balance scale?
29. Suppose you have 40 blue socks and 40 brown socks in a drawer. How many socks must you take from the drawer (without looking) to be sure of getting a pair of the same color?
30. Five books of five different colors are placed on a shelf. The orange book is between the gray and pink books, and these three books are consecutive. The gold book is not leftmost on the shelf and the pink book is not rightmost on the shelf. The brown book is separated from the pink book by two books. If the gold book is not next to the brown book, what is the complete ordering of the books?
31. If a clock takes 5 seconds to strike 5:00 (chiming five times), how long does it take to strike 10:00 (chiming ten times)? Assume each chime occurs simultaneously (takes no time).
32. One day in the maternity ward, the name tags for four girl babies become mixed up.

    a. In how many different ways could two babies be tagged correctly and two babies be tagged incorrectly?

    b. In how many different ways could three of the babies be tagged correctly and one baby be tagged incorrectly?

33–40: Projects: Real-World Problems. Consider the following complex problems, which do not have a single straightforward solution. Describe how you would approach the problem (without carrying out the process). List the assumptions and information needed to carry out your solution. Assess the uncertainty in these assumptions and data. Determine whether you believe the problem can be solved and whether a particular solution might generate controversy.

33. You are asked to calculate the cost of installing enough bike racks on campus to solve a bicycle parking problem.
34. What is the cost and what are the risks involved in installing enough battery-charging stations (for cell phones and other small electronic devices) to serve the students on your campus?
35. You want to know whether having a nationally ranked football program means more or less money for academic programs at your university.
36. A city imposes a law that requires all drivers to turn off their engines while stopped at traffic lights. Estimate the percent reduction in fuel use that this law could produce. What disadvantages might result from the law?
37. You decide that, in the interest of protecting the environment, you will convert your home heating and hot water system to solar power. How much will this conversion cost or save over the next 10 years?
38. Suppose that China and India decide to use their extensive coal reserves to supply energy to their populations at the same per capita level as in the United States. How much carbon dioxide would be added to the atmosphere?
39. Suppose that a city added new bus routes and handed out free bus passes. How many people would give up driving in favor of the bus? How much money, overall, would this cost or save the city?
40. You are planning to have a child. How much will it cost to raise the child over the next 18 years, assuming you provide for the child in the way you think best?

IN YOUR WORLD

41. **Textbook Analysis.** Although research shows that most adults today have difficulty with "story problems," we might hope that the next generation will have less difficulty. Find a current textbook in mathematics that is used in middle schools. Read through the "story problems" in the textbook. Write an analysis of the problems and conclude with an opinion as to whether the problems make mathematics meaningful.
42. **Multiple Solutions.** Find an example of a real problem for which, because of insufficient data, we cannot distinguish between two or more possible solutions. The problem might come from a news report or from your own experiences. What additional data would be useful?
43. **Multiple Methods.** Find an example of a real problem that could potentially be solved by two or more competing methods. The problem might come from a news report or from your own experiences. Describe each method. Which one do you think is better? Why?
44. **Novel Solution.** Find a news report concerning a problem in business or science that was solved by a surprising method. Describe the method and explain why it was useful.

Chapter 2 Summary

UNIT	KEY TERMS	KEY IDEAS AND SKILLS
2A	Understand-Solve-Explain (p. 71) units (p. 73) unit analysis (p. 73) conversion factor (p. 79) currency (p. 82)	**Apply** the three-step problem-solving process: (p. 72) 1. Understand the problem. 2. Solve the problem. 3. Explain your result. **Understand** the key words *per, of, square*, and *cube* and the key symbol the hyphen. (p. 74) **Apply** unit analysis to check answers and help solve problems. (p. 77) **Write** conversion factors in three equivalent forms. (p. 79) **Understand** why unit conversions mean multiplying by 1. (p. 79) **Apply** unit conversions with units raised to powers. (p. 81)
2B	U.S. customary system (USCS) (p. 89) metric system (SI) (p. 91) Celsius, Fahrenheit, Kelvin, absolute zero (p. 95) energy units: Calorie, joule, kilowatt-hour (p. 96) power units: 1 watt = 1 joule/s (p. 97) density (p. 98) concentration (p. 98) blood alcohol content (BAC) (p. 99)	**Know** the metric prefixes. (p. 91) **Understand** and convert basic USCS and metric units. (p. 93) **Understand** temperatures in °F, °C, and Kelvin, and how to convert between them. (p. 95) **Understand** the difference between energy and power: Power is the *rate* at which energy is used. (p. 97) **Apply** energy units in unit analysis and problem solving. (p. 97) **Apply** the concept of density to materials, population, and information. (p. 98) **Apply** the concept of concentration to medical dosage problems, air and water pollution, and blood alcohol content. (p. 100)
2C		**Remember** eight hints for problem solving: (p. 107) 1. There may be more than one answer. 2. There may be more than one method. 3. Use appropriate tools. 4. Consider simpler, similar problems. 5. Consider equivalent problems with simpler solutions. 6. Approximations can be useful. 7. Try alternative patterns of thought. 8. Do not spin your wheels.

3 NUMBERS IN THE REAL WORLD

Life is filled with numbers that may at first seem incomprehensible: a population measured in billions of people, national budgets measured in trillions of dollars, and distances ranging from nanometers to light-years. Moreover, numbers come in many forms, including percentages and index numbers (such as the Consumer Price Index), and nearly always involve uncertainties. In this chapter, we will discuss the use and interpretation of numbers we encounter in our daily lives.

Q

Today's nuclear power plants use a process called *nuclear fission*, in which large atoms (such as those of uranium or plutonium) are split apart. Suppose, instead, we had the ability to generate power through *nuclear fusion*—which combines hydrogen atoms to make the harmless gas helium—using hydrogen extracted from ordinary water as the fuel. If you had a portable fusion power plant and hooked it up to the faucet of your kitchen sink, how much power could you generate from the hydrogen in the water flowing through it?

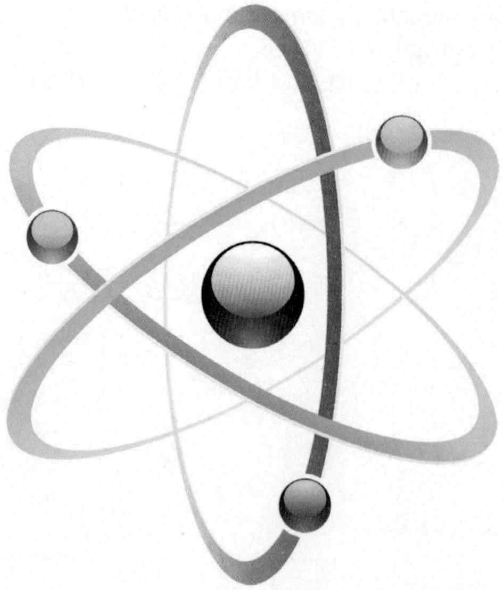

- **A** Enough to provide for all the electricity, heat, and air conditioning you use in your home or apartment
- **B** Enough to provide for the energy needs of 10 homes or apartments
- **C** Enough to provide for the energy needs of 100 homes or apartments
- **D** Enough to provide for the energy needs of 1000 homes or apartments
- **E** Enough to provide for all the energy needs of the entire United States

The concept of number is the obvious distinction between the beast and man. Thanks to number, the cry becomes song, noise acquires rhythm, the spring is transformed into a dance, force becomes dynamic, and outlines figures.

—Joseph de Maistre, 19th-century French philosopher

A

This question is an example of what we call an "order of magnitude" question, meaning that instead of asking us to look for an exact answer, it asks for only a general sense of the size of the answer. In this case, each answer choice differs from the previous one by at least a factor of 10. Knowing something only within a factor of 10 might seem like knowing very little, but it can often be quite meaningful. For example, a business will operate very differently if it estimates its customer base at 1000 people than if it estimates it at 100 or 10,000. In the same way, each answer choice to our fusion question would give a very different sense of the ways in which fusion might be useful.

So how do you figure out the answer? You could guess, but surveys of other students have found that very few guess correctly. The better approach is to calculate, and in this case you need only two pieces of readily available data to find the solution. Think about how you'd approach the problem, and when you're ready, check the solution that appears in Unit 3B, Example 5 (p. 145).

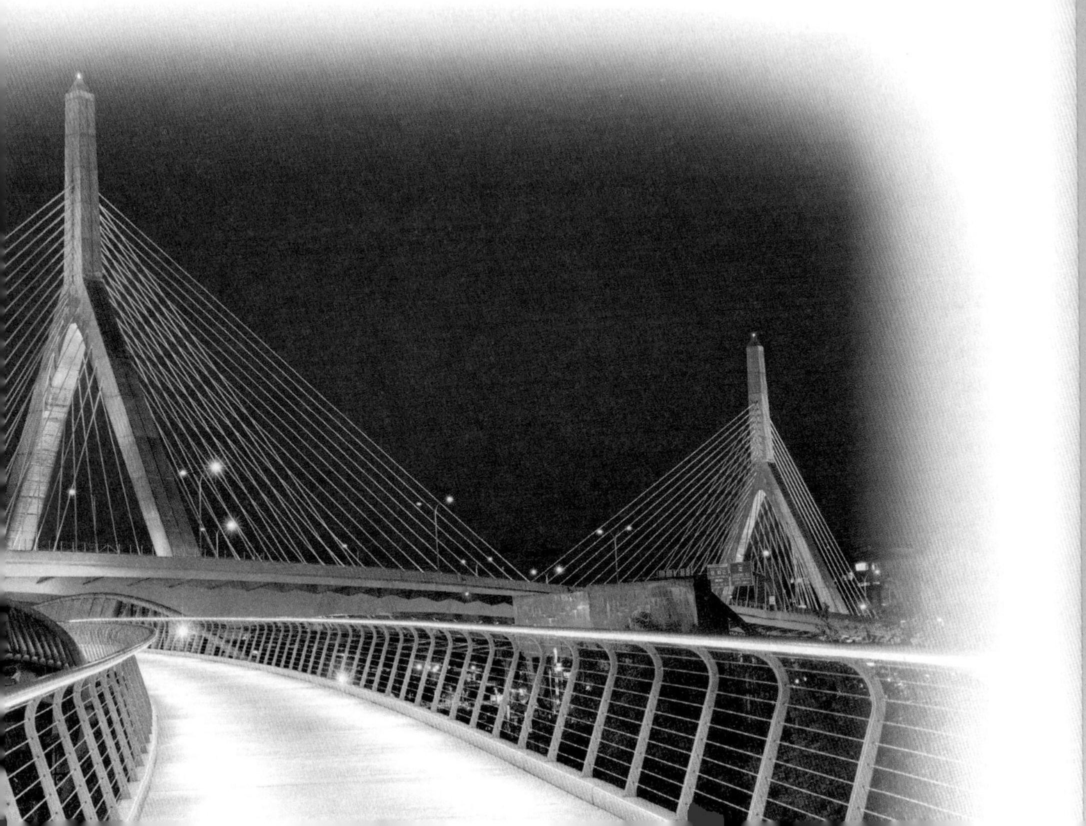

UNIT 3A

% Uses and Abuses of Percentages: Become familiar with subtle uses and abuses of percentages.

UNIT 3B

Putting Numbers in Perspective: Develop techniques for giving perspective to the many large and small numbers we encounter in daily life.

UNIT 3C

Dealing with Uncertainty: Understand the types of errors that affect measured numbers and explore ways of dealing with the inevitable uncertainty of numbers in the daily news.

UNIT 3D

Index Numbers: The CPI and Beyond: Study the role of index numbers, particularly the Consumer Price Index (CPI), in modern life.

UNIT 3E

How Numbers Can Deceive: Polygraphs, Mammograms, and More: Explore how numbers can be deceiving unless we interpret them carefully.

ACTIVITY

Big Numbers

Use this activity to gain a sense of the kinds of problems this chapter will enable you to analyze. Additional activities are available online in MyLab Math.

As a warm-up for thinking about the many ways in which we use numbers in the real world, let's investigate some large numbers that play key roles in all our lives. Use the Web to find the numbers you need to answer the following questions. If possible, work with two or three other students, with each of you looking up all the numbers. You may find that you and your colleagues find different results; if that happens, discuss how each of you found your numbers, why they differ, and which estimate is likely to be the best one.

1. What are the current populations of the United States and the world? What fraction of the world's population lives in the United States?
2. What is the federal government's deficit expected to be for this fiscal year? How much does that amount represent for each person in the United States?
3. What is the current federal debt of the United States? How much does the debt represent for each person in the United States? (*Note:* See Unit 4F if you are unsure of the difference between *deficit* and *debt*.)

4. How much gasoline is used in the United States each year? What is the average use per person, and what is the average cost per person?
5. Find the total annual budget and the total enrollment for your college or university. Then divide the budget by the enrollment to determine the average cost of educating each student. How does this number compare to what you actually pay in tuition? Can you explain why the two numbers are different?
6. How many videos are currently posted on YouTube? How many views does the average video receive? How many views does the most popular video receive?

UNIT 3A Uses and Abuses of Percentages

News reports frequently express quantitative information with percentages. Unfortunately, while percentages themselves are rather basic—they are just an alternative form of fractions—they are often used in very subtle ways. For example, consider the following statement about a change in e-cigarette usage from 2014 to 2015:

> *The rate of e-cigarette usage among middle school students was up 36 percent, to 5.3 percent.*

The percentages in this statement are used correctly, but the phrase "up 36 percent, to 5.3 percent" is not easy to interpret. In this unit, we will investigate this statement (see Example 11), along with other subtle uses and abuses of percentages.

Three Ways of Using Percentages

Consider the following statements from news reports:

- A total of 13,000 newspaper employees, 2.6% of the newspaper work force, lost their jobs.
- The company's stock fell 15% last week, to $44.25.
- The advanced battery lasts 125% longer than the standard one, but costs 200% more.

On close examination, each statement uses a percentage in a different way. The first uses a percentage to express a *fraction* of the total work force. The second uses a percentage to describe a *change* in stock price. The third uses percentages to *compare* the performance and costs of batteries.

Using Percentages as Fractions

Percent is just a fancy way of saying "divided by 100," so $P\%$ simply means $P/100$. For example, 5.3% means 5.3/100, or 0.053. Therefore, if 5.3% of middle school students use e-cigarettes among a group of 100,000 middle school students, then the number using e-cigarettes is 5.3% *of* 100,000, or

$$5.3\% \times 100{,}000 = 0.053 \times 100{,}000 = 5300$$

Notice that the word *of* told us to multiply. We've found that if the rate of e-cigarette usage among middle school students is 5.3%, then there are about 5300 e-cigarette users among 100,000 of those students.

Brief Review Percentages

The words *per cent* mean "per 100" or "divided by 100." We use the symbol % as shorthand for *per cent*. For example, we read 50% as "50 percent" and its meaning is

$$50\% = \frac{50}{100} = 0.5$$

More generally, for a number P,

$$P\% = \frac{P}{100}$$

For example:

$$100\% = \frac{100}{100} = 1 \qquad 200\% = \frac{200}{100} = 2$$

$$350\% = \frac{350}{100} = 3.5$$

Note that multiplying a number by 100% does not change its value, because 100% is just another way of writing 1. For example, multiplying 1.25 by 100% gives

$$1.25 \times 100\% = 125\%$$

That is, 125% is just another way of writing 1.25. These basic ideas lead to the following rules for converting between percentages and decimals or common fractions.

- **To convert a percentage to a common fraction:** Replace the % symbol with division by 100; simplify the fraction if necessary.

  Example: $25\% = \frac{25}{100} = \frac{1}{4}$

- **To convert a percentage to a decimal:** Drop the % symbol and divide by 100 (equivalent to moving the decimal point two places to the left).

  Example: $25\% = \frac{25}{100} = 0.25$

- **To convert a decimal to a percentage:** Multiply by 100 (equivalent to moving the decimal point two places to the right) and insert the % symbol.

  Example: $0.43 = \frac{43}{100} = 43\%$

- **To convert a common fraction to a percentage:** First convert the common fraction to decimal form, using a calculator if necessary; then convert the decimal to a percentage.

  Example: $\frac{1}{5} = 0.2 = 20\%$

▶ Now try Exercises 15–30.

EXAMPLE 1 Presidential Survey

An opinion poll finds that 35% of 1069 people surveyed said that the President is doing a good job. How many said the President is doing a good job?

Solution This percentage is used as a fraction and *of* indicates multiplication, so we multiply 35% by the 1069 respondents:

$$35\% \times 1069 = 0.35 \times 1069 = 374.15 \approx 374$$

We've found that about 374 of the 1069 people surveyed said the President is doing a good job. Note that we rounded the answer to the nearest whole number, because it wouldn't make sense to have a fraction of a person responding to the survey.

▶ Now try Exercises 37–42.

Using Percentages to Describe Change

Percentages are often used to describe how a quantity changes with time. As an example, suppose the population of a town has risen from 10,000 a decade ago to 15,000 today. We can express the change in population in two basic ways:

- Because the population rose by 5000 people (from 10,000 to 15,000), we say that the **absolute change** in the population was 5000 people.
- Because the increase of 5000 people was 50% of the starting population of 10,000, we say that the **relative change** in the population was 50%, or 0.5.

In general, calculating an absolute or relative change always involves two numbers: a starting number, or **reference value**, and a **new value**. Once we identify these two values, we can calculate the absolute and relative change with the formulas in the following box. Note that the absolute and relative changes are *positive* if the new value is greater than the reference value and *negative* if the new value is less than the reference value.

Absolute and Relative Change

The **absolute change** describes the actual increase or decrease from a reference value to a new value:

$$\text{absolute change} = \text{new value} - \text{reference value}$$

The **relative change** is the size of the absolute change in comparison to the reference value and can be expressed as a percentage:

$$\text{relative change} = \frac{\text{new value} - \text{reference value}}{\text{reference value}} \times 100\%$$

EXAMPLE 2 Stock Price Rise

During a 6-month period, Lunar Industry's stock doubled in price from \$7 to \$14. What were the absolute and relative changes in the stock price?

Solution The reference value is the starting stock price of \$7 and the new value is the later stock price of \$14, so the absolute change is

$$\text{absolute change} = \text{new value} - \text{reference value}$$

$$\$14 - \$7 = \$7$$

The relative change is

$$\text{relative change} = \frac{\text{new value} - \text{reference value}}{\text{reference value}} \times 100\% = \frac{\$14 - \$7}{\$7} \times 100\% = 100\%$$

The doubling of Lunar Industry's stock price from \$7 to \$14 represented an absolute change of \$7 and a relative change of 100%.

▲ **CAUTION!** Note that when a quantity *doubles* in value, it has increased by 100% from its prior value (*not* by 200%, as many people guess). ▲

▶ Now try Exercises 43–44.

Think About It Choose and explain the correct answer: If a population triples, then it has increased by ________. (a) 100% (b) 200% (c) 300% (d) 400%

EXAMPLE 3 World Population Growth

Estimated world population increased from 2.7 billion in 1953 to 7.5 billion in 2018. Describe the absolute and relative change in world population over this 65-year period.

Solution The reference value is the 1953 population estimate of 2.7 billion, and the new value is the 2018 population estimate of 7.5 billion. We use these values in the absolute and relative change formulas:

$$\begin{aligned}\text{absolute change} &= \text{new value} - \text{reference value} \\ &= 7.5 \text{ billion} - 2.7 \text{ billion} = 4.8 \text{ billion} \\ \text{relative change} &= \frac{\text{new value} - \text{reference value}}{\text{reference value}} \times 100\% \\ &= \frac{7.5 \text{ billion} - 2.7 \text{ billion}}{2.7 \text{ billion}} \times 100\% \approx 178\%\end{aligned}$$

Estimated world population increased by 4.8 billion people, or by 178%, from 1953 to 2018.

▶ Now try Exercises 45–46.

BY THE WAY

If you are an 18-year-old college student, world population today is *larger* than it was when you were born by about 1.4 billion people—or by more than four times the current population of the United States.

EXAMPLE 4 Depreciating a Computer

You bought a new laptop computer three years ago for \$1000. Today, it is worth only \$300. Describe the absolute and relative change in the laptop's value.

Solution The reference value is the original price of \$1000, and the new value is its current worth of \$300. The absolute and relative changes in the laptop's value are

$$\begin{aligned}\text{absolute change} &= \text{new value} - \text{reference value} \\ &= \$300 - \$1000 = -\$700 \\ \text{relative change} &= \frac{\text{new value} - \text{reference value}}{\text{reference value}} \times 100\% \\ &= \frac{\$300 - \$1000}{\$1000} \times 100\% = -70\%\end{aligned}$$

The absolute change in the laptop's value is −\$700 and the relative change is −70%. The negative signs tell us that the laptop's monetary value has *decreased* with time.

▶ Now try Exercises 47–48.

Using Percentages for Comparisons

Percentages are also commonly used to compare two numbers. Suppose we want to compare the price of a \$50,000 Mercedes to the price of a \$40,000 Lexus. The difference between the Mercedes and Lexus prices is

$$\$50{,}000 - \$40{,}000 = \$10{,}000$$

That is, the Mercedes costs \$10,000 *more than* the Lexus. We can also express this difference as a percentage of the Lexus price:

$$\frac{\$10{,}000}{\$40{,}000} = 0.25 = 25\%$$

In relative terms, the Mercedes costs 25% *more than* the Lexus.

Because we are *comparing to* the Lexus price, we say that the Lexus price is the **reference value**. *Notice that the reference value follows the word "than."* The Mercedes price is the **compared value**. We can now define the absolute and relative *difference* between the quantities, much as we defined absolute and relative change. The absolute and relative differences are *positive* if the compared value is greater than the reference value and *negative* if the compared value is less than the reference value.

Absolute and Relative Difference

The **absolute difference** is the actual difference between the compared value and the reference value:

$$\text{absolute difference} = \text{compared value} - \text{reference value}$$

The **relative difference** describes the size of the absolute difference in comparison to the reference value and can be expressed as a percentage:

$$\text{relative difference} = \frac{\text{compared value} - \text{reference value}}{\text{reference value}} \times 100\%$$

That's essentially all there is to comparisons, except for one subtlety: There's no particular reason why we chose the Lexus price as the reference value. We can just as easily use the Mercedes price as our reference value. In that case, the Lexus price is the compared value and the absolute difference is

$$\text{Lexus price} - \text{Mercedes price} = \$40{,}000 - \$50{,}000 = -\$10{,}000$$

The negative sign tells us that the Lexus costs \$10,000 *less than* the Mercedes. Again, note that the reference value follows *than*. The relative difference is

$$\frac{\text{Lexus price} - \text{Mercedes price}}{\text{Mercedes price}} \times 100\% = \frac{-\$10{,}000}{\$50{,}000} \times 100\% = -20\%$$

The negative sign indicates that the Lexus costs 20% *less than* the Mercedes.

We now have two ways to express the relative difference in the car prices:

- The Mercedes costs 25% more than the Lexus; or
- the Lexus costs 20% less than the Mercedes.

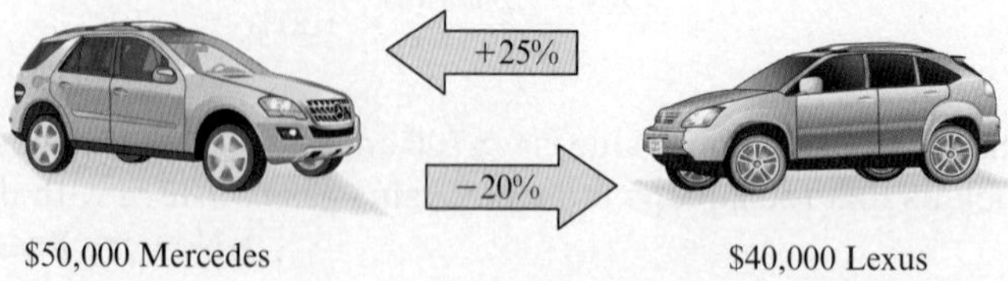

Both statements are correct, yet they contain *different* percentages. That is why it is so important to keep careful track of the reference and compared values.

For those who like formulas, we can write the following rule: If quantity A is $P\%$ *more* than quantity B, then quantity B is $\frac{100P}{100 + P}$ percent *less* than quantity A.

EXAMPLE 5 Income Comparison

Recent data showed that California ranked first among the 50 states in average income, at about \$68,900 per person, and West Virginia ranked last at \$46,600 per person.

a. How much lower is average income in West Virginia than in California?
b. How much higher is average income in California than in West Virginia?

Answer both questions in both absolute and relative terms.

Solution

a. We are asked how much lower income is in West Virginia than in California. Remembering that the reference value follows the word *than*, we use the California income as the reference value and the West Virginia income as the compared value:

$$\begin{aligned}\text{absolute difference} &= \text{compared value} - \text{reference value}\\ &= \$46{,}600 - \$68{,}900 = -\$22{,}300\\ \text{relative difference} &= \frac{\text{compared value} - \text{reference value}}{\text{reference value}} \times 100\%\\ &= \frac{\$46{,}600 - \$68{,}900}{\$68{,}900} \times 100\% \approx -32.4\%\end{aligned}$$

The negative signs tell us that average income in West Virginia is *less than* that in California by \$22,300, or by about 32.4%.

b. This time West Virginia follows *than*, so we use the West Virginia income as the reference value and the California income as the compared value:

$$\begin{aligned}\text{absolute difference} &= \text{compared value} - \text{reference value}\\ &= \$68{,}900 - \$46{,}600 = \$22{,}300\\ \text{relative difference} &= \frac{\text{compared value} - \text{reference value}}{\text{reference value}} \times 100\%\\ &= \frac{\$68{,}900 - \$46{,}600}{\$46{,}600} \times 100\% \approx 47.9\%\end{aligned}$$

Average income in California is *more than* that in West Virginia by \$22,300, or by about 47.9%.

▲ **CAUTION!** Note that while the absolute differences in parts (a) and (b) are the same except for the sign (positive or negative) no matter which value you choose as the reference value, the relative differences are not the same. That is why you must conclude with a statement that makes your choices of the compared and reference values clear. ▲

▶ Now try Exercises 49–52.

Of versus *More Than*

Consider a population that triples in size from 200 to 600. There are two equivalent ways to state this change with percentages:

- Using *more than:* The new population is 200% *more than* the original population. Here we are stating the *relative change* in the population:

$$\text{relative change} = \frac{\text{new value} - \text{reference value}}{\text{reference value}} \times 100\%$$

$$= \frac{600 - 200}{200} \times 100\% = 200\%$$

- Using *of:* The new population is 300% *of* the original population, which means it is three times the size of the original population. Here we are looking at the *ratio* of the new population to the original population:

$$\frac{\text{new population}}{\text{original population}} = \frac{600}{200} = 3.00 = 300\%$$

Notice that the percentages in the *of* and *more than* statements are related by 300% = 100% + 200%. This leads to the following general relationships.

Of versus More Than (or Less Than)

- If the new or compared value is P% *more than* the reference value, it is $(100 + P)\%$ *of* the reference value.
- If the new or compared value is P% *less than* the reference value, it is $(100 - P)\%$ *of* the reference value.

Think About It Use the relative change formula to confirm that a population that grows from 200 to 600 increases in size by 200%.

EXAMPLE 6 Income Difference

Carol earns 200% more than William. What is Carol's income as a percentage *of* William's? How many times as large as William's income is Carol's?

Solution We use the rule that P% *more than* means $(100 + P)\%$ *of*. Because Carol's income is 200% more than William's, we set $P = 200$. Therefore, Carol's income is $(100 + 200)\% = 300\%$ *of* William's income. Because 300% = 3, Carol earns 3 times as much as William.

▲ **CAUTION!** Unfortunately, people often use the words "more than" in a way that is not technically correct. This example shows that because Carol earns 200% *more than* William, she earns 3 times *as much as* he does. If you were instead to say that she earns "3 times more than" he does (as many people do), your statement would be incorrect because it would imply 300% more, rather than 200% more. ▲

▶ Now try Exercises 53–56.

EXAMPLE 7 Sale!

A store is having a "25% off" sale. How does an item's sale price compare to its original price?

Solution The phrase "25% off" means that an item's sale price is 25% *less than* its original price. The sale price is $(100 - 25)\% = 75\%$ *of* the original price. For example, if the original price is \$100, the sale price is \$75. ▶ Now try Exercises 57–60.

Think About It One store advertises "30% off everything!" Another store advertises "Sale prices are 30% of original prices!" Which store is having the bigger sale? Explain.

Percentages of Percentages

Percentage changes or comparisons can be particularly confusing when the values *themselves* are percentages. Suppose a bank increases its mortgage interest rate from 3% to 4%. It's tempting to say that the interest rate increased by 1%, but this statement is ambiguous at best. The interest rate increased by 1 *percentage point,* but the relative change in the interest rate was

$$\frac{\text{new value} - \text{reference value}}{\text{reference value}} \times 100\% = \frac{4\% - 3\%}{3\%} \times 100\% = 33\%$$

In other words, you can say that the bank raised the interest rate by 33%, even though the actual rate increased by only 1 percentage point. More specifically, the *absolute change* in the interest rate was 1 percentage point, while the *relative change* in the interest rate was 33%.

Percentage Points versus Percent (or %)

When you see a change or difference expressed in *percentage points,* you can assume it is an *absolute* change or difference. If it is expressed with the % sign or the word *percent,* it should be a *relative* change or difference.

Brief Review What Is a Ratio?

Suppose we want to compare two quantities, such as the prices of a BMW that costs \$80,000 and a Honda that costs \$20,000. We can, of course, find the absolute or relative difference in the quantities. But another way to make the comparison is to compute the *ratio* of the two quantities. In this case, the two quantities are \$80,000 and \$20,000, so their ratio is

$$\frac{\$80{,}000}{\$20{,}000} = \frac{4}{1} = 4$$

Notice the cancellation of the unit *dollars* on the top and bottom of the fraction. This is a general characteristic of ratios: Because comparisons make sense only when we compare quantities with the same units, ratios always end up without units. Also note that we can state this result in several equivalent ways:

- The ratio of the BMW price to the Honda price is 4 to 1. (The ratio can also be stated as 4, but "4 to 1" is more common.)
- The BMW price is 4 times the price of the Honda.
- The ratio of the Honda price to the BMW price is 1 to 4, which we can also state as $\frac{1}{4}$, 0.25, or 25%.

Example: Earth's average density is about 5.5 grams per cubic centimeter, and the average density of Saturn is about 0.7 gram per cubic centimeter. What is the ratio of their densities?

Solution: We divide Earth's density by Saturn's density:

$$\frac{\text{average density of Earth}}{\text{average density of Saturn}} = \frac{5.5\ \text{g}/\cancel{\text{cm}^3}}{0.7\ \text{g}/\cancel{\text{cm}^3}} \approx 8$$

The ratio of Earth's average density to Saturn's average density is about 8 to 1. Alternatively, we can say that Earth's density is about 8 times that of Saturn or that Saturn's density is about $\frac{1}{8}$ that of Earth. Again, notice how the units canceled, leaving the ratio without units.

▶ Now try Exercises 31–36.

EXAMPLE 8 Newspaper Readership Declines

According to Pew Research Center polls, the percentage of adults who regularly read a daily newspaper fell from about 54% in 2004 to about 38% in 2017. Describe this change in newspaper readership.

Solution The drop in readership from 54% to 38% represents a decline of $54 - 38 = 16$ percentage points. This is the *absolute change* in the percentage of adults who read a daily newspaper. The *relative change* in readership is

$$\frac{\text{new value} - \text{reference value}}{\text{reference value}} \times 100\% = \frac{38\% - 54\%}{54\%} \times 100\% \approx -30\%$$

The negative sign indicates a *decrease* in readership. We say that regular newspaper readership dropped by about 30% in relative terms or by 16 percentage points in absolute terms.

▲ **CAUTION!** Notice that while the drop from 54% to 38% represents a decline of 16 percentage points, it is *not* correct to call it a 16% decline. (As we just found, a drop from 54% to 38% represents a relative decline of 30%.) ▲

▶ Now try Exercises 61–64.

EXAMPLE 9 Care in Wording

If you can't convince them, confuse them.
—Harry S. Truman

Assume that 40% of the registered voters in Carson City are Republicans. Read the following questions carefully, and give the most appropriate answers.

a. The percentage of voters registered as Republicans is 25% higher in Freetown than in Carson City. What percentage of the registered voters in Freetown are Republicans?
b. The percentage of voters registered as Republicans is 25 percentage points higher in Freetown than in Carson City. What percentage of the registered voters in Freetown are Republicans?

Solution

a. We are told that the percentage of registered Republicans in Carson City is 40% and that the percentage of registered Republicans in Freetown is 25% higher. We interpret the 25% as a relative difference. Because 25% of 40% is 10% $(0.25 \times 0.40 = 0.10)$, we add this value to the Carson City percentage to find that the percentage of registered Republicans in Freetown is $40\% + 10\% = 50\%$.
b. We interpret 25 *percentage points* as an absolute difference, so we add this value to the percentage of Republicans in Carson City. The percentage of registered Republicans in Freetown is $40\% + 25\% = 65\%$.

▶ Now try Exercises 65–66.

Solving Percentage Problems

Consider this statement:

Retail prices are 25% more than wholesale prices.

If you knew a wholesale price, how would you calculate the retail price? One way to begin is by translating the *more than* statement into an *of* statement. The statement becomes

Retail prices are $(100 + 25)\% = 125\%$ *of wholesale prices.*

Replacing *of* with multiplication, we write this statement mathematically as

$$\text{retail price} = 125\% \times \text{wholesale price}$$

With this equation, we can find a retail price from a wholesale price. For example, if the wholesale price is \$10, the retail price is $125\% \times \$10 = 1.25 \times \$10 = \$12.50$.

We can rearrange the same equation to find a wholesale price from a retail price. To do so, we first divide both sides of the equation by 125%:

$$\frac{\text{retail price}}{125\%} = \frac{125\% \times \text{wholesale price}}{125\%}$$

We then interchange the left and right sides to write the result:

$$\text{wholesale price} = \frac{\text{retail price}}{125\%}$$

For example, if the retail price is \$15, the wholesale price is $\$15/1.25 = \$12$. We can now generalize our results.

Solving Percentage Problems

Given that some final (or compared) value is $P\%$ *more than* an initial (or reference) value, calculations will be easier if you first convert the *more than* statement to an *of* statement, which you can write as multiplication:

$$\text{final value} = (100 + P)\% \times \text{initial value}$$

Use the equation in the above form if you are given P and the initial value. If you are given the final value and want to find the initial value, then rearrange the above equation:

$$\text{initial value} = \frac{\text{final value}}{(100 + P)\%}$$

If the final value is *less than* the initial value, then use $(100 - P)$ instead of $(100 + P)$ in the above equations.

EXAMPLE 10 Tax Calculations

a. You purchase a t-shirt with a labeled (pre-tax) price of \$17. The local sales tax rate is 5%. What is your final cost (including tax)?

b. Your receipt shows that you paid \$19.26 for a phone case, tax included. The local sales tax rate is 7%. What was the labeled (pre-tax) price of the case?

c. You are eligible for a 15% student discount for basketball tickets that otherwise cost \$55. How much will you pay?

BY THE WAY

Percentages are often called *rates*. For example, a sales tax of 6% is called a *tax rate* of 6%. If 5.3% of middle school students use e-cigarettes, we say that the *e-cigarette usage rate* among middle school students is 5.3%.

Solution

a. To be sure the process is clear, let's apply our Understand-Solve-Explain strategy (p. 72).

Understand: We are asked to find the final cost of the t-shirt, and we are given its labeled price of \$17 and the sales tax rate of 5%. One way to think about this is to recognize that the final cost will therefore be 5% *more than* the labeled price, which is the same as saying that the final cost will be 105% *of* the labeled price.

Solve: We write *of* as multiplication and substitute the labeled price of \$17:

$$\text{final cost} = 105\% \times \text{labeled price} = 1.05 \times \$17 = \$17.85$$

Explain: We've found that with a labeled price of \$17 and a tax rate of 5%, the final cost for the t-shirt is \$17.85.

b. We use the same process as in part (a) but without writing out all the details. We know that with a tax rate of 7%, the final cost of the phone case is 7% *more than* its labeled price, which means

$$\text{final cost} = (100 + 7)\% \times \text{labeled price} = 107\% \times \text{labeled price}$$

In this case, we are given the final cost of \$19.26 and asked to find the labeled price, so we proceed as follows:

Starting equation:	$\text{final cost} = 107\% \times \text{labeled price}$
Divide both sides by 107%:	$\text{labeled price} = \dfrac{\text{final cost}}{107\%}$
Replace 107% by 1.07, substitute final cost of \$19.26, and calculate:	$\text{labeled price} = \dfrac{\text{final cost}}{1.07} = \dfrac{\$19.26}{1.07} = \$18.00$

The labeled price of the phone case was \$18.00. Note that a good way to check this answer is to confirm that adding a 7% sales tax to a price of \$18.00 returns the given receipt value: $1.07 \times \$18.00 = \$19.26$.

c. This question is very similar to the one in part (a) except that because you are getting a 15% discount, the final ticket price is *less than* the initial ticket price. We therefore *subtract* 15% from 100% (alternatively, you can think of it as adding $P = -15\%$ to 100%):

Starting equation:	$\text{final price} = (100 - P)\% \times \text{initial price}$
Substitute given values:	$\text{final price} = (100 - 15)\% \times \$55 = 85\% \times \$55$
Replace 85% by 0.85 and calculate:	$\text{final price} = 0.85 \times \$55 = \$46.75$

Your discounted ticket price will be \$46.75. ▶ **Now try Exercises 67–68.**

EXAMPLE 11 Up 36%, to 5.3%

Consider the statement from the beginning of this unit:

The rate of e-cigarette usage among middle school students was up 36 percent, to 5.3 percent.

What was the rate of e-cigarette usage among middle school students before this rise?

Solution Here, "5.3 percent" is telling us that 5.3% of middle school students now use e-cigarettes. And "36 percent" expresses the relative change from the previous rate to the new rate, telling us that the new rate is 36% *more than* the previous rate. Changing from a *more than* statement to an *of* statement, and writing *of* as multiplication, we have

$$\text{new rate} = (100 + 36)\% \times \text{previous rate} = 136\% \times \text{previous rate}$$

We solve for the previous rate by dividing both sides by 136% = 1.36:

$$\text{previous rate} = \frac{\text{new rate}}{136\%} = \frac{\text{new rate}}{1.36}$$

Now we find the previous rate by substituting 5.3% for the new rate:

$$\text{previous rate} = \frac{5.3\%}{1.36} \approx 3.9\%$$

The previous e-cigarette usage rate for middle school students was 3.9%. Note that, in this case, we left the numerator as a percentage in the above equation (rather than converting it to decimal form) so that our answer would be a percentage.

▶ Now try Exercises 69–70.

Abuses of Percentages

Because percentages can be so subtle, many people misuse them—sometimes inadvertently and sometimes deliberately. In the rest of this unit, we investigate a few common abuses of percentages.

Beware of Shifting Reference Values

Consider the following situation: Because of losses experienced by your employer, you agree to accept a temporary 10% pay cut. Your employer promises to give you a 10% pay raise after six months. Will that pay raise restore your original salary?

We can answer this question by assuming some arbitrary number for your original weekly pay, such as \$500. A 10% pay cut means that your pay will decrease by 10% of \$500, or \$50, so your weekly pay after the cut is

$$\$500 - \$50 = \$450$$

The subsequent raise increases your pay by 10% of \$450, or \$45, making your weekly pay

$$\$450 + \$45 = \$495$$

Notice that the 10% pay cut followed by the 10% pay raise leaves you with *less* than your original salary. This result arises because the reference value for the calculations shifted during the problem: It was \$500 in the first calculation and \$450 in the second.

EXAMPLE 12 Shifting Investment Value

A stockbroker offers the following defense to angry investors: "I admit that the value of your investments fell 60% during my first year on the job. This year, however, their value has increased by 75%, so you are now 15% ahead!" Evaluate the stockbroker's defense.

Solution Imagine that you began with an investment of \$1000. During the first year, your investment lost 60% of its value, or \$600, leaving you with \$400. During the second year, your investment gained 75% of \$400, or $0.75 \times \$400 = \$300$. Therefore, at the end of the second year, your investment was worth $\$400 + \$300 = \$700$, which is still *less* than your original investment of \$1000 and certainly not a 15% gain overall. We can trace the problem with the stockbroker's defense to a shifting reference value: It was \$1000 for the first calculation and \$400 for the second.

▶ Now try Exercises 71–72.

EXAMPLE 13 Tax Cuts

A politician promises, "If elected, I will cut your taxes by 20% for each of the first three years of my term, for a total cut of 60%." Evaluate the promise.

Solution The politician neglected the effects of shifting reference values. A cut of 20% in each of three years will *not* make an overall cut of 60%. To see what really happens, suppose that you currently pay \$1000 in taxes. The following table shows how your taxes change over the three years.

Year	Tax paid in previous year	20% of previous year's tax	New tax this year
1	\$1000	\$200	\$800
2	\$800	\$160	\$640
3	\$640	\$128	\$512

Over three years, your taxes decline by \$1000 − \$512 = \$488. Because this is only 48.8% of \$1000, over the three years your tax bill declines by 48.8% not by 60%.

▸ Now try Exercises 73–74.

Less Than Nothing

BY THE WAY

The light bulb example related here was used as the title story in the book *200% of Nothing* by A. K. Dewdney (Wiley, 1993). The book contains many other interesting stories of misuses of numbers.

We often see numbers that represent large "more than" percentages. For example, a price of \$40 is 300% more than a price of \$10. However, in most cases it is not possible to have a "less than" percentage that is greater than 100%. To see why, consider an advertisement claiming that replacing standard light bulbs with energy-efficient ones would use "200 percent less energy." If you think about it, you'll realize that such a savings is impossible. If the new light bulbs used 100% less energy, they'd be using no energy at all. The only way they could use 200% less would be if they actually *produced* energy. Clearly, whoever wrote the advertisement made a mistake.

EXAMPLE 14 Impossible Sale

A store advertises that it will take "150% off" the price of all merchandise. What should happen when you go to the checkout to buy a \$500 item?

Solution If the price were 100% off, the item would be free. So if the price is 150% off, the store should *pay you* half the item's cost, or \$250. More likely, the store manager did not understand percentages.

▸ Now try Exercises 75–80.

Think About It Can an athlete give 110% effort? Can a glass of juice have 110% of the minimum daily requirement for vitamin C? Explain.

Don't Average Percentages

Technical Note

We usually think of the average of two numbers as their sum divided by 2. Technically, this type of average is called a *mean.* We'll discuss other definitions of *average* in Unit 6A.

Suppose you got 70% of the questions correct on a midterm exam and 90% correct on the final exam. Can we conclude that you answered 80% of all the questions correctly? It might be tempting to say yes—after all, 80% is the average (mean) of 70% and 90%. But it would be wrong, unless both tests had the same number of questions.

To see why, let's suppose that the midterm had 10 questions and the final had 100 questions, for a total of 110 questions. Your 70% score on the midterm means you answered 7 questions correctly, while your 90% score on the final means you answered 90 questions correctly. Therefore, on the two exams combined, you answered 97 out of 110 questions correctly, which is 88.2% (because $97/110 = 0.882$). This is much higher than the 80% "average" of the two individual exam percentages.

This example carries a very important lesson: As a general rule, *you should never average percentages.*

EXAMPLE 15 Batting Average

In baseball, a player's batting average represents the percentage of at-bats in which he got a hit. For example, a batting average of .350 means that the player got a hit 35% of the times he batted. Suppose a player had a batting average of .200 during the first half of the season and .400 during the second half of the season. Can we conclude that his batting average for the entire season was .300 (the average of .200 and .400)? Why or why not? Give an example that illustrates your reasoning.

Solution No. For example, suppose the player had 300 at-bats during the first half of the season and 200 at-bats during the second half, for a total of 500 at-bats. His first-half batting average of .200 means he got hits in 20% of his 300 at-bats, or $0.2 \times 300 = 60$ hits. His second-half batting average of .400 means he got hits in 40% of his 200 at-bats, or $0.4 \times 200 = 80$ hits. For the season, he got a total of $60 + 80 = 140$ hits in his 500 at-bats, so his season batting average was $140/500 = 28\%$, or .280—not the .300 found by averaging his first-half and second-half batting percentages. (In fact, the only case in which his season average would be .300 is if he had precisely the same number of at-bats in each half of the season.) ▶ Now try Exercises 81–82.

Quick Quiz 3A

Choose the best answer to each of the following questions. Explain your reasoning with one or more complete sentences.

1. The price of a meal at a four-star restaurant is 200% more than the $100 it was a couple of decades ago. The current price of such a meal is

   a. $200. b. $300. c. $400.

2. The population of a town increases from 50,000 to 75,000. What are the absolute and relative changes in the population?

   a. absolute change = 25,000; relative change = 25%

   b. absolute change = 25,000; relative change = 50%

   c. absolute change = 25,000; relative change = −25%

3. Suppose the value of a home changed by −20% over the past five years. This means that

   a. a mistake was made in the calculation, because relative change cannot be negative.

   b. the house increased in value over the past five years.

   c. the house decreased in value over the past five years.

4. Emily scored 50% higher on the SAT than Joshua did. This means that

   a. Joshua's score was 50% lower than Emily's.

   b. Joshua's score was half of Emily's.

   c. Joshua's score was two-thirds of Emily's.

5. The price of a movie ticket increased from $10 to $12. This means that the new price is

   a. 20% of the old price.

   b. 80% of the old price.

   c. 120% of the old price.

6. Your receipt shows that you paid $47.96 for a new shirt, including sales tax. The sales tax rate is 9%. The amount you paid in sales tax was

   a. $\$47.96 \times 0.09$. b. $\dfrac{\$47.96}{1.09}$. c. $\$47.96 - \dfrac{\$47.96}{1.09}$.

7. Consider this statement: "The interest rate on auto loans has increased 50% over the past decade and now stands at 6%." What can you conclude?

   a. The interest rate a decade ago was 4%.

   b. The interest rate a decade ago was 3%.

   c. The interest rate a decade ago was 9%.

8. A friend has a textbook that originally cost $150. The friend says that you can have it for 100% less than he paid for it. Your price will be

   a. $50. b. $75. c. $0 (free).

9. You currently earn $1000 per month, but you are expecting your earnings to increase 10% per year. This means that in five years you expect to be earning

   a. less than $1500 per month.

   b. exactly $1500 per month.

   c. more than $1500 per month.

10. During high school, Elise won 30% of the swim races she entered. During college, Elise won 20% of the swim races she entered. We can conclude that, in high school and college combined, Elise won

    a. 25% of the races she entered.

    b. more than 20% but less than 30% of the races she entered.

    c. more than 26% but less than 28% of the races she entered.

Exercises 3A

REVIEW QUESTIONS

1. Describe the three basic uses of percentages. Give a sample statement that uses a percentage in each of the three ways.
2. Distinguish between absolute and relative change. Give an example that illustrates how we calculate a relative change.
3. Distinguish between absolute and relative difference. Give an example that illustrates how we calculate a relative difference.
4. Explain the difference between the key words *of* and *more than* when dealing with percentages. How are their meanings related?
5. Explain the difference between the terms *percent* (%) and *percentage points*. Give an example of how they can differ for the same situation.
6. Give an example to explain why, in general, it is not legitimate to average percentages.

DOES IT MAKE SENSE?

Decide whether each of the following statements makes sense (or is clearly true) or does not make sense (or is clearly false). Explain your reasoning.

7. In many European countries, the percentage (relative) change in population has been negative in recent decades.
8. The price of tuition has tripled since my parents went to school—that's a 200% increase in price!
9. I've decreased my caloric intake by 125%, which has helped me lose weight.
10. If you earn 20% more than I do, then I must earn 20% less than you do.
11. If the tax rate increases by 10 percentage points every year, in a decade we'll be paying everything we earn to taxes.
12. We found that these rare cancers were 700% more common in children living near the toxic landfill than in the general population.
13. The rate of return on our fund increased by 50%, to 15%.
14. My bank increased the interest rate on my savings account 100%, from 2% to 4%.

BASIC SKILLS & CONCEPTS

15–30: Fractions, Decimals, Percentages. Express the following numbers in three forms: as a reduced fraction, as a decimal, and as a percentage. (Refer to the Brief Review on p. 123.)

15. 2/5	16. 30%	17. 0.20	18. 0.85
19. 150%	20. 2/3	21. 4/9	22. 1.25
23. 5/8	24. 44%	25. 69%	26. 4.25
27. 7/5	28. 121%	29. 4/3	30. 0.666...

31–36: Review of Ratios. Compare the following pairs of numbers *A* and *B* in three ways. (Refer to the Brief Review on p. 129.)

a. Find the ratio of *A* to *B*.
b. Find the ratio of *B* to *A*.
c. Complete the sentence: *A* is ________ percent of *B*.

31. $A = 52{,}252$ is the number of deaths due to AIDS in the United States in 1995, and $B = 12{,}333$ is the number of deaths due to AIDS in the United States in 2014.
32. $A = 40{,}229{,}000$ is the number of Americans over age 65 in 2010, and $B = 88{,}458{,}000$ is the projected number of Americans over age 65 in 2050.
33. $A = 1.6$ million is the 2017 population of Philadelphia, and $B = 2.1$ million is the 2017 population of Houston.
34. $A = 472$ is the average mathematics SAT score in Maine, and $B = 523$ is the average mathematics SAT score in Vermont.
35. $A = 69.50$ million is the number of votes received by Barack Obama in 2008, and $B = 62.98$ million is the number of votes received by Donald Trump in 2016.
36. $A = 90.8\%$ is the 2015 high school graduation rate in Iowa (first in the nation), and $B = 68.6\%$ is the high school graduation rate in New Mexico (last in the nation).

37–42: Percentages as Fractions. In the following statements, express the first number as a percentage of the second number.

37. Donald Trump won 302 electoral votes in 2016, out of a total of 538.
38. Apple's 2015 revenue was \$233 million, compared to Walmart's \$482 million.
39. The full-time year-round median salary for U.S. men in 2015 was \$50,383, and the full-time year-round median salary for U.S. women in 2015 was \$39,621.
40. In 2015, there were 61 million Americans under the age of 15, and a total of 321 million Americans.
41. The 2015 population of California was approximately 39.1 million. The 2010 population was approximately 33.9 million.
42. The United States has an estimated 7000 nuclear weapons out of an estimated total of 15,350 nuclear weapons worldwide (as of 2016).
43. **Salary Comparisons.** Clint's salary increased from \$25,000 to \$35,000 over a three-year period. Helen's salary increased from \$30,000 to \$42,000 over the same period. Whose salary increased more in absolute terms? In relative terms? Explain.
44. **Population Comparison.** Between the 2010 U.S. Census and 2015, the estimated population of El Paso, Texas, increased from approximately 649,000 to 681,000. During the same period, the population of Chandler, Arizona, increased from approximately 236,000 to 260,000. Which city had the greater absolute change in population? Which city had the greater relative change in population?

45–48: Percentage Change. Find the absolute change and the relative change in the following cases.

45. The number of refugees in the world increased from 8.7 million in 2005 to 16.1 million in 2015.

46. The average student loan debt after a bachelor's degree increased from $18,550 in 2004 to $28,950 in 2014.

47. The U.S. per capita consumption of beef decreased from 67.8 pounds in 2000 to 55.4 pounds in 2016.

48. The number of commercial U.S. oldies radio stations decreased from 729 in 2006 to 351 in 2016.

49–52: Percentage Comparisons. Complete the following sentences.

49. The life expectancy (at birth) of American women (81.2 years) is ________ percent greater than the life expectancy of men (76.4 years).

50. The life expectancy (at birth) of all Americans in 2014 (78.8 years) is ________ percent greater than the life expectancy in 1900 (47.3 years).

51. In 2015, the median age at first marriage of U.S. women (27.1 years) was ________ percent less than the median age at first marriage of U.S. men (29.2 years).

52. The marriage rate in the United States in 2014 (6.9 marriages per 1000 people) is ________ percent less than the marriage rate in 1980 (10.6 per 1000 people).

53–56: *Of* versus *More Than*. Fill in the blanks in the following statements.

53. The population of Michigan is 63% greater than the population of Missouri, so Michigan's population is ______ % of Missouri's population.

54. The area of Norway is 24% more than the area of Colorado, so Norway's area is ______ % of Colorado's area.

55. The population of Hawaii is 52% less than the population of Arkansas, so the population of Hawaii is ______ % of the population of Arkansas.

56. The net worth of Warren Buffett is 18.9% less than the net worth of Bill Gates, so Warren Buffett's net worth is ______ % of Bill Gates's net worth.

57–60: Prices and Sales. Fill in the blanks in the following statements.

57. The wholesale price of a toaster is 30% less than the retail price. Therefore, the wholesale price is ________ times the retail price.

58. A store is having a 50% off sale. Therefore, the original price of an item is ________ times as much as the sale price.

59. The original retail price of a TV is 20% more than its on-sale price. Therefore, the retail price is ________ times the on-sale price.

60. A store is having a quarter-off (25% off) sale. The sale price for an item with a regular price of $120 is ________.

61–64: Percentages of Percentages. Describe each of the following changes in two ways: as an absolute change in terms of percentage points and as a relative change in terms of a percentage.

61. The percentage of Republicans in the House of Representatives increased from 40.9% in 2010 to 55.3% in 2016.

62. The percentage of Icelandic people accessing the Internet increased from 44.5% in 2000 to 98.2% in 2015 (highest percentage in the world).

63. Alcohol use by high school seniors decreased from 50.0% in 2000 to 35.3% in 2015.

64. The percentage of cars in the world manufactured in the United States decreased from 75.5% in 1950 to 13.3% in 2015.

65. **Care in Wording.** Assume that 30% of city employees in Carson City ride the bus to work. Consider the following two statements:

- The percentage of city employees who ride the bus to work is 10% higher in Freetown than in Carson City.
- The percentage of city employees who ride the bus to work is 10 percentage points higher in Freetown than in Carson City.

For each case, state the percentage of city employees in Freetown who ride the bus to work. Briefly explain why the two statements have different meanings.

66. **Ambiguous News.** The average annual precipitation on Mt. Washington, in New Hampshire, is 90 inches. During one particularly wet year, different news reports carried the following statements.

- The precipitation this year is 200% of normal.
- The precipitation this year is 200% above normal.

Do the two statements have the same meaning? That is, how many inches of precipitation is implied by each one? Explain.

67–70: Solving Percentage Problems. Solve the following percentage problems.

67. The total (after-tax) cost of a laptop computer is $1278.24. The local sales tax rate is 7.6%. What is the retail (pre-tax) price?

68. A store buys lawnmowers from the manufacturer and marks up the price by 40% to arrive at a retail price. Fill in the blank: The wholesale price is ________ times the retail price.

69. In 2015, alcohol use among high school seniors was down from the previous year by 5.6%, to 35.3%. What was the usage rate for the previous year (2014)?

70. Between 2000 and 2015, the number of twin births in the United States increased by 13%, to approximately 135,000. About how many twin births were there in 2000?

71–74: Shifting Reference Value. State whether the following statements are true or false, and explain why. If a statement is false, state the true change.

71. If the national unemployment rate decreased at an annual rate of 2% per year for three consecutive years, then the rate decreased by 6% over the three-year period.

72. You receive a pay raise of 6%, followed later by a pay cut of 6%. After the two changes in pay, your salary is unchanged.

73. If sales in your pizza restaurant increase by 11% one year and decrease by 3% the following year, your sales are up by 8% over two years.

74. A high school reports that its students' SAT scores were down by 10% in one year. The next year, however, SAT scores rose by 20%. The high school principal announces, "Overall, test scores have improved by 10% over the past two years."

75–80: Is It Possible? Determine whether the following claims could be true. Explain your answer.

75. Anna bought a surfboard online and saved 125%.
76. Scott is 200% taller than his son.
77. Average hotel room rates have increased 100% in the last 20 years.
78. Through hard training, Renee improved her 10-kilometer running time by 100%.
79. Your computer has 200% more storage than mine.
80. Your computer has 200% less storage than mine.
81. **Average Percentages.** Suppose you have an 80% average in a class going into the final exam. You receive a 90% grade on the final exam. In general, does it follow that your average for the entire course is 85%? Explain.
82. **Average Percentages.** A basketball player with a career free throw shooting average of 80% hit all his free throws (100% average) in a single game. Does it follow that his new free throw shooting average is $(80\% + 100\%)/2 = 90\%$? Explain.

FURTHER APPLICATIONS

83–86: Analyzing Percentage Statements. Assuming the given information is accurate, determine whether the following statements are true. Provide an explanation.

83. Forty percent of the students at an orientation are women and 20% of the women are engineering students. Therefore, $40\% \times 20\% = 8\%$ of the students at the orientation are female engineering students.

84. Sixty percent of the students at an orientation are men and 30% of the students at the orientation are arts majors. Therefore, $60\% \times 30\% = 18\%$ of the students at the orientation are male arts majors.
85. Fifty percent of the cars in the rental pool have Bluetooth capability and 20% of the cars have GPS. Therefore, $50\% + 20\% = 70\%$ of the cars have Bluetooth or GPS.
86. Thirty percent of commuters take a train into the city and 55% of commuters drive into the city. Therefore, $30\% + 55\% = 85\%$ of commuters either take the train or drive into the city.

87–90: Solving Percentage Problems. Solve the following percentage problems.

87. The 4550 men in the arena comprised 85% of the people in the arena. How many people were in the arena?
88. In 2016, 84.0% of households with televisions subscribed to cable, an increase of 7.8% over 2000. What percentage of households with televisions subscribed to cable in 2000?
89. With 2.4 million digital subscribers, *Game Informer Magazine* (ranked first) has 1150% more digital subscribers than *National Geographic* (ranked third). How many digital subscribers does *National Geographic* have?
90. The number of daily evening newspapers in the United States has decreased by 46.5% to 389 newspapers between 2000 and 2015. How many daily evening newspapers were there in 2000?

91–94: Percentages in the News. Answer the question that follows each quote from a news source.

91. "Since 2008, Jackson Family Wines has reduced its annual water usage by 31 percent." By what fraction was the 2008 water usage reduced?
92. "Concessions in Brooklyn more than doubled during the period, to 15.4 percent from 6.6 percent." What was the percentage increase in concessions during the period?
93. "The unemployment rate has risen more than a percentage point, to 8.5% in February from 7.1% last November." What is the relative change in the unemployment rate expressed as a percentage?
94. "The American Booksellers Association, which represents independent booksellers, includes 1500 businesses at 2500 locations. Twenty years ago, it represented 4700 businesses at 5500 locations." What are the percentage changes in the numbers of businesses and locations represented?
95. **Stock Market Losses.**

    a. The largest single-day *point* loss of the Dow Jones Industrial Average occurred on September 29, 2008, when the market lost 778 points and closed at 10,365. What was the percentage change?

    b. The largest single-day *percentage* loss of the Dow Jones Industrial Average occurred on October 19, 1987, when the market closed down 22.6% at 1739. What was the point change?

IN YOUR WORLD

96. **Percentages.** Find three recent news reports that quote percentages. In each case, describe the use of the percentage (as a fraction, to describe change, or for comparison) and explain its context.
97. **Percentage Change.** Find a recent news report that quotes a percentage change. Describe the meaning of the change.
98. **Abuse of Percentages.** Find a news article or report in which the use of a percentage is either suspicious or wrong. If possible, clarify or correct the statement.

UNIT 3B Putting Numbers in Perspective

We hear numbers in the millions, billions, or trillions nearly every day, in contexts such as government spending or memory storage on phones and computers. Yet relatively few people understand what these large numbers really mean. In this unit, we will study several techniques for putting large (or small) numbers into a perspective that gives them real meaning.

Writing Large and Small Numbers

Working with large and small numbers is much easier when we write them in a special format known as **scientific notation**. We express numbers in this format by writing a number *between* 1 and 10 multiplied by a power of 10. (See the Brief Review on p. 140 for a review of powers of 10.) For example, a billion is 10 to the ninth power, or 10^9, so we write 6 billion in scientific notation as 6×10^9. Similarly, we write 420 in scientific notation as 4.2×10^2, and 0.67 as 6.7×10^{-1}.

BY THE WAY

In the United States, a *billion* is a thousand million, or 10^9, and a *trillion* is a thousand billion, or 10^{12}. But in Great Britain and Germany, a billion is a million million, or 10^{12}, and a trillion is 10^{18}. We will use only the U.S. meanings in this book.

Definition

Scientific notation is a format in which a number is expressed as a number *between* 1 and 10 multiplied by a power of 10.

Scientific notation makes it easy to write large and small numbers. We must be careful, however, not to let this ease of writing deceive us. For example, it's so easy to write the number 10^{80} that we might think it's not all that big—but it is larger than the total number of atoms in the known universe.

EXAMPLE 1 Numbers in Scientific Notation

Rewrite each of the following statements using scientific notation.

a. Total spending in the new federal budget is \$4,200,000,000,000.
b. The diameter of a hydrogen nucleus is about 0.000000000000001 meter.

Solution Notice how much easier it is to read the numbers with scientific notation.

a. Total spending in the new federal budget is $\$4.2 \times 10^{12}$, or \$4.2 trillion.
b. The diameter of a hydrogen nucleus is about 1×10^{-15} meter.

▶ Now try Exercises 23–26.

Approximations with Scientific Notation

Scientific notation can help us *approximate* answers without a calculator. For example, we can quickly approximate the answer to 5795×326 by rounding 5795 to 6000 and 326 to 300. Writing the rounded numbers in scientific notation, we then see that

$$5795 \times 326 \approx (6 \times 10^3) \times (3 \times 10^2) = 18 \times 10^5 = 1{,}800{,}000$$

Notice that this approximation is a fairly good estimate of the exact answer, which you can confirm to be 1,889,170.

Brief Review: Working with Scientific Notation

Converting to Scientific Notation

To convert a number from ordinary notation to scientific notation:

Step 1. Move the decimal point to come after the *first* nonzero digit.

Step 2. For the power of 10, use the number of places the decimal point moves; the power is *positive* if the decimal point moves to the left and *negative* if it moves to the right.

Examples:

$$3042 \xrightarrow[\text{3 places to left}]{\text{decimal moves}} 3.042 \times 10^3$$

$$0.00012 \xrightarrow[\text{4 places to right}]{\text{decimal moves}} 1.2 \times 10^{-4}$$

$$226 \times 10^2 \xrightarrow[\text{2 places to left}]{\text{decimal moves}} (2.26 \times 10^2) \times 10^2 = 2.26 \times 10^4$$

Converting from Scientific Notation

To convert a number from scientific notation to ordinary notation:

Step 1. The power of 10 indicates how many places to move the decimal point; move it to the *right* if the power of 10 is positive and to the *left* if it is negative.

Step 2. If moving the decimal point creates any open places, fill them with zeros.

Examples:

$$4.01 \times 10^2 \xrightarrow[\text{2 places to right}]{\text{move decimal}} 401$$

$$3.6 \times 10^6 \xrightarrow[\text{6 places to right}]{\text{move decimal}} 3{,}600{,}000$$

$$5.7 \times 10^{-3} \xrightarrow[\text{3 places to left}]{\text{move decimal}} 0.0057$$

Multiplying or Dividing with Scientific Notation

Multiplying or dividing numbers expressed in scientific notation simply requires operating on the powers of 10 and the other parts of the numbers separately.

Examples:

$$(6 \times 10^2) \times (4 \times 10^5) = (6 \times 4) \times (10^2 \times 10^5) = 24 \times 10^7 = 2.4 \times 10^8$$

$$\frac{4.2 \times 10^{-2}}{8.4 \times 10^{-5}} = \frac{4.2}{8.4} \times \frac{10^{-2}}{10^{-5}} = 0.5 \times 10^{-2-(-5)} = 0.5 \times 10^3 = 5 \times 10^2$$

Note that, in both examples, we first found an answer in which the number multiplied by a power of 10 was *not* between 1 and 10. We then followed the process for converting the final answer into scientific notation.

Addition and Subtraction with Scientific Notation

In general, we must write numbers in ordinary notation before adding or subtracting.

Examples:

$$(3 \times 10^6) + (5 \times 10^2) = 3{,}000{,}000 + 500 = 3{,}000{,}500 = 3.0005 \times 10^6$$

$$(4.6 \times 10^9) - (5 \times 10^8) = 4{,}600{,}000{,}000 - 500{,}000{,}000 = 4{,}100{,}000{,}000 = 4.1 \times 10^9$$

When both numbers have the *same* power of 10, we can factor out the power of 10 first.

Examples:

$$(7 \times 10^{10}) + (4 \times 10^{10}) = (7 + 4) \times 10^{10} = 11 \times 10^{10} = 1.1 \times 10^{11}$$

$$(2.3 \times 10^{-22}) - (1.6 \times 10^{-22}) = (2.3 - 1.6) \times 10^{-22} = 0.7 \times 10^{-22} = 7.0 \times 10^{-23}$$

▸ Now try Exercises 15–22.

EXAMPLE 2 Checking Answers with Approximations

You and a friend are trying to estimate how much garbage New York City residents produce every day. New York City's population is about 8.6 million, and you find online that an average resident produces about 2.4 pounds, or 0.0012 ton, of garbage each day. Your friend quickly presses calculator buttons and tells you that the total daily garbage production for the city is about 225 tons. Without using your calculator, determine whether this answer is reasonable.

Solution First, you recognize that the answer can be found simply by multiplying the population by the garbage production per person:

$$8.6 \text{ million } \cancel{\text{persons}} \times 0.0012 \frac{\text{ton}}{\cancel{\text{person}}}$$

You can make a quick estimate by writing 8.6 million as 8.6×10^6, which is nearly 10^7, and writing 0.0012 as 1.2×10^{-3}, which is close to 10^{-3}. Therefore, the product should be approximately

$$10^7 \times 10^{-3} = 10^{7-3} = 10^4 = 10{,}000$$

Clearly, your friend's answer of 225 tons is too small. Notice that this approximation technique provided a useful check, even without providing an exact answer.

▶ Now try Exercises 27–28.

Giving Meaning to Numbers

We are now ready to move toward our goal of putting numbers in perspective. As with problem solving (see Chapter 2), there's no single recipe for gaining perspective, but a few simple techniques can be helpful. Here we present three such techniques: estimation, comparisons, and scaling.

Perspective Through Estimation

How high is 1000 feet? For most people, the quantity "1000 feet" has little meaning by itself. However, we can give the number some perspective by estimating what it would mean for a tall building. For example, we can expect each story in a tall building to be about 10 feet from floor to ceiling, which means that 1000 feet is the approximate height of a 100-story building.

Keep in mind that estimates are not meant to be exact. For example, the 105-story One World Trade Center has a roof height of 1368 feet, or about 13 feet per story. Therefore, our estimate of 10 feet per story was off by about 3 feet, or 30%. Nevertheless, if we start with no idea about what 1000 feet looks like, picturing a 100-story building gives us a reasonable sense of what it means.

Some estimates are useful even if they only get us within a very broad range of an exact value. For example, you can infer a lot about the character of a town or city simply by knowing whether its population is in the ten thousands or the millions—the former is the size of a small college town and the latter is a huge city. Estimates that give only a broad range of values, such as "in the millions," are called **order of magnitude** estimates.

BY THE WAY

One World Trade Center's roof height is identical to that of the original World Trade Center's North Tower, which was destroyed in the 9/11/2001 terrorist attacks. Above the roof, a spire and antenna give One World Trade Center a total height of 1776 feet, chosen to symbolize the 1776 Declaration of Independence.

Definition

An **order of magnitude** estimate specifies only a broad range of values, usually within one or two powers of ten, such as "in the ten thousands" or "in the millions."

Technical Note

In science, the term *order of magnitude* refers specifically to powers of 10. For example, 10^{23} is said to be five orders of magnitude larger than 10^{18}, because it is 10^5 times larger ($10^{23} = 10^{18} \times 10^5$).

In ordinary language, we usually indicate an order of magnitude estimate by actually using the word *order.* For example, we might say that the population of the United States is "on the order of 300 million," by which we mean it is nearer to 300 million than to, say, 200 million or 400 million. Note that the context determines how we interpret an order of magnitude estimate. When astronomers say that the number of stars in a galaxy is "of order 100 billion," they mean it could be anywhere within about a factor of 10 of this number—that is, between about 10 billion and 1 trillion. This is a much wider range than the range implied when we say "on the order of 300 million" for the population of the United States, but it is appropriate to the context.

EXAMPLE 3 Order of Magnitude of Ice Cream Spending

Make an order of magnitude estimate of total annual spending on ice cream in the United States.

Solution We can calculate total annual ice cream spending by multiplying the amount the average person spends each year by the total population. We can find the per-person annual spending by multiplying the average cost of a serving of ice cream by the average number of servings a person eats in a year. Putting these ideas together, we can find the total annual ice cream spending with the following equation:

$$\underbrace{\text{total annual spending}}_{\text{units of } \frac{\$}{\text{yr}}} = \underbrace{\text{servings per person per year}}_{\text{units of } \frac{\cancel{\text{servings}}}{\cancel{\text{person}} \times \text{yr}}} \times \underbrace{\text{price per serving}}_{\text{units of } \frac{\$}{\cancel{\text{serving}}}} \times \underbrace{\text{population}}_{\text{units of } \cancel{\text{persons}}}$$

Notice how the units work out, telling us that our equation will indeed give us the units of dollars per year that we want for the answer.

We can reasonably guess that an average person has something like 50 servings of ice cream per year (which means about one a week). The price of ice cream is of order \$1 per serving (as opposed to, say, 10¢ or \$10). The U.S. population is of order 300 million (3×10^8) people. Using those numbers, a reasonable estimate is

$$\begin{aligned} \text{total annual spending} &= 50 \frac{\cancel{\text{servings}}}{\cancel{\text{person}} \times \text{yr}} \times \frac{\$1}{\cancel{\text{serving}}} \times (3 \times 10^8 \ \cancel{\text{persons}}) \\ &= \frac{\$1.5 \times 10^{10}}{1 \text{ yr}} \end{aligned}$$

▲ **CAUTION!** Make sure you can get this answer with your calculator. If your answer differs from \$1.5 x 10^{10}, then you probably used the key labeled "EE" (or "E" or "exp") incorrectly; the correct usage is explained in the Using Technology box on the next page. ▲

Total annual spending on ice cream in the United States is of order $\$1.5 \times 10^{10}$, or \$15 billion, per year. The actual number might be a few times more or less than this; such as between about \$5 billion and \$50 billion. Still this estimate has told us a lot—before we started, we had no idea how much Americans spend on ice cream, and now we know that they spend *billions* of dollars on it each year.

▶ Now try Exercises 29–38.

USING TECHNOLOGY

Scientific Notation

Calculators and computers usually use a special format for numbers in scientific notation. Consider the number 3.5 million, which we write in scientific notation as 3.5×10^6. On a calculator or computer, this number will usually be written in a format like 3.5E6. The E is meant to stand for "exponent," but it's easier to think of it as follows:

Calculator/computer uses "E" to mean "times 10 to the power that follows."

$3.5 \times 10^6 \longrightarrow$ 3.5E6

the correct way to write 3.5 million in scientific notation — the same number on a calculator or computer

Be sure to notice that "E" stands for "times 10 to the power that follows," which means that you do not need to enter the 10. Remembering this fact will help you avoid common mistakes. For example:

- To enter a power of 10, such as the number 1 million, or 10^6, you must put it in scientific notation (1×10^6) by entering 1E6 on your calculator or computer. Many people mistakenly enter 10E6, but that really means $10 \times 10^6 = 10^7$.
- Another common mistake is entering a number like 3.5×10^6 as $3.5 \times$ 10E6, but that actually means $3.5 \times 10 \times 10^6 = 10^7$.

Standard Calculators Most calculators have a key labeled either "E," "EE," or "exp" for entering the power of 10 (the "exponent"). For example, you enter the number 3.5×10^6 with the key sequence

[3.5] (EE) [6]

Microsoft Excel Type the letter E as above when you enter a number in scientific notation into a spreadsheet cell. For example, enter the number 3.5×10^6 by typing "=3.5E6" into the cell, as shown in the screen shot on the left below.

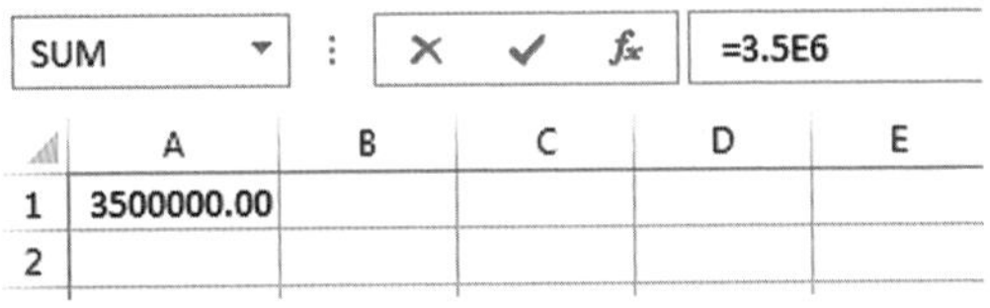

Microsoft Excel 2016, Windows 10, Microsoft Corporation.

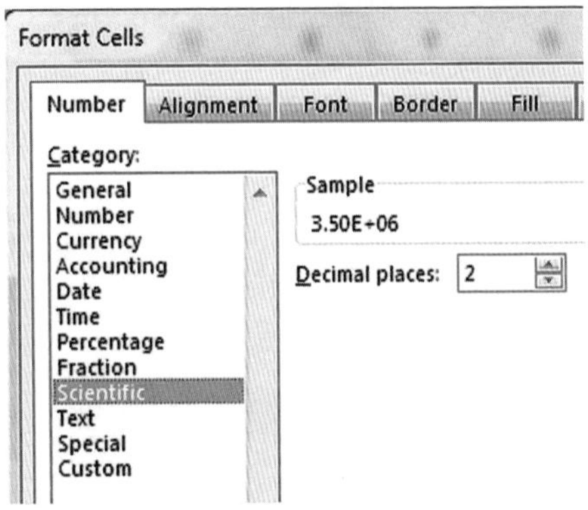

Note that, by default, Excel will format the number without scientific notation in the cell. To change that option, select "Scientific" as the Number format for the cell, as indicated in the screen shot on the right above. Excel then shows the exponent as "+06," with the "+" indicating that it is a positive exponent.

Google Google has a built-in calculator, which you can use simply by typing your calculations into the search box and then hitting Return. The screen shot below illustrates how to multiply $3.5 \times 10^6 \times 7$; note the use of * for multiplication and "e" for the power of 10.

Think About It Suppose you work for a company that distributes ice cream to stores throughout the United States. Market research tells you that a $25 million advertising campaign could give you an additional 5% of the total U.S. market for ice cream. Based on the estimate in Example 3, would the advertising campaign be worth its cost? Explain.

Perspective Through Comparisons

A second general way to put numbers in perspective is by making comparisons. Consider $100 billion, a number that you may find in the news almost any day. It's easy to say, but how big is it? Let's think of it in terms of counting. Suppose you were asked to count $100 billion in one-dollar bills. How long would it take? Clearly, if we assume you can count 1 bill each second, it would take 100 billion (10^{11}) seconds, which we can put in perspective by converting to years with a chain of conversions:

$$10^{11}\,\cancel{\text{s}} \times \left(\frac{1\ \cancel{\text{min}}}{60\ \cancel{\text{s}}}\right) \times \left(\frac{1\ \cancel{\text{hr}}}{60\ \cancel{\text{min}}}\right) \times \left(\frac{1\ \cancel{\text{day}}}{24\ \cancel{\text{hr}}}\right) \times \left(\frac{1\ \text{yr}}{365\ \cancel{\text{days}}}\right) \approx 3171\ \text{yr}$$

In other words, you would need *more than three thousand years* to count $100 billion in $1 bills (at a rate of 1 bill per second). And that assumes that you never take a break: no sleeping, no eating, and absolutely no dying!

Comparisons are particularly useful in dealing with units that are relatively unfamiliar, such as energy units. Table 3.1 lists various energies that you can use in comparisons. For example, we immediately see that U.S. annual energy consumption is roughly $\frac{1}{6}$ of world annual energy consumption.

EXAMPLE 4 U.S. vs. World Energy Consumption

Compare the U.S. population to the world population and U.S. energy consumption to world energy consumption. What does this tell you about energy usage by Americans?

Solution The world population is of order 7 billion (7×10^9), and the U.S. population is of order 300 million (3×10^8). Comparing the two quantities, we find

$$\frac{\text{U.S. population}}{\text{world population}} \approx \frac{3 \times 10^8}{7 \times 10^9} = \frac{3}{7} \times 10^{-1} \approx 0.043$$

The U.S. population is only about 4% of the world's population but, as shown in Table 3.1, the United States uses about 1/5.7 or about 18%, of the world's energy. That is, Americans use roughly four times as much energy per person as the world average.

▶ Now try Exercises 39–42.

TABLE 3.1 Selected Energy Comparisons

Item	Energy (joules)
Energy released by metabolism of 1 average candy bar	1×10^6
Energy needed for 1 hour of running (adult)	4×10^6
Energy released by burning 1 liter of oil	1.2×10^7
Electrical energy used in an average home daily	5×10^7
Energy released by burning 1 kilogram of coal	1.6×10^9
Energy released by fission of 1 kilogram of uranium-235	5.6×10^{13}
Energy released by fusion of hydrogen in 1 liter of water	6.9×10^{13}
U.S. annual energy consumption	1.0×10^{20}
World annual energy consumption	5.7×10^{20}
Annual energy generation of Sun	1×10^{34}

EXAMPLE 5 Fusion Power

We are now ready to return to our chapter-opening question (p. 120): If you had a fusion power plant and hooked it up to the faucet of your kitchen sink, how much power could you generate from the hydrogen in the water flowing through it?

Solution Let's write out our Understand-Solve-Explain process for this one.

Understand: We recall that power is the *rate* at which energy is used, that is, energy divided by time. We are given that the source of the power will be fusion of hydrogen in the water flowing through a faucet, and Table 3.1 gives the amount of energy that could be generated by fusion of the hydrogen in 1 liter of water. We can therefore apply unit analysis to this problem, and recognize that we get an answer with the correct units by multiplying the flow rate through the faucet (in liters per second) by the energy that fusion could release from each liter of water:

$$\underbrace{\text{total power}}_{\text{units of } \frac{\text{joules}}{\text{s}}} = \underbrace{(\text{water flow rate})}_{\text{units of } \frac{\text{liters}}{\text{s}}} \times \underbrace{(\text{energy from each liter of water})}_{\text{units of } \frac{\text{joules}}{\text{liter}}}$$

We are not given the flow rate, but that is easy to measure. For example, if you place a 1-liter pitcher under a typical kitchen faucet, you'll find that it fills in about 20 seconds. Therefore, the flow rate is about 1/20, or 0.05, liter per second.

Solve: We have all the information needed to solve the problem, so we put the data in the equation and calculate:

$$\begin{aligned}\text{total power} &= (\text{water flow rate}) \times (\text{energy from each liter of water}) \\ &= 0.05 \frac{\cancel{\text{L}}}{\text{s}} \times 6.9 \times 10^{13} \frac{\text{joule}}{\cancel{\text{L}}} \\ &= 3.45 \times 10^{12} \frac{\text{joule}}{\text{s}}\end{aligned}$$

Explain: We have found that a fusion generator could in principle produce a power of 3.45 trillion joules per second, or 3.45 trillion watts (3.45 terawatts), from your kitchen faucet. To make this answer meaningful, we need to compare it to something understandable. Because Table 3.1 gives us U.S. and world energy usage per year, let's convert the fusion power we found in joules per second into joules per year, which we do with a chain of unit conversions as follows:

$$3.45 \times 10^{12} \frac{\text{joule}}{\cancel{\text{s}}} \times \frac{60\ \cancel{\text{s}}}{1\ \cancel{\text{min}}} \times \frac{60\ \cancel{\text{min}}}{1\ \cancel{\text{hr}}} \times \frac{24\ \cancel{\text{hr}}}{1\ \cancel{\text{day}}} \times \frac{365\ \cancel{\text{day}}}{1\ \text{yr}} \approx 1.1 \times 10^{20} \frac{\text{joule}}{\text{yr}}$$

Notice that our result is larger than the U.S. annual energy consumption of 1.0×10^{20} joules. Therefore, the correct answer to the chapter-opening question is E: If we could do fusion with all the hydrogen in the water flowing from a typical kitchen faucet, we could produce enough energy to meet *all* U.S. energy needs. That is, a single fusion power plant hooked up to your kitchen faucet would produce enough energy that we would no longer need to use any other energy source, including oil, coal, gas, hydroelectric, wind, and nuclear fission.

▶ Now try Exercises 43–46.

Think About It Fusion power plants not only could generate immense energy but also would be safer and cleaner (for the amount of energy generated) than any other known energy technology. Unfortunately, decades of effort have not yet succeeded in producing a viable commercial nuclear fusion power technology. The U.S. government currently spends about $400 million per year on fusion research. Do you think this is the right spending level? How do you think the availability of fusion energy would change our world?

BY THE WAY

Nuclear *fission* means splitting large atomic nuclei into smaller ones, while nuclear *fusion* means combining small nuclei into larger ones. Current nuclear power plants use fission of uranium or plutonium atoms. The Sun generates energy by fusion, as do thermonuclear weapons (also known as *hydrogen bombs*).

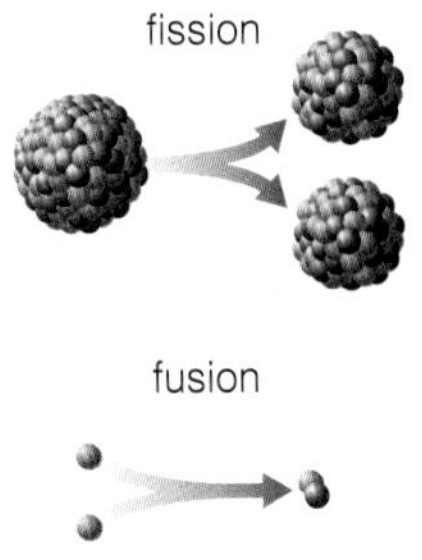

BY THE WAY

Scientists working on fusion power usually use a form of hydrogen called *deuterium*. (Ordinary hydrogen nuclei contain just a single proton; the deuterium nucleus also contains a neutron.) About 1 in 6400 hydrogen atoms is deuterium, so the flow rate needed to power the United States would be 6400 times higher than that calculated in Example 5—about the flow rate of a small creek.

Perspective Through Scaling

A third general technique for giving meaning to numbers uses *scaling* or scale models. You are probably familiar with three common ways of expressing scales used on maps:

- *Verbally:* A scale can be described in words such as "One centimeter represents one kilometer" or, more simply, as "1 cm = 1 km." This scale means that 1 cm on the map represents 1 km of actual distance.
- *Graphically:* A marked miniruler on a map can show the scale visually (Figure 3.1).
- *As a ratio:* We can state the ratio of distance on a map to actual distance. For example, there are 100,000 centimeters in a kilometer (because there are 100 centimeters in 1 meter and 1000 meters in 1 kilometer), so a scale where 1 centimeter represents 1 kilometer can be described as a **scale ratio** of 1 to 100,000 (or 1/100,000).

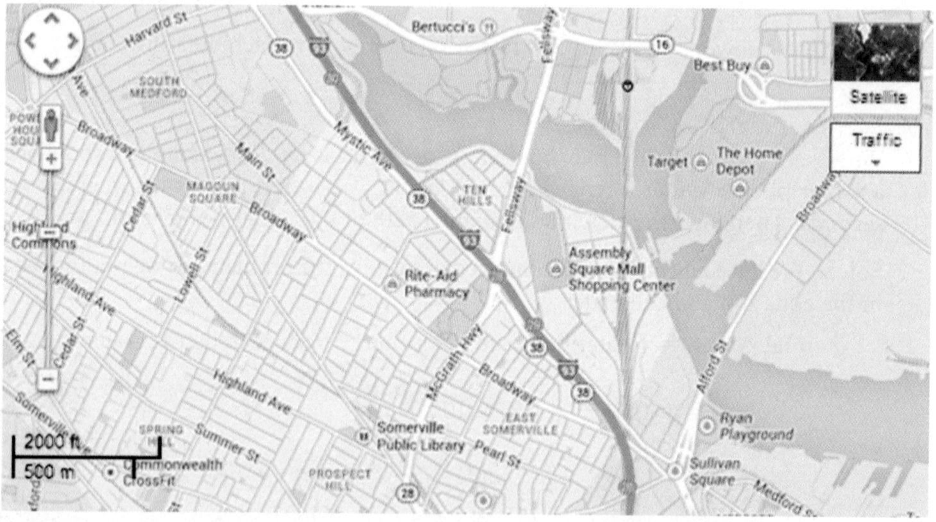

FIGURE 3.1 The miniruler at the lower left acts as the map scale. In this case, the length of the upper segment represents 2000 feet in the city, while the slightly shorter lower segment represents 500 meters.

Scales have many uses other than on maps. Architects and engineers build scale models to visualize their plans. *Timelines* represent the scale of time; a given distance along a timeline represents a certain number of years. *Time-lapse photography* and computer simulations allow us to represent large blocks of time in short periods. For example, in a 4-second time-lapse video clip that shows 24 hours of weather, each second of the video represents 6 hours of real time. Similarly, a computer simulation of continental drift might represent 1 billion years of geographical change in just 1 minute.

EXAMPLE 6 Scale Ratio

A city map states, "One inch represents one mile." What is the scale ratio for this map?

Solution We find the ratio by converting 1 mile into inches:

$$1\ \cancel{\text{mi}} \times 5280\ \frac{\cancel{\text{ft}}}{\cancel{\text{mi}}} \times 12\ \frac{\text{in}}{\cancel{\text{ft}}} = 63{,}360\ \text{in}$$

One inch on the map represents 1 actual mile, or 63,360 inches. Therefore, the scale ratio for this map is 1 to 63,360, meaning that actual distances are 63,360 times the corresponding distances on the map. Note that scale ratios never have units.

▸ Now try Exercises 47–50.

EXAMPLE 7 Earth and Sun

The distance from the Earth to the Sun is about 150 million kilometers. The diameter of the Sun is about 1.4 million kilometers, and the equatorial diameter of the Earth is about 12,760 kilometers. Put these numbers in perspective by using a scale model of the solar system with a 1 to 10 billion scale.

Solution The scale ratio tells us that actual sizes and distances are 10 billion (10^{10}) times as large as the model sizes and distances, so we find scaled sizes and distances by dividing the actual values by 10^{10}. For the Earth–Sun distance, we find

$$\begin{aligned}\text{scaled Earth–Sun distance} &= \frac{\text{actual distance}}{10^{10}}\\ &= \frac{1.5 \times 10^{8}\text{ km}}{10^{10}}\\ &= 1.5 \times 10^{-2}\text{ km} \times 10^{3}\frac{\text{m}}{\text{km}} = 15\text{ m}\end{aligned}$$

Note that, in the last step, we converted the distance from kilometers to meters because it is easier to understand "15 meters" than "0.015 kilometer." We find the scaled Sun and Earth diameters similarly, this time converting the units to centimeters and millimeters, respectively:

$$\begin{aligned}\text{scaled Sun diameter} &= \frac{\text{actual Sun diameter}}{10^{10}}\\ &= \frac{1.4 \times 10^{6}\text{ km}}{10^{10}}\\ &= 1.4 \times 10^{-4}\text{ km} \times 10^{5}\frac{\text{cm}}{\text{km}} = 14\text{ cm}\end{aligned}$$

$$\begin{aligned}\text{scaled Earth diameter} &= \frac{\text{actual Earth diameter}}{10^{10}}\\ &= \frac{1.276 \times 10^{4}\text{ km}}{10^{10}}\\ &= 1.276 \times 10^{-6}\text{ km} \times 10^{6}\frac{\text{mm}}{\text{km}} = 1.276\text{ mm}\end{aligned}$$

The model Sun, at 14 centimeters in diameter, is roughly the size of a grapefruit. The model Earth, at about 1.3 millimeters in diameter, is about the size of the ball point in a pen, and the distance between them is 15 meters. ▶ Now try Exercise 51.

BY THE WAY

The Voyage scale model solar system on the National Mall in Washington, DC, uses the 1 to 10 billion scale described in Example 7. The photo below shows a boy touching the model Sun. Visible to the left are the pedestals that hold the inner planets Mercury, Venus, Earth, and Mars.

Think About It Find a grapefruit or similar-sized ball and the ball point from a pen. Set them 15 meters apart to represent the Sun and Earth with a 1 to 10 billion scale. How does a scale model like this one make it easier to understand our solar system? Discuss.

EXAMPLE 8 Distances to the Stars

The distance from the Earth to the nearest stars besides the Sun (the three stars of the Alpha Centauri system) is about 4.3 light-years. On the 1 to 10 billion scale of Example 7, how far are these stars from the Earth? (*Note:* Recall that 1 light-year $\approx 9.46 \times 10^{12}$ km.)

Solution We are given the distance to the nearest stars in light-years, which we can convert to kilometers:

$$4.3\text{ light-years} \times \frac{9.46 \times 10^{12}\text{ km}}{1\text{ light-year}} \approx 4.1 \times 10^{13}\text{ km}$$

We now divide by 10 billion to find the distance on the 1 to 10 billion scale:

$$\text{scaled distance} = \frac{\text{actual distance}}{10^{10}} \approx \frac{4.1 \times 10^{13}\ \text{km}}{10^{10}}$$
$$= 4.1 \times 10^{3}\ \text{km} = 4100\ \text{km}$$

The distance to even the nearest stars on this scale is more than 4000 kilometers, or approximately the distance across the United States. ▶ Now try Exercise 52.

Think About It Suppose that an Earth-like planet is orbiting a nearby star. Based on the results of Examples 7 and 8, discuss the challenge of trying to detect such a planet. (Despite the challenge, scientists have already discovered thousands of planets around other stars, including many that are Earth-size or smaller.)

BY THE WAY

According to modern science, Earth and our solar system formed just over $4\frac{1}{2}$ billion years ago from the collapse of a large cloud of interstellar gas. Powerful telescopes allow us to observe similar clouds in which stars are forming today.

EXAMPLE 9 Timeline

Human civilization, at least since the time of ancient Egypt, is on the order of 5000 years old. The age of the Earth is on the order of 5 billion years. Suppose we use the length of a football field, or about 100 meters, as a timeline to represent the age of the Earth. If we put the birth of the Earth at the start of the timeline, how far from the line's end would human civilization begin?

Solution First, we compare the 5000-year history of human civilization to the 5 billion-year age of the Earth:

$$\frac{5000\ \text{yr}}{5\ \text{billion yr}} = \frac{5 \times 10^{3}\ \cancel{\text{yr}}}{5 \times 10^{9}\ \cancel{\text{yr}}} = 10^{-6}$$

That is, 5000 years is about 10^{-6}, or one *millionth,* of the age of the Earth. One millionth of a 100-meter (10^2 m) timeline is

$$10^{-6} \times 10^{2}\ \text{m} = 10^{-6+2}\ \text{m} = 10^{-4}\ \cancel{\text{m}} \times \frac{10^{3}\ \text{mm}}{1\ \cancel{\text{m}}} = 0.1\ \text{mm}$$

On a timeline where the Earth's history stretches the length of a football field, human civilization shows up only in the final tenth of a millimeter.

▶ Now try Exercises 53–54.

Putting It All Together: Case Studies

We've studied several techniques for putting numbers in perspective, but in many cases we gain more perspective by using two or more techniques together. Sometimes we need to apply a bit of creativity to think of ways to make sense of a particular number. The following case studies illustrate a few more of the many ways of putting numbers in perspective. As you study them, ask whether the numbers now have more meaning and how else you might give meaning to the numbers.

CASE STUDY
How Big Is a University?

Consider a university with 25,000 students. The number 25,000 is small compared to many of the numbers we've dealt with in this chapter, but it still takes thought to put it in perspective. Here's one way: Imagine that the new university president wants to get to know the students. She proposes to meet for lunch with groups of 5 students at a time. Is it possible for her to have lunch with *all* the students?

To answer, let's assume she holds the lunch meetings 5 days per week. With 5 students at each lunch, she will meet $5 \times 5 = 25$ students per week. Since there are 25,000 students at the school, it would take her 1000 weeks to have lunch with everyone. If we now assume that she has the lunches for 50 weeks each year (out of the total of 52 weeks in a year), it would take her $1000 \text{ weeks} \div 50 \text{ weeks/yr} = 20$ years to meet all the students. But after only 4 years, most of the 25,000 students would have graduated and been replaced by a new group of 25,000 students. Therefore, it would *not* be possible to meet all the students at these small group lunches. The lesson from this example is that while 25,000 people might not sound like much, you could not get to know all of them in 4 years—or even a lifetime.

Think About It Today, a typical congressional representative has more than 700,000 constituents. Is it possible for a representative to campaign by going door to door to meet everyone in his or her district? Explain.

CASE STUDY
What Is a Billion Dollars?

One way to put \$1 billion in perspective is to ask a question like "How many people can you employ with \$1 billion per year?" Let's suppose that employees receive salary and benefits worth an average of \$100,000 and that it costs a business an additional \$100,000 per year in overhead for each employee (costs for office space, computer services, and so on). The total cost of an employee is therefore \$200,000, so \$1 billion would allow a business to hire

$$\frac{\$1 \text{ billion}}{\$200{,}000 \text{ per employee}} = \frac{\$10^9}{\$2 \times 10^5/\text{employee}} = 5 \times 10^3 \text{ employees}$$

One billion dollars per year could support a work force of some 5000 employees.

Another way to put \$1 billion in perspective also points out how different numbers can be, even when they sound similar (like million, billion, and trillion). Suppose you become a sports star and earn a salary of \$1 million per year. How long would it take you to earn a billion dollars? We simply divide \$1 billion by your salary of \$1 million/year:

$$\frac{\$1 \text{ billion}}{\$1 \text{ million/yr}} = \frac{\$10^9}{\$10^6/\text{yr}} = 10^3\text{yr} = 1000 \text{ yr}$$

Even at a salary of \$1 million per year, earning a billion dollars would take a thousand years.

A billion here, a billion there; soon you're talking real money.

—Attributed to former Illinois Senator Everett Dirksen

CASE STUDY
The Scale of the Atom

We and our planet are made from *atoms*, which consist of a nucleus (made from *protons* and *neutrons*) surrounded by a "cloud" of *electrons*. A typical atom has a diameter of about 10^{-10} meter (as defined by its electron cloud), while its nucleus is about 10^{-15} meter in diameter. Can we put these numbers in perspective?

Let's begin with the atom itself. Because its diameter is 10^{-10} meter, or one *ten-billionth* of a meter, we could fit 10 billion (10^{10}) atoms in a line along a meter stick. A centimeter is $1/100$ (or $1/10^2$) of a meter, so we could fit $10^{10}/10^2 = 10^8$, or 100 million, atoms along a 1-centimeter line. Therefore, if we could shrink people down

BY THE WAY

The size of an atom is determined by the size of its electron cloud, but most of the mass of an atom is contained in its nucleus. Because the nucleus is so tiny compared to the atom itself, we have the surprising fact that atoms consist mostly of empty space!

to the size of atoms, the roughly 300 million people in the United States could fit along a line only about 3 centimeters—just over an inch—long.

Next, let's compare the diameter of the atom to the diameter of its nucleus:

$$\frac{\text{diameter of atom}}{\text{diameter of nucleus}} = \frac{10^{-10}\ \cancel{\text{m}}}{10^{-15}\ \cancel{\text{m}}} = 10^{-10-(-15)} = 10^{5}$$

The atom itself is about 10^5, or 100,000, times as large as its nucleus. That is, if we made a scale model of an atom in which the nucleus were the size of a marble (1 cm), the atom would have a diameter of about 100,000 centimeters, which is 1 kilometer, or more than half a mile.

CASE STUDY
Until the Sun Dies

We can hope that life will flourish on Earth until the Sun dies, which astronomers estimate will be in about 5 billion (5×10^9) years. How long is 5 billion years? First, let's compare the Sun's remaining lifetime to a long human lifetime of 100 years:

$$\frac{5 \times 10^9\ \cancel{\text{yr}}}{100\ \cancel{\text{yr}}} = 5 \times 10^7$$

The Sun's remaining lifetime is equivalent to about 5×10^7, or 50 million, human lifetimes. We can add perspective by dividing a human lifetime of 100 years by 50 million, then converting the result from years to minutes:

$$\frac{100\text{ yr}}{5 \times 10^7} = 2 \times 10^{-6}\ \cancel{\text{yr}} \times \frac{365\ \cancel{\text{days}}}{1\ \cancel{\text{yr}}} \times \frac{24\ \cancel{\text{hr}}}{1\ \cancel{\text{day}}} \times \frac{60\text{ min}}{1\ \cancel{\text{hr}}} \approx 1\text{ min}$$

Therefore, a human lifetime in comparison to the Sun's remaining life is roughly the same as *one minute* in comparison to a human lifetime.

How about human creations? The Egyptian pyramids are often described as "eternal." But they are slowly eroding because of wind, rain, air pollution, and the impact of tourists. All traces of them will have vanished within a few million years. If we estimate the lifetime of the pyramids as 5 million (5×10^6) years, then

$$\frac{\text{Sun's remaining lifetime}}{\text{lifetime of pyramids}} \approx \frac{5 \times 10^9\ \cancel{\text{yr}}}{5 \times 10^6\ \cancel{\text{yr}}} = 10^3 = 1000$$

A few million years may seem like a long time, but the Sun's remaining lifetime is a thousand times as long, an idea captured in a somewhat different way in the following famous poem:

I met a traveller from an antique land
Who said: Two vast and trunkless legs of stone
Stand in the desert . . . Near them, on the sand,
Half sunk, a shattered visage lies, whose frown,
And wrinkled lip, and sneer of cold command,
Tell that its sculptor well those passions read
Which yet survive, stamped on these lifeless things,
The hand that mocked them, and the heart that fed.
And on the pedestal these words appear:

"My name is Ozymandias, king of kings:
Look on my works, ye Mighty, and despair!"
Nothing beside remains. Round the decay
Of that colossal wreck, boundless and bare,
The lone and level sands stretch far away.

—Percy Bysshe Shelley, "Ozymandias"

Quick Quiz 3B

Choose the best answer to each of the following questions. Explain your reasoning with one or more complete sentences.

1. The number 300,000,000 is the same as

   a. 3×10^7 b. 3×10^8 c. 3×10^9

2. You are multiplying 1277 times 14,385. You expect the answer to be a number between 1 and 10 times

   a. 10^3. b. 10^5. c. 10^7.

3. Fill in the blank: $\$10^{10}$ is __________ times as much money as $\$10^6$.

   a. four

   b. one thousand

   c. ten thousand

4. You are asked to estimate the total amount of gasoline that will be used by all Americans this year. As an order of magnitude estimate, you determine that the answer should be somewhere around

   a. 1 million gallons.

   b. 100 million gallons.

   c. 100 billion gallons.

5. You are wondering how many dollar bills you'd need to lay end to end to stretch the 400,000-kilometer distance from Earth to the Moon. You find a website that says the answer is 8 million dollar bills. With a quick estimate, you conclude that this answer is

   a. a reasonable order of magnitude estimate.

   b. way too large (the actual number would be much less than 8 million).

   c. way too small (the actual number would be much more than 8 million).

6. You are given some data and asked to calculate how long the Sun can continue to shine before it dies. Your answer is 1.2×10^{-10} years. Based on this answer, you conclude that

   a. the Sun will die within just a few centuries.

   b. the death of the Sun is something we do not need to worry about for a very long time.

   c. your answer is wrong.

7. You are looking at a map with a scale of 1 inch = 100 miles. If two towns are separated on the map by 3.5 inches, the actual distance between them is

   a. 100 miles. b. 350 miles. c. 350 inches.

8. An NFL quarterback is offered a new contract paying him \$20 million per year. How long would it take him at this rate to earn \$1 billion?

   a. 5 years b. 50 years c. 100 years

9. You are running for mayor this year in a city with 300,000 households. You decide to campaign by going door to door and spending a few minutes talking with members of every household in the town. This plan

   a. will require about three months of your time.

   b. will require about a year of your time.

   c. is unfeasible, given how long it would take.

10. A lottery ticket on which the odds of winning are 1 in 1 million is given to every person in attendance at a college football game. The most likely result is that there will be

    a. no winners in the stadium.

    b. one winner in the stadium.

    c. many winners in the stadium.

Exercises 3B

REVIEW QUESTIONS

1. Briefly describe *scientific notation*. How is it useful for writing large and small numbers? How is it useful for making approximations?

2. Explain how we can use estimation to put numbers in perspective. Give an example.

3. What is an *order of magnitude estimate*? Explain why such an estimate can be useful even though it may be as much as 10 times too large or too small.

4. Explain how we can use comparisons to put numbers in perspective. Give an example.

5. Describe three common ways of expressing the scale of a map or model. How would you show a scale of 1 cm = 100 km graphically? How would you describe it as a ratio?

6. Explain how we can use scaling to put numbers in perspective. Give an example.

7. Suppose that the Sun were the size of a grapefruit. How big and how far away would the Earth be on this scale? How far would the nearest stars (besides the Sun) be?

8. Describe several ways of putting each of the following in perspective: the size of a large university; $1 billion; the size of an atom; the Sun's remaining lifetime.

DOES IT MAKE SENSE?

Decide whether each of the following statements makes sense (or is clearly true) or does not make sense (or is clearly false). Explain your reasoning.

9. I read a book that had 10^5 words in it.

10. I've seen about 10^{50} commercials on TV.

11. I work in an office building that is 300 feet high.

12. In total, Americans spend about a billion dollars per year on housing costs (rent and home mortgage payments).

13. A popular local restaurant serves 5 million dinners each year.

14. The CEO of the company earned more money last year than the company's 500 lowest-paid employees combined.

BASIC SKILLS & CONCEPTS

15–20: Review of Scientific Notation. In the following exercises, use the skills covered in the Brief Review on p. 140.

15. Convert each of the following numbers from scientific to ordinary notation, and write its name.

Example: $2 \times 10^3 = 2000 =$ two thousand

a. 6×10^4 b. 3×10^5 c. 3.4×10^5
d. 2×10^{-3} e. 6.7×10^{-2} f. 3×10^{-6}

16. Convert each of the following numbers from scientific to ordinary notation, and write its name.

a. 9×10^7 b. 1.1×10^3 c. 2.3×10^{-3}
d. 4×10^{-6} e. 1.23×10^8 f. 2.34×10^{-2}

17. Write each of the following numbers in scientific notation.

a. 468 b. 126,547 c. 0.04
d. 9736.23 e. 12.56 f. 0.8642

18. Write each of the following numbers in scientific notation.

a. 3578 b. 984.35 c. 0.0058
d. 624.87 e. 0.0003005 f. 98.180004

19. Do the following operations without a calculator, and show your work clearly. Be sure to express answers in scientific notation. You may round your answers to one decimal place (as in 3.2×10^5).

a. $(3 \times 10^3) \times (2 \times 10^2)$ b. $(4 \times 10^5) \times (2 \times 10^5)$
c. $(3 \times 10^3) + (2 \times 10^2)$ d. $(6 \times 10^{10}) \div (3 \times 10^5)$

20. Do the following operations without a calculator, and show your work clearly. Be sure to express answers in scientific notation. You may round your answers to one decimal place (as in 3.2×10^5).

a. $(3 \times 10^4) \times (3 \times 10^8)$ b. $(3.2 \times 10^5) \times (2 \times 10^4)$
c. $(5 \times 10^3) + (5 \times 10^2)$ d. $(9 \times 10^{13}) \div (3 \times 10^{10})$

21–22: They Don't Look That Different! Compare the numbers in each pair, and give the factor by which the numbers differ. *Example:* 10^6 is 10^2, or 100, times as large as 10^4.

21. a. 10^{24}, 10^{18}
b. 10^{17}, 10^{27}
c. 1 trillion, 1 million

22. a. 250 million, 5 billion
b. 6×10^3, 3×10^{-3}
c. 10^{-8}, 2×10^{-13}

23–26: Using Scientific Notation. Rewrite the following statements using a number in scientific notation.

23. My hard drive has a capacity of 1.2 terabytes. (Recall that *tera-* means "trillion.")

24. The number of different eight-character passwords that can be made with 26 letters and 10 numerals is approximately 2.8 trillion.

25. The diameter of a typical atom is about 0.5 nanometer. (Recall that *nano-* means "billionth.")

26. A large power plant generates 2.2 gigawatts of electrical power. (Recall that *giga-* means "billion.")

27–28: Approximation with Scientific Notation. Estimate the following quantities without using a calculator. Then find a more precise result, using a calculator if necessary. Discuss whether your approximation technique worked.

27. a. $260{,}000 \times 200$
b. 5.1 million $\times$ 1.9 thousand
c. $600{,}000 \div 3100$

28. a. 5.6 billion $\div$ 200
b. 30 million $\div$ 140,000
c. $9000 \times 54{,}986$

29–32: Perspective Through Estimation. Use estimation to make the following comparisons. Discuss your conclusion.

29. Which is greater: the amount you spend in a month on coffee or the amount you spend in a month on gasoline?

30. Could a person ride a bike across the United States (New York to California) in a month? If not, about how long would it take?

31. Could an average person lift the weight of $200 in quarters?

32. Which is greater: the amount you spend in a year on transportation or the amount you spend in a year on food?

33–38: Order of Magnitude Estimates. Make order of magnitude estimates of the following quantities. Explain the assumptions you use in your estimates.

33. Total annual spending on pizza in the United States

34. Total annual spending on movie tickets in the United States

35. The amount of (non-alcoholic) fluids an average person drinks in a year

36. The total amount of soda consumed in the United States in a year

37. The amount of gasoline an average car uses in a year

38. Total annual gasoline usage in the United States

39–46: Energy Comparisons. Use Table 3.1 to answer the following questions.

39. How many average candy bars would you have to eat to supply the energy needed for 8 hours of running?

40. How many liters of oil are required to supply the electrical energy needs of an average home for a year?

41. Compare the energy released by burning 1 kilogram of coal to that released by fission of 1 kilogram of uranium-235.

42. Compare the energy released by burning 1 liter of oil to that released by fusion of the hydrogen in 1 liter of water.

43. If you could generate energy by fusing the hydrogen in water, how much water would you need to generate the electrical energy used daily by 10,000 typical homes?

44. If you could generate energy by fusing the hydrogen in water, how much water would you need to supply all the energy currently consumed worldwide in 1 *day*?

45. How many kilograms of uranium would be required to supply the energy needs of the United States for 1 day using fission?

46. Suppose that we could somehow capture all the energy released by the Sun for just 1 second. Would this energy be enough to supply U.S. energy needs for a year? Explain.

47–50: Scale Ratios. Find the scale ratios for the following maps.

47. 1 centimeter on the map represents 20 kilometers.

48. 1 inch on the map represents 5 miles.

49. 1 cm(map) = 500 km (actual)

50. 1 in(map) = 10 km (actual)

51. **Scale Model Solar System.** The following table gives size and distance data for the planets. Calculate the scaled size and distance for each planet using a 1 to 10 billion scale model solar system. Give your results in table form. Then write one or two paragraphs that describe your findings in words and give perspective to the size of our solar system.

Planet	Diameter	Average distance from Sun
Mercury	4880 km	57.9 million km
Venus	12,100 km	108.2 million km
Earth	12,760 km	149.6 million km
Mars	6790 km	227.9 million km
Jupiter	143,000 km	778.3 million km
Saturn	120,000 km	1427 million km
Uranus	52,000 km	2870 million km
Neptune	48,400 km	4497 million km

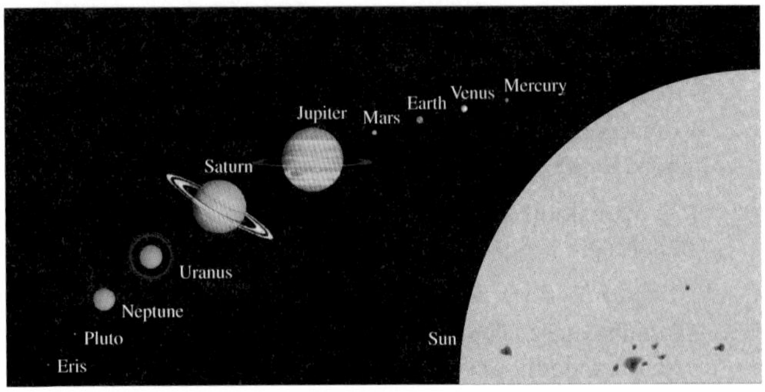

52. **Interstellar Travel.** The fastest spaceships launched to date are traveling away from Earth at speeds of about 50,000 kilometers per hour. How long would such a spaceship take to reach Alpha Centauri? (*Hint:* See Example 8.) Based on your answer, write one or two paragraphs discussing whether interstellar travel is a realistic possibility today.

53. **Universal Timeline.** According to modern science, Earth is about 4.5 billion years old and written human history extends back about 10,000 years. Suppose you represent the entire history of Earth with a 100-meter-long timeline, with the birth of Earth on one end and today at the other end.

a. What distance represents 500 million years?

b. How far from the end of the timeline does written human history begin?

54. **Universal Clock.** According to modern science, Earth is about 4.5 billion years old and written human history extends back about 10,000 years. Suppose you represent the entire history of Earth by 12 hours on a clock, with the birth of Earth at the stroke of midnight and today at the stroke of noon.

a. How much time on the clock represents 500 million years?

b. How long before noon does written human history begin?

FURTHER APPLICATIONS

55–62: Making Numbers Understandable. Restate the following facts as indicated.

55. There were approximately 2.69 million deaths in the United States in 2015. Express this quantity in deaths per minute.

56. There were approximately 2.2 million marriages in the United States in 2015. Express this quantity in marriages per hour.

57. Approximately 101.5 million passengers passed through the world's busiest airport, Atlanta's Hartsfield-Jackson Airport, in 2015. Express this quantity in passengers per hour.

58. Americans spent approximately $277 billion on new automobiles in 2015. Express this amount in dollars per person. (Use a population of 321 million.)

59. At the end of 2016, the U.S. gross debt stood at about $20 trillion. Express this amount in dollars per person. (Use a population of 321 million.)

60. The 2014 prison population in the United States was approximately 1.56 million. How many stadiums with a capacity of 60,000 would this population fill?

61. Walmart's reported revenue for 2015 was $482 billion. How many $30,000 cars would this amount buy per minute?

62. How many football fields (with dimensions 300 feet by 160 feet) are needed to accommodate 50,000 students if each student has 4 square feet to stand?

63. **Cells in the Human Body.** Estimates of the number of cells in the human body vary over an order of magnitude. Indeed, the precise number varies from one individual to another and depends on whether you count bacterial cells. Here is one way to make an estimate.

a. Assume that an average cell has a diameter of 6 micrometers (6×10^{-6} meter), which means it has a volume of 100 cubic micrometers. How many cells are there in a cubic centimeter?

b. Estimate the number of cells in a liter, using the fact that a cubic centimeter equals a milliliter.

c. Estimate the number of cells in a 70-kilogram (154-pound) person, assuming that the human body is 100% water (actually it is about 60–70% water) and that 1 liter of water weighs 1 kilogram.

64. **Emissions.** For every gallon of gasoline burned by an automobile, approximately 10.2 kilograms of carbon dioxide are emitted into the atmosphere. Estimate the total amount of carbon dioxide added to the atmosphere by all automobile travel in the United States over the past year.

65. **The Amazing Amazon.** An issue of *National Geographic* contained the following statement:

> *Dropping less than two inches per mile after emerging from the Andes, the Amazon drains a sixth of the world's runoff into the ocean. One day's discharge at its mouth—4.5 trillion gallons—could supply all U.S. households for five months.*

Based on this statement, determine how much water an average U.S. household uses each month. Does this answer seem reasonable? Explain any estimates you make.

66. **Wood for Energy?** A total of about 180,000 terawatts of solar power reaches Earth's surface, of which about 0.06% is used by plants for photosynthesis. Of the energy that goes to photosynthesis, about 1% ends up stored in plant matter (including wood). (Recall that 1 watt = 1 joule/s; 1 terawatt = 10^{12} watts.)

a. Calculate the total amount of energy that becomes stored in plant matter each second.

b. Suppose that power stations generated electricity by burning plant matter. If all the energy stored in plants could be converted to electricity, what average power, in terawatts, would be possible? Would it be enough to meet world electricity demand, which is of order 10 terawatts?

c. Based on your answer to part (b), can you draw any conclusions about why humans depend on fossil fuels, such as oil and coal, which are the remains of plants that died long ago? Explain.

67. **Stellar Corpses: White Dwarfs and Neutron Stars.** A few billion years from now, after exhausting its nuclear engine, the Sun will become a type of remnant star called a *white dwarf.* The white dwarf will still have nearly the same mass (about 2×10^{30} kg) as the Sun does today, but its radius will be only about that of Earth (about 6400 km).

a. Calculate the average density of the white dwarf in units of kilograms per cubic centimeter.

b. What is the mass of a teaspoon of material from the white dwarf? (*Hint:* A teaspoon is about 4 cubic centimeters.) Compare this mass to the mass of something familiar (for example, a person, a car, a tank).

c. A *neutron star* is a type of stellar remnant compressed to even greater densities than a white dwarf. Suppose that a neutron star has a mass 1.4 times the mass of the Sun but a radius of only 10 kilometers. What is its density? Compare the mass of 1 cubic centimeter of neutron star material to the total mass of Mt. Everest (about 5×10^{10} kg).

68. **Until the Sun Dies.** It took 65 million years from the time the dinosaurs were wiped out by an asteroid impact until humans arrived on the scene. Today, we have technology that could wipe out all humans if we do not use it wisely. Suppose we wipe ourselves out, and it then takes 65 million years for the next intelligent species to arise on Earth. Then suppose the same thing happens to them, with another intelligent species arising 65 million years later. If this process could continue until the Sun dies in about 5 billion years, how many more times could intelligent species arise on Earth at 65-million-year intervals?

69. **Personal Consumption.** The Bureau of Economic Analysis estimates that in 2015, personal consumption expenditures of Americans totaled $12.3 trillion. The major categories of these expenditures were durable goods ($1.36 trillion; for example, cars, furniture, recreational equipment), non-durable goods ($2.66 trillion; for example, food, clothing, fuel), and services ($8.3 trillion; for example, health care, education, transportation).

a. What was the approximate *annual* per capita spending on personal consumption? Assume a population of 321 million.

b. What was the approximate *daily* per capita spending on personal consumption?

c. On average, about what percentage of personal consumption spending was devoted to services? Is this figure consistent with your own spending?

d. Spending on health care was estimated to be $2.1 trillion. About what percentage of all personal consumption spending was devoted to health care?

e. In 2000, the total spending on personal consumption was $6.8 trillion, while health care spending was $918 billion. Compare the percentage increase in total spending and health care spending between 2000 and 2015.

70–73: Scaling Problems. Scaling techniques can be used to estimate physical quantities. To estimate a large quantity, you might measure a representative small sample and find the total quantity by "scaling up." To estimate a small quantity, you might measure several of the small quantities together and "scale down." In each of the following, describe your estimation technique and answer the questions.

Example: How thick is a sheet of a paper?

Solution: One way to estimate the thickness of a sheet of paper is to measure the thickness of a ream (500 sheets) of paper. A particular ream was 7.5 centimeters thick. Thus, a sheet of paper from this ream was 7.5 cm ÷ 500 = 0.015 cm, or 0.15 millimeter, thick.

70. How much does a sheet of paper weigh?
71. How thick is a penny? A nickel? A dime? A quarter? Would you rather have your height stacked in pennies, nickels, dimes, or quarters? Explain.
72. How much does a grain of sand weigh? How many grains of sand are in a typical playground sand box?
73. How many stars are visible in the sky on the clearest, darkest nights in your home town?

IN YOUR WORLD

74. **Energy Comparisons.** Using data available from the Energy Information Administration website, choose a few measures of U.S. or world energy consumption or production. Make comparisons that put these numbers in perspective.
75. **Nuclear Fusion.** Learn about the current state of research into building commercially viable fusion power plants. What obstacles must still be overcome? Do you think fusion power will be a reality in your lifetime? Explain.
76. **Scale Model Solar System.** Visit the website for the Voyage scale model solar system on the National Mall in Washington, DC. Write a brief report on what you learn.
77. **Richest People.** Find the net worth of the world's three richest people. Put these monetary values in perspective through any techniques you wish.
78. **Large Numbers.** Search today's news for as many numbers larger than 100,000 as you can find. Briefly explain the context within which each large number is used.
79. **Perspective in the News.** Find an example in the recent news in which a reporter uses a technique to put a number in perspective. Describe the example. Do you think the technique is effective? Can you think of a better way to put the number in perspective? Explain.
80. **Putting Numbers in Perspective.** Find at least two examples of very large or small numbers in recent news reports. Use a technique of your choosing to put each number in perspective in a way that you believe most people would find meaningful.

TECHNOLOGY EXERCISE

81. **Scientific Notation with Technology.** Use a calculator, Excel, or a search engine to do the following calculations.

    a. Find the distance that light travels in a year at a speed of 186,000 miles/second (a distance called a *light-year*).

    b. Evaluate $\frac{52 \cdot 51 \cdot 50 \cdot 49 \cdot 48}{5 \cdot 4 \cdot 3 \cdot 2 \cdot 1}$, the number of different 5-card hands that can be dealt from a 52-card deck of cards.

    c. Annual worldwide emissions of carbon dioxide were estimated to be 37 billion metric tons in 2015. Express this quantity in per capita terms (metric tons per person in the world). Assume a world population of 7.5 billion.

    d. Earth's mass is 6.0×10^{24} kilograms. Its volume is 1.1×10^{12} cubic kilometers. Find the density of Earth (mass/volume) in units of grams per cubic centimeter. (For comparison, the density of water is 1 gram per cubic centimeter.)

    e. The universe is estimated to be about 14 billion years old. What is its age in seconds?

UNIT 3C Dealing with Uncertainty

In early 2001, after the U.S. federal budget had shown a *surplus* (the federal government took in more money than it spent) for four straight years, economists projected that continuing surpluses would add up to a total of $5.6 trillion over the next 10 years, spurring political arguments about how to spend the windfall. In reality, the projected surplus not only vanished, but the next 10 years saw the federal government go some $9 trillion further into debt. In other words, the difference between the projection and the reality turned out to be almost $15 trillion, or almost $50,000 for every man, woman, and child in the United States. How could the projection have been so wrong?

The answer is that like all estimates, the projection was only as good as the assumptions that went into it, and these assumptions included highly uncertain predictions about the future of the economy, future tax rates, and future spending. In all fairness, the economists who made the projection were aware of these uncertainties, but the news media and politicians tended to report the surplus projection as an indisputable fact.

". . . practically no one knows what they're talking about when it comes to numbers in the newspapers. And that's because we're always quoting other people who don't know what they're talking about like politicians and stock-market analysts."
—Molly Ivins (1944–2007), syndicated (U.S.) columnist

This story of vanishing trillions holds an important lesson. Many of the numbers we encounter in daily life are far less certain than we are told, and we can be severely misled unless we learn to examine and interpret uncertainties for ourselves. In this unit, we discuss ways of dealing with the inevitable uncertainty of numbers we encounter in daily life.

Significant Digits

Suppose you measure your weight to be 132 pounds on a scale that can be read only to the nearest pound. Saying that you weigh 132.00 pounds would be misleading, because it would incorrectly imply that you know your weight to the nearest *hundredth* of a pound, rather than to the nearest pound. In other words, the measurements 132 pounds and 132.00 pounds do *not* mean the same thing.

The digits in a number that represent actual measurements are called **significant digits**. For example, 132 pounds has 3 significant digits and implies a measurement to the nearest pound, while 132.00 pounds has 5 significant digits and implies a measurement to the nearest hundredth of a pound.

Note that zeros are significant when they represent actual measurements, but not when they serve only to locate the decimal point. We assume that the zeros in 132.00 pounds are significant because there is no reason to include them except to represent an actual measurement. In contrast, we assume that the zeros in 600 centimeters are *not* significant, because they serve only to tell us that the decimal point comes to their right. Rewriting 600 centimeters as 6 meters makes it easier to see that only the 6 is a significant digit.

The only subtlety in counting significant digits arises when we cannot be sure whether zeros are truly significant. For example, suppose your professor states that there are 200 students in your class. Without further information, you have no way to know whether she means exactly 200 students or roughly 200. We can avoid this kind of ambiguity by writing numbers in scientific notation. In that case, zeros appear only when they are significant. For example, an enrollment of 2×10^2 implies a measurement to the nearest hundred students, while 2.00×10^2 implies exactly 200 students.

BY THE WAY

The speed of light provides a good example of the difficulty with zeros. When we state it as 300,000 km/s, the standard rules suggest that it has only one significant digit. However, its measured value to 9 significant digits is 299,792.458 km/s, which means the first two zeros actually are significant, because in scientific notation we would round the value to 3.00×10^5 km/s. (If you look back at Example 1 in Unit 2A, this is why we rounded the answer for a light-year to 3 significant digits.)

SUMMARY When Are Digits Significant?

Type of Digit	Significance
Nonzero digits	Always significant
Zeros that follow a nonzero digit *and* lie to the right of the decimal point (as in 4.20 or 3.00)	Always significant
Zeros *between* nonzero digits (as in 4002 or 3.06) or other significant zeros (such as the first zero in 30.0)	Always significant
Zeros to the *left* of the first nonzero digit (as in 0.006 or 0.00052)	Never significant
Zeros to the *right* of the last nonzero digit but before the decimal point (as in 40,000 or 210)	Not significant unless stated otherwise

EXAMPLE 1 Counting Significant Digits

State the number of significant digits and the implied meaning of the following numbers.

a. a time of 11.90 seconds
b. a length of 0.000067 meter
c. a weight of 0.0030 gram
d. a population reported as 240,000
e. a population reported as 2.40×10^5

Solution

a. The number 11.90 seconds has 4 significant digits and implies a measurement to the nearest 0.01 second.
b. The number 0.000067 meter has 2 significant digits and implies a measurement to the nearest 0.000001 meter. Note that we can rewrite this number as 67 micrometers, showing clearly that it has only 2 significant digits.
c. The number 0.0030 has 2 significant digits. The leading zeros are not significant because they serve only as placeholders, as we can see by rewriting the number as 3.0 milligrams. The final zero is significant because there is no reason to include it unless it was measured.
d. We assume that the zeros in 240,000 people are not significant. Therefore, the number has 2 significant digits and implies a measurement to the nearest 10,000 people.
e. The number 2.40×10^5 has 3 significant digits. Although this number means 240,000, the scientific notation shows that the first zero is significant, so it implies a measurement to the nearest 1000 people. ▸ Now try Exercises 15–26.

EXAMPLE 2 Rounding with Significant Digits

For each of the following operations, give your answer with the specified number of significant digits.

a. 7.7 mm × 9.92 mm; give your answer with 2 significant digits.
b. 240,000 × 72,106; give your answer with 4 significant digits.

Solution

a. $7.7 \text{ mm} \times 9.92 \text{ mm} = 76.384 \text{ mm}^2$. Because we are asked to give the answer with 2 significant digits, we round to 76 mm^2.
b. $240{,}000 \times 72{,}106 = 1.730544 \times 10^{10}$. Because we are asked to give the answer with 4 significant digits, we round to 1.731×10^{10}. ▸ Now try Exercises 27–32.

Understanding Errors

We are now ready to deal with errors themselves. We'll begin by describing two types of errors. Then we'll discuss how to quantify the sizes of errors and how to report final results to account for errors.

Brief Review Rounding

The basic process of rounding numbers takes just two steps.

Step 1. Decide which decimal place (e.g., tens, ones, tenths, or hundredths) is the smallest that should be kept.

Step 2. Look at the number in the next place to the *right* (for example, if rounding to tenths, look at hundredths). If the value in the next place is *less than 5*, round *down*; if it is *5 or greater*, round *up*.

For example, the number 382.2593 is given to the nearest ten-thousandth. It can be rounded in the following ways:

382.2593 rounded to the nearest thousandth is 382.259.
382.2593 rounded to the nearest hundredth is 382.26.
382.2593 rounded to the nearest tenth is 382.3.
382.2593 rounded to the nearest one is 382.
382.2593 rounded to the nearest ten is 380.
382.2593 rounded to the nearest hundred is 400.

(Some statisticians use a more complex rounding rule if the value in the next column is exactly 5: They round up if the last digit being kept is odd and down if it is even. We won't use that rule in this book.) ▸ Now try Exercises 13–14.

Types of Error: Random and Systematic

Broadly speaking, measurement errors fall into two categories: random errors and systematic errors. An example will illustrate the difference.

Suppose you work in a pediatric clinic and use a digital scale to weigh babies. If you've ever worked with babies, you know that they usually don't lie quietly when put on a scale. Their wriggling and kicking tends to shake the scale, making the readout jump around. For the case shown in Figure 3.2a, you could equally well record the baby's weight as anything between about 14.5 and 15.0 pounds. We say that the shaking of the scale introduces a **random error**, because any particular measurement may be either too high or too low.

BY THE WAY

A systematic error in which a scale's measurements differ consistently from the true values is also called a *calibration error.* You can test the calibration of a scale by putting known weights on it, such as weights that are 2, 5, 10, and 20 pounds, and making sure that the scale gives the expected readings.

Now suppose you have weighed babies all day. At the end of the day, you notice that the scale reads 1.2 pounds when there is nothing on it (Figure 3.2b). If you assume that this problem had been present all day, then every measurement you made was high by 1.2 pounds. This type of error is called a **systematic error**, because it is caused by an error in the measurement *system*—that is, an error that consistently (systematically) affects all measurements.

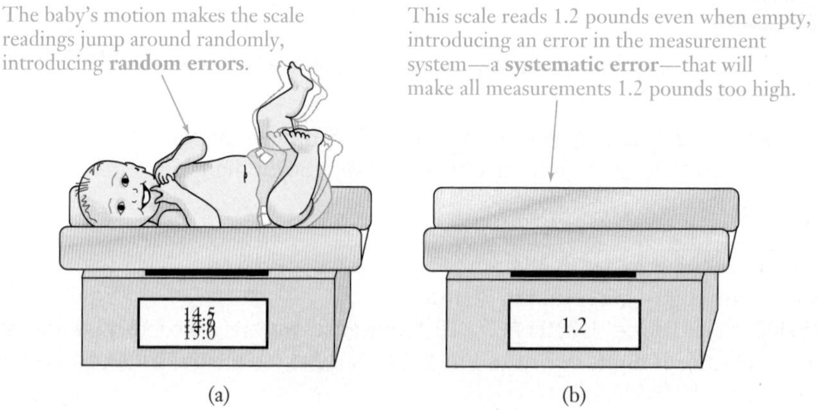
The baby's motion makes the scale readings jump around randomly, introducing random errors.
14.5
15.0
This scale reads 1.2 pounds even when empty, introducing an error in the measurement system—a systematic error—that will make all measurements 1.2 pounds too high.
1.2
(a)
(b)

FIGURE 3.2

> **Two Types of Measurement Error**
>
> **Random errors** occur because of random and inherently unpredictable events in the measurement process.
>
> **Systematic errors** occur when there is a problem in the measurement system that affects all measurements in the same way, such as making them all too low or too high by the same amount.

If you discover a systematic error, you can go back and adjust the affected measurements. In contrast, the unpredictable nature of random errors makes it impossible to correct for them. However, you can minimize the effects of random errors by making many measurements and averaging them. For example, if you measure the baby's weight ten times, your measurements will probably be too high in some cases and too low in others. You can therefore get a better estimate of the baby's true weight by averaging the ten individual measurements.

Think About It Go to a website (such as time.gov) that gives the precise current time. How far off is your clock or watch? Describe possible sources of random and systematic errors in your time-keeping.

EXAMPLE 3 Errors in Global Warming Data

Scientists studying global warming need to know how the average temperature of the entire Earth, or the *global average temperature,* has changed with time. Consider two difficulties (among many others) in trying to interpret historical temperature data from the early 20th century: (1) temperatures were measured with simple thermometers and the data were recorded by hand; and (2) most temperature measurements were recorded in or near urban areas, which tend to be warmer than surrounding rural areas because of heat released by human activity. Discuss whether each of these two difficulties produces random or systematic errors, and consider the implications of these errors.

Solution The first difficulty involves *random errors* because people undoubtedly made occasional errors in reading the thermometers, in calibrating the thermometers, and in recording temperature readings. There is no way to predict whether any individual reading is correct, too high, or too low. However, if there are multiple readings for the same region on the same day, averaging these readings can minimize the effects of the random errors.

The second difficulty involves a *systematic error* because the excess heat in urban areas always causes the temperature reading to be higher than it would be otherwise. By studying and understanding this systematic error, researchers can correct the temperature data for this error.

▶ Now try Exercises 33–38.

BY THE WAY

The fact that urban areas tend to be warmer than they would be in the absence of human activity is often called the *urban heat island effect.* Major causes of it include heat released by burning fuel in automobiles, homes, and industry, and the fact that pavement and buildings tend to retain heat from sunlight.

EXAMPLE 4 The Census

The Constitution of the United States mandates a census of the population every 10 years. The United States Census Bureau conducts the census by attempting to survey every individual and family in the United States. Suggest several sources of both random and systematic error in the census.

Solution Random errors can occur if surveys are improperly filled out or if Census Bureau employees make errors when they enter the data from the surveys. These errors are random because some of them will mean counting too many people and others will mean counting too few.

Systematic errors can arise from problems in the survey process. For example, the difficulty of counting the homeless and the fact that undocumented aliens may try to hide their presence both tend to lead to undercounts of the population. Other systematic errors tend to cause overcounts. These include college students being counted both at home and at their school residences, and children of divorced parents being counted in both households.

Although there is no way to determine the overall effects of random errors, statistical studies can estimate the effects of systematic errors. For example, the 2010 census found a population of about 308.7 million people, and follow-up statistical studies found that overcounts and undercounts were almost evenly matched, suggesting that this was an accurate estimate of the actual population at the time.

▶ Now try Exercises 39–42.

BY THE WAY

Between census years, the U.S. Census Bureau *estimates* population based on statistical data for the birth rate, death rate, and immigration rate. The Bureau estimates that the U.S. population passed 325 million in 2017 and is growing at a rate of about 1 person every 15 seconds, or more than 2 million people per year.

Estimated Dates of U.S. Population Milestones
(populations in millions)

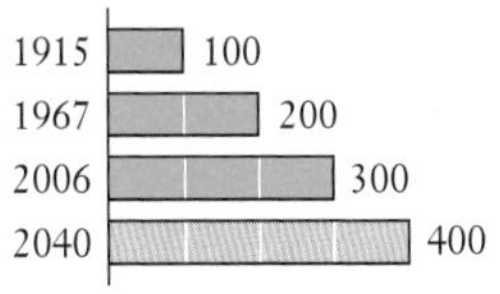

Size of Errors: Absolute vs. Relative

Besides wanting to know whether an error is random or systematic, we often want to know whether an error is big enough to be of concern or small enough to be unimportant. An example should clarify the idea.

Suppose you go to the store and buy what you think is 6 pounds of hamburger, but because the store's scale is poorly calibrated, you actually get only 4 pounds. You'd probably be upset by this 2-pound error. Now suppose you are buying hamburger for a town barbeque and you order 3000 pounds of hamburger, but you actually receive only 2998 pounds. You are short by the same 2 pounds as before, but in this case the error probably doesn't seem very important.

In more technical language, the 2-pound error in both cases is an **absolute error**—it describes how far the claimed or measured value lies from the true value. A **relative error** compares the size of an absolute error to the true value. The relative error for the first case is fairly large because the absolute error of 2 pounds is half the true weight of 4 pounds; we say that the relative error is 2/4, or 50%. In contrast, the relative error for the second case is the absolute error of 2 pounds divided by the true hamburger weight of 2998 pounds, which is only $2/2998 \approx 0.00067$, or 0.067%.

Absolute and Relative Error

The **absolute error** describes how far a measured (or claimed) value lies from the true value:

$$\text{absolute error} = \text{measured value} - \text{true value}$$

The **relative error** compares the size of the absolute error to the true value and is often expressed as a percentage:

$$\text{relative error} = \frac{\text{measured value} - \text{true value}}{\text{true value}} \times 100$$

The absolute and relative errors are *positive* when the measured or claimed value is greater than the true value and *negative* when the measured or claimed value is less than the true value. Note the similarity between the ideas of absolute and relative error and those of absolute and relative change or difference (see Unit 3A).

EXAMPLE 5 Absolute and Relative Error

Find the absolute and relative error in each case.

a. Your true weight is 125 pounds, but a scale says you weigh 130 pounds.
b. The government claims that a program costs \$49.0 billion, but an audit shows that the true cost is \$50.0 billion.

Solution

a. The measured value is the scale reading of 130 pounds, and the true value is your true weight of 125 pounds. The absolute and relative errors are

$$\begin{aligned}\text{absolute error} &= \text{measured value} - \text{true value}\\ &= 130\text{ lb} - 125\text{ lb} = 5\text{ lb}\\ \text{relative error} &= \frac{\text{measured value} - \text{true value}}{\text{true value}} \times 100,\\ &= \frac{5\text{ lb}}{125\text{ lb}} \times 100\% = 4\%\end{aligned}$$

The measured weight is too high by 5 pounds, or 4%.

b. We treat the claimed cost of \$49.0 billion as the measured value. The true value is the true cost of \$50.0 billion. The absolute and relative errors are

$$\text{absolute error} = \$49.0\text{ billion} - \$50.0\text{ billion} = -\$1.0\text{ billion}$$

$$\text{relative error} = \frac{\$49.0\text{ billion} - \$50.0\text{ billion}}{\$50.0\text{ billion}} \times 100\% = -2\%$$

The claimed cost is too low by \$1.0 billion, or 2%. ▶ **Now try Exercises 43–50.**

Describing Results: Accuracy and Precision

Two key ideas about any reported value are its *accuracy* and its *precision*. Although these terms are often used interchangeably in English, they are not the same thing.

The goal of any measurement is to obtain a value that is as close as possible to the *true value*. **Accuracy** describes how close the measured value lies to the true value. **Precision** describes the amount of detail in the measurement. For example, suppose a census says that the population of your hometown is 72,453 but the true population is 96,000. The census value of 72,453 is quite precise because it seems to tell us the exact count, but it is not very accurate because it is nearly 25% smaller than the actual population of 96,000. Note that accuracy is usually defined by relative error rather than absolute error. For example, if a company projects sales of $7.30 billion and true sales turn out to be $7.32 billion, we say the projection was quite accurate because it had a relative error of less than 1%, even though the absolute error of $0.02 billion represents $20 million.

BY THE WAY

In 1999, NASA lost the $160 million *Mars Climate Orbiter* when engineers sent it very precise computer instructions in English units (pounds) and the spacecraft software interpreted them in metric units (kilograms). In other words, the loss occurred because the very precise instructions were actually quite inaccurate!

Definitions

Accuracy describes how closely a measurement approximates a true value. An accurate measurement has a small relative error.

Precision describes the amount of detail in a measurement.

EXAMPLE 6 Accuracy and Precision in Your Weight

Suppose your true weight is 102.4 pounds. The scale at the doctor's office, which can be read only to the nearest quarter pound, says that you weigh $102\frac{1}{4}$ pounds. The scale at the gym, which gives a digital readout to the nearest 0.1 pound, says that you weigh 100.7 pounds. Which scale is more *precise*? Which is more *accurate*?

Solution The scale at the gym is more *precise*, because it gives your weight to the nearest tenth of a pound while the doctor's scale gives your weight only to the nearest quarter pound. However, the scale at the doctor's office is more *accurate*, because its value is closer to your true weight. ▶ Now try Exercises 51–54.

Think About It In Example 6, we need to know your *true* weight to determine which scale is more accurate. But how would you know your true weight? Can you ever be *sure* that you know your true weight? Explain.

CASE STUDY
Does the Census Measure the True Population?

Upon completing the 2010 Census, the U.S. Census Bureau reported a population of 308,745,538 (on April 1, 2010), thereby implying an exact count of everyone living in the United States. Unfortunately, such a precise count could not possibly be as accurate as it seems to imply.

Even in principle, the only way to get an exact count of the number of people living in the United States would be to count everyone *instantaneously*. Otherwise, the count would be off because, for example, an average of about eight births and four deaths occur every minute in the United States.

In fact, the census is conducted over a period of many months, so the actual population on a given date could not be known. Moreover, the census results are affected by both random and systematic errors (see Example 4 on p. 159). A more honest report of the population would use much less precision—for example, stating the population as "about 310 million." In fairness to the Census Bureau, their detailed reports explain the uncertainties in the population count, but these uncertainties are rarely mentioned by the media.

Summary: Dealing with Errors

Mistakes are the portals of discovery.

—James Joyce

Let's briefly summarize what we have discussed about measurement and errors.

- Errors can occur in many ways, but generally can be classified as one of two basic types: random errors or systematic errors.
- Whatever the source of an error, its size can be described in two different ways: as an absolute error or as a relative error.
- Once a measurement is reported, we can evaluate it in terms of its accuracy and its precision.

Combining Measured Numbers

Suppose you live in a city with a population of 400,000. One day, your best friend moves to your city to share your apartment. What is the population of your city now?

You might be tempted to add your friend to the city's population, making the new population 400,001. However, the number 400,000 has only 1 significant digit, implying that the population is known only to the nearest 100,000 people. The number 400,001 has 6 significant digits, which implies that you know the population *exactly.* Clearly, your friend's move cannot change the fact that the population is known only to the nearest 100,000. The population is still 400,000, despite the addition of your friend.

As this example illustrates, we must be very careful when we combine measured numbers. Otherwise, we may state answers with more certainty than they deserve. In scientific or statistical work, researchers conduct careful analyses to determine how to combine numbers properly. For most purposes, however, we can use two simple rounding rules.

USING TECHNOLOGY

Rounding in Excel

You can specify the precision to which numbers are rounded in Excel. Open the "Format Cells," dialog box. You will then see an option to format numbers in the cell, which allows you to choose how many decimal places are displayed.

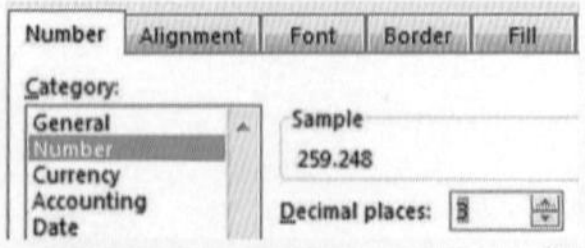

Microsoft Excel 2016, Windows 10, Microsoft Corporation.

Combining Measured Numbers

Rounding rule for addition or subtraction: Round your answer to the same precision as the *least precise* number in the problem.

Rounding rule for multiplication or division: Round your answer to the same number of *significant digits* as the measurement with the *fewest significant digits.*

Note: To avoid errors, you should do the rounding only *after* completing all the operations, not during intermediate steps.

Remember that these rounding rules tell you the *most* precision that can be justified. In many cases, the justified precision may actually be less than these rules imply.

EXAMPLE 7 Combining Measured Numbers

a. A book written 30 years ago states that the oldest Mayan ruins are 2000 years old. How old are they now?

b. The government in a town of 82,000 people plans to spend \$41.5 million this year. Assuming all this money must come from taxes, what average amount must the city collect from each resident?

Solution

a. Because the book is 30 years old, we might be tempted to add 30 years to 2000 years to get 2030 years for the age of the ruins. However, 2000 years is the less precise of the two numbers: It is precise only to the nearest 1000 years, while 30 years is precise to the nearest 10 years. Therefore, the answer also should be precise only to the nearest 1000 years:

$$\underbrace{2000\text{ yr}}_{\text{precise to nearest 1000}} + \underbrace{30\text{ yr}}_{\text{precise to nearest 10}} = \underbrace{2030\text{ yr}}_{\text{must round to nearest 1000}} \approx \underbrace{2000\text{ yr}}_{\text{correct final answer}}$$

Given the precision of the age of the ruins, they are still 2000 years old, despite the 30-year age of the book.

b. We find the average tax by dividing the \$41.5 million, which has 3 significant digits, by the population of 82,000, which has 2 significant digits. The population has the fewest significant digits, so the answer should be rounded to match its 2 significant digits.

$$\underbrace{\$41{,}500{,}000}_{\text{3 significant digits}} \div \underbrace{82{,}000\text{ persons}}_{\text{2 significant digits}} = \underbrace{\$506.10\text{ per person}}_{\text{must round to 2 significant digits}} \approx \underbrace{\$510\text{ per person}}_{\text{correct final answer}}$$

The average resident must pay about \$510 in taxes. ▶ Now try Exercises 55–62.

Quick Quiz 3C

Choose the best answer to each of the following questions. Explain your reasoning with one or more complete sentences.

1. The \$5.6 trillion surplus that government economists had projected in 2001 never materialized because the people who made this projection
   a. didn't understand how the government budget works.
   b. based it on assumptions about the future economy that turned out to be untrue.
   c. made basic errors in their calculations.
2. Under the standard rules for counting significant digits, which of the following numbers has the *most* significant digits?
   a. 5.0×10^{-1} b. 5×10^{3} c. 500,000
3. Under the standard rules for counting significant digits, which of the following numbers has the *most* significant digits?
   a. 1.02 b. 1.020 c. 0.000020
4. You are trying to measure the outside temperature at a particular time. You are likely to get better results if you average readings on three different thermometers rather than using just a single thermometer, because the averaging will
   a. reduce the effects of any random errors.
   b. eliminate the effects of systematic errors.
   c. increase the precision of your measurements.
5. You are trying to measure the outside temperature at a particular time. If you use three thermometers and place all three in direct sunlight, the sunlight is likely to cause your measurements to suffer from
   a. random errors.
   b. systematic errors.
   c. a decrease in precision.
6. A testing service makes an error that causes all the SAT scores for several thousand students to be low by 50 points. This is an example of
   a. poor precision.
   b. scores being affected by random errors.
   c. scores being affected by a systematic error.
7. A testing service makes an error that causes all the SAT scores for several thousand students to be low by 50 points. Which statement is true?
   a. All the scores had the same relative error, but the absolute errors varied.
   b. All the scores had the same absolute error, but the relative errors varied.
   c. Both the absolute and relative errors were the same in all cases.
8. A digital scale shows that you weigh 112.7 pounds, but you actually weigh 146 pounds. Which statement is true?
   a. The scale was fairly precise but not very accurate.
   b. The scale was fairly accurate but not very precise.
   c. The scale had a small absolute error but a large relative error.
9. At a particular moment, the U.S. National Debt Clock says that the federal debt is \$20,958,652,995,023.45 This reading is
   a. very precise but not necessarily accurate.
   b. very accurate but not necessarily precise.
   c. both very precise and very accurate.
10. Your car gets 29 miles per gallon and has a gas tank that can hold 10.0 gallons of gas. According to the standard rules for combining measured numbers, the distance this car can go on a full tank of gas is
    a. 290 miles. b. 290.0 miles. c. 300 miles.

Exercises 3C

REVIEW QUESTIONS

1. What are *significant digits*? How can you tell whether zeros are significant?
2. Distinguish between *random errors* and *systematic errors*. How can we minimize the effects of random errors? How can we account for the effect of a systematic error?
3. Distinguish between the *absolute error* and the *relative error* in a measurement. Give an example in which the absolute error is large but the relative error is small and another example in which the absolute error is small but the relative error is large.
4. Distinguish between *accuracy* and *precision*. Give an example of a measurement that is precise but inaccurate and another example of a measurement that is accurate but imprecise.
5. Why can it be misleading to give measurements with more precision than is justified by the measurement process?
6. State the rounding rules for adding and subtracting measured numbers and for multiplying and dividing measured numbers. Give examples of the use of these rules.

DOES IT MAKE SENSE?

Decide whether each of the following statements makes sense (or is clearly true) or does not make sense (or is clearly false). Explain your reasoning.

7. Next year's federal deficit will be $443.45 billion.
8. In many developing nations, official estimates of the population may be off by 10% or more.
9. My height is 5 feet, 6.3980 inches.
10. Wilma used her paces to measure the dimensions of her backyard to the nearest inch.
11. More precision is useless if the measurement is inaccurate.
12. A $2 million error is a lot of money, but it represents only 0.1% of the company's revenue.

BASIC SKILLS & CONCEPTS

13–14: Review of Rounding. In the following exercises, use the skills covered in the Brief Review on p. 157.

13. Round the following numbers to the nearest whole number.

a. 6.34	b. 98.245	c. 0.34
d. 356.678	e. 12,784.1	f. 3.499
g. 7386.5	h. −15.9	i. −14.1

14. Round the following numbers to the nearest tenth and nearest ten.

a. 682.48	b. 354.499	c. 2323.51
d. 987.654	e. −65.47	f. 128.55
g. −35.78	h. 678.5	i. 0.024

15–26: Counting Significant Digits. State the number of significant digits and the implied precision of the following numbers.

15. 8234	16. 923.12
17. 800	18. 0.003
19. 7.435 mi	20. 350,000 yr
21. 1.2×10^4 s	22. 0.004502 m
23. 328.4536 lb	24. 0.000005 kg
25. 1657.3 km/s	26. 2.123×10^{12} mi

27–32: Rounding with Significant Digits. Carry out the indicated operations and give your answer with the specified number of significant digits.

27. 23×12.4; 3 significant digits
28. 386×43.2; 5 significant digits
29. $988 \div 10.3$; 2 significant digits
30. $345 \div 0.36$; 5 significant digits
31. $(1.82 \times 10^3) \times (6.5 \times 10^{-2})$; 3 significant digits
32. $(7.345 \times 10^5) \times (8.424 \times 10^{-5})$; 5 significant digits

33–38: Sources of Error. Describe possible sources of random and systematic errors in the following measurements.

33. A bird census in which watchers count the birds in a 10-acre preserve over a one-day period
34. The amount of money that surveyed people say they donated to charity in the past year
35. The annual incomes of 200 people obtained from their tax returns
36. Lap times in the Indianapolis 500 auto race
37. Speeds of cars recorded by a police officer using a radar gun
38. The number of votes cast for two candidates in a U.S. Senate race
39. **Tax Audit.** A tax auditor reviewing a tax return looks for several kinds of problems, including (1) mistakes made in entering or calculating numbers on the tax return and (2) places where the taxpayer reported income dishonestly. Discuss whether each problem involves random or systematic errors.
40. **Zika Epidemic.** Researchers studying the progression of an epidemic disease such as infection by the Zika virus need to know how many people are suffering from the disease. Two of the many problems they face in this research are (1) some people who are suffering from Zika infection are misdiagnosed as having other diseases, and vice versa, and (2) many cases of Zika infection occur in poor or remote areas where medical diagnosis may not be available. Discuss whether each problem involves random or systematic errors. Can you think of other measurement difficulties that may affect this research?
41. **Safe Air Travel.** Before taking off, a pilot is supposed to set the aircraft altimeter to the elevation of the airport. A pilot leaves San Diego (altitude 17 feet) with her altimeter set to 517 feet.

Explain how this affects the altimeter readings throughout the flight. What kind of error is this?

42. **Cutting Lumber.** A lumber yard employee cuts 30 three-foot boards by measuring a three-foot length with a tape measure and then making a cut (30 times). Later, careful measurements show that all the boards are either slightly more than or less than 37 inches in length. What kind of measurement errors are involved in this case?

43–50: Absolute and Relative Errors. Find the absolute and relative errors in the following situations.

43. Your true height is 1.73 meters (5 feet 8 inches), but a nurse in your doctor's office measures your height as 1.76 meters.

44. The label on a bag of concrete says "60 pounds," but the true weight is only 58 pounds.

45. Your bike speedometer reads 26 miles per hour when you are actually traveling 24 miles per hour.

46. You buy a sweater for \$60.45 including tax, but your credit card statement shows a charge of \$70.45.

47. You measure the length of a flower garden to be ten 3-foot paces. A tape measure gives the length as 33.5 feet.

48. The actual distance to Minot is 46.7 miles, but your odometer reads 45.8 miles.

49. Your actual body temperature is 98.4°F, but your thermometer gives a reading of 97.9°F.

50. The vote count in a precinct is 1876. It is later discovered that three votes were overlooked.

51–54: Accuracy and Precision. For each pair of measurements, state which one is more accurate and which one is more precise.

51. Your true height is 70.50 inches. A tape measure that can be read to the nearest $\frac{1}{8}$ inch gives your height as $70\frac{3}{8}$ inches. A new laser device at the doctor's office that gives readings to the nearest 0.05 inch gives your height as 70.90 inches.

52. Your true height is 62.50 inches. A tape measure that can be read to the nearest $\frac{1}{8}$ inch gives your height as $62\frac{5}{8}$ inches. A new laser device at the doctor's office that gives readings to the nearest 0.05 inch gives your height as 62.50 inches.

53. Your weight is 52.55 kilograms. A scale at a health clinic that gives weight measurements to the nearest half kilogram gives your weight as 53 kilograms. A digital scale at the gym that gives readings to the nearest 0.01 kilogram gives your weight as 52.88 kilograms.

54. Your weight is 52.55 kilograms. A scale at a health clinic that gives weight measurements to the nearest half kilogram gives your weight as $52\frac{1}{2}$ kilograms. A digital scale at the gym that gives readings to the nearest 0.01 kilogram gives your weight as 51.48 kilograms.

55–62: Combining Numbers. Use the appropriate rounding rules to do the following calculations. Express the result with the correct precision or correct number of significant digits.

55. Add 0.6 pound to 136 pounds to find your weight before you went for a run.

56. Multiply 12 ft^3 by 62.4 lb/ft^3.

57. Divide 163 miles by 2.3 hour.

58. You drive 357.8 miles on a 13.5-gallon tank of gasoline. What is your gas mileage?

59. To her birth weight of 8 lb 4 oz, your baby adds 14.6 oz. What is her new weight in ounces?

60. At the hardware store, you buy a 55-kilogram bag of sand. You also buy 1.25 kilograms of nails. What is the total weight of your purchases?

61. What is the per capita cost of a \$2.8 million library in a city with 120,400 people?

62. Garden mulch costs \$46/yd^3. How much does it cost to fill your 1.25-yd^3 truck with mulch?

FURTHER APPLICATIONS

63–70: Believable Facts? Discuss possible sources of error in the following measurements. Then state whether you think the measurement is believable, given the precision with which it is stated.

63. The most visited theme park in the world is Tokyo Disneyland, which had 16,600,000 visitors in 2015.

64. In 2015, car dealers in the United States sold 5,611,411 domestic cars and 1,913,612 imported cars.

65. The U.S. Census Bureau reported the 2015 U.S. population to be 321,418,820.

66. The most common last name in the U.S. population is *Smith*, which is the last name of 0.881% of the population.

67. Asia has a land area of 30,875,906 square kilometers.

68. According to the United Nations, 36,700,000 people were living with HIV/AIDS worldwide in 2015.

69. The New York Public Library has 20,889,337 items of printed material.

70. The U.S. magazine with the greatest circulation in 2016 was *AARP The Magazine*, with 23,144,225 copies mailed out.

71. **Propagation of Error.** Suppose you want to cut 20 identical boards of length 4 feet. The procedure is to measure and cut the first board, then use the first board to measure and cut the second board, then use the second board to measure and cut the third board, and so on.

a. What are the possible lengths of the 20th board, if each time you cut a board there is a maximum error of $\pm\frac{1}{4}$ inch?

b. What are the possible lengths of the 20th board, if each time you cut a board there is a maximum error of $\pm 0.5\%$?

72. **Analyzing a Calculation.** According to 2015 estimates by the U.S. Census Bureau, the U.S. population was 321,418,820. According to the U.S. Geological Survey, the land area of the United States is 3,531,905 mi^2. Dividing the population by the area, we find that the population density of the country is 91.00437866 people/mi^2.

a. Discuss the possible sources of error in this calculation.

b. State the population and land area with the precision you think are justified.

c. Give an estimate of the population density that you think is reasonable.

IN YOUR WORLD

73. **Random and Systematic Errors.** Find a recent news report that gives a measured quantity (for example, a report of population, average income, or the number of homeless people). Write a short description of how the quantity was measured, and describe any likely sources of either random or systematic errors. Overall, do you think that the reported measurement was accurate?

74. **Absolute and Relative Errors.** Find a recent news report that describes some mistake in a measured, estimated, or projected number (for example, a budget projection that turned out to be incorrect). In words, describe the size of the error in terms of both absolute error and relative error.

75. **Accuracy and Precision.** Find a recent news article that causes you to question accuracy or precision. For example, the article might report a figure with more precision than you think is justified, or it may cite a figure that you know is inaccurate. Write a short summary of the article, and explain why you question its accuracy or precision.

76. **Uncertainty in the News.** Look for news articles from the past week to find at least two numbers from each of the following: national/international news, local news, sports news, and business news. In each case, describe the number and its context, and discuss any uncertainty that you think is associated with the number.

77. **The 2020 Census.** Depending on whether it has already happened, investigate plans or results for the 2020 census. How do the chosen methods compare to those from past censuses? Discuss and form opinions about any controversies concerning how the data are presented and used.

UNIT 3D Index Numbers: The CPI and Beyond

You've probably heard about **index numbers** such as the Consumer Price Index, the Producer Price Index, and the Consumer Confidence Index. Index numbers are common because they provide a simple way to compare measurements made at different times or in different places. In this unit, we'll investigate the meaning and use of index numbers, focusing especially on the Consumer Price Index (CPI).

What Is an Index Number?

Let's start with an example using gasoline prices. Table 3.2 shows the average price of gasoline in the United States at 10-year intervals from 1965 to 2015. Suppose that, instead of just knowing these prices, we want to know how each of them compares to the 1985 price. One way to do this would be to express each year's price as a percentage of the 1985 price. For example, dividing the 1975 price by the 1985 price, we find that the 1975 price was 47.5% of the 1985 price:

$$\frac{\text{1975 price}}{\text{1985 price}} = \frac{\$0.57}{\$1.20} = 0.475 = 47.5\%$$

Proceeding similarly for each of the other years, we can calculate all the prices as percentages of the 1985 price. The third column of Table 3.2 shows the results. Note that the percentage for 1985 is 100%, because we chose the 1985 price as the reference value.

Now look at the last column of Table 3.2. It is identical to the third column, except we dropped the % signs. This simple change converts each number from a percentage to a *price index,* which is one type of index number. The statement "1985 = 100" in the column heading shows that the reference value is the 1985 price. In this case, there's no

TABLE 3.2 Average Gasoline Prices (per gallon)

Year	Price	Price as a Percentage of 1985 Price	Price Index (1985 = 100)
1965	\$0.31	25.8%	25.8
1975	\$0.57	47.5%	47.5
1985	\$1.20	100.0%	100.0
1995	\$1.21	100.8%	100.8
2005	\$2.31	192.5%	192.5
2015	\$2.52	210.0%	210.0

Source: U.S. Department of Energy; prices are the year-long average for all grades of gasoline.

BY THE WAY

The 10-year increments in Table 3.2 hide substantial price variations that occurred within each decade. For example, while the 2015 average gasoline price was only slightly higher than the 2005 price, actual prices during that period ranged from below \$1.90 to nearly \$4.20 per gallon.

particular advantage to stating the comparison in terms of index numbers rather than percentages. However, as we'll see shortly, it's helpful to use index numbers in cases where many factors are being considered simultaneously.

Index Numbers

An **index number** provides a simple way to compare measurements made at different times or in different places. The value at one particular time (or place) is chosen as the *reference value.* The index number for any other time (or place) is

$$\text{index number} = \frac{\text{value}}{\text{reference value}} \times 100$$

Note: There is no particular rule for rounding index numbers, but in this book we will generally round them to the nearest tenth.

EXAMPLE 1 Finding an Index Number

Suppose the cost of gasoline today is \$2.70 per gallon. Using the 1985 price as the reference value, find the price index for gasoline today.

Solution We use the 1985 price of \$1.20 per gallon (see Table 3.2) as the reference value to find the price index for the \$2.70 gasoline price today:

$$\text{index number} = \frac{\text{current price}}{\text{1985 price}} \times 100 = \frac{\$2.70}{\$1.20} \times 100 = 225.0$$

The index number for a current price of \$2.70 is 225.0, which means the price is 225% of the 1985 price. ▶ Now try Exercises 11–12.

BY THE WAY

The term *index* is commonly used for almost any kind of number that provides a useful comparison, even when the numbers are not standard index numbers. For example, body mass index (BMI) provides a way of comparing people by height and weight, but is defined without any reference value. Specifically, body mass index is defined as weight (in kilograms) divided by height (in meters) squared.

Think About It Find the actual price of gasoline today at a nearby gas station. What is the gasoline price index for today's price, with the 1985 price as the reference value?

Making Comparisons with Index Numbers

The primary purpose of index numbers is to facilitate comparisons. For example, suppose we want to know how much more expensive gas was in 2015 than in 1985. Because Table 3.2 uses the 1985 price as the reference value and shows that the price index for 2015 was 210.0, we conclude that the price of gasoline in 2015 was 210.0% of the 1985 price, or 2.10 times the 1985 price.

We can also make comparisons when neither value is the reference value. For example, suppose we want to know how much more expensive gas was in 1995 than in 1965. We find the answer by dividing the index numbers for the two years:

$$\frac{\text{index number for 1995}}{\text{index number for 1965}} = \frac{100.8}{25.8} \approx 3.91$$

The 1995 price was about 3.91 times the 1965 price, or 391% *of* the 1965 price. In other words, the amount of gas that cost \$1.00 in 1965 cost \$3.91 in 1995.

EXAMPLE 2 Using the Gas Price Index

Use Table 3.2 to answer the following questions:

a. Suppose it cost \$16.00 to fill your gas tank in 1985. How much did it cost to buy the same amount of gas in 2015?

b. Suppose it cost \$20.00 to fill your gas tank in 2005. How much did it cost to buy the same amount of gas in 1965?

Solution

a. Table 3.2 shows that the price index (1985 = 100) for 2015 was 210, which means that the price of gasoline in 2015 was 210% of the 1985 price. So the 2015 cost of gas that cost \$16.00 in 1985 was

$$210\% \times \$16.00 = 2.1 \times \$16.00 = \$33.60$$

b. Table 3.2 shows that the price index (1985 = 100) for 2005 was 192.5 and the index for 1965 was 25.8. Dividing the index numbers, we find that the cost of the same amount of gasoline in 1965 was this fraction of the cost in 2005:

$$\frac{\text{index number for 1965}}{\text{index number for 2005}} = \frac{25.8}{192.5} = 0.1340$$

Therefore, gas that cost \$20.00 in 2005 cost 0.1340 × \$20.00 = \$2.68 in 1965.

▶ Now try Exercises 13–14.

Changing the Reference Value

There's no particular reason why we chose the 1985 price as the reference value in Table 3.2, and it is easy to recast the table with a different reference value. For example, let's compute the gasoline price index with the 2005 price as the reference value.

The first two columns of Table 3.2 do not change because they give actual years and prices of gasoline. But this time, the reference value is the 2005 price of \$2.31. We use this value to find the index numbers for the other years. For example, the index number for the 1965 gasoline price is

$$\frac{\text{1965 price}}{\text{2005 price}} \times 100 = \frac{\$0.31}{\$2.31} \times 100 = 13.4$$

This price index tells us that 1965 gasoline prices (per gallon) were about 13.4% of 2005 prices. As we should expect, the index number is *not* the same as it was when we used 1985 as the reference year. Table 3.3 shows the index numbers with both 1985 and 2005 as the reference year. Both sets of numbers are equally valid, but meaningful only when we know the reference year.

Think About It For practice in finding index numbers, use the index number formula to confirm all the values in the last column of Table 3.3.

TABLE 3.3 The Gasoline Price Index with Two Different Reference Years

Year	Price	Price Index (1985 = 100)	Price Index (2005 = 100)
1965	\$0.31	25.8	13.4
1975	\$0.57	47.5	24.7
1985	\$1.20	100.0	51.9
1995	\$1.21	100.8	52.4
2005	\$2.31	192.5	100.0
2015	\$2.52	210.0	109.1

EXAMPLE 3 2005 Index

Using the 2005 price as the reference value, find the price index if today's gasoline price is \$2.70 per gallon. Compare your answer to the answer for the same gasoline price in Example 1, where 1985 was the reference year.

Solution With the 2005 price as the reference value, the index number for a current price of \$2.70 is

$$\text{index number} = \frac{\text{current price}}{\text{2005 price}} \times 100 = \frac{\$2.70}{\$2.31} \times 100 = 116.9$$

As we should expect, this price index is different from the value of 225.0 found in Example 1, where we used 1985 as the reference year. Both numbers are valid, but meaningful only when we know the reference value. Specifically, the 2005 price index tells us that the current price of \$2.70 is 116.9% of the 2005 price, while the 1985 price index tells us that the current price is 225.0% of the 1985 price.

▶ Now try Exercises 15–16.

The Consumer Price Index

We've seen that the price of gasoline has risen substantially with time. Most other prices and wages have also risen, a phenomenon we call **inflation**. (Prices and wages occasionally decline with time, which is *deflation*.) Changes in actual prices therefore are not very meaningful unless we compare them to the overall rate of inflation, which is measured by the **Consumer Price Index (CPI)**.

The Consumer Price Index is computed and reported monthly by the U.S. Bureau of Labor Statistics. It represents an average of prices for a sample of goods, services, and housing. The monthly sample consists of more than 60,000 items. The details of the data collection and index calculation are fairly complex, but the CPI itself is a simple index number. Table 3.4 shows the average annual CPI over a 40-year period. Currently, the reference value for the CPI is an average of prices during the period 1982–1984, which is why the table says "1982 − 1984 = 100."

The Consumer Price Index

The **Consumer Price Index (CPI)** is based on an average of prices for a sample of more than 60,000 goods, services, and housing costs. It is computed and reported monthly.

The CPI allows us to compare overall prices at different times. For example, to find out how much higher typical prices were in 2015 than in 1995, we compute the ratio of

Technical Note

The government measures two consumer price indices: The CPI-U is based on products thought to reflect the purchasing habits of all urban consumers, and the CPI-W is based on the purchasing habits only of wage earners. Table 3.4 shows the CPI-U.

TABLE 3.4 Average Annual Consumer Price Index (1982–1984 = 100)

Year	CPI	Year	CPI	Year	CPI	Year	CPI
1977	60.6	1987	113.6	1997	160.5	2007	207.3
1978	65.2	1988	118.3	1998	163.0	2008	215.3
1979	72.6	1989	124.0	1999	166.6	2009	214.5
1980	82.4	1990	130.7	2000	172.2	2010	218.1
1981	90.9	1991	136.2	2001	177.1	2011	224.9
1982	96.5	1992	140.3	2002	179.9	2012	229.6
1983	99.6	1993	144.5	2003	184.0	2013	233.0
1984	103.9	1994	148.2	2004	188.9	2014	236.7
1985	107.6	1995	152.4	2005	195.3	2015	237.0
1986	109.6	1996	156.9	2006	201.6	2016	240.0

the CPIs for the two years, using the shorthand notation CPI_{2015} to represent the CPI for 2015 (and similar notation for other years):

$$\frac{CPI_{2015}}{CPI_{1995}} = \frac{237.0}{152.4} = 1.555$$

Based on this CPI ratio, typical prices in 2015 were about 1.555 times those in 1995. For example, a typical item that cost \$1000 in 1995 would have cost \$1555 in 2010.

Think About It Look up the most recent monthly value of the CPI. How does it compare to the 2016 CPI?

EXAMPLE 4 Cost of Living

Suppose you needed \$30,000 to maintain a particular standard of living in 2010. How much would you have needed in 2016 to have the same standard of living? Assume that the average price of your typical purchases has risen at the same rate as the Consumer Price Index (CPI).

Solution We compare prices for two different years by comparing the CPIs for those years:

$$\frac{CPI_{2016}}{CPI_{2010}} = \frac{240.0}{218.1} = 1.100$$

The index numbers tell us that typical prices in 2016 were 1.100 times those in 2010. Therefore, if you needed \$30,000 to maintain your standard of living in 2010, then in 2016 you would have needed 1.100 × \$30,000 = \$33,000 to have the same standard of living. ▶ **Now try Exercises 17–18.**

The Rate of Inflation

The **rate of inflation** refers to the relative change in the CPI from one year to the next. For example, the inflation rate from 1978 to 1979 was the relative increase in the CPI between those two years:

$$\text{inflation rate 1978 to 1979} = \frac{CPI_{1979} - CPI_{1978}}{CPI_{1978}}$$

$$= \frac{72.6 - 65.2}{65.2} = 0.113 = 11.3\%$$

As measured by the CPI, the rate of inflation from 1978 to 1979 was 11.3%.

The Rate of Inflation

The **rate of inflation** from one year to the next is the relative change in the Consumer Price Index.

If I had to populate an asylum with people certified insane, I'd just pick 'em from all those who claim to understand inflation.
—Will Rogers

EXAMPLE 5 Inflation Comparison

Find the inflation rate from 2015 to 2016. How does it compare to the inflation rate in the late 1970s?

Solution We apply the relative change formula to find the inflation rate from 2015 to 2016:

$$\frac{\text{CPI}_{2016} - \text{CPI}_{2015}}{\text{CPI}_{2015}} = \frac{240.0 - 237.0}{237.0} = 0.01266 \approx 1.3\%$$

This recent inflation rate of about 1.3% is much lower than the inflation rates in the late 1970s.

▶ Now try Exercises 19–20.

Adjusting Prices for Inflation

Because prices of most commodities tend to rise over time, we can't compare the prices fairly unless we take into account the increase we would have expected from inflation. For example, the CPI data in Table 3.4 tell us that typical prices rose from 1985 to 2015 by a factor of:

$$\frac{\text{CPI}_{2015}}{\text{CPI}_{1985}} = \frac{237.0}{107.6} \approx 2.20$$

In the language of economics, we say that \$1.00 in "1985 dollars" was equivalent to \$2.20 in "2015 dollars."

Now consider what happened to gasoline prices. Table 3.2 shows that the 1985 price of gasoline was \$1.20 per gallon. Therefore, if this price had risen with inflation, in 2015 it would have been 2.20 × \$1.20 = \$2.64 per gallon. Because Table 3.2 shows that the actual 2015 gasoline price was only \$2.52, we say that the 2015 price of gasoline was *lower than* the 1985 price in "real terms," meaning prices adjusted for inflation. Figure 3.3 shows more than six decades of annual price data for both actual gasoline prices and prices adjusted to 2016 dollars. Notice that, in real terms, the 2016 average annual price was not much above the lowest prices over the period shown.

Think About It Suppose that adjusted prices in Figure 3.3 were adjusted to 1986 dollars rather than 2016 dollars. Where would the two curves cross in that case? Would the general trends look the same or different? Explain.

USING TECHNOLOGY

The Inflation Calculator

The U.S. Bureau of Labor Statistics (BLS) provides an online inflation calculator that allows you to adjust prices for any pair of years (based on the CPI). Search for "inflation calculator" on the BLS website.

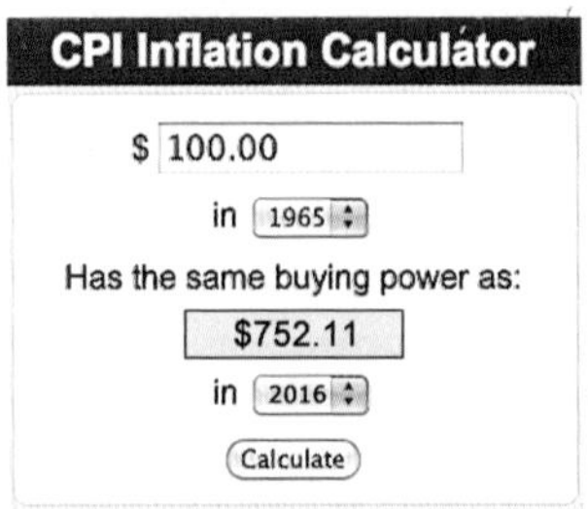

Adjusting Prices for Inflation

Given a price in dollars for year X ($\$_X$) the equivalent price in dollars for year Y ($\$_Y$) is

$$\text{price in } \$_Y = (\text{price in } \$_X) \times \frac{\text{CPI}_Y}{\text{CPI}_X}$$

where X and Y represent years, such as 1996 and 2016.

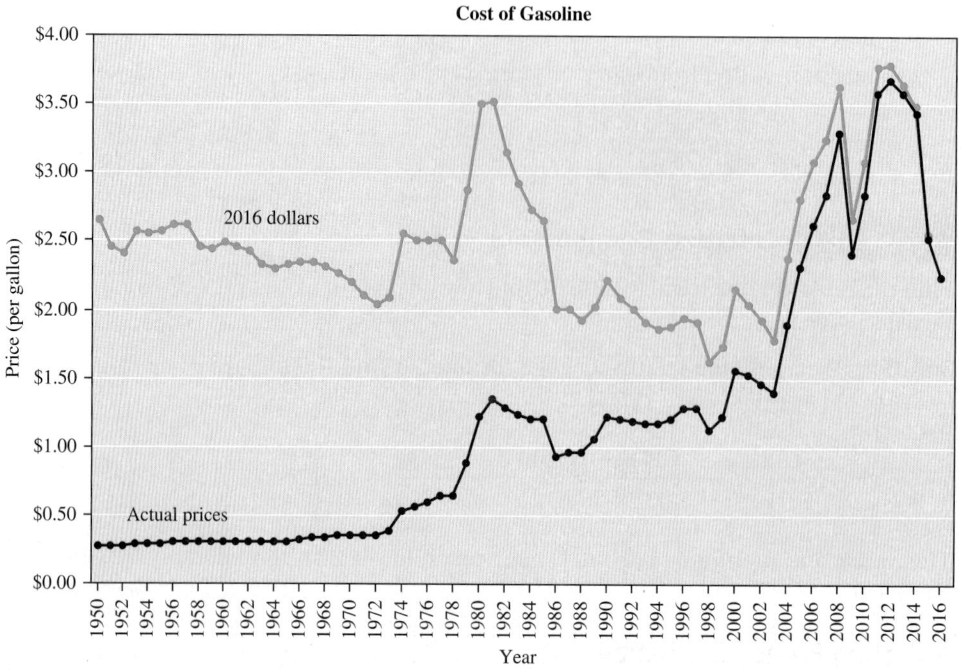

FIGURE 3.3 Gasoline prices (annual average), 1950–2016. Note that, because we use 2016 dollars for the adjusted prices, the actual and adjusted prices are the same for that year. *Source:* U.S. Energy Information Administration.

BY THE WAY

Salaries of professional athletes were once kept low because the players were not allowed to offer their skills in the free market ("free agency"). That changed after star baseball player Curt Flood filed suit against Major League Baseball in 1970. Flood lost his suit when the Supreme Court ruled in favor of baseball management in 1972, but the process he set in motion (toward free agency) was unstoppable.

EXAMPLE 6 Baseball Salaries

In 1987, the mean (average) salary for Major League baseball players was $412,000. In 2016, it was $4,380,000. Compare the two average salaries in "2016 dollars" ($\$_{2016}$). How does the rise in baseball salaries compare to the overall rise in prices due to inflation?

Solution We convert the 1987 salary of $412,000 into 2016 dollars:

$$\text{salary in } \$_{2016} = (\text{salary in } \$_{1987}) \times \frac{\text{CPI}_{2016}}{\text{CPI}_{1987}}$$

$$= \$412{,}000 \times \frac{240.0}{113.6}$$

$$\approx \$870{,}000$$

Based on changes in the CPI, a salary of $412,000 in 1987 was equivalent to a salary of about $870,000 in 2016. In other words, the actual 2016 average salary ($4,380,000) of a Major League player was about $\frac{\$4{,}380{,}000}{\$870{,}000} \approx 5$ times what it would have been if the average salary had risen only with inflation. We conclude that in "real terms" (terms adjusted for inflation), average baseball salaries rose by approximately a factor of 5 from 1987 to 2016.

▶ Now try Exercises 21–22.

IN YOUR WORLD

The Chained CPI and the Federal Budget

The Consumer Price Index plays a significant role in the federal budget, affecting both revenues and spending. For example:

- Tax rates depend on the CPI. Each year, the government raises the levels of income at which different income tax rates take effect (tax brackets) by the amount that the CPI changes. This annual adjustment is supposed to protect you from the effects of inflation. If the brackets did not change, then people with no change in their standard of living would gradually move to higher tax brackets.
- Payments for Social Security and other government benefit programs are adjusted upward each year by the amount the CPI changes. These increases are supposed to make sure that benefits increase enough that recipients can maintain the same standard of living.

But what if the CPI overstates the real change in the cost of living? In that case, people with a certain standard of living would effectively pay less in taxes over time (because the tax brackets would rise faster than the cost of living), while people receiving benefit payments would effectively get benefit increases over time. Clearly, this combination of lower taxes and higher benefit payments would tend to worsen the federal deficit.

In fact, most economists believe that the standard CPI *does* overstate the true effects of inflation, because of at least two systematic errors that make the CPI rise faster than the real cost of living. First, from one month to the next, the CPI is based on changes in the prices of particular items at particular stores. However, if the price of an item rises at one store, consumers often buy it for less at another store, and if the price rises at all stores, consumers may substitute a similar but lower-priced item (such as a different brand of the same product). This "price substitution" effect means that consumers don't find their actual costs rising as much as the CPI indicates. Second, the CPI tracks prices of "typical" items purchased by consumers at any given time, but it does not account for changes or improvements in these items with time. For example, the data used in computing this year's CPI may show that a typical cell phone is 10% more expensive than a typical cell phone a few years ago, but these data do not account for the fact that today's cell phones have more power. The data therefore overstate the effect of the cell phone price rise, because they don't account for the fact that consumers get more for their money today.

These ideas suggest that the federal government's budget picture could be improved by linking changes in tax rates and cost-of-living adjustments to a value that more accurately reflects inflation than the current CPI. In fact, the government already computes something called a "chained CPI" that is designed specifically for this purpose. The figure below shows how both revenue and spending would be different if the government tied changes in tax rates and benefit amounts to the chained CPI rather than the standard CPI. As this book goes to press in 2017, politicians are considering making this change, though the fact that it means more tax revenue and lower benefit payments means a lot of people are against it. To learn the current status of the debate, search for "chained CPI."

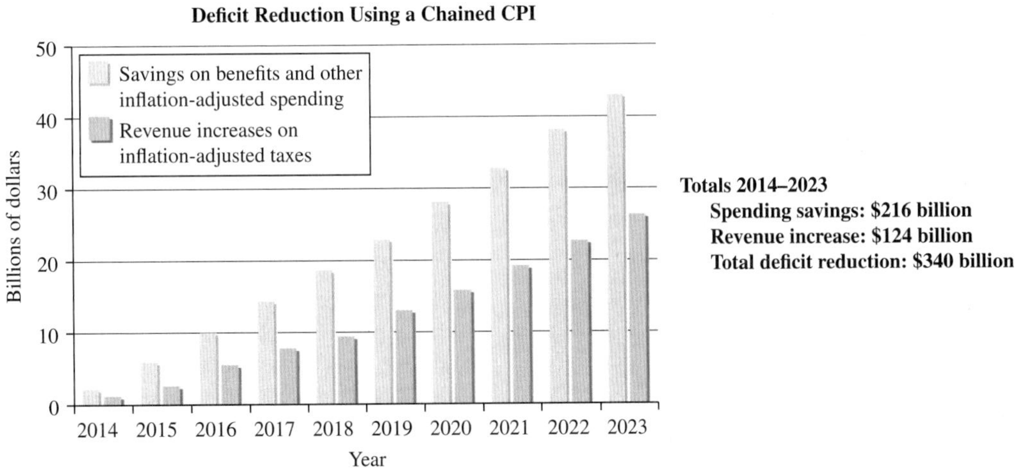

Projected revenue increases and spending savings if benefit increases and tax bracket changes were tied to the chained CPI instead of the standard CPI (CPI-U). Graph is based on the assumption that the changes took effect under policies as they stood in 2014. Data from the Congressional Budget Office.

EXAMPLE 7 Computer Prices

A typical smart phone that cost \$700 in 2016 had a computing power roughly equivalent to that of a supercomputer that sold for \$30 million in 1985. If computer prices had risen with inflation, how much would the 1985 supercomputer have cost in 2016? What does this tell us about the cost of computing power?

Solution We convert the 1985 price of \$30 million into 2016 dollars:

$$\text{price in } \$_{2016} = (\text{price in } \$_{1985}) \times \frac{\text{CPI}_{2016}}{\text{CPI}_{1985}}$$

$$= \$30 \text{ million} \times \frac{240.0}{107.6}$$

$$\approx \$67 \text{ million}$$

If computer prices had risen with inflation, the computing power of the 1985 supercomputer would have cost about \$67 million in 2016. Instead, it cost only \$700—which is barely 1/100,000 as much. In real terms, the price of computing power has fallen tremendously with time.

▶ Now try Exercises 23–26.

Other Index Numbers

The Consumer Price Index is only one of many index numbers you'll see in news reports. Some are also price indices, such as the Producer Price Index (PPI), which measures the prices that producers (manufacturers) pay for the goods they purchase (rather than the prices that consumers pay). Other indices attempt to measure more qualitative variables. For example, the Consumer Confidence Index is based on a survey designed to measure consumer attitudes so that businesses can gauge whether people are likely to be spending or saving. New indices are created frequently by groups attempting to provide simple comparisons.

Quick Quiz 3D

Choose the best answer to each of the following questions. Explain your reasoning with one or more complete sentences.

1. Look at the gasoline price index in Table 3.2. What does the 2005 price index of 192.5 tell us?

   a. Gas in 2005 cost 192.5 times as much as gas in 1985.

   b. Gas in 2005 cost 1.925 times as much as gas in 1985.

   c. Gas in 2005 cost 192.5¢ per gallon.

2. Consider a gasoline price index with 1985 = 100 chosen as the reference. If the price of gas today is \$2.55 do you need any additional information to compute the index number for today's price?

   a. No.

   b. Yes; you need to know the current CPI.

   c. Yes; you need to know the price of gas in 1985.

3. The Consumer Price Index (CPI) is designed to

   a. tell us the current cost of living for an average person.

   b. provide a fair comparison of how prices change with time.

   c. describe how the cost of gasoline has changed with time.

4. As shown in Table 3.4, the CPI was 163.0 for 1998 and 184.0 for 2003. This tells us that typical prices in 2003 were

   a. 21¢ higher than prices in 1998.

   b. 21% higher than prices in 1998.

   c. 184.0/163.0 times prices in 1998.

5. The CPI is currently published with a reference value of 100 for the years 1982–1984. Suppose the CPI were recalculated with 1995 as the reference year. Then the CPI for 2008 would be

   a. the same (215.3) as it is with 1982–1984 as the reference period.

   b. higher than 215.3.

   c. lower than 215.3.

6. Suppose we created a price index for computers, remembering that computer prices have fallen with time. If we used 1995 = 100 as the reference value for the computer price index, the price index today would be

   a. still about 100.

   b. much less than 100.

   c. much more than 100.

7. Over the past three decades, the cost of college has increased at a much greater rate than the CPI. This tells us that for the average family

   a. college has become more difficult to afford.

   b. college has become easier to afford.

   c. a college education is more valuable today than ever before.

8. Suppose your salary has been rising at a greater rate than the CPI. In principle, this should mean that

   a. your standard of living has improved.

   b. your standard of living has declined.

   c. you must now be working more hours.

9. Study Figure 3.3. Adjusted for inflation, the lowest gasoline prices during the period shown occurred during the years

   a. 1950–1952. b. 1980–1982. c. 1998–1999.

10. Assume that, from 1990 to 2015, housing prices in San Diego tripled. If we created a housing price index for San Diego with 1990 = 100 as the reference value, the index for 2015 would be

   a. 3. b. 130. c. 300.

Exercises 3D

REVIEW QUESTIONS

1. What is an *index number*? Briefly describe how index numbers are calculated and what they mean.
2. What is the *Consumer Price Index (CPI)*? How is it supposed to be related to inflation?
3. In making price comparisons, why is it important to adjust prices for the effects of inflation? Briefly describe how we use the CPI to adjust prices.
4. List a few other uses of index numbers besides the CPI. Why is it important to understand an index before deciding whether to trust it?

DOES IT MAKE SENSE?

Decide whether each of the following statements makes sense (or is clearly true) or does not make sense (or is clearly false). Explain your reasoning.

5. The price per gallon of gasoline has risen from only a quarter in 1918 to about $2.55 today, thereby making it much more difficult for the poor to afford fuel for their cars.
6. My salary has remained the same for the past 7 years, but my standard of living has fallen.
7. Benjamin Franklin said, "A penny saved is a penny earned," but if he were alive today, he would be talking about a dollar rather than a penny.
8. The prices of cars have risen steadily, but when the prices are adjusted for inflation, cars are cheaper today than they were a couple of decades ago.
9. When we chart today's price of milk in 1995 dollars, we find that it has become slightly more expensive, but when we chart it in 1975 dollars, we find that it has become cheaper.
10. The Consumer Price Index is a nice theoretical idea, but it has no impact on me, as a student on financial aid.

BASIC SKILLS & CONCEPTS

11–16: Gasoline Price Index. Use Table 3.2 to answer the following questions.

11. Suppose the current price of gasoline is $2.55. Find the current price index, using the 1985 price as the reference value.
12. Suppose the current price of gasoline is $2.30. Find the current price index, using the 1985 price as the reference value.
13. If it cost $12 to fill a gas tank in 1985, how much would it have cost to fill the same tank in 2015?
14. If it cost $15 to fill a gas tank in 1995, how much would it have cost to fill the same tank in 2015?
15. If it cost $10 to fill a gas tank in 1975, what fraction of the same tank could you fill with $10 in 2015?
16. Revise the gasoline price indices in Table 3.2 using the 1995 price as the reference value.

17–26: Understanding the CPI. Use Table 3.4 to answer the following questions. Assume that all prices have risen at the same rate as the CPI.

17. If someone needed $25,000 to maintain a certain standard of living in 1986, how much would be needed to maintain the same standard of living in 2016?
18. If someone needed $45,000 to maintain a certain standard of living in 1996, how much would be needed to maintain the same standard of living in 2016?
19. What was the overall increase in prices due to inflation as a percentage from 1995 to 2005?
20. What was the overall increase in prices due to inflation as a percentage from 2000 to 2016?
21. A box of macaroni and cheese cost $0.25 in 1977. What was its price in 2010 dollars?
22. A car cost $1500 in 1980. What was its price in 2013 dollars?
23. If a movie ticket cost $10 in 2016, what was its price in 1980 dollars?

24. If a ski lift ticket cost $100 in 2016, what was its price in 1985 dollars?

25. What was the purchasing power of $1 in 1977 in terms of 2010 dollars?

26. What was the purchasing power of $1 in 1979 in terms of 2016 dollars?

FURTHER APPLICATIONS

27–30: Housing Price Index. Realtors use an index to compare housing prices in major cities. The housing price index values for several cities are given in the table below. Use the following formula to answer the questions.

$$\underset{(\text{other town})}{\text{price}} = \underset{(\text{your town})}{\text{price}} \times \frac{\text{index of other town}}{\text{index in your town}}$$

City	Index	City	Index
Atlanta	100	Las Vegas	115
Boston	146	Los Angeles	190
Chicago	105	Miami	161
Dallas	121	Phoenix	122
Denver	136	San Francisco	172

27. For a house valued at $400,000 in Phoenix, find the price of a comparable house in Miami.

28. For a house valued at $600,000 in Boston, find the price of a comparable house in Los Angeles.

29. For a house valued at $300,000 in Dallas, find the price of a comparable house in San Francisco.

30. For a house valued at $1.2 million in San Francisco, find the price of a comparable house in Denver.

31. **Health Care Spending.** Total spending on health care in the United States rose from $85 billion in 1977 to $3.2 trillion in 2015. Compare the relative change in health care spending to the overall rate of inflation as measured by the Consumer Price Index.

32. **Airfare.** According to the U.S. Bureau of Transportation Statistics, the average domestic airfare (round trips, no special discounts or fees) increased from $292 for 1995 to $349 for 2016.

a. Calculate the relative change in the average domestic airfare from 1995 to 2016, and compare this change to the overall rate of inflation as measured by the Consumer Price Index.

b. In "2016 constant dollars," the average domestic airfare for 1995 was $460. Does this mean that airfares rose more or less than expected with inflation based on the Consumer Price Index? Explain.

33. **Private College Cost.** According to the College Board, the average tuition and fees at private 4-year colleges and universities increased from $8396 in 1990 to $33,480 for the 2016–2017 school year (per full-time equivalent student). Calculate the relative change in private college cost over this time period, and compare it to the overall rate of inflation as measured by the CPI.

34. **Public College Cost.** According to the College Board, the average in-state tuition and fees at public 4-year colleges and universities increased from $1780 in 1990 to $9650 for the 2016–2017 school year (per full-time equivalent student). Calculate the relative change in public college cost over this time period, and compare it to the overall rate of inflation as measured by the CPI.

35–42: Federal Minimum Wage. Use the following table, which shows the federal minimum wage over the past 70 years, to answer the following questions.

Federal Minimum Wage (for Years in Which a Change Occurred)

Year	Actual dollars	1996 dollars	Year	Actual dollars	1996 dollars
1938	$0.25	$2.78	1979	$2.90	$6.27
1939	$0.30	$3.39	1981	$3.35	$5.78
1945	$0.40	$3.49	1990	$3.50	$4.56
1950	$0.75	$4.88	1991	$4.25	$4.90
1956	$1.00	$5.77	1996	$4.75	$4.75
1961	$1.25	$6.41	1997	$5.15	$5.03
1967	$1.40	$6.58	2007	$5.85	$4.42
1968	$1.60	$7.21	2008	$6.55	$4.77
1974	$2.00	$6.37	2009	$7.25	$5.12
1976	$2.30	$6.34	2016*	$7.25	$4.74
1978	$2.65	$6.38			

*There has been no change in the federal minimum wage since 2009, but 2016 is included for easy reference; 1996 dollars based on CPI-U. *Source:* U. S. Department of Labor.

35. According to this table, how much is $1 in 1938 dollars worth in 1996 dollars?

36. According to this table, how much is $3 in 1956 dollars worth in 1996 dollars?

37. Explain why the minimum wage for 1996 is the same in actual and 1996 dollars.

38. Explain why the 2009 minimum wage in actual dollars is greater than the 2009 minimum wage in 1996 dollars.

39. Use Table 3.4 to convert the 1979 minimum wage from actual dollars to 1996 dollars. Is the result consistent with the entry for 1979 in the minimum wage table above?

40. In terms of purchasing power, would you rather have earned the minimum wage in 1968 or 2009? Explain.

41. How high would the minimum wage need to have been in 2016 to match the highest inflation-adjusted value shown in the table (that is, the highest value in 1996 dollars)? How does that compare to the actual minimum wage in 2016? How does it compare to the minimum wage of $15 that many labor activists have been advocating?

42. You are listening to an argument in which Paul claims that the minimum wage has never been higher, because it has

been rising for the past 70 years. Paula counters that the minimum wage actually needs to be increased, because it has been decreasing almost consistently since 1968. Based on the data in the minimum wage table above, write a one-paragraph explanation of each argument. Which argument do you think is stronger? Why?

43. **Fan Cost Index.** The cost of attending a Major League baseball game is summarized by the Fan Cost Index (FCI), which according to its originators is the price of four adult average-price tickets, two small draft beers, four small soft drinks, four regular hot dogs, parking, two programs, and two caps. The following table shows the 2016 FCI for several Major League teams and the Major League average.

Team	FCI
Boston	\$360.66
New York (Yankees)	\$337.20
Chicago (Cubs)	\$312.32
Colorado	\$193.96
San Diego	\$182.82
Arizona	\$132.10
Major League average	\$219.53

a. The FCI values are given in dollars; so is the FCI really an "index"? Explain.

b. Consider an index in which the reference value is the 2016 Major League FCI average; that is, set this value equal to 100. Then revise the above table so that all values are expressed in terms of this index.

44. **Price of Gold.** The price of gold (end-of-year closing price in dollars per troy ounce) is shown in the table below.

Year	1986	1996	2006	2016
Price	\$391	\$369	\$636	\$1152

a. The prices shown in the table are *not* adjusted for inflation. Revise the above table to express all prices in terms of 2016 dollars.

b. If you bought 3 ounces of gold in 1986 and sold it in 1996, would you have seen a profit (adjusted for inflation)? Explain.

c. If you bought 3 ounces of gold in 2006 and sold it in 2016, would you have seen a profit (adjusted for inflation)? Explain.

45. **Economic Freedom Index.** The Heritage Foundation compiles a "global economic freedom index," a composite measure of the support that 186 different countries provide for economic and business growth. Among the factors that comprise the index are business freedom, trade freedom, government spending, and government regulation. Visit the website for the global economic freedom index to answer the following questions.

a. According to the index, what are the only five "fully free" countries (scores between 80 and 100)?

b. Of the "fully free" countries, which ones have an increasing index?

c. Where is the United States on the list? To what do you attribute its (relatively low) ranking?

IN YOUR WORLD

46. **Consumer Price Index.** Find a recent news report that includes a reference to the Consumer Price Index. Briefly describe how the Consumer Price Index is important in the story.

47. **Chained CPI?** Find arguments on both sides of the question of whether the standard CPI overstates inflation and should be replaced with the chained CPI. Write a short summary of the arguments. Then state and defend your own opinion as to whether the change should be made.

48. **Producer Price Index.** Go to the Producer Price Index (PPI) home page. Read the overview and recent news releases. Write a short summary describing the purpose of the PPI and how it is different from the CPI. Also summarize any important recent trends in the PPI.

49. **Consumer Confidence Index.** Use a search engine to find recent news about the Consumer Confidence Index. After studying the news, write a short summary of what the Consumer Confidence Index attempts to measure and describe any recent trends in the index.

50. **Human Development Index.** The United Nations Development Programme regularly releases its Human Development Report. A closely watched finding of this report is the Human Development Index (HDI), which measures the overall achievements of a country in three basic dimensions of human development: life expectancy, educational attainment, and adjusted income. Find the most recent copy of this report, and investigate exactly how the HDI is defined and computed.

TECHNOLOGY EXERCISES

51–56: Inflation Calculator. Use the Bureau of Labor Statistics inflation calculator to complete the following sentences. Round your answers to the nearest dollar. *Note:* If the calculator asks you to choose a month, choose January.

51. \$100 in 1980 has the same buying power as \$ ________ in 2009.

52. \$10 in 2009 has the same buying power as \$ ________ in 1920.

53. \$25 in 1930 has the same buying power as \$ ________ in 2009.

54. \$1000 in 2008 has the same buying power as \$ ________ in 1915.

55. Suppose that Y is the year in which you were born and that \$$A$ in year Y is worth \$$B$ today. Is A greater than or less than B?

56. Suppose that Y is the year in which you were born and that \$$C$ today is worth \$$D$ in year Y. Is C greater than or less than D?

UNIT 3E

How Numbers Can Deceive: Polygraphs, Mammograms, and More

The government administers polygraph tests ("lie detectors") to new applicants for sensitive security jobs. The polygraph tests are reputed to be 90% accurate. That is, they supposedly catch 90% of the people who lie during their interviews and validate 90% of the people who are truthful. Most people therefore guess that only 10% of the people who fail their polygraph test have been falsely accused of lying. In fact, the actual percentage of false accusations can be *much* higher—more than 90% in some cases. How can this be?

We'll discuss the answer soon, but the moral of this story should already be clear: Numbers may not lie, but they can be deceiving if we do not interpret them carefully. In this unit, we'll discuss several common ways in which numbers can deceive.

Better in Each Case, but Worse Overall

Suppose a pharmaceutical company creates a new treatment for acne. To decide whether the new treatment is better than an old treatment, the company gives the old treatment to 90 patients and the new treatment to 110 patients. Table 3.5 summarizes the results after four weeks of treatment, broken down according to which treatment was given and whether the patient's acne was mild or severe. If you study the table carefully, you will notice the following key facts:

TABLE 3.5 Results of Acne Treatments

	Mild Acne		Severe Acne	
	Cured	Not Cured	Cured	Not Cured
Old Treatment	2	8	40	40
New Treatment	30	60	12	8

Among patients with mild acne:

- 10 received the old treatment and 2 were cured, for a 20% cure rate.
- 90 received the new treatment and 30 were cured, for a 33% cure rate.

Among patients with severe acne:

- 80 received the old treatment and 40 were cured, for a 50% cure rate.
- 20 received the new treatment and 12 were cured, for a 60% cure rate.

Notice that the new treatment had a higher cure rate both for patients with mild acne (33% for the new treatment versus 20% for the old) and for patients with severe acne (60% for the new treatment versus 50% for the old). Is it therefore fair for the company to claim that its new treatment is better than the old treatment?

At first, this might seem to make sense. But instead of looking at the two groups of patients separately, let's look at the overall results:

Old treatment: A total of 90 patients received the old treatment and 42 were cured (2 out of 10 with mild acne and 40 out of 80 with severe acne), for an overall cure rate of $42/90 = 46.7\%$.

New treatment: A total of 110 patients received the new treatment and 42 were cured (30 out of 90 with mild acne and 12 out of 20 with severe acne), for an overall cure rate of $42/110 = 38.2\%$.

HISTORICAL NOTE

The general case in which a data set gives different results for each of several group comparisons than it does when the groups are taken together is known as *Simpson's paradox*, so named because it was described by Edward Simpson in 1951. However, the same idea was proposed around 1900 by Scottish statistician George Yule.

Overall, the old treatment had the higher cure rate, despite the fact that the new treatment had a higher cure rate for both mild and severe acne cases.

This example illustrates that it is possible for something to appear better in each of two or more group comparisons but still be worse overall. If you look carefully, you'll see that this occurs because of the way in which the overall results are divided into unequally sized groups (in this case, mild acne patients and severe acne patients).

EXAMPLE 1 Who Played Better?

Table 3.6 gives the shooting performance of two players in the two halves of a basketball game. Sheryl had a higher shooting percentage both in the first half (40% to 25%) and in the second half (75% to 70%). Does this mean that Sheryl had the better shooting percentage for the game?

TABLE 3.6 Basketball Shots

	First Half			Second Half		
Player	Baskets	Attempts	Percentage	Baskets	Attempts	Percentage
Sheryl	4	10	40%	3	4	75%
Candace	1	4	25%	7	10	70%

Solution No, and we can see why by looking at the overall game statistics. Sheryl made a total of 7 baskets (4 in the first half and 3 in the second half) on 14 shots (10 in the first half and 4 in the second half), for an overall shooting percentage of $7/14 = 50\%$. Candace made a total of 8 baskets on 14 shots, for an overall shooting percentage of $8/14 = 57.1\%$. Even though Sheryl had a higher shooting percentage in both halves, Candace had a better overall shooting percentage for the game.

▶ **Now try Exercises 11–12.**

EXAMPLE 2 Smoking Does Not Make You Live Longer

In the early 1970s, a medical study in England involved many adult residents from a district called Wickham. Twenty years later, a follow-up study looked at the survival rates of the people from the original study. The follow-up study found the following surprising results (D. R. Appleton, J. M. French, and M. P. Vanderpump, *American Statistician*, Vol. 50, 1996, pp. 340–341):

- Among the adult smokers, 24% died during the 20 years since the original study.
- Among the adult nonsmokers, 31% died during the 20 years since the original study.

Do these results suggest that smoking can make you live longer?

Solution No, because the given results don't tell us the ages of the smokers and nonsmokers. It turned out that, in the original study, the nonsmokers were older on average than the smokers. The higher death rate among nonsmokers simply reflected the fact that death rates tend to increase with age. When the results were broken into age groups, they showed that for any given age group, nonsmokers had a higher 20-year survival rate than smokers. That is, 55-year-old nonsmokers were more likely to reach age 75 than 55-year-old smokers, and so on. Rather than suggesting that smoking prolongs life, a careful study of the data showed just the opposite. ▶ **Now try Exercises 13–16.**

Does a Positive Mammogram Mean Cancer?

The American Cancer Society recommends an annual mammogram to screen for breast cancer in all women between the ages of 45 and 54, and in many cases for younger and older women as well. These screenings are typically presumed to be about 85% accurate for both positive and negative results, meaning that they will correctly give a positive result in 85% of cases in which breast cancer is present and correctly give a negative result in 85% of cases in which there is no cancer. (In reality, the accuracy is usually different for positive results than for negative results, but we will assume it is the same for both.) Now suppose that you are a woman who has just had a mammogram, and the result has come back positive for cancer. How concerned should you be?

BY THE WAY

The accuracy of breast cancer screening is rapidly improving; newer technologies, including digital mammograms and ultrasounds, appear to achieve accuracies near 98%. The most definitive test for cancer is a biopsy, though even biopsies can miss cancers if they are not taken with sufficient care. If you have negative tests but are still concerned about an abnormality, ask for a second opinion.

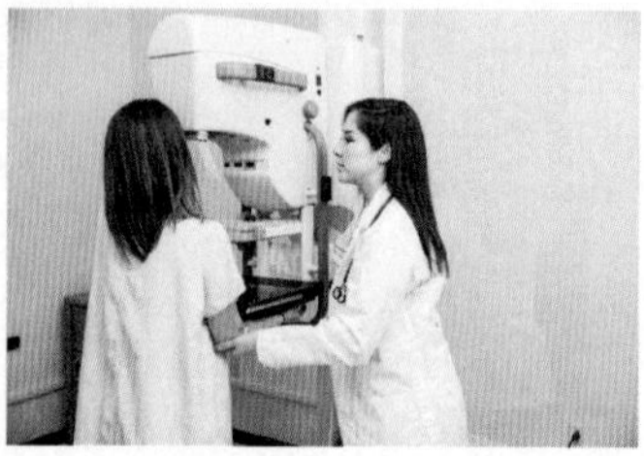

Given the 85% accuracy, most people guess that the positive result means an 85% chance that you have cancer. But in fact, the chance is much lower. To see why, let's assume that the overall breast cancer incidence rate among the screened population is 1%. That is, 1% of all women who undergo mammogram screening actually have breast cancer at the time of the screening. (The actual incidence rate varies among women of different ethnic groups and with various genetic factors; we've chosen the 1% value to keep the numbers simple.)

Let's use these numbers to consider a study in which mammograms are given to 10,000 women. With a 1% incidence rate, $1\% \times 10{,}000 = 100$ of the women will have breast cancer and the remaining 9900 women will not. In that case, the 85% accuracy of the screening will lead to the following results:

Among the 100 women with cancer:

- The mammogram will correctly identify the cancer in 85% of these 100 women, or 85 women. These cases are called **true positives**.
- The remaining 15 women will have a negative mammogram result, even though they have cancer. These cases are **false negatives**.

Among the 9900 women with no cancer:

- The mammogram will correctly give a negative result for 85% of these 9900 women, representing a total of $85\% \times 9900 = 8415$ women. These cases are **true negatives**.
- The remaining $9900 - 8415 = 1485$ women will have a positive mammogram result, even though they do not have cancer. These cases are **false positives**.

Table 3.7 summarizes these results. Notice that, overall, the mammogram screening gives a positive result for 85 women who have cancer and for 1485 women who do *not* have cancer, for a total of $85 + 1485 = 1570$ positive results. Because only 85 of these are true positives (the rest are false positives), the percentage of women with positive results who actually have cancer is only $85/1570 = 0.054$, or 5.4%. In conclusion, even if your screening comes back with a positive result, it's more likely that you don't have cancer than that you do.

TABLE 3.7 Results for a (Hypothetical) Sample of 10,000 Mammograms

	Cancer Present	No Cancer	Total
Positive Mammogram	85 true positives	1485 false positives	1570
Negative Mammogram	15 false negatives	8415 true negatives	8430
Total	100	9900	10,000

EXAMPLE 3 False Negatives

Based on the numbers in Table 3.7, what is the percentage of women with negative mammogram results who actually have cancer (false negatives)?

Solution For the 10,000 cases summarized in Table 3.7, the mammograms are negative for 15 women with cancer and for 8415 women without cancer. The total number of negative results is $15 + 8415 = 8430$. The percentage of women with false negatives is $15/8430 = 0.0018 = 0.18\%$, or slightly less than 2 in 1000. In other words, the chance that a woman with a negative mammogram has cancer is very small.

▶ Now try Exercises 17–18.

Polygraphs and Drug Tests

We're now ready to return to the question asked at the beginning of this unit, about how a 90% accurate polygraph test can lead to a surprising number of false accusations. The explanation is very similar to that found in the case of the mammograms.

Suppose the government gives the polygraph test to 1000 applicants for sensitive security jobs. Further suppose that 990 of these 1000 people tell the truth throughout their polygraph interview, while only 10 people lie. For a test that is 90% accurate, we find the following results:

- Of the 10 people who lie, the polygraph correctly identifies 90%, meaning that 9 fail the test (they are identified as liars) and 1 passes.
- Of the 990 people who tell the truth, the polygraph correctly identifies 90%, meaning that $90\% \times 990 = 891$ truthful people pass the test and the other $10\% \times 990 = 99$ people fail the test.

Figure 3.4 summarizes these results. The total number of people who fail the test is $9 + 99 = 108$. Of these, only 9 were actually liars; the other 99 were falsely accused of lying. That is, 99 out of 108, or $99/108 = 91.7\%$, of the people who fail the test were actually telling the truth.

The percentage of people who are falsely accused in any real situation depends on both the accuracy of the test and the proportion of people who are lying. Nevertheless, for the numbers given here, we have an astounding result: Assuming the government rejects applicants who fail the polygraph test, then almost 92% of the rejected applicants were actually being truthful and may have been highly qualified for the jobs.

BY THE WAY

A polygraph, often called a "lie detector," measures a variety of bodily functions including heart rate, skin temperature, and blood pressure. Polygraph operators look for subtle changes in these functions that typically occur when people lie. However, polygraph results have never been allowed as evidence in criminal proceedings, because 90% accuracy is far too low for justice. In addition, studies show that polygraphs are easily fooled by people who train to beat them.

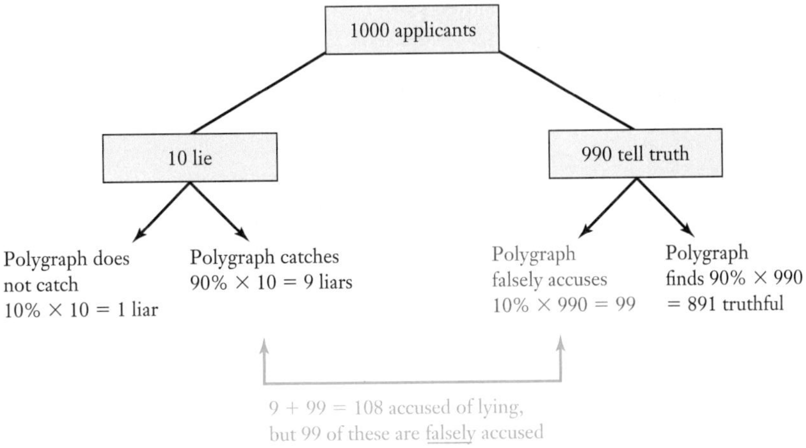

FIGURE 3.4 A tree diagram summarizes results of a 90% accurate polygraph test for 1000 people, of whom only 10 are lying.

Think About It Imagine that you are falsely accused of a crime. The police suggest that, if you are truly innocent, you should agree to take a polygraph test. Would you do it? Why or why not?

EXAMPLE 4 High School Drug Testing

All athletes participating in a regional high school track-and-field championship must provide a urine sample for a drug test. Those who test positive for drugs are eliminated from the meet and suspended from competition for the following year. Studies show that, at the laboratory selected, the drug tests are 95% accurate. Assume that 4% of the athletes actually use drugs. What fraction of the athletes who fail the test are falsely accused and therefore suspended without cause?

Solution The easiest way to answer this question is by using some sample numbers. If there are 1000 athletes in the meet and 4% of them use drugs, then 40 athletes use drugs and 960 athletes do not. In that case, the 95% accurate drug test should return the following results:

- 95% of the 40 athletes who use drugs, or $0.95 \times 40 = 38$ athletes, test positive. The other 2 athletes who use drugs test negative.
- 95% of the 960 athletes who do not use drugs test negative, but 5% of these 960 athletes test positive. The number of athletes who fail despite not using drugs is $0.05 \times 960 = 48$.

The total number of athletes who test positive is $38 + 48 = 86$. But 48 of these athletes, or $48/86 = 56\%$, are actually nonusers. Despite the 95% accuracy of the drug test, more than half of the suspended students are innocent of drug use.

▶ Now try Exercises 19–20.

Political Math

Another type of deception occurs when two sides in a debate argue with different sets of numbers. Some classic cases arise in the politics of taxes.

Consider the two charts shown in Figure 3.5. Both purport to show effects of the tax rate cuts enacted under President Bush in 2001, which remained in effect

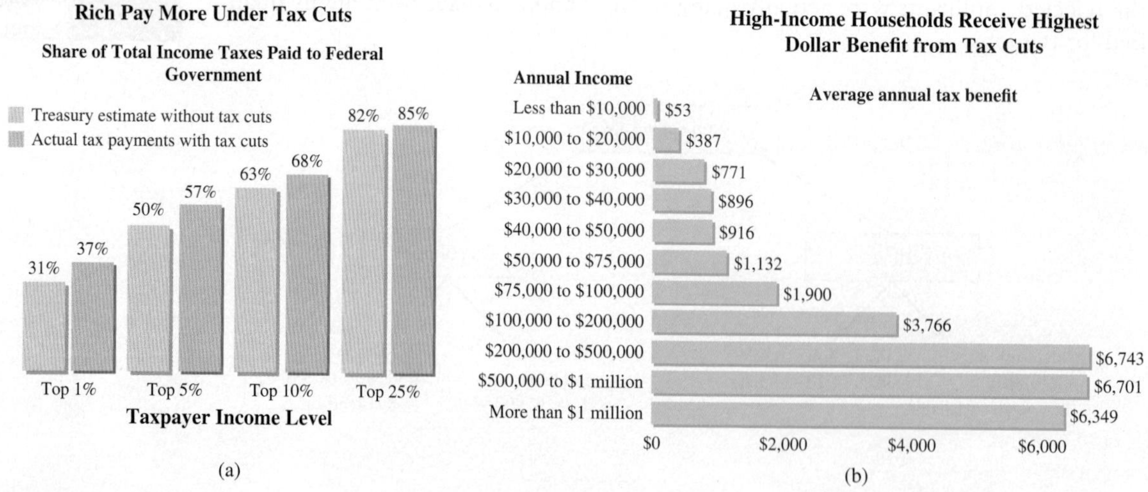

FIGURE 3.5 Both charts purport to show effects on higher-income households of tax cuts that were in effect from 2001 through 2012, but they were designed to support opposing conclusions. (a) Adapted from a graph published in *The American—The Journal of the American Enterprise Institute,* based on data from the U.S. Department of the Treasury. (b) Adapted from a graph published by the Center on Budget and Policy Priorities, based on data from the Congressional Joint Committee on Taxation.

through 2012. The chart in Figure 3.5a, created by people who would like to see new tax cuts, indicates that the rich ended up paying more under the Bush-era tax cuts than they would have otherwise. Figure 3.5b, created by opponents of new tax cuts, shows that the rich received far more benefit from the Bush-era tax cuts than lower-income taxpayers. The two charts therefore seem contradictory, because the first seems to indicate that the rich paid more while the second seems to indicate that they paid less.

Let us not seek the Republican answer or the Democratic answer, but the right answer.
—John F. Kennedy

Which story is right? In fact, both of the graphs are accurate and show data from reputable sources. The seemingly opposing claims arise from the way in which each group chose its data. The tax cut supporters show the *percentage of total taxes* that the rich paid with the cuts and what they would have paid without them. The title stating that the "rich pay more" therefore means that the tax cuts led them to pay a higher percentage of total taxes. However, if total tax revenue also was lower than it would have been without the cuts (as it was), a higher percentage of total taxes could still mean lower absolute dollars. The opponents of the tax cut show these absolute dollar savings, which show that the largest benefits went to the rich.

Which side was being more fair? Neither, really. The supporters have deliberately focused on a relative change (percentage) in order to mask the absolute change, which would be less favorable to their position. The opponents focused on the absolute change, but neglected to mention the fact that the wealthy pay most of the taxes. Unfortunately, this type of "selective truth" is very common when it comes to numbers, especially those tied up in politics.

EXAMPLE 5 A Cut or an Increase?

Suppose federal government spending for a popular education program was \$100 million this year, and that Congress has proposed to increase this amount to \$102 million for next year. Explain how this can cause lobbyists who oppose the program to complain about the size of the increase, while lobbyists who support the program will complain that it is being cut. Assume that the Consumer Price Index is expected to rise by 3% over the next year.

Politicians ... simply cook figures to suit their purpose, use obscure measures of economic performance, and indulge in horrendous examples of chart abuse, all in the name of disguising unpalatable truths.
—A. K. Dewdney, *200% of Nothing*

Solution Spending *is* rising, since it is going from \$100 million to \$102 million, so the opponents of the program can legitimately complain about this increase. However, if the Consumer Price Index rises by 3%, then this year's \$100 million is equivalent to \$103 million next year, which means an increase to \$102 million is not enough of an increase to keep pace with inflation. For this reason, the supporters can also legitimately claim that the funding is being cut in "constant dollars," because \$102 million next year will not buy as much as \$100 million this year. ▶ Now try Exercises 21–22.

Quick Quiz 3E

Choose the best answer to each of the following questions. Explain your reasoning with one or more complete sentences.

1. Study Table 3.5. What does the number 8 in the lower right cell mean?

   a. 8 people with severe acne were not cured by the new treatment.

   b. 8 people with severe acne were cured by the new treatment.

   c. 8% of the people with severe acne were not cured by the new treatment.

2. Study Table 3.5. Which statement is *not* supported by the table?

   a. The new treatment cured a higher percentage of the people with any kind of acne than did the old treatment.

   b. The new treatment cured a higher percentage of the people with mild acne than did the old treatment.

   c. The new treatment cured a higher percentage of the people with severe acne than did the old treatment.

3. During their first year in college, Derek's GPA was 3.4 and Terry's was 3.2. During their second year, Derek's GPA was 3.6 and Terry's was 3.5. Which of the following statements is *not* necessarily true?

   a. Derek had a higher GPA than Terry during their first year and again during their second year.

   b. Derek's overall GPA for the two-year period was higher than Terry's.

   c. Derek's overall GPA for the two-year period was somewhere between 3.4 and 3.6.

4. A *false negative* in a cancer screening test means that

   a. a person tested negative but actually has cancer.

   b. a person tested negative and does not have cancer.

   c. a person tested positive but does not actually have cancer.

5. A *false positive* in a drug test for steroids means that

   a. a person tested negative but actually took steroids.

   b. a person tested positive and actually took steroids.

   c. a person tested positive but did not actually take steroids.

6. Study Table 3.7. The total number of women who did *not* have cancer was

   a. 100. b. 8415. c. 9900.

7. Study Table 3.7. The total number of women whose tests gave *incorrect* results was

   a. 100. b. 1500. c. 1570.

8. Suppose that a home pregnancy test is 99% accurate. Which statement is *not* necessarily true?

   a. The test will give correct results to 99% of the women.

   b. Among the women who are pregnant, only 1% will get a result saying they're not pregnant.

   c. Among the women who test negative, 99% are not pregnant.

9. Study the graph in Figure 3.5a. Which of the following is *not* true according to this graph?

   a. Taxpayers in the top 1% income level paid a greater percentage of total federal income tax revenue than they would have without the tax cuts.

   b. Taxpayers in the top 1% income level paid 37% of total federal income tax revenue.

   c. Taxpayers in the top 1% income level paid more money in income taxes than they would have without the tax cuts.

10. Study the graph in Figure 3.5b. Which of the following is *not* true according to this graph?

    a. Taxpayers in all income brackets shown on the graph paid less in taxes than they would have without the tax cuts.

    b. Taxpayers earning $75,000 to $100,000 paid an average of $1900 in income tax.

    c. Taxpayers earning more than $1 million paid an average of $6349 less than they would have without the tax cuts.

Exercises 3E

REVIEW QUESTIONS

1. Professional athletes are routinely barred from participation for using banned substances. For such a test, what is a *false positive*? What is a *false negative*? What is *a true positive*? What is a *true negative*?
2. Briefly explain why a positive result on a screening test for a disease does not necessarily mean that a patient has the disease.
3. Explain how it is possible for an accurate polygraph or drug test to result in a large proportion of false accusations.
4. Give an example explaining how politicians on both sides of an issue can use numbers to support their case without lying.

DOES IT MAKE SENSE?

Decide whether each of the following statements makes sense (or is clearly true) or does not make sense (or is clearly false). Explain your reasoning.

5. The new drug lowered blood pressure more than the old drug for both the men and the women in the study, but the old drug was more effective overall.
6. Our total class score is based only on homework and exams, and I have better scores than you on both homework and exams. Therefore, I have a better overall score than you.
7. Baggage screening machines are 98% accurate in identifying bags that contain banned materials. Therefore, if the screening shows a bag contains banned materials, then it almost certainly does.
8. The polygraph test showed that the suspect was lying when he claimed to be innocent, so he must be guilty.
9. The Republicans claim the tax cut benefits everyone equally, but the Democrats say it favors the rich. Clearly, one side must be lying.
10. The agency suffered a real cut in its annual budget, even though it got a higher dollar amount this year than last.

BASIC SKILLS & CONCEPTS

11. **Batting Percentages.** The table below shows the batting records of two baseball players in the first half (first 81 games) and last half of a season.

Player	Hits	At-bats	Batting average
First half			
Josh	50	150	.333
Jude	10	50	.200
Second half			
Josh	35	70	.500
Jude	70	150	.467

a. Which player had the higher batting average in the first half of the season?

b. Which player had the higher batting average in the second half of the season?

c. Which player had the higher overall batting average for the season?

d. Briefly explain why the results might seem surprising or paradoxical.

12. **Jeter and Justice.** The following table shows the number of hits (H), number of at-bats (AB), and batting average (AVG = H/AB) for Major Leaguers Derek Jeter and David Justice in 1995 and 1996.

	1995	1996
Jeter	12 H 48 AB AVG = .250	183 H 582 AB AVG = .314
Justice	104 H 411 AB AVG = .253	45 H 140 AB AVG = .321

Source: Based on Ken Ross, *A Mathematician at the Ballpark: Odds and Probabilities for Baseball Fans*, Pi Press, 2004.

a. Which player had the higher batting average in both 1995 and 1996?

b. Compute the batting average for each player for the two years combined.

c. Which player had the higher combined batting average for 1995 and 1996?

d. Briefly explain why the results might seem surprising or paradoxical.

13. **Test Scores.** The table below shows eighth-grade mathematics test scores in Nebraska and New Jersey. The scores are separated according to the race of the students. Also shown are the state averages for all races.

	White	Nonwhite	Average for all races
Nebraska	281	250	277
New Jersey	283	252	272

Source: Based on National Assessment of Educational Progress scores for 1992, from *Chance*, Spring 1999.

a. Which state had the higher scores in both racial categories? Which state had the higher overall average across both racial categories?

b. Explain how a state could score lower in both categories and still have a higher overall average.

c. The following table gives the actual percentages of whites and nonwhites in each state. Use these percentages to verify that the overall average test score in Nebraska is 277, as claimed in the first table.

	White	Nonwhite
Nebraska	87%	13%
New Jersey	66%	34%

d. Use the racial percentages to verify that the overall average test score in New Jersey is 272, as claimed in the first table.

e. Briefly explain why the results might seem surprising or paradoxical.

14. **Test Scores.** Consider the following table comparing the grade averages and mathematics SAT scores of high school students in 1988 and 1998.

	Percentage of students		SAT score		
Grade average	**1988**	**1998**	**1988**	**1998**	**Change**
A+	4%	7%	632	629	−3
A	11%	15%	586	582	−4
A−	13%	16%	556	554	−2
B	53%	48%	490	487	−3
C	19%	14%	431	428	−3
Overall average			504	514	+10

Source: Based on *Chance*, Vol. 12, No. 2, 1999, from data in *New York Times*, September 2, 1999.

a. In general terms, how did the SAT scores of the students in the five grade categories change between 1988 and 1998?

b. How did the overall average SAT score change between 1988 and 1998?

c. Briefly explain why the results might seem surprising or paradoxical.

15. **Tuberculosis Deaths.** The following table shows deaths due to tuberculosis (TB) in New York City and Richmond, Virginia, in 1910.

Race	Population	TB deaths
New York		
White	4,675,000	8400
Nonwhite	92,000	500
Total	4,767,000	8900
Richmond		
White	81,000	130
Nonwhite	47,000	160
Total	128,000	290

Source: Based on Cohen and Nagel, *An Introduction to Logic and Scientific Method*, Harcourt, Brace and World, 1934.

a. Compute the death rates for whites, nonwhites, and all residents in New York City.

b. Compute the death rates for whites, nonwhites, and all residents in Richmond.

c. Briefly explain why the results might seem surprising or paradoxical.

16. **Weight Training.** Two cross-country running teams participated in a (hypothetical) study in which a fraction of each team used weight training to supplement a running workout. The remaining runners did not use weight training. At the end of the season, the mean improvement in race times (in seconds) was recorded in the following table.

	Mean time improvement (in seconds)		
	With weight training	Without weight training	Team average
Gazelles	10	2	6.0
Cheetahs	9	1	6.2

Describe how the results recorded in the table might seem surprising or paradoxical. Resolve the paradox by finding the percentage of each team that used weight training.

17. **More Accurate Mammograms.** Like Table 3.7, the following table is based on the assumption that the incidence rate of breast cancer is 1%. However, this table assumes that mammogram screening is 90% accurate (versus 85% accurate, as assumed in Table 3.7).

	Cancer present	No cancer	Total
Positive mammogram	90 true positives	990 false positives	1080
Negative mammogram	10 false negatives	8910 true negatives	8920
Total	100	9900	10,000

a. Verify that the numbers in the table are correct. Show your work.

b. Suppose a patient has a positive mammogram. What is the chance that she has cancer?

c. What is the chance that a patient with a positive mammogram does not have cancer?

d. Suppose a patient has a negative mammogram. What is the chance that she does have cancer?

18. **Disease Test.** Suppose a test for a disease is 90% accurate for those who have the disease (true positives) and 90% accurate for those who do not have the disease (true negatives). Within a sample of 2000 patients, the incidence rate of the disease is the national average, which is 2%.

a. Verify that the entries in the following table agree with the information given and that the overall incidence rate is 2%. Explain.

	Disease	No disease	Total
Test positive	36	196	232
Test negative	4	1764	1768
Total	40	1960	2000

b. Of those with the disease, what percentage test positive?

c. Of those who test positive, what percentage have the disease? Compare this result to that from part (b) and explain why they are different.

d. Of those who test positive, what percentage do *not* have the disease?

19. **Performance Enhancement.** Suppose that a test for performance-enhancing drugs is 90% accurate (it will correctly detect 90% of people who use such drugs and it will correctly detect 90% of people who do not). Suppose that, 2% of the 2000 athletes in a major meet have actually taken performance enhancing drugs, and all athletes are given the drug test.

a. Verify that the entries in the table below agree with the information given. Explain each entry.

	Users	Non-users	Total
Test finds drug use	36	196	232
Test finds no drug use	4	1764	1768
Total	40	1960	2000

b. How many athletes in total were accused of using performance-enhancing drugs? Of these, how many were using and how many were not? What percentage of those accused of using were falsely accused?

c. How many athletes in total are cleared of using performance-enhancing drugs? Of these, how many were actually using the drugs? What percentage does this represent of those cleared?

20. **Real Polygraph Test.** The results in the table below are from experiments conducted by researchers Charles R. Honts (Boise State University) and Gordon H. Barland (Department of Defense Polygraph Institute). In each case, it was known whether the subject lied, so the table indicates when the polygraph test was correct.

	Did the subject actually lie?	
	No	Yes
Polygraph test indicated that the subject *lied*.	15	42
Polygraph test indicated that the subject did *not lie*.	32	9

a. How many subjects did the test find to be lying? Of those, how many were actually lying and how many were telling the truth? What percentage of those who were found to be lying were not actually lying?

b. How many subjects did the test find to be telling the truth? Of those, how many were actually telling the truth? What percentage of those who were found to be telling the truth were actually truthful?

21. **Political Math.** Suppose federal government spending for a popular housing program was $1 billion this year and Congress has proposed increasing spending for the program to $1.01 billion for next year. Assume that the Consumer Price Index is expected to rise by 3% over the next year. Those who support the program complain that the program is being cut. Those who oppose the program complain that the program is being increased. Explain each position.

22. **A Tax Cut.** According to a professional analysis of one proposed federal tax cut (by the accounting firm Deloitte and Touche), the cut would have resulted in the following income tax savings: a single person with a household income of $41,000 would save $211 in taxes; a single person with a household income of $530,000 would save $12,838 in taxes; a married couple with two children and a household income of $41,000 would save $1208 in taxes; and a married couple with two children and a household income of $530,000 would save $13,442 in taxes.

a. Find the absolute difference in savings between a single person earning $41,000 and a single person earning $530,000. Then express the savings as a percentage of earnings for each person.

b. Find the absolute difference in savings between a married couple with two children earning $41,000 and a married couple with two children earning $530,000. Then express the savings as a percentage of earnings for each couple.

c. Write a paragraph either defending or disputing the position that the tax cut helps lower-income people.

FURTHER APPLICATIONS

23. **Basketball Records.** Consider the following hypothetical basketball records for Spelman and Morehouse Colleges.

	Spelman College	Morehouse College
Home games	10 wins, 19 losses	9 wins, 19 losses
Away games	12 wins, 4 losses	56 wins, 20 losses

a. Give numerical evidence to support the claim that Spelman College has a better team than Morehouse College.

b. Give numerical evidence to support the claim that Morehouse College has a better team than Spelman College.

c. Which claim do you think makes more sense? Why?

24. **Better Drug.** Two drugs, A and B, were tested on a total of 2000 patients, half of whom were women and half of whom were men. Drug A was given to 900 patients, and Drug B to 1100 patients. The results are presented in the table below.

	Women	Men
Drug A	5 of 100 cured	400 of 800 cured
Drug B	101 of 900 cured	196 of 200 cured

a. Give numerical evidence to support the claim that Drug B is more effective than Drug A.

b. Give numerical evidence to support the claim that Drug A is more effective than Drug B.

c. Which claim do you think makes more sense? Why?

25. **HIV Risks.** The New York State Department of Health estimates a 10% rate of HIV infection among the "at risk" population and a 0.3% rate for the general population. Tests for HIV are 95% accurate in detecting both true negatives and true positives. Random selection of 5000 "at risk" people and 20,000 people from the general population results in the following table.

	"At risk" population	
	Test positive	Test negative
Infected	475	25
Not infected	225	4275
	General population	
	Test positive	Test negative
Infected	57	3
Not infected	997	18,943

a. Verify that incidence rates for the general and "at risk" populations are 0.3% and 10%, respectively. Also, verify that detection rates for the general and "at risk" populations are 95%.

b. Consider a patient in the "at risk" category. Of those with HIV, what percentage test positive? Of those who test positive, what percentage have HIV? Explain why these two percentages are different.

c. Suppose a person in the "at risk" category tests positive for HIV. As a doctor using this table, how would you describe the patient's chance of actually having HIV? Compare this figure to the overall rate of HIV in the "at risk" population.

d. Consider the general population. Of those with HIV, what percentage test positive? Of those who test positive, what percentage have HIV? Explain why these two percentages are different.

e. Suppose a person in the general population tests positive for HIV. As a doctor using this table, how would you describe the patient's chance of actually having HIV? Compare this figure to the overall incidence rate of HIV.

26. **Sensitivity and Specificity.** Consider the following table showing the results of mammogram screenings for 64,810 women.

	Cancer present	No cancer	Total
Test positive	132	983	1115
Test negative	45	63,650	63,695
Total	177	64,633	64,810

Source: Boston University School of Public Health.

a. According to these data, what is the incidence rate of breast cancer among the women who had the mammograms?

b. What percentage of women who tested positive had cancer?

c. What percentage of women who tested negative had cancer?

d. The **sensitivity** of a screening test is the accuracy of the test among people with the disease. In this case, what percentage of women with the disease tested positive?

e. The **specificity** of a screening test is the accuracy of the test among people without the disease. In this case, what percentage of women without the disease tested negative?

27. **Airline Arrivals.** The following table shows real arrival data for two airlines in five cities (airline names have been changed).

	Excelsior Airlines		Paradise Airlines	
Destination	**Percentage on-time**	**Number of arrivals**	**Percentage on-time**	**Number of arrivals**
Los Angeles	88.9%	559	85.6%	811
Phoenix	94.8%	233	92.1%	5255
San Diego	91.4%	232	85.5%	448
San Francisco	83.1%	605	71.3%	449
Seattle	85.8%	2146	76.7%	262
Total		3775		7225

Source: Data from *Technical Review*.

a. Which airline has the higher percentage of on-time flights to the five cities?

b. Compute the percentage of on-time flights for the two airlines over all five cities.

c. Explain the apparent inconsistency in these results.

IN YOUR WORLD

28. **Polygraph Arguments.** Visit websites devoted to either opposing or supporting the use of polygraph tests. Summarize the arguments on both sides, specifically noting the role that false negative rates play in the discussion.

29. **Drug Testing.** Explore the issue of drug testing either in the workplace or for athletic competitions. Discuss the legality of drug testing in these settings and the accuracy of the tests that are commonly conducted.

30. **Cancer Screening.** Investigate recommendations concerning routine screening for some type of cancer (for example, breast cancer, prostate cancer, colon cancer). Explain how the accuracy of the screening test is measured. How is the test useful? How can its results be misleading?

31. **Tax Change.** Find information about a recently proposed tax cut or increase, and find the arguments about fairness being used by both proponents and opponents of the change. Discuss the numerical data cited by each side. Which side do you think has the stronger argument? Why?

Chapter 3 Summary

UNIT	KEY TERMS	KEY IDEAS AND SKILLS
3A	absolute change (p. 124) relative change (p. 124) reference value (p. 124) absolute difference (p. 126) relative difference (p. 126)	**Absolute change** = new value − reference value (p. 124) **Relative change** = $\frac{\text{new value} - \text{reference value}}{\text{reference value}} \times 100\%$ (p. 124) **Absolute difference** = compared value − reference value (p. 126) **Relative difference** = $\frac{\text{compared value} - \text{reference value}}{\text{reference value}} \times 100\%$ (p. 126) **Understand** the *of* versus *more than* rule. (p. 128) **Understand** that percentage points mean something different from percent (%). (p. 129) **Solve** percentage problems. (p. 131) **Identify** common abuses of percentages. (p. 133)
3B	scientific notation (p. 139) order of magnitude (p. 141)	**Write and interpret** numbers in scientific notation. (p. 140) **Put numbers in perspective** through: • estimation, including order of magnitude estimates (p. 141) • comparisons (p. 144) • scaling (p. 146)
3C	significant digits (p. 156) random error (p. 158) systematic error (p. 158) absolute error (p. 160) relative error (p. 160) accuracy (p. 161) precision (p. 161)	**Distinguish** significant digits from nonsignificant zeros. (p. 156) **Identify and distinguish** between random and systematic errors. (p. 158) **Absolute error** = measured value − true value (p. 160) **Relative error** = $\frac{\text{measured value} - \text{true value}}{\text{true value}} \times 100\%$ (p. 160) **Distinguish** between accuracy and precision. (p. 161) **Apply rounding rules** for combining approximate numbers. (p. 162) Addition/subtraction: Round to the precision of the *least precise* number in the problem. (p. 162) Multiplication/division: Round to the same number of significant digits as in the number in the problem with the *fewest significant digits.* (p. 162)
3D	index number (p. 166) inflation (p. 169) Consumer Price Index (CPI) (p. 169) rate of inflation (p. 171)	**Understand index numbers** and how they are useful for comparisons. (p. 167) **Index number** = $\frac{\text{value}}{\text{reference value}} \times 100$ (p. 167) **Understand** how the CPI is used to measure inflation. (p. 170) **Adjust** prices for inflation with the CPI: (p. 171) $\text{price in } \$_Y = (\text{price in } \$_X) \times \frac{\text{CPI}_Y}{\text{CPI}_X}$
3E	false positive (p. 180) true positive (p. 180) false negative (p. 180) true negative (p. 180)	**Understand and give examples** of how individual trials may indicate a different result than the combined trials. (p. 180) **Understand and give examples** of how mammograms and polygraphs can lead to surprising results. (p. 181) **Understand and give examples** of how Democrats and Republicans can make different claims about the same data, even if neither side is lying. (p. 182)

4 MANAGING MONEY

Managing your personal finances is a complex task in the modern world. If you are like most Americans, you already have a bank account and at least one credit card. You may also have student loans, a home mortgage, and various investment accounts. In this chapter, we'll discuss key issues in personal financial management, including budgeting, savings, loans, taxes, and investments. We'll also explore how the federal government manages its money, which affects all of us.

Q

You're a high school graduate and figure you could use an extra $1 million. Which of the following is the best strategy for getting it?

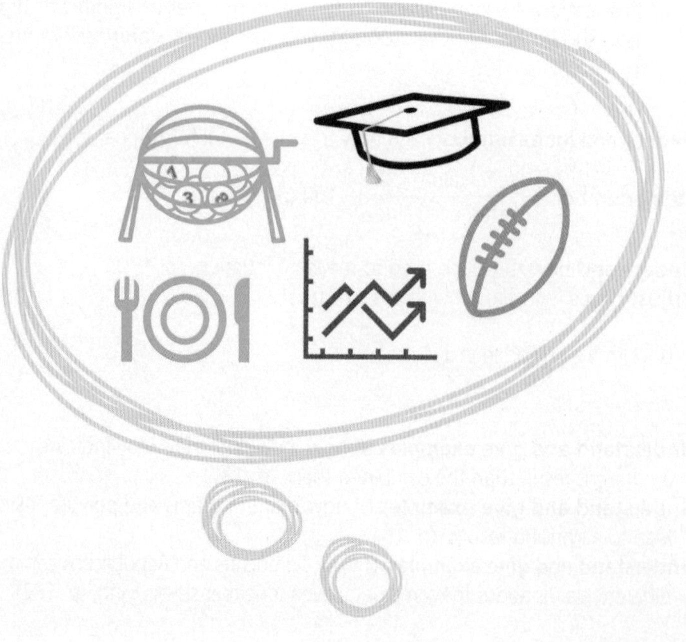

- (A) Wait for the lottery to have an unusually large prize, then buy a lot of tickets.
- (B) Develop your athletic skills in hopes of becoming a professional athlete.
- (C) Go to college.
- (D) Invest in the stock market.
- (E) Get a restaurant job in hopes that you can move up through the ranks to management.

> A fool and his money are soon parted.
>
> —English proverb

A A good way to approach this question is by starting with options you can rule out. The most obvious wrong answer is probably A: Your chances of winning the lottery are very small, and nearly all lottery players lose far more money than they ever win back. Choice B may be enticing if you happen to be a gifted athlete, but a few statistics will show that it's still a long shot. For example, the National Collegiate Athletic Association (NCAA) reports that only about 3 in 10,000 senior boys on high school basketball teams and 8 in 10,000 on football teams will end up being drafted by a professional team. For all sports combined, there are only about 10,000 professional athletes in the United States, or less than 1 in 30,000 Americans, which means your odds of becoming one aren't much different from the abysmal odds of winning the lottery.

We're left with choices C, D, and E. Choice D, investing in the stock market, can be a good long-term strategy, but few investors earn $1 million even over a lifetime, and poor choices or bad luck can produce losses. The management track of choice E could work, but nearly all of the good management jobs go to college graduates, leaving C as the correct answer. It's true: The average college graduate will earn more than $1 million more over a career than the average high school graduate. To learn exactly how, see Example 7 in Unit 4A (p. 197).

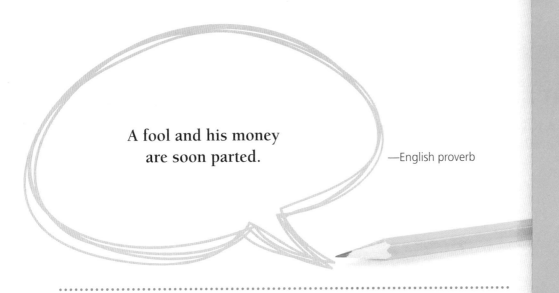

UNIT 4A

Taking Control of Your Finances: Review the basics of personal budgeting.

UNIT 4B

The Power of Compounding: Explore the basic principles of compound interest.

UNIT 4C

Savings Plans and Investments: Calculate the future value of savings plans and study investments in stocks and bonds.

UNIT 4D

Loan Payments, Credit Cards, and Mortgages: Understand the mathematics of loan payments, including those for student loans, credit cards, and mortgages.

UNIT 4E

Income Taxes: Explore the mathematics of income taxes and the political issues that surround them.

UNIT 4F

Understanding the Federal Budget: Examine the federal budget process and related political issues.

Student Loans

Use this activity to gain a sense of the kinds of problems this chapter will enable you to study. Additional activities are available online in MyLab Math.

If you are like most college students in the United States, you have received money from at least one student loan, and you may need more student loans in the future. Even if you don't personally have a student loan, many of your friends probably do. But student loans don't always have the same cost. For example, two students who each borrow $10,000 won't necessarily owe the same amounts at graduation, and they may not have the same interest rate or monthly payments once they take a job. Moreover, rules, fees, and penalties also differ for student loans. Because these loans are such an important part of most students' lives, let's use them to begin thinking about the financial topics that will be covered in this chapter.

1. Do you have a student loan? If you do, did you really need it to attend college, or did you have other options for paying for your college education? If you don't have a student loan, how have you managed to avoid having one, and do you think you will need one in the future?
2. If you have one or more student loans, look up their terms. What are the interest rates? When will you have to start paying back the loans? How long will you have to make payments, and what will your monthly payments be? If you don't have a student loan, answer these questions for a loan that you could in principle take out.
3. Suppose that you need to borrow $10,000 to pay for your next year of college. Use resources available through your college or on the Web to investigate your options for a new $10,000 student loan. Which option seems the best, and why?
4. Whether or not you have any student loans, college is expensive. Do you think that it is worth its cost? How do you expect to benefit from your college education?

UNIT 4A Taking Control of Your Finances

Money isn't everything, but it certainly has a great influence on our lives. Most people would like to have more money, and there's no doubt that more money allows you to do things that simply aren't possible with less. However, when it comes to personal happiness, studies show that the amount of money you have is less important than having your personal finances under control. Even high-income people who lose control of their finances tend to suffer from financial stress, which in turn leads to higher divorce rates and other difficulties in personal relationships, higher rates of depression, and a variety of other ailments. In contrast, people who manage their money well are more likely to say they are happy, even when they are not particularly wealthy. So if you want to attain happiness—along with any financial goals you might have—the first step is to make sure you understand *your* personal finances enough to keep them well under control.

Take Control

If you're reading this text, you are probably a college student, in which case you are probably facing financial challenges that you never had to deal with before. If you are a recent high school graduate, this may be the first time you are responsible for your own financial well-being. If you are coming back to school after years in the work force or as a parent, you now must juggle the cost of college along with all the other financial challenges of daily life.

The key to success in meeting these financial challenges is to make sure you always control them, rather than letting them control you. The first step in gaining control is to keep track of your finances. For example, you should always know your bank account balance, so that you never need to worry about overdrafts or having your debit card rejected. Similarly, you should know what you are spending on your credit card, and whether it's going to be possible for you to pay off the card at the end of the month or if your spending will dig you deeper into debt. And, of course, you should spend money wisely and at a level that you can afford.

There are lots of books and websites aimed at helping you control your finances, but in the end they all come back to the same basic idea: You need to know how much money you have and how much money you spend and then find a way to live within your means. If you can do that, as summarized in the following box, you have a good chance at financial success and happiness.

Controlling Your Finances

- Know your bank balance. Avoid bouncing a check or having your debit card rejected.
- Know what you spend; in particular, keep track of your debit and credit card spending.
- Don't buy on impulse. Think first; then buy only if you are sure the purchase makes sense for you.
- Make a budget, and don't overspend it.

EXAMPLE 1 Latte Money

Calvin isn't rich, but he gets by, and he loves sitting down for a latte at the college coffee shop. With tax and tip, he usually spends \$5 on his large latte. He gets at least one a day (on average), and about every three days he has a second one. He figures it's not such a big indulgence. Is it?

Solution One latte a day means 365 lattes per year. A second one every third day adds about $365/3 = 121$ more lattes (rounding down). That comes to $365 + 121 = 486$ lattes a year. At \$5 apiece, they have a total cost of

$$486 \times \$5 = \$2430$$

Calvin's latte habit is costing him more than \$2400 per year. That might not be much if he's financially well off. But it's more than two months of rent for an average college student; it's enough to allow Calvin to take a friend out for a \$100 dinner twice a month; and it's enough that, if he saved it, with interest he could easily build a savings balance of more than \$25,000 over the next 10 years. ▸ Now try Exercises 13–20.

EXAMPLE 2 Credit Card Interest

Cassidy has begun keeping her spending under better control, but she still can't fully pay off her credit card balance. She maintains an average monthly balance of about \$1100, and her card charges a 24% annual interest rate, which it bills at a rate of 2% per month. How much is she spending on credit card interest?

BY THE WAY

Some credit card companies set minimum payments that cover only interest—in which case you will *never* pay off your credit card if you make only the minimum payments. To avoid this problem, be sure you always pay at least something toward the principal. You can use the "2 by 24" rule: If your annual interest rate is 24%, be sure you are paying more than 2% of the balance each month. Better yet, pay off the entire balance so you won't be paying such exorbitant interest rates.

Solution Her average monthly interest is 2% of the \$1100 average balance, which is

$$0.02 \times \$1100 = \$22$$

Multiplying by 12 months in a year gives her annual interest payment:

$$12 \times \$22 = \$264$$

Interest alone is costing Cassidy more than \$260 per year—a significant amount for someone living on a tight budget. Clearly, she'd be a lot better off if she could find a way to pay off that credit card balance quickly and end those interest payments.

▶ Now try Exercises 21–24.

Master Budget Basics

As you can see from Examples 1 and 2, one of the keys to deciding what you can afford is knowing your personal budget. Making a **budget** means keeping track of how much money you have coming in and going out and then deciding what adjustments you need to make. The following box summarizes the four basic steps in making a budget.

A Four-Step Budget-Making Process

Step 1. Determine your average monthly income. Be sure to include an average monthly amount for any income you do not receive monthly (such as once-per-year payments).

Step 2. Determine your average monthly expenses. Be sure to include an average amount for expenses that don't recur monthly, such as expenses for tuition, books, vacations, insurance, and holiday gifts.

Step 3. Determine your net monthly **cash flow** by subtracting your total expenses from your total income.

Step 4. Make adjustments as needed.

For most people, the most difficult part of the budget process is making sure not to leave anything out of the list of monthly expenses. A good technique is to keep careful track of your expenses for a few months. For example, carry a small note pad with you and write down everything you spend, or use a personal budgeting app that will work with your phone or tablet. And don't forget your occasional expenses, or else you may severely underestimate your average monthly costs.

Once you've made your income and expenses lists for Steps 1 and 2, doing Step 3 is just arithmetic: Subtracting your monthly expenses from your monthly income gives you your overall monthly cash flow. If your cash flow is positive, you will have money left over at the end of each month, which you can use for savings. If your cash flow is negative, you have a problem: You'll need to find a way to balance your budget, either by earning more, spending less, using savings, or taking out a loan.

EXAMPLE 3 College Expenses

In addition to your monthly expenses, you have the following college expenses that you pay twice a year: \$3500 for your tuition each semester, \$750 in student fees each semester, and \$800 for textbooks each semester. How should you handle these expenses in computing your monthly budget?

Solution Because you pay these expenses twice a year, the total amount you pay over a whole year is

$$2 \times (\$3500 + \$750 + \$800) = \$10{,}100$$

To average this total expense for the year on a monthly basis, we divide it by 12:

$$\$10{,}100 \div 12 \approx \$842$$

Your average monthly college expenses for tuition, fees, and textbooks come to a little less than \$850, so you should put \$850 per month into your monthly expenses list.

▶ Now try Exercises 25–30.

EXAMPLE 4 College Student Budget

Brianna is creating a budget. The expenses she pays monthly are \$700 for rent, \$120 for gas for her car, \$140 for health insurance, \$75 for auto insurance, \$25 for renters insurance, \$110 for her cell phone, \$100 for utilities, about \$300 for groceries, and about \$250 for entertainment, including eating out. In addition, over the entire year she spends \$12,000 for college expenses, about \$1000 on gifts for family and friends, about \$1500 for vacations at spring and winter break, about \$800 on clothes, and \$600 in gifts to charity. Her income consists of a monthly, after-tax paycheck of about \$1600 and a \$3000 scholarship that she received at the beginning of the school year. Find her total monthly cash flow.

Solution Let's use the four-step budget-making process.

Step 1: Determine average monthly income. Brianna's direct monthly income is her \$1600 monthly paycheck. In addition, her \$3000 scholarship means an average scholarship income of \$3000/12 = \$250 per month. Therefore, her total average monthly income is

$$\$1600 + \$250 = \$1850$$

Step 2: Determine average monthly expenses. First, we add up Brianna's monthly expenses:

$$\$700 + \$120 + \$140 + \$75 + \$25 + \$110 + \$100 + \$300 + \$250 = \$1820$$

Next we consider her annual expenses, which total

$$\$12{,}000 + \$1000 + \$1500 + \$800 + \$600 = \$15{,}900$$

We divide the annual expenses by 12 to find their monthly average: \$15,900/12 = \$1325 per month. Therefore, her total average monthly expenses, including both those paid monthly and those paid annually, are

$$\$1820 + \$1325 = \$3145$$

Step 3: Determine monthly cash flow. We find Brianna's cash flow by subtracting her expenses from her income:

$$\begin{aligned}\text{monthly cash flow} &= \text{monthly income} - \text{monthly expenses}\\ &= \$1850 - \$3145\\ &= -\$1295\end{aligned}$$

Her monthly cash flow is about −\$1300. The fact that it is negative means she is spending more than she is taking in by about \$1300 per month, or about \$1300 × 12 = \$15,600 per year.

Step 4: Make adjustments as needed. Brianna clearly has a budget problem. Unless she can find a way to earn more or spend less, she will have to cover her excess expenditure either by drawing on past savings (her own or her family's) or by going into debt. It seems clear she needs to make adjustments, though they will require difficult choices on her part.

▶ Now try Exercises 31–34.

BY THE WAY

The cost of a college education is significantly more than what students pay in tuition and fees. On average, tuition and fees cover about two-thirds of the total cost at private colleges and universities, one-third of the cost at public four-year institutions, and one-fifth of the cost at two-year public colleges. The rest is covered by taxpayers, alumni donations, grants, and other revenue sources.

Think About It Look carefully at the list of expenses for Brianna in Example 4. What would you recommend she do to try to get her budget into better balance? How does her list of expenses compare to your own, and does the example hold any lessons for you? Explain.

Adjust Your Budget

If you're like most people, a careful analysis of your budget will prove very surprising. For example, many people find that they are spending a lot more in certain categories than they had imagined and that the items they thought were causing their biggest difficulties are small compared to other items. Once you evaluate your current budget, you'll almost certainly want to make adjustments to improve your cash flow for the future.

There are no set rules for adjusting your budget, so you'll need to use your critical thinking skills to come up with a plan that makes sense for you. If your finances are complicated—for example, if you are a returning college student who is juggling a job and family while attending school—you might benefit from consulting a financial advisor or reading a few books about financial planning.

BY THE WAY

Spending patterns have shifted a great deal over time. A century ago, the average American family spent 43% of its income on food and 23% on housing. Today, food accounts for only 13% of the average family's spending, while housing accounts for 33%. Notice that the *combined* percentage for food and housing has declined from 66% to 46% over the past century, implying that families now spend significantly higher percentages of their income on other items, including leisure activities.

You might also find it helpful to evaluate your own spending against average spending patterns. For example, if you are spending a higher percentage of your money on entertainment than the average person, you might want to consider finding lower-cost entertainment options. Figure 4.1 summarizes the average spending patterns for people of different ages in the United States.

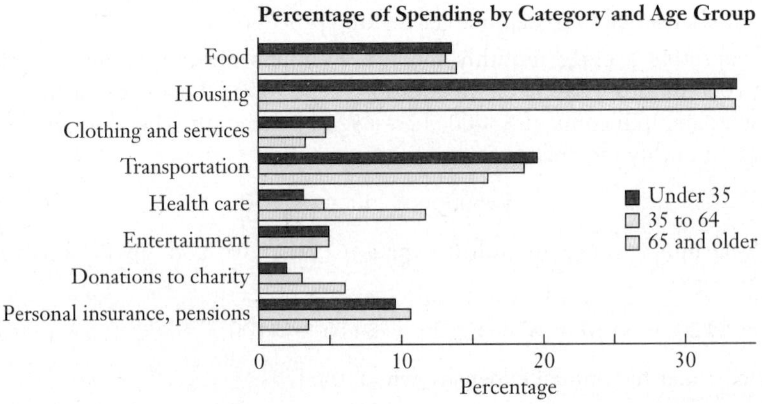

FIGURE 4.1 Average spending patterns by age group in the United States. Note: The data show spending per "consumer unit," which is defined to be either a single person or a family sharing a household. *Source:* U.S. Bureau of Labor Statistics.

EXAMPLE 5 Affordable Rent?

You've worked up a budget and find that you have $1500 per month available for all your personal expenses combined. According to the spending averages in Figure 4.1, how much should you be spending on rent?

Solution Figure 4.1 shows that the percentage of spending for housing varies very little across age groups; it is close to 1/3, or 33%, for all the groups. Based on this average and your available budget, your rent should be about 33% of $1500, or $500 per month. That's lower than rents for apartments in most college towns, which means you face a choice: Either you can put a higher proportion of your income toward rent—in which case you'll have less left over for other types of expenditures—or you can seek a way of keeping rent down, such as finding a roommate. ▶ Now try Exercises 35–40.

Look at the Long Term

Figuring out your monthly budget is a crucial step in taking control of your personal finances, but it is only the beginning. Once you have understood your budget, you need to start looking at longer-term financial issues. The general principle is always the same:

Before making any major expenditure or investment, be sure you figure out how it will affect your finances over the long term.

EXAMPLE 6 Cost of a Car

Jorge commutes both to his job and to school, driving a total of about 250 miles per week. His current car is fully paid off, but it's getting old. He spends about $1800 per year on repairs, and the car gets only about 18 miles per gallon. He's thinking about buying a new hybrid that will cost $25,000; it gets 54 miles per gallon and should be maintenance-free aside from oil changes over the next 5 years. Should he do it?

Solution To figure out whether buying the new car makes sense, Jorge needs to consider many factors. Let's start with gas. His 250 miles per week of driving means about $250\ \text{mi/wk} \times 52\ \text{wk/yr} = 13{,}000$ miles per year of driving. His current car gets 18 miles per gallon, which means he needs about 720 gallons of gas:

$$\frac{13{,}000\ \text{mi}}{18\dfrac{\text{mi}}{\text{gal}}} \approx 720\ \text{gal}$$

If we assume that gas costs $3 per gallon, this comes to $720 \times \$3 = \$2160$ per year. Notice that the 54-miles-per-gallon gas mileage for the new car is three times the 18-miles-per-gallon gas mileage for his current car, so gasoline for the new car would cost only $1/3$ as much, or about $720. He would therefore save $\$2160 - \$720 = \$1440$ each year on gas. He would also save the $1800 per year that he's currently spending on repairs, making his total annual savings about $\$1440 + \$1800 = \$3240$.

Over 5 years, Jorge's total savings on gasoline and repairs would come to about $\$3240/\text{yr} \times 5\ \text{yr} = \$16{,}200$. Although this is less than the $25,000 he would spend on the new car, the savings are starting to look pretty good, and they will get better if he keeps the new car for more than 5 years or if he can sell it for a decent price at the end of 5 years. On the other hand, if he has to take out a loan to buy the new car, his interest payments will add an extra expense; insurance for the new car may cost more as well. What would *you* do in this situation? ▶ **Now try Exercises 41–46.**

EXAMPLE 7 The Value of a College Degree

In 2017, the average (median) salary for a full-time worker aged 25 or older who was a college graduate with a bachelor's degree (or higher) was about $67,000 per year, while the average (median) salary for a full-time worker with only a high school diploma was about $37,000 per year. Based on these data, how much more does the college graduate earn over a typical 40-year career? (*Source*: Data from U.S. Bureau of Labor Statistics.)

Solution The difference in the median incomes is

$$\$67{,}000 - \$37{,}000 = \$30{,}000$$

Therefore, over a 40-year career, the total difference is

$$40\ \text{yr} \times \frac{\$30{,}000}{\text{yr}} = \$1{,}200{,}000$$

Based on these data, the average college graduate can expect to earn about *$1.2 million more* over a career than the average high school graduate. Although this amount does not include the cost of college (or the "lost" earnings for not working while in college), it is still clear that going to college usually pays off in the long term. In addition, the unemployment rate for college graduates has generally been less than the unemployment rate for high school graduates (see Figure 5.11b in Unit 5D). ▶ **Now try Exercises 47–50.**

BY THE WAY
Remember that the earnings data in Example 7 are overall averages, and there are differences among college graduates based on majors and individual factors. The highest earnings generally go to students who major in high-demand fields such as mathematics, science, or engineering. And no matter what your major, on average you'll earn more if you study more and get better grades.

EXAMPLE 8 Cost of a College Class

Across all institutions, the average cost of a three-credit college class is approximately \$1500. Suppose that, between class time, commute time, and study time, the average class requires about 10 hours per week of your time. Assuming that you could have had a job paying \$10 per hour, what is the net cost of the class compared to working? Is it a worthwhile expense?

Solution A typical college semester lasts 14 weeks, so your "lost" work wages for the time you spend on the class come to

$$14 \text{ wk} \times \frac{10 \text{ hr}}{\text{wk}} \times \frac{\$10}{\text{hr}} = \$1400$$

We find your net cost for the class by adding this amount to the \$1500 that the class itself costs. The result is \$2900. Whether this expense is worthwhile is subjective, but remember that the average college graduate earns about \$1.2 million more over a career than the average high school graduate. And also remember that, on average, students who do better in college do even better in their career earnings. ▶ Now try Exercises 47–48.

Think About It Following up on Example 8, suppose that you are having difficulty in a particular class, but believe that you could raise your grade by cutting back on your work hours to allow more time for studying. How would you decide whether you should do this? Explain.

Insurance—Protection Against Unexpected Crises

No matter how well you plan your budget, an unexpected crisis can throw everything into disarray. The purpose of insurance is to help you mitigate the effects of such crises. The many types of insurance include the following: Health insurance is designed to help cover medical expenses; auto insurance may cover damage and liability related to car crashes; homeowners insurance may cover damage and liability related to your home; renters insurance may cover damage to your belongings; and life insurance may pay benefits to your family in the event of your death.

Some forms of insurance are mandatory. For example, you are required to have auto insurance if you own a car, and you are generally required to have homeowners insurance if you have a mortgage on your home. Other forms of insurance are optional, and you must carefully evaluate whether these are worth their costs. For example, life insurance can be very important if you support a family, but is probably unnecessary if you are young and single.

There are many factors to consider when buying insurance, but the bottom line is always a tradeoff between the costs of an insurance policy, its potential benefits, and the risks you believe you have (see Unit 7D for a more general discussion of risk). The three most common types of insurance costs are summarized in the box on the next page.

The benefit and risk sides are often very complicated, and you should research any insurance policy carefully before making a purchase decision. For example, be sure you consider all the following factors:

- *What does the policy cover?* Insurance nearly always has limitations. For example, health insurance usually covers treatment only from particular doctors and hospitals (those included are often said to be "in the network" for your policy), and policies can differ greatly in what types of preventive care they cover, whether they cover mental health treatments, and so on. Similarly, homeowners or renters insurance may cover damages from some disasters (such as fire) but not others (such as flood or earthquake). Before you purchase any policy, be sure you know exactly what is covered and what is not.

Insurance Costs

The overall cost of an insurance policy usually involves some combination of the following:

The **premium** is the amount you pay to purchase the policy. Premiums are often paid once or twice a year, though sometimes you may pay them more often.

A **deductible** is the amount you are personally responsible for before the insurance company will pay anything. For example, if your auto insurance policy has a $1000 deductible per incident, this means that you will pay the first $1000 for any damages from any crash.

A **co-payment** usually applies to health insurance and is the amount you pay each time you use a particular service that is covered by the insurance policy. For example, if you have $50 co-payment for office visits, you will pay $50 each time you visit a doctor's office on an insured visit.

BY THE WAY

In the United States, one of the great recent political controversies has concerned health insurance. Health insurance was made mandatory under the Affordable Care Act of 2010 (often called "Obamacare"), and this law also prevented exclusions for pre-existing conditions and mandated numerous minimums in terms of what would be covered. However, Congress and President Trump were trying to replace this law when this text was written in 2017, so be sure you understand any enacted changes and the impact they may have on you.

- *What are the policy's maximum benefits?* An insurance policy usually also specifies maximums that it will pay. For example, a renters insurance policy may cover only up to $1000 in losses, so if you lose more than that, you'll be responsible for the additional amount.
- *What exceptions lead to a lack of coverage?* Insurance policies often have exceptions that may leave you without coverage in some situations. For example, some health insurance policies may not cover "pre-existing conditions," meaning health problems that you had at the time you purchased the policy; many homeowners or renters insurance policies do not cover natural disasters such as earthquakes or floods; and life insurance policies usually will not pay benefits to your family if you die by suicide or as a result of engaging in risky activities.
- *What is your potential cost if you **don't** purchase coverage?* Once you understand the insurance policy options available to you, ask yourself what risk you take if you *don't* purchase insurance. For example, what could happen if you don't purchase health insurance and then have an unexpected medical problem or accident?

Only if you understand the costs and potential benefits of various types of insurance can you make an informed decision about whether to buy a particular type and which policy to choose.

EXAMPLE 9 Emergency Room Visit

Suppose you have an accident and end up in the emergency room, receiving a $7000 bill for your treatment. Fortunately, you have health insurance, but your policy has a $1000 annual deductible, a $250 co-payment for emergency room visits, and pays only 80% of the remaining balance. How much will you pay out of pocket for the emergency room visit? Assume that you haven't had any other medical expenses in the current year.

Solution Your total payment has three parts:

- The $1000 deductible, which you will pay in full since you have not already paid any of it earlier in the year
- The $250 co-payment for an emergency room visit
- Your share of the remaining balance. The total bill is $7000, but you've already paid $1250 (the $1000 deductible plus the $250 co-payment). Therefore, the remaining balance is $\$7000 - \$1250 = \$5750$. The insurance company pays 80% of this, so you owe the other 20%, which is $0.2 \times \$5750 = \$1150$.

Your total out-of-pocket cost is $\$1000 + \$250 + \$1150 = \$2400$.

▶ Now try Exercises 53–54.

EXAMPLE 10 Deductible Tradeoff

Suppose you are choosing between two health insurance policies. A low-deductible policy has premiums of \$250 per month and has a \$1000 annual deductible. A high-deductible policy has premiums of only \$150 per month but has a \$6000 annual deductible. Assume that the two plans are otherwise identical. How would you evaluate the cost difference between the policies?

Solution We analyze the differences in the plans:

- The low-deductible policy has higher monthly premiums than the high-deductible policy by $\$250 - \$150 = \$100$ per month, or \$1200 per year. Therefore, if you don't use any health care services, the low-deductible policy costs you \$1200 more than the higher deductible policy.
- With the low-deductible policy, you'll pay the first \$1000 of health care costs you incur during a year, but no more for covered services.
- With the high-deductible policy, you'll pay the first \$6000 of health care costs you incur during a year, which means you could potentially incur costs up to \$5000 higher than you would with the low-deductible policy.

Putting these ideas together, the trade-off is clear. If you are fortunate and don't end up with any major health care expenses, the high-deductible policy will save you \$1200 in premiums. However, if you end up with significant health care expenses, the deductible difference could mean as much as \$5000 more out of your pocket with the high-deductible policy. In other words, you must decide whether it's worth saving \$1200 in premiums and facing the risk of an extra \$5000 in out-of-pocket expenses. ▶ Now try Exercises 55–56.

Base Financial Goals on Solid Understanding

It's rare for a financial question to have a clear "best" answer for everyone. Instead, your decisions depend on your current circumstances, your goals for the future, and some unavoidable uncertainty. The key to financial success is to approach all your financial decisions with a clear understanding of the available choices.

In the rest of this chapter, we'll study several crucial topics in finance to help you build the understanding needed to reach your financial goals. To prepare yourself for this study, it's worth taking a few moments to think about the impact that each of these topics will have on your financial life:

- Achieving your financial goals will almost certainly require that you build up savings over time. Although it may be difficult to save while you are in college, ultimately you must find a way to save. You will also need to understand how savings work and how to choose appropriate savings plans and investments. These are the topics of Units 4B and 4C.
- You will probably need to borrow money at various points in your life. You may already have credit cards, or you may be using student loans to help pay for college. In the future, you are likely to need a loan to make a large purchase, such as a car or a home. Because borrowing is very expensive, it's critical that you understand the basic mathematics of loans so you can make wise choices; this is the topic of Unit 4D.
- Many of our financial decisions have consequences for the taxes we pay. Sometimes, these tax consequences can be large enough to influence our decisions. For example, the fact that interest on house payments is sometimes tax deductible while rent is not may influence your decision to rent or buy. While no one can expect to understand tax law fully, it's important to have at least a basic understanding of how taxes are computed and how they can affect your financial decisions; this is the topic of Unit 4E.

- Finally, we do not live in isolation, and our personal finances are inevitably intertwined with those of the government. For example, when politicians allow the government to run deficits, it means that future politicians will have to collect more tax dollars from you or your children. Unit 4F focuses on the federal budget and explains what it may mean for your future.

Quick Quiz 4A

Choose the best answer to each of the following questions. Explain your reasoning with one or more complete sentences.

1. By evaluating your monthly budget, you can learn how to
   a. keep your personal spending under control.
   b. make better investments.
   c. earn more money.
2. The two things you must keep track of in order to understand your budget are
   a. your income and your spending.
   b. your wages and your bank interest.
   c. your wages and your credit card debt.
3. A *negative* monthly cash flow means that
   a. your investments are losing value.
   b. you are spending more money than you are taking in.
   c. you are taking in more money than you are spending.
4. When you are making your monthly budget, how should you handle your once-a-year expenses for holiday gifts in December?
   a. Ignore them.
   b. Include them only in your calculation for December's budget.
   c. Divide them by 12 and include them as a monthly expense.
5. For the average person, the single biggest category of expense is
   a. food. b. housing. c. entertainment.
6. According to Figure 4.1, which of the following expenses tends to increase the most as a person ages?
   a. housing b. transportation c. health care
7. Which of the following is *necessary* if you want to make monthly contributions to savings?
   a. You must have a positive monthly cash flow.
   b. You must be spending less than 20% of your income on food and clothing.
   c. You must not owe money on any loans.
8. Sandy's automobile insurance policy has an annual deductible of \$500. Her bill for a recent collision repair, which was her first insurance claim of the year, was \$850. How much did Sandy pay out-of-pocket for the repair?
   a. \$500 b. \$350
   c. \$0 (Her insurance covered the entire bill.)
9. Suppose you have a health insurance policy with an annual premium of \$4800, an annual deductible of \$1000, and co-payments of \$25 for visits to doctors' offices. Suppose you go through the year with no medical bills at all. What is your total cost for the year?
   a. \$4800 b. \$5800 c. \$5825
10. Thomas receives a bill for \$2700 for an out-patient procedure at a local clinic. His health insurance has a \$1500 deductible and a \$200 co-payment for the procedure. The insurance company pays 80% of the remaining balance. How much of the cost does the insurance company cover?
    a. \$2700 b. \$1500 c. \$800

Exercises 4A

REVIEW QUESTIONS

1. Why is it so important to understand your personal finances? What types of problems are more common among people who do not have their finances under control?
2. List four crucial things you should do if you want to keep your finances under control, and describe how you can achieve each one.
3. What is a *budget*? What is *cash flow*? Describe the four-step process of figuring out your monthly budget.
4. Summarize how average spending patterns change with age. How can comparing your own spending to average spending patterns help you evaluate your budget?
5. Distinguish among *premiums*, *deductibles*, and *co-payments* as costs of insurance polices. Then list a few factors you should consider when evaluating the benefits of an insurance policy.
6. What items should you include when calculating how much it costs you to attend college? How can you decide whether this is a worthwhile expense?

DOES IT MAKE SENSE?

Decide whether each of the following statements makes sense (or is clearly true) or does not make sense (or is clearly false). Explain your reasoning.

7. When I figured out my monthly budget, I included only my rent and my spending on gasoline, because nothing else could possibly add up to much.
8. My monthly cash flow was −$150, which explained why my credit card debt kept rising.
9. My vacation cost a total of $1800, which I entered into my monthly budget as $150 per month.
10. Emma and Emily are good friends who do everything together, spending the same amounts on eating out, entertainment, and other leisure activities. Yet Emma has a negative monthly cash flow while Emily's is positive, because Emily has more income.
11. Brandon discovered that his daily routine of buying a slice of pizza and a soda at lunch was costing him more than $15,000 per year.
12. I bought the health insurance policy with the lowest monthly premiums, because that's sure to be the best option for my long-term financial success.

BASIC SKILLS & CONCEPTS

13–20: Extravagant Spending? Compute the total cost per year of the following pairs of expenses. Then complete this sentence for each pair: On an *annual* basis, the first expense is ______% of the second expense.

13. Maria spends $25 every week on coffee and $150 per month on food.
14. Jeremy buys *The New York Times* from a newsstand for $2.50 a day (skipping Sundays) and spends $30 per week on gasoline for his car. (Assume that there are 52 weeks in a year.)
15. Suzanne's cell phone bill is $85 per month, and she spends $200 per year on student health insurance.
16. Homer spends $10 per week on lottery tickets and $600 per month for rent.
17. Sheryl spends $12 every week on cigarettes and spends $40 a month on dry cleaning.
18. Ted goes to a club or concert every two weeks and spends an average of $60 each time; he spends $500 a year on car insurance.
19. Vern drinks three 6-packs of beer each week at a cost of $9 each and spends $800 per year on his textbooks.
20. Sandy fills the gas tank of her car an average of once every two weeks at a cost of $35 per tank; her cable TV/Internet service costs $80 per month.

21–24: Interest Payments. Find the *monthly* interest payment in each of the following situations. When an annual rate is given, assume that monthly interest rates are 1/12 of annual interest rates.

21. You maintain an average balance of $750 on your credit card, which carries an 18% annual interest rate.
22. Brooke's credit card has an annual interest rate of 21% on her unpaid balance, which averages $900.
23. Sam bought a new TV for $2800. He made a down payment of $400 and then financed the balance through the store. Unfortunately, he was unable to make the first monthly payment and now pays 3% interest per month on the balance (while he watches his TV).
24. Veronica owes a clothing store $700, but until she makes a payment, she pays 9% interest per month.

25–30: Prorating Expenses. Prorate the following expenses and find the corresponding *monthly* expense.

25. During one year, Riley pays $6500 for tuition and fees, plus $600 for textbooks, for each of two semesters.
26. During one year, Saul takes 15 credit-hours for each of three quarters. Tuition and fees amount to $600 per credit-hour. Textbooks average $350 per quarter.
27. Talib pays a semiannual premium of $750 for automobile insurance, a monthly premium of $150 for health insurance, and an annual premium of $500 for life insurance.
28. Xan pays $500 per month in rent, a semiannual car insurance premium of $800, and an annual health club membership fee of $900.
29. In filing his income tax, Finn reported annual contributions of $250 to a public radio station, $300 to a public TV station, $150 to a local food bank, and $450 to other charitable organizations.
30. Melinda spends an average of $35 per week on gasoline and $50 every three months on a daily newspaper.

31–34: Net Cash Flow. The following tables show expenses and income for various individuals. Find each person's net monthly cash flow (it may be negative or positive). Assume that amounts shown for salaries and wages are after taxes and that 1 month = 4 weeks.

31.

Income	Expenses
Part-time job: $650/month	Rent: $500/month
College fund from grandparents: $400/month	Groceries: $60/week
Scholarship: $6000/year	Tuition and fees: $3600 twice a year
	Incidentals: $120/week

32.

Income	Expenses
Part-time job: $1200/month	Rent: $600/month
Student loan: $7000/year	Groceries: $70/week
Scholarship: $8000/year	Tuition and fees: $7500/year
	Health insurance: $40/month
	Entertainment: $200/month
	Phone: $65/month

33.

Income	Expenses
Salary: $1900/month	Rent: $800/month
	Groceries: $90/week
	Utilities: $125/month
	Health insurance: $150/month
	Car insurance: $420 twice a year
	Gasoline: $25/week
	Miscellaneous: $400/month

34.

Income	Expenses
Salary: $36,000/year	House payments: $800/month
Pottery sales: $200/ month	Groceries: $150/week
	Household expenses: $400/month
	Health insurance: $250/month
	Car insurance: $600 twice a year
	Savings plan: $200/month
	Donations: $600/year
	Miscellaneous: $800/month

35–40: Budget Allocations. Determine whether the following spending patterns are equal to, above, or below the national averages given in Figure 4.1. Assume that amounts given for salaries and wages are after taxes.

35. A single 30-year-old woman with a monthly salary of $3600 spends $900 per month on rent.

36. A couple under the age of 30 has a combined household income of $4400 per month and spends $400 per month on entertainment.

37. A single 42-year-old man with a monthly salary of $4200 spends $200 per month on health care.

38. A 32-year-old couple with a combined household income of $45,500 per year spends $700 per month on transportation.

39. A retired couple (over 65 years old) with a fixed monthly income of $5200 spends $800 per month on health care.

40. A family with a 45-year-old wage earner has an annual household income of $50,000 and spends $2500 per month on housing.

41–46: Making Decisions. Consider the following situations, which each involve two options. Determine which option is less expensive. Are there unstated factors that might affect your decision?

41. You currently drive 250 miles per week in a car that gets 21 miles per gallon of gas. You are considering buying a new fuel-efficient car for $16,000 (after the trade-in allowance on your current car), which gets 45 miles per gallon. Insurance premiums for the new and old cars are $800 and $400 per year, respectively. You anticipate spending $1500 per year on repairs to the old car and having no repairs on the new car. Assume that gas costs $3.50 per gallon. Over a 5-year period, is it less expensive to keep your old car or buy the new car?

42. You currently drive 300 miles per week in a car that gets 15 miles per gallon of gas. You are considering buying a new fuel-efficient car for $12,000 (after the trade-in allowance on your current car), which gets 50 miles per gallon. Insurance premiums for the new and old cars are $800 and $600 per year, respectively. You anticipate spending $1200 per year on repairs to the old car and having no repairs on the new car. Assume that gas costs $3.50 per gallon. Over a 5-year period, is it less expensive to keep your old car or buy the new car?

43. You must decide whether to buy a new car for $22,000 or lease that car for a 3-year period. Under the terms of the lease, you make a down payment of $1000 and have monthly payments of $250. At the end of 3 years, the leased car has a residual value (the amount you will pay if you choose to buy the car at the end of the lease period) of $10,000. Assume that if you buy the car, you can sell it at the end of 3 years for the same amount as the residual value. Is it less expensive to buy or to lease?

44. You must decide whether to buy a new car for $22,000 or lease that car for a 4-year period. Under the terms of the lease, you make a down payment of $1000 and have monthly payments of $300. At the end of 4 years, the leased car has a residual value (the amount you will pay if you choose to buy the car at the end of the lease period) of $8000. Assume that if you buy the car, you can sell it at the end of 4 years for the same amount as the residual value. Is it less expensive to buy or to lease?

45. You have a choice between going to an in-state college where you will pay $4000 per year for tuition and an out-of-state college where the tuition is $6500 per year. The cost of living is much higher at the in-state college; you can expect to pay $700 per month in rent there, compared to $450 per month at the other college. Assuming all other factors are equal, which is the less expensive choice on an annual (12-month) basis?

46. If you stay in your home town, you can go to Concord College at a reduced tuition of $3000 per year and pay $800 per month in rent. Or you can leave home, go to Versalia College with a $10,000 scholarship (per year), pay $16,000 per year in tuition, and pay $350 per month to live in a dormitory. You will pay $2000 per year to travel back and forth from Versalia College. Assuming all other factors are equal, which is the less expensive choice on an annual (12-month) basis?

47–50. Value of Education. The following table shows median monthly earnings (in 2015) for women and men by their highest achieved level of education.

	Educational Attainment			
	High school graduate	Associate's degree	Bachelor's degree	Advanced degree
Women	$6936	$7932	$11,580	$14,220
Men	$9012	$10,464	$14,988	$19,560

Source: U.S. Bureau of Labor Statistics.

47. Assuming the difference shown in the table remains constant over a 40-year career, approximately how much more does a man with a bachelor's degree earn than a man with only a high school education?

48. Assuming the difference shown in the table remains constant over a 40-year career, approximately how much less does a woman with a bachelor's degree earn than a woman with an advanced degree?

49. As a percentage, how much more does a man with a bachelor's degree earn than a woman with a bachelor's degree? Assuming that difference remains constant over a 40-year career, how much more does the man earn than the woman?

50. As a percentage, how much more does a man with an advanced degree earn than a woman with an advanced degree? Assuming that difference remains constant over a 40-year career, how much more does the man earn than the woman?

51–52. Choices. Consider the following pairs of options and answer the questions that follow.

51. You could take a 15-week, three-credit college course, which requires 10 hours per week of your time and costs $500 per credit-hour in tuition. Or you could work during those hours at a job paying $10 per hour. What is the net cost of the class

compared to working? Based on your answer and the fact that the average college graduate earns nearly $28,000 per year more than the average high school graduate, write a few sentences giving your opinion as to whether the college course is a worthwhile expense.

52. You could have a part-time job (20 hours per week) that pays $15 per hour, or you could have a full-time job (40 hours per week) that pays $12 per hour. Because of the extra free time, you will spend $150 per week more on entertainment with the part-time job than with the full-time job. After accounting for the extra entertainment expense, how much more is your weekly cash flow with the full-time job than with the part-time job? Neglect taxes and other expenses.

53. Assume your automobile insurance policy has a semiannual (twice a year) premium of $650 and carries a $1000 annual deductible. In one particularly unfortunate year, you have collision repair bills of $1120 and $1660.

a. What is your total out-of-pocket expense for insurance and repairs for the year?

b. What would your total expense have been if you did not have the insurance policy?

54. According to one estimate, the average cost of having a baby (prenatal care, routine delivery, and post-partum care) in the United States is $8800. Suppose you have health insurance with an annual premium of $2100, a deductible of $1500 (for the year), and a one-time co-payment of $250 for services relating to the birth of a baby. What is the out-of-pocket cost of having a baby?

55. You have a choice of two health insurance plans:

- Plan A (low deductible): Annual premium is $5400, deductible is $400 (for the year), and there are no co-payments. The insurance plan covers 100% of all expenses after the deductible is met.
- Plan B (high deductible): Annual premium is $1500, deductible is $2000 (for the year), and the co-payments are $100 for an emergency room visit, $100 for radiology procedures such as CT scans, and $30 for an office visit. The insurance covers 80% of all expenses after the deductible is met and your co-payments are subtracted.

Find the out-of-pocket expenses under each plan assuming that, in one year, you have one emergency room visit that costs $3500, one annual check-up that costs $300, and one CT scan that costs $700.

56. **Health Care Choices.** You have a choice of two health insurance plans with the following terms. Assume that both plans cover 100% of expenses after the deductible and co-payments.

Plan A	Plan B
Monthly premium: $300	Monthly premium: $700
Annual deductible: $5000	Annual deductible: $1500
Office visit co-payment: $25	Office visit co-payment: $25
Emergency room co-payment: $500	Emergency room co-payment: $200
Surgical operations co-payment: $250	Surgical operations: no co-payment

Suppose that during a one-year period your family received the following medical bills.

Service	Total cost (before insurance)
Jan. 23: Emergency room	$800
Feb. 14: Office visit	$100
Apr. 13: Surgery	$1400
June 14: Surgery	$7500
July 1: Office visit	$100
Sept. 23: Emergency room	$1200

a. Determine your annual health care expenses if you have Plan A.

b. Determine your annual health care expenses if you have Plan B.

c. Determine your health care expenses for the year if you have no health insurance. Does having no health insurance create any other risks for you? Explain.

FURTHER APPLICATIONS

57. **Laundry Upgrade.** Suppose that you currently own a clothes dryer that costs $25 per month to operate (for electricity usage). A new, more efficient dryer costs $620 and has an estimated operating cost of $15 per month. How long will it take for the new dryer to pay for itself?

58. **Solar Payback Period.** Julie is considering installing solar photovoltaic panels on the roof of her house. Her monthly electricity bills currently average $85. The cost of installing a photovoltaic system is $18,400; however, she expects a 40% reduction in this cost due to tax credits and local rebates. Assuming all of her electrical needs are met by the new system and neglecting possible revenue when the system puts electricity back into the grid, what is the approximate payback period for the photovoltaic system?

59. **Insurance Deductibles.** In each of the following cases, determine how much you would pay with and without the insurance policy.

a. You have a car insurance policy with a $500 deductible (per claim) for collision damage. During a two-year period, you file claims for $450 and $925. The annual premium for the policy is $550.

b. You have a car insurance policy with a $200 deductible (per claim) for collision damage. During a two-year period, you file claims for $450 and $1200. The annual premium for the policy is $650.

c. You have a car insurance policy with a $1000 deductible (per claim) for collision damage. During a two-year period, you file claims for $200 and $1500. The annual premium for the policy is $300.

d. Explain why lower insurance premiums go with higher deductibles.

60. **The Real Cost of Car Ownership.** Suppose you take out a $0 down payment loan to buy a car. Your monthly payments on the loan are $260. In addition to the loan payments, you must pay auto insurance with a semiannual (twice a year)

premium of $480, an annual registration fee of $240, an estimated $50 per month for gas and maintenance, and $30 per month for a campus parking permit.

a. What are the monthly costs of owning the car? What percentage of the total monthly cost is the loan payment?

b. Suppose that, as part of your student fees, you can buy a bus pass that covers all your travel to and from campus. The cost of the bus pass is $90 for each of two semesters. How much would you save per month during a 9-month school year with the bus pass?

61. **Car Leases.** Consider the following three lease options for a new car. Determine which lease is least expensive, assuming that you buy the car when the lease expires. (The residual value is the car's price at that time.) Discuss other factors that might affect your decision.

- Plan A: $1000 down payment, $400 per month for 2 years, residual value = $10,000
- Plan B: $500 down payment, $250 per month for 3 years, residual value = $9,500
- Plan C: $0 down payment, $175 per month for 4 years, residual value = $8,000

62. **Health Costs.** Assume that you have a (relatively simple) health insurance plan with the following provisions:

- Annual deductible is $500. Insurance company pays 100% of all costs after deductible and co-payments.
- Office visits require a co-payment of $25.
- Emergency room visits have a $200 co-payment.
- Surgical operations have a $1000 co-payment.
- You pay a monthly premium of $350.

During a one-year period, your family has the following expenses.

Service	Total cost (before insurance)
Feb. 18: Office visit	$100
Mar. 26: Emergency room	$580
Apr. 23: Office visit	$100
May 14: Surgery	$6500
July 1: Office visit	$100
Sept. 23: Emergency room	$840

a. Determine your health care expenses for the year with the insurance policy.

b. Determine your health care expenses for the year if you did not have the insurance policy.

63–66: Personal Finances. The following exercises involve analyses of your personal income and expenses. (*Note*: If you do not wish to reveal your personal finances in work that you turn in, you may use hypothetical values for income and expenses, but be sure to state that you are doing so and clearly explain the assumptions behind your values, such as whether they are for an average college student at your school, a single parent taking a college class, or some other hypothetical student.)

63. **Daily Expenditures.** Keep a list of everything you spend money on during one entire day. Categorize each expenditure, and then make a table with one column for the categories and one column for the expenditures. Add a third column in which you compute how much you'd spend in a year if you spent the same amount every day.

64. **Weekly Expenditures.** Repeat Exercise 63, but this time make the list for a full week of spending rather than just one day.

65. **Prorated Expenditures.** Make a list of all the major expenses you have each year that you do not pay on a monthly basis, such as college expenses, holiday expenses, and vacation expenses. For each item, estimate the amount you spend in a year, and then determine the prorated amount that you should use when you determine your monthly budget.

66. **Monthly Cash Flow.** Create your complete monthly budget, listing all sources of income and all expenditures, and use it to determine your net monthly cash flow. Be sure to include small but frequent expenditures and prorated amounts for large expenditures. Explain any assumptions you make in creating your budget. When the budget is complete, write a paragraph or two explaining what you learned about your own spending patterns and what adjustments you may need to make to your budget.

IN YOUR WORLD

67. **Personal Budgets.** Many apps and websites provide personal budget advice and worksheets. Find one to help you organize your budget for at least two months. Is it effective in helping you plan your finances? Discuss how tracking your budget led to insights that you would not have had otherwise.

68. **Your Health Insurance.** What kind of health insurance do you have? Analyze its costs and benefits, then compare them to what you would have with at least two other health insurance options that are available to you. Based on your findings, decide whether you should keep or change your current health insurance.

69. **U.S. Health Insurance.** What is the current status of federal law in the United States regarding health insurance? Has the law changed since the Affordable Care Act was passed in 2010? If so, explain how. Also discuss any efforts to make changes to current laws.

70. **Personal Bankruptcies.** The rate of personal bankruptcies has risen in recent years. Find at least three news articles on the subject, document the increase in bankruptcies, and explain the primary reasons for the increase.

71. **Consumer Debt.** Find data on the increase in consumer (credit card) debt in the United States. Based on your findings, do you think the extent of consumer debt is (a) a crisis, (b) a significant occurrence but nothing to worry about, or (c) a good thing? Justify your conclusion.

72. **U.S. Savings Rate.** When it comes to saving disposable income, Americans have a remarkably low savings rate. Find sources that compare the savings rates of Asian and European countries to that of the United States. Discuss your observations and put your own savings habits on the scale.

UNIT 4B The Power of Compounding

BY THE WAY

As a compromise, the New College administrator suggested reducing the annual interest rate from 4% to 2%, in which case the college was owed only about $8.9 million. This, he said, would help with a modernization project at the College. The royal family has not yet paid.

On July 18, 1461, King Edward IV of England borrowed the modern equivalent of $384 from New College of Oxford. The King soon paid back $160, but never repaid the remaining $224. The debt was forgotten for 535 years. Upon its rediscovery in 1996, a New College administrator wrote to the Queen of England asking for repayment, with interest. Assuming an interest rate of 4% per year, the administrator calculated that the college was owed $290 billion.

Unfortunately for New College, there was no clear record of a promise to repay the debt with interest, and even if there were, it might be difficult to demand payment of a debt that had been forgotten for more than 500 years. But this example still illustrates what is sometimes called the "power of compounding": the remarkable way that money grows when interest continues to accumulate year after year.

Simple vs. Compound Interest

Imagine that you deposit $1000 in Honest John's Money Holding Service, which promises to pay 5% interest each year. At the end of the first year, Honest John's sends you a check for

$$5\% \times \$1000 = 0.05 \times \$1000 = \$50$$

Because you receive a check for your interest, your balance with Honest John's remains $1000, so you get the same $50 payment at the end of the second year, and the same again for the third year. Your total interest for the three years is

$$3 \times \$50 = \$150$$

Therefore, your original investment of $1000 grew to $1150. Honest John's method of payment represents **simple interest**, in which interest is paid only on your initial investment. More generally, the amount of money on which interest is paid is called the **principal**.

Now, suppose that you place the $1000 in a bank account that pays the same 5% interest once a year. But instead of paying you the interest directly, the bank adds the interest to your account. At the end of the first year, the bank deposits $5\% \times \$1000 = \$50$ interest into your account, raising your balance to $1050. At the end of the second year, the bank again pays you 5% interest. This time, however, the 5% interest is paid on the balance of $1050, so it amounts to

$$5\% \times \$1050 = 0.05 \times \$1050 = \$52.50$$

Adding this $52.50 raises your balance to

$$\$1050 + \$52.50 = \$1102.50$$

This is the new balance on which your 5% interest is computed at the end of the third year, so your third interest payment is

$$5\% \times \$1102.50 = 0.05 \times \$1102.50 = \$55.13$$

Therefore, your balance at the end of the third year is

$$\$1102.50 + \$55.13 = \$1157.63$$

Despite identical interest rates, you end up with $7.63 more if you use the bank instead of Honest John's. The difference comes about because the bank pays you *interest on the interest* as well as on the original principal. This type of interest payment is called **compound interest**.

Definitions

The **principal** in financial formulas is the balance on which interest is paid.

Simple interest is interest paid only on the original investment and not on any interest added at later dates.

Compound interest is interest paid both on the original investment and on all interest that has been added to the original investment.

EXAMPLE 1 Savings Bond

While banks almost always pay compound interest, bond issuers usually pay simple interest. Suppose you invest \$1000 in a savings bond that pays simple interest of 10% per year. How much total interest will you receive in 5 years? If the bond paid compound interest, would you receive more or less total interest? Explain.

Brief Review Powers and Roots

We have already reviewed powers of 10 (see the Brief Review, p. 92). Now, for our work with financial formulas, we review powers and roots more generally.

Basics of Powers

A number raised to the *n*th power is that number multiplied by itself *n* times (*n* is called an *exponent*). For example:

$$2^1 = 2 \qquad 2^2 = 2 \times 2 = 4 \qquad 2^3 = 2 \times 2 \times 2 = 8$$

A number raised to the *zero* power is defined to be 1. For example:

$$2^0 = 1$$

Negative powers are the reciprocals of the corresponding positive powers. For example:

$$5^{-2} = \frac{1}{5^2} = \frac{1}{5 \times 5} = \frac{1}{25} \qquad 2^{-3} = \frac{1}{2^3} = \frac{1}{2 \times 2 \times 2} = \frac{1}{8}$$

Power Rules

In the following rules, *x* represents a number being raised to a power, and *n* and *m* are exponents. Note that these rules work only when all powers apply to the same number *a*.

- To multiply powers of the same number, *add* the exponents:

$$a^n \times a^m = a^{n+m}$$

$$\textit{Example: } 2^3 \times 2^2 = 2^{3+2} = 2^5 = 32$$

- To divide powers of the same number, *subtract* the exponents:

$$\frac{a^n}{a^m} = a^{n-m}$$

$$\textit{Example: } \frac{5^3}{5^2} = 5^{3-2} = 5^1 = 5$$

- When a power is raised to another power, *multiply* the exponents:

$$(a^n)^m = a^{n \times m}$$

$$\textit{Example: } (2^2)^3 = 2^{2\times 3} = 2^6 = 64$$

Basics of Roots

Finding a root is the reverse of raising a number to a power. Second roots, or square roots, are written with a number under the root symbol $\sqrt{\ }$. More generally, we indicate an *n*th root by writing a number under the symbol $\sqrt[n]{\ }$. For example:

$$\sqrt{4} = 2 \text{ because } 2^2 = 2 \times 2 = 4$$

$$\sqrt[3]{27} = 3 \text{ because } 3^3 = 3 \times 3 \times 3 = 27$$

$$\sqrt[4]{16} = 2 \text{ because } 2^4 = 2 \times 2 \times 2 \times 2 = 16$$

$$\sqrt[6]{1{,}000{,}000} = 10 \text{ because } 10^6 = 1{,}000{,}000$$

Roots as Fractional Powers

The *n*th root of a number is the same as the number raised to the $1/n$ power. That is,

$$x^{1/n} = \sqrt[n]{x}$$

For example:

$$64^{1/3} = \sqrt[3]{64} = 4$$

$$1{,}000{,}000^{1/6} = \sqrt[6]{1{,}000{,}000} = 10$$

▶ **Now try Exercises 15–26.**

Solution With simple interest, every year you receive the same interest payment: $10\% \times \$1000 = \$100$. Therefore, you receive a total of \$500 in interest over 5 years. With compound interest, you receive *more* than \$500 in interest because the interest each year is calculated on your growing balance rather than on your original investment. For example, because your first interest payment of \$100 raises your balance to \$1100, your next compound interest payment is $10\% \times \$1100 = \$110$, which is more than the simple interest payment of \$100. For the same interest rate, compound interest always raises your balance faster than simple interest. ▶ **Now try Exercises 51–54.**

The Compound Interest Formula

Let's return to King Edward's debt to the New College. We can calculate the amount owed to the college by pretending that the \$224 the king failed to pay back promptly was deposited into an interest-bearing account for 535 years. Let's assume, as the New College administrator did, that the interest rate was 4% per year. For each year, we can calculate the interest and the new balance with interest. The first three columns of Table 4.1 show these calculations for 4 years.

To find the total balance, we could continue the calculations to 535 years. Fortunately, there's a much easier way. The 4% annual interest rate means that each end-of-year balance is 104% of, or 1.04 times, the previous year's balance. As shown in the last column of Table 4.1, we can get each balance in a single step as follows:

- The balance at the end of 1 year is the initial deposit of \$224 times 1.04:

$$\$224 \times 1.04 = \$232.96$$

- The balance at the end of 2 years is the 1-year balance times 1.04, which is equal to the initial deposit times $(1.04)^2$:

$$\$224 \times 1.04 \times 1.04 = \$224 \times (1.04)^2 = \$242.28$$

- The balance at the end of 3 years is the 2-year balance times 1.04, which is equal to the initial deposit times $(1.04)^3$:

$$\$224 \times 1.04 \times 1.04 \times 1.04 = \$224 \times (1.04)^3 = \$251.97$$

Continuing the pattern, we find that the balance after Y years is the initial deposit times 1.04 raised to the Yth power. For example, the balance after $Y = 10$ years is

$$\$224 \times (1.04)^{10} = \$331.57$$

We can generalize this result by looking carefully at the previous equation. Notice that \$224 is the initial amount, which we will refer to as the **starting principal**, or P. The 1.04 is 1 plus the interest rate of 4%, or 0.04. The exponent 10 is the number of times that the interest has been compounded. Let's write the equation again, adding these identifiers and turning it around to put the result on the left:

$$\underbrace{\$331.57}_{\text{accumulated balance, } A} = \underbrace{\$224}_{\text{starting principal, } P} \times \underbrace{(1.04)}_{1 + \text{interest rate}}{}^{10} \leftarrow \text{number of compounding periods}$$

USING TECHNOLOGY

Powers

Most calculators have a special key labeled y^x or ^ for raising a number to any power. Most spreadsheets and computer programs use the ^ symbol. For example, to calculate 1.04^{535}, you enter 1.04^535. The screen shot below shows the calculation in Excel, with the answer in cell A1.

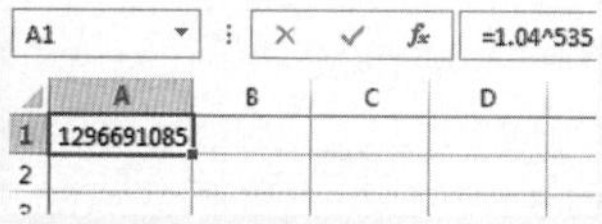

Microsoft Excel 2016, Windows 10, Microsoft Corporation.

When interest is compounded just once a year, as it is in this case, the interest rate is called the **annual percentage rate**, or **APR**. The number of compounding periods is

TABLE 4.1 Calculating Compound Interest (starting principal, P = \$224; annual interest rate, APR = 4%)

After N Years	Interest	Balance	Single-Step Calculation of Balance
1 year	$4\% \times \$224 = \$8.96$	$\$224 + \$8.96 = \$232.96$	$\$224 \times 1.04 = \$232.96$
2 years	$4\% \times \$232.96 = \$9.32$	$\$232.96 + \$9.32 = \$242.28$	$\$224 \times (1.04)^2 = \$242.28$
3 years	$4\% \times \$242.28 = \$9.69$	$\$242.28 + \$9.69 = \$251.97$	$\$224 \times (1.04)^3 = \$251.97$
4 years	$4\% \times \$251.97 = \$10.08$	$\$251.97 + \$10.08 = \$262.05$	$\$224 \times (1.04)^4 = \$262.05$

then simply the number of years Y over which the principal earns interest. We therefore obtain the following general formula for interest compounded once a year.

The Compound Interest Formula (for Interest Paid Once a Year)

$$A = P \times (1 + \text{APR})^Y$$

where
A = accumulated balance after Y years
P = starting principal
APR = annual percentage rate (as a decimal)
Y = number of years

Notes: (1) The starting principal, P, is often called the **present value** (PV), because we usually begin a calculation with the amount of money in an account at present. (2) The accumulated balance, A, is often called the **future value** (FV), because it is the amount that will be accumulated at some time in the future. (3) When using this formula, you must express the APR as a decimal rather than as a percentage.

Technical Note
For the more general case in which the interest rate is not necessarily compounded on an annual (APR) basis, the compound interest formula is written

$$A = P \times (1 + i)^N$$

where i is the interest rate and N is the total number of compounding periods.

In the New College case, the annual interest rate is APR = 4% = 0.04, and interest is paid over a total of 535 years. The accumulated balance after Y = 535 years is

$$\begin{aligned} A &= P \times (1 + \text{APR})^Y \\ &= \$224 \times (1 + 0.04)^{535} \\ &= \$224 \times (1.04)^{535} \\ &= \$224 \times 1{,}296{,}691{,}085 \\ &\approx \$2.9 \times 10^{11} = \$290 \text{ billion} \end{aligned}$$

As the administrator claimed, a 4% interest rate for 535 years would make the original \$224 debt grow to \$290 billion.

EXAMPLE 2 Simple and Compound Interest

You invest \$100 in two accounts that each pay an interest rate of 10% per year, but one pays simple interest and the other pays compound interest. Make a table to show the growth of each account over a 5-year period. Use the compound interest formula to verify the amount shown in the table after 5 years for the compound interest account.

Solution The simple interest is the same amount each year: 10% × \$100 = \$10. The compound interest grows from year to year, because it is paid on the accumulated interest as well as on the starting principal. Table 4.2 summarizes the calculations.

TABLE 4.2 Calculations for Example 2 (starting principal, P = \$100; annual interest rate, APR = 10%)

End of Year	Simple Interest Account: Interest Paid	Simple Interest Account: Old Balance + Interest = New Balance	Compound Interest Account: Interest Paid	Compound Interest Account: Old Balance + Interest = New Balance
1	10% × \$100 = \$10	\$100 + \$10 = \$110	10% × \$100 = \$10	\$100 + \$10 = \$110
2	10% × \$100 = \$10	\$110 + \$10 = \$120	10% × \$110 = \$11	\$110 + \$11 = \$121
3	10% × \$100 = \$10	\$120 + \$10 = \$130	10% × \$121 = \$12.10	\$121 + \$12.10 = \$133.10
4	10% × \$100 = \$10	\$130 + \$10 = \$140	10% × \$133.10 = \$13.31	\$133.10 + \$13.31 = \$146.41
5	10% × \$100 = \$10	\$140 + \$10 = \$150	10% × \$146.41 = \$14.64	\$146.41 + \$14.64 = \$161.05

To verify the final entry in the table with the compound interest formula, we use the starting principal $P = \$100$ and the annual interest rate $\text{APR} = 10\% = 0.1$, with interest paid for $Y = 5$ years. The accumulated balance A is

$$\begin{aligned} A &= P \times (1 + \text{APR})^Y \\ &= \$100 \times (1 + 0.1)^5 \\ &= \$100 \times 1.1^5 \\ &= \$100 \times 1.6105 \\ &= \$161.05 \end{aligned}$$

▲ **CAUTION!** As shown here, be sure you enter the APR into the interest formula as a decimal (0.1) rather than as a percentage (10%). ▲

This result agrees with the amount in the table. Overall, the account paying compound interest builds to \$161.05, while the simple interest account reaches only \$150, even though both pay interest at the same 10% rate. Although the 10% interest rate that we assumed here is quite high compared to what most banks pay, the basic point should be clear: For the same interest rate, compound interest is always better for the investor than simple interest.

▶ **Now try Exercises 55–56.**

Compound Interest as Exponential Growth

The New College case demonstrates the remarkable way in which money can grow with compound interest. Figure 4.2 shows how the value of the New College debt rises during the first 100 years, assuming a starting value of \$224 and an interest rate of 4% per year. Note that while the value rises slowly at first, it rapidly accelerates; so in later years the value grows by much more each year than it did during earlier years.

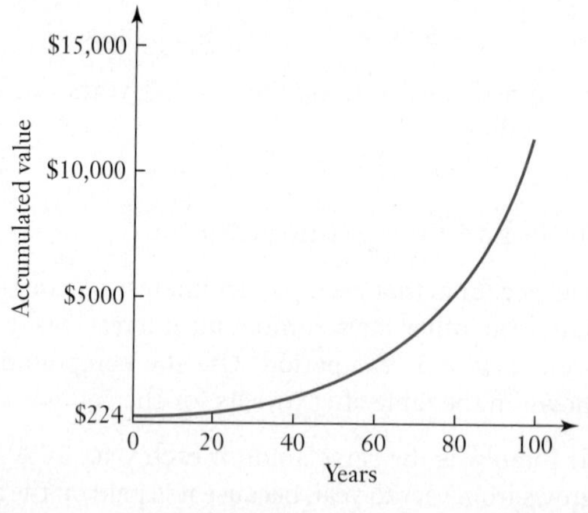

FIGURE 4.2 The value of the debt in the New College case during the first 100 years, at an interest rate of 4% per year. Note that the value rises much more rapidly in later years than in earlier years—a hallmark of exponential growth.

This rapid growth is a hallmark of what we generally call *exponential growth*. You can see how exponential growth gets its name by looking again at the general compound interest formula:

$$A = P \times (1 + \text{APR})^Y$$

Because the starting principal P and the interest rate APR have fixed values for any particular compound interest calculation, the growth of the accumulated value A depends only on Y (the number of times interest has been paid), which appears as an *exponent* in the calculation.

Exponential growth is one of the most important topics in mathematics, with applications that include population growth, resource depletion, and radioactivity. We will

USING TECHNOLOGY

The Compound Interest Formula

Standard Calculators You can do compound interest calculations on any calculator that has a key for raising numbers to powers (y^x or ∧). The only "trick" is making sure you follow the standard **order of operations**:

1. Parentheses: Do terms in parentheses first.
2. Exponents: Do powers and roots next.
3. Multiplication and Division: Work from left to right.
4. Addition and Subtraction: Work from left to right.

You can remember the order of operations with the mnemonic "Please Excuse My Dear Aunt Sally."

Let's apply this order of operations to the compound interest problem from Example 2, in which we have $P = \$100$, $\text{APR} = 0.1$, and $Y = 5$ years.

General Procedure	Our Example	Calculator Steps	Output
$A = P \times (1 + \text{APR})^Y$	$A = 100 \times (1 + 0.1)^5$	Step 1 1 + 0.1 =	1.1
1. parentheses	1. parentheses	Step 2 ∧ 5 =	1.61051
2. exponent	2. exponent	Step 3 × 100 =	161.051
3. multiply	3. multiply		

Note: Do not round answers in intermediate steps; only the final answer should be rounded to the nearest cent.

Excel Use the built-in function FV (for *future value*) for compound interest calculations in Excel. The table below explains the inputs that go in the parentheses of the FV function.

Input	Description	Our Example
rate	The interest rate for each compounding period	Because the interest is compounded once a year, the interest rate is the annual rate, $\text{APR} = 0.1$.
nper	The total number of compounding periods	For interest compounded once a year, the total number of compounding periods is the number of years, $Y = 5$.
pmt	The amount of any payment made each month	No payment is being made monthly in our example, so we enter 0.
pv	The present value, equivalent to the starting principal P	We use the starting principal, $P = 100$.
type	An optional input related to whether monthly payments are made at the beginning (type = 0) or end (type = 1) of a month	The input type does not apply in this case because there is no monthly payment, so we do not include it.

The screen shot on the left below shows the use of the FV function for our sample calculation. Note: You could get the final result by typing values directly into the FV function, but as shown in the screen shot on the right below, it is better to show your work. Here we put variable names in Column A and values in Column B, using the FV function in cell B5. Besides making your work clearer, this approach makes it easy to do "what if" scenarios, such as changing the interest rate or number of years.

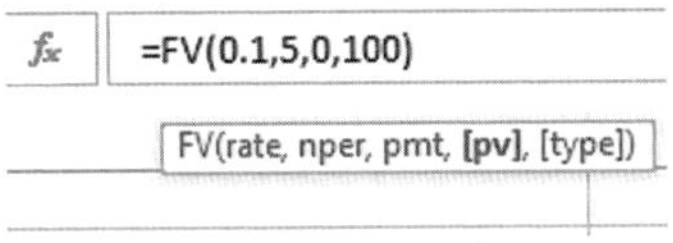

Microsoft Excel 2016, Windows 10, Microsoft Corporation.

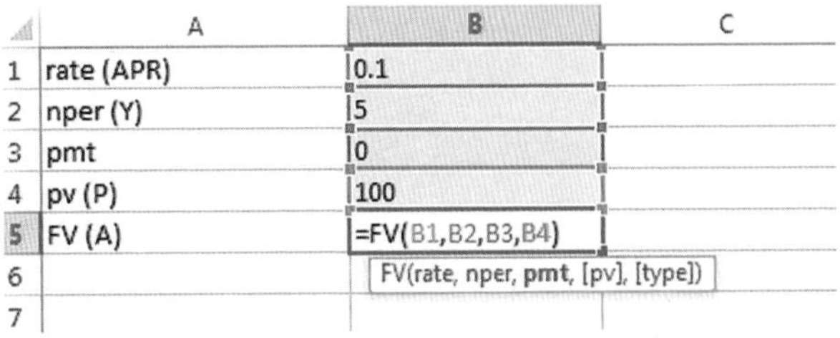

study exponential growth in much more detail in Chapter 8. In this chapter, we focus only on its applications in finance.

EXAMPLE 3 New College Debt at 2%

If the interest rate is 2%, calculate the amount due to New College using:

a. simple interest. **b.** compound interest.

Solution

a. The following steps show the calculation with simple interest for a starting principal $P = \$224$ and an annual interest rate of 2%:

The simple interest due each year is 2% of the starting principal:	$2\% \times \$224 = 0.02 \times \$224 = \$4.48$
Over 535 years, the total interest due is:	$535 \times \$4.48 = \$2396.80$
The total due after 535 years is the starting principal plus the interest:	$\$224 + \$2396.80 = \$2620.80$

With simple interest, the payoff amount after 535 years is \$2620.80.

b. To find the amount due with compound interest, we set the annual interest rate to APR = 2% = 0.02 and the number of years to $Y = 535$. Then we use the formula for compound interest paid once a year:

$$\begin{aligned} A = P \times (1 + \text{APR})^Y &= \$224 \times (1 + 0.02)^{535} \\ &= \$224 \times (1.02)^{535} \\ &\approx \$224 \times 39{,}911 \\ &\approx \$8.94 \times 10^6 \end{aligned}$$

The amount due with compound interest is about \$8.94 million—far higher than the amount due with simple interest. ▶ **Now try Exercises 57–58.**

Effects of Interest Rate Changes

Notice the remarkable effects of small changes in the compound interest rate. In Example 3, we found that a 2% compound interest rate leads to a payoff amount of \$8.94 million after 535 years. Earlier, we found that a 4% interest rate for the same 535 years leads to a payoff amount of \$290 billion—which is more than 30,000 times as large as \$8.94 million. In other words, doubling the interest rate produces far more than a doubling of the accumulated balance. Figure 4.3 contrasts the values of the New College debt during the first 100 years at interest rates of 2% and 4% Note that the rate change doesn't make much difference for the first few years, but over time the higher rate yields a much greater accumulated value.

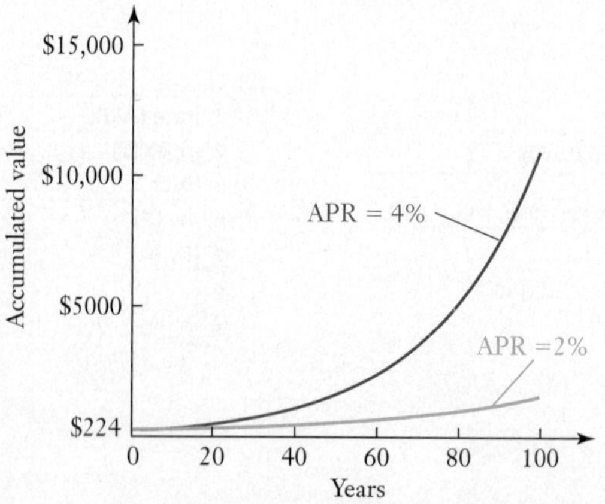

FIGURE 4.3 Comparison of the value of the debt in the New College case during the first 100 years at interest rates of 2% and 4%.

Think About It Suppose the interest rate for the New College debt were 3%. Without calculating, do you think the value after 535 years would be halfway between the values at 2% and 4% or closer to one or the other of these values? Now, check your guess by calculating the value at 3%. What happens at an interest rate of 6%? Briefly discuss why small changes in the interest rate can lead to large changes in the accumulated value.

EXAMPLE 4 Mattress Investments

Your grandfather put \$100 under his mattress 50 years ago. If he had instead invested it in a bank account paying 3.5% interest compounded yearly (roughly the average U.S. rate of inflation during that period), how much would it be worth now?

Solution The starting principal is $P = \$100$. The annual percentage rate is APR $= 3.5\% = 0.035$. The number of years is $Y = 50$. So the accumulated balance would be

$$\begin{aligned} A = P \times (1 + \text{APR})^Y &= \$100 \times (1 + 0.035)^{50} \\ &= \$100 \times (1.035)^{50} \\ &= \$558.49 \end{aligned}$$

Invested at a rate of 3.5%, the \$100 would be worth over \$550 today. Unfortunately, the money was put under a mattress, so it still has a face value of only \$100.

▶ Now try Exercises 59–62.

Compound Interest Paid More Than Once a Year

Suppose you could put \$1000 into an investment that pays compound interest at an annual percentage rate of APR = 8%. If the interest is paid all at once at the end of a year, you'll receive interest of

$$8\% \times \$1000 = 0.08 \times \$1000 = \$80$$

Therefore, your year-end balance will be \$1000 + \$80 = \$1080.

Now, assume instead that the investment pays interest *quarterly*, or four times a year (once every 3 months). The quarterly interest rate is one-fourth of the annual interest rate:

$$\text{quarterly interest rate} = \frac{\text{APR}}{4} = \frac{8\%}{4} = 2\% = 0.02$$

Table 4.3 shows how quarterly compounding affects the \$1000 starting principal during the first year.

TABLE 4.3 Quarterly Interest Payments ($P = \$1000$, APR = 8%)

After *N* Quarters	Interest Paid	New Balance
1st quarter (3 months)	2% × \$1000 = \$20	\$1000 + \$20 = \$1020
2nd quarter (6 months)	2% × \$1020 = \$20.40	\$1020 + \$20.40 = \$1040.40
3rd quarter (9 months)	2% × \$1040.40 = \$20.81	\$1040.40 + \$20.81 = \$1061.21
4th quarter (1 full year)	2% × \$1061.21 = \$21.22	\$1061.21 + \$21.22 = \$1082.43

The year-end balance with quarterly compounding (\$1082.43) is *greater* than the year-end balance with interest paid just once a year (\$1080). That is, when interest is compounded more than once a year, the balance increases by *more* than the APR in 1 year.

We can find the same results with the compound interest formula. Remember that the basic form of the compound interest formula is

$$A = P \times (1 + \text{interest rate})^{\text{number of compoundings}}$$

where A is the accumulated balance and P is the starting principal. In our current case, the starting principal is $P = \$1000$, the quarterly payments have an interest rate of $\text{APR}/4 = 0.02$, and in one year the interest is paid four times. Therefore, the accumulated balance at the end of one year is

$$A = P \times (1 + \text{interest rate})^{\substack{\text{number of} \\ \text{compoundings}}} = \$1000 \times (1 + 0.02)^4 = \$1082.43$$

Generalizing, if interest is paid n times per year, the interest rate at each payment is APR/n. The total number of times that interest is paid after Y years is nY. We therefore find the following formula for interest paid more than once each year.

Compound Interest Formula for Interest Paid *n* Times Per Year

$$A = P\left(1 + \frac{\text{APR}}{n}\right)^{(nY)}$$

where

A = accumulated balance after Y years

P = starting principal

APR = annual percentage rate (as a decimal)

n = number of compounding periods per year

Y = number of years

Note that Y is not necessarily an integer; for example, a calculation for a period of 6 months would use $Y = 0.5$.

Think About It Confirm that substituting $n = 1$ into the formula for interest paid n times per year gives you the formula for interest paid once a year. Explain why this should be true.

EXAMPLE 5 Monthly Compounding at 3%

You deposit \$5000 in a bank account that pays an APR of 3% and compounds interest monthly. How much money will you have after 5 years? Compare this amount to the amount you'd have if interest were paid only once each year.

Solution The starting principal is $P = \$5000$ and the interest rate is $\text{APR} = 0.03$. Monthly compounding means that interest is paid $n = 12$ times a year, and we are considering a period of $Y = 5$ years. We put these values into the compound interest formula to find the accumulated balance, A:

$$\begin{aligned} A = P \times \left(1 + \frac{\text{APR}}{n}\right)^{(nY)} &= \$5000 \times \left(1 + \frac{0.03}{12}\right)^{(12 \times 5)} \\ &= \$5000 \times (1.0025)^{60} \\ &= \$5808.08 \end{aligned}$$

For interest paid only once each year, we find the balance after 5 years by using the formula for compound interest paid once a year:

$$\begin{aligned} A = P \times (1 + \text{APR})^Y &= \$5000 \times (1 + 0.03)^5 \\ &= \$5000 \times (1.03)^5 \\ &= \$5796.37 \end{aligned}$$

After 5 years, monthly compounding gives you a balance of \$5808.08 while annual compounding gives you a balance of \$5796.37. That is, monthly compounding earns $\$5808.08 - \$5796.37 = \$11.71$ more, even though the APR is the same in both cases.

▶ Now try Exercises 63–70.

USING TECHNOLOGY

The Compound Interest Formula for Interest Paid More Than Once a Year

Standard Calculators The procedure when interest is paid more than once a year is essentially the same as that for the basic compound interest formula (see Using Technology, p. 211), except you enter APR/n instead of APR and nY instead of Y. Let's apply the procedure to Example 5, in which $P = \$5000$, APR $= 0.03$, $n = 12$, and $Y = 5$ years.

General Procedure	Our Example	Calculator Steps*	Output
$A = P \times \left(1 + \frac{APR}{n}\right)^{(nY)}$ 1. parentheses 2. exponent 3. multiply	$A = 5000 \times \left(1 + \frac{0.03}{12}\right)^{(12\times5)}$ 1. parentheses 2. exponent 3. multiply	Step 1 1 (+) 0.03 (÷) 12 (=)	1.0025
		Step 2 (∧) (() 12 (×) 5 ()) (=)	1.1616...
		Step 3 (×) \$5000 (=)	5808.08

*If your calculator does not have parentheses keys, then find the exponent ($nY = 12 \times 5$) before you begin, and keep track of it on paper or in the calculator's memory.

Excel Use the built-in function FV just as for the basic compound interest formula (p. 211), *except*

- because *rate* is the interest rate *for each compounding period*, in this case use the monthly interest rate APR/n = 0.03/12.
- because *nper* is the total number of compounding periods, in this case use $nY = 12 \times 5$. (Note that Excel uses an asterisk * for multiplication.)

The screen shot on the left below shows the direct entry of the FV function for our example. However, it's best to show your work by referencing clearly labeled cells. In this case, we start with cells for APR, n, and Y, because these are the variables used in the compound interest formula in this text. These are then referenced to create the inputs for the FV function. You should create your own Excel worksheet to confirm that you get the result from Example 5 ($A = \$5808.08$).

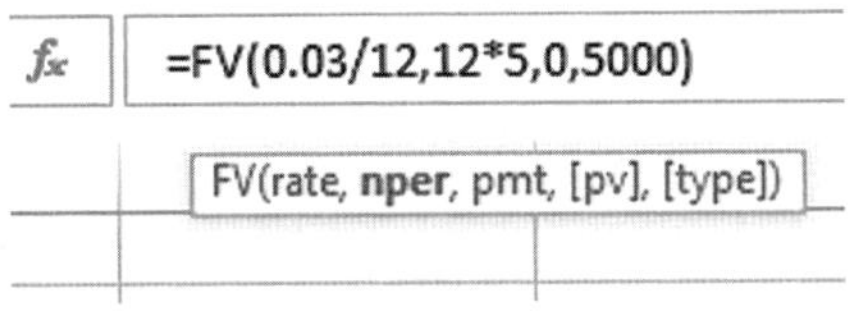

Microsoft Excel 2016, Windows 10, Microsoft Corporation.

	A	B
1	APR	0.03
2	n (monthly compounding)	12
3	Y (number of years)	5
4	**rate** (for each compounding period = APR/n)	=B1/B2
5	**nper** (total number of compounding periods = nY)	=B2*B3
6	**pmt** (no monthly payment in this case)	0
7	**pv** (present value = starting principal P)	5000
8	**FV** (future value = accumulated balance A)	=FV(B4,B5,B6,B7)
9		

USING TECHNOLOGY

APY in Excel

The Excel function EFFECT returns the APY (the *effective yield*) from the APR and the number of compounding periods per year (n); the format is EFFECT(APR, n).

Example: With APR = 0.08 and $n = 4$, you find the APY (0.08243) by entering

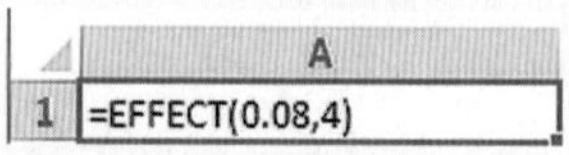

For cases where you know the APY and need to find the APR, use the NOMINAL function.

Example: With APR = 0.08243 and $n = 4$, (quarterly compounding), you find the APR (0.08) by entering

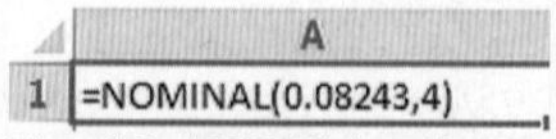

Microsoft Excel 2016, Windows 10, Microsoft Corporation.

Technical Note

Most banks divide the APR by 360, rather than 365, when calculating the interest rate and APY for daily compounding. Therefore, the results found here may not agree exactly with actual interest amounts paid by a bank.

Annual Percentage Yield (APY)

We've seen that in one year, money grows by *more* than the APR when interest is compounded more than once a year. For example, we found that with quarterly compounding and an 8% APR, a starting principal of \$1000 increases to \$1082.43 in one year. This represents a relative increase of 8.243%:

$$\text{relative increase} = \frac{\text{absolute increase}}{\text{starting principal}} = \frac{\$82.43}{\$1000} = 0.08243 = 8.243\%$$

This relative increase over one year is called the **annual percentage yield (APY)**. Note that it depends only on the annual interest rate (APR) and the number of compounding periods, not on the starting principal.

Definition

The **annual percentage yield (APY)**—also called the *effective yield*, or simply the *yield*—is the actual percentage by which a balance increases in one year. It is equal to the APR if interest is compounded annually. It is greater than the APR if interest is compounded more than once a year.

Banks usually list both the annual percentage rate (APR) and the annual percentage yield (APY). However, the APY is what your money really earns and is the more important number when you are comparing interest rates. Banks are required by law to state the APY on interest-bearing accounts.

EXAMPLE 6 More Compounding Means a Higher Yield

You deposit \$1000 into an account with APR = 8%. Find the annual percentage yield with *monthly* compounding and with *daily* compounding.

Solution The easiest way to find the annual percentage yield is by finding the balance at the end of 1 year. We have $P = \$1000$, APR = 8% = 0.08, and $Y = 1$ year.

With monthly compounding: We set $n = 12$, so after 1 year the balance is

$$\begin{aligned} A = P \times \left(1 + \frac{\text{APR}}{n}\right)^{(nY)} &= \$1000 \times \left(1 + \frac{0.08}{12}\right)^{(12\times 1)} \\ &= \$1000 \times (1.006666667)^{12} \\ &= \$1083.00 \end{aligned}$$

Your balance will increase by \$83.00, so the annual percentage yield is

$$\text{APY} = \text{relative increase in 1 year} = \frac{\$83.00}{\$1000} = 0.083 = 8.3\%$$

With monthly compounding, the annual percentage yield is 8.3% (which is more than the APR).

With daily compounding: We set $n = 365$, so after 1 year the balance is

$$\begin{aligned} A = P \times \left(1 + \frac{\text{APR}}{n}\right)^{(nY)} &= \$1000 \times \left(1 + \frac{0.08}{365}\right)^{(365\times 1)} \\ &= \$1000 \times (1.000219178)^{365} \\ &= \$1083.28 \end{aligned}$$

Your balance will increase by \$83.28, so the annual percentage yield is

$$\text{APY} = \text{relative increase in 1 year} = \frac{\$83.28}{\$1000} = 0.08328 = 8.32\%$$

Notice that the annual percentage yield is slightly higher with daily compounding than with monthly compounding.
▶ Now try Exercises 71–74.

Continuous Compounding

Suppose that interest were compounded more often than daily—say, every second or every trillionth of a second. How would this affect the annual percentage yield?

Let's examine what we've found so far for APR = 8%. If interest is compounded annually (once a year), the annual percentage yield is simply APY = APR = 8%. With quarterly compounding, we found APY = 8.243%. With monthly compounding, we found APY = 8.300%. With daily compounding, we found APY = 8.328%. Clearly, more frequent compounding means a higher APY (for a given APR).

However, notice that the change gets smaller as the frequency of compounding increases. For example, changing from annual compounding ($n = 1$) to quarterly compounding ($n = 4$ increases the APY quite a bit, from 8% to 8.243%. In contrast, going from monthly ($n = 12$) to daily ($n = 365$) compounding increases the APY only slightly, from 8.300% to 8.328%. In fact, the APY can't get much larger than it already is for daily compounding.

Table 4.4 shows the APY for various compounding periods, and Figure 4.4 is a graph of those results. Note that the APY does *not* grow indefinitely. Instead, it approaches a *limit* that is very close to the value of 8.3287068% found for $n = 1$ billion. In fact, even if we could compound *infinitely many times* per year, the annual percentage yield would not exceed 8.3287068%. Compounding infinitely many times per year is called **continuous compounding**. It represents the best possible compounding for a particular APR. With continuous compounding, the compound interest formula takes the following special form.

TABLE 4.4 Annual Yield (APY) for APR = 8% with Various Numbers of Compounding Periods (n)

n	APY	n	APY
1	8.0000000%	1000	8.3283601%
4	8.2432160%	10,000	8.3286721%
12	8.2999507%	1,000,000	8.3287064%
365	8.3277572%	10,000,000	8.3287067%
500	8.3280135%	1,000,000,000	8.3287068%

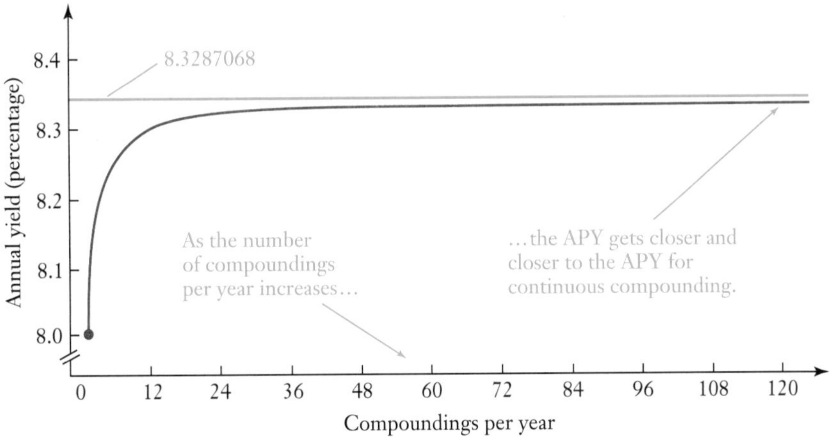

FIGURE 4.4 The annual percentage yield (APY) for APR = 8% and different numbers of compounding periods. Note that it approaches a limit, which represents the yield obtained with continuous compounding.

USING TECHNOLOGY

Powers of e

You can find powers of e in Excel with the function EXP(power). *Example:* To find $e^{0.8}$, enter

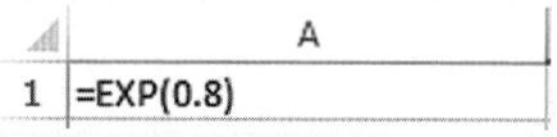

Microsoft Excel 2016, Windows 10, Microsoft Corporation.

In Google, simply type e and use ∧ to raise it to a power.

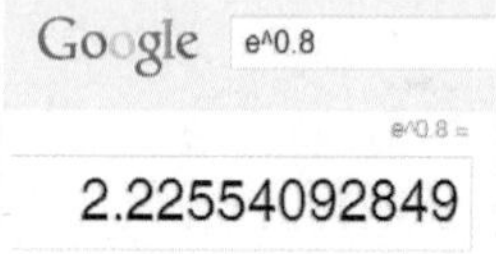

Compound Interest Formula for Continuous Compounding

$$A = P \times e^{(\text{APR} \times Y)}$$

where

A = accumulated balance after Y years

P = starting principal

APR = annual percentage rate (as a decimal)

Y = number of years

The number e is a special irrational number with a value of $e \approx 2.71828$. You can compute e to a power with the e^x key on your calculator.

Think About It Look for the e^x key on your calculator. Use it to enter e^1 and thereby verify that $e \approx 2.71828$. Bonus: Also verify the value in Excel and with Google.

HISTORICAL NOTE

Like the number π that arises so frequently in mathematics, the number e is one of the universal mathematical constants. It appears in countless applications, most importantly to describe exponential growth and decay processes. The notation e was proposed in 1727 by the Swiss mathematician Leonhard Euler (pronounced "Oiler"). Like π, the number e is not only an irrational number, but also an example of what is called a *transcendental number*.

EXAMPLE 7 Continuous Compounding

You deposit $100 in an account with an APR of 8% and continuous compounding. How much will you have after 10 years?

Solution We have $P = \$100$, APR $= 8\% = 0.08$, and $Y = 10$ years of continuous compounding. The accumulated balance after 10 years is

$$A = P \times e^{(\text{APR} \times Y)} = \$100 \times e^{(0.08 \times 10)} = \$100 \times e^{0.8} = \$222.55$$

▲ **CAUTION!** Be sure you can get the above answer by using the number e correctly with your calculator or computer. ▲

Your balance will be $222.55 after 10 years.

▶ Now try Exercises 75–80.

Brief Review Four Basic Rules of Algebra

A major goal of algebra is to solve equations for some variable. Here we review four basic rules that are often useful in this process.

Four Basic Rules

The following rules can always be used:

1. You can interchange the left and right sides of an equation. That is, if $x = y$, it is also true that $y = x$.
2. You can add or subtract the same quantity on both sides of an equation.
3. You can multiply or divide both sides of an equation by the same quantity, as long as you do not multiply or divide by zero.
4. You can raise both sides of an equation to the same power or take the same root on both sides (which is equivalent to raising both sides to the same fractional power).

Examples—Adding and Subtracting

Example: Solve the equation $x - 9 = 3$ for x.
Solution: We isolate x by adding 9 to both sides:

$$x - 9 + 9 = 3 + 9 \quad \rightarrow \quad x = 12$$

Example: Solve the equation $y + 6 = 2y$ for y.
Solution: We put all terms with y on the right side by subtracting y from both sides:

$$y + 6 - y = 2y - y \quad \rightarrow \quad 6 = y$$

Interchanging the left and right sides gives us the answer: $y = 6$.
Example: Solve the equation $8q - 17 = p + 4q - 2$ for p.
Solution: We isolate p by subtracting $4q$ from both sides while also adding 2 to both sides:

$$8q - 17 - 4q + 2 = p + 4q - 2 - 4q + 2$$
$$\downarrow$$
$$4q - 15 = p$$

Interchanging the left and right sides, we get the answer: $p = 4q - 15$.

Examples—Multiplying and Dividing

Example: Solve the equation $4x = 24$ for x.

Planning Ahead with Compound Interest

Suppose you have a new baby and want to make sure that you'll have \$100,000 for his or her college education. Assuming that your baby will start college in 18 years, how much money should you deposit now?

If we know the interest rate, we can answer this question with a "backward" compound interest calculation. We start with the amount, A, needed after 18 years and then calculate the necessary starting principal, P. The following example illustrates the calculation.

BY THE WAY

The process of finding the amount (present value) that must be deposited today to yield some particular future amount is called *discounting* by financial planners.

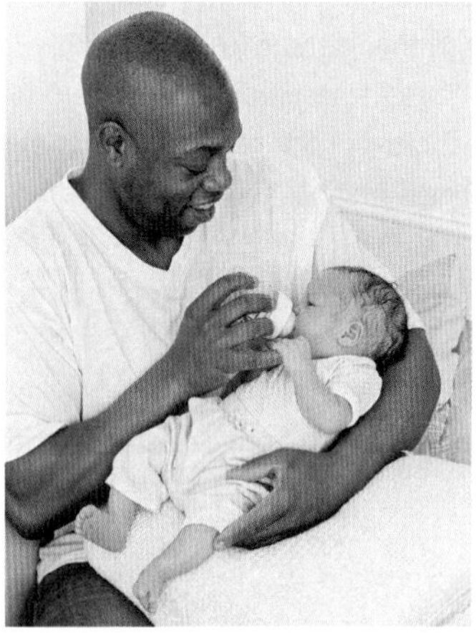

EXAMPLE 8 College Fund at 3%, Compounded Monthly

Suppose you invest money in an account with an interest rate of APR = 3%, compounded monthly, and leave it there for the next 18 years. How much would you have to deposit now to realize \$100,000 after 18 years?

Solution We know the interest rate (APR = 0.03), the number of years of compounding ($Y = 18$), the amount desired after 18 years ($A = \$100{,}000$), and $n = 12$ for monthly compounding. To find the starting principal (P) that must be deposited now, we solve the compound interest formula for P, then substitute the given values.

Start with the compound interest formula for interest paid more than once a year:	$A = P \times \left(1 + \frac{\text{APR}}{n}\right)^{nY}$
Interchange the left and right sides and divide both sides by $\left(1 + \frac{\text{APR}}{n}\right)^{nY}$:	$P = \frac{A}{\left(1 + \frac{\text{APR}}{n}\right)^{nY}}$
Substitute the values APR = 0.03, $Y = 18$, $A = \$100{,}000$, and $n = 12$:	$P = \frac{\$100{,}000}{\left(1 + \frac{0.03}{12}\right)^{12 \times 18}} = \frac{\$100{,}000}{(1.0025)^{216}} = \$58{,}314.11$

Solution: We isolate x by dividing both sides by 4:

$$\frac{4x}{4} = \frac{24}{4} \quad \longrightarrow \quad x = 6$$

Example: Solve the equation $y = 3x + 9$ for x.

Solution: We isolate the term containing x by subtracting 9 from both sides:

$$y - 9 = 3x + 9 - 9 \quad \longrightarrow \quad y - 9 = 3x$$

Next, we divide both sides by 3 to isolate x, then interchange the two sides to write the final answer:

$$\frac{y-9}{3} = \frac{3x}{3} \quad \longrightarrow \quad x = \frac{y-9}{3}$$

Example: Solve the equation $\frac{3z}{4} - 2 = 10$ for z.

Solution: First, we isolate the term containing z by adding 2 to both sides:

$$\frac{3z}{4} - 2 + 2 = 10 + 2 \quad \longrightarrow \quad \frac{3z}{4} = 12$$

Now we multiply both sides by $\frac{4}{3}$:

$$\frac{3z}{4} \times \frac{4}{3} = 12 \times \frac{4}{3} \quad \longrightarrow \quad z = 16$$

Examples—Powers and Roots

Example: Find the positive solution of the equation $x^4 = 16$.

Solution: We solve for x by raising both sides to the fourth power:

$$(x^4)^{1/4} = 16^{1/4}$$

This leaves x on the left side [from the rule $(a^n)^m = a^{n \times m}$], and on the right side the $\frac{1}{4}$ power is the same as the fourth root:

$$x^{4 \times 1/4} = 16^{1/4} \quad \longrightarrow \quad x = \sqrt[4]{16} = 2$$

The positive solution to the equation is $x = 2$. (Another solution is $x = -2$, but we will generally ignore negative solutions in this text.) ▸ **Now try Exercises 27–50.**

IN YOUR WORLD

Effects of Low Interest Rates

We've seen that compounding can make money grow substantially over time even with annual interest rates that are relatively low by historical standards, such as 3%. But in recent years, most bank interest rates have fallen even lower. The graph below shows how typical bank interest rates have varied over the past several decades. Although these rates were sometimes over 10% and the average (median) for the period has been about 5.6%, the rates have approached zero in recent years. Moreover, interest rates in the past were often higher than the rate of inflation, which meant the real value of money in bank accounts rose with time. In recent years, however, interest rates have generally been *lower* than the inflation rate, which means the money loses real value over time.

Besides making it difficult to keep pace with inflation, these low rates can have devastating effects on retirees who had hoped to live off the interest from a lifetime of savings. For example, consider a retired couple who saved $500,000 for retirement. At an interest rate of 5%, they would earn $25,000 per year in interest and could use this money to live on without reducing their $500,000 principal at all. But at the 0.05% that was typical for bank accounts in 2017, the interest would be only $250 per year. If the couple needed $25,000 to live, they would have to draw down their principal by about this amount each year, in which case their retirement account would be empty in about 20 years, possibly leaving them dependent solely on Social Security. We'll discuss the future of Social Security in Unit 4F, but this case already shows why it is an issue that draws highly emotional responses. Alternatively, the couple might try to find better returns in other types of investments, but as we'll discuss in Unit 4C, the search for better returns always comes with additional risk of losses.

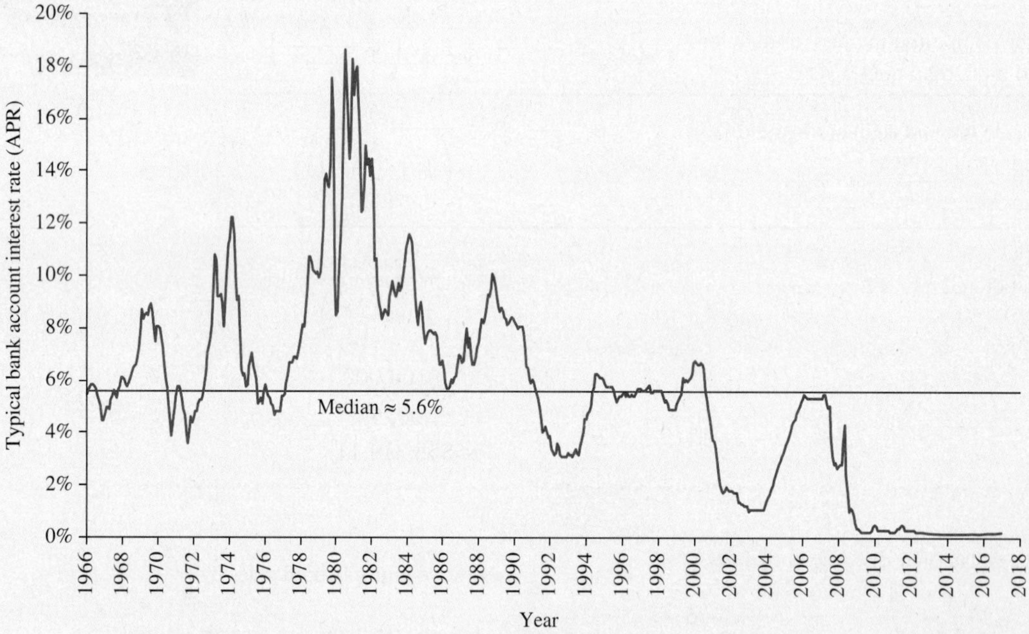

Typical bank interest rates since 1966. Based on monthly averages for 3-month CDs. *Source:* Department of the Treasury.

Depositing $58,314 now will yield the desired $100,000 in 18 years—assuming that the 3% APR doesn't change and that you make no withdrawals or additional deposits. You can easily check this answer: Applying the compound interest formula with a starting balance of $58,314.11 gives an accumulated balance of $100,000 after 18 years.

▶ **Now try Exercises 81–88.**

Think About It Aside from long-term government bonds, it is extremely difficult to find investments with a constant interest rate for 18 years. Nevertheless, financial planners often make such assumptions when exploring investment options. Explain why such calculations can be useful, despite the fact that you can't be sure of a steady interest rate.

Quick Quiz 4B

Choose the best answer to each of the following questions. Explain your reasoning with one or more complete sentences.

1. Consider two investments, one earning simple interest and one earning compound interest. If both start with the same initial deposit (and you make no other deposits or withdrawals) and earn the same annual interest rate, after 2 years the account with simple interest will have

   a. a greater balance than the account with compound interest.

   b. a smaller balance than the account with compound interest.

   c. the same balance as the account with compound interest.

2. An account with interest compounded annually and an APR of 6% increases in value each year by a factor of

   a. 1.06. b. 1.6. c. 1.006.

3. After 5 years, an investment with interest compounded annually and an APR of 5.5% increases in value by a factor of

   a. 1.55^5 b. 5×1.055 c. 1.055^5

4. An account with an APR of 4% and quarterly compounding increases in value every 3 months by

   a. 1%. b. 1/4%. c. 4%.

5. With the same deposit, APR, and length of time, an investment with monthly compounding yields

   a. a greater balance than an account with daily compounding.

   b. a smaller balance than an account with quarterly compounding.

   c. a greater balance than an account with annual compounding.

6. The annual percentage rate (APR) is always

   a. greater than the APY.

   b. less than or equal to the APY.

   c. the same as the APY.

7. Consider two accounts earning compound interest, one with an APR of 4%, the other with an APR of 2%, and both with the same initial deposit (and no further deposits or withdrawals). After 20 years, how much more interest will the account with APR = 4% have earned than the account with APR = 2%?

   a. less than twice as much

   b. exactly twice as much

   c. more than twice as much

8. If you deposit \$250 in an account with an APR of 6% and continuous compounding, the balance after 2 years is

   a. $\$250 \times e^{0.12}$ b. $\$250 \times e^2$ c. $\$250 \times (1 + 0.06)^2$

9. Suppose you use the compound interest formula to calculate how much you must deposit into a college fund today if you want it to grow in value to \$20,000 in 10 years. The calculation assumes that

   a. the average APR remains constant for 10 years.

   b. the fund has continuous compounding.

   c. the fund earns simple interest rather than compound interest.

10. A bank account with compound interest exhibits what we call

    a. linear growth. b. simple growth. c. exponential growth.

Exercises 4B

REVIEW QUESTIONS

1. What is the difference between *simple interest* and *compound interest*? Why do you end up with more money with compound interest?
2. Explain how New College could claim that a debt of \$224 from 535 years ago grew to be worth \$290 billion. How does this show the "power of compounding"?
3. Explain why the term APR/n appears in the compound interest formula for interest paid n times a year.
4. State the compound interest formula for interest paid once a year. Define *annual percentage rate (APR)* and Y.
5. State the compound interest formula for interest paid more than once a year. Define all the variables.
6. What is an *annual percentage yield (APY)*? Explain why, for a given APR, the APY is higher if the interest is compounded more frequently.
7. What is *continuous compounding*? How does the APY for continuous compounding compare to the APY for, say, daily compounding? Explain the formula for continuous compounding.
8. Give an example of a situation in which you might want to solve the compound interest formula to find the amount P that must be invested now to yield a particular amount A in the future.

DOES IT MAKE SENSE?

Decide whether each of the following statements makes sense (or is clearly true) or does not make sense (or is clearly false). Explain your reasoning.

9. Simple Bank offers simple interest at 4.5% per year, which is clearly a better deal than the 4.5% compound interest rate at Complex Bank.

10. Both banks were paying the same annual percentage rate (APR), but one had a higher annual percentage yield than the other (APY).

11. The bank that pays the highest annual percentage rate (APR) is always the best deal.

12. No bank could afford to pay interest every trillionth of a second because, with compounding, it would soon owe everyone infinite dollars.

13. My bank paid an annual interest rate (APR) of 5.0%, but at the end of the year, my account balance had grown by 5.1%.

14. If you deposit $10,000 in an investment account today, it can double in value to $20,000 in just a couple of decades even at a relatively low interest rate (say, 4%).

BASIC SKILLS & CONCEPTS

15–26: Review of Powers. Use the skills covered in the Brief Review on p. 207 to evaluate or simplify the following expressions.

15. 3^2
16. 3^4
17. 2^5
18. 3^{-2}
19. $25^{1/2}$
20. $81^{1/2}$
21. $64^{-1/3}$
22. $2^3 \times 2^5$
23. $3^4 \div 3^2$
24. $6^2 \times 6^{-2}$
25. $25^{1/2} \div 25^{-1/2}$
26. $3^3 + 2^3$

27–50: Algebra Review. Use the skills covered in the Brief Review on p. 219 to solve the following equations.

27. $x - 4 = 16$
28. $y + 4 = 7$
29. $z - 10 = 6$
30. $2x = 8$
31. $4y = 16$
32. $4y + 2 = 18$
33. $5z - 1 = 19$
34. $1 - 6y = 13$
35. $3x - 4 = 2x + 6$
36. $5 - 4s = 6s - 5$
37. $3a + 4 = 6 + 4a$
38. $3n - 16 = 53$
39. $6q - 20 = 60 + 4q$
40. $5w - 5 = 3w - 25$
41. $t/4 + 5 = 25$
42. $2x/3 + 4 = 2x$
43. $x^2 = 49$
44. $y^3 = 27$
45. $(x - 4)^2 = 36$
46. $p^{1/3} = 3$
47. $(t/3)^2 = 16$
48. $w^2 + 2 = 27$
49. $u^9 = 512$
50. $v^3 + 4 = 68$

51–54: Simple Interest. Calculate the amount of money you will have in each account after 5 years, assuming that the account earns simple interest.

51. You deposit $800 in an account with an annual interest rate of 5%.

52. You deposit $1500 in an account with an annual interest rate of 4%.

53. You deposit $3200 in an account with an annual interest rate of 3.5%.

54. You deposit $1800 in an account with an annual interest rate of 3.8%.

55–56: Simple vs. Compound Interest. Complete the following tables, which show the performance of two investments over a 5-year period. Round all figures to the nearest dollar.

55. Suzanne deposits $3000 in an account that earns simple interest at an annual rate of 2.5%. Derek deposits $3000 in an account that earns compound interest at an annual rate of 2.5%.

End of year	Suzanne's annual interest	Suzanne's balance	Derek's annual interest	Derek's balance
1				
2				
3				
4				
5				

56. Ariel deposits $5000 in an account that earns simple interest at an annual rate of 3%. Travis deposits $5000 in an account that earns compound interest at an annual rate of 3%.

End of year	Ariel's annual interest	Ariel's balance	Travis's annual interest	Travis's balance
1				
2				
3				
4				
5				

57–62: Compound Interest. Use the compound interest formula to compute the balance in each account after the stated period of time, assuming that interest is compounded annually.

57. $5000 is invested at an APR of 4% for 10 years.

58. $20,000 is invested at an APR of 3.5% for 20 years.

59. $15,000 is invested at an APR of 3.2% for 25 years.

60. $3000 is invested at an APR of 1.8% for 12 years.

61. $10,000 is invested at an APR of 3.7% for 12 years.

62. $40,000 is invested at an APR of 2.8% for 30 years.

63–70: Compounding More Than Once a Year. Use the appropriate compound interest formula to compute the balance in each account after the stated period of time.

63. $5000 is invested for 10 years with an APR of 2% and quarterly compounding.

64. $4000 is invested for 5 years with an APR of 3% and daily compounding.

65. $25,000 is invested for 5 years with an APR of 3% and daily compounding.

66. $10,000 is invested for 5 years with an APR of 2.75% and monthly compounding.

67. $4000 is invested for 20 years with an APR of 6% and monthly compounding.

68. $30,000 is invested for 15 years with an APR of 4.5% and daily compounding.

69. $25,000 is invested for 30 years with an APR of 3.7% and quarterly compounding.

70. $15,000 is invested for 15 years with an APR of 4.2% and monthly compounding.

71–74: Annual Percentage Yield (APY). Find the annual percentage yield (to the nearest 0.01%) in each case.

71. A bank offers an APR of 4.1% compounded daily.

72. A bank offers an APR of 3.2% compounded monthly.

73. A bank offers an APR of 1.23% compounded monthly.

74. A bank offers an APR of 2.25% compounded quarterly.

75–80: Continuous Compounding. Use the formula for continuous compounding to compute the balance in each account after 1, 5, and 20 years. Also, find the APY for each account.

75. A $5000 deposit in an account with an APR of 4.5%

76. A $2000 deposit in an account with an APR of 3.1%

77. A $7000 deposit in an account with an APR of 4.5%

78. A $3000 deposit in an account with an APR of 7.5%

79. A $3000 deposit in an account with an APR of 6%

80. A $500 deposit in an account with an APR of 2.7%

81–84: Planning Ahead. How much must you deposit today into each of the following accounts in order to have $25,000 in 8 years for a down payment on a house? Assume that no additional deposits are made.

81. An account with annual compounding and an APR of 6%

82. An account with quarterly compounding and an APR of 4.5%

83. An account with monthly compounding and an APR of 6%

84. An account with daily compounding and an APR of 4%

85–88: College Fund. How much must you deposit today into each of the following accounts in order to have a $120,000 college fund in 15 years? Assume that no additional deposits are made.

85. An APR of 5.5%, compounded annually

86. An APR of 5.5%, compounded daily

87. An APR of 2.85%, compounded quarterly

88. An APR of 3.5%, compounded monthly

FURTHER APPLICATIONS

89–90: Small Rate Differences. The following pairs of investment plans are identical except for a small difference in interest rates. Compute the balances in both accounts after 10 and 30 years. Discuss the difference.

89. Chang invests $500 in an account that earns 3.5% compounded annually. Kio invests $500 in a different account that earns 3.75% compounded annually.

90. José invests $1500 in an account that earns 5.6% compounded annually. Marta invests $1500 in a different account that earns 5.7% compounded annually.

91. **Comparing Annual Yields.** Consider an account with an APR of 5.3% Find the APY with quarterly compounding, monthly compounding, and daily compounding. Comment on how changing the compounding period affects the annual yield.

92. **Comparing Annual Yields.** Consider an account with an APR of 5%. Find the APY with quarterly compounding, monthly compounding, and daily compounding. Comment on how changing the compounding period affects the annual yield.

93. **Rates of Compounding.** Compare the accumulated balance in two accounts that both start with an initial deposit of $1000. Both accounts have an APR of 5.5%, but one account compounds interest annually while the other account compounds interest daily. Make a table that shows the interest earned each year and the accumulated balance in both accounts for the first 10 years. Compare the balance in the accounts, in percentage terms, after 10 years. Round all figures to the nearest dollar.

94. **Understanding Annual Percentage Yield (APY).**

a. Explain why APR and APY are the same with annual compounding.

b. Explain why APR and APY are different with daily compounding.

c. Does APY depend on the starting principal, *P*? Why or why not?

d. How does APY depend on the number of compoundings during a year, *n*? Explain.

95. **Comparing Investment Plans.** Rosa invests $3000 in an account with an APR of 4% and annual compounding. Julian invests $2500 in an account with an APR of 5% and annual compounding.

a. Compute the balance in each account after 5 and 20 years.

b. Determine, for each account and for the periods of 5 and 20 years, the percentage of the balance that is interest.

c. Comment on the effect of interest rates and patience.

96. **Comparing Investment Plans.** Paula invests $4000 in an account with an APR of 4.8% and continuous compounding. Petra invests $3600 in an account with an APR of 5.6% and continuous compounding.

a. Compute the balances in the accounts after 5 and 20 years.

b. Determine, for each account and for the periods of 5 and 20 years, the percentage of the balance that is interest.

c. Comment on the effect of interest rates and patience.

97. **Retirement Fund.** Suppose you want to accumulate $120,000 for your retirement in 30 years. You have two choices: Plan A is to make a single deposit into an account with annual compounding and an APR of 5%. Plan B is to make a single deposit into an account with continuous compounding and an APR of 4.8%. How much do you need to deposit in each account in order to reach the goal?

98. **Your Bank Account.** Find the current APR, the compounding period, and the claimed APY for your personal savings account (or pick a rate from a nearby bank if you don't have an account).

a. Calculate the APY on your account. Does your calculation agree with the APY claimed by the bank? Explain.

b. Suppose you receive a gift of $10,000 and place it in your account. If the interest rate never changes, how much will you have in 10 years?

c. Suppose you could find another bank that offers savings accounts with an APR that is 2 percentage points higher than yours, with the same compounding period. With the \$10,000 deposit, how much would you have after 10 years? Briefly discuss how this result compares to the result from part (b).

99–101: Finding Time Periods. Use a calculator and possibly some trial and error to answer the following questions.

99. How long will it take your initial deposit to triple at an APR of 6% compounded annually?

100. How long will it take your initial deposit to grow by 50% at an APR of 7% compounded annually?

101. You deposit \$1000 in an account that pays an APR of 7% compounded annually. How long will it take for your balance to reach \$100,000?

102. **Continuous Compounding.** Explore continuous compounding by answering the following questions.

a. For an APR of 12%, make a table similar to Table 4.4 in which you display the APY for $n = 4, 12, 365, 500, 1000$.

b. Find the APY for continuous compounding at an APR of 12%.

c. Show the results of parts (a) and (b) on a graph similar to Figure 4.4.

d. In words, compare the APY with continuous compounding to the APY with other types of compounding.

e. You deposit \$500 in an account with an APR of 12%. With continuous compounding, how much money will you have at the end of 1 year? At the end of 5 years?

103. **Philanthropy.** Charles Feeney is an investor who made billions of dollars but decided to give it all away (mostly anonymously) to causes such as education, human rights, and public health. In 2012, having donated over \$6 billion, Mr. Feeney still had \$1.5 billion remaining that he wished to donate before he died.

a. Suppose Mr. Feeney set up an *endowment* (an account that provides ongoing funds from interest only) with his \$1.5 billion and earned interest at an annual rate of 6% compounded monthly. How much could he give away each year *in interest alone* (leaving the balance unchanged)?

b. Repeat part (a) but with an interest rate of 5% compounded monthly. Would the account generate enough interest to make 75 annual donations of \$1 million each? Explain.

c. In fact, Mr. Feeney did not set up an endowment with his last \$1.5 billion, but instead donated all of this remaining money over a period of 5 years. (He kept \$2 million for him and his wife to live on for the rest of their lives, but you can ignore this, as well as any interest earned during the 5-year period, in your calculations.) On average, how much did he give away each week during those 5 years?

104. **Retirement Fund.** A retired couple plans to supplement their Social Security with interest earned by a \$120,000 retirement fund.

a. If the fund compounds interest monthly at an annual rate of 6%, which the couple takes out and spends, how much interest is generated each month?

b. Suppose the annual interest rate suddenly drops to 3%. What is the resulting interest payment each month?

c. Estimate the annual interest rate needed to generate \$900 each month in interest.

IN YOUR WORLD

105. **Rate Comparisons.** Find a website that compares interest rates available for ordinary savings accounts at different banks. What is the range of rates currently being offered? What is the best deal? How does your own bank account compare?

106. **Bank Advertisement.** Find two bank advertisements that refer to compound interest rates. Explain the terms in each advertisement. Which bank offers the better deal? Explain.

107. **Power of Compounding.** In an advertisement or article about an investment plan, find a description of how money has grown (or will grow) over a period of many years. Discuss whether the description is correct.

TECHNOLOGY EXERCISES

108. **Evaluating Powers.** Use a calculator or Excel to evaluate the following expressions.

a. 6^{12}

b. 1.01^{40}

c. 20×1.05^{16}

d. 4^{-5}

e. 1.08^{-20}

109. **Compound Interest with Excel: Annual Compounding.** Use the future value (FV) function in Excel to compute the balance in each of the following accounts.

a. An account with annual compounding, an APR of 10%, and an initial deposit of \$100, after 5 years

b. An account with annual compounding, an APR of 2%, and an initial deposit of \$224, after 535 years

110. **Compound Interest with Excel: Dependence on Parameters.** Suppose you deposit \$500 in an account with an APR of 3% and annual compounding. As explained in Using Technology on page 211, fill the cells on an Excel spreadsheet as follows:

	A	B
1	rate (APR)	value
2	nper (Y)	value
3	pmt	0
4	pv (P)	value
5	FV (A)	= FV(B1, B2, B3, B4)

By changing the input values, answer the following questions.

a. What is the balance after 20 years?

b. If you double the APR in part (a), will the balance be double, more than double, or less than double the balance in part (a)?

c. If you double the number of years in part (a), will the balance be double, more than double, or less than double the balance in part (a)?

d. If you double the amount of the deposit in part (a), will the balance be double, more than double, or less than double the balance in part (a)?

111. **Compound Interest with Excel: Multiple Compoundings per Year.** Use the future value (FV) function in Excel to compute the balance in each of the following accounts.

a. An account with monthly compounding, an APR of 3%, and an initial deposit of $5000, after 5 years

b. An account with monthly compounding, an APR of 4.5%, and an initial deposit of $800, after 30 years

c. An account with daily compounding, an APR of 3.75%, and an initial deposit of $1000, after 50 years

112. **Effective Yield.** Use the effective yield function (EFFECT) in Excel to compute the APY (or effective yield) for each of the following accounts.

a. An account with quarterly compounding and an APR of 4%

b. An account with monthly compounding and an APR of 4%

c. An account with daily compounding and an APR of 4%

d. Based on the results of parts (a), (b), and (c), estimate the APY for an account with continuous compounding (where the number of compoundings becomes very large).

113. **Exponential Function.** Use a calculator, Excel, or Google to evaluate the following quantities.

a. $e^{3.2}$

b. $e^{0.065}$

c. The APY (effective yield) of an account with continuous compounding and an APR of 4%

UNIT 4C Savings Plans and Investments

Suppose you want to save money, perhaps for retirement or for your child's college expenses. You could deposit a lump sum of money today and let it grow through the power of compound interest. But what if you don't have a large lump sum to start such an account?

For most people, a more realistic way to save is by depositing smaller amounts on a regular basis. For example, you might put $50 a month into savings. Such long-term **savings plans** are so popular that many have special names—and some get special tax treatment—including Individual Retirement Accounts (IRAs), 401(k) plans, and 529 college savings plans.

BY THE WAY

Financial planners call any series of equal, regular payments an *annuity.* Savings plans are a type of annuity, as are loans that you pay with equal monthly payments.

The Savings Plan Formula

We can use an example to see how savings plans work. Suppose you deposit $100 into a savings plan at the end of each month. To keep the numbers simple, suppose that your plan pays interest monthly at an annual rate of APR = 12%, or 1% per month.

- You begin with $0 in the account. At the end of month 1, you make the first deposit of $100.
- At the end of month 2, you receive the monthly interest on the $100 already in the account, which is 1% × $100 = $1. In addition, you make your monthly deposit of $100. Your balance at the end of month 2 is

$$\underbrace{\$100}_{\text{prior balance}} + \underbrace{\$1.00}_{\text{interest}} + \underbrace{\$100}_{\text{new deposit}} = \$201.00$$

- At the end of month 3, you receive 1% interest on the $201 already in the account, or 1% × $201 = $2.01. Adding your monthly deposit of $100, you have a balance at the end of month 3 of

$$\underbrace{\$201.00}_{\text{prior balance}} + \underbrace{\$2.01}_{\text{interest}} + \underbrace{\$100}_{\text{new deposit}} = \$303.01$$

Table 4.5 continues these calculations through 6 months. In principle, we could extend this table indefinitely—but it would take a lot of work. Fortunately, there's a much easier way: the **savings plan formula** (see the box on the next page).

TABLE 4.5 Savings Plan Calculations (\$100 monthly deposits; APR = 12%, or 1% per month)

End of ...	Prior Balance	Interest on Prior Balance	End-of-Month Deposit	New Balance
Month 1	\$0	\$0	\$100	\$100
Month 2	\$100	1% × \$100 = \$1	\$100	\$201
Month 3	\$201	1% × \$201 = \$2.01	\$100	\$303.01
Month 4	\$303.01	1% × \$303.01 = \$3.03	\$100	\$406.04
Month 5	\$406.04	1% × \$406.04 = \$4.06	\$100	\$510.10
Month 6	\$510.10	1% × \$510.10 = \$5.10	\$100	\$615.20

Note: The last column shows the new balance at the end of each month, which is the sum of the prior balance, the interest, and the end-of-month deposit.

Technical Note

This version of the savings plan formula assumes that the same periods are used for the regular deposits (payments) and compounding. For example, if payments are made monthly, interest is also calculated and paid monthly.

Savings Plan Formula (Regular Payments)

$$A = \text{PMT} \times \frac{\left[\left(1 + \frac{\text{APR}}{n}\right)^{(nY)} - 1\right]}{\left(\frac{\text{APR}}{n}\right)}$$

where

A = accumulated savings plan balance
PMT = regular payment (deposit) amount
APR = annual percentage rate (as a decimal)
n = number of payment periods per year
Y = number of years

As with compound interest, the accumulated balance (A) is often called the *future value* (FV); the *present value* is the starting principal (P), which is \$0 because we assume the account has no balance before the payments begin.

EXAMPLE 1 Using the Savings Plan Formula

Use the savings plan formula to calculate the balance after 6 months for an APR of 12% and monthly payments of \$100.

Solution We have monthly payments of PMT = \$100, APR = 0.12, n = 12 because the payments are made monthly, and $Y = \frac{1}{2}$ because 6 months is a half-year. Using the savings plan formula, we can find the balance after 6 months:

$$A = \text{PMT} \times \frac{\left[\left(1 + \frac{\text{APR}}{n}\right)^{(nY)} - 1\right]}{\left(\frac{\text{APR}}{n}\right)} = \$100 \times \frac{\left[\left(1 + \frac{0.12}{12}\right)^{(12\times 1/2)} - 1\right]}{\left(\frac{0.12}{12}\right)}$$

$$= \$100 \times \frac{\left[(1.01)^{6} - 1\right]}{0.01} = \$615.20$$

Note that this answer agrees with the value in Table 4.5. ▸ Now try Exercises 15–18.

EXAMPLE 2 Retirement Plan

At age 30, Michelle starts an IRA to save for retirement by depositing \$100 at the end of each month. If she can count on an APR of 6%, how much will she have when she retires 35 years later at age 65? Compare the IRA's accumulated value to her total deposits over this time period.

MATHEMATICAL INSIGHT

Derivation of the Savings Plan Formula

We can derive the savings plan formula by looking at the example in Table 4.5 in a different way. Instead of calculating the balance at the end of each month (as in Table 4.5), we calculate how the value of each individual payment (deposit) and its interest changes by the end of month 6.

The first \$100 payment was made at the *end* of month 1. Therefore, by the end of month 6, this first payment has collected interest for $6 - 1 = 5$ months (at the end of months 2, 3, 4, 5, and 6). Using the general form of the compound interest formula (Unit 4B), with payment amount PMT = \$100 and *monthly* interest rate $i = 0.01$, the value of the first payment after $n = 5$ interest payments is

$$\text{PMT} \times (1 + i)^5 = \$100 \times 1.01^5$$

Similarly, the second \$100 payment has earned interest for $6 - 2 = 4$ months, so its value at the end of month 6 is

$$\text{PMT} \times (1 + i)^4 = \$100 \times 1.01^4$$

The table below continues the calculations. Note that the total for the second column agrees with the result found in Table 4.5. The last column shows how the compound interest formula applies in general to each individual payment.

End-of-month payment	Value after month 6	Value generalized for *N* months
1	$\$100 \times 1.01^5$	$\text{PMT} \times (1 + i)^{N-1}$
2	$\$100 \times 1.01^4$	$\text{PMT} \times (1 + i)^{N-2}$
3	$\$100 \times 1.01^3$	⋮
4	$\$100 \times 1.01^2$	
5	$\$100 \times 1.01$	$\text{PMT} \times (1 + i)^1$
6	\$100	PMT
Total	\$615.20	(sum of terms above)

The sum of the terms in the last column is the accumulated balance A for any savings plan after N months:

$$\begin{aligned} A = {} & \text{PMT} \\ & + \text{PMT} \times (1 + i)^1 \\ & + \cdots \\ & + \text{PMT} \times (1 + i)^{N-1} \end{aligned}$$

We can simplify this formula with the algebra shown in Equation 1 below, in which we first multiply both sides by $(1 + i)^1$ and then subtract the original equation from the new equation. Note that all but two terms cancel on the right, leaving us with

$$A(1 + i) - A = PMT(1 + i)^N - \text{PMT}$$

The left side of this equation simplifies to Ai (because $A(1 + i) - A = A + Ai - A = Ai$), and we can factor out PMT to write the right side as $\text{PMT} \times \left[(1 + i)^N - 1\right]$. The full equation is now

$$Ai = \text{PMT} \times \left[(1 + i)^N - 1\right]$$

Dividing both sides by i gives us the savings plan formula:

$$A = \text{PMT} \times \frac{\left[(1 + i)^N - 1\right]}{i}$$

This is the same as the savings plan formula given in the text if you substitute $i = \text{APR}/n$ for the interest rate per period and $N = nY$ for the number of payments (where n is the number of payments per year and Y is the number of years).

Equation 1:

$$\begin{aligned} A(1 + i) &= \text{PMT}(1 + i)^1 + \cdots + \text{PMT}(1 + i)^{N-1} + \text{PMT}(1 + i)^N \\ -A &= -[\text{PMT} + \text{PMT}(1 + i)^1 + \cdots + \text{PMT}(1 + i)^{N-1}] \\ \hline A(1 + i) - A &= -\text{PMT} + \text{PMT}(1 + i)^N \\ &= \text{PMT}(1 + i)^N - \text{PMT} \end{aligned}$$

Technical Note

A savings plan in which payments are made at the end of each month is called an *ordinary annuity*. A plan in which payments are made at the beginning of each period is called an *annuity due*. In both cases, the accumulated amount, *A*, at some future date is called the *future value* of the annuity. The formulas in this unit apply only to ordinary annuities.

Solution We use the savings plan formula with payments of PMT = \$100, an interest rate of APR = 0.06, and $n = 12$ for monthly deposits. The balance after $Y = 35$ years is

$$A = \text{PMT} \times \frac{\left[\left(1 + \frac{\text{APR}}{n}\right)^{(nY)} - 1\right]}{\left(\frac{\text{APR}}{n}\right)} = \$100 \times \frac{\left[\left(1 + \frac{0.06}{12}\right)^{(12\times35)} - 1\right]}{\left(\frac{0.06}{12}\right)}$$

$$= \$100 \times \frac{\left[(1.005)^{420} - 1\right]}{0.005}$$

$$= \$142{,}471.03$$

Because 35 years is 420 months ($35 \times 12 = 420$), the total amount of Michelle's deposits over 35 years is

$$420 \text{ months} \times \frac{\$100}{\text{months}} = \$42{,}000$$

She will deposit a total of \$42,000 over 35 years. However, thanks to compounding, her IRA will have a balance of more than \$142,000—*more than three times* the amount of her contributions. ▶ **Now try Exercises 19–22.**

Planning Ahead with Savings Plans

Most people start savings plans with a particular goal in mind, such as saving enough for retirement or enough to buy a new car in a couple of years. For planning ahead, the important question is this: Given a financial goal (the total amount, *A*, desired after a certain number of years), what regular payments are needed to reach the goal? The following two examples show how the calculations work.

EXAMPLE 3 College Savings Plan at 3%

When your child is born, you set a goal of building a \$100,000 college fund in 18 years by making regular, end-of-month deposits. Assuming an APR of 3%, calculate how much you should deposit monthly. How much of the final value comes from actual deposits and how much from interest?

Solution The goal is to accumulate $A = \$100{,}000$ over $Y = 18$ years. The interest rate is APR = 0.03, and monthly payments mean $n = 12$. The goal is to calculate the required monthly payment, PMT. We therefore need to solve the savings plan formula for PMT and then substitute the given values for *A*, APR, *n*, and *Y*. The following steps show the calculation.

Start with the savings plan formula:

$$A = \text{PMT} \times \frac{\left[\left(1 + \frac{\text{APR}}{n}\right)^{(nY)} - 1\right]}{\left(\frac{\text{APR}}{n}\right)}$$

Multiply both sides by $\left(\frac{\text{APR}}{n}\right)$ and divide both sides by $\left[\left(1+\frac{\text{APR}}{n}\right)^{(nY)}-1\right]$:

$$A \times \frac{\left(\frac{\text{APR}}{n}\right)}{\left[\left(1+\frac{\text{APR}}{n}\right)^{(nY)}-1\right]} = \text{PMT} \times \frac{\cancel{\left[\left(1+\frac{\text{APR}}{n}\right)^{(nY)}-1\right]}}{\cancel{\left(\frac{\text{APR}}{n}\right)}} \times \frac{\cancel{\left(\frac{\text{APR}}{n}\right)}}{\cancel{\left[\left(1+\frac{\text{APR}}{n}\right)^{(nY)}-1\right]}}$$

Interchange the left and right sides:

$$\text{PMT} = A \times \frac{\left(\frac{\text{APR}}{n}\right)}{\left[\left(1+\frac{\text{APR}}{n}\right)^{(nY)}-1\right]}$$

Substitute the values APR = 0.03, n = 12, Y = 18, and A = \$100,000:

$$\text{PMT} = \$100{,}000 \times \frac{\left(\frac{0.03}{12}\right)}{\left[\left(1+\frac{0.03}{12}\right)^{(12\times18)}-1\right]}$$

$$= \$100{,}000 \times \frac{0.0025}{\left[(1.0025)^{216}-1\right]}$$

$$= \$349.72$$

Assuming a constant APR of 3%, monthly payments of \$349.72 will give you \$100,000 after 18 years. During that time, you will deposit a total of

$$18\ \cancel{\text{yr}} \times \frac{12\ \cancel{\text{mo}}}{\cancel{\text{yr}}} \times \frac{\$349.72}{\cancel{\text{mo}}} \approx \$75{,}540$$

Just over three-fourths of the \$100,000 comes from your actual deposits; the other one-fourth is the result of compound interest. ▶ Now try Exercises 23–26.

Think About It Compare the result of Example 3 to that of Example 8 in Unit 4B. Notice that both have the same interest rate of 3% and both have a value of \$100,000 at the end of 18 years, but one does it with a savings plan and the other by starting with a large, lump-sum deposit. Discuss the pros and cons of each approach. How would *you* decide which approach to use?

EXAMPLE 4 A Comfortable Retirement

You would like to retire 25 years from now and have a retirement fund from which you can draw an income of \$50,000 per year—forever! How can you do it? Assume a constant APR of 7%.

Solution You can achieve your goal by building a retirement fund that is large enough to earn \$50,000 per year *from interest alone.* In that case, you can withdraw the interest for your living expenses while leaving the principal untouched. The principal will then continue to earn the same \$50,000 interest year after year (assuming there is no change in interest rates).

USING TECHNOLOGY

The Savings Plan Formula

Standard Calculators As with compound interest calculations, the only "trick" to using the savings plan formula on a standard calculator is following the correct order of operations. The following calculation shows the correct order for the numbers from Example 2, in which PMT = \$100, APR = 0.06, n = 12, and Y = 35 years; as always, be sure that you *do not round any answers until the end of the calculation*.

General Procedure	Our Example	Calculator Steps*	Output
$A = \text{PMT} \times \dfrac{\left[\left(1+\dfrac{\text{APR}}{n}\right)^{(nY)} - 1\right]}{\left(\dfrac{\text{APR}}{n}\right)}$ 1. inner parentheses; 2. exponent; 3. outer parentheses; 4. divide; 5. multiply	$A = 100 \times \dfrac{\left[\left(1+\dfrac{0.06}{12}\right)^{(12\times 35)} - 1\right]}{\left(\dfrac{0.06}{12}\right)}$ 1. inner parentheses; 2. exponent; 3. outer parentheses; 4. divide; 5. multiply	Step 1 1 (+) .006 (÷) 12 (=)	1.005
		Step 2 (∧) (( ) 12 (×) 35 ( )) (=)	8.12355...
		Step 3 (−) 1 (=)	7.12355...
		Step 4 (÷) (( ) 0.06 (÷) 12 ( )) (=)	1424.71029
		Step 5 (×) 100 (=)	142471.029

*If your calculator does not have parentheses keys, do those steps before you start and keep track of the results on paper or in the calculator's memory.

Excel We use the built-in function FV, as we did for the compound interest formula (see Using Technology in Unit 4B, pp. 211 and 215). In this case, the inputs are as shown in the screen shot at the right.

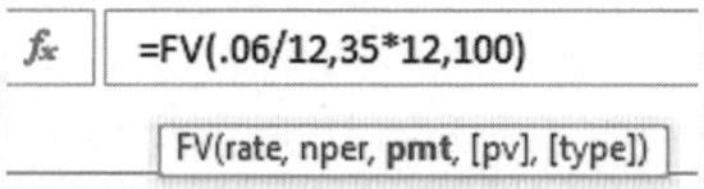

Microsoft Excel 2016, Windows 10, Microsoft Corporation.

Note:

- *rate* is the interest rate for each compounding period, which in this case is the monthly interest rate, APR/n = 0.06/12.
- *nper* is the total number of compounding periods, which in this case is nY = 12 × 35.
- *pmt* is the monthly payment of \$100.
- *pv* is left blank, because there is no starting principal (present value) in this case.
- *type* is also left blank, which indicates that type = 0. This is the default value; it indicates monthly payments made at the *end* of each period (month), as is the case for all the examples in this text. For the rarer case in which payments are made at the beginning of each payment period, you would set type = 1.

You may find it easier to see the inputs if you use the dialog box that comes up when you choose the FV function (the method for choosing the function varies by Excel version), which will look something like this:

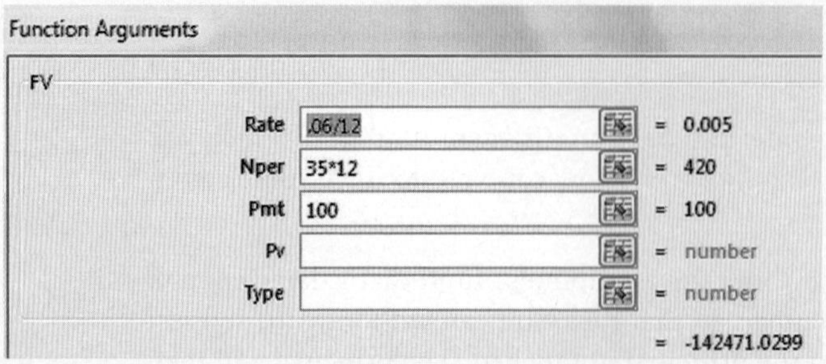

You may wonder why the result (shown at the lower right) is negative. This is an artifact of the way Excel's financial functions handle cash flow. Positive amounts are considered *inflows* of money and negative amounts are *outflows*. In this case, the positive monthly payments of \$100 are inflows into the savings plan. The future value is then negative because it is the amount that you will eventually draw *out* of the savings plan to pay for the home, college, or whatever else you were saving for. Note that, aside from this subtlety, the result is the same as we found in Example 2.

What balance do you need to earn \$50,000 annually from interest? Since we are assuming an APR of 7%, the \$50,000 must be 7% = 0.07 of the total balance. That is,

$$\$50{,}000 = 0.07 \times (\text{total balance})$$

Dividing both sides by 0.07, we find

$$\text{total balance} = \frac{\$50{,}000}{0.07} = \$714{,}286$$

In other words, with a 7% APR, a balance of about \$715,000 allows you to withdraw \$50,000 per year without ever reducing the principal.

Let's assume you will try to accumulate this balance of $A = \$715{,}000$ by making regular monthly deposits into a savings plan. We have APR = 0.07, $n = 12$ (for monthly deposits), and $Y = 25$ years. As in Example 3, we calculate the required monthly deposits by using the savings plan formula solved for PMT.

$$\text{PMT} = \frac{A \times \frac{\text{APR}}{n}}{\left[\left(1 + \frac{\text{APR}}{n}\right)^{(nY)} - 1\right]} = \frac{\$715{,}000 \times \frac{0.07}{12}}{\left[\left(1 + \frac{0.07}{12}\right)^{(12\times 25)} - 1\right]}$$

$$= \frac{\$715{,}000 \times 0.0058333}{\left[(1.0058333)^{300} - 1\right]}$$

$$= \$882.64$$

If you deposit about \$883 per month over the next 25 years, you will achieve your retirement goal—assuming you can count on a 7% APR (which is high by historical standards). Although saving almost \$900 per month is a lot to ask, it can be easier than it sounds thanks to special tax treatment for retirement plans (see Unit 4E). ▶ Now try Exercises 27–28.

BY THE WAY

An account that provides a permanent source of income without reducing its principal is called an *endowment.* Many charitable foundations use endowments. They spend each year's interest (or a portion of the interest) on their charitable activities, leaving the principal untouched to earn interest in future years.

Total and Annual Return

In the examples so far, we've assumed that you get a constant interest rate for a long period of time. In reality, interest rates usually vary over time. Consider a case in which you initially deposit \$1000 and it grows to \$1500 in 5 years. Although the interest rate may have varied during the 5 years, we can still describe the change in both total and annual terms.

Your **total return** is the percentage change in the investment value over the 5-year period:

$$\text{total return} = \frac{\text{new value} - \text{starting principal}}{\text{starting principal}} \times 100\%$$

$$= \frac{\$1500 - \$1000}{\$1000} \times 100\% = 50\%$$

The total return on this investment is 50% over 5 years.

Your **annual return** is the *average annual rate* at which your money grew over the 5 years. That is, it is the constant annual percentage yield (APY) that would give the same result in 5 years. One way to determine this annual return is through trial and error. If you test APY = 8.5% = 0.085 with a starting principal $P = \$1000$ and number of years $Y = 5$, you'll find that the principal grows to approximately $A = \$1500$:

$$A = P \times (1 + \text{APY})^{Y} = \$1000 \times (1 + 0.085)^{5} = \$1503.66$$

You can find a more exact answer using the annual return formula in the following box.

Total and Annual Return

Consider an investment that grows from an original principal P to a later accumulated balance A. The **total return** is the percentage change in the investment value:

$$\text{total return} = \frac{(A - P)}{P} \times 100\%$$

The **annual return** is the annual percentage yield (APY) that would give the same overall growth over Y years. The formula is

$$\text{annual return} = \left(\frac{A}{P}\right)^{(1/Y)} - 1$$

This formula gives the annual return as a decimal; multiply by 100% to express it as a percentage. (See Exercise 70 to derive this formula.)

USING TECHNOLOGY

Fractional Powers (Roots)

Recall that raising a number to a fractional power such as $1/Y$ is the same as taking the Yth root. Some calculators have a key labeled $x^{1/y}$ or $\sqrt[y]{x}$, but it is often easier to use the exponent key (∧) and parentheses to enter the calculation directly.

Example: Calculate

$$\sqrt[4]{2.8} = 2.8^{1/4}$$

by pressing

2.8 ∧ (1 ÷ 4) =.

In Excel, use the ∧ symbol to raise to a power, with the fractional power in parentheses, as shown in the screen shot below.

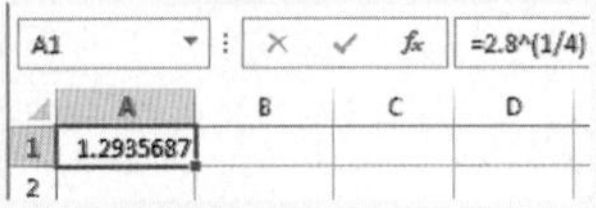

Microsoft Excel 2016, Windows 10, Microsoft Corporation.

EXAMPLE 5 Mutual Fund Gain

You invest \$3000 in the Clearwater mutual fund. Over 4 years, your investment grows in value to \$8400. What are your total and annual returns for the 4-year period?

Solution You have a starting principal of $P = \$3000$ and an accumulated value of $A = \$8400$ after $Y = 4$ years. Your total and annual returns are

$$\text{total return} = \frac{(A - P)}{P} \times 100\% = \frac{(\$8400 - \$3000)}{\$3000} \times 100\% = 180\%$$

$$\text{annual return} = \left(\frac{A}{P}\right)^{1/Y} - 1 = \left(\frac{\$8400}{\$3000}\right)^{1/4} - 1$$

$$= \sqrt[4]{2.8} - 1 \approx 0.294 = 29.4\%$$

Your total return is 180%, meaning that the value of your investment after 4 years is 1.8 times its original value. Your annual return is approximately 0.294, or 29.4%, meaning that your investment has grown by an average of 29.4% each year. ▶ **Now try Exercises 29–32.**

EXAMPLE 6 Investment Loss

You purchased shares in NewWeb.com for \$2000. Three years later, you sold them for \$1100. What were your total return and annual return on this investment?

Solution You had a starting principal of $P = \$2000$ and an accumulated value of $A = \$1100$ after $Y = 3$ years. Your total and annual returns were

$$\text{total return} = \frac{(A - P)}{P} \times 100\% = \frac{(\$1100 - \$2000)}{\$2000} \times 100\% = -45\%$$

$$\text{annual return} = \left(\frac{A}{P}\right)^{1/Y} - 1 = \left(\frac{\$1100}{\$2000}\right)^{(1/3)} - 1 = \sqrt[3]{0.55} - 1 = -0.18$$

Your total return was −45%, meaning that your investment lost 45% of its original value. Your annual return was −0.18, or −18%, meaning that your investment lost an average of 18% of its value each year. ▶ **Now try Exercises 33–36.**

Types of Investments

By combining what we've covered about savings plans with the ideas of total and annual return, we can now study investment options. Most investments fall into one of the three basic categories described in the following box.

Three Basic Types of Investments

Stock (or *equity*) gives you a share of ownership in a company. You invest by purchasing shares of the stock, and the only way to get your money out is to sell the stock. Because stock prices change with time, the sale may give you either a gain or a loss on your original investment.

A **bond** (or *debt*) represents a promise of future cash. Bonds are usually issued by either a government or a corporation. The issuer pays you simple interest (as opposed to compound interest) and promises to pay back your initial investment plus interest at some later date.

Cash investments include money you deposit into bank accounts, certificates of deposit (CD), and U.S. Treasury bills. Cash investments generally earn interest.

BY THE WAY

There are many other types of investments besides the basic three, such as rental properties, precious metals, commodities, futures, and derivatives. These investments generally are more complex and often have higher risk than the basic three.

There are two basic ways to invest in any of these categories: (1) You can invest directly, which means buying individual investments yourself (often through a broker). (2) You can invest indirectly by purchasing shares in a **mutual fund**, where a professional fund manager invests your money along with the money of others participating in the fund.

Investment Considerations: Liquidity, Risk, and Return

No matter what type of investment you make, you should evaluate the investment in terms of three general considerations.

- **Liquidity**: How difficult is it to take out your money? An investment from which you can withdraw money easily, such as an ordinary bank account, is said to be **liquid**. The liquidity of an investment like real estate is much lower because real estate can be difficult to sell.
- **Risk**: Is your investment principal at **risk**? The safest investments are federally insured bank accounts and U.S. Treasury bills—there's virtually no risk of losing the principal you've invested. Stocks and bonds are much riskier because they can drop in value, in which case you may lose part or all of your principal.
- **Return**: How much **return** (total or annual) can you expect on your investment? A higher return means you earn more money. In general, low-risk investments offer relatively low returns, while high-risk investments offer the prospects of higher returns—along with the possibility of losing your principal.

BY THE WAY

The U.S. Treasury issues bills, notes, and bonds. *Treasury bills* are essentially cash investments that are highly liquid and very safe. *Treasury notes* are essentially bonds with 2- to 10-year terms. *Treasury bonds* have terms longer than 10 years.

Historical Returns

One of the most difficult tasks of investing is trying to balance risk and return. Although there is no way to predict the future, historical trends offer at least some guidance. To study historical trends, financial analysts generally look at an *index* that describes the overall performance of some category of investment. The best-known index is the **Dow Jones Industrial Average (DJIA)**, which reflects the average prices of the stocks of 30 large companies. (The 30 companies are chosen by the editors of the *Wall Street Journal*.) Figure 4.5 shows historical data for the DJIA.

BY THE WAY

The DJIA is the most famous stock index, but others that track larger numbers of stocks may give a better picture of the overall market. These include the *Standard and Poor's 500* (S&P 500), which tracks 500 large-company stocks; the *Russell 2000*, which tracks 2000 small-company stocks; and the *NASDAQ composite*, which tracks 100 large-company stocks listed on the NASDAQ exchange.

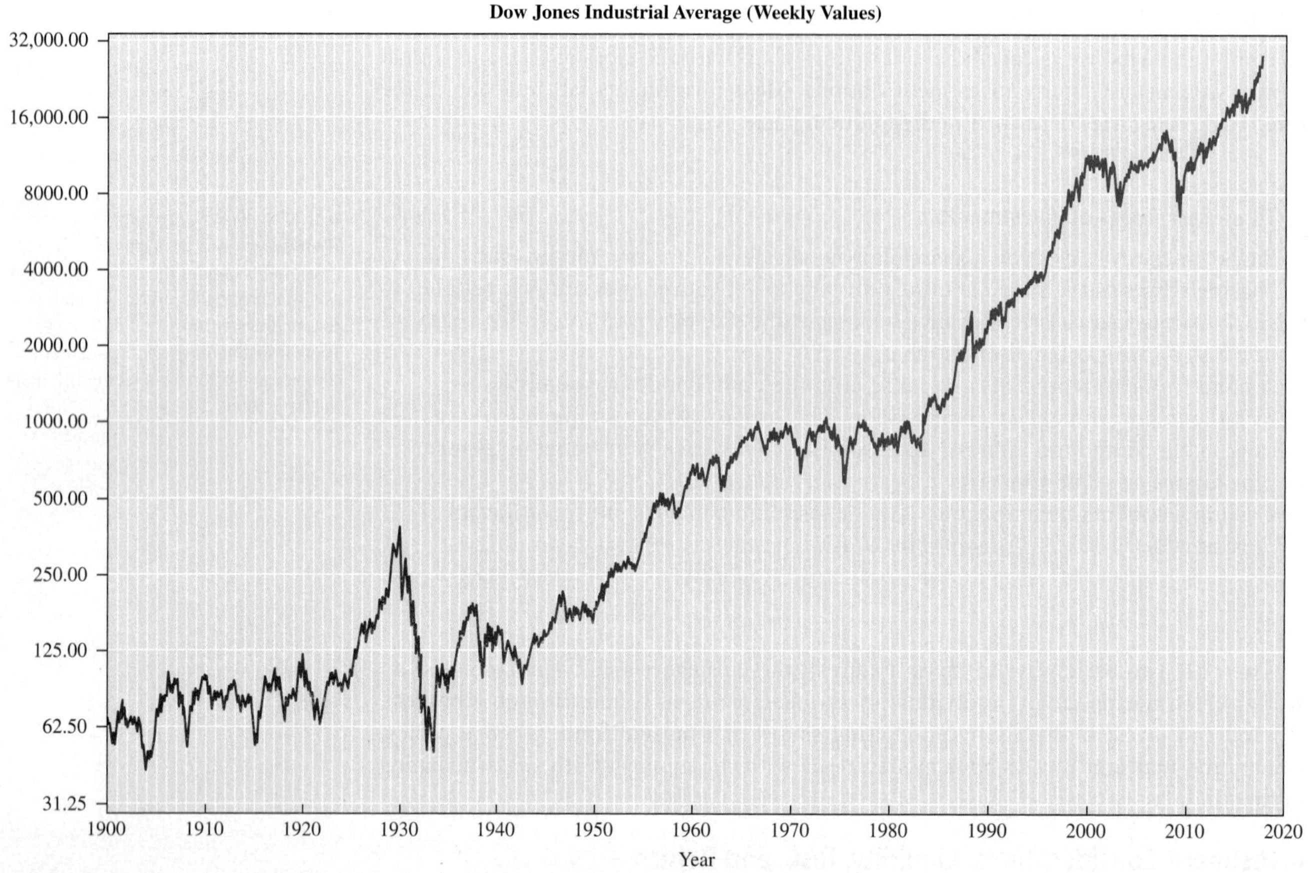

FIGURE 4.5 Historical values of the Dow Jones Industrial Average, 1900 through mid-2017. Note that the numbers on the vertical axis double with each equivalent increment in height; this type of exponential graph makes it easier to see the up-and-down trends that occurred when the value of the DJIA was low compared to its value today.
Sources: Dow Jones & Company; StockCharts.com.

TABLE 4.6

Historical Returns by Category, 1900–2016

Category	Average Annual Return
Stocks	6.4%
Bonds	2.0%
Cash	0.8%

Source: Credit Suisse Global Investment Returns Yearbook 2017.

Notice that the long-term trend shown in Figure 4.5 looks quite good. In fact, stocks have historically proven to be a far better long-term investment than bonds or cash (Table 4.6). Over shorter time periods, however, stocks can be risky. For example, if just before the crash of 1929 you had invested in a mutual fund that tracked the DJIA, you would have had to wait some 25 years before the fund returned to its pre-crash value.

Think About It Find today's closing value for the DJIA. Suppose you had invested in a mutual fund that tracked the DJIA on March 1, 2017, a day on which the DJIA reached 21,116 (a record high at the time). Would your investment be worth more or less today? By how much?

EXAMPLE 7 Historical Returns

Suppose your great-great-grandmother invested \$100 at the end of 1900 in each of three funds that tracked the averages of stocks, bonds, and cash, respectively. Assuming that her investments grew at the rates given in Table 4.6, approximately how much would each investment have been worth at the end of 2016?

Solution We find the value of each investment with the compound interest formula (for interest compounded once a year), setting the interest rate (APR) to the average annual return for each category. In each case, the starting principal is $P = \$100$ and $Y = 116$ (the number of years from the end of 1900 to the end of 2016).

$$\text{Stocks (annual return} = 0.064\text{): } A = P \times (1 + \text{APR})^Y$$
$$= \$100 \times (1 + 0.064)^{116} \approx \$133{,}000$$
$$\text{Bonds (annual return} = 0.020\text{): } A = P \times (1 + \text{APR})^Y$$
$$= \$100 \times (1 + 0.020)^{116} \approx \$994$$
$$\text{Cash (annual return} = 0.008\text{): } A = P \times (1 + \text{APR})^Y$$
$$= \$100 \times (1 + 0.008)^{116} \approx \$252$$

Notice the enormous difference in the growth of $100 for these types of investments. Of course, the fact that stocks have been the long-term investment of choice in the past is no guarantee that they will remain the best long-term investment for the future.

▶ Now try Exercises 37–38.

Think About It Typically, financial planners recommend that younger people invest a larger proportion of their money in stocks and less in cash, while recommending the opposite to people who are retired or nearing retirement. Do you think this is good advice? Why or why not?

Financial Data

If you decide to invest, you can track your investments online. Let's look briefly at what you must know to understand commonly published data about stocks, bonds, and mutual funds.

Stocks

In general, there are two ways to make money on stocks:

- You can make money if you sell a stock for more than you paid for it, in which case you have a **capital gain** on the sale of the stock. Of course, you also can lose money on a stock (a *capital loss*) if you sell shares for less than you paid for them or if the company goes into bankruptcy.
- You can make money while you own the stock if the corporation distributes part or all of its profits to stockholders as **dividends**. Each share of stock is paid the same dividend, so the amount of money you receive depends on the number of shares you own.

Before you invest in any stock, you should check its current stock quote; Figure 4.6 explains the key data that you'll find in a typical stock quote. In addition, you should learn more about the company by studying its annual report and visiting its website. You can also get independent research reports from many investment services (usually for a fee) or by working with a stockbroker (to whom you pay commissions when you buy or sell stock).

BY THE WAY

A *corporation* is a legal entity created to conduct a business. Ownership is held through shares of stock. For example, owning 1% of a company's stock means owning 1% of the company. Shares of stock in *privately held* corporations are owned only by a limited group of people. Shares of stock in *publicly held* corporations are traded on a public exchange, such as the New York Stock Exchange or the NASDAQ, and anyone can own them.

EXAMPLE 8 Understanding a Stock Quote

Answer the following questions by assuming that Figure 4.6 shows an actual Microsoft stock quote that you found online today.

a. What is the symbol for Microsoft stock?
b. What was the price per share at the start of the day?
c. Based on the current price, what is the total value of the shares that have been traded so far today?
d. What fraction of all Microsoft shares have been traded so far today?
e. Suppose you own 100 shares of Microsoft. Based on the current price and dividend yield, what are your shares currently worth and what total dividend should you expect to receive this year?
f. How much profit did Microsoft earn per share in the past year?
g. How much total profit did Microsoft earn in the past year?

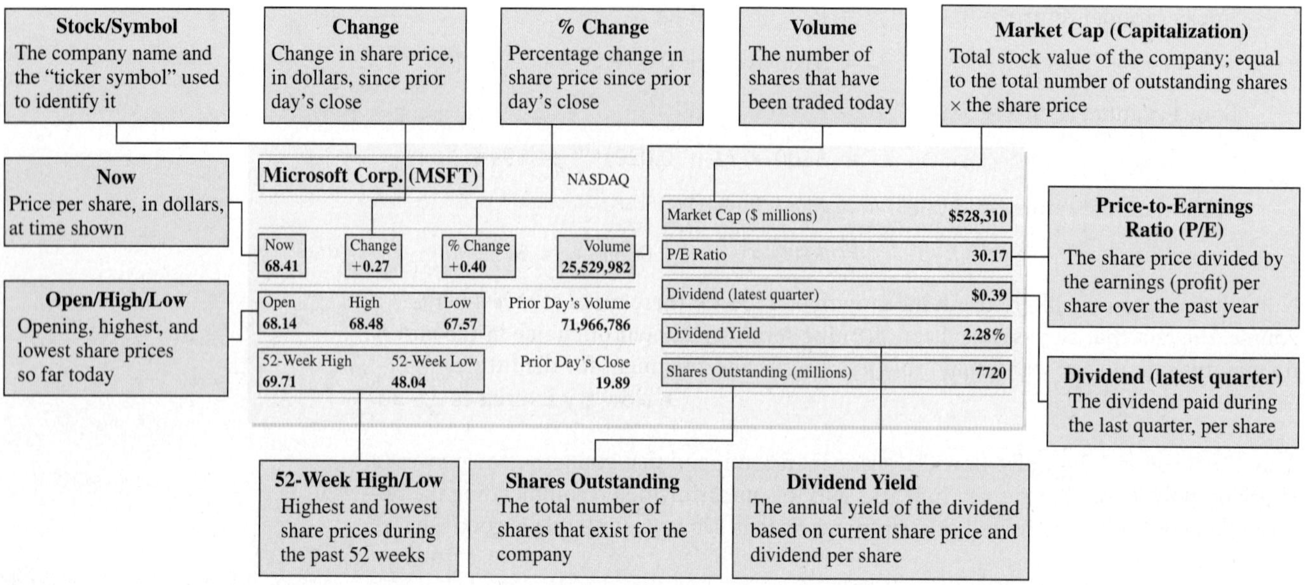

FIGURE 4.6 Explanation of the key data that you'll find in an online stock quote.

Solution

a. As shown at the top of the quote, Microsoft's stock symbol is MSFT.

b. The "Open" value is the price at the start of the day, which was \$68.14.

c. The volume shows that 25,529,982 shares of Microsoft stock were traded today. At the current price of \$68.41 per share, the value of these shares is

$$25{,}529{,}982 \text{ shares} \times \$68.41/\text{share} \approx \$1{,}747{,}000{,}000$$

The total value of shares traded today is about \$1.747 billion.

d. We divide the 25,529,982 shares traded today by the total number of shares outstanding, which is quoted as 7720 million, or 7,720,000,000, to find that about 0.0033, or 0.33%, of all shares have traded today.

e. At the current price, your 100 shares are worth $100 \times \$68.41 = \$6841$. The dividend yield is 2.28%, so at that rate you should earn $\$6841 \times 0.0228 = \$155.97$ in dividend payments this year.

f. The P/E ratio of 30.17 indicates that Microsoft's share price is 30.17 times its earnings per share for the past year. Therefore, its earnings (or profit) per share were

$$\text{earnings per share} = \frac{\text{share price}}{\text{P/E ratio}} = \frac{\$68.41}{30.17} = \$2.27$$

g. From part (f) we know that Microsoft earned \$2.27 per share in profits. Multiplying this by the total number of outstanding shares, we find that Microsoft's total profit for the year was about $\$2.27 \times 7720$ million $\approx \$17{,}524$ million, or \$17.524 billion. ▶ **Now try Exercises 39–46.**

BY THE WAY

Historically, most gains from stocks have come from increases in stock prices, rather than from dividends. Stocks in companies that pay consistently high dividends are called *income stocks* because they provide ongoing income to stockholders. Stocks in companies that reinvest most profits in hopes of growing larger are called *growth stocks*. Of course, a so-called "growth stock" can still decline or the company can go out of business.

Think About It Find today's stock quote for Microsoft. How has it changed since the quote in Figure 4.6, which was from mid-2017? What does this change suggest about how Microsoft has been doing as a company?

Building a Portfolio

Before you bought a new television for a few hundred dollars, you'd probably do a fair amount of research to make sure that you were getting a good buy. You should be even more diligent when making investments that may determine your entire financial future.

The best way to plan your savings is to learn about investments by reading financial news and some of the many books and websites devoted to finance. You may also want to consult a professional financial planner (but be sure it is one who follows a "fiduciary rule," meaning that the planner puts the client's interests ahead of his or her own). With this background, you will be prepared to create a personal financial *portfolio* (set of investments) that meets your needs.

Most financial advisors recommend that you create a *diversified* portfolio—that is, a portfolio with a mixture of low-risk and high-risk investments. No single mixture is right for everyone. Your portfolio should balance risk and return in a way that is appropriate for your situation. For example, if you are young and retirement is far in the future, you may be willing to have a relatively risky portfolio that offers the hope of high returns. In contrast, if you are already retired, you may want a low-risk portfolio that promises a safe and steady stream of income.

No matter how you structure your portfolio, the key to achieving your financial goals is making sure that you save enough money. You can use the tools in this unit to help you determine what is "enough." Make a reasonable estimate of the annual return you can expect from your overall portfolio. Use this annual return as the interest rate in the savings plan formula, and calculate how much you must invest each month or each year to meet your goals (see Examples 3 and 4). Then make sure you actually put this money in your investment plan. If you need further motivation, consider this: Every \$100 you spend today is gone. However, even at a fairly low (by historical standards) annual return of 4%, every \$100 you invest today will be worth \$148 in 10 years, \$219 in 20 years, and \$711 in 50 years.

Bonds

Most bonds that are issued have three main characteristics:

- The **face value** (or *par value*) of the bond is the price you must pay the issuer to buy it at the time it is issued.
- The **coupon rate** of the bond is the *simple interest rate* that the issuer promises to pay. For example, a coupon rate of 8% on a bond with a face value of \$1000 means that the issuer will pay you interest of $8\% \times \$1000 = \$80$ each year.
- The **maturity date** of the bond is the date on which the issuer promises to repay the face value of the bond.

Bonds would be simple if that were the end of the story. However, bonds can also be bought and sold after they are issued, in what is called the *secondary bond market*. For example, suppose you own a bond with a \$1000 face value and a coupon rate of 8%. Further suppose that new bonds with the same level of risk and same time to maturity are issued with a coupon rate of 9%. In that case, no one would pay \$1000 for your bond because the new bonds offer a higher interest rate. However, you may be able to sell your bond at a *discount*—that is, for less than its face value. In contrast, suppose that new bonds are issued with a coupon rate of 7%. In that case, buyers will prefer your 8% bond to the new bonds and therefore may pay a *premium* for your bond—a price greater than its face value.

Consider a case in which you buy a bond with a face value of \$1000 and a coupon rate of 8% for only \$800. The bond issuer will still pay simple interest of 8% of \$1000, or \$80 per year. However, because you paid only \$800 for the bond, *your* return for each year is

$$\frac{\text{amount you earn}}{\text{amount you paid}} = \frac{\$80}{\$800} = 0.1 = 10\%$$

BY THE WAY

A company that needs cash can raise it either by issuing new shares of stock or by issuing bonds. Issuing new shares of stock reduces the ownership fraction represented by each share and hence can depress the value of the shares. Issuing bonds obligates the company to pay interest to bondholders. Companies must balance these factors in deciding whether to raise cash through bond issues or stock offerings.

BY THE WAY

Bonds are graded in terms of risk by independent rating services. Bonds with a AAA rating are presumed to have the lowest risk and bonds with a D rating have the highest risk. Unfortunately, during the financial crisis that began in 2007, many bond ratings turned out to have been overstated.

More generally, the **current yield** of a bond is defined as the amount of interest it pays each year divided by the bond's current price (*not* its face value).

Current Yield of a Bond

$$\text{current yield} = \frac{\text{annual interest payment}}{\text{current price of bond}}$$

A bond selling at a discount from its face value has a current yield that is higher than its coupon rate. The reverse is also true: A bond selling at a premium over its face value has a current yield that is lower than its coupon rate. These facts lead to the rule that *bond prices and yields move in opposite directions.*

Bond prices are usually quoted in *points,* which means percentage of face value. Most bonds have a face value of \$1000. Thus, for example, a bond that closes at 102 points is selling for 102% × \$1000 = \$1020.

EXAMPLE 9 **Bond Interest**

The closing price of a U.S. Treasury bond with a face value of \$1000 is quoted as 105.97 points, for a current yield of 3.7%. If you buy this bond, how much annual interest will you receive?

Solution The 105.97 points means the bond is selling for 105.97% of its face value or

$$105.97\% \times \$1000 = \$1059.70$$

This is the current price of the bond. We are also given its current yield of 3.7%, so we can solve the current yield formula to find the annual interest payment:

Start with the current yield formula:	$\text{current yield} = \dfrac{\text{annual interest}}{\text{current price}}$
Multiply both sides by *current price:*	$\text{current yield} \times \text{current price} = \dfrac{\text{annual interest}}{\cancel{\text{current price}}} \times \cancel{\text{current price}}$
Simplify and interchange the left and right sides:	$\text{annual interest} = \text{current yield} \times \text{current price}$
Use the given current yield (3.7% = 0.037) and the value found above for current price	$\text{annual interest} = 0.037 \times \$1059.70 = \$39.21$

The annual interest payment on this bond is \$39.21. ▶ **Now try Exercises 47–54.**

Mutual Funds

When you buy shares in a mutual fund, the fund manager takes care of the day-to-day decisions about when to buy and sell individual stocks or bonds in the fund. Therefore, in comparing mutual funds, the most important factors are the fees charged for investing and how well the fund performs. Figure 4.7 shows a sample mutual fund quote, in this case for a Vanguard fund that is designed to track the S&P 500 stock index. The quote makes it easy to see the past performance of the fund, which you can compare to that of other funds. Of course, as stated in every mutual fund prospectus, *past performance is no guarantee of future results.*

Most mutual fund tables do not show the fees charged. For that, you must call or check the website of the company offering the mutual fund. Because fees are generally withdrawn automatically from your mutual fund account, they can have a big impact on your

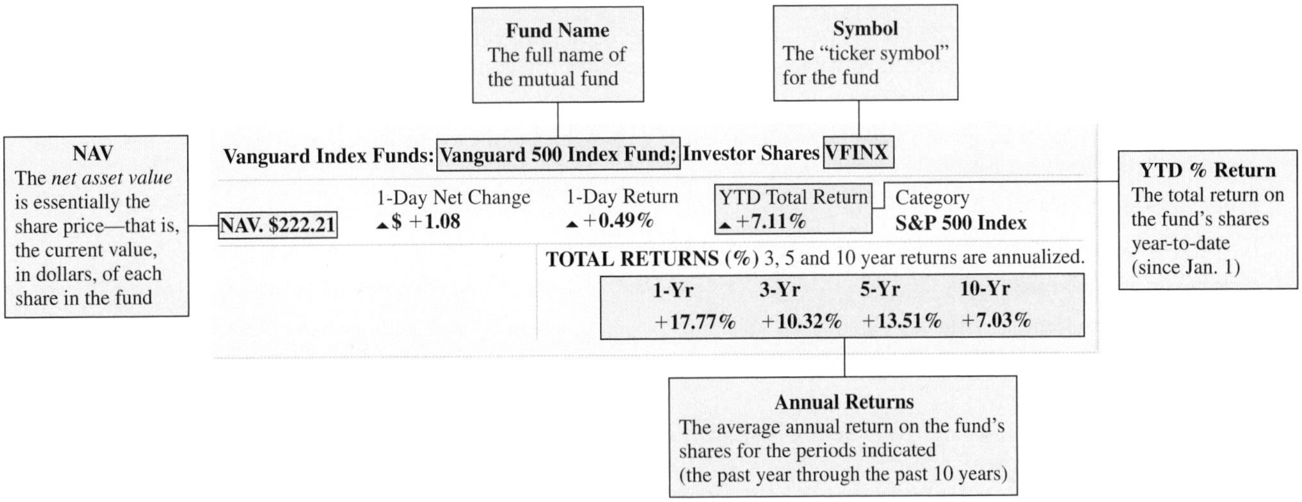

FIGURE 4.7 Explanation of the key data that you'll find in an online mutual fund quote.

long-term gains. For example, if you invest \$1000 in a fund that charges a 5% annual fee, only \$950 is actually invested. Over many years, this can significantly reduce your total return.

EXAMPLE 10 Understanding a Mutual Fund Quote

Answer the following questions by assuming that Figure 4.7 shows an actual Vanguard 500 mutual fund quote that you found online today.

a. Suppose you decide to invest \$3000 in this fund today. How many shares will you be able to buy?

b. Suppose you had invested \$3000 in this fund 3 years ago. How much would your investment be worth now?

c. Suppose you had invested \$3000 in this fund 10 years ago. How much would your investment be worth now?

Solution

a. To find the number of shares you can buy, divide your investment of \$3000 by the current share price, which is the NAV (net asset value) of \$222.21:

$$\frac{\$3000}{\$222.21} \approx 13.5$$

Your \$3000 investment will buy 13.5 shares in the fund.

b. The annual return for the past 3 years was 10.32% = 0.1032. We use this value as the APR in the compound interest formula, with a term of $Y = 3$ years and a starting principal of $P = \$3000$:

$$A = P \times (1 + \text{APR})^Y = \$3000 \times (1 + 0.1032)^3 \approx \$4028$$

Your investment would have increased by more than \$1000 in value to about \$4028.

c. The annual return for the past 10 years was 7.03% = 0.0703. We use this value as the APR in the compound interest formula, with a term of $Y = 10$ years and a starting principal of $P = \$3000$:

$$A = P \times (1 + \text{APR})^Y = \$3000 \times (1 + 0.0703)^{10} \approx \$5918$$

Over 10 years, your \$3000 investment would have nearly doubled in value, to \$5918.

▶ Now try Exercises 55–56.

BY THE WAY

Mutual funds collect fees in two ways. Some funds charge a commission, or *load,* when you buy or sell shares. Funds that do not charge commissions are called *no-load* funds. Nearly all funds charge an *annual fee,* which is usually a percentage of your investment's value. In general, fees are higher for funds that require more research on the part of the fund manager.

Think About It Find today's quote for the Vanguard 500 Index Fund. How does its recent performance compare to what is shown in the quote in Figure 4.7?

Quick Quiz 4C

Choose the best answer to each of the following questions. Explain your reasoning with one or more complete sentences.

1. In the savings plan formula, assuming all other variables are constant, the accumulated balance in the savings account

   a. decreases as n increases.

   b. increases as APR increases.

   c. decreases as Y increases.

2. In the savings plan formula, assuming all other variables are constant, the accumulated balance in the savings account

   a. decreases as n increases.

   b. increases as PMT increases.

   c. decreases as Y increases.

3. The *total return* on a 5-year investment is

   a. the value of the investment after 5 years.

   b. the difference between the final and initial values of the investment.

   c. the relative change in the value of the investment.

4. The *annual return* on a 5-year investment is

   a. the average of the amounts that you earned in each of the 5 years.

   b. the annual percentage yield that gives the same increase in the value of the investment.

   c. the amount you earned in the best of the 5 years.

5. Suppose you deposited $200 per month into a savings plan for 5 years and at the end of that period your balance was $22,200. The amount you earned in interest was

   a. $10,200. b. $20,200.

   c. impossible to compute without knowing the APR.

6. Which of the following sets of characteristics describes the best investment?

   a. low risk, high liquidity, and high return

   b. high risk, low liquidity, and high return

   c. low risk, high liquidity, and low return

7. Company A has 1 million shares outstanding and a share price of $10. Company B has 10 million shares outstanding and a share price of $9. Company C has 100,000 shares outstanding and a share price of $100. Which company has the greatest market capitalization?

   a. Company A b. Company B

   c. Company C

8. Excalibur's P/E ratio of 75 tells you that

   a. its current share price is 75 times its earnings per share over the past year.

   b. its current share price is 75 times the total value of the company if it were sold.

   c. it offers an annual dividend that is 1/75 of its current share price.

9. The price you pay for a bond with a face value of $5000 selling at 103 points is

   a. $5300. b. $5150. c. $5103.

10. The 1-year return on a mutual fund

    a. must be greater than the 3-year return.

    b. must be less than the 3-year return.

    c. could be greater than or less than the 3-year return.

Exercises 4C

REVIEW QUESTIONS

1. What is a *savings plan*? Explain the savings plan formula.
2. Give an example of a situation in which you might want to solve the savings plan formula to find the regular payment, PMT, required to achieve some goal.
3. Distinguish between the *total return* and the *annual return* on an investment. How do you calculate the annual return? Give an example.
4. Briefly describe the three basic types of investments: stocks, bonds, and cash. How can you invest in these types directly? How can you invest in them indirectly through a mutual fund?
5. Explain what we mean by an investment's *liquidity, risk*, and *return*. How are risk and return usually related?
6. Contrast the historical returns for different types of investments. How do financial indexes, such as the DJIA, help keep track of historical returns?
7. Define the *face value, coupon rate*, and *maturity date* of a bond. What does it mean to buy a bond at a *premium*? At a *discount*? How can you calculate the current yield of a bond?
8. Briefly describe the meaning of key data values given in online quotes for stocks and mutual funds.

DOES IT MAKE SENSE?

Decide whether each of the following statements makes sense (or is clearly true) or does not make sense (or is clearly false). Explain your reasoning.

9. If interest rates stay at 4% APR and I continue to make my monthly $25 deposits into my retirement plan, I should be able to retire in 30 years with a comfortable income.
10. My financial advisor showed me that I could reach my retirement goal with deposits of $200 per month and an average annual return of 7%. But I don't want to deposit that much of

my paycheck, so I'm going to reach the same goal by getting an average annual return of 15% instead.

11. I'm putting all my savings into stocks because stocks always outperform other types of investment over the long term.

12. I'm hoping to withdraw money to buy my first house soon, so I need to put it into an investment that is fairly liquid.

13. I bought a fund advertised on the Web that says it uses a secret investment strategy to get an annual return twice that of stocks, with no risk at all.

14. I'm already retired, so I need low-risk investments. That's why I put most of my money in U.S. Treasury bills, notes, and bonds.

BASIC SKILLS & CONCEPTS

15–18: Savings Plan Formula. Assume monthly deposits and monthly compounding for the following savings plans.

15. Find the savings plan balance after 12 months with an APR of 3% and monthly payments of $75.

16. Find the savings plan balance after 5 years with an APR of 2.5% and monthly payments of $200.

17. Find the savings plan balance after 3 years with an APR of 4% and monthly payments of $200.

18. Find the savings plan balance after 24 months with an APR of 5% and monthly payments of $125.

19–22: Investment Plans. Use the savings plan formula to answer the following questions.

19. At age 25, you set up an IRA (individual retirement account) with an APR of 5%. At the end of each month, you deposit $150 in the account. How much will the IRA contain when you retire at age 65? Compare that amount to the total deposits made over the time period.

20. A friend has an IRA with an APR of 6.25% She started the IRA at age 25 and deposits $100 per month. How much will her IRA contain when she retires at age 65? Compare that amount to the total deposits made over the time period.

21. You put $300 per month in an investment plan that pays an APR of 3.5%. How much money will you have after 18 years? Compare this amount to the total deposits made over the time period.

22. You put $200 per month in an investment plan that pays an APR of 4.5%. How much money will you have after 18 years? Compare that amount to the total deposits made over the time period.

23–26: Planning for the Future. Use the savings plan formula to answer the following questions.

23. Your goal is to create a college fund for your child. Suppose you find a fund that offers an APR of 5% How much should you deposit monthly to accumulate $170,000 in 15 years?

24. At age 35, you start saving for retirement. If your investment plan pays an APR of 6% and you want to have $1 million when you retire in 30 years, how much should you deposit monthly?

25. You want to purchase a new car in 3 years and expect the car to cost $30,000. Your bank offers a plan with a guaranteed APR of 5.5% if you make regular monthly deposits. How much should you deposit each month to end up with $30,000 in 3 years?

26. At age 20 when you graduate, you start saving for retirement. If your investment plan pays an APR of 4.5% and you want to have $2.5 million when you retire in 45 years, how much should you deposit monthly?

27. **Comfortable Retirement.** Suppose you are 30 years old and would like to retire at age 60, having accumulated a retirement fund from which you could draw an income of $100,000 per year—forever! How much would you need to deposit each month to do this? Assume a constant APR of 6%.

28. **Very Comfortable Retirement.** Suppose you are 25 years old and would like to retire at age 65, having accumulated a retirement fund from which you could draw an income of $200,000 per year—forever! How much would you need to deposit each month to do this? Assume a constant APR of 6%.

29–36: Total and Annual Returns. Compute the total and annual returns on each of the following investments.

29. Five years after buying 100 shares of XYZ stock for $60 per share, you sell them for $9400.

30. You pay $8000 for a municipal bond. When it matures after 20 years, you receive $12,500.

31. Twenty years after purchasing shares in a mutual fund for $6500, you sell them for $11,300.

32. Three years after buying 200 shares of XYZ stock for $25 per share, you sell them for $8500.

33. Three years after paying $3500 for shares in a startup company, you sell them for $2000 (at a loss).

34. Five years after paying $5000 for shares in a new company, you sell them for $3000 (at a loss).

35. Ten years after purchasing shares in a mutual fund for $7500, you sell them for $12,600.

36. Ten years after purchasing shares in a mutual fund for $10,000, you sell them for $2200 (at a loss).

37. **Historical Returns.** Suppose that at the end of 1900 your great-great-grandfather invested $300 in each of three funds that tracked the averages of stocks, bonds, and cash, respectively. Assuming that these investments grew at the rates given in Table 4.6, approximately how much would each have been worth at the end of 2016?

38. **Historical Returns.** Suppose that at the end of 1900 your great-great-grandmother made the following investments: $10 in stocks, $75 in bonds, and $500 in cash. Assuming that these investments grew at the rates given in Table 4.6, approximately how much would each have been worth at the end of 2016?

39–40: Reading Stock Tables.

39. Answer the following questions, assuming that the stock quote in Figure 4.8 is one that you found online today.

Intel Corporation (INTC)				Market Cap ($ millions)	$167,860
Last 35.48	Change +0.24	% Change +0.71%	Volume 10,368,000	P/E Ratio	15.4
Open 35.15	High 35.53	Low 35.13		Dividend (latest quarter)	$0.26
				Dividend Yield	3.07%
52-Week High 38.45	52-Week Low 29.50			Shares Outstanding (millions)	4709

FIGURE 4.8

a. What is the symbol for Intel stock?

b. What was the price per share at the end of the day yesterday?

c. Based on the current price, what is the total value of the shares that have been traded so far today?

d. What percentage of all Intel shares have been traded so far today?

e. Suppose you own 100 shares of Intel stock. Based on the current price and dividend yield, what total dividend should you expect to receive this year?

f. What were the earnings per share for Intel?

g. How much total profit did Intel earn in the past year?

40. Answer the following questions, assuming that the stock quote in Figure 4.9 is one that you found online today.

Walmart Stores (WMT)				Market Cap ($ millions)	$246,680
Last 78.81	Change +1.28	% Change 1.65%	Volume 10,650,000	P/E Ratio	17.85
Open 77.97	High 79.44	Low 77.77		Dividend (latest quarter)	$0.46
				Dividend Yield	2.63%
52-Week High 77.66	52-Week Low 65.28			Shares Outstanding (millions)	3031

FIGURE 4.9

a. What is the symbol for Walmart stock?

b. What was the price per share at the end of the day yesterday?

c. Based on the current price, what is the total value of the shares that have been traded so far today?

d. What percentage of all Walmart shares have been traded so far today?

e. Suppose you own 100 shares of Walmart stock. Based on the current price and dividend yield, what total dividend should you expect to receive this year?

f. What were the earnings per share for Walmart?

g. How much total profit did Walmart earn in the past year?

41–44: Price-to-Earning Ratio. For each stock described below, answer the following questions.

a. How much were earnings per share?

b. Does the stock seem overpriced, underpriced, or priced about right, given that historically P/E ratios range from 12 to 14?

41. Costco closed at $171.60 per share with a P/E ratio of 20.84.

42. General Mills closed at $52.65 per share with a P/E ratio of 16.14.

43. IBM closed at $151.98 per share with a P/E ratio of 12.49.

44. Alphabet (parent company of Google) closed at $954.65 per share with a P/E ratio of 30.77.

45. **Stock Look-Up.** Find a quote for Exxon Mobil on the Web, and locate the closing stock price and the current P/E ratio. How much were the earnings per share during the previous year? Does the stock seem overpriced? Explain.

46. **Stock Look-Up.** Find a quote for Wells Fargo and Company on the Web, and locate the closing stock price and the current P/E ratio. How much were the earnings per share during the previous year? Does the stock seem overpriced? Explain.

47–50: Bond Yields. Compute the current yield on each of the following bonds.

47. A $1000 U.S. Treasury bond with a coupon rate of 2.0% that has a market value of $950

48. A $1000 U.S. Treasury bond with a coupon rate of 2.5% that has a market value of $1050

49. A $1000 U.S. Treasury bond with a coupon rate of 5.5% that has a market value of $1100

50. A $10,000 U.S. Treasury bond with a coupon rate of 3.0% that has a market value of $9500

51–54: Bond Interest. Compute the annual interest you would earn on each of the following bonds.

51. A $1000 U.S. Treasury bond with a current yield of 3.9% that is quoted at 105 points

52. A $1000 U.S. Treasury bond with a current yield of 1.5% that is quoted at 98 points

53. A $1000 U.S. Treasury bond with a current yield of 6.2% that is quoted at 114.3 points

54. A $10,000 U.S. Treasury bond with a current yield of 3.6% that is quoted at 102.5 points

55. **Mutual Fund Growth.** Answer the following questions, assuming that the mutual fund quote in Figure 4.10 is one that you found online today.

Vanguard Limited-Term Tax-Exempt Fund (VMLTX)				
NAV. $10.99	1-Day Net Change ▲$0.01		1-Day Return ▲+0.10%	
	TOTAL RETURNS (%) 3, 5 and 10 year returns are annualized.			
	YTD	1-Yr	5-Yr	10-Yr
Fund	1.97%	0.69%	1.25%	2.45%

FIGURE 4.10

a. Suppose you invest $5000 in this fund today. How many shares will you buy?

b. Suppose you had invested $5000 in this fund 5 years ago. How much would your investment be worth now?

c. Suppose you had invested $5000 in this fund 10 years ago. How much would your investment be worth now?

56. **Mutual Fund Growth.** Answer the following questions, assuming that the mutual fund quote in Figure 4.11 is one that you found online today.

Vanguard Long-Term Bond Index (VBLTX)				
NAV. $13.93	1-Day Net Change ▲$ 0.00	1-Day Return ▲+0.01%		
TOTAL RETURNS (%) 3, 5 and 10 year returns are annualized.				
	YTD	1-Yr	5-Yr	10-Yr
Fund	4.62%	2.63%	3.98%	7.16%

FIGURE 4.11

a. Suppose you invest $2500 in this fund today. How many shares will you buy?

b. Suppose you had invested $2500 in this fund 5 years ago. How much would your investment be worth now?

c. Suppose you had invested $2500 in this fund 10 years ago. How much would your investment be worth now?

FURTHER APPLICATIONS

57–60: Who Comes Out Ahead? Consider the following pairs of savings plans. Compare the balances in each plan after 10 years. In each case, which person deposited more money in the plan? Which of the two investment strategies do you believe was better? Assume that the compounding and payment periods are the same for each pair.

57. Yolanda deposits $200 per month in an account with an APR of 5%; Zach deposits $2400 at the end of each year in an account with an APR of 5%.

58. Polly deposits $50 per month in an account with an APR of 6%; Quint deposits $40 per month in an account with an APR of 6.5%.

59. Juan deposits $400 per month in an account with an APR of 6%; Maria deposits $5000 at the end of each year in an account with an APR of 6.5%.

60. George deposits $40 per month in an account with an APR of 7%; Harvey deposits $150 per quarter in an account with an APR of 7.5%.

61–64: Will It Work? Suppose you want to accumulate $50,000 for a college fund over the next 15 years. Determine whether the following investment plans will allow you to reach your goal. Assume that the compounding and payment periods are the same for all the accounts.

61. You deposit $50 per month into an account with an APR of 7%.

62. You deposit $75 per month into an account with an APR of 7%.

63. You deposit $100 per month into an account with an APR of 6%.

64. You deposit $200 per month into an account with an APR of 5%.

65. **Total Return on Stock.** Suppose you bought XYZ stock 1 year ago for $5.80 per share and sell it at $8.25. You also pay a commission of $0.25 per share on your sale. What is the total return on your investment?

66. **Total Return on Stock.** Suppose you bought XYZ stock 1 year ago for $46.00 per share and sell it at $8.25. You also pay a commission of $0.25 per share on your sale. What is the total return on your investment?

67. **Death and the Maven (A True Story).** In December 1995, 101-year-old Anne Scheiber died and left $22 million to Yeshiva University. This fortune was accumulated through shrewd and patient investment of a $5000 nest egg over the course of 50 years. In turning $5000 into $22 million, what were Scheiber's total and annual returns? How did her annual return compare to the average annual return for stocks (see Table 4.6)?

68. **Cell Phones to Dollars.** According to one study, the average monthly cell phone bill in the United States is $70 (up 31% since 2009). If a 20-year old student with an average bill gives up her cell phone and each month invests the $70 she would have spent on her phone bill in a savings plan that averages a 4% annual return, how much will she have saved by the time she is 65?

69. **Get Started Early!** Mitch and Bill are both age 75. When Mitch was 25 years old, he began depositing $1000 per year into a savings account. He made deposits for the first 10 years, at which point he was forced to stop making deposits. However, he left his money in the account, where it continued to earn interest for the next 40 years. Bill didn't start saving until he was 45 years old, but for the next 30 years he made annual deposits of $1000. Assume that both accounts earned an average annual return of 5% (compounded once a year).

a. How much money does Mitch have in his account at age 75?

b. How much money does Bill have in his account at age 75?

c. Compare the amounts of money that Mitch and Bill deposited into their accounts.

d. Write a paragraph summarizing your conclusions about this parable.

70. **Deriving the Annual Return Formula.** If you deposit $$P$ in an account with an annual percentage yield of APY, after Y years the balance in the account is $A = P \times (1 + \text{APY})^Y$. Use the rules of algebra (see the Brief Review on p. 219 in Unit 4B) to solve this formula for APY; the result should be the annual return formula given in the text. *Hint:* Start by dividing both sides by P; then isolate the term with APY by raising both sides to the $1/Y$ power; then subtract the same value from both sides to find APY.

IN YOUR WORLD

71. **Advertised Investment.** Find an advertisement for an investment plan. Describe some of the cited benefits of the plan. Using what you learned in this unit, identify at least one possible drawback of the plan.

72. **Investment Tracking.** Choose three stocks, three bonds, and three mutual funds that you think would make good investments. Imagine that you invest $1000 in each of these nine investments. Use the Web to track the value of your investment portfolio over the next 5 weeks. Based on the portfolio value at the end, find your return for the 5-week period. Which investments did best, and which did worst?

73. **Company Research.** Choose one of the 30 companies in the DJIA, and carry out research on that company as if you were a prospective investor. You should consider the following questions: How has the company performed over the last year? 5 years? 10 years? Does the company offer dividends? How do you interpret its P/E ratio? Overall, do you think the company's stock is a good investment? Why or why not?

74. **Financial Websites.** Visit one of the many financial news and advice websites. Describe the services offered by the website. Explain whether, as an active or prospective investor, you find the website useful.

75. **Online Brokers.** Visit the websites of at least two online brokers. How do their services differ? Compare the commissions charged by the brokers.

76. **Personal Investment Options.** Does your employer offer you the option of enrolling in a savings or retirement plan? If so, describe the available options and discuss the advantages and disadvantages of each.

TECHNOLOGY EXERCISES

77. **Savings Plan Formula with Excel.** Use the future value (FV) function in Excel to answer the following questions.

a. What is the balance in an account after 25 years with monthly deposits of \$100 and an APR of 4%?

b. If you double the APR in part (a), is the balance double, more than double, or less than double the balance in part (a)?

c. If you double the number of years in part (a), is the balance double, more than double, or less than double the balance in part (a)?

78. **Savings Plan Formula with Excel.** Abe deposits \$50 each month for 40 years in an account with an APR of 5.5%. Beatrice deposits \$100 each month for 20 years in an account with an APR of 5.5%.

a. Verify that Abe and Beatrice deposit the same amount of money during the stated periods of time. How much do they deposit?

b. Compute the accumulated balance in each account and explain the results.

79. **Computing Roots.** Use a calculator or Excel to compute the following quantities.

a. $2.8^{1/4}$

b. $120^{1/3}$

c. The annual return on an account in which an initial investment of \$250 increases to \$1850 over a period of 15 years

UNIT 4D Loan Payments, Credit Cards, and Mortgages

Do you have a credit card? Have you taken out student loans or a loan for a car? Do you own a house? Chances are that you owe money because of at least one of these things. If so, you not only have to pay back the money you borrowed but also have to pay interest on the money that you owe. In this unit, we study the basic mathematics of loans.

Loan Basics

Suppose you borrow \$1200 at the annual interest rate of APR = 12%, or 1% per month. At the end of the first month, you owe interest in the amount of

$$1\% \times \$1200 = \$12$$

If you paid *only* this \$12 in interest, you'd still owe \$1200. That is, the total amount of the loan, called the **principal**, would still be \$1200. In that case, you'd owe the same \$12 in interest the next month. In fact, if you paid only the interest each month, the loan would never be paid off and you'd pay \$12 per month forever.

If you hope to make progress in paying off the loan, you need to pay back part of the principal as well as the interest. For example, suppose that you paid \$200 toward the principal each month, plus the current interest. At the end of the first month, you'd pay \$200 toward principal *plus* \$12 for the 1% interest you owe, making a total payment of \$212. Because you've paid \$200 toward principal, your new loan principal would be \$1200 − \$200 = \$1000.

At the end of the second month, you'd again pay \$200 toward principal and 1% interest. But this time the interest is on the \$1000 that you still owe. Your interest payment therefore would be 1% × \$1000 = \$10, making your total payment \$210. Table 4.7 shows the calculations for the 6 months until the loan is paid off.

Loan Basics

For any loan, the **principal** is the amount of money owed at any particular time. Interest is charged on the principal of a loan. To pay off a loan, you must gradually pay down the principal. The **loan term** is the time you have to pay back the loan in full.

TABLE 4.7 Payments and Principal for a \$1200 Loan with Principal Paid Off at \$200/Month

End of...	Prior Principal	Interest on Prior Principal	Payment Toward Principal	Total Payment	New Principal
Month 1	\$1200	1% × \$1200 = \$12	\$200	\$212	\$1000
Month 2	\$1000	1% × \$1000 = \$10	\$200	\$210	\$800
Month 3	\$800	1% × \$800 = \$8	\$200	\$208	\$600
Month 4	\$600	1% × \$600 = \$6	\$200	\$206	\$400
Month 5	\$400	1% × \$400 = \$4	\$200	\$204	\$200
Month 6	\$200	1% × \$200 = \$2	\$200	\$202	\$0

Installment Loans

Table 4.7 shows your total payment decreasing from month to month because of the declining amount of interest that you owe. There's nothing inherently wrong with this method of paying off a loan, but most people prefer to pay the *same* total amount each month because it makes it easier to plan a budget. A loan that you pay off with equal regular payments is called an **installment loan** (or *amortized loan*).

Suppose you wanted to pay off your \$1200 loan with 6 equal monthly payments. How much should you pay each month? Because the payments in Table 4.7 vary between \$202 and \$212, it's clear that the equal monthly payments must lie somewhere in this range. The exact amount is not obvious, but we can calculate it with the **loan payment formula**.

BY THE WAY
About two-thirds of all college students earning bachelor's degrees have at least one student loan, with an average (mean) debt of about \$37,000 at graduation.

Loan Payment Formula (Installment Loans)

$$\text{PMT} = \frac{P \times \left(\frac{\text{APR}}{n}\right)}{\left[1 - \left(1 + \frac{\text{APR}}{n}\right)^{(-nY)}\right]}$$

where

PMT = regular payment amount

P = starting loan principal (amount borrowed)

APR = annual percentage rate

n = number of payment periods per year

Y = loan term in years

In our current example, the starting loan principal is $P = \$1200$, the annual interest rate is APR $= 12\%$, the loan term is $Y = \frac{1}{2}$ year (6 months), and monthly payments mean $n = 12$. The loan payment formula gives

$$\text{PMT} = \frac{P \times \left(\frac{\text{APR}}{n}\right)}{\left[1 - \left(1 + \frac{\text{APR}}{n}\right)^{(-nY)}\right]} = \frac{\$1200 \times \left(\frac{0.12}{12}\right)}{\left[1 - \left(1 + \frac{0.12}{12}\right)^{(-12\times 1/2)}\right]}$$

$$= \frac{\$1200 \times (0.01)}{\left[1 - (1 + 0.01)^{-6}\right]} = \frac{\$12}{1 - 0.942045235} = \$207.06$$

The monthly payments would be \$207.06; note that, as we expected, the monthly payment is between \$202 and \$212.

Note that, because the principal of an installment loan is paid down while the payments remain the same, the following two features apply to all installment loans:

- The interest due each month gradually decreases.
- The amount paid toward principal each month gradually increases.

Early in the loan term, the portion going toward interest is relatively high and the portion going toward principal is relatively low. As the term proceeds, this pattern gradually reverses, and toward the end of the loan term most of the payments go to principal and relatively little to interest.

EXAMPLE 1 Student Loan

Suppose you have a student loan totaling \$7500 when you graduate from college. The interest rate is APR $= 9\%$, and the loan term is 10 years. What are your monthly payments? How much will you pay over the lifetime of the loan? What is the total interest you will pay on the loan?

Technical Note

Because we assume the compounding period is the same as the payment period and because we round payments to the nearest cent, the calculated payments may differ slightly from actual payments.

Solution The starting loan principal is $P = \$7500$, the interest rate is APR $= 0.09$, the loan term is $Y = 10$ years, and $n = 12$ for monthly payments. We use the loan payment formula to find the monthly payments:

$$\text{PMT} = \frac{P \times \left(\frac{\text{APR}}{n}\right)}{\left[1 - \left(1 + \frac{\text{APR}}{n}\right)^{(-nY)}\right]} = \frac{\$7500 \times \left(\frac{0.09}{12}\right)}{\left[1 - \left(1 + \frac{0.09}{12}\right)^{(-12\times 10)}\right]}$$

$$= \frac{\$7500 \times (0.0075)}{\left[1 - (1 + 0.0075)^{-120}\right]}$$

$$= \frac{\$56.25}{\left[1 - 0.407937305\right]}$$

$$= \$95.01$$

Your monthly payments are \$95.01. Over the 10-year term, your total payments will be

$$10 \text{ yr} \times 12 \frac{\text{mo}}{\text{yr}} \times \frac{\$95.01}{\text{mo}} = \$11{,}401.20$$

Of this amount, \$7500 pays off the principal. The rest, or $\$11{,}401 - \$7500 = \$3901$, represents interest payments. **Now try Exercises 13–24.**

USING TECHNOLOGY

The Loan Payment Formula (Installment Loans)

Standard Calculators As with compound interest and savings plan calculations, you can do loan payment calculations with a standard calculator, as long as you follow the correct order of operations. The following calculation shows one correct way to handle the numbers from Example 1, in which $P = \$7500$, $\text{APR} = 0.09$, $n = 12$, and $Y = 10$ years; as always, *do not round any answers until the end of the calculation.*

General Procedure	Our Example	Calculator Steps*		Output
$\text{PMT} = \dfrac{P \times \left(\dfrac{\text{APR}}{n}\right)}{\left[1 - \left(1 + \dfrac{\text{APR}}{n}\right)^{(-nY)}\right]}$	$\text{PMT} = \dfrac{\$7500 \times \left(\dfrac{0.09}{12}\right)}{\left[1 - \left(1 + \dfrac{0.09}{12}\right)^{(-12\times 10)}\right]}$	Step 1	1 (+) 0.09 (÷) 12 (=)	1.0075
1. Parentheses	1. Parentheses	Step 2	(∧) (( ) (−) 12 (×) 10 ( )) (=)	0.407937305
2. Exponent	2. Exponent	Step 3	1 (−) 0.407937305 (=)	0.592062695
3. Compute denominator	3. Compute denominator	Step 4	7500 (×) 0.09 (÷) 12 (÷) 0.592062695 (=)	95.0068...
4. Compute numerator and divide by denominator	4. Compute numerator and divide by denominator			

*If your calculator does not have parentheses keys, then you will need to keep track of the results of Steps 1–3 either on paper or in the calculator's memory. Do not round intermediate results.

Excel Use the built-in function PMT to compute payments for installment loans. The inputs are similar to those we used earlier with the FV function and are shown in the screen shot on the left below. You may find it easier to see the inputs if you use the dialog box that appears when you choose the PMT function from the Insert menu; that box resemble the screen shot on the right.

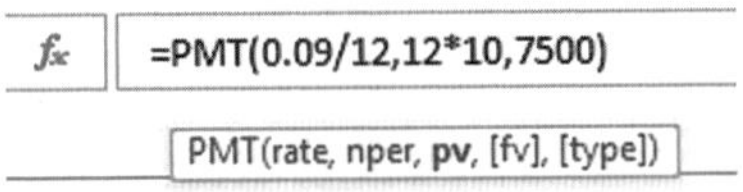

Microsoft Excel 2016, Windows 10, Microsoft Corporation.

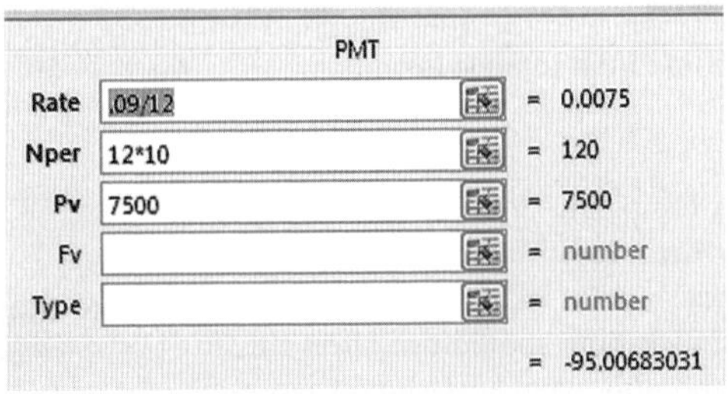

Note that the variables are similar to those we used earlier with the FV function:

- *rate* is the interest rate for each payment period, which in this case is the monthly interest rate: $\text{APR}/n = 0.09/12$.
- *nper* is the total number of payment periods, which in this case is $nY = 12 \times 10$.
- *pv* is the present value, which is the starting loan principal of \$7500.
- *fv* is the future value, which is \$0 because the goal is to pay off the loan completely; note that it is 0 by default, so we can leave fv blank.
- *type* is also left blank, because we use the default value (type = 0) to indicate monthly payments made at the *end* of each period (month).

BY THE WAY

A table of principal and interest payments over the life of a loan is called an *amortization schedule*. Most banks will provide an amortization schedule for any loan you are considering.

EXAMPLE 2 Principal and Interest Payments

For the loan in Example 1, calculate the portions of your payments that go to principal and to interest during the first 3 months.

Solution The monthly interest rate is APR/12 = 0.09/12 = 0.0075. For a \$7500 starting principal, the interest due at the end of the first month is

$$0.0075 \times \$7500 = \$56.25$$

Because your monthly payment (calculated in Example 1) is \$95.01 and the interest is \$56.25, the remaining \$95.01 − \$56.25 = \$38.76 goes to principal. Therefore, after your first payment, your new loan principal is

$$\$7500 - \$38.76 = \$7461.24$$

Table 4.8 continues the calculations for months 2 and 3. Note that, as expected, the interest payment gradually decreases and the payment toward principal gradually increases. Also note that, for these first 3 months of a 10-year loan, more than half of each payment goes toward interest. We could continue this table through the life of the loan (see Using Technology on p. 249), but it's much easier to use software that has built-in functions for finding the principal and interest portions of loan payments.

MATHEMATICAL INSIGHT

Derivation of the Loan Payment Formula

Suppose you borrow a starting principal P for a loan term of N months at a monthly interest rate i. In most real cases, you would make monthly payments on this loan. However, suppose the lender did not want monthly payments, but instead wanted you to pay back the loan with compound interest in a lump sum at the end of the loan term. We can find this lump sum amount with the general compound interest formula:

$$A = P \times (1 + i)^N$$

In financial terms, this lump sum amount, A, is called the *future value* of your loan. (The *present value* is the original loan amount, P.) From the lender's point of view, allowing you to spread your payments out over time should not affect this future value, so your total monthly payments should represent the same future value, A. We already have a formula for determining the future value with monthly payments—it is the general form of the savings plan formula from Unit 4C:

$$A = \text{PMT} \times \frac{\left[(1 + i)^N - 1\right]}{i}$$

We now have two different expressions for A, so we set them equal:

$$\text{PMT} \times \frac{\left[(1 + i)^N - 1\right]}{i} = P \times (1 + i)^N$$

To find the loan payment formula, we need to solve this equation for PMT. We first divide both sides by the fraction on the left; you should confirm that the equation then becomes

$$\text{PMT} \times \frac{P \times (1 + i)^N \times i}{\left[(1 + i)^N - 1\right]}$$

Next, we divide both the numerator and the denominator of the fraction on the right by $(1 + i)^N$:

$$\text{PMT} = \frac{\dfrac{P \times \cancel{(1 + i)^N} \times i}{\cancel{(1 + i)^N}}}{\dfrac{\left[(1 + i)^N - 1\right]}{(1 + i)^N}}$$

The numerator simplifies to $P \times i$. To simplify the denominator, we expand it and then write its second term with a negative exponent as follows:

$$\frac{\left[(1 + i)^N - 1\right]}{(1 + i)^N} = \frac{\cancel{(1 + i)^N}}{\cancel{(1 + i)^N}} - \frac{1}{(1 + i)^N}$$

$$= 1 - (1 + i)^{-N}$$

Substituting the simplified terms for the numerator and the denominator, we find the loan payment formula:

$$\text{PMT} = \frac{P \times i}{1 - (1 + i)^{-N}}$$

To put the loan payment formula in the form given in the text, we substitute $i = \text{APR}/n$ for the interest rate per period and $N = nY$ for the total number of payments (where n is the number of payments per year and Y is the number of years).

TABLE 4.8 Interest and Principal Portions of Payments on a \$7500 Loan (10-year term, APR = 9%)

End of...	Interest = 0.0075 × Balance	Payment Toward Principal	New Principal
Month 1	0.0075 × \$7500 = \$56.25	\$95.01 − \$56.25 = \$38.76	\$7500 − \$38.76 = \$7461.24
Month 2	0.0075 × \$7461.24 = \$55.96	\$95.01 − \$55.96 = \$39.05	\$7461.24 − \$39.05 = \$7422.19
Month 3	0.0075 × \$7422.19 = \$55.67	\$95.01 − \$55.67 = \$39.34	\$7422.19 − \$39.34 = \$7382.85

▸ Now try Exercises 25–26.

Think About It Many people mistakenly guess that a 9% interest rate means that only 9% of their loan payments go to interest, but Example 2 shows that the actual portion going to interest can be much higher, especially during the early period of the loan term. Explain why this is the case and how the portion of the payments going toward interest will change with time.

USING TECHNOLOGY

Principal and Interest Portions of Loan Payments

You can use Excel to make a table of principal and interest amounts for loan payments like the one shown in Table 4.8. The most direct way of doing this is shown in the following screen shot.

	A	B	C	D	E
1					
2		Month	Interest	Principal	Balance
3					7500
4		1	=0.09*E3/12	=95.01-C4	=E3-D4
5		2	=0.09*E4/12	=95.01-C5	=E4-D5
6		3	=0.09*E5/12	=95.01-C6	=E5-D6

To understand how the calculation works, note the following:

- The table begins with the starting balance of \$7500 in cell E3.
- Cell C4 calculates the interest for month 1, which is the balance from cell E3 times APR/*n*, which in this case is 0.09/12.
- Cell D4 calculates the principal for month 1 by subtracting the interest payment (cell C4) from the monthly payment of \$95.01.
- Cell E4 calculates the new balance after month 1 by subtracting the principal payment (cell D4) from the prior balance (cell E3).
- Each successive row then repeats the same pattern.

You should make this table for yourself in Excel and confirm that it gives the values shown in Table 4.8.

The only difficulty with this approach is that finding the interest and principal amounts in, say, month 25 requires 25 rows in your table. If you only want to know the interest and principal amounts for a particular month, you can use Excel's functions IPMT and PPMT, respectively, as shown in the screen shots below for the 25th month; note that the variable *per* is the period for which you want to know the payments, which is 25 in this case. As with the PMT function (see Using Technology, p. 247), the *fv* and *type* inputs are left blank.

```
fx  =IPMT(0.09/12,25, 120,7500)
    IPMT(rate, per, nper, pv, [fv], [type])
```

```
fx  =PPMT(0.09/12,25, 120,7500)
    PPMT(rate, per, nper, pv, [fv], [type])
```

Microsoft Excel 2016, Windows 10, Microsoft Corporation.

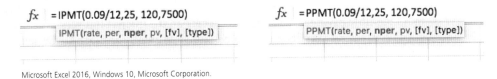

Choices of Rate and Term

You'll usually have several choices of interest rate and loan term when taking out a loan. For example, a bank might offer a 3-year car loan at 8%, a 4-year car loan at 9%, and a 5-year car loan at 10%. You'll pay less total interest with the shortest-term, lowest-rate loan, but this loan will have the highest monthly payments. You'll have to evaluate your choices and make the decision that is best for your personal situation.

EXAMPLE 3 Choice of Auto Loans

You need a $6000 loan to buy a used car. Your bank offers a 3-year loan at 8%, a 4-year loan at 9%, and a 5-year loan at 10%. Calculate your monthly payments and total interest over the loan term with each option.

Solution We are looking for payments on an installment loan, so we use the loan payment formula:

$$\text{PMT} = \frac{P \times \left(\frac{\text{APR}}{n}\right)}{\left[1 - \left(1 + \frac{\text{APR}}{n}\right)^{(-nY)}\right]}$$

BY THE WAY

You should always watch out for financial scams, especially when borrowing money. Keep in mind what is sometimes called the first rule of finance: *If it sounds too good to be true, it probably is!*

The three columns in the following table show the calculations for the three choices of APR and loan term. Note that all three have the same amount borrowed ($P = \$6000$) and $n = 12$ (for monthly payments). As we should expect, a longer-term loan leads to lower monthly payments but higher total interest.

	$Y = 3$ years (36 months), APR $= 0.08$	$Y = 4$ years (48 months), APR $= 0.09$	$Y = 5$ years (60 months), APR $= 0.10$
Use the loan payment formula to find the monthly payment:	$\text{PMT} = \frac{\$6000 \times \left(\frac{0.08}{12}\right)}{\left[1 - \left(1 + \frac{0.08}{12}\right)^{(-12 \times 3)}\right]}$ $= \$188.02$	$\text{PMT} = \frac{\$6000 \times \left(\frac{0.09}{12}\right)}{\left[1 - \left(1 + \frac{0.09}{12}\right)^{(-12 \times 4)}\right]}$ $= \$149.31$	$\text{PMT} = \frac{\$6000 \times \left(\frac{0.10}{12}\right)}{\left[1 - \left(1 + \frac{0.10}{12}\right)^{(-12 \times 5)}\right]}$ $= \$127.48$
Multiply the monthly payment by the loan term in months to find the total payments:	$188.02/month × 36 months = $6768.72	$149.31/month × 48 months = $7166.88	$127.48/month × 60 months = $7648.80
Subtract the principal from the total payments to find the total interest:	$6768.72 − $6000 = $768.72	$7166.88 − $6000 = $1166.88	$7648.80 − $6000 = $1648.80

▶ Now try Exercises 27–28.

Think About It Consider your own current financial situation. If you needed a $6000 car loan, which option from Example 3 would you choose? Why?

Credit Cards

Credit card loans differ from installment loans in that you are not required to pay off your balance in any set period of time. Instead, you are required to make only a minimum monthly payment that generally covers all the interest but very little principal. As a result, it takes a very long time to pay off your credit card loan if you make only the minimum payments. If you wish to pay off your loan in a particular amount of time, you should use the loan payment formula to calculate the necessary payments.

IN YOUR WORLD

Avoiding Credit Card Trouble

Used properly, credit cards offer many conveniences. They are safer and easier to carry than cash, they offer monthly statements that list everything charged to the card, they can be used as ID for things like car rentals, and in some cases they offer rewards that can be beneficial, such as small rebates or airline miles. But credit cards are also easy to abuse, and many people get into financial trouble as a result. A few simple guidelines can help you avoid credit card trouble:

- Use only one credit card. People who accumulate balances on several cards often lose track of their overall debt.
- If possible, pay off your balance in full each month. If you can do this, be sure that your credit card offers an interest-free "grace period" of at least 25 to 30 days on purchases so you will not have to pay any interest.
- If you can't pay your balance fully each month, be very careful not to accumulate too much debt. Most credit cards have very high interest rates compared to other types of loans, making it easy to get into financial trouble if you get overextended using credit cards.

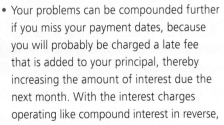

- Your problems can be compounded further if you miss your payment dates, because you will probably be charged a late fee that is added to your principal, thereby increasing the amount of interest due the next month. With the interest charges operating like compound interest in reverse, failure to pay on time can put a person into an ever-deepening financial hole.
- Shop around for the best credit card deals. If you expect to carry a balance, compare interest rates and annual fees. If you will pay off your balance each month, compare annual fees and benefits, such as cash back or airline miles.
- Watch out for *teaser rates* that try to entice you to get a new credit card by offering a low starting interest rate. The interest rate usually becomes much higher after just a few months, so these are rarely a good deal.
- Never use your credit card for a cash advance except in an emergency, because nearly all credit cards charge both fees and high interest rates for cash advances. Moreover, most credit cards charge interest immediately on cash advances, even if there is a grace period on purchases.
- If you own a home, consider replacing a common credit card with a home equity credit line. You'll generally get a lower interest rate, and the interest may be tax deductible.
- If you have trouble controlling your spending, use cash rather than a credit card. Studies show that most people spend significantly less when paying cash then when paying with a credit card (or debit card).
- If you find yourself in a deepening financial hole, consult a financial advisor right away. A good place to start is with the National Foundation for Credit Counseling (www.nfcc.org). The longer you wait, the worse off you'll be in the long run.

EXAMPLE 4 Credit Card Debt

Suppose you have a credit card balance of $2300 with an annual interest rate of 21%. You decide to pay off your balance over 1 year. How much will you need to pay each month? Assume that you make no further purchases with the credit card.

Solution The amount borrowed is $P = \$2300$, the interest rate is $\text{APR} = 0.21$, and you make $n = 12$ payments per year. Because you want to pay off the loan in 1 year, we set $Y = 1$. The required payments are

$$\text{PMT} = \frac{P \times \left(\frac{\text{APR}}{n}\right)}{\left[1 - \left(1 + \frac{\text{APR}}{n}\right)^{(-nY)}\right]} = \frac{\$2300 \times \left(\frac{0.21}{12}\right)}{\left[1 - \left(1 + \frac{0.21}{12}\right)^{(-12\times 1)}\right]} = \$214.16$$

You must pay $214.16 per month to pay off the balance in 1 year.

▲ **CAUTION!** Be sure to note that the credit card's annual interest rate of 21% is much higher than the interest rate on most other loans, which is a major reason why it is easy to get into financial trouble by spending too much using credit cards. ▲

▸ Now try Exercises 29–32.

BY THE WAY

About three-fourths of adult Americans have at least one credit card. Among those who carry a balance on their cards, the average (mean) credit card debt is about $16,000, and the average annual interest rate is about 17%—far higher than the interest rate on most other consumer loans.

Think About It Continuing Example 4, suppose you can get a personal loan at a bank at an annual interest rate of 10%. Should you take this loan and use it to pay off your $2300 credit card debt? Discuss the pros and cons.

EXAMPLE 5 A Deepening Hole

Paul has gotten into credit card trouble. He has a balance of $9500 and just lost his job. His credit card company charges interest at a rate of APR = 21%, compounded daily. Suppose the credit card company allows him to suspend his payments until he finds a new job—but continues to charge interest. If it takes him a year to find a new job, how much will he owe when he starts that job?

Solution Because Paul is not making payments during the year, this is not a loan payment problem. Instead, it is a compound interest problem, in which Paul's balance of $9500 grows at an annual rate of 21%, compounded daily. We use the compound interest formula with $P = \$9500$, $\text{APR} = 0.21$, $Y = 1$ year, and $n = 365$ (for daily compounding). At the end of the year, his loan balance will be

$$\begin{aligned} A &= P \times \left(1 + \frac{\text{APR}}{n}\right)^{(nY)} \\ &= \$9500 \times \left(1 + \frac{0.21}{365}\right)^{(365 \times 1)} \\ &= \$11{,}719.23 \end{aligned}$$

During his year of unemployment, interest alone will make Paul's credit card balance grow from $9500 to more than $11,700, an increase of more than $2200. Clearly, this increase will only make it more difficult for Paul to get back on his financial feet.

▶ Now try Exercises 33–36.

Mortgages

BY THE WAY The curious word *mortgage* comes from Latin and old French. It literally means "dead pledge."

One of the most popular types of installment loans is a home **mortgage**, which is designed specifically to help you buy a home. Mortgage interest rates generally are lower than interest rates on other types of loans because your home itself serves as a payment guarantee. If you fail to make your payments, the lender (usually a bank or mortgage company) can take possession of your home, through the process called *foreclosure,* and sell it to recover some or all of the amount loaned to you.

There are several considerations in getting a home mortgage. First, the lender may require a **down payment**, typically 10% to 20% of the purchase price. Then the lender will loan you the rest of the money needed to purchase the home. Most lenders also charge fees, or **closing costs**, at the time you take out a loan. Closing costs can be substantial and may vary significantly between lenders, so you should be sure that you understand them. In general, there are two types of closing costs:

- Direct fees, such as fees for getting the home appraised and checking your credit history, for which the lender charges a fixed dollar amount. These fees typically range from a few hundred to a couple of thousand dollars.
- Fees charged as **points**, where each point is 1% of the loan amount. Many lenders divide points into two categories: an *origination fee* that is charged on all loans and *discount points* that vary for loans with different rates. For example, a lender might charge an origination fee of 1 point (1%) on all loans, then offer you a choice of interest rates depending on how many discount points you are willing to pay. Despite their different names, there is no essential difference between an origination fee and discount points.

As always, you should watch out for any fine print that may affect the cost of your loan. For example, make sure that there are no *prepayment penalties* if you decide to pay off your loan early. Most people pay off mortgages early, either because they sell the home or because they decide to refinance the loan to get a better interest rate or to change their monthly payments.

BY THE WAY

A mortgage is said to be *underwater* if its loan balance is greater than the value of the house. The burst of the housing bubble (see the Chapter 1 Activity, p. 4) left many homeowners with underwater mortgages.

Definitions

A home **mortgage** is an installment loan designed specifically to finance the purchase of a home.

A **down payment** is the amount of money you must pay up front in order to be given a mortgage or other loan.

Closing costs are fees you must pay in order to be given the loan. They may include a variety of direct costs, or fees charged as **points**, where each point is 1% of the loan amount. In most cases, lenders are required to give you a clear assessment of closing costs before you sign for the loan.

Fixed Rate Mortgages

The simplest type of home-buying loan is a **fixed rate mortgage**, which includes a guarantee that the interest rate will not change over the life of the loan. Most fixed rate mortgages have a term of either 15 or 30 years, with lower interest rates on the shorter-term loans.

EXAMPLE 6 Fixed Rate Payment Options

You need a loan of \$100,000 to buy your new home. The bank offers either a 30-year loan at an APR of 5% or a 15-year loan at 4.5% Compare your monthly payments and total loan cost under the two options. Assume that the closing costs are the same in both cases and therefore do not affect your choice.

Solution Mortgages are installment loans, so we use the loan payment formula:

$$\text{PMT} = \frac{P \times \left(\frac{\text{APR}}{n}\right)}{\left[1 - \left(1 + \frac{\text{APR}}{n}\right)^{(-nY)}\right]}$$

For both cases, we have $P = \$100{,}000$ and monthly payments, for which $n = 12$; the table below shows the calculations for the two different interest rates and loan terms.

	$Y = 30$ years, $\text{APR} = 0.05$	$Y = 15$ years, $\text{APR} = 0.045$
Use the loan payment formula to find the monthly payment:	$\text{PMT} = \frac{\$100{,}000 \times \left(\frac{0.05}{12}\right)}{\left[1 - \left(1 + \frac{0.05}{12}\right)^{(-12\times 30)}\right]} = \$536.82$	$\text{PMT} = \frac{\$100{,}000 \times \left(\frac{0.045}{12}\right)}{\left[1 - \left(1 + \frac{0.045}{12}\right)^{(-12\times 15)}\right]} = \$764.99$
Multiply the monthly payment by the loan term in months to find the total payments:	$30 \text{ yr} \times 12 \frac{\text{mo}}{\text{yr}} \times \frac{\$536.82}{\text{mo}} \approx \$193{,}255$	$15 \text{ yr} \times 12 \frac{\text{mo}}{\text{yr}} \times \frac{\$764.99}{\text{mo}} \approx \$137{,}698$

Note that the monthly payments on the 15-year loan are higher by about $\$765 - \$537 = \$228$. However, the 15-year loan saves you about $\$193{,}255 - \$137{,}698 = \$55{,}557$ in total payments. That is, the 15-year loan saves you a lot in the long run, but it's a good choice only if you are confident that you can afford the higher monthly payments over the next 15 years. (See Example 8 for an alternative payment strategy.)

▶ Now try Exercises 37–40.

Think About It Do a quick Web search to find today's average interest rate for 15-year and 30-year fixed mortgage loans. How would the payments in Example 6 differ with the current rates?

EXAMPLE 7 Closing Costs

Great Bank offers a \$100,000, 30-year, 5% fixed rate loan with closing costs of \$500 plus 1 point. Big Bank offers a lower rate of 4.75% on a 30-year loan of the same amount, but with closing costs of \$1000 plus 2 points. Evaluate the two options.

Solution In Example 6, we calculated the monthly payments on the 5% loan to be \$536.82. At the lower 4.75% rate, the monthly payments are

$$\text{PMT} = \frac{P \times \left(\frac{\text{APR}}{n}\right)}{\left[1 - \left(1 + \frac{\text{APR}}{n}\right)^{(-nY)}\right]} = \frac{\$100{,}000 \times \left(\frac{0.0475}{12}\right)}{\left[1 - \left(1 + \frac{0.0475}{12}\right)^{(-12 \times 30)}\right]} = \$521.65$$

You would save about \$15 per month with Big Bank's lower interest rate. Now we must consider the difference in closing costs. Big Bank charges you an extra \$500 plus an extra 1 point (or 1%), which is \$1000 on a \$100,000 loan. Therefore, the choice comes down to this: Big Bank would cost you an extra \$1500 up front, but save you about \$15 per month in payments. We divide to find the time it would take to recoup the extra \$1500:

$$\frac{\$1500}{\$15/\text{mo}} = 100 \text{ mo} = 8\tfrac{1}{3} \text{ yr}$$

It would take you more than 8 years to save the extra \$1500 that Big Bank charges up front. Unless you are sure that you will be staying in the house (and keeping the same loan) for much more than 8 years, you probably should go with the lower closing costs at Great Bank, even though your monthly payments will be slightly higher.

▶ Now try Exercises 41–44.

Prepayment Strategies

Because of the long loan term, the early payments on a mortgage go mostly to interest. For example, Figure 4.12 shows the portions of each payment going to principal and interest for a 30-year, \$100,000 loan at 5%. As we found in Example 6 and you can see from the areas of the two regions in Figure 4.12, the total interest paid on this mortgage is nearly as much as the total principal.

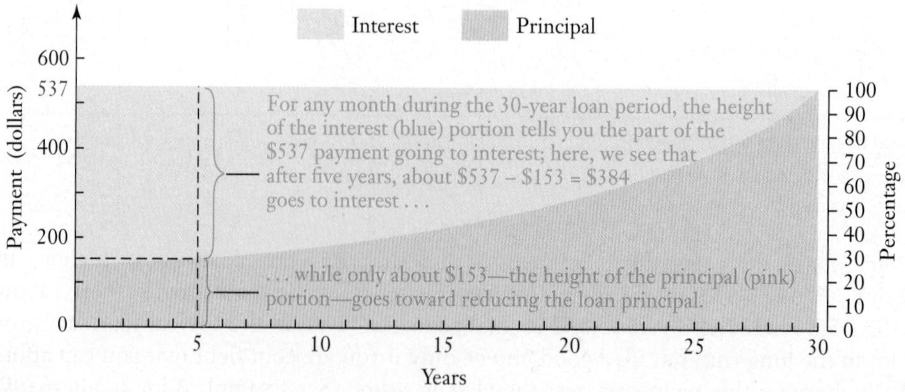

FIGURE 4.12 Portions of monthly payments going to principal and interest over the life of a 30-year, \$100,000 loan at 5%. (The calculations leading to this graph can be done by making a principal and interest table as described in Using Technology on p. 249.)

IN YOUR WORLD

Choosing or Refinancing a Loan

As you've seen throughout this unit, the interest rate and the loan term are the primary factors in determining loan payments. But several other factors need to be considered when you take out or refinance any loan (*refinancing* means replacing an existing loan with a new loan):

- Be sure that you understand the loan. For example, is the interest rate fixed or variable? What is the term of the loan? When are payments due?
- Is a down payment required? If so, how will you afford it? If not, could you get a better interest rate by offering to make a down payment?
- What fees and closing costs will be charged? Be sure you identify *all* closing costs, including origination fees and discount points, as different lenders may quote their fees differently.
- Watch out for fine print that may make the loan more expensive than it appears. Be especially wary of prepayment penalties, because you may later decide to pay the loan off early or to refinance it at a better interest rate.
- Refinancing at a lower interest rate can save you money, but it is not always a good idea. Be sure to consider these two additional factors when deciding whether to refinance a loan:
  1. How long will it take before the lower interest rate makes up for the fees and closing costs you must pay to refinance? As a general rule, you should not refinance a loan for which these costs take more than about 2 to 3 years to recoup, unless you are convinced that you will be holding the new loan for a much longer time.
  2. Remember that refinancing "resets the clock" on a loan. For example, suppose you have been paying off a 10-year student loan for 4 years. If you keep this loan, you will pay it off in 6 more years. But if you refinance with a new 10-year loan, you will have payments for 10 years starting from now. So even if refinancing reduces your monthly payments, it may not be worth it because you will be making payments for 10 more years instead of only 6 more years.
- Most important, regardless of what a bank or loan broker might say, be sure that *you* feel confident that you will be able to afford your loan payments throughout the life of the loan, even in the event that you temporarily lose your job or have other unexpected financial difficulties.

You can therefore save a lot if you can reduce your interest payments. As long as there are no prepayment penalties, one way to do this is to pay extra toward the principal, particularly early in the term. For example, suppose you pay an extra \$100 toward the principal in the first monthly payment of your \$100,000 loan. That is, instead of paying the required \$537 (see Example 6), you pay \$637. Because you've reduced your loan balance by \$100, you will save the compounded value of this \$100 over the rest of the 30-year loan term—which is nearly \$450. In other words, paying an extra \$100 in the first month saves you about \$450 in interest over the 30 years.

BY THE WAY

Mortgage rates have varied substantially with time. In the 1980s, average U.S. rates for new mortgages (30-year fixed) were almost always above 10%, peaking at more than 18% in 1981. During 2016, average rates dipped to record lows near 3%.

EXAMPLE 8 An Alternative Strategy

An alternative strategy to the mortgage options in Example 6 is to take the 30-year loan at 5%, but try to pay it off in 15 years by making larger payments than are required. To carry out this plan, how much would you have to pay each month? Discuss the pros and cons of this strategy.

Solution To adjust for paying off the 5% loan in 15 years, we set APR $= 0.05$ and $Y = 15$; we still have $P = \$100{,}000$ and $n = 12$. The monthly payments are

$$\text{PMT} = \frac{P \times \left(\frac{\text{APR}}{n}\right)}{\left[1 - \left(1 + \frac{\text{APR}}{n}\right)^{(-nY)}\right]} = \frac{\$100{,}000 \times \left(\frac{0.05}{12}\right)}{\left[1 - \left(1 + \frac{0.05}{12}\right)^{(-12 \times 15)}\right]} = \$790.79$$

BY THE WAY

The 2008 financial crisis occurred in part because mortgage lenders were able to collect fees for making loans and then sell the loans to other financial institutions, which created an incentive to make risky loans because the risk was transferred to new owners. The new owners of the loans, in turn, often bundled the loans and used "reinsurers," who supposedly insured the loans against default, but then defaulted themselves. You can learn more from many books and from movies including *The Big Short* (2015) and *Inside Job* (2010).

In Example 6, we found that the 30-year loan requires payments of about \$537, so paying off this loan in 15 years requires paying more than the minimum by about \$791 − \$537 = \$254 per month.

Note that this payment is also about \$26 per month *more* than the payment of \$765 required with the 15-year loan (see Example 6), because the 15-year loan has a lower interest rate. Clearly, if you *know* you're going to pay off the loan in 15 years, you should take the lower-interest 15-year loan. However, taking the 30-year loan has one advantage: Because your required monthly payments are only \$537, you can always drop back to this level in some months if you can't afford the extra needed to pay off the loan in 15 years. ▶ Now try Exercises 45–46.

Think About It Even if you can afford extra mortgage payments, some financial advisors suggest instead using that extra money to invest in stocks, bonds, or other investments. Why might this suggestion make sense? Can you think of counterarguments? Overall, what would *you* do if you had an extra \$100 per month: use it to pay off a loan early or invest it?

Adjustable Rate Mortgages

A fixed rate mortgage is advantageous for you because your monthly payments never change. However, it poses a risk to the lender. Imagine that you take out a fixed, 30-year loan of \$100,000 from Great Bank at a 4% interest rate. Initially, the loan may seem like a good deal for Great Bank. But suppose that, 2 years later, prevailing interest rates have risen to 5%. If Great Bank still had the \$100,000 that it lent to you, it could lend it out to someone else at this higher rate. Instead, it's stuck with the 4% rate that you are paying. In effect, Great Bank loses potential future income if prevailing rates rise and you have a fixed rate loan.

Lenders can lessen the risk of rising interest rates by charging higher rates for longer-term loans. That is why rates generally are higher for 30-year loans than for 15-year loans. But an even lower-risk strategy for the lender is an **adjustable rate mortgage (ARM)**, in which the interest rate you pay changes whenever prevailing rates change. Because of the reduced long-term risk to lenders, ARMs generally have much lower initial interest rates than fixed rate loans. For example, a bank offering a 4% rate on a fixed 30-year loan might offer an ARM that begins at 3%. Most ARMs guarantee their starting interest rate for the first 6 months or 1 year, but after that period the interest rate will move up or down to reflect prevailing rates. Most ARMs also include a *rate cap* that cannot be exceeded. For example, if your ARM begins at an interest rate of 3%, you may be promised that your interest rate can never go higher than a rate cap of 8%. Making a decision between a fixed rate loan and an ARM can be one of the most important financial decisions of your life.

BY THE WAY

Watch out for "teaser" rates on adjustable rate mortgages. Under normal circumstances, your rate on an ARM rises only if prevailing interest rates rise. However, some lenders offer low teaser rates—rates below the prevailing rates—for the first few months of an ARM. While teaser rates may be attractive, the longer-term policies of the ARM are far more important.

EXAMPLE 9 Rate Approximations for ARMs

You have a choice between a 30-year fixed rate loan at 4% and an ARM with a first-year rate of 3%. Neglecting compounding and changes in principal, estimate your monthly savings with the ARM during the first year on a \$100,000 loan. Suppose that the ARM rate rises to 5% by the third year. How will your payments be affected?

Solution Because mortgage payments mostly go to interest in the early years of a loan, we can make approximations by assuming that the principal remains unchanged. For the 4% fixed rate loan, the interest on the \$100,000 loan for the first year will be approximately $4\% \times \$100{,}000 = \$4000$. With the 3% ARM, your first-year interest will be approximately $3\% \times \$100{,}000 = \$3000$. The ARM will save you about \$1000 in interest during the first year, which means a monthly savings of about $\$1000 \div 12 \approx \$83$.

By the third year, when rates reach 5%, the situation is reversed. The rate on the ARM is now 1 percentage point above the rate on the fixed rate loan. Instead of saving

$83 per month, you'd be paying $83 per month *more* on the ARM than on the 4% fixed rate loan. Moreover, if interest rates remain high on the ARM, you will continue to make these high payments for many years to come. Therefore, while ARMs reduce risk for the lender, they add risk for the borrower.

▶ Now try Exercises 47–48.

Think About It In recent years, another type of mortgage loan became popular: the *interest only loan*, in which you pay only interest and pay nothing toward principal. Most financial experts advise against these loans, because your principal never gets paid off, and many think they played a role in creating the housing bubble. Can you think of any circumstances under which such a loan might make sense for a home buyer? Explain.

Quick Quiz 4D

Choose the best answer to each of the following questions. Explain your reasoning with one or more complete sentences.

1. In the loan payment formula, assuming all other variables are constant, the monthly payment
   a. increases as P increases.
   b. increases as APR decreases.
   c. increases as Y increases.
2. With the same APR and amount borrowed, a 15-year loan will have
   a. a higher monthly payment than a 30-year loan.
   b. a lower monthly payment than a 30-year loan.
   c. a payment that could be greater or less than that of a 30-year loan.
3. With the same term and amount borrowed, a loan with a higher APR will have
   a. a lower monthly payment than a loan with a lower APR.
   b. a higher monthly payment than a loan with a lower APR.
   c. a payment that could be greater or less than that of a loan with a lower APR.
4. In the early years of a 30-year mortgage loan,
   a. most of the monthly payment goes to the principal.
   b. most of the monthly payment goes to interest.
   c. equal amounts go to principal and interest.
5. If you make monthly payments of $1000 on a 10-year loan, your total payments over the life of the loan amount to
   a. $1000. b. $100,000. c. $120,000.
6. Credit card loans are different from installment loans in that
   a. credit card loans do not require regular (monthly) payments.
   b. credit card loans do not have an APR.
   c. credit card loans do not have a set loan term.
7. A loan of $200,000 that carries a 2-point origination fee requires an advance payment of
   a. $2000. b. $40,000. c. $4000.
8. A $120,000 loan with $500 in closing costs plus 1 point requires an advance payment of
   a. $1500. b. $1700. c. $500.
9. You are currently paying off a student loan with an interest rate of 9% and a monthly payment of $450. You are offered the chance to refinance the remaining balance with a new 10-year loan with an interest rate of 8%, which will give you a significantly lower monthly payment. Refinancing in this way
   a. is always a good idea.
   b. is a good idea if it lowers your monthly payment by at least $100.
   c. may or may not be a good idea, depending on closing costs and how many years are remaining in your current loan term.
10. Consider two mortgage loans with the same amount borrowed and the same APR. Loan 1 is fixed for 15 years, and Loan 2 is fixed for 30 years. Which statement is true?
    a. Loan 1 will have higher monthly payments, but you'll pay less total interest over the life of the loan.
    b. Loan 1 will have lower monthly payments, and you'll pay less total interest over the life of the loan.
    c. Both loans will have the same monthly payments, but you'll pay less total interest with Loan 1.

Exercises 4D

REVIEW QUESTIONS

1. Suppose you pay only the interest on a loan. Will the loan ever be paid off? Why or why not?
2. What is an *installment loan*? Explain the meaning and use of the loan payment formula.
3. Explain, in general terms, how the portions of loan payments going to principal and interest change over the life of the loan.

4. Suppose that you need a loan of $100,000 and are offered a choice of a 3-year loan at 5% interest or a 5-year loan at 6% interest. Discuss the pros and cons of each choice.

5. How do credit card loans differ from ordinary installment loans? Why are credit card loans particularly dangerous?

6. What is a *mortgage*? What is a down payment on a mortgage? Explain how closing costs, including points, can affect the choice of a mortgage.

DOES IT MAKE SENSE?

Decide whether each of the following statements makes sense (or is clearly true) or does not make sense (or is clearly false). Explain your reasoning.

7. The interest rate on my student loan is only 6%, yet more than half of my payments are currently going toward interest rather than principal.

8. My student loans were all 20-year loans at interest rates of 7% or above, so when my bank offered me a 20-year loan at 6%, I took it and used it to pay off the student loans.

9. I make only the minimum required payments on my credit card balance each month, because that way I'll have more of my own money to keep.

10. I carry a large credit card balance, and I had a credit card that charged an annual interest rate of 12%. So when I found another credit card that promised a 3% interest rate for the first 3 months, it was obvious that I should switch to this new card.

11. I had a choice between a fixed rate mortgage at 4% and an adjustable rate mortgage that started at 2.5% for the first year with a maximum increase of 1.5 percentage points a year. I took the adjustable rate, because I'm planning to move within three years.

12. Fixed rate loans with 15-year terms have lower interest rates than loans with 30-year terms, so it always makes sense to take the 15-year loan.

BASIC SKILLS & CONCEPTS

13–14: Loan Terminology. Consider the following loans.

a. Identify the amount borrowed, the annual interest rate, the number of payments per year, the loan term, and the payment amount.

b. How many total payments does the loan require? What is the total amount paid over the full term of the loan?

c. Of the total amount paid, what percentage is for the principal and what percentage is for interest?

13. You borrowed $120,000 at an APR of 6%, which you are paying off with monthly payments of $1013 for 15 years.

14. You borrowed $15,000 at an APR of 9%, which you are paying off with monthly payments of $190 for 10 years.

15–24: Loan Payments. Consider the following loans.

a. Calculate the monthly payment.

b. Determine the total amount paid over the term of the loan.

c. Of the total amount paid, what percentage is for the principal and what percentage is for interest?

15. A student loan of $100,000 at a fixed APR of 6% for 20 years

16. A student loan of $24,000 at a fixed APR of 7% for 10 years

17. A home mortgage of $400,000 with a fixed APR of 3.5% for 30 years

18. A home mortgage of $150,000 with a fixed APR of 4% for 15 years

19. A home mortgage of $100,000 with a fixed APR of 3% for 15 years

20. A home mortgage of $200,000 with a fixed APR of 4% for 30 years

21. You borrow $10,000 for a period of 3 years at a fixed APR of 8%.

22. You borrow $10,000 for a period of 5 years at a fixed APR of 6.5%.

23. You borrow $150,000 for a period of 15 years at a fixed APR of 5%.

24. You borrow $100,000 for a period of 30 years at a fixed APR of 5.5%.

25–26: Principal and Interest. For each of the following loans, make a table (as in Example 2) showing the amounts of each monthly payment that go toward principal and interest for the first 3 months of the loan.

25. A home mortgage of $150,000 with a fixed APR of 4% for 30 years

26. A student loan of $24,000 at a fixed APR of 8% for 15 years

27. **Choosing a Personal Loan.** You need to borrow $15,000 to buy a car, and you determine that you can afford monthly payments of $325. The bank offers three choices: a 3-year loan at 7% APR, a 4-year loan at 7.5% APR, or a 5-year loan at 8% APR. Which loan best meets your needs? Explain your reasoning.

28. **Choosing a Personal Loan.** You need to borrow $4000 to pay off your credit cards and you can afford monthly payments of $150. The bank offers three choices: a 2-year loan at 8% APR, a 3-year loan at 9% APR, or a 4-year loan at 10% APR. Which loan best meets your needs? Explain your reasoning.

29–32: Credit Card Debt. Suppose that on January 1 you have a balance of $10,000 on each of the following credit cards, which you want to pay off in the given amount of time. Assume that you make no additional charges on the card after January 1.

a. Calculate your monthly payments.

b. When the card is paid off, how much will you have paid since January 1?

c. What percentage of the total payment found in part (b) is for interest?

29. The credit card APR is 18%, and you want to pay off the balance in 1 year.

30. The credit card APR is 20%, and you want to pay off the balance in 2 years.

31. The credit card APR is 21%, and you want to pay off the balance in 3 years.

32. The credit card APR is 22%, and you want to pay off the balance in 1 year.

33. **Credit Card Debt.** Assume you have a balance of \$1200 on a credit card with an APR of 18%, or 1.5% per month. You start making monthly payments of \$200, but at the same time you charge an additional \$75 per month to the credit card. Assume that interest for a given month is based on the balance for the previous month. The following table shows how you can calculate your monthly balance.

Month	Payment	Expenses	Interest	New balance
0	—	—	—	\$1200
1	\$200	\$75	1.5% × \$1200 = \$18	\$1200 − \$200 + \$75 + \$18 = \$1093
2	\$200	\$75		
3	\$200	\$75		

Complete and extend the table to show your balance at the end of each month until the debt is paid off. How long does it take to pay off the credit card debt?

34. **Credit Card Debt.** Repeat the table in Exercise 33, but this time assume that you make monthly payments of \$300. Extend the table as long as necessary until your debt is paid off. How long does it take to pay off your debt?

35. **Credit Card Woes.** The following table shows the expenses and payments for 8 months on a credit card account with an initial balance of \$300. Assume that the interest rate is 1.5% per month (18% APR) and that interest for a given month is charged on the balance from the previous month. Complete the table. After 8 months, what is the balance on the credit card? Comment on the effect of the interest and the initial balance, in light of the fact that for 7 of the 8 months expenses never exceeded payments.

Month	Payment	Expenses	Interest	New balance
0	—	—	—	\$300
1	\$300	\$175	1.5% × \$300 = \$4.50	\$179.50
2	\$150	\$150		
3	\$400	\$350		
4	\$500	\$450		
5	0	\$100		
6	\$100	\$100		
7	\$200	\$150		
8	\$100	\$80		

36. **Teaser Rate.** You have a total credit card debt of \$4000. You receive an offer to transfer this debt to a new card with an introductory APR of 6% for the first 6 months. After that, the rate becomes 24%.

a. What is the monthly interest payment on \$4000 during the first 6 months? (Assume you pay nothing toward principal and don't charge any further amounts on the new card.)

b. What is the monthly interest payment on \$4000 *after* the first 6 months? Comment on the change from the teaser rate.

37–40: Comparing Loan Options. Compare the monthly payment and total payment for the following pairs of loan options. Assume that both loans are fixed rate and have the same closing costs. Discuss the pros and cons of each loan.

37. You need a \$400,000 loan.
Option 1: a 30-year loan at an APR of 8%
Option 2: a 15-year loan at an APR of 7.5%

38. You need a \$150,000 loan.
Option 1: a 30-year loan at an APR of 8%
Option 2: a 15-year loan at an APR of 7%

39. You need a \$60,000 loan.
Option 1: a 30-year loan at an APR of 7.15%
Option 2: a 15-year loan at an APR of 6.75%

40. You need a \$180,000 loan.
Option 1: a 30-year loan at an APR of 7.25%
Option 2: a 15-year loan at an APR of 6.8%

41–44: Closing Costs. Consider the following pairs of choices for a \$120,000 mortgage loan. Calculate the monthly payment and total closing costs for each choice. Explain which loan you would choose and why.

41. Choice 1: 30-year fixed rate mortgage at 4% with closing costs of \$1200 and no points
Choice 2: 30-year fixed rate mortgage at 3.5% with closing costs of \$1200 and 2 points

42. Choice 1: 30-year fixed rate mortgage at 4% with no closing costs and no points
Choice 2: 30-year fixed rate mortgage at 3% with closing costs of \$1200 and 4 points

43. Choice 1: 30-year fixed rate mortgage at 4.5% with closing costs of \$1200 and 1 point
Choice 2: 30-year fixed rate mortgage at 4.25% with closing costs of \$1200 and 3 points

44. Choice 1: 30-year fixed rate mortgage at 3.5% with closing costs of \$1000 and no points
Choice 2: 30-year fixed rate mortgage at 3% with closing costs of \$1500 and 4 points

45. **Accelerated Loan Payment.** Suppose you have a student loan of \$30,000 with an APR of 9% for 20 years.

a. What are your required monthly payments?

b. Suppose you would like to pay the loan off in 10 years instead of 20. What monthly payments will you need to make?

c. Compare the total amounts you'll pay over the loan term if you pay the loan off in 20 years versus 10 years.

46. **Accelerated Loan Payment.** Suppose you have a student loan of \$60,000 with an APR of 8% for 25 years.

a. What are your required monthly payments?

b. Suppose you would like to pay the loan off in 15 years instead of 25. What monthly payments will you need to make?

c. Compare the total amounts you'll pay over the loan term if you pay the loan off in 25 years versus 15 years.

47. **ARM Rate Approximations.** You have a choice between a 30-year fixed rate mortgage at 4.5% and an adjustable rate mortgage (ARM) with a first-year rate of 3%. Estimate your monthly savings with the ARM during the first year for a \$150,000 loan. Suppose that the ARM rate rises to 6.5% at the start of the third year. Approximately how much extra will you then be paying over what you would have paid if you had taken the fixed rate loan? Neglect compounding and changes in principal when making your estimates.

48. **ARM Rate Approximations.** You have a choice between a 30-year fixed rate mortgage at 4% and an ARM with a first-year rate of 2.5%. Estimate your monthly savings with the ARM during the first year for a \$150,000 loan. Suppose that the ARM rate rises to 5.75% at the start of the second year. Approximately how much extra will you then be paying over what you would have paid if you had taken the fixed rate loan? Neglect compounding and changes in principal when making your estimates.

FURTHER APPLICATIONS

49. **How Much House Can You Afford?** You can afford monthly payments of \$500. If current mortgage rates are 3.75% for a 30-year fixed rate loan, how much can you afford to borrow? If you are required to make a 20% down payment and you have the cash on hand to do it, how expensive a home can you afford? (*Hint:* You will need to solve the loan payment formula for P.)

50. **Refinancing.** Suppose you take out a 30-year \$200,000 mortgage with an APR of 6%. You make payments for 5 years (60 monthly payments) and then consider refinancing the original loan. The new loan would have a term of 20 years, have an APR of 5.5%, and be in the amount of the unpaid balance on the original loan. (The amount you borrow on the new loan would be used to pay off the balance on the original loan.) The administrative cost of taking out the second loan would be \$2000.

a. What are the monthly payments on the original loan?

b. A short calculation shows that the unpaid balance on the original loan after 5 years is \$186,046, which would become the amount of the second loan. What would the monthly payments be on the second loan?

c. What would be the total amount you would pay if you continued with the original 30-year loan without refinancing?

d. What would be the total amount you would pay with the refinancing plan?

e. Compare the two options and decide which one you would choose. What other factors should be considered in making the decision?

51. **Student Loan Consolidation.** Suppose you have the following three student loans: \$10,000 with an APR of 6.5% for 15 years, \$15,000 with an APR of 7% for 20 years, and \$12,500 with an APR of 7.5% for 10 years.

a. Calculate the monthly payment for each loan individually.

b. Calculate the total of your monthly payments during the life of all three loans (combined).

c. A bank offers to consolidate your three loans into a single loan with an APR of 6.5% and a loan term of 15 years. What will your monthly payments be in that case? What will your total monthly payments be over the 15 years? Discuss the pros and cons of accepting this loan consolidation.

52. **Bad Deals: Car-Title Lenders.** Some businesses called "car-title lenders" offer quick cash loans in exchange for holding the title to your car as collateral (you lose your car if you fail to pay off the loan). In many states, these lenders operate under pawnbroker laws that allow them to charge a fee that is a percentage of the unpaid balance. Suppose you need \$2000 in cash, and a car-title lender offers you a loan at an interest rate of 2% per month *plus* a monthly fee of 20% of the unpaid balance.

a. How much will you owe in interest and fees on your \$2000 loan at the end of the first month?

b. Suppose that you pay only the interest and fees each month. How much will you pay over the course of a full year?

c. Suppose that you instead obtain a loan of \$2000 from a bank with a term of 3 years and an APR of 10%. What are your monthly payments in that case? Compare these to the payments to the car-title lender.

53. **Payday Loans.** *Payday loans* (also called *cash advance loans*) are short-term, high-interest loans marketed as a way for people to borrow money for a short period of time until their next paycheck arrives. However, like many other short-term loans, they should be used with extreme caution; indeed, they are outlawed or highly regulated in some states. Payday loans typically carry a fixed fee of \$15 to \$30 per \$100 borrowed, and lenders sometimes demand

access to your checking account to make sure you will pay the fees.

a. Consider a payday loan that has a fixed fee of \$20 per \$100 borrowed that is due in 2 weeks (when your paycheck arrives). If you borrow \$500, what is the total fee (not including your \$500 principal) you owe the lender at the end of 2 weeks?

b. What is the interest rate for that 2-week period?

c. What is the equivalent annual percentage rate (APR) for this loan? (Assume that there are 52 weeks in 1 year.) How does this rate compare to other common types of loans, such as credit cards or mortgages?

54. **13 Payments.** Suppose you want to borrow \$100,000, and you find a bank offering a 20-year term for a loan of that amount, with an APR of 6%.

a. What are your monthly payments?

b. Instead of making 12 payments per year, you save enough money to make a 13th payment each year, of the same amount as your regular monthly payment found in part (a). How long will it take to pay off the loan?

55. **Other Than Monthly Payments.** Suppose you want to borrow \$100,000 and you find a bank offering a 20-year term for a loan of that amount, with an APR of 6%.

a. Find your regular payments if you make them yearly, monthly, biweekly (every 2 weeks), or weekly, that is, for $n = 1, 12, 26, 52$.

b. Compute the total payment for each case in part (a).

c. Compare the total payments computed in part (b). Discuss the pros and cons of the payment plans.

IN YOUR WORLD

56. **Choosing a Mortgage.** Imagine that you work for an accounting firm and a client has told you that he is buying a house and needs a loan of \$120,000. His monthly income is \$4000, and he is single with no children. He has \$14,000 in savings that can be used for a down payment. Find the current rates available from local banks for both fixed rate mortgages and adjustable rate mortgages (ARMs). Analyze the offerings, and summarize orally or in writing the best options for your client, along with the pros and cons of each option.

57. **Credit Card Statement.** Look carefully at the terms of financing explained on your most recent credit card statement. Explain all the important terms, including the interest rates that apply, annual fees, and grace periods.

58. **Credit Card Comparisons.** Visit a website that gives comparisons between credit cards. Briefly explain the factors that are considered in the comparisons. How does your own credit card compare to other credit cards? Based on this comparison, do you think you would be better off with a different credit card?

59. **Home Financing.** Visit a website that offers online home financing. Describe the options offered, and discuss the advantages and disadvantages of each option.

60. **Online Car Purchase.** Find a car online that you might want to buy. Find a loan that you would qualify for, and calculate your monthly payments and total payment over the life of the loan. Next, suppose that you start a savings plan instead of buying the car, depositing the same amounts that would have gone to car payments. Estimate how much you would have in your savings plan by the time you graduate from college. Explain your assumptions.

61. **Student Financial Aid.** There are many websites that offer student loans. Visit a website that offers such loans, and describe the terms of a particular loan. Discuss the advantages and disadvantages of financing a student loan online rather than through a bank or through your university or college.

62. **Scholarship Scams.** The Federal Trade Commission keeps track of many financial scams related to college scholarships. Read about two different types of scams, and report on how they work and how they hurt people who are taken in by them.

63. **Financial Scams.** Many websites keep track of current financial scams. Visit some of these sites, and report on one scam that has already hurt a lot of people. Describe how the scam works and how it hurts those who are taken in by it.

TECHNOLOGY EXERCISES

64. **Loan Payments Using Excel.** Use the PMT function in Excel to answer the following questions.

a. Find the monthly payment for a \$7500 loan with an APR of 9% and a loan term of 10 years.

b. Describe the change in the monthly payment for the loan in part (a) if the loan term is doubled.

c. Describe the change in the monthly payment for the loan in part (a) if the APR is halved.

65. **Loan Table in Excel.** Consider the loan in Exercise 64 (a \$7500 loan with an APR of 9% and a loan term of 10 years).

a. Follow the guidelines in Using Technology on p. 249 to construct a table showing the interest payment and loan balance after each month. Verify that, with monthly payments of \$95.01, the loan balance reaches \$0 after 120 months.

b. How much interest is paid in the first month of the loan? How much is paid toward the principal in the first month of the loan?

c. How much interest is paid in the last month of the loan? How much is paid toward the principal in the last month of the loan?

UNIT 4E Income Taxes

In this world, nothing is certain but death and taxes.
—Benjamin Franklin

There are many types of taxes, including sales tax, gasoline tax, and property tax. But for most Americans, the largest tax burden comes from federal taxes on wages and other income. These taxes also represent one of the most contentious political issues of our time. In this unit, we explore a few of the many aspects of federal income taxes.

The hardest thing in the world to understand is the income tax.
—Albert Einstein

Income Tax Basics

It's quite possible that *no one* fully understands federal income taxes. The complete tax code is many thousands of pages long, and it changes almost every year as Congress enacts new tax laws. Nevertheless, the basic ideas behind federal income taxes are relatively simple, and most of the arcane rules apply only to relatively small segments of the population. As a result, most people can not only file their own taxes but also understand them well enough to make intelligent decisions about both personal finances and political tax questions.

Figure 4.13 summarizes the steps in a basic income tax calculation. We'll go through the steps in this flow chart, defining terms as we go.

- The process begins with your **gross income**, which is *all* your income for the year, including wages, tips, profits from a business, interest or dividends from investments, and any other income you receive.
- Some gross income is not taxed (in the year it is received), such as contributions to individual retirement accounts (IRAs) and other tax-deferred savings plans. These untaxed portions of gross income are called *adjustments*. Subtracting your adjustments from your gross income gives your **adjusted gross income**.
- Most people are entitled to certain **exemptions** and **deductions**—amounts that are subtracted from adjusted gross income before taxes are calculated. Once you subtract your exemptions and deductions, you are left with your **taxable income**.
- A tax table or tax rate computation allows you to determine how much tax you owe on your taxable income. However, you may not actually have to pay this much tax if you are entitled to any **tax credits**, such as the child tax credit that many parents can claim. Subtracting the amount of any tax credits gives your **total tax**.
- Finally, most people have already paid part or all of their tax bill during the year, either through *withholding* (by an employer) or by paying quarterly *estimated taxes* (for those who are self-employed). You subtract the taxes that you've already paid to determine how much you still owe. If you have paid more than you owe, then you should receive a **tax refund**.

HISTORICAL NOTE

An income tax was first levied in the United States in 1862 (during the Civil War), but was abandoned a few years later. The 16th Amendment to the Constitution, ratified in 1913, gave the federal government full authority to levy an income tax.

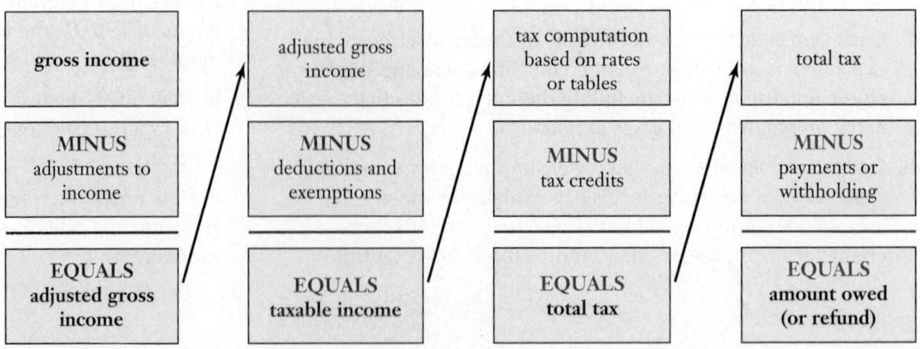

FIGURE 4.13 Flow chart showing the basic steps in calculating income tax.

EXAMPLE 1 Income on Tax Forms

Karen earned wages of \$38,600, received \$750 in interest from a savings account, and contributed \$1200 to a tax-deferred retirement plan. She is entitled to a personal exemption of \$4050 and had deductions totaling \$6350. Find her gross income, adjusted gross income, and taxable income.

Solution Karen's gross income is the sum of all her income, which means the sum of her wages and her interest:

$$\text{gross income} = \$38{,}600 + \$750 = \$39{,}350$$

Her \$1200 contribution to a tax-deferred retirement plan counts as an adjustment to her gross income, so her adjusted gross income (AGI) is

$$\text{AGI} = \text{gross income} - \text{adjustments} = \$39{,}350 - \$1200 = \$38{,}150$$

To find her taxable income, we subtract her exemptions and deductions:

$$\begin{aligned}\text{taxable income} &= \text{AGI} - \text{exemptions} - \text{deductions} \\ &= \$38{,}150 - \$4050 - \$6350 = \$27{,}750\end{aligned}$$

Her taxable income is \$27,750. ▶ Now try Exercises 19–22.

Filing Status

Tax calculations depend on your **filing status**, such as single or married. Most people fall into one of four categories for filing status:

- *Single* applies if you are unmarried, divorced, or legally separated.
- *Married filing jointly* applies if you are married *and* you and your spouse file a single tax return. (In some cases, this category also applies to widows or widowers.)
- *Married filing separately* applies if you are married *and* you and your spouse file two separate tax returns.
- *Head of household* applies if you are unmarried and are paying more than half the cost of supporting a dependent child or parent.

We will use these four categories in the rest of our discussion.

BY THE WAY

U.S. federal income taxes are collected by the *Internal Revenue Service* (IRS), which is part of the U.S. Department of the Treasury. Most people file federal taxes by completing a tax form, such as Form 1040, 1040A, or 1040EZ.

Exemptions and Deductions

Both exemptions and deductions are subtracted from your adjusted gross income. However, they are calculated differently, which is why they have different names.

Exemptions are a fixed amount per person (\$4050 in 2017 for most people). You can claim the amount of an exemption for yourself and each of your *dependents* (for example, children whom you support).

Deductions vary from one person to another. The most common deductions include interest on home mortgages, contributions to charities, and taxes you've paid to other agencies (such as state income taxes and local property taxes). However, you don't necessarily have to add up all your deductions. When you file your taxes, you have two options for deductions, of which you should choose the larger one, since it will reduce your taxes more:

- You can choose a **standard deduction**, the amount of which depends on your filing status.
- You can choose **itemized deductions**, in which case you add up all the individual deductions to which you are entitled.

Note that you get *either* the standard deduction *or* itemized deductions, not both.

While exemptions and deductions are usually easy to calculate, there's a complication for high-income taxpayers: Those with incomes above certain thresholds (for example, $261,500 for a single person in 2017) are subject to "phase out" rules that reduce the dollar value of both exemptions and deductions. In addition, many middle- to high-income taxpayers are subject to the *alternative minimum tax* (AMT), which can also limit the value of deductions. The rules for these changes can be quite complex, and we will ignore them for the calculations in this chapter.

EXAMPLE 2 Should You Itemize?

Suppose you have the following deductible expenditures: $4500 for interest on a home mortgage, $900 for contributions to charity, and $250 for state income taxes. Your filing status entitles you to a standard deduction of $6350. Should you itemize your deductions or take the standard deduction?

Solution The total of your deductible expenditures is

$$\$4500 + \$900 + \$250 = \$5650$$

If you itemize your deductions, you can subtract $5650 when finding your taxable income. But if you take the standard deduction, you can subtract $6350. You are better off taking the standard deduction.

▶ Now try Exercises 23–28.

Tax Rates

For ordinary income (as opposed to dividends and capital gains, which we'll discuss later), the United States has a **progressive income tax**, meaning that people with higher taxable incomes pay at a higher tax *rate*. The system works by assigning different **marginal tax rates** to different income ranges (or *margins*). For example, suppose you are single and your taxable income is $25,000. Under 2017 tax rates, you would pay 10% tax on the first $9325 and 15% tax on the remaining $15,675. In this case, we say that your *marginal rate* is 15% or that you are in the 15% *tax bracket*. For each major filing status, Table 4.9 shows the marginal tax rate, standard deduction, and exemptions for 2017. If you are calculating taxes for a year other than 2017, you must get an updated tax rate table.

BY THE WAY

Congress can change the tax laws at any time, and when this text was written in mid-2017, both Congress and the President had proposed significant changes to the federal tax system, some of which may be implemented by the time you are reading this chapter.

TABLE 4.9 2017 Marginal Tax Rates, Standard Deductions, and Exemptions*

Tax Rate**	Single	Married Filing Jointly	Married Filing Separately	Head of Household
10%	up to $9325	up to $18,650	up to $9325	up to $13,350
15%	up to $37,950	up to $75,900	up to $37,950	up to $50,800
25%	up to $91,900	up to $153,100	up to $76,550	up to $131,200
28%	up to $191,650	up to $233,350	up to $116,675	up to $212,500
33%	up to $416,700	up to $416,700	up to $208,350	up to $416,700
35%	up to $418,400	up to $470,700	up to $235,350	up to $444,550
39.6%	above $418,400	above $470,700	above $235,350	above $444,550
Standard deduction	$6350	$12,700	$6350	$9350
Exemption (per person)	$4050	$4050	$4050	$4050

* This table ignores (i) exemption and deduction phase-outs that apply to high-income taxpayers; (ii) the alternative minimum tax (AMT) that affects many middle- and high-income taxpayers; (iii) potential changes in tax law made after this text was printed that may have changed the values given in this table for 2017.

** Each higher marginal rate begins where the prior one leaves off. For example, for a single person, the 15% marginal rate affects income starting at $9325, which is where the 10% rate leaves off, and continuing up to $37,950.

Think About It Find a table of marginal tax rates for the current year. Have the tax brackets (10%, 15%, ..., 39.6%) changed since 2017? Have the income thresholds for each tax bracket changed? Briefly summarize the changes you notice.

EXAMPLE 3 Marginal Tax Computations

Using 2017 rates, calculate the tax owed by each of the following people. Assume that they all take the standard deduction and neglect any tax credits.

There is no such thing as a good tax.

—Winston Churchill

I like paying taxes. With them I buy civilization.

—Justice Oliver Wendell Holmes

a. Deirdre is single with no dependents. Her adjusted gross income is \$90,000.

b. Robert is a head of household taking care of two dependent children. His adjusted gross income is \$90,000.

c. Jessica and Frank are married with no dependents. They file jointly. They each have \$90,000 in adjusted gross income, or a total adjusted gross income of \$180,000.

Solution

a. First, we must find Deirdre's taxable income. She is entitled to a personal exemption of \$4050 and a standard deduction of \$6350. We subtract these amounts from her adjusted gross income to find her taxable income:

$$\text{taxable income} = \$90{,}000 - \$4050 - \$6350 = \$79{,}600$$

Now we calculate her taxes using the rates in the "Single" column in Table 4.9. She is in the 25% tax bracket because her taxable income is above \$37,950 but below the 28% threshold of \$91,900. Therefore, she owes 10% on the first \$9325 of her taxable income, 15% on her taxable income above \$9325 but below \$37,950, and 25% on her taxable income above \$37,950.

$$\underbrace{(10\% \times \$9325)}_{\substack{\text{10\% marginal rate on first}\\ \text{\$9325 of taxable income}}} + \underbrace{(15\% \times [\$37{,}950 - \$9325])}_{\substack{\text{15\% marginal rate on taxable income}\\ \text{between \$9325 and \$37,950}}} + \underbrace{(25\% \times [\$79{,}600 - \$37{,}950])}_{\substack{\text{25\% marginal rate on taxable}\\ \text{income above \$37,950}}}$$

$$= \$932.50 + \$4293.75 + \$10{,}412.50$$

$$= \$15{,}638.75$$

Rounded to the nearest dollar, Deirdre's tax is \$15,639.

b. Robert is entitled to *three* exemptions of \$4050 each—one for himself and one for each of his two children. As a head of household, he is also entitled to a standard deduction of \$9350. We subtract these amounts from his adjusted gross income to find his taxable income:

$$\text{taxable income} = \$90{,}000 - (3 \times \$4050) - \$9350 = \$68{,}500$$

We calculate Robert's taxes using the "Head of Household" rates. His taxable income of \$68,500 puts him in the 25% tax bracket, so his tax is

$$\underbrace{(10\% \times \$13{,}350)}_{\substack{\text{10\% marginal rate on first}\\ \text{\$13,350 of taxable income}}} + \underbrace{(15\% \times [\$50{,}800 - \$13{,}350])}_{\substack{\text{15\% marginal rate on taxable income}\\ \text{between \$13,350 and \$50,800}}} + \underbrace{(25\% \times [\$68{,}500 - \$50{,}800])}_{\substack{\text{25\% marginal rate on taxable}\\ \text{income above \$50,800}}}$$

$$= \$1335.00 + \$5617.50 + \$4425.00$$

$$= \$11{,}377.50$$

Rounded to the nearest dollar, Robert's tax is \$11,378.

c. Jessica and Frank are each entitled to one exemption of \$4050. Because they are married filing jointly, their standard deduction is \$12,700. We subtract these amounts from their combined adjusted gross income to find their taxable income:

$$\text{taxable income} = \$180{,}000 - (2 \times \$4050) - \$12{,}700 = \$159{,}200$$

BY THE WAY

In part (c) of Example 3, Jessica and Frank each earned the same amount as Deirdre, in part (a), but they each paid the equivalent of $15,730 which is more than the $15,639 that Deirdre paid. This feature of the tax code, whereby people pay more when married than they would if they were single, is called the *marriage penalty*. Not all couples are affected the same way; some even get a marriage benefit instead, especially if one spouse earns significantly more than the other.

We calculate their taxes using the "Married Filing Jointly" rates. Their taxable income of $159,200 puts them in the 28% tax bracket, so their tax is

$$\underbrace{(10\% \times \$18{,}650)}_{\substack{\text{10\% marginal rate on first}\\ \text{\$18,650 of taxable income}}} + \underbrace{(15\% \times [\$75{,}900 - \$18{,}650])}_{\substack{\text{15\% marginal rate on taxable income}\\ \text{between \$18,650 and \$75,900}}}$$

$$+ \underbrace{(25\% \times [\$153{,}100 - \$75{,}900])}_{\substack{\text{25\% marginal rate on taxable income}\\ \text{between \$75,900 and \$153,100}}} + \underbrace{(28\% \times [\$159{,}200 - \$153{,}100])}_{\substack{\text{28\% marginal rate on taxable}\\ \text{income above \$153,100}}}$$

$$= \$1865.00 + \$8587.50 + \$19{,}300.00 + \$1708.00$$

$$= \$31{,}460.50$$

Rounding down, Jessica and Frank's combined tax is $31,460, equivalent to $15,730 each.

▶ **Now try Exercises 29–36.**

Think About It All four individuals in Example 3 have the same $90,000 in adjusted gross income, yet they each pay a different amount in taxes. Do you believe the differences are fair? Why or why not? (Bonus: Could their *gross incomes* have differed even though their adjusted gross incomes were the same? Explain.)

Tax Credits and Deductions

Tax credits and tax deductions may sound similar, but they are very different. Suppose you are in the 15% tax bracket. A tax *credit* of $500 reduces your total tax bill by the full $500. In contrast, a tax *deduction* of $500 reduces your taxable income by $500, which means it saves you only $15\% \times \$500 = \$75$ in taxes. As a rule, tax credits are more valuable than tax deductions.

Congress authorizes tax credits for only specific situations, such as a (maximum) $1000 tax credit for each child (as of 2017). In contrast, for those who itemize, deductions depend on how you spend your money. The most valuable deduction for most people who itemize is the **mortgage interest tax deduction**, which allows you to deduct the *interest* (but not the principal) you pay on a home mortgage. Many people also get substantial deductions from donating money to charities and from deducting state and local taxes.

EXAMPLE 4 Tax Credits vs. Tax Deductions

Suppose you are in the 28% tax bracket. How much does a $1000 tax credit save you? How much does a $1000 charitable contribution (which is tax deductible) save you? Answer these questions both for the case in which you itemize deductions and for the case in which you take the standard deduction.

Solution The entire $1000 tax credit is deducted from your tax bill and therefore saves you a full $1000, whether you itemize deductions or take the standard deduction. The tax value of the $1000 charitable contribution depends on whether you itemize:

BY THE WAY

As of 2017, fewer than one-third of all taxpayers itemize their deductions. The rest take the standard deduction and hence get no additional benefit from the deductibility of things like mortgage interest and charitable contributions.

- If you itemize your deductions, the $1000 charitable contribution reduces your taxable income by $1000. Therefore, if you are in the 28% tax bracket, it will save you $0.28 \times \$1000 = \$280$ in taxes.
- If you do not itemize (because your total itemized deductions are less than the standard deduction), then the $1000 contribution will save you nothing at all.

▶ **Now try Exercises 37–42.**

EXAMPLE 5 Rent or Own?

Suppose you are in the 28% tax bracket and you itemize your deductions. You are trying to decide whether to rent an apartment or buy a house. The apartment rents for $1400 per month. You've investigated your loan options, and you've determined that if you buy the house, your monthly mortgage payments will be \$1600, of which an average of \$1250 goes toward interest during the first year. Compare the monthly rent to the mortgage payment. Is it cheaper to rent the apartment or buy the house? Assume that you have enough other deductions to itemize them rather than taking the standard deduction.

Solution The monthly cost of the apartment is \$1400 in rent. For the house, however, we must take into account the value of the mortgage deduction. The monthly interest of \$1250 is tax deductible. Because you are in the 28% tax bracket, this deduction saves you $28\% \times \$1250 = \$350$. As a result, the true monthly cost of the mortgage is the payment minus the tax savings, or

$$\$1600 - \$350 = \$1250$$

Despite the fact that the mortgage payment is \$200 higher than the rent, its true cost to you is \$150 per month less because of the tax savings from the mortgage interest deduction. Of course, as a homeowner, you will have other costs, such as for maintenance and repairs that you may not have to pay if you rent. Moreover, we have assumed that you are itemizing deductions; if your total deductions were below (or not too far above) the standard deduction, then the deduction for the interest would not provide you with the benefit calculated here. ▸ **Now try Exercises 43–44.**

Think About It Aside from the lower monthly cost, what other factors would affect your decision about whether to rent or buy in Example 5?

EXAMPLE 6 Varying Value of Deductions

Drew is in the 15% marginal tax bracket. Marian is in the 35% marginal tax bracket. They each itemize their deductions. They each donate \$5000 to charity. Compare their true costs for the charitable donation.

Solution The \$5000 contribution to charity is tax deductible. Because Drew is in the 15% tax bracket, this contribution saves him $15\% \times \$5000 = \$750$ in taxes. Therefore, the true cost of his contribution is the contributed amount of \$5000 minus his tax savings of \$750, or \$4250. For Marian, who is in the 35% tax bracket, the contribution saves $35\% \times \$5000 = \$1750$ in taxes. Therefore, the true cost of her contribution is $\$5000 - \$1750 = \$3250$. The true cost of the donation is considerably lower for Marian because she is in a higher tax bracket. (*Note*: In fact, Drew would probably get no tax benefit from his contribution, because very few people in the 15% tax bracket itemize their deductions.) ▸ **Now try Exercises 45–46.**

BY THE WAY
Americans donate an average of 4.7% of income to charity, which is far higher than the average charitable giving of people in any other industrialized nation.

Think About It As shown in Example 6, tax deductions are more valuable to people in higher tax brackets. Some people argue that this is unfair because it means that tax deductions save more money for richer people than for poorer people. Others argue that it is fair, because richer people pay a higher tax rate in the first place. What do you think? Defend your opinion.

Social Security and Medicare Taxes

In addition to being subject to taxes computed with the marginal rates, some income is subject to Social Security and Medicare taxes, which are collected under the obscure name of **FICA** (Federal Insurance Contribution Act). Taxes collected under FICA are used to pay Social Security and Medicare benefits, primarily to people who are retired.

FICA taxes are calculated on *all* wages, tips, and self-employment income; you may not subtract any adjustments, exemptions, or deductions when calculating these taxes. FICA does not apply to income from sources such as interest, dividends, or profits from sales of stock. FICA taxes are paid by both employers and employees in equal shares. For 2017, the FICA tax rates were

- 7.65% on the first $127,200 of income from wages, with the employer matching this 7.65%
- 1.45% on any income from wages in excess of $127,200, with the employer matching this 1.45%

Individuals who are self-employed must pay both the employee and the employer shares of FICA taxes. As a result, self-employed individuals pay FICA taxes on their actual income at *double* the rates paid by individuals who are not self-employed.

BY THE WAY
The additional Medicare tax was implemented as part of the Affordable Care Act passed in 2010. In late 2017, Congress was considering whether to repeal this tax.

An additional Medicare tax applies (at the time this text was written in mid-2017) to taxpayers with "modified adjusted gross income" (which can differ from adjusted gross income if you have tax-exempt or tax-deductible income) above $200,000 for individuals (single or head of household) or above $250,000 for married couples. Above these thresholds, the additional Medicare tax is

- 3.8% on most income that is not otherwise subject to FICA taxes, such as income from investments and dividends
- 0.9% surcharge on ordinary income

Note that the 0.9% surcharge is in addition to the 1.45% employer and employee shares of FICA taxes that go to Medicare, so its overall effect is to make the total Medicare rate above the thresholds the same 3.8% as it is for investments (because $2 \times 1.45\% + 0.9\% = 3.8\%$).

EXAMPLE 7 FICA Taxes

In 2017, Jude earned $26,000 in wages and tips from her job waiting tables. Calculate her FICA taxes and her total tax bill including marginal taxes. What is the overall tax rate on her gross income, including both FICA and income taxes? Assume that she is single and takes the standard deduction.

Solution Jude's entire income of $26,000 is subject to the 7.65% FICA tax rate:

$$\text{FICA tax} = 7.65\% \times \$26{,}000 = \$1989$$

Now we must find her income tax. We get her taxable income by subtracting her $4050 personal exemption and $6350 standard deduction:

$$\text{taxable income} = \$26{,}000 - \$4050 - \$6350 = \$15{,}600$$

BY THE WAY
Most Americans pay more in FICA tax than in ordinary income tax, if you include the portion of FICA taxes paid by employers.

From Table 4.9, her income tax is 10% on the first $9325 of her taxable income and 15% on the remaining amount of $\$15{,}600 - \$9325 = \$6275$. Therefore, her income tax is $(10\% \times \$9325) + (15\% \times \$6275) = \$1874$ (rounded up). Her total tax, including both FICA and income tax, is

$$\text{total tax} = \text{FICA} + \text{income tax} = \$1989 + \$1874 = \$3863$$

Her overall tax rate, including both FICA and income tax, is

$$\frac{\text{total tax}}{\text{gross income}} = \frac{\$3863}{\$26{,}000} \approx 0.149$$

Jude's overall tax rate is about 14.9%. Note that she pays more in FICA tax than in income tax. ▶ Now try Exercises 47–52.

Dividends and Capital Gains

Not all income is created equal, at least not in the eyes of the tax collector! In particular, dividends (from stocks) and capital gains—profits from the sale of stock or other property—get special tax treatment. **Capital gains** are divided into two subcategories.

Short-term capital gains are profits on items sold within 12 months of their purchase; they are taxed at the same rates as ordinary income (the rates in Table 4.9), except that they are not subject to FICA taxes. **Long-term capital gains** are profits on items held for more than 12 months before being sold, and most long-term capital gains and dividends are taxed at lower rates than other income such as wages and interest earnings.

As of 2017, the tax rates for long-term capital gains and dividends were

- 0% for income in the 10% and 15% tax brackets
- 15% for income in all higher tax brackets *except* the highest 39.6% bracket
- 20% for income in the 39.6% tax bracket

In addition, as noted earlier, capital gains and dividends are subject to the 3.8% Medicare tax if your income is above the thresholds where this tax kicks in (and the tax remained in place as the law stood when this text was written).

BY THE WAY

Capital gains have not always been taxed at lower rates. For example, the tax reform law signed by President Reagan in 1986 applied the same rates to both ordinary income and capital gains. Supporters of lower capital gains rates argue that they help the economy by encouraging investment in new businesses and products that involve risk on the part of the investor.

EXAMPLE 8 Dividend and Capital Gains Income

In 2017, Serena was single and lived off an inheritance. Her gross income consisted solely of \$90,000 in dividends and long-term capital gains. She had no adjustments to her gross income, but had \$12,000 in itemized deductions and a personal exemption of \$4050. How much tax does she owe? What is her overall tax rate?

Solution She owes no FICA tax because her income is not from wages. She had no adjustments to her gross income, so we find her taxable income by subtracting her itemized deductions and personal exemption:

$$\text{taxable income} = \$90{,}000 - \$12{,}000 - \$4050 = \$73{,}950$$

Because her income is all dividends and long-term capital gains, she pays tax at the special rates for these types of income. The 0% rate for dividends and long-term capital gains applies to the income on which she would have been taxed at 10% or 15% if it had been ordinary income, which from Table 4.9 means her first \$37,950 of income. The rest of her income is taxed at the special 15% capital gains rate. Her total tax is

$$\underbrace{(0\% \times \$37{,}950)}_{\text{0\% capital gains rate}} + \underbrace{(15\% \times [\$73{,}950 - \$37{,}950])}_{\text{15\% capital gains rate}} = \$0 + \$5400 = \$5400$$

Her overall tax rate is

$$\frac{\text{total tax}}{\text{gross income}} = \frac{\$5400}{\$90{,}000} = 0.06$$

Serena's overall tax rate is 6%. ▶ Now try Exercises 53–54.

Think About It Note that Serena in Example 8 had a gross income more than triple that of Jude in Example 7. Compare their tax payments and overall tax rates. Who pays more tax? Who pays at a higher tax rate? Explain.

Tax-Deferred Income

The tax code tries to encourage long-term savings by allowing you to *defer* income taxes on contributions to certain types of savings plans, called **tax-deferred savings plans**. Money that you deposit into such savings plans is not taxed for the year in which you make the deposit. Instead, it will be taxed in the future when you make withdrawals from the plan.

Tax-deferred savings plans go by a variety of names, such as *individual retirement accounts* (IRAs), *qualified retirement plans* (QRPs), *401(k) plans*, and more. All are subject to strict rules. For example, you generally are not allowed to withdraw money from any

BY THE WAY

With tax-deferred savings, you will eventually pay tax on the money when you withdraw it. With *tax-exempt* investments, you never have to pay tax on the earnings. For example, some government bonds are tax-exempt, and Roth IRAs are individual retirement accounts for which you pay taxes on money you deposit now, but your earnings are tax-exempt when you withdraw them.

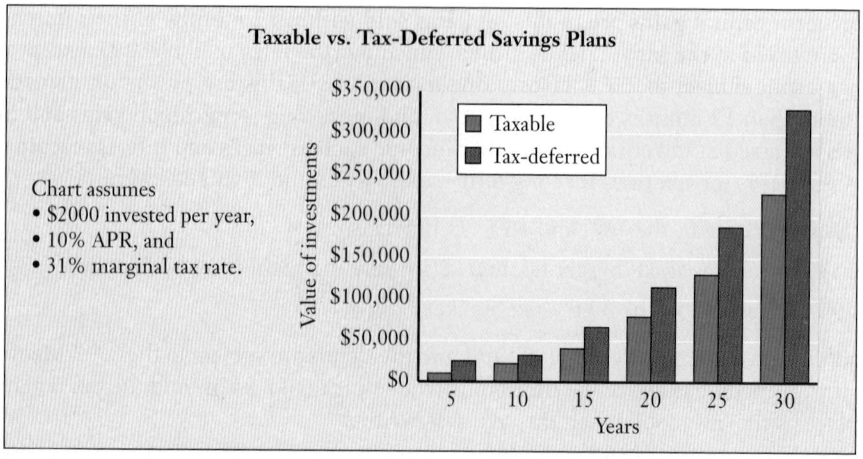

FIGURE 4.14 Comparison of the growth in value of a tax-deferred savings plan and an ordinary savings plan, assuming that tax on the interest is paid from the plan in the latter case. Notice the substantial advantage of the tax-deferred plan, assuming all else is equal.

of these plans until you reach age $59\frac{1}{2}$. Anyone can set up a tax-deferred savings plan, and *you should*, regardless of your current age. Why? Because they offer two key advantages in saving for your long-term future.

First, contributions to tax-deferred savings plans count as *adjustments* to your present gross income and are not part of your taxable income. As a result, the contributions cost you less than contributions to savings plans without special tax treatment. For example, suppose you are in the 28% marginal tax bracket. If you deposit $100 in an ordinary savings account, your tax bill is unchanged and you have $100 less to spend on other things. But if you deposit $100 in a tax-deferred savings account, you do not have to pay tax on that $100. With your 28% marginal rate, you therefore save $28 in taxes, so the amount you have to spend on other things decreases by only $100 − $28 = $72.

The second advantage of tax-deferred savings plans is that their earnings are also tax deferred. With an ordinary savings plan, you must pay taxes on the earnings each year, which effectively reduces your earnings. With a tax-deferred savings plan, *all* of the earnings accumulate from one year to the next. Over many years, this tax saving can make the value of tax-deferred savings accounts rise much more quickly than that of ordinary savings accounts (Figure 4.14).

EXAMPLE 9 Tax-Deferred Savings Plan

Suppose you are single, have a taxable income of $65,000, and make monthly payments of $500 to a tax-deferred savings plan. How do the tax-deferred contributions affect your monthly take-home pay?

Solution Table 4.9 shows that your marginal tax rate is 25%. Each $500 contribution to a tax-deferred savings plan therefore reduces your tax bill by

$$25\% \times \$500 = \$125$$

In other words, $500 goes into your tax-deferred savings account each month, but your monthly paychecks go down by only $500 − $125 = $375. The special tax treatment makes it significantly easier for you to afford the monthly contributions needed to build your retirement fund.

▸ **Now try Exercises 55–58.**

Think About It The complexity of the tax code makes many people wish for a simpler system, and as this text was being written, Congress and the Trump administration had both proposed major overhauls of the federal tax system. Have any such proposals been implemented? Do they make the kinds of changes that *you* think would be helpful?

Quick Quiz 4E

Choose the best answer to each of the following questions. Explain your reasoning with one or more complete sentences.

1. The total amount of income you receive is called your
   a. gross income.
   b. net income.
   c. taxable income.
2. If your taxable income puts you in the 25% marginal tax bracket,
   a. your tax is 25% of your taxable income.
   b. your tax is 25% of your gross income.
   c. your tax is 25% of only a portion of your income; the rest is taxed at a lower rate.
3. Suppose you are in the 25% marginal tax bracket. Then a *tax credit* of $1000 will reduce your tax bill by
   a. $1000.
   b. $150.
   c. $500.
4. Suppose you are in the 15% marginal tax bracket and earn $25,000. Then a *tax deduction* of $1000 will reduce your tax bill by
   a. $1000.
   b. $150.
   c. $500.
5. Suppose that in the past year your only deductible expenses were $5000 in mortgage interest and $2000 in charitable contributions. If you are entitled to a standard deduction of $6350, then the maximum total deduction you can claim is
   a. $6350.
   b. $7000.
   c. $13,350.
6. Assume you are in the 25% tax bracket and you are entitled to a standard deduction of $6350. If you have no other deductible expenses, by how much will a $1000 charitable contribution reduce your tax bill?
   a. $0
   b. $250
   c. $1000
7. What is the FICA tax?
   a. a tax on investment income
   b. another name for the marginal tax rate system
   c. a tax collected primarily to fund Social Security and Medicare
8. Based on the FICA rates for 2017, which of the following people pays the highest *percentage* of his or her income in FICA taxes?
   a. Joe, whose income consists of $12,000 from his job at Burger Joint
   b. Kim, whose income is $175,000 in wages from her job as an aeronautical engineer
   c. David, whose income is $1,000,000 in capital gains from investments
9. Jerome, Jenny, and Jacqueline all have the same *taxable income*, but Jerome's income is entirely from wages at his job, Jenny's income is a combination of wages and short-term capital gains, and Jacqueline's income is all from dividends and long-term capital gains. If you count both income taxes and FICA, how do their tax bills compare?
   a. They all pay the same amount in taxes.
   b. Jerome pays the most, Jenny the second most, and Jacqueline the least.
   c. Jacqueline pays the most, Jenny the second most, and Jerome the least.
10. When you put money into a tax-deferred retirement plan,
    a. you never have to pay tax on this money.
    b. you pay tax on this money now, but not when you withdraw it later.
    c. you do not pay tax on this money now, but you pay tax on money you withdraw from the plan later.

Exercises 4E

REVIEW QUESTIONS

1. Explain the basic process of calculating income taxes, as shown in Figure 4.13. What is the difference between gross income, adjusted gross income, and taxable income?
2. What is meant by *filing status*? How does it affect tax calculations?
3. What are *exemptions* and *deductions*? How should you choose between taking the standard deduction and itemizing deductions?
4. What is meant by a *progressive income tax*? Explain the use of marginal tax rates in calculating taxes. What is meant by a *tax bracket*?
5. What is the difference between a *tax deduction* and a *tax credit*? Why is a tax credit more valuable?
6. Explain how a deduction, such as the mortgage interest tax deduction, can save you money. Why do deductions benefit people in different tax brackets differently?
7. What are *FICA taxes*? What type of income is subject to FICA taxes?
8. How are dividends and capital gains treated differently than other income by the tax code?
9. Explain how you can benefit from a tax-deferred savings plan.
10. Why do tax-deferred savings plans tend to grow in value faster than ordinary savings plans?

DOES IT MAKE SENSE?

Decide whether each of the following statements makes sense (or is clearly true) or does not make sense (or is clearly false). Explain your reasoning. Assume 2017 tax rates and policies.

11. We're both single with no children and both have the same total (gross) income, so we must both pay the same amount in taxes.
12. The $1000 child tax credit sounds like a good idea, but it doesn't help me because I take the standard deduction rather than itemizing deductions.
13. When I calculated carefully, I found that it was cheaper for me to buy a house than to continue renting, even though my rent payments were lower than my new mortgage payments.
14. My husband and I paid $5000 in mortgage interest this year, but we did not get a tax benefit from it because we took the standard deduction.
15. Bob and Sue were planning to get married in December of this year, but they postponed their wedding until January when they found it would save them money in taxes.
16. Regina's income last year was $10 million, all from long-term capital gains. Phil had the same income from his salary (wages) as a professional athlete. Despite their equal incomes, Phil paid nearly twice as much in federal taxes.
17. I'm self-employed and earned a total of only $10,000 last year, yet I still had to pay 15.3% of my income in taxes.
18. I started contributing $400 each month to my tax-deferred savings plan, but my take-home pay declined by only $300.

BASIC SKILLS & CONCEPTS

19–22: Income on Tax Forms. Find the gross income, adjusted gross income, and taxable income for each of the following individuals.

19. Antonio earned wages of $47,200, received $2400 in interest from a savings account, and contributed $3500 to a tax-deferred retirement plan. He was entitled to a personal exemption of $4050 and took the standard deduction of $6350.
20. Marie earned wages of $28,400, received $95 in interest from a savings account, was entitled to a personal exemption of $4050, and took the standard deduction of $6350.
21. Isabella earned wages of $88,750, received $4900 in interest from a savings account, and contributed $6200 to a tax-deferred retirement plan. She was entitled to a personal exemption of $4050 and had itemized deductions totaling $9050.
22. Lebron earned wages of $3,452,000, received $54,200 in interest from savings, and contributed $30,000 to a tax-deferred retirement plan. He was not allowed to claim a personal exemption (because of his high income) but did have itemized deductions totaling $674,500.

23–24: To Itemize or Not? Decide whether you should itemize your deductions or take the standard deduction in the following cases.

23. Your deductible expenditures are $8600 for interest on a home mortgage, $2700 for contributions to charity, and $645 for state income taxes. Your filing status entitles you to a standard deduction of $12,700.
24. Your deductible expenditures are $3700 for contributions to charity and $760 for state income taxes. Your filing status entitles you to a standard deduction of $6350.

25–28: Income Calculations. Compute the gross income, adjusted gross income, and taxable income in each of the following situations. Use the exemptions and deductions in Table 4.9. Explain how you decided whether to itemize deductions or use the standard deduction.

25. Suzanne is single and earned wages of $33,200. She received $350 in interest from a savings account. She contributed $500 to a tax-deferred retirement plan. She had $450 in itemized deductions from charitable contributions.
26. Malcolm is single and earned wages of $23,700. He had $4500 in itemized deductions from interest on a mortgage.
27. Wanda is married, but she and her husband filed separately. Her salary was $33,400, and she earned $500 in interest. She had $1500 in itemized deductions and claimed three exemptions, for herself and two children.
28. Emily and Juan are married and filed jointly. Their combined wages were $75,300. They earned $2000 from a rental property they own, and they received $1650 in interest. They claimed four exemptions, for themselves and two children. They contributed $3240 to their tax-deferred retirement plans, and their itemized deductions totaled $9610.

29–36: Marginal Tax Calculations. Use the marginal tax rates in Table 4.9 to compute the tax owed in each of the following situations.

29. Gene is single and had a taxable income of $35,400.
30. Sarah and Marco are married and filing jointly, with a taxable income of $87,500.
31. Bobbi is married but filing separately, with a taxable income of $77,300.
32. Abraham is single with a taxable income of $23,800.
33. Paul is a head of household with a taxable income of $89,300. He is entitled to a $1000 tax credit.
34. Pat is a head of household with a taxable income of $57,000. She is entitled to a $1000 tax credit.
35. Winona and Jim are married and filing jointly, with a taxable income of $105,500. They are entitled to a $2000 tax credit.
36. Chris is married but filing separately, with a taxable income of $127,500.

37–42: Tax Credits and Tax Deductions. Determine how much the following individuals or couples will save in taxes if they use the specified tax credits or deductions.

37. Midori and Tremaine are in the 28% tax bracket and take the standard deduction. How much will their tax bill be reduced if they qualify for a $500 tax credit?
38. Vanessa is in the 35% tax bracket and takes the standard deduction. How much will her tax bill be reduced if she qualifies for a $1500 tax credit?

39. Rosa is in the 15% tax bracket and takes the standard deduction. How much will her tax bill be reduced if she makes a \$1000 contribution to charity?

40. Shiro is in the 15% tax bracket and itemizes his deductions. How much will his tax bill be reduced if he makes a \$1000 contribution to charity?

41. Sebastian is in the 28% tax bracket and itemizes his deductions. How much will his tax bill be reduced if he makes a \$1000 contribution to charity?

42. Santana is in the 39.6% tax bracket and itemizes her deductions. How much will her tax bill be reduced if she makes a \$1000 contribution to charity?

43–44: Rent or Own? Consider the following choices between paying rent and making house payments. Including the amount saved by taking the mortgage interest deduction, determine whether the monthly rent is greater than or less than the monthly house payments during the first year. In both cases, assume that you itemize deductions.

43. You are in the 33% tax bracket. Your apartment rents for \$1600 per month. Your monthly mortgage payments would be \$2000, of which an average of \$1800 per month would go toward interest during the first year.

44. You are in the 28% tax bracket. Your apartment rents for \$600 per month. Your monthly mortgage payments would be \$675, of which an average of \$600 per month would go toward interest during the first year.

45. **Varying Value of Deductions.** Maria is in the 33% tax bracket. Steve is in the 15% tax bracket. They each itemize their deductions and pay \$10,000 in mortgage interest during the year. Compare their true costs for mortgage interest. How does your answer change if, like most people in the 15% bracket, Steve does not itemize?

46. **Varying Value of Deductions.** Yolanna is in the 35% tax bracket. Alia is in the 10% tax bracket. They each itemize their deductions, and they each donate \$4000 to charity. Compare their true costs for charitable donations. How does your answer change if, like most people in the 10% bracket, Alia does not itemize?

47–52: FICA Taxes. For each of the following individuals, calculate the FICA taxes and income taxes to obtain the total tax owed. Then find the overall tax rate on the gross income, including both FICA and income taxes. Assume that all the individuals are single and take the standard deduction. Use the tax rates in Table 4.9.

47. Luis earned \$28,000 from wages as a computer programmer.

48. Carla earned a salary of \$34,500 and received \$750 in interest.

49. Jack earned a salary of \$44,800 and received \$1250 in interest.

50. Alejandro earned a salary of \$130,200 and received \$4450 in interest.

51. Brittany earned \$48,200 in wages and tips. She had no other income.

52. Larae earned \$21,200 in wages and tips. She had no other income.

53–54: Dividends and Capital Gains. Calculate the total tax (FICA and income taxes) owed by each individual in the following pairs. Compare their overall tax rates. Assume that all the individuals are single and take the standard deduction. Use the tax rates in Table 4.9 and the special rates for dividends and capital gains given in the text.

53. Pierre earned \$120,000 in wages. Katarina earned \$120,000, all from dividends and long-term capital gains.

54. Deion earned \$60,000 in wages. Josephina earned \$60,000, all from dividends and long-term capital gains.

55–58: Tax-Deferred Savings Plans. Calculate the change in your monthly take-home pay when you make each tax-deferred contribution.

55. You are single and have a taxable income of \$18,000. You make monthly contributions of \$400 to a tax-deferred savings plan.

56. You are single and have a taxable income of \$45,000. You make monthly contributions of \$600 to a tax-deferred savings plan.

57. You are married filing jointly and have a taxable income of \$90,000. You make monthly contributions of \$800 to a tax-deferred savings plan.

58. You are married filing jointly and have a taxable income of \$200,000. You make monthly contributions of \$800 to a tax-deferred savings plan.

FURTHER APPLICATIONS

59–62: Marriage Penalty. Consider the following couples, who are engaged to be married. Calculate their income tax in two ways: (1) if they delay their marriage until next year so they can file their tax returns as individuals at the single tax rate this year and (2) if they marry before the end of this year and file a joint return. Assume that each person takes one exemption and the standard deduction. Use the tax rates in Table 4.9. Do the couples face a marriage penalty if they marry before the end of the year? (Married rates apply for the entire year, regardless of when during the year the marriage took place.)

59. Gabriella and Roberto have adjusted gross incomes of \$96,400 and \$82,600, respectively.

60. Joan and Paul have adjusted gross incomes of \$32,500 and \$29,400, respectively.

61. Mia and Steve each have an adjusted gross income of \$185,000.

62. Lisa has an adjusted gross income of \$85,000, and Patrick is a student with no income.

63. **Different Rates for Different People.** Example 3 showed the marginal income tax calculations for three sets of people: Deirdre, Robert, and the married couple Jessica and Frank.

    a. Calculate the FICA tax owed by each of the three sets, assuming that the given adjusted gross incomes came from ordinary wages.

    b. Calculate the total tax (the marginal tax from Example 3 plus FICA tax) owed by each of the three sets.

c. Calculate the overall tax rate for each set as a percentage of the adjusted gross income.

d. Compare the tax rates for the three sets to each other and to Serena from Example 8. Who pays at the highest rate? Who pays at the lowest?

e. Briefly explain why the five individuals have different tax rates, even though each had the same adjusted gross income of $90,000.

64. **Warren Buffett and His Secretary.** In 2012, when the maximum tax rate on long-term capital gains and dividends was 15%, billionaire Warren Buffett famously argued for a higher rate by noting that his 15% tax rate was lower than his secretary's. In 2017, the maximum rate on long-term capital gains and dividends was 23.8% (including both the 20% capital gains rate and the 3.8% Medicare tax). For this exercise, assume that Buffet paid at this maximum rate, though in reality his rate would have been lower due to allowed deductions for charitable contributions and other items. Then compare his rates from both 2012 and 2017 to the rates for each of the following individuals (using only the 2017 rates), including both marginal tax rates and FICA taxes. In all cases, assume that the individuals are single and take the standard deduction.

a. Buffett's secretary, who reportedly earned a salary of $200,000 per year

b. A waitress earning $22,000 per year

c. A teacher earning $56,000 per year

d. A self-employed businesswoman earning $110,000 per year

65–66. Who Pays Income Taxes? When it comes to income taxes, one of the key political debates concerns whether the overall system is fair. For Exercises 65–66, use the following data collected from the IRS in 2014 by the Pew Research Center.

Federal Individual Income Tax Data

Adjusted gross income	Percentage of federal income tax returns filed	Percentage of total federal income tax paid
Less than $15,000	24.3%	0.1%
$15,000 to $29,999	20.4%	1.4%
$30,000 to $49,999	17.6%	4.1%
$50,000 to $99,999	21.7%	14.9%
$100,000 to $199,999	11.8%	21.9%
$200,000 to $249,999	1.5%	5.9%
$250,000 and above	2.7%	51.6%

Source: Pew Research Center.

65. a. Consider people with adjusted gross incomes of less than $50,000. What percentage of federal tax returns were filed by this income group, and what percentage of total federal income taxes did this income group pay?

b. Consider people with adjusted gross incomes of more than $250,000. What percentage of federal tax returns were filed by this income group, and what percentage of total federal income taxes did this income group pay?

c. Regardless of your personal beliefs, use your answers from parts (a) and (b) to write a one-paragraph argument claiming that the distribution of income taxes is unfair to the rich.

d. Regardless of your personal beliefs, use your answers from parts (a) and (b) to write a one-paragraph argument claiming that the distribution of income taxes is unfair to the poor.

66. a. Consider people with adjusted gross incomes of less than $100,000. What percentage of federal tax returns were filed by this income group, and what percentage of total federal income taxes did this income group pay?

b. Consider people with adjusted gross incomes of more than $100,000. What percentage of federal tax returns were filed by this income group, and what percentage of total federal income taxes did this income group pay?

c. Do you think your answers to parts (a) and (b) indicate that the current distribution of income taxes is fair or unfair? Defend your opinion.

d. As discussed in this unit, working Americans pay FICA taxes in addition to income taxes; in fact, the U.S. Treasury Department estimates that all but the top-earning 20% of taxpayers pay more in FICA taxes than in federal income taxes. Does this fact alter your opinion from part (c)? Write one or two paragraphs discussing the overall fairness of the federal income and FICA tax system.

IN YOUR WORLD

67. **Tax Simplification Plans.** Use the Web to investigate a recent proposal to simplify federal tax laws and filing procedures. What are the advantages and disadvantages of the simplification plan, and who supports it?

68. **Fairness Issues.** Choose some part of the federal tax code that has issues of fairness associated with it (for example, capital gains rates, the marriage penalty, or the alternative minimum tax [AMT]). Use the Web to investigate the current status of the item you've chosen. Have new laws been passed that affect it? What are the advantages and disadvantages of recent or proposed changes, and who supports the changes? Summarize your own opinion about whether current tax law is unfair in this respect, and, if it is, describe what you think should be done about it.

69. **Consumption Tax.** Some people have proposed that the income tax be replaced by a consumption tax. What is a consumption tax? Find Web reports discussing the pros and cons of a consumption tax, and state your own opinion about whether it should be considered.

70. **Your Tax Return.** Briefly describe your own experiences with filing a federal income tax return. Do you file your own returns? If so, do you use a computer software package or a professional tax advisor? Will you change your filing method in the future? Why or why not?

UNIT 4F Understanding the Federal Budget

So far in this chapter, we have discussed issues of financial management that affect us directly as individuals. But we are also affected by the way our government manages its finances. In this unit, we will discuss a few of the basic concepts needed to understand the federal budget and some of the major issues behind the federal deficit and debt.

Budget Basics

In principle, the federal budget works much like your personal budget (Unit 4A) or the budget of a small business. All budgets have **receipts**, or income, and **outlays**, or expenses. **Net income** is the difference between receipts and outlays; recall that we called this *cash flow* for a personal budget. When receipts exceed outlays, net income is positive and the budget has a **surplus** (profit). When outlays exceed receipts, net income is negative and the budget has a **deficit** (loss).

Definitions

Receipts, or income, represent money that has been collected.

Outlays, or expenses, represent money that has been spent.

Net income = receipts − outlays.

If net income is positive, the budget has a **surplus**.

If net income is negative, the budget has a **deficit**.

Note that a deficit means that more money was spent than was collected, which is possible only if you (or a business or the government) cover this excess spending with money either borrowed or withdrawn from savings.

EXAMPLE 1 Personal Budget

Suppose your gross income last year was \$40,000. Your expenditures were \$20,000 for rent and food, \$2000 for interest on your credit cards and student loans, \$6000 for car expenses, and \$9000 for entertainment and miscellaneous expenses. You also paid \$8000 in taxes. Did you have a deficit or a surplus?

Solution The total of your outlays, including tax, was

$$\$20{,}000 + \$2000 + \$6000 + \$9000 + \$8000 = \$45{,}000$$

Because your outlays were greater than your income of \$40,000 by \$5000, your personal budget had a \$5000 deficit. Therefore, you must have either withdrawn \$5000 from savings or borrowed that amount to cover your deficit. ▶ Now try Exercises 15–16.

Deficit and Debt

The terms *deficit* and *debt* are easy to confuse, but there's an important difference between them. Your *deficit* is the shortfall in your budget (or the budget of a business or the government) for any single year. Your **debt** is the total amount of money that you are obligated to repay—often accumulated over many years. In other words, each year's deficit adds more to the total debt.

Debt vs. Deficit

A **deficit** represents money that is borrowed (or taken from savings) during a *single* year.

The **debt** is the total amount of money owed to lenders, which may result from accumulating deficits over many years.

A Small-Business Analogy

Before we focus on the federal budget, let's investigate the simpler books of an imaginary company with not-so-imaginary problems. Table 4.10 summarizes four years of budgets for the Wonderful Widget Company, which started with a clean slate at the beginning of 2014.

The first column shows that during 2014, the company had receipts of \$854,000 and total outlays of \$1,000,000. The company's net income was

$$\$854{,}000 - \$1{,}000{,}000 = -\$146{,}000$$

The negative sign tells us that the company had a *deficit* of \$146,000. The company had to borrow money to cover this deficit and ended the year with a debt of \$146,000. The debt is shown as a negative number because it represents money owed to someone else.

In 2015, receipts increased to \$908,000, while outlays increased to \$1,082,000. These outlays included a \$12,000 interest payment on the debt from 2014. The deficit for 2015 was

$$\$908{,}000 - \$1{,}082{,}000 = -\$174{,}000$$

and the company had to borrow \$174,000 to cover this deficit. Further, it had no money with which to pay off the debt from 2014. Therefore, the accumulated debt at the end of 2015 was

$$\$146{,}000 + \$174{,}000 = \$320{,}000$$

Here is the key point: Because the company again failed to balance its budget in its second year, its total debt continued to grow. As a result, its interest payment in 2016 increased to \$26,000.

In 2017, the company's owners decided to change strategy. They froze operating expenses and employee benefits (at 2016 levels) and actually *cut* security expenses.

TABLE 4.10 Budget Summary for the Wonderful Widget Company (in thousands of dollars)

	2014	2015	2016	2017
Total Receipts	854	908	950	990
Outlays				
Operating Expenses	525	550	600	600
Employee Benefits	200	220	250	250
Security	275	300	320	300
Interest on Debt	0	12	26	47
Total Outlays	1000	1082	1196	1197
Surplus or Deficit	−146	−174	−246	−207
Debt (Accumulated)	**−146**	**−320**	**−566**	**−773**

However, the interest payment rose substantially because of the rising debt. Despite the attempts to curtail outlays and despite another increase in receipts, the company still ran a deficit in 2017 and the total debt continued to grow.

Think About It Suppose you were a loan officer for a bank in 2018, when the Wonderful Widget Company applied for a further loan to cover its increasing debt. Would you lend it money? If so, would you attach any special conditions to the loan? Explain.

EXAMPLE 2 Growing Interest Payments

Consider the data in Table 4.10 for the Wonderful Widget Company. Assume that the \$47,000 interest payment in 2017 was for the 2016 debt of \$566,000. What was the annual interest rate? If the interest rate remained the same, what would be the required payment in 2018 for the debt at the end of 2017? What would the payment be if the interest rate rose by 2 percentage points?

Solution Paying \$47,000 interest on a debt of \$566,000 means an interest rate of

$$\frac{\$47,000}{\$566,000} = 0.083$$

The interest rate was 8.3% (fairly typical for corporate bonds). At the end of 2017, the debt stood at \$773,000. At the same interest rate, the next interest payment would be

$$0.083 \times \$773,000 = \$64,159$$

If the interest rate rose by 2 percentage points, to 10.3%, the next interest payment would be

$$0.103 \times \$773,000 = \$79,619$$

A 2-percentage-point change in the interest rate increases the interest payment by more than \$15,000. ▶ Now try Exercises 17–18.

The Federal Budget

The Wonderful Widget Company example shows that a succession of deficits leads to a rising debt. The increasing interest payments on that debt, in turn, make it even easier to run deficits in the future. The company's story is a mild version of what has happened to the U.S. budget.

A national debt, if it is not excessive, will be to us a national blessing.
—Alexander Hamilton, 1781

Trends in the Federal Deficit and Debt

Figure 4.15a shows the federal government's net income from 1975 through 2017. Note that the budget has been in deficit nearly every year, with the notable exception of the years 1998 to 2001. Figure 4.15b shows how the debt has steadily climbed as a result.

There can be no freedom or beauty about a home life that depends on borrowing and debt.
—Henrik Ibsen, 1879

There's great debate about how much of a problem these deficits and the growing debt pose for our future, and we'll discuss this issue shortly. But it's worth noting that the current deficits represent a reversal of what appeared to be a very positive trend in the 1990s. After the deficit reached nearly \$300 billion in 1992—a record high at the time—a combination of tax increases and a strong economy led to deficit reductions each year, until the deficit became a surplus in 1998. By 2000, the surplus had grown so large that economists projected that the government would fully pay off the national debt by 2013. Of course, that did not happen, because the government budget returned to deficit in 2002 and has remained in deficit ever since.

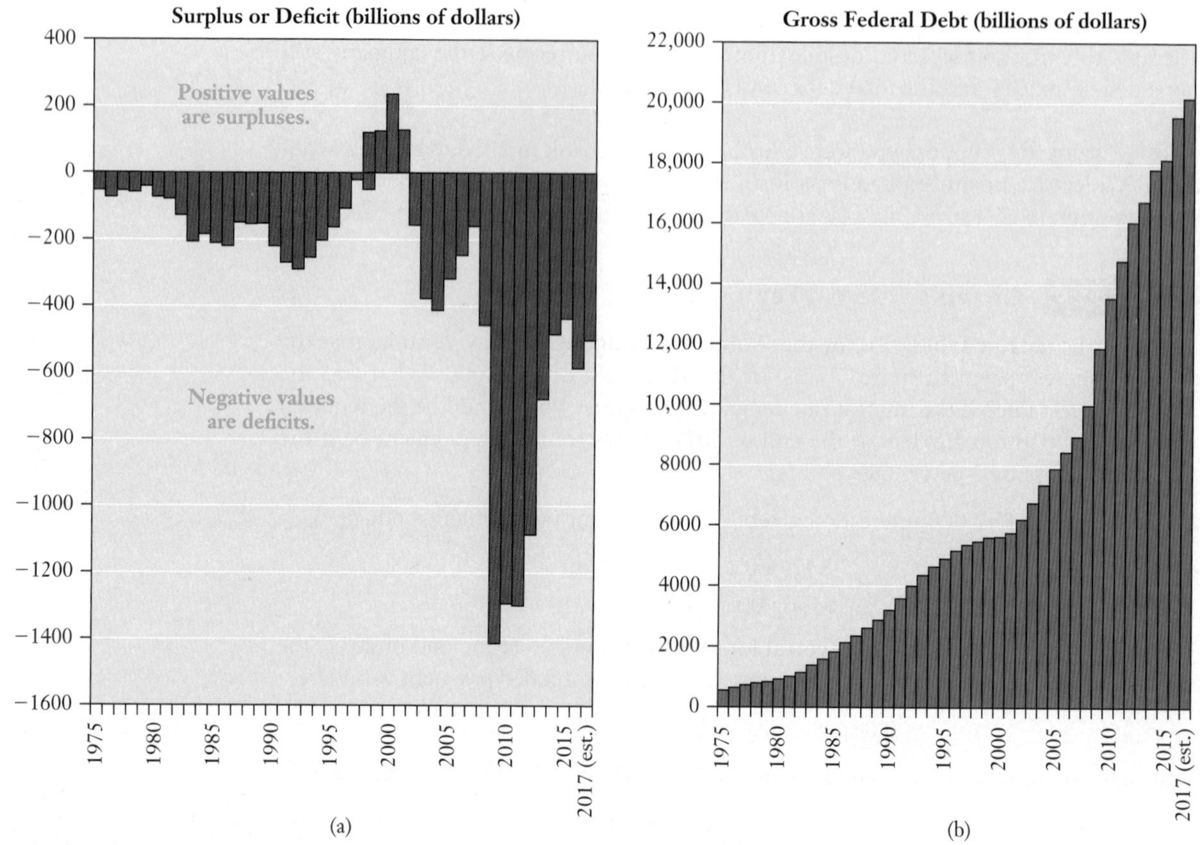

FIGURE 4.15 (a) Annual deficits or surpluses 1975–2017. (b) Accumulated gross federal debt for the same period. Data are based on fiscal years, which end on September 30. *Source:* U.S. Office of Management and Budget.

EXAMPLE 3 The Federal Debt

The federal debt at the end of 2017 was approximately \$20 trillion. If this debt were divided evenly among the roughly 325 million citizens of the United States, how much would *you* owe?

Solution This question is easiest to answer by putting the numbers in scientific notation. We divide the debt of \$20 trillion (\$$2.0 \times 10^{13}$) by the population of 325 million (3.25×10^8):

$$\frac{\$2.0 \times 10^{13}}{3.25 \times 10^8 \text{ persons}} \approx \$6.2 \times 10^4/\text{person} = \$62{,}000/\text{person}$$

Your personal share of the total federal debt is approximately \$62,000.

▶ Now try Exercises 19–20.

Think About It How does your share of the national debt compare to personal debts that you owe? Explain.

Revenues or Spending?

The federal government has run large deficits in recent years because its spending (through outlays) is much greater than its revenue (through receipts). Much of the political debate over deficits revolves around the question of whether they are caused by too little revenue or too much spending.

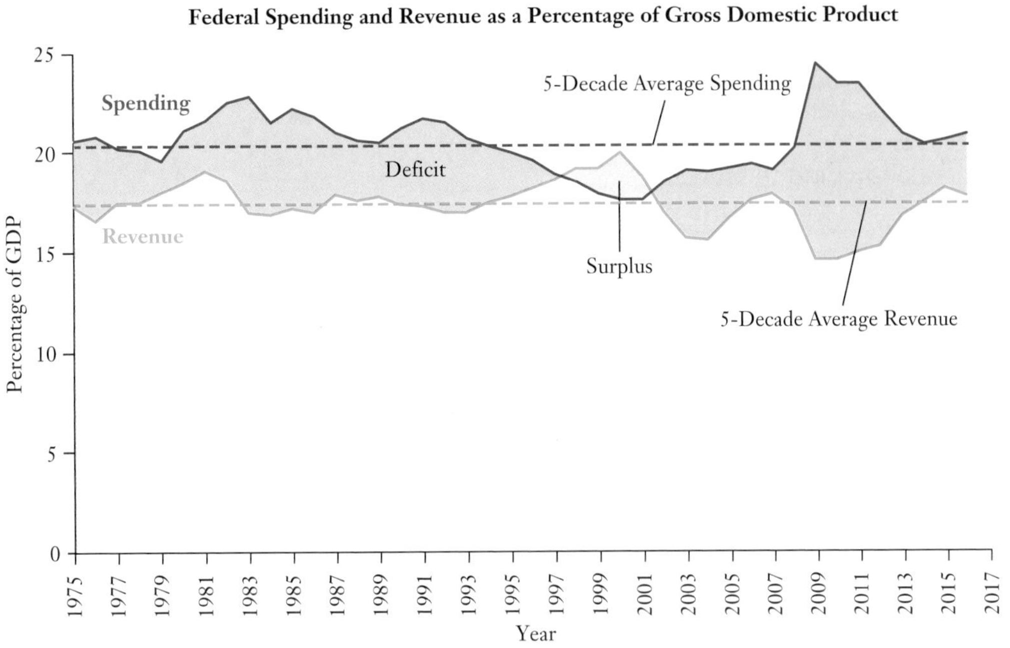

FIGURE 4.16 Federal spending and revenue as a percentage of GDP. The gap between the two is a deficit when spending is higher and a surplus when revenue is higher. Dashed lines indicate averages over the past five decades. Notice how the differences between spending and revenue in this graph compare to the actual dollar amounts for deficits and surpluses shown in Figure 4.15a. *Source:* Congressional Budget Office.

Figure 4.16 shows more than 40 years of spending and revenue as a percentage of the **gross domestic product (GDP)**, which is the most commonly used measure of the overall size of the national economy. Economists generally consider numbers for federal revenue and spending to be more meaningful when expressed as a percentage of GDP than in absolute dollars, for much the same reason that your own spending depends on your income. As the figure shows, the main reason the government has run deficits for most of the past five decades is the fact that spending has been higher than revenues, by an average of almost 3% of GDP.

Think About It As Figure 4.16 makes clear, our options for reducing the federal deficit are (a) cut spending only, (b) raise revenue only, or (c) do some combination of the two. What option do you favor? Defend your opinion.

EXAMPLE 4 Deficit and GDP

Economists also commonly consider deficit and debt numbers as percentages of gross domestic product (GDP). For 2016, the GDP was about \$18.6 trillion, the total deficit was \$587 billion (or \$0.587 trillion), and the end-of-year debt was about \$19.5 trillion. Find the deficit and debt as percentages of GDP.

Solution We simply divide the deficit or debt in dollars by the GDP and convert to a percentage:

$$\text{deficit as \% of GDP} = \frac{\text{total deficit}}{\text{GDP}} \times 100\% = \frac{\$0.587 \text{ trillion}}{\$18.6 \text{ trillion}} \times 100\% \approx 3.2\%$$

$$\text{debt as \% of GDP} = \frac{\text{total debt}}{\text{GDP}} \times 100\% = \frac{\$19.5 \text{ trillion}}{\$18.6 \text{ trillion}} \times 100\% \approx 105\%$$

Technical Note

The government also collects revenues from a few "business-like" activities, such as charging entrance fees at national parks. However, for historical reasons, these revenues are subtracted from outlays instead of being added to receipts when the government publishes its budget. Although this method of accounting may seem odd, it does not affect overall calculations of the surplus or deficit.

BY THE WAY

Under the U.S. Constitution, only Congress can authorize the borrowing of money by the federal government. Congress therefore legislates a "debt ceiling" that authorizes borrowing up to some pre-approved level; to avoid the risk of default on the debt, Congress must increase this ceiling whenever the federal debt approaches it. In recent years, Congress has several times had difficulty passing increases in the debt ceiling; in 2011, the delay lasted long enough to cause the first-ever downgrade of the U.S. government's credit rating.

For comparison, the 2016 deficit value was only slightly above the five-decade average of 2.9% of GDP. However, the 105% value for debt as a percentage of GDP was the highest this number had reached since World War II, and continuing deficits may send it higher still in coming years. ▶ Now try Exercises 21–24.

Following the Money

To get a better understanding of why there is such a large gap between spending and revenue, we need to look more carefully at how the federal government gets its receipts and spends its outlays. The pie charts in Figure 4.17 show the major sources of revenue and major categories of spending.

On the revenue side, notice that more than 80% of the federal government's receipts come from the combination of individual income taxes and the FICA taxes collected primarily for Social Security and Medicare (Unit 4E). Another 9% come from corporate income taxes. The rest comes from *excise taxes*—which include taxes on alcohol, tobacco, and gasoline—and a variety of "other" categories that include gift taxes and fines collected by the government.

The spending side is a little more complex, because current law treats some of the categories shown in the pie differently from others. In particular, the spending categories fall into two major groups:

- **Mandatory outlays** are expenses that are paid automatically *unless* Congress acts to change them. Most of the mandatory outlays are for "entitlements" such as Social Security, Medicare, and other payments to individuals. (They are called *entitlements* because the law specifically states the conditions under which individuals are entitled to them.) Interest on the debt is also considered a mandatory outlay, because it must be paid to prevent the government from being in default on its loans.
- **Discretionary outlays** are the ones that Congress must vote on each year and that the President must then sign into law. In the spending pie chart, discretionary expenses are subdivided into those that affect national defense and security and all the rest ("Non-defense discretionary"), a broad category that includes education,

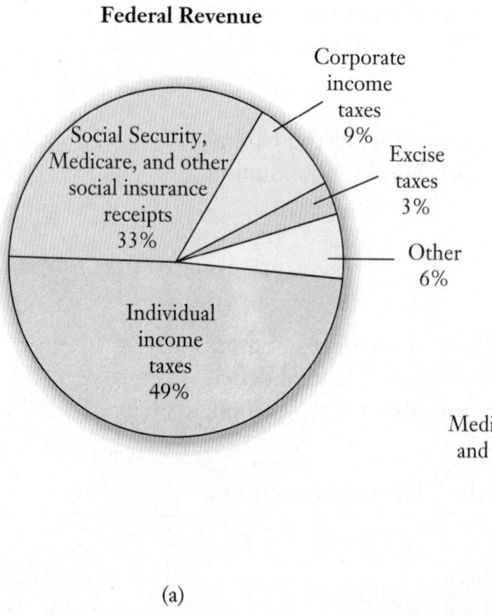

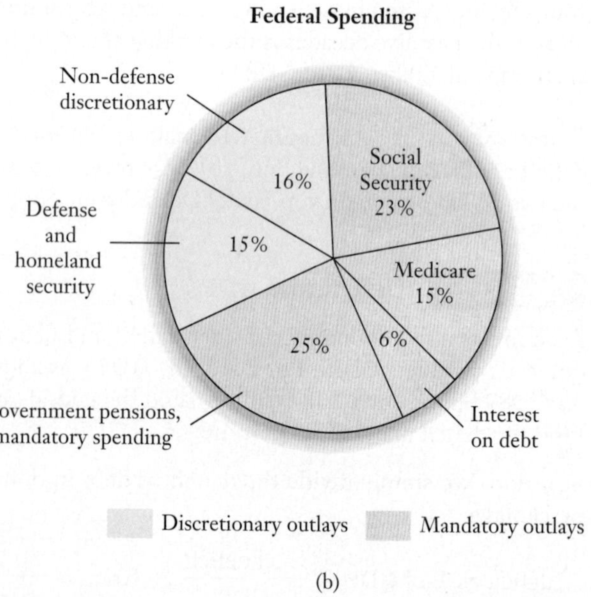

FIGURE 4.17 Approximate makeup of (a) federal revenue and (b) federal spending for fiscal year 2016. For spending, note that all categories *except* "Defense and homeland security" and "Non-defense discretionary" are considered mandatory.
Source: Congressional Budget Office.

roads and transportation, agriculture, food and drug safety, consumer protection, housing, the space program, energy development, scientific research, international aid, and virtually every other government program you've ever heard of.

Notice that only the two blue wedges in the pie chart are discretionary expenses, and they total to only 31% of the total spending. The practical effect of this fact is that Congress generally exerts control only over this relatively small portion of the full budget (because politically it is easier for Congress to leave the mandatory outlays alone).

Technical Note

As discussed later, the government tracks both the total (gross) debt and the publicly held (net) debt. The $330 billion interest that Example 5 cites for 2016 is the interest on the total debt. Most news reports, and the wedge in Figure 4.17b, are based on the lower amount of interest paid on the publicly held debt.

EXAMPLE 5 Interest on the Debt

For 2016, interest on the debt totaled $330 billion, and the total debt at the time was about $19.5 trillion. What was the annual interest rate paid by the government? Suppose instead that interest rates paid by the government had been at their historical average of about 3%. How would that have affected the interest payment in 2016? Comment on the meaning of your results. Useful data: Total 2016 government spending on all education (including student loans), training, and social services was about $114 billion; total spending for NASA was about $18.5 billion.

Solution To make the calculation easier, note that $19.5 trillion is the same as $19,500 billion (because 1 trillion = 1,000 billion). Therefore, paying $330 billion interest on a debt of $19.5 trillion means an interest rate of

$$\frac{\$330 \text{ billion}}{\$19{,}500 \text{ billion}} \approx 0.017 = 1.7\%$$

If the interest rate had instead been 3%, the interest payment would have been

$$0.03 \times \$19{,}500 \text{ billion} = \$585 \text{ billion}$$

This is $585 billion − $330 billion = $255 billion more than the actual interest payment.

Notice that the interest payment of $330 billion was already nearly triple what the government spent on all education, training, and social services combined, and about 18 times what it spent on NASA. The extra $255 billion that would have been needed at the historical average interest rate would have made the budget picture far worse.

▶ Now try Exercises 25–26.

Think About It Pick a government program that you think is worthwhile, and look up what the government spent on it in 2016. How did the spending for your program compare to the total interest on the debt?

Future Projections

The fact that most government spending is in the mandatory category creates a difficult problem for anyone hoping to reduce the deficit or balance the federal budget. Because the debt is still growing, the mandatory interest payments are virtually certain to rise (they could fall only if interest rates moved lower), and Example 5 shows that they may rise much more if interest rates move back toward their historical averages. The rest of the mandatory expenses are the entitlements that go to individuals, and politicians have found it very difficult to reduce these expenses.

Figure 4.18 shows how different categories of spending were projected to rise under the law as it stood at the end of 2016. The following are among the many lessons you might take away from this chart:

- With these spending levels, the only way to balance the federal budget would be with a substantial increase in tax revenues.

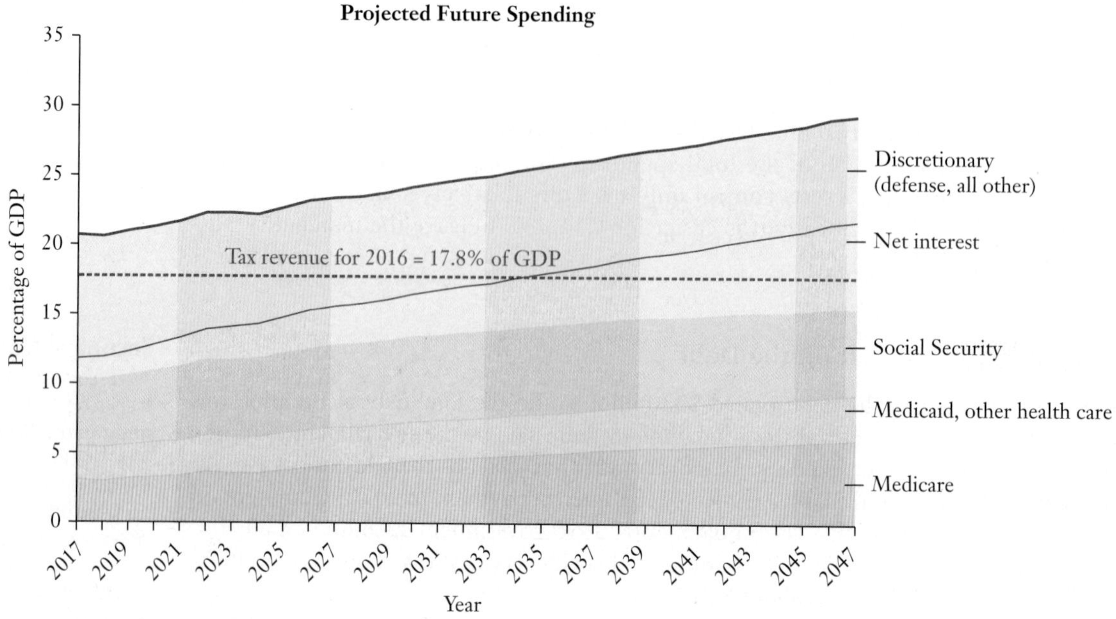

FIGURE 4.18 Projected federal government spending by category as a percentage of GDP, assuming federal laws as they stood at the end of 2016. *Source:* Congressional Budget Office.

- At current tax levels, the budget would remain in deficit *even if discretionary expenses were reduced to zero*—and remember that discretionary spending includes national defense and many other critical programs.

The key point should be clear: The only way to get future deficits under control is to either reduce mandatory spending on entitlements, raise taxes, or do some combination of the two. This lesson is well known to politicians of all stripes, and numerous high-level commissions have made this point repeatedly.

Technical Note
In principle, there is a third way (besides cutting spending or raising taxes) that future deficits could be brought under control: economic growth that raises the GDP enough so that spending would become a much smaller percentage of GDP. However, economists consider such growth to be extremely unlikely.

Think About It Have there been any major changes to entitlement spending since 2017 that would change the projections shown in Figure 4.18? If so, what are they? If not, are any proposals for such changes currently being considered?

EXAMPLE 6 Discretionary Squeeze

As shown in Figure 4.18, mandatory spending on Social Security and Medicare is expected to grow substantially as more people retire in coming decades. Suppose that the government decided to hold total spending steady, but as projected, the proportions of spending going to Social Security and Medicare rise from their combined 38% in 2016 to 43% over the next couple decades. As a percentage of total outlays, how much would discretionary spending have to decrease to cover this increase in Social Security and Medicare? Comment on how this scenario would affect discretionary programs and Congress's power to control the deficit. Assume the budget proportions shown in Figure 4.17b.

Solution If the percentage going to Social Security and Medicare rises by 5 percentage points, then the proportion of spending for all other programs would have to drop by 5 percentage points for the total to remain 100%. Figure 4.17b shows that discretionary spending made up a total of 31% of the 2016 budget (15% for defense and 16% for other discretionary programs). Therefore, if all the money for the increase in Social Security and Medicare spending came from the discretionary category, discretionary spending would fall from 31% to 26% of total outlays. Notice that, as a percentage, this is a change in discretionary spending of

$$\frac{\text{new value} - \text{old value}}{\text{old value}} \times 100\% = \frac{26\% - 31\%}{31\%} \times 100\% \approx -16\%$$

In other words, to make up for the increased entitlement spending without increasing total spending, Congress would have to cut discretionary programs by about 16%—and remember that discretionary spending includes national defense as well as politically popular programs for education, transportation, energy, science, and more. Again, we see that there's little hope of reducing future budget deficits unless Congress enacts some combination of significant reductions in entitlement spending or increases in taxes.

▶ Now try Exercises 27–32.

Benefits are popular. Paying for benefits is extremely unpopular.
—John Danforth, former U.S. Senator (Republican, Missouri)

Think About It Among politicians, one popular proposal for reducing future entitlement spending is to leave it untouched (or nearly untouched) for current retirees, but to reduce it substantially for future retirees. Why do you think politicians like this plan? How would it affect *you*?

Strange Numbers: Publicly Held and Gross Debt

Take another look at Figure 4.15, and you may notice something rather strange: Even in the years when the government ran a surplus (1998–2001), the debt still continued to increase. More generally, if you look at the numbers in detail, you'll find that the debt tends to rise from one year to the next by *more* than the amount of the deficit for the year. To understand why this occurs, we must investigate government accounting in a little more detail.

Financing the Debt

Remember that whenever you run a deficit, you must cover it either by withdrawing from savings or by borrowing money. The federal government does both. It withdraws money from its "savings," and it borrows money from people and institutions willing to lend to it.

Let's consider borrowing first. The government borrows money by selling Treasury bills, notes, and bonds (see Unit 4C) to the public. If you buy one of these Treasury issues, you are effectively lending the government money that it promises to pay back with interest. By the end of 2016, the government had borrowed a total of about \$14 trillion through the sale of Treasury issues. This debt, which the government must eventually pay back to those who hold the Treasury issues, is called the **publicly held debt** (sometimes called the *net debt* or the *marketable debt*).

BY THE WAY
Nearly half of the U.S. publicly held debt is owed to foreign individuals and banks, and about half of this amount is owed to the combination of the two largest holders, China and Japan.

The government's "savings" consist of special accounts, called **trust funds**, which are supposed to help the government meet its future obligations to mandatory spending programs. The biggest trust fund by far is for Social Security. Over the past few decades, the government collected much more money in Social Security taxes (through FICA) than it paid out in Social Security benefits. Legally, the government is required to invest this excess money in the Social Security trust fund so that it would be there when needed for future retirees. But there's a catch: Before the government borrows from the public to finance a deficit, it first tries to cover the deficit by borrowing from its own trust funds.

In fact, the government has to date borrowed every penny it ever deposited into the Social Security trust fund, and the same is true for other trust funds. In other words, there is *no actual money* in any of the trust funds, including Social Security. Instead, they are filled with the equivalent of a stack of IOUs—more technically, with Treasury bills—representing the government's promise to return the money it has borrowed from itself, with interest.

As of the end of 2016, the government's debt to its own trust funds was more than \$5 trillion. Adding this amount to the publicly held debt of \$14 trillion, we get a **gross debt** of more than \$19 trillion. This is the total debt shown in Figure 4.15b, and it

The trust fund more accurately represents a stack of IOUs to be presented to future generations for payment, rather than a build-up of resources to fund future benefits.
—John Hambor, former research director for the Social Security Administration

represents the total amount that the government is obligated to eventually repay from government receipts other than those collected for Social Security and other trust funds.

Two Kinds of National Debt

The **publicly held debt** (or *net debt*) represents money the government must repay to individuals and institutions that bought Treasury issues.

The **gross debt** includes both the publicly held debt and money that the government owes to its own trust funds, such as the Social Security trust fund.

On-Budget and Off-Budget: Effects of Social Security

To illustrate how trust funds affect the two kinds of debt, consider 2001, when the federal government ran a $128 billion surplus (see Figure 4.15a), meaning that the government really did collect $128 billion more than it spent. The government used this surplus to buy back some of the Treasury notes and bonds it had sold to the public, which reduced the publicly held debt.

However, the government also collected excess Social Security taxes, which legally had to be deposited in the Social Security trust fund. In addition, the government owed the trust fund interest for all the money it had borrowed from the trust fund in the past. When we add both the excess Social Security taxes and the owed interest, it turns out that the government was *supposed to* deposit $161 billion in the Social Security trust fund in 2001. But the government had already spent the $161 billion (on programs other than Social Security), leaving no cash available to deposit in the trust fund. The government therefore "deposited" $161 billion worth of IOUs (in the form of Treasury bills) in the trust fund, adding to the stack of IOUs already there from the past. Because IOUs represent loans, the government effectively borrowed $161 billion from the Social Security trust fund. When we subtract this borrowed amount from the $128 billion surplus, the government's income for 2001 becomes

$$\underbrace{\$128 \text{ billion}}_{\text{unified net income}} - \underbrace{\$161 \text{ billion}}_{\text{off-budget net income}} = \underbrace{-\$33 \text{ billion}}_{\text{on-budget net income}}$$

BY THE WAY
If you want complete details of budgets at both federal and other (state and local) levels, explore the website USAfacts.org, launched in 2017.

With the Social Security trust fund included, the $128 billion surplus turns into a $33 billion deficit! In government-speak, the actual net income ($128 billion in 2001) is called the **unified net income**. Social Security is said to be **off-budget**, so its $161 billion deficit is the "off-budget net income." The difference after subtracting this amount is the **on-budget** net income.

Although Social Security is the only major expenditure that is legally considered off-budget, other trust funds also represent future repayment obligations. Because the government borrowed from all these other trust funds as well, the gross debt rose in 2001 by considerably more than the $33 billion on-budget deficit. In fact, promises made to trust funds caused the gross debt—which is the debt that must actually be repaid in the future—to rise by $141 billion in 2001.

Unified Budget, On-Budget, and Off-Budget

The **unified budget** represents all federal revenues and spending. For accounting purposes, the government divides the unified budget into two parts:

- The portion of the unified budget that concerns Social Security is considered **off-budget**.
- The rest of the unified budget is considered **on-budget**.

Therefore, the following relationship holds:

$$\text{unified net income} - \text{off-budget net income} = \text{on-budget net income}$$

EXAMPLE 7 On- and Off-Budget

The federal government had a unified net income of −$587 billion (a deficit) in 2016. However, the government also collected $34 billion more in Social Security revenue than it paid out in Social Security benefits. What do we call this excess $34 billion of Social Security revenue, and what happened to it? What was the government's on-budget deficit for 2016? Explain.

Solution The excess $34 billion of Social Security revenue represents the *off-budget* net income (a surplus) for 2016, because it is counted separately from the rest of the budget (that's what makes it "off" budget). By law, this $34 billion had to be added to the Social Security trust fund. Unfortunately, it had already been spent (on programs other than Social Security), so the government instead added $34 billion worth of IOUs (Treasury bills) to the trust fund. Because this $34 billion worth of IOUs will have to be repaid eventually, it is included in the calculation of the on-budget deficit:

$$\underbrace{-\$587\text{ billion}}_{\text{unified net income}} - \underbrace{\$34\text{ billion}}_{\text{off-budget net income}} = \underbrace{-\$621\text{ billion}}_{\text{on-budget net income}}$$

In other words, the amount by which the government actually overspent in 2016 was the on-budget deficit of $621 billion. Moreover, because of the government's other trust funds and other accounting details, the gross debt rose by even more than this amount.

▶ Now try Exercises 33–34.

The Future of Social Security

Imagine that you decide to set up a retirement savings plan that will allow you to retire comfortably at age 65. Using the savings plan formula (see Unit 4C), you determine that you can achieve your retirement goal by making monthly deposits of $250 into your retirement plan. So you start the plan by making your first $250 deposit.

However, the very next day, you decide you want a new TV and find yourself $250 short of what you need. You therefore decide to "borrow" back the $250 you just deposited into your retirement plan. Because you don't want to fall behind on your retirement savings, you write yourself an IOU promising to put the $250 back. Moreover, recognizing that you would have earned interest on the $250 if you'd left it in the account, at the end of the month you write yourself an additional IOU to replace this lost interest.

Month after month and year after year, you continue in the same way, always diligently depositing your $250, but then withdrawing it so you can spend it on something else, and replacing it with IOUs for the withdrawn money and the lost interest. When you finally reach age 65, your retirement plan will contain IOUs that say you owe yourself enough money to retire—but there will be *no actual money* in your account. Obviously, you won't be able to live off the IOUs you wrote to yourself.

This method of "saving" for retirement may sound silly, but it essentially describes the Social Security trust fund. The government has been diligently depositing the excess money collected through FICA into the Social Security trust fund, then immediately withdrawing it for other purposes while replacing it with Treasury bills that are nothing more than IOUs.

It has been easy to ignore these accounting tricks while the IOUs keep piling up, but that has been possible only because revenues from Social Security taxes have exceeded spending on Social Security benefits. This trend is expected to reverse as more people retire in coming years, which means the government will need to start redeeming the IOUs that it has written to itself.

To see the problem vividly, consider the year 2034, which is approximately when the government's "intermediate" projections (meaning those that are neither especially optimistic nor especially pessimistic) say the Social Security trust fund will run dry. That

BY THE WAY

Social Security benefits differ from private retirement benefits in at least two major ways. First, Social Security benefits are guaranteed. Private retirement accounts may rise or fall in value, thereby changing how much you can afford to withdraw during retirement; in contrast, Social Security promises a particular monthly payment in any circumstances. Second, Social Security benefits are paid as long as you live, but cannot be passed on to your heirs. In contrast, money in a private retirement account can be passed on through your will.

Technical Note
Projections are made in current dollars, so it is not necessary to adjust projected numbers for future inflation.

year, projected Social Security payments will be about \$600 billion more than collections from Social Security taxes, which means the government will be redeeming its last \$600 billion in IOUs from the Social Security trust fund. But since the government owes this money to itself, it will have to find some other source for this \$600 billion. Generally speaking, the government could find this money through some combination of the following three options: (1) it could cut spending on discretionary programs; (2) it could borrow the money from the public by selling more debt (in the form of Treasury bills, notes, and bonds); or (3) it could raise other taxes.

You may notice that any of these three possibilities could have a dramatic impact on *you*. The \$600 billion is nearly as much as the total amount of *all* non-defense discretionary spending, so the Social Security shortfall could not be covered even with major cuts to discretionary programs. Borrowing the money would increase the long-term deficit and debt, and the third option would mean you'd be taxed much more heavily than you are today. But unless our politicians act to alleviate these problems before they occur, you will be forced to face the consequences.

BY THE WAY
Social Security is sometimes called the "third rail" of politics, because politicians who try to touch Social Security often lose their next election. The term comes from the New York City subway system, where the trains run on two rails and the third rail carries electricity at very high voltage. Touching the third rail generally causes instant death.

Think About It Some proposals for solving the Social Security problem call for converting part or all of the program to private savings accounts. Do you think this would be a good idea? Defend your opinion.

EXAMPLE 8 Tax Increase

In 2016, individual income taxes made up about 49% of total government receipts of about \$3.3 trillion. Suppose that the government needed to raise an additional \$600 billion through individual income taxes. How much would taxes have to increase? Neglect any economic problems that the tax increase might cause.

Solution Individual income taxes accounted for 49% of \$3.3 trillion, or about $0.49 \times \$3.3$ trillion $\approx$ \$1.6 trillion in government revenue. Raising an additional \$600 billion (\$0.6 trillion) would bring this total to \$1.6 trillion + \$0.6 trillion = \$2.2 trillion. In percentage terms, this amount represents a tax increase of

$$\frac{\text{new value} - \text{old value}}{\text{old value}} \times 100\% = \frac{\$2.2\text{ trillion} - \$1.6\text{ trillion}}{\$1.6\text{ trillion}} \times 100\% \approx 38\%$$

In other words, on average everyone's income taxes would have to increase by nearly 40% in order to generate this additional revenue. ▶ **Now try Exercises 35–36.**

Quick Quiz 4F

Choose the best answer to each of the following questions. Explain your reasoning with one or more complete sentences.

1. In 2018, Bigprofit.com had \$1 million more in outlays than in receipts, bringing the total amount it owed lenders to \$7 million. We say that at the end of 2018 Bigprofit.com had
   a. a deficit of \$7 million and a debt of \$1 million.
   b. a deficit of \$1 million and a debt of \$7 million.
   c. a surplus of \$1 million and a deficit of \$7 million.
2. If the U.S. government decided to pay off the federal debt by asking for an equal contribution from all U.S. citizens, you'd be asked to pay approximately
   a. \$620. b. \$6200. c. \$62,000.
3. Compared to the historical average over the past 50 years, tax revenues as a percentage of gross domestic product (GDP) for the most recent years (2015 and 2016) shown in Figure 4.16 are
   a. about average.
   b. more than 2 percentage points below average.
   c. more than 2 percentage points above average.
4. In terms of the U.S. budget, what do we mean by *discretionary* outlays?
   a. money that the government spends on things that aren't really important
   b. money that the government spends on programs that Congress must authorize every year
   c. programs funded by FICA taxes

5. Which of the following expenses is *not* considered a mandatory outlay in the U.S. federal budget?
   a. national defense
   b. interest on the debt
   c. Medicare
6. Currently, the majority of government spending goes to
   a. mandatory expenses.
   b. national defense.
   c. science and education.
7. Suppose the government collects \$50 billion more in Social Security taxes than it pays out in Social Security benefits. Under current law, what happens to this "extra" \$50 billion?
   a. It is physically deposited into a bank that holds it to be used for future Social Security benefits.
   b. It is used to fund other government programs.
   c. It is returned in the form of rebates to those who paid the excess taxes.
8. If the government were able to pay off the *publicly held debt*, who would receive the money?
   a. The money would be distributed among all U.S citizens.
   b. The money would go to holders of Treasury bills, notes, and bonds.
   c. The money would go to future retirees through the Social Security trust fund.
9. Which of the following best describes the *total* amount of money the federal government has obligated itself to pay back in the future?
   a. the publicly held debt
   b. the gross debt
   c. the off-budget debt
10. By the year 2036, the government is expected to owe several hundred billion dollars more in Social Security benefits each year than it will collect in Social Security taxes. Although all options for covering this shortfall might be politically difficult, which of the following is *not* an option even in principle?
   a. The shortfall could be covered by tax increases.
   b. The shortfall could be covered by additional borrowing from the public.
   c. The shortfall could be covered by reducing the spending on education grants.

Exercises 4F

REVIEW QUESTIONS

1. Define *receipts, outlays, net income, surplus*, and *deficit* as they apply to annual budgets.
2. What is the difference between a deficit and a debt? How large is the federal debt?
3. Explain why years of running deficits makes it increasingly difficult to get a budget into balance.
4. What is the *gross domestic product (GDP)*, and why do economists often look at budget numbers as a percentage of GDP?
5. Briefly summarize the makeup of federal receipts and federal outlays. Distinguish between mandatory outlays and discretionary outlays.
6. How does the federal government finance its debt? Distinguish between the publicly held debt and the gross debt.
7. Briefly describe the Social Security trust fund. What's in it? What problems may this cause in the future?
8. Distinguish between an off-budget deficit (or surplus) and an on-budget deficit (or surplus). What is the unified deficit (or surplus)?

DOES IT MAKE SENSE?

Decide whether each of the following statements makes sense (or is clearly true) or does not make sense (or is clearly false). Explain your reasoning.

9. My share of the federal government's debt is greater than the cost of a new car.
10. My share of the federal government's annual *interest* payments on the federal debt is greater than the cost of a new car.
11. Because Social Security is off-budget, we could cut Social Security taxes with no impact on the rest of the federal government.
12. The government collected more money than it spent this year, but its gross debt still increased.
13. Because Social Security is an entitlement program and is funded by mandatory government spending, I know it will be there when I retire in 40 years.
14. The federal deficit can easily be eliminated through cuts to discretionary spending.

BASIC SKILLS & CONCEPTS

15. **Personal Budget Basics.** Suppose your after-tax annual income is \$38,000. Your annual expenses are \$12,000 for rent, \$6000 for food and household expenses, \$1200 for interest on credit cards, and \$8500 for entertainment, travel, and other.
   a. Do you have a surplus or a deficit? Explain.
   b. Next year, you expect to get a 3% raise. You think you can keep your expenses unchanged, with one exception: You plan to spend \$8500 on a car. Explain the effect of this purchase on your budget.
   c. As in part (b), assume that you will get a 3% raise for next year. If you can limit your expenses to a 1% increase (over the prior year), could you afford \$7500 in tuition and fees without going into debt?

16. **Personal Budget Basics.** Suppose your after-tax income is \$28,000. Your annual expenses are \$8000 for rent, \$4500 for food and household expenses, \$1600 for interest on credit cards, and \$10,400 for entertainment, travel, and other.

a. Do you have a surplus or a deficit? Explain.

b. Next year, you expect to get a 2% raise, but plan to keep your expenses unchanged. Will you be able to pay off \$5200 in credit card debt? Explain.

c. As in part (b), assume that you will get a 2% raise for next year. If you can limit your expenses to a 1% increase, could you afford \$3500 for a wedding and honeymoon without going into debt?

17. **The Wonderful Widget Company's Future.** Extending the budget summary of the Wonderful Widget Company (Table 4.10), assume that, for 2018, total receipts are \$1,050,000, operating expenses are \$600,000, employee benefits are \$200,000, and security costs are \$250,000.

a. Based on the accumulated debt at the end of 2017, calculate the 2018 interest payment. Assume an interest rate of 8.2%.

b. Calculate the total outlays for 2018, the year-end surplus or deficit, and the year-end accumulated debt.

c. Based on the accumulated debt at the end of 2018, calculate the 2019 interest payment, again assuming an 8.2% interest rate.

d. Assume that in 2019 the company has receipts of \$1,100,000, holds operating expenses and employee benefits to their 2018 levels, and spends no money on security. Calculate the total outlays for 2019, the year-end surplus or deficit, and the year-end accumulated debt.

e. Imagine that you are the CFO (chief financial officer) of the Wonderful Widget Company at the end of 2019. Write a three-paragraph statement to shareholders about the company's future prospects.

18. **The Wonderful Widget Company's Future.** Extending the budget summary of the Widget Company (Table 4.10), assume that, for 2018, total receipts are \$975,000, operating expenses are \$850,000, employee benefits are \$290,000, and security costs are \$210,000.

a. Based on the accumulated debt at the end of 2017, calculate the 2018 interest payment. Assume an interest rate of 8.2%.

b. Calculate the total outlays for 2018, the year-end surplus or deficit, and the year-end accumulated debt.

c. Based on the accumulated debt at the end of 2018, calculate the 2019 interest payment, again assuming an 8.2% interest rate.

d. Assume that in 2019 the company has receipts of \$1,050,000, holds operating expenses and employee benefits to their 2018 levels, and spends no money on security. Calculate the total outlays for 2019, the year-end surplus or deficit, and the year-end accumulated debt.

e. Imagine that you are the CFO (chief financial officer) of the Wonderful Widget Company at the end of 2019. Write a three-paragraph statement to shareholders about the company's future prospects.

19. **Per-Worker Debt.** Suppose that in 2017 the government had decided to pay off the \$20 trillion federal debt with a one-time charge distributed equally among the approximately 160 million people in the civilian work force. How much would each worker have been charged?

20. **Per-Family Debt.** Suppose that in 2017 the government had decided to pay off the \$20 trillion federal debt with a one-time charge distributed equally among the approximately 118 million U.S. households. How much would each household be charged?

21–24. Deficit, Debt, and GDP. Consider the following table showing past and projected federal revenue, spending, and GDP. All figures are in billions of dollars and rounded to the nearest billion.

Year	Surplus or deficit	Debt	GDP
2000	236	5600	10,100
2010	−1294	13,600	14,800
2020 (projected)	−540	22,500	21,900

21. Find the deficit or surplus and debt as a percentage of GDP in 2000 and 2010. Comment on the changes.

22. Find the percent change in the *debt* between 2000 and 2010. Then find the percent change *in the debt as a percentage of GDP* between 2000 and 2010.

23. Find the deficit as a percentage of GDP in 2010 and 2020 (projected). What is the percentage change during this period?

24. Find the debt as a percentage of GDP in 2010 and 2020 (projected). What is the percentage change during this period?

25–26. Interest Payments.

25. Suppose that the federal debt in 2020 is \$22.5 trillion. If interest rates remain at the 2016 level of 1.7%, find the annual interest payment on the debt. How much does the interest payment change with a 0.5-point increase in the interest rate?

26. Suppose that the federal debt in 2022 is \$24 trillion. If interest rates remain at the 2016 level of 1.7%, find the annual interest payment on the debt. How much does the interest payment change with a 1-point increase in the interest rate?

27–32: Budget Analysis. Assume that the composition of federal outlays and receipts shown in Figure 4.17 remained the same in 2017. Also assume that total revenue for 2017 was \$3.31 trillion and the total spending was \$3.89 trillion.

27. About how much 2017 revenue (in dollars) came from individual income taxes? About how much came from the category "Social Security, Medicare, and other social insurance receipts"?

28. About how much spending (in dollars) in 2017 was discretionary? About how much spending was non-discretionary?

29. Suppose that in 2018, total spending increases by 1.6% and total revenue increases by 2.0% from the 2017 levels given above. Would the 2018 deficit be larger, smaller, or the same as the 2017 deficit? Explain.

30. Suppose that in 2018, total spending and total revenue both increase by 2% from the 2017 levels given above. Would the 2018 deficit be larger, smaller, or the same as the 2017 deficit? Explain.

31. a. About how much was spent (in dollars) on Social Security and Medicare combined in 2017?

b. About how much was spent (in dollars) on all discretionary expenses (combined) in 2017?

c. Suppose that over a period of several years, overall spending remained at the 2017 level of $3.89 trillion, but the share of Social Security and Medicare spending increased from 38% to 40% of total spending. Assuming that all the needed cuts came from discretionary spending, how much would this spending decrease? Give your answer both in dollars and as a percentage.

32. a. About how much was spent on *non-defense* discretionary outlays in 2017?

b. Suppose that combined spending on Social Security and Medicare increased by 10% from its 2017 level (as found in Exercise 31a), but total spending remained unchanged. Assuming that all the needed cuts came from non-defense discretionary spending, how much would this spending decrease? Give your answer both in dollars and as a percentage.

33. **On- and Off-Budget.** Suppose that in a recent year the government has a unified net income of $40 billion, but was supposed to deposit $180 billion in the Social Security trust fund. What was the on-budget surplus or deficit? Explain.

34. **On- and Off-Budget.** Suppose that in a recent year the government has a unified net income of $40 billion, but was supposed to deposit $205 billion in the Social Security trust fund. What was the on-budget deficit? Explain.

35. **Social Security Finances.** Suppose the year is 2025, and the government needs to pay out $350 billion more in Social Security benefits than it collects in Social Security taxes. Briefly discuss the options for getting this money.

36. **Social Security Finances.** Suppose the year is 2025, and the government needs to pay out $525 billion more in Social Security benefits than it collects in Social Security taxes. Briefly discuss the options for getting this money.

FURTHER APPLICATIONS

37. **Counting the Federal Debt.** Suppose you began counting the approximately $20 trillion of the 2017 federal debt, $1 at a time. If you could count $1 each second, how long would it take to complete the count? Specify the answer in years.

38. **Paving with the Federal Debt.** Suppose you began covering the ground with $1 bills. If you had the approximately $20 trillion of the 2017 federal debt in $1 bills, how much total area could you cover? Compare this area to the total land area of the United States, which is about 10 million square kilometers. (*Hint:* Assume that a dollar bill has an area of 100 square centimeters.)

39. **Rising Debt.** Suppose the federal debt increases at an annual rate of 1% per year. Use the compound interest formula (Unit 4B) to determine the size of the debt in 10 years and in 50 years. Assume that the current size of the debt (the principal for the compound interest formula) is $20 trillion.

40. **Rising Debt.** Suppose the federal debt increases at an annual rate of 2% per year. Use the compound interest formula (Unit 4B) to determine the size of the debt in 10 years and in 50 years. Assume that the current size of the debt (the principal for the compound interest formula) is $20 trillion.

41. **Retiring the Public Debt.** Consider the 2017 publicly held debt of about $14.5 trillion. Use the loan payment formula (Unit 4D) to determine the annual payments needed to pay this debt off in 10 years. Assume an annual interest rate of 3%.

42. **Retiring the Public Debt.** Consider the 2017 publicly held debt of about $14.5 trillion. Use the loan payment formula (Unit 4D) to determine the annual payments needed to pay this debt off in 15 years. Assume an annual interest rate of 2%.

43. **National Debt Lottery.** Imagine that, through some political or economic miracle, the gross debt stopped rising after 2017 and remained at $20 trillion. To retire the gross debt, the government decided to have a national lottery. Suppose that every U.S. citizen bought a $1 lottery ticket every week, thereby generating about $325 million in weekly lottery revenue. Because lotteries typically use half their revenue for prizes and lottery operations, assume that $163 million would go toward debt reduction each week. How long would it take to retire the debt through this lottery?

44. **National Debt Lottery.** Suppose the government hopes to pay off the 2017 gross debt of about $20 trillion with a national lottery. For the debt to be paid off in 50 years, how much would each citizen have to spend on lottery tickets each year? Assume that half of the lottery revenue goes toward debt reduction and that there are 325 million citizens.

IN YOUR WORLD

45. **Political Action.** This unit outlined numerous budgetary problems facing the U.S. government, as they stood at the time the text was written in 2017. Has there been any significant political action to deal with any of these problems? Learn what, if anything, has changed since 2017; then write a one-page position paper outlining your own recommendations for the future.

46. **Debt Problem.** How serious a problem is the gross debt? Find arguments on both sides of this question. Summarize the arguments, and state your own opinion.

47. **Social Security Solutions.** Research various proposals for solving the budgetary problems of Social Security. Choose one proposal that you think is worthwhile, and write a one- to two-page report summarizing it and describing why you think it is a good idea.

48. **Medicare Solutions.** Like Social Security, Medicare is projected to represent a growing share of federal spending as the population ages and health care costs rise. Find one or more articles that detail problems and potential solutions for Medicare. Write a short summary of the issues, and state your own opinion of what should be done.

49. **Back to Surplus?** Research how the government managed to turn from deficits to surpluses from 1998 to 2001. What changed to put the budget back in deficit after that? Do you think the government could find a way to return to surpluses again in the coming decade? Summarize your findings and opinion by writing a one-page memo titled "Back to Surplus."

Chapter 4 Summary

UNIT	KEY TERMS	KEY IDEAS AND SKILLS
4A	budget (p. 194) cash flow (p. 194) insurance costs premium (p. 199) deductible (p. 199) co-payment (p. 199)	**Understand** the importance of controlling your finances. (p. 193) **Know** how to make a budget. (p. 194) **Be aware of** factors that help determine whether your spending patterns make sense for your situation. (p. 196) **Understand** the costs and potential benefits of various types of insurance. (p. 198)
4B	principal (p. 206) simple interest (p. 206) compound interest (p. 206) annual percentage rate (APR) (p. 208) variable definitions: *A, P, N, n, Y* (pp. 209, 214) annual percentage yield (APY) (p. 216)	**General form of the compound interest formula:** (p. 209) $A = P \times (1 + i)^N$ **Compound interest formula for interest paid once a year:** (p. 209) $A = P \times (1 + \text{APR})^Y$ **Compound interest formula for interest paid *n* times a year:** (p. 214) $A = P\left(1 + \frac{\text{APR}}{n}\right)^{(nY)}$ **Compound interest formula for continuous compounding:** (p. 218) $A = P \times e^{(\text{APR} \times Y)}$ **Know** when and how to apply these formulas. (pp. 208–219)
4C	savings plan (p. 225) total return (p. 231) annual return (p. 231) mutual fund (p. 233) investment considerations liquidity (p. 233) risk (p. 233) return (p. 233) bond characteristics face value (p. 237) coupon rate (p. 237) maturity rate (p. 237) current yield (p. 238)	**Savings plan formula:** (p. 226) $A = \text{PMT} \times \frac{\left[\left(1 + \frac{\text{APR}}{n}\right)^{(nY)} - 1\right]}{\left(\frac{\text{APR}}{n}\right)}$ **Return on investments:** (p. 232) $\text{total return} = \frac{(A - P)}{P}$ $\text{annual return} = \left(\frac{A}{P}\right)^{(1/Y)} - 1$ **Understand** types of investments: stocks, bonds, cash. (p. 233) **Read** online quotes for stocks, bonds, and mutual funds. (pp. 236, 239) **Remember** important principles of investing: (p. 237) Higher returns usually involve higher risk. High commissions and fees can dramatically lower returns. Build an appropriately diversified portfolio.
4D	installment loan (p. 245) mortgage (p. 252) down payment (p. 252) closing costs (p. 252) points (p. 252) fixed rate mortgage (p. 253) adjustable rate mortgage (p. 256)	**Loan payment formula:** (p. 245) $\text{PMT} = \frac{P \times \left(\frac{\text{APR}}{n}\right)}{\left[1 - \left(1 + \frac{\text{APR}}{n}\right)^{(-nY)}\right]}$ **Understand** the uses and dangers of credit cards. (p. 251) **Understand** considerations in choosing a mortgage. (p. 252) **Understand** strategies for early payment of loans. (p. 254)

UNIT	KEY TERMS	KEY IDEAS AND SKILLS
4E	gross income (p. 262) adjusted gross income (p. 262) exemptions, deductions, tax credits (p. 262) taxable income (p. 262) filing status (p. 263) progressive income tax (p. 264) marginal tax rates (p. 264) FICA (p. 267) capital gains (p. 268)	**Define** different types of income as they apply to taxes. (p. 262) **Use** tax rate tables to calculate taxes. (p. 264) **Distinguish between** tax credits and tax deductions. (p. 266) **Calculate** FICA taxes. (p. 268) **Be aware of** special tax rates for dividends and capital gains. (p. 268) **Understand** the benefits of tax-deferred savings plans. (p. 269)
4F	receipts, outlays (p. 275) net income (p. 275) surplus (p. 275) deficit (p. 275) debt (p. 275) mandatory outlays (p. 280) discretionary outlays (p. 280) publicly held debt (p. 283) gross debt (p. 283) on-budget, off-budget (p. 284) unified budget (p. 284)	**Distinguish between** a deficit and a debt. (p. 276) **Understand** basic principles of the federal budget. (p. 277) **Distinguish between** publicly held debt and gross debt. (p. 284) **Be familiar with** major issues concerning the future of Social Security. (p. 284)

5 STATISTICAL REASONING

Is your drinking water safe? Do most people approve of the President's economic plan? How much is the cost of health care rising? These questions and thousands more like them can be answered only through statistical studies. Indeed, statistical information appears in the news every day, making the ability to understand and reason with statistics crucial to modern life.

Q

You want to know whether Americans generally support or oppose a new health care plan. Which of the following approaches is most likely to give you an accurate result?

- **A** Short interviews with 1000 randomly selected Americans, conducted by a professional polling organization such as Gallup or the Pew Research Center
- **B** In-depth interviews with 10,000 Americans selected at random from among those who have been hospitalized in the past year, conducted by a professional polling organization
- **C** Short interviews with 100,000 Americans, chosen by asking 100 randomly selected people at each of 1000 grocery stores at 9 a.m. on a particular Monday morning, conducted by volunteers working for a citizens' group
- **D** A poll in which more than 2 million people register their opinions online, conducted by a television news channel
- **E** A special election held nationwide, in which all registered voters have an opportunity to answer a question about their opinion on the health care plan

Statistical thinking will one day be as necessary for efficient citizenship as the ability to read and write.

—H. G. Wells, as paraphrased by Samuel Wilks

A Many people guess that the special election would be the best bet, because it would include the largest number of people, and that the online poll with 2 million people would be the second best. In fact, neither of those is likely to produce very meaningful results, and the best choice by far is choice A.

This fact is quite astonishing when you think about it. A survey of 1000 people means asking only about 1 out of every 325,000 Americans; to put this in perspective, this is like choosing just a single individual from about six football stadiums full of people. But well-conducted polls and surveys can indeed produce excellent results; they can even give us quantitative measures of how confident we should be in their results. To understand how this is possible, you first need to understand the basic concepts of statistics that we discuss in this chapter. And to learn exactly why A is correct and the other choices are not, see Unit 5B, Example 3.

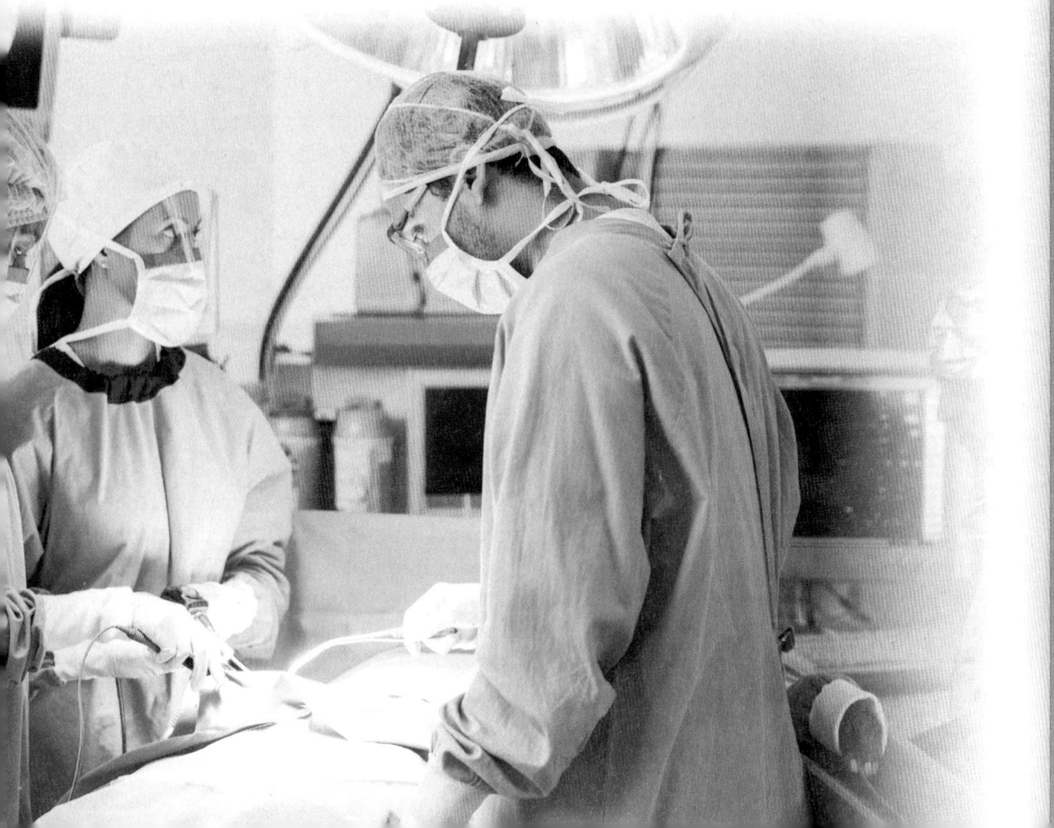

UNIT 5A

Fundamentals of Statistics: Understand how statistical studies are conducted, with emphasis on the importance of sampling.

UNIT 5B

Should You Believe a Statistical Study? Be familiar with eight useful guidelines for evaluating statistical claims.

UNIT 5C

Statistical Tables and Graphs: Interpret and create basic tables and graphs, including frequency tables, bar graphs, pie charts, histograms, and line charts.

UNIT 5D

Graphics in the Media: Interpret and explore common types of media graphics.

UNIT 5E

Correlation and Causality: Investigate correlations and learn how to decide whether a correlation is the result of causality.

Cell Phones and Driving

Use this activity to gain a sense of the kinds of problems this chapter will enable you to analyze. Additional activities are available online in MyLab Math.

Is it safe to use a cell phone while driving? The science of statistics provides a way to approach this question, and the results of many studies indicate that the answer is no. The National Safety Council estimates that approximately 1.6 million car crashes each year (more than a quarter of the total) are caused by some type of distraction, most commonly the use of a cell phone for talking or texting. In fact, some studies suggest that merely talking on a cell phone makes you as dangerous as a drunk driver. As preparation for your study of statistics in this chapter, work individually or in groups to research the issues raised in the following questions. Discuss your findings.

1. Think about the physical process of using a cell phone (either talking or texting) while driving, and list possible reasons why it could be distracting and cause a crash.
2. When the link between cell phone use and crashes was first discovered, many people thought the problem could be solved by mandating that only hands-free cell phone systems be allowed in cars, and many localities, states, and nations enacted laws allowing drivers to use only hands-free systems. However, more recent studies show that hands-free systems are nearly as dangerous as regular cell phones—and that talking on a hands-free system is much more dangerous than talking to a passenger sitting next to you. Why don't hands-free systems eliminate the danger of cell phone use while driving?
3. The fact that many crashes involve cell phone use does not necessarily prove that the use of cell phones *caused* the crashes. What kinds of studies might prove that cell phone use is the cause of crashes? How could such studies be conducted? Look for results of actual studies of this issue.
4. Find some actual data that shed light on the issue of cell phones and driving. Explain what the data show. Do you think the data are summarized clearly, or could they have been displayed in a better way?
5. Have you personally ever been involved in a crash or a close call in which you think a cell phone played a role? If so, how confident are you that the cell phone use was responsible?
6. Statistical studies are most useful when they lead to intelligent action. Given the apparent link between cell phone use and car crashes, what do *you* think should be done about the issue? Defend your opinion.

UNIT 5A Fundamentals of Statistics

The field of statistics plays a major role in modern society. It's used to determine whether a new drug is effective in treating cancer. It's involved when agricultural inspectors check the safety of the food supply. It's used in every opinion poll and survey. In business, it's used for market research. Sports statistics are part of daily conversation for millions of people. Indeed, you'll be hard-pressed to think of a topic that is not linked in some way to statistics.

But what *is* (or *are*) statistics? When stated in the singular, *statistics* is the *science* that helps us understand how to collect, organize, and interpret numbers or other information about some topic; we refer to the numbers or other pieces of information as *data*. When the term is plural, *statistics* are the actual data that describe some characteristic. For example, if there are 30 students in your class and they range in age from 17 to 64, the numbers "30 students," "17 years," and "64 years" are statistics that describe your class.

BY THE WAY

You'll sometimes hear the word *data* used as a singular synonym for *information*, but technically the word *data* is plural. One piece of information is called a *datum*, and two or more pieces are called *data*.

Two Definitions of *Statistics*

- Statistics is the *science* of collecting, organizing, and interpreting data.
- Statistics are the *data* (numbers or other pieces of information) that describe or summarize something.

How Statistics Works

Statistical studies are conducted in many different ways and for many different purposes, but they all share a few characteristics. To get the basic ideas, consider the *Nielsen ratings*, which are used to estimate the numbers of people watching various television shows.

Suppose the Nielsen ratings tell you that *The Big Bang Theory* was last week's most popular show, with 22 million viewers. You probably know that no one counted all 22 million people. But you may be surprised to learn that the Nielsen ratings are based on data from only about 5000 homes. To understand how Nielsen can draw a conclusion about millions of Americans from only a few thousand homes, we need to investigate the principles behind statistical research.

Nielsen's goal is to draw conclusions about the viewing habits of all Americans. In the language of statistics, we say that Nielsen is interested in the **population** of all Americans. The characteristics of this population that Nielsen seeks to learn—such as the number of people watching each television show—are called **population parameters**. Note that, although we usually think of a population as a group of people, in statistics a population can be any kind of group—people, animals, or things. For example, in a study of college costs, the population might be *all colleges and universities*, and the population parameters might include the amounts charged for tuition, fees, and housing.

Nielsen cannot realistically study the entire population of all Americans, so the company instead studies a much smaller **sample** consisting of about 5000 homes. The individual measurements that Nielsen collects from the sample, such as who is watching each show at each time, constitute the *raw data*. Nielsen then consolidates these raw data into a set of numbers that characterize the sample, such as the percentage of young male viewers watching *The Big Bang Theory*. These numbers are called **sample statistics**.

BY THE WAY

Arthur C. Nielsen founded his company and invented market research in 1923. He began producing ratings for radio programs in 1942 and added television ratings in the 1960s. Nielsen's company now also tracks usage of many other types of media.

Definitions

The **population** in a statistical study is the *complete* set of people or things being studied.

The **sample** is the subset of the population from which the raw data are actually obtained.

Population parameters are specific numbers of interest to researchers that describe certain characteristics of the population.

Sample statistics are numbers describing characteristics of the sample, found by consolidating or summarizing the raw data collected from the sample.

EXAMPLE 1 Population and Sample

For each of the following cases, describe the population, sample, population parameters, and sample statistics.

a. Agricultural inspectors for Jefferson County measure the levels of residue from three common pesticides on 25 ears of corn from each of the 104 corn-producing farms in the county.

b. Anthropologists estimate the average brain size (skull size) of early Neanderthals in Europe by studying skulls found at three sites in southern Europe.

Solution

a. The population the inspectors are interested in consists of all ears of corn grown in the county. The sample consists of the 25 ears from each of the 104 farms (a total of $25 \times 104 = 2600$ ears of corn). The population parameters are the average levels of residue from the three pesticides on *all* corn grown in the county. The sample statistics are the average levels of residue measured on the ears of corn in the sample.

b. The population the anthropologists are interested in consists of all early Neanderthals in Europe. The sample consists of the relatively few individual Neanderthals whose skulls were found at the three sites. The population parameter of interest is the average brain size of all Neanderthals, and the sample statistic is the average brain size of the individuals in the sample. ▶ Now try Exercises 15–20.

The Process of a Statistical Study

BY THE WAY

Statisticians often divide their subject into two major branches. *Descriptive statistics* is the branch that deals with *describing* data using tables, graphs, or sample statistics. *Inferential statistics* is the branch that deals with *inferring* (or estimating) population characteristics from sample data.

Because Nielsen does not study the entire population of all Americans, it cannot measure any population parameters. Instead, the company tries to infer reasonable values for population parameters from the sample statistics (which it *does* measure).

The process of inference is simple in principle, though it must be carried out with great care. For example, suppose Nielsen finds that 7% of the people in its sample watched *The Big Bang Theory*. If this sample accurately represents the entire population of all Americans, then Nielsen can infer that *approximately* 7% of all Americans watched the show. In other words, the sample statistic of 7% is used as an estimate for the population parameter. (By using certain other statistical techniques that we'll discuss in Unit 6D, Nielsen can also estimate the uncertainty in the inferred population parameters.)

Once Nielsen has estimates of the population parameters, it can draw general conclusions about what Americans were watching. The process used by Nielsen Media Research is similar to that used in many statistical studies, summarized in the box below.

Basic Steps in a Statistical Study

1. State the goals of your study. That is, determine the population you want to study and exactly what you'd like to learn about it.
2. Choose a representative sample from the population.
3. Collect raw data from the sample and summarize these data by finding sample statistics of interest.
4. Use the sample statistics to infer the population parameters.
5. Draw conclusions: Determine what you learned and how it addresses your goals.

Figure 5.1 summarizes the general relationships among a population, a sample, the sample statistics, and the population parameters.

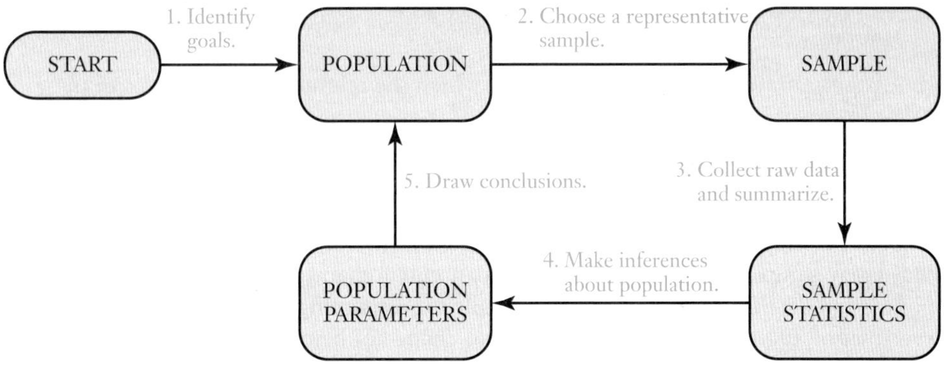

FIGURE 5.1
Elements of a statistical study.

EXAMPLE 2 Unemployment Survey

Each month, the U.S. Labor Department surveys 60,000 households to determine characteristics of the U.S. work force. One population parameter of interest is the U.S. unemployment rate, defined as the percentage of people who are unemployed among all those who are either employed or actively seeking employment. Describe how the five basic steps of a statistical study apply to this research.

Solution The steps apply as follows.

Step 1. The goal of the research is to learn about employment (or unemployment) within the population of all Americans who are either employed or actively seeking employment.

Step 2. The Labor Department chooses a sample consisting of people employed or seeking employment from 60,000 households.

Step 3. The Labor Department asks questions of the people in the sample, and their responses constitute the raw data for the research. The department then consolidates these data into sample statistics, such as the percentage of people in the sample who are unemployed.

Step 4. Based on the sample statistics, the Labor Department makes estimates of the corresponding population parameters, such as the unemployment rate for the entire United States.

Step 5. The Labor Department draws conclusions based on the population parameters and other information. For example, it might use the current and past unemployment rates to draw conclusions about whether jobs have been created or lost.

▶ Now try Exercises 21–26.

BY THE WAY

By the Labor Department definition, someone who is not working is not necessarily unemployed. For example, stay-at-home moms and dads are not counted among the unemployed unless they are actively trying to find a job, and people who had been trying to find work but gave up in frustration are not counted as unemployed.

Choosing a Sample

Choosing a sample may be the most important step in any statistical study. If the sample fairly represents the entire population, then it's reasonable to make inferences from the sample to the population. But if the sample is *not* representative, then there's little hope of drawing accurate conclusions about the population.

To see why, suppose you decide to estimate the average height and weight of male students at a large university by measuring the heights and weights of a sample of 100 students. A sample consisting only of members of the football team would not be reliable, because these athletes tend to be larger than most other men. In contrast, suppose you select your sample with a computer program that randomly draws student numbers from the entire male population at the university. In this case, the 100 students in your sample are likely to be representative of the population. You can therefore expect that the average height and weight of students in the sample are reasonable estimates of the averages for all male students.

Definition

A **representative sample** is a sample in which the relevant characteristics of the sample members are generally the same as those of the population.

USING TECHNOLOGY

Random Numbers

Many calculators have a key that generates random numbers, and a Web search will turn up many random number generators. In Excel, the RAND function generates uniformly distributed random numbers between 0 and 1.

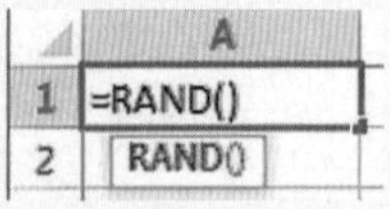

Alternatively, the Excel function RANDBETWEEN generates a random integer between any two given numbers; the screen shot below shows how to generate random numbers between 1 and 25.

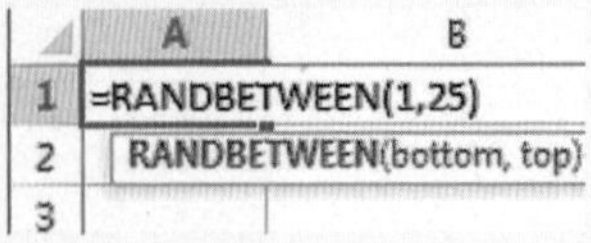

A sample drawn with a computer program that selects students at random is an example of a **simple random sample**. More technically, simple random sampling means that every sample of a particular size has an equal chance of being selected from the population. In the case of the student sample for estimating average height and weight, every set of 100 male students has an equal chance of being selected from the population of male students.

Simple random sampling is not always practical or necessary, so other sampling techniques are sometimes used. The following box summarizes four of the most common sampling techniques, and Figure 5.2 illustrates the ideas.

Common Sampling Methods

Simple random sampling: We choose a sample of items in such a way that every sample of the same size has an equal chance of being selected.

Systematic sampling: We use a simple system to choose the sample, such as selecting every 10th or every 50th member of the population.

Convenience sampling: We choose a sample that is convenient to select, such as people who happen to be in the same classroom.

Stratified sampling: We use this method when we are concerned about differences among subgroups, or *strata,* within a population. We first identify the subgroups and then draw a simple random sample within each subgroup. The total sample consists of all the samples from the individual subgroups.

Regardless of what type of sampling is used, always keep the following key ideas in mind:

- No matter how a sample is chosen, the study can be successful only if the sample is representative of the population.
- Sample size is important, because a large well-chosen sample has a better chance of being representative than a small one. However, the selection process is even more important: A small well-chosen sample is likely to give better results than a large poorly chosen sample.
- Even if a sample is chosen in the best possible way, we can never be *sure* that it is representative of the population. We can only conclude that it has a strong likelihood of being representative.

EXAMPLE 3 Sampling Methods

Identify the type of sampling used in each of the following cases and comment on whether the sample is likely to be representative of the population.

a. You are conducting a survey of students in a dormitory. You choose your sample by knocking on the door of every 10th room.

b. To survey opinions on a possible property tax increase, a research firm randomly draws the addresses of 150 homeowners from a public list of all homeowners.

c. Agricultural inspectors for Jefferson County check the levels of residue from three common pesticides on 25 ears of corn from each of the 104 corn-producing farms in the county.

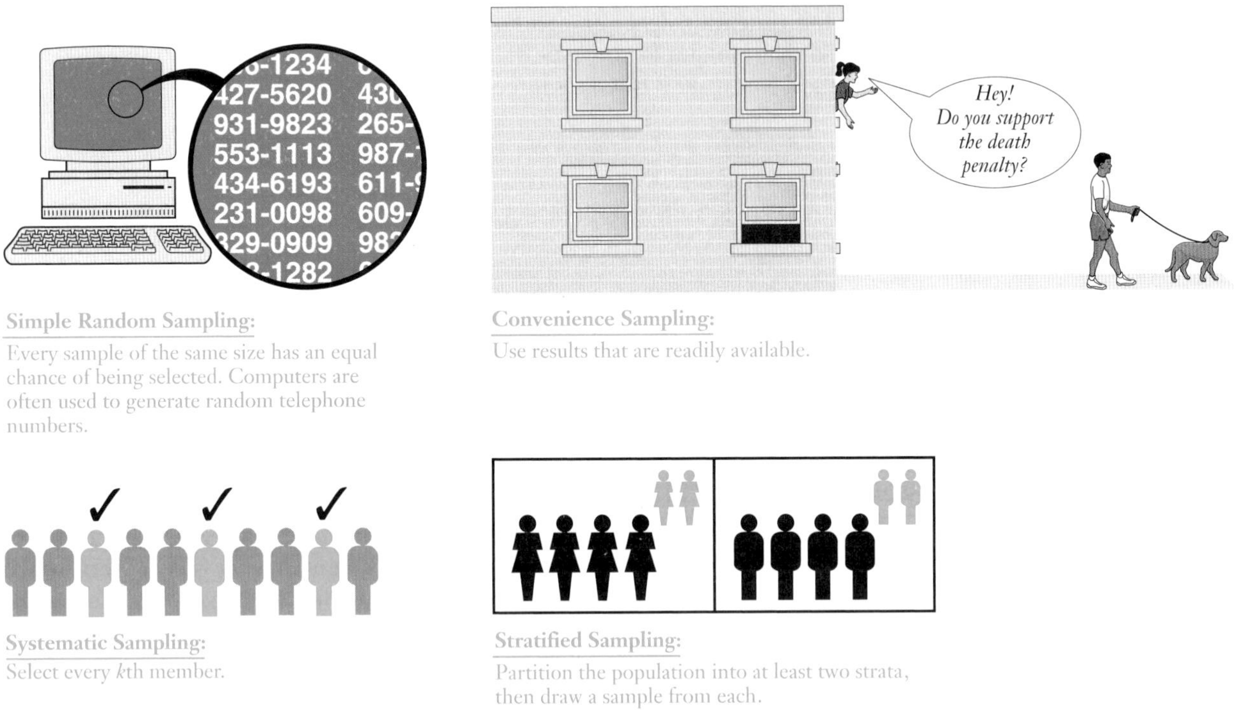
427-5620
931-9823
553-1113
434-6193
231-0098
Simple Random Sampling:
Every sample of the same size has an equal chance of being selected. Computers are often used to generate random telephone numbers.
Hey!
Do you support the death penalty?
Convenience Sampling:
Use results that are readily available.
Systematic Sampling:
Select every *k*th member.
Stratified Sampling:
Partition the population into at least two strata, then draw a sample from each.

FIGURE 5.2 Common sampling techniques.

d. Anthropologists determine the average brain size of early Neanderthals in Europe by studying skulls found at three sites in southern Europe.

BY THE WAY
Neanderthals lived between about 100,000 and 30,000 years ago in Eurasia and northern Africa. They were physiologically distinct from *Homo sapiens* (modern humans), and skull measurements suggest that Neanderthals had larger brains. Although Neanderthals went extinct, genetic evidence indicates that they interbred with *Homo sapiens* and that 1%–4% of modern human DNA originated with Neanderthals.

Solution

a. Choosing every 10th room makes this a systematic sample. The sample may be representative, as long as students were randomly assigned to rooms.

b. The records presumably list all homeowners, so drawing randomly from this list produces a simple random sample. It has a good chance of being representative of the population.

c. Each farm may have different pesticide use, so the inspectors consider corn from each farm as a subgroup (stratum) of the full population. By checking 25 ears of corn from each of the 104 farms, the inspectors are using stratified sampling. If the ears are collected randomly on each farm, each set of 25 is likely to be representative of its farm.

d. By studying skulls found at selected sites, the anthropologists are using a convenience sample. They have little choice, because only a few skulls remain from the many Neanderthals who once lived in Europe. However, it seems reasonable to assume that these skulls are representative of the larger population.

▶ Now try Exercises 27–34.

Watching Out for Bias

Imagine that someone tried to estimate the average weight of all men at a college using a sample consisting only of football players. As noted earlier, we do not expect this sample to be representative of the population; instead, we say that this sample is *biased* because the football players in the sample are likely to differ in average weight from the population of all men at the college. More generally, the term **bias** refers to any problem in the design or conduct of a statistical study that tends to favor certain results.

Bias can arise in many different ways. For example, a researcher may be biased if he or she has a personal stake in the outcome of the study. In that case, the researcher might distort (intentionally or unintentionally) the true meaning of the data. You should always be on the lookout for any type of bias that may affect the results or interpretation of a statistical study. We'll discuss sources of bias further in Unit 5B.

Definition

A statistical study suffers from **bias** if its design or conduct tends to favor certain results.

Think About It Explain why concerns about potential bias lead television networks to use Nielsen (an independent company) to measure ratings rather than doing it themselves.

Types of Statistical Study

HISTORICAL NOTE

Statistics originated with the collection of census and tax data, which are affairs of state. That is why the word *state* is at the root of the word *statistics*.

Broadly speaking, most statistical studies fall into one of two categories: observational studies and experiments. Nielsen's research on television viewing is an **observational study** because it is designed to *observe* the television-viewing behavior of the people in its 5000 sample homes. Note that an observational study may involve activities that go beyond the usual definition of *observing*. Measuring people's weights requires interacting with them, as in asking them to stand on a scale. But in statistics, we consider these measurements to be observations because the interactions do not change people's weights. Similarly, an opinion poll in which researchers conduct in-depth interviews is considered observational as long as the researchers attempt only to learn people's opinions, not to change them.

In contrast, consider a medical study designed to test whether large doses of vitamin C prevent colds. To conduct this study, the researchers must ask some people in the sample to take large doses of vitamin C. This type of statistical study is called an **experiment**, because some participants receive a treatment (in this case, vitamin C) that they would not otherwise receive.

Two Basic Types of Statistical Study

1. In an **observational study**, researchers observe or measure characteristics of the sample members but do not attempt to influence or modify those characteristics.
2. In an **experiment**, researchers apply a treatment to some or all of the sample members and then observe the effects of the treatment.

It is difficult to determine whether an experimental treatment works unless you compare groups that receive the treatment to groups that don't. In the vitamin C study, for example, researchers might create two groups of people: a **treatment group** that takes large doses of vitamin C and a **control group** that does not take vitamin C. The researchers can then look for differences in the numbers of colds among people in the two groups. Having a control group is usually crucial to interpreting the results of experiments.

With proper treatment, a cold can be cured in a week. Left to itself, it may linger for seven days.

—A medical folk saying

In an experiment, it is important for the treatment and control groups to be alike in all respects except for the treatment. For example, if the treatment group consisted of active people with good diets and the control group consisted of sedentary people with poor diets, we could not attribute any differences in colds to vitamin C alone. To avoid this type of problem, assignments to the control and treatment groups must be done randomly.

Treatment and Control Groups

The **treatment group** in an experiment is the group of sample members who receive the treatment being tested.

The **control group** in an experiment is the group of sample members who do *not* receive the treatment being tested.

It is important for the treatment and control groups to be selected randomly and to be alike in all respects except for the treatment.

Think About It Consider a computer function that generates random numbers between 0 and 100. How could you use this function to assign participants in an experiment to the treatment and control groups?

The Placebo Effect and Blinding

When an experiment involves people, bias can arise simply because subjects know they are part of the experiment. For example, if a treatment group of people taking vitamin C gets fewer colds than a control group of people who don't, we can't be sure that the vitamin C was responsible; it might be that people in the treatment group stayed healthier because they *believed* that vitamin C works. This is possible because stress and other psychological factors have been shown to affect resistance to colds. Cases like this, in which people seem to improve because they believe the treatment is helpful, are said to suffer bias from a **placebo effect**. (The word *placebo* comes from the Latin "to please.")

To distinguish between results caused by a placebo effect and results due to a treatment, researchers generally try to make sure that the participants do not know whether they are part of the treatment or control group. The researchers accomplish this by giving people in the control group a **placebo**—something that looks or feels just like the treatment being tested, but lacks its active ingredients.

BY THE WAY

The placebo effect can be surprisingly powerful. In some studies, up to 75% of participants receiving a placebo actually improved. Nevertheless, different researchers disagree about the strength and precise origins of the placebo effect.

Definitions

A **placebo** lacks the active ingredients of the treatment being tested in a study, but looks or feels enough like the treatment so that participants cannot distinguish whether they are receiving the placebo or the real treatment.

The **placebo effect** refers to the situation in which patients improve simply because they believe they are receiving a useful treatment.

In statistical terminology, the practice of keeping people in the dark about who is in the treatment group and who is in the control group is called **blinding**. A **single-blind** experiment is one in which the participants don't know which group they belong to, but the experimenters (the people administering the treatment) do know. Using a placebo is one way to create a single-blind experiment. Sometimes, a single-blind experiment can still be unreliable if the experimenters can subtly influence outcomes. For example, in an experiment that involves interviews, the experimenters might speak differently to people who received the real treatment than to those who received the placebo. This type of problem can be avoided by making the experiment **double-blind**, which means neither the participants nor the experimenters know who belongs to each group. (Of course, someone must keep track of the two groups in order to evaluate the results at the end. In typical double-blind experiments, researchers hire experimenters to make any necessary contact with the participants.)

Blinding in Experiments

An experiment is **single-blind** if the participants do not know whether they are members of the treatment group or members of the control group, but the experimenters do know.

An experiment is **double-blind** if neither the participants nor the experimenters (people administering the treatment) know who belongs to the treatment group and who belongs to the control group.

EXAMPLE 4 What's Wrong with This Experiment?

For each of the experiments described below, identify any problems and explain how the problems could have been avoided.

a. A chiropractor performs adjustments on 25 patients with back pain. Afterward, 18 patients say they feel better. The chiropractor concludes that the adjustments are an effective treatment.

b. A new drug for a type of attention disorder is supposed to make children less disruptive. Randomly selected children suffering from the disorder are divided into treatment and control groups. Those in the control group receive a placebo that looks just like the real drug. The experiment is single-blind. Experimenters interview the children individually to decide whether they became more polite.

Solution

a. The 25 patients who receive adjustments represent a treatment group, but this study lacks a control group. The patients may be feeling better because of a placebo effect rather than any real effect of the adjustments. The chiropractor might have improved his study by hiring an actor to do a fake adjustment (one that feels like a real manipulation, but doesn't conform to chiropractic guidelines) on a control group. Then he could have compared the results in the two groups to see whether a placebo effect was involved.

b. Because the experimenters know which children received the real drug, during the interviews they may inadvertently speak differently to these children or interpret the children's behavior differently. The experiment should have been double-blind; then the experimenters conducting the interviews would not have known which children received the real drug and which children received the placebo.

▶ Now try Exercises 35–40.

Retrospective Studies

Sometimes it may be impractical or unethical to conduct an experiment. For example, suppose we want to study how marijuana use during pregnancy affects newborn babies. Because it is already known that marijuana can be harmful during pregnancy, it would be unethical to divide a sample of pregnant mothers randomly into two groups and then force the members of one group to use marijuana. However, we may be able to conduct a **retrospective study** (also called a *case-control study*), in which the participants naturally fall into two groups. In this example, the **cases** consist of mothers who used marijuana (by choice) during a past pregnancy, and the **controls** consist of mothers who did *not* use marijuana.

A retrospective study is observational because the researchers do not change the behavior of the participants, but it resembles an experiment because the cases effectively represent a treatment group and the controls represent a control group. (Sometimes, researchers plan ahead to look for case and control groups in the future, in which case the study is called *prospective* rather than retrospective.)

Definitions

A **retrospective study** (or *case-control study*) is an observational study that uses data from the past, such as official records or past interviews, and in which the sample naturally divides into a group of **cases** who engaged in the behavior under study and a group of **controls** who did not.

EXAMPLE 5 Which Type of Study?

For each of the following questions, what type of statistical study is most likely to lead to an answer? Why?

a. What is the average income of stockbrokers?
b. Do seat belts save lives?
c. Can lifting weights improve runners' times in 10-kilometer races?
d. Can a new herbal remedy reduce the severity of colds?

Solution

a. An *observational study* can tell us the average income of stockbrokers. We need only to survey (observe) the brokers.
b. It would be unethical to do an experiment in which some people were told to wear seat belts and others were told *not* to wear them. Instead, we can conduct a *retrospective study*. People who wore seat belts in crashes represent the cases and people who did not wear them are the controls. By comparing the death rates among the cases and controls, we can learn whether seat belts save lives. (They do.)
c. We need an *experiment* to determine whether lifting weights can improve runners' 10K times. One group of runners will be put on a weight-lifting program, and a control group will be asked to stay away from weights. We must try to ensure that all other aspects of their training are similar. Then we can see whether the runners in the weight-lifting group improve their times more than those in the control group. Note that we cannot use blinding in this experiment because there is no way to prevent participants from knowing whether they are lifting weights.
d. We should use a *double-blind experiment*, in which some participants get the actual remedy while others get a placebo. We need double-blind conditions because the severity of a cold may be affected by mood or other factors that experimenters might inadvertently influence.

▶ Now try Exercises 41–46.

Surveys and Opinion Polls

Surveys and opinion polls are observational studies and are among the most common types of statistical study. Survey and poll results should generally include a number called the **margin of error**, which is used as follows.

Suppose a poll finds that 46% of the public supports the President, with a margin of error of 3 percentage points. The 46% is a sample statistic; that is, 46% of the people in a sample said they support the President. The margin of error helps us understand how well this sample statistic is likely to approximate the true population parameter (in this case, the percentage of *all* Americans who support the President). By adding and subtracting the margin of error from the sample statistic, we find a range of values, or a **confidence interval**, that is *likely* to contain the population parameter. In this case, we add and subtract 3 percentage points to find a confidence interval from 43% to 49%.

BY THE WAY

Politicians and marketers often pretend they are trying to conduct a true opinion poll or survey when, in fact, they are deliberately trying to get particular results. These types of surveys are called *push polls* because they try to "push" people's opinions.

Definition

The **margin of error** in a statistical study is used to describe a **confidence interval** that is likely to contain the true population parameter. We find this interval by subtracting and adding the margin of error from the sample statistic obtained in the study. That is, as shown on the number line below, the confidence interval extends from (sample statistic − margin of error) to (sample statistic + margin of error).

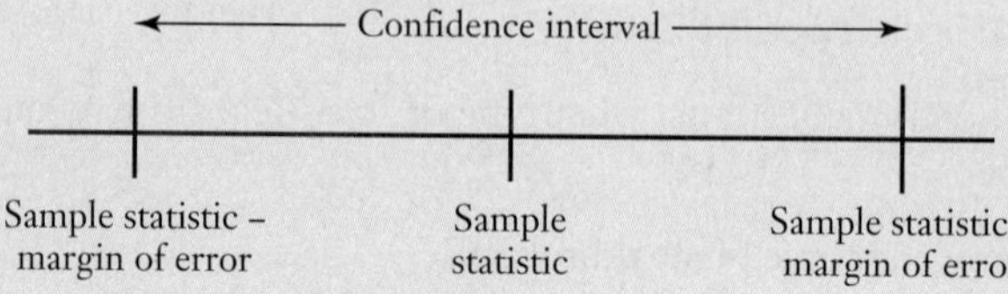

How confident can we be in a poll result? Unless we are told otherwise, we assume that the margin of error is defined to give us 95% confidence that the confidence interval contains the population parameter. We'll discuss the precise meaning of "95% confidence" in Unit 6D, but for now you can think of it as follows: If the poll were repeated 20 times with 20 different samples, 19 of the 20 polls (that is, 95% of the polls) would have a confidence interval that contains the true population parameter.

EXAMPLE 6 Close Election

An election eve poll finds that 52% of surveyed voters plan to vote for Smith, and she needs a majority (more than 50%) to win without a runoff. The margin of error in the poll is 3 percentage points. Should we assume she will win?

Solution We subtract and add the margin of error of 3 percentage points to find a confidence interval

$$\text{from } 52\% - 3\% = 49\% \quad \text{to} \quad 52\% + 3\% = 55\%$$

We can be 95% confident that the actual percentage of people planning to vote for her is between 49% and 55%. Because this confidence interval leaves open the possibility of both a majority and less than a majority, this election is too close to call.

▶ **Now try Exercises 47–50.**

Think About It In Example 6, suppose the poll found 65% of voters planning to vote for the candidate. Should she be confident of a win?

Quick Quiz 5A

Choose the best answer to each of the following questions. Explain your reasoning with one or more complete sentences.

1. You conduct a poll in which you randomly select 1000 registered voters from Texas and ask if they approve of the job their governor is doing. The *population* for this study is
   a. all registered voters in the state of Texas.
   b. the 1000 people that you interview.
   c. the governor of Texas.
2. Results of the poll described in Question 1 would most likely suffer from *bias* if you chose the participants from
   a. all registered voters in Texas.
   b. all people with a Texas driver's license.
   c. people who donated money to the governor's campaign.

3. When we say that a sample is *representative* of the population, we mean that
   a. the results found for the sample are similar to those we would find for the entire population.
   b. the sample is very large.
   c. the sample was chosen in the best possible way.
4. A poll concerning support for mass transit is conducted by interviewing people in 100 large cities, with 25 people randomly selected to be interviewed in each of those cities. What type of sampling was used?
   a. simple random sampling.
   b. systematic sampling.
   c. stratified sampling.
5. Consider an experiment designed to test whether cash incentives improve school attendance. The researcher chooses two groups of 100 high school students. She offers one group $10 for every week of perfect attendance. She tells the other group that they are part of an experiment but does not give them any incentive. The students who do *not* receive an incentive represent
   a. the treatment group.
   b. the control group.
   c. the observation group.
6. The experiment described in Question 5 is
   a. single-blind. b. double-blind. c. not blinded at all.
7. The purpose of a *placebo* is
   a. to prevent participants from knowing whether they belong to the treatment group or the control group.
   b. to distinguish between the cases and the controls in a retrospective study.
   c. to determine whether diseases can be cured without any treatment.
8. An experiment is single-blind if
   a. it lacks a treatment group.
   b. it lacks a control group.
   c. the participants do not know whether they belong to the treatment or the control group.
9. Poll X predicts that Powell will receive 49% of the vote, while Poll Y predicts that he will receive 52% of the vote. Both polls have a margin of error of 3 percentage points. What can you conclude?
   a. One of the two polls must have been conducted poorly.
   b. The two polls are consistent with each other.
   c. Powell will receive 50.5% of the vote.
10. A survey reveals that 12% of Americans believe Elvis is still alive, with a margin of error of 4 percentage points. The confidence interval for this poll is
   a. from 10% to 14%. b. from 8% to 16%.
   c. from 4% to 20%.

Exercises 5A

REVIEW QUESTIONS

1. Why do we say that the term *statistics* has two meanings? Give both meanings.
2. Define the terms *population, sample, population parameter,* and *sample statistics* as they apply to statistical studies.
3. Describe the five basic steps in a statistical study, and give an example of their application.
4. Why is it so important that a statistical study use a representative sample? Briefly describe four common sampling methods.
5. What is *bias*? How can it affect a statistical study? Give examples of several forms of bias.
6. Describe and contrast *observational studies* and *experiments*. What do we mean by the *treatment group* and the *control group* in an experiment? What do we mean by *cases* and *controls* in a *retrospective study*?
7. What is a *placebo*? Describe the *placebo effect*, and explain how it can make experiments difficult to interpret. How can making an experiment *single-blind* or *double-blind* help?
8. What is meant by the *margin of error* in a survey or opinion poll? How is this number used to identify a *confidence interval*?

DOES IT MAKE SENSE?

Decide whether each of the following statements makes sense (or is clearly true) or does not make sense (or is clearly false). Explain your reasoning.

9. In my statistical study, I used a sample that was larger than the population.
10. I followed all the guidelines for sample selection carefully, but my sample still did not reflect the characteristics of the population.
11. I wanted to test the effects of vitamin C on colds, so I gave the treatment group vitamin C and gave the control group vitamin D.
12. I don't believe the results of the experiment because the results were based on interviews and the study was not double-blind.
13. A pollster plans to improve survey results by conducting polls in which the margin of error is zero.

14. By choosing my sample carefully, I can make a good estimate of the average height of Americans by measuring the heights of only 500 people.

BASIC SKILLS & CONCEPTS

15–20: Population and Sample. For the following studies, identify the population, sample, population parameters, and sample statistics.

15. In order to gauge public opinion on the President's plan to contain Iran's nuclear program, the Pew Research Center surveyed 1001 Americans by telephone.

16. An AP/CBS telephone poll of 998 randomly selected Americans revealed that 6 in 10 of the respondents believe there has been progress in finding a cure for cancer in the last 30 years.

17. Astronomers typically determine the distance to a galaxy (a huge collection of billions of stars) by measuring the distances to just a few stars within it and taking the mean (average) of these distance measurements.

18. In a test of the effectiveness of garlic for lowering cholesterol, 47 adult subjects were treated with garlic in a processed tablet form. Cholesterol levels were measured before and after the treatment. The average (mean) change in the subjects' LDL cholesterol level was 3.2 mg/dL)

19. In an Accountemps survey of 150 senior executives, 47% said that the most common job interview mistake is to have little or no knowledge of the company where the applicant is being interviewed.

20. The Higher Education Research Institute conducts an annual study of attitudes of incoming college students by surveying approximately 241,000 first-year students at 340 colleges and universities. There are approximately 1.4 million first-year college students in this country.

21–26: Steps in a Study. Describe how you would apply the five basic steps of a statistical study to the following issues.

21. You want to determine what percentage of high school seniors regularly use a cell phone while driving.

22. A supermarket manager wants to determine whether the variety of products in her store meets customers' needs.

23. You want to know the percentage of American college students who attend home basketball games at their college.

24. You want to know the typical percentage of the bill that is left as a tip in restaurants.

25. You want to know the average lifetime of golden retrievers.

26. You want to know the average number of years required for college students to receive a bachelor's degree.

27. **Representative Sample?** You want to determine the average number of classes skipped by first-year students at a small college during a particular semester. State whether each of the following samples is likely to be representative and explain why or why not.

    Sample 1: 100 first-year students who belong to a sorority or fraternity

    Sample 2: 100 first-year students who play a varsity sport

    Sample 3: The first 100 first-year students whom you meet at the student union

    Sample 4: 100 first-year students taking honors humanities courses

28. **Representative Sample?** You want to determine the typical dietary habits of students at a college. State whether each of the following samples is likely to be representative and explain why or why not.

    Sample 1: Students in a single sorority

    Sample 2: Students majoring in public health

    Sample 3: Students who participate in intercollegiate sports

    Sample 4: Students enrolled in a required history class

29–34: Identify the Sampling Method. Identify the sampling method (simple random sampling, systematic sampling, convenience sampling, or stratified sampling) in the following studies.

29. An IRS (Internal Revenue Service) auditor randomly selects for audits 100 single taxpayers in each of the filing tax brackets.

30. *US Weekly* magazine predicts Grammy Award winners by polling readers who voluntarily respond by e-mail.

31. A study of the use of antidepressants selects 50 participants between the ages of 20 and 29, 50 participants between the ages of 30 and 39, and 50 participants between the ages of 40 and 49.

32. Every 100th cast iron pipe that comes off an assembly line is tested for strength.

33. Student ID numbers are randomly selected by a computer for a survey of student opinions on college athletics.

34. A taste test for chips and salsa is conducted at the entrance to a supermarket.

35–40: Type of Study. Determine whether each of the following studies described is observational or an experiment. If the study is an experiment, identify the control and treatment groups, and discuss whether single- or double-blinding is necessary. If the study is observational, state whether it is a retrospective study, and if so, identify the cases and controls.

35. A study at the University of Southern California separated 108 volunteers into groups, based on psychological tests designed to determine how often they lied and cheated. Those with a tendency to lie had different brain structures than those who did not lie (*British Journal of Psychiatry*).

36. A National Cancer Institute study of 716 melanoma patients and 1014 cancer-free patients matched by age, sex, and race found that those having a single large mole had twice the risk of melanoma. Having 10 or more moles was associated with a 12 times greater risk of melanoma (*Journal of the American Medical Association*).

37. A study of 2002 runners looked for relationships between specific types of running injuries and variables such as height, weight, body mass index, age, and running history (*British Journal of Sports Medicine*).

38. In a study of the effects of magnets on back pain, some subjects were treated with magnets while others were given nonmagnetic devices with a similar appearance (*Journal of the American Medical Association*). (The magnets did not appear to be effective in treating back pain.)

39. A study by Brown University social scientists found that 18.6% of first-year women students surveyed at an upstate New York college experienced sexual assault during that first year (*Journal of Adolescent Health*).

40. Using a survey of 1502 Americans, the Pew Research Center determined that 85% of Americans believe that the country is more politically divided than in the past and that those divisions will persist.

41–46: What Type of Study? What type of statistical study is most likely to lead to an answer to the following questions? If the study is an experiment, identify the control and treatment groups, and discuss whether single- or double-blinding is necessary. If the study is observational, state whether it is a retrospective study, and if so, identify the cases and controls.

41. Which of two available treatments is more effective in protecting trees from emerald ash borers?

42. Which of eight airlines has the highest customer satisfaction?

43. Over a period of many years, which National Basketball Association teams with high-altitude home courts have better records?

44. Is deep massage effective at relieving lower back pain?

45. Does taking a daily multivitamin reduce the incidence of strokes?

46. Do horoscopes based on birth dates give accurate predictions?

47–50: Margin of Error. The following summaries of statistical studies give a sample statistic and a margin of error. Find the confidence interval and answer any additional questions.

47. A poll is conducted the day before an election for state senator. There are two candidates running. The poll shows that 53% of the voters surveyed favor the Republican candidate, with a margin of error of 2.5 percentage points. Should the Republican be confident that she'll have a victory party? Why or why not?

48. A Gallup poll found that 36% of Americans favor a law banning the manufacture, sale, and possession of semi-automatic (assault) rifles (down from 60% in 2000). The margin of error was 4 percentage points. Is it reasonable to claim that one-third of Americans support such a law? Explain.

49. A national survey by the Pew Research Center of 1521 respondents reached on land lines and cell phones found that 46% of adults favored legalized abortion. The margin of error was 3 percentage points. Is it reasonable to claim that a majority of Americans opposed legalized abortion (at the time of the survey)? Explain.

50. In a survey of 1002 people, 701 (which is 70%) said that they voted in the most recent presidential election (based on data from ICR Research Group). The margin of error for the survey was 3 percentage points. However, actual voting records show that only 61% of all eligible voters actually did vote. Does this necessarily imply that people lied when they answered the survey? Explain.

FURTHER APPLICATIONS

51. **Effectiveness of a New Drug.** As part of the U.S. Food and Drug Administration's approval process, a new rheumatoid arthritis drug was compared to a placebo. The randomized, double-blind study with 482 patients showed that 41% of those given the new drug had a decrease in symptoms, while 19% of those given the placebo experienced improvement (*Western Journal of Medicine*).

a. Which patients were in the treatment group and the control group?

b. Do the results appear to offer evidence that the new drug was effective? Why or why not?

c. Do the results appear to indicate that a placebo effect was present in these trials? Explain.

d. If you were on the panel deciding whether to approve the new drug, how would you vote based on this study? Explain your reasoning.

52–57: Real Studies. Consider the following statistical studies.

a. Identify the population and the population parameter of interest.

b. Describe the sample and sample statistic for the study.

c. Identify the type of study.

d. Discuss what additional facts you would like to know before you believed the study or acted on the results of the study.

52. A study done at the Center for AIDS and STD at the University of Washington tracked the survival rates of 17,517 asymptomatic North American patients with HIV who started drug therapy at different points in the progression of the infection. It was discovered that asymptomatic patients who postponed antiretroviral treatment until their disease was more advanced faced a higher risk of dying than those who had initiated drug treatment earlier (*New England Journal of Medicine*).

53. A Fox News poll of 900 registered voters found that 19% of Americans "regift" (give gifts that they received as gifts). Women (21%) are more likely to regift than men (16%), and the results are nearly independent of income. The margin of error was 3 percentage points.

54. The phenomenon of *stereotype threat* occurs when people, often women or minorities, are reminded before taking a test that the test will be used to evaluate intellectual or academic abilities. Studies show that test performance declines

significantly when such reminders are given compared to performance when they are not given (*Psychological Science*, 15, 829–836).

55. A Fox News poll carried out by phone interviews of 1006 registered American voters (434 land lines and 572 cell phones) found that 84% of the voters thought that fake news was hurting the country (61% "very concerned" and 23% "somewhat concerned") and 79% felt confident that they could identify fake news when they read it.

56. A study of acupuncture used in the relief of chronic non-specific low back pain compared acupuncture treatments to "usual care" in a sample of 241 patients. The researchers reported "modest benefit to health" for acupuncture patients and a slightly higher cost than usual care treatments (*The BMJ* [formerly *British Medical Journal*]).

57. In nearly every year since 1974, the General Social Survey has asked this question of Americans:

> *The U.S. Supreme Court has ruled that no state or local government may require the reading of the Lord's Prayer or Bible verses in public schools. Do you approve or disapprove of the court ruling?*

In 1974, 30.8% of respondents approved of the ruling and 66.1% disapproved. In 2014, 39.1% of respondents approved of the ruling and 56.7% disapproved.

58. **The General Social Survey.** The General Social Survey (GSS), established in 1972 by the National Opinion Research Center (NORC) at the University of Chicago, has tracked trends in opinions and attitude for over 40 years. Visit the GSS site (gss.norc.org), go to "GSS Data Explorer" and then "Explore GSS Data," enter a keyword or phrase, and choose a survey topic. You can then see the survey question and data for all the years in which that question has been asked.

a. Use the GSS site to find survey results for the question "Do you believe in life after death?" How have the percentages of people who answered "Yes," "No," and "Don't know" to this question changed between 1973 and 2014?

b. Use the GSS site to find survey results for the question "Do you think the use of marijuana should be made legal or not?" How has the percentage of people who answered "Yes" to this question changed between 1973 and 2014?

c. Find a survey question that interests you. Report on the results you find and compare them with your expected results.

IN YOUR WORLD

59. **Statistics in the News.** Select three news stories from the past week that involve statistics in some way. For each case, write one or two paragraphs describing the role of statistics in the story.

60. **Statistics in Your Major.** Write two to three paragraphs describing the ways in which you think the science of statistics is important in your major field of study. (If you have not chosen a major, answer this question for a major that you are considering.)

61. **Statistics in Sports.** Home field (or home court or home ice) advantage is real. Find at least two of many analyses of home field advantage and report on your findings. In what sport (baseball, football, basketball, or soccer) is home field advantage the greatest? What is the most plausible explanation for home field advantage?

62. **Sample and Population.** Find a report in today's news concerning any type of statistical study. What is the population being studied? What is the sample? Why do you think the sample was chosen as it was?

63. **Poor Sampling.** Find a news article about a study that attempts to describe some characteristic of a population, but that you believe involved poor sampling (for example, a sample that was too small or not representative of the population under study). Describe the population, the sample, and what you think was wrong with the sample. Briefly discuss how you think the poor sampling affected the study results.

64. **Good Sampling.** Find a recent news article that describes a statistical study in which the sample was well chosen. Describe the population, the sample, and why you think the sample was a good one.

65. **Margin of Error.** Find a report of a recent survey or poll. Interpret the sample statistic and margin of error quoted for the survey or poll.

TECHNOLOGY EXERCISES

Answer the following questions using procedures described in the Using Technology boxes in this unit or with **StatCrunch** (available in MyLab Math).

66–67. Random Numbers. Use a calculator, Excel, or StatCrunch to generate the random numbers for the following exercises.

66. Generate and write down each list of numbers:

a. Ten random numbers between 0 and 1

b. Ten random numbers between 0 and 10

c. Ten random numbers between 1 and 2

d. Ten random numbers between 10 and 20

67. a. Generate ten random numbers between 0 and 1. What is the mean of the numbers (to find this, divide the sum of the numbers by 10)?

b. Generate three more lists of ten random numbers between 0 and 1 and find the mean for each list. How do the four means (including the one for your first set) compare?

c. Without carrying out the calculation, what number do you think the average of 1000 random numbers between 0 and 1 is near? Explain.

UNIT 5B Should You Believe a Statistical Study?

Most statistical research is carried out with integrity and care. Nevertheless, bias can arise in many ways, making it important to examine reports of statistical research carefully. In this unit, we discuss eight guidelines that can help you answer the question "Should I believe a statistical study?" The following box summarizes the guidelines.

Eight Guidelines for Evaluating a Statistical Study

1. *Get a Big Picture View of the Study.* You should understand the goal of the study, the population that was under study, and whether the study was observational or an experiment.
2. *Consider the Source.* Look for potential sources of bias on the part of the researchers.
3. *Look for Bias in the Sample.* Decide whether the sampling method was likely to produce a representative sample.
4. *Look for Problems in Defining or Measuring the Variables of Interest.* Ambiguity in the variables can make it difficult to interpret reported results.
5. *Beware of Confounding Variables.* If the study neglected potential confounding variables, its results may not be valid.
6. *Consider the Setting and Wording in Surveys.* Look for anything that might tend to produce inaccurate or dishonest responses.
7. *Check That Results Are Presented Fairly.* Check whether the study supports the conclusions that are presented in the media.
8. *Stand Back and Consider the Conclusions.* Evaluate whether the study achieved its goals. If so, do the conclusions make sense and have practical significance?

Guideline 1: Get a Big Picture View of the Study

Before evaluating the details of a statistical study, we must know what it is about. A good starting point for gaining a big picture view is to answer these basic questions:

- What was the goal of the study?
- What was the population under study? Was the population clearly and appropriately defined?
- Was the study observational or an experiment? If it was an experiment, was it single- or double-blind, and were the treatment and control groups properly randomized? Given the goal, was the type of study appropriate?

EXAMPLE 1 Appropriate Type of Study?

Imagine the following (hypothetical) news report: "Researchers gave 100 participants their astrological horoscopes and asked whether the horoscopes appeared to be accurate; 85% of the participants answered *yes* (the horoscopes were accurate). The researchers concluded that horoscopes are valid most of the time." Analyze this study according to Guideline 1.

Solution The goal of the study was to determine the validity of horoscopes. Based on the news report, it appears that the study was *observational:* The researchers

BY THE WAY
Surveys show that nearly half of Americans believe their horoscopes. However, in controlled experiments, the predictions of horoscopes come true no more often than would be expected by chance.

simply asked the participants about the accuracy of the horoscopes. However, because the accuracy of a horoscope is somewhat subjective, this study should have been a controlled experiment in which some people were given their actual horoscope and others were given a fake horoscope. Then the researchers could have looked for differences between the two groups. Moreover, because researchers could easily influence the results by how they questioned the participants, the experiment should have been double-blind. In summary, the type of study was inappropriate for the goal and its results are meaningless.

▶ Now try Exercises 9–10.

Guideline 2: Consider the Source

Statistical studies are supposed to be objective, but the people who carry them out and fund them may be biased. Always be sure to consider the source of a study and evaluate the potential for biases that might invalidate the study's conclusions.

EXAMPLE 2 Is Smoking Healthy?

By 1963, enough research on the health dangers of smoking had accumulated that the Surgeon General of the United States publicly announced that smoking is bad for health. Research done since that time has built further support for this claim. However, while the vast majority of studies show that smoking is unhealthy, a few past studies found no dangers from smoking, and perhaps even health *benefits*. These studies generally were carried out by the Tobacco Research Institute, funded by the tobacco companies. Analyze the Tobacco Research Institute studies according to Guideline 2.

Solution Tobacco companies have a financial interest in minimizing the dangers of smoking. Because the studies carried out at the Tobacco Research Institute were funded by the tobacco companies, there may have been pressure on the researchers to produce results to the companies' liking. This *potential* for bias does not mean that the research *was* biased, but the fact that it contradicted virtually all other research on the subject gave reason for concern.

▶ Now try Exercises 11–12.

BY THE WAY
After decades of arguing to the contrary, in 1999 the Philip Morris Company—the world's largest seller of tobacco products—publicly acknowledged that smoking causes lung cancer, heart disease, emphysema, and other serious diseases. Shortly thereafter, Philip Morris changed its name to Altria.

Guideline 3: Look for Bias in the Sample

Look for bias that may prevent the sample from being representative of the population. The following two forms of bias are particularly common in sample selection.

Bias in Choosing a Sample

Selection bias (or a *selection effect*) occurs whenever researchers *select* their sample in a way that would tend to make it unrepresentative of the population. For example, a pre-election poll that surveys only registered Republicans has selection bias because it is unlikely to reflect the opinions of all voters.

Participation bias occurs whenever people *choose* whether to participate. For example, if participants must take action to participate in a survey rather than being selected at random, those who feel more strongly about the survey issue are more likely to participate. (Surveys or polls in which people choose whether to participate are often called *self-selected* or *voluntary response surveys*.)

CASE STUDY
The 1936 *Literary Digest* Poll

The *Literary Digest,* a popular magazine of the 1930s, successfully predicted the outcomes of several elections using large polls. In 1936, editors of the *Literary Digest* conducted a particularly large poll in advance of the presidential election. They randomly chose a sample of 10 million people from various lists, including names in telephone books and rosters of country clubs. They mailed a postcard "ballot" to each of these 10 million people. About 2.4 million people returned the postcard ballots. Based on the returned ballots, the editors of the *Literary Digest* predicted that Alf Landon would win the presidency by a margin of 57% to 43% over the incumbent Franklin D. Roosevelt. Instead, Roosevelt won with 62% of the popular vote. How did such a large survey go so wrong?

The sample suffered from both selection bias and participation bias. The selection bias arose because the *Literary Digest* chose its 10 million names in ways that favored affluent people. For example, selecting names from telephone books favored those who could afford telephones, which were relatively expensive in 1936. Similarly, country club members are usually wealthy. The selection bias favored the Republican Landon because affluent voters of the 1930s tended to vote for Republican candidates.

The participation bias arose because return of the postcard ballots was voluntary, so people who felt strongly about the election were more likely to return their postcard ballots. This bias also tended to favor Landon because he was the challenger—people who did not like President Roosevelt could express their desire for change by returning the postcards. Together, the two forms of bias made the sample results useless, despite the large number of people surveyed.

HISTORICAL NOTE

A young pollster named George Gallup conducted his own survey prior to the 1936 presidential election. Using interviews with only 3000 randomly selected people at a time, he correctly predicted the outcome of the election and even gained insights into how the *Literary Digest* poll went wrong. Gallup went on to establish a very successful polling organization.

EXAMPLE 3 Comparing Polling Techniques

Look back at the chapter-opening question (p. 292), which offered five possible choices for a survey on Americans' views of a new health care plan. We already know that the correct answer is A, the carefully conducted poll of only 1000 Americans. Explain why each of the other choices would *not* be expected to produce representative results. Then explain why A is the correct choice.

Solution The problems with choices B through E all involve various forms of bias. Let's start with E, the special election. While an election with *large* voter turnout is arguably the best possible survey of public opinion, special elections tend to draw small turnouts, and a special election held solely to gauge public opinion would likely draw even fewer voters than one with real stakes. The special election therefore suffers from participation bias, and those who do vote will tend to be people with the strongest opinions. Choice D, the online poll with 2 million participants, also suffers from participation bias because people choose whether to participate in online polls. The results of choices D and E are no more likely to be accurate than the prediction of the *Literary Digest* poll in 1936.

Choices C and B both suffer from selection bias. Choice C (the interviews conducted at grocery stores at 9 a.m. on a Monday) is essentially a convenience sample (sampling people who were shopping at the time), but it is not likely to represent the population because Monday-morning interviews at grocery stores will tend to overrepresent stay-at-home parents and underrepresent people who work a standard business week. In addition, this survey used volunteer interviewers, which may be a concern because interviewers generally need training to ensure that they are objective. Choice B uses professional interviewers but selects participants from people who had been hospitalized in the past year. This sample introduces selection bias, likely weighted toward older

BY THE WAY

Self-selection has become a crisis for legitimate pollsters. Decades ago, most people selected for a poll agreed to participate, and as recently as the late 1990s, the response rates were above 35%. Today, however, response rates for most polls have fallen below 10%. This may explain why recent pre-election polls have often proved unreliable.

and less healthy people. Moreover, recently hospitalized people could have a different view of the health care system than those in the general population.

Therefore, despite including far fewer people than any of the other options, choice A is likely to produce the most representative result. In fact, using techniques discussed in Unit 6D, the carefully conducted poll of 1000 Americans turns out to have a margin of error of less than about 4 percentage points, which is quite good despite the seemingly small poll size.

▶ Now try Exercises 13–14.

Guideline 4: Look for Problems in Defining or Measuring the Variables of Interest

Statistical studies usually attempt to measure *something,* and we call the things being measured the **variables of interest** in the study. The term *variable* simply refers to an item or quantity that can vary or take on different values. For example, variables in the Nielsen ratings include *show being watched* and *number of viewers.*

Definition

A **variable** is any item or quantity that can vary or take on different values.

The **variables of interest** in a statistical study are the items or quantities that the study seeks to measure.

Results of a statistical study may be especially difficult to interpret if the variables being studied are difficult to define or measure. For example, imagine trying to conduct a study of how exercise affects resting heart rates. The variables of interest would be *amount of exercise* and *resting heart rate*, but both variables are difficult to define and measure. For example, does *amount of exercise* include time spent walking to class? Unless the definitions are clearly specified, the results of a study will be difficult to evaluate.

Think About It What are the challenges of defining and measuring *resting heart rate?*

EXAMPLE 4 Can Money Buy Love?

A Roper survey of randomly selected individuals who are among the wealthiest 1% of Americans found that these people would pay an average of $487,000 for "true love," $407,000 for "great intellect," $285,000 for "talent," and $259,000 for "eternal youth." Analyze this result according to Guideline 4.

Solution The variables in this study are difficult to define. How, for example, do you define "true love"? And does it mean true love for a day, a lifetime, or something else? Similarly, does the ability to balance a spoon on your nose constitute "talent"? Because the variables are so poorly defined, it's likely that different people interpreted them differently, which would make the survey results meaningless.

▶ Now try Exercise 15.

EXAMPLE 5 Illegal Drug Supply

A commonly quoted statistic is that law enforcement authorities succeed in stopping only about 10% of the illegal drugs entering the United States. Should you believe this statistic?

Solution There are essentially two variables in the study: *quantity of illegal drugs intercepted* and *quantity of illegal drugs NOT intercepted.* It should be relatively easy to measure the quantity of illegal drugs that law enforcement officials intercept. However, because the drugs are illegal, it's unlikely that anyone is reporting the quantity of drugs that are *not* intercepted. While there may be data (such as from arrests, overdoses, etc.) that could be carefully evaluated to construct an estimate of the illegal drug supply, a number as specific as 10% seems more likely to be one that a police officer was quoted as calling "PFA," for "pulled from the air."

▶ Now try Exercise 16.

Guideline 5: Beware of Confounding Variables

Variables that are *not intended* to be part of the study can make it difficult to interpret results properly. Such variables are often called **confounding variables**, because they confound (confuse) a study's results.

It's not easy to discover confounding variables. Sometimes they are discovered years after a study was completed, and sometimes they are not discovered at all. But some confounding variables are fairly obvious and can be discovered simply by thinking hard about factors that may have influenced a study's results.

EXAMPLE 6 Radon and Lung Cancer

Radon is a radioactive gas produced by natural processes (the decay of uranium) in the ground. The gas can leach into buildings through the foundation and can accumulate in relatively high concentrations if doors and windows are closed. Imagine a study that seeks to determine whether radon gas causes lung cancer by comparing the lung cancer rate in Colorado, where radon gas is fairly common, with the lung cancer rate in Hong Kong, where radon gas is less common. Suppose the study finds that the lung cancer rates in the two sites are nearly the same. Is it fair to conclude that radon is *not* a significant cause of lung cancer?

Solution The variables under study are *amount of radon* and *lung cancer rate*. However, radon gas is not the only possible cause of lung cancer. For example, smoking is a known cause of lung cancer, and the smoking rate in Hong Kong is much higher than the smoking rate in Colorado. *Smoking rate* is therefore a confounding variable in this study, so any conclusions about radon and lung cancer will be invalid unless the researchers also take the smoking rate into account. In fact, careful studies show that radon gas *can* cause lung cancer, and the U.S. Environmental Protection Agency (EPA) recommends taking steps to prevent radon from building up indoors.

▶ Now try Exercises 17–18.

BY THE WAY

Many hardware stores sell simple kits that you can use to test whether radon gas is accumulating in your home. If it is, the problem can be eliminated by installing a "radon mitigation" system, which usually consists of a fan that blows the radon out from under the house before it can get in.

Guideline 6: Consider the Setting and Wording in Surveys

Even when a survey is conducted with proper sampling and with clearly defined terms and questions, you should watch out for problems with the setting or wording that might produce inaccurate or dishonest responses. Dishonest responses are particularly likely when the survey concerns sensitive subjects, such as personal habits or income. For example, the question "Do you cheat on your income taxes?" is unlikely to elicit honest answers from those who cheat, especially if the setting does not guarantee complete confidentiality.

Sometimes just the order of the words in a question can affect the outcome. A poll conducted in Germany asked people in two randomly selected groups one of the following two questions:

- *Would you say that traffic contributes more or less to air pollution than industry does?*
- *Would you say that industry contributes more or less to air pollution than traffic does?*

The only difference is the order of the words *traffic* and *industry*, but this difference dramatically changed the results: In the group asked the first question, 45% answered traffic and 32% answered industry. In the group asked the second question, only 24% answered traffic while 57% answered industry.

BY THE WAY

People are more likely to choose the item that comes first in a survey because of what psychologists call the *availability error*—the tendency to make judgments based on what option is perceived as most *available*. Professional polling organizations take great care to avoid this problem; for example, they may pose a question with two answer choices in one order to half the people in the sample and in the opposite order to the other half.

IN YOUR WORLD

The Gun Debate: Defensive Gun Use

It's hard to escape the debate over guns in America, and both sides of the debate often quote statistics that can be very difficult to verify or interpret. So in forming your own opinions on gun issues, it's important to understand both the meaning and uncertainty of the statistics you hear. As an example, let's consider statistics on what is called defensive gun use, or DGU for short, which is the use of a gun in self-defense or defense of others (to save a life or prevent a crime). Researchers have been trying for at least 20 years to determine the annual number of DGUs, but different researchers have come up with estimates varying over a huge range (from about 80,000 to 2.5 million DGUs per year). If the number of DGUs is large, as gun rights advocates claim, it justifies self-defense as a reason to carry a firearm. If the number of DGUs is small, or if DGUs often have tragic consequences, as gun control supporters claim, then these statistics offer a reason to restrict firearms.

Why do the statistics vary over such a wide range? We can understand some of the reasons by applying the guidelines of this unit. Applying Guideline 1, we recognize that the only plausible way to estimate the number of DGUs is with an observational study in which gun owners are surveyed or police or news media records of DGUs are examined. This fact quickly raises issues of bias in the sample (Guideline 3): For surveys, it may be difficult to identify a representative sample of gun owners, while police and news media records usually depend on voluntary reporting. Guideline 4 comes into play in defining the variable of interest: deciding what counts as a DGU. For example in self-reporting of DGUs, different people may decide differently whether an incident in which a gun was present or brandished (rather than fired) should be counted, or whether a criminal's use of a gun in self-defense should count as a DGU. Confounding variables (Guideline 5) can arise in determining who is on the defense in situations in which both parties have guns, and the survey question itself will introduce the possibility of inaccurate or dishonest responses (Guideline 6). For example, the contentious nature of the topic might lead those who support gun rights to exaggerate their use of guns for self-defense. And even those who are trying to be honest may give inaccurate responses because of a common cognitive effect called *telescoping*, in which people incorrectly recall the timing of a past event; in this case, telescoping may cause a respondent to claim that a DGU occurred during the last year when it actually occurred further in the past.

These and other difficulties explain the wide range of estimates of DGUs, which brings us to our final two guidelines. You need to look carefully at any results of surveys on gun issues that you read about and decide whether you think they have been presented fairly (Guideline 7). Then you should step back and carefully consider the conclusions (Guideline 8), to decide whether they affect your own opinions about guns.

EXAMPLE 7 **Do You Want a Tax Cut?**

The Republican National Committee commissioned a poll to find out whether Americans supported its proposed tax cuts. Asked "Do you favor a tax cut?" a large majority answered yes. Should we conclude that Americans supported the proposal?

Solution A question like "Do you favor a tax cut?" is biased because it does not give other options (much like the fallacy of *limited choice* discussed in Unit 1A). In fact, other polls conducted at the same time showed a similarly large majority expressing great concern about federal deficits. Indeed, support for the tax cuts was far lower when the question was asked by independent organizations in the form "Would you favor a tax cut even if it increased the federal deficit?"

▶ Now try Exercises 19–20.

Guideline 7: Check That Results Are Presented Fairly

Even when a statistical study is done well, it may be misrepresented in graphs or concluding statements. Researchers may occasionally misinterpret the results of their own studies or jump to conclusions that are not supported by the results, particularly when they have personal biases toward certain interpretations. In other cases, reporters may misinterpret a survey or jump to unwarranted conclusions that make a story seem more spectacular. Misleading graphs are an especially common problem (see Unit 5D). In general, you should look for inconsistencies between the interpretation of a study (in pictures and words) and any actual data given with it.

EXAMPLE 8 Does the School Board Need a Statistics Lesson?

The school board in Boulder, Colorado, created a hubbub when it announced that 28% of Boulder school children were reading "below grade level" and hence concluded that methods of teaching reading needed to be changed. The announcement was based on reading tests on which 28% of Boulder school children scored below the national average for their grade. Do these data support the board's conclusion?

Solution The fact that 28% of Boulder children scored below the national average for their grade implied that 72% scored at or above the national average. Therefore, the school board's ominous statement about students reading "below grade level" made sense only if "grade level" meant the national average score for a particular grade. But this interpretation of "grade level" would imply that half the students in the nation are always below grade level, no matter how high the scores. Teaching methods may have needed improvement, but these data did not justify that conclusion.

▸ Now try Exercises 21–22.

Guideline 8: Stand Back and Consider the Conclusions

Finally, even if a study seems reasonable based on consideration of the first seven guidelines, you should stand back and consider the conclusions. Ask yourself questions such as these:

- Did the study achieve its goals?
- Do the conclusions make sense?
- Can you rule out alternative explanations for the results?
- If the conclusions do make sense, do they have any practical significance?

EXAMPLE 9 Practical Significance

An experiment is conducted in which the weight losses of people who try a new "Fast Diet Supplement" are compared to the weight losses of a control group of people who try to lose weight in other ways. After eight weeks, the results show that the treatment group lost an average of one-half pound more than the control group. Assuming that it has no dangerous side effects, does this study suggest that the Fast Diet Supplement is a good treatment for people wanting to lose weight?

An extraordinary claim requires extraordinary proof.
—Marcello Truzzi

Solution Compared to the average person's body weight, an additional weight loss of one-half pound hardly matters at all. So while the results of this study may be accurate and interesting, they don't seem to have much practical significance.

▸ Now try Exercises 23–26.

Quick Quiz 5B

Choose the best answer to each of the following questions. Explain your reasoning with one or more complete sentences.

1. You read about an issue that was the subject of an observational study when it clearly should have been investigated with a double-blind experiment. The results from the observational study are therefore

   a. still valid, but a little less reliable.

   b. valid, but only if you first correct for the fact that the wrong type of study was done.

   c. essentially meaningless.

2. A study conducted by the oil company British Petroleum shows that there was no lasting damage from a large oil spill in Gulf of Mexico. This conclusion

   a. is definitely invalid, because the study was biased.

   b. may be correct, but the potential for bias means that you should look very closely at how the conclusion was reached.

   c. could be correct if it falls within the confidence interval of the study.

3. Consider a study designed to learn about the social networks of first-year college students in the United States, in which researchers randomly interviewed students living in on-campus dormitories. The way this sample was chosen means the study will suffer from

   a. selection bias.

   b. participation bias.

   c. confounding variables.

4. The show *The Voice* selects winners based on votes cast by anyone who wants to vote. This means that the winner

   a. is the person most Americans want to win.

   b. may or may not be the person most Americans want to win, because the voting is subject to participation bias.

   c. may or may not be the person most Americans want to win, because the voting should have been double-blind.

5. Consider an experiment in which you measure the weights of 6-year-olds. The variable of interest in this study is

   a. the size of the sample.

   b. the weights of 6-year-olds.

   c. the ages of the children under study.

6. Consider a survey in which 1000 people are asked "How often do you go to the dentist during a year?" The variable of interest in this study is

   a. the number of annual visits to the dentist.

   b. the 1000-person size of the sample.

   c. the quality of dental care.

7. Imagine that a survey of randomly selected people finds that people who used sunscreen were *more* likely to have been sunburned in the past year. Which explanation for this result seems most likely?

   a. Sunscreen is useless.

   b. The people in the study all used sunscreen that had passed its expiration date.

   c. People who use sunscreen are more likely to spend time in the sun.

8. You want to know whether people prefer Smith or Jones for mayor, and you are considering two possible ways to word the survey question. Wording X is "Do you prefer Smith or Jones for mayor?" Wording Y is "Do you prefer Jones or Smith for mayor?" (That is, the names are reversed in the two wordings.) The best approach is to

   a. use Wording X for everyone.

   b. use the same wording for everyone—it doesn't matter whether it is Wording X or Wording Y.

   c. use Wording X for half the people and Wording Y for the other half

9. A *self-selected survey* is one in which

   a. the people being surveyed decide which question to answer.

   b. people decide for themselves whether to be part of the survey.

   c. the people who design the survey are also the survey participants.

10. If a statistical study is carefully conducted in every possible way, then

    a. its results must be correct.

    b. we can have confidence in its results, but it is still possible that they are not correct.

    c. we say that the study is perfectly biased.

Exercises 5B

REVIEW QUESTIONS

1. Briefly describe each of the eight guidelines for critically evaluating statistical studies. Give an example to which each guideline applies.
2. Describe and contrast *selection bias* and *participation bias* in sampling. Give an example of each.
3. What do we mean by *variables of interest* in a study?
4. What are *confounding variables*, and what problems can they cause?

DOES IT MAKE SENSE?

Decide whether each of the following statements makes sense (or is clearly true) or does not make sense (or is clearly false). Explain your reasoning.

5. More than 1 million people sent texts in response to a TV survey question, so this survey result is more valid than a survey of a random sample of 200 people.
6. The survey of religious beliefs suffered from selection bias because the questionnaires were handed out only at Catholic churches.
7. My experiment proved that vitamin C reduces the severity of colds, because I controlled the experiment for every confounding variable.
8. Everyone who jogs for exercise should try the new training regimen, because careful studies suggest it can increase your speed by up to 1%.

BASIC SKILLS AND CONCEPTS

9–20: Should You Believe This Study? Based solely on the information given, do you have reason to question the results or the methods used in the following hypothetical studies? Explain your reasoning.

9. A survey of hourly wages of fast-food workers in a large city samples 20 workers from each of ten fast-food restaurants.
10. An experimental, double-blind study investigates whether people who skip lunch are more likely to feel tired in the afternoon.
11. A study by the liberal Center for American Progress is designed to assess a new Republican budget plan.
12. A study financed by a major pharmaceutical company is intended to determine whether its new high blood pressure drug is more effective than similar drugs of competing companies.
13. A TV talk show host asks the TV audience to text "1" if they support and "2" if they oppose a law requiring background checks for all purchases of firearms.
14. A state Democratic Party polls its members to determine whether its candidate for governor is likely to win against the Republican candidate.
15. Researchers design five survey questions to determine whether Europeans are more optimistic than Americans.
16. A government study is designed to determine the percentage of taxpayers who understate their income, based on people who had their tax returns audited.
17. Researchers conclude that body mass index is the single most important factor in predicting the performance of world-class swimmers.
18. In a study of obesity among children, researchers monitor the eating and exercise habits of the participating children, carefully recording everything they eat and all their activity.
19. Sociologists studying alcohol abuse circulate a questionnaire asking each respondent "Did you drink to excess in the past week?"
20. To gauge public opinion on whether there should be a constitutional amendment to ban capital punishment, a survey asked people, "Do you support legalized murder?"

21–26: Should You Believe This Claim? Based solely on the information given about the following hypothetical studies, decide whether you would believe the stated claim. Justify your conclusion.

21. An educational research group that tracks tuition rates finds that tuition at a particular small college is 50% more than it was 10 years ago.
22. A new diet program claims that 200 randomly selected participants lost up to 15 pounds in six weeks and that the program works for anyone with enough discipline.
23. Citing a higher incidence of deaths due to binge drinking among first-year students, the college president claims that banning drinking in student housing will save lives.
24. A spokesperson for a Major League baseball team claims that the average attendance at home games is 45,236, up 12% over the previous season.
25. The local Chamber of Commerce claims that the average number of employees among all businesses in town is 12.5.
26. A car manufacturer claims that its newest model of low-emission diesel car has 20% lower emissions than competing models.

FURTHER APPLICATIONS

27–34: Bias. Identify at least one potential source of bias in the following studies or claims. Explain why the bias would or would not affect your view of the study or claim.

27. The White House claimed that 1.5 million people attended the 2016 inauguration.
28. Based on a survey of 2718 people, the National Opinion Research Center concluded that 32% of Americans *always* make a special effort to sort and recycle glass, cans, plastic, or papers, and 24% *often* make such an effort.
29. Grocery shoppers on a Saturday morning are invited to taste-test orange juice and vote for their favorite brand.
30. An article in *Journal of Nutrition* noted that chocolate is rich in flavonoids. The article reports that "regular consumption of foods rich in flavonoids may reduce the risk of coronary heart disease." The study received funding from Mars, Inc., the candy company, and the Chocolate Manufacturers Association.

31. Based on a Pew Research Center survey of 35,000 American adults, the percentages of four-year college degree holders among Hindus, atheists, Muslims, Catholics, and all American adults are 77%, 43%, 39%, 27%, and 26%, respectively.

32. According to a *New York Times*/CBS News poll, 60% of baseball fans are bothered by steroid use by players and 44% say that those who used steroids should not be allowed in the Baseball Hall of Fame.

33. Based on interviews with about 175,000 adults, a Gallup poll found that Mississippi is the most religious state (59% claimed to be very religious) and Vermont is the least religious (21% claimed to very religious).

34. A study (published in the *Canadian Medical Association Journal*) of 20 nations discovered that Germany has the highest average (mean) number of annual visits to a doctor (8.5), while Finland has the fewest (3.2).

35. **It's All in the Wording.** Princeton Survey Research Associates did a study for *Newsweek* magazine illustrating the effects of wording in a survey. Two questions were asked:

- *Do you personally believe that abortion is wrong?*
- *Whatever your own personal view of abortion, do you favor or oppose a woman in this country having the choice to have an abortion with the advice of her doctor?*

To the first question, 57% of the respondents replied yes and 36% responded no. In response to the second question, 69% of the respondents favored allowing women to have the choice and 24% opposed allowing women to have the choice. Discuss why the two questions produced seemingly contradictory results. How could the results of the questions be used selectively by various groups?

36. **It's All in the Wording.** The Pew Research Center reports the following survey results for two questions on the same topic.

- When asked "Do you favor or oppose taking military action in Iraq to end Saddam Hussein's rule?" 68% favored military action while 25% opposed it.
- When asked "Do you favor or oppose taking military action in Iraq to end Saddam Hussein's rule even if it meant that U.S. forces might suffer thousands of casualties?" 43% favored military action and 48% opposed it.

Discuss why the two questions produced such different results. Do you feel that either set of results is accurate? If not, how would you ask the question to obtain more accurate results?

37–42: Stat-Bites. Much like sound bites of news stories, statistical studies are often reduced to one- or two-sentence "stat-bites." For the following stat-bites taken from various news sources, discuss what crucial information is missing and what more you would want to know before acting on the study.

37. *USA Today* reports that more than 60% of adults avoid visits to the dentist because of fear.

38. A Fox News poll reveals that of 77% Americans say "Merry Christmas" rather than "Happy Holidays."

39. CNN reports on a Zagat survey of America's top restaurants, which found that "only nine restaurants achieved a rare 29 out of a possible 30 rating and none of those restaurants is in the Big Apple."

40. According to a Netflix study, 48% of American couples are "cheaters": They promise each other to watch a film together, but one of them first watches it alone.

41. According to *USA Today*, 26% of Americans rate the potato as their favorite vegetable, making it the most popular vegetable.

42. Thirty percent of newborns in India would qualify for intensive care if they were born in the United States.

43–44: Accurate Headlines? Consider the following headlines, each followed by a brief summary of a study. Discuss whether the headline accurately represents the study.

43. Headline: "Drugs shown in 98 percent of movies"
Story summary: A "government study" claims that drug use, drinking, or smoking was depicted in 98% of the top movie rentals (Associated Press).

44. Headline: "Sex more important than jobs"
Story summary: A survey found that 82% of 500 people interviewed by phone ranked a satisfying sex life as important or very important, while 79% ranked job satisfaction as important or very important (Associated Press).

45. **What Is the Question?** Discuss the differences between the following questions, each of which could be the basis for a statistical study.

- What percentage of Internet dates lead to marriage?
- What percentage of marriages begin with Internet dates?

46. **Exercise and Dementia.** A recent study in the *Annals of Internal Medicine* was summarized by the Associated Press, in part, as follows:

> *The study followed 1740 people aged 65 and older who showed no signs of dementia at the outset. The participants' health was evaluated every two years for six years. Out of the original pool, 1185 were later found to be free of dementia, 77 percent of whom reported exercising three or more times a week; 158 people showed signs of dementia, only 67 percent of whom said they exercised that much. The rest either died or withdrew from the study.*

a. How many people completed the study?

b. Fill in the following two-way table (with numbers of individuals), using the figures given in the above passage:

	Exercise	No exercise	Total
Dementia			
No dementia			
Total			

c. Draw a Venn diagram with two overlapping circles to illustrate the data.

47. **Illegal Voting?** The "Cooperative Congressional Election Study," published in 2014, claimed to find evidence of voting by noncitizens. The conclusion was based largely on a 2008 survey in which approximately 38,000 registered voters were asked both

whether they voted and whether they were citizens. A total of 339 of those surveyed reported being noncitizens, and a total of 48 of these people also said they voted.

a. Based on the survey, what percentage of noncitizens claim to have voted?

b. One difficulty with any survey is *response error*, in which, for example, people accidentally check the wrong box. Suppose that the response error rate for this survey was only 0.1%, meaning that 99.9% of those surveyed answered the survey questions accurately. How many people would have answered the citizenship question incorrectly? (Note: Data suggest that most surveys have response error rates significantly higher than 0.1%.)

c. Assume that your result from part (b) represents citizens who accidentally said they were noncitizens when they were citizens, and that all these people voted. If all other results from this survey were accurate (though this is unlikely), how would this one set of errors change the number of noncitizens who voted? How large a response error could have accounted for *all* the noncitizen voting found in the survey?

d. The 2008 survey was repeated in 2010, with some (but not all) of the same people asked the same questions about citizenship status and voting that year. There were indeed changes in responses to the citizenship question among those who participated in the survey both times, suggesting response errors. In addition, a total of 85 people claimed to be noncitizens in both the 2008 and 2010 surveys, and among these *zero* reported having voted. How does this result support the claims of those people who say the study was flawed and that, in fact, it offered no evidence of noncitizen voting?

IN YOUR WORLD

48. **Polling Organization.** Go to the website for a major professional polling organization. Study results from a recent poll, and evaluate the poll according to the guidelines in this unit.

49. **Applying the Guidelines.** Find a recent news report about a statistical study on a topic that you find interesting. Write a short critique of the study, in which you apply each of the eight guidelines given in this unit. (Some of the guidelines may not apply to the particular study you are analyzing. In that case, explain why the guideline is not applicable.)

50. **Believable Results.** Find a recent news report about a statistical study whose results you believe are meaningful and important. In one page or less, summarize the study and explain why you find its results believable.

51. **Unbelievable Results.** Find a recent news report about a statistical study whose results you *don't* believe are meaningful or important. In one page or less, summarize the study and why you don't believe its claims.

UNIT 5C Statistical Tables and Graphs

Whether you look at a news magazine article, a corporate annual report, or a government study, you are almost sure to see tables and graphs of statistical data. Some of these tables and graphs are simple; others can be quite complex. Some make it easy to understand the data; others may be confusing or even misleading. In this unit, we'll investigate some of the basic principles behind tables and graphs, preparing for coverage of more complex graphics in Unit 5D.

Frequency Tables

A teacher makes the following list of the grades she gave her 25 students on an essay:

A C C B C D C C F D C C C B B A B D B A A B F C B

This list contains all the grades, but it isn't easy to read. A better way to display these data is with a **frequency table**—a table showing the number of times, or **frequency**, that each grade appears (Table 5.1). The five possible grades are called the **categories** for the table.

TABLE 5.1

Grade	Frequency
A	4
B	7
C	9
D	3
F	2
Total	**25**

Frequency Tables

A basic **frequency table** has two columns:

- The first column lists all the **categories** of data.
- The second column lists the **frequency** for each category, which is the number of data values in the category.

There are two common variations on the frequency. The **relative frequency** for a category expresses its frequency as a fraction or percentage of the total. For example, 4 of the 25 students received A grades, so the relative frequency for A grades is 4/25, or 16%. The total relative frequency must always be 1, or 100% (though rounding may sometimes cause the relative frequencies in a table or chart to add up to slightly more or less than 100%). The **cumulative frequency** is the number of responses in a particular category and all preceding categories. For example, the cumulative frequency for grades of C and above is 20, because 20 students received grades of either A, B, or C.

Definitions

The **relative frequency** of any category is the fraction (or percentage) of the data values that fall in that category:

$$\text{relative frequency} = \frac{\text{frequency in category}}{\text{total frequency}}$$

The **cumulative frequency** of any category is the number of data values in that category and all preceding categories.

EXAMPLE 1 Relative and Cumulative Frequency

Add to Table 5.1 columns showing the relative and cumulative frequencies.

Solution Table 5.2 shows the new columns and calculations.

TABLE 5.2

Grade	Frequency	Relative Frequency	Cumulative Frequency
A	4	4/25 = 16%	4
B	7	7/25 = 28%	7 + 4 = 11
C	9	9/25 = 36%	9 + 7 + 4 = 20
D	3	3/25 = 12%	3 + 9 + 7 + 4 = 23
F	2	2/25 = 8%	2 + 3 + 9 + 7 + 4 = 25
Totals	**25**	**1 = 100%**	**25**

▶ Now try Exercises 13–14.

Think About It Briefly explain *why* the total relative frequency should always be 1, or 100%.

Data Types

Essay grades are generally subjective, because different teachers might score the same essay differently. We say that the grade categories A through F are **qualitative**, because they represent *qualities* such as bad or good. In contrast, scores on a multiple-choice exam are **quantitative**, because they represent an actual count (or measurement) of the number of correct answers. As we'll see shortly, distinguishing between qualitative and quantitative data can be useful in creating tables or graphs.

Data Types

Qualitative data consist of values that can be placed into nonnumerical categories.

Quantitative data represent counts or measurements.

EXAMPLE 2 Data Types

Classify each of the following types of data as either qualitative or quantitative.

a. Brand names of shoes in a consumer survey
b. Heights of students
c. Audience ratings of a film on a scale of 1 to 5, where 5 means excellent

Solution

a. Brand names are nonnumerical categories, so they are qualitative data.
b. Heights are measurements, so they are quantitative data.
c. Although the film rating categories involve numbers, the numbers represent subjective opinions about a film, not counts or measurements. These data are therefore qualitative, despite being stated as numbers. ▶ Now try Exercises 15–22.

Binning Data

When we deal with quantitative data categories, it's often useful to group, or **bin**, the data into categories that cover a range of possible values. For example, in a table of income levels, it might be useful to create bins covering ranges from \$0 to \$19,999, \$20,000 to \$39,999, and so on. The frequency of each bin would then be the number of people with incomes in that bin.

USING TECHNOLOGY

Frequency Tables in Excel

Excel is easy to use for creating statistical tables and calculations. The following steps show how to create the frequency table for Example 1. The screenshot on the left shows the Excel table with the formulas, and the one on the right shows the results of the formulas.

1. Create columns for the grade and frequency data, then enter the data; the screen shots show these data in columns B and C. At the bottom of column C, in cell C8, use the SUM function to compute the total frequency.
2. Compute the relative frequency (column D) by dividing each frequency in column C by the total frequency from cell C8. Enter the formula for the first row (=C3/\$C\$8) and use the editing option "Fill Down" to put the correct formulas in the remaining rows. Note: When using Fill Down, you must include the dollar signs in front of C and 8 to make the reference to cell C8 an *absolute* cell reference. Without these dollar signs, using Fill Down would make the cell reference shift down (becoming C9, C10, etc.) in each row, which would be incorrect in this case.
3. Cumulative frequency (column E) is the total of all the frequencies up to a given category. The first row shows "=C3" because cell C3 contains the frequency for A grades. The next row (=E3 + C4) starts with the value in the prior row (cell E3) and adds the frequency for B grades (cell C4). The pattern continues for the remaining rows, which you can fill with the Fill Down option.

	A	B	C	D	E
1					
2		Grade	Frequency	Relative Frequency	Cumulative Frequency
3		A	4	=C3/\$C\$8	=C3
4		B	7	=C4/\$C\$8	=E3+C4
5		C	9	=C5/\$C\$8	=E4+C5
6		D	3	=C6/\$C\$8	=E5+C6
7		F	2	=C7/\$C\$8	=E6+C7
8		Total	=SUM(C3:C7)	=SUM(D3:D7)	=C8

	A	B	C	D	E
1					
2		Grade	Frequency	Relative Frequency	Cumulative Frequency
3		A	4	16%	4
4		B	7	28%	11
5		C	9	36%	20
6		D	3	12%	23
7		F	2	8%	25
8		Total	25	100%	25

EXAMPLE 3 **Binned Exam Scores**

Consider the following set of 20 scores on a 100-point exam:

76 80 78 76 94 75 98 77 84 88 81 72 91 72 74 86 79 88 72 75

Determine appropriate bins, and construct a frequency table. Include columns for relative frequency and cumulative frequency, and interpret the cumulative frequency for this case.

Solution The scores range from 72 to 98, so one way to group the data is with 5-point bins. The first bin includes scores from 95 to 99, the second bin includes scores from 90 to 94, and so on. Note that there is no overlap between bins and each bin has the same width. We then count the frequency (the number of scores) in each bin. For example, only 1 score is in bin 95 to 99 (the high score of 98) and 2 scores are in bin 90 to 94 (the scores of 91 and 94). Table 5.3 shows the complete frequency table. In this case, we interpret the cumulative frequency of any bin to be the total number of scores in *or above* that bin. For example, the cumulative frequency of 6 for the bin 85 to 89 means that 6 scores were 85 or higher.

TABLE 5.3 **Frequency Table for Binned Exam Scores**

Scores	Frequency	Relative Frequency	Cumulative Frequency
95 to 99	1	$1/20 = 0.05$	1
90 to 94	2	$2/20 = 0.10$	$2 + 1 = 3$
85 to 89	3	$3/20 = 0.15$	$3 + 2 + 1 = 6$
80 to 84	3	$3/20 = 0.15$	$3 + 3 + 2 + 1 = 9$
75 to 79	7	$7/20 = 0.35$	$7 + 3 + 3 + 2 + 1 = 16$
70 to 74	4	$4/20 = 0.20$	$4 + 7 + 3 + 3 + 2 + 1 = 20$
Totals	**20**	**1**	**20**

▶ Now try Exercises 23–24.

Bar Graphs and Pie Charts

Bar graphs and pie charts are commonly used to display qualitative data. You are probably familiar with both, but let's review the basic ideas.

A **bar graph** uses a set of bars to represent the frequency (or relative frequency) of each category: the higher the frequency, the longer the bar. The bars can be either vertical or horizontal. Figure 5.3 shows a vertical bar graph based on the essay grade data in Table 5.1. Note that the graph shows both frequency and relative frequency: Frequency is marked along the left axis and relative frequency along the right. Also note the importance of clear labeling: Without proper labels, a graph is meaningless.

Important Labels for Graphs

Title/caption: The graph should have a title or caption (or both) that explains what is being shown and, if applicable, lists the source of the data.

Vertical scale and title: Numbers along the vertical axis should clearly indicate the scale. The numbers should line up with the tick marks—the marks along the axis that precisely locate the numerical values. Include an axis title that describes the variable that the numbers represent.

Horizontal scale and title: The categories should be clearly indicated along the horizontal axis; tick marks are not necessary for qualitative data but should be used with quantitative data. Include an axis title that describes the variable that the categories represent.

Legend: If multiple data sets are displayed on a single graph, include a legend or key to identify the individual data sets.

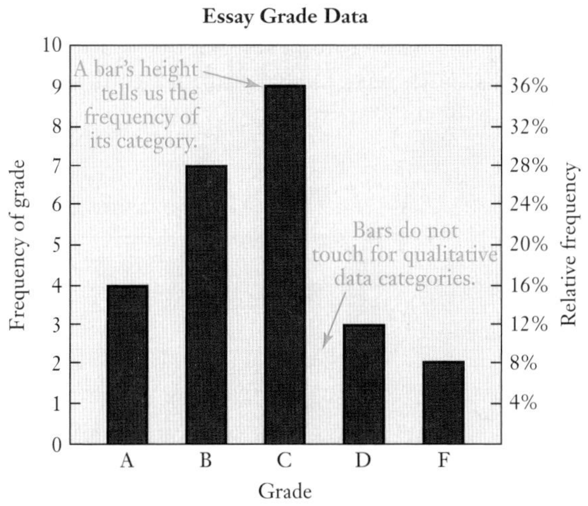

FIGURE 5.3 Bar graph for the essay grade data in Table 5.1.

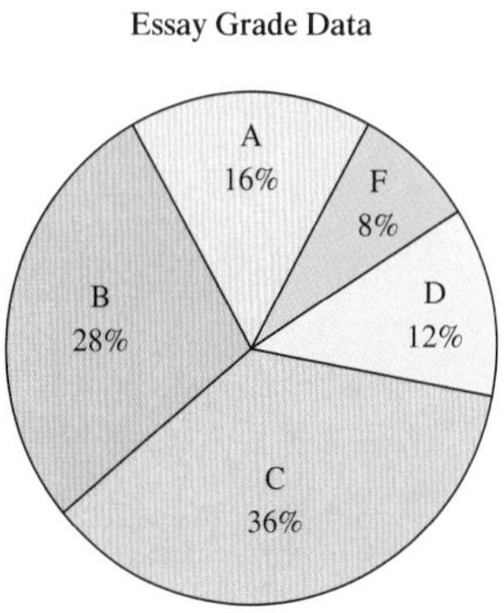

FIGURE 5.4 Pie chart for the essay grade data in Table 5.1.

Pie charts are used primarily for relative frequencies, because the total pie must always represent the total relative frequency of 100%. Figure 5.4 shows a pie chart for the essay grade data. The size of each wedge is proportional to the relative frequency of the category it represents. In other words, each wedge spans an angle given by the following formula:

$$\text{wedge angle} = \text{relative frequency} \times 360°$$

For example, the wedge for A grades in Figure 5.4 spans an angle of $0.16 \times 360° = 57.6°$.

EXAMPLE 4 Carbon Dioxide Emissions

Carbon dioxide (CO_2) is released into the atmosphere primarily by the combustion of fossil fuels (oil, coal, natural gas). Table 5.4 lists the eight countries that emit the most carbon dioxide. Make bar graphs for the total emissions and the emissions per person. Put the bars in descending order of size.

TABLE 5.4 The World's Eight Leading Emitters of Carbon Dioxide (2015)

Country	Total CO_2 Emissions (millions of metric tons)	Per-Person CO_2 Emissions (metric tons)
China	10,357	7.5
United States	5414	16.9
India	2274	1.7
Russia	1617	11.3
Japan	1237	9.8
Germany	798	10.0
Iran	648	8.2
Saudi Arabia	601	19.4

Source: Global Carbon Atlas (data from Boden et al., 2016; UNFCCC, 2016; BP, 2016).

BY THE WAY

By U.S. Department of Energy estimates, China first surpassed the United States as the leading emitter of carbon dioxide in 2005. As recently as 1990, the United States emitted nearly twice as much total carbon dioxide as China.

Solution The categories are the countries, and the frequencies are the data values. The total emissions are given in units of "millions of metric tons," and the highest value in these units is 10,357; therefore, a range of 0 to 12,000 makes a good

HISTORICAL NOTE

A bar graph with the bars in descending order is often called a Pareto chart, after Italian economist Vilfredo Pareto (1848–1923).

choice for the vertical scale. The per-person emissions are given in metric tons, and the highest value is 19.4 for Saudi Arabia; therefore, a range of 0 to 20 works well. Figure 5.5 shows the two bar graphs, with bars placed in order of descending height.

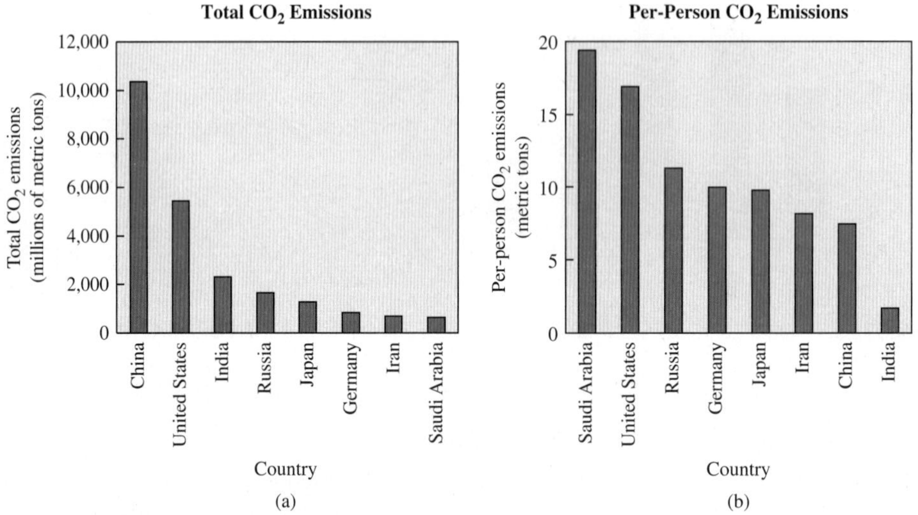

FIGURE 5.5 Bar graphs for (a) total carbon dioxide emissions by country and (b) per-person carbon dioxide emissions by country.

▶ Now try Exercises 25–26.

Think About It Most people around the world aspire to a standard of living like that in the United States. Suppose that to achieve this standard, the rest of the world's per-person carbon dioxide emissions rose to the same level as that in the United States. What consequences might this have for the world? Defend your opinion.

EXAMPLE 5 Simple Pie Chart

Among the registered voters in Rochester County, 25% are Democrats, 25% are Republicans, and 50% are Independents. Construct a pie chart to represent the party affiliations.

Solution Because Democrats and Republicans each represent 25% of the voters, the wedges for Republicans and Democrats each occupy 25%, or one-fourth, of the pie. Independents represent half of the voters, so their wedge occupies the remaining half of the pie. Figure 5.6 shows the result. As always, note the importance of clear labeling.

▶ Now try Exercises 27–28.

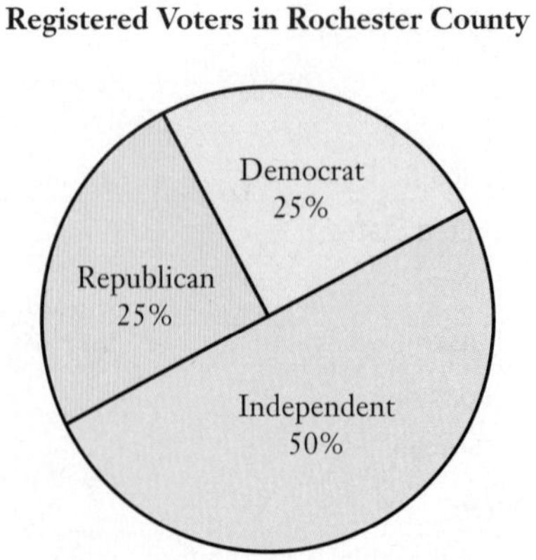

FIGURE 5.6 Party affiliations of registered voters in Rochester County.

USING TECHNOLOGY

Bar Graphs and Pie Charts in Excel

Excel can be used to make many types of statistical graphs. Let's start with a bar graph for the grade data in Table 5.1. The basic process is as follows, though the details vary with different versions of Excel.

1. Starting with the frequency table created in the Using Technology box on p. 321, select the grade letters (column B) and the frequencies (column C).
2. Choose a chart type from the Insert menu; here we chose "2-D Column," which is Excel's name for a basic bar graph. The screen shot below shows the process and result.
3. You can customize the labels on the bar graph. In most versions of Excel, a right click on the bar graph will allow you to change the axes and other labels; some versions also offer dialog boxes for changing the labels.

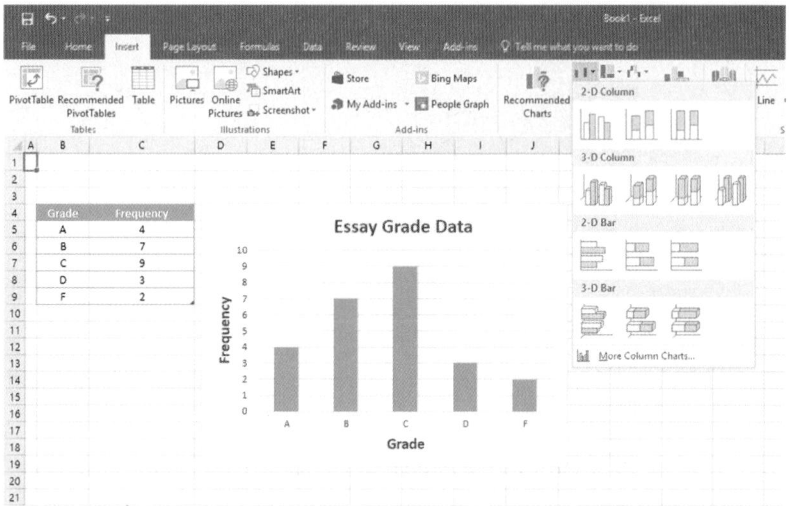

Making a pie chart is similar to making a bar graph, except for the following:

- For a pie chart, you will probably want to select the relative frequencies rather than the frequencies (though both will work); it may be helpful to cut and paste these data so they are next to the letter grades.
- Choose "2-D Pie" rather than "2-D Column" as the chart type from the Insert menu.
- Excel offers options for labeling, colors, and other decorative features. A right click on the pie chart will allow you to change the chart type and data labels. The screen shot below shows some of the options in one version of Excel; other versions have similar tools.

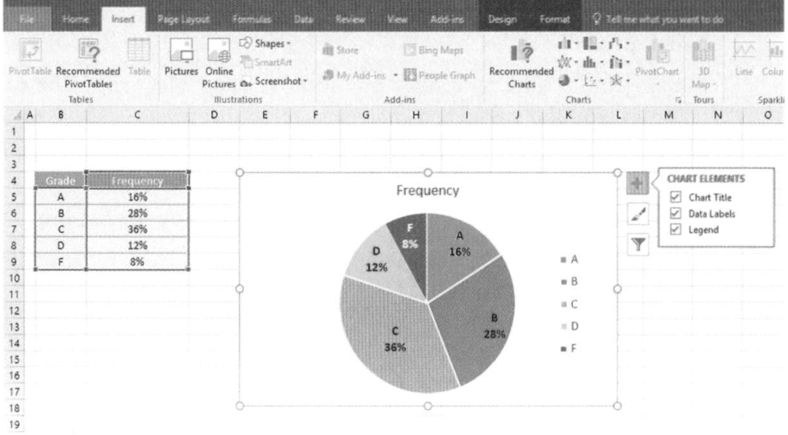

Microsoft Excel 2016, Windows 10, Microsoft Corporation.

Histograms and Line Charts

Technical Note

You may find varying definitions for the terms *histogram* and *bar graph*. In this text, a bar graph is any graph that uses bars, and histograms are bar graphs used for quantitative data categories.

For quantitative data categories, the two most common types of graphs are *histograms* and *line charts*. A **histogram** is a bar graph in which the data categories are quantitative. The bars on a histogram must follow the natural order of the numerical categories, and the widths of the bars have a specific meaning. Figure 5.7a shows a histogram for the binned exam data of Table 5.3. Notice that the width of each bar represents 5 points on the exam. The bars in the histogram touch each other because there are no gaps between the categories.

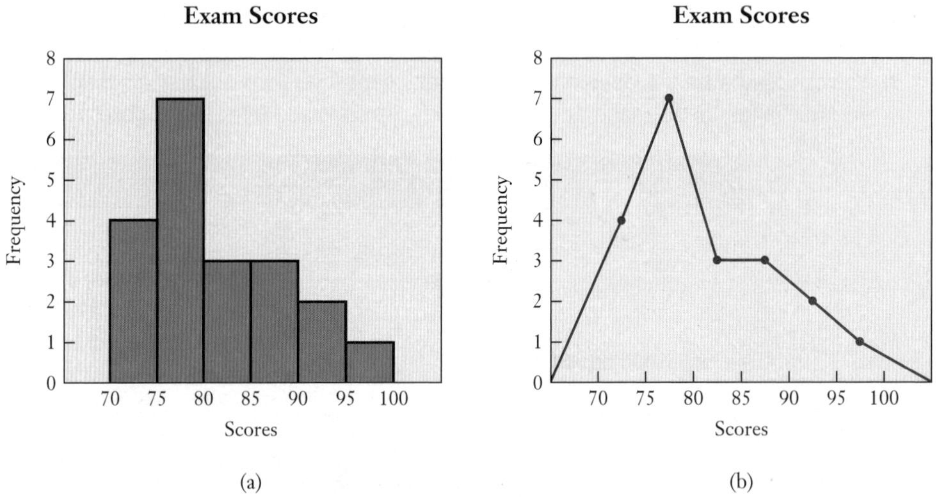

FIGURE 5.7 (a) Histogram for the data in Table 5.3. (b) Line chart for the same data.

Figure 5.7b shows a **line chart** for the same data. To make the line chart, we use a dot (instead of a bar) to represent the frequency of each data category; that is, the dots go at the tops of the bars on the corresponding histogram. Because the data are binned into 5-point bins, we place a dot at the center of each bin. For example, the dot for the data category 70–75 goes at 72.5 along the horizontal axis. After the dots are placed, we connect them with straight line segments. To make the graph look complete, we connect the points at the far left and far right back down to a frequency of zero.

Line charts and histograms are often used to show how some variable changes with time. Because these graphs have *time* on the horizontal axis, they are often called **time-series graphs**.

Definitions

A **histogram** is a bar graph for quantitative data categories. The bars have a natural order, and the bar widths have a specific meaning.

A **line chart** shows the data value for each category as a dot, with the dots connected by lines. For each dot, the horizontal position is the *center* of the bin it represents and the vertical position is the data value for the bin.

A **time-series graph** is a line chart or histogram in which the horizontal axis represents time.

EXAMPLE 6 Oscar-Winning Female Actors

Table 5.5 shows the ages (at the time when they won the award) of all Academy Award–winning female actors through 2017. Make a histogram and a line chart to display these data. Discuss the results.

Solution The data are quantitative and organized in 10-year bins. Figure 5.8 shows the data as both a histogram and a line chart; the line chart overlays the histogram so you can see how the two diagrams compare. Note that the histogram bars touch one another because there are no gaps between the categories.

USING TECHNOLOGY

Line Charts in Excel

To use Excel to create a line chart, as shown in the screen shot below for the binned data from Example 3 (p. 322), follow these steps:

1. To get the dots in the centers of the bins for the scores, enter the center point of each bin in column B. Then enter the frequencies in column C.
2. Select the scores and frequencies, then choose the chart type "Scatter" with the option for connecting points with straight lines.
3. Use the chart options to improve design, labels, and more.

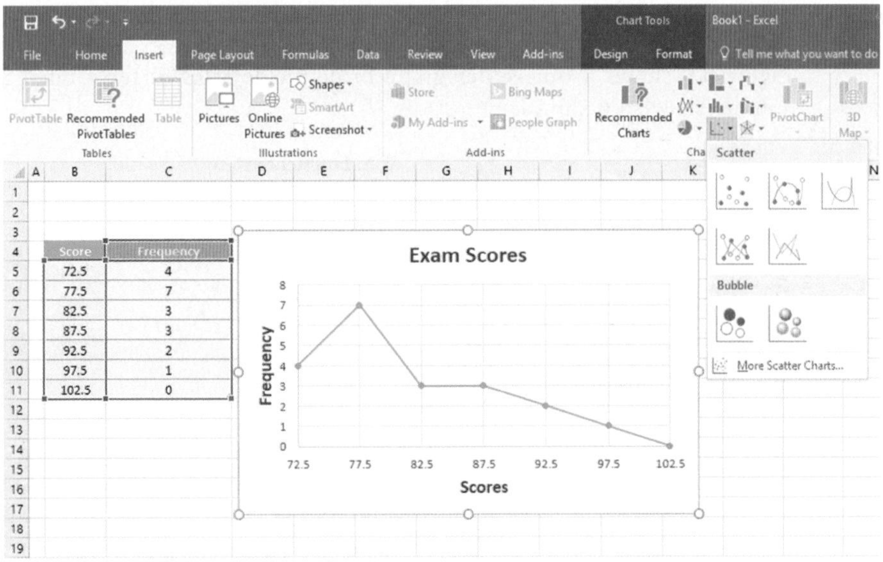

Microsoft Excel 2016, Windows 10, Microsoft Corporation.

Alternatively, you can create a line chart in Excel by choosing the "Line" option for chart type. In that case, select *only* the frequencies when you begin the graphing process; then, in the "Source Data" dialog box, choose "Series" and select the scores (column B) as the "X values." The resulting graph should look the same as that created with the "Scatter" option, except the data points will not have dots.

Note: Generating histograms in Excel requires the use of add-ins, such as the Data Analysis add-in that can be installed with some versions of Excel.

TABLE 5.5 Ages of Academy Award–Winning Female Actors at Time of Award (through 2017)

Age	Number of Female Actors
20–29	32
30–39	34
40–49	14
50–59	2
60–69	6
70–79	1
80–89	1

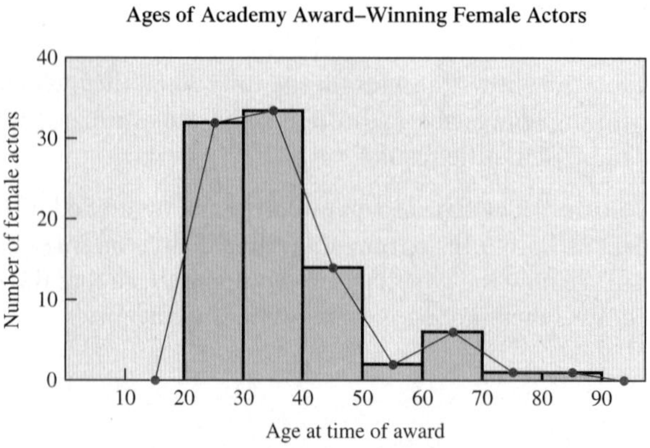

FIGURE 5.8 Histogram and line chart for ages of Academy Award–winning female actors through 2017.

▲ CAUTION! Notice that the line chart overlay in Figure 5.8 connects to zero on either side of the histogram. This is important in a case like this, because we can presume that age categories 10–19 and 90–99 have data values of zero, even though these are not shown in Table 5.5. ▲

The data show that most female actors win the award at a fairly young age, which stands in contrast to the older ages of most male winners of Best Actor (see Exercise 29). Many female actors believe this difference arises because Hollywood producers rarely make movies that feature older women in strong character roles.

▶ Now try Exercises 29–30.

EXAMPLE 7 Time-Series Graph

Figure 5.9 shows a time-series graph of homicide rates in the United States. Briefly summarize what it shows.

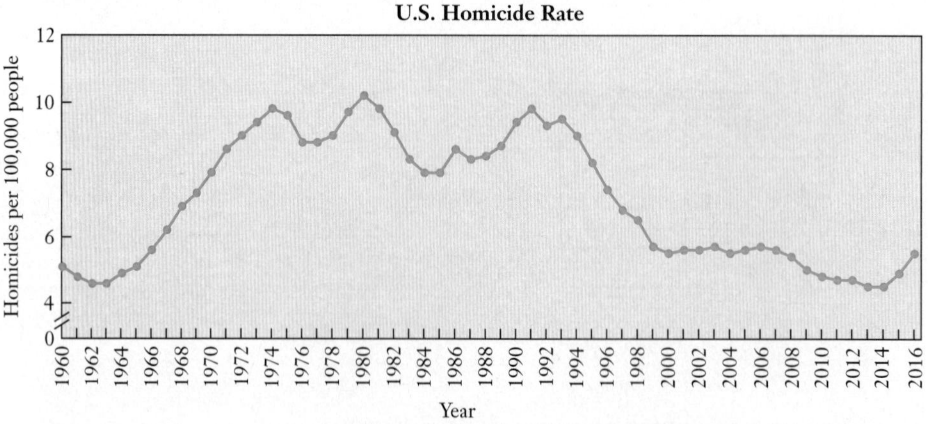

FIGURE 5.9 U.S. homicide rate per 100,000 people. *Source:* FBI Uniform Crime Reports.

Solution The graph shows how the homicide rate (deaths per 100,000 people) has changed since 1960. We see that the homicide rate rose dramatically—more than doubling—from a minimum around 1962 to a first peak around 1974. It then remained high, with some variations, through about 1993. After 1993, it fell dramatically through the year 2000, then stayed nearly constant for a few years and dropped again from 2008 through 2012. However, the rate has since increased, though it is still much lower than it was a few decades ago.

▶ Now try Exercises 31–32.

Quick QUIZ 5C

Choose the best answer to each of the following questions. Explain your reasoning with one or more complete sentences.

1. In a class of 200 students, 50 students received a grade of B. What was the *relative* frequency of a B grade?
   a. 25 b. 0.25
   c. It cannot be calculated with the information given.
2. For the class described in Question 1, what was the *cumulative* frequency of a grade of B or above?
   a. 25 b. 0.25
   c. It cannot be calculated with the information given.
3. Which of the following is an example of *qualitative* data?
   a. height in inches b. ratings of movies
   c. ticket prices at theaters
4. The *sizes* of the wedges in a pie chart tell you
   a. the number of categories in the pie chart.
   b. the frequencies of the categories in the pie chart.
   c. the relative frequencies of the categories in the pie chart.
5. You have a table listing ten tourist attractions and their annual numbers of visitors. Which type of display would be most appropriate for these data?
   a. a bar graph b. a pie chart c. a line chart
6. In the table of tourist attractions and visitors from Question 5, where should you put the names of the ten tourist attractions?
   a. They should be in the title of the display.
   b. They should be in alphabetical order along the vertical axis.
   c. They should be listed along the horizontal axis.
7. You have a list of the GPAs of 100 college graduates, precise to the nearest 0.001. You want to make a frequency table for these data. A good first step would be to
   a. group all the data into bins with widths of 0.2 of a grade point.
   b. draw a pie chart for the 100 individual GPAs.
   c. count how many people have identical GPAs.
8. You have a list of the average gasoline price for each month during the past year. Which type of display would be most appropriate for these data?
   a. a bar graph
   b. a pie chart
   c. a line chart
9. A *histogram* is
   a. a graph that shows how some quantity has changed through history.
   b. a graph that shows cumulative frequencies.
   c. a bar graph for quantitative data.
10. You have a histogram and you want to convert it into a line chart. A good first step would be to
    a. make a list of all the categories in alphabetical order.
    b. place a dot at the top of each bar, in the center of the bar.
    c. calculate all the relative frequencies that you can read from the histogram.

Exercises 5C

REVIEW QUESTIONS

1. What is a *frequency table*? Explain what we mean by the *categories* and *frequencies*, and how we can use them to calculate *relative frequencies* and *cumulative frequencies*.
2. What is the distinction between *qualitative* data and *quantitative* data? Give a few examples of each.
3. What is the purpose of *binning*? Give an example in which binning is useful.
4. What two types of graphs are most common when the categories are qualitative data? Describe the construction of each.
5. Describe the importance of labeling on a graph, and briefly discuss the kinds of labels that should be included on graphs.
6. What two types of graphs are most common when the categories are quantitative data? Describe the construction of each.

DOES IT MAKE SENSE?

Decide whether each of the following statements makes sense (or is clearly true) or does not make sense (or is clearly false). Explain your reasoning.

7. I made a frequency table with two columns, one headed with *State* and one with *State Capitol*.
8. The relative frequency of B grades in our class was 0.3.
9. The cumulative frequency of C grades in our class of 30 students was 40.
10. Your bar graph must be wrong, because you have 10 bars but there were only 5 data categories.
11. Your pie chart must be wrong, because when I added the percentages on your wedges, they totaled 124%.
12. I rearranged the bars on my histogram so that the tallest bar came first.

BASIC SKILLS & CONCEPTS

13–14: Frequency Tables. Make frequency tables for the following data sets. Include columns for relative frequency and cumulative frequency.

13. Final grades of 30 students in a math class:
A A A A A A B B B B B B C C C C C C C C C C D D D D D F F F

14. A website that reviews recent movies lists 5 five-star films (the highest rating), 15 four-star films, 15 three-star films, 10 two-star films, and 5 one-star films.

15–22: Qualitative versus Quantitative. Determine whether the following variables are qualitative or quantitative, and explain why.

15. The eye color of students in a class

16. The responses of customers on a satisfaction survey that used a scale from 0 = terrible to 5 = fantastic

17. Home prices in a small town

18. Daily snowfall (in inches) during January in Syracuse, New York

19. The flavors of ice cream sold at a delicatessen

20. The breeds of all cats at an animal rescue shelter

21. The annual salaries of NBA basketball players

22. The gold medal count of each team in the 2016 Olympics

23–24: Binned Frequency Tables. Use the given bin sizes to make a frequency table for the following data set:

89	67	78	75	64	70	83	95	69	84
77	88	98	90	92	68	86	79	60	96

Include columns for relative frequency and cumulative frequency.

23. Use 5-point bins (95 to 99, 90 to 94, etc.).

24. Use 10-point bins (90 to 99, 80 to 89, etc.).

25. **Most Populous Countries.** The following table gives the populations of the five most populous countries (2016 estimates). Make a bar graph for these data, with the bars in descending order.

Country	Population (millions)
China	1374
India	1267
United States	321
Indonesia	258
Brazil	206

26. **Beef Production.** The following table shows beef production of the six largest beef producers in the world (2016 data from USDA). Make a bar graph for these data, with the bars in descending order.

Beef producer	Amount of beef (million metric tons)
United States	11.4
Brazil	9.3
European Union	7.9
China	6.9
India	4.3
Argentina	2.6

27–28: Pie Charts. Construct pie charts for the following data sets.

27. The percentage of American mothers aged 40 to 44 with various numbers of children (2014 data from Pew Research Center)

Number of children	Percentage of mothers
1	22%
2	41%
3	24%
4 or more	13%

28. The percentages of people who said each category was the most important resource in finding their current job (2015 data from Pew Research Center)

Resource for job hunt	Percentage ranking resource as "most important"
Online resources	34%
Close family/friend connection	20%
Professional connection	17%
Acquaintance or friend-of-friend	7%
Job fair, conference, or special event	5%
Employment agency	5%
Other	12%

29. **Oscar-Winning Male Actors.** The following frequency table shows ages of Academy Award–winning male actors through 2017 in 10-year age bins. Draw a histogram to display the binned data.

Ages of Academy Award–Winning Male Actors at Time of Award (through 2017)

Age	Number of actors
20–29	1
30–39	31
40–49	38
50–59	13
60–69	6
70–79	1

30. **ACT Test Takers.** The following table shows binned data for the percentages of high school graduates who took the ACT test in 2016 among the 50 states and the District of Columbia. Make a histogram of the data.

ACT Test Takers by State

Percentage of high school graduates	Number of states
≤20%	2
21–40%	14
41–60%	6
61–80%	6
81–100%	23

31. **Cell Phone Subscriptions.** The following table shows the numbers of cell phone subscriptions (in millions) in the United States for various years. Construct a time-series graph for the data. Does the graph show straight-line growth (linear growth)? Or is the growth faster than linear? (The number of subscriptions can be larger than the population because some people have more than one subscription, such as for personal and business cell phones.)

Year	Number (millions)	Year	Number (millions)
1997	55	2007	255
1999	86	2009	286
2001	128	2011	316
2003	159	2013	336
2005	208	2015	378

32. **Death Rates.** Figure 5.10 is a time-series graph showing overall death rates (from all causes) in the United States for 1900–2015. The spike in 1919 was due to a worldwide epidemic of influenza. Write a few sentences summarizing the overall trend, describing how much the death rate changed over this period of time and putting the 1919 spike into context in terms of its impact on the population.

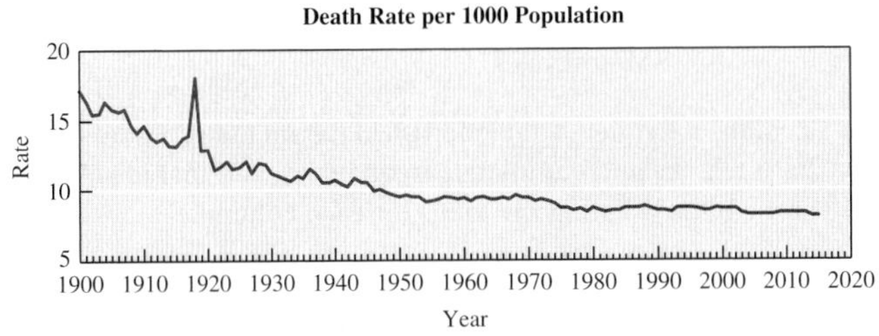

FIGURE 5.10 *Source:* National Center for Health Statistics.

FURTHER APPLICATIONS

33. **U.S. Electrical Energy.** The following table gives the percentage of electrical energy generated in the United States from various sources (2015 data). Display the data as both a bar graph and pie chart. Then write a short paragraph discussing the pros and cons of each display in terms of interpretation.

Energy source	Percentage of total energy generated
Coal	33%
Natural gas	33%
Nuclear power	20%
Renewables, non-hydropower	7%
Hydropower	6%
Other	1%

Source: U.S. Energy Information Administration.

34. **Marriage Data.** The following table shows the percentages of adult American men and women who have been married never, exactly once, exactly twice, and three or more times. Display the data for men and women in separate pie charts and write a short paragraph explaining any noticeable trends or patterns.

Number of marriages	Percentage of women	Percentage of men
0	27%	33%
1	58%	52%
2	12%	12%
3 or more	3%	3%

35. **Ages of Nobel Prize Winners.** The following frequency table categorizes Nobel Prize winners in literature (through 2016) by their age at the time they received the award. Display the data in a histogram and write a short paragraph explaining any noticeable trends or patterns.

Age (years)	Number of winners	Age (years)	Number of winners
<50	9	70–74	18
50–54	10	75–79	15
55–59	18	80–84	4
60–64	18	85–89	2
65–69	19		

36. **Movie Theater Admissions.** The following table shows data for U.S. movie theater admissions (per week). Display the data as a time-series graph and write a short paragraph explaining any noticeable trends or patterns.

Year	Admissions/week (millions)	Year	Admissions/week (millions)
1945	79.0	1985	20.3
1955	39.9	1995	23.3
1965	19.8	2005	26.5
1975	19.9	2015	25.4

37. **Student Religious Affiliations.** The following table gives the stated religious affiliations of approximately 140,000 first-year college students in 2016. Display the data with a pie chart and write a short paragraph discussing what it shows.

Religion	Percentage of sample
Protestant	37.5%
Roman Catholic	24.3%
Jewish	2.7%
Other	5.9%
None	29.5%

Source: UCLA Higher Education Research Institute.

38. **No Religious Affiliation Trends.** The following table gives the percentage of first-year college students who answered "None" when asked about their religious affiliation in selected years. Display the data as a time-series graph and

write a short paragraph explaining any noticeable trends or patterns.

Year	Percentage claiming no religious affiliation
1985	9.4%
1990	12.3%
1995	14.0%
2000	14.9%
2005	17.4%
2010	23.0%
2015	29.5%

Source: UCLA Higher Education Research Institute.

39. **Immigrant Data.** The following table gives the percentage of the U.S. population that was foreign-born in selected years. Display the data as a time-series graph and write a short paragraph explaining any noticeable trends or patterns.

Year	Percent foreign-born	Year	Percent foreign-born
1940	8.8%	1990	8.0%
1950	6.9%	2000	10.4%
1960	5.4%	2010	12.2%
1970	4.7%	2015	13.3%
1980	6.2%		

40. **Ages of Presidents.** The following table gives the order of the Presidents of the United States and the ages at their inauguration. Display these data with a bar graph using one bar for each president. Write a paragraph describing significant features of the data.

Order	1	2	3	4	5	6	7	8	9
Age	57	61	57	57	58	57	61	54	68
Order	10	11	12	13	14	15	16	17	18
Age	51	49	64	50	48	65	52	56	46
Order	19	20	21	22	23	24	25	26	27
Age	54	49	51	47	55	55	54	42	51
Order	28	29	30	31	32	33	34	35	36
Age	56	55	51	54	51	60	62	43	55
Order	37	38	39	40	41	42	43	44	45
Age	56	61	52	69	64	46	54	47	70

IN YOUR WORLD

41. **Frequency Tables.** Find a recent news article that includes some type of frequency table. Briefly describe the table, and explain how it adds to the news report. Do you think the table was constructed in the best possible way for the article? If so, why? If not, what would you have done differently?

42. **Bar Graph.** Find a recent news article that includes a bar graph with qualitative data categories. Briefly explain what the graph shows, and discuss whether it helps make the point of the news article.

43. **Pie Chart.** Find a recent news article that includes a pie chart. Briefly discuss the effectiveness of the pie chart. For example, would it be better if the data were displayed in a bar graph rather than a pie chart? Could the pie chart be improved in other ways?

44. **Histogram.** Find a recent news article that includes a histogram. Briefly explain what the histogram shows, and discuss whether it helps make the point of the news article. Are the labels clear? Is the histogram a time-series graph? Explain.

45. **Line Chart.** Find a recent news article that includes a line chart. Briefly explain what the line chart shows, and discuss whether it helps make the point of the news article. Are the labels clear? Is the line chart a time-series graph? Explain.

TECHNOLOGY EXERCISES

Answer the following questions using procedures described in the Using Technology boxes in this unit or with **StatCrunch** (available in MyLab Math).

46. **Parking Lot Data.** Use Excel or StatCrunch with the following vehicle counts observed during a survey of a student parking lot.

Category of vehicle	Frequency
American cars	30
Japanese cars	25
English cars	5
Other European cars	12
Motorcycles	8

a. Make a frequency table for the data. Include columns for both relative and cumulative frequencies, and include a row at the bottom listing the totals for each column.

b. Make a bar graph that displays these data.

c. Make a pie chart for these data.

47. **U.S. Population in Poverty.** Use Excel or StatCrunch with the following data on the percentage of the U.S. population below the poverty level in various years from 1960 through 2015.

Year	Percentage below poverty level
1960	22.2%
1970	12.6%
1980	13.0%
1990	13.5%
2000	11.3%
2010	15.1%
2015	13.5%

a. Make a line chart that displays these data.

b. Make a bar graph for these data.

48. **StatCrunch Data Entry.** To gain practice entering data in StatCrunch, go to the StatCrunch work space and in the first column (var1), enter 20 whole numbers between 1 and 10.

a. Make a table showing the frequency, relative frequency, cumulative frequency, and cumulative relative frequency of the numbers 1, 2, 3, …, 10. (Go to the Stat menu and choose "Tables" and then "Frequency.")

b. Make a bar graph of the data set.

c. Make a pie chart of the data set and check that the legend agrees with the frequency table from part (a).

49. **NFL Data.** Open the StatCrunch shared data set called *NFL Players 2016* (choose "Explore" and then "Data" and type the data set name in the search box).

a. Make a bar graph of the ages of the players.

b. Make a pie chart of the age distribution for all players.

c. Now make a pie chart of the age distribution for each position on a team. (Go to the Graph menu and choose "Pie Chart" and then "With Data"; then select "Age" for the column and "Position" for "Group B.") How many pie charts are produced? Why?

50. **StatCrunch Project.** Choose one data set available in StatCrunch that is of interest to you and that lends itself to making a bar graph or pie chart.

a. State the data set you've chosen and write a sentence about why it is of interest to you.

b. Make a bar graph or pie chart that displays the data distribution effectively.

UNIT 5D Graphics in the Media

The basic graphs we have studied so far are only the beginning of the many ways to depict data visually. In this unit, we will explore some of the more complex types of graphics that are common in the media, along with a few cautions to keep in mind when interpreting graphics.

Graphics Beyond the Basics

Many graphical displays of data go beyond the basic types discussed in Unit 5C. Here we explore a few of the types that are most common in the news media.

Multiple Bar Graphs and Line Charts

A **multiple bar graph** is a simple extension of a regular bar graph. It has two or more sets of bars that allow comparison of two or more data sets. All the data sets must have the same categories so they can be displayed on the same graph. Figure 5.11a is a multiple bar graph with two sets of bars, one for men and one for women. (The *median* is a type of average that we discuss in Unit 6A.)

The data categories for Figure 5.11a (levels of educational attainment by gender) are qualitative, which makes a bar graph the best choice for display. In cases for which data categories are quantitative, a **multiple line chart** is often a better choice. Figure 5.11b shows

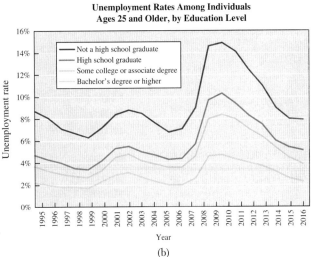

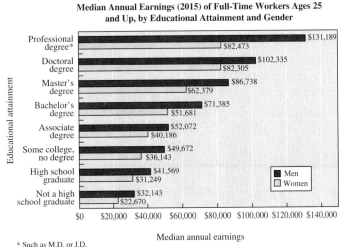

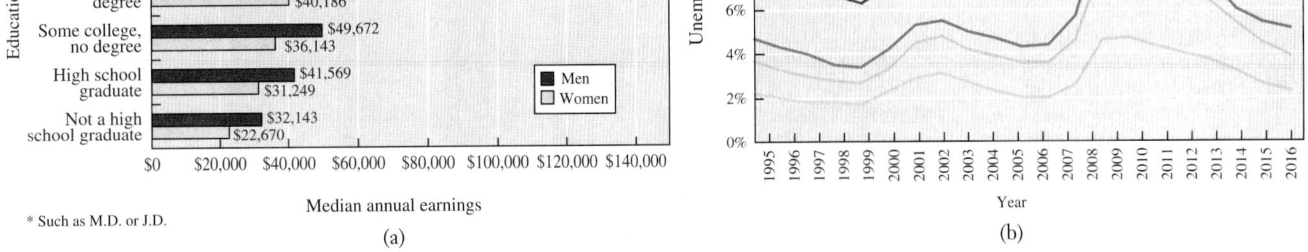

FIGURE 5.11 (a) A multiple bar graph showing the relationship between earnings and educational attainment and gender (2015 data). (b) A multiple line chart showing how the unemployment rate has varied with time for people with different levels of educational attainment. *Sources:* (a) U.S. Census Bureau; (b) Bureau of Labor Statistics.

time-series data using four different lines for four different data sets, each based on a different level of educational attainment. The data are quantitative in this case because the categories (on the horizontal axis) are years and the data values are unemployment rates, both of which are measured quantitatively.

EXAMPLE 1 Education Pays

What general messages are revealed by the graphs in Figure 5.11? Comment on how the use of the multiple bar and line graphics helps convey these messages.

Solution Figure 5.11a conveys at least two clear messages. First, by looking at the bars across all the categories, we see that people with greater education have significantly higher median incomes, confirming our earlier finding (see the chapter-opening question for Chapter 4) that education is a good financial investment, at least on average. The second message conveyed by the graph is that for equivalent levels of educational attainment, women still earn much less than men. Figure 5.11b shows another added value of education: Unemployment is significantly lower for more highly educated people.

Multiple bars or lines work well for these graphs because they allow easy comparisons. For example, if the bar graphs for men and women were shown separately, it would be much more difficult to see that women earn less than men with the same education. Similarly, if the unemployment line charts were shown separately, our eyes would be drawn more to the trends with time than to the crucial differences in the unemployment rates for people with different levels of education.

▶ Now try Exercises 13–16.

EXAMPLE 2 Gender Differences in Mathematics

The Program for International Student Assessment (PISA), administered through the Organization for Economic Cooperation and Development (OECD), tracks the educational performance of students around the world by administering standardized tests to samples of students from different countries every three years. Figure 5.12 shows recent mathematics results from six selected countries, using two sets of bars: one for boys and one for girls. The first set of bars shows that in the United States, boys outperform girls on the mathematics test, a fact sometimes used to argue that boys are inherently better than girls at mathematics. Based on the overall graph, comment on the validity of this argument.

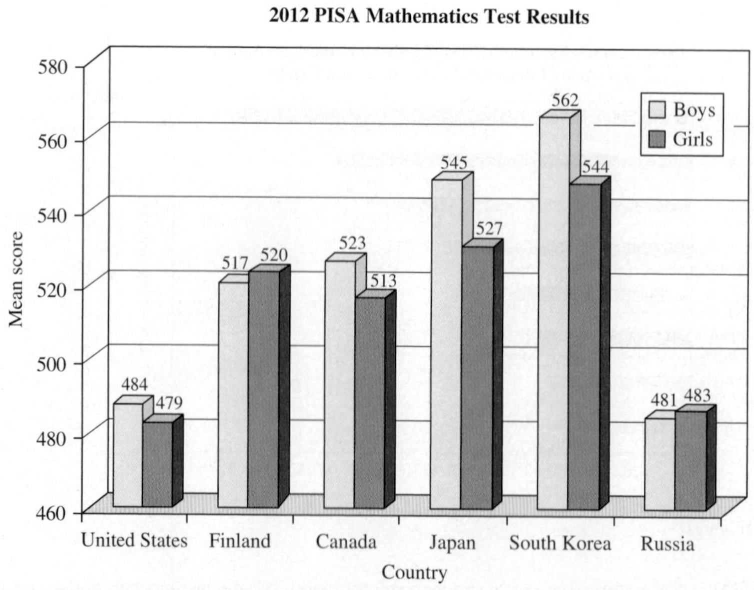

FIGURE 5.12 A double bar graph showing results for boys and girls on the 2012 PISA mathematics test in six selected countries. *Source:* Data from the Organization for Economic Cooperation and Development.

Solution The "gender gap" in which boys outperform girls in the United States also appears in the scores of some other countries—but not all. For example, both Finland and Russia show girls outperforming boys. Moreover, the scores of *girls* in Finland, Canada, Japan, and South Korea are all significantly higher than those of *boys* in the United States. These facts do not support a claim of inherent differences in mathematical ability between girls and boys, and are consistent with the alternative hypothesis that differences in cultural factors or educational practices may be responsible.

▲ **CAUTION!** Notice that the vertical scale in Figure 5.12 does not start from zero, which tends to exaggerate the differences between the bars. See Example 9 for more detail. ▲

▶ Now try Exercises 17–18.

Stack Plots

Another way to show two or more related data sets simultaneously is with a **stack plot**, which shows different data sets stacked upon one another. The following two examples show that data can be stacked in both bar graphs and line charts.

EXAMPLE 3 College Costs

Figure 5.13 shows a stack plot using stacked bars laid out horizontally. Briefly explain how to interpret each bar. What can you conclude about the primary reason that different types of college have significantly different total costs?

Solution Each bar is divided into sections, which are color-coded according to the key at the upper right. For example, the top bar shows that the total budget for the average commuter student at a two-year public college is $17,000 and the light blue segment in that bar shows that this average student spends $8060 on room and board.

The differences in the total bar lengths highlight the differences in total costs among the types of institution. By looking at the color-coded segments within the bars, we see that the primary reason for these overall differences comes from differences in tuition. Other costs are more similar across the different types of institution.

▶ Now try Exercises 19–20.

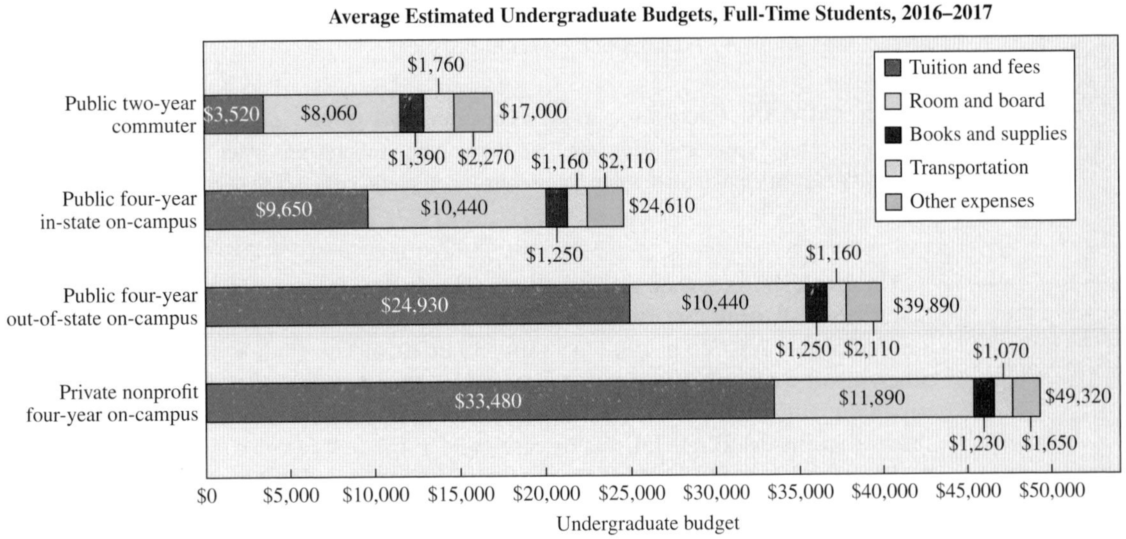

FIGURE 5.13 This stack plot uses stacked horizontal bars to show the breakdown of average student budgets at different types of institutions. *Source:* The College Board, "Trends in College Pricing 2016."

Think About It What type of college do you attend? How does your own budget compare to the average budget for students at similar institutions?

EXAMPLE 4 Trends in Health Data

Figure 5.14 shows how mortality due to four serious diseases has changed since 1900, as measured in deaths from these diseases per 100,000 people. What was the death rate for cardiovascular disease in 1980? Discuss the general trends visible on this graph.

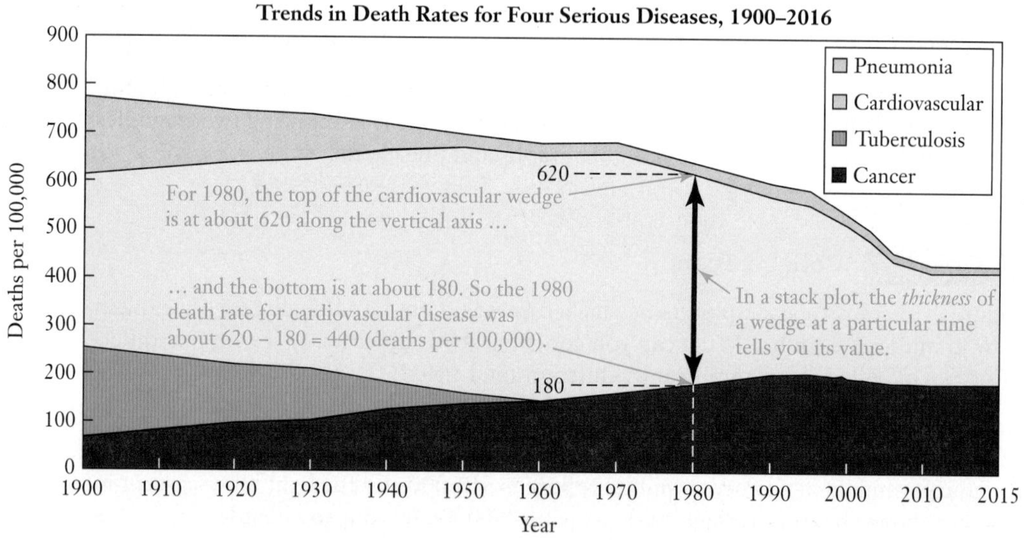

FIGURE 5.14 A stack plot showing trends in death rates from four diseases. *Source:* Centers for Disease Control and Prevention.

Solution Each disease has its own color-coded region, or wedge, identified in the key at the upper right. The *thickness* of any wedge at a particular time tells you its value at that time. For 1980, the cardiovascular wedge extends from about 180 to 620 on the vertical axis, so its thickness is about 440. Therefore, the death rate in 1980 for cardiovascular disease was about 440 deaths per 100,000 people.

▲ **CAUTION!** Figure 5.14 also shows one drawback to stack plots: When data values for a category become very small, as they do for tuberculosis after about 1960, the stack plot can make them seem to disappear completely, even if they continue at a very low level. ▲

The graph shows several general trends. The downward slope of the top wedge shows that the overall death rate due to these four diseases decreased substantially, from nearly 800 deaths per 100,000 in 1900 to about 435 in 2015. The gradual disappearance of the tuberculosis wedge shows that this disease was once a major killer, but has been nearly wiped out since 1960. The widening and then narrowing of the cardiovascular wedge shows that the death rate for this disease grew in the first few decades shown, but then declined (and has held fairly steady for about the past decade). The cancer wedge shows that the cancer death rate rose steadily until the mid-1990s, but has dropped somewhat since then.

▶ **Now try Exercises 21–24.**

BY THE WAY

The overall U.S. death rate from all causes declined fairly steadily for decades—a fact reflected in rising life expectancy (see Unit 7D). However, the trend unexpectedly reversed in 2015 and 2016, when the overall death rate rose, primarily as a result of increases in deaths from drug overdoses, car crashes, and suicide.

Graphs of Geographical Data

We are often interested in geographical patterns in data. Figure 5.15 shows one common way of displaying **geographical data**. In this case, the map shows trends in energy use per capita (per person) in the 50 states (and the District of Columbia). The actual data values are shown in small print with each state, while the color coding reflects the binned categories listed in the key.

The display in Figure 5.15 works well because each state is associated with a unique energy usage per person. For data that vary continuously across geographical areas, a **contour map** is more convenient. Figure 5.16 shows a contour map of temperatures over the United States at a particular time. Each of the *contours* (curvy lines) connects locations with the same temperature. For example, the temperature was 50°F everywhere along the contour labeled 50°F and the temperature was 60°F everywhere along the contour labeled 60°F. Between these two contours, the temperature was between

USING TECHNOLOGY

Graphs with Multiple Data Sets

Excel provides a variety of options for making multiple bar graphs, multiple line charts, stack plots, and more. The screen shot below shows some of the data and an Excel graph that went into making Figure 5.14. Excel provides several ways of creating a desired graph, but here's one process:

1. Enter all the data. Notice that the disease categories are in row 2, and the years are in column A. The data cells show the death rates per 100,000.
2. Select all the data, then choose "Chart" from the Insert menu to bring up the chart options.
3. Choose your chart option. For this particular screen shot, we chose the chart type "Stacked Area."
4. From this point, getting to a final graphic like Figure 5.14 is just a matter of choosing labeling options and other decorative features. Most of these features can be added within Excel by right-clicking on the chart elements, though sometimes it is easier to add them by importing the Excel graph into an art- or photo-editing program.

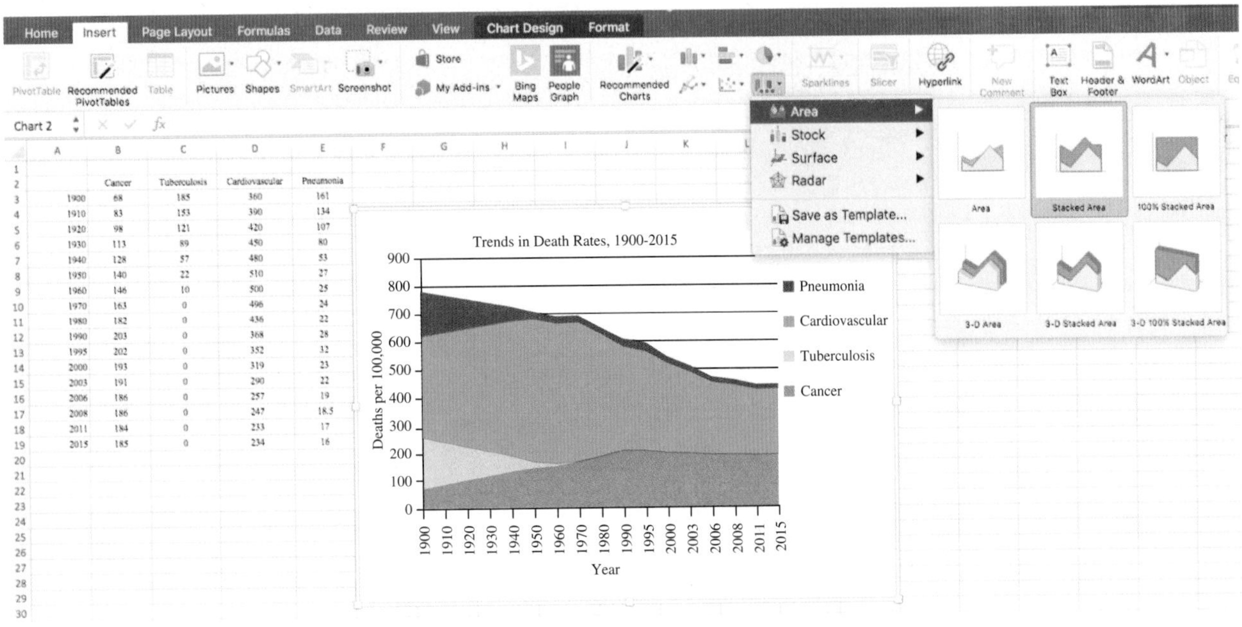

Microsoft Excel 2016, Windows 10, Microsoft Corporation.

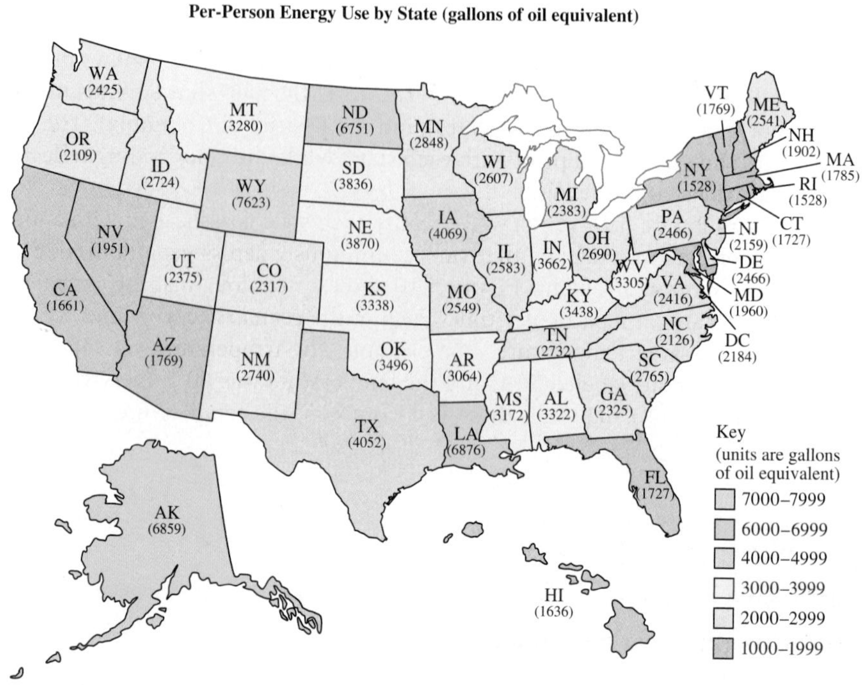

FIGURE 5.15 Geographical data can be displayed with a color-coded map. These data show per-person energy usage by state, in units of "gallons of oil equivalent"; that is, the data represent the amount of oil that each person would use if all the energy were generated by burning oil. *Source:* U.S. Energy Information Administration, State Energy Data System (2013 data).

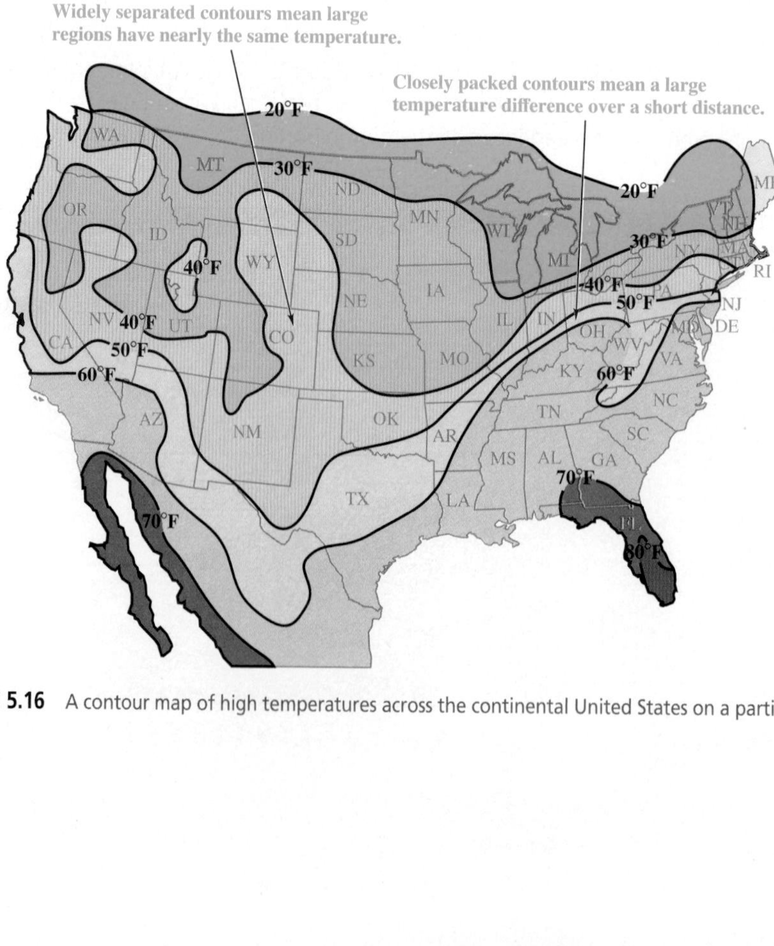

FIGURE 5.16 A contour map of high temperatures across the continental United States on a particular day.

50°F and 60°F. Note that more closely spaced contours (such as those over Ohio) mean that temperature was varying more rapidly with distance. To make the graph easier to read, the regions between adjacent contours are color-coded.

EXAMPLE 5 Interpreting Geographical Data

Use Figures 5.15 and 5.16 to answer the following questions.

a. What geographical characteristics are common to states with the lowest energy usage per person?
b. Were there any temperatures above 80°F in the United States on the day shown in Figure 5.16? If so, where?

Solution

a. The color coding shows that the states in the lowest category of energy use per person are all either warm-weather states (CA, AZ, FL, NV, and HI) or states in the more densely populated regions of the northeast (NY, NH, CT, MA, and RI).
b. The 80°F contour passes through southern Florida, so the parts of Florida south of this contour had temperatures above 80°F. ▶ Now try Exercises 25–28.

Think About It Look for a current weather map. How are the temperature contours shown? Interpret the temperature data.

The greatest value of a picture is when it forces us to notice what we never expected to see.
—John Tukey

Three-Dimensional Graphics

Today, computer software makes it easy to give almost any graph a three-dimensional appearance. For example, the double bar graph in Figure 5.12 looks good with its three-dimensional appearance, but it could show the same data without the added third dimension of depth. In other words, its three-dimensional effects are purely cosmetic.

In contrast, each of the three axes in Figure 5.17 carries distinct information, making it a true three-dimensional graph. Notice that the bars for 2015 are similar to the ones shown in Figure 5.5 for the top five emitters of carbon dioxide. The added dimension in Figure 5.17 is *time*, so the three dimensions are countries, carbon dioxide emissions, and time.

The advantage of three-dimensional graphics is that they allow us to show richer data sets. The drawback is that, as evident in Figure 5.17, it can be difficult to read the data precisely. Three-dimensional graphics therefore tend to work best when they are interactive (online), so that figures can be rotated or viewed from different perspectives.

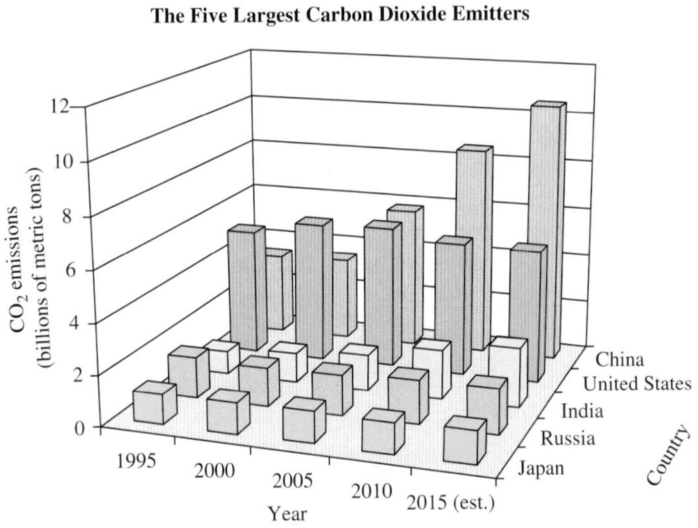

FIGURE 5.17 This graph shows true three-dimensional data. The three dimensions are the countries, their total carbon dioxide emissions, and time. *Source:* World Bank.

EXAMPLE 6 Three-Dimensional Carbon Dioxide Emissions

Based on Figure 5.17, about when did China surpass the United States in total carbon dioxide emissions?

Solution By looking along the axis listing the countries, we see that blue bars represent the United States and purple bars represent China. For both 1995 and 2000, the U.S. bars are clearly taller than the bars for China, indicating that the United States had more total emissions. However, for 2005 the bars of the two countries are about equal, and by 2010 China's bar is clearly taller than the bar for the United States. Therefore, China surpassed the United States in total emissions sometime between 2005 and 2010, and probably closer to 2005. ▶ Now try Exercises 29–30.

Infographics

The graphic types we have studied so far are common and fairly easy to create. But the availability of sophisticated software has made more complex graphics increasingly common. One common type, known as an **infographic** (short for "information graphic"), generally presents a large, interrelated set of information and data in a visual way, with the goal of making it easy to understand and interpret all the information.

Figure 5.18 shows an infographic summarizing survey data collected from first-year college students. Notice that it contains a combination of data values (such as percentages) and words and icons designed to help explain what is going on. This is certainly a case where a picture is worth far more than a thousand words. Many online infographics go even further, adding interactive or animated features that add to their wealth of information.

EXAMPLE 7 First-Year Student Survey

Answer the following questions based on Figure 5.18.

a. What percentage of students feel that their coursework is relevant to their lives?
b. Can you draw any conclusions about whether exercise contributes to a successful balance of academics and extracurricular activities?

Solution

a. The statement near the top indicates that 58% of first-year students feel that their coursework is relevant to their lives.
b. The fact that 56% exercise and 79% feel they have a successful balance means there must be some overlap between the two categories. However, without further information, we cannot draw any conclusions about the extent of the overlap between the categories or whether there is any cause and effect involved.

▶ Now try Exercises 31–32.

Think About It Spend a few minutes studying Figure 5.18. Which of its features do you think are particularly effective? What features might be subject to misinterpretation? What, if anything, would you have done differently if you had created the graphic?

A Few Cautions About Graphics

As we have seen, graphics can offer clear and meaningful summaries of statistical data. However, even well-made graphics can be misleading if we are not careful in interpreting them, and poorly made graphics are almost always misleading. Moreover, some people use graphics in deliberately misleading ways. Here we discuss a few of the more common ways in which graphics can lead us astray.

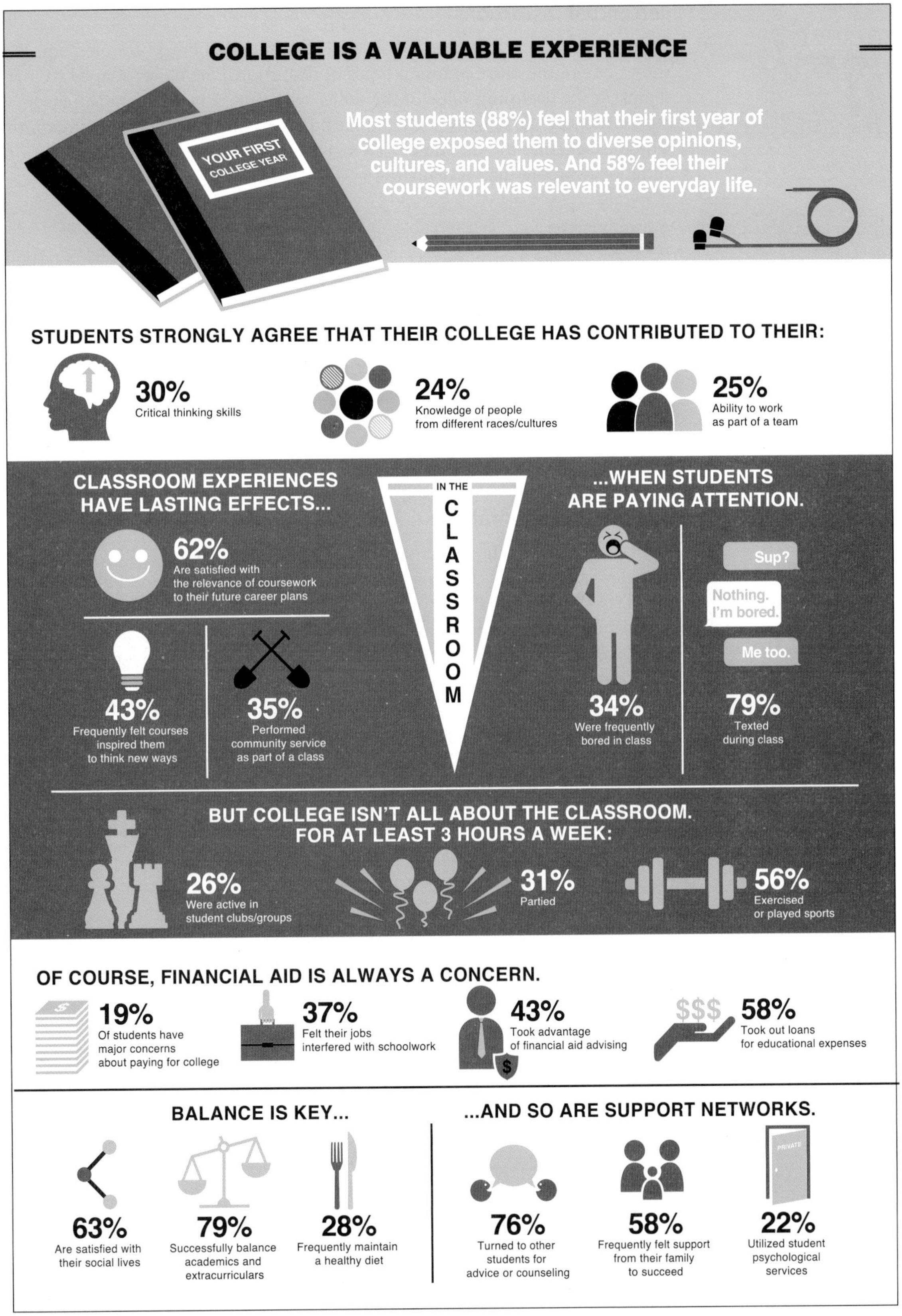

FIGURE 5.18 This infographic summarizes a great deal of information from a survey of first-year college students.
Source: Higher Education Research Institute, UCLA.

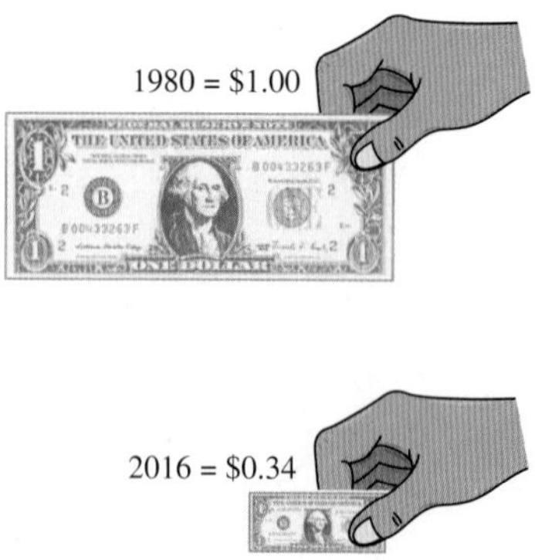

FIGURE 5.19 The lengths of the dollars are proportional to their spending power, but our eyes are drawn to the areas of the bills, which exaggerate the difference. This type of distortion is so common that 19th-century German researchers gave it its own name, which translates roughly as "the old goosing up the effect by squaring the eyeball trick."

Perceptual Distortions

Many graphics are drawn in a way that distorts our perception of them. Figure 5.19 shows one of the most common types of distortion. The dollar-shaped bars are used to represent the declining value of the dollar over time. The problem is that the values are represented by the *lengths* of the dollar bills, but our eyes tend to focus on their *areas*, which exaggerate the difference. This gives the perception that the value of the dollar shrank even more than it really did.

EXAMPLE 8 Area Distortion

How does the reduction in the *areas* of the dollar bills in Figure 5.19 compare to the reduction in their lengths?

Solution The labels on the figure indicate that the 2016 dollar was worth \$0.34 in 1980 dollars, so the 2016 dollar bill was drawn with a length 0.34 times that of the 1980 dollar bill. This means the width is also shorter by this same factor, so the area of the 2016 dollar bill is $0.34^2 \approx 0.12$ that of the 1980 dollar bill. In other words, while the actual 2016 value of the dollar was 34% of the 1980 value, the area change makes it appear that it was only about 12% of that value. ▸ **Now try Exercises 33–34.**

Axes That Exaggerate

Figure 5.20a shows the percentage of college students since 1910 who were women. At first glance, it appears that this percentage grew by a huge margin after about 1950. But the vertical axis scale does not begin at zero and does not end at 100%. The increase is still substantial but looks far less dramatic if we redraw the graph with the vertical axis covering the full range of 0 to 100% (Figure 5.20b). From a mathematical point of view, leaving out the zero point on a scale is perfectly honest and can make it easier to see small-scale trends in the data. Nevertheless, as Figure 5.20 shows, it can be visually deceptive if you don't study the scale carefully.

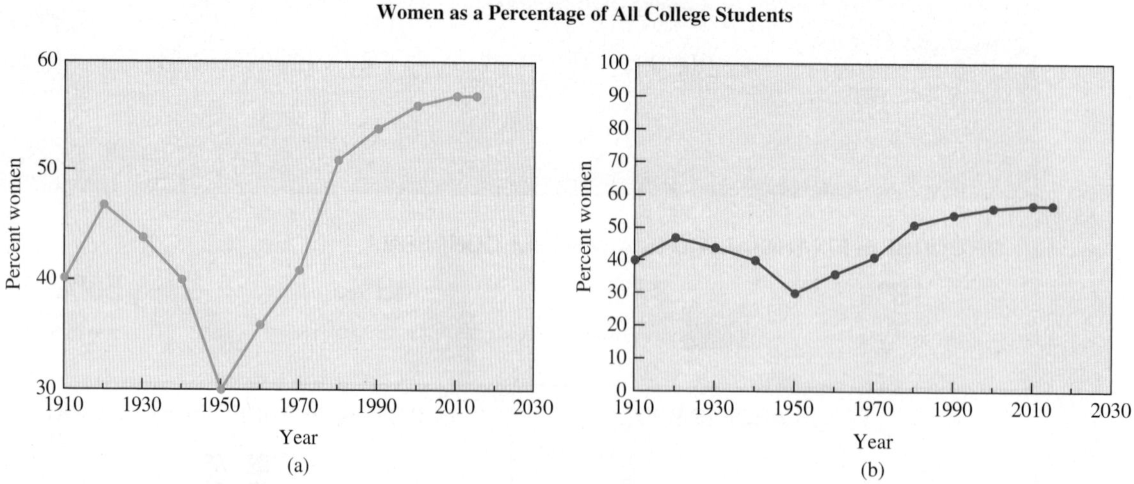

FIGURE 5.20 Both graphs show the same data, but they look very different because their vertical scales have different ranges. *Source:* National Center for Education Statistics.

EXAMPLE 9 Exaggerated Differences

Look back at Figure 5.12. The actual mean scores on PISA mathematics test for U.S. boys and girls were 484 and 479, respectively. How does this actual difference compare to the difference in the heights of the bars shown?

Solution We can find the relative difference between the boys score of 484 and the girls score of 479 with the formula from Unit 3A:

$$\text{relative difference} = \frac{\text{compared value} - \text{reference value}}{\text{reference value}} \times 100\% = \frac{484 - 479}{479} \times 100\% \approx 1\%$$

The boys' score is only about 1% higher than the girls' score. However, because the vertical axis starts from a score of 460, the bar for the boys has a height corresponding to $484 - 460 = 24$ points, while the bar for girls corresponds to $479 - 460 = 19$ points. This makes the relative difference appear to be $(24 - 19)/19 \approx 0.26$, or 26%, which greatly exaggerates the actual difference. (The graph was drawn with a nonzero starting point in order to call attention to the small differences, which would be virtually unnoticeable if the vertical axis started from zero.) ▶ Now try Exercises 35–36.

Nonlinear Scales

It's also important to watch for and carefully interpret graphs that use nonlinear scales, meaning scales in which each increment does not always represent the same change in value. Consider Figure 5.21a, which shows how the speeds of the fastest computers have increased with time. At first glance, it appears that speeds have been increasing almost linearly. For example, it might look as if the speed increased by about the same amount from 1990 to 2000 as it did from 1970 to 1980. However, notice that each tick mark on the vertical axis in Figure 5.21a represents a *tenfold* increase in speed. We then see that in terms of millions of calculations per second, computer speed grew from about 10 in 1970 to 1000 in 1980 and from about 10,000 in 1990 to 1 million in 2000. This type of scale is called an **exponential scale** (or *logarithmic scale*), because it grows by powers (in this case, powers of 10) and powers are indicated with *exponents*. It is always possible to convert an exponential scale back to a linear scale, as shown in Figure 5.21b.

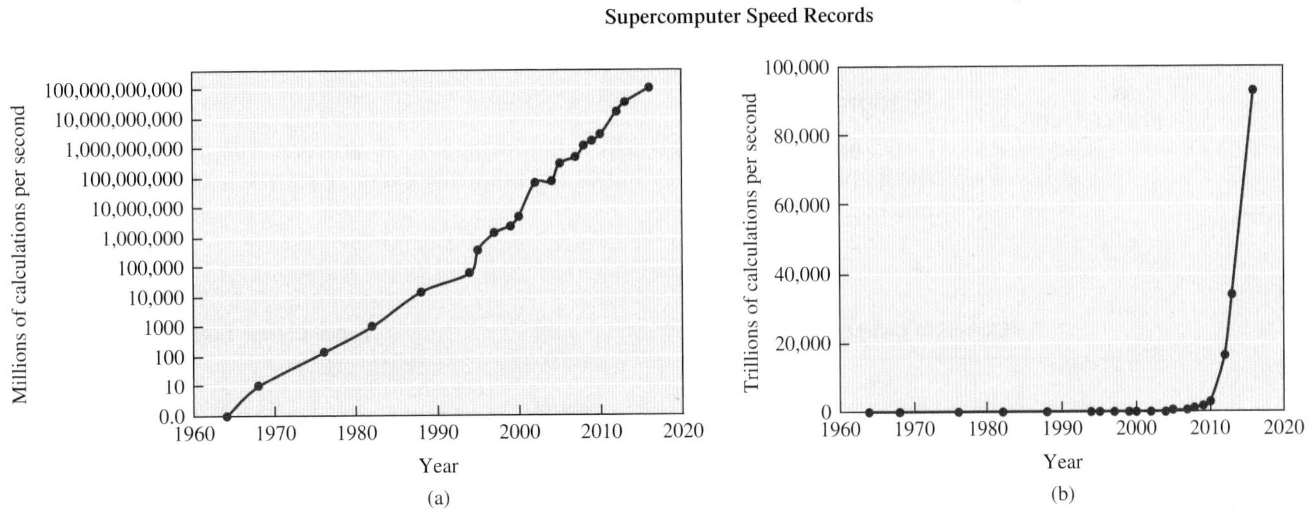

FIGURE 5.21 Both graphs show the same data, but the graph on the left uses an exponential scale.

EXAMPLE 10 Computer Speed

Which graph in Figure 5.21 can you use to determine how many times faster the record supercomputer speed was in 2000 than in 1970? Use your answer to comment more generally on when exponential scales are useful.

Solution Figure 5.21a shows clearly that in the units used for the vertical axis (millions of calculations per second) the record supercomputer speed increased from about

BY THE WAY
In 1965, Intel founder Gordon E. Moore predicted that advances in technology would allow computer chips to double in power roughly every two years. This idea is now called *Moore's law,* and it has held fairly true ever since Moore first stated it.

10 in 1970 to about 10,000,000 in 2000, which represents an increase by a factor of about 1 million. In contrast, Figure 5.21b makes it impossible to determine the speed change from 1970 to 2000, because both data points are visually indistinguishable from zero. This illustrates the general idea that exponential scales are useful whenever data vary over a huge range of values. ▶ Now try Exercises 37–38.

Percentage Change Graphs

Is college tuition getting more or less expensive? If you didn't look carefully, Figure 5.22a might lead you to conclude that after peaking in the mid-2000s, the tuition at public colleges generally decreased. But look more closely and you'll see that the vertical axis on this graph represents the *percentage change* in tuition. The decrease in data values therefore tells us only that tuition was increasing by *smaller* amounts, not that it was decreasing. Actual tuitions are shown in Figure 5.22b, which makes it clear that they increased every year.

Graphs that show percentage change are very common; you'll find them in the financial news almost every day. Although they are perfectly honest, you can be misled unless you interpret them with great care.

EXAMPLE 11 College Costs

For the time period shown in Figure 5.22, in what year did private college tuition increase the most? In what year was the tuition highest?

Solution We can answer the first question with Figure 5.22a, because it shows percentage change. We see that the highest percentage change for private college tuition occurred in the academic year 2001–2002, when it increased by about 8%. To answer the question about when tuition was highest, we must refer to Figure 5.22b. We see that tuition has risen every year, so the highest tuition came in the most recent year shown, which is 2016–2017, when private college tuition was about $33,500.

▶ Now try Exercises 39–40.

Get your facts first, and then you can distort them as much as you please.
—Mark Twain

Pictographs

Pictographs are graphs embellished with additional artwork. The artwork may make the graph more appealing, but as the following example shows, it can also distract or mislead.

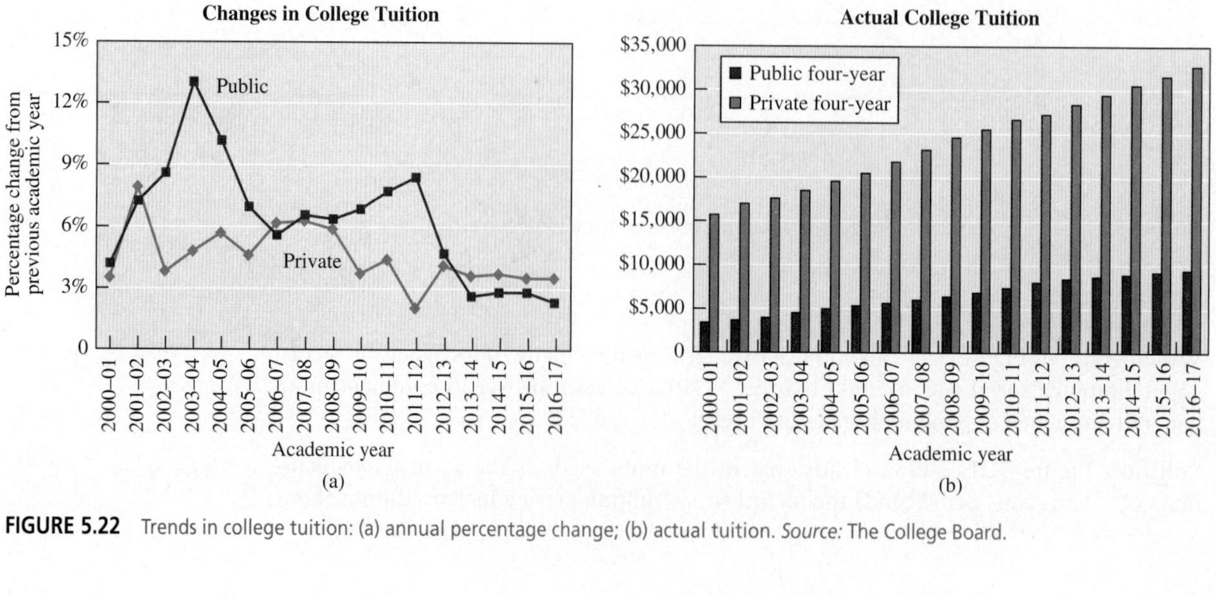

FIGURE 5.22 Trends in college tuition: (a) annual percentage change; (b) actual tuition. *Source:* The College Board.

EXAMPLE 12 World Population Art

Figure 5.23 is a pictograph based on the past and projected growth of world population from 1804 to 2040 (future values are based on United Nations intermediate-case projections). Discuss which aspects of the pictograph represent real data and which are statistically meaningless. Also discuss any ways in which the data are potentially misleading.

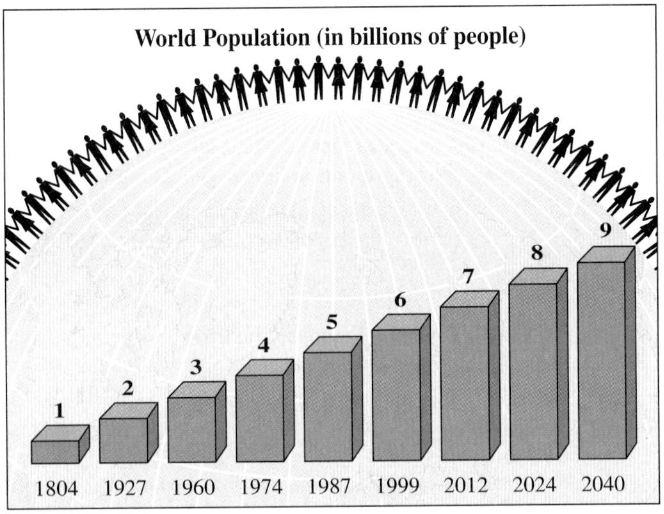

FIGURE 5.23 A pictograph of world population growth. *Source:* United Nations Population Division, future projections based on intermediate-case assumptions.

Solution The bar graph along the bottom shows real data, but the rest of the artwork (the globe and the hand-holding people) is purely decorative and statistically meaningless. For the bar graph, if you measure with a ruler, you'll find that the bar heights are reasonable representations of the populations for the indicated years. For example, the bar for 4 billion people is twice as tall as the bar for 2 billion people, as it should be with a linear scale. However, the graph is potentially misleading in the way the bars are spaced along the horizontal axis. Because the bars are arranged with equal space between them, a quick glance makes it appear that the population has risen linearly with time. In fact, close inspection shows that the time intervals between bars differ greatly, thereby masking the fact that recent increases of 1 billion people have occurred in much less time than they did in the past. For example, it took 123 years (from 1804 to 1927) for the population to increase from 1 billion to 2 billion, but only 12 years (from 1987 to 1999) to increase from 5 billion to 6 billion. ▶ Now try Exercises 41–42.

BY THE WAY

If world population continued doubling at the same rate as it did in the late 20th century, it would reach 34 billion by 2100, 192 billion by 2200, and by about 2650 it would have become so large that the human population would not fit on the Earth, even if everyone stood elbow to elbow everywhere.

Quick Quiz 5D

Choose the best answer to each of the following questions. Explain your reasoning with one or more complete sentences.

1. Based on Figure 5.12, which statement is correct?

   a. Boys' scores in the United States are greater than girls' scores in Canada.

   b. Boys' scores in the United States are nearly equal to girls' scores in Canada.

   c. Boys' scores in the United States are less than girls' scores in Canada.

2. Consider Figure 5.13. Notice that the red segment of the second bar from the top starts farther right than the red segment of the top bar. This fact tells us that

   a. books and supplies cost more at public 4-year institutions than at public 2-year institutions.

   b. room and board cost more at public 4-year institutions than at public 2-year institutions.

   c. tuition and fees plus room and board are higher at public 4-year institutions than at public 2-year institutions.

3. Consider Figure 5.14. What was the death rate for pneumonia in 1940?

   a. Approximately 60 deaths per 100,000

   b. Approximately 750 deaths per 100,000

   c. Approximately 640 deaths per 100,000

4. Consider Figure 5.15. According to this graph, what is per capita energy use in Missouri (MO)?

   a. between 2000 and 2999 gallons of oil equivalent

   b. between 3000 and 3999 gallons of oil equivalent

   c. more than 4000 gallons of oil equivalent

5. Consider Figure 5.16. On the day shown, the temperature in Des Moines, Iowa (IA) was

   a. 30°F. b. 40°F.

   c. between 30°F and 40°F.

6. Suppose you are given a contour map showing elevation (altitude) for the state of Vermont. The region with the most closely spaced contours represents

   a. the highest altitude. b. the lowest altitude.

   c. the steepest terrain.

7. Which of the following best describes the perceptual distortion that occurs in Figure 5.19?

   a. The meaningful data are represented by the dollar bill lengths, but our eyes tend to focus on their areas.

   b. Given that the dollar bill for 2016 is shown smaller, the hand should also have been shown smaller.

   c. The change in the dollar bill sizes is deceptive, because actual dollar bills were the same size in 2016 and in 1980.

8. Consider Figure 5.20a. The way the graph is drawn

   a. makes it completely invalid.

   b. makes the changes from one decade to the next appear larger than they really were.

   c. makes it more difficult to see the upward and downward trends that have occurred over time.

9. Consider Figure 5.21a. Moving one tick mark up the vertical axis represents an increase in computer speed of

   a. 1 billion calculations per second.

   b. a factor of 2.

   c. a factor of 10.

10. Consider Figure 5.22a. In years where the graph slopes downward with time

   a. college tuition decreased.

   b. college tuition rose, but more slowly than in the prior years.

   c. college tuition rose, but the new tuition represented a lower proportion of the average person's income.

Exercises 5D

REVIEW QUESTIONS

1. Briefly describe the construction and use of multiple bar graphs, multiple line charts, and stack plots.
2. What are *geographical data*? Briefly describe at least two ways to display geographical data. Be sure to explain the meaning of contours on a contour map.
3. What are three-dimensional graphics? Explain the difference between graphics that only appear three-dimensional and those that show truly three-dimensional data.
4. What are *infographics*, and what is their goal?
5. Describe how perceptual distortions and scales that do not go all the way to zero can each be misleading. Why are graphics with these features sometimes useful?
6. What is an *exponential scale*? When is an exponential scale useful?
7. Explain how a graph that shows percentage change can show descending bars (or a descending line) even when the variable of interest is increasing.
8. What is a *pictograph*? How can a pictograph enhance a graph? How can it make a graph misleading?

DOES IT MAKE SENSE?

Decide whether each of the following statements makes sense (or is clearly true) or does not make sense (or is clearly false). Explain your reasoning.

9. My bar chart contains more information than yours, because my bars are three-dimensional.
10. I used an exponential scale because the data values for my categories ranged from 7 to 450,000.
11. There's been only a very slight rise in our stock price over the past few months, but I wanted to make it look dramatic so I started the vertical scale from the lowest price rather than from zero.
12. A graph showing the yearly rate of change in the number of computer users has a slight downward trend, even though the actual number of users is rising.

BASIC SKILLS & CONCEPTS

13–16: Educational Value. Use the graphs in Figure 5.11 to answer the following questions.

13. a. What is the percentage difference in earnings for men with a bachelor's degree compared to women with a bachelor's degree?

b. What is the percentage difference in earnings for men with a professional degree (such as an M.D. or J.D.) compared to women with a professional degree?

14. a. What is the percentage difference in earnings for women with a professional degree (such as an M.D. or J.D.) compared to women with a bachelor's degree?

b. What is the percentage difference in earnings for men with a professional degree compared to men with a bachelor's degree?

c. Does there appear to be a gender difference between men and women with respect to the increased earnings they would receive with a professional degree compared to a bachelor's degree? Explain.

15. a. On average, people spend about 45 years in the work force before retiring. How much more would the average male college graduate (bachelor's degree) earn during these 45 years than the average male high school graduate?

b. How much more would the average female college graduate with a professional degree earn over a 45-year career than the average female college graduate with only a bachelor's degree?

16. a. Briefly describe how the unemployment rate has changed with time over the period shown in Figure 5.11b. Has the trend in differences between educational attainment level remained consistent during this time? Explain.

b. For 2016, approximately how much more likely was a high school dropout to be unemployed than a worker with a bachelor's degree? State your answer as a multiplicative factor (that is, in the form "*x* times more likely").

17. **Gender and Mathematics.** Consider the data displayed in Figure 5.12.

a. Which group scored higher on the PISA test, boys in Canada or girls in South Korea? Estimate the test scores of each group.

b. Which group scored higher on the PISA test, boys in Finland or girls in Japan? Estimate the test scores of each group.

18. **Gender and Science.** The following table gives data for boys and girls in six countries who took the PISA science test in 2012. Make a double bar graph for these data, with the vertical axis running from 460 to 560.

PISA Science Test Results (2012)

Country	Mean (boys)	Mean (girls)
United States	497	498
Finland	537	554
Canada	527	524
Japan	552	541
South Korea	539	536
Germany	524	524

19–20: College Costs Stack Plot. Use Figure 5.13 to answer the following questions.

19. a. Which cost category varies the least among the different types of colleges? Can you explain why this category varies so little?

b. Ignoring the "other expenses" category, the general trend is for all categories to cost more as you look down the chart from 2-year public colleges to 4-year private colleges. However, one category is an exception to this trend. Which category? Can you explain why?

20. Use the data shown in the figure to construct a bar graph for tuition only at the four different types of colleges, using *vertical* bars.

21–22. Disease Stack Plot. Use Figure 5.14 to answer the following questions.

21. a. For each of the four diseases individually, state whether its death rate increased, decreased, or remained nearly constant between 1900 and 1950.

b. About when did the death rate for cardiovascular disease reach its highest value between 1900 and 2015, and what was the value?

c. What was the approximate death rate due to cancer in 2000?

d. Based on the trends in the graph, speculate on which of these four diseases will be responsible for the most deaths in 2050. Explain.

22. a. For each of the four diseases individually, state whether its death rate increased, decreased, or remained nearly constant between 1950 and 2015.

b. Compare the death rates due to cancer in 1980 and 2015.

c. In about what year shown on the graph was the death rate due to cardiovascular disease at about 320 deaths per 100,000?

d. In 2015, there were 3.0 cases (not deaths) of tuberculosis per 100,000. Could 3.0 deaths per 100,000 be represented in Figure 5.14? Explain.

23. Figure 5.24 shows a stack plot of federal government spending categories as a percentage of the total spending, with data through 2016.

a. Briefly explain how to interpret each wedge. What were the approximate percentages of the federal budget spent on Medicare in 1976 and in 2016?

b. In approximately what years was spending on national defense the greatest, as a percentage of the budget?

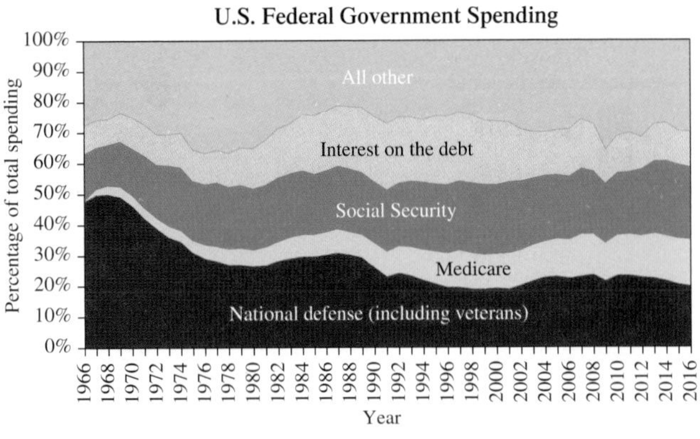

FIGURE 5.24 *Source:* Office of Management and Budget.

c. As a percentage of the budget, how has spending for Social Security changed since 1966?

d. Can you use this graph to draw any conclusions about how *total* government spending has changed with time? Why or why not?

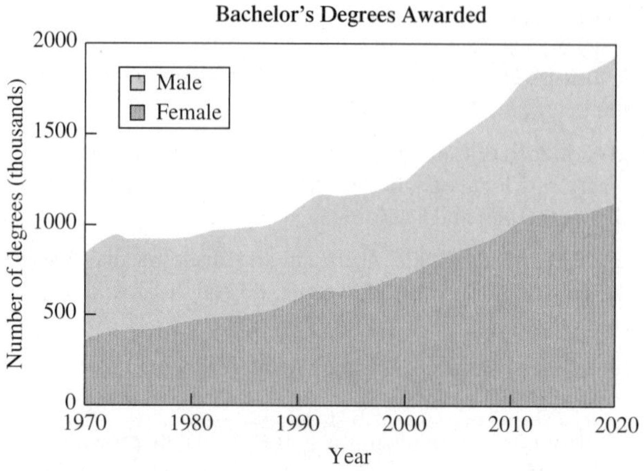

FIGURE 5.25 *Source:* National Center for Education Statistics.

24. **College Degrees.** Figure 5.25 shows the numbers of bachelor's degrees awarded to males and females since 1970. (The last few years are projections.)

a. Estimate the numbers of bachelor's degrees awarded to males and to females (separately) in 2020.

b. About when were the numbers of bachelor's degrees equal for males and females?

c. During what decade did the *total* number of degrees awarded increase the most?

d. Do you think the stack plot is an effective way to display these data? Briefly discuss other ways that might have been used instead.

25–26: Melanoma Mortality. Figure 5.26 shows the female mortality rates for melanoma (a malignant form of skin cancer) across the counties of the continental United States. Use the figure to answer the following questions.

25. Are there broad regions where melanoma mortality is more common than others? Which ones, and what do they have in common?

26. Are there any locations that stand out as unusual and that might therefore warrant special study? Explain.

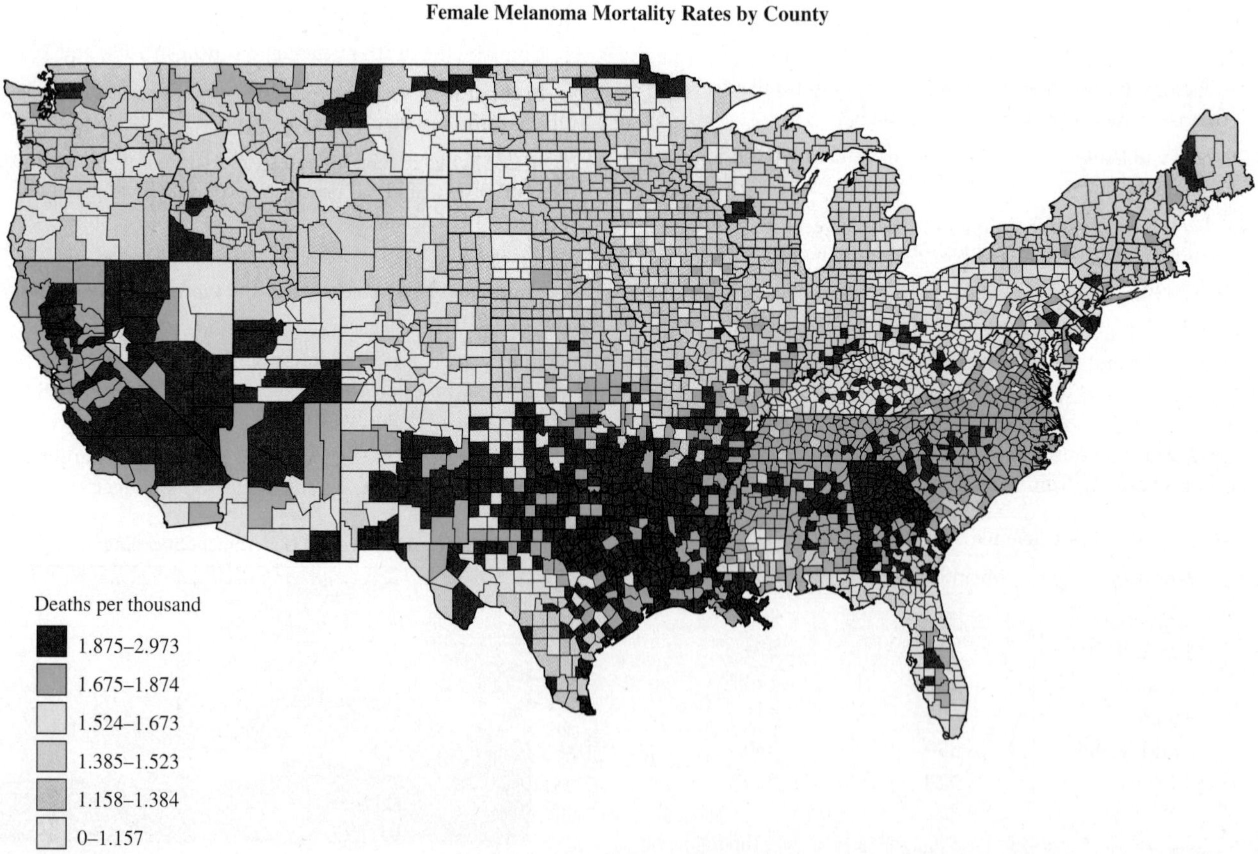

FIGURE 5.26 *Source:* "Female Melanoma Mortality Rates by County," Karen Kafadar, University of Colorado-Denver. Reprinted with permission.

27–28: Contour Map. Figure 5.27 shows an elevation contour map with six points (A through F) marked on it. Assume that points A and B correspond to mountain summits and that the contour lines represent altitude changes of 40 feet. Use the figure to answer the following questions.

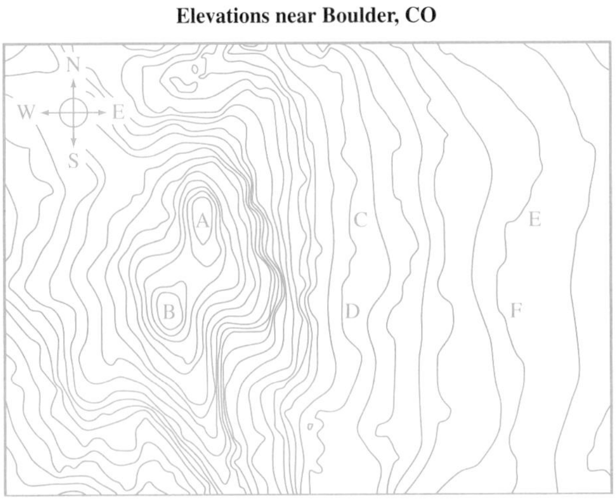

FIGURE 5.27

27. a. If you walk from A to C, do you walk uphill or downhill? Explain how you know.

b. Does your elevation change more in walking from B to D or from D to F? Explain how you know.

28. a. If you walk directly from E to F, does your elevation increase, decrease, or remain the same? Explain how you know.

b. What is your net elevation change if you walk from A to C to D to A? Explain how you know.

29–30: 3-D Bars. Figure 5.28 shows the age distribution of the U.S. population from 2010 through 2050 (with projections for the later years). Use the figure to answer the following questions.

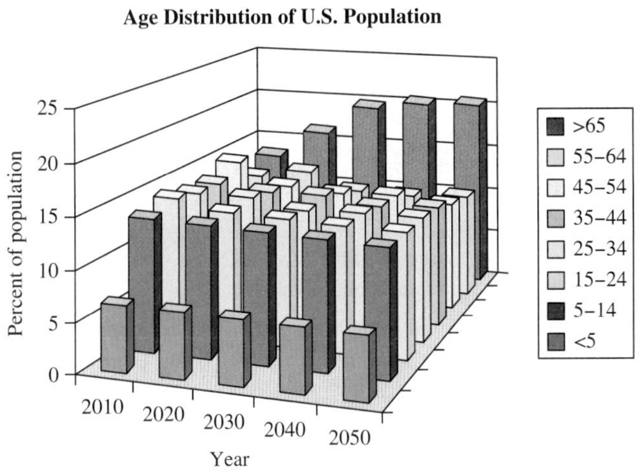

FIGURE 5.28

29. a. About what percentage of the population was over age 65 in 2010?

b. About what percentage of the population is projected to be over age 65 in 2050?

c. Describe the projected change in the 45–54 age group between 2010 and 2050.

30. a. In about what year did (will) 45- to 54-year-olds comprise the largest percentage of the population?

b. In about what year did (will) the population of over 65-year-olds peak as a percentage of the population?

c. Discuss any significant trends that you see in the data.

31–32: Infographic. Use Figure 5.18 to answer the following questions.

31. a. Within a group of 500 typical first-year college students, approximately how many have loans to help pay for their education?

b. Can you conclude that some students who are satisfied with their coursework also text in class? Explain.

32. a. What percentage of students do *not* party at least 3 hours per week?

b. The graphic shows two different percentages for categories that mention relevance of coursework. Is this a contradiction? Why or why not?

33. **Volume Distortion.** Figure 5.29 depicts the amounts of daily oil consumption (in millions of barrels) in the United States and Japan (2015 data). Does the illustration accurately depict the data? Why or why not?

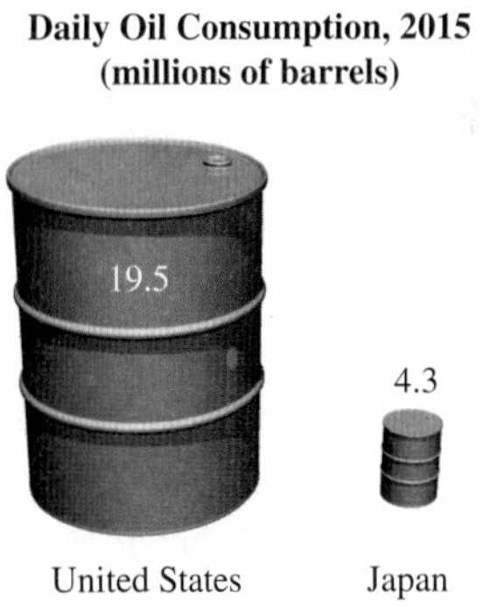

FIGURE 5.29

34. **Three-Dimensional Pies.** The pie charts in Figure 5.30 represent the percentages of Americans in three age categories in 1990 and 2050 (projected). Briefly explain how the three-dimensional effects create a perceptual distortion in this case. Why would flat pies (without the three-dimensional effects) give a more accurate representation of the data?

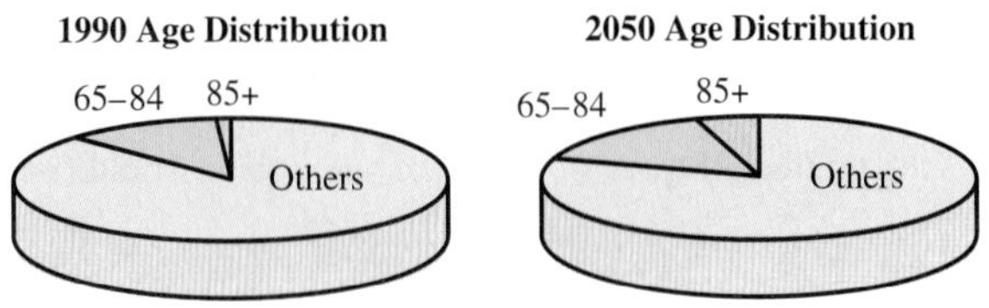

FIGURE 5.30 *Source:* U.S. Census Bureau.

35. **Refugees to Europe.** Figure 5.31 shows the numbers of refugees arriving in Europe from three countries of origin in 2016.

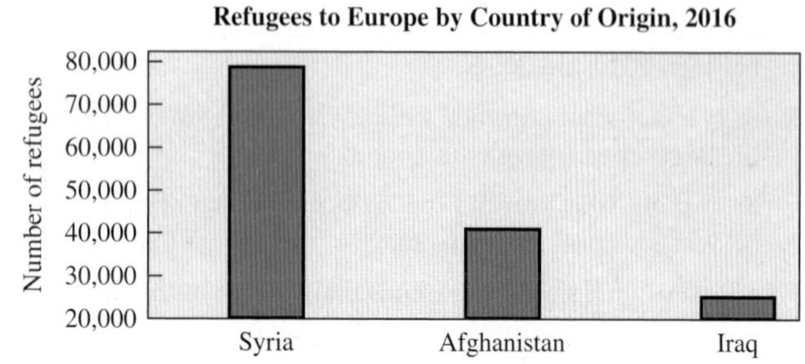

FIGURE 5.31 *Source:* United Nations High Commission for Refugees.

a. What is the ratio of refugees from Iraq to refugees from Syria? What is the ratio of the lengths of the corresponding bars in the graph?

b. Based on your answer to part (a), do you consider the graph to be deceptive? Explain.

c. Redraw the bar graph with a starting point of zero on the vertical axis.

36. **Life Expectancy.** Figure 5.32 shows the life expectancy of Americans at birth since 1950.

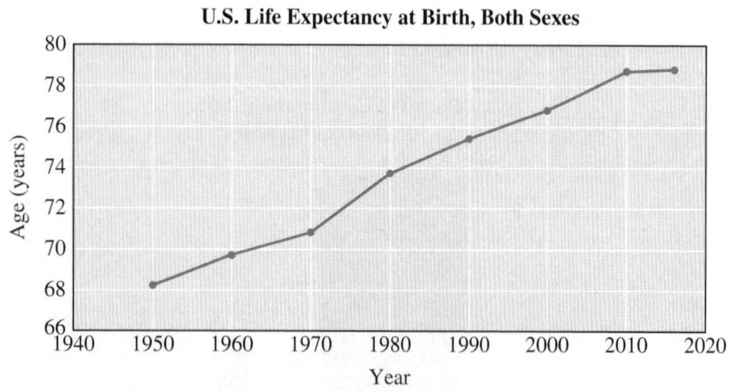

FIGURE 5.32 *Source:* National Center for Health Statistics.

a. How much has life expectancy increased (in absolute and relative terms) since 1950?

b. Based on your answer to part (a), do you consider the graph to be deceptive? Explain.

c. Redraw the graph with a starting point of zero on the vertical axis. Which version of the graph (the original below or your redrawn version) do you consider more useful for showing trends in life expectancy? Why?

37. **Moore's Law.** *Moore's law* refers to a prediction made in 1965 by Intel cofounder Gordon Moore, who suggested that the number of transistors on computer chips would rise exponentially with time. With some variation, the prediction has held true for decades. Use the following table to answer the questions below.

Year	Transistors	Year	Transistors
1971	2300	1993	31,000,000
1974	4500	1997	7,500,000
1978	29,000	2000	42,000,000
1982	55,000	2003	411,000,000
1985	275,000	2011	1,160,000,000
1989	1,180,000	2015	7,200,000,000

a. Construct a time-series graph of these data, using a uniform scale on both axes.

b. Construct an exponential graph of these data in which the subdivisions on the vertical axis are 1, 10, 100, 1000, and so on.

c. Which of your two graphs is more useful for these data? Explain.

38. **Cell Phone Subscriptions.** The following table shows the number of cell phone subscriptions in the United States for selected years. (The number of subscriptions can be larger than the population because some people have more than one subscription, such as for personal and business cell phones.) Display the data using both an ordinary vertical scale and an exponential vertical scale. Which graph is more useful? Why? (*Hint:* For the exponential scale, use tick marks at 1 million, 10 million, and 100 million.)

Year	Subscriptions (millions)	Year	Subscriptions (millions)
1990	5	2002	141
1995	34	2003	159
1997	55	2007	255
1998	69	2010	303
1999	86	2012	325
2000	110	2015	378
2001	128		

39. **College Tuition.** Answer the following questions for the time period shown in Figure 5.22.

a. In what academic year did public college tuition rise by the largest percentage, and about what was that percentage?

b. In the year found in part (a), about what was the percentage increase in private college tuition?

c. In the year found in part (a), which had the larger increase in actual tuition (in dollars): public or private colleges? Explain.

40. **Percentage Change in the CPI.** Figure 5.33 shows the percentage change in the CPI over recent years.

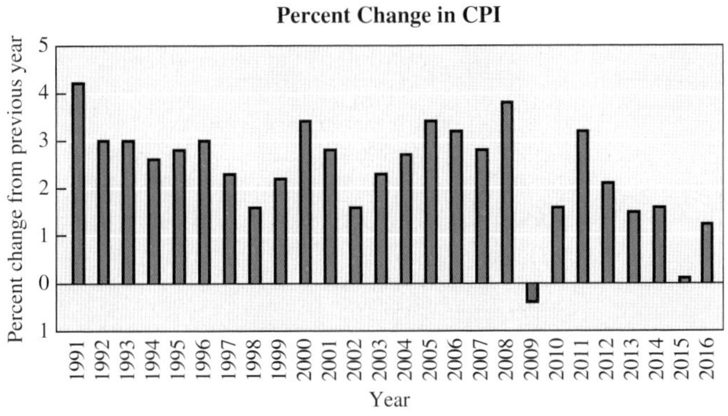

FIGURE 5.33

a. In what year (of those displayed) was the change in the CPI the greatest?

b. In what year was the actual CPI the highest? How do you know?

c. Briefly explain the meaning of the red bar for 2009.

41. **World Population.** Recast the population data in Figure 5.23 with a linear horizontal axis. What trends are clear in your new graph that are not clear in the original? Explain.

42. **Find a Pictograph.** Find a pictograph (*USA Today* frequently includes them, and there are other possible sources), and briefly discuss which aspects of it represent real data, which are statistically meaningless, and ways in which the pictograph is potentially misleading.

FURTHER APPLICATIONS

43. **Net Grain Production.** Net grain production is the difference between the amount of grain a country produces and the amount of grain its residents consume. It is positive if the country produces more than it consumes and negative if the country consumes more than it produces. Figure 5.34 shows the net grain production of four countries in 1990 and projected for 2030.

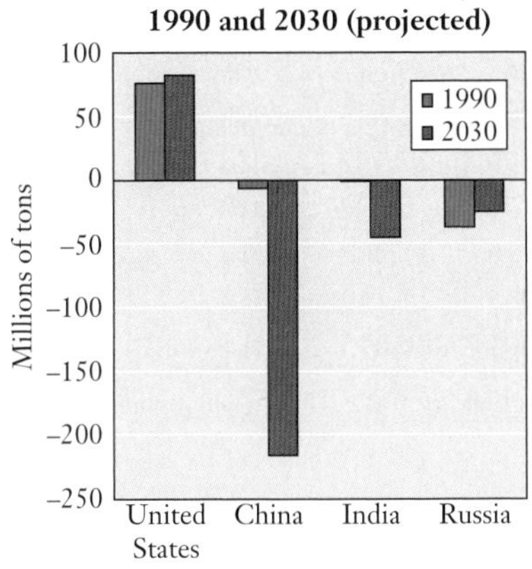

FIGURE 5.34

a. Which of the four countries had to import grain to meet consumption needs in 1990?

b. Which of the four countries are expected to import grain to meet consumption needs in 2030?

c. Given that India and China are the world's two most populous countries, what does this graph tell you about how world agriculture must change between now and 2030?

44. **Marriage and Divorce Rates.** The graph in Figure 5.35 depicts U.S. marriage and divorce rates for selected years. The marriage rates are depicted by the blue bars, and the divorce rates are depicted by the red bars. The rates are given as number of marriages or divorces per 1000 people in the population.

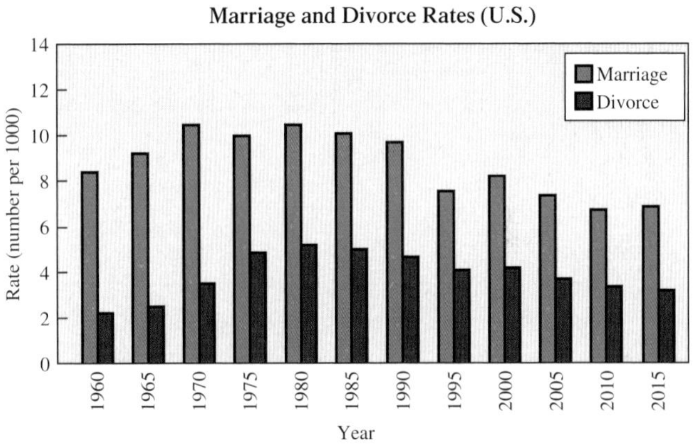

FIGURE 5.35 *Source:* Department of Health and Human Services

a. Why do these data consist of marriage and divorce *rates* rather than total numbers of marriages and divorces? Comment on any trends that you observe in these rates, and give plausible historical and sociological explanations for these trends.

b. Redraw the graph as a multiple line chart (one line for marriage rate and one for divorce rate). Briefly discuss the advantages and disadvantages of the two different representations of this particular data set.

45. **Double Horizontal Scale.** The graph in Figure 5.36 shows *simultaneously* the number of births in the United States during two time periods: 1946–1964 and 1977–1994.

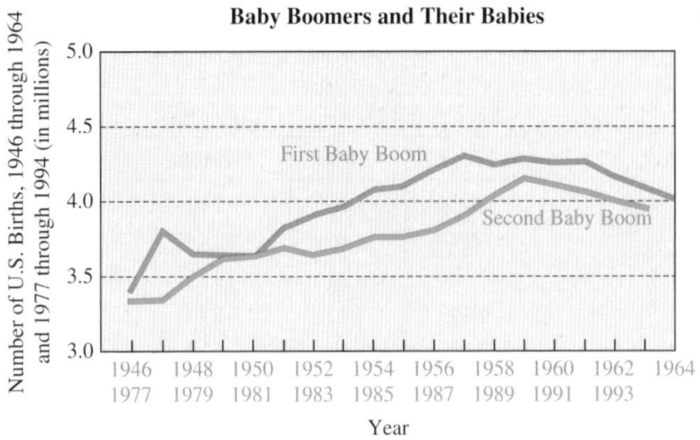

FIGURE 5.36 *Source:* Based on data from the National Center for Health Statistics.

a. When did the first baby boom peak?

b. When did the second baby boom peak?

c. Why do you think the designer of this display chose to superimpose the two time intervals, rather than use a single time scale from 1946 through 1994?

46. **Seasonal Effects on Schizophrenia?** The graph in Figure 5.37 shows data on the relative risk of schizophrenia among people born in different months.

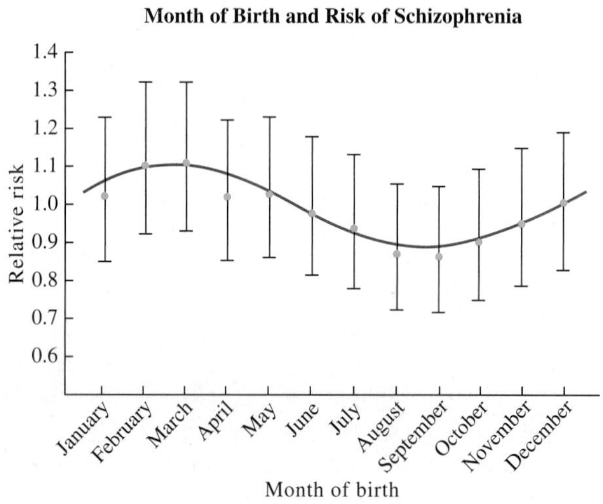

FIGURE 5.37 *Source: New England Journal of Medicine.*

a. Note that the scale of the vertical axis does not start at zero. Sketch the same risk curve using an axis that starts at zero. Comment on the effect of this change.

b. Each value of the relative risk is shown with a dot at its most likely value and with an "error bar" indicating the range in which the data value probably lies. The study concludes that "the risk was also significantly associated with the season of birth." Given the size of the error bars, does this claim appear justified? (Is it possible to draw a flat line that passes through all of the error bars?)

47–52: Creating Graphics. Make a graphical display of the following data sets, choosing any type of display that you feel is appropriate. Explain your choice of display and discuss any interesting features in the data.

47. The following table shows the number of bachelor's degrees (in thousands) conferred on men and women in U.S. colleges and universities in selected years (and projected for 2020).

Bachelor's Degrees (thousands)

Year	Men	Women
1960	260	140
1970	460	335
1980	480	475
1990	490	530
2000	510	710
2010	700	920
2020	810	1050

Source: National Center for Education Statistics.

48. The following table shows the percentages of American men and women who had never been married in selected years since 1960.

Percentages of Adults Who Never Married

Year	Women	Men
1960	19.0%	25.3%
1980	22.5%	29.6%
2000	25.1%	31.3%
2015	28.9%	34.8%

Source: U.S. Census Bureau.

49. The following table gives the median age of the U.S. population in selected years. (Half the population is below the median age and half is above the median age.)

Year	U.S. median age (years)
1920	25.3
1930	26.5
1940	29.0
1950	30.2
1960	29.5
1970	28.1
1980	30.0
1990	32.8
2000	35.3
2010	37.2
2015	37.8

Source: U.S. Census Bureau.

50. The following table gives the total number of automobile fatalities and the number of fatalities in which alcohol was involved for selected years.

Automobile Fatalities

Year	Total	Alcohol involved	Year	Total	Alcohol involved
1982	43,945	26,173	2000	41,945	17,380
1984	44,257	24,762	2002	42,815	17,419
1986	46,087	25,017	2004	42,643	16,919
1988	47,087	23,833	2006	42,708	15,829
1990	44,599	22,587	2008	34,017	11,773
1992	39,250	18,290	2010	32,885	10,200
1994	40,716	17,308	2012	33,561	10,322
1996	42,065	17,749	2014	32,675	9967
1998	41,501	16,673	2016 (est.)	37,200	10,300

Source: National Highway Traffic Safety Administration.

51. The following table gives the number of U.S. daily newspapers and their total circulation (in millions) for selected years since 1920.

Year	Number of daily newspapers	Circulation (millions)
1920	2042	27.8
1930	1942	39.6
1940	1878	41.1
1950	1772	53.9
1960	1763	58.8
1970	1748	62.1
1980	1747	62.2
1990	1611	62.3
2000	1485	56.1
2010	1390	45.2
2015	1350	34.9

Source: Editor & Publisher International Yearbook.

52. The following table lists U.S. labor force participation rates (as percentages) for women with children, binned in two categories based on the age of the woman's youngest child.

Percentages of Mothers in the Labor Force		
Year	Youngest child aged 6 to 17 years	Youngest child under 6 years of age
1980	64.3%	46.8%
1985	69.9%	53.5%
1990	74.7%	58.2%
1995	76.4%	62.3%
2000	79.0%	65.3%
2005	76.9%	62.6%
2010	76.5%	63.9%

Source: Bureau of Labor Statistics.

53. **Gender Wage Gap.** The following table gives the average (median) weekly incomes of American workers (men and women) in several age categories.

a. Choose an appropriate graph to display these data. Then write a short paragraph explaining any noticeable trends or patterns.

b. For each age category, compute the gender wage gap: the ratio of the women's to men's earnings. Then write a short paragraph explaining any noticeable patterns.

	Weekly Income	
Age	Men	Women
16–24	\$505	\$470
25–34	\$791	\$710
35–44	\$1024	\$845
45–54	\$1063	\$829
55–64	\$1054	\$795
65 and over	\$1032	\$736

Source: U.S. Bureau of Labor Statistics.

54. **Senate Infographic.** The Senate of the 115th Congress includes 52 Republicans, 44 Democrats, and 2 Independents; 79 men and 21 women; 3 African Americans, 4 Hispanics, and 3 Asian Americans; 1 openly gay member; 1 Buddhist and 8 Jewish members; 7 members in their 80s (at the start of the session in 2017), 14 in their 70s, 40 in their 60s, 25 in their 50s, 13 in their 40s, and 1 in his 30s. Devise an infographic that best illustrates these data.

IN YOUR WORLD

55. **News Graphics.** Find a recent news report that shows a multiple bar graph or stack plot. Comment on the effectiveness of the display. Could another type of graphic have been used to depict the same data?

56. **Geographical Data.** Find an example of a graph of geographical data in a recent news report. Comment on the effectiveness of the display. Could another type of graphic have been used to depict the same data?

57. **Three-Dimensional Effects.** Find an example of a three-dimensional display in a recent news report. Are the data three-dimensional or are the three-dimensional effects cosmetic? Comment on the effectiveness of the display. Could another type of graphic have been used to depict the same data?

58. **Graphic Confusion.** Find an example in a recent news report of a graph that is misleading in one of the ways discussed in this unit. Explain what makes the graph misleading, and describe how it could have been drawn more honestly.

59. **Outstanding News Graph.** Find a graph from a recent news report that, in your opinion, does a truly outstanding job of displaying data visually. Discuss what the graph shows, and explain why you think it is so outstanding.

60. **Interactive Infographics.** Find a few examples of infographics, especially those that are interactive or animated. Good sources include gapminder.org, dailyinfographic.com, and many news sites (*The New York Times* often features them). Choose one that you particularly like, and write a short summary of what you like about it.

TECHNOLOGY EXERCISES

Answer the following questions using procedures described in the Using Technology boxes in this unit or with **StatCrunch** (available in MyLab Math).

61–66. Graphs with Multiple Data Sets. Consider the following data on various causes of deaths in selected years since 1980.

	(a) Motor vehicle	(b) Falling	(c) Poisoning	(d) Choking	(e) Drowning	(f) Total
1980	53,172	13,294	4331	3249	7257	105,718
1990	46,814	12,313	5803	3303	4685	91,983
2000	43,354	13,322	12,757	4313	3482	97,900
2010	35,332	26,009	33,041	4570	3782	120,859
2015	35,398	31,959	42,032	4816	3406	136,053

Sources: National Safety Council; National Center for Health Statistics.

61. Use Excel or StatCrunch to create a multiple line chart showing the number of deaths due to each of the five causes with a separate line for each cause.

62. Use Excel or StatCrunch to create a stack plot (similar to Figure 5.14) showing the *number* of deaths due to each of the five causes with a separate slice for each cause.

63. Use Excel or StatCrunch to create a stack plot (similar to Figure 5.14) showing the *percentage* of deaths due to each of the five causes with a separate slice for each cause.

64. Use Excel or StatCrunch to create a stacked bar graph (similar to Figure 5.13) showing the *number* of deaths due to each of the five causes with a separate bar for each year.

65. Use Excel or StatCrunch to create a stacked bar graph (similar to Figure 5.13) showing the *percentage* of deaths due to each of the five causes with a separate bar for each year.

66. Use Excel or StatCrunch to create a three-dimensional bar graph (similar to Figure 5.17) to show: the five causes of death in the table, time (the years shown in the table), and the number of deaths due to each cause for each year.

67. **Arctic Sea Ice.** Open the StatCrunch shared data set called *Arctic Sea Ice Volume* (choose "Explore" then "Data" and then type the data set name in the search box).

a. Make a multiple line chart for the March and September data for the years 1979 through 2013.

b. Discuss the trend in the data over the years covered and compare the graphs for March and September.

68. **StatCrunch Project.** Choose one data set available in StatCrunch that is of interest to you and that lends itself to making one of the graphic types discussed in this unit.

a. State the data set you've chosen along with a sentence about why it is of interest to you.

b. Make a graphical display of the data set. Explain why you chose the type of graphic you did, and how it is useful to understanding the data.

UNIT 5E Correlation and Causality

A major goal of many statistical studies is to determine whether one factor *causes* another. For example, does smoking cause lung cancer? In this unit, we will discuss how statistics can be used to search for *correlations* that might suggest a cause-and-effect relationship. Then we'll explore the more difficult task of establishing causality.

Seeking Correlation

What does it mean when we say that smoking *causes* lung cancer? It certainly does *not* mean that you'll get lung cancer if you smoke a single cigarette. It does not even mean that you'll get lung cancer if you smoke heavily for many years, because some heavy smokers do not get lung cancer. Rather, it is a *statistical* statement meaning that you are *much more likely* to get lung cancer if you smoke than if you don't smoke.

How did researchers learn that smoking causes lung cancer? The process began with informal observations, as doctors noticed that a surprisingly high proportion of their patients with lung cancer were smokers. These observations led to carefully conducted studies in which researchers compared lung cancer rates among smokers and nonsmokers. These studies showed clearly that heavier smokers were more likely to get lung cancer. In more formal terms, we say that there is a **correlation** between the variables *amount of smoking* and *incidence of lung cancer.* A correlation is a special type of relationship between variables in which a rise or fall in one goes along with a corresponding rise or fall in the other.

BY THE WAY

Smoking is linked to many serious diseases besides lung cancer, including heart disease and emphysema. Smoking is also linked with less lethal health conditions, such as premature skin wrinkling and sexual impotence.

Definition

A **correlation** exists between two variables when higher values of one variable are consistently associated with higher values of another variable or when higher values of one variable are consistently associated with lower values of another variable.

Here are a few other examples of correlations:

- There is a correlation between the variables *height* and *weight* for people. That is, taller people tend to weigh more than shorter people.
- There is a correlation between the variables *demand for apples* and *price of apples*. That is, demand tends to decrease as prices increase.
- There is a correlation between *practice time* and *skill* among piano players. That is, those who practice more tend to be more skilled.

Establishing a correlation between two variables does *not* mean that a change in one variable *causes* a change in the other. The correlation between smoking and lung cancer did not by itself prove that smoking causes lung cancer. We could imagine, for example, that some gene predisposes a person both to smoking and to lung cancer. Nevertheless, identifying the correlation was the crucial first step in learning that smoking causes lung cancer.

Think About It Suppose there actually was a gene that made people prone to both smoking and lung cancer. Explain why we would still find a strong correlation between smoking and lung cancer in that case, but would not be able to say that smoking caused lung cancer.

Scatterplots

Table 5.6 shows the production cost and U.S. gross receipts (total revenue from ticket sales) for the 15 biggest-budget movies of all time (through 2016). Movie executives presumably hope there is a favorable correlation between the production budget and the receipts. That is, they hope that spending more to produce a movie will result in higher box office receipts. But is there such a correlation? We can look for a correlation by making a **scatterplot** showing the relationship between the variables *production cost* and *gross receipts*.

TABLE 5.6 Biggest-Budget Movies (through 2016)

Movie	Production Cost (millions of dollars)	U.S. Gross Receipts (millions of dollars)
Pirates of the Caribbean: On Stranger Tides (2011)	378	241
Star Wars: Episode VII (2015)	306	937
Pirates of the Caribbean: At World's End (2007)	300	309
Spectre (2015)	300	200
Avengers: Age of Ultron (2015)	280	459
The Lone Ranger (2013)	275	89
The Dark Knight Rises (2012)	275	448
John Carter (2012)	264	73
Tangled (2010)	260	201
Spider-Man 3 (2007)	258	337
Harry Potter and the Half-Blood Prince (2009)	250	302
The Hobbit: Battle of the Five Armies (2014)	250	255
Captain America: Civil War (2016)	250	408
Batman v Superman (2016)	250	330
Avatar (2009)	237	761

Note: Budgets are estimates, which may differ significantly from different sources (because studios do not usually release precise figures). Gross receipts are for United States only; worldwide receipts are often substantially higher. These figures are not adjusted for inflation.

Definition

A **scatterplot** is a graph in which each point represents the values of two variables.

Technical Note

We often have some reason to think that one variable depends at least in part on the other. In the case of Figure 5.38, we might guess that gross receipts should depend on the production cost. We therefore call production cost the *explanatory variable* and gross receipts the *response variable*, because the production cost might help explain the gross receipts. The explanatory variable is usually plotted on the horizontal axis and the response variable on the vertical axis.

The following procedure describes how to make a scatterplot of the data in Table 5.6, which is shown in Figure 5.38:

1. We assign one variable to each axis and label each axis with values that comfortably fit the data. Here we assign *production cost* to the horizontal axis and *gross receipts* to the vertical axis. We choose a range of \$200 to \$400 million for the production cost axis and \$0 to \$1000 million (\$1 billion) for the gross receipts axis.
2. For each movie in Table 5.6, we plot a *single point* at the horizontal position corresponding to its production cost and the vertical position corresponding to its gross receipts. For example, the point for the movie *Tangled* goes at the position \$260 million on the horizontal axis and \$201 million on the vertical axis. The dashed lines on Figure 5.38 show how we locate this point.
3. (Optional) If we wish, we can label data points, as is done for selected points in Figure 5.38.

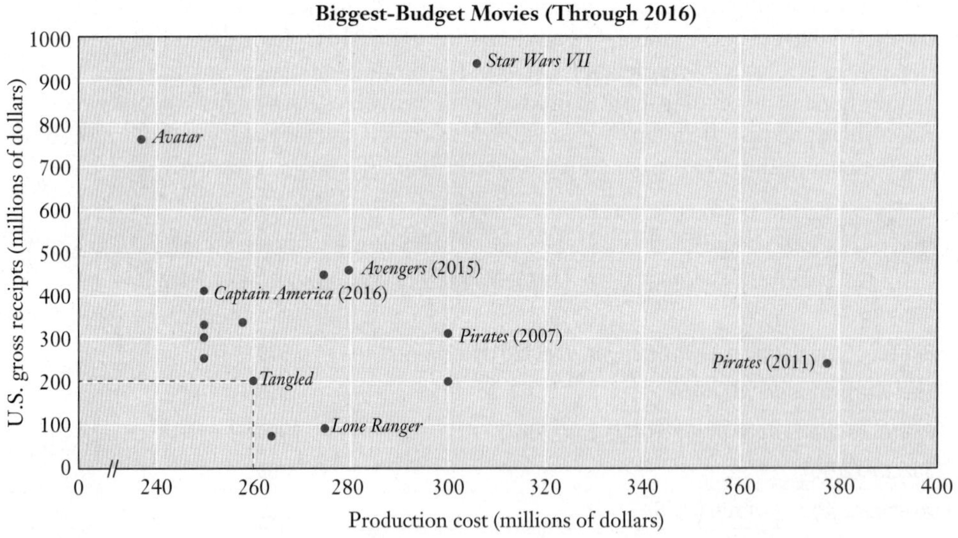

FIGURE 5.38 Scatterplot for the data in Table 5.6.

Think About It By studying Table 5.6, associate each of the unlabeled data points in Figure 5.38 with a particular movie.

Types of Correlation

Look carefully at the scatterplot for movies in Figure 5.38. The dots seem to be scattered about with no apparent pattern. In other words, at least for these big-budget movies, there appears to be little or no correlation between the amount of money spent producing the movie and the amount of money it earned in gross receipts.

Now consider the scatterplot in Figure 5.39, which shows the weights (in carats) and retail prices of 23 diamonds. Here the dots show a clear upward trend, indicating that larger diamonds generally cost more. The correlation is not perfect. For example, the heaviest diamond is not the most expensive. But the overall trend seems fairly clear.

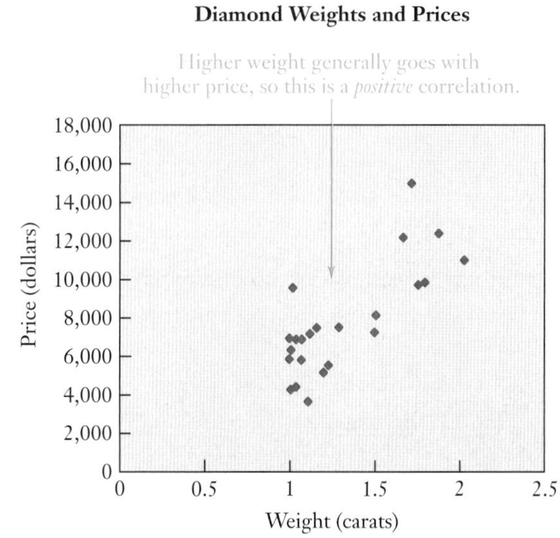

FIGURE 5.39 A scatterplot for diamond weights and prices.

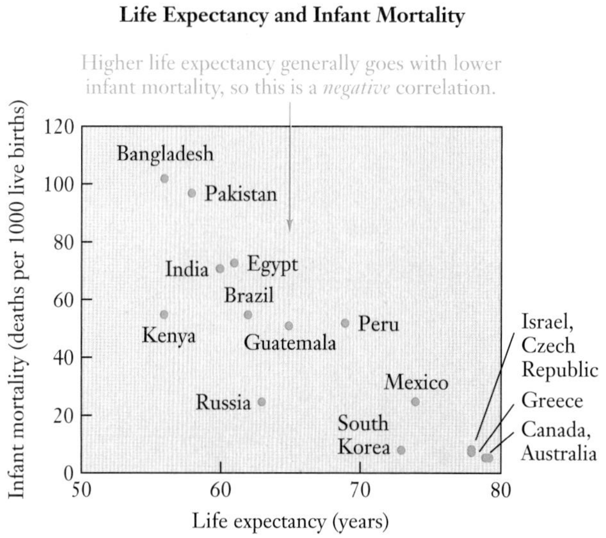

FIGURE 5.40 A scatterplot for life expectancy and infant mortality.

Because the prices tend to increase with the weights, we say that Figure 5.39 shows a **positive correlation**.

In contrast, Figure 5.40 shows a scatterplot for the variables *life expectancy* and *infant mortality* in 16 countries. We again see a clear trend, but this time it is a **negative correlation**: Countries with *higher* life expectancy tend to have *lower* infant mortality.

Besides stating whether a correlation exists, we can also discuss its strength. The more closely the data follow the general trend, the stronger is the correlation.

Relationships Between Two Data Variables

Positive correlation: Both variables tend to increase (or decrease) together.

Negative correlation: The two variables tend to change in opposite directions, with one increasing while the other decreases.

No correlation: There is no apparent relationship between the two variables.

Strength of a correlation: The more closely two variables follow the general trend, the stronger the correlation (which may be either positive or negative). In a perfect correlation, all data points lie on a straight line.

BY THE WAY

In statistics, the *correlation coefficient* provides a quantitative measure of the strength of a correlation. It is defined to be 1 for a perfect (meaning all data points lie on a single straight line) positive correlation, −1 for a perfect negative correlation, and 0 for no correlation.

EXAMPLE 1 Inflation and Unemployment

Prior to the 1990s, most economists assumed that the unemployment rate and the inflation rate were negatively correlated. That is, when unemployment goes down, inflation goes up, and vice versa. Table 5.7 shows unemployment and inflation data for the period 1990–2016. Make a scatterplot for these data. Based on your diagram, does it appear that the data support the historical claim of a link between the unemployment and inflation rates?

Solution We make the scatterplot by plotting the variable *unemployment rate* on the horizontal axis and the variable *inflation rate* on the vertical axis. To make the graph easy to read, we use values ranging from about 3.5% to 10% for the unemployment rate and from 0 to 6% for the inflation rate. Figure 5.41 shows the result. To the eye, there does not appear to be any obvious correlation between the two variables. (A calculation confirms that there is nearly zero correlation.) These data do *not* support the historical claim of a negative correlation between the unemployment and inflation rates.

TABLE 5.7 **U.S. Unemployment and Inflation Rates**

Year	Unemployment Rate (%)	Inflation Rate (%)	Year	Unemployment Rate (%)	Inflation Rate (%)
1990	5.6	5.4	2004	5.5	2.7
1991	6.8	4.2	2005	5.1	3.4
1992	7.5	3.0	2006	4.6	3.3
1993	6.9	3.0	2007	4.6	2.8
1994	6.1	2.6	2008	5.8	3.8
1995	5.6	2.8	2009	9.3	0*
1996	5.4	3.0	2010	9.6	1.6
1997	4.9	2.3	2011	8.9	3.2
1998	4.6	2.3	2012	8.1	2.1
1999	4.3	2.2	2013	7.4	1.5
2000	4.0	3.4	2014	6.2	1.6
2001	4.2	1.8	2015	5.3	0.0
2002	5.8	1.6	2016	4.9	1.3
2003	6.0	2.3			

*The 2009 inflation rate was negative (-0.4%), but we have set it to zero here.

Source: U.S. Bureau of Labor Statistics.

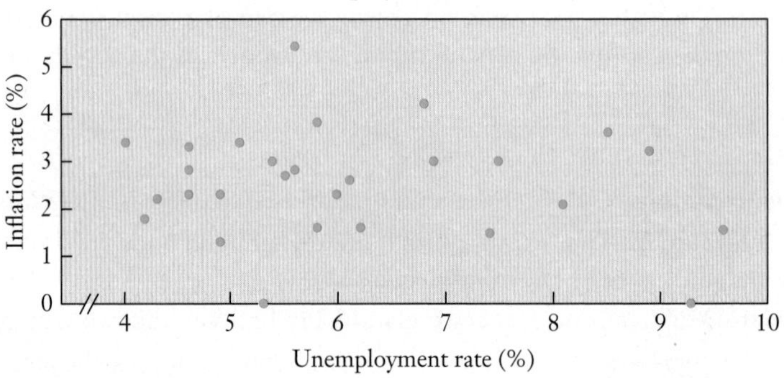

FIGURE 5.41 Scatterplot for the data in Table 5.7.

▶ **Now try Exercises 13–14.**

EXAMPLE 2 Accuracy of Weather Forecasts

The scatterplots in Figure 5.42 show two weeks of data comparing the actual high temperature for the day with the temperature predicted by the same-day forecast (left diagram) and the three-day forecast (right diagram). Discuss the types of correlation on each diagram.

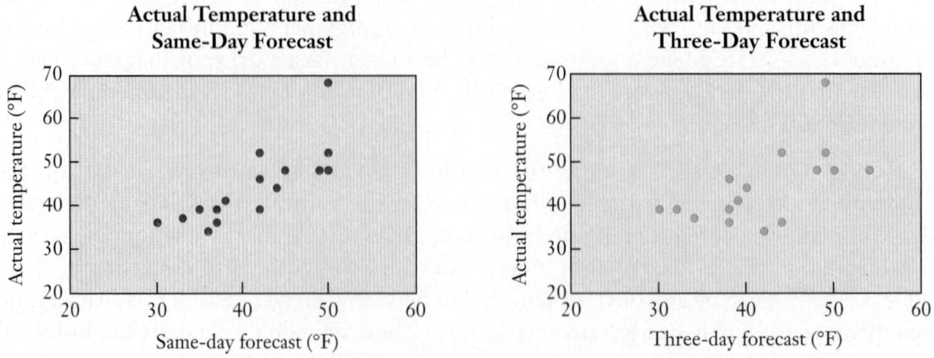

FIGURE 5.42 Comparison of actual high temperatures with same-day and three-day forecasts.

Solution Both scatterplots show a general trend in which higher predicted temperatures mean higher actual temperatures. That is, both show positive correlations. However, the points in the left diagram lie more nearly on a straight line, indicating a stronger correlation than in the right diagram. This makes sense, because we expect weather forecasts to be more accurate on the same day than three days in advance.

▶ Now try Exercises 15–16.

Possible Explanations for a Correlation

We began by stating that correlations can help us search for cause-and-effect relationships. But we've already seen that causality is not the only possible explanation for a correlation. For example, the predicted temperatures on the horizontal axis of

USING TECHNOLOGY

Scatterplots in Excel

Excel makes it easy to generate scatterplots; the screen shot below shows the process for Figure 5.39.

1. Enter the data, which are shown in columns B (weight) and C (price).
2. Select the columns for the two variables on the scatterplot, which in this case are columns B and C.
3. Click the Insert tab and choose "XY Scatter" as the chart type, with no connecting lines. You can then use "Chart Options" (which comes up with a right click in the graph) to customize the design, axis range, labels, and more.
4. (Optional) The straight line on the graph, called a *best-fit line*, is added by right-clicking on any data point and selecting the option "Add Trendline." In the Format Trendline menu, be sure to choose the option "Linear" for the trendline. You'll also find options to display the two items shown in the upper left of the graph: the equation of the line and a value called R^2, which is the square of the correlation coefficient: The correlation coefficient is defined so that values close to -1 indicate a strong negative correlation and values close to $+1$ indicate a strong positive correlation; values near 0 indicate a weak or no correlation. (Values of R^2 are always positive, and values close to 1 indicate a strong correlation while values close to 0 indicate a weak or no correlation.)

Note: You can also calculate the correlation coefficient directly with the built-in function CORREL.

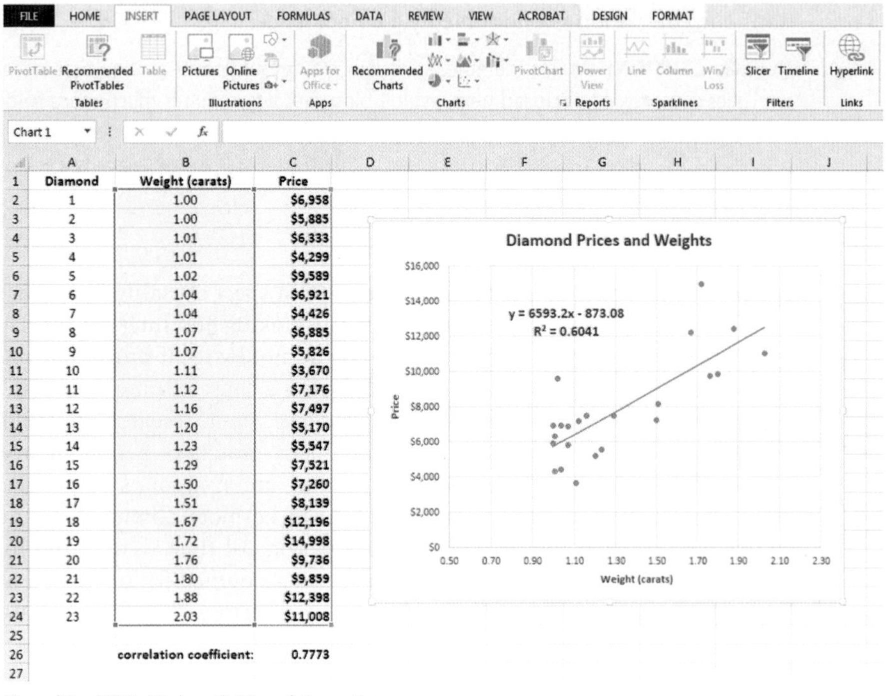

Diamond	Weight (carats)	Price
1	1.00	$6,958
2	1.00	$5,885
3	1.01	$6,333
4	1.01	$4,299
5	1.02	$9,589
6	1.04	$6,921
7	1.04	$4,426
8	1.07	$6,885
9	1.07	$5,826
10	1.11	$3,670
11	1.12	$7,176
12	1.16	$7,497
13	1.20	$5,170
14	1.23	$5,547
15	1.29	$7,521
16	1.50	$7,260
17	1.51	$8,139
18	1.67	$12,196
19	1.72	$14,998
20	1.76	$9,736
21	1.80	$9,859
22	1.88	$12,398
23	2.03	$11,008
	correlation coefficient:	0.7773

Microsoft Excel 2016, Windows 10, Microsoft Corporation.

Figure 5.42 certainly do not *cause* the actual temperatures on the vertical axis. The following box summarizes three possible explanations for a correlation.

Possible Explanations for a Correlation

1. The correlation may be a *coincidence*.
2. Both variables might be directly influenced by some *common underlying cause*.
3. One of the correlated variables may actually be a *cause* of the other. Note that, even in this case, it may be only one of several causes.

EXAMPLE 3 Explanation for a Correlation

Consider the correlation between infant mortality and life expectancy in Figure 5.40. Which of the three possible explanations for a correlation most likely applies? Explain.

Solution The negative correlation between infant mortality and life expectancy is an example of a common underlying cause. Both variables depend on an underlying variable that we might call *quality of health care*. In countries where health care is better in general, infant mortality is lower and life expectancy is higher.

▶ Now try Exercises 17–18.

EXAMPLE 4 How to Get Rich in the Stock Market (Maybe)

Consider the following real correlation: The stock market tends to rise for the rest of the year (overall) when a team from the old, pre-1970 NFL wins the Super Bowl and tends to fall otherwise. This correlation successfully matched 28 of the first 32 Super Bowls to the performance of the stock market, which made the "Super Bowl Indicator" a far more reliable predictor of the market's trend than any professional stockbroker during the same period. In fact, as of 2017, the Super Bowl Indicator had still been correct 40 out of 50 times—an impressive 80% success rate. Should you therefore make a decision about whether to invest in the stock market based on the NFL origins of the most recent Super Bowl winner?

Solution Despite the strength of the correlation, it seems inconceivable that the origin of the winning team could actually *cause* the stock market to move in a particular direction. The correlation is undoubtedly a coincidence.

▲ **CAUTION!** As this example points out, you should never assume that correlation implies causality; as discussed next, establishing causality requires much more evidence than correlation alone. ▲

▶ Now try Exercises 19–24.

Establishing Causality

The truth is rarely pure and never simple.

—Oscar Wilde

Suppose you have discovered a correlation and suspect causality. How can you test your suspicion? Let's return to the relationship of smoking and lung cancer. The strong correlation between smoking and lung cancer did not by itself prove that smoking causes lung cancer. So how was smoking established as a cause of lung cancer?

The answer involved several lines of evidence. First, researchers found correlations between smoking and lung cancer among many groups of people: women, men, and people of different races and cultures. Second, among groups of people that seemed otherwise identical, lung cancer was found to be more common in smokers. Third, people who smoked more and for longer periods of time were found to have higher rates of lung cancer. Fourth, when researchers accounted for other potential causes of lung cancer (such as exposure to radon gas or asbestos), they found that almost all the remaining lung cancer cases occurred among smokers (or people exposed to second-hand smoke).

These four lines of evidence made a strong case, but still did not rule out the possibility that some other factor, such as genetics, predisposes people both to smoking and to lung cancer. However, two additional lines of evidence made this possibility highly unlikely. One line of evidence came from animal experiments. In controlled experiments, animals were divided into randomly chosen treatment and control groups. The experiments still found a correlation between inhalation of cigarette smoke and lung cancer, which effectively ruled out a genetic factor, at least in the animals. The final line of evidence came from biologists studying cell cultures (that is, small samples of human lung tissue). The biologists discovered the basic process by which ingredients in cigarette smoke can create cancer-causing mutations. This process does not appear to depend in any way on specific genetic factors, making it all but certain that lung cancer is caused by smoking and not by any preexisting genetic factor. The fact that second-hand smoke exposure is also associated with lung cancer further argues against a genetic factor (because second-hand smoke affects nonsmokers) but is consistent with the idea that ingredients in cigarette smoke create cancer-causing mutations.

The following box summarizes these ideas about establishing causality. Generally speaking, the case for causality is stronger when more of these guidelines are met.

Guidelines for Establishing Causality

If you suspect that a particular variable (the suspected cause) is causing some effect:

1. Look for situations in which the effect is correlated with the suspected cause even while other factors vary.
2. Among groups that differ only in the presence or absence of the suspected cause, check that the effect is similarly present or absent.
3. Look for evidence that larger amounts of the suspected cause produce larger amounts of the effect.
4. If the effect might be produced by other potential causes (besides the suspected cause), make sure that the effect still remains after accounting for these other potential causes.
5. If possible, test the suspected cause with an experiment. If the experiment cannot be performed with humans for ethical reasons, consider doing the experiment with animals, cell cultures, or computer models.
6. Try to determine the physical mechanism by which the suspected cause produces the effect.

BY THE WAY

The first four guidelines in the box are called *Mill's methods,* after the English philosopher and economist John Stuart Mill (1806–1873). Mill was a leading scholar of his time and an early advocate of women's right to vote.

Think About It There's a great deal of controversy concerning whether animal experiments are ethical. What is your opinion of animal experiments? Defend your opinion.

CASE STUDY
Air Bags and Children

By the mid-1990s, passenger-side air bags had become commonplace in cars. Statistical studies showed that the air bags saved many lives in moderate- to high-speed collisions. But a disturbing pattern also appeared. In at least some cases, young children, especially infants and toddlers in child car seats, were killed by air bags in low-speed collisions.

BY THE WAY

Based on the findings of many studies, the government now recommends that child car seats *never* be used on the front seat and that children under age 12 sit in the back seat if possible.

At first, many safety advocates found it difficult to believe that air bags could be the cause of the deaths. But the observational evidence became stronger, meeting the first four guidelines for establishing causality. For example, the greater risk to infants in child car seats fit Guideline 3, because it indicated that being closer to the air bags increased the risk of death. (A child car seat sits on top of the built-in seat, thereby putting a child closer to the air bags than the child would be otherwise.)

To seal the case, safety experts undertook experiments using dummies. They found that children, because of their small size, often sit where they could be easily hurt by the explosive opening of an air bag. The experiments also showed that an air bag could impact a child car seat hard enough to cause death, thereby revealing the physical mechanism by which the deaths occurred.

CASE STUDY
Cell Phones and Distracted Driving

The issue of cell phones and driving has followed a trajectory similar to that of air bags and children. The correlation between cell phone use and car crashes was also thought to be coincidental at first, because talking on a cell phone doesn't seem to be so different from talking to a passenger, and the latter did not appear to be correlated with crash rates. As evidence for the correlation accumulated, the idea of coincidence became harder and harder to accept, so researchers began to look for other explanations, and one seemed obvious on the surface: Most cell phone users were holding the cell phones in their hands, which seemed likely to mean less control while driving. This fact led many states and countries to pass laws requiring hands-free devices in cars. However, further studies showed that there was still a correlation between cell phone use and crashes, even when the devices were being used hands-free.

At this point, researchers began to question the assumption that talking on a cell phone is just like talking to a passenger. Brain scans conducted during simulated driving sessions soon showed that talking on a cell phone activates different areas of the brain than talking with someone sitting next to you. This offered a potential explanation for why talking on a cell phone might have different effects than talking to a passenger. Follow-up studies provided additional evidence for the idea that cell phone use is a source of distraction that can cause crashes with or without hands-free devices. Most shockingly, some studies have found that the distraction caused by talking on a cell phone while driving—even with a hands-free device—can make you as dangerous as a drunk driver. Texting or using other computing devices while driving is even worse, because it causes the same type of distraction and also forces your eyes off the road. The National Safety Council now estimates that approximately 1.6 million car crashes each year, more than a quarter of the total, are caused by some type of distraction.

BY THE WAY

Despite the media attention given to the serious problems caused by distracted driving, a 2013 survey found that 70% of Americans admitted to chatting on their cell phone while driving during the 30 days before the survey, and about 30% admitted to having sent text messages while driving during the same period.

Confidence in Causality

The six guidelines offer a way to examine the strength of a case for causality, but often we must make decisions before causality is established. We therefore need some way to decide how confident we are in a cause-and-effect relationship.

Unfortunately, while many areas of mathematics and statistics offer accepted techniques quantifying possible errors or uncertainties, there are no such quantitative measures for questions of causality. However, another area of study has dealt with practical problems of causality for hundreds of years: our legal system. You may be familiar with the following three broad ways of expressing a legal level of confidence.

BY THE WAY
For criminal trials, the Supreme Court endorsed this guidance from Justice Ginsburg: "Proof beyond a reasonable doubt is proof that leaves you firmly convinced of the defendant's guilt. There are very few things in this world that we know with absolute certainty, and in criminal cases the law does not require proof that overcomes every possible doubt. If, based on your consideration of the evidence, you are firmly convinced that the defendant is guilty of the crime charged, you must find him guilty. If on the other hand, you think there is a real possibility that he is not guilty, you must give him the benefit of the doubt and find him not guilty."

Broad Levels of Confidence in Causality

Possible cause: We have discovered a correlation, but cannot yet determine whether the correlation implies causality. In the legal system, possible cause (such as thinking that a particular suspect possibly caused a particular crime) is often the reason for starting an investigation.

Probable cause: We have good reason to suspect that the correlation involves cause, perhaps because some of the guidelines for establishing causality are satisfied. In the legal system, probable cause is the general standard for getting a judge to grant a warrant for a search or wiretap.

Cause beyond reasonable doubt: We have found a physical model that is so successful in explaining how one thing causes another that it seems unreasonable to doubt the causality. In the legal system, cause beyond reasonable doubt is the usual standard for conviction. It generally demands that the prosecution show how and why (essentially the physical model) the suspect committed the crime. Note that beyond *reasonable* doubt does *not* mean beyond *all* doubt.

While these broad levels of confidence may seem vague, they give us at least some common language for discussing confidence in causality. If you study law, you will learn much more about the subtleties of interpreting these terms.

Quick Quiz 5E

Choose the best answer to each of the following questions. Explain your reasoning with one or more complete sentences.

1. When we say that X is correlated with Y, we mean that
   a. X causes Y.
   b. increasing values of X are consistently associated with increasing values of Y.
   c. increasing values of X are consistently associated with either increasing or decreasing values of Y.
2. Consider Figure 5.40. According to this diagram, life expectancy in Brazil is about
   a. 22 years.
   b. 62 years.
   c. 58 years.
3. If the points on a scatterplot fall on a nearly straight line sloping downward, the two variables have
   a. a strong negative correlation.
   b. no correlation.
   c. a weak positive correlation.
4. If the points on a scatterplot fall into a broad swath that slopes upward, the two variables have
   a. a strong negative correlation.
   b. a weak positive correlation.
   c. no correlation.
5. When can you rule out the possibility that changes to variable X *cause* changes to variable Y?
   a. when there is no correlation between X and Y
   b. when there is a negative correlation between X and Y
   c. when a scatterplot of the two variables shows points lying in a straight line
6. What type of correlation would you expect between exercise and body mass index (a measure of how much body fat a person has relative to his or her height and weight)?
   a. none
   b. positive: more exercise would be consistently associated with higher body mass index

c. negative: more exercise would be consistently associated with lower body mass index

7. You have found a higher rate of birth defects among babies born to women exposed to second-hand smoke. Which of the following should you also expect to find if the second-hand smoke *caused* the birth defects?

a. higher rates of these birth defects are correlated with exposure to greater amounts of smoke

b. these types of birth defects occur *only* in babies whose mothers were exposed to smoke and never to any other babies

c. the types of birth defects in these babies are more debilitating than other types of birth defects

8. Based on the data in Figure 5.39, about how much (on average) would you expect a 0.5-carat diamond to cost?

a. $2000 b. $7000 c. $12,000

9. Which of the following statements best describes the correlation between accidents and texting while driving?

a. It is a coincidence.

b. There is a common underlying cause.

c. Texting while driving is a likely cause of accidents.

10. A finding by a jury that a person is guilty "beyond reasonable doubt" is supposed to mean that

a. the person is definitely guilty.

b. all 12 members of the jury believed that there was more than a 50% chance that the person was guilty.

c. any reasonable person would conclude that the evidence was sufficient to establish guilt.

Exercises 5E

REVIEW QUESTIONS

1. What is a *correlation*? Give three examples of pairs of variables that are correlated.
2. What is a *scatterplot*, and how is one made? How can we use a scatterplot to look for a correlation?
3. Define and distinguish among *positive correlation*, *negative correlation*, and *no correlation*. How do we determine the strength of a correlation?
4. Describe the three general categories of explanation for a correlation. Give an example of each.
5. Briefly describe each of the six guidelines presented in this unit for establishing causality. Give an example of the application of each guideline.
6. Briefly describe three levels of confidence in causality, and explain how they can be useful when we do not have absolute proof of causality.

DOES IT MAKE SENSE?

Decide whether each of the following statements makes sense (or is clearly true) or does not make sense (or is clearly false). Explain your reasoning.

7. There is a strong negative correlation between the price of tickets and the number of tickets sold. This suggests that if we want to sell a lot of tickets, we should lower the price.
8. There is a strong positive correlation between the amount of time spent studying and grades in mathematics classes. This suggests that if you want to get a good grade, you should spend more time studying.
9. I found a nearly perfect positive correlation between variable *A* and variable *B* and therefore was able to conclude that an increase in variable *A* causes an increase in variable *B*.
10. I found a nearly perfect negative correlation between variable *C* and variable *D* and therefore was able to conclude that an increase in variable *C* causes a decrease in variable *D*.
11. I had originally suspected that an increase in variable *E* would cause a decrease in variable *F*, but I no longer believe this because I found no correlation between the two variables.
12. If causality has been established beyond reasonable doubt, then we can be 100% confident that the causality is real.

BASIC SKILLS & CONCEPTS

13–16: Interpreting Scatterplots. Consider the following scatterplots.

a. State whether the diagram shows a positive correlation, a negative correlation, or no correlation. If there is a positive or negative correlation, is it strong or weak?

b. Summarize any conclusions that you can draw from the diagram.

13.

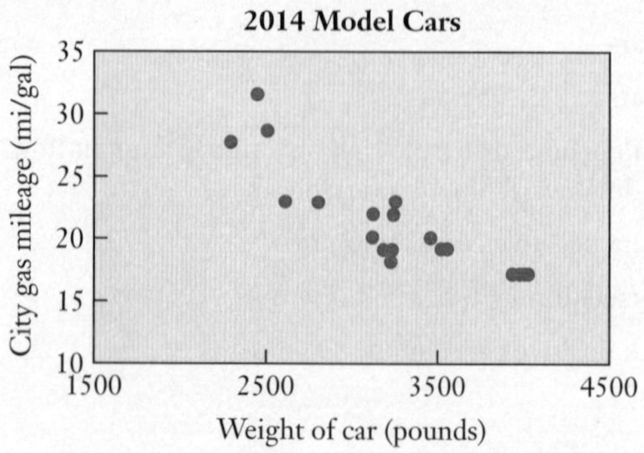

14. **U.S. Presidential Elections, 1964–2016**

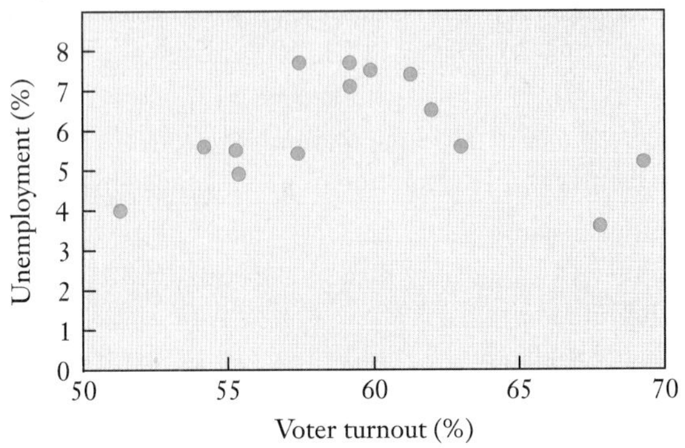

15. **Charitable Giving (11 states) as Percentage of Adjusted Gross Income (AGI)**

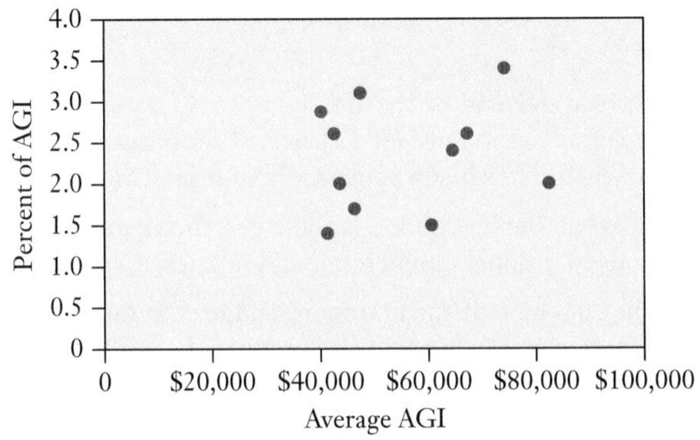

16. **U.S. Farms, 1950–2015**

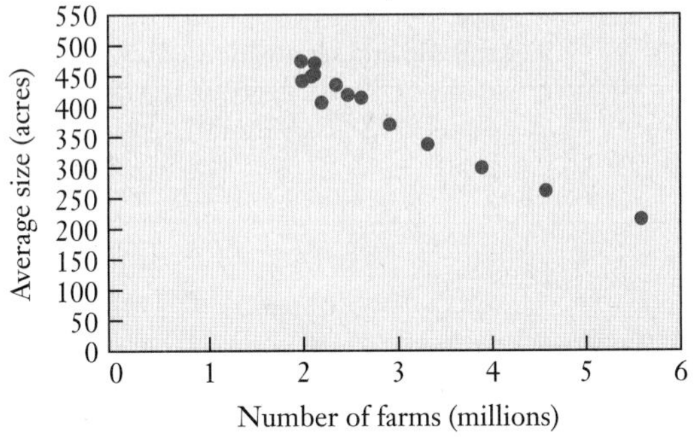

17–24: Types of Correlation. For the following pairs of variables, state the units that might be used to measure each variable. Then state whether you believe that the two variables are correlated. If you believe they are correlated, state whether the correlation is positive or negative. Explain your reasoning.

17. For cities: latitude and average high temperature in June
18. For people: weight and how frequently a person attends rock concerts
19. For people: age and time spent on social networking sites
20. For a mountain hike: altitude and air temperature along the route
21. For the 50 states: the year a state joined the Union and the land area of the state
22. For people: height and waist size
23. For countries: fertility rate of women and life expectancy
24. For school districts: average property tax in a school district and average number of computers in each school

FURTHER APPLICATIONS

25–30: Making Scatterplots. Consider the following data sets.

a. Make a scatterplot for the data.

b. State whether the two variables appear to be correlated, and if so, state whether the correlation is positive, negative, strong, or weak.

c. Suggest a reason for the correlation or lack of correlation. If you suspect causality, discuss what further evidence you need to establish it.

25. The following table gives number of home runs and batting average for baseball's MVPs (Most Valuable Players) for 2009–2016 (NL = National League, AL = American League). There is no entry for the AL 2011 or NL 2013 winners because they were pitchers.

MVP	Home runs	Batting average
Albert Pujols (2009 NL)	47	.327
Joe Mauer (2009 AL)	28	.365
Joey Votto (2010 NL)	37	.324
Josh Hamilton (2010 AL)	32	.359
Ryan Braun (2011 NL)	33	.332
Buster Posey (2012 NL)	24	.336
Miguel Cabrera (2012 AL)	44	.330
Andrew McCutchen (2013 NL)	21	.317
Miguel Cabrera (2013 AL)	44	.348
Mike Trout (2014 AL)	36	.287
Bryce Harper (2015 NL)	42	.330
Josh Donaldson (2015 AL)	41	.297
Kris Bryant (2016 NL)	39	.292
Mike Trout (2016 AL)	29	.315

26. The following table gives per capita personal income and percent of the population below the poverty level for ten states in 2015.

State	Per capita personal income (dollars)	Percentage of population below poverty level
California	52,651	13.9%
Colorado	50,410	9.9%
Illinois	49,471	10.9%
Iowa	44,971	10.4%
Minnesota	50,541	7.8%
Montana	41,280	11.9%
Nevada	42,185	13.0%
New Hampshire	54,817	7.3%
Utah	39,045	9.3%
West Virginia	37,047	14.5%

Source: U.S. Department of Commerce.

27. The following table gives the average hours of traditional (network and cable) television watched per week in five age categories (data for 2016). For the age categories, use the data points 6, 15, 21, 40, 65.

Age	Weekly TV hours
2–11	21.1
12–17	15.5
18–24	17.3
25–54	31
>55	49

Source: Nielsen Media Research.

28. The following table gives the average teacher salary and the expenditure on public education per pupil (both in dollars) for ten states in 2015.

State	Average teacher salary (dollars)	Per-pupil expenditure (dollars)
Alabama	48,611	9,185
Alaska	66,755	20,117
Arizona	45,406	7,461
Connecticut	71,709	17,759
Massachusetts	75,398	16,594
North Dakota	50,025	8,518
Oregon	58,811	11,127
Texas	50,713	8,826
Utah	45,848	7,711
Wyoming	57,414	16,318

Source: National Center for Education Statistics.

29. The following table gives the literacy rate and the infant mortality rate (per 1000 births) for ten countries.

Country	Literacy rate (%)	Infant mortality rate (per 1000 births)
Afghanistan	38	66
Austria	98	3
Burundi	86	54
Colombia	95	14
Ethiopia	49	41
Germany	99	3
Liberia	48	53
New Zealand	99	5
Turkey	95	12
United States	99	6

Sources: UNESCO for literacy rates, 2015; UNICEF and WHO for infant mortality, 2015.

30. The following table gives the 2016 population of eight moderately populated nations and the projected percentage population change by 2050.

Country	2016 population (millions)	Projected population change by 2050
Afghanistan	33.3	91.6%
France	66.8	3.16%
Japan	126.7	−15.4%
Saudi Arabia	28.1	43.0%
Sudan	36.7	61.0%
Germany	80.7	−11.4%
Australia	23	26.1%
Thailand	68.2	−3.2%

Source: Estimates from the United Nations.

31. **Federal Aid and Graduation Rates.** Figure 5.43 shows the high school graduation rates for 13 selected states and the percentage of each state's budget coming from federal aid.

a. Based on the scatterplot, is there evidence that increased federal aid produces higher graduation rates? Explain.

b. The data include the minimum and maximum values in both categories (federal aid and graduation rates). Does the graph accurately reflect the true range of graduation rates across these states? Does the graph accurately reflect the true range of federal aid across these states? Explain.

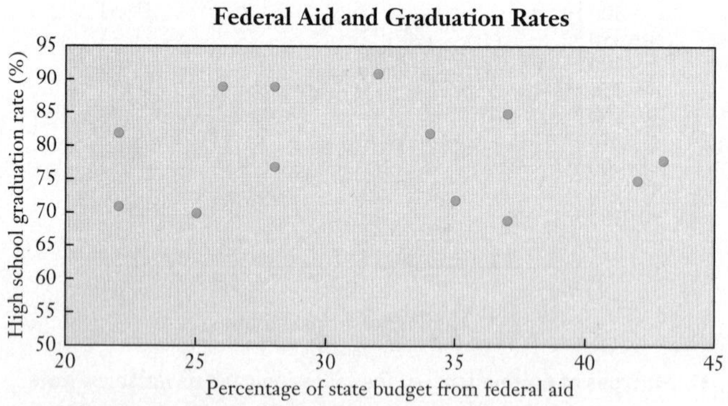

FIGURE 5.43 *Sources:* Based on data from National Center for Education Statistics; Tax Foundation.

32–37: Correlation and Causality. Consider the following statements about a correlation. In each case, state the correlation clearly (for example, there is a positive correlation between variable A and variable B). Then state whether the correlation is most likely due to coincidence, a common underlying cause, or a direct cause. Explain your answer.

32. In a large resort city, the number of burglaries increased as the number of hotel rooms increased.

33. Over the past three decades, the number of miles of freeways in Los Angeles has grown and traffic congestion has worsened.

34. Over the past 30 years, data show that the number of games won by Major League baseball teams increased with team payroll (*Source*: fivethirtyeight.com).

35. In the past three decades, levels of atmospheric carbon dioxide have increased while the number of pirates worldwide has decreased.

36. Automobile gas mileage decreases with tire pressure.

37. Over a period of 20 years, the number of bartenders and ministers in a city both increase.

38. **Identifying Cause: Headaches.** You are trying to identify the cause of late-afternoon headaches that plague you several days each week. For each of the following tests and observations, explain which of the six guidelines for establishing causality you used and what you concluded.

- The headaches occur only on days when you go to work.
- If you stop drinking Coke at lunch on days when you go to work, the headaches persist.
- In the summer, the headaches occur less frequently if you open the windows of your office slightly. They occur even less often if you open the windows of your office fully.

Having made all these observations, what reasonable conclusion can you reach about the cause of the headaches?

39. **Smoking and Lung Cancer.** There is a strong correlation between tobacco smoking and incidence of lung cancer, and most physicians believe that tobacco smoking causes lung cancer. However, not everyone who smokes gets lung cancer. Briefly describe how smoking could cause cancer when not all smokers get cancer.

40. **Longevity of Orchestra Conductors.** A famous study in *Forum on Medicine* (1978) concluded that the mean lifetime of conductors of major orchestras was 73.4 years, about 5 years longer than that of all American males at the time. The author claimed that a life of music *causes* a longer life. Evaluate the claim of causality and propose other explanations for the longer life expectancy of conductors.

41. **High-Voltage Power Lines.** Suppose that people living near a particular high-voltage power line have a higher incidence of cancer than people living farther from the power line. Can you conclude that the high-voltage power line is the cause of the elevated cancer rate? If not, what other explanations might there be for it? What other types of research would you like to see before you conclude that high-voltage power lines cause cancer?

42. **Soccer and Birthdays.** A recent study revealed that the best soccer players in the world tend to have birthdays in the earlier months of the year. Is this a coincidence, or can you find a plausible explanation?

IN YOUR WORLD

43. **Global Warming.** Learn about the evidence linking human activity to global warming; one good place to start is with the "Global Warming Primer" written by an author of this book (go to globalwarmingprimer.com). Based on what you learn, what level of causality do you assign to the idea that human activity causes global warming? How do *you* think we should deal with the problem of global warming?

44. **Success in the NFL.** Find last season's NFL team statistics. Make a table showing the following for each team: number of wins, average yards gained on offense per game, and average yards allowed on defense per game. Make scatterplots to explore the correlations between offense and wins and between defense and wins. Discuss your findings. Do you think that there are other team statistics that would yield stronger correlations with the number of wins?

45. **Correlations in the News.** Find a recent news report that describes some type of correlation. Describe the correlation. Does the article give any sense of the strength of the correlation? Does it suggest that the correlation reflects any underlying causality? Briefly discuss whether you believe the implications the article makes with respect to the correlation.

46. **Causation in the News.** Find a recent news report in which a statistical study has led to a conclusion of causation. Describe the study and the claimed causation. Do you think the claim of causation is legitimate? Explain.

47. **Legal Causation.** Find a news report concerning an ongoing legal case, either civil or criminal, in which establishing causality is important to the outcome. Briefly describe the issue of causation in the case and how the ability to establish or refute causality will influence the outcome of the case.

TECHNOLOGY EXERCISES

Answer the following questions using procedures described in the Using Technology boxes in this unit or with **StatCrunch** (available in MyLab Math).

48. **Making a Scatterplot.** Consider the following data on violent crime rate (measured in violent crimes reported per 100,000 people in the population) and state expenditure per pupil in the public schools in ten different states.

State	Violent crime rate	Per-pupil expenditure (dollars)
California	396	11,145
Connecticut	237	17,759
Florida	540	9223
Maine	128	8957
Mississippi	279	8779
New Jersey	261	20,925
Ohio	285	11,530
Tennessee	608	8809
Utah	216	7711
Wisconsin	290	11,424

Sources: National Center for Education Statistics, 2015; FBI, 2014.

a. Make a scatterplot of the data.

b. Do the variables in the data set appear to be correlated? Express the correlation in words.

c. If you are using Excel, add a best-fit line (also called a trend line) to the scatterplot.

49. **Exercise and TV.** Open the StatCrunch shared data set called *ExerciseHours* (choose "Explore" then "Data" and then type the data set name in the search box), which gives various characteristics of 50 college students (gender, handedness, exercise hours per week, TV hours per week, pulse, and more).

a. Make a scatterplot of exercise hours and TV hours. Comment of the correlation that you see in the plot. Is it positive or negative? Strong or weak?

b. (Optional) The correlation coefficient is briefly explained in the By the Way on p. 357. Compute the correlation coefficient for exercise hours and TV hours. (Go to "Stat" then "Summary stats" and "Correlation"). Does the correlation coefficient confirm your observation in part (a)?

50. **StatCrunch Project.** Choose one data set available in StatCrunch that is of interest to you and that lends itself to making a scatterplot because you suspect a correlation.

a. State the data set you've chosen along with a sentence about why you expect a correlation.

b. Make a scatterplot for your chosen data, and add a best-fit line. Does it show a correlation as expected? Explain.

Chapter 5 Summary

UNIT	KEY TERMS	KEY IDEAS AND SKILLS
5A	statistics (p. 295) population (p. 295) sample (p. 295) population parameter (p. 295) sample statistics (p. 295) bias (p. 300) observational study (p. 300) experiment (p. 300) placebo, placebo effect (p. 301) blinding (single-blind, double-blind) (p. 301) retrospective study (p. 302) margin of error (p. 303) confidence interval (p. 303)	**Know** two meanings of *statistics*—statistics *is* a science and statistics *are* data. (p. 295) **Understand and interpret** the five basic steps in a statistical study. (p. 296) **Understand** the importance of a representative sample. (p. 297) **Be familiar** with four common sampling methods: (p. 298) simple random sampling systematic sampling convenience sampling stratified sampling **Distinguish between** observational studies and experiments; also recognize observational retrospective studies. (p. 300) **Understand** the placebo effect and the importance of blinding in experiments. (p. 301) **Find** a confidence interval from a margin of error: (p. 303) from (sample statistic − margin of error) to (sample statistic + margin of error)
5B	selection bias (p. 310) participation bias (p. 310) variable (in a statistical study) (p. 312)	**Understand and apply** eight guidelines for evaluating a statistical study. (p. 309)
5C	frequency table (p. 319) categories (p. 319) frequency (p. 319) relative frequency (p. 320) cumulative frequency (p. 320) data types qualitative (p. 320) quantitative (p. 320) bar graph (p. 322) pie chart (p. 323) histogram (p. 326) line chart (p. 326) time-series graph (p. 326)	**Interpret and create** frequency tables. (p. 319) **Interpret and create** bar graphs and pie charts. (p. 322) **Interpret and create** histograms and line charts. (p. 326)
5D	multiple bar graph (p. 333) stack plot (p. 335) geographical data (p. 337) contour map (p. 337) infographic (p. 340)	**Interpret** multiple bar graphs, stack plots, contour maps, and other media graphs. (p. 333) **Distinguish between** true three-dimensional data and graphs that have a three-dimensional look simply for cosmetic reasons. (p. 339) **Be aware of** common cautions about graphs. (p. 340)
5E	correlation (p. 354) scatterplot (p. 355) causality (p. 261)	**Distinguish between** correlation and causality. (p. 354) **Create and interpret** scatterplots and use them to identify correlations: (p. 355) positive, negative, or no correlation strength of correlation **Know** three possible explanations for a correlation: (p. 359) coincidence common underlying cause true cause **Understand and apply** six guidelines for establishing causality. (p. 360)

6 PUTTING STATISTICS TO WORK

In Chapter 5, we discussed some of the many ways in which we see statistics used in the media, such as in opinion polls, tables and graphs, summaries of statistical studies, and attempts to establish cause and effect. Now, we are ready to do some actual statistical calculations. The tools discussed in this chapter will give you a strong introduction to the powerful science of statistics.

Q

Consider the following two claims about whether the government should do more to help the poor and the middle class:

Claim 1: Adjusted for inflation, income has been relatively stagnant for more than three decades, with average household income in the United States rising less than 10% over that time period.

Claim 2: Adjusted for inflation, income has risen significantly over recent decades, with average household income in the United States rising nearly 30% over that time period.

Which of the following statements about these claims is correct?

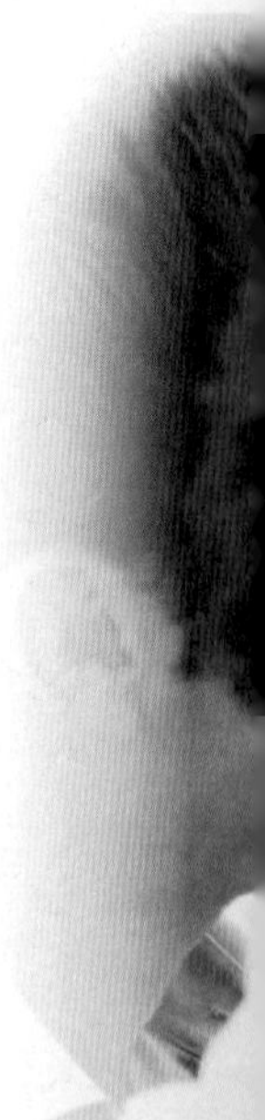

- **A** Claim 1 is true and Claim 2 is false.
- **B** Claim 1 is false and Claim 2 is true.
- **C** The two claims are contradictory, so one of them must be false.
- **D** Neither claim is true, but we can assume that the true change in average household income is between the values (10% and 30%) given in the two claims.
- **E** Both claims are true.

> Some people hate the very name of statistics, but I find them full of beauty and interest. Whenever they are not brutalized, but delicately handled by the higher methods, and are warily interpreted, their power of dealing with complicated phenomena is extraordinary.
>
> —Sir Francis Galton (1822–1911)

A Here's a hint: The first claim tends to be made by liberals and the second one by conservatives. You can probably see why. Liberals want to argue that standards of living for average Americans are stagnant, because it bolsters their arguments for stronger government policies to help income growth, especially among the poor and middle class. Conservatives want to argue that average incomes have been rising, because that would suggest that government intervention is unnecessary.

The question of which claim is true might therefore seem to be a question about which side is more truthful. But here's the surprising answer: E. Although the claims may sound contradictory, *they are both true*. The reason is that there is more than one way to define the term "average." To learn exactly how both claims can be true, see Example 8 in Unit 6A.

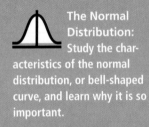

UNIT 6A

Characterizing Data: Investigate ways of describing data by their average value—including mean, median, and mode—and their distribution.

UNIT 6B

Measures of Variation: Explore common measures of the variation, or spread, in a data set: the range, the five-number summary, and the standard deviation.

UNIT 6C

The Normal Distribution: Study the characteristics of the normal distribution, or bell-shaped curve, and learn why it is so important.

UNIT 6D

Statistical Inference: Explore three crucial ideas used when *inferring* a conclusion about a population from results for a sample: statistical significance, the margin of error, and hypothesis testing.

Are We Smarter Than Our Parents?

Use this activity to gain a sense of the kinds of problems this chapter will enable you to analyze. Additional activities are available online in MyLab Math.

Kids tend to think that they're smarter than their parents, but is it possible that they're right? The most common way to measure "intelligence" is with an IQ test, which consists of questions intended to measure innate abilities. Each child's score is compared to the scores of other children taking the same test, and the final result—the child's IQ—is reported using a scale that is *defined* so that the average score is 100. In other words, even though the questions change from year to year (and there are multiple versions of IQ tests), the average IQ is always 100. Moreover, the distribution of IQ scores is scaled to fall in a particular pattern known as a bell-shaped curve, or *normal distribution*. We'll discuss the normal distribution in Unit 6C; for now, note just two facts about the IQ distribution:

- Roughly 3% of all people score an IQ above 130; psychologists describe these scores as "intellectually very superior."
- Roughly 3% of all people score an IQ below 70, which is considered "intellectually deficient."

In the early 1980s, a political science professor named Dr. James Flynn began to look at the results from individual questions given to children on IQ tests (as opposed to the scaled IQ scores). Knowing that the tests and questions changed over time, he focused on those questions that did *not* change (or that were easily compared), which gave him a way to compare the results from different times. To his surprise, he found that the actual test scores had been rising ever since the tests were first given many decades earlier, at a rate equivalent to about 0.3 IQ point per year. This trend—now called *the Flynn effect*—still continues, so the rise in scores over the 80 years between 1930 and 2010 was approximately $80 \times 0.3 = 24$ points. Figure 6.A shows this effect. The average IQ score in 2010 was still 100, since that is how a score of 100 is defined. However, if we put an average child's performance in 2010 on the IQ scale from 1930, we'd find that the child would have an IQ of 124. Overall, on the 1930 scale, nearly half of today's children would rate as "intellectually superior" and virtually none would have been "intellectually deficient."

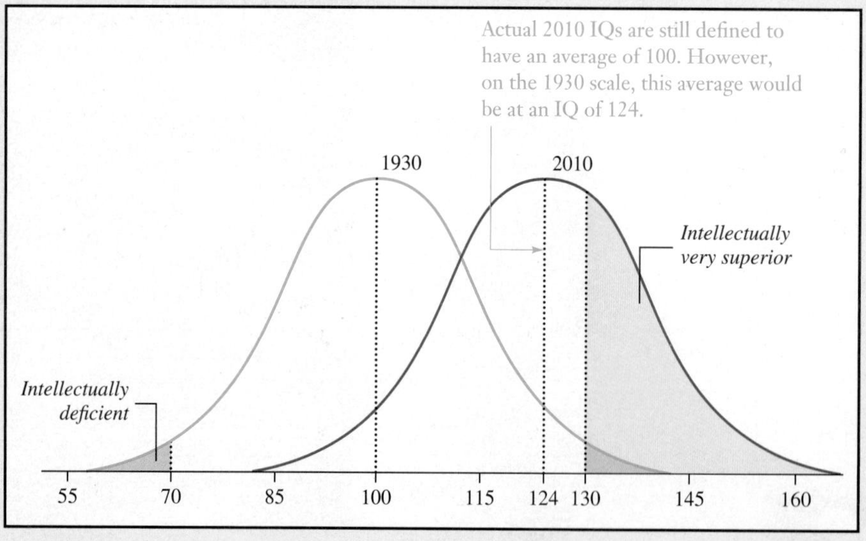
Actual 2010 IQs are still defined to have an average of 100. However, on the 1930 scale, this average would be at an IQ of 124.
1930
2010
Intellectually very superior
Intellectually deficient
55
70
85
100
115
124
130
145
160

FIGURE 6.A Although the average IQ score is defined to be 100, comparison of IQ tests taken at different times shows that actual scores have been rising at a rate of about 0.3 IQ point per year. This graph shows how IQ scores in 2010 compare to IQ scores from 1930 if they were scored on the 1930 scale. Notice that nearly half of children in 2010 would rate "intellectually very superior" on the 1930 test, while virtually none would rate "intellectually deficient."

The Flynn effect seems to imply that today's children are indeed smarter than their parents and grandparents, at least as measured by IQ scores. Of course, that also leads to questions about whether IQ is truly a measure of innate intelligence. To explore the Flynn effect further, work in small groups to answer the following questions.

1. The horizontal axis in Figure 6.A is labeled with IQs on the 1930 scale, showing that 2010 IQ scores are the equivalent of scores 24 points higher than those from 1930. This fact also implies that a certain IQ score in 1930 would translate into a lower score in 2010. Add a second set of labels to the horizontal axis showing the 2010 scale, on which the peak of the 2010 curve represents an IQ of 100. Use this scale to determine what a person who scored an IQ of 100 in 1930 would have scored in 2010. Approximately what fraction of the 1930 population (of IQ test takers) would now be considered "intellectually deficient" on the 2010 scale? Do you think that this fraction accurately represents intellectual capacities of people of that time? Explain.
2. Einstein never actually took an IQ test, but let's suppose he took one in 1910. Psychologists have estimated that Einstein's intellect put him in the top 0.01% of the population at that time, which means he would have scored about 160 in 1910. However, as you saw in Question 1, the Flynn effect means this score would be the equivalent of a lower score today. Assuming that the scores have changed at the rate of 0.3 point per year, what would Einstein's 160 from 1910 be on the 2010 IQ scale? On the 2020 scale? Explain.
3. Based on your answer to Question 2 and the data given in the bulleted list above, about what percentage of today's children have an IQ as high as Einstein? Do you think there really are this many "Einsteins" among us today? Defend your opinion.
4. There is little doubt among social scientists that the Flynn effect is real, meaning that IQ test scores really have been rising. However, there is great debate about what it means. As already discussed, one possible explanation for the effect is that children today are smarter than children of the past. Make a list of other possible explanations for the Flynn effect, briefly explaining each one.
5. Here is an additional fact about the Flynn effect: IQ tests consist of a variety of different question types. Some questions test knowledge, such as vocabulary or arithmetic, while others test conceptual or abstract thinking. By examining results on different question types, social scientists have learned that nearly all the gains of the Flynn effect have occurred on conceptual and abstract questions; results for knowledge questions have barely changed over the decades. Reconsider your list from Question 4 in light of this fact. Does it cause you to change your list of alternative explanations? Explain.
6. For each of the alternative explanations now on your list, suggest a way of testing whether it is correct. Briefly list the data you would need to collect in order to test it, and discuss how you might in principle go about getting these data.
7. Overall, do you think that kids today really are smarter than their parents and grandparents, or do you favor a different explanation for the Flynn effect? Write a short summary of the explanation your group favors.

UNIT 6A Characterizing Data

Frequency tables and graphs (see Unit 5C) show us how one or more data variables are distributed over chosen categories, so we say that they describe the **distribution** of the variable(s). For example, the variable for essay grades is the letter grade (A, B, C, D, or F), so a distribution of essay grades shows how many students received each grade.

Definition

The **distribution** of a variable refers to the way its values are spread over all possible values. We can display a distribution with a table or graph.

In many cases, however, we are less interested in the complete distribution than in a few descriptive terms that summarize it. In this unit, we'll study how we can characterize a data distribution by its center, or average, and by its shape.

What Is Average?

The term *average* is used so often that you may be surprised that it does not always have the same meaning. Let's begin by studying three common ways of characterizing the *center* of a data distribution, all of which are sometimes called the "average."

Mean, Median, and Mode

Table 6.1 shows the number of movies (original and sequels or prequels) in each of five popular science fiction film franchises. What is the average number of films in these franchises? One way to answer this question is to compute the **mean**, which we find by dividing the total number of films by the number of franchises (five):

$$\text{mean} = \frac{6 + 8 + 13 + 8 + 5}{5} = \frac{40}{5} = 8.0$$

Technical Note

There are no hard and fast rules for precision or rounding when calculating the mean, but in this book we will round values to *one more* decimal place than is found in the original list of data. For example, Table 6.1 gives the data values as whole numbers, so we state their mean to the nearest tenth.

We say that these five franchises have a mean of 8.0 films. More generally, we find the mean of any data set by summing all the data values and then dividing by the number of data values. The mean is what most people think of as the "average." It represents the balance point for a data distribution, as shown in Figure 6.1.

We could also describe the "average" number of films with another measure of center: the **median**, or middle value, of the data set. To find a median, we first arrange the data values in ascending (or descending) order, repeating data values that appear more than once. If the number of values is odd, there is exactly one value in the middle of the list, and this value is the median. If the number of values

TABLE 6.1 Five Science Fiction Film Franchises

Title	Number of Movies in Series (as of 2017)
Alien	6
Planet of the Apes	8
Star Trek	13
Star Wars	8
Terminator	5

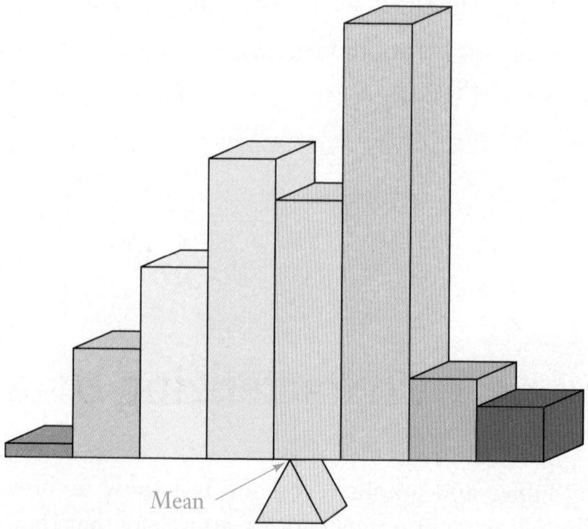

FIGURE 6.1 A histogram made from blocks would balance at the position of its mean.

is even, there are two values in the middle of the list, and the median is the number that lies halfway between them. For Table 6.1, putting the data in ascending order gives the list

$$5, 6, 8, 8, 13$$

The median number of films among these franchises is 8, because 8 is the middle number in the list.

The **mode** is the most common value (or group of values) in a data set. In the case of the film franchises, the mode is 8 because this value occurs twice in the data set, while the other values occur once. A data set may have one mode, more than one mode, or no mode. Sometimes the mode refers to a group of closely spaced values rather than a single value.

Measures of Center in a Distribution

The **mean** is what we most commonly call the average value. It is found as follows:

$$\text{mean} = \frac{\text{sum of all values}}{\text{total number of values}}$$

The **median** is the middle value in the sorted data set, or halfway between the two middle values if the number of values is even.

The **mode** is the most common value (or group of values) in a distribution.

EXAMPLE 1 Price Data

Eight grocery stores sell the PR energy bar for the following prices:

\$1.09 \$1.29 \$1.29 \$1.35 \$1.39 \$1.49 \$1.59 \$1.79

Find the mean, median, and mode for these prices.

Solution The mean price is \$1.41:

$$\text{mean} = \frac{\begin{pmatrix} \$1.09 + \$1.29 + \$1.29 + \$1.35 \\ + \$1.39 + \$1.49 + \$1.59 + \$1.79 \end{pmatrix}}{8} = \$1.41$$

To find the median, we first sort the data in ascending order:

$$\underbrace{\$1.09 \quad \$1.29 \quad \$1.29}_{\text{3 values below}} \quad \underbrace{\$1.35 \quad \$1.39}_{\text{2 middle values}} \quad \underbrace{\$1.49 \quad \$1.59 \quad \$1.79}_{\text{3 values above}}$$

Because there are eight prices (an even number), there are two values in the middle of the list: \$1.35 and \$1.39. Therefore, the median lies halfway between these two values, which we calculate by adding them and dividing by 2:

$$\text{median} = \frac{\$1.35 + \$1.39}{2} = \$1.37$$

The median price is \$1.37. The mode is \$1.29, because this price occurs more times (twice) than any other price. ▶ Now try Exercises 13–18.

Effects of Outliers

To explore the differences among the mean, median, and mode, imagine that the five graduating seniors on a men's college basketball team receive the following first-year contract offers to play in the National Basketball Association (zero indicates that the player did not receive a contract offer):

$$0 \quad 0 \quad 0 \quad 0 \quad \$10{,}000{,}000$$

The mean contract offer is

$$\text{mean} = \frac{0 + 0 + 0 + 0 + \$10{,}000{,}000}{5} = \$2{,}000{,}000$$

Is it therefore fair to say that the *average* senior on this basketball team received a $2 million contract offer?

Not really. The problem is that the single player receiving the large offer makes the mean much larger than it would be otherwise. If we ignore this one player and look only at the other four, the mean contract offer is zero. Because the single value of $10,000,000 is so extreme compared to the others, we call it an **outlier** (or *outlying value*). As our example shows, an outlier can pull the mean significantly upward (or downward), thereby making the mean unrepresentative of the data set as a whole.

While the outlier pulls the mean contract offer upward, it has no effect on the *median* contract offer, which remains zero for the five players. In general, the value of an outlier has little or no effect on either the median or mode, because outliers don't lie in the middle of a data set and are not common values. Table 6.2 summarizes the characteristics of the mean, median, and mode, including the effects of outliers on each measure.

BY THE WAY

A survey once found that geography majors at the University of North Carolina had a far higher mean starting salary than geography majors from other schools. The reason for the high mean turned out to be a single outlier—the basketball superstar and geography major named Michael Jordan, shown here in 1999 when he received the *Sports Illustrated* "Athlete of the Century" award.

Definition

An **outlier** in a data set is a data value that is much higher or much lower than almost all other values. An outlier can change the mean of a data set but generally does not affect the median or mode.

USING TECHNOLOGY

Mean, Median, and Mode in Excel

Excel provides the built-in function AVERAGE for calculating a mean and separate functions MEDIAN and MODE for finding those statistics. The screen shot below shows the use of these functions for the data from Example 1. Column B shows the data, column E shows the functions, and column F shows the results. Note that the results agree with those calculated in Example 1.

	A	B	C	D	E	F
1	Data	1.09		Mean	=AVERAGE(B1:B8)	1.41
2		1.29		Median	=MEDIAN(B1:B8)	1.37
3		1.29		Mode	=MODE(B1:B8)	1.29
4		1.35				
5		1.39				
6		1.49				
7		1.59				
8		1.79				

Microsoft Excel 2016, Windows 10, Microsoft Corporation.

TABLE 6.2 Comparison of Mean, Median, and Mode

Measure	Definition	How Common?	Existence	Takes Every Value into Account?	Affected by Outliers?	Advantages
Mean	$\frac{\text{sum of all values}}{\text{total number of values}}$	Most familiar "average"	Always exists	Yes	Yes	Commonly understood; works well with many statistical methods
Median	Middle value (of an ordered data set)	Common	Always exists	No (aside from counting the total number of values)	No	When there are outliers, may be more representative of an "average" than the mean
Mode	Most frequent value	Sometimes used	May be no mode, one mode, or more than one mode	No	No	Most appropriate for qualitative data

Think About It Is it fair to use the median as the *average* contract offer for the five players? Why or why not?

Deciding how to deal with outliers is one of the more important issues in statistics. Sometimes, as in our basketball example, an outlier is a legitimate value that must be understood in order to interpret the mean and median properly. Other times, outliers may indicate mistakes in a data set. Deciding when outliers are important and when they are mistakes can be very difficult.

EXAMPLE 2 Mistake?

A track coach wants to determine an appropriate heart rate for her athletes during their workouts. She chooses five of her best runners and asks them to wear heart rate monitors during a workout. In the middle of the workout, she reads the following heart rates for the five athletes: 130, 135, 140, 145, 325. Which is a better measure of the average in this case: the mean or the median? Why?

Solution Four of the five values are fairly close together and seem reasonable for mid-workout heart rates. The high value of 325 is an outlier. This outlier seems likely to be a mistake (perhaps caused by a faulty heart rate monitor), because anyone with such a high heart rate would be in cardiac arrest. If the coach uses the mean as the average, she will be including this outlier—which means she will be including any mistake made when it was recorded. If she uses the median as the average, she'll have a more reasonable value, because the median won't be affected by the outlier.

▶ Now try Exercises 19–20.

"Average" Confusion

The different meanings of "average" can lead to confusion. Sometimes this confusion arises because we are not told whether the "average" is the mean or the median, and other times because we are not given enough information about how the average was computed. The following two examples illustrate such situations.

EXAMPLE 3 Wage Dispute

A newspaper surveys wages for assembly workers in regional high-tech companies and reports an average of $42 per hour. The workers at one large firm immediately request a pay raise, claiming that they work as hard as employees at other companies but their average wage is only $36. The management rejects their request, telling them that they are *overpaid* because their average wage, in fact, is $48. Can both sides be right? Explain.

Figures won't lie, but liars will figure.

—Charles H. Grosvenor

Solution Both sides can be right if they are using different definitions of *average*. In this case, the workers may be using the median while management is using the mean. For example, imagine there are only five workers at the firm and their wages are \$36, \$36, \$36, \$36, and \$96. The median of these five wages is \$36 (as the workers claim), but the mean is \$48 (as management claims). ▸ Now try Exercises 21–22.

EXAMPLE 4 Which Mean?

All 100 first-year students at a small college take three courses in the Core Studies Program. Two courses are taught in large lectures, with all 100 students in a single class. The third course is taught in 10 classes of 10 students each. Students and administrators get into an argument about whether classes are too large. The students claim that the mean size of their Core Studies classes is 70. The administrators claim that the mean class size is only 25 students. Can both sides be right? Explain.

Solution The students have calculated the mean size of the classes in which each student is personally enrolled. Each student is taking two classes with enrollments of 100 and one class with an enrollment of 10, so the mean size of each student's classes is

$$\frac{\text{total enrollment in student's classes}}{\text{number of classes student is taking}} = \frac{100 + 100 + 10}{3} = 70$$

The administrators have calculated the mean enrollment in all classes. There are two classes with 100 students and 10 classes with 10 students, making a total enrollment of 300 students in 12 classes. The mean enrollment per class is

$$\frac{\text{total enrollment}}{\text{number of classes}} = \frac{300}{12} = 25$$

The two claims about the mean are both correct, but very different, because the students and administrators used different means. The students calculated the mean *class size per student*, while the administrators calculated the mean number of *students per class*.

▸ Now try Exercises 23–26.

Think About It In Example 4, could the administrators redistribute faculty assignments so that all classes had 25 students each? How? Discuss the advantages and disadvantages of such a change.

Shapes of Distributions

We next turn our attention to describing the overall shape of a distribution. We can *see* the complete shape of a distribution on a graph. Because we are interested in the *general* shapes of distributions, it's easier to focus on smooth curves that fit the data, rather than dealing with the actual data. Figure 6.2 shows three examples of curves that approximate data distributions, two in which the distributions are shown as histograms and one in which the distribution is shown as a line chart. In each case, the smooth curve makes a good approximation to the original distribution.

Number of Modes

One simple way to describe the shape of a distribution is by its number of modes, or peaks. The distributions in Figure 6.2a and Figure 6.2c have one mode, so we say that they are *unimodal*, or single-peaked. The distribution in Figure 6.2b has two modes, even though the second peak is lower than the first; it is a *bimodal* distribution. Other distributions can have no modes (they are called *uniform distributions*) or more than two modes.

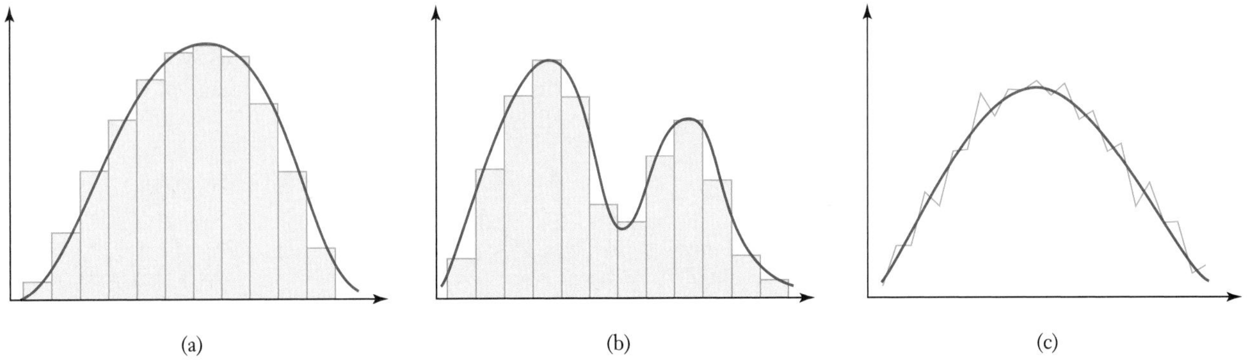

FIGURE 6.2 The smooth curves approximate the actual shapes of the distributions. Note that (a) and (c) are single-peaked (or unimodal), while (b) is double-peaked (or bimodal).

EXAMPLE 5 Number of Modes

How many peaks would you expect for each of the following distributions? Why? Make a rough sketch for each distribution, with clearly labeled axes.

a. Heights of 1000 randomly selected adult women
b. Hours spent watching football on TV in January for 1000 randomly selected adult Americans
c. Weekly sales throughout the year at a retail clothing store for children
d. The numbers of people with particular last digits (0 through 9) in their Social Security numbers

Solution Figure 6.3 shows sketches of the distributions.

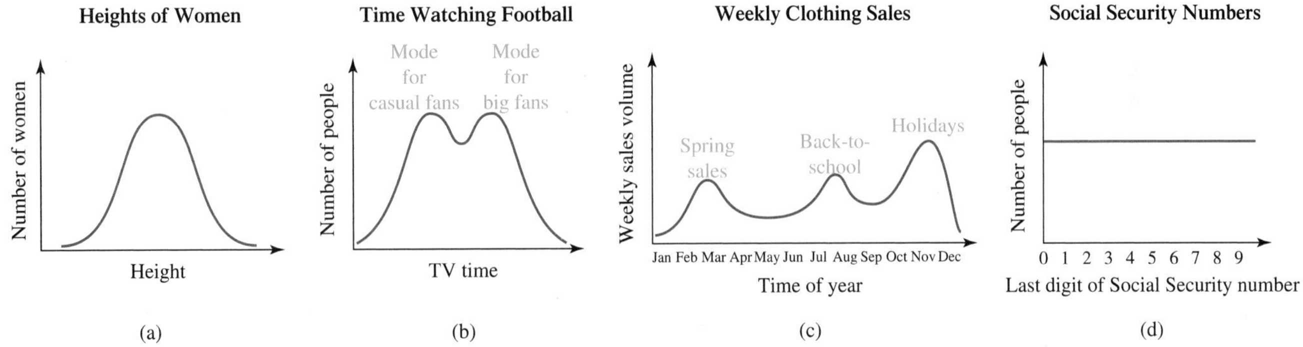

FIGURE 6.3 Sketches for Example 5.

a. The distribution of heights of women is single-peaked (unimodal) because many women have a height at or near the mean height, with fewer and fewer women having heights much greater or less than the mean.
b. The distribution of times spent watching football on TV for 1000 randomly selected adult Americans is likely to be bimodal (two modes). One mode represents the most common watching time of big fans and the other that of more casual fans.
c. The distribution of weekly sales throughout the year at a retail clothing store for children is likely to have several modes. For example, it will probably have a mode in spring for sales of summer clothing, a mode in late summer for back-to-school sales, and another mode in winter for holiday sales.
d. The last digits of Social Security numbers are essentially random, so the number of people with each different last digit (0 through 9) should be about the same. That is, about 10% of all Social Security numbers end in 0, 10% end in 1, and so on. It is therefore a uniform distribution with no mode. ▶ Now try Exercises 27–34, parts (a) and (b).

Symmetry or Skewness

A second way to describe the shape of a distribution is in terms of its symmetry (or lack thereof). A distribution is **symmetric** if its left half is a mirror image of its right half. The distributions in Figure 6.4 are all symmetric.

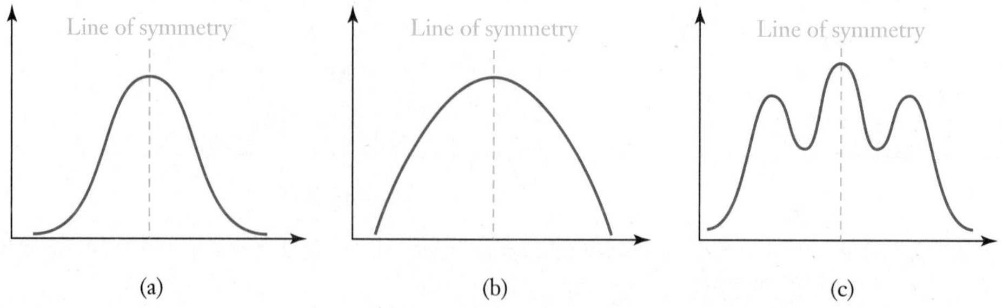

FIGURE 6.4 These distributions are all symmetric because their left halves are mirror images of their right halves. Note that (a) and (b) are single-peaked (unimodal) while (c) is triple-peaked (trimodal).

A distribution that is not symmetric must have values that tend to be more spread out on one side than the other. In this case, we say that the distribution is **skewed**. Figure 6.5a shows a distribution in which the values are more spread out on the left, meaning that some values are outliers at low values. Such a distribution is **left-skewed** (or *negatively* skewed), because it looks as if it has a tail that has been pulled toward the left. Figure 6.5b shows a distribution in which outlier values form a tail to the right (high values), making it **right-skewed** (or *positively* skewed).

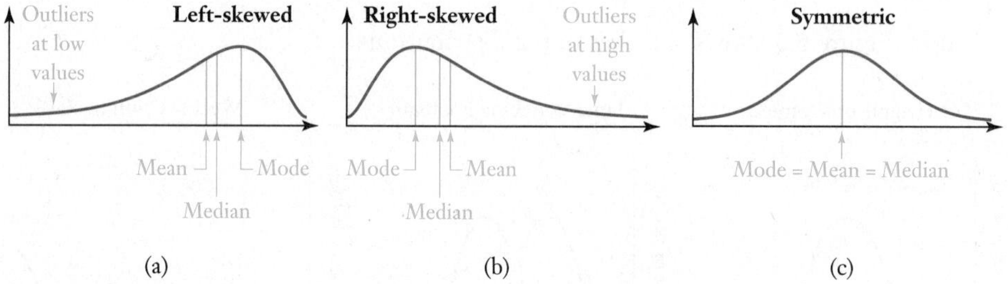

FIGURE 6.5 (a) Skewed to the left (left-skewed): The mean and median are less than the mode. (b) Skewed to the right (right-skewed): The mean and median are greater than the mode. (c) Symmetric distribution: The mean, median, and mode are the same.

Figure 6.5 also shows how skewness affects the relative positions of the mean, median, and mode. By definition, the mode is the peak in a single-peaked distribution. In a left-skewed distribution, both the mean and median lie to the left of the mode, with values less than the mode. In addition, outliers at the low end of the data set make the mean less than the median. Similarly, in a right-skewed distribution, the mean and median lie to the right of the mode, and the outliers at the high end of the data set make the mean greater than the median. When the distribution is symmetric and single-peaked, both the mean and the median are equal to the mode.

Symmetry and Skewness

A single-peaked distribution is **symmetric** if its left half is a mirror image of its right half.

A single-peaked distribution is **left-skewed** if its values are more spread out on the left side of the mode.

A single-peaked distribution is **right-skewed** if its values are more spread out on the right side of the mode.

Think About It Which is a better measure of the "average" (or of the *center* of the distribution) for a skewed distribution: the median or the mean? Why?

EXAMPLE 6 Skewness

For each of the following situations, state whether you expect the distribution to be symmetric, left-skewed, or right-skewed. Explain.

a. Heights of a sample of 100 women
b. Number of books read during the school year by fifth-graders
c. Speeds of cars on a road where a visible patrol car is using radar to detect speeders

Solution

a. The distribution of heights of women is symmetric, because roughly equal numbers of women are shorter and taller than the mean height and extremes of height are rare on either side of the mean.
b. The distribution of the number of books read is right-skewed. Most fifth-grade children read a moderate number of books during the school year, but a few voracious readers will read far more than most other students. These students will therefore be outliers with high values for the number of books read, creating a tail on the right side of the distribution.
c. Drivers usually slow down when they are aware of a patrol car looking for speeders. Few if any drivers will be exceeding the speed limit, but some drivers tend to slow to speeds well below the limit. Therefore, the distribution of speeds is left-skewed, with a mode near the speed limit but with a few cars going well below the speed limit.

▶ Now try Exercises 27–34, part (c).

BY THE WAY

Speed kills. On average in the United States, someone is killed in an auto accident about every 13 minutes. About one-third of these fatalities involve a speeding driver.

Think About It In ordinary English, the term *skewed* is often used to mean that something is distorted or depicted in an unfair way. How is this use of *skewed* related to its meaning in statistics?

Variation

A third way to describe a distribution is by its **variation**, which is a measure of how much the data values are spread out. A distribution in which most data values are clustered together has low variation. As shown in Figure 6.6a, such a distribution has a fairly sharp peak. The variation is higher when the data are distributed more widely around the center, which makes the peak broader. Figure 6.6b shows a distribution with moderate variation, and Figure 6.6c shows a distribution with high variation. We'll discuss methods for quantitatively describing variation in Unit 6B.

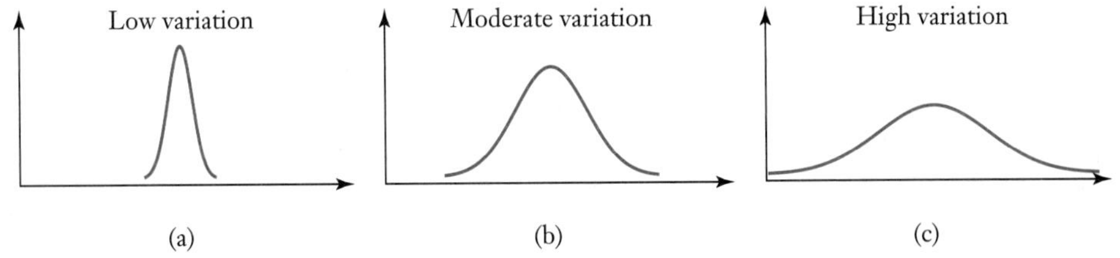

FIGURE 6.6 From left to right, these three distributions have increasing variation.

Definition

Variation describes how widely data values are spread out from the center of a data set.

EXAMPLE 7 Variation in Marathon Times

How would you expect the variation to differ between times in an Olympic marathon and times in the New York Marathon? Explain.

Solution An Olympic marathon involves only elite runners, whose times are likely to be clustered relatively near world-record times. The New York Marathon includes runners of all abilities, whose times are spread over a very wide range (from near the world record to many hours). Therefore, we expect much greater variation among the times in the New York Marathon than in an Olympic marathon.

▶ Now try Exercises 27–34, part (d).

EXAMPLE 8 Changes in Household Income

We now return to the chapter-opening question (p. 370), which offered two seemingly different claims about changes in household income over the past three decades. One claim stated that the average income had risen less than 10%, while the other stated that it had risen nearly 30%. Explain how both statements can be true and what this fact tells you about the shape of the distribution.

Solution The claims can both be true because the first one refers to the *median* while the second refers to the *mean*. The fact that the mean is significantly higher than the median tells us that the distribution of household incomes is right-skewed (see Figure 6.5b). This makes sense because most households are middle-class, so the mode of the household income distribution is a middle-class income. But a small number of very high-income households pull the mean to a considerably higher value than either the mode or the median, stretching the distribution to the right (high-income) side. The fact that some incomes are far higher than others also tells us that the distribution has a relatively large variation, at least if we measure variation in terms of the total range of values.

▶ Now try Exercises 35–36.

Quick Quiz 6A

Choose the best answer to each of the following questions. Explain your reasoning with one or more complete sentences.

1. You want to find the *median* weight of 51 pumpkins. What do you need to do?

   a. List the pumpkin weights in increasing order and find the weight of the pumpkin in the middle.

   b. Find the total weight of all the pumpkins and divide by the number of pumpkins.

   c. Count the pumpkins that have exactly the same weight.

2. On an astronomy exam, 20 students score below 79 and 25 students score above 79. The median score is

   a. 79. b. greater than 79. c. less than 79.

3. One hundred students take a chemistry exam. All but two of the students score between 50 and 70 points, but one student gets a 21 and one student gets a 98. The scores of 21 and 98 represent

   a. mean scores. b. skewed scores. c. outliers.

4. Twenty students take a political science exam. Eighteen score between 70 and 75, and two score 100. What can you conclude?

   a. The mean is higher than the median.

   b. The median is higher than the mean.

   c. The median is higher than the mode.

5. A survey asks students to state how many sodas they drink per week. The results show a mean of 12 sodas per week and a median of 8 sodas per week. What can you conclude?

   a. Something must have been done wrong in computing the mean and median.

   b. Most students drink between 8 and 12 sodas per week.

   c. At least one student drinks more than 16 sodas per week.

6. Among professional actors, a small number of superstars earn much more money than most others. A distribution of actor salaries is therefore

   a. symmetric.

   b. left-skewed, with outliers at low values.

   c. right-skewed, with outliers at high values.

7. The distribution of wages at a company is right-skewed with outliers at high values. Assuming you would like a high wage, you would hope that your wage was closer to the

   a. mean. b. median. c. mode.

8. Compared to a distribution with a broad central peak, a distribution with a sharp central peak

   a. has low variation.

   b. has high variation.

   c. is symmetric.

9. If you compared the distribution of weights of 20 elite female gymnasts to the distribution of weights of 20 randomly selected women, you would expect the variation of the weights to be

   a. greater for the gymnasts.

   b. greater for the randomly selected women.

   c. the same for both groups.

10. The mayor of a town is considering a run for governor. She conducts a poll asking registered voters to rate their likelihood of voting for her on a scale of 1 to 5, where 1 means "definitely would not vote for her" and 5 means "definitely would vote for her." The most encouraging result would be

   a. low median, high variation.

   b. high median, low variation.

   c. high median, high variation.

Exercises 6A

REVIEW QUESTIONS

1. Define and distinguish among *mean*, *median*, and *mode*.
2. What are *outliers*? Describe the effects of outliers on the mean, median, and mode.
3. Briefly describe at least two possible sources of confusion about the "average."
4. Give simple examples of a *unimodal* (single-peaked) distribution and a *bimodal* (double-peaked) distribution.
5. What do we mean when we say that a distribution is *symmetric*? Give simple examples of a symmetric distribution, a *left-skewed* distribution, and a *right-skewed* distribution.
6. What do we mean by the *variation* of a distribution? Give simple examples of distributions with different amounts of variation.

DOES IT MAKE SENSE?

Decide whether each of the following statements makes sense (or is clearly true) or does not make sense (or is clearly false). Explain your reasoning.

7. In my data set of 10 exam scores, in which no two people got the same score, the mean was equal to the score of the person with the third highest grade.
8. In my data set of 10 exam scores, in which no two people got the same score, the median was equal to the score of the person with the third highest grade.
9. I asked the renters of 15 apartments in my neighborhood how much they paid per month. One apartment had a much higher rent than all the others, and this outlier caused the mean rent to be higher than the median rent.
10. Two extremely tall people skewed the distribution of heights to the smaller values.
11. The distribution of grades was left-skewed, but the mean, median, and mode were all the same.
12. There's much more variation in the ages of the general population than in the ages of students in my college extension course, but both distributions have the same mean.

BASIC SKILLS & CONCEPTS

13–18: Mean, Median, and Mode. Compute the mean, median, and mode of the following data sets.

13. Number of words on nine randomly selected pages of *War and Peace*:

    350 360 345 340 355 375 363 345 358

14. Body temperature (in degrees Fahrenheit) of randomly selected normal and healthy adults:

    98.6 98.6 98.0 98.0 99.0
    98.4 98.4 98.4 98.4 98.6

15. Blood alcohol content of drivers involved in fatal crashes (data from the U.S. Department of Justice):

    0.27 0.17 0.17 0.16 0.13 0.24
    0.29 0.24 0.14 0.16 0.12 0.16

16. Number of games played in 12 matches of a recent U.S. Open Tennis Championships:

    48 52 55 45 50 42 52 40 39 43 46 35

17. Actual times (in seconds) recorded when statistics students participated in an experiment to test their ability to determine when 1 minute (60 seconds) had passed:

    53 52 75 62 68 58 49 49

18. Weights (in pounds) of randomly selected 50-pound bags of dog food:

48.5 49.6 48.4 49.9 50.2 50.3 47.9 48.3 49.2 49.8

19. **Outlier Coke.** The contents of cans of Coca-Cola vary slightly in weight. Here are the measured weights of seven cans, in pounds:

0.8161 0.8194 0.8165 0.8176 0.7901 0.8143 0.8126

Find the mean and median of these weights. Which, if any, of these weights would you consider to be an outlier? What are the mean and median weights if the outlier is excluded?

20. **Margin of Victory.** The following data give the margin of victory in the Super Bowl for 2003–2017.

27 3 3 11 12 3 4 14 6 4 3 35 4 14 6

a. Find the mean and median margin of victory.

b. Identify the outlier(s) in the data set. If you eliminate the outlier(s), what are the new mean and median?

21–26: Appropriate Average. For each of the following distributions, decide whether you expect the mean, median, or mode to give the best representation of the center of the distribution, and explain why.

21. Per capita earnings in New York City
22. Age at first marriage for women in the United States
23. Number of times people change jobs during their careers
24. Daily snowfall in Omaha in January
25. The individual weights of oranges in a 5-pound bag
26. Scores on a very difficult test on which a perfect score is 200 points

27–34: Describing Distributions. Consider the following distributions.

a. How many peaks would you expect the distribution to have? Explain.

b. Make a sketch of the distribution.

c. Would you expect the distribution to be symmetric, left-skewed, or right-skewed? Explain.

d. Would you expect the variation of the distribution to be small, moderate, or large? Explain.

27. The exam scores for 50 students when 5 students scored between 90 and 100, 10 students scored between 80 and 89, 20 students scored between 70 and 79, and 15 students scored below 70, with their scores spread out to a low of 40
28. The weights of all people who use an ice rink that is open to figure skaters in the morning and to hockey players in the afternoon
29. The annual snowfall amounts in 50 randomly selected American cities
30. The numbers of people whose street address ends in 0, 1, 2, 3, 4, 5, 6, 7, 8, and 9 (the data set has 10 values)
31. The monthly sales of anti-freeze over a one-year period at a store in Anchorage, Alaska (the distribution will look best if you start from July rather than January)
32. The weights of cars at a dealership at which about half of the inventory consists of compact cars and half of the inventory consists of sport utility vehicles
33. The prices of 20-pound bags of 20 different brands of dog food
34. The ages of patrons who visit an amusement park

35–36: Understanding Distributions. For the given exam results, briefly describe the shape and variation of the distribution.

35. Exam results for 100 students: median 75, mean 75, low score 60, high score 90
36. Exam results for 100 students: median 60, mean 70, low score 50, high score 85

FURTHER APPLICATIONS

37–40: Smooth Distributions. For each histogram, draw a smooth curve that captures its important features. Then classify the distribution according to its number of peaks, symmetry or skewness, and variation.

37. The histogram in Figure 6.7 shows times between eruptions of Old Faithful geyser in Yellowstone National Park.

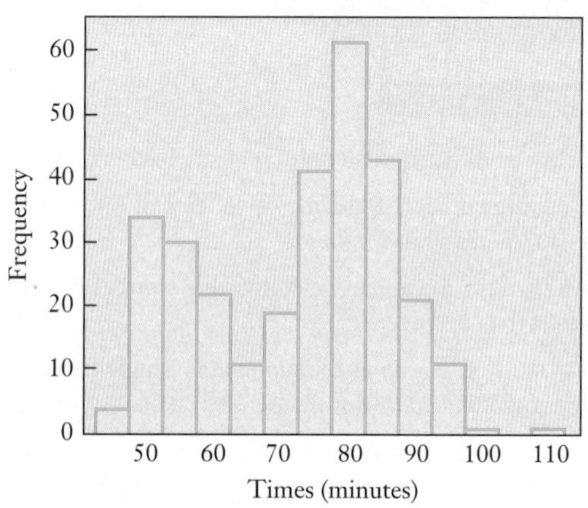

FIGURE 6.7 *Source:* Hand et al., *Handbook of Small Data Sets.*

38. The histogram in Figure 6.8 shows the F-scale measurements of 490 tornadoes from recent years. (The F-scale runs from 0 to 5; a higher number indicates a stronger tornado.)

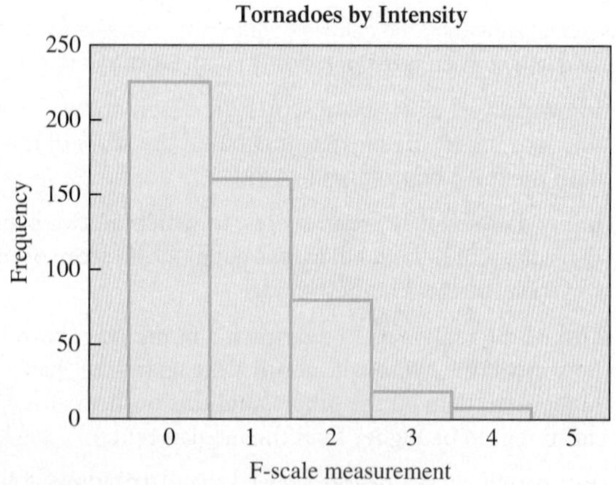

FIGURE 6.8 *Source:* National Weather Service.

39. The histogram in Figure 6.9 shows the circumference (in centimeters) of the upper arm of 300 randomly selected adults.

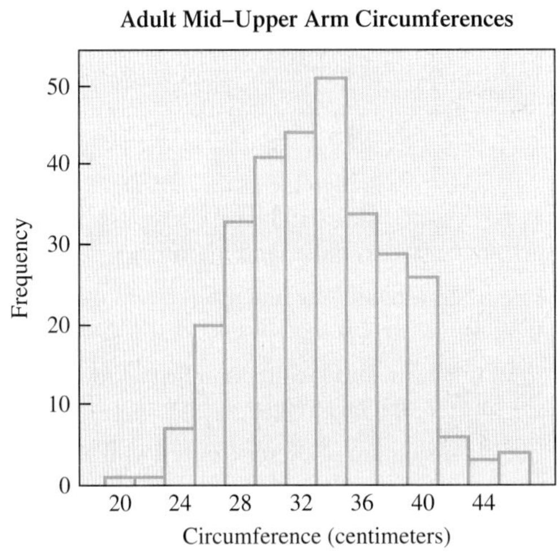

FIGURE 6.9

40. Figure 6.10 is a histogram of digits drawn in California's Daily Four lottery.

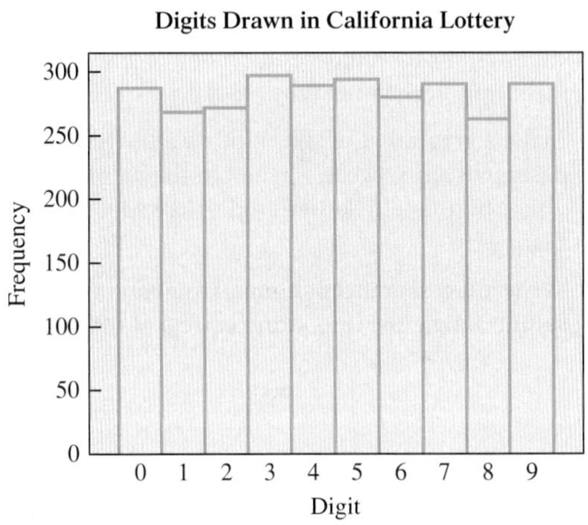

FIGURE 6.10

41. **Family Income.** Suppose you study family income in a random sample of 300 families. You find that the mean family income is \$55,000; the median is \$45,000; and the highest and lowest incomes are \$250,000 and \$2400, respectively.

a. Draw a rough sketch of the income distribution, with clearly labeled axes. Describe the distribution as symmetric, left-skewed, or right-skewed.

b. How many families in the sample earned less than \$45,000? Explain how you know.

c. Based on the given information, can you determine how many families earned more than \$55,000? Why or why not?

42. **Airline Delays.** Suppose you are a scheduler for a major airline, analyzing a distribution of flight delays (departures) during one day at an airport. The distribution of delay times has the following properties: The mean delay is 10 minutes, the median delay is 6 minutes, the mode of the distribution is 0 minutes, and the maximum delay is 25 minutes.

a. What is the minimum delay in this distribution? Draw a rough sketch of the distribution with clearly labeled axes. Is the distribution skewed (and if so, which way) or symmetric?

b. What percentage of flights were delayed less than 6 minutes?

c. What percentage of flights were delayed more than 6 minutes?

d. Discuss the pros and cons of using the mean and the median to present the average flight delay of your airline to the public.

43–44: Weighted Means. We often deal with *weighted means*, in which different data values carry different *weights* in the calculation of the mean. For example, if the final exam counts for 50% of your final grade and 2 midterms each count for 25%, then you must assign weights of 50% and 25% to the final and midterms, respectively, before computing the mean score for the term. Apply the idea of weighted means in the following exercises.

43. **GPA.** A student has completed the 15 credits in the table below during one semester. Grades are weighted as follows: A = 4.0, A− = 3.7, B+ = 3.4, B = 3.0, B− = 2.7, C+ = 2.4, C = 2.0.

a. Find the student's GPA for the semester.

b. What grade would the student need in French to raise her GPA over 3.5?

Course	Credits	Grade
Statistics	4	A
Eastern Religions	3	B+
French	4	C+
Geology	3	B−
Geology Lab	1	A−

44. **Class Grade.** Ryan is taking an advanced psychology class in which the midterm and final exams are worth 35% each and homework is worth 30% of the final grade. On a 100-point scale, his midterm exam score was 85.5, his homework average score was 94.1, and his final exam score was 88.5.

a. On a 100-point scale, what is Ryan's overall average for the class?

b. Ryan was hoping to get an A in the class, which requires an overall score of 93.5 or higher. Could he have scored high enough on the final exam to get an A in the class?

IN YOUR WORLD

45. **Salary Data.** Find salary data for a career you are considering. What are the mean and median salaries for this career? How do these salaries compare to those of other careers that interest you?

46. **Tax Statistics.** The IRS website provides statistics collected from tax returns on income, refunds, and much more. Choose a set of tax statistics and study the distribution. Describe the distribution in words, and discuss anything you learn that is relevant to national tax policies.

47. **Education Statistics.** Visit the website of the National Center for Education Statistics and view the most recent copy of the *Digest of Educational Statistics*. Choose one of the many tables or graphs in the report that display a distribution. Describe the variables that are presented, discuss the effectiveness of the presentation, and interpret the results.

48. **Averages in the News.** Find three recent news reports that refer to some type of average. In each case, explain whether the average is a mean, a median, or some other type of average.

49. **Daily Averages.** Cite three examples of averages that you deal with in your own life (such as grade point average or batting average). In each case, explain whether the average is a mean, a median, or some other type of average. Briefly describe how the average is useful to you.

50. **Distributions in the News.** Find three recent examples in the news of distributions shown as histograms or line charts. Over each distribution, draw a smooth curve that captures its general features. Then classify the distribution according to its number of peaks, symmetry or skewness, and variation.

TECHNOLOGY EXERCISES

Answer the following questions using procedures described in the Using Technology boxes in this unit or with **StatCrunch** (available in MyLab Math).

51. **Comparing Averages.** The five starting players on two basketball teams (Team A and Team B) have the following weights in pounds.

A: {160, 155, 125, 115, 115} B: {145, 140, 135, 125, 110}

Use a calculator, Excel, or StatCrunch to answer the following questions.

a. Compute the mean, median, and mode of the starting player weights for each team. Then decide if or how each coach could make a claim to have the heavier team.

b. Is the mean weight of the two teams combined equal to the mean of the two mean weights found for the teams individually? Explain.

c. Is the median weight of the two teams combined equal to the median of the two median weights found for the teams individually? Explain.

52. **Summary Statistics in StatCrunch.** Load the StatCrunch data set *ExerciseHours*, which gives various characteristics of 50 college students (gender, handedness, exercise hours per week, TV hours per week, pulse, and more).

a. Find the mean and median number of hours the students in the sample exercised per week.

b. Find the mean and median number of hours the students in the sample watched TV per week.

c. Make a histogram of exercise hours with 5-hour bins. Is the distribution left-skewed, right-skewed, or symmetric?

d. What percentage of students in the sample are left-handed?

53. **StatCrunch Project.** Choose a data set available in StatCrunch that is of interest to you; in this case, you might wish to choose a relatively small and simple one.

a. Identify the data set you've chosen, and describe in a sentence why it is of interest to you.

b. Find the mean, median, and mode for the data set, and explain how each is relevant.

c. Make a graph that displays the data distribution effectively, choosing an appropriate way to bin the data if necessary. Is the distribution left-skewed, right-skewed, or symmetric? Explain why.

d. Write a few sentences summarizing how your data analysis helps illuminate the important aspects of the data set.

UNIT 6B Measures of Variation

In Unit 6A, we saw how to describe variation qualitatively, by looking at the shape of a distribution. We now turn to quantitative measures of variation.

Why Variation Matters

Imagine customers waiting in line for tellers at two different banks. Customers at Big Bank can enter any one of three different lines leading to three different tellers. Best Bank also has three tellers, but all customers wait in a single line and are called to the next available teller. The following values are waiting times, in minutes, for eleven customers at each bank. The times are arranged in ascending order.

Big Bank (three lines):	4.1	5.2	5.6	6.2	6.7	7.2	7.7	7.7	8.5	9.3	11.0
Best Bank (one line):	6.6	6.7	6.7	6.9	7.1	7.2	7.3	7.4	7.7	7.8	7.8

You'll probably find more unhappy customers at Big Bank than at Best Bank, but it's *not* because the average wait is any longer. In fact, you can verify that the mean and median waiting times are 7.2 minutes at both banks. The difference in customer satisfaction comes from the *variation* in waiting times at the two banks. The waiting times at Big Bank vary over a fairly wide range, so a few customers have long waits and are likely to become annoyed. In contrast, the variation of the waiting times at Best Bank is small, so customers probably feel that they are being treated roughly equally. The histograms in Figure 6.11 show the differences in variation visually.

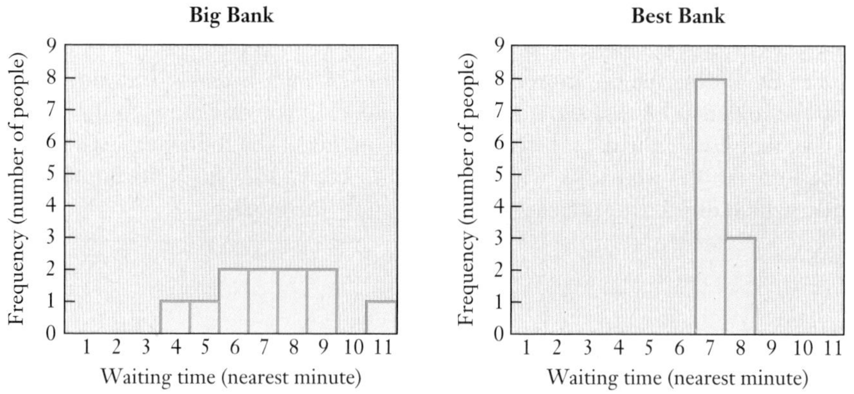

FIGURE 6.11 Histograms for the waiting times at Big Bank and Best Bank, shown with data binned to the nearest minute.

BY THE WAY
Understanding how lines form (known as *queuing*) is important not only with people but also with data. Major corporations often employ statisticians to help them make sure that data stream smoothly and without bottlenecks when flowing among servers and over the Internet.

Think About It Explain *why* Big Bank, with three separate lines, has a greater variation in waiting times than Best Bank. Then consider several places where you commonly wait in lines feeding to multiple clerks, such as the grocery store, a financial aid office, a concert ticket outlet, or a fast food restaurant. Do these places use a single customer line or multiple lines? If multiple lines are used, do you think a single line would be better? Explain. ▶ **Now try Exercises 13–14.**

Range

The simplest way to describe the variation of a data set is to compute its **range**, defined as the difference between the highest (maximum) and lowest (minimum) values. For example, the waiting times for Big Bank vary from 4.1 to 11.0 minutes, so the range is $11.0 - 4.1 = 6.9$ minutes. The waiting times for Best Bank vary from 6.6 to 7.8 minutes, so the range is $7.8 - 6.6 = 1.2$ minutes. The range for Big Bank is much larger, reflecting its data set's greater variation.

Definition

The **range** of a data set is the difference between its highest and lowest data values:

$$\text{range} = \text{highest value (max)} - \text{lowest value (min)}$$

Although the range is easy to compute and can be useful, it occasionally can be misleading, as the next example shows.

EXAMPLE 1 Misleading Range

Consider the following two sets of quiz scores for nine students. Which set has the greater range? Would you also say that this set has the greater variation?

Quiz 1:	1	10	10	10	10	10	10	10	10
Quiz 2:	2	3	4	5	6	7	8	9	10

Solution The range for Quiz 1 is 10 − 1 = 9 points, which is greater than the range for Quiz 2 of 10 − 2 = 8 points. However, aside from a single low score (an outlier), Quiz 1 has no variation at all because every other student got a 10. In contrast, no two students got the same score on Quiz 2, and the scores are spread throughout the list of possible scores. Quiz 2 therefore has greater variation in scores even though Quiz 1 has the greater range. ▸ **Now try Exercises 15–18, part (a).**

Quartiles and the Five-Number Summary

A better way to describe variation is to consider a few intermediate data values in addition to the high and low values. A common approach involves looking at the **quartiles**, or values that divide the data distribution into quarters. The following list repeats the waiting times at the two banks, with the quartiles shown in bold. Note that the middle quartile, which divides the data set in half, is simply the median.

			lower quartile ↓			median ↓			upper quartile ↓		
Big Bank:	4.1	5.2	**5.6**	6.2	6.7	**7.2**	7.7	7.7	**8.5**	9.3	11.0
Best Bank:	6.6	6.7	**6.7**	6.9	7.1	**7.2**	7.3	7.4	**7.7**	7.8	7.8

Technical Note
Statisticians do not universally agree on the precise methods for calculating quartiles, and different procedures can result in slightly different values.

Definitions

The **lower quartile** (or *first quartile*) divides the lowest fourth of a data set from the upper three-fourths. It is the median of the data values in the *lower half* of a data set. (Exclude the middle value in the data set if the number of data points is odd.)

The **middle quartile** (or *second quartile*) is the median of the entire data set.

The **upper quartile** (or *third quartile*) divides the lowest three-fourths of a data set from the upper fourth. It is the median of the data values in the *upper half* of a data set. (Exclude the middle value in the data set if the number of data points is odd.)

Once we know the quartiles, we can describe a distribution with a **five-number summary** consisting of the lowest value, the lower quartile, the median, the upper quartile, and the highest value. For the waiting times at the two banks, the five-number summaries are

Big Bank		*Best Bank*	
lowest value =	4.1	lowest value =	6.6
lower quartile =	5.6	lower quartile =	6.7
median =	7.2	median =	7.2
upper quartile =	8.5	upper quartile =	7.7
highest value =	11.0	highest value =	7.8

The Five-Number Summary

The **five-number summary** for a data set consists of the following five numbers:

lowest value lower quartile median upper quartile highest value

We can display the five-number summary with a **boxplot** (or *box-and-whisker plot*), using the procedure described in the following box. Figure 6.12 shows the boxplots for the bank waiting times. Both the box and the "whiskers" are broader for Big Bank than for Best Bank, indicating that the waiting times have greater variation at Big Bank.

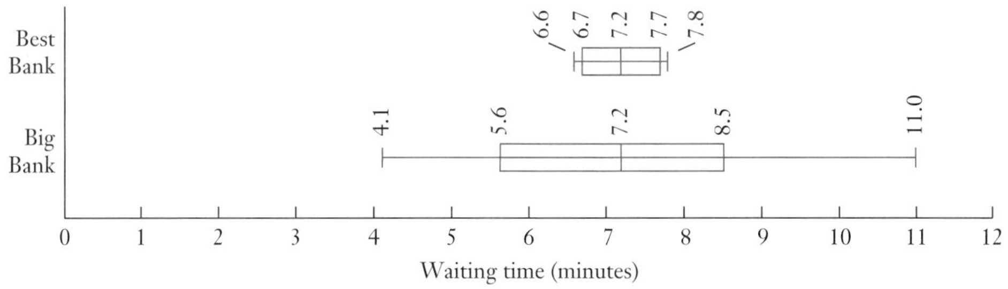

FIGURE 6.12 Boxplots for the waiting times at Big Bank and Best Bank.

Drawing a Boxplot

Step 1. Draw a number line that spans all the values in the data set.

Step 2. Enclose the values from the lower to the upper quartile in a box. (The thickness of the box has no meaning.)

Step 3. Draw a vertical line through the box at the median.

Step 4. Add whiskers (horizontal lines with short vertical segments at their ends) extending to the lowest and highest values.

HISTORICAL NOTE

The boxplot was invented by American statistician John Tukey (1915–2000), widely regarded as one of the most accomplished statisticians of the 20th century. Tukey invented many powerful statistical methods and was a proponent of a thriving area of statistics called *exploratory data analysis (EDA).*

EXAMPLE 2 Race Times

Consider the following two sets of twenty 100-meter running times (in seconds):

Set 1:

9.92	9.97	9.99	10.01	10.06	10.07	10.08	10.10	10.13	10.13
10.14	10.15	10.17	10.17	10.18	10.21	10.24	10.26	10.31	10.38

Set 2:

9.89	9.90	9.98	10.05	10.35	10.41	10.54	10.76	10.93	10.98
11.05	11.21	11.30	11.46	11.55	11.76	11.81	11.85	11.87	12.00

Compare the variation in the two data sets with five-number summaries and boxplots.

Solution Each set has 20 data points, so the median lies halfway between the 10th and 11th running times. The lower quartile is the median of the *lower half* of the times, which lies halfway between the 5th and 6th times. The upper quartile is the median of the *upper half* of the times, which lies halfway between the 15th and 16th times. You should confirm that the five-number summaries are as follows.

First Set of Running Times		*Second Set of Running Times*	
lowest value =	9.92	lowest value =	9.89
lower quartile =	10.065	lower quartile =	10.385
median =	10.135	median =	11.015
upper quartile =	10.195	upper quartile =	11.655
highest value =	10.38	highest value =	12.00

Figure 6.13 shows the corresponding boxplots on the same scale, clearly illustrating the greater variation in the second set of times.

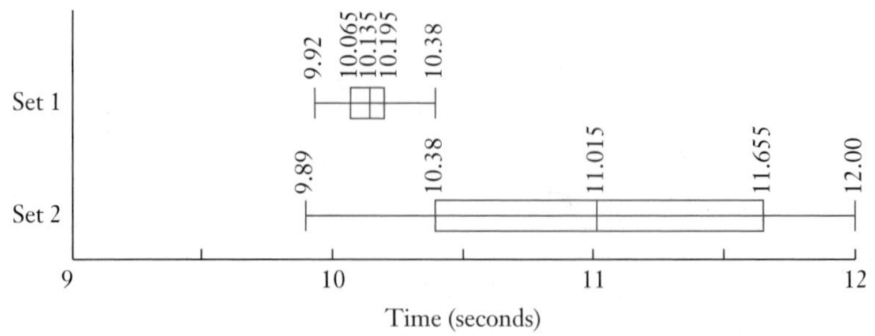

FIGURE 6.13

▶ Now try Exercises 15–18, part (b).

Standard Deviation

The five-number summary characterizes a data set's variation well, but statisticians often prefer to describe variation with a single number. The single number most commonly used to describe variation is called the **standard deviation**, which is a measure of how far data values are spread around the mean of a data set.

To understand the idea, consider the data sets of waiting times at banks, for which the mean waiting time was 7.2 minutes at both Big Bank and Best Bank. A waiting time of 8.2 minutes then has a **deviation** from the mean of 8.2 minutes − 7.2 minutes = 1.0 minute, because it is 1.0 minute greater than the mean. Similarly, a waiting time of 5.2 minutes has a deviation from the mean of 5.2 minutes − 7.2 minutes = −2.0 minutes (*negative* 2 minutes), because it is 2.0 minutes *less* than the mean.

The purpose of the standard deviation is to represent an average of the *sizes* of the individual deviations, without regard to their signs (positive or negative). To do that, we square the deviations before adding them up, then take the square root at the end of the calculation. The following box summarizes the complete procedure for calculating the standard deviation.

Technical Notes

(1) The standard deviation formula given here, in which we divide the sum of the squared deviations by the total number of data values *minus* 1, is technically valid only for data from *samples*. When dealing with populations, we do not subtract 1 in Step 4. In this book, we will use only the standard deviation formula for samples.

(2) The result of Step 4, which is the square of the standard deviation, is called the *variance.*

Calculating the Standard Deviation

Step 1. Compute the mean of the data set. Then find the deviation from the mean for every data value by subtracting the mean from the data value. That is, for every data value:

$$\text{deviation from mean} = \text{data value} - \text{mean}$$

Step 2. Find the *squares* (second power) of all the deviations from the mean.

Step 3. Add all the squares of the deviations from the mean.

Step 4. Divide this sum by the total number of data values *minus* 1.

Step 5. The standard deviation is the square root of this quotient.

All of these steps are summarized by the following standard deviation formula:

$$\text{standard deviation} = \sqrt{\frac{\text{sum of (deviations from the mean)}^2}{\text{total number of data values} - 1}}$$

Note that, because we square the deviations in Step 2 and take the square root in Step 5, the units of the standard deviation are the same as the units of the data values. For example, if the data values have units of minutes, the standard deviation also has units of minutes.

EXAMPLE 3 Calculating Standard Deviation

Calculate the standard deviations for the waiting times at Big Bank and Best Bank.

Solution We follow the five steps to calculate the standard deviations. Table 6.3 shows how to organize the work in the first three steps. The first column for each bank lists the waiting times (in minutes). The second column lists the deviations from the mean (Step 1), which is 7.2 minutes for both banks. The third column lists the squares of the deviations (Step 2). We add all the squared deviations to find the sum at the bottom of the third column (Step 3). For Step 4, we divide the sums from Step 3 by the total number of data values *minus* 1. Because there are 11 data values, we divide by 10:

$$\textit{Big Bank:} \quad \frac{38.46}{10} = 3.846$$

$$\textit{Best Bank:} \quad \frac{1.98}{10} = 0.198$$

▲ **CAUTION!** For Step 4, don't forget to divide the sums by the total number of data values *minus* 1. ▲

Finally, Step 5 tells us that the standard deviations are the square roots of the numbers from Step 4:

$$\textit{Big Bank:} \quad \text{standard deviation} = \sqrt{3.846} \approx 1.96 \text{ minutes}$$

$$\textit{Best Bank:} \quad \text{standard deviation} = \sqrt{0.198} \approx 0.44 \text{ minute}$$

We conclude that the standard deviation of the waiting times is 1.96 minutes at Big Bank and 0.44 minute at Best Bank. As we expected, the waiting times show greater variation at Big Bank, which is why the lines at Big Bank annoy more customers than do those at Best Bank. ▶ Now try Exercises 15–18, part (c).

Think About It Look closely at the individual deviations in Table 6.3. Do the standard deviations for the two data sets seem like reasonable "averages" for the deviations? Explain.

TABLE 6.3 Calculating Standard Deviation

BIG BANK			BEST BANK		
Time	Deviation (Time − Mean)	(Deviation)^2	Time	Deviation (Time − Mean)	(Deviation)^2
4.1	4.1 − 7.2 = −3.1	$(-3.1)^2 = 9.61$	6.6	6.6 − 7.2 = −0.6	$(-0.6)^2 = 0.36$
5.2	5.2 − 7.2 = −2.0	$(-2.0)^2 = 4.00$	6.7	6.7 − 7.2 = −0.5	$(-0.5)^2 = 0.25$
5.6	5.6 − 7.2 = −1.6	$(-1.6)^2 = 2.56$	6.7	6.7 − 7.2 = −0.5	$(-0.5)^2 = 0.25$
6.2	6.2 − 7.2 = −1.0	$(-1.0)^2 = 1.00$	6.9	6.9 − 7.2 = −0.3	$(-0.3)^2 = 0.09$
6.7	6.7 − 7.2 = −0.5	$(-0.5)^2 = 0.25$	7.1	7.1 − 7.2 = −0.1	$(-0.1)^2 = 0.01$
7.2	7.2 − 7.2 = 0.0	$(0.0)^2 = 0.0$	7.2	7.2 − 7.2 = 0.0	$(0.0)^2 = 0.0$
7.7	7.7 − 7.2 = 0.5	$(0.5)^2 = 0.25$	7.3	7.3 − 7.2 = 0.1	$(0.1)^2 = 0.01$
7.7	7.7 − 7.2 = 0.5	$(0.5)^2 = 0.25$	7.4	7.4 − 7.2 = 0.2	$(0.2)^2 = 0.04$
8.5	8.5 − 7.2 = 1.3	$(1.3)^2 = 1.69$	7.7	7.7 − 7.2 = 0.5	$(0.5)^2 = 0.25$
9.3	9.3 − 7.2 = 2.1	$(2.1)^2 = 4.41$	7.8	7.8 − 7.2 = 0.6	$(0.6)^2 = 0.36$
11.0	11.0 − 7.2 = 3.8	$(3.8)^2 = 14.44$	7.8	7.8 − 7.2 = 0.6	$(0.6)^2 = 0.36$
Mean = 7.2		**Sum = 38.46**	**Mean = 7.2**		**Sum = 1.98**

Interpreting the Standard Deviation

A good way to develop a deeper understanding of the standard deviation is to consider an approximation called the **range rule of thumb**, summarized in the following box.

BY THE WAY

Another interpretation of the standard deviation uses a mathematical rule called *Chebyshev's theorem*. It states that, for any data set, at least 75% of all data values lie within 2 standard deviations of the mean, and at least 89% of all data values lie within 3 standard deviations of the mean.

The Range Rule of Thumb

The standard deviation is *approximately* related to the range of a distribution by the **range rule of thumb**:

$$\text{standard deviation} \approx \frac{\text{range}}{4}$$

If we know the range of a data set (range = highest − lowest), we can use this rule to estimate the standard deviation. Alternatively, if we know the standard deviation for a data set, we estimate the lowest and highest values as follows:

$$\text{lowest value} \approx \text{mean} - (2 \times \text{standard deviation})$$
$$\text{highest value} \approx \text{mean} + (2 \times \text{standard deviation})$$

The range rule of thumb does not work well when the highest or lowest value is an outlier.

The range rule of thumb works reasonably well only for data sets in which values are distributed fairly evenly. You must therefore use judgment in deciding whether the range rule of thumb is applicable in a particular case, and in all cases remember that the range rule of thumb yields rough approximations, not exact results.

USING TECHNOLOGY

Standard Deviation in Excel

The built-in Excel function STDEV automates the calculation of a standard deviation—all you have to do is enter the data and then use the function. The screen shot below shows the process for the waiting times at Big Bank, with the functions shown in column E and the results in column F. Note that the results agree with those found for Big Bank in Example 3.

	A	B	C	D	E	F
1		**Big Bank**				
2	**Data**	4.1		**Mean**	=AVERAGE(B2:B12)	7.2
3		5.2		**St. Dev.**	=STDEV(B2:B12)	1.96
4		5.6				
5		6.2				
6		6.7				
7		7.2				
8		7.7				
9		7.7				
10		8.5				
11		9.3				
12		11				

Microsoft Excel 2016, Windows 10, Microsoft Corporation.

EXAMPLE 4 Using the Range Rule of Thumb

Use the range rule of thumb to estimate the standard deviations for the waiting times at Big Bank and Best Bank. Compare the estimates to the actual standard deviations found in Example 3.

Solution The waiting times for Big Bank vary from 4.1 to 11.0 minutes, which means a range of $11.0 - 4.1 = 6.9$ minutes. The waiting times for Best Bank vary from 6.6 to 7.8 minutes, for a range of $7.8 - 6.6 = 1.2$ minutes. The range rule of thumb gives the following estimates for the standard deviations:

$$\textit{Big Bank:}\qquad \text{standard deviation} \approx \frac{6.9}{4} \approx 1.7$$

$$\textit{Best Bank:}\qquad \text{standard deviation} \approx \frac{1.2}{4} \approx 0.3$$

The actual standard deviations calculated in Example 3 are 1.96 and 0.44, respectively. For these two cases, the estimates from the range rule of thumb slightly underestimate the actual standard deviations. Nevertheless, the estimates put us in the right ballpark, showing that the rule is useful. ▶ Now try Exercises 15–18, part (d).

EXAMPLE 5 Estimating a Range

Studies of the gas mileage of a Prius under varying driving conditions show that it gets a mean of 45 miles per gallon with a standard deviation of 4 miles per gallon. Estimate the minimum and maximum gas mileage that you can expect under ordinary driving conditions.

Solution From the range rule of thumb, the lowest and highest values for gas mileage are approximately

$$\begin{aligned}\text{lowest value} &\approx \text{mean} - (2 \times \text{standard deviation})\\ &= 45 - (2 \times 4)\\ &= 37\end{aligned}$$

$$\begin{aligned}\text{highest value} &\approx \text{mean} + (2 \times \text{standard deviation})\\ &= 45 + (2 \times 4)\\ &= 53\end{aligned}$$

The range of gas mileage for the Prius extends roughly from 37 to 53 miles per gallon. ▶ Now try Exercises 15–18, part (e).

BY THE WAY

Technologies such as catalytic converters have helped reduce the amounts of many pollutants emitted by cars (per mile driven), but burning less gasoline is the only way to reduce carbon dioxide emissions that cause global warming. This is a major reason why auto manufacturers are developing high-mileage hybrid vehicles and zero-emission vehicles that run on electricity or fuel cells.

Quick Quiz 6B

Choose the best answer to each of the following questions. Explain your reasoning with one or more complete sentences.

1. The lowest score on an exam was 62, the median score was 75, and the high score was 96. The range was
   a. 34. b. 62. c. 75.
2. Which of the following is *not* part of a five-number summary?
   a. the lowest value b. the median c. the mean
3. The lower quartile for hourly wages at a coffee shop is \$11.25 and the upper quartile is \$13.75. What can you conclude?
   a. Half the workers earn between \$11.25 and 13.75.
   b. The median is \$12.50.
   c. The range is \$2.50.

4. Is it possible for a distribution to have a mean that is lower than its lower quartile?
   a. Yes; this is always the case.
   b. Yes, but only if there are outliers at low values.
   c. No.
5. Suppose you are given the mean and just one data value from a distribution. What can you calculate?
   a. the range
   b. the deviation for the single data value
   c. the standard deviation
6. The standard deviation is best described as a measure of
   a. the average values in a data set.
   b. the spread of the data values around the mean.
   c. the median of a data set.
7. What type of data distribution has a *negative* standard deviation?
   a. a distribution in which most of the values are negative numbers
   b. a distribution in which most of the values lie below the mean
   c. None—the standard deviation cannot be negative.
8. For any distribution, it is always true that
   a. the range is at least as large as the standard deviation.
   b. the standard deviation is at least as large as the range.
   c. the range is always at least twice the mean.
9. Which data set would you expect to have the highest standard deviation?
   a. heights (lengths) of newborn infants
   b. heights of all elementary school children
   c. heights of first-grade boys
10. Professors Smith, Jones, and Garcia all received the same mean grade of 2.7 (a B–) on their student evaluations last semester. The standard deviations were 0.2 for Smith, 0.5 for Jones, and 1.1 for Garcia. Which professor received grades above the mean from the most students?
   a. Smith b. Jones c. Garcia

Exercises 6B

REVIEW QUESTIONS

1. Consider two grocery stores at which the mean time waiting in line is the same but the variation is different. At which store would you expect the customers to have more complaints about the waiting time? Explain.
2. Describe how we define and calculate the *range* of a distribution.
3. What are the *quartiles* of a distribution? How do we find them?
4. Define the five-number summary, and explain how to depict it visually with a boxplot.
5. Describe the process of calculating a standard deviation. Give a simple example of this calculation (such as calculating the standard deviation of the numbers 2, 3, 4, 4, and 6). What is the standard deviation if all of the sample values are the same?
6. Briefly describe the use of the range rule of thumb for interpreting the standard deviation. What are its limitations?

DOES IT MAKE SENSE?

Decide whether each of the following statements makes sense (or is clearly true) or does not make sense (or is clearly false). Explain your reasoning.

7. The distributions of scores on two exams had the same range, so they had the same median.
8. The highest exam score was in the upper quartile of the distribution.
9. For the 30 students who took the test, the high score was 80, the median was 74, and the low score was 40.
10. I examined the data carefully, and the range was greater than the standard deviation.
11. The standard deviation for the heights of a group of 5-year-old children is smaller than the standard deviation for the heights of a group of children who range in age from 3 to 15.
12. The mean gas mileage of the compact cars we tested was 34 miles per gallon, with a standard deviation of 5 gallons.

BASIC SKILLS & CONCEPTS

13. **Big Bank Verification.** Find the mean and median for the waiting times at Big Bank given in the beginning of this unit. Show your work clearly, and verify that both are 7.2 minutes.
14. **Best Bank Verification.** Find the mean and median for the waiting times at Best Bank given in the beginning of this unit. Show your work clearly, and verify that both are 7.2 minutes.

15–18: Comparing Variations. Consider the following data sets.

a. Find the mean, median, and range for each of the two data sets.

b. Give the five-number summary and draw a boxplot for each of the data sets.

c. Find the standard deviation for each of the data sets.

d. Apply the range rule of thumb to estimate the standard deviation of each of the data sets. How well does the rule work in each case? Briefly discuss why it does or does not work well.

e. Based on all your results, compare and discuss the two data sets in terms of their center and variation.

15. The table below gives the 2016 Cost of Living Index (COLI) for six cities with a relatively high cost of living and six cities with a relatively low cost of living. The index is based on housing, utilities, grocery items, transportation, health care, and miscellaneous goods and services; 100 represents the national average.

High-cost cities		Low-cost cities	
Manhattan, NY	228	Memphis, TN	83
San Francisco, CA	177	Tupelo, MS	83
Washington DC	149	Ashland, OH	82
Oakland, CA	149	Kalamazoo, MI	80
Boston, MA	148	Richmond, IN	80
Seattle, WA	145	McAllen, TX	76

Source: Council for Community and Economic Research.

16. The table below gives the average sales tax rate (state plus local) in six east coast states and six western states.

East coast states		Western states	
Florida	6.80%	California	8.25%
Massachusetts	6.25%	Arkansas	1.76%
Maryland	6.00%	Oregon	0%
New Hampshire	0%	Washington	8.92%
New York	8.49%	Arizona	8.25%
Rhode Island	7.00%	Utah	6.76%

Source: Tax Foundation.

17. The table below shows the fraction of games won (to the nearest thousandth) by six Major League Baseball teams in the American and National Leagues for the 2016 season. The lists include the teams with the best and worst win-loss records in both leagues.

National League	0.420	0.463	0.484	0.537	0.586	0.640
American League	0.364	0.420	0.500	0.549	0.584	0.586

18. The following data sets give the approximate lengths of Beethoven's nine symphonies and Mahler's nine symphonies (in minutes).

Beethoven	28	36	50	33	30	40	38	26	68
Mahler	52	85	94	50	72	72	80	90	80

FURTHER APPLICATIONS

19–20: Understanding Variation. The following exercises give four data sets consisting of seven numbers.

a. Make a histogram for each set.

b. Give the five-number summary and draw a boxplot for each set.

c. Compute the standard deviation for each set.

d. Based on your results, briefly explain how the standard deviation provides a useful single-number summary of the variation in these data sets.

19. The following sets of numbers all have a mean of 9:

$$\{9, 9, 9, 9, 9, 9, 9\}, \{8, 8, 9, 9, 9, 10, 10\},$$
$$\{8, 8, 8, 9, 10, 10, 10\}, \{6, 6, 6, 9, 12, 12, 12\}$$

20. The following sets of numbers all have a mean of 6:

$$\{6, 6, 6, 6, 6, 6, 6\}, \{5, 5, 6, 6, 6, 7, 7\},$$
$$\{5, 5, 5, 6, 7, 7, 7\}, \{3, 3, 3, 6, 9, 9, 9\}$$

21. **Pizza Deliveries.** After recording the pizza delivery times for two different pizza shops, you conclude that one pizza shop has a mean delivery time of 30 minutes with a standard deviation of 3 minutes. The other shop has a mean delivery time of 29 minutes with a standard deviation of 12 minutes. Interpret these figures. If you liked the pizzas from both shops equally well, which one would you order from? Why?

22. **Airline Arrival Times.** Two airlines have data on the arrival times of their flights. An arrival time of +2 minutes means the flight arrived 2 minutes early. An arrival time of −5 minutes means the flight arrived 5 minutes late. Skyview Airlines has a mean arrival time of 0.5 minute with a standard deviation of 9.6 minutes. SkyHigh Airlines has a mean arrival time of −5 minutes with a standard deviation of 4.0 minutes. Explain the meaning of these figures and why they would affect your choice of airline.

23. **Portfolio Standard Deviation.** The book *Investments* by Zvi Bodie, Alex Kane, and Alan Marcus claims that the returns for investment portfolios with a single stock have a standard deviation of 0.55, while the returns for portfolios with 32 stocks have a standard deviation of 0.325. Explain how the standard deviation measures the risk in these two types of portfolios.

24. **Defect Rates.** Two factories each produce 1000 computer chips per day. In Factory A, the mean number of defective chips per day is 3 with a standard deviation of 2.5. In Factory B, the mean number of defective chips per day is 4 with a standard deviation of 0.5. Use these figures to either defend or discredit the claim that Factory A has a more reliable manufacturing process.

25. **Ice Cream Deviations.** Each night you total the day's sales and the total volume of ice cream sold in your shop. You notice that when an employee named Ben works, the mean price of the ice cream sold is $2.30 per pint with a standard deviation of $0.05. On nights when an employee named Jerry

works, the mean price of the ice cream sold is $2.25 per pint with a standard deviation of $0.35. Which employee likely receives more complaints that his servings are too small? Explain.

26. **Vet Data.** A small animal veterinarian reviews her records for the day and notes that she has seen eight dogs and eight cats with the following weights (in pounds):

Dogs: 13, 23, 37, 45, 55, 63, 76, 102

Cats: 4, 4, 5, 9, 10, 15, 19, 22

a. Before analyzing these data sets, make a conjecture about which set has the larger mean, median, and standard deviation. Explain your reasoning.

b. Compute the mean and standard deviation of each set.

27. **Olympic 100-Meter Times.** The table below shows the times (in seconds) of all finishers in the finals of the men's 100-meter race in the 2000, 2008, and 2016 summer Olympics.

2000: 9.87, 9.99, 10.04, 10.08, 10.09, 10.13, 10.17

2008: 9.69, 9.89, 9.91, 9.93, 9.95, 9.97, 10.01, 10.03

2016: 9.81, 9.89, 9.91, 9.93, 9.94, 9.96, 10.04, 10.06

a. Find the mean and standard deviation of each data set.

b. Is there evidence that during this 16-year period runners have gotten faster, either individually or as a group?

28. **Commute Times.** Jack and Juan live next to each other and commute to the same office. During one week, the two men record the following commute times (in minutes). Find the mean and standard deviation of each data set, and give a plausible explanation for the differences you observe.

Jack: 32, 38, 39, 42, 45

Juan: 28, 29, 30, 30, 31

29. **Quality Control.** An auto transmission manufacturer receives ball bearings from two different suppliers. The ball bearings must have a specified diameter of 16.30 millimeters (mm) with a tolerance of ± 0.1 mm. Recent shipments from the two suppliers had ball bearings with the following diameters.

Supplier A: 16.25, 16.27, 16.29, 16.31, 16.34, 16.37, 16.41

Supplier B: 16.19, 16.22, 16.28, 16.34, 16.39, 16.42, 16.44

a. Find the mean and standard deviation of each data set.

b. Draw a boxplot for each data set, and mark the tolerance on each boxplot.

c. What percentage of ball bearings from each supplier meet the specifications?

IN YOUR WORLD

30. **Web Data Sets.** Go to any website that gives data sets, such as the Census Bureau, the U.S. Energy Information Administration, or the National Center for Heath Statistics. Choose three data sets that you find interesting. In each case, describe the data distribution and discuss the variation in the distribution. How is the variation important in understanding the data?

31. **Ranges in the News.** Find two examples of data distributions in recent news reports; they may be given either as tables or as graphs. In each case, state the range of the distribution and explain its meaning in the context of the news report.

32. **Summarizing a News Data Set.** Find an example of a data distribution given in the form of a table in a recent news report. Make a five-number summary and a boxplot for the distribution.

33. **Range Rule in the News.** For each of the two data distributions you found in Exercise 31, estimate the standard deviation by applying the range rule of thumb. Discuss whether you think the estimate is valid in each case.

TECHNOLOGY EXERCISES

Answer the following questions using procedures described in the Using Technology boxes in this unit or with **StatCrunch** (available in MyLab Math).

34. **Computing Standard Deviation.** The following data set gives the time intervals in minutes between eruptions of Old Faithful geyser in Yellowstone National Park.

98 92 95 87 96 90 65 92 95 93 98 94

a. Use a calculator, Excel, or StatCrunch to compute the mean, median, and standard deviation of the data.

b. Suppose Old Faithful missed the second eruption in the set above, giving the following new data set:

190 95 87 96 90 65 92 95 93 98 94

How does this change affect the mean and standard deviation computed in part (a)?

c. Suppose we add to the original data set a 13th time interval with a value equal to the mean of that data set (that is, the mean of the 12 time intervals shown at the beginning of this exercise). Does this added data value change the mean, median, or standard deviation of the original data set?

35. **Variation in StatCrunch.** Load the StatCrunch data set *ExerciseHours*, which gives various characteristics of 50 college students (gender, handedness, exercise hours per week, TV hours per week, pulse, and more).

a. Find the range and standard deviation of the hours spent exercising per week.

b. Find the range and standard deviation of the hours spent watching TV.

c. Make a boxplot of the exercise data set grouped by year. The result will be separate boxplots for first-year students, sophomores, juniors, and seniors. Comment on the results.

d. Make a boxplot of the data for TV hours per week grouped by handedness. The result will be separate boxplots for right- and left-handed students. Comment on the results.

36. **StatCrunch Project.** Choose a data set available in StatCrunch that is of interest to you; in this case, you might wish to choose a relatively small and simple one.

  a. Identify the data set you've chosen, and describe in a sentence why it is of interest to you.

  b. Find the mean, median, range, and standard deviation for the data set.

  c. Make a graph that displays the variation clearly. Explain why you chose that type of graph (boxplot, histogram, etc.).

  d. Write a few sentences summarizing what the mean, median, and variation in this data set tell you about the distribution.

UNIT 6C The Normal Distribution

Figure 6.14 shows two examples of what is often called a *bell-shaped curve*, but is more formally called a **normal distribution**. The normal distribution arises so frequently that we give it special attention in statistics.

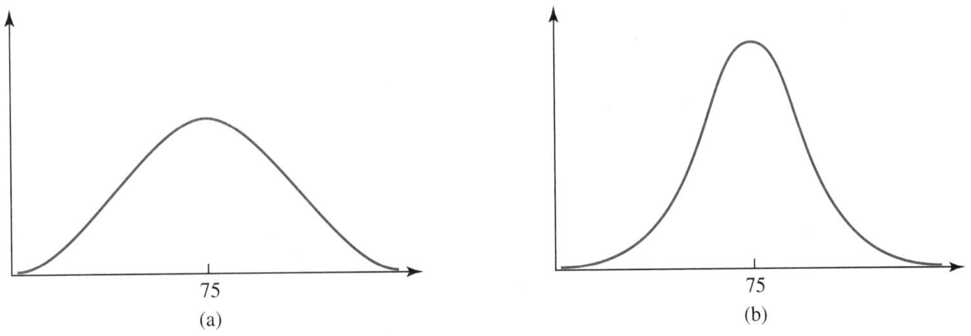

FIGURE 6.14 Both distributions are normal and have the same mean of 75, but the distribution in part (a) has a larger standard deviation.

Normal distributions always have the same characteristic bell shape and can be characterized by just two numbers: their mean and their standard deviation (see Unit 6B). The mean determines the location of the bell's peak, while the standard deviation determines the bell's width. Notice that the normal distributions in Figure 6.14 both have the same mean of 75, but the distribution in part (a) is wider because it has a greater standard deviation.

HISTORICAL NOTE

Using data taken from French and Scottish soldiers, the Belgian social scientist Adolphe Quetelet realized in the 1830s that human characteristics such as height and chest circumference are normally distributed. This observation led him to coin the term "the average man."

Definition

The **normal distribution** is a symmetric, bell-shaped distribution with a single peak. Its peak corresponds to the mean, median, and mode of the distribution. Its variation is characterized by the standard deviation of the distribution.

EXAMPLE 1 The Normal Shape

Figure 6.15 shows two distributions: (a) a famous data set of the chest sizes of 5738 Scottish militiamen collected in about 1846 and (b) the distribution of the population densities of the 50 states. Is either distribution a normal distribution? Explain.

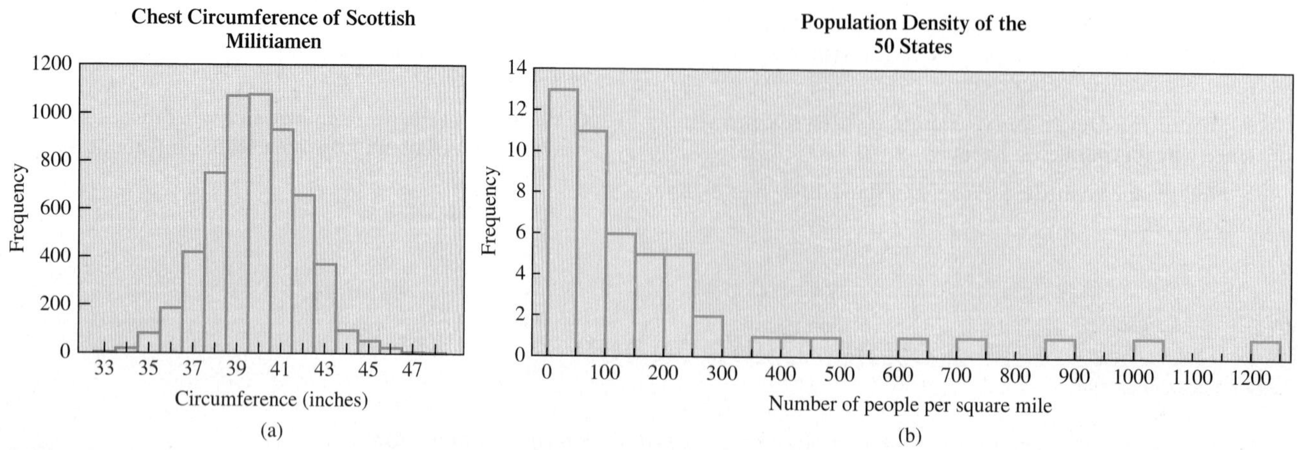

FIGURE 6.15

Solution The distribution of chest sizes in Figure 6.15a is nearly symmetric, with a mean between 39 and 40 inches. Values far from the mean are less common, and the distribution is approximately bell-shaped, so it is approximately a normal distribution. The distribution in Figure 6.15b shows that most states have low population densities, but a few have much higher densities. This fact makes the distribution right-skewed, so it is not a normal distribution. ▶ Now try Exercises 11–12.

What Is Normal?

We can better appreciate the importance of the normal distribution if we understand why it is so common. Consider a human characteristic such as height. Most men or women have heights clustered near the mean height (for their sex), so a data set of heights has a peak at the mean height. But as we consider heights increasingly far from the mean on either side, we find fewer and fewer people. This "tailing off" of heights far from the mean produces two symmetric tails on either side of the mean. As a result, the overall height distribution closely approximates a normal distribution.

On a deeper level, any quantity that is the result of *many* factors is likely to follow a normal distribution. Adult heights are the result of many genetic and environmental factors. Scores on SAT tests or IQ tests tend to be normally distributed because each test score is determined from many individual test questions. Sports statistics, such as batting averages, tend to be normally distributed because they involve many people with many different levels of skill. More generally, we can expect a data set to have a nearly normal distribution if it meets the following conditions.

Conditions for a Normal Distribution

A data set that satisfies the following four criteria is likely to have a nearly normal distribution:

1. Most data values are clustered near the mean, giving the distribution a well-defined single peak.
2. Data values are spread evenly around the mean, making the distribution symmetric.
3. Larger deviations from the mean become increasingly rare, producing the symmetric tapering tails of the distribution.
4. Individual data values result from a combination of many different factors, such as genetic and environmental factors.

EXAMPLE 2 Is It a Normal Distribution?

Which of the following variables would you expect to have a normal or nearly normal distribution?

a. Scores on a very easy test
b. Foot lengths of a random sample of adult women

Solution

a. Tests have a maximum score (100%) that limits the size of the data values. If a test is very easy, the mean will be high and many scores will be near the maximum. The fewer lower scores can be spread out well below the mean. We therefore expect the distribution of scores to be left-skewed and not normal.

b. Foot length is a human trait determined by many genetic and environmental factors. We therefore expect lengths of women's feet to cluster near a mean and become less common farther from the mean in both directions, giving the distribution the bell shape of a normal distribution. ▶ Now try Exercises 13–18.

The Standard Deviation in Normal Distributions

Because all normal distributions have the same bell shape, knowing the mean and standard deviation of a distribution tells us a lot about where the data values lie. Recall from Unit 6B that the standard deviation gives a measure of the "average" spread of data values around the mean. A simple rule, often called the **68-95-99.7 rule**, gives approximate guidelines for the percentages of data values that lie within 1, 2, and 3 standard deviations of the mean for any normal distribution. The following box states the rule in words, and Figure 6.16 shows it graphically.

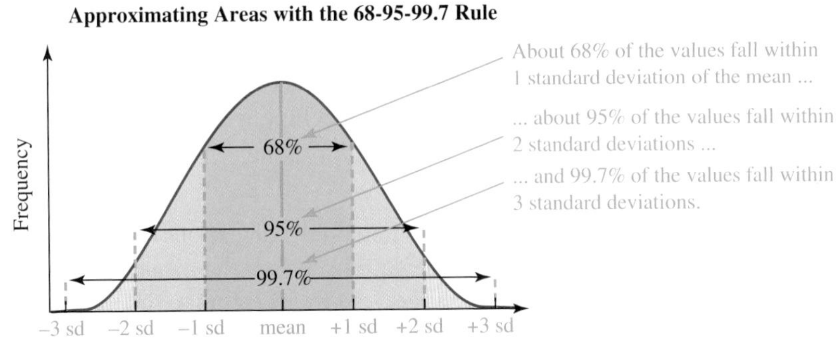

FIGURE 6.16 A normal distribution illustrating the 68-95-99.7 rule. Notice that the horizontal axis is marked with the position of the mean and the positions corresponding to 1, 2, and 3 standard deviations above and below the mean; "sd" is used as an abbreviation for "standard deviation."

BY THE WAY
Although we've abbreviated "standard deviation" as "sd" in Figure 6.16, in statistical work the more common abbreviation is the Greek letter σ (sigma). The mean is commonly abbreviated with the Greek letter μ (mu).

The 68-95-99.7 Rule for a Normal Distribution

- About 68% (more precisely, 68.3%), or just over two-thirds, of the data points fall within 1 standard deviation of the mean.
- About 95% (more precisely, 95.4%) of the data points fall within 2 standard deviations of the mean.
- About 99.7% of the data points fall within 3 standard deviations of the mean.

EXAMPLE 3 SAT Scores

The tests that make up the individual parts of the SAT are designed so that their scores are normally distributed with a mean of 500 and a standard deviation of 100 (and a range from 200 to 800). Interpret this statement.

Solution From the 68-95-99.7 rule, about 68% of students have scores within 1 standard deviation (100 points) of the mean of 500 points; that is, about 68% of the students score between 400 and 600. About 95% of the students score within 2 standard deviations (200 points) of the mean, or between 300 and 700. And about 99.7% of the students score within 3 standard deviations (300 points) of the mean, or between 200 and 800. Figure 6.17 illustrates this interpretation graphically. Note that the horizontal axis shows both actual scores and the distance from the mean in standard deviations.

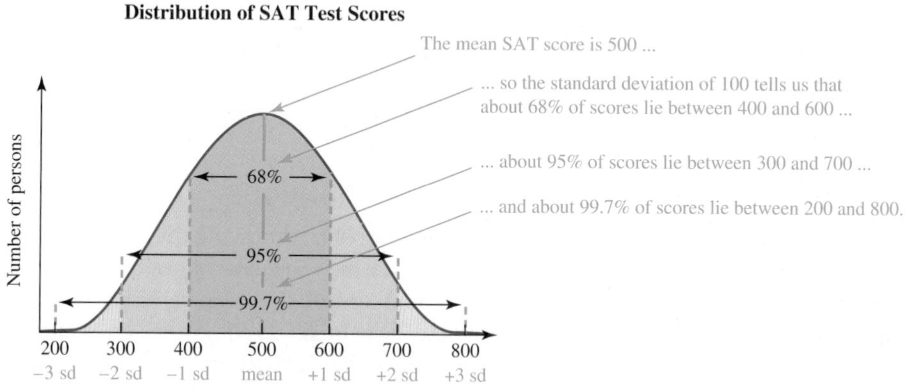
Distribution of SAT Test Scores
The mean SAT score is 500 ...
... so the standard deviation of 100 tells us that about 68% of scores lie between 400 and 600 ...
... about 95% of scores lie between 300 and 700 ...
... and about 99.7% of scores lie between 200 and 800.
Number of persons
68%
95%
99.7%
200
300
400
500
600
700
800
−3 sd
−2 sd
−1 sd
mean
+1 sd
+2 sd
+3 sd

FIGURE 6.17 The normal distribution of scores on individual parts of the SAT.

▶ Now try Exercise 19.

EXAMPLE 4 Detecting Counterfeits

Vending machines can be adjusted to reject coins above and below certain weights. The weights of legal U.S. quarters are normally distributed with a mean of 5.67 grams and a standard deviation of 0.0700 gram. If a vending machine is adjusted to reject quarters that weigh more than 5.81 grams and less than 5.53 grams, what percentage of legal quarters will be rejected by the machine?

Solution A weight of 5.81 is 0.14 gram, or 2 standard deviations, above the mean. A weight of 5.53 is 0.14 gram, or 2 standard deviations, below the mean. Therefore, by accepting only quarters within the weight range 5.53 to 5.81 grams, the machine accepts quarters that are within 2 standard deviations of the mean and rejects those that are more than 2 standard deviations from the mean. By the 68-95-99.7 rule, about 95% of legal quarters will be accepted and about 5% of legal quarters will be rejected.

▶ Now try Exercise 20.

Applying the 68-95-99.7 Rule

We can apply the 68-95-99.7 rule to find approximate answers to many questions about the frequencies (or relative frequencies) of data values in a normal distribution. Consider an exam taken by 1000 students for which the scores are normally distributed with a mean of 75 and a standard deviation of 7. Suppose we want to know how many students scored above 82, which is 7 points, or 1 standard deviation, above the mean of 75. The 68-95-99.7 rule tells us that about 68% of the scores are *within* 1 standard deviation of the mean. Therefore, about 100% − 68% = 32% of the scores are *more than* 1 standard deviation from the mean. Half of this 32%, or 16%, of the scores are more than 1 standard deviation *below* the mean; the other 16% of the scores are more than 1 standard deviation *above* the mean (Figure 6.18a). We conclude that about 16% of 1000 students, or 160 students, scored above 82.

Similarly, suppose we want to know how many students scored below 61, which is 14 points, or 2 standard deviations, below the mean of 75. The 68-95-99.7 rule tells us that about 95% of the scores are *within* 2 standard deviations of the mean, so about 5% of the scores are *more than* 2 standard deviations from the mean. Half of this 5%, or 2.5%, of the scores are more than 2 standard deviations *below* the mean (Figure 6.18b), so we conclude that about 2.5% of 1000 students, or 25 students, scored below 61.

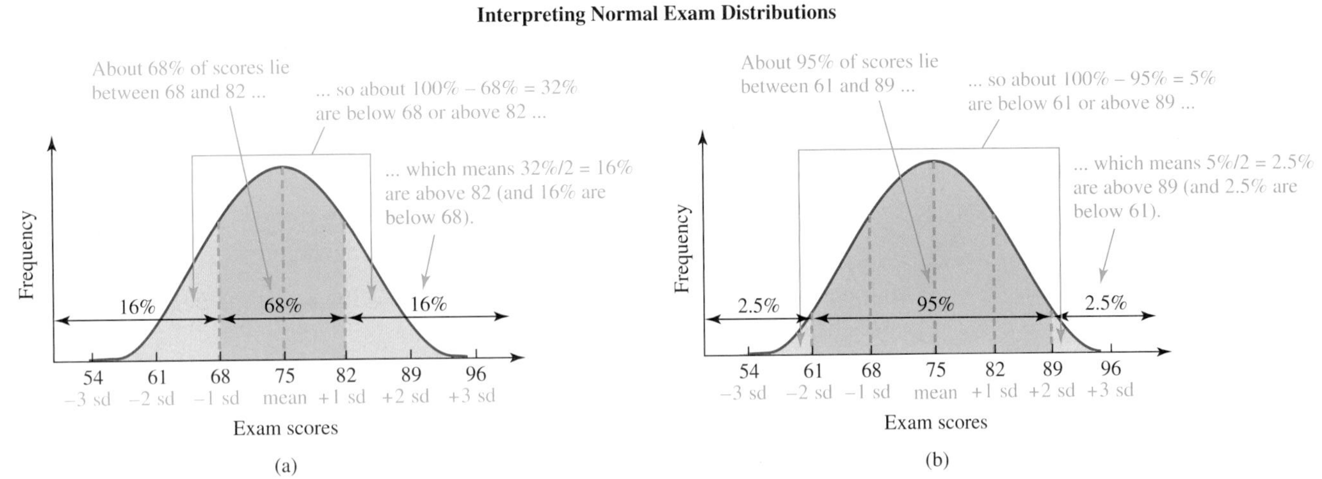

FIGURE 6.18 A normal distribution of test scores with a mean of 75 and a standard deviation of 7. (a) 68% of the scores lie within 1 standard deviation of the mean. (b) 95% of the scores lie within 2 standard deviations of the mean.

EXAMPLE 5 Normal Auto Prices

A survey finds that the prices paid for two-year-old Ford Fusion cars are normally distributed with a mean of \$17,500 and a standard deviation of \$500. Consider a sample of 10,000 people who bought two-year-old Ford Fusions.

a. How many people paid between \$17,000 and \$18,000?
b. How many paid less than \$17,000?
c. How many paid more than \$19,000?

Solution

a. Given the mean of \$17,500 and standard deviation of \$500, prices between \$17,000 and \$18,000 are within 1 standard deviation of the mean. From the 68-95-99.7 rule, about 68% of the 10,000 car buyers, or 6800 people, paid prices in this range.

b. The remaining 32%, or 3200 people, paid prices more than 1 standard deviation from the mean, or outside the \$17,000 to \$18,000 range. Half of these people, or 1600 people, paid prices below \$17,000 (while the other half paid prices above \$18,000).

c. A price of \$19,000 is \$1500 above the mean of \$17,500. Because the standard deviation is \$500, this price of \$19,000 is 3 standard deviations above the mean. From the 68-95-99.7 rule, about 99.7% of the car buyers paid prices *within* 3 standard deviations of the mean, leaving 0.3% paying either more or less. Half of the 0.3%, or 0.15%, paid a price *more than* 3 standard deviations above the mean. Out of 10,000 people, 0.15% represents 15 people. That is, approximately 15 people paid more than \$19,000.

▶ Now try Exercises 21–24.

Standard Scores

The 68-95-99.7 rule applies to data values that are exactly 1, 2, or 3 standard deviations from the mean. For other cases, we can generalize this rule if we know precisely how many standard deviations from the mean a particular data value lies. The number of standard deviations a data value lies above or below the mean is called its **standard score** (or *z-score*), often abbreviated by the letter z. For example:

- The standard score of the mean is $z = 0$ because the mean is 0 standard deviations from the mean.
- The standard score of a data value 1.5 standard deviations above the mean is $z = 1.5$.
- The standard score of a data value 2.4 standard deviations below the mean is $z = -2.4$.

The following box explains the computation of standard scores.

Computing Standard Scores

The number of standard deviations a data value lies above or below the mean is called its **standard score** (or *z-score*), defined by

$$z = \text{standard score} = \frac{\text{data value} - \text{mean}}{\text{standard deviation}}$$

The standard score is positive for data values above the mean and negative for data values below the mean.

EXAMPLE 6 Standard IQ Scores

BY THE WAY

IQ stands for "intelligence quotient." The term was coined by French psychologist Alfred Binet (1857–1911), who also created the first IQ test. He gave the test to children of different ages and calculated IQ by dividing a child's "mental age" by his or her actual age and multiplying the result by 100. For example, a 5-year-old who scored as well as an average 6-year-old was said to have a "mental age" of 6 and an IQ of $6 \div 5 \times 100 = 120$. Binet did not believe that IQ was innate and created his test with the goal of identifying children who could increase their intelligence if they were given additional help in school.

The Stanford-Binet IQ test is scaled so that its scores are normally distributed with a mean of 100 and a standard deviation of 15. Find the standard scores for IQ scores of 85, 100, and 125.

Solution We calculate the standard scores for these IQ scores by using the standard score formula with a mean of 100 and standard deviation of 15.

$$\textit{Standard score for } 85: \quad z = \frac{85 - 100}{15} = -1.00$$

$$\textit{Standard score for } 100: \quad z = \frac{100 - 100}{15} = 0.00$$

$$\textit{Standard score for } 125: \quad z = \frac{125 - 100}{15} = 1.67$$

We interpret these standard scores as follows: 85 is 1.00 standard deviation *below* the mean, 100 is equal to the mean, and 125 is 1.67 standard deviations *above* the mean. Figure 6.19 shows these values on the distribution of IQ scores.

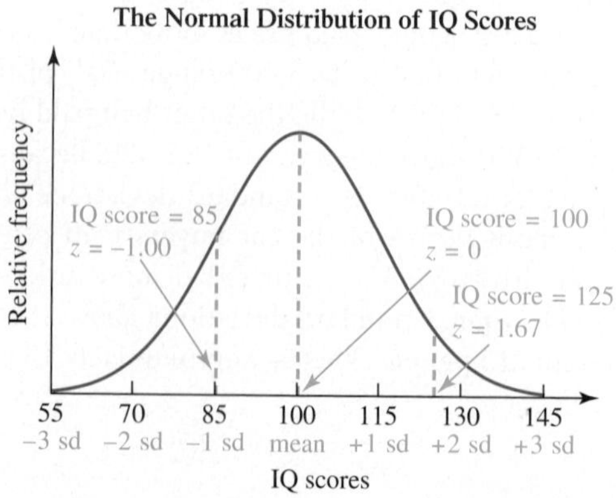

FIGURE 6.19

▸ Now try Exercises 25–28.

Standard Scores and Percentiles

Once we know the standard score of a data value, the properties of the normal distribution allow us to find the data value's **percentile** in the distribution. You are probably familiar with the idea of percentiles. For example, if you scored in the 45th percentile on the SAT, 45% of the SAT scores were lower than yours.

Percentiles

The **percentile** of a data value is the percentage of all data values in a data set that are *less than or equal to* it.

We can convert standard scores to percentiles with computer software or a *standard score table*, such as Table 6.4. For each of many standard scores in a normal distribution, the table gives the percentage of values in the distribution *less than or equal to* that value. For example, Table 6.4 shows that 55.96% of the values in a normal distribution have a standard score less than or equal to 0.15. In other words, if we are looking for an integer percentile, a data value with a standard score of 0.15 lies in the 55th percentile (nearly in the 56th).

Think About It Is it possible for someone to score above the 100th percentile on a standardized test? Why or why not?

TABLE 6.4 Standard Scores and Percentiles for a Normal Distribution

z-score	Percentile	z-score	Percentile	z-score	Percentile	z-score	Percentile
−3.5	0.02	−1.0	15.87	0.0	50.00	1.1	86.43
−3.0	0.13	−0.95	17.11	0.05	51.99	1.2	88.49
−2.9	0.19	−0.90	18.41	0.10	53.98	1.3	90.32
−2.8	0.26	−0.85	19.77	0.15	55.96	1.4	91.92
−2.7	0.35	−0.80	21.19	0.20	57.93	1.5	93.32
−2.6	0.47	−0.75	22.66	0.25	59.87	1.6	94.52
−2.5	0.62	−0.70	24.20	0.30	61.79	1.7	95.54
−2.4	0.82	−0.65	25.78	0.35	63.68	1.8	96.41
−2.3	1.07	−0.60	27.43	0.40	65.54	1.9	97.13
−2.2	1.39	−0.55	29.12	0.45	67.36	2.0	97.72
−2.1	1.79	−0.50	30.85	0.50	69.15	2.1	98.21
−2.0	2.28	−0.45	32.64	0.55	70.88	2.2	98.61
−1.9	2.87	−0.40	34.46	0.60	72.57	2.3	98.93
−1.8	3.59	−0.35	36.32	0.65	74.22	2.4	99.18
−1.7	4.46	−0.30	38.21	0.70	75.80	2.5	99.38
−1.6	5.48	−0.25	40.13	0.75	77.34	2.6	99.53
−1.5	6.68	−0.20	42.07	0.80	78.81	2.7	99.65
−1.4	8.08	−0.15	44.04	0.85	80.23	2.8	99.74
−1.3	9.68	−0.10	46.02	0.90	81.59	2.9	99.81
−1.2	11.51	−0.05	48.01	0.95	82.89	3.0	99.87
−1.1	13.57	0.0	50.00	1.0	84.13	3.5	99.98

EXAMPLE 7 Cholesterol Levels

Cholesterol levels in men 18 to 24 years of age are normally distributed with a mean of 178 and a standard deviation of 41.

a. In what percentile is a 20-year-old man with a cholesterol level of 190?
b. What cholesterol level corresponds to the 90th percentile, the level at which treatment may be necessary?

Solution

a. The standard score for a cholesterol level of 190 is

$$z = \frac{\text{data value} - \text{mean}}{\text{standard deviation}} = \frac{190 - 178}{41} \approx 0.29$$

Table 6.4 shows that a standard score of 0.29 corresponds to almost the 62nd percentile.

b. Table 6.4 shows that 90.32% of all data values have a standard score less than 1.3. That is, the 90th percentile is about 1.3 standard deviations above the mean. The standard deviation is 41, so 1.3 standard deviations is $1.3 \times 41 = 53.3$. Adding this value to the mean, we find that the 90th percentile begins at about $178 + 53 = 231$. A person with a cholesterol level above 231 may need treatment.

▶ **Now try Exercises 29–30.**

EXAMPLE 8 Women in the Army

The heights of American women ages 18 to 24 are normally distributed with a mean of 65 inches and a standard deviation of 2.5 inches. In order to serve in the U.S. Army, women must be between 58 inches and 80 inches tall. What percentage of American women are ineligible to serve based on their height?

Solution The standard scores for the army's minimum and maximum heights of 58 inches and 80 inches are

$$\textit{For 58 inches: } z = \frac{\text{data value} - \text{mean}}{\text{standard deviation}} = \frac{58 - 65}{2.5} = -2.8$$

$$\textit{For 80 inches: } z = \frac{\text{data value} - \text{mean}}{\text{standard deviation}} = \frac{80 - 65}{2.5} = 6.0$$

Table 6.4 shows that a standard score of −2.8 corresponds to the 0.26 percentile. A standard score of 6.0 does not appear in the table, which means it is above the 99.98th percentile (since that is the highest percentile shown in the table). Therefore, 0.26% of all American women are too short to serve in the army and fewer than 0.02% of all American women are too tall to serve in the army. Altogether, fewer than about 0.26% + 0.02% = 0.28% of women, or about one out of every 400 women, are ineligible to serve in the army based on their height.

▶ **Now try Exercises 31–32.**

Think About It Note that the army standards reject a larger percentage of short women than tall women. Why do you think this is the case? Is it fair?

USING TECHNOLOGY

Standard Scores and Percentiles in Excel

You can use built-in functions of Excel to find both standard scores and percentiles.

- The function STANDARDIZE returns a data value's standard score, based on the data value and the mean and standard deviation for the data set. Enter the function in the form "=STANDARDIZE(data value, mean, standard deviation)."
- The function NORM.DIST returns a data value's percentile in a normal distribution. Enter the function in the form "=NORM.DIST(data value, mean, standard deviation, TRUE)," where the fourth input (TRUE) is necessary to get a percentile result.

The screen shot below shows the calculation of the standard score and percentile for the data in Example 6a; the functions reference the cells (B1, B2, B3) containing the data value, mean, and standard deviation, respectively, and column C shows the results. Note that the percentile is returned as a fraction, so you need to multiply that number by 100 to express the percentile as a percentage.

	A	B	C
1	**data value**	190	
2	**mean**	178	
3	**standard deviation**	41	
4	**standard score**	=STANDARDIZE(B1,B2,B3)	0.29
5	**percentile**	=NORMDIST(B1,B2,B3,TRUE)	0.62

Microsoft Excel 2016, Windows 10, Microsoft Corporation.

Quick Quiz 6C

Choose the best answer to each of the following questions. Explain your reasoning with one or more complete sentences.

1. Graphs of normal distributions
   a. always look exactly the same.
   b. always have the same characteristic bell shape.
   c. can have any shape as long as they have a sharp central peak.
2. In a normal distribution, the mean
   a. is equal to the median.
   b. is greater than the median.
   c. can be greater or less than the median.
3. In a normal distribution, data values farther from the mean are
   a. less common than data values close to the mean.
   b. more common than data values close to the mean.
   c. equally as common as data values close to the mean.
4. Consider wages at a fast food restaurant where most of the workers earn the minimum wage. Would you expect the wages of all workers at this restaurant to have a normal distribution?
   a. Yes, because wage distributions are always normal.
   b. No, because the minimum wage is not enough money to live on.
   c. No, because a normal distribution is symmetric, but no one earns less than the minimum wage.
5. In a normal distribution, about two-thirds of the data values fall within
   a. 1 standard deviation of the mean.
   b. 2 standard deviations of the mean.
   c. 3 standard deviations of the mean.
6. Suppose a car driven under different conditions gets a mean gas mileage of 40 miles per gallon with a standard deviation of 3 miles per gallon. If you drive this car many times, the gas mileage on 95% of the trips will be between
   a. 37 and 43 miles per gallon.
   b. 34 and 46 miles per gallon.
   c. 31 and 49 miles per gallon.
7. Consider again the car described in Question 6. On about what percentage of the trips will the gas mileage be less than 37 miles per gallon?
   a. 16% b. 2.5% c. 68%
8. Consider an exam with a normal distribution of scores with a mean of 75 and a standard deviation of 6. If you get a 66 on the exam, your standard score (z-score) is
   a. 66. b. -9. c. -1.5.

9. A friend tells you that his IQ is in the 102nd percentile. You can conclude that
   a. he is smarter than 102% of all people.
   b. he is smarter than 2% of all people.
   c. he doesn't understand percentiles.

10. The height of a particular 7-year-old girl has a standard score of −0.50 among the heights of all 7-year-old girls. From Table 6.4, you know that
   a. she is 0.5 inch shorter than average for her age.
   b. she is taller than about 69% of all 7-year-old girls.
   c. she is shorter than about 69% of all 7-year-old girls.

Exercises 6C

REVIEW QUESTIONS

1. What is a *normal distribution*? Briefly describe the conditions that are likely to lead to a normal distribution. Give an example of a data set that meets these conditions, and explain how it meets them.
2. What is the *68-95-99.7 rule* for normal distributions? Explain how it can be used to answer questions about frequencies of data values in a normal distribution.
3. What is a *standard score*? How do you find the standard score for a particular data value?
4. What is a *percentile*? Describe how Table 6.4 allows you to relate standard scores and percentiles.

DOES IT MAKE SENSE?

Decide whether each of the following statements makes sense (or is clearly true) or does not make sense (or is clearly false). Explain your reasoning.

5. The heights of male basketball players at Kentucky College are normally distributed with a mean of 6 feet 8 inches and a standard deviation of 4 inches.
6. The weights of babies born at Belmont Hospital are normally distributed with a mean of 6.8 pounds and a standard deviation of 1.2 pounds.

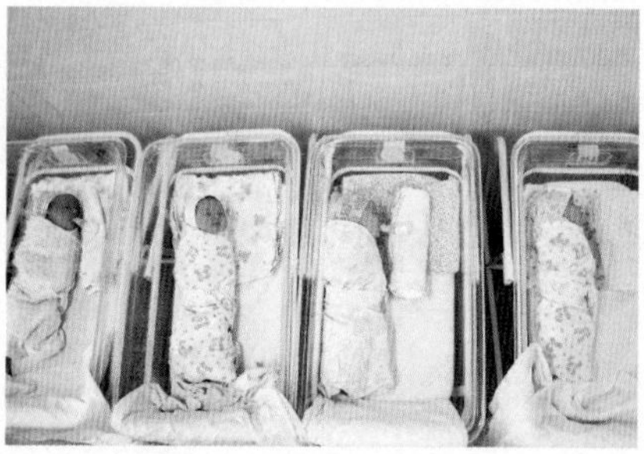

7. The weights of babies born at Belmont Hospital are normally distributed with a mean of 6.8 pounds and a standard deviation of 7 pounds.
8. On yesterday's mathematics exam, the standard score was 75.
9. My professor graded the final exam on a curve, and she gave a grade of A to anyone who had a standard score of 2 or more.
10. Jack is in the 50th percentile for height, so he is of median height.

BASIC SKILLS & CONCEPTS

11–12: Normal Shape. Consider the following sets of three distributions, all of which are drawn to the same scale. Identify the two distributions in each set that are normal. Of the two normal distributions, which one has the larger variation?

11.

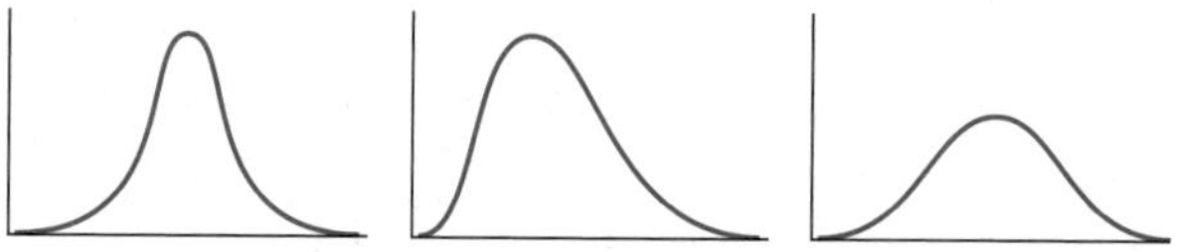

(a) (b) (c)

12.

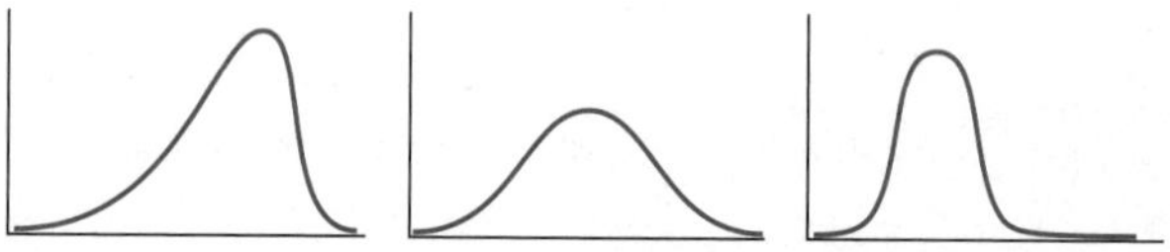

(a) (b) (c)

13–18: Normal Distributions. State, with an explanation, whether you would expect the following data sets to be normally distributed.

13. The delay in departure of trains from a station (note that trains, buses, and airplanes cannot leave early)
14. The weights of bags of flour labeled "25 pounds"
15. The weights of 200 new Ford-250 pickup trucks
16. The heights of 100 high jumps of a top-ranked athlete during a season
17. Scores awarded a large class on an easy economics exam
18. The last digit of the Social Security number of 1000 randomly selected people
19. **The 68-95-99.7 Rule.** A set of test scores is normally distributed with a mean of 100 and a standard deviation of 20. Use the 68-95-99.7 rule to find the percentage of scores in each of the following categories.
   a. greater than 100
   b. greater than 120
   c. less than 80
   d. less than 140
   e. less than 60
   f. less than 120
   g. greater than 80
   h. between 80 and 120

20. **The 68-95-99.7 Rule.** The resting heart rates for a sample of individuals are normally distributed with a mean of 70 and a standard deviation of 15. Use the 68-95-99.7 rule to find the percentage of heart rates in each of the following categories.

a. greater than 85
b. less than 40
c. less than 85
d. less than 100
e. greater than 70
f. less than 55
g. greater than 40
h. between 55 and 100

21–28: Psychology Exam. The scores on a psychology exam were normally distributed with a mean of 67 and a standard deviation of 8.

21. About what percentage of scores were less than 59?
22. About what percentage of scores were greater than 83?
23. A failing grade on the exam was anything 2 or more standard deviations below the mean. What was the cutoff for a failing score? Approximately what percentage of the students failed?
24. If 200 students took the exam, approximately how many students scored above 59?
25. What is the standard score for an exam score of 67?
26. What is the standard score for an exam score of 71?
27. What is the standard score for an exam score of 79?
28. What is the standard score for an exam score of 55?

29–30: Standard Scores and Percentiles. Use Table 6.4 to find the standard score and percentile of the following data values.

29. a. A data value 1 standard deviation below the mean
b. A data value 0.5 standard deviation below the mean
c. A data value 1.5 standard deviations above the mean
30. a. A data value 0.5 standard deviation above the mean
b. A data value 2 standard deviations above the mean
c. A data value 1.2 standard deviations below the mean

31–32: Percentiles. Use Table 6.4 to find the approximate standard score of the following data values. Then state the approximate number of standard deviations that the value lies above or below the mean.

31. a. A data value in the 30th percentile
b. A data value in the 70th percentile
c. A data value in the 55th percentile
32. a. A data value in the 20th percentile
b. A data value in the 45th percentile
c. A data value in the 94th percentile

FURTHER APPLICATIONS

33–36: Pregnancy Length. Actual lengths of pregnancy terms are nearly normally distributed about a mean pregnancy length (of about 38 to 39 weeks) with a standard deviation of 15 days.

33. About what percentage of births would be expected to occur within 15 days of the mean pregnancy length?
34. About what percentage of births would be expected to occur within 1 month of the mean pregnancy length?
35. About what percentage of births would be expected to occur more than 15 days after the mean pregnancy length?
36. About what percentage of births would be expected to occur more than 30 days after the mean pregnancy length? (*Note:* In practice, doctors will induce labor before this time, because of potential danger to the baby.)
37. **Heights.** According to data from the National Health Survey, the heights of all adult American women are normally distributed with a mean of 63.6 inches and a standard deviation of 2.5 inches. Give the standard score and approximate percentile for a woman with each of the following heights.

a. 64 inches
b. 61 inches
c. 59.8 inches
d. 64.8 inches

38. **Body Mass Index (BMI).** The body mass indexes of American men between ages 30 and 50 are normally distributed with a mean of 26.2 and a standard deviation of 4.7.

a. Determine the standard score and the percentile of a BMI of 25.

b. Determine the standard score and the percentile of a BMI of 28.

c. Several health organizations have declared that men with a BMI of 25 or greater are overweight and that men with a BMI of 30 or greater are obese. What percentage of American men are overweight? Obese?

39. **Is It Likely?** Suppose you read that the average height of a class of 45 eighth-graders is 55 inches with a standard deviation of 40 inches. Is this likely? Explain.
40. **Is It Likely?** The exam scores in your statistics course have a mean of 70 and a standard deviation of 50. Is this likely? Explain.

41–47: GRE Scores. Scores on the verbal section of the Graduate Record Exam (GRE) have a mean of 150 and a standard deviation of 8.5. Scores on the quantitative section of the exam have a mean of 152 and a standard deviation of 8.9. Assume the scores are normally distributed.

41. Suppose a graduate school requires (among other qualifications) that applicants score at or above the 90th percentile on both exams. What verbal and quantitative scores are required?
42. What percentage of students taking the verbal exam score above 160?
43. What percentage of students taking the quantitative exam score above 160?
44. Students intending to study engineering in graduate school have a mean score of 159 on the quantitative exam and a mean score of 149 on the verbal exam.

a. Find the percentile of 159 on the quantitative exam.

b. Find the percentile of 149 on the verbal exam.

45. Students intending to study arts and humanities in graduate school have a mean score of 150 on the quantitative exam and a mean score of 157 on the verbal exam.

a. Find the percentile of 150 on the quantitative exam.

b. Find the percentile of 157 on the verbal exam.

46. Scores on the quantitative exam range between 130 and 170. What are the percentiles of these scores?

47. Scores on the verbal exam range between 130 and 170. What are the percentiles of these scores?

IN YOUR WORLD

48. **Normal Distributions.** Many data sets described in the news have nearly normal distributions, though the article may not say so explicitly. In news reports, find two data sets that you suspect have nearly normal distributions. Explain your reasoning.

49. **Normal Demonstration.** Do a Web search on the keywords "normal distribution," and find an animated demonstration of the normal distribution. Describe how the demonstration works and the useful features that you observed.

TECHNOLOGY EXERCISES

Answer the following questions using procedures described in the Using Technology boxes in this unit or with **StatCrunch** (available in MyLab Math).

50. **Heights of American Men.** The heights of American adult men are normally distributed with a mean of 69.7 inches and a standard deviation of 2.7 inches. Use a calculator, Excel, or StatCrunch to answer the following questions.

    a. What are the standard score and percentile of a height of 72 inches?

    b. What are the standard score and percentile of a height of 65 inches?

    c. What percentage of the American adult male population is taller than 6′4″?

    d. What percentage of the American adult male population is shorter than 5′4″?

    e. Assume that the American population consists of 12 million adult men. Approximately how many men are at least as tall as the basketball player Lebron James, who is 6′ 8″ tall?

51. **Normal Distributions in StatCrunch.** Go to the StatCrunch work space and generate approximate data for a normal distribution by choosing "Data" and then "Simulate" and "Normal." You can generate 100 data values by entering 100 for "Rows" and 1 for "Columns." Then enter values for the mean (choose 0) and standard deviation (choose 3).

    a. Once the data values are displayed, make a bar graph of them showing relative frequency. A window will appear to ask if you want the data binned first. Respond "OK." Comment on the shape of the distribution.

    b. You will get a better picture of the distribution if you bin the data manually. Go to "Data" and "Bin" and then enter –4 for "Start at" and 0.5 for "Binwidth." A new column of binned data appears in the work space. Make a bar graph of the binned data.

    c. Repeat parts (a) and (b) with fewer than and more than 100 data values. Comment on how well the resulting data sets approximate a normal distribution.

52. **StatCrunch Project.** Choose a data set available in StatCrunch that you suspect may have a nearly normal distribution.

    a. Identify the data set you've chosen, and write a sentence explaining why you think it may have a normal or nearly normal distribution.

    b. Make a graph that displays the data distribution clearly. Explain why you chose the type of graph you did.

    c. Is the distribution nearly normal as you expected? Write a few sentences summarizing what you've learned from the distribution, and explaining why it is or is not normal.

UNIT 6D Statistical Inference

The ideas we have discussed so far in this chapter apply to describing data collected from a *sample*. However, as we discussed in Unit 5A, the goal of most statistical studies is to learn something about an entire population. The heart of most statistical studies therefore lies with *inferring* a conclusion about a population from results for a sample.

The process of statistical inference can be complex, but the basic ideas are easy to understand. In this unit, we will briefly explore three important ideas: *statistical significance*, the *margin of error* (with a little more detail than in Unit 5A), and *hypothesis testing*.

Statistical Significance

Suppose you are trying to test whether a coin is fair; that is, whether it is equally likely to land heads up or tails up. If you toss the coin 100 times and get 52 heads and 48 tails, should you conclude that the coin is unfair? No. We *expect* some variation from one sample of 100 tosses to another, and therefore should not be surprised to observe a small deviation from a perfect 50–50 split between heads and tails. In other words, we expect small deviations to occur *by chance*.

In contrast, suppose you toss a coin 100 times and get 20 heads and 80 tails. This is such a substantial deviation from a 50–50 split that you'd conclude that the coin is unfair. To see why, look at Figure 6.20, which shows the results expected if we were to repeat the 100 tosses of a fair coin many times. Notice that the relative frequencies of 20 or fewer heads are extremely low; in fact, a precise calculation shows that 20 or fewer heads should be expected in less than 1 in 1 billion sets of 100 tosses of a fair coin. Therefore, while it is *possible* that you observed a very rare set of 100 tosses, it's far more likely that the coin is unfair. When the difference between what is observed and what is expected is too great to be explained by chance alone, we say the difference is **statistically significant**.

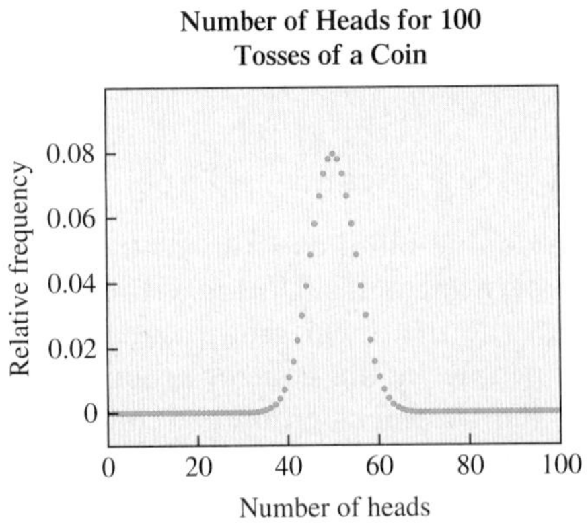

FIGURE 6.20 This graph shows the relative frequency of heads when a coin is repeatedly tossed 100 times. For example, the peak shows the relative frequency of exactly 50 heads to be about 0.08, which means you should expect to get exactly 50 heads in 100 tosses only about 8% of the time.

Definition

A set of measurements or observations in a statistical study is said to be **statistically significant** if it is unlikely to have occurred by chance.

EXAMPLE 1 Significant Events?

Are the following events statistically significant? Explain.

a. The team with the worst win–loss record in basketball wins one game against the team with the best record.

b. In terms of the global average temperature, the 16 years from 2001 to 2016 were 16 of the 17 hottest years on record, using records that go back to 1880.

Solution

a. One win is *not* statistically significant because although we expect a team with a poor win–loss record to lose most of its games, we also expect it to win occasionally, even against the best teams.

b. Having 16 of the 17 hottest years on record occur in a single 16-year period is statistically significant. Indeed, having such a streak of hot years is so unlikely to have occurred by chance alone that it provides significant evidence of a warming Earth.

▶ Now try Exercises 15–18.

BY THE WAY

Scientists are confident that global warming is *caused* by human activity because causality has been established according to all six of the guidelines introduced in Unit 5E. In particular, the physical mechanism is well understood: Human activity, such as the burning of fossil fuels, releases greenhouse gases (especially carbon dioxide) into the atmosphere, and these gases trap heat and energy through the greenhouse effect, which has been measured in the laboratory and verified through studies of other planets.

From Sample to Population

We can now apply the idea of statistical significance to making an inference from a sample to a population. Suppose that in a poll of 1000 randomly selected people, 51% support the President. A week later, in another poll with a different random sample of 1000 people, only 49% support the President. Should we conclude that the opinions of all Americans changed during the one week between the polls?

You can probably guess that the answer is *no*. Because each poll involved a *sample* of only 1000 people, we shouldn't expect its result to be an *exact* match for the entire population of Americans. Because the two polls found only slightly different percentages (51% versus 49%), it's quite possible that the percentage of all Americans supporting the President did not change at all.

In terms of statistical significance, the change in the poll results from 51% to 49% is *not* statistically significant. Therefore, we should not infer any real change in the population based on these polls. In contrast, if the poll results had fallen from 70% one week to 30% the next, we would conclude that the change *was* statistically significant and that the opinions of Americans really did change in just one week.

EXAMPLE 2 Statistical Significance in Experiments

A researcher conducts a double-blind experiment that tests whether a new herbal formula is effective in preventing colds. During a three-month period, the 100 randomly selected people in the treatment group take the herbal formula while the 100 randomly selected people in the control group take a placebo. The results show that 30 people in the treatment group get colds, compared to 32 people in the control group. Can we conclude that the herbal formula is effective in preventing colds?

Solution Whether a person gets a cold during any three-month period depends on many unpredictable factors. Therefore, we should not expect the number of people with colds in any two groups of 100 people to be exactly the same. In this case, the difference between 30% of the treatment group and 32% of the control group getting colds is small enough to be explainable by chance. That is, the difference is not statistically significant, and we should not conclude that the treatment made any difference at all.

▶ Now try Exercises 19–20.

Quantifying Statistical Significance

In Example 2, it's fairly obvious that the difference between 30% and 32% in the two groups is small enough to have occurred by chance and therefore is not statistically significant. But suppose that 22% of the people in the treatment group had colds compared to 34% of the people in the control group. Is this difference enough to be statistically significant? To answer this question, we need a quantitative definition for statistical significance. Quantitatively, statistical significance is defined as follows.

Quantifying Statistical Significance

- If the likelihood (probability) of an outcome as extreme as the one observed is 0.05 (1 in 20, or 5%) or less, the difference is statistically significant *at the 0.05 level.*
- If the likelihood (probability) of an outcome as extreme as the one observed is 0.01 (1 in 100, or 1%) or less, the difference is statistically significant *at the 0.01 level.*

For the outcome of 22% colds in the treatment group and 34% in the control group, the difference turns out to be significant at the 0.05 level (but not at the 0.01 level). This means that if the treatment had no effect at all, we could still expect to see these results in up to about 1 out of 20 similar experiments *just by chance alone*. You can probably see that some caution is in order when working with statistical significance. While significance at the 0.05 level provides some evidence that the treatment works, and significance at the 0.01 level provides even stronger evidence, neither level is conclusive.

BY THE WAY In case you are wondering how we know that the difference between 22% colds in the treatment group and 34% in the control group (for groups of 100 people) is statistically significant at the 0.05 level but not at the 0.01 level, the answer is that you can calculate it with something called the *chi-square statistic*, which you will study if you take a statistics course.

Think About It Suppose an experiment finds that people taking a new herbal remedy get fewer colds than people taking a placebo. The results are statistically significant at the 0.01 level. Has the experiment *proven* that the herbal remedy works? Explain.

EXAMPLE 3 Polio Vaccine Significance

In 1954, a large experiment was conducted to test the effectiveness of a new vaccine for polio created by Dr. Jonas Salk (1914–1995). A sample of 400,000 children was chosen from the population of all children in the United States. Half of these 400,000 children received an injection of the Salk vaccine. The other half received an injection of a placebo containing only salt water. Among the children receiving the Salk vaccine, 33 contracted polio, compared to 115 polio cases among the children who did not get the Salk vaccine. Calculations show that the probability of this difference between the groups occurring by chance is less than 0.01. Discuss the implications of this result.

Solution The results of the polio vaccine test are statistically significant at the 0.01 level, meaning that if the vaccine had no effect, the observed results would occur by chance in fewer than 1 in 100 similar experiments. Therefore, we can be confident that the vaccine was responsible for the fewer cases of polio in the treatment group. (In fact, the probability of the Salk results occurring by chance is *much* less than 0.01, so researchers were quite convinced that the vaccine worked.) As a result of this conclusion, the vaccine was immediately put into wide use. ▶ Now try Exercises 21–24.

BY THE WAY If you had been a parent in the 1940s or 1950s, polio would have been one of your greatest fears. Each year during this long polio epidemic, thousands of young children were paralyzed by the disease. Thanks to the polio vaccine, the disease is now rare and a global vaccination campaign seeks to completely eradicate it. In 2016 there were only 42 new cases of polio worldwide, the fewest ever.

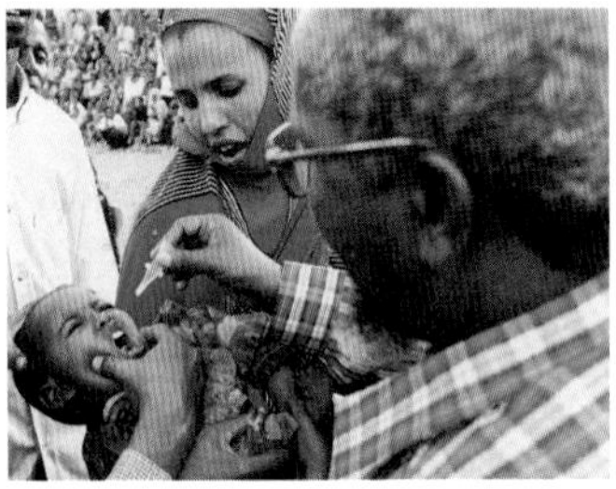

Margin of Error and Confidence Intervals

Recall that in Unit 5A, we briefly discussed how a *margin of error* allows us to define a *confidence interval* for a statistical study. Now we are ready to explore these concepts in greater detail.

Suppose you want to find out what the 1500 students at a high school think about their education by asking "Are you satisfied with the education you are receiving?" Let's further suppose that *if* you could ask all 1500 students, you would find that 900 of the 1500 responses, or 60%, would be "Yes." This 60% proportion of "Yes" responses is a *population parameter*, because it represents the responses of the entire population of the school.

Now suppose that, as with most surveys, you decide it's impractical to question all 1500 students, so you instead choose a random sample of only 25 students to survey. The responses of this sample are as follows, with each Y (for Yes) or N (for No) representing one student's answer to the question:

YNYYNYYNYNYYNYYYNNYYYNNYY

Counting carefully, you find that 16 of the 25 responses, or 64%, are Y. Because it was measured from a sample, 64% is a *sample statistic*, and as we should expect, it is *not* exactly equal to the true population parameter of 60%. If we choose a different sample of 25 students, we might get a different result. For example, suppose another sample produces the following responses:

YYNNNYYNYYYNNYNYYNYNYNYYN

This time, 14 of the 25 responses, or 56%, are Y.

Technical Note In this discussion, we deal with percentages, or *proportions*. In other situations (such as test scores or height measurements), we may be interested in a *mean* rather than a proportion. The general ideas are the same, but the specific details and formulas are different for means and proportions.

Now, imagine that you could repeat the process of drawing a random sample of 25 students hundreds or thousands of times. Because the true proportion of Y responses in the population is 60%, most of the samples would have a proportion of Ys close to 60%. Samples with proportions far from 60% would be relatively rare, and samples with proportions higher than 60% would occur about as often as samples with proportions lower than 60%. If you look back at Unit 6C, you'll see that we have just described three of the four conditions for a *normal distribution*.

In fact, a mathematical theorem, called the *central limit theorem*, states that the distribution of proportions from many samples of the same size is approximately a normal distribution. The theorem also says that the mean of this distribution is the true proportion for the entire population (the population parameter). Figure 6.21 shows this distribution. It is called a **sampling distribution** because it consists of proportions from many individual samples.

The central limit theorem tells us not only that the sampling distribution is approximately normal with the true population proportion as its mean, but also that its standard deviation is approximately $1/(2\sqrt{n})$, where n is the size of the samples. Two standard deviations is twice this value, or $1/\sqrt{n}$, and the 68-95-99.7 rule (see Unit 6C) tells us that about 95% of all possible samples have proportions within 2 standard deviations of the true population proportion. We therefore use this value as the approximate margin of error for a 95% **confidence interval**.

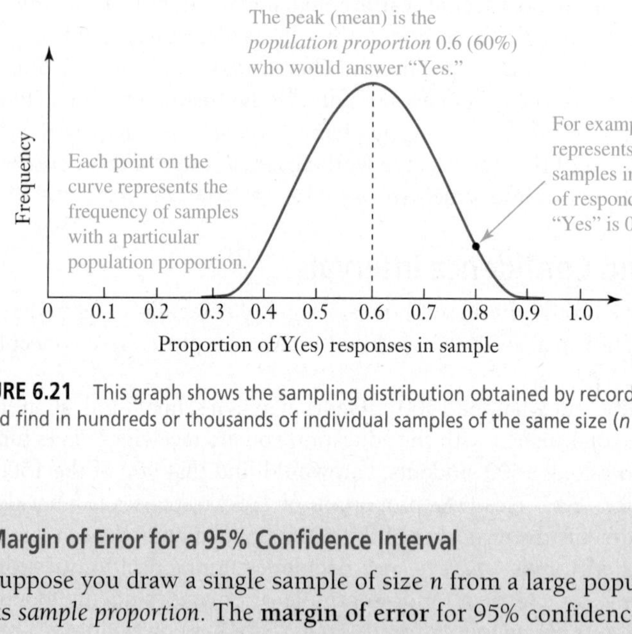
Expected Distribution of Sample Proportions
The peak (mean) is the population proportion 0.6 (60%) who would answer "Yes."
Frequency
Each point on the curve represents the frequency of samples with a particular population proportion.
For example, this point represents the frequency of samples in which the proportion of respondents who would answer "Yes" is 0.8 (80%).
0
0.1
0.2
0.3
0.4
0.5
0.6
0.7
0.8
0.9
1.0
Proportion of Y(es) responses in sample

FIGURE 6.21 This graph shows the sampling distribution obtained by recording the proportions you would find in hundreds or thousands of individual samples of the same size ($n = 25$ in this case).

Technical Note

The precise formula for the margin of error is

$$E = 1.96\sqrt{\frac{\hat{p}(1-\hat{p})}{n}}$$

where $\hat{p}$ is the proportion found for the sample. This formula requires $n\hat{p} \geq 5$ and $n(1-\hat{p}) \geq 5$, conditions that are usually met in practice.

Margin of Error for a 95% Confidence Interval

Suppose you draw a single sample of size n from a large population and measure its *sample proportion*. The **margin of error** for 95% confidence is

$$\text{margin of error} \approx \frac{1}{\sqrt{n}}$$

When we find a 95% confidence interval, we usually say that we can be "95% confident" that it contains the true value of the population parameter. A more precise interpretation is as follows (Figure 6.22): If we took a large number of samples (all of the same size) and found the confidence interval for each of them, 95% of those confidence intervals would contain the true population parameter, while 5% would not contain the true population parameter.

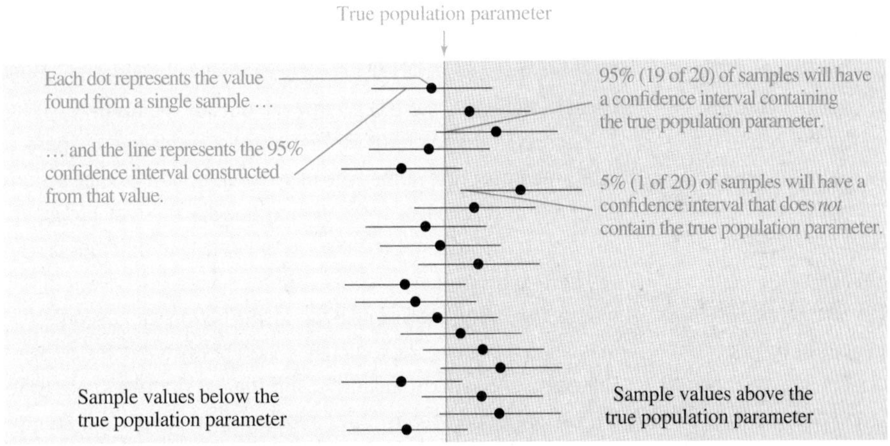

FIGURE 6.22 This figure illustrates the idea that, on average, we would find that 95% (or 19 out of 20) of any set of samples will have a 95% confidence interval that contains the true population parameter.

EXAMPLE 4 Poll Margins

Find the margin of error and 95% confidence interval for the following surveys.

a. A survey of 500 people finds that 52% plan to vote for Smith for governor.
b. A survey of 1500 people finds that 87% support stricter penalties for child abuse.

Solution

a. The proportion measured for the sample is 52%, or 0.52. For a sample size of $n = 500$ people, the margin of error is approximately

$$\frac{1}{\sqrt{n}} = \frac{1}{\sqrt{500}} \approx 0.045$$

Adding and subtracting 0.045, or 4.5%, from the sample proportion of 52% produces a 95% confidence interval from 47.5% to 56.5%. We can be 95% confident that the true proportion of people who plan to vote for Smith lies in this interval.

b. The proportion measured for the sample is 87%, or 0.87. For a sample size of $n = 1500$ people, the margin of error is approximately

$$\frac{1}{\sqrt{n}} = \frac{1}{\sqrt{1500}} \approx 0.026$$

Adding and subtracting 0.026, or 2.6%, from the sample proportion of 87% produces a 95% confidence interval from 84.4% to 89.6%. We can be 95% confident that the true proportion of people who support the stricter penalties lies in this interval.

▶ Now try Exercises 25–28.

EXAMPLE 5 Unemployment Rate

Suppose the Bureau of Labor Statistics finds 3420 unemployed people in a sample of $n = 60{,}000$ people. Estimate the population unemployment rate and give a 95% confidence interval.

Solution The sample proportion is the unemployment rate for the sample:

$$\frac{3420}{60{,}000} = 0.057$$

IN YOUR WORLD

Is Polling Reliable?

You're probably aware that the 2016 presidential election proved to be a huge surprise to pollsters: Dozens of pre-election polls predicted that Hillary Clinton would win, but Donald Trump was elected President. This result left pollsters—and the public—wondering whether polls can be trusted at all. This is a crucial question, because we count on reliable polling to gauge public opinion. So let's look at both what went right and what went wrong with the polling data in 2016.

What went right: In an analysis of the final set of pre-election polls, Real Clear Politics found that most polls predicted a Clinton win by between 1 and 6 percentage points, and that her mean advantage among the polls they tracked was 3.2 percentage points. In fact, Clinton won the national popular vote by a margin of 2.1 percentage points. This means that the national popular vote results were within about 1 percentage point of the general predictions, which puts them well within the typical margin of error for opinion polls (usually 3 to 4 percentage points, depending on the sample size).

What went wrong: Where the polls failed was in predicting electoral college results, which are tallied state by state. More specifically, the polls incorrectly forecast a Clinton win in several key "swing states"—states that were expected to be close—suggesting that there were systematic errors in these swing state polls. No one knows exactly why these polls proved incorrect, but at least three issues probably played a role.

First, many pollsters conduct their polls with random dialing of phone numbers, but response rates (the percentage of people who answer the phone and agree to answer the poll questions) can be very low, sometimes under 10%. As we discussed in Unit 5B, this can introduce bias that can make the sample unrepresentative. For example, Trump tended to have his greatest support among working-class individuals, but these individuals may have been less likely to be home to answer a phone call on a land line, simply because they were more likely to work outside the home.

Second, to try to adjust for bias introduced by low response rates, most polls use some type of *weighting*; that is, they assign extra weight to the responses of individuals from small groups that are likely to be underrepresented in the data. But different pollsters use different weighting processes, and because the weighted groups have very small sample sizes, small errors can become magnified. For example, the pre-election polls conducted by the University of Southern California and the *Los Angeles Times* (USC/LAT) in 2016 consistently found much greater support for Trump among African Americans than other polls, and *The New York Times* traced the difference to a single African American supporter of Trump who happened to be included in this poll (which was repeated several times) with a response weighted about 30 times as much as the average respondent. Clearly, weighting decisions like this could cause inaccurate predictions from polls in closely contested states.

Third, pollsters can only ask people whether they *plan* to vote, so they can miss out on voter enthusiasm. For example, if Trump supporters were more enthusiastic than Clinton supporters, then they may have been more likely to actually cast a vote, in which case the polls would have underestimated Trump's share of the vote.

To return to the crucial question of whether polls are reliable, we are left with good news and bad news. The good news is that the polls actually performed fairly well overall, suggesting that you can consider polls to be reasonably reliable as a *general* gauge of public opinion. The bad news is that if you want to answer a very specific question, such as who will win a close election in a particular state, the 2016 results suggest that there are too many uncertainties to give a reliable prediction.

This proportion is likely to be close to the true population unemployment rate. The margin of error is approximately

$$\frac{1}{\sqrt{60{,}000}} \approx 0.004$$

We add and subtract the margin of error of 0.004 from the sample proportion of 0.057, yielding a 95% confidence interval from 0.053 to 0.061, or 5.3% to 6.1%. We can be 95% confident that the interval from 5.3% to 6.1% contains the true unemployment rate for the population.

▶ **Now try Exercises 29–32.**

BY THE WAY

The actual margin of error for the Bureau of Labor Statistics unemployment survey is smaller than that found here because the bureau surveys 60,000 *households*, which contain more than 60,000 people.

Think About It We've seen that the margin of error is most commonly defined for a 95% confidence interval, as opposed to some higher level of confidence such as 99% or 99.99%. Do you think this level of confidence is enough for most surveys? Can you think of cases for which it is not enough?

Hypothesis Testing

Advertisers make claims about their products. Schools make claims about the quality of education they offer. Lawyers make claims about a suspect's guilt or innocence. Medical diagnoses are claims about the presence or absence of disease. Pharmaceutical companies make claims about the effectiveness of their drugs. But how can we decide whether any of these claims are true? Statistics offers an answer, through a powerful set of techniques for testing the validity of claims. The techniques go by the name of **hypothesis testing**, and they play a fundamental role in almost every aspect of modern life.

To illustrate the ideas behind hypothesis testing, let's consider the case of a product, called Gender Choice, that was once claimed to increase a woman's chance of giving birth to a baby girl from the usual chance of about 50%. How could we test whether this claim is true?

One way would be to study a random sample of babies born to women who used the Gender Choice product. If the product does *not* work, we would expect about half of these babies to be girls. If it does work, we would expect significantly more than half to be girls. In other words, we need to compare the actual proportion of baby girls in the sample to the proportion expected if the product does not work. This means we are dealing with two claims, or *hypotheses*, about the product:

- The **null hypothesis** is the claim that Gender Choice does *not* work and the proportion of baby girls born to women who used the product is unchanged from its normal value of about 50%.
- The **alternative hypothesis** is the claim that Gender Choice *does* work and the proportion of baby girls born to women who used the product is significantly greater than 50%.

More generally, the null hypothesis should make a specific claim about the value of a population parameter (generally either a population mean or a population proportion). In this case, the null hypothesis makes the specific claim that the proportion of baby girls (the population parameter) is about equal to 50%, as expected under normal conditions without intervention.

BY THE WAY
Gender Choice was a real product, but the U.S. Food and Drug Administration (FDA) stopped its sale because there was no evidence that it worked as claimed. However, newer technologies have proven successful in allowing gender selection, raising ethical questions about whether or when such selection should be an option for parents.

Null and Alternative Hypotheses

The **null hypothesis** is the starting assumption for a hypothesis test. Although other forms are possible, in this book we consider only null hypotheses that claim a specific value for a population parameter:

$$\text{null hypothesis: population parameter} = \text{claimed value}$$

An **alternative hypothesis** is a claim that the population parameter has a value *different* from that claimed by the null hypothesis.

BY THE WAY
The word *null* comes from the Latin *nullus*, meaning "nothing." A null hypothesis often states that there is no special effect or difference. The null hypothesis is often denoted by H_0 (pronounced "H naught"). The alternative hypothesis is often denoted by H_a.

EXAMPLE 6 Dog Hypotheses

A dog food company claims that its special diet mix will allow Labrador retrievers to live to a mean age of 15.0 years (significantly longer than the overall mean longevity for the breed). A consumer group says that this claim is overstated and that the actual mean will be less than the company's claim. State the null and alternative hypotheses for a hypothesis test.

BY THE WAY

Different dog breeds have different average lifetimes. The longest-lived dogs are miniature pinschers and poodles, which live to average ages just under 15 years. Larger dog breeds tend to live shorter lives; for example, Great Danes live to an average of only about 8 years. Labrador retrievers live to an average of about $12\frac{1}{2}$ years.

Solution The null hypothesis is the company's specific claim that dogs eating the special diet will live to a mean age of 15.0 years. The alternative hypothesis is the consumer group's claim that the actual mean age will not be 15.0 years but less than that. In summary:

null hypothesis: mean age = 15.0 years
alternative hypothesis: mean age < 15.0 years

▶ **Now try Exercises 33–38, part (a).**

Outcomes of a Hypothesis Test

A hypothesis test always begins with the assumption that the null hypothesis is true. We then test to see whether the data give us reason to think otherwise. As a result, there generally are only two possible outcomes to any hypothesis test, summarized in the following box.

Two Possible Outcomes of a Hypothesis Test

- *Reject* the null hypothesis, because we have evidence in support of the alternative hypothesis.
- Do *not reject* the null hypothesis, because we lack sufficient evidence to support the alternative hypothesis.

Note that "accepting the null hypothesis" is *not* a possible outcome, because the null hypothesis is the starting assumption. The hypothesis test may give us reason to reject that starting assumption, but it cannot by itself give us reason to conclude that the starting assumption is true.

Think About It Our legal principles maintain that criminal suspects are innocent until proven guilty. At the end of a criminal trial, juries can render only two possible verdicts: guilty or not guilty. Putting these ideas into the language of hypothesis testing, state the null hypothesis for a criminal trial. How do the two possible verdicts correspond to the two possible outcomes of a hypothesis test?

EXAMPLE 7 Gender Choice Outcomes

Describe the two possible outcomes of a hypothesis test concerning the Gender Choice product.

Solution As already discussed, the null hypothesis states that the proportion of girl babies is the expected 50%. The alternative hypothesis in this case states that the proportion is greater than 50%. There are two possible outcomes:

1. We can reject the null hypothesis and accept the alternative hypothesis. In this case, we conclude that the proportion of girl babies is greater than 50% for people using Gender Choice and that the product actually works.
2. We can decide not to reject the null hypothesis. This means we have no grounds for doubting the null hypothesis, but it does *not* prove that the null hypothesis is true. For Gender Choice, it means that, while we lack evidence that the product works, we have not proven that it does *not* work.

▶ **Now try Exercises 33–38, part (b).**

The Null Hypothesis: To Reject or Not to Reject

Suppose you study a sample of 100 babies born to women using the Gender Choice product and find that 64 of the babies are girls. How do you decide whether this constitutes evidence for rejecting the null hypothesis? The answer comes down to statistical significance.

If the null hypothesis is true, then the proportion of baby girls among the *population* of Gender Choice users is about 50%. However, it is still possible that a single *sample* of 100 babies could contain 64 girls, because we don't expect every sample to match the population. We therefore would like to know the likelihood that a single random sample of 100 babies would have 64 or more girls, under the assumption that the actual population proportion of girls is 50%. Using statistical methods that we will not cover here, this particular likelihood (probability) turns out to be about 3 in 1000, or 0.003. Because 0.003 is less than 0.01, observing 64 or more baby girls would be statistically significant at the 0.01 level. In other words, we would have good reason to reject the null hypothesis, which would in turn support the alternative hypothesis that Gender Choice works.

Hypothesis Test Decisions

We decide the outcome of a hypothesis test by comparing the actual sample statistic (mean or proportion) to the statistic expected if the null hypothesis is true.

- If the likelihood (probability) of a sample statistic at least as extreme as the observed statistic is less than 1 in 100 (or 0.01), the result is statistically significant at the 0.01 level. This significance level offers strong evidence for rejecting the null hypothesis (and accepting the alternative hypothesis).
- If the likelihood (probability) of a sample statistic at least as extreme as the observed statistic is less than 1 in 20 (or 0.05), the result is statistically significant at the 0.05 level. This significance level offers moderate evidence for rejecting the null hypothesis.
- If the likelihood (probability) of a sample statistic at least as extreme as the observed statistic is greater than 1 in 20, the result is not statistically significant and therefore does not provide sufficient grounds for rejecting the null hypothesis.

EXAMPLE 8 Birth Weight Significance

A county health official believes that the mean birth weight of male babies at a local hospital is greater than the national average of 3.39 kilograms. A random sample of 145 male babies born at that hospital has a mean birth weight of 3.61 kilograms. Assuming that the mean birth weight of all male babies born at the hospital is the national average of 3.39 kilograms, a calculation shows that the likelihood of selecting a sample with a mean birth weight of at least 3.61 kilograms is 0.032. Formulate the null and alternative hypotheses. Then discuss whether the sample provides evidence for rejecting or not rejecting the null hypothesis.

Solution The null hypothesis is the claim that the mean birth weight of all male babies born at this hospital is the national average of 3.39 kilograms. The alternative hypothesis is the claim of the health official—that the hospital mean is higher than the national average.

$$\text{null hypothesis: mean birth weight} = 3.39 \text{ kilograms}$$
$$\text{alternative hypothesis: mean birth weight} > 3.39 \text{ kilograms}$$

Technical Note
In statistics, the actual probability of obtaining a particular sample result (under the assumption that the null hypothesis is true) is called a *P*-value. The *P*-value for the sample in Example 8 is the given probability of 0.032.

We are given that, *if* the null hypothesis is true, the likelihood of observing a sample with a mean of at least 3.61 kilograms is 0.032, or 32 in 1000. This is less than 0.05 but greater than 0.01, so the result is statistically significant at the 0.05 level but not at the 0.01 level. Statistical significance at the 0.05 level provides moderate evidence for rejecting the null hypothesis, in which case the county official's claim would be supported.

▶ Now try Exercises 39–44.

Think About It The idea that a hypothesis test cannot lead us to accept the null hypothesis is an example of the old dictum that "absence of evidence is not evidence of absence." As an illustration of this idea, explain why it would be nearly impossible to prove that some legendary animal (such as Bigfoot or the Loch Ness Monster) does *not* exist.

Quick Quiz 6D

Choose the best answer to each of the following questions. Explain your reasoning with one or more complete sentences.

1. Suppose you toss a coin 100 times and get 38 heads. Compared to what you would expect for a fair coin, based on Figure 6.20, this result is

   a. not statistically significant.

   b. statistically significant at the 0.05 level.

   c. statistically significant at the 0.01 level.

2. Researchers are testing a new cancer treatment. To determine whether the treatment offers a statistically significant improvement in survival rates, the researchers must

   a. give the treatment to at least 1000 people.

   b. compare the survival rate with the new treatment to that with other treatments or no treatment.

   c. determine the physiological mechanism through which the new treatment works.

3. Working with equal sample sizes, researchers compared three new cold remedies to a placebo. Which remedy is most likely to be most effective?

   a. the one that gave results statistically significant at the 0.01 level

   b. the one that gave results statistically significant at the 0.05 level

   c. the one that gave results that were not statistically significant

4. Researchers conduct 20 different experiments to test the effectiveness of a new weight loss pill compared to the effectiveness of a placebo. For 19 of the 20 experiments, the results are not statistically significant. However, one of the experiments shows that the pill works better than the placebo with statistical significance at the 0.05 level. A fair report of the combined results from all 20 experiments would say that

   a. the treatment works effectively about 5% of the time.

   b. we can have 95% confidence that the treatment works for at least some people.

   c. there is no evidence that the treatment works.

5. A poll finds that 35% of the people surveyed approve of the President's handling of the economy, with a margin of error (for 95% confidence) of 3%. The 95% confidence interval for this poll is

   a. 32% to 35%.

   b. 32% to 38%.

   c. 33.5% to 36.5%.

6. Suppose the poll described in Question 5 were repeated one month later (with a new sample) and 33% of the people surveyed approve of the President's handling of the economy. Which is the correct conclusion about the percentage of the population that approves of the President's handling of the economy?

   a. The percentage of the population approving the handling has definitely dropped during the month between surveys.

   b. We can be 95% confident that the percentage of the population approving the handling has dropped during the month between surveys.

   c. The two polls results do not differ enough to allow us to draw a conclusion about any change.

7. Consider a survey with a margin of error of 4%. If you want to reduce the margin of error to 2%, you should repeat the survey with

   a. half as many people.

   b. twice as many people.

   c. four times as many people.

8. You want to test your hypothesis that local gas stations are charging much more than the national average price for gasoline. A good *null hypothesis* for this test would be

   a. local gas prices equal the national average gas price.

   b. local gas prices are higher than the national average gas price.

   c. local gas prices show much more variation than national gas prices.

9. You carry out a test of the hypothesis described in Question 8. If the results show that you *cannot* reject the null hypothesis, then

a. local gas prices are actually the same as the national average.

b. local gas prices are actually below the national average.

c. you cannot be sure that local prices are the same as the national average, but there is insufficient evidence for a claim that local gas prices are above the national average.

10. Suppose you again carry out a test of the hypothesis described in Question 8, but this time your results show that prices are above the national average, and by an amount that you might find by chance in only 1 out of 100 similar tests. A good description of this result would be that

a. you've proven that local gas prices are above the national average.

b. you've found a result that is statistically significant at the 0.01 level.

c. there's a 99% chance that local prices are at least 1% above the national average.

Exercises 6D

REVIEW QUESTIONS

1. What is *statistical inference*? Why is it important?
2. Explain the meaning of *statistical significance*. What does it mean for a result to be statistically significant at the 0.05 level? At the 0.01 level?
3. How does the idea of statistical significance apply to the question of whether results from a sample can be generalized to conclusions about a population? Explain.
4. Explain why you should usually expect that a proportion found in a single sample will be close to, but not exactly match, the true proportion in a population.
5. Briefly describe the use of the formula for margin of error. Give an example in which you interpret the margin of error in terms of 95% confidence.
6. What is the purpose of a *hypothesis test*? How do we formulate the *null hypothesis* and *alternative hypothesis* for a test?
7. What are the two possible conclusions of a hypothesis test? Explain, and also explain why accepting the null hypothesis is *not* a possible outcome.
8. Briefly discuss how the idea of statistical significance helps us decide whether to reject or not reject a null hypothesis.

DOES IT MAKE SENSE?

Decide whether each of the following statements makes sense (or is clearly true) or does not make sense (or is clearly false). Explain your reasoning.

9. The study found that more people were cured by the new drug than by the old drug, but the result was not statistically significant.
10. There can no longer be any doubt that our product, the Magic Diet Pill, really works. We financed a study in which the pill led to more weight loss than a placebo, with results significant at the 0.05 level.
11. Both agencies conducted their surveys carefully, both asked the same question and found about the same proportion of "yes" responses, and both interviewed the same number of people. However, the margin of error was smaller for Agency A's survey than for Agency B's survey.
12. If you want to reduce the margin of error in your pre-election survey, you should use a larger sample.
13. We began the study with the null hypothesis stating that the toxic waste dump had no effect on disease rates among the residents nearby. The alternative hypothesis therefore stated that the toxic waste dump did have an effect on disease rates among the residents nearby.
14. We began the study with the null hypothesis stating that the toxic waste dump had no effect on the residents nearby, and our study proved this hypothesis to be true.

BASIC SKILLS & CONCEPTS

15–20: Subjective Significance. For each of the following events, state whether you think the difference between what occurred and what you would expect by chance is statistically significant. Explain.

15. In 120 rolls of a standard six-sided die, you get 2 sixes.
16. In 200 tosses of a fair coin, you get 103 heads.
17. An airline with a 95% on-time departure rate has 19 out of 400 flights with late departures.
18. A basketball player with a 89% free-throw percentage misses 20 free throws in a row.
19. Ten winners of the Power Ball lottery in a row bought their tickets at the same 7-Eleven store.
20. Of the 40 students in your mathematics class, 35 have the same last name.
21. **Human Body Temperature.** A study by University of Maryland researchers measured the body temperatures of 106 individuals. The mean for the sample was 98.20°F. The accepted value for human body temperature is 98.60°F. If we assume that the mean body temperature is actually 98.60°F, the probability of getting a sample with a mean of 98.20°F or less turns out to be less than 1 in 1 million. Is this result significant at the 0.05 level? At the 0.01 level? Would it be reasonable to

conclude that the accepted value for human body temperature is wrong? Explain.

22. **Seat Belts and Children.** In a study of children injured in automobile crashes (*American Journal of Public Health*, Vol. 82, No. 3), those wearing seat belts had a mean stay of 0.83 day in an intensive care unit. Those not wearing seat belts had a mean stay of 1.39 days. The probability of this difference in means occurring by chance turns out to be less than 1 in 10,000. Is this result significant at the 0.05 level? At the 0.01 level? Would it be reasonable to conclude that seat belts reduce the severity of injuries? Explain.

23. **SAT Preparation.** A study of 75 students who took an SAT preparation course (*American Education Research Journal*, Vol. 19, No. 3) concluded that the mean improvement in SAT scores was 0.6 point. If we assume that the preparation course has no effect, the probability of getting a mean improvement of 0.6 point by chance is 0.08, or 8 in 100. Discuss whether this preparation course results in statistically significant improvement.

24. **Weight by Age.** A National Health Survey determined that the mean weight of a sample of 804 men ages 25 to 34 was 176 pounds, while the mean weight of a sample of 1657 men ages 65 to 74 was 164 pounds. The difference is significant at the 0.01 level. Interpret this result.

25–32: Margin of Error. Find the margin of error and the 95% confidence interval for the following studies. Briefly interpret the 95% confidence interval.

25. According to a Gallup poll of 1012 people, about one-third (32%) of Americans keep a dog for protection.

26. A recent survey of 65,000 households by the U.S. Department of Labor reported an unemployment rate of 4.3%.

27. A 2016 survey of 2010 adults by the Pew Research Center found that 83% of those surveyed support background checks for private and gun show sales.

28. A 2017 Gallup poll of 1035 adults found that 37% of those surveyed believe the United Nations is doing a good job solving global problems.

29. A 2017 NBC poll found that 53% of the 1000 people surveyed believe that Congress should investigate possible Russian interference in the 2016 general election.

30. Using a sample consisting of 2192 adults, a 2016 Harris poll found that "two-thirds (66%) of adults say they oppose laws allowing businesses to refuse service to LGBT persons because of [business owners'] religious objections." What is the margin of error and confidence interval for this poll? Does the 66% represent people who favor or oppose the rights of LGTB people in this case? Explain.

31. Surveying 1549 adults, the Pew Research Center reported that "an overwhelming majority of Americans (82%) support requiring all healthy schoolchildren to be vaccinated for measles, mumps and rubella."

32. A 2017 CBS poll of 1257 adults concluded that 79% of American adults believe that if a wall is built on the Mexican border, the United States will pay for it.

33–38: Formulating Hypotheses. Consider the following claims related to statistical studies.

a. State the null and alternative hypotheses for a hypothesis test.

b. Describe the two possible outcomes of the test, using the context of the given situation.

33. A college president claims that the six-year graduation rate at her four-year college is higher than the national average of 60%.

34. The Food and Drug Administration claims that the amount of vitamin C in tablets produced by a company is less than the advertised 500 milligrams (mg).

35. The governor claims that the median high school teacher salary in her state is above the national median of $57,900.

36. The state hydrologist claims that the total precipitation for the year just ending is less than the average of 27.5 inches recorded over the prior ten years.

37. Federal Aviation Administration officials claim that the percentage of flights delayed more than 15 minutes at Denver International Airport is greater than the national average of 15%.

38. A high school principal claims that the cigarette smoking rate in his school is less than the national average of 9.3%.

39–44: Hypothesis Tests. The following exercises describe the results of a hypothesis test in terms of the chance of obtaining a particular sample. In each case, use the given context to formulate the null and alternative hypotheses. Then discuss whether the sample provides evidence for rejecting or not rejecting the null hypothesis.

39. The owner of a car rental company claims that the mean annual mileage for the population of all cars in his fleet is more than 11,725 miles (which is the mean annual mileage for all cars in the United States). A random sample of $n = 225$ cars from his fleet has a mean annual mileage of 12,000 miles. Assuming that the mean annual mileage for all cars in his fleet is 11,725 miles, the likelihood of selecting a random sample with a mean annual mileage of 12,000 miles or more is 0.01.

40. A Senate candidate claims that a majority of voters support her. A poll of 400 voters finds that the proportion of voters who support the candidate is 0.51 (51%). Assuming that the proportion of people in the population who support her is $p = 0.5$, the likelihood of selecting a sample in which the proportion is 0.51 or more is 0.345.

41. A hospital administrator finds that the mean hospital stay for a sample of 81 women after childbirth is 2.3 days. She claims that the mean stay at her hospital is greater than the national average of 2.1 days. Assuming that the average at her hospital is the same as the national average, the likelihood of observing a sample with a mean of 2.3 days or more is 0.17.

42. Each baseball in a random sample of 40 new balls is dropped onto a concrete surface and the bounce height is recorded. The heights have a mean of 92.67 inches (based on data from *USA Today*). A statistically literate manager claims that the mean bounce height for all baseballs is 90.10 inches and the likelihood of observing a sample mean of 92.67 inches is 0.035.

43. A marketing company claims that the mean household income of HDTV owners across the population is greater than $50,000. A random sample of 1700 households with HDTVs shows that the mean household income is $51,182. Assuming that the true mean is $50,000, the likelihood of selecting a sample with a mean income of $51,182 or more is 0.007.

44. A Roper poll used a sample of 100 randomly selected car owners. Within the sample, the mean time for ownership of a single car was 7.01 years. Assuming that the actual mean ownership time is 7.5 years, the likelihood of selecting a sample with a mean time of ownership of 7.01 years or less is 0.10.

FURTHER APPLICATIONS

45. **Nielsen Ratings.** Suppose Nielsen Media Research reports that 12.8% of all households with TVs watched CBS's *NCIS* during a particular week. Interpret this result with a 95% confidence interval, assuming a sample size of 5000 households.

46. **Male Birth Weight.** The mean birth weight of male babies born to 121 mothers taking a vitamin supplement is 3.67 kilograms (based on New York Department of Health data). The national average birth weight of male babies is 3.39 kilograms. Discuss the significance of the sample observation given that the probability of observing such a sample is estimated to be 0.0015.

47. **Better Margin of Error.** Suppose you want to decrease the margin of error by a factor of 2—for example, from 4% to 2%. By what factor must you increase the sample size? Explain.

48. **Better Margin of Error.** Suppose you want to decrease the margin of error by a factor of 10—for example, from 1% to 0.1%. By what factor must you increase the sample size? Explain.

49. **Gallup Description.** The Gallup Organization includes the following explanation after its summary of an opinion poll. Are the figures in the explanation (sample size and margin of error) consistent? Comment on the provision that is made about "error and bias."

> *The results reported here are based on telephone interviews with a randomly selected national sample of 1,019 adults 18 years and older. For results based on the whole sample, one can say with 95 percent confidence that the maximum error attributable to sampling and other random effects is plus or minus 3 percentage points. In addition to sampling error, question wording and practical difficulties in conducting surveys can introduce error or bias into the findings of public opinion polls.*

IN YOUR WORLD

50. **Recent Polls.** Visit the websites of polling organizations, and gather the results of a particular poll. Describe the details of the actual polling procedure. Interpret the results in terms of a 95% confidence interval. What conclusions can you draw? Explain.

51. **Genetically Modified Foods.** Use the Web to learn about current studies of whether genetically modified foods are safe for human consumption and for the environment. Choose a study that interests you. Describe how it is an example of a hypothesis test, and discuss its results and conclusions.

52. **Statistical Significance.** Find a recent news report in which the idea of statistical significance is used. Write a one-page summary of the study and the result that is considered to be statistically significant. Be sure to discuss whether you believe the result and why.

53. **Margin of Error.** Find three recent surveys that quote a result as a proportion (or percentage) along with a margin of error. Interpret each result in terms of a 95% confidence interval. Based on the sample size in each case, calculate

the margin of error with the formula given in this unit. Does your calculation agree with the stated margin of error? Explain.

54. **Hypothesis Testing.** Find a news report describing a statistical study that involved a hypothesis test. (News reports rarely state explicitly that a hypothesis test was used, but you should be able to decide from the context whether a hypothesis test was involved.) State the null and alternative hypotheses for the study, and describe how a conclusion was reached.

TECHNOLOGY EXERCISE

Use **StatCrunch** (available in MyLab Math) for the following question.

55. **Confidence Interval.** Go to StatCrunch and choose "Explore" and "Data"; then select the category "Surveys" and choose one data set that is of interest to you.

    a. Identify the data set you've chosen and write a sentence or two explaining the data it contains.

    b. Construct a relevant 95% confidence interval for some aspect of the data. Explain your results.

Chapter 6 Summary

UNIT	KEY TERMS	KEY IDEAS AND SKILLS
6A	distribution (p. 373) mean (p. 374) median (p. 374) mode (p. 375) outlier (p. 376) symmetric distribution (p. 380) skewed (left or right) distribution (p. 380) variation (p. 381)	**Measures of center in a distribution:** (p. 374) $\text{Mean} = \dfrac{\text{sum of all values}}{\text{total number of values}}$ Median is the middle value. Mode is the most common value. **Understand** effects of outliers. (p. 376) **Be aware of** potential confusion with the term *average*. (p. 377) **Characterize** the shape of a distribution by • number of peaks (p. 378) • symmetry (symmetric, left-skewed, or right-skewed) (p. 380) • variation (p. 381)
6B	range (p. 387) quartiles (p. 388) five-number summary (p. 388) boxplot (p. 388) standard deviation (p. 390) range rule of thumb (p. 392)	**Understand and calculate** measures of variation: • range = highest value − lowest value (p. 387) • five-number summary (lowest value, lower quartile, median, upper quartile, highest value) (p. 388) • $\text{standard deviation} = \sqrt{\dfrac{\text{sum of (deviations from the mean)}^2}{\text{total number of data values} - 1}}$ (p. 390) **Interpret** a standard deviation with the range rule of thumb. (p. 392)
6C	normal distribution (p. 397) 68-95-99.7 rule (p. 399) standard score (p. 401) percentile (p. 403)	**Describe** a normal distribution and list the conditions under which it can be expected. (p. 397) **Understand and apply** the 68-95-99.7 rule for normal distributions. (p. 399) **Use** a standard score table for normal distributions. (p. 401)
6D	statistical significance (p. 408) sampling distribution (p. 412) confidence interval (p. 412) margin of error (p. 412) hypothesis testing (p. 415) null hypothesis (p. 415) alternative hypothesis (p. 415)	**Understand** the concept of statistical significance, including significance at the 0.05 level and significance at the 0.01 level. (p. 408) **Understand** how a margin of error arises from a sampling distribution. (p. 411) **Understand and conduct** a basic hypothesis test and make decisions based on the test. (p. 415)

7 PROBABILITY: LIVING WITH THE ODDS

Probability is involved in virtually every decision we make. Sometimes the role of probability is clear, as in deciding whether to buy a lottery ticket or whether to plan a picnic based on the probability of rain. In other cases, probability guides decisions on a deeper level. For example, you might choose a particular college because you believe that it is most likely to meet your personal needs. In this chapter, we will look at just how practical and powerful probability can be in our everyday lives.

Q

Suppose there are 25 students in your math class. What is the probability at least two students in the class have the same birthday (such as February 5 or July 22)?

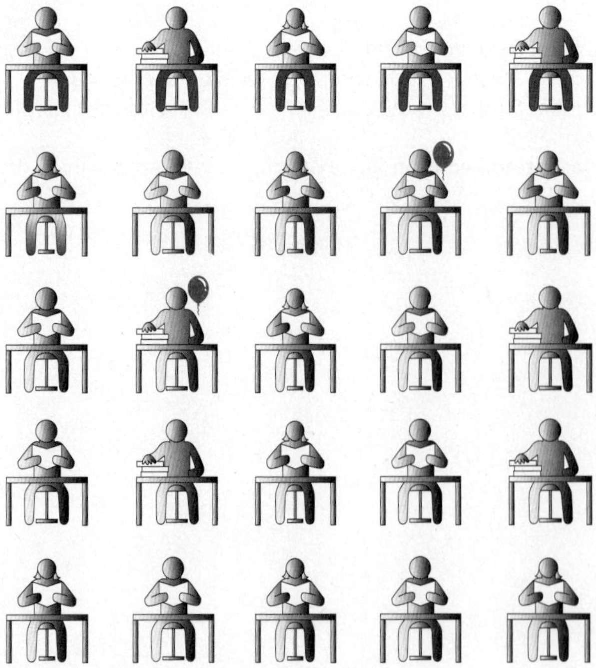

- **A** About 0.01 (or 1 in 100)
- **B** About 0.25 (or 1 in 4)
- **C** About 0.6 (or 3 in 5)
- **D** Exactly $\frac{2}{365}$
- **E** Exactly $\frac{25}{365}$

Probability is the very guide of life.

—Cicero (106–43 B.C.E.)

A This question may not be particularly "important," in that knowing the answer is unlikely to affect your daily life. However, it illustrates an important fact about probability: Research shows that we tend to have poor intuition about probability, and even experts are sometimes wrong about probabilities until they do the detailed calculations. This fact leads many people to underestimate or overestimate probabilities by large margins, which may explain why we often engage in risky activities (such as talking on a cell phone while driving) while fearing low-risk ones (such as flying on a commercial airline). For this reason, the study of probability can be very useful to the decision making we do throughout our lives.

For the birthday question, most people guess that the probability is fairly low. After all, there are only 25 students in the class, while there are 365 possible birthdays in a year. Indeed, if we asked the probability that another student in the class had *your* birthday, it would be close to (but not exactly) $\frac{25}{365}$, which is less than 0.07, meaning that you'd find someone with your birthday in only about 7 out of 100 randomly selected groups of 25 people. However, the question does not ask about *your* birthday, but rather about *any* two students having the same birthday—and this probability turns out to be about 0.57, meaning that there is a better than 50–50 chance that two people in any randomly selected group of 25 people will have the same birthday. You'll find the calculation for this surprising answer in Example 8 in Unit 7E.

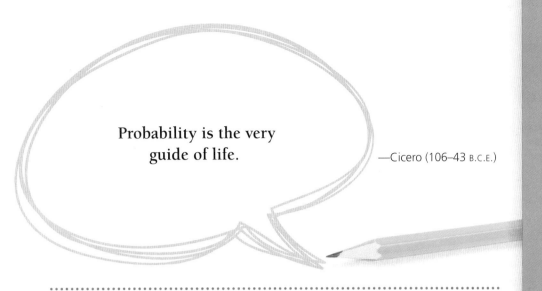

UNIT 7A

Fundamentals of Probability: Explore basic concepts of probability and learn three methods for determining probabilities: theoretical, relative frequency, and subjective.

UNIT 7B

Combining Probabilities: Learn basic rules for adding and multiplying probabilities.

UNIT 7C

The Law of Large Numbers: Explore the law of large numbers and its application to lotteries, insurance, and more.

UNIT 7D

Assessing Risk: Use probability to describe risk in connection with travel, diseases, and life expectancy.

UNIT 7E

Counting and Probability: Learn about permutations and combinations and apply them to probability and to exploring coincidences.

ACTIVITY Lotteries

Use this activity to gain a sense of the kinds of problems this chapter will enable you to analyze. Additional activities are available online in MyLab Math.

Lotteries are big business and a major revenue source for many governments. As of 2017, all but seven states in the United States have some type of lottery, and total spending on these lotteries is approximately $75 billion per year. About one-fourth of that money ends up as state revenue, with the rest going to prizes and expenses. Lotteries present many lessons in probability, and lottery statistics fuel great debate over whether lotteries are an appropriate way for governments to generate revenue. Working individually or in groups, consider the following questions.

1. Given a U.S. population of 325 million, about how much does the average American spend on lotteries each year? How much does the average *adult* spend on the lottery? (About 23% of the population is under age 18.)
2. Surveys indicate that fewer than half of all American adults play the lottery in any given year. What is the average spending per adult lottery player? Does this result surprise you, or is it consistent with what you would expect based on the spending of your friends or family members who play the lottery?
3. Does your state have a lottery? If so, look up the various games available and the probabilities of winning different prizes. If not, look up the multi-state Powerball game. Consider the fact that lottery operators have spent millions of dollars researching how to encourage people to play the lottery as much as possible. Discuss how this research is applied in the selection of the games and probabilities for your state (or to Powerball).

4. Find advertisements for your state lottery (or Powerball). Do the advertisements give a fair assessment of the probabilities you found in question 3? Do they seem deceptive in any way? Explain.
5. Research suggests that lower-income people tend to play lotteries more often than higher-income people. Search for statistics concerning the percentage of income spent on the lottery by people in different income groups. Is it fair to say, as critics allege, that lotteries are a form of regressive taxation? Defend your opinion.
6. Based on your answers to the preceding questions and any additional data you can find, discuss the various moral dimensions of lotteries. For example: Is it morally responsible for governments to sponsor a form of gambling? For governments that need more revenue, are lotteries a good alternative to higher taxes? Are lotteries unfair to the poor or uneducated?

UNIT 7A Fundamentals of Probability

Probability plays such a huge role in our lives that it is essential to understand how it works. In this unit, we'll discuss fundamental concepts required for probability calculations.

Let's begin by considering a toss of two coins. Figure 7.1 shows that there are four different ways the coins can fall. Each of these four ways is a different **outcome** of the

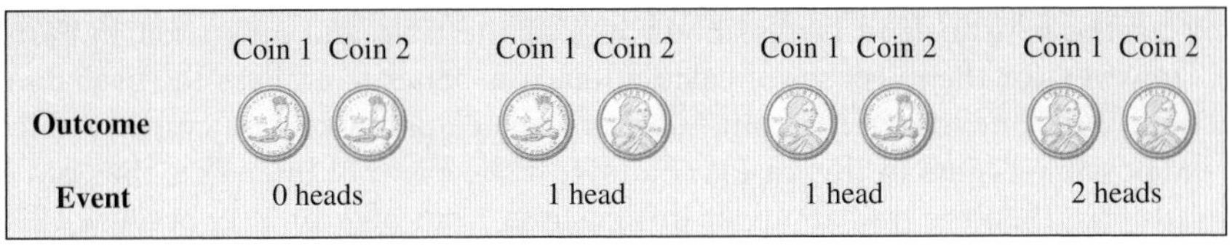

FIGURE 7.1 The four possible outcomes for a toss of two coins. The two middle outcomes both represent the same event of 1 head.

coin toss. For convenience, we abbreviate tails with T and heads with H, so that we can write the four possible outcomes as follows:

- TT: Coin 1 and Coin 2 both land tails.
- TH: Coin 1 lands tails and Coin 2 lands heads.
- HT: Coin 1 lands heads and Coin 2 lands tails.
- HH: Coin 1 and Coin 2 both land heads.

Note that, when we consider the possible outcomes, the order matters. That is, TH is a different outcome from HT.

Now, suppose we are interested only in the number of heads. Because the two middle outcomes (TH and HT) each have exactly 1 head, we say that these two outcomes represent the same **event**. That is, an event describes one or more possible outcomes that all have the same property of interest—in this case, the same number of heads. Figure 7.1 also shows that if we are counting the number of heads on two coins, the four possible outcomes represent only three different events: 0 heads (or 0 H), 1 head (or 1 H), and 2 heads (or 2 H).

Definitions

Outcomes are the most basic possible results of observations or experiments. For example, if you toss two coins, one possible outcome is HT and another possible outcome is TH.

An **event** consists of one or more outcomes that share a property of interest. For example, if you toss two coins and count the number of heads, the outcomes HT and TH both represent the same *event* of 1 head (and 1 tail).

EXAMPLE 1 Family Outcomes and Events

Consider families with two children. List all the possible *outcomes* for the birth order of boys and girls. If we are only interested in the total number of boys in the families, what are the possible *events*?

Solution Using B to represent a boy and G to represent a girl, there are four different possible birth orders (outcomes): BB, BG, GB, and GG. Because we are asked to consider only the total number of boys, the possible events with two children are: 0 boys, 1 boy, and 2 boys. The event 0 boys (0 B) consists of the single outcome GG, the event 1 boy (1 B) consists of the outcomes GB and BG, and the event 2 boys (2 B) consists of the single outcome BB. ▶ Now try Exercises 17–18.

Expressing Probabilities

Consider a single coin and assume that it is "fair," meaning that it is equally likely to land heads (H) or tails (T). In everyday language, we say that the chance of the coin landing heads on a single toss is "50–50," which means that over many tosses we expect the coin to land heads about 50% of the time and tails about 50% of the time. However, for the purposes of calculation, it is better to express the probability as a fraction. With one toss of the coin, there are two equally likely outcomes (which are also the possible events in this case): H and T. If the event of interest is the coin landing heads (H), then the probability of this event is 1 out of 2, or 1/2, because H is 1 of the 2 equally likely outcomes (H and T). Using the letter *P* to represent probability, we express the probability of heads as follows:

$$P(\text{H}) = \frac{1}{2} = 0.5$$

We read this statement as "The probability of heads equals one-half, or 0.5." More generally, we use the notation *P*(event) to mean the probability of any event. We often denote events by letters or symbols, as we did in using H for "heads."

We can use the same notation when the event of interest for the single coin is that it lands tails, which also has a probability of 1/2:

$$P(\text{T}) = \frac{1}{2} = 0.5$$

BY THE WAY

In very rare cases, a coin can land on its edge. However, left to itself, the coin will eventually fall over, which is why heads and tails are considered the only possible outcomes. Still, you might think of exceptions under special circumstances. For example, if you toss a coin in the International Space Station, it will float weightlessly along with everything else.

Because no event besides heads (H) or tails (T) is possible, it is *certain* that we will toss either a *head or* a *tail*. That is, the probability of an event in which the coin lands as *either heads or tails* is 2 out of 2, or:

$$P(\text{coin lands either heads or tails}) = \frac{2}{2} = 1$$

Generalizing the coin toss example, probabilities always have values between 0 and 1. A probability of 0 means an event is impossible, and a probability of 1 means an event is certain to occur. In between the extremes of 0 and 1, larger fractions correspond to more likely events. Figure 7.2 shows the scale of probability values, along with common expressions of likelihood.

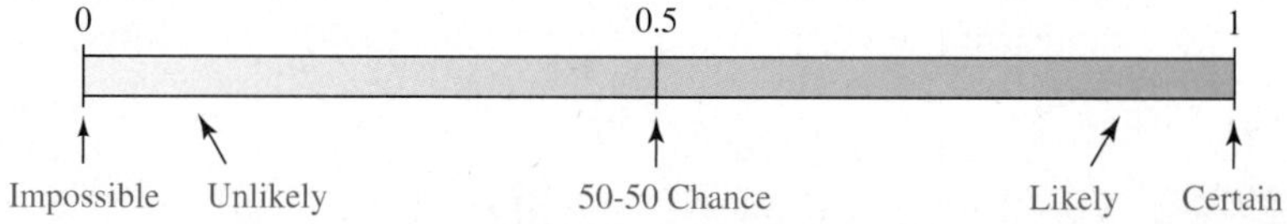

FIGURE 7.2 The scale shows various degrees of certainty as expressed by probabilities.

Expressing Probability

The probability of an event, expressed as *P*(event), is always between 0 and 1 (inclusive). A probability of 0 means the event is impossible and a probability of 1 means the event is certain.

Note on rounding: If possible, express a probability as an *exact* fraction or decimal; otherwise, round the result, usually to three significant digits, as in 0.00457.

Think About It Place each of the following events on the scale in Figure 7.2, and explain how you chose where to put them: (a) the event of the Sun being above the horizon in the daytime; (b) the event of being in two places at the same time; (c) the event of being hit by a bus; (d) the event of getting an A in your math class.

Theoretical Probabilities

There are three basic methods for finding probabilities, which we will call the *theoretical method*, the *relative frequency method*, and the *subjective method*. We'll begin with the theoretical method.

When we say that the probability of heads for a coin toss is $1/2$, we are assuming that the coin is fair and is equally likely to land heads or tails. In essence, the probability is based on a *theory* of how the coin behaves, so we say that the probability of $1/2$ comes from the **theoretical method**. When all outcomes are equally likely, we can use the following procedure to calculate theoretical probabilities.

BY THE WAY

Theoretical methods are also called *a priori* methods. The words *a priori* are Latin for "before the fact" or "before experience."

Theoretical Method for Equally Likely Outcomes

Step 1. Count the total number of possible outcomes.

Step 2. Among all the possible outcomes, count the number of ways the event of interest, A, can occur.

Step 3. Determine the probability, $P(A)$, from

$$P(A) = \frac{\text{number of ways } A \text{ can occur}}{\text{total number of possible outcomes}}$$

To see how the procedure works, look back at Figure 7.1, which shows that there are four possible outcomes for a two-coin toss (HH, HT, TH, and TT) and each is equally likely. However, if we are interested in the event of 2 heads, there is only one way (HH) that it can occur. Therefore, its probability is

$$P(2 \text{ heads}) = \frac{\text{number of ways 2 heads can occur}}{\text{total number of possible outcomes}} = \frac{1}{4}$$

EXAMPLE 2 Coins and Dice

Apply the theoretical method to find the probability of:

a. exactly one head when you toss two coins.
b. getting a 3 when you roll a 6-sided die.

Solution

a. There are four possible and equally likely outcomes for tossing two coins: HH, HT, TH, and TT. Two of these four outcomes represent the event of exactly one head (TH and HT). Therefore, the probability of exactly one head is

$$P(1 \text{ head}) = \frac{\text{number of ways 1 head can occur}}{\text{total number of possible outcomes}} = \frac{2}{4} = \frac{1}{2}$$

When tossing two coins together, the probability of getting exactly one head is $1/2$, meaning that in many tosses we expect to get exactly one head (and one tail) about half of the time.

b. Figure 7.3 shows the six possible outcomes for rolling a single 6-sided die. Each outcome is equally likely and only one outcome is rolling a 3, so the probability of rolling a 3 is

$$P(\text{rolling a } 3) = \frac{\text{number of ways the outcome 3 can occur}}{\text{total number of possible outcomes}} = \frac{1}{6}$$

Technical Note

Except when stated otherwise, we assume that dice are fair, six-sided cubes with face values of 1 through 6.

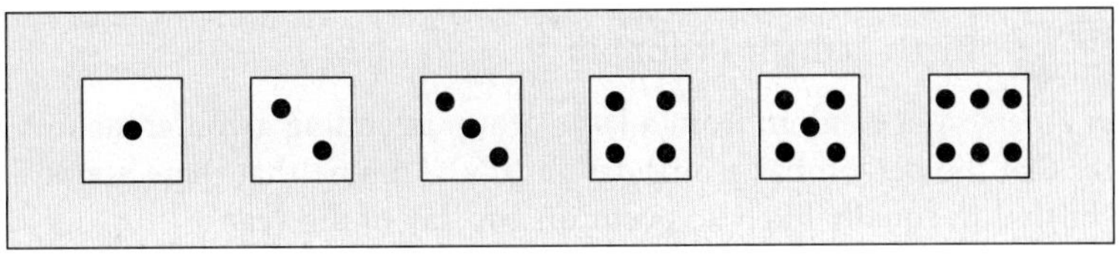

FIGURE 7.3 The six possible outcomes for a roll of one die.

The probability of rolling a 3 on a 6-sided die is $1/6$; that is, if you roll the die many times, you should see a 3 about $1/6$ of the time. ▶ Now try Exercises 19–20.

EXAMPLE 3 Playing Card Probabilities

Figure 7.4 shows the 52 playing cards in a standard deck. There are four *suits*: hearts, spades, diamonds, and clubs. Each suit has cards labeled with the numbers 2 through 10, plus a jack, queen, king, and ace (for a total of 13 cards in each suit). Notice that hearts and diamonds are red, while spades and clubs are black. If you draw one card at random from a standard deck, what is the probability that it is a spade?

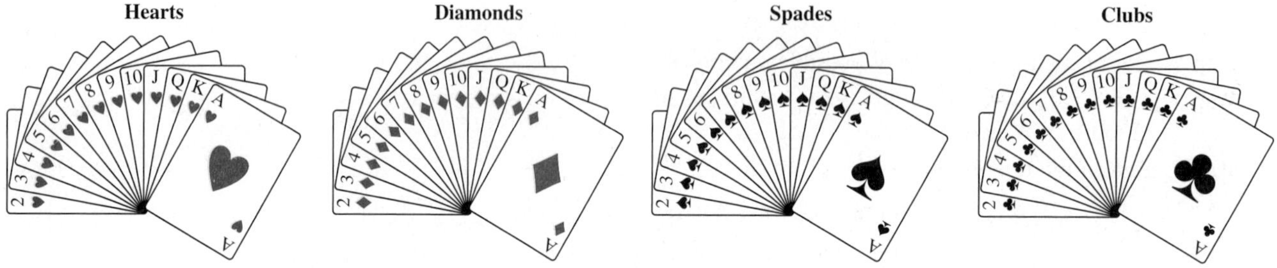

FIGURE 7.4 Playing cards in a standard 52-card deck.

Solution We can use the three-step theoretical method, because each of the 52 cards is equally likely to be drawn.

Step 1. Each card represents a possible outcome, so there are 52 possible outcomes.

Step 2. The event of interest is drawing a spade, and there are 13 spades in the deck.

Step 3. The probability that a randomly drawn card is a spade is

$$P(\text{spade}) = \frac{\text{number of outcomes that are spades}}{\text{total number of possible outcomes}} = \frac{13}{52} = \frac{1}{4}$$

▶ Now try Exercises 21–22.

EXAMPLE 4 Two Girls and a Boy

What is the probability that a randomly selected family with three children has two girls and one boy? Assume boys and girls are equally likely.

Solution We apply the three-step theoretical method.

Step 1. There are two possible outcomes for each birth: boy (B) or girl (G). For a family with three children, the total number of possible outcomes (birth orders) is $2 \times 2 \times 2 = 8$ (see the Brief Review on p. 431). The eight possible outcomes are the birth orders BBB, BBG, BGB, BGG, GBB, GBG, GGB, and GGG.

Step 2. Of these eight possible outcomes, three have two girls and one boy: BGG, GBG, and GGB.

Step 3. The probability that a family with three children has two girls and one boy is

$$P(2\text{ girls}, 1\text{ boy}) = \frac{\text{number of outcomes with 2 girls, 1 boy}}{\text{total number of outcomes}} = \frac{3}{8} = 0.375$$

To summarize, the probability that a family with three children has two girls and one boy is $3/8$, or 0.375. ▶ Now try Exercises 23–24.

BY THE WAY
In reality, birthdays are not quite randomly distributed throughout the year. After accounting for the different lengths of months, January has the lowest birth rate and June and July have the highest birth rates.

EXAMPLE 5 Birth Month Probability

You select a person at random from a large group attending a conference. What is the probability that the person has a birthday in July? Assume that there are 365 days in a year and births are equally likely to occur on any day of the year.

Solution Because we assume that all possible birthdays are equally likely, we can apply the three-step theoretical method.

Step 1. Each possible birthday represents an outcome, so there are 365 possible outcomes.

Step 2. July has 31 days, so 31 of the 365 total outcomes represent the event of a July birthday.

Step 3. The probability that a randomly selected person has a July birthday is

$$P(\text{July birthday}) = \frac{\text{number of days that are a July birthday}}{\text{total number of possible birthdays}} = \frac{31}{365} \approx 0.0849$$

which is slightly more than 1 in 12.

▶ **Now try Exercises 25–28.**

Brief Review The Multiplication Principle

Suppose we toss two coins and want to know the total number of possible outcomes. The toss of the first coin has two possible outcomes: heads (H) or tails (T). The toss of the second coin also has the possible outcomes H and T. Figure 7.5a shows that, as a result, there are a total of $2 \times 2 = 4$ possible outcomes for the two coins together. Similarly, Figure 7.5b shows that if we toss three coins, there are $2 \times 2 \times 2 = 8$ possible outcomes. We can generalize these examples with the following rule, often called the **multiplication principle**.

> **The Multiplication Principle**
>
> Suppose there are M possible outcomes for one process and N possible outcomes for a second process. The total number of possible outcomes for the two processes combined is $M \times N$.

This idea can be extended to any number of processes. For example, if a third process has R possible outcomes, the total number of possible outcomes for the three processes combined is $M \times N \times R$.

Example: How many outcomes are possible if you roll two fair dice?

Solution: The first die has six possible outcomes (see Figure 7.3). The second die also has six possible outcomes. The total number of possible outcomes for two dice is $6 \times 6 = 36$.

Example: A restaurant menu offers two choices for an appetizer, five choices for a main course, and three choices for a dessert. How many different three-course (appetizer, main course, dessert) meals are possible?

Solution: There are two possible outcomes when selecting an appetizer, five possible outcomes when selecting a main course, and three possible outcomes when selecting dessert. The total number of possible outcomes is $2 \times 5 \times 3 = 30$; that is, 30 different three-course meals are possible.

Example: A college offers 12 natural science classes, 15 social science classes, 10 English classes, and 8 fine arts classes. You must take one class from each category to meet your core requirements. How many possible ways are there to meet the core requirements?

Solution: Each selection of a course from a category represents a distinct process. To find the total number of ways to meet the requirements, we multiply the number of choices in each category: $12 \times 15 \times 10 \times 8 = 14{,}400$. There are 14,400 different ways to choose your courses to meet the requirements.

▶ **Now try Exercises 13–16.**

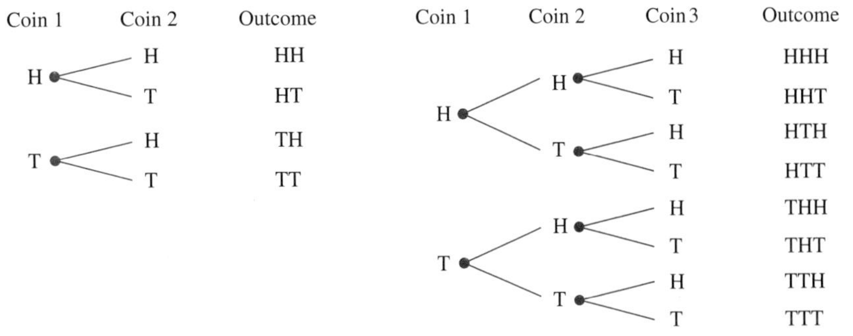

(a) (b)

FIGURE 7.5 Tree diagrams showing the possible outcomes of tossing (a) two coins and (b) three coins.

He that leaves nothing to chance will do few things ill, but will do very few things.
—George Savile Halifax

Think About It With the same assumptions as in Example 5, what is the probability of selecting a person with a July 4 birthday? Explain.

Relative Frequency Probabilities

A second way to determine probabilities is to *approximate* the probability of an event A by making many observations and counting the number of times event A occurs. This approach is called the **relative frequency method** (or *empirical method*). For example, if we observe that it rains an average of 100 days per year, we might say that the probability of rain on a randomly selected day is $100/365$. We apply this method as follows.

Relative Frequency Method

Step 1. Repeat or observe a process many times and count the number of times the event of interest, A, occurs.

Step 2. Estimate $P(A)$ using this formula:

$$P(A) = \frac{\text{number of times } A \text{ occurred}}{\text{total number of observations}}$$

BY THE WAY

Climate change makes it much more difficult to predict the frequency of future extreme weather events from past records. For example, by using computer simulations to predict storm frequencies in a warming world, researchers from MIT and Princeton found that what past records show to be "500-year floods" may occur as frequently as every 25 years (on average) by the end of this century.

EXAMPLE 6 500-Year Flood

Geological records indicate that a river has crested above a particular high flood level 4 times in the past 2000 years. Using the relative frequency method, what is the probability that the river will crest above this flood level next year?

Solution Based on the data, the relative frequency probability of the river cresting above flood stage in any single year is

$$\frac{\text{number of years with flood}}{\text{total number of years}} = \frac{4}{2000} = \frac{1}{500}$$

Because a flood of this magnitude occurs on average once every 500 years, it is called a "500-year flood." The probability of having a flood of this magnitude in any given year is $1/500$, or 0.002.

▲ **CAUTION!** We should *not* expect a "500-year flood" to occur exactly every 500 years; it is simply a way of expressing that such a flood has a $1/500$ chance of occurring in any single year. ▲

▶ Now try Exercises 29–30.

EXAMPLE 7 Fair Coin Test

Suppose you toss two coins 100 times and observe the following results:

- 0 heads occurs 22 times.
- 1 head occurs 51 times.
- 2 heads occurs 27 times.

Compare the relative frequency probabilities to the theoretical probabilities. Do you have reason to suspect that the coins are unfair?

Solution Let's start with the theoretical probabilities for the different numbers of heads. Of the four possible outcomes for a two-coin toss shown in Figure 7.1, only one has 0 heads, so the theoretical probability for this outcome is $P(0 \text{ heads}) = 1/4$, or 0.25. Similarly, the theoretical probability for 2 heads is $P(2 \text{ heads}) = 0.25$. And as we found in Example 2 (part a), the theoretical probability for exactly 1 head is $P(1 \text{ head}) = 2/4$, or 0.5.

We next find the relative frequency probabilities by dividing the number of times each event occurred by the total of 100 tosses

Relative frequency of 0 heads: $\dfrac{\text{number of times 0 heads occurs}}{\text{total number of coin tosses}} = \dfrac{22}{100} = 0.22$

Relative frequency of 1 head: $\dfrac{\text{number of times 1 head occurs}}{\text{total number of coin tosses}} = \dfrac{51}{100} = 0.51$

Relative frequency of 2 heads: $\dfrac{\text{number of times 2 heads occurs}}{\text{total number of coin tosses}} = \dfrac{27}{100} = 0.27$

Comparing these relative frequency probabilities to the theoretical probabilities, we see that the values are fairly close (for example, 0.51 versus 0.5 for 1 head). Therefore, this particular set of 100 tosses does *not* provide evidence that the coins are unfair. (A more precise calculation confirms that results similar to this one are common with fair coins.)

▶ Now try Exercises 31–32.

Think About It How does the conclusion of Example 7 resemble the idea of a *hypothesis test*, discussed in Unit 6D?

Subjective Probabilities

The third method for determining probabilities is to estimate a **subjective probability** using experience or intuition. For example, you could make a subjective estimate of the probability that a friend will be married within the next year or the probability that a good grade in your math class will help you get the job you want.

Three Approaches to Finding Probability

A **theoretical probability** is based on the assumption that all outcomes are equally likely. It is calculated by dividing the number of ways an event can occur by the total number of possible outcomes.

A **relative frequency probability** is based on observations or experiments. It is the relative frequency of the event of interest.

A **subjective probability** is an estimate based on experience or intuition.

BY THE WAY
Another approach to finding probabilities, called the *Monte Carlo method*, uses computer simulations. This method essentially finds relative frequency probabilities, but the experiments are computer simulations rather than actual experiments in a lab.

EXAMPLE 8 Which Method?

Identify the method that resulted in the following statements.

a. I'm certain that you'll be happy with this car.
b. Based on housing data, the chance that someone will move to a new residence during the coming year is about 1 in 8.
c. The probability of rolling a 7 with a 12-sided die is 1/12.

Solution

a. This is a subjective probability, which is based on the opinion of the speaker.
b. This is a relative frequency probability, because it is based on data showing that about 1/8 of the population moves to a new residence during a given year.
c. This is a theoretical probability, because it is based on the assumption that a fair 12-sided die is equally likely to land on any of its 12 sides.

▶ Now try Exercises 33–34.

Probability of an Event *Not* Occurring

Suppose we are interested in the probability that a particular event or outcome does *not* occur. For example, consider the probability of a giving a wrong answer on a multiple-choice question with five possible answers. The probability of answering correctly with a random guess is $1/5$, so the probability of *not* answering correctly is $4/5$. Notice that the sum of the two probabilities is 1, because the answer must be either right or wrong. We can generalize this idea.

Probability of an Event *Not* Occurring

If the probability of an event A is $P(A)$, we write the probability that event A does *not* occur as $P(\text{not } A)$. Because an event either does or does not occur, the sum of these probabilities is $P(A) + P(\text{not } A) = 1$. Therefore, the probability that event A does *not* occur is

$$P(\text{not } A) = 1 - P(A)$$

EXAMPLE 9 Not Two Girls

What is the probability that a randomly chosen family with three children does *not* have two girls and one boy? Assume boys and girls are equally likely.

Solution In Example 4, we found that the probability of two girls and one boy in a family with three children is $3/8$, or 0.375. Therefore, the probability that a family with three children does *not* have two girls and one boy is $1 - 3/8 = 5/8$, or 0.625.

▶ Now try Exercises 35–38.

Probability Distributions

Consider again the tossing of two coins together, for which there are four possible outcomes (see Figure 7.1) representing three unique events: 0 heads, 1 head, and 2 heads. We've already found the probabilities for these events, so we can summarize them with a **probability distribution** that displays the probability for *every possible event* of interest (in this case, events involving heads). Table 7.1 shows the probability distribution for tossing two coins as a table, and Figure 7.6 shows it as a histogram. Keep in mind that the sum of all the probabilities in a probability distribution must be 1, because it is certain that one of the events will occur.

TABLE 7.1 Probability Distribution for Tossing Two Coins

Event	Probability
2 heads, 0 tails	0.25
1 head, 1 tail	0.50
0 heads, 2 tails	0.25
Total	1

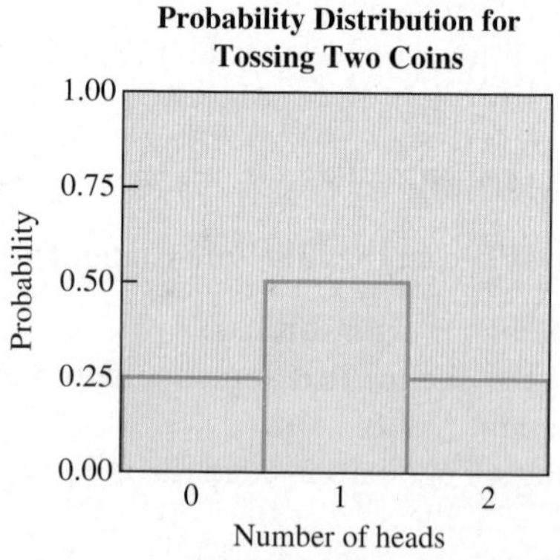

FIGURE 7.6 Histogram showing the probability distribution for tossing two coins.

Making a Probability Distribution

A **probability distribution** represents the probabilities of all possible events of interest. To make a table of a probability distribution:

Step 1. List all possible *outcomes*.

Step 2. Identify outcomes that represent the same *event*. Find the probability of each event.

Step 3. Make a table or figure that displays all the probabilities. The sum of all the probabilities must be 1.

EXAMPLE 10 Tossing Three Coins

Make a probability distribution for the number of heads that occur when three coins are tossed simultaneously.

Solution We follow the process described in the box above.

Step 1. As shown in Figure 7.5 (in the Brief Review, p. 431) there are eight possible outcomes: HHH, HHT, HTH, HTT, THH, THT, TTH, and TTT.

Step 2. The events of interest are the numbers of heads, so the eight outcomes represent four possible events: 0 heads, 1 head, 2 heads, and 3 heads. Notice that only one outcome represents the event of 3 heads (and 0 tails), so its probability is $1/8$; the same is true for the event of 0 heads (and 3 tails). The remaining events, 1 head (and 2 tails) and 2 heads (and 1 tail), each occur three times, so their probabilities are each $3/8$.

Step 3. Table 7.2 shows the probability distribution, with the four events listed in the left column and their probabilities in the right column. Notice that the sum of the probabilities is 1, as it must be. We can also show the distribution with a graph (Figure 7.7).

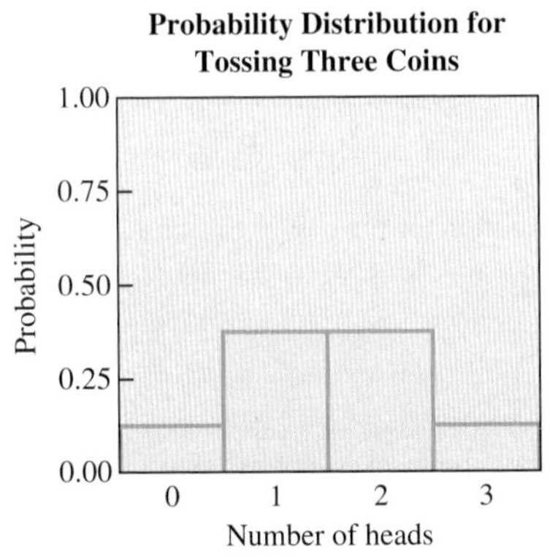

FIGURE 7.7 Histogram showing the probability distribution for tossing three coins.

TABLE 7.2 Probability Distribution for Tossing Three Coins

Result	Probability
3 heads (0 tails)	1/8
2 heads (1 tail)	3/8
1 head (2 tails)	3/8
0 heads (3 tails)	1/8
Total	**1**

▶ Now try Exercises 39–40.

Think About It How many different *outcomes* are possible when you toss four coins? If you are interested in the number of heads, how many different *events* are possible?

The false ideas prevalent among all classes of the community respecting chance and luck illustrate the truth that common consent argues almost of necessity of error.

—Richard Proctor, *Chance and Luck*

EXAMPLE 11 Two Dice Distribution

Make a probability distribution for the sum of the dice when two fair dice are rolled together. What is the most probable sum?

Solution There are six ways for each die to land (see Figure 7.3), so there are $6 \times 6 = 36$ possible outcomes for a single roll of two dice. Table 7.3 shows all 36 outcomes by listing one die along the rows and the other along the columns, with each cell showing the sum for the two dice.

TABLE 7.3 Outcomes and Sums for Rolling Two Dice

	1	2	3	4	5	6
1	$1 + 1 = 2$	$1 + 2 = 3$	$1 + 3 = 4$	$1 + 4 = 5$	$1 + 5 = 6$	$1 + 6 = 7$
2	$2 + 1 = 3$	$2 + 2 = 4$	$2 + 3 = 5$	$2 + 4 = 6$	$2 + 5 = 7$	$2 + 6 = 8$
3	$3 + 1 = 4$	$3 + 2 = 5$	$3 + 3 = 6$	$3 + 4 = 7$	$3 + 5 = 8$	$3 + 6 = 9$
4	$4 + 1 = 5$	$4 + 2 = 6$	$4 + 3 = 7$	$4 + 4 = 8$	$4 + 5 = 9$	$4 + 6 = 10$
5	$5 + 1 = 6$	$5 + 2 = 7$	$5 + 3 = 8$	$5 + 4 = 9$	$5 + 5 = 10$	$5 + 6 = 11$
6	$6 + 1 = 7$	$6 + 2 = 8$	$6 + 3 = 9$	$6 + 4 = 10$	$6 + 5 = 11$	$6 + 6 = 12$

TABLE 7.4 Probability Distribution for the Sum of Two Dice

Event (Sum)	Probability
2	1/36
3	2/36
4	3/36
5	4/36
6	5/36
7	6/36
8	5/36
9	4/36
10	3/36
11	2/36
12	1/36
Total	1

Notice that the possible sums, which are the *events* of interest in this case, are 2 through 12. We find the probability of each sum by counting the number of times it occurs and dividing by the total of 36 possible outcomes. For example, the five highlighted outcomes in the table have a sum of 8, so the probability of a sum of 8 is 5/36. Table 7.4 and Figure 7.8 show the probability distribution. The most likely event is a sum of 7, which occurs for 6 of the outcomes and therefore has a probability of 6/36, or 1/6.

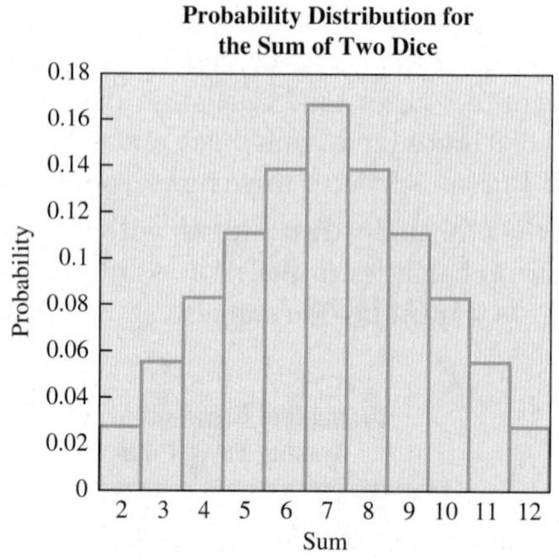

FIGURE 7.8 Probability distribution for the sum of two dice.

▶ Now try Exercises 41–42.

Stating the Odds

You've probably noticed that there are several ways to express the idea of *probability*. For example, saying that the *chance* (or *likelihood*) of getting heads on a coin toss is 1 in 2, or 1/2, is equivalent to saying that the *probability* is 1/2. Another common term for probability is **odds**, but strictly speaking the two terms have different meanings. Odds are commonly stated as the *ratio* of the probability that a particular event will occur to the probability that it will not occur.

Definition

The **odds for** an event *A* are given by

$$\text{odds for event } A = \frac{P(A)}{P(\text{not } A)}$$

The **odds against** an event *A* are given by

$$\text{odds against event } A = \frac{P(\text{not } A)}{P(A)}$$

Note: In gambling, the term *odds on* generally means "odds against."

Although odds are defined as a fraction, we usually state them by reducing the fraction to simplest terms and reading it as a ratio. For example, the probability of rolling a 4 on a single die is 1/6 and the probability of *not* rolling a 4 is 5/6. Therefore, the *odds for* rolling a 4 are

$$\frac{P(4)}{P(\text{not } 4)} = \frac{1/6}{5/6} = \frac{1}{5}$$

We usually state these *odds for* rolling a 4 as *1 to 5*. Taking the reciprocal, we can say the *odds against* rolling a 4 are *5 to 1*.

In gambling, the *odds on* (meaning the "odds against") usually expresses how much you can gain with a win for each dollar you bet. For example, suppose the odds on a particular horse in a horse race are 3 to 1. Then for each \$1 you bet on this horse, you will gain \$3 if the horse wins. If you place a \$2 bet and the horse wins, you will gain $3 \times \$2 = \$6$. (You will also receive back the money you bet, so when you collect on your \$2 bet, you will receive the \$6 gain plus your original \$2, for a total of \$8.) Of course, as with any form of gambling, you are more likely to lose than to win.

EXAMPLE 12 Two-Coin Odds

What are the odds for getting two heads when tossing two coins together? What are the odds against it?

Solution As we found earlier, the probability of two heads when tossing two coins is 1/4. That is, $P(2 \text{ heads}) = 1/4$. Therefore, the probability of *not* getting two heads is $P(\text{not 2 heads}) = 1 - 1/4 = 3/4$. So the *odds for* getting two heads are

$$\text{odds for event of 2 heads} = \frac{P(2 \text{ heads})}{P(\text{not 2 heads})} = \frac{1/4}{3/4} = \frac{1}{3}$$

The odds for getting two heads in two coin tosses are 1 to 3. We take the reciprocal of the *odds for* to find the *odds against*, so the odds against getting two heads when tossing two coins are 3 to 1.

▶ Now try Exercises 43–46.

BY THE WAY

What's a blue moon? Many people now say that a blue moon is the second full moon in a calendar month. But the editors of *Sky and Telescope* discovered that this definition came from a mistake in a 1946 issue of their magazine. No one knows the original definition of *blue moon*, but the phrase "once in a blue moon" has come to mean an extremely low probability of occurrence.

EXAMPLE 13 Horse Race Payoff

At a horse race, the odds on Blue Moon are given as 7 to 2. If you bet \$10 and Blue Moon wins, how much will you gain?

Solution The 7 to 2 odds mean that, for each \$2 you bet on Blue Moon, you gain \$7 if you win. A \$10 bet is equivalent to five \$2 bets, so you gain $5 \times \$7 = \$35$. You also get your \$10 back, so you will receive \$45 when you collect on your winning ticket.

▶ Now try Exercises 47–48.

Quick Quiz 7A

Choose the best answer to each of the following questions. Explain your reasoning with one or more complete sentences.

1. Suppose you toss a coin three times. In terms of the number of heads, which of the following outcomes represents the same event as an outcome of tails, heads, tails (THT)?

   a. THH

   b. TTH

   c. TTT

2. During the course of the basketball season, Lisa made 80 out of 100 free throws. When we say that her probability of making a free throw is 0.8, what type of probability are we stating?

   a. a theoretical probability

   b. a relative frequency probability

   c. a subjective probability

3. A box contains 20 fruits, but only 4 of them are oranges. When we say that the probability of drawing one of the oranges at random is 0.2, what type of probability are we stating?

   a. a theoretical probability

   b. a relative frequency probability

   c. a subjective probability

4. Suppose the probability of winning a certain prize in the lottery is 0.001. What is the probability of *not* winning?

   a. 0.001

   b. 1 + 0.001

   c. 1 − 0.001

5. When you toss one fair coin, the probability of heads is 1/2. Assuming all the coins are fair, this means that

   a. if you toss two coins, you'll get 1 head and 1 tail.

   b. if you toss 100 coins, you'll get 50 heads and 50 tails.

   c. if you toss 1000 coins, the number of heads will be close to, but not necessarily exactly, 500.

6. On a roll of two dice, Serena bets that the sum will be 6, and Mackenzie bets that the sum will be 9. Who has a higher probability of winning the bet? (*Hint:* See Table 7.4.)

   a. Serena

   b. Mackenzie

   c. Both have an equal probability of winning.

7. Suppose you toss four 6-sided dice. How many *outcomes* are possible?

   a. 6

   b. $4 \times 4 \times 4 \times 4 \times 4 \times 4$

   c. $6 \times 6 \times 6 \times 6$

8. Suppose you toss three 6-sided dice. How many different sums are possible?

   a. 11

   b. 16

   c. $6 \times 6 \times 6$

9. You are playing five-card poker with a deck of 52 cards. If you make a probability distribution showing the individual probabilities of all possible hands, the sum of all the individual probabilities will be

   a. 1.

   b. 5.

   c. 52.

10. The odds on (or odds against) TripleTreat winning the Kentucky Derby are 4 to 1. This means that whoever set the odds thinks that TripleTreat's probability of winning is

    a. 1 in 5, or 1/5.

    b. 1 in 4, or 1/4.

    c. 4 in 5, or 4/5.

Exercises 7A

REVIEW QUESTIONS

1. Distinguish between an *outcome* and an *event* in dealing with probabilities. Give an example in which the same event can occur through two or more outcomes.
2. What does it mean when we write *P*(event)? What is the possible range of values for *P*(event), and why?
3. Briefly describe the *theoretical, relative frequency*, and *subjective methods* for finding probabilities. Give an example of each.
4. How is the probability of an event *not* occurring related to the probability that it *does* occur? Why?
5. What is a *probability distribution*? Explain how to make a table or histogram of a probability distribution.
6. Explain the common usage of the term *odds* and its usage in gambling.

DOES IT MAKE SENSE?

Decide whether each of the following statements makes sense (or is clearly true) or does not make sense (or is clearly false). Explain your reasoning.

7. When I toss four coins, there are six different outcomes that represent the event of 2 heads and 2 tails.
8. The probability that my sister will get into the college of her choice is 3.7.
9. I estimate that the probability of my getting married in the next 3 years is 0.7.
10. Because either there is life on Mars or there is not, the probability of life on Mars is 0.5.
11. The probability that Jonas will win the race is 0.6 and the probability that he will not win is 0.5.
12. Based on data showing that we've had snow on Christmas in 27 of the past 100 years, the probability of snow on Christmas this year is 0.27.

BASIC SKILLS & CONCEPTS

13–16: Review of the Multiplication Principle. Use the skills covered in the Brief Review on p. 431 to answer the following questions.

13. How many different choices of car do you have if a particular model comes in 12 colors and 4 styles (sedan, station wagon, SUV, or hatchback)?
14. A local ski shop sells nine types of skis, eight types of bindings, and twelve types of boots. How many different ski/binding/boot packages are available?
15. A restaurant has a special menu that features two choices of salad, eight choices of entree, and six choices of dessert. How many different three-course meals could you order?
16. Four states each elect a delegate to a regional commission. Within each state there are two candidates running for the delegate position. How many different commissions are possible?
17. **Double-Header Outcomes and Events.** Suppose the New York Yankees play a double-header (two games on the same day). Using W for a win and L for a loss, list all the possible

outcomes for the double-header. If we are only interested in the total number of games the Yankees win, what are the possible *events* for the two games?

18. **Weather Outcomes and Events.** Suppose we describe the weather as either sunny (S) or cloudy (C). List all the possible *outcomes* for the weather on three consecutive days. If we are only interested in the number of sunny days, what are the possible *events* for the two consecutive days?

19–28: Theoretical Probabilities. Use the theoretical method to determine the probability of the following outcomes and events. State any assumptions that you make.

19. Tossing two coins and getting either no tails or one tail
20. Rolling a single die and getting an even number (2, 4, or 6)
21. Drawing a king from a standard deck of cards
22. Drawing a red card (heart or diamond) from a standard deck of cards
23. Randomly selecting a two-child family with two boys
24. Randomly selecting a three-child family with exactly two boys
25. A randomly selected person has a birthday in April.
26. A randomly selected person has a birthday in a month beginning with J.
27. Sharing a birthday with another person when you both have birthdays in December
28. A randomly selected person was born on a Sunday.

29–32: Relative Frequency Probabilities. Use the relative frequency method to answer the following questions.

29. The local weather forecast has been accurate for 15 of the past 27 days. Based on this fact, what is the relative frequency probability that the forecast for tomorrow will be accurate?
30. Halfway through the season, a soccer player has made 12 penalty kicks in 18 attempts. Based on her performance to date, what is the relative frequency probability that she will make her next penalty kick?
31. You toss a coin 100 times and get only 15 heads. Do you have reason to suspect the coin is unfair? Explain.
32. You toss two coins 100 times and get 0 heads 23 times, 1 head 51 times, and 2 heads 26 times. Do you have reason to suspect the coins are unfair? Explain

33–34: Which Type of Probability? State which method (theoretical, relative frequency, or subjective) should be used to answer the following questions.

33. What is the probability of being dealt a pair of aces in a five-card hand?
34. What is the probability of being injured in an automobile accident in the next year?

35–38: Event *Not* Occurring. Determine the probability of the following events. State any assumptions that you use.

35. What is the probability of *not* rolling a 1 with a fair die?
36. What is the probability of *not* tossing 3 heads with three fair coins?
37. What is the probability that a 69% free-throw shooter will miss his next free throw?
38. What is the probability that the next person you meet was *not* born in the spring (April, May, June)? (Assume 365 days in a year.)

39–42: Probability Distributions. Make a probability distribution for the given set of events. You may use either a table or a graph (or both).

39. The number of boys in families with three children
40. The number of heads when four fair coins are tossed
41. The sums that result when two fair 4-sided dice (tetrahedrons) are tossed
42. The number of girls in families with four children

43–46: Odds. Use the definition given in the text to find both the *odds for* and the *odds against* the following events.

43. Rolling a fair die and getting a 5
44. Flipping two fair coins and getting 2 heads
45. Rolling two fair dice and getting a double 6
46. Randomly drawing a heart from a standard deck of cards

47–48: Gambling Odds. Use the definition of *odds* in betting to find the following odds.

47. The odds on (against) your bet are 3 to 5. If you bet $20 and win, how much will you gain?
48. The odds on (against) your bet are 6 to 5. If you bet $20 and win, how much will you gain?

FURTHER APPLICATIONS

49–66: Computing Probabilities. Decide which method (theoretical, relative frequency, or subjective) is appropriate, and compute or estimate the following probabilities. If you use the subjective method, explain your reasoning.

49. Drawing a red face card (red jack, red queen, or red king) from a standard deck of cards
50. Rolling a 12-sided die and getting an even number
51. Drawing an even-numbered card (2, 4, 6, 8, or 10) from a standard deck of cards
52. Rolling three dice and getting the same number on all three dice
53. Randomly meeting someone born between midnight and 2:00 a.m.
54. Randomly meeting someone born in a month beginning with a vowel (assume 365 days in a year)
55. Randomly meeting someone with a phone number that ends in a 0 or 1
56. Randomly meeting someone born on a weekend (Saturday or Sunday)
57. Randomly selecting a three-child family with three girls
58. Randomly selecting a red tie from a drawer that holds five blue ties, six red ties, and seven green ties
59. Randomly meeting someone whose Social Security number ends in the same digit as yours

60. The German men's soccer team winning the next World Cup

61. A baseball player with a .250 batting average will get a hit at his next at-bat

62. Tossing a fair coin and a fair die in the air and seeing them land with a head and a 6

63. Correctly guessing a stranger's birth date knowing only that she was born on an odd-numbered day of June

64. Not having a 50-year flood this year

65. Not hearing your favorite group on a random shuffle of your playlist that has 889 songs, of which 127 are by your favorite group

66. Rolling a sum of 8 when you roll two dice

67–68: Marble Probability Distributions

67. Suppose you reach in and draw two marbles at random from a bag containing 10 white marbles (W), 10 black marbles (B), and 10 red marbles (R).

a. List all possible outcomes of this process (for example, RR, BW, and so on).

b. Make a table or graph showing the probability distribution for the events of 0, 1, and 2 black marbles.

68. Suppose you reach in and draw three marbles at random from a bag containing 10 white marbles (W), 10 black marbles (B), and 10 red marbles (R).

a. List all possible outcomes of this process (for example, RRR, BWR).

b. Make a table or graph showing the probability distribution for the events of 0, 1, 2, and 3 black marbles.

69–72: More Counting. Answer the following counting questions.

69. The local paint store offers 28 basic colors, each of which can be prepared with one of 4 different textures. How many different paint combinations are available?

70. You are required to take five courses, one each in humanities, sociology, science, math, and music. You have a choice of four humanities courses, three sociology courses, five science courses, two math courses, and three music courses. How many different sets of five courses are possible?

71. In designing your new home entertainment center, you have a choice of seven different TVs, nine different amplifiers, and eleven different speaker sets. How many different systems could you design?

72. The car model you are considering comes with or without leather seats, with or without a sun roof, with or without window tinting, and in eight stock colors. How many different versions of the car are available?

73. **Gender Politics.** The following table gives the gender and political party of the 100 delegates at a political convention. Suppose you encounter a delegate at random.

	Women	Men
Republican	21	28
Democrat	25	16
Independent	6	4

a. What is the probability that you meet a man?

b. What is the probability that you meet a Democrat?

c. What is the probability that you do not meet an Independent?

d. What is the probability that you meet a male Republican?

e. What is the probability that you meet someone who is not a female Republican?

74. **Senior Citizens.** In 2015, there were 48 million people over 65 years of age in the total U.S. population of 321 million. By 2050, it is estimated that there will be 88 million people over 65 years of age in a total U.S. population of 400 million. Would your chances of meeting a person over 65 at random be greater in 2015 or 2050? Explain.

75. **Marriage Status.** The following table gives percentages of all American women and men in various marriage categories.

Status	Women	Men
Married	51.2%	53.7%
Never married	28.9%	34.8%
Divorced	11.3%	8.9%
Widowed	8.6%	2.6%

Source: Current Population Surveys, U.S. Census Bureau.

a. What is the probability that a randomly encountered American woman is married?

b. What is the probability that a randomly encountered American man is not married?

c. Can you determine the overall probability that an American person is widowed from these data alone? Explain.

76. **Deceptive Odds.** Suppose event *A* has a 0.99 probability of occurring and event *B* has a 0.96 probability of occurring—both high probabilities. Compute the odds for event *A* and the odds for event *B*. Comment on the relative difference between the odds for the two events compared to the relative difference between the probabilities. How are the odds deceptive in this case?

77. **Project: Thumb Tack Probabilities.** Find a standard thumb tack, and practice tossing it onto a flat surface. Notice that there are two different outcomes: The tack can land with its point facing down or with its point facing up.

a. Toss the tack 50 times, and record the outcomes.

b. Give the relative frequency probabilities of the two outcomes based on these results.

c. If possible, ask several other people to repeat the process. How well do your probabilities agree?

78. **Project: Three-Coin Toss.** Toss three coins (at once) 50 times and record the outcomes in terms of the number of heads. Based on your observations, give the relative frequency probability of each result. Do the relative frequency results agree with the theoretical probabilities? Explain.

IN YOUR WORLD

79. **Blood Groups.** The four major blood groups are designated A, B, AB, and O. Within each group there are two Rh types: positive

and negative. Find data on the relative frequency of blood groups, including the Rh types. Make a table showing the probability of meeting someone in each of the eight blood groups.

80. **Rare Accidents.** Find data that give the relative frequency probabilities of various types of rare accidents (such as being killed by lightning, by a shark bite, or by a falling airplane part). Are the relative frequency probabilities consistent with your intuition?

81. **Probability in the News.** Find a news article or research report that makes use of a probability. Interpret the probability, and discuss whether it is theoretical, relative frequency, or subjective.

82. **Probability in Your Life.** Describe a recent instance from your own life when you used probability to make a decision. What type of probability did you use? How did it help you make the decision?

83. **Gambling Odds.** Find an advertisement for a gambling establishment or lottery in which the term *odds* is used, and interpret its meaning.

TECHNOLOGY EXERCISES

Answer the following questions using procedures described in the Using Technology boxes in this unit or with **StatCrunch** (available in MyLab Math).

84. **Coin Toss Simulation.** You can use StatCrunch to generate simulations of coin tosses as follows: In the StatCrunch work space (select "Open StatCrunch" on the main menu), choose "Applets" then "Simulation" and then "Coin Flipping." Then enter the number of coins you wish to toss in your simulation and choose "Compute!" An empty graph will appear, with options along the top for the number of runs in your simulation.

    a. Choose 30 for your number of coins; then when the graph appears, choose 1 run and observe the number of heads in this run. Then choose 5 runs to see the results. Repeat this process and watch how the frequencies change for different numbers of heads. Are the frequencies behaving as you expect as you increase the number of runs? Explain.

    b. Start over with 30 coins, but now choose 1000 runs. Attach a screen shot of your result, and briefly explain what it shows and why it makes sense in light of what you found in part (a).

    c. Repeat part (b), but with 50 coins in each of the 1000 runs. Attach a picture of your result, and briefly explain how it compares to your result from part (b).

85. **Dice Simulation.** Follow the method used in Exercise 84 except choose "Dice Rolling" rather than "Coin Flipping" for your StatCrunch simulation.

    a. Choose 2 as the number of dice, so that the event of interest is the sum of two dice. Watch how the result changes from that for 1 roll as you add additional rolls, 5 rolls at a time. Briefly describe the pattern you notice.

    b. Start over and do 1000 rolls of the 2 dice. Attach a screen shot of your result, and briefly describe how it compares to the theoretical probabilities shown in Figure 7.8.

UNIT 7B Combining Probabilities

In 1654, a French con man known as the Chevalier de Méré (his given name was Antoine Gombaud) bet other gamblers even money that he could roll at least one 6 in four rolls of a standard 6-sided die. The Chevalier made a profit on this bet, although he had calculated his chance of winning incorrectly.

Because the probability of rolling a 6 in one roll of a die is $1/6$, the Chevalier falsely reasoned that his chances would be four times as great with four rolls, or $4/6 = 0.67$. In fact, the chance of rolling a 6 within four tries is only 0.52 (as we will show shortly). Nevertheless, because a probability of 0.52 means winning 52% of the time (and losing only 48% of the time), the Chevalier profited over the long run.

As gamblers caught on to this "con game," they were no longer willing to play. The Chevalier therefore proposed a new game in which he rolled *two* dice and bet that he could throw a *double-6* within 24 rolls. He knew that, on any single roll of the two dice, the probability of a double-6 is $1/36$ (see Example 11 in Unit 7A). He then used his faulty reasoning to guess that, if he rolled the pair of dice 24 times, his odds of rolling at least one double-6 would be $24/36 = 0.67$.

To his dismay, the Chevalier began losing money. To understand his losses, he turned to the mathematician Blaise Pascal, who, in turn, began a correspondence about probability with the mathematician Pierre de Fermat. Pascal and Fermat quickly recognized the error in the Chevalier's reasoning and developed methods for properly combining probabilities. In this unit, we will study some of these methods.

HISTORICAL NOTE

Blaise Pascal (1623–1662) was a precocious Parisian who made fundamental contributions to mathematics and physics. He was deeply religious, and his work *Pensées* is an important treatise of western philosophy. Pascal died at the age of 39. Pierre de Fermat (1601–1665) was a lawyer, but his passion was mathematics. He spent his life corresponding with mathematicians and preoccupied with difficult problems. One of his most famous claims, known as *Fermat's last theorem*, was not proved until 1994—more than 350 years after he stated it.

EXAMPLE 1 Flaw in the Chevalier's Thinking

Consider the Chevalier's first bet, in which he guessed that the probability of rolling at least one 6 in four rolls would be four times the single-roll probability of 1/6, or 4/6. If we extended the same logic, what would we find for the probability of rolling at least one 6 in five rolls and in six rolls? Explain how this extension proves that his logic was incorrect.

Solution By the Chevalier's logic, the probability of rolling at least one 6 in five rolls would be $5 \times 1/6 = 5/6$, and the probability of rolling at least one 6 in six rolls would be $6 \times 1/6 = 6/6$, or 1. Because a probability of 1 means certainty (100% chance of the event occurring), his logic implies that you would always roll at least one 6 every time you roll a die six times. But this is not true, as it is possible to roll a die six times without rolling a 6 once. Therefore, the Chevalier's logic was flawed. (The correct calculation is shown at the end of this unit.) ▶ Now try Exercises 11–12.

Think About It Find a 6-sided die, roll it six times, and write down the number of times you roll a 6. Repeat your experiment at least 10 more times. How many times did you get at least one 6? How many times did you *not* get at least one 6? What do your results say about the Chevalier's logic?

And Probabilities

Example 1 shows why the Chevalier's method for combining probabilities was incorrect, but what is the correct method? Let's start with the case in which we want to know the probability that two events will both occur.

BY THE WAY

Whether events are independent is not always obvious. Consider a set of basketball free throws by a single player. Some people argue that the probability of success is the same for each free throw. Others suggest that the psychological impact of success or failure on one free throw affects the player's chances of making the next one. Despite many statistical studies of basketball free throws, no one has clearly established which view is correct.

Independent Events

We begin with probabilities for **independent events**, where the occurrence of one event does not affect the probability of the other event occurring. For example if you toss two coins, the outcome for one coin is independent of the outcome for the other. Similarly, if you roll a die twice, the outcome of the first roll does not affect the outcome of the second.

Suppose you toss two fair dice and want to know the probability that *both* will come up 4 (a "double-4"). One way to find the probability is to consider the two tossed dice as a *single* toss of two dice. As shown in Table 7.3, double-4 is 1 of 36 possible outcomes, so its probability is 1/36.

Alternatively, we can consider the two dice individually. That is, we are looking for the probability that one die comes up 4 *and* the other die also comes up 4. We therefore say that we are looking for an ***and* probability** (or *joint probability*). We calculate this probability by recognizing that for each die, the probability of a 4 is 1/6, so we can find the probability that the first die *and* the second die show a 4 by multiplying the individual probabilities together:

$$P(\text{double-4}) = P(4 \text{ on one die}) \times P(4 \text{ on other die}) = \frac{1}{6} \times \frac{1}{6} = \frac{1}{36}$$

It is possible to prove that this method always works for two independent events, and that it can be extended to additional events.

And Probability: Independent Events

Two events are **independent** if the occurrence of one event does not affect the probability of the other event. Consider two independent events A and B with probabilities $P(A)$ and $P(B)$. The ***and* probability** that A and B both occur is

$$P(A \text{ and } B) = P(A) \times P(B)$$

This principle can be extended to any number of independent events. For example, the probability of A, B, and a third independent event C occurring together is

$$P(A \text{ and } B \text{ and } C) = P(A) \times P(B) \times P(C)$$

EXAMPLE 2 Three Coins

Suppose you toss three fair coins. What is the probability of getting three tails?

Solution Because coin tosses are independent, we multiply the probabilities of getting tails on each individual coin:

$$P(3 \text{ tails}) = \underbrace{P(\text{tails})}_{\text{coin 1}} \times \underbrace{P(\text{tails})}_{\text{coin 2}} \times \underbrace{P(\text{tails})}_{\text{coin 3}} = \frac{1}{2} \times \frac{1}{2} \times \frac{1}{2} = \frac{1}{8}$$

The probability that three tossed coins all land on tails is $1/8$ (which we also found in Example 10 of Unit 7A using a different method). ▶ Now try Exercises 13–14.

EXAMPLE 3 Three Dice

What is the probability of rolling three 6's in a row with a single fair die?

Solution Because die rolls are independent and the probability of a 6 on a single roll is $1/6$, we multiply:

$$\begin{aligned} P(\text{three 6's in a row}) &= P(6 \text{ on first roll}) \times P(6 \text{ on second roll}) \times P(6 \text{ on third roll}) \\ &= \frac{1}{6} \times \frac{1}{6} \times \frac{1}{6} = \frac{1}{216} \end{aligned}$$

The probability of rolling three 6's in a row is 1 in 216. ▶ Now try Exercises 15–16.

EXAMPLE 4 Consecutive Floods

Find the probability that a 100-year flood (a flood with a 0.01 probability of occurring in a given year) will hit the same city in two consecutive years. Assume that a flood in one year does not affect the likelihood of a flood in the next year.

Solution We are told to assume that floods in different years are independent events. Therefore, we find the probability of 100-year floods in two consecutive years by multiplying the probability for each individual year:

$$\begin{aligned} P(2 \text{ consecutive flood years}) &= P(\text{flood in year 1}) \times P(\text{flood in year 2}) \\ &= 0.01 \times 0.01 = 0.0001 \end{aligned}$$

The probability of a 100-year flood in two consecutive years is 1 in 10,000. While events with such a low probability can certainly occur, remember that this calculation *assumes* that the chance of a flood is 0.01 in any year. If a 100-year flood were to occur in two consecutive years, we might wonder whether the 0.01 probability of a single flood was correct.

▶ Now try Exercises 17–18.

Dependent Events

Suppose you take pieces of candy randomly from a box that initially contains five chocolates and five caramels. Clearly, the probability of getting a chocolate on your first draw is 5/10, or 1/2. Now, suppose you get a chocolate on the first draw and eat it. What is the probability of getting another chocolate on your second draw?

Because you've already eaten one chocolate, the box now contains only nine pieces of candy, of which four are chocolate. Therefore, the probability of getting a chocolate on the second draw is 4 out of 9, or 4/9 (Figure 7.9), which is *not* the same as the 1/2 probability for the first draw. Because the candy selected on your first draw affects the probability for your second draw, we say that the events are **dependent events**.

CHOCOLATE AND CARAMEL CANDY SAMPLER

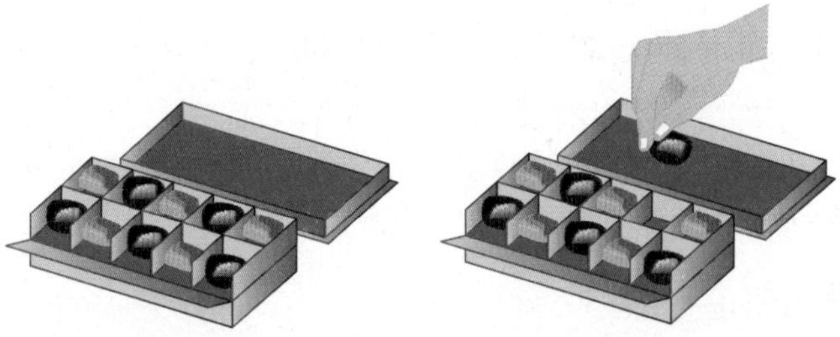

5/10 of the candies are chocolates (a) | 4/9 of the candies are chocolates (b)

FIGURE 7.9 A box of candy contains five chocolates and five caramels. (a) If you choose one candy at random, the probability of getting a chocolate is 5/10, or 1/2. (b) If the first candy is a chocolate and you eat it, then there are only four chocolates left among the nine remaining candies. Therefore, the probability of getting another chocolate on the second draw is 4/9.

Calculating the probability for dependent events still involves multiplying the individual probabilities, but we must take into account how prior events affect subsequent events. In the case of the candy box, we find the probability of getting two chocolates in a row by multiplying the 1/2 probability on the first draw by the 4/9 probability on the second draw:

$$\begin{aligned} P(2\text{ chocolates}) &= \underbrace{P(\text{chocolate})}_{\text{first draw}} \times \underbrace{P(\text{chocolate given chocolate on first draw})}_{\text{second draw}} \\ &= \frac{1}{2} \times \frac{4}{9} = \frac{2}{9} \end{aligned}$$

The probability of drawing two chocolates in a row is 2/9. We can generalize this example to any set of dependent events.

Think About It Suppose that instead of eating the first candy you draw from the box in Figure 7.9a, you put it back. Without doing any calculations, decide whether the probability of drawing two chocolates in a row when you replace the candy in the box is larger or smaller than the probability of drawing two chocolates in a row when you eat the first candy. Explain your reasoning.

***And* Probability: Dependent Events**

Two events are **dependent** if the occurrence of one event affects the probability of the other event. The ***and* probability** that dependent events A and B both occur is

$$P(A \text{ and } B) = P(A) \times P(B \text{ given } A)$$

where "$P(B \text{ given } A)$" means the probability of event B given the occurrence of event A. This principle can be extended to any number of dependent events. For example, the *and* probability of three dependent events A, B, and C is

$$P(A \text{ and } B \text{ and } C) = P(A) \times P(B \text{ given } A) \times P(C \text{ given } A \text{ and } B)$$

BY THE WAY

$P(B \text{ given } A)$, often written more compactly as $P(B|A)$, is called a *conditional probability*. Methods for dealing with conditional probabilities were first explored by the Reverend Thomas Bayes (1702–1761). His work also dealt with how both subjective beliefs and evidence can affect probabilities, and his theory on this has become so important—particularly in fields such as economics and law—that it is now called *Bayesian statistics* in his honor.

EXAMPLE 5 Bingo

The game of Bingo involves drawing labeled buttons from a bin at random, without replacing those drawn. There are 75 buttons, 15 for each of the letters B, I, N, G, and O. What is the probability of drawing two B buttons in the first two selections?

Solution Bingo involves dependent events because selected buttons are not replaced, so removing a button changes the contents of the bin. The probability of drawing a B on the first draw is 15/75. If this occurs, the bin has 74 buttons remaining, of which 14 are the letter B. Therefore, the probability of drawing a B button on the second draw is 14/74. The probability of drawing two B buttons in the first two selections is

$$P(\text{B and B}) = \underbrace{P(\text{B})}_{\text{first draw}} \times \underbrace{P(\text{B given B on first draw})}_{\text{second draw}} = \frac{15}{75} \times \frac{14}{74} \approx 0.0378$$

The probability of drawing a B button on the first two selections is 0.0378, or just under 4 in 100.

▶ Now try Exercises 19–20.

EXAMPLE 6 Jury Selection

A three-person jury must be selected at random from a pool that has 6 men and 6 women. What is the probability of selecting an all-male jury?

Solution Because there are 6 men in the pool of 12 people, the probability that the first juror is male is 6/12. If the first juror is male, the remaining pool has 5 men among 11 people. Therefore, the probability that the second juror is also male is 5/11. If both the first two jurors are male, the pool is left with 4 men among 10 people, making the probability of a third male juror 4/10. The combined probability is

$$\begin{aligned} P(3 \text{ men}) &= \underbrace{P(\text{man})}_{\text{juror 1}} \times \underbrace{P(\text{man given male juror 1})}_{\text{juror 2}} \times \underbrace{P(\text{man given male jurors 1 and 2})}_{\text{juror 3}} \\ &= \frac{6}{12} \times \frac{5}{11} \times \frac{4}{10} = \frac{120}{1320} \approx 0.091 \end{aligned}$$

The probability of selecting an all-male jury is about 0.09, or 9 in 100.

▶ Now try Exercises 21–22.

BY THE WAY

In 1968, the famed baby doctor Benjamin Spock (whose books on child care sold more than 50 million copies) was convicted by an all-male jury of conspiracy to encourage draft resistance during the Vietnam War, even though he and his alleged co-conspirators had never met. The defense argued that a jury with women would have been more sympathetic. The conviction was overturned on appeal, though primarily for other reasons.

Either/Or Probabilities

Suppose we want to know the probability that *either* of two events occurs, rather than the probability that both events occur. In that case, we are looking for an *either/or*

probability, such as the probability that either a #2 bus or a #9 bus will be the next to arrive at your stop. As with *and* probabilities, there are two cases to consider; we call them *overlapping* and *non-overlapping* events.

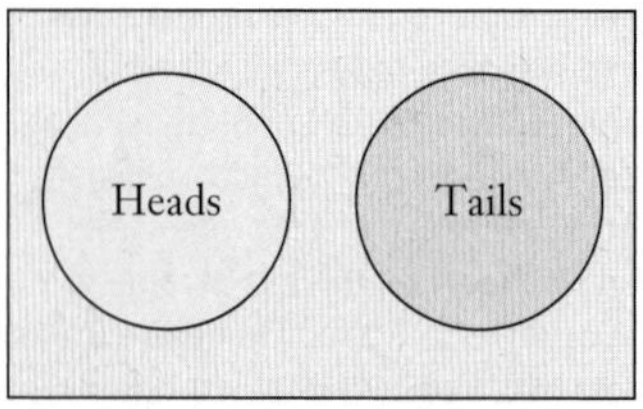

FIGURE 7.10 Venn diagram for non-overlapping events.

Non-overlapping Events

A coin can land either heads *or* tails, but it can't land on both heads *and* tails at the same time. When two events cannot occur together, they are said to be **non-overlapping** (or *mutually exclusive*). We can represent non-overlapping events with a Venn diagram (see Unit 1C) in which the circles represent the events and the circles do not overlap. For example, Figure 7.10 shows the Venn diagram for the possible events for a single coin toss. We can calculate *either/or* probabilities for non-overlapping events with the following rule.

Either/Or Probability: Non-overlapping Events

Two events are **non-overlapping** if they cannot occur together. If A and B are non-overlapping events, the probability that either A or B occurs is

$$P(A \text{ or } B) = P(A) + P(B)$$

This principle can be extended to any number of non-overlapping events. For example, the probability that either event A, event B, or event C occurs is

$$P(A \text{ or } B \text{ or } C) = P(A) + P(B) + P(C)$$

EXAMPLE 7 Either/Or Dice

Suppose you roll a single die. What is the probability of rolling either a 2 or a 3?

Solution The events of rolling a 2 and rolling a 3 are non-overlapping because a single die can land only one way. Each probability is 1/6 (because there are 6 ways for the die to land), so the combined probability is

$$P(2 \text{ or } 3) = P(2) + P(3) = \frac{1}{6} + \frac{1}{6} = \frac{2}{6} = \frac{1}{3}$$

The probability of rolling either a 2 or a 3 is 1/3. ▶ Now try Exercises 23–24.

Think About It Use the either/or rule for non-overlapping events to find the probability that a fair coin will land either heads or tails. Is the answer what you expect? Explain.

Overlapping Events

Suppose you have a standard deck of 52 cards (see Figure 7.4) and want to know the probability of drawing either a queen or a club. Because there are 4 queens in the deck, the probability of drawing a queen is 4/52. Similarly, because there are 13 clubs in the deck, the probability of drawing a club is 13/52. The sum of the two probabilities is

$$\frac{4}{52} + \frac{13}{52} = \frac{17}{52}$$

However, this is *not* the probability of drawing a queen *or* a club. The Venn diagram in Figure 7.11 shows why. One circle represents the 4 queens in the deck and the other represents the 13 clubs. The circles overlap because one card—the queen of clubs—is both a queen *and* a club. That is, the events of drawing a queen and drawing a club are **overlapping events** because both can occur together. If we simply add the individual probabilities, as above, the queen of clubs gets counted twice (once as a queen and once as a club). We must therefore correct the calculation by subtracting the double-counted probability, which is the 1/52 probability of drawing the queen of clubs (because this card is just 1 of the 52 cards in the deck). This gives us

$$P(\text{queen or club}) = P(\text{queen}) + P(\text{club}) - P(\text{queen and club})$$
$$= \frac{4}{52} + \frac{13}{52} - \frac{1}{52} = \frac{16}{52}$$

The probability of drawing a queen or a club is 16/52. You can also see this result from the Venn diagram, which shows 16 cards (out of 52 total) that are either a queen or a club (or both).

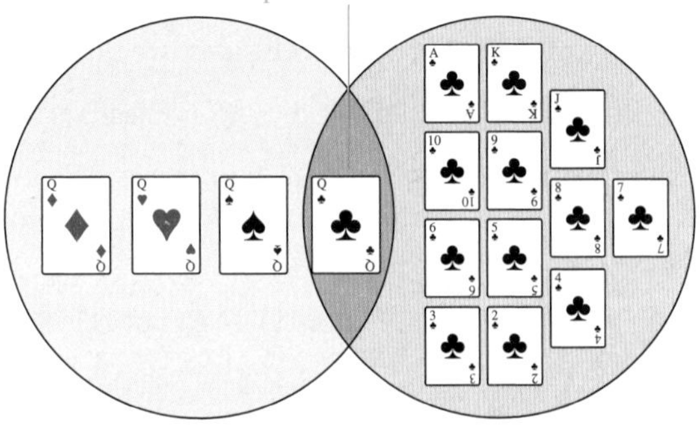

FIGURE 7.11 Venn diagram for overlapping events. One circle represents the queens in a deck of cards and the other represents the clubs. The overlap region contains the queen of clubs, which belongs in both circles.

Either/Or Probability: Overlapping Events

Two events are **overlapping** if they *can* occur together. If A and B are overlapping events, the probability that either A or B occurs is

$$P(A \text{ or } B) = P(A) + P(B) - P(A \text{ and } B)$$

EXAMPLE 8 Democrats and Women

There are eight people in a room: two Democratic men, two Republican men, two Democratic women, and two Republican women. If you select one person at random from this room, what is the probability that you will select *either* a woman or a Democrat?

The principal means for ascertaining truth—induction and analogy—are based on probabilities; so that the entire system of human knowledge is connected with the theory of probability.

—Pierre Simon, Marquis de Laplace (1819)

Solution The probability of selecting a Democrat is 1/2, because half of the people in the room are Democrats. Similarly, half of the people are women, so the probability of selecting a woman is also 1/2. We cannot simply add these probabilities, because two of the eight people in the room are both Democrats and women. The probability of choosing one of these two Democratic women is 2/8, or 1/4. Therefore, the probability of selecting either a woman or a Democrat is

$$P(\text{woman or Democrat}) = P(\text{woman}) + P(\text{Democrat}) - P(\text{woman and Democrat})$$
$$= \frac{1}{2} + \frac{1}{2} - \frac{1}{4} = \frac{3}{4}$$

The probability that you will select either a woman or a Democrat is 3/4, or 0.75.

▶ Now try Exercises 25–28.

The *At Least Once* Rule

Suppose you toss a coin four times. What is the probability of getting *at least one* head? One way to solve the problem is by recognizing that there are four different ways to get at least one head (H): by getting 1 head, 2 heads, 3 heads, or 4 heads. We can therefore treat this as an either/or probability and add the individual probabilities:

$$P(\text{at least one head in 4 tosses}) = P(1\text{ H}) + P(2\text{ H}) + P(3\text{ H}) + P(4\text{ H})$$

However, to complete this calculation, we'd first need to find the four individual probabilities in the above equation.

Fortunately, there is an easier way. Because you either *do* get at least one head or *do not* get at least one head (which means no heads) in the four coin tosses, the sum of these two probabilities is 1:

$$P(\text{at least one head}) + P(\text{no heads}) = 1$$

Subtracting the probability of no heads from both sides, we find

$$P(\text{at least one head}) = 1 - P(\text{no heads})$$

It's easy to find the probability of no heads in four tosses: We know the probability of not getting heads (which means getting tails) on one toss is $P(\text{not H}) = 1/2$, so the probability of no heads in four tosses is

$$\begin{aligned} P(\text{no heads}) &= \underbrace{P(\text{not H})}_{\text{toss 1}} \times \underbrace{P(\text{not H})}_{\text{toss 2}} \times \underbrace{P(\text{not H})}_{\text{toss 3}} \times \underbrace{P(\text{not H})}_{\text{toss 4}} \\ &= [P(\text{not H})]^4 = \left(\frac{1}{2}\right)^4 = \frac{1}{16} \end{aligned}$$

Therefore, the probability of *at least one* head in 4 coin tosses is

$$P(\text{at least one head}) = 1 - P(\text{no heads}) = 1 - \frac{1}{16} = \frac{15}{16}$$

The following box generalizes this idea.

> **The *At Least Once* Rule (for Independent Events)**
>
> Suppose the probability of an event A occurring in one trial is $P(A)$. If all trials are independent, the probability that event A occurs *at least once* in n trials is
>
> $$\begin{aligned} P(\text{at least one event } A \text{ in } n \text{ trials}) &= 1 - P(\text{no } A \text{ in } n \text{ trials}) \\ &= 1 - [P(\text{not } A)]^n \end{aligned}$$

EXAMPLE 9 **At Least One Head with Three Coins**

Use the *at least once* rule to find the probability of getting at least one head when you toss three coins.

Solution In this case, event A represents heads (H) on one toss. We apply the *at least once* rule for $n = 3$ tosses:

$$P(\text{at least one H in 3 tosses}) = 1 - P(\text{no H in 3 tosses}) = 1 - [P(\text{not H})]^3$$

The probability of getting no heads in one toss is $1/2$, so our final result is

$$\begin{aligned} P(\text{at least one H in 3 tosses}) &= 1 - [P(\text{not H in 1 toss})]^3 \\ &= 1 - \left(\frac{1}{2}\right)^3 = 1 - \frac{1}{8} = \frac{7}{8} \end{aligned}$$

The probability of getting at least one head when you toss three coins is 7/8. Notice that you can see this visually in Figure 7.5, which shows that all but one (TTT) of the 8 possible outcomes contain at least one head. ▸ Now try Exercises 29–30.

EXAMPLE 10 100-Year Flood

Find the probability that a region will experience *at least one* 100-year flood (a flood that has a 0.01 chance of occurring in any given year) during the next 100 years. Assume flood events are independent from year to year.

Solution Because there's a 0.01 probability of a 100-year flood in any one year, the probability that such a flood will *not* occur in any one year is $1 - 0.01 = 0.99$. The *at least once* rule gives us the probability of at least one of these floods in 100 years:

$$\begin{aligned} P(\text{at least one flood in 100 years}) &= 1 - [P(\text{no flood in 1 year})]^{100} \\ &= 1 - [0.99]^{100} \approx 0.634 \end{aligned}$$

The probability that a 100-year flood will occur at least once in the next 100 years is 0.634, or almost 2 out of 3. ▸ Now try Exercises 29–30.

Think About It Suppose a region has *not* experienced a 100-year flood in 100 years. Should you be surprised? Would this fact make you think the region is "due" for a flood? Explain.

EXAMPLE 11 Lottery Chances

You purchase 10 lottery tickets, for which the probability of winning some type of prize on a single ticket is 1 in 10. What is the probability that you will have at least one winning ticket among the 10 tickets?

Lottery: a tax on people who are bad at math.
—Message circulated on the Internet

Solution Because the probability of winning on any one ticket is 0.1 (and independent of winning on other tickets), the probability of *not* winning with one ticket is $1 - 0.1 = 0.9$. Therefore, the probability of winning at least once with ten tickets is

$$\begin{aligned} P(\text{at least one winner in 10 tickets}) &= 1 - [P(\text{not winning on 1 ticket})]^{10} \\ &= 1 - [0.9]^{10} \approx 0.651 \end{aligned}$$

The probability of at least one winning ticket among 10 is 0.651. This also means that the probability of all ten tickets being losers is $1 - 0.651 = 0.349$. In other words, you stand a better than 1 in 3 chance of losing on all ten tickets you bought.
▸ Now try Exercises 33–34.

HISTORICAL NOTE

As early as 3600 B.C.E., rounded bones called *astragali* were used much like dice for games of chance in the Middle East. Our familiar cubical dice appeared about 2000 B.C.E. in Egypt and in China. Playing cards were invented in China in the 10th century and brought to Europe in the 14th century.

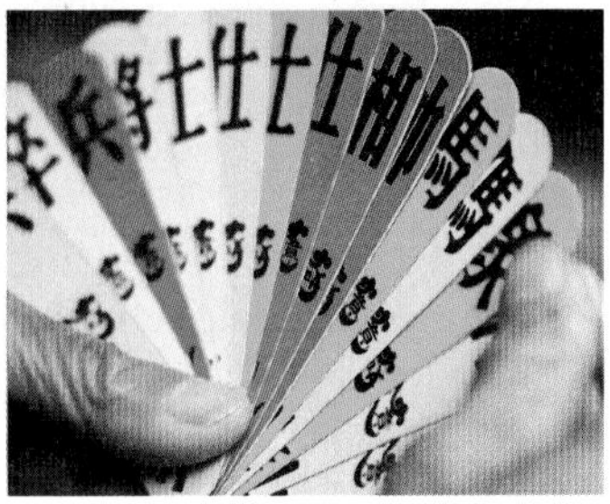

Return to the Chevalier

We now return to the story of the Chevalier de Méré. Recall that, in his first game, he bet that he could get at least one 6 in four rolls of a die. We can use the *at least once* rule to calculate his probability of winning. The probability of a 6 on any single roll is 1/6, so the probability of *not 6* is 5/6. Therefore, the probability of getting at least one 6 in four rolls is

$$\begin{aligned} P(\text{at least one 6 in 4 rolls}) &= 1 - [P(\text{not 6 in 1 roll})]^4 \\ &= 1 - \left[\frac{5}{6}\right]^4 \approx 0.518 \end{aligned}$$

The Chevalier's probability of winning in the first game was 0.518. Although this is considerably lower than the 4/6 (0.667) probability he had guessed, it is still better than even odds, so he was likely to come out ahead if he played the game many times.

In his second game, he bet that he could roll a double-6 in 24 rolls of two dice. The probability of a double-6 on any single roll is $1/36$, so the probability of *not* getting a double-6 on a single roll is $35/36$. Therefore, the probability of getting at least one double-6 in 24 rolls is

$$P(\text{at least one double-6 in 24 rolls}) = 1 - [P(\text{not double-6 in 1 roll})]^{24}$$
$$= 1 - \left[\frac{35}{36}\right]^{24} \approx 0.491$$

The Chevalier's probability of winning in his second game was 0.491, which is slightly less than even odds. Therefore, he was likely to lose more games than he won.

Think About It Comment on how the difference between a 52% and a 49% chance of winning can be the difference between rags and riches in gambling. How can casinos exploit this idea?

Quick Quiz 7B

Choose the best answer to each of the following questions. Explain your reasoning with one or more complete sentences.

1. The probability of rolling two dice and getting a double-6 is 1 in 36. Suppose you roll two dice *twice*. Which statement is *not* true?

   a. The probability of getting a double-6 is $1/36$ for each individual roll.

   b. The probability of getting a double-6 both times is $1/36 \times 1/36$.

   c. The probability of getting a double-6 on at least one of the two rolls is $2/36$.

2. The rule $P(A \text{ and } B) = P(A) \times P(B)$ holds

   a. in all cases.

   b. only if it is possible for both *A* and *B* to occur together (simultaneously).

   c. only if an outcome of *A* on one trial does not affect the probability of an outcome of *B* on the next.

3. In which of the following cases are the events *dependent*?

   a. the probability of an Olympic diver scoring 8 or above on two consecutive dives

   b. the probability of selecting a red M&M from a bag, eating it, and then selecting another red M&M

   c. the probability of two people in your class having a birth date of October 31

4. A box of candy contains five dark chocolates and five white chocolates. If you pick randomly and eat each candy after choosing it, what is the probability of choosing three dark chocolates in a row?

   a. $1/2 \times 1/2 \times 1/2$ b. $5/10 \times 4/10 \times 3/10$

   c. $1/2 \times 4/9 \times 3/8$

5. The events of being born on a Monday and being born in June are

   a. overlapping. b. mutually exclusive.

   c. independent.

6. You roll two dice. Based on the probabilities shown in Table 7.4, which of the following has a probability *greater* than 0.5?

   a. rolling a sum of 2 or 3 or 4 or 5

   b. rolling a sum of 2 or 3 or 4 or 10 or 11 or 12

   c. rolling a sum of 5 or 6 or 7 or 8

7. You roll two dice twice. Based on the probabilities shown in Table 7.4, what is the probability that you'll get a sum of 3 on the first roll and a sum of 4 on the second roll?

   a. $2/36 \times 3/36$ b. $2/36 + 3/36$

   c. $(2/36 \times 3/36)^2$

8. You toss two coins 10 times, and you want to know the probability of getting two heads at least once in the 10 trials. To get the answer in the easiest possible way, what is the first thing you should calculate?

   a. the probability of getting two heads exactly once during the 10 trials

   b. the probability of *not* getting two heads in a single trial

   c. the probability of getting two heads on both of the first two trials

9. You purchase 10 lottery tickets for which the probability of winning some prize on a single ticket is 1 in 50 (or 0.02). Your probability of having at least one winner among your 10 tickets is

   a. 10 in 50, or 1 in 5. b. 0.02^{10}.

   c. $1 - 0.98^{10}$.

10. One in 10 people on campus has blond hair. In 20 random encounters, what is the probability of meeting at least one blond-haired person?

    a. $1 - 0.1^{20}$

    b. $1 - 0.9^{20}$

    c. 0.9^{20}

Exercises 7B

REVIEW QUESTIONS

1. How did the gambling habits of the Chevalier de Méré help launch the mathematical study of probability?
2. Give an example in which we would be interested in an *and* probability. How do we determine whether the events are independent or dependent? Give examples of each case, and explain how we calculate the probabilities.
3. Give an example in which we would be interested in an *either/or* probability. How do we determine whether the events are overlapping or non-overlapping? Give examples of each case, and explain how we calculate the probabilities.
4. What is the *at least once* rule? Explain how the *at least once* rule can be used to find the correct probabilities in the games of the Chevalier de Méré.

DOES IT MAKE SENSE?

Decide whether each of the following statements makes sense (or is clearly true) or does not make sense (or is clearly false). Explain your reasoning.

5. The probability of getting heads *and* tails when you toss a coin is 0, but the probability of getting heads *or* tails is 1.
6. If you toss a coin and get heads three times in a row, you're due to get tails on the next toss.
7. The probability of drawing an ace or a spade from a deck of cards is the same as the probability of drawing the ace of spades.
8. I can't believe you chose the lottery numbers 1-2-3-4-5-6. Getting six numbers in a row is much less likely than getting other random numbers.
9. My chance of getting a 5 on a roll of one die is $1/6$, so my chance of getting at least one 5 when I roll six dice is $6/6 = 1$
10. To find the probability that at least one of my 25 lottery tickets is a winner, I calculated the probability that none of my tickets is a winner and subtracted it from 1.

BASIC SKILLS & CONCEPTS

11. **Chevalier's Logic with Coins.** The Chevalier de Méré's faulty logic would suggest that since the probability of heads on a coin toss is $1/2$, the probability of at least one head in two coin tosses is $2 \times 1/2 = 2/2 = 1$. Find two coins, toss them together, and record your results as either 0 heads, 1 head, or 2 heads. Repeat this experiment 10 times (or more if needed to get a toss with no heads), recording your result each time. What fraction of the time did you find at least one head? How does this show the flaw in the Chevalier's logic?
12. **Chevalier's Logic with Two Die.** The Chevalier de Méré's faulty logic led him to conclude that the probability of getting a double-6 within 24 rolls of two 6-sided dice is 0.67. Find two dice and roll them until you roll a double-6; record the number of rolls it took to get the double-6. Repeat this experiment at least 10 times. What fraction of the time did you roll a double-6 within 24 rolls? Based on your results, what can you conclude about the Chevalier's reasoning?

13–22: *And* Probabilities. Determine whether the events described in each exercise are independent or dependent. Then find the *and* probability of the events.

13. Getting five heads when you toss five (fair) coins simultaneously
14. Tossing a coin four times in a row and getting HTHT, in that order
15. Rolling a fair die three time and getting in order 1, 2, 3
16. The next five births at a hospital all being boys
17. Discovering that your three best friends were all born on a Sunday
18. Drawing three aces in a row from a standard deck of cards when the drawn card is returned to the deck each time
19. Randomly drawing and immediately eating three red M&Ms in a row from a bag that contains 10 red M&Ms out of 30 M&Ms total
20. Selecting a pair of red socks on three successive days from a drawer that originally has 5 pairs of red socks and 10 pairs of black socks
21. Selecting an all-male five-person jury from a pool of 10 men and 10 women
22. Randomly selecting an all-American four-person committee from a pool of 8 Canadians and 12 Americans

23–28: *Either/Or* Probabilities. Determine whether the events described in each exercise are overlapping or non-overlapping. Then find the *either/or* probability of the events.

23. Getting a sum of either 2, 3, 4, or 5 on a roll of two dice
24. Getting a sum of either 5 or 9 on a roll of two dice
25. Drawing either a black ace or a red king on one draw from a regular deck of cards
26. Drawing either a 10 or a heart from a regular deck of cards
27. Randomly choosing a black sock or a small sock from a drawer in which half of the socks are black and half are white, and half of each color are small and half are large.
28. Randomly meeting a four-child family with either exactly one or exactly two boy children

29–34: *At Least Once* Problems. Use the *at least once* rule to find the probabilities of the following events.

29. Getting at least one head when tossing three fair coins
30. Getting at least one tail when tossing five fair coins
31. Getting rain at least once in three days when the probability of rain on each single day is 0.3

32. Getting at least one 50-year flood in the next ten years
33. Purchasing 15 lottery tickets and having at least one winner when the probability of winning is 0.01 on a single ticket
34. Rolling at least one 6 in three rolls of a fair die

FURTHER APPLICATIONS

35–55: Assorted Probabilities. Use the method of your choice to determine the following probabilities.

35. Randomly choosing either a standard red die or a standard green die, rolling the die once, and rolling a green even number
36. Randomly choosing either a standard red die or a standard green die, rolling the die once, and rolling a green number or an even number
37. Getting at least one even number in five rolls of a single die
38. Spinning two winners in a row with a wheel of fortune on which the winner is one of 36 equally likely outcomes
39. Drawing four spades in a row from a standard deck of cards when the drawn card is not returned to the deck each time
40. Being dealt three black cards off the top of a standard deck of well-shuffled cards
41. Drawing either a face card (jack, queen, or king) or a club from a standard deck of cards
42. Drawing either a 6, 7, or 8 from a standard deck of cards
43. Getting a green light at a busy intersection at least once in five times through the intersection, given that the light in your direction is green 4/10 of the time
44. Selecting at random a blue long-sleeved shirt from a closet in which one-third of the shirts are short-sleeved and two-thirds are long-sleeved, and half of the long-sleeved shirts are blue
45. Drawing at least one ace when you draw a card from a standard deck 6 times (replacing the card each time you draw)
46. On a roulette wheel on which there are 38 equally likely slots, 18 red, 18 black, and 2 green, spinning three green numbers in a row
47. Purchasing five winning lottery tickets in a row when each ticket has a 1 in 8 chance of being a winner
48. Four rainy days in a row when the forecast calls for a "40% chance of rain" each day
49. Meeting at least one left-handed person in eight random encounters on campus when the incidence rate of left-handedness is 11% (11 in 100 people are left-handed)
50. Getting at least one parking ticket among four occasions when you do not pay the parking meter, when the chance of getting a ticket when not paying is 0.15
51. Randomly meeting either a woman or a Republican in a group composed of 30 Democratic men, 20 Republican men, 50 Democratic women, and 60 Republican women
52. Randomly meeting either a man or an American in a group composed of 25 French women, 15 French men, 30 American women, and 20 American men
53. Randomly meeting a three-child family with either all boy or all girl children
54. Meeting at least one person with flu in 8 random encounters on campus when the infection rate is 5% (5 in 100 people have flu)
55. Getting at least one false test for strep throat in four trials when the probability of a false report is 1 in 50 (0.02)
56. **How Many Rolls?** At least how many times do you have to roll a fair die to be sure that the probability of rolling at least one 6 is greater than 9 in 10 (0.9)?
57. **Probability and Court Cases.** The data in the following table show the outcomes of guilty and not-guilty pleas in 1028 criminal court cases.

	Guilty plea	Not-guilty plea
Sent to prison	392	58
Not sent to prison	564	14

Source: Brereton and Casper, *Law and Society Review*, Vol. 16, No. 7.

a. What is the probability that a randomly selected defendant either pled guilty or was sent to prison?

b. What is the probability that a randomly selected defendant either pled not guilty or was not sent to prison?

58. **Testing a Drug.** A new cold medication was tested by giving 100 people the drug and 100 people a placebo. A control group consisted of 100 people who were given no treatment. The number of people in each group who showed improvement is shown in the table below.

	Cold drug	Placebo	Control	Total
Improvement	70	55	20	145
No improvement	30	45	80	155
Total	100	100	100	300

a. What is the probability that a randomly selected person in the study either was given the placebo or was in the control group?

b. What is the probability that a randomly selected person improved?

c. What is the probability that a randomly selected person was given the drug and improved?

d. What is the probability that a randomly selected person who improved was given the drug?

e. What is the probability that a randomly selected person who was given the drug improved?

f. Based on these data, does the drug appear to be effective? Explain.

59. **Polling Calls.** A telephone pollster has a list of names and telephone numbers for 45 voters, 20 of whom are listed as registered Democrats and 25 of whom are listed as registered Republicans. Calls are made in random order. Suppose you want to find the probability that the first two calls are to Republicans.

a. Are the two calls independent or dependent events? Explain.

b. If you treat them as dependent events, what is the probability that the first two calls are to Republicans?

c. If you treat them as independent events, what is the probability that the first two calls are to Republicans?

d. Compare the results of parts (b) and (c).

60. **Dominant and Recessive Genes.** Many traits are controlled by a dominant gene, *A*, and a recessive gene, *a*. A child gets two genes, one from each parent. Suppose a child's parents each have the gene combination *Aa*; that is, each parent is equally likely to pass either the *A* or the *a* gene to the child. Make a table showing the probability distribution for the child's possible gene combinations, which are *AA*, *Aa*, and *aa*. If the combinations *AA* and *Aa* both result in the same dominant trait (say, brown hair) and *aa* results in the recessive trait (say, blond hair), what is the probability that a child will have the dominant trait? What is the probability that the child will have the recessive trait?

61. **Better Bet for the Chevalier.** Suppose that the Chevalier de Méré had bet that he could roll a double-6 in 25 rolls, rather than 24. In that case, what would have been his probability of winning? Had he made this bet, would he still have lost over time? Explain.

62. **Lottery Odds.** The probability of a \$2 winner in a particular state lottery is 1 in 20, the probability of a \$5 winner is 1 in 50, and the probability of a \$10 winner is 1 in 200.

a. What is the probability of getting either a \$2, \$5, or \$10 winner? Compare this to the probability of getting only a \$2 winner.

b. If you buy 50 lottery tickets, what is the probability that you will get at least one \$5 winner?

c. If you buy 100 lottery tickets, what is the probability that you will get at least one \$10 winner?

63. **Miami Hurricanes.** Studies of the Florida Everglades show that, historically, the Miami region is hit by a hurricane about every 20 years.

a. Based on the historical record, what is the empirical probability that Miami will be hit by a hurricane next year?

b. What is the probability that Miami will be hit by hurricanes in two consecutive years?

c. What is the probability that Miami will be hit by at least one hurricane in the next 10 years?

64. **Field Goals.** Jake is a college football placekicker who successfully kicks two-thirds of field goals that are less than 30 yards, one-third of field goals that are between 30 and 45 yards, and no field goals longer than 45 yards. In a single game, Jake attempts three field goals of lengths 20, 25, and 35 yards.

a. What is the probability that all the field goals are successful?

b. What is the probability that at least one field goal is successful?

c. What is the probability that exactly one field goal is successful?

d. In a single game, Jake attempts three field goals of lengths 20, 25, and 55 yards. What is the probability that all the field goals are successful?

65. **One-and-One Free Throws.** A basketball player is fouled in a one-and-one situation, meaning she is awarded two free throws only if she makes the first free throw. Her season average for free throws is 0.7 (70%). Using her season average as a relative frequency probability, find:

a. the probability that she will miss her first free throw.

b. the probability that she will make her first free throw and miss her second.

c. the probability that she will make both free throws. Which outcome is most likely?

IN YOUR WORLD

66. **Lottery Chances.** Find a lottery website, and study the probabilities of winning. Based on your findings, what is the probability of winning at least once if you play 10 times? Explain any assumptions you make.

67. **Combined Probability in the News.** Find a recent news report in which two or more probabilities were combined in some way. Describe how and why they were combined.

68. **Combined Probability in Your Life.** Cite a recent situation or decision in your life that involved an *and* probability or an *either/or* probability. How did the probability affect the situation or decision?

UNIT 7C The Law of Large Numbers

If you toss a coin once, you cannot predict exactly how it will land; you can only state that the probability of a head is 0.5. If you toss the coin 100 times, you still cannot predict precisely how many heads will occur. However, you can reasonably expect to get heads *close to* 50% of the time (see Figure 6.20). If you toss the coin 1000 times, you can expect the proportion of heads to be even closer to 50%. In general, the more times you toss the coin, the closer the percentage of heads will be to exactly 50%. The idea that large numbers of events may show some pattern even when individual events are unpredictable is called the **law of large numbers** (or the *law of averages*).

The Law of Large Numbers

Consider an event A with probability $P(A)$ in a single trial. The **law of large numbers** holds that:

- For a large number of trials, the proportion in which event A occurs will be close to the probability $P(A)$.
- The larger the number of trials, the closer the proportion should be to $P(A)$.

This law holds as long as each trial is independent of prior trials, so that an individual trial always has the same probability, $P(A)$.

We can illustrate the law of large numbers with a die-rolling experiment. The probability of a 1 on a single roll is $P(1) = 1/6 \approx 0.167$. To avoid the tedium of rolling the die many times, we can use a computer to *simulate* random rolls of the die. Figure 7.12 shows a computer simulation of rolling a single die 5000 times. The horizontal axis gives the number of rolls, and the height of the curve at any point gives the proportion of 1s up to that point. Although the curve bounces around when the number of rolls is small, the proportion of 1s approaches the probability 0.167, in agreement with the law of large numbers.

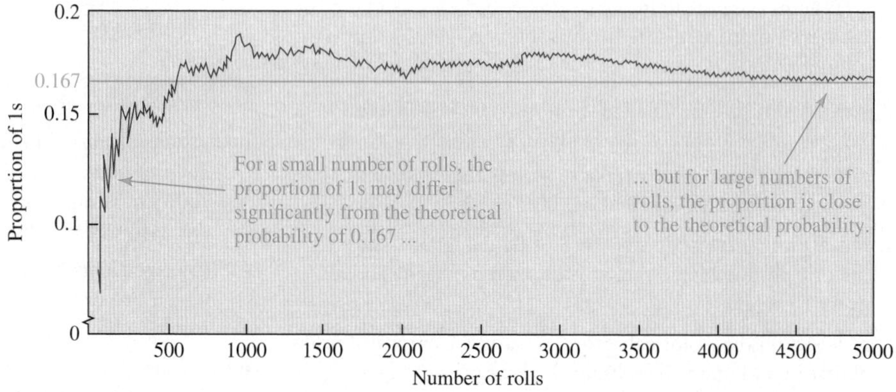

FIGURE 7.12 Results of computer simulation of rolling a die. As the number of rolls increases, the proportion of 1s gets close to the $1/6 \approx 0.167$ probability of getting a 1 on a single roll.

Think About It Suppose you did another computer simulation of 5000 coin tosses and made a graph like that in Figure 7.12. In what ways might you expect the new graph to differ from Figure 7.12? In what ways would you expect it to look similar? Explain.

EXAMPLE 1 Roulette

As shown at the left, a roulette wheel has 38 numbers: 18 black numbers, 18 red numbers, and the numbers 0 and 00 in green. Assume that all possible outcomes (the 38 numbers) have equal probability.

a. What is the probability of getting a red number on any spin?
b. If patrons in a casino spin the wheel 100,000 times, about how many times will a red number be the outcome?

Solution

a. The theoretical probability of getting a red number on any spin is

$$P(A) = \frac{\text{number of ways red can occur}}{\text{total number of outcomes}} = \frac{18}{38} = 0.474$$

b. The law of large numbers tells us that as the game is played more and more times, the proportion of times that the wheel shows a red number should get closer to 0.474. In 100,000 tries, the result should be a red number close to 47.4% of the time, or about 47,400 times.

▶ Now try Exercises 13–14.

Expected Value

Suppose the InsureAll Company sells a special type of insurance that pays the policyholder \$100,000 in the event that he or she has to quit a job because of serious illness. Based on data from past claims, the relative frequency probability that a policyholder will make such a claim is 1 in 500. Should the insurance company expect to earn a profit if it sells the policies for \$250 each?

If InsureAll sells only a few policies, the profit or loss is unpredictable. For example, selling 100 policies for \$250 each generates revenue of $100 \times \$250 = \$25{,}000$. If none of the 100 policyholders files a claim, InsureAll makes a significant profit. However, if InsureAll must pay a \$100,000 claim to even one policyholder, it will face a huge loss.

In contrast, if InsureAll sells a large number of policies, the law of large numbers tells us that the proportion of policies for which claims must be paid should be very close to the 1 in 500 probability for a single policy. For example, if the company sells 1 million policies, it should expect that the number of policyholders making the \$100,000 claim will be close to

$$\underbrace{1{,}000{,}000}_{\text{number of policies}} \times \underbrace{\frac{1}{500}}_{\text{probability of \$100,000 claim}} = 2000$$

Paying these 2000 claims will cost

$$2000 \times \$100{,}000 = \$200 \text{ million}$$

This cost is an *average* of \$200 for each of the 1 million policies, which means that if the policies sell for \$250 each, the company should expect to earn an average of $\$250 - \$200 = \$50$ per policy. We call this average the **expected value** for each policy; note that it is "expected" only if the company sells a large number of policies.

We can find the same expected value with a more formal procedure. The insurance example involves two distinct events, each with a particular *probability* and *value* for the company:

1. In the event that a person buys a policy, the value to the company is the \$250 price of the policy. The probability of this event is 1 because everyone who buys a policy pays the \$250.
2. In the event that a person is paid for a claim, the value to the company is $-\$100{,}000$; the value is negative because the company loses money. The probability of this event is $1/500$.

HISTORICAL NOTE

The insurance industry's roots lie in birth and death data compiled by John Graunt in London in the 1660s. The data were improved by Edmund Halley (for whom Halley's Comet is named), and they led to a thriving insurance industry. In 1687, Edward Lloyd opened a coffee house in London that specialized in insuring almost all risks, including maritime travel and trade. Lloyd's of London remains a major insurer today.

Technical Note

There are alternative ways to calculate expected value. In the insurance example, we could instead define event 1 as a policy with no claim and event 2 as a policy on which the company pays a $100,000 claim. Event 1 has a value to the company of $250 (the sales price of a policy) and a probability of 499/500. Event 2 has a value to the company of $250 (the policy price) *minus* the $100,000 paid out with a claim, or −$99,750, and a probability of 1/500. Notice that this method gives the same expected value of $50:

$$\left(\$250 \times \frac{499}{500}\right) + \left(-\$99{,}750 \times \frac{1}{500}\right) = \$50$$

Some statisticians prefer this alternative method because the sum of the event probabilities is 1.

We now multiply the value of each event by its probability and add the results to find the expected value of each insurance policy:

$$\text{expected value} = \underbrace{\$250}_{\substack{\text{value of}\\ \text{policy sale}}} \times \underbrace{1}_{\substack{\text{probability of}\\ \text{earning \$250 on sale}}} + \underbrace{(-\$100{,}000)}_{\substack{\text{value of}\\ \text{claim}}} \times \underbrace{\frac{1}{500}}_{\substack{\text{probability of}\\ \text{paying claim}}}$$

$$= \$250 - \$200 = \$50$$

This expected profit of $50 per policy is the same answer we found earlier. Note that it amounts to a profit of $50 million on sales of 1 million policies.

Think About It Should InsureAll expect a profit of $50,000 on 1000 policies? Explain.

Expected Value

Consider two events, each with its own value and probability. The **expected value** based on these two events is

$$\text{expected value} = \left(\begin{matrix}\text{value of}\\ \text{event 1}\end{matrix}\right) \times \left(\begin{matrix}\text{probability}\\ \text{of event 1}\end{matrix}\right) + \left(\begin{matrix}\text{value of}\\ \text{event 2}\end{matrix}\right) \times \left(\begin{matrix}\text{probability}\\ \text{of event 2}\end{matrix}\right)$$

This formula can be extended to any number of events by including more terms in the sum.

EXAMPLE 2 Lottery Expectations

Suppose that $1 lottery tickets have the following probabilities and values: 1 in 5 to win a free ticket (worth $1), 1 in 100 to win $5, 1 in 100,000 to win $1000, and 1 in 10 million to win $1 million. What is the expected value of a lottery ticket? Discuss the implications. (Note: Winners do *not* get back the $1 they spend on the ticket.)

Solution An easy way to proceed is to make a table like that shown below, listing all the relevant events with their values and probabilities. We are calculating the expected value of a lottery ticket for a person who buys one ticket; the ticket price therefore has a negative value because it costs that person money, while the values of winnings are positive.

Event	*Value*	*Probability*	*Value × Probability*
Ticket purchase	−$1	1	$-\$1 \times 1 = -\$1$
Win free ticket	$1	$\frac{1}{5}$	$\$1 \times \frac{1}{5} = \$0.20$
Win $5	$5	$\frac{1}{100}$	$\$5 \times \frac{1}{100} = \$0.05$
Win $1000	$1000	$\frac{1}{100{,}000}$	$\$1000 \times \frac{1}{100{,}000} = \$0.01$
Win $1 million	$1,000,000	$\frac{1}{10{,}000{,}000}$	$\$1{,}000{,}000 \times \frac{1}{10{,}000{,}000} = \$0.10$
			Sum: − $0.64

The expected value is the sum of all the products Value × Probability, which the final column of the table shows to be −$0.64. In other words, averaged over many tickets, a lottery player should expect to lose 64¢ for each $1 lottery ticket. If a person buys, say, 1000 tickets, he or she should expect to *lose* about 1000 × $0.64 = $640.

▶ Now try Exercises 15–18.

EXAMPLE 3 Art Auction

You are at an art auction, trying to decide whether to bid \$50,000 for a particular painting. You believe that you have a 1/2 probability of reselling the painting to a client in New York for \$70,000 and a 1/4 probability of reselling the painting to a client in San Francisco for \$80,000. Otherwise, you will have to keep the painting—an option that has benefits, but not monetary ones. What is the expected value of a \$50,000 bid?

Solution There are three relevant events to consider:

Event 1. With probability 1, you will pay \$50,000 to buy the painting. This option represents a monetary loss to you, so it has a negative value.

Event 2. With probability 1/2, you will sell the painting for \$70,000. This option is a gain, so it has a positive value.

Event 3. With probability 1/4, you will sell the painting for \$80,000. This option is also a gain, so it has a positive value.

Therefore, the expected value of a \$50,000 bid is

$$\begin{aligned}\text{expected value} &= \begin{pmatrix}\text{value of}\\ \text{event 1}\end{pmatrix} \times \begin{pmatrix}\text{prob. of}\\ \text{event 1}\end{pmatrix} + \begin{pmatrix}\text{value of}\\ \text{event 2}\end{pmatrix} \times \begin{pmatrix}\text{prob. of}\\ \text{event 2}\end{pmatrix}\\ &\quad + \begin{pmatrix}\text{value of}\\ \text{event 3}\end{pmatrix} \times \begin{pmatrix}\text{prob. of}\\ \text{event 3}\end{pmatrix}\\ &= (-\$50{,}000 \times 1) + \left(\$70{,}000 \times \frac{1}{2}\right) + \left(\$80{,}000 \times \frac{1}{4}\right)\\ &= -\$50{,}000 + \$35{,}000 + \$20{,}000 = \$5000\end{aligned}$$

With a \$50,000 bid, you could expect to gain \$5000 by the purchase.

▲ **CAUTION!** For this example, remember that expected value refers to expected gain or loss over *many* such purchases, so you should *not* expect to earn \$5000 in this single case. ▲

▶ Now try Exercises 19–22.

BY THE WAY

Edvard Munch's 1895 painting *The Scream* sold at auction in 2012 for \$120 million, making it the third-most expensive painting ever sold at auction, as of 2017.

The Scream (1893), Edvard Munch. Munch Museum, Oslo. © 2013 ARS, NY

The Gambler's Fallacy

Consider a simple coin toss game in which you win \$1 if the coin lands heads and lose \$1 if it lands tails. Suppose you toss the coin 100 times and get 45 heads and 55 tails, putting you \$10 in the hole. Are you "due" for a streak of better luck?

You probably recognize that the answer is *no*. Each coin toss is independent of previous tosses, so your past bad luck has no bearing on your future chances. However, many gamblers—especially compulsive gamblers—guess just the opposite. They believe that when their luck has been bad, it's due for a change. This mistaken belief is often called the **gambler's fallacy** (or the *gambler's ruin*).

Definition

The **gambler's fallacy** is the mistaken belief that a streak of bad luck makes a person "due" for a streak of good luck (or that a streak of good luck will continue).

The value of our expectations always signifies something in the middle between the best we can hope for and the worst we can fear.
—Jacob Bernoulli, 17th-century mathematician

One reason people succumb to the gambler's fallacy is a misunderstanding of the law of large numbers. In the coin toss game, the law of large numbers tells us that the proportion of heads tends to be closer to 50% for larger numbers of games. But this does *not* mean that you are likely to recover early losses. The following example shows why.

HISTORICAL NOTE

Russian writer Fyodor Dostoyevsky was also a compulsive gambler who played at roulette throughout Europe in the 19th century. He once reported that he "won 10,000 francs at first, . . . but I got carried away and dropped half my winnings." His gambling experience formed the basis of his novel *The Gambler*.

EXAMPLE 4 Continued Losses

You are playing the coin toss game, in which you win \$1 for heads and lose \$1 for tails. After 100 tosses, you are \$10 in the hole because you have 45 heads and 55 tails. You continue playing until you've tossed the coin 1000 times, at which point you've gotten 480 heads and 520 tails. Does the result agree with the law of large numbers? Have you gained back any of your losses? Explain.

Solution The proportion of heads in your first 100 tosses was 45%. After 1000 tosses, the proportion of heads has increased to 480 out of 1000, or 48%, which agrees with the law of large numbers because the proportion is now closer to 50%. However, after 1000 tosses, you've won \$480 (for the 480 heads) while losing \$520 (for the 520 tails), for a net loss of \$40. In other words, your losses have actually *increased* from \$10 to \$40, despite the fact that the proportion of heads grew closer to 50%.

▶ Now try Exercises 23–26.

Streaks

Another aspect of the gambler's fallacy involves expectations about streaks. Suppose you toss a coin six times and see the outcomes HHHHHH (all heads). Then you toss it six more times and see the outcomes HTTHTH. Most people would say that the latter outcome is "natural" while the streak of all heads is surprising. But, in fact, both outcomes are equally likely. The total number of possible outcomes for six coins is $2 \times 2 \times 2 \times 2 \times 2 \times 2 = 64$ so every individual outcome has the same probability of $1/64$.

Moreover, suppose you just tossed six heads and had to bet on the outcome of the next toss. You might think that, given the run of heads, a tail is "due" on the next toss. But the probability of a head or a tail on the next toss is still 0.50; the coin has no memory of previous tosses.

BY THE WAY

The craps table in a casino looks quite elaborate, but the only thing that counts in the game itself is the result from each roll of the dice.

EXAMPLE 5 Hot Hand at the Craps Table

The popular casino game of craps involves rolling dice. Suppose you are playing craps and suddenly find yourself with a "hot hand" in which you roll a winner on ten consecutive bets. Is your hand really "hot"? Should you increase your bet because you are on a hot streak? Assume that there is a 0.486 probability of winning on a single play (the best odds available in craps).

Solution First, let's find the probability of winning ten times in a row. With a probability of 0.486 on a single play, the probability of ten straight wins is

$$(0.486)^{10} \approx 0.000735$$

Your streak of ten straight wins has a probability of only about 7 in 10,000, which might make you think that your hand really is "hot." However, look around the casino. If it is a large casino, several hundred people may be playing at the craps tables, and there may be *tens of thousands* of individual rolls each night. It is almost inevitable that someone will have a "hot streak" of ten straight wins. Indeed, even longer hot streaks are likely during any given night. Your apparent "hot hand" is a mere coincidence and your probability of winning on your next bet is still only 0.486. Increasing your bet based on a mere coincidence would be foolish.

▶ Now try Exercises 27–28.

EXAMPLE 6 Planning for Rain

A farmer knows that at this time of year in his part of the country, the probability of rain on a given day is 0.5. It hasn't rained in 10 days, and he needs to decide whether to start irrigating. Is he justified in postponing irrigation because his region is due for a rainy day?

Solution The 10-day dry spell is unexpected, and, like a gambler, the farmer is having a "losing streak." However, if we assume that weather events are independent from one day to the next (which is often but not always the case), then the probability of rain is still 0.5 on any given day, so the region is not "due" for rain. ▶ Now try Exercises 29–30.

Luck is a mighty queer thing. All you know about it for certain is that it's bound to change.

—Bret Harte, "The Outcasts of Poker Flat"

The House Edge

A casino makes money because games are set up so that the expected earnings of patrons are negative (losses). Because the casino earns whatever patrons lose, the casino's earnings are positive. The amount that the casino, or *house*, can expect to earn per dollar bet is called the **house edge**. That is, the house edge is the expected value *to the casino* of a particular bet.

Definition

Gambling houses (casinos) generally have games set up so that the house always has a higher probability of winning than patrons. For any particular game, the **house edge** is the expected value *to the house* of each individual bet.

The house edge varies from game to game. It tends to be greatest in games where big winnings are possible, such as slot machines. It tends to be least in games where strategy can improve a patron's odds, such as blackjack. But the house edge is always present, and gamblers who believe they can beat the house regularly are sadly succumbing to the gambler's fallacy.

EXAMPLE 7 The House Edge in Roulette

The game of roulette is usually set up so that betting on *red* is a 1 to 1 bet. That is, you win the same amount of money as you bet if any red number comes up. Betting on a single number is a 35 to 1 bet. That is, you win 35 times as much as you bet if your single number comes up. Find the house edge and the amount of money the casino should expect to earn if patrons wager a total of \$1 million for (a) a bet on red; and (b) a bet on a single number.

Solution

a. The house edge is the casino's expected earnings on the bet. The probability of a red number is 18/38 (see Example 1), so the probability of *not red* is 20/38. If a patron bets \$1 on red, the casino pays \$1 if the wheel stops on red and earns \$1 if it does not, so the value for the casino of *red* is −\$1 and the value of *not red* is \$1. Therefore, the house edge when patrons bet on red is

$$\left(\underbrace{-\$1}_{\text{value to casino of red}} \times \underbrace{\frac{18}{38}}_{\text{probability of red}}\right) + \left(\underbrace{\$1}_{\text{value to casino of not red}} \times \underbrace{\frac{20}{38}}_{\text{probability of not red}}\right) \approx \$0.053$$

The casino can expect to gain \$0.053 per dollar gambled on red. If patrons bet \$1 million on red, the casino can expect to earn about \$1 million × 0.053 = \$53,000.

b. The probability of winning a bet on a single number is 1/38 because there are 38 numbers on the roulette wheel. The 35 to 1 payoff offered by the casino on this bet means the casino pays the patron 35 times the amount of the bet, so its value to the casino is −\$35. The probability that the bet does not win is 37/38. The casino gains

BY THE WAY

The French mathematician and philosopher Blaise Pascal (1623–1662) used expected value to argue for belief in God. He used *p* to represent the probability that God exists and assigned values to the consequences of belief and nonbelief. He claimed that if God exists, the consequence of nonbelief would be infinitely bad (negative). He therefore concluded that no matter how small *p* might be, the expected value of belief in God is positive and the expected value of nonbelief is infinitely negative. Of course, other philosophers have disputed Pascal's analysis.

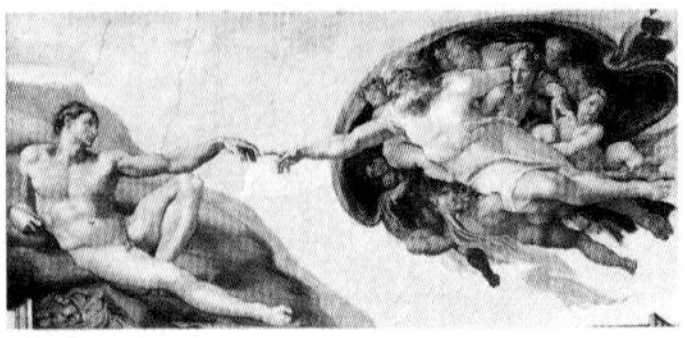

the patron's bet in this case, so its value is \$1. The house edge when patrons bet on a single number is

$$\underbrace{\left(-\$35\right.}_{\text{value to casino of hitting the number}} \times \underbrace{\left.\frac{1}{38}\right)}_{\text{probability of hitting the number}} + \underbrace{\left(\$1\right.}_{\text{value to casino of not hitting the number}} \times \underbrace{\left.\frac{37}{38}\right)}_{\text{probability of not hitting the number}} \approx \$0.053$$

This is the same house edge as for the bet on red. Again, if patrons bet \$1 million on single numbers, the house can expect to earn \$53,000.

▶ Now try Exercises 31–32.

Think About It Given the house edge, is it possible for a casino to lose money on its gambling results? Explain.

Quick Quiz 7C

Choose the best answer to each of the following questions. Explain your reasoning with one or more complete sentences.

1. Suppose that the probability of a hurricane striking Florida in any single year is 1 in 10 and that this probability has been the same for the past 100 years. Which of the following is implied by the law of large numbers?

   a. If no hurricanes have hit Florida in the past 10 years, then the probability of a hurricane hitting next year is greater than 0.1.

   b. Florida has been hit by close to (but not necessarily exactly) 10 hurricanes in the past 100 years.

   c. Florida has been hit by exactly 10 hurricanes in the past 100 years.

2. Consider a lottery with 100 million tickets in which each ticket has a unique number. Each ticket is sold for \$1, and one ticket is drawn for a single prize of \$75 million (and no other prizes). The expected value of a single ticket is

   a. \$1. b. \$75. c. −\$0.25.

3. Consider the lottery described in Question 2. If you were to spend \$1 million to purchase 1 million lottery tickets, the most likely result would be that

   a. you would be the lottery winner.

   b. you would win back \$750,000 of your \$1 million.

   c. you would lose your entire \$1 million.

4. You are betting on a game in which each bet has an expected value of −\$0.40 This means that

   a. you will win \$0.40 every time you play.

   b. you will lose \$0.40 every time you play.

   c. if you play the game many times, on average you will lose about \$0.40 per game.

5. An insurance company knows that the average cost to build a home in a new California subdivision is \$100,000 and that in any particular year there is a 1 in 50 chance of a wildfire destroying all the homes in the subdivision. Based on these data and assuming the insurance company wants a positive expected value when it sells policies, what is the minimum that the company must charge for fire insurance policies in this subdivision?

   a. \$50 per year b. \$2000 per year

   c. \$100,000 per year

6. You know a shortcut to work that uses side streets instead of the freeway. Most of the time, the shortcut takes 5 minutes off the driving time by freeway. However, about once in every ten trips, an accident blocks traffic on the shortcut, with the result that the trip takes 20 minutes longer than it would by freeway. In terms of the amount of time the shortcut saves, what is the expected value of the shortcut?

   a. $(5 \text{ minutes} \times 0.9) + (-20 \text{ minutes} \times 0.1)$

   b. $(5 \text{ minutes} \times 0.1) + (-25 \text{ minutes} \times 0.1)$

   c. 5 minutes

7. Cameron is betting on a game in which the probability of winning is 1 in 4. He's lost ten games in a row, so he decides to double his bet on the eleventh game. This strategy shows

   a. good logic, because he's due for a win.

   b. good logic, because he's more likely to win when the bet is larger.

   c. poor logic, as he has a 75% chance of losing the double bet.

8. Cameron is betting on a game in which the probability of winning is 1 in 4. He's *won* ten games in a row, so he decides to double his bet on the eleventh game. This strategy shows

   a. good logic, because he's having a good day and will probably win again.

   b. poor logic, because on such a good day he should bet much more.

   c. poor logic, as he has a 75% chance of losing the double bet.

9. A $1 slot machine at a casino is set so that it returns 97% of all the money put into it in the form of winnings, with most of the winnings in the form of huge but low-probability jackpots. What is your probability of winning when you put $1 in this slot machine?

   a. 0.03   b. 0.97

   c. It cannot be calculated from the given data, but it is certainly quite low.

10. Consider the slot machine described in Question 9. Which statement is *true* if patrons have put $10 million into this slot machine?

    a. The casino's profit has been close to $300,000.

    b. 97% of the patrons have been winners.

    c. 3% of the patrons have been winners.

Exercises 7C

REVIEW QUESTIONS

1. Explain the meaning of the *law of large numbers*. Does this law say anything about what will happen in a single observation or experiment? Why or why not?
2. In 10 tosses of a fair coin, should you be surprised to see 6 heads? In 1000 tosses of a fair coin, should you be surprised to see 600 heads? Explain in terms of the law of large numbers.
3. What is an *expected value*, and how is it computed? Should we always expect to get the expected value? Why or why not?
4. What is the *gambler's fallacy*? Give an example.
5. Explain why the probability is the same for any particular set of outcomes for 10 tosses of a fair coin. How does this idea affect our thinking about streaks?
6. What is the *house edge* in gambling? Explain how it virtually guarantees that a casino will win more money than it loses.

DOES IT MAKE SENSE?

Decide whether each of the following statements makes sense (or is clearly true) or does not make sense (or is clearly false). Explain your reasoning.

7. The expected value to me of each raffle ticket I purchased is −$0.85.
8. The expected value of each insurance policy our company sells is $150, so if we sell 10 more policies, our profits will increase by $1500.
9. If you toss a coin four times, it's much more likely to land in the order HTHT than HHHH (where H stands for heads and T for tails).
10. I haven't won in my last 25 pulls on the slot machine, so I'm due to win on the next pull.
11. I haven't won in my last 25 pulls on the slot machine, so I must be having a bad day and I'm sure to lose if I play again.
12. I've lost $750 so far today on roulette. I'm going to play a little longer so that I can reduce my losses to $500.

BASIC SKILLS & CONCEPTS

13. **Understanding the Law of Large Numbers.** Suppose you toss a fair coin 10,000 times. Should you expect to get exactly 5000 heads? Why or why not? What does the law of large numbers tell you about the results you are likely to get?
14. **Speedy Driver.** Suppose a man who has a habit of driving fast has never had a speeding ticket. What does it mean to say that "the law of averages [large numbers] will catch up with him"? Is it true? Explain.

15–18: Expected Value in Games. Find the expected value (to you) of the described game. Would you expect to win or lose money in 1 game? In 100 games? Explain.

15. You are given 5 to 1 odds against tossing three heads with three coins, meaning you win $5 if you succeed and you lose $1 if you fail.
16. You are given 9 to 1 odds against tossing three heads with three coins, meaning you win $9 if you succeed and you lose $1 if you fail.
17. You are given 3 to 1 odds against rolling two even numbers with the roll of two fair dice, meaning you win $3 if you succeed and you lose $1 if you fail.
18. You are given 7 to 1 odds against rolling a double number (for example, two 1s or two 2s) with the roll of two fair dice, meaning you win $7 if you succeed and you lose $1 if you fail.

19–20: Insurance Claims. Find the expected value (to the company) per policy sold. If the company sells 10,000 policies, what is the expected profit or loss? Explain.

19. An insurance policy sells for $300. Based on past data an average of 1 in 100 policyholders will file a $10,000 claim, an average of 1 in 250 policyholders will file a $25,000 claim, and an average of 1 in 500 policyholders will file a $50,000 claim.
20. An insurance policy sells for $600. Based on past data, an average of 1 in 50 policyholders will file a $5000 claim, an average of 1 in 100 policyholders will file a $10,000 claim, and an average of 1 in 200 policyholders will file a $30,000 claim.

21. **Expected Wait.** Suppose you arrive at a bus stop randomly, so all arrival times are equally likely. The bus arrives regularly every 20 minutes without delay (say, on the hour, 20 minutes and 40 minutes past the hour). What is the expected value of your waiting time? Explain.

22. **Expected Wait.** A bus arrives at a bus stop at noon, 12:20, and 1:00. You arrive at the bus stop at random times between noon and 1:00 every day, so all arrival times are equally likely.

 a. What is the probability that you will arrive at the bus stop between noon and 12:20? What is your mean waiting time in that case? (Exercise 21 may be helpful.)

 b. What is the probability that you will arrive at the bus stop between 12:20 and 1:00? What is your mean waiting time in that case?

 c. Overall, what is your expected waiting time for the bus?

 d. Would your expected waiting time be longer or shorter if the bus arrived at equally spaced intervals (say, noon, 12:30, 1:00)? Explain.

23. **Gambler's Fallacy and Coins.** Suppose you play a coin toss game in which you win $1 if a head appears and lose $1 if a tail appears. In the first 100 coin tosses, heads comes up 46 times and tails comes up 54 times.

 a. What percentage of times has heads come up in the first 100 tosses? What is your net gain or loss at this point?

 b. Suppose you toss the coin 200 more times (a total of 300 tosses), and at that point heads has come up 47% of the time. Is this increase in the percentage of heads consistent with the law of large numbers? Explain. What is your net gain or loss at this point?

 c. How many heads would you need in the next 100 tosses in order to break even after 400 tosses? Is this likely to occur?

 d. Suppose that, still behind after 400 tosses, you decide to keep playing because you are "due" for a winning streak. Explain how this belief would illustrate the gambler's fallacy.

24. **Gambler's Fallacy and Dice.** Suppose you roll a die with the following rules: You win $1 if the die comes up with an even number and you lose $1 if it comes up odd.

 a. Suppose you get 45 even numbers in your first 100 rolls. How much money have you won or lost?

 b. On the second 100 rolls, your luck improves and you roll 47 even numbers. How much money have you won or lost over 200 rolls?

 c. Your luck continues to improve, and you roll 148 even numbers in your next 300 rolls. How much money have you won or lost over your total of 500 rolls?

 d. How many even numbers would you have to roll in the next 100 rolls to break even? Is this likely? Explain.

 e. What were the percentages of even numbers after 100, 200, and 500 rolls? Explain how this illustrates the law of large numbers, even while your losses increased.

25. **Can You Catch Up?** Suppose you toss a fair coin 100 times, getting 42 heads and 58 tails, which is 16 more tails than heads.

 a. Explain why, on your next toss, the *difference* between the numbers of heads and tails is as likely to grow to 17 as it is to shrink to 15.

 b. Extend your explanation from part (a) to explain why, if you toss the coin 1000 more times, the final difference between the numbers of heads and tails is as likely to be larger than 16 as it is to be smaller than 16.

 c. Suppose you are betting on heads with each coin toss. After the first 100 tosses, you are well on the losing side. Explain why, if you continue to bet, you will most likely remain on the losing side. How is this answer related to the gambler's fallacy?

26. **Baseball Batting Averages.** Based on his record of the last five seasons, Sal expects to have a .300 batting average (meaning that he averages 3 hits for every 10 times at bat) for the current season.

 a. During the first month of the season, Sal is at bat 80 times and gets 20 hits. What is his batting average for this month?

 b. During the next month, Sal gets 22 hits in 80 times at bat. What is his batting average for the second month and for the season so far?

 c. During the third month, Sal breaks through and gets 27 hits in 80 times at bat. What is his batting average for the third month and for the season so far?

 d. How many hits must Sal get in 80 times at bat during the fourth and final month to reach a .300 batting average for the season?

27. **Lottery Draw.** Consider a lottery game in which six balls are drawn randomly from a set of balls numbered 1 through 42. One week, the winning combination consists of balls numbered 5, 12, 23, 32, 36, and 41. The next week, the winning balls are numbered 1, 2, 3, 4, 5, and 6. Is the second winning set more or less likely than or just as likely as the first? Explain.

28. **Coin Streak.** You toss a coin 1000 times and record the results. You then look through your results and find that at one point you got 10 heads in a row. What is the probability of getting 10 heads in 10 tosses? Should you be surprised to find this "streak" in your set of 1000 tosses? Explain.

29. **Hot Shooter.** Maria plays basketball and is a 70% free throw shooter (meaning on average she hits 7 of ten 10 free throws). During a 25-game season, Maria had a game in which she attempted five free throws and hit all five of them. Do you think she had a "hot" day, or was it just chance? Explain your reasoning.

30. **Royal Flush Luck.** The probability of being dealt a royal flush (ace, king, queen, jack, and 10 of the same suit) in a five-card poker hand is about 1 in 650,000. According to the Las Vegas Tourist Bureau, approximately 70,000 people gamble in Las Vegas every day, and those who play poker are dealt many hands during a day.

a. Is it likely that you will be dealt a royal flush?

b. Should you be surprised that someone in Las Vegas is dealt a royal flush during any given day?

31. **House Edge in Blackjack.** In a large casino, the house wins on its blackjack tables with a probability of 50.7%. All bets at blackjack are 1 to 1: If you win, you gain the amount you bet; if you lose, you lose the amount you bet.

a. If you bet $1 on each hand, what is the expected value to you of a single game? What is the house edge?

b. If you played 100 games of blackjack in an evening, betting $1 on each hand, how much should you expect to win or lose? Explain.

c. If you played 100 games of blackjack in an evening, betting $5 on each hand, how much should you expect to win or lose? Explain.

d. If patrons bet $1,000,000 on blackjack in one evening, how much should the casino expect to earn? Explain.

32. **Profitable Casino.** Averaged over all games and all bets placed, the house edge of a particular casino is $0.055 per dollar gambled. If a total of $100 million is wagered in the casino over the course of a year, what is the casino's total gain? Explain.

FURTHER APPLICATIONS

33–34: Powerball. The table below gives prizes and probabilities of winning (on a single $1 ticket) for the Multi-State Powerball lottery.

Prize	Probability
Jackpot	1 in 292,201,338
$1,000,000	1 in 11,688,053
$50,000	1 in 913,129
$100	1 in 36,525
$100	1 in 14,494
$7	1 in 580
$7	1 in 701
$4	1 in 92
$4	1 in 38

33. Find the expected value of the winnings for a single lottery ticket if the jackpot is $30 million. How much can you expect to win or lose each year if you buy 10 lottery tickets per week? Should you actually expect to win or lose this amount? Explain.

34. Find the expected value of the winnings for a single lottery ticket if the jackpot is $45 million. How much can you expect to win or lose each year if you buy 20 lottery tickets per week? Should you actually expect to win or lose this amount?

35–36: Mega Millions. The table below gives prizes and probabilities of winning (on a single $1 ticket) for the Mega Millions lottery.

Prize	Probability
Jackpot	1 in 258,890,850
$1,000,000	1 in 18,492,203
$5000	1 in 739,688
$500	1 in 52,835
$50	1 in 10,720
$5	1 in 766
$5	1 in 473
$2	1 in 56
$1	1 in 21

35. Find the expected value of the winnings for a single lottery ticket if the jackpot is $15 million. How much can you expect to win or lose if you buy 100 lottery tickets? Should you actually expect to win or lose this amount? Explain.

36. Find the expected value of the winnings for a single lottery ticket if the jackpot is $40 million. How much can you expect to win or lose if you buy 500 lottery tickets? Should you actually expect to win or lose this amount? Explain.

37. **Extra Points in Football.** National Football League teams have the option of trying to score either 1 or 2 extra points after a touchdown. A team can get 1 point by kicking the ball through the goal posts or 2 points by running or passing the ball across the goal line. A team makes zero points if either attempt fails. The kicking distance required for a one-point conversion was increased starting with the 2015 season.

a. Over 22 seasons before the kicking distance was increased, 23,325 one-point conversions were made out of 23,684 attempts. Based on the relative frequency, what was the probability of making a one-point conversion? What was the expected value for one-point conversions in these seasons?

b. In the first two seasons (2015 and 2016) after the kicking distance was increased, 2265 one-point conversions were made out of 2412 attempts. Based on the relative frequency, what was the probability of making a one-point conversion? What was the expected value for one-point conversions in these seasons?

c. Over 10 recent NFL seasons, 344 two-point conversions were made out of 718 attempts. Based on the relative frequency, what was the probability of making a two-point conversion? What was the expected value for two-point conversions in these seasons?

d. Based on these results, which option generally made more sense before the kicking distance was increased? After it was increased? Bonus: If you are a football fan, explain why there might be cases in which it would make sense for a team to choose the option with the lower expected value.

38. **Roulette.** When you bet $5 on the number 7 in roulette at a typical casino, you have a 37/38 probability of losing $5 and a 1/38 probability of making a net gain of $175 (after paying your bet). If you bet $5 on an odd number, the probability of losing $5 is 20/38 and the probability of a net gain of $5 is 18/38.

a. What is the expected value of betting $5 on the number 7?

b. What is the expected value of betting $5 on an odd number?

c. Which is the best option: bet on 7, bet on an odd number, or don't bet?

39. **Household Size.** It is estimated that 57% of Americans live in households with 1 or 2 people, 32% live in households with 3 or 4 people, and 11% live in households with 5 or more people. Using 1.5 for the expected number of people in households with 1 or 2 people, 3.5 for households with 3 or 4 people, and 6 for households of 5 or more people, find the expected number of people in an American household. How is this related to the mean household size?

IN YOUR WORLD

40. **Law of Large Numbers.** Describe a news story in which the law of large numbers is mentioned. Is the term used in an accurate way? Explain.

41. **Personal Law of Large Numbers.** Describe a situation in which you personally have made use of the law of large numbers, either correctly or incorrectly. Why did you use the law of large numbers in this situation? Was it helpful?

42. **The Gambler's Fallacy.** Describe a situation in which you or someone you know has fallen victim to the gambler's fallacy. How should the situation have been dealt with?

43. **The Morality of Gambling.** Republican Senator Richard Lugar (Indiana) attacked state-supported lotteries and the trend toward increasing legalization of gambling, saying "the spread of gambling is a measure of the moral erosion taking place in our country.... It says that if you play enough, you can hit the jackpot and be freed from the discipline of self-support through a job or the long commitment to ongoing education." Do a Web search on "gambling and morality" to find other opinions on this issue. Write a short summary of your findings and your own opinion, which you should defend clearly.

TECHNOLOGY EXERCISE

Answer the following question with **StatCrunch** (available in MyLab Math).

44. **Die Simulation.** Use StatCrunch to generate simulations of rolling a single die as follows: In the StatCrunch work space, choose "Applets" then "Simulation" and then "Dice Rolling." Then enter 1 for the number of dice to roll and click on "Compute!" An empty graph will appear with options along the top for the number of runs in your simulation.

a. Choose 5 runs to see the results for 5 rolls of the single die. Then continue by choosing 5 additional runs (multiple times) until you have a total of 30 runs on your graph. Are the frequencies close to what you expect? Explain.

b. Repeat the experiment with 50, 100, and 1000 die rolls. Are the results consistent with the law of large numbers? Explain.

UNIT 7D Assessing Risk

A smooth-talking but honest salesman comes to you, offering a new product:

I can't reveal the details yet, but you will love this product! It will improve your life in more ways than you can count. Its only downside is that it will eventually kill everyone who uses it. Will you buy one?

Not likely! After all, could any product be so great that you would die for it? A few weeks later, the salesman shows up again:

No one was buying, so we've made some improvements. Your chance of being killed by the product is now only 1 in 10. Ready to buy?

Despite the improvement, most people still would send the salesman home and wait for his inevitable return:

Okay, this time we've really perfected it. We've made it so safe that it would take about 20 years for it to kill as many Americans as live in San Francisco (though admittedly it will injure another 4 million each year). Still, you'll love it, and it can be yours for only about $30,000.

You may be surprised to realize that most people jump at this offer. The product is, after all, the automobile. It does indeed improve our lives in many ways, and the median new car price is more than $30,000. However, given nearly 40,000 annual fatalities (and more than 4 million injuries requiring medical attention) due to car crashes in the United States, it kills the equivalent of the population of San Francisco (roughly 800,000) in about 20 years. The National Safety Council estimates that these deaths and injuries cost the United States more than $400 billion per year.

As this example shows, we frequently make tradeoffs between benefits and risks. In this unit, we'll see how ideas of probability can help us quantify risk, thereby allowing us to make informed decisions about such tradeoffs. ▶ Now try Exercises 9–12.

BY THE WAY

Car crashes are the leading cause of death among young people between 6 and 25 years of age. In the United States, vehicle occupants account for about 85% of car crash fatalities, while pedestrians and cyclists account for the other 15%.

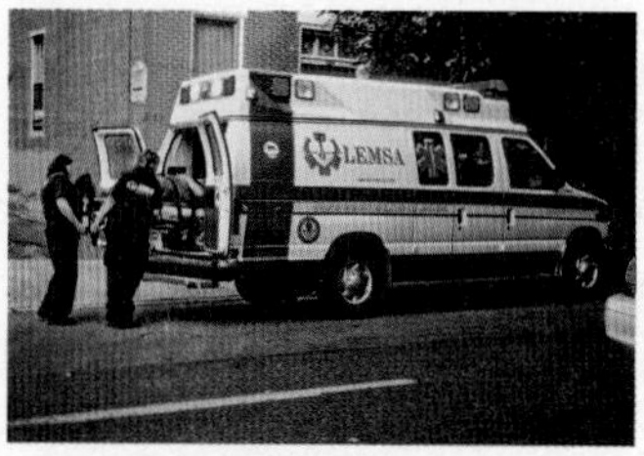

IN YOUR WORLD

Terrorism, Risk, and Human Psychology

The very word *terrorism* tells you that its intent is to provoke fear among ordinary citizens. However, careful research shows that one reason terrorism provokes fear is because the ways in which we perceive risks are often disconnected from actual statistics.

In the case of terrorism, a Cato Institute study found that a total of 3432 murders were committed by terrorists on U.S. soil from 1975 through 2015, including the 2983 murders that occurred in the attacks on 9/11/2001. This works out to an average of about 85 terrorist-caused deaths per year during that period, and only 11 deaths per year if you exclude the 9/11 attacks. Using the latter number, we find that nearly *4000 times* as many Americans die each year in car crashes and nearly *3000 times* as many die due to injury by firearms (unrelated to terrorism). Indeed, the Consumer Product Safety Commission found that the average American is about as likely to be crushed to death by a piece of furniture as to die in a terrorist attack.

Why, then, do surveys consistently show that terrorism ranks near the top of our list of fears? Researchers have identified numerous aspects of human psychology that contribute to the answer. For example, we tend to be more fearful in situations in which we lack control, which is one reason why many people fear flying (where control rests with the pilots) more than driving. The seemingly random nature of terrorist attacks plays to this fear, since we have no control over when or where they will occur.

A second factor in fear of terrorism is its potential to harm a large number of people at once. This same factor explains why a commercial airline crash that kills, say, 350 people anywhere in the world is major news, even though the *daily* global death toll from car crashes is 10 times as large.

A third factor arises from the fact that terrorist attacks inevitably make the news, leading to what psychologists call an *availability error*: We fear terrorism more than many everyday threats that pose large risks because the media coverage makes it more available to our minds.

We may not be able to change the way our brains are wired to react to the fears evoked by terrorism, but we can control the way we respond to these fears. On a political level, for example, we should be careful to consider the tradeoffs involved when we decide to commit resources or implement policies that aim to alleviate one risk over other risks that may be larger. On an individual level, consider the lesson embodied in the ironic fact that people who drive long distances because of a fear of flying actually increase their likelihood of being killed during their trip. Fear can be useful when channeled appropriately, but decisions should also be based on rational evaluation of real data and consequences.

Risk and Travel

Are you safer in a small car or a sport utility vehicle? Are cars today safer than those 30 years ago? If you need to travel across the country, are you safer flying or driving? To answer these and many similar questions, we must quantify the risk involved in travel. We can then make decisions appropriate for our own personal circumstances.

Travel risk is often expressed in terms of an **accident rate** or **death rate**. For example, suppose an annual accident rate is 750 accidents per 100,000 people. This means that, within an average group of 100,000 people, 750 will have an accident over the period of a year. The statement is in essence a relative frequency probability: It tells us that the probability of a person being involved in an accident (in one year) is 750 in 100,000, or 0.0075.

This concept of travel risk is straightforward, but we must still interpret the numbers with care. For example, travel risks are sometimes stated *per 100,000 people*, as above, but other times they are stated *per trip* or *per mile*. If we use death rates *per trip* to compare the risks of flying and driving, we neglect the fact that airplane trips are typically much longer than automobile trips. Similarly, if we use accident rates *per person*, we neglect the fact that most automobile accidents involve only minor injuries.

EXAMPLE 1 Is Driving Getting Safer?

Figure 7.13 shows the number of automobile fatalities and the total number of vehicle-miles driven (among all Americans) for each year over a period of more than four decades. In terms of the death rate *per 100 million vehicle-miles* driven, how has the risk of driving changed over this period?

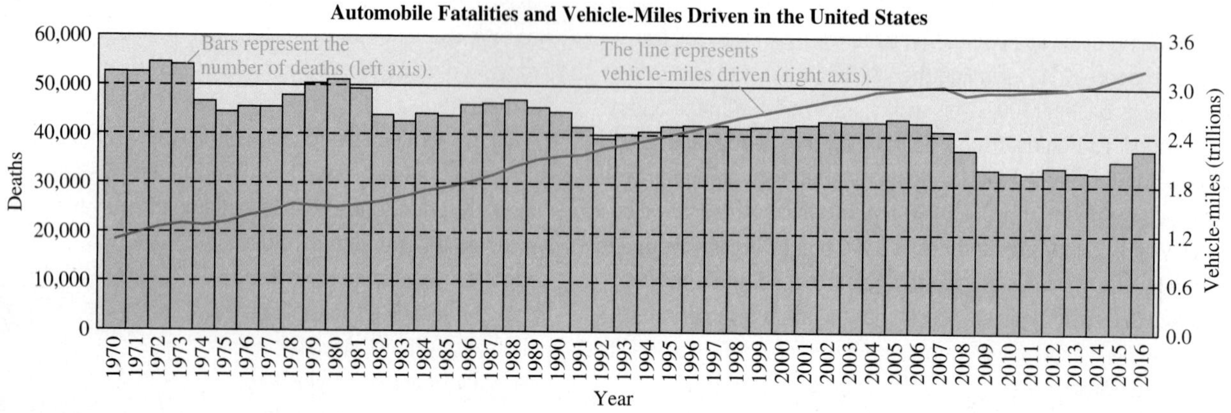

FIGURE 7.13 This graph shows annual automobile fatalities (bars) and vehicle-miles driven (line) in the United States. *Source:* National Transportation Safety Board.

Solution We start by reading approximate numbers from the graph. For 1970, the bar shows that there were about 52,000 deaths, while the blue curve (which uses the values on the right side of the chart) shows about 1.1 trillion vehicle-miles driven. For 2016, the bar shows about 37,000 deaths, while the blue curve shows about 3.2 trillion miles vehicle-driven. We can compute the death rates per mile by dividing the number of deaths by the number of vehicle-miles driven in each year. However, because we are asked to find the death rates per 100 million vehicle-miles, we also do a unit conversion in which we use the fact that 100 million $= 10^8$. The calculations are as follows:

$$\textit{1970:}\quad \frac{52{,}000 \text{ deaths}}{1.1 \times 10^{12} \text{ miles}} \times \frac{10^8 \text{ miles}}{100 \text{ million miles}} \approx 4.7 \text{ deaths per 100 million miles}$$

$$\textit{2016:}\quad \frac{37{,}000 \text{ deaths}}{3.2 \times 10^{12} \text{ miles}} \times \frac{10^8 \text{ miles}}{100 \text{ million miles}} \approx 1.2 \text{ deaths per 100 million miles}$$

From 1970 to 2016, the death rate for automobile crashes fell from about 4.7 to 1.2 deaths per 100 million vehicle-miles; you should confirm with the relative change formula (see Unit 3A) that this represents a nearly 75% decrease. Most researchers attribute the lower death rate to better automobile design and safety features such as shoulder belts and air bags. However, notice also that after a long general decline, the number of automobile fatalities increased substantially in 2015 and 2016, which researchers attribute largely to distracted driving, along with increased numbers of people driving under the influence of marijuana.

▶ Now try Exercises 13–14.

Technical Note

Different organizations count fatalities due to car crashes differently. For example, the National Safety Council counts deaths arising from crashes in both traffic and non-traffic situations (such as in driveways or parking lots) and deaths that occur as long as a year after a crash, while the National Transportation Safety Board counts only traffic deaths occurring within 30 days after a crash. The National Safety Council therefore reports higher numbers of fatalities.

EXAMPLE 2 Which Is Safer: Flying or Driving?

For the five-year period 2012–2016, U.S. commercial airlines flew about 8 billion miles per year, and had a total of only 5 crash-related fatalities, or an average (mean) of 1 death per year. Use these numbers to calculate the death rate per mile of air travel. Compare the risk of flying to the risk of driving.

Solution Using the given annual data, the death rate per mile of air travel is

$$\frac{1 \text{ death}}{8 \times 10^9 \text{ miles}} = 1.25 \times 10^{-10} \text{ death per mile}$$

This rate, which we can state equivalently as about 1 death per 10 billion miles flown, is only about 1/100, or 1%, of the 2016 rate of 1.2 deaths per 100 million miles for driving. Based on these data, flying is much less risky than driving, on a per mile basis.

▶ Now try Exercises 15–16.

BY THE WAY

All 5 of the fatalities cited in Example 2 occurred in 2013. Aside from that, there were *no* other commercial airline fatalities in the United States for the years 2010 through 2016 (the most recent year for which data were available when this book went to press).

Think About It Do you know anyone with a fear of flying? Do you think the above statistics should influence that fear? Defend your opinion.

Vital Statistics

Data concerning births and deaths of citizens, called *vital statistics* (because the Latin *vita* means "life"), are important to understanding risk-benefit tradeoffs. For example, insurance companies use vital statistics to assess risks and set rates. Health professionals study vital statistics to assess medical progress and decide where research resources should be concentrated. Demographers use birth and death rates to predict future population trends.

One important set of vital statistics covers causes of death. Table 7.5 presents sample data for the United States. A more detailed table might, for example, categorize the data by age, sex, and race. Vital statistics like those in Table 7.5 are often expressed in terms of deaths per person or deaths per 100,000 people, which makes it easier to compare the rates for different years and for different states or countries.

TABLE 7.5 Leading Causes of Death in the United States (2015)

Cause	Deaths	Cause	Deaths
Heart disease	614,300	Alzheimer's disease	93,500
Cancer	591,700	Diabetes	76,500
Chronic respiratory diseases	147,100	Pneumonia/Influenza	55,200
Accidents (including car crashes)	136,100	Kidney disease	48,100
Stroke	133,100	Suicide	42,800

Source: Centers for Disease Control and Prevention.

BY THE WAY

Experts estimate that more than 90% of all suicide deaths could be prevented with counseling or treatment, making it one of the most preventable causes of death.

EXAMPLE 3 Interpreting Vital Statistics

Assume an approximate U.S. population of 325 million. Use Table 7.5 to find and compare death rates per person and per 100,000 people for accidents and cancer.

Solution We find the death rates per person by dividing the numbers of deaths by the total population of 325 million:

$$\textit{Accidents/crashes:} \quad \frac{136{,}100 \text{ deaths}}{325{,}000{,}000 \text{ people}} \approx 0.00042 \text{ death per person}$$

$$\textit{Cancer:} \quad \frac{591{,}700 \text{ deaths}}{325{,}000{,}000 \text{ people}} \approx 0.0018 \text{ death per person}$$

We can convert these numbers to deaths per 100,000 people by multiplying the per person rates by 100,000. We find that accidents are responsible for about 42 deaths per 100,000 people, while cancer is responsible for about 180 deaths per 100,000 people.

▶ Now try Exercises 17–20.

BY THE WAY

College students face particularly serious health risks from alcohol consumption. The National Institutes of Health estimates that alcohol contributes to 1400 deaths (many through car crashes in which alcohol was a factor), 500,000 injuries, and 70,000 cases of sexual assault among U.S. college students each year.

Think About It Table 7.5 suggests that the probability of death from heart disease is much higher than the probability of death by accident, but these data include all age groups. How do you think the risk of death from heart disease or from an accident would differ between young people and old people? Explain.

Life Expectancy

One of the most commonly cited vital statistics is *life expectancy*, which is often used to compare overall health at different times or in different countries. The idea will be clearer if we start by looking at death rates. Figure 7.14a shows the U.S. death rate (deaths per 1000 people) for people of different ages. Note that there is an elevated risk of death shortly after birth, after which the death rate drops to very low levels. At about 15 years of age, the death rate begins a gradual rise.

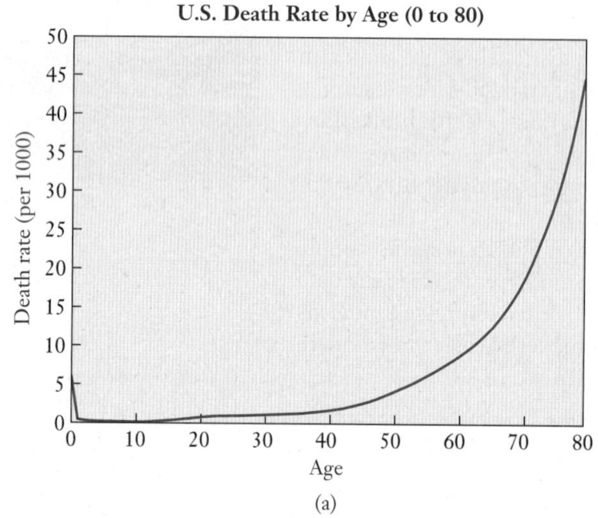

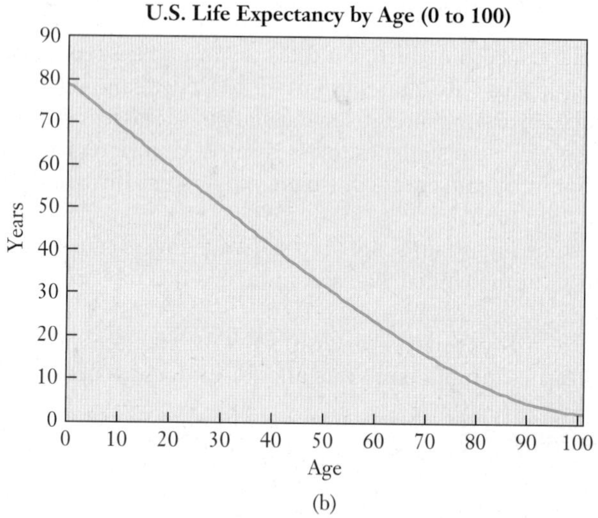

FIGURE 7.14 (a) The U.S. death rate for different ages. (b) Life expectancy for different ages. *Source:* Centers for Disease Control and Prevention (2012 data, released 2016).

Figure 7.14b shows the **life expectancy** of Americans (men and women combined) of different ages, defined as the number of additional years a person of a given age can expect to live on average. As we would expect, life expectancy is higher for younger people because, on average, they have longer left to live. At birth, the life expectancy of Americans today is about 79 years (76 years for men and 81 years for women).

Definition

Life expectancy is the number of additional years a person of a given age can expect to live on average. *Life expectancy at birth* is the life expectancy for a newborn infant.

BY THE WAY

Life expectancies vary widely around the world. According to the *CIA World Factbook*, as of 2015 Monaco had the highest life expectancy at 89.5 years, while Chad had the lowest at 49.8 years. The United States ranked 43rd in life expectancy, even though it ranks first in health care spending per person by a large margin.

Life expectancies have been rising with time because of improvements in medical science and public health (Figure 7.15). This fact means that life expectancies must be interpreted carefully, because they are calculated based on *current* death rates. For example, when we say that the life expectancy of an infant born today is 79 years, we mean that the average baby born today will live to age 79 *if there are no future changes* in medical science or public health. In fact, most people expect that life expectancies will continue to increase as public health and medical treatments continue to improve. Therefore, while life expectancy provides a useful measure of current overall health, it should not be considered a *prediction* of future life spans.

EXAMPLE 4 Life Expectancies by Age

Using Figure 7.14b, find the life expectancies of a 20-year-old and a 60-year-old American in 2016. Are the numbers consistent? Explain.

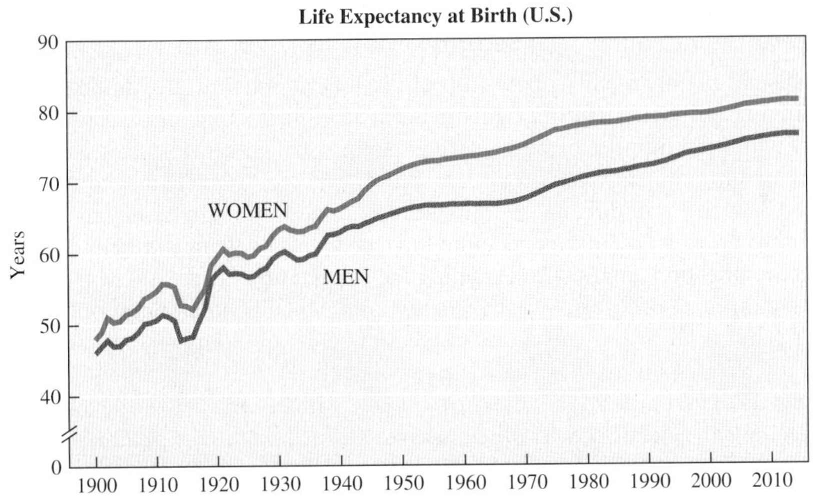

FIGURE 7.15 Changes in U.S. life expectancy at birth since the beginning of the 20th century. *Sources:* Centers for Disease Control and Prevention; National Center for Health Science Statistics.

The reports of my death are greatly exaggerated.
—Mark Twain, from London in cable to the Associated Press

Solution The graph shows that the life expectancy at age 20 is about 60 years and at age 60 is about 23 years. This means that an average 20-year-old American can expect to live about 60 more years, to age 80. An average 60-year-old can expect to live about 23 more years, to age 83.

It might at first seem strange that 60-year-olds have a longer average life span than 20-year-olds (83 years versus 80 years). But remember that life expectancies are based on current data. If there were no changes in medicine or public health, a 60-year-old would have a greater probability of reaching age 83 than a 20-year-old simply because he or she has already made it to age 60. However, if medicine and public health continue to improve, today's 20-year-olds may live to older ages than today's 60-year-olds.

▶ **Now try Exercises 21–24.**

EXAMPLE 5 Will You Live to Be 110?

Use Figure 7.15 to estimate the rise in U.S. life expectancy at birth during the 20th century. Suppose that life expectancies rise by the same amount during this century. What will life expectancies at birth be for children born in the United States in the year 2100? What are the implications to your own life expectancy? Explain.

Solution Figure 7.15 shows that, for women, life expectancy at birth rose from about 48 years in 1900 to about 80 years in 2000, an increase of $80 - 48 = 32$ years. For men, life expectancy at birth rose from about 46 years in 1900 to about 74 years in 2000, an increase of $74 - 46 = 28$ years. If the same increase occurs in this century, then in 2100:

- Life expectancy for women will be 32 years greater than 80 years, or 112 years.
- Life expectancy for men will be 28 years greater than 74, or 102 years.

In other words, if life expectancy at birth in the United States increases as much in the 21st century as it did in the 20th, it will reach 112 for women and 102 for men. Note that, although this example deals with life expectancy at birth, life expectancies for all ages (not just at birth) tend to rise together with time (see Example 4), so that elderly people in 2100 will be living to ages similar to those indicated by the life expectancy of newborns. Therefore, if life expectancies continue their trend from the 20th century, there is a good chance that today's high school and college students will live to average ages over 100 for men and 110 for women.

▶ **Now try Exercises 25–26.**

Think About It Do you think life expectancy will really rise to over 100 by the end of this century? Why or why not?

CASE STUDY
Life Expectancy and Social Security

Because of the changing age makeup of the U.S. population, the number of retirees qualifying for Social Security benefits is expected to become much larger in the future, while the number of wage earners paying Social Security taxes is expected to grow much more slowly. As a result, one of the biggest challenges to the future of Social Security is finding a way to make sure there is enough money to pay benefits to future retirees. We discussed some aspects of this challenge in Unit 4F, but a further complication comes with changing life expectancy.

Any estimate of future Social Security costs depends on assumptions about how long people will live, because unless there are corresponding changes in retirement age, longer lifespans result in more years collecting Social Security benefits. Unfortunately, these assumptions are difficult to make with confidence. For example, recent projections by Social Security officials have assumed that life expectancy at birth will rise only slightly during the rest of this century. However, as discussed in Example 5, it seems reasonable to imagine that life expectancy might increase by a fairly large amount. If that turns out to be the case, it will be great for public health, but will greatly exacerbate the budgetary problems facing Social Security (and Medicare).

Think About It Compare the life expectancies of men and women shown in Figure 7.15. Do they have any implications for Social Security or other social policy? For insurance rates? Discuss.

Quick Quiz 7D

Choose the best answer to each of the following questions. Explain your reasoning with one or more complete sentences.

1. Given almost 40,000 annual fatalities due to car crashes in the United States, the average number of Americans killed in car crashes each day is approximately

   a. 10 b. 100 c. 1000

2. Based on Figure 7.13, over the past 40 years, the number of vehicle-miles driven has generally

   a. decreased with time. b. held steady with time.

   c. increased with time.

3. Based on the data in Figure 7.13, what can you conclude about how the death rate per vehicle-mile driven has changed over the past few decades?

   a. The death rate per vehicle-mile has varied up and down with no overall trend.

   b. The death rate per vehicle-mile has generally decreased with time.

   c. The death rate per vehicle-mile has generally increased with time.

4. You are told the number of deaths due to AIDS in South Africa in a given year. To calculate the death rate per person, what else do you need to know?

   a. the number of people infected with AIDS in South Africa

   b. the amount of time the victims had the disease

   c. the population of South Africa

5. Based on Table 7.5 and a U.S. population of 325 million, the death rate from diabetes *per 100,000 people* is

   a. 76,500. b. $\dfrac{76{,}500}{325\text{ million}}$.

   c. $\dfrac{76{,}500}{325\text{ million}} \times 100{,}000$.

6. Based on Figure 7.14a, besides the elderly, the group of Americans facing the greatest risk of death is

   a. infants. b. teenagers.

   c. men in their mid-30s.

7. Which statement about life expectancy is true?

   a. It is the same for everyone in the United States.

   b. It tells you how long you will live.

   c. It decreases as you get older.

8. The life expectancy at birth of an American girl today is about 81 years. Does this mean that we should expect girls born today to live an average of 81 years?

   a. Yes; that's what we mean by life expectancy.

   b. No, because their life expectancies will get lower as they get older.

   c. No, because this life expectancy assumes no change in medical science or public health, both of which are almost certain to change.

9. Based on Figure 7.15, which statement is *not* true?

a. Americans today live, on average, more than 50% longer than Americans in 1900.

b. The life expectancies of men are rapidly catching up with the life expectancies of women.

c. If the increase in life expectancy during 2000–2100 is the same as during 1900–2000, the life expectancy of men will exceed 100 by 2100.

10. If life expectancies continued to rise as they did in the 20th century, how would the increase affect current projections about the future of Social Security?

a. It would mean current projections underestimate how much Social Security will pay in benefits.

b. It would mean current projections underestimate how much revenue Social Security will receive from Social Security taxes.

c. It would have no effect, because budget planners already assume that life expectancies will increase at the same rate as in the 20th century.

Exercises 7D

REVIEW QUESTIONS

1. Briefly explain why quantifying risk is important to decision making.
2. Give an example of a rate used to measure risk in travel. Why are rates more useful than total numbers of deaths or accidents in measuring risk? How are such rates similar to probabilities?
3. What are *vital statistics*? How are they usually described? Give a few examples.
4. Explain the meaning of the term *life expectancy*. How does life expectancy change with age? How is it affected by changes in the overall health of a population?

DOES IT MAKE SENSE?

Decide whether each of the following statements makes sense (or is clearly true) or does not make sense (or is clearly false). Explain your reasoning.

5. No one can succeed in selling a product that will kill thousands of people per year.
6. Fewer people are killed in motorcycle accidents than in automobile accidents, so motorcycles must be safer than cars.
7. Your life expectancy is the major factor in determining how long you live.
8. A 60-year-old has a shorter life expectancy than a 20-year-old.

BASIC SKILLS & CONCEPTS

9–12: Data in Perspective. Use these data to help answer the following questions:

- U.S. data (2016 estimates): car crash fatalities = 40,200; car crash injuries = 4.6 million; cost to society = \$430 billion. (*Source*: National Safety Council.)
- Global data (estimated): car crash fatalities ≈ 1.3 million; car crash injuries ≈ 50 million; car crash fatalities with victims under age 25 ≈ 400,000. (*Source*: Association for Safe International Road Travel.)

9. About how often (once every x minutes or x seconds) on average is a person killed in a car crash in the United States? How often is a person injured in a car crash?
10. Repeat Exercise 9 for the global data.
11. Use the given data on the societal cost of car crashes and the approximate U.S. population of 325 million to calculate the average social cost per person.
12. Globally, how many young people (under age 25) die in car crashes each day on average?
13. **Twenty-Year Trend in Automobile Safety.** Use the table to answer the following questions.

Year	U.S. population (millions)	Traffic fatalities	Licensed drivers (millions)	Vehicle-miles (trillions)
1995	263	41,817	177	2.4
2015	321	35,092	218	3.1

Source: National Transportation Safety Board.

a. Express the 1995 and 2015 fatality rates in deaths per 100 million vehicle-miles traveled.

b. Express the 1995 and 2015 fatality rates in deaths per 100,000 population.

c. Express the 1995 and 2015 fatality rates in deaths per 100,000 licensed drivers.

d. Are the changes in the three fatality rates from 1995 to 2015 that you found in parts (a)–(c) consistent with one another? Briefly comment on the general conclusions they suggest.

14. **Recent Trend in Automobile Traffic Safety.** Use the table to answer the following questions.

Year	Traffic and non-traffic fatalities	Vehicle-miles (trillions)
2014	35,398	3.03
2016	40,200	3.22

Source: National Safety Council; 2016 data are provisional.

a. Find the percentage increase in fatalities from 2014 to 2016. Is this change consistent with the general long-term trend in automobile fatalities? Explain.

b. Find the fatality rate in deaths per 100 million vehicle-miles traveled for 2014 and 2016. Can the increase in vehicle-miles explain the percentage increase in fatalities that you found in part (a)? Explain.

c. Suggest other possible explanations for the increase in fatalities from 2014 to 2016, and briefly discuss any evidence backing those explanations.

15–16. General Aviation Safety. The table shows the numbers of accidents, fatalities, hours flown, and miles flown for U.S. airlines with scheduled service in selected years.

Year	Accidents	Fatalities	Hours flown (millions)	Miles flown (billions)
2000	49	89	16.7	7.1
2005	34	22	18.7	7.8
2010	28	0	17.2	7.3
2015	27	0	17.4	7.6

Source: National Transportation Safety Board.

15. a. Compute the accident rate per million hours flown in 2000 and 2015. By this measure, has travel on U.S. airlines become safer?

b. Compute the accident rate per billion miles flown in 2000 and 2015. By this measure, has travel on U.S. airlines become safer?

16. a. Compute the fatality rate per million hours flown in 2000 and 2015. By this measure, has travel on U.S. airlines become safer?

b. Compute the fatality rate per billion miles flown in 2000 and 2015. By this measure, has travel on U.S. airlines become safer?

17–20: Causes of Death. Use Table 7.5, and assume a U.S. population of 325 million.

17. What is the relative frequency probability of death due to diabetes for an American during a single year? How much greater is the risk of death due to diabetes than the risk of death due to kidney disease?

18. How much greater is the risk of death from an accident than the risk of death from kidney disease?

19. If you lived in a typical U.S. city of 500,000, how many people would you expect to die of a stroke in a year?

20. If you lived in a typical U.S. city of 500,000, how many people would you expect to die of chronic respiratory diseases in a year?

21–24: Mortality Rates. Use the graphs in Figure 7.14 to answer the following questions.

21. Estimate the death rate for 60- to 65-year-olds. Assuming there were about 14.2 million 60- to 65-year-olds, how many people in this bracket could be expected to die in a year?

22. Estimate the death rate for 25- to 35-year-olds. Assuming there were about 44 million 25- to 35-year-olds, how many people in this age bracket could be expected to die in a year?

23. Suppose that a life insurance company insures 1 million 50-year-old people in a given year. (Assume a death rate of 5 per 1000 people.) The cost of the premium is $200 per year, and the death benefit is $50,000 What is the expected profit or loss for the insurance company?

24. Suppose that a life insurance company insures 5000 40-year-old people in a given year. (Assume a death rate of 3 per 1000 people.) The cost of the premium is $200 per year, and the death benefit is $50,000. How much can the company expect to gain (or lose) in a year?

25–26: Life in the Next Century. Example 5 assumed that the life expectancy between 2000 and 2100 would increase by the same absolute amount as it did between 1900 and 2000. For the following exercises, suppose instead that life expectancy between 2000 and 2100 will increase by the same *percentage* as it did between 1900 and 2000.

25. What would the life expectancy for women be in 2100? Do you think this calculation gives a more or less realistic estimate than Example 5 of life expectancy in 2100? Explain.

26. What would the life expectancy for men be in 2100? Do you think this calculation gives a more or less realistic estimate of life expectancy in 2100? Explain.

FURTHER APPLICATIONS

27. **High/Low U.S. Birth Rates.** The highest and lowest birth *rates* in the United States in 2015 were in Utah and New Hampshire, respectively. Utah reported 51,154 births with a population of about 3.0 million people. New Hampshire reported 12,302 births with a population of about 1.3 million people. Use these data to answer the following questions.

a. On average, how many people were born each day of the year in Utah?

b. On average, how many people were born each day of the year in New Hampshire?

c. What was the birth rate in Utah in births per 100,000 people?

d. What was the birth rate in New Hampshire in births per 100,000 people?

28. **High/Low U.S. Death Rates.** In 2015, the highest and lowest death rates in the United States were in West Virginia and Hawaii, respectively. West Virginia reported 22,186 deaths with a population of about 1.8 million. Hawaii reported 10,767 deaths with a population of 1.4 million.

a. Compute the death rates in West Virginia and Hawaii in deaths per 100,000 people.

b. How many people die each day in West Virginia?

c. How many people die each day in Hawaii?

29. **U.S. Birth and Death Rates.** In 2015, the U.S. population was about 321 million. The overall birth rate was 12.5 births per 1000 people, and the overall death rate was 8.4 deaths per 1000 people.

a. Approximately how many births were there in the United States in 2015?

b. About how many deaths were there in the United States in that year?

c. Based on births and deaths alone (i.e., not counting immigration), about how much did the U.S. population increase during 2015?

d. Suppose that during 2015 the U.S. population actually increased by 3.0 million. Based on this fact and your results from part (c), estimate how many people immigrated to the United States. What proportion of the overall population growth was due to immigration?

30. **Multiple Births.** In 2015, there were approximately 4 million births in the United States, of which 135,000 were twin births (that is, 2 × 135,000 babies), 4200 were triplet births, and 246 were quadruplet births. Assume that births of five or more babies in one delivery are negligible in number. *Note*: For parts (b) and (c), remember that the total number of babies is larger than the total number of births, because each twin birth represents 2 babies, each triplet birth represents 3 babies, and each quadruplet birth represents 4 babies.

a. If you were a woman who gave birth in 2015, what was the approximate probability that you had more than one baby? Assume that multiple births are randomly distributed in the population.

b. What is the approximate probability that a randomly selected newborn in 2015 was a twin?

c. What is the approximate probability that a randomly selected newborn in 2015 was a triplet *or* a quadruplet?

31. **Aging Population.** The table shows the total U.S. population and the number of Americans over age 65 for various years since 1950.

Year	1950	1960	1970	1980	1990	2000	2010	2015
U.S. population (millions)	151	179	203	227	249	281	309	321
Americans over age 65 (millions)	12.7	17.2	20.9	26.1	31.9	34.9	40.4	47.7

a. Make a table showing the percentage of Americans over age 65 between 1950 and 2015.

b. What was the percentage increase in the U.S. population between 1950 and 2015?

c. What was the percentage increase in the over-65 population between 1950 and 2015?

d. In 2015, there were approximately 169 million Americans of working age (25–65 years old). Estimate the number of people who were actually working and contributing to Social Security in 2015, and compare this number to the over-65 population. Explain any assumptions in your estimate.

e. How do these results affect the future of programs such as Social Security and Medicare?

32. **Psychology of Expected Values.** The psychologists Amos Tversky and Daniel Kahneman studied the psychology of risk by posing survey questions related to expected value. Consider the following two questions.

Question 1: Which option would you rather take, A or B?

Option A: 100% chance of gaining $250

Option B: 25% chance of gaining $1000; 75% chance of gaining nothing

Question 2: Which option would you rather take, C or D?

Option C: 100% chance of losing $750

Option D: 75% chance of losing $1000; 25% chance of losing nothing

a. Before doing calculations, answer both questions based on your intuition.

b. For each question, find the expected value of each option.

c. Tversky and Kahneman discovered that for Question 1, a majority of people chose option A, while for Question 2, a majority of people chose option D. Are these majority responses consistent with the expected values? Explain.

d. Do you think these results are relevant to the human tendency to misjudge risks (as discussed in In Your World on p. 465)? If so, explain why. If not, explain why not.

IN YOUR WORLD

33. **Car Safety in Perspective.** The opening paragraphs of this unit suggest one way of putting automobile safety in perspective, but there are many others. For example, law professor and judge Guido Calabresi uses an analogy to frame the issue: He asks students to imagine a deity offering a gift that will improve their lives but at the price of having about 1000 randomly selected people put to death weekly. Explain how this analogy relates to U.S. car crash statistics, then create your own example to put either U.S. or global car crash data into perspective. (You may wish to use the statistics offered in Exercises 9–12.)

34. **Improving Car Safety.** Find a recent news report discussing some suggestion for improving automobile safety, such as laws or technological solutions to limit distracted driving or driving while intoxicated, or self-driving technology that will bring potential safety benefits. Summarize any statistics given in the report, and discuss your overall opinion of the suggestion.

35. **Vital Statistics in the News.** Find a recent news report that gives new data about any aspect of vital statistics. Summarize the report and the statistics, and discuss any personal or social implications of the new data.

36. **Life Expectancy in the News.** Find a recent news report that discusses life expectancies in any context. Summarize the report, and discuss the social implications of the given life expectancy data.

37. **Global Life Expectancy.** Find data about how life expectancies compare among different nations and why they differ. Choose one nation or region with a relatively low life expectancy, and write a short report about why the life expectancy is low and what the prospects are for increasing it in the near future.

38. **Visualizing Life Expectancy.** Go to Gapminder.org, and find the site's interactive graphic of income per person versus life expectancy (titled "Wealth & Health of Nations"). Play the animation, and study it carefully. Draw at least two general conclusions from it, and briefly summarize your conclusions and describe how they are shown in the graphic.

39. **Risk in the News.** Find a recent news report discussing some current fear among the public. Is the level of fear warranted? Evaluate by giving data on the risks related to this fear in comparison to other known risks.

UNIT 7E Counting and Probability

In previous units, we have considered theoretical probabilities in which we could count the possible outcomes either with simple diagrams or by applying the multiplication principle (see the Brief Review on p. 431). In this unit, we'll introduce a few counting tools that will enable us to consider a wider range of probability problems. We'll also explore how probability relates to our interpretations of coincidences.

Arrangements with Repetition

Suppose a state has license plates that display seven numerals (Figure 7.16). How many different seven-number license plates are possible? There are 10 possibilities (the numerals 0 through 9) for the first position and the same 10 possibilities for the second position, so there are $10 \times 10 = 100$ possibilities for the first two positions combined. Similarly, because there are also 10 possibilities for the third position, there are $10 \times 10 \times 10 = 1000$ possibilities for the first three positions. For all seven positions, each with 10 possibilities, the total number of possibilities is

$$10 \times 10 \times 10 \times 10 \times 10 \times 10 \times 10 = 10^7$$

FIGURE 7.16 For each of the seven positions on the license plate, there are 10 choices, resulting in 10^7 possible license plates.

With seven numerals, there are 10 million different possible license plates.

The license plate problem involves making selections from a *single* group of choices—in this case, the numerals 0–9. However, the same numerals may be selected over and over again. This type of counting problem, in which we repeatedly select from the same group of choices, is called **arrangements with repetition**.

> **Arrangements with Repetition**
>
> If we make r selections from a group of n choices, a total of $\underbrace{n \times n \times \cdots \times n}_{r \text{ times}} = n^r$ different arrangements are possible.

EXAMPLE 1 Arrangements with Repetition

a. How many seven-character license plates are possible if both numerals *and* uppercase letters can be used in any order?

b. How many six-character passwords can be made by combining lowercase letters, uppercase letters, numerals, and the characters @, $, !, *, and &?

Solution

a. With 10 numerals and 26 letters in the alphabet, there are 36 choices for each of the seven positions. We are therefore selecting $r = 7$ characters with $n = 36$ choices for each character, so the number of seven-character license plates is

$$n^r = 36^7 = 78{,}364{,}164{,}096$$

There are more than 78 billion possible seven-character license plates.

b. Each password character can be selected from the set of 26 lowercase letters, 26 uppercase letters, 10 numerals, and the five characters @, $, !, *, and &, making a total of 67 characters to choose from. For six-character passwords, we select $r = 6$ characters with $n = 67$ choices for each character. The total number of possible passwords is

$$n^r = 67^6 = 90{,}458{,}382{,}169$$

More than 90 billion passwords are possible with these choices.

▶ Now try Exercises 23–24.

Think About It Compare the numbers of possibilities in Example 1 to the human population. Will a state using seven-character license plates ever run out of possibilities? Do you think a six-character password is secure against hackers? Explain.

Permutations

Suppose you coach a team of four swimmers. How many different ways can you order these swimmers in a four-person relay team?

You can choose any of the four swimmers for the first leg. Once you've chosen the first swimmer, you have three swimmers left to choose from for the second leg. Therefore, the number of possible choices for the first two legs combined is $4 \times 3 = 12$. You then have two swimmers to choose from for the third leg, and only one choice for the last leg. The total number of possible arrangements of the four swimmers for the relay is

$$4 \times 3 \times 2 \times 1 = 24$$

Figure 7.17 shows all 24 possible relay orders with a tree diagram in which the four swimmers are identified as A, B, C, and D. The first row shows the four possible choices for the first leg. The second row shows the three possible second-leg choices for each of the first-leg choices. The third row shows the remaining two choices for the third leg, and the fourth row shows the single remaining choice for the last leg in each case. Counting across the bottom row, we see the total of 24 possibilities. Each of the 24 different relay orders is called a **permutation** of the four letters A, B, C, and D.

BY THE WAY The word *permute* (root of *permutation*) comes from a Latin term meaning "to change throughout."

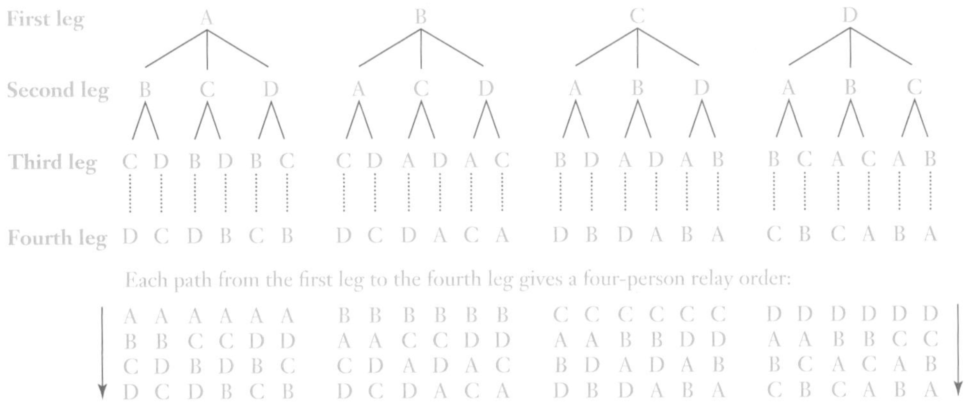

FIGURE 7.17 Tree diagram for the relay order possibilities with four swimmers, A, B, C, and D. There are four choices for the first leg. Each possible choice for the first leg leads to three remaining choices for the second leg. After the swimmers have been chosen for the first two legs, two choices remain for the third leg. Once the first three legs have been chosen, only one swimmer remains for the fourth leg.

The relay example involves selection from a single group (the four swimmers), in which each member of the group can be selected only once and the *order of arrangement* matters. For example, the relay order ABCD is different from the order DCBA. For any similar case, the following box summarizes how we calculate the total number of possible permutations.

Permutations

We are dealing with **permutations** whenever all selections come from a single group of items, no item may be selected more than once, and the *order of arrangement matters* (for example, ABC is considered to be different from CBA). If every member of the group is selected in each possible permutation then the total number of permutations possible with a group of n items is $n!$ (read as "n factorial), where

$$n! = n \times (n - 1) \times \cdots \times 2 \times 1$$

USING TECHNOLOGY

Factorials

Many calculators have a special key for factorials, usually labeled "$n!$" For example, to calculate $6!$ ($= 720$), press ⑥ $n!$ =. Excel has a built-in function FACT for factorials (see the screen shot below), and you can use Google to calculate a factorial simply by typing the expression (such as "6!") into the search box.

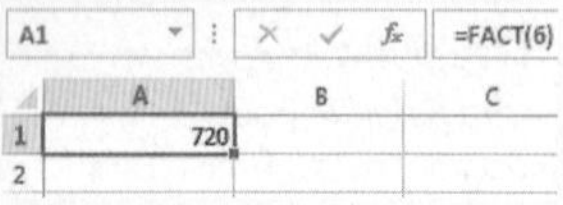

Microsoft Excel, Windows 10, Microsoft Corporation.

EXAMPLE 2 Class Schedules

A middle school principal needs to schedule six different classes—algebra, English, history, Spanish, science, and gym—in six different time periods. How many different class schedules are possible?

Solution This is a case of permutations because all selections come from the same group of classes, each class is scheduled only once, and the order of the classes matters (because a schedule that begins with Spanish is different from one that begins with gym). The principal may choose any of the six classes for the first period, leaving five choices for the second period. After the first two periods are filled, there are four choices left for the third period. Similarly, there are three choices for the fourth period, two choices for the fifth period, and only one choice for the sixth period. Altogether, the number of possible schedules is

$$6! = 6 \times 5 \times 4 \times 3 \times 2 \times 1 = 720$$

The principal can schedule the six classes in 720 different ways.

▶ Now try Exercises 25–26.

The Permutations Formula

Suppose you coach a team of ten swimmers, from whom you must put together a four-person relay team. How many possibilities do you have in this case?

Brief Review Factorials

Products of the form $4 \times 3 \times 2 \times 1$ come up so frequently in counting problems that they have a special name. Whenever a positive integer n is multiplied by all the preceding positive integers, the result is called **n factorial** and is denoted $n!$ (the exclamation mark is read as "factorial"). For example:

$$1! = 1$$
$$2! = 2 \times 1 = 2$$
$$3! = 3 \times 2 \times 1 = 6$$
$$4! = 4 \times 3 \times 2 \times 1 = 24$$
$$5! = 5 \times 4 \times 3 \times 2 \times 1 = 120$$

In general,

$$n! = n \times (n-1) \times (n-2) \times \cdots \times 2 \times 1$$

Note that $n!$ grows rapidly with n. For example, $20! \approx 2.4 \times 10^{18}$, $40! \approx 8.2 \times 10^{47}$, and $60! \approx 8.3 \times 10^{81}$. In fact, factorials quickly become so large that many calculators cannot handle them above $n = 69$ (if the calculator limit is 10^{100}). Note also that, by definition, $0! = 1$.

Example: Calculate each of the following *without* using the factorial key on your calculator.

a. $\dfrac{6!}{4!}$ b. $\dfrac{25!}{22!}$ c. $\dfrac{200!}{199!}$

Solution:

a. We can write out the entire calculation, but it is easier if we simplify it by canceling like terms in the numerator and denominator:

$$\frac{6!}{4!} = \frac{6 \times 5 \times \cancel{4} \times \cancel{3} \times \cancel{2} \times \cancel{1}}{\cancel{4} \times \cancel{3} \times \cancel{2} \times \cancel{1}} = \frac{6 \times 5 \times \cancel{4!}}{\cancel{4!}}$$
$$= 6 \times 5 = 30$$

b. This calculation is easier if we recognize that $25! = 25 \times 24 \times 23 \times 22!$, giving us

$$\frac{25!}{22!} = \frac{25 \times 24 \times 23 \times \cancel{22!}}{\cancel{22!}} = 25 \times 24 \times 23 = 13{,}800$$

c. If you tried 200! on your calculator, you would probably get an error because it is too large for most calculators to handle. But this calculation is easy if we recognize that $200! = 200 \times 199!$, which gives

$$\frac{200!}{199!} = \frac{200 \times \cancel{199!}}{\cancel{199!}} = 200$$

▶ Now try Exercises 11–22.

This time, you can choose one of ten swimmers for the first leg of the relay. Once you make the first choice, you have nine choices for the second leg, eight choices for the third leg, and seven choices for the fourth leg (Figure 7.18). The total number of relay possibilities is

$$10 \times 9 \times 8 \times 7 = 5040$$

Each of the 5040 teams represents a different *permutation* because each relay swimmer is selected from the same group of ten swimmers, no swimmer can swim more than once, and the order of the swimmers in the relay is important. However, in this case the relay teams make use of only four of the total of ten swimmers. That is, the number of possible relay teams made from "ten swimmers selected four at a time" is 5040.

We can find a general formula for permutations by writing the product $10 \times 9 \times 8 \times 7$ in a slightly different way:

$$10 \times 9 \times 8 \times 7 = \frac{10 \times 9 \times 8 \times 7 \times \not{6}!}{\not{6}!} = \frac{10!}{(10-4)!}$$

Note that, in the final formula on the right,

- the number 10 in the numerator is the number of swimmers from which the coach may choose the four relay participants, and
- the number $10 - 4 = 6$ in the denominator is the number of swimmers who do *not* swim in the relay.

Permutations have their own special notation: We read ${}_{10}P_4$ as "the number of permutations of ten items selected four at a time." Using this compact notation, we have

$${}_{10}P_4 = \frac{10!}{(10-4)!} = 5040$$

Generalizing, we have the following permutations formula.

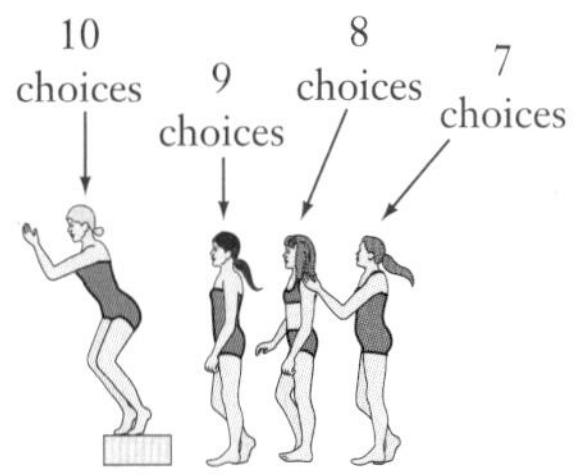

FIGURE 7.18 With ten swimmers from which to choose a four-person relay team, you have ten choices for the first position, nine for the second position, eight for the third position, and seven for the fourth position. This results in $10 \times 9 \times 8 \times 7 = 5040$ possible relay teams.

The Permutations Formula

Permutations occur whenever all selections come from a single group of items, no item may be selected more than once, and the *order of arrangement matters* (for example, ABC is considered different from CBA). If we make r selections from a group of n items, the number of permutations is

$${}_nP_r = \frac{n!}{(n-r)!} = \underbrace{n \times (n-1) \times (n-2) \times \cdots \times (n-r+1)}_{\text{There will be a total of } r \text{ terms multiplied together.}}$$

where ${}_nP_r$ is read as "the number of permutations of n items taken r at a time."

USING TECHNOLOGY

Permutations

Many calculators have a special key for permutations, usually labeled "${}_nP_r$." For example, to find the number of permutations of $n = 10$ items taken $r = 4$ at a time, press (10) (${}_nP_r$) (4) (=). In Excel, use the built-in function PERMUT, as shown in the screen shot below.

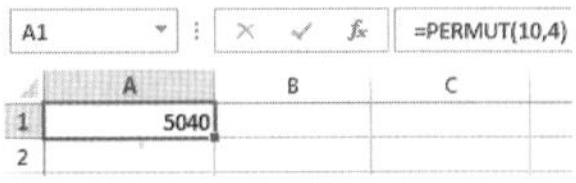

Microsoft Excel, Windows 10, Microsoft Corporation.

EXAMPLE 3 Leadership Election

A city has 12 candidates running for three leadership positions. The top vote-getter will become the mayor, the second-place vote-getter will become the deputy mayor, and the third-place vote-getter will become the treasurer. How many outcomes are possible for the three leadership positions?

Solution The voters choose $r = 3$ leaders from a group of $n = 12$ candidates. The order in which leaders are chosen matters because each of the three leadership positions is different. Therefore, we are looking for the number of permutations of 12 candidates taken 3 at a time:

$$_{12}P_3 = \frac{12!}{(12-3)!} = \frac{12!}{9!} = \frac{12 \times 11 \times 10 \times \cancel{9!}}{\cancel{9!}} = 12 \times 11 \times 10 = 1320$$

There are 1320 possible outcomes for the three leadership positions.

▶ Now try Exercises 27–28.

EXAMPLE 4 Batting Orders

A Little League manager has 15 children on her team. How many ways can she form a 9-player batting order?

Solution The manager forms a batting order by selecting $r = 9$ players from the roster of $n = 15$ players. Each batter is selected only once and the order matters, so the number of possible batting orders is the number of permutations of 15 players taken 9 at a time:

$$_{15}P_9 = \frac{15!}{(15-9)!} = \frac{15!}{6!}$$
$$= 15 \times 14 \times 13 \times 12 \times 11 \times 10 \times 9 \times 8 \times 7 = 1{,}816{,}214{,}400$$

Nearly 2 *billion* batting orders are possible for a baseball team with a roster of 15 players.

▶ Now try Exercises 29–30.

Think About It Major League baseball teams carry a roster of 25 players. Without actually calculating, describe how the calculation of the number of possible batting orders would be different for a Major League team than for the Little League team in Example 4.

Combinations

A committee is a group of important individuals who singly can do nothing, but who together can agree that nothing can be done.

—Fred Allen, comedian

Suppose that the five members of a city council decide they need a three-person committee to study the impact of a new shopping center. The council members are Ursula, Vern, Wendy, Yolanda, and Zeke (U, V, W, Y, and Z, for short). How many three-person committees can be formed from the five council members?

This is *not* a permutation problem because we are interested only in the makeup of the committee, *not* the order in which the names are listed. For example, if we were dealing with permutations, we would consider ZWU to be different from WZU. In this case, however, both arrangements are equivalent because they both represent the same committee members (Zeke, Wendy, and Ursula). Problems of this type are called **combinations** problems.

One way to count the possible committees is to list them all. Going in alphabetical order, we begin by listing all three-person committees that include Ursula:

UVW UVY UVZ UWY UWZ UYZ

Next, we list the committees that include Vern but not Ursula:

VWY VWZ VYZ

This leaves only one committee that includes neither Ursula nor Vern:

WYZ

We've found a total of ten three-person committees that can be made from the five council members.

Of course, listing all the possible committees will become tedious if the numbers are larger. Fortunately, we can get the same answer with an alternative strategy. First, we find the number of *permutations* of the $n = 5$ council members selected $r = 3$ at a time:

$$_5P_3 = \frac{5!}{(5-3)!} = \frac{5!}{2!} = 5 \times 4 \times 3 = 60$$

That is, there are 60 possible permutations when we choose three people from a group of five. However, because order matters for permutations but does not matter for committees, the number of permutations is an *overcount* of the actual number of different committees. More specifically, any three-person committee can be listed in $3! = 3 \times 2 \times 1 = 6$ different orders. For example, the committee consisting of Zeke, Yolanda, and Wendy has six different permutations:

ZYW ZWY YZW YWZ WZY WYZ

Because each three-person committee is counted $3! = 6$ times by the permutations formula, this formula gives us six times the actual number of committees. We must therefore divide the number of permutations by 3! to find the number of committees:

$$\frac{_5P_3}{3!} = \frac{60}{3!} = \frac{60}{3 \times 2 \times 1} = 10$$

This is the same result we obtained by listing the three-person committees. The permutations part of the equation is $_nP_r$ where $n = 5$ and $r = 3$. We then divided this term by $r! = 3!$ to correct for the overcounting. Generalizing, we find the following formula for combinations.

USING TECHNOLOGY

Combinations

Many calculators have a special key for combinations, usually labeled "$_nC_r$." For example, to find the number of combinations of $n = 10$ items taken $r = 4$ at a time, press 10 $_nC_r$ 4 =. In Excel, use the built-in function COMBIN, as shown in the screen shot below.

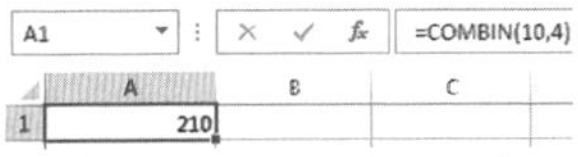

Microsoft Excel, Windows 10, Microsoft Corporation.

Combinations

Combinations occur whenever all selections come from a single group of items, no item may be selected more than once, and the order of arrangement does *not* matter (for example, ABC is considered the same as CBA). If we make r selections from a group of n items, the number of possible combinations is

$$_nC_r = \frac{_nP_r}{r!} = \frac{n!}{(n-r)! \times r!}$$

where $_nC_r$ is read as "the number of combinations of n items taken r at a time."

EXAMPLE 5 Ice Cream Combinations

Suppose that you select 3 different flavors of ice cream in a shop that carries 12 flavors. How many flavor combinations are possible?

Solution We are looking for the number of combinations of $n = 12$ flavors selected $r = 3$ at a time. From the combinations formula, the number of flavor combinations is

$$_{12}C_3 = \frac{12!}{(12-3)! \times 3!} = \frac{12!}{9! \times 3!} = \frac{12 \times 11 \times 10 \times 9!}{9! \times 3!} = \frac{12 \times 11 \times 10}{3 \times 2 \times 1}$$

$$= \frac{1320}{6} = 220$$

There are 220 different three-flavor combinations possible from the 12 flavors.

▶ Now try Exercises 31–32.

EXAMPLE 6 Poker Hands

How many different five-card poker hands can be dealt from a standard deck of 52 cards? What is the probability of one particular hand, such as a *royal flush* of hearts (a hand consisting of the ace, king, queen, jack, and 10 of hearts)?

Solution This is a combinations problem, because the order of the cards in a hand does not matter. The number of combinations of $n = 52$ cards drawn $r = 5$ at a time is

$$_{52}C_5 = \frac{52!}{(52-5)! \times 5!} = \frac{52!}{47! \times 5!} = \frac{52 \times 51 \times 50 \times 49 \times 48 \times 47!}{47! \times 5 \times 4 \times 3 \times 2 \times 1} = 2{,}598{,}960$$

About 2.6 million hands are possible in five-card poker. Any particular hand, such as the royal flush of hearts, therefore has a probability of about 1 in 2.6 million.

▶ Now try Exercises 33–34.

Think About It Consider the following unimpressive poker hand: 5 of clubs, 7 of hearts, 3 of diamonds, 2 of spades, 9 of hearts. How does the probability of this hand, which has no value in poker, compare to the probability of a royal flush of spades, which is the highest-value hand in poker? Explain.

EXAMPLE 7 (Not) Winning the Lottery

Suppose you play a lottery in which the winner is chosen by drawing 6 balls at random from a drum containing 52 numbered balls (numbered 1 through 52). What is the probability that your 6 numbers will match the 6 winning numbers?

HISTORICAL NOTE

New Hampshire began the first legalized lottery in the United States in 1964. Since then, all but seven states have adopted lotteries as a source of revenue. Lotteries have existed in various forms and places for centuries, including the Roman Empire (1st century) and Italy in the Middle Ages.

Solution The order in which the balls are drawn does not matter, so finding the total number of possible outcomes is a combinations problem. The number of combinations for $n = 52$ balls selected $r = 6$ at a time is

$$_{52}C_6 = \frac{52!}{(52-6)! \times 6!} = \frac{52!}{46! \times 6!} = 20{,}358{,}520$$

There are more than 20 million possible combinations of 6 numbers, so the chance that your 6 numbers will match the 6 winning numbers is less than 1 in 20 million.

▶ Now try Exercises 35–40.

Probability and Coincidence

The word *coincidence* literally refers to one or more incidents happening together, such as to events happening in the same place or at the same time. However, in everyday language, we usually call something a coincidence only when it seems surprising in some way. These types of coincidences come in many different forms. You may be surprised to find that two people at your dinner party have the same birthday or that a friend's mother met your mother on a trip to China. You might flip a coin and get heads ten times in a row or have a dream that seems to predict an event that later occurs. It is tempting to chalk these cases up to something more mysterious than mere coincidence. But the laws of probability dictate that many coincidences are bound to happen, even though the particular form of a coincidence is unpredictable.

Coincidences Are Bound to Happen

Although a *particular* coincidence may be highly unlikely, *some* similar coincidence may be extremely likely or even certain to occur. In general, this means that coincidences should be expected to occur, with their likelihoods dictated by the laws of probability.

As a simple example, consider the big prize in a multi-state lottery. The prize will eventually be won, which means that it is certain (probability = 1) that *someone* will win the lottery. However, with many millions of people playing (and many of them playing multiple times), it is extremely unlikely the winner will be *you*.

The general difference between *some* coincidence and a *particular* coincidence brings us back to the question that opened this chapter (p. 424). Most people expect the probability of finding two people with the same birthday in a group of only 25 people to be low. However, while it would be relatively rare to find another person with *your* birthday in such a group, the probability of finding *some* pair of people with the same birthday is almost 3 in 5. Using techniques developed throughout this chapter, the following example now shows you the calculations needed to find the precise probabilities for these two cases.

BY THE WAY

Studies show that most people playing lotteries vastly overestimate their chances of winning. Here's one good way to think about it: Imagine playing a lottery with 1 million players and 1 winner. Because 999,999 out of the 1 million players will not win, there is a 99.9999% chance that the following statement will be true: "Someone will win, but it will *not* be you."

EXAMPLE 8 Birthday Coincidence

Suppose there are 25 students in your class. Find each of the following probabilities. Assume that there are 365 days in a year.

a. The probability that at least one person in the class has the same birthday as *you*.

b. The probability that *some* pair of students in the class of 25 shares the *same* birthday. Compare this result to the result from part (a).

Solution

a. This is an *at least once* question (see Unit 7B). We begin by recognizing that, with 365 days in a year, the probability that any particular student has your birthday is 1/365. Therefore, the probability that a particular student does not have your birthday is 364/365. From the *at least once* rule, the probability that at least one of the 24 other students has your birthday is

$$\begin{aligned} P(\text{at least one with your birthday}) &= 1 - [P(\text{not your birthday})]^{24} \\ &= 1 - \left[\frac{364}{365}\right]^{24} \\ &= 1 - 0.936 = 0.064 \end{aligned}$$

The probability that at least one other student shares your birthday is only 0.064, which is about 6%, or 1 in 16.

b. This question differs from the question in part (a) because the pair sharing the same birthday does not have to include you; it can be *any* pair of two students. But it is still an *at least once* question, because we are looking for the probability of a shared birthday among at least one pair of students. We can therefore find the answer as follows:

$$P(\text{at least one pair of shared birthdays}) = 1 - P(\text{no shared birthdays})$$

The easiest way to find the probability of no shared birthdays is by calculating the probability that all 25 students have *different* birthdays. We begin with the simpler, similar problem of considering just two students from the class. The first student has a birthday on 1 of the 365 days of the year, so the probability that the second student has a different birthday is 364/365. Now we add a third student. If the first two students have different birthdays, then 2 of the 365 days are already "taken." Therefore, the probability that the third student's birthday falls on a third different day is 363/365, since 363 days are not yet "taken" for birthdays. Putting the probabilities together, we find that the chance that the first three students all have different birthdays is

$$\underbrace{\frac{364}{365}}_{\substack{\text{probability that first two students} \\ \text{have different birthdays}}} \times \underbrace{\frac{363}{365}}_{\substack{\text{probability that third student} \\ \text{also has different birthday}}}$$

Similarly, if the first three students all have different birthdays, the probability that a fourth student has a different birthday from any of the first three is 362/365. And so on. Finally, if the first 24 students all have different birthdays, then 24 of the 365 days are "taken," leaving 365 − 24 = 341 possible different birthdays for the 25th student. Overall, the probability that all 25 students have different birthdays is

$$\frac{364}{365} \times \frac{363}{365} \times \cdots \times \frac{341}{365} = \frac{364 \times 363 \times \cdots \times 341}{365^{24}}$$

Although we could write the product in the numerator more compactly as 364!/340!, the factorials are too large for most calculators to handle. Therefore, the better way to proceed is by computing the numerator and denominator separately. You should confirm that the result is

$$\frac{364 \times 363 \times \cdots \times 341}{365^{24}} \approx \frac{1.348 \times 10^{61}}{3.126 \times 10^{61}} \approx 0.431$$

This is the probability that there are no shared birthdays among the 25 students in the class. Therefore, the probability that there is at least one pair of shared birthdays is

$$\begin{aligned} P(\text{at least one pair of shared birthdays}) &= 1 - P(\text{no shared birthdays}) \\ &\approx 1 - 0.431 = 0.569 \end{aligned}$$

There's a probability of 0.569, which is a chance of about 57%, or almost 3 in 5, that at least two of the 25 students have the same birthday! Most people are very surprised at how much higher this probability is than the 0.064 probability that at least one of the students shares *your* birthday, but it illustrates the fact that *some* coincidence is far more likely than a *particular* coincidence. ▶ **Now try Exercises 41–42.**

Quick Quiz 7E

Choose the best answer to each of the following questions. Explain your reasoning with one or more complete sentences.

1. You are asked to create a 5-character password, and each character may be any of the 26 letters of the alphabet or the 10 numerals 0 through 9. How many different passwords are possible?

   a. 36 × 5 b. 5^{36} c. 36^5

2. A waitress has four different entrees for the four people sitting at a table, but has forgotten which person ordered which entree. In how many different ways could she serve the entrees?

   a. 4 b. 4^4 c. 24

3. A teacher has 28 students, and 5 of them will be chosen to participate in a play that has 5 distinct roles. Which of the following questions requires calculating *permutations*?

   a. How many different sets of 5 children can be selected for the 5 roles?

   b. How many different choices are possible for each role in the play?

   c. Once the 5 children have been chosen, how many different ways can their roles be arranged?

4. The number of permutations of 12 objects grouped 5 at a time is written

   a. ${}_5P_{12}$ b. ${}_{12}P_5$ c. ${}_{12}P_{12}$

5. A soccer coach has a team of 15 children and 7 of the children are on the field at one time. Which number is *largest?*

   a. the number of combinations of 7 children that can be chosen from the 15

   b. the number of permutations of 7 children that can be chosen from the 15

   c. the number of different ways that 7 children can be arranged among the 7 positions on the field

6. One term in the denominator of the combinations formula is $(n - r)!$. Suppose you are trying to determine the number of different possible 4-person teams that can be put together from a group of 9 people. In this case, the term $(n - r)!$ means

   a. 4 × 3 × 2 × 1.

   b. 9 × 8 × 7 × 6 × 5.

   c. 5 × 4 × 3 × 2 × 1.

7. The number of different 4-person teams (order does not matter) that can be put together from a group of 9 people is

a. 9!.

b. $9 \times 8 \times 7 \times 6$.

c. $\frac{9 \times 8 \times 7 \times 6}{4 \times 3 \times 2 \times 1}$.

8. One person in a stadium filled with 100,000 people is chosen at random to win a free airline ticket. The probability that *someone* will win the ticket

a. is 1 in 100,000.

b. is 1.

c. depends on how the winner is chosen.

9. One person in a stadium filled with 100,000 people is chosen at random to win a free pair of airline tickets. What is the probability that it will *not* be you?

a. 1 in 100,000

b. 0.99

c. 0.99999

10. There are 365 possible birthdays in a year. In a class of 25 students, the chance of finding 2 students with the same birthday is

a. 25/365.

b. $2 \times 25/365$.

c. greater than 0.5.

Exercises 7E

REVIEW QUESTIONS

1. What are *arrangements with repetition*? Give an example of a situation in which the n^r formula gives the number of possible arrangements.
2. What do we mean by *permutations*? Explain the meaning of each of the terms in the permutations formula. Give an example of its use.
3. What do we mean by *combinations*? Explain the meaning of each of the terms in the combinations formula. Give an example of its use.
4. Explain what we mean when we say that *some* outcome is much more likely than a similar particular outcome. How does this idea affect our perception of coincidences?

DOES IT MAKE SENSE?

Decide whether each of the following statements makes sense (or is clearly true) or does not make sense (or is clearly false). Explain your reasoning.

5. I used the permutations formula to determine how many possible relay orders we could make with the 10 girls on our swim team.
6. I used the combinations formula to determine how many different five-card poker hands are possible.
7. The number of different possible batting orders for 9 players on a 25-person baseball team is so large that there's no hope of trying them all out.
8. It must be my lucky day, because the five-card poker hand I got had only about a 1 in 2.5 million chance of being dealt to me.
9. The probability that two people in a randomly selected group will have the same last name is much higher than the probability that someone will have the same last name as I do.
10. Someone wins the lottery every week, so I figure that if I keep playing eventually I will be the one who wins.

BASIC SKILLS & CONCEPTS

11–22: Review of Factorials. Use the skills covered in the Brief Review on p. 476 to evaluate the following quantities *without* using the factorial key on your calculator (you may use the multiplication key). Show your work.

11. $6!$

12. $12!$

13. $\frac{5!}{3!}$

14. $\frac{10!}{8!}$

15. $\frac{12!}{4!3!}$

16. $\frac{9!}{4!2!}$

17. $\frac{11!}{3!(11-3)!}$

18. $\frac{30!}{29!}$

19. $\frac{8!}{3!(8-3)!}$

20. $\frac{30!}{28!}$

21. $\frac{6!\,8!}{4!\,5!}$

22. $\frac{15!}{2!\,13!}$

23–40: Counting Methods. Answer the following questions using the appropriate counting technique: arrangements with repetition, permutations, or combinations. Be sure to explain why the particular counting technique applies to the problem.

23. How many different seven-digit phone numbers can be formed?
24. How many different seven-character passwords can be formed from the lowercase letters of the alphabet?
25. How many different five-character passwords can be formed from the lowercase letters of the alphabet if repetition is not allowed?

26. How many ways can the ten performances at a piano recital be ordered?

27. A city council with eight members must elect a four-person executive committee consisting of a mayor, deputy mayor, secretary, and treasurer. How many executive committees are possible?

28. A city council with nine members must appoint a leadership team consisting of mayor, treasurer, and secretary. How many teams are possible?

29. The President must assign ambassadors to six different foreign embassies. From a pool of ten candidates, how many different diplomatic teams can she form?

30. How many anagrams (rearrangements) of the letters ILOVEMATH can you make?

31. Suppose you have 12 music playlists from which you must choose 6 for an upcoming party. If you are not particular about the order in which the music is played, how many 6-playlist sets are possible?

32. How many different 5-card hands can be dealt from a 48-card pinochle deck?

33. How many 6-person lineups can be formed from a 10-player volleyball roster? (Every player plays every position.)

34. A dog shelter is giving away 12 different dogs, but you have room for only 3 of them. How many different dog families could you have?

35. How many license plates can be made with seven characters, *XXX–YYYY*, where *X* is a letter of the alphabet and *Y* is a numeral 0–9?

36. How many different groups of 7 balls can be drawn from a barrel containing balls numbered 1–32?

37. How many different birth orders with respect to gender are possible in a family with six children? (For example, BBBGGG and BGBGGB are different orders.)

38. How many different telephone numbers of the form *aaa-bbb-cccc* can be formed if the area code *aaa* can only begin with the numbers 2 through 7 and the exchange *bbb* cannot begin with 0?

39. How many different three-letter "words" can be formed from the genetic alphabet ACGT?

40. The debate club has 12 members, but only 4 can compete at the next meet. How many 4-person teams are possible?

41–42: Birthday Coincidences. Suppose you are part of a group of people at a dinner party. For the situations given, find the probability that at least one of the other guests has *your* birthday and the probability that *some* pair of guests shares the same birthday. Discuss your results. (Assume 365 days in a year.)

41. You are one of 12 people at the party.

42. You are one of 20 people at the party.

FURTHER APPLICATIONS

43. **Ice Cream Shop.** Josh and John's Ice Cream Shop offers 20 different flavors of ice cream and 8 different toppings. Answer the following questions by using the appropriate counting technique (multiplication principle, arrangements with repetitions, permutations, or combinations). Explain why you chose the particular counting technique.

a. How many different sundaes can you create using one of the ice cream flavors and one of the toppings?

b. How many different triple cones can you create from the 20 flavors if the same flavor may be used more than once? Assume that you specify which flavor goes on the bottom, middle, and top.

c. Using the 20 flavors, how many different triple cones can you create with 3 *different* flavors if you specify which flavor goes on the bottom, middle, and top?

d. Using the 20 flavors, how many different triple cones can you create with 3 *different* flavors if you don't care about the order of the flavors on the cone?

44. **Telephone Numbers.** A ten-digit phone number in the United States consists of a three-digit area code followed by a three-digit *exchange* followed by a four-digit number.

a. The first digit of the area code cannot be 0 or 1. The first digit of the exchange cannot be 0 or 1. How many different ten-digit phone numbers can be formed? Can a city with 2 million telephone numbers be served by a single area code? Explain.

b. How many exchanges are needed to serve an area code shared by 80,000 people? Explain.

45. **Pizza Hype.** Luigi's Pizza Parlor advertises 56 different three-topping pizzas. How many individual toppings does Luigi actually use? Ramona's Pizzeria advertises 36 different two-topping pizzas. How many toppings does Ramona actually use? (*Hint:* In these problems, you are given the total number of combinations, and you must find the number of toppings that are used.)

46. **ZIP Codes.** The U.S. Postal Service uses both five-digit and nine-digit ZIP codes.

a. How many five-digit ZIP codes are available to the U.S. Postal Service?

b. For a U.S. population of 300 million people, what is the average number of people per five-digit ZIP code if all possible ZIP codes are used? Explain.

c. How many nine-digit ZIP codes are available to the U.S. Postal Service? Could everyone in the United States have his or her own personal nine-digit ZIP code? Explain.

47–54: Counting and Probability. Find the probability of the given event.

47. Choosing six lottery numbers that match six randomly selected balls when the balls are numbered 1 through 32

48. Choosing five lottery numbers that match five randomly selected balls when the balls are numbered 1 through 40

49. Being dealt a 10, jack, queen, king, and ace, all of the same suit from a standard 52-card deck

50. Guessing the top three winners (in order) from a group of 16 finalists in a soccer tournament

51. Guessing the top four winners (in any order) from a group of 12 finalists in a spelling bee

52. Randomly selecting 3 students from Utah out of a group of 12 students, 5 of whom are from Utah

53. Being dealt 5 cards from a standard 52-card deck and getting four of a kind (for example, four aces)

54. Being in the first half of the program when you are one of ten performers whose order of performance is randomly selected

55. **Hot Streaks.** Suppose that 2000 people are all playing a game for which the chance of winning is 48%.

a. Assuming everyone plays exactly five games, what is the probability of one person winning five games in a row? On average, how many of the 2000 people could be expected to have a "hot streak" of five games?

b. Assuming everyone plays exactly ten games, what is the probability of one person winning ten games in a row? On average, how many of the 2000 people could be expected to have a "hot streak" of ten games?

56. **Joe DiMaggio's Record.** One of the longest-standing records in sports is the 56-game hitting streak of baseball player Joe DiMaggio. Assume that a player on a long "hot streak" is batting .400, which is about the best that anyone ever hits over a period of 50 or more games. (A batting average of .400 means the batter gets a hit 40% of the time. Typically, only a handful of players each year hit that well for a period as long as 56 games.)

a. What is the probability that a player batting .400 will get at least one hit in four at-bats?

b. Use the result in part (a) to calculate the probability of a .400 hitter getting a hit in one stretch of 56 consecutive games, assuming four at-bats per game.

c. Suppose that, instead of batting .400, a player has a more ordinary average of .300. In that case, what is the probability of the player getting a hit in one stretch of 56 consecutive games, assuming four at-bats per game?

d. Considering the results in parts (b) and (c) and the fact that baseball has been played for about 100 years, are you surprised that *someone* set a record of hitting in 56 consecutive games? Explain clearly.

57. **Mega Millions.** The lottery game Mega Millions® is played in over 40 states and requires $1 per ticket to play. To win the jackpot, you must correctly pick five unique numbers from balls numbered 1 through 75 (order doesn't matter) *and* correctly pick the Megaball number, which is chosen from balls numbered 1 through 15. What is the probability of winning the jackpot?

58. **Coin Streaks.** Toss a coin 100 times, and record your results (heads or tails) in order. What was your longest streak of consecutive heads or tails? Calculate the probability of this streak by itself. For example, if your longest streak was four heads, what is the probability of four heads in four tosses? Should you be surprised that you got such a streak? Explain.

IN YOUR WORLD

59. **Lottery Chances.** Find an article or advertisement about a lottery in which choosing a winner involves combinations. Calculate the probability of winning. Does your calculation agree with the advertised probability? Explain.

60. **Amazing Coincidence.** Discuss a seemingly amazing coincidence that you've read about or experienced yourself. Based on ideas of probability, decide whether the coincidence was bound to happen to someone. Was it really as amazing as it seemed? Explain.

61. **The "Monty Hall Problem."** A famous controversy about a probability question erupted over an item in the "Ask Marilyn" column in *Parade* magazine, written by Marilyn Vos Savant. The problem her column addressed was loosely based on a TV game show called *Let's Make a Deal*, hosted by Monty Hall, and hence is known as the "Monty Hall Problem." Here's the question:

> *Suppose that you're on a game show and you're given the choice of three doors: Behind one door is a car; behind the other two doors are goats. You pick a door, say No. 1, and the host, who knows what's behind the doors, opens another door, say No. 3, which has a goat. He then says to you, "Do you want to change your pick to door No. 2?" Is it to your advantage to switch your choice?*

Marilyn answered that the probability of winning was higher if the contestant switched. This answer generated a huge number of letters, including a few from mathematicians, claiming that she was wrong. Marilyn answered with the following logic. When you first pick door No. 1, the chance that you picked the one with the car is 1/3. The probability that you chose a door with a goat is 2/3. When the host opens door No. 3 to reveal a goat, it does not change the 1/3 probability that you picked the right door in the first place. Thus, as only one other door remains, the probability that it contains the car is 2/3. Visit a few of the many websites devoted to the Monty Hall Problem to gain some understanding of its subtleties. Do you agree with Marilyn's logic? If so, try to explain it in your own words. If not, present an alternative approach.

TECHNOLOGY EXERCISES

Answer the following questions using procedures described in the Using Technology boxes in this unit or with **StatCrunch** (available in MyLab Math).

62. **Comparing Factorials to Powers.**

a. Use Excel or StatCrunch to complete the following table.

n	$n!$	10^n
1		
2		
3		
4		
5		
6		
7		
8		
9		
10		

b. Your table should show that $10! < 10^{10}$. Extend the table to determine the smallest integer n such that $n! > 10^n$.

63. **Computing Lottery Probabilities.** Use Excel or StatCrunch to do the following calculations.

a. Compute the number of ways in which 5 lottery balls can be selected from a pool of 44 balls. What is the probability of winning such a lottery (by matching all five numbers)?

b. Compute the number of ways in which 6 lottery balls can be selected from a pool of 40 balls. What is the probability of winning such a lottery (by matching all six numbers)?

64. **Poker Probability.** The following formula gives the probability of being dealt a full house (three cards of one kind and two cards of another kind, such as 99933) in five-card poker:

$$P(\text{full house}) = \frac{{}_{13}C_1 \times {}_4C_3 \times {}_{12}C_1 \times {}_4C_2}{{}_{52}C_5}$$

Use Excel or StatCrunch to evaluate this probability.

65. **The Birthday Problem.** StatCrunch can be used to do random simulations of the birthday problem. In the StatCrunch work space, choose "Applets" then "Simulation" and then "Birthday Problem."

a. Simulate the birthday problem for 5 different classes, each with a class size of 30 students. In how many classes does exactly one pair of students share a birthday? In how many classes are there no shared birthdays?

b. Now do the simulation with 1000 classes. In what percentage of the 1000 classes does exactly one pair of students share a birthday? In what percentage of the 1000 classes are there no shared birthdays? In what percentage of the 1000 classes is there more than one pair of shared birthdays?

c. In Example 8 of this unit, we found that the probability of a shared birthday in a class of 25 students is about 0.57. Does this seem consistent with the results of the simulations you did with 30 students? Explain.

Chapter 7 Summary

UNIT	KEY TERMS	KEY IDEAS AND SKILLS
7A	outcome (p. 426) event (p. 427) theoretical probability (p. 429) multiplication principle (p. 431) relative frequency probability (p. 432) subjective probability (p. 433) probability distribution (p. 434) odds (p. 436)	**Distinguish among** theoretical, empirical, and subjective probabilities. (p. 429) **Determine** theoretical probabilities: (p. 429) $P(A) = \dfrac{\text{number of ways } A \text{ can occur}}{\text{total number of outcomes}}$ **Make** a probability distribution. (p. 434)
7B	independent events (p. 442) *and* probability (p. 442) dependent events (p. 444) *either/or* probability (p. 445) non-overlapping events (p. 446) overlapping events (p. 446)	***And* probability, independent events:** (p. 442) $P(A \text{ and } B) = P(A) \times P(B)$ ***And* probability, dependent events:** (p. 444) $P(A \text{ and } B) = P(A) \times P(B \text{ given } A)$ ***Either/or* probability, non-overlapping events:** (p. 445) $P(A \text{ or } B) = P(A) + P(B)$ ***Either/or* probability, overlapping events:** (p. 446) $P(A \text{ or } B) = P(A) + P(B) - P(A \text{ and } B)$ ***At least once* rule:** (p. 448) $P(\text{at least one event } A \text{ in } n \text{ trials}) = 1 - P(\text{no } A \text{ in } n \text{ trials}) = 1 - [P(\text{not } A)]^n$
7C	law of large numbers (p. 454) expected value (p. 456) gambler's fallacy (p. 457) house edge (p. 459)	**Understand and apply** the law of large numbers. (p. 453) **Calculate and interpret** expected values: (p. 455) $\text{expected value} = \begin{pmatrix}\text{event 1}\\ \text{value}\end{pmatrix} \times \begin{pmatrix}\text{event 1}\\ \text{probability}\end{pmatrix} + \begin{pmatrix}\text{event 2}\\ \text{value}\end{pmatrix} \times \begin{pmatrix}\text{event 2}\\ \text{probability}\end{pmatrix}$
7D	accident rate (p. 465) death rate (p. 465) life expectancy (p. 468)	**Measure** risk in terms of accident or death rates. (p. 464) **Understand and interpret** vital statistics and life expectancy. (p. 467)
7E	arrangements with repetition (p. 474) permutations (p. 475) combinations (p. 478)	**Arrangements with repetition:** (p. 474) For r selections from a group of n choices, the number of arrangements is n^r. **Permutations:** (p. 475) For r selections from a group of n items ${}_nP_r = \dfrac{n!}{(n-r)!}$ **Combinations:** (p. 478) For r selections from a group of n items, ${}_nC_r = \dfrac{{}_nP_r}{r!} = \dfrac{n!}{(n-r)! \times r!}$ **Coincidences are bound to happen**—understand why this is true and what the implications are for probability. (p. 480)

8 EXPONENTIAL ASTONISHMENT

World population is currently growing by more than 80 million people per year—enough to populate an entire new United States in only about 4 years. A growing population means new challenges for our species, which we can meet only if we understand this growth. In this chapter, we will investigate the mathematical laws of growth—specifically, exponential growth. We will focus on what we call exponential astonishment: the intuition-defying reality of exponential growth. Although we will emphasize population growth, we will also study many other important topics, including the decay of waste from nuclear power plants, the depletion of natural resources, and the environmental effects of acid rain.

Q

You place a single, microscopic bacterium into a nutrient-filled bottle. The bacterium grows rapidly, and after 1 minute it divides, so there are now two bacteria in the bottle. These bacteria grow and divide at the same rate, so after 2 minutes there are four bacteria in the bottle, after 3 minutes there are eight bacteria, and so on. Suppose that after 1 hour of this growth pattern, the bacteria fill a 1-liter bottle. If they continued to grow at the same rate, how many bottles would they fill at the end of a second hour?

A one
B two
C three
D four
E a million trillion

> **The greatest shortcoming of the human race is our inability to understand the exponential function.**
>
> —Albert A. Bartlett

A You should not need to guess, so be sure you have a good reason for your chosen answer before you read on. OK ... if you've made your choice, we can get right to the point. The correct answer is E. In fact, by the end of the second hour, there would be enough bacteria to cover the entire surface of the Earth—both land and oceans—in a layer about 2 meters deep. Clearly, this type of growth by doublings can produce astonishing results.

This idea is important, because growth by doublings is very common. It is a characteristic of any quantity that exhibits what we call *exponential growth*, in which growth occurs by the same percentage every fixed time period (such as each minute, month, or year). Such growth occurs with compound interest, cancerous tumors, human population growth, and much more. In this chapter, you'll find many examples of the surprising consequences of exponential growth and decay. And to find out how the answer to the multiple-choice question above is obtained, see Example 2 in Unit 8A.

UNIT 8A

Growth: Linear vs. Exponential: Distinguish between linear growth and exponential growth, and explore the remarkable effects of the repeated doublings that characterize exponential growth.

UNIT 8B

Doubling Time and Half-Life: Find and understand the doubling time for exponential growth and the half-life for exponential decay.

UNIT 8C

Real Population Growth: Study the reasons for and the limits to real population growth, which is neither strictly linear nor strictly exponential.

UNIT 8D

Logarithmic Scales: Earthquakes, Sounds, and Acids: Explore the meaning and use of three important *logarithmic scales:* the magnitude scale for earthquakes, the decibel scale for sound, and the pH scale for acidity.

ACTIVITY

Towers of Hanoi

Use this activity to gain a sense of the kinds of problems this chapter will enable you to analyze. Additional activities are available online in MyLab Math.

You can explore key ideas of exponential growth with the game called *Towers of Hanoi* (shown in the photo), which consists of three pegs and a set of disks of varying sizes, each with a hole in its center so that it can be moved from peg to peg. The game begins with all the disks stacked on one peg in order of decreasing size. The object of the game is to move the entire stack of disks to a different peg, following two rules:

Rule 1. Only one disk can be moved at a time.
Rule 2. A larger disk can never be placed on top of a smaller disk.

You can easily find or make a version of the Towers of Hanoi game; many websites have online simulations of the game, or you can make your own version by cutting disks out of cardboard. Play the game with seven disks, looking for the most efficient strategy for moving the disks. Once you have found the best strategy, answer the following questions.

1. Briefly describe or demonstrate the strategy that most efficiently moves the disks from one peg to another.

2. You can view the game as a series of goals. The first goal is to end up with 1 disk on another peg, the second goal is to end up with 2 disks on another peg, and so on, until all the disks are on another peg. The first goal requires only one move: taking the smallest (top) disk and moving it to a different peg. The second goal then requires two more moves: first moving the second-smallest disk to the empty peg and then putting the smallest disk on top of it. Continue the game with the most efficient strategy for moving the disks, and complete the following table as you proceed.

Goal	Moves required for this step	Total moves in game so far
1 disk on another peg	1	1
2 disks on another peg	2	$1 + 2 = 3$
3 disks on another peg		
4 disks on another peg		
5 disks on another peg		
6 disks on another peg		
7 disks on another peg		

3. Look at the patterns in the table. Find general formulas for the second and third columns after n steps. Confirm that your formulas give the correct results for all the entries in the table. (*Hint:* If you are using the most efficient strategy, both formulas will involve powers of 2, with n appearing in the exponent.)

4. Use the formula for the total moves in the game (column 3) to predict the total number of moves required to complete the game with 10 disks (rather than 7).

5. The game is related to a Hindu legend claiming that, at the beginning of the world, the Brahma put three large diamond needles on a brass slab in

the Great Temple and placed 64 disks of solid gold on one needle; the disks were arranged in order of decreasing size, just like the disks in the Towers of Hanoi game. Working in shifts, day and night, the temple priests moved the golden disks according to the two rules of the game. How many total moves are required to move the entire set of 64 disks to another needle?

6. The legend holds that, upon completion of the task of moving all 64 disks, the temple will crumble and the world will come to an end. Assume that the priests can move very fast, so they move one disk each second. Based on your answer to Question 5, how many years would it take to move the entire stack of 64 disks? If the legend is true, do we have anything to worry about right now? (Useful data: Scientists estimate the current age of the universe to be about 14 billion years.)

7. The way in which the number of moves required for each step increases is an example of exponential growth—the topic of this chapter. Briefly comment on what this game illustrates about the nature of exponential growth.

UNIT 8A Growth: Linear vs. Exponential

Imagine two communities, Straightown and Powertown, each with an initial population of 10,000 people (Figure 8.1). Straightown grows at a constant rate of 500 people per year, so its population reaches 10,500 after 1 year, 11,000 after 2 years, 11,500 after 3 years, and so on. Powertown grows at a constant rate of 5 *percent* per year. Because 5% of 10,000 is 500, Powertown's population also reaches 10,500 after 1 year. In the second year, however, Powertown's population increases by 5% of 10,500, which is 525, to 11,025. In the third year, Powertown's population increases by 5% of 11,025, or by 551 people. Figure 8.1 contrasts the populations of the two towns over a period of several decades. Note that Powertown's population rises ever more steeply and quickly outpaces Straightown's.

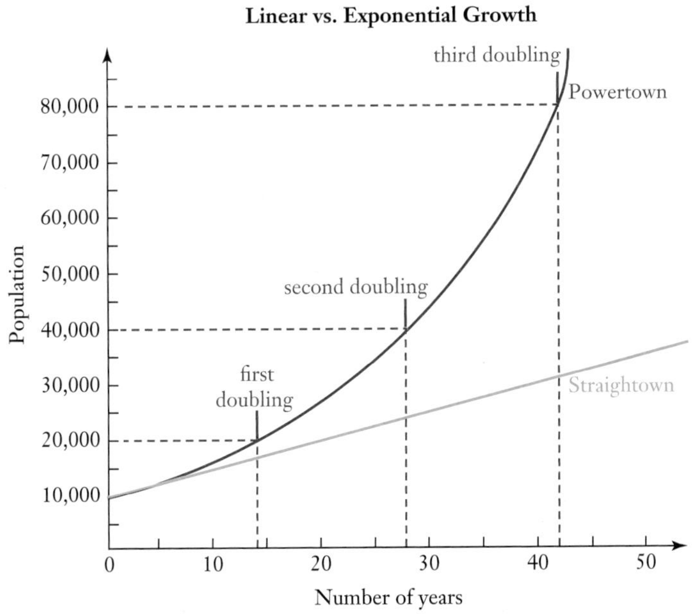

FIGURE 8.1 Straightown grows linearly, while Powertown grows exponentially.

Straightown and Powertown illustrate two fundamentally different types of growth. Straightown grows by the same *absolute* amount—500 people—each year, which is characteristic of **linear growth**. In contrast, Powertown grows by the same *relative* amount—5%—each year, which is characteristic of **exponential growth**.

Two Basic Growth Patterns

Linear growth occurs when a quantity grows by the same *absolute* amount in each unit of time.

Exponential growth occurs when a quantity grows by the same *relative* amount—that is, by the same *percentage*—in each unit of time.

The terms *linear* and *exponential* can also be applied to quantities that decrease with time. In those cases, *linear decay* means a quantity decreases by the same absolute amount in each unit of time, and *exponential decay* means a quantity decreases by the same relative (percentage) amount in each unit of time.

EXAMPLE 1 Linear or Exponential?

For each of the following situations, state whether the growth (or decay) is linear or exponential, and answer the question about it.

a. The number of students at Wilson High School has increased by 50 in each of the past 4 years. If the student population 4 years ago was 750, what is it today?
b. The price of milk has been rising 3% per year. If the price of a gallon of milk was \$4 a year ago, what is it now?
c. Tax law allows you to depreciate the value of your equipment by \$200 per year. If you purchased the equipment 3 years ago for \$1000, what is its depreciated value today?
d. The memory capacity of state-of-the-art computer storage devices has been doubling approximately every 2 years. If a company's top-of-the-line drive holds 16 terabytes today, what will it hold in 6 years?
e. The price of high-definition TV sets has been falling by about 25% per year. If the price is \$1000 today, what can you expect it to be in 2 years?

Solution

a. The number of students increased by the same absolute amount each year, so this is *linear growth*. Because the student population increased by 50 students per year, in 4 years it grew by $4 \times 50 = 200$ students, from 750 to 950.
b. The price rises by the same percent each year, so this is *exponential growth*. If the price was \$4 a year ago, it increased by $0.03 \times \$4 = \$0.12$, making the price \$4.12.
c. The equipment value decreases by the same absolute amount each year, so this is *linear decay*. In 3 years, the value decreases by $3 \times \$200 = \$600$, so the value decreases from \$1000 to \$400.
d. A doubling is the same as a 100% increase, so the 2-year doubling time represents *exponential growth*. With a doubling every 2 years, the capacity will double three times in 6 years: from 16 terabytes to 32 terabytes after 2 years, from 32 to 64 terabytes after 4 years, and from 64 to 128 terabytes after 6 years.
e. The price decreases by the same percentage each year, so this is *exponential decay*. From \$1000 today, the price will fall by 25%, or $0.25 \times \$1000 = \$250$, in 1 year. Therefore, next year's price will be \$750. The following year, the price will again fall by 25%, or $0.25 \times \$750 = \$187.50$, so the price after 2 years will be $\$750 - \$187.50 = \$562.50$.

▶ **Now try Exercises 9–16.**

The Impact of Doublings

Look again at the graph of Powertown's population in Figure 8.1. After about 14 years, the original population has doubled to 20,000. In the next 14 years, it doubles again to 40,000. It then doubles again, to 80,000, 14 years after that. This type of repeated *doubling*, in which each doubling occurs in the same amount of time, is a hallmark of exponential growth.

The time it takes for each doubling depends on the rate of the exponential growth. In Unit 8B, we'll see how the doubling time depends on the percentage growth rate. Here we'll explore three parables that show how doublings make exponential growth so different from linear growth.

Parable 1: From Hero to Headless in 64 Easy Steps

Legend has it that, when chess was invented in ancient times, a king was so enchanted that he said to the inventor, "Name your reward."

"If you please, king, put one grain of wheat on the first square of my chessboard," said the inventor. "Then place two grains on the second square, four grains on the third square, eight grains on the fourth square, and so on." The king gladly agreed, thinking the man a fool for asking for a few grains of wheat when he could have had gold or jewels. But let's see how it adds up for the 64 squares on a chessboard.

Table 8.1 shows the calculations. Each square gets twice as many grains as the previous square, so the number of grains on any square is a power of 2. The third column shows the total number of grains up to each point, and the last column gives a simple formula for the total number of grains.

TABLE 8.1

Square	Grains on This Square	Total Grains Thus Far	Formula for Total Grains
1	$1 = 2^0$	1	$2^1 - 1$
2	$2 = 2^1$	$1 + 2 = 3$	$2^2 - 1$
3	$4 = 2^2$	$3 + 4 = 7$	$2^3 - 1$
4	$8 = 2^3$	$7 + 8 = 15$	$2^4 - 1$
5	$16 = 2^4$	$15 + 16 = 31$	$2^5 - 1$
⋮	⋮	⋮	⋮
64	2^{63}		$2^{64} - 1$

From the pattern in the last column, we see that the grand total for all 64 squares is $2^{64} - 1$ grains. How much wheat is this? With a calculator, you can confirm that $2^{64} = 1.8 \times 10^{19}$, or about 18 *billion billion*. Not only would it be impossible to fit so many grains on a chessboard, but this number is larger than the total number of grains of wheat harvested in all human history. The king never finished paying the inventor and, according to legend, instead had him beheaded. ▶ **Now try Exercises 17–20.**

Parable 2: The Magic Penny

One lucky day, you meet a leprechaun who promises to give you fantastic wealth, but hands you only a penny before disappearing. You head home and place the penny under your pillow. The next morning, to your surprise, you find two pennies under your pillow. The following morning, you find four pennies, and the morning after that, eight pennies. Apparently, the leprechaun gave you a *magic* penny: While you sleep, each magic penny turns into *two* magic pennies. Table 8.2 shows your growing wealth. Note that

TABLE 8.2

Day	Amount Under Pillow
0	$\$0.01 \times 2^0 = \$0.01$
1	$\$0.01 \times 2^1 = \$0.02$
2	$\$0.01 \times 2^2 = \$0.04$
3	$\$0.01 \times 2^3 = \$0.08$
4	$\$0.01 \times 2^4 = \$0.16$
⋮	⋮
T	$\$0.01 \times 2^t$

"day 0" is the day you met the leprechaun. Generalizing from the first four rows of the table, the amount under your pillow after t days is

$$\$0.01 \times 2^t$$

We can use this formula to figure out how long it will be until you have fantastic wealth. After $t = 9$ days, you'll have $\$0.01 \times 2^9 = \$5.12$, which is barely enough to buy lunch. But by the end of a month, or $t = 30$ days, you'll have $\$0.01 \times 2^{30} = \$10{,}737{,}418.24$. That is, you'll be a multi-millionaire within a month, and you'll need a much larger pillow! In fact, if your magic pennies keep doubling, by the end of just 51 days, you'll have $\$0.01 \times 2^{51} \approx \$22.5$ trillion, which is about enough to pay off the national debt of the United States.

▶ Now try Exercises 21–24.

Parable 3: Bacteria in a Bottle

For our third parable, we return to the topic explored in the chapter-opening question (on p. 488). Suppose you place a single bacterium in a bottle at 11:00 a.m. It grows and at 11:01 divides into two bacteria. These two bacteria each grow and at 11:02 divide into four bacteria, which grow and at 11:03 divide into eight bacteria, and so on.

Now, suppose the bacteria continue to double every minute, and the bottle is full at 12:00. You may already realize that the number of bacteria at this point must be 2^{60} (because they doubled every minute for 60 minutes), but the important fact is that we have a bacterial disaster on our hands: Because the bacteria have filled the bottle, the entire bacterial colony is doomed. Let's examine this disaster in greater detail by asking a few questions about the demise of the colony.

BY THE WAY

The bacteria in a bottle parable was developed by University of Colorado Professor Albert A. Bartlett (1923–2013), who delivered more than 1740 lectures on the lessons of exponential growth around the country over a period of 40 years.

Question 1: The disaster occurred because the bottle was completely full at 12:00. When was the bottle *half*-full?

Answer: Because it took 1 hour to fill the bottle, many people guess that it was half-full after a half-hour, or at 11:30. However, because the bacteria *double* in number every minute, they must also have doubled during the last minute, which means the bottle went from being half-full to full during the final minute. That is, the bottle was half-full at 11:59, just 1 minute before the disaster.

Question 2: Imagine that *you* are a mathematically sophisticated bacterium, and at 11:56 you recognize the impending disaster. You immediately jump on your soapbox and warn that unless your fellow bacteria slow their growth dramatically, the end is just 4 minutes away. Will anyone believe you?

Answer: Note that the question is *not* whether you are correct, because you are: The bottle will indeed be full in just 4 minutes. Rather, the question is whether others who have not done the calculations will believe you. As we've already seen, the bottle would be half-full at 11:59. Continuing to work backward through the doublings each minute, we find that it would be $\frac{1}{4}$ full at 11:58, $\frac{1}{8}$ full at 11:57, and $\frac{1}{16}$ full at 11:56. Therefore, if your fellow bacteria look around the bottle at 11:56, they'll see that only $\frac{1}{16}$ of the bottle's space has been used. In other words, $\frac{15}{16}$ of the bottle's space is unused, which means the amount of unused space is *15 times* the amount of used space. You are asking your fellow bacteria to believe that, in just the next 4 minutes, they'll fill 15 times as much space as they did in their entire 56-minute history. Unless they do the mathematics for themselves, they are unlikely to take your warnings seriously. Figure 8.2 shows the situation graphically. Note that the bottle remains nearly empty for most of the 60 minutes, but the continued doublings fill it rapidly in the final 4 minutes.

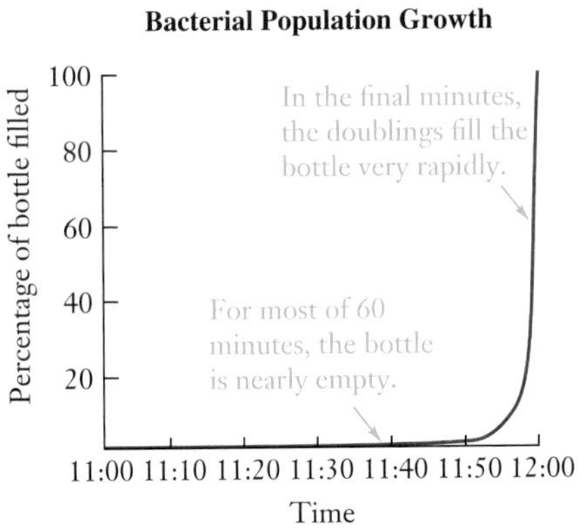

FIGURE 8.2 The population of the bacteria in the bottle.

Question 3: It's 11:59 and, with the bottle half-full, your fellow bacteria are finally taking your warnings seriously. They quickly start a space program, sending little bacterial spaceships out into the lab in search of new bottles. Thankfully, they discover three more bottles (making a total of four, including the one already occupied). Working quickly, they initiate a mass migration by packing bacteria onto spaceships and sending them to the new bottles. They successfully distribute the population evenly among the four bottles, just in time to avert the disaster. Given that they now have four bottles rather than just one, how much time have they gained for their civilization?

Answer: Because it took 1 hour to fill one bottle, you might guess that it would take 4 hours to fill four bottles. But remember that the bacterial population continues to *double* each minute. If there are enough bacteria to fill one bottle at 12:00, there will be enough to fill two bottles by 12:01 and four bottles by 12:02. The discovery of three new bottles gives the bacteria only 2 additional minutes.

Question 4: Suppose the bacteria continue their space program, constantly looking for more bottles. Is there any hope that further discoveries will allow the colony to continue its exponential growth?

Answer: Let's do some calculations. After n minutes, the bacterial population is 2^n. For example, it is $2^0 = 1$ when the first bacterium starts the colony at 11:00, $2^1 = 2$ at 11:01, $2^2 = 4$ at 11:02, and so on. There are 2^{60} bacteria when the first bottle fills at 12:00, and 2^{62} bacteria when four bottles are full at 12:02. Suppose that, somehow, the bacteria managed to keep doubling every minute until 1:00. By that time, the number of bacteria would be 2^{120} because it has been 120 minutes since the colony began. Now, we must figure out how much space they'd require for this population.

The smallest bacteria measure approximately 10^{-7}m (0.1 micrometer) across. If we assume that the bacteria are roughly cube-shaped, the volume of a single bacterium is

$$(10^{-7}\text{ m})^3 = 10^{-21}\text{ m}^3$$

Therefore, the colony of 2^{120} bacteria would occupy a total volume of

$$2^{120} \times 10^{-21}\text{ m}^3 \approx 1.3 \times 10^{15}\text{ m}^3$$

With this volume, the bacteria would cover the entire surface of the Earth in a layer more than 2 meters deep! (See Exercise 27 to calculate this result for yourself.)

In fact, if the doublings continued for just $5\frac{1}{2}$ hours, the volume of bacteria would exceed the volume of the entire universe (see Exercise 28). Needless to say, this cannot happen. The exponential growth of the colony cannot possibly continue for long, no matter what technological advances might be imagined.

Facts do not cease to exist because they are ignored.

—Aldous Huxley

EXAMPLE 2 **Number of Bottles**

How many bottles would the bacteria fill at the end of the second hour?

Solution For this calculation, we start with the fact that the bacteria have filled 1 bottle at the end of the first hour (at 12:00). As they continue to double, they fill $2^1 = 2$ bottles at 12:01, $2^2 = 4$ bottles at 12:02, and so on. In other words, during the second hour, the number of bottles filled is 2^m, where m is the number of minutes that have passed since 12:00. Because there are 60 minutes in the second hour, the number of bottles at the end of the second hour is 2^{60}. With a calculator, you will find that

$$2^{60} \approx 1.15 \times 10^{18}$$

At the end of the second hour, the bacteria would fill approximately 10^{18} bottles. Using the rules for working with powers of 10 (see the Brief Review in Chapter 2, on p. 92), we can write $10^{18} = 10^6 \times 10^{12}$. We recognize that $10^6 = 1$ million and $10^{12} = 1$ trillion. Therefore, 10^{18} is a *million trillion*—which is the correct answer for the chapter-opening question on p. 488.

▶ **Now try Exercises 25–28.**

Think About It Some people have suggested that we could find room for an exponentially growing human population by colonizing other planets in our solar system. Is this possible?

Doubling Lessons

The three parables reveal at least two key lessons about the repeated doublings that arise with exponential growth. First, if you look back at Table 8.1, you'll notice that the number of grains on each square is nearly equal to the total number of grains on all previous squares combined. For example, the 16 grains on the fifth square are 1 more than the total of 15 grains on the first four squares combined.

Second, all three parables show quantities growing to impossible proportions. We cannot possibly fit all the wheat harvested in world history on a chessboard, we cannot fit $22 trillion worth of pennies under your pillow, and a colony of bacteria could not keep growing until it filled the universe. The following box summarizes the two lessons.

Key Facts About Exponential Growth

- Exponential growth leads to repeated doublings. With each doubling, the size of the most recent doubling is approximately equal to the *sum* of all preceding doublings.
- Exponential growth cannot continue indefinitely. After only a relatively small number of doublings, exponentially growing quantities reach impossible proportions.

Quick Quiz 8A

Choose the best answer to each of the following questions. Explain your reasoning with one or more complete sentences.

1. A town's population increases in one year from 100,000 to 120,000. If the population is growing *linearly,* at a steady rate, then at the end of a second year it will be

   a. 120,000. b. 140,000. c. 144,000.

2. A town's population increases in one year from 100,000 to 120,000. If the population is growing *exponentially* at a steady rate, then at the end of a second year it will be

   a. 120,000. b. 140,000. c. 144,000.

3. The balance owed on your credit card doubles from $1000 to $2000 in 6 months. If your balance is growing *exponentially*, how much longer will it be until it reaches $8000?

a. 6 months b. 12 months c. 18 months

4. The number of songs in your music library has increased from 200 to 400 in 3 months. If the number of songs is increasing *linearly*, how much longer will it be until you have 800 songs?

a. 3 months b. 6 months c. $1\frac{1}{2}$ months

5. Which of the following is an example of *exponential decay*?

a. The population of a rural community decreases by 100 people per year.

b. The price of gasoline decreases by $0.02 per week.

c. Government support for education decreases by 1% per year.

6. On a chessboard with 64 squares, you place 1 penny on the first square, 2 pennies on the second square, 4 pennies on the third square, and so on. If you could follow this pattern to fill the entire board, about how much money would you need in total?

a. about $1.28

b. about $500,000

c. about 10,000 times as much as the current U.S. federal debt

7. At 11:00 you place a single bacterium in a bottle, and at 11:01 it divides into 2 bacteria, which at 11:02 divide into 4 bacteria, and so on. How many bacteria will be in the bottle at 11:30?

a. 2×30 b. 2^{30} c. 2×10^{30}

8. Consider the bacterial population described in Question 7. How many more bacteria are in the bottle at 11:31 than at 11:30?

a. 30 b. 2^{30} c. 2×10^{30}

9. Consider the bacterial population described in Question 7. If the bacteria occupy a volume of 1 cubic meter at 12:02 and continue their exponential growth, when will they occupy a volume of 2 cubic meters?

a. 12:03 b. 12:04 c. 1:02

10. Which of the following is *not* true of any exponentially growing population?

a. With every doubling, the population increase is nearly equal to the total increase from all previous doublings.

b. The steady growth makes it easy to see any impending crisis long before the crisis becomes severe.

c. The exponential growth must eventually stop.

Exercises 8A

REVIEW QUESTIONS

1. Describe the basic differences between *linear growth* and *exponential growth*.
2. Briefly explain how repeated doublings characterize exponential growth. Describe the impact of doublings, using the wheat on the chessboard or magic penny parable.
3. Briefly summarize the story of the bacteria in the bottle. Be sure to explain the answers to the four questions asked in the text, and describe why the answers are surprising.
4. Explain the meaning of the two key facts about exponential growth given at the end of this unit. Then create your own example of exponential growth and describe the influence and impact of repeated doubling.

DOES IT MAKE SENSE?

Decide whether each of the following statements makes sense (or is clearly true) or does not make sense (or is clearly false). Explain your reasoning.

5. Money in a bank account earning compound interest at an annual percentage rate of 1.2% is an example of exponential growth.
6. Suppose you had a magic bank account in which your balance doubled each day. If you started with just $1, you'd be a multi-millionaire in less than a month.
7. A small town that grows exponentially can become a large city in just a few decades.
8. Human population has been growing exponentially for a few centuries, and we can expect this trend to continue forever in the future.

BASIC SKILLS & CONCEPTS

9–16: Linear or Exponential? State whether the growth (or decay) is linear or exponential, and answer the associated question.

9. The population of MeadowView is increasing at a rate of 300 people per year. If the population is 2400 today, what will it be in 4 years?
10. The population of Winesburg is increasing at a rate of 10% per year. If the population is 100,000 today, what will it be in 3 years?
11. During an episode of hyperinflation that occurred in Venezuela in 2016, the price of food increased at a rate of about 40% per month. If a food item cost 1000 bolívars (the country's unit of currency) at the beginning of this period, what was its price 4 months later?
12. The price of a gallon of gasoline is increasing by 4¢ per week. If the price is $3.20 per gallon today, what will it be in 10 weeks?

13. The price of computer memory is decreasing at a rate of 15% per year. If a memory chip costs \$50 today, what will it cost in 3 years?

14. The value of your car is decreasing by 10% per year. If the car is worth \$10,000 today, what will it be worth in 2 years?

15. The value of your house is increasing by \$2000 per year. If it is worth \$100,000 today, what will it be worth in 5 years?

16. The value of your house is increasing by 2% per year. If it is worth \$100,000 today, what will it be worth in 3 years?

17–20: Chessboard Parable. Use the chessboard parable presented in the text. Assume that each grain of wheat weighs 1/7000 pound.

17. How many grains of wheat should be placed on square 16 of the chessboard? Find the total number of grains and their total weight (in pounds) at this point.

18. How many grains of wheat should be placed on square 32 of the chessboard? Find the total number of grains and their total weight (in pounds) at this point.

19. What is the total weight of all the wheat when the chessboard is full?

20. The total world harvest of all grains (wheat, rice, and corn) in 2016 was about 2.5 billion tons. How does this total compare to the weight of the wheat on the chessboard? (1 ton = 2000 pounds)

21–24: Magic Penny Parable. Use the magic penny parable presented in the text.

21. How much money would you have after 21 days?

22. Suppose that you stacked the pennies after 21 days. How high would the stack rise, in kilometers (km)? (*Hint*: Find a few pennies and a ruler.)

23. How many days would elapse before you had a total of more than \$1 billion? (*Hint*: Proceed by trial and error.)

24. Suppose that you could keep making a single stack of the pennies. After how many days would the stack be long enough to reach the nearest star (besides the Sun), which is about 4.3 light-years (4.0×10^{13} km) away? (*Hint*: Proceed by trial and error.)

25–28: Bacteria in a Bottle Parable. Use the bacteria parable presented in the text.

25. How many bacteria are in the bottle at 11:50? What fraction of the bottle is full at that time?

26. How many bacteria are in the bottle at 11:15? What fraction of the bottle is full at that time?

27. **Knee-Deep in Bacteria.** The total surface area of Earth is about 5.1×10^{14} m^2. Assume that the bacteria continued their doublings (as described in the text) for a total of 2 hours, at which point they were distributed uniformly over Earth's surface. How deep would the bacterial layer be? Would it be knee-deep, more than knee-deep, or less than knee-deep? (*Hint*: You can find the approximate depth by dividing the bacteria volume by Earth's surface area.)

28. **Bacterial Universe.** Suppose the bacteria in the parable continued to double their population every minute. How long would it take until their volume exceeded the total volume of the observable universe, which is about 10^{79} m^3? (*Hint*: Proceed by trial and error.)

FURTHER APPLICATIONS

29. **Human Doubling.** Human population in the year 2000 was about 6 billion and was increasing with a doubling time of 50 years. Suppose population continued this growth pattern from the year 2000 into the future.

a. Extend the following table, showing the population at 50-year intervals under this scenario, until you reach the year 3000. Use scientific notation, as shown.

Year	Population
2000	6×10^9
2050	$12 \times 10^9 = 1.2 \times 10^{10}$
2100	$24 \times 10^9 = 2.4 \times 10^{10}$
⋮	⋮

b. The total surface area of Earth is about 5.1×10^{14} m^2. Assuming that people could occupy all this area (in reality, most of it is ocean), approximately when would people be so crowded that every person would have only 1 m^2 of space?

c. Suppose that, when we take into account the area needed to grow food and to find other resources, each person actually requires about 10^4 m^2 of area to survive. About when would we reach this limit?

d. Suppose that we learn to colonize other planets and moons in our solar system. The total surface area of the worlds in our solar system that could potentially be colonized (not counting gas planets such as Jupiter) is roughly five times the surface area of Earth. Under the assumptions of part (c), could humanity fit in our solar system in the year 3000? Explain.

30. **Doubling Time vs. Initial Amount.**

a. Would you rather start with one penny (\$0.01) and double your wealth every day or start with one dime (\$0.10) and double your wealth every 5 days (assuming you want to get rich)? Explain.

b. Would you rather start with one penny (\$0.01) and double your wealth every day or start with \$1000 and double your wealth every 2 days (assuming you want to get rich in the long run)? Explain.

c. Which is more important in determining how fast exponential growth occurs: the doubling time or the initial amount? Explain.

31. **Facebook Users.** The table shows the number of monthly active users of Facebook (the number of people who use Facebook at least once a month) over a 3-year period.

	Month			
	Dec. 2013	Dec. 2014	Dec. 2015	Dec. 2016
Monthly active users (millions)	1228	1393	1591	1860
Absolute change over previous year (millions)	—			
Percent change over previous year	—			

a. Fill in the middle row of the table showing the absolute change in the number of active monthly users.

b. Fill in the bottom row of the table showing the percent change in the number of active monthly users.

c. Does the increase in Facebook users appear to be linear or exponential? Is it reasonable for investors to expect this growth pattern to continue for decades? Justify your answers.

IN YOUR WORLD

32. **Linear Growth.** Identify at least two news stories that describe a quantity undergoing *linear* growth or decay. Describe the growth or decay process in each.

33. **Exponential Growth.** Identify at least two news stories that describe a quantity undergoing *exponential* growth or decay. Describe the growth or decay process in each.

34. **Computing Power.** Choose an aspect of computing power (such as processor speed or memory chip capacity) and investigate its growth over time. Is the growth pattern closer to linear or exponential? Do you think the past pattern of growth can continue in the future? Explain.

35. **Web Growth.** Investigate the growth of the Web itself, in terms of both number of users and number of Web pages. Has the growth been linear or exponential? How do you think the growth will change in the future? Explain.

UNIT 8B Doubling Time and Half-Life

Exponential growth leads to repeated *doublings* and exponential decay leads to repeated *halvings*. However, in most cases of exponential growth or decay, we are given a *rate* of growth or decay—usually as a percentage—rather than the time required for doubling or halving. In this unit, we'll convert between growth (or decay) rates and doubling (or halving) times.

Doubling Time

The time required for each doubling in exponential growth is called the **doubling time**. For example, the doubling time for the magic penny (see Unit 8A) was 1 day, because your wealth doubled each day. The doubling time for the bacteria in the bottle was 1 minute.

Given the doubling time, we can easily calculate the value of a quantity at any time. Consider an initial population of 10,000 that grows with a doubling time of 10 years:

- In 10 years, or one doubling time, the population increases by a factor of 2, to a new population of $2 \times 10{,}000 = 20{,}000$.
- In 20 years, or two doubling times, the population increases by a factor of $2^2 = 4$, to a new population of $4 \times 10{,}000 = 40{,}000$.
- In 30 years, or three doubling times, the population increases by a factor of $2^3 = 8$, to a new population of $8 \times 10{,}000 = 80{,}000$.

To write a general formula, let's use t for the amount of time that has passed and T_{double} for the doubling time. Note that after $t = 30$ years with a doubling time of $T_{\text{double}} = 10$ years, there have been $t/T_{\text{double}} = 30/10 = 3$ doublings. Generalizing, the number of doublings after a time t is t/T_{double}. That is, the size of the population after time t is the initial population times $2^{t/T_{\text{double}}}$.

Calculations with the Doubling Time

After a time t, an exponentially growing quantity with a doubling time of T_{double} increases in size by a factor of $2^{t/T_{double}}$. The new value of the growing quantity is related to its initial value (at $t = 0$) by

$$\text{new value} = \text{initial value} \times 2^{t/T_{double}}$$

Think About It Consider an initial population of 10,000 that grows with a doubling time of 10 years. Confirm that the above formula gives a population of 80,000 after 30 years, as we found earlier. What does the formula predict for the population after 50 years?

EXAMPLE 1 Doubling with Compound Interest

Compound interest (Unit 4B) produces exponential growth because an interest-bearing account grows by the same percentage each year. Suppose your bank account has a doubling time of 21 years. By what factor does your balance increase in 50 years?

Solution The doubling time is $T_{double} = 21$ years, so after $t = 50$ years your balance increases by a factor of

$$2^{t/T_{double}} = 2^{50\text{ yr}/21\text{ yr}} = 2^{2.3810} \approx 5.21$$

▲ **CAUTION!** Before you calculate with the doubling time formula, be sure that the units of t and T_{double} are the same, as they are in this example (both given in years). If they are not given with the same units, you'll first need to convert one of them to the units of the other. ▲

For example, if you start with a balance of \$1000, in 50 years it will grow to about \$1000 × 5.21 = \$5,210.

▶ **Now try Exercises 25–32.**

EXAMPLE 2 World Population Growth

World population doubled from 3 billion in 1960 to 6 billion in 2000. Suppose that world population continued to grow (after 2000) with the same doubling time of 40 years. What would the population be in 2050? In 2200?

Solution The doubling time is $T_{double} = 40$ years. If we let $t = 0$ represent 2000, the year 2050 is $t = 50$ years later. If the 2000 population of 6 billion is used as the initial value, the population in 2050 would be

$$\begin{aligned}\text{new value} &= \text{initial value} \times 2^{t/T_{double}}\\ &= 6\text{ billion} \times 2^{50\text{ yr}/40\text{ yr}}\\ &= 6\text{ billion} \times 2^{1.2} \approx 14.3\text{ billion}\end{aligned}$$

By 2200, which is $t = 200$ years after 2000, the population would reach

$$\begin{aligned}\text{new value} &= \text{initial value} \times 2^{t/T_{double}}\\ &= 6\text{ billion} \times 2^{200\text{ yr}/40\text{ yr}}\\ &= 6\text{ billion} \times 2^{5} = 192\text{ billion}\end{aligned}$$

If world population continued to grow at the same rate it did between 1960 and 2000, it would be on track to reach 14 billion by 2050 and 192 billion by 2200.

▶ **Now try Exercises 33–34.**

Think About It Do you think that it's really possible for the human population on Earth to reach 192 billion? Why or why not?

The Approximate Doubling Time Formula

Consider an ecological study of a prairie dog community. The community contains 100 prairie dogs when the study begins, and researchers determine that the population is increasing at a rate of 10% per month. That is, each month the population grows to 110% *of*, or 1.1 times, its previous value (see the "*of* versus *more than*" rule in Unit 3A). Table 8.3 tracks the population growth (rounded to the nearest whole number).

TABLE 8.3 Growth of a Prairie Dog Community

Month	Population	Month	Population
0	100	8	$(1.1)^8 \times 100 = 214$
1	$(1.1)^1 \times 100 = 110$	9	$(1.1)^9 \times 100 = 236$
2	$(1.1)^2 \times 100 = 121$	10	$(1.1)^{10} \times 100 = 259$
3	$(1.1)^3 \times 100 = 133$	11	$(1.1)^{11} \times 100 = 285$
4	$(1.1)^4 \times 100 = 146$	12	$(1.1)^{12} \times 100 = 314$
5	$(1.1)^5 \times 100 = 161$	13	$(1.1)^{13} \times 100 = 345$
6	$(1.1)^6 \times 100 = 177$	14	$(1.1)^{14} \times 100 = 380$
7	$(1.1)^7 \times 100 = 195$	15	$(1.1)^{15} \times 100 = 418$

Note that the population nearly doubles (to 195) after 7 months, then nearly doubles again (to 380) after 14 months. This roughly 7-month doubling time is related to the 10% growth rate as follows:

$$\text{doubling time} \approx \frac{70}{\text{percentage growth rate}} = \frac{70}{10/\text{month}} = 7 \text{ months}$$

This formula, in which the doubling time is approximately 70 divided by the percentage growth rate, works whenever the growth rate is relatively small (less than about 15%). It is often called the **rule of 70**.

Approximate Doubling Time Formula (Rule of 70)

For a quantity growing exponentially at a rate of P% per time period, the doubling time is *approximately*

$$T_{\text{double}} \approx \frac{70}{P}$$

This approximation works best for small growth rates and breaks down for growth rates over about 15%.

EXAMPLE 3 Population Doubling Time

World population reached 7.5 billion in 2017 and was growing at a rate of about 1.1% per year. What is the approximate doubling time at this growth rate? If this growth rate were to continue, what would world population be in 2050? Compare this number to the result in Example 2.

Solution We can use the approximate doubling time formula because the growth rate is much less than 15%. The percentage growth rate of 1.1% per year means we set $P = 1.1/\text{year}$.

$$T_{\text{double}} \approx \frac{70}{P} = \frac{70}{1.1/\text{year}} \approx 64 \text{ years}$$

The doubling time is approximately 64 years. The year 2050 is $t = 33$ years after 2017, so the population in 2050 would be

$$\begin{aligned} \text{new value} &= \text{initial value} \times 2^{t/T_{\text{double}}} \\ &= 7.5 \text{ billion} \times 2^{33\text{ yr}/64\text{ yr}} \\ &= 7.5 \text{ billion} \times 2^{0.5156} \approx 10.7 \text{ billion} \end{aligned}$$

At a 1.1% annual growth rate starting from the 2017 population of 7.5 billion, world population would reach about 10.7 billion in 2050. This is more than 3 billion fewer people than predicted in Example 2, reflecting the fact that population growth has slowed.

▶ Now try Exercises 35–36.

Think About It United Nations intermediate projections suggest that world population will be 9.7 billion in 2050. For these projections, what assumption is being made about the population growth rate between now and 2050 compared to its current growth rate? Do you think this assumption is valid? Why or why not?

EXAMPLE 4 Solving the Doubling Time Formula

World population doubled in the 40 years from 1960 to 2000. What was the average percentage growth rate during this period? Contrast this growth rate with the 2017 growth rate of 1.1% per year.

Solution We answer the question by solving the approximate doubling time formula for P. Multiplying both sides of the formula by P and dividing both sides by T_{double}, we have

$$P \approx \frac{70}{T_{\text{double}}}$$

Substituting $T_{\text{double}} = 40$ years, we find

$$P \approx \frac{70}{T_{\text{double}}} = \frac{70}{40 \text{ years}} = 1.75/\text{year}$$

The average population growth rate between 1960 and 2000 was about $P\% = 1.75\%$ per year. This is significantly higher than the 2017 growth rate of 1.1% per year.

▶ Now try Exercises 37–40.

Exponential Decay and Half-Life

Exponential *decay* occurs whenever a quantity *decreases* by the same percentage in every fixed time period (for example, by 20% every year). In that case, the value of the quantity repeatedly decreases to half its value, with each halving occurring in a time called the **half-life**.

You may have heard half-life applied to radioactive materials such as uranium or plutonium. For example, radioactive plutonium-239 (Pu-239) has a half-life of about 24,000 years. To understand the meaning of the half-life, suppose that 100 pounds of Pu-239 is deposited at a nuclear waste site. The plutonium gradually decays into other substances as follows:

- In 24,000 years, or one half-life, the amount of Pu-239 declines to $\frac{1}{2}$ of its original value, or to $\frac{1}{2} \times 100$ pounds $= 50$ pounds.
- In 48,000 years, or two half-lives, the amount of Pu-239 declines to $\left(\frac{1}{2}\right)^2 = \frac{1}{4}$ of its original value, or to $\frac{1}{4} \times 100$ pounds $= 25$ pounds.
- In 72,000 years, or three half-lives, the amount of Pu-239 declines to $\left(\frac{1}{2}\right)^3 = \frac{1}{8}$ of its original value, or to $\frac{1}{8} \times 100$ pounds $= 12.5$ pounds.

BY THE WAY

Plutonium-239 is the chemical element plutonium in a form (or *isotope*) with atomic mass number 239. Atomic mass number is the total number of protons and neutrons in the nucleus. Because all plutonium nuclei have 94 protons, Pu-239 nuclei have $239 - 94 = 145$ neutrons.

We can generalize this idea much as we did earlier for the doubling time. A single halving reduces a quantity by a factor of $\frac{1}{2}$, two halvings reduce it by a factor of $\left(\frac{1}{2}\right)^2$, and three halvings reduce it by a factor of $\left(\frac{1}{2}\right)^3$. If we let t be the amount of time that has passed and T_{half} be the half-life, then the number of halvings after a time t is t/T_{half}. That is, the quantity after time t is the original quantity times this factor of $\left(\frac{1}{2}\right)^{t/T_{half}}$.

Calculations with the Half-Life

After a time t, an exponentially decaying quantity with a half-life of T_{half} decreases in size by a factor of $\left(\frac{1}{2}\right)^{t/T_{half}}$. The new value of the decaying quantity is related to its initial value (at $t = 0$) by

$$\text{new value} = \text{initial value} \times \left(\frac{1}{2}\right)^{t/T_{half}}$$

EXAMPLE 5 Carbon-14 Decay

Radioactive carbon-14 has a half-life of about 5700 years. It collects in organisms only while they are alive. Once they are dead, it only decays. What fraction of the carbon-14 in an animal bone still remains 1000 years after the animal has died?

Solution The half-life is $T_{half} = 5700$ years, so the fraction of the initial amount of carbon-14 remaining after $t = 1000$ years is

$$\left(\frac{1}{2}\right)^{t/T_{half}} = \left(\frac{1}{2}\right)^{1000\text{ yr}/5700\text{ yr}} \approx 0.885$$

For example, if the bone originally contained 1 kilogram of carbon-14, the amount remaining after 1000 years is approximately 0.885 kilogram. This steady decay means that we can use measurements of the amount of carbon-14 to determine the age of bones found at archaeological sites, as we'll discuss in Unit 9C.

▲ **CAUTION!** Before you calculate with the half-life formula, be sure that the units of t and T_{half} are the same, as they are in this example (both given in years). If they are not given with the same units, you'll first need to convert one of them to the units of the other. ▲

▶ Now try Exercises 41–44.

BY THE WAY

Ordinary carbon is carbon-12, which is stable (not radioactive). Carbon-14 is produced in the Earth's atmosphere by high-energy particles coming from the Sun. It mixes with ordinary carbon and therefore becomes incorporated into living tissue through respiration (breathing).

EXAMPLE 6 Plutonium after 100,000 Years

Suppose that 100 pounds of Pu-239 is deposited at a nuclear waste site. How much of it will still be present in 100,000 years?

Solution The half-life of Pu-239 is $T_{half} = 24{,}000$ years. Given an initial amount of 100 pounds, the amount remaining after $t = 100{,}000$ years is

$$\text{new value} = \text{initial value} \times \left(\frac{1}{2}\right)^{t/T_{half}} = 100\text{ pounds} \times \left(\frac{1}{2}\right)^{100{,}000\text{ yr}/24{,}000\text{ yr}}$$
$$\approx 5.6\text{ pounds}$$

About 5.6 pounds of the original 100 pounds of Pu-239 will still be present in 100,000 years.

▶ Now try Exercises 45–48.

HISTORICAL NOTE

The atomic bomb that devastated Nagasaki, Japan, during World War II generated its destructive power by fission of plutonium-239. (The Hiroshima bomb used uranium-235.)

Think About It Plutonium, which is not found naturally on the Earth, is made in nuclear reactors for use both as fuel for nuclear power plants and in nuclear weapons. Based on its half-life, explain why the safe disposal of Pu-239 poses a challenge.

The Approximate Half-Life Formula

The approximate doubling time formula (the rule of 70) we found earlier works equally well for exponential decay if we replace the doubling time with the half-life and the percentage growth rate with the percentage decay rate.

Technical Note

Some texts consider P to be negative for exponential decay, in which case the half-life is approximated by $70/|P|$, where $|P|$ is the absolute value of P.

Approximate Half-Life Formula

For a quantity decaying exponentially at a rate of $P\%$ per time period, the half-life is *approximately*

$$T_{\text{half}} \approx \frac{70}{P}$$

This approximation works best for small decay rates and breaks down for decay rates over about 15%.

EXAMPLE 7 Devaluation of Currency

Suppose that inflation causes the value of the Russian ruble to fall at a rate of 12% per year (relative to the dollar). At this rate, approximately how long does it take for the ruble to lose half its value?

Solution We can use the approximate half-life formula because the decay rate is less than 15%. The 12% decay rate means we set $P = 12/\text{year}$.

$$T_{\text{half}} \approx \frac{70}{P} = \frac{70}{12/\text{year}} \approx 5.8 \text{ years}$$

The half-life is a little less than 6 years, meaning that the ruble loses half its value (against the dollar) in 6 years. ▶ **Now try Exercises 49–52.**

Exact Formulas for Doubling Time and Half-Life

The approximate doubling time and half-life formulas are useful because they are easy to remember. However, for more precise work, or for rates of growth or decay where the approximate formulas break down, we need the exact formulas given below. (We will see how these formulas are derived in Unit 9C.) These formulas use the **fractional growth rate**, defined as $r = P/100$, with r positive for growth and negative for decay. For example, if the percentage growth rate is 5% per year, the fractional growth rate is $r = 0.05$ per year. For a 5% *decay* rate per year, the fractional growth rate is $r = -0.05$ per year. The formulas also use logarithms, which are reviewed in the next Brief Review (on p. 506).

BY THE WAY

Don't let logarithms intimidate you! As the Brief Review (on p. 506) explains, they are not difficult, despite their odd-sounding name. Just remember that "$\log_{10} x$" simply means "the power of 10 that makes x." For example, if you are asked to evaluate $\log_{10} 100$, remember that this means "the power of 10 that makes 100," and you know this is 2 because $10^2 = 100$. You may find it helpful to mentally replace "log" with "the power of," so that "$\log_{10} x$" becomes "the power of 10 (that makes x)".

Exact Doubling Time Formula

For an exponentially growing quantity with a fractional growth rate r, the doubling time is

$$T_{\text{double}} = \frac{\log_{10} 2}{\log_{10}(1 + r)}$$

Note: The units of time used for T_{double} and r must be the same. For example, if the fractional growth rate is 0.05 per *month*, then the doubling time also will be measured in months.

Exact Half-Life Formula

For an exponentially decaying quantity, in which the fractional decay rate r is negative $(r < 0)$, the half-life is

$$T_{\text{half}} = -\frac{\log_{10} 2}{\log_{10}(1 + r)}$$

Note: The units of time used for T_{half} and r must be the same.

Note that because r is positive in the doubling time formula and negative in the half-life formula, both formulas will yield positive values.

EXAMPLE 8 Large Growth Rate

A population of rats is growing at a rate of 80% per month. Find the exact doubling time for this growth rate, and compare it to the doubling time found with the approximate doubling time formula.

Solution The growth rate of 80% per month means $P = 80/\text{month}$ or $r = 0.8/\text{month}$. The doubling time is

$$T_{\text{double}} = \frac{\log_{10} 2}{\log_{10}(1 + 0.8)} = \frac{0.301030}{\log_{10}(1.8)} = \frac{0.301030}{0.255273} \approx 1.18 \text{ months}$$

▲ **CAUTION!** Be sure to check that you get the above answer with your calculator. If you don't, then you may be using the wrong calculator key to find base 10 logs. ▲

The doubling time is about 1.2 months. This answer should make sense: With the population growing by 80% in a month, we expect it to take a little over 1 month to grow by 100% (which is a doubling). In contrast, the approximate doubling time formula predicts a doubling time of $70/P = 70/80 = 0.875$ month, which is less than 1 month. We see that the approximate formula does not work well for large growth rates. ▶ Now try Exercises 53–54.

EXAMPLE 9 Ruble Revisited

Suppose the Russian ruble is falling in value against the dollar at 12% per year. Using the exact half-life formula, determine how long it takes the ruble to lose half its value. Compare your answer to the approximate answer found in Example 7.

Solution The percentage decay rate is $P = 12\%/\text{yr}$. Because this is a rate of *decay*, we set the fractional growth rate to $r = -0.12/\text{year}$. The half-life is

$$T_{\text{half}} = -\frac{\log_{10} 2}{\log_{10}(1 - 0.12)} = -\frac{0.301030}{-0.055517} \approx 5.42 \text{ years}$$

The ruble loses half its value against the dollar in about 5.4 years. This result is only about 0.4 year less than the 5.8 years obtained with the approximate formula. We see that the approximate formula is reasonably accurate for the 12% decay rate.

▶ Now try Exercises 55–56.

USING TECHNOLOGY

Logarithms

Most calculators have a key for computing common (base 10) logarithms, usually but not always labeled "log." To be sure you are using the correct key, check that your calculator returns $\log_{10} 10 = 1$.

In Excel, use the built-in function LOG10, as shown in the screen shot below. (You can also use the LOG function, which works for any base but uses base 10 by default.)

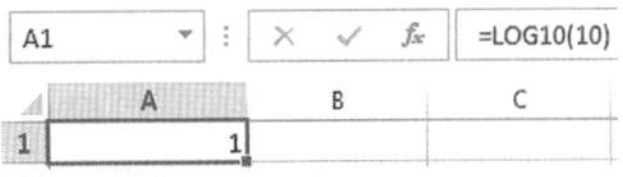

Microsoft Excel, Windows 10, Microsoft Corporation.

Brief Review Logarithms

A **logarithm** (or **log**, for short) is a power or exponent. In this book, we focus on base 10 logs, also called **common logs**, which are defined as follows:

$\log_{10} x$ means *the power of 10 that makes x.*

For example:

$$\log_{10} 1000 = 3 \quad \text{because } 10^3 = 1000$$
$$\log_{10} 10{,}000{,}000 = 7 \quad \text{because } 10^7 = 10{,}000{,}000$$
$$\log_{10} 1 = 0 \quad \text{because } 10^0 = 1$$
$$\log_{10} 0.1 = -1 \quad \text{because } 10^{-1} = 0.1$$
$$\log_{10} 30 \approx 1.477 \quad \text{because } 10^{1.477} \approx 30$$

Four important rules follow directly from the definition of a logarithm.

Rule 1. Taking the logarithm of a power of 10 gives the power. That is,

$$\log_{10} 10^x = x$$

Rule 2. Raising 10 to a power that is the logarithm of a number gives the number. That is,

$$10^{\log_{10} x} = x \quad (x > 0)$$

Rule 3. Because powers of 10 are multiplied by adding their exponents, we have the *addition rule* for logarithms:

$$\log_{10} xy = \log_{10} x + \log_{10} y \quad (x > 0 \text{ and } y > 0)$$

Rule 4. We can "bring down" an exponent within a logarithm by applying the *power rule* for logarithms:

$$\log_{10} a^x = x \times \log_{10} a \quad (a > 0)$$

Most calculators have a key to compute $\log_{10}$ for any positive number. You should find this key on your calculator and use it to verify that $\log_{10} 1000 = 3$ and $\log_{10} 2 \approx 0.301030$.

Example: Given that $\log_{10} 2 \approx 0.301030$, find each of the following:

a. $\log_{10} 8$

b. $10^{\log_{10} 2}$

c. $\log_{10} 200$

Solution:

a. We notice that $8 = 2^3$. Therefore, from Rule 4,

$$\log_{10} 8 = \log_{10} 2^3$$
$$= 3 \times \log^{10} 2 \approx 3 \times 0.301030 = 0.90309$$

b. From Rule 2,

$$10^{\log_{10} 2} = 2$$

c. Notice that $200 = 2 \times 100 = 2 \times 10^2$. Therefore, from Rule 3,

$$\log_{10} 200 = \log_{10}(2 \times 10^2) = \log_{10} 2 + \log_{10} 10^2$$

From Rule 1, we know that $\log_{10} 10^2 = 2$, so

$$\log_{10} 200 = \log_{10}(2 \times 10^2) = \log_{10} 2 + \log_{10} 10^2$$
$$\approx 0.301030 + 2 = 2.301030$$

Example: Someone tells you that $\log_{10} 600 = 5.778$. Should you believe this?

Solution: Because 600 is between 100 and 1000, $\log_{10} 600$ must be between $\log_{10} 100$ and $\log_{10} 1000$. From Rule 1, we find that $\log_{10} 100 = \log_{10} 10^2 = 2$ and $\log_{10} 1000 = \log_{10} 10^3 = 3$. Therefore, $\log_{10} 600$ must be between 2 and 3, and the claimed value of 5.778 must be wrong.

▶ Now try Exercises 13–24.

Quick Quiz 8B

Choose the best answer to each of the following questions. Explain your reasoning with one or more complete sentences.

1. Suppose the value of an investment doubles every 8 years. By what factor will its value rise in 30 years?

   a. $2^{30/8}$ b. $2^{8/30}$ c. 8×30

2. Suppose your salary increases at a rate of 4.5% per year. Then your salary will double in approximately

   a. 4.5 years. b. $\frac{70}{4.5}$ years. c. $\frac{4.5}{70}$ years.

3. Which of the following is *not* a good approximation of a doubling time?

   a. Inflation running at 35% per year will cause prices to double in about 2 years.

   b. A town growing at 2% per year will double its population in about 35 years.

   c. A bank account balance growing at 7% per year will double in about 10 years.

4. A town's population doubles in 23 years. Its percentage growth rate is approximately

a. 23% per year. b. $\frac{70}{23}$ per year. c. $\frac{23}{70}$ per year.

5. Radioactive tritium (hydrogen-3) has a half-life of about 12 years, which means that if you start with 1 kg of tritium, 0.5 kg will decay during the first 12 years. How much will decay during the next 12 years?

a. $\frac{12}{0.5}$ kg b. 0.5 kg c. 0.25 kg

6. Radioactive uranium-235 has a half-life of about 700 million years. Suppose a rock is 2.8 billion years old. What fraction of the rock's original uranium-235 still remains?

a. $\frac{1}{2}$ b. $\frac{1}{16}$ c. $\frac{1}{700}$

7. The population of an endangered species decreases at a rate of 10% per year. Approximately how long will it take the population to decrease to half of its current value?

a. 7 years b. 10 years c. 70 years

8. $\log_{10} 10^8 =$

a. 100,000,000 b. 108 c. 8

9. A rural population decreases at a rate of 20% per decade. If you wish to calculate this population's exact half-life, you should set the fractional growth rate per decade to

a. $r = 20$.

b. $r = 0.2$.

c. $r = -0.2$.

10. A new company's revenues increase at 15% per year. The doubling time for the revenues is

a. $\frac{\log_{10} 2}{\log_{10} 1.15}$ years.

b. $\frac{\log_{10} 2}{\log_{10} 0.85}$ years.

c. $\frac{\log_{10} (1 + 0.15)}{\log_{10} 2}$ years.

Exercises 8B

REVIEW QUESTIONS

1. What is a *doubling time*? Suppose a population has a doubling time of 25 years. By what factor will it grow in 25 years? In 50 years? In 100 years?
2. Given a doubling time, explain how you calculate the value of an exponentially growing quantity at any time *t*.
3. State the approximate doubling time formula and the conditions under which it works well. Give an example.
4. What is a *half-life*? Suppose a radioactive substance has a half-life of 1000 years. What fraction will be left after 1000 years? After 2000 years? After 4000 years?
5. Given a half-life, explain how you calculate the value of an exponentially decaying quantity at any time *t*.
6. State the approximate half-life formula and the conditions under which it works well. Give an example.
7. Briefly describe the exact doubling time and half-life formulas. Explain all their terms.
8. Give an example in which it is important to use the exact doubling time or half-life formula, rather than the approximate formula. Explain why the approximate formula does not work well in this case.

DOES IT MAKE SENSE?

Decide whether each of the following statements makes sense (or is clearly true) or does not make sense (or is clearly false). Explain your reasoning.

9. Our town is growing with a doubling time of 25 years, so its population will triple in 50 years.
10. Our town is growing at a rate of 7% per year, so its population will double about every 10 years.
11. A toxic chemical decays with a half-life of 10 years, so half of it will be gone 10 years from now and all the rest will be gone 20 years from now.
12. The half-life of plutonium-239 is about 24,000 years, so we can expect some of the plutonium produced in recent decades to still be around 100,000 years from now.

BASIC SKILLS & CONCEPTS

13–22: Logarithms. Refer to the Brief Review on p. 506. Determine whether each statement is true or false *without* doing any calculations. Explain your reasoning.

13. $10^{0.928}$ is between 1 and 10.
14. $10^{3.334}$ is between 500 and 1000.
15. $10^{-5.2}$ is between $-100{,}000$ and $-1{,}000{,}000$.
16. $10^{-2.67}$ is between 0.001 and 0.01.
17. $\log_{10} \pi$ is between 3 and 4.
18. $\log_{10} 96$ is between 3 and 4.
19. $\log_{10} 1{,}600{,}000$ is between 16 and 17.
20. $\log_{10}(8 \times 10^9)$ is between 9 and 10.
21. $\log_{10}\left(\frac{1}{4}\right)$ is between -1 and 0.
22. $\log_{10} 0.00045$ is between 5 and 6.

23–24. Logarithm Calculations

23. Using the approximation $\log_{10} 2 \approx 0.301$, evaluate each of the following expressions *without* a calculator.

a. $\log_{10} 16$ b. $\log_{10} 20{,}000$
c. $\log_{10} 0.05$ d. $\log_{10} 128$
e. $\log_{10} 0.02$ f. $\log_{10} (1/32)$

24. Using the approximation $\log_{10} 5 \approx 0.699$, evaluate each of the following expressions *without* a calculator.

a. $\log_{10} 500$ b. $\log_{10} 25$
c. $\log_{10} (1/125)$ d. $\log_{10} 0.2$
e. $\log_{10} 0.05$ f. $\log_{10} 625$

25–32: Doubling Time. Each exercise gives a doubling time for an exponentially growing quantity. Answer the question(s) about the growth of each quantity.

25. The doubling time of a population of fruit flies is 12 hours. By what factor does the population increase in 36 hours? In 1 week?

26. The doubling time of a bank account balance is 16 years. By what factor does it grow in 32 years? In 96 years?

27. The doubling time of a city's population is 20 years. How long does it take for the population to quadruple?

28. Prices are rising with a doubling time of 3 months. By what factor do prices increase in a year?

29. The initial population of a town is 15,000, and it grows with a doubling time of 10 years. What will the population be in 12 years? In 24 years?

30. The initial population of a town is 5000, and it grows with a doubling time of 8 years. What will the population be in 12 years? In 28 years?

31. The number of cells in a tumor doubles every 1.5 months. If the tumor begins as a single cell, how many cells will there be after 20 months? After 3 years?

32. The number of cells in a tumor doubles every 5 months. If the tumor begins with a single cell, how many cells will there be after 3 years? After 4 years?

33–34: World Population. In 2017, estimated world population was 7.5 billion. Use the given doubling time to predict the world population in 2027, 2067, and 2117.

33. Assume a doubling time of 40 years.

34. Assume a doubling time of 60 years.

35. **Rabbits.** A community of rabbits begins with an initial population of 100 and grows by 7% per month. Make a table, similar to Table 8.3, that shows the population for each of the next 15 months. Based on the table, find the doubling time of the population and briefly discuss how well the approximate doubling time formula works for this case.

36. **Mice.** A community of mice begins with an initial population of 1000 and grows 20% per month. Make a table, similar to Table 8.3, that shows the population for each of the next 15 months. Based on the table, find the doubling time of the population and briefly discuss how well the approximate doubling time formula works for this case.

37–40: Approximate Doubling Time Formula. Use the approximate doubling time formula (rule of 70). Discuss whether the formula is valid for the case described.

37. The Consumer Price Index is increasing at a rate of 3.2% per year. What is its doubling time? By what factor will prices increase in 3 years?

38. A city's population is growing at a rate of 2.5% per year. What is its doubling time? By what factor will the population increase in 50 years?

39. Gasoline prices are rising at a rate of 1.2% per month. What is their doubling time? By what factor will prices increase in 1 year? In 8 years?

40. Oil consumption is increasing at a rate of 0.9% per year. What is its doubling time? By what factor will oil consumption increase in a decade?

41–48: Half-Life. Each exercise gives a half-life for an exponentially decaying quantity. Answer the questions about the quantity's decay.

41. The half-life of a radioactive substance is 40 years. If you start with some amount of this substance, what fraction of that amount will remain in 80 years? In 120 years?

42. The half-life of a radioactive substance is 300 years. If you start with some amount of this substance, what fraction of that amount will remain in 120 years? In 2500 years?

43. The half-life of a drug in the bloodstream is 16 hours. What fraction of the original drug dose remains in the bloodstream in 24 hours? In 72 hours?

44. The half-life of a drug in the bloodstream is 4 hours. What fraction of the original drug dose remains in the bloodstream in 24 hours? In 48 hours?

45. The current population of a threatened animal species is 1 million, but it is declining with a half-life of 24 years. How many animals will be left in 30 years? In 70 years?

46. The current population of a threatened animal species is 1 million, but it is declining with a half-life of 25 years. How many animals will be left in 30 years? In 70 years?

47. Cobalt-56 has a half-life of 77 days. If you start with 1 kilogram of cobalt-56, how much will remain after 100 days? After 200 days?

48. Radium-226 is a metal with a half-life of 1600 years. If you start with 1 kilogram of radium-226, how much will remain after 1000 years? After 10,000 years?

49–52: Approximate Half-Life Formula. Use the approximate half-life formula. Discuss whether the formula is valid for the case described.

49. Urban encroachment is causing the area of a forest to decline at a rate of 6% per year. What is the approximate half-life of the forest? Based on this approximate half-life, about what fraction of the forest will remain in 50 years?

50. A clean-up project is reducing the concentration of a pollutant in the water supply, with a 3.5% decrease per week. What

is the approximate half-life of the concentration of the pollutant? Based on this approximate half-life, about what fraction of the original amount of the pollutant will remain when the project ends after 1 year (52 weeks)?

51. The Great African Elephant Census, completed in 2016, found a total population of about 350,000 African elephants, and concluded that the population was decreasing at a rate of about 8% per year, primarily due to poaching. What is the approximate half-life for the population? Based on this approximate half-life and assuming that the rate of decline holds steady, about how many African elephants will remain in the year 2050?

52. The production of a gold mine decreases by 5% per year. What is the approximate half-life for the production decline? Based on this approximate half-life and assuming that the mine's current annual production is 5000 kilograms, about what will its production be in 10 years?

53–56: Exact Formulas. Compare the doubling times found with the approximate and exact doubling time formulas. Then use the exact doubling time formula to answer the given question.

53. Inflation is causing prices to rise at a rate of 6% per year. For an item that costs $600 today, what will the price be in 4 years?

54. Hyperinflation is driving up prices at a rate of 80% per month. For an item that costs $1000 today, what will the price be in 1 year?

55. A nation of 100 million people is growing at a rate of 3.5% per year. What will the nation's population be in 30 years?

56. A family of 100 termites invades your house, and its population increases at a rate of 20% per week. How many termites will be in your house after 1 year (52 weeks)?

FURTHER APPLICATIONS

57. **Plutonium on Earth.** Scientists think that Earth once had naturally existing plutonium-239. Suppose Earth had 10 trillion tons of Pu-239 when it formed. Given plutonium's half-life of 24,000 years and Earth's current age of 4.6 billion years, how much would remain today? Use your answer to explain why plutonium is not found naturally on Earth today.

58. **Nuclear Weapons.** Thermonuclear weapons use tritium for their nuclear reactions. Tritium is a radioactive form of hydrogen (containing 1 proton and 2 neutrons) with a half-life of about 12 years. Suppose a nuclear weapon contains 1 kilogram of tritium. How much will remain in 50 years? Use your answer to explain why thermonuclear weapons require regular maintenance.

59. **Fossil Fuel Emissions.** Over the past century, global emissions of carbon dioxide from the burning of fossil fuels have risen at an average rate of about 4% per year. If emissions continue to increase at this rate, about how much higher will total emissions be in 2050 than they were in 2015?

60. **Yucca Mountain.** The U.S. government spent nearly $10 billion planning and developing a nuclear waste facility at Yucca Mountain (Nevada), though the project was cancelled in 2011. The intent had been for the facility to store up to 77,000 metric tons of nuclear waste safely for at least 1 million years. Suppose it had been successful and stored the maximum amount of waste in the form of plutonium-239 with a half-life of 24,000 years. How much plutonium would have remained after 1 million years?

61. **Crime Rate.** The homicide rate decreases at a rate of 3% per year in a city that had 800 homicides in the most recent year. At this rate, in how many years will the number of homicides reach 400 in a year?

62. **Drug Metabolism.** A particular antibiotic is eliminated from the blood with a half-life of 16 hours. How many hours does it take for a 100-milligram dose to decrease to 30 milligrams in the blood? To 1 milligram in the blood?

63. **Atmospheric Pressure.** The pressure of Earth's atmosphere at sea level is approximately 1000 millibars (mb), and it decreases by a factor of 2 for every 7 kilometers of increase in altitude.

   a. If you live at an elevation of 1 kilometer (roughly 3300 feet), what is the atmospheric pressure?

   b. What is the atmospheric pressure at the top of Mount Everest (elevation of 8848 meters)?

   c. By approximately what percentage does atmospheric pressure decrease every kilometer?

IN YOUR WORLD

64. **Doubling Time.** Find a news story that gives an exponential growth rate. Find the approximate doubling time from the growth rate, and discuss the implications of the growth.

65. **Radioactive Half-Life.** Find a news story that discusses some type of radioactive material. If it is not given, look up the half-life of the material. Discuss the implications for disposal of the material.

66. **World Population Growth.** Find data on world population growth over the past 50 years. Estimate the population growth rate over each decade. Compute the associated doubling times. Write a two-paragraph statement on the trends that you observe.

67. **Growth/Decay Project.** Choose some quantity that is of interest to you that you suspect may either be growing or decaying exponentially with time. Explain your choice and why you suspect exponential change. Then find actual data for your quantity over a period of a least 10 years, and discuss whether the data support your hypothesis that it is growing or decaying exponentially. Explain clearly.

TECHNOLOGY EXERCISES

68. **Logarithms I.** Use a calculator or Excel to find each of the logarithms in Exercise 23. Give answers to 6 decimal places.

69. **Logarithms II.** Use a calculator or Excel to find each of the logarithms in Exercise 24. Give answers to 6 decimal places.

UNIT 8C Real Population Growth

Perhaps the most important application of exponential growth concerns human population. From the time of the earliest humans more than 2 million years ago until about 10,000 years ago, human population probably never exceeded 10 million. The advent of agriculture brought about more rapid population growth. Human population is estimated to have reached about 250 million by C.E. 1, and it continued growing slowly to about 500 million by 1650.

Exponential growth set in with the Industrial Revolution. Our rapidly developing ability to grow food and exploit natural resources allowed us to build more homes for more people. Meanwhile, improvements in medicine and health science lowered death rates dramatically. In fact, world population began growing at a rate exceeding that of steady exponential growth, in which the doubling time would have remained constant. Population doubled from 500 million to 1 billion in the 150 years from 1650 to 1800. It then doubled again, to 2 billion, by 1922, only about 120 years. The next doubling, to 4 billion, was complete by 1974, a doubling time of only 52 years. World population is projected to reach 8 billion by 2024, which will mean the most recent doubling occurred in just 50 years. Figure 8.3 shows the estimated human population over the past 12,000 years.

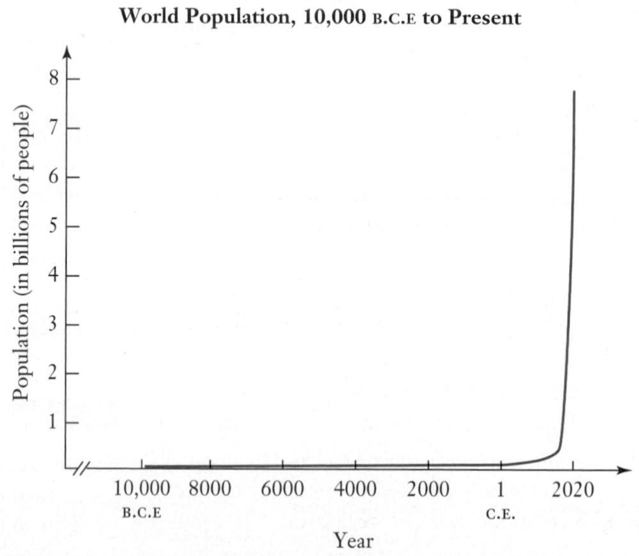

FIGURE 8.3 World population. *Source:* Data from United Nations; 2020 value is a projection.

To put current world population growth in perspective, consider the following facts:

- Every 4 years, the world adds nearly as many people as the total population of the United States.
- Each month, world population increases by nearly the equivalent of the population of Switzerland.
- While you study during the next hour and a half, world population will increase by about 10,000 people.

Projections of future population growth have large uncertainties, and demographers often state low, intermediate, and high projections based on differing assumptions about future growth rates. As of 2017, the United Nations intermediate projections predict that world population will reach 9.7 billion by 2050 and 11 billion by 2100. In the United States, where population is growing more slowly than the world average, the population is still expected to increase by about *100 million* people over the next 45 years. Fortunately, while the absolute increase in population remains huge, these numbers indicate that the growth rate is declining. This fact makes many researchers suspect that the growth rate will continue its downward trend, thereby preventing a population catastrophe.

BY THE WAY

Most of the projected population growth (9 out of 10 people) will take place in the developing regions. By 2050, more than 900 million people are expected to be added to Asia alone.

EXAMPLE 1 Varying Growth Rate

The average annual growth rate for world population since 1650 has been about 0.7%. However, the annual rate has varied significantly. It peaked at about 2.1% during the 1960s and is currently (as of 2017) about 1.1%. Find the approximate doubling time for each of these growth rates. Use each doubling time to predict world population in 2060, based on a 2017 population of 7.5 billion.

Solution Using the approximate doubling time formula (Unit 8B), we find the doubling times for the three rates:

$$\text{For } 0.7\%: \quad T_{\text{double}} \approx \frac{70}{P} = \frac{70}{0.7/\text{year}} = 100 \text{ years}$$

$$\text{For } 2.1\%: \quad T_{\text{double}} \approx \frac{70}{P} = \frac{70}{2.1/\text{year}} \approx 33 \text{ years}$$

$$\text{For } 1.1\%: \quad T_{\text{double}} \approx \frac{70}{P} = \frac{70}{1.1/\text{year}} \approx 64 \text{ years}$$

To predict approximate world population in 2060, we use the formula

$$\text{new value} = \text{initial value} \times 2^{t/T_{\text{double}}}$$

We set the initial value to 7.5 billion and note that 2060 is $t = 43$ years after 2017:

$$\text{For } 0.7\%: \quad \text{2060 population} \approx 7.5 \text{ billion} \times 2^{43\text{yr}/100\text{yr}} \approx 10.1 \text{ billion}$$

$$\text{For } 2.1\%: \quad \text{2060 population} \approx 7.5 \text{ billion} \times 2^{43\text{yr}/33\text{yr}} \approx 18.5 \text{ billion}$$

$$\text{For } 1.1\%: \quad \text{2060 population} \approx 7.5 \text{ billion} \times 2^{43\text{yr}/64\text{yr}} \approx 11.9 \text{ billion}$$

Notice the large differences in the predicted population for the different growth rates. Clearly, decisions we make today to affect the growth rate will have major implications for human population in the future. ▶ Now try Exercises 13–16.

BY THE WAY

For individual countries, the growth rate depends on immigration and emigration as well as on birth and death rates. In the United States, immigration comprises about half the overall growth rate. Of course, immigration and emigration do not affect *world* population, because no one is moving to or from Earth.

What Determines the Growth Rate?

The world population growth rate is simply the difference between the birth rate and the death rate. For example, for 2017 the global averages were about 19 births per 1000 people and 8 deaths per 1000 people per year. Therefore, the population growth rate was about

$$\frac{19}{1000} - \frac{8}{1000} = \frac{11}{1000} = 0.011 = 1.1\%$$

Overall Growth Rate

The world population **growth rate** is the difference between the *birth rate* and the *death rate*:

$$\text{growth rate} = \text{birth rate} - \text{death rate}$$

Interestingly, birth rates have dropped rapidly throughout the world during the past 60 years—the same period that has seen the largest population growth in history. Indeed, worldwide birth rates have never been lower than they are today. Today's rapid population growth comes from the fact that death rates have fallen even more dramatically.

BY THE WAY

Although Example 2 looks at annual birth rates, an alternative way to look at changes in fertility is by the number of children the average woman bears in her lifetime. Before the 20th century, the average woman gave birth to more than 6 children during her lifetime, and nearly half of all children did not survive to adulthood. Today, on a worldwide basis, the average woman gives birth to 2.5 children. Demographers estimate that the population will level out if the worldwide fertility rate falls to between 2.0 and 2.1 children.

EXAMPLE 2 Birth and Death Rates

In 1950, the worldwide birth rate was 37 births per 1000 people and the worldwide death rate was 19 deaths per 1000 people. By 1975, the birth rate had fallen to 28 births per 1000 people and the death rate to 11 deaths per 1000 people. Contrast the overall population growth rates in 1950 and 1975.

Solution In 1950, the overall population growth rate was

$$\frac{37}{1000} - \frac{19}{1000} = \frac{18}{1000} = 0.018 = 1.8\%$$

In 1975, the overall growth rate was

$$\frac{28}{1000} - \frac{11}{1000} = \frac{17}{1000} = 0.017 = 1.7\%$$

Despite a significant fall in the worldwide birth rate during the 25-year period, the population growth rate barely changed because the death rate fell almost as much.

▶ **Now try Exercises 17–20.**

Think About It Suppose that medical science finds a way to extend human lifespans significantly. How would this affect the population growth rate?

Carrying Capacity and Real Growth Models

As we saw in Unit 8A, exponential growth cannot continue indefinitely. Indeed, human population cannot continue to grow much longer at its current rate, because we'd be elbow to elbow over the entire Earth in just a few centuries. Theoretical models of population growth therefore assume that human population is ultimately limited by the **carrying capacity** of Earth—the number of people that Earth can support.

Definition

For any particular species in a given environment, the **carrying capacity** is the maximum sustainable population. That is, it is the largest population the environment can support for extended periods of time.

Two important models for populations approaching the carrying capacity are (1) a gradual leveling off, known as *logistic growth*, and (2) a rapid increase followed by a rapid decrease, known as *overshoot and collapse*. Let's investigate each model.

Logistic Growth

A logistic growth model assumes that the population growth rate gradually decreases as the population approaches the carrying capacity. For example, if the carrying capacity is 12 billion people, a logistic growth model assumes that the population growth rate decreases as this number is approached. The growth rate falls to zero as the carrying capacity is approached, allowing the population to remain steady at that level thereafter.

Logistic Growth

When the population is small relative to the carrying capacity, **logistic growth** is exponential with a fractional growth rate close to the base growth rate, r. As the population approaches the carrying capacity, the logistic growth rate approaches zero. The fractional logistic growth rate at any particular time depends on the population at that time, the carrying capacity, and the base growth rate r:

$$\text{logistic growth rate} = r \times \left(1 - \frac{\text{population}}{\text{carrying capacity}}\right)$$

Figure 8.4 contrasts logistic and exponential growth for the same base growth rate r. In the exponential case, the growth rate equals r at all times. In the logistic case, the growth rate starts out equal to r, so the logistic curve and the exponential curve look the same at early times. As time progresses, the logistic growth rate decreases and approaches zero, and the population levels off at the carrying capacity.

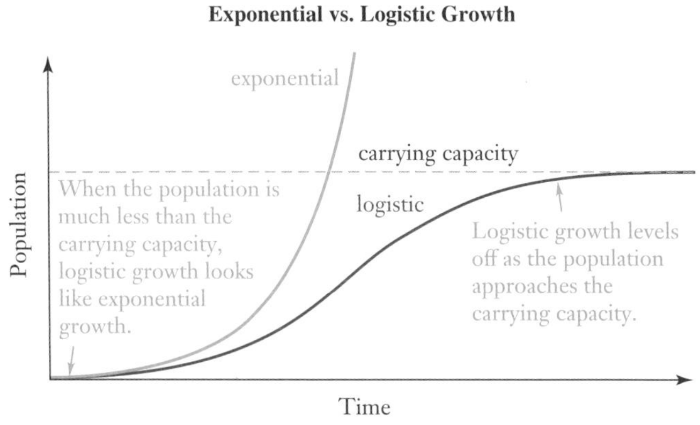
Exponential vs. Logistic Growth
exponential
carrying capacity
logistic
When the population is much less than the carrying capacity, logistic growth looks like exponential growth.
Logistic growth levels off as the population approaches the carrying capacity.
Population
Time

FIGURE 8.4 This graph contrasts exponential growth with logistic growth for the same base growth rate r.

EXAMPLE 3 Are We Growing Logistically?

The worldwide population growth rate has been slowing since around 1960, when it was about 2.1% and the population was about 3 billion. Assume that these values for the growth rate and population represent one point in time on a logistic growth curve with a carrying capacity of 12 billion. Does this model successfully predict the 2017 growth rate of 1.1% given the 2017 population of 7.5 billion? Explain.

Solution We begin by using the 1960 data to find the base growth rate r in a logistic model that uses the 1960 data values. You should confirm that solving the logistic growth rate formula for r gives

$$r = \frac{\text{logistic growth rate (in 1960)}}{\left(1 - \frac{\text{population(in 1960)}}{\text{carrying capacity}}\right)}$$

Substituting the 1960 growth rate of 2.1% = 0.021, the 1960 population of 3 billion, and a carrying capacity of 12 billion, we find

$$r = \frac{0.021}{\left(1 - \frac{3 \text{ billion}}{12 \text{ billion}}\right)} = \frac{0.021}{(1 - 0.25)} \approx 0.028 = 2.8\%$$

We now use this value of the base growth rate r to predict the growth rate for the 2017 population of about 7.5 billion:

$$\text{2017 growth rate} = 0.028 \times \left(1 - \frac{7.5 \text{ billion}}{12 \text{ billion}}\right) \approx 0.011$$

This logistic model successfully predicts the 2017 growth rate of 1.1%. Therefore, it is reasonable to say that the worldwide population has been growing logistically since about 1960. If population growth continues to follow this logistic pattern, then the growth rate will continue to decline and the population will gradually level out at about 12 billion. However, human population has *not* been showing logistic growth over longer periods. The base growth rate we found for our logistic model, $r = 0.028$, implies that the actual population growth rate should have started at 2.8% long ago and gradually declined to 2.1% by 1960. In fact, the 1960 growth rate was an all-time peak. In summary, while we see evidence that population growth has followed a logistic trend since about 1960, it has not followed this trend over longer periods. Therefore, it is still too soon to conclude that the logistic trend will continue in the future. ▶ Now try Exercises 21–22.

BY THE WAY

The overshoot and collapse model characterizes many predator-prey populations. The population of a predator increases rapidly, causing the prey population to collapse. Once the prey population collapses, the predator population must also collapse because of lack of food. Once the predator population collapses, the prey population can begin to recover—as long as it has not gone extinct.

Overshoot and Collapse

A logistic model assumes that the growth rate automatically adjusts as the population approaches the carrying capacity. However, because of the astonishing rate of exponential growth, real populations often increase beyond the carrying capacity in a relatively short period of time. This phenomenon is called **overshoot**.

When a population overshoots the carrying capacity of its environment, a decrease in the population is inevitable. If the overshoot is substantial, the decrease can be rapid and severe—a phenomenon known as **collapse**. Figure 8.5 contrasts logistic growth with overshoot and collapse.

Think About It The concept of carrying capacity can be applied to any localized environment. Consider the decline of past civilizations such as the ancient Greeks, Romans, Mayans, and Ancestral Puebloans. Does an overshoot and collapse model describe the fall of any of these or other civilizations? Explain.

What Is the Carrying Capacity?

Given that human population cannot grow exponentially forever, logistic growth is clearly preferable to overshoot and collapse. Logistic growth means a sustainable future population, while overshoot and collapse might mean the end of our civilization.

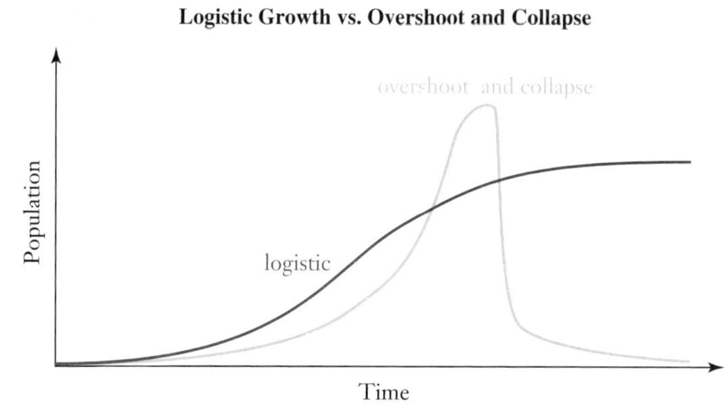

FIGURE 8.5 This graph contrasts logistic growth with overshoot and collapse.

The most fundamental question about population growth therefore concerns the carrying capacity. Example 3 suggested that we are currently following a logistic growth pattern *if* the carrying capacity is 12 billion. In that case, we are on a path to long-term population stability. However, if the carrying capacity is lower than 12 billion—or if it is higher and the growth rate goes back up—then we could face a situation of overshoot and collapse. Unfortunately, any estimate of carrying capacity is subject to great uncertainty, for at least four important reasons:

- The carrying capacity depends on consumption of resources such as energy. However, different countries consume at very different rates. For example, the carrying capacity is much lower if we assume that the growing population will consume energy at the U.S. average rate rather than the Japanese average rate (which is about half the U.S. rate).
- The carrying capacity depends on assumptions about the environmental impact of the average person. A larger average impact on the environment means a lower carrying capacity.
- The carrying capacity can change with both human technology and the environment. For example, estimates of carrying capacity typically consider the availability of freshwater. However, if we can develop new sources of energy (such as fusion), nearly unlimited amounts of freshwater may be obtained through the desalinization of seawater. Conversely, global warming is altering the environment and might reduce our ability to grow food, thereby lowering the carrying capacity.
- Even if we could account for the many individual factors in the carrying capacity (such as food production, energy, and pollution), the Earth is such a complex system that precisely predicting the carrying capacity may well be impossible. For example, no one can predict whether or how much the loss of rain forest species affects the carrying capacity.

The history of attempts to guess the carrying capacity of the Earth is full of incorrect predictions. Among the most famous was that made by English economist Thomas Malthus (1766–1834). In a 1798 paper entitled *An Essay on the Principle of Population as It Affects the Future Improvement of Society,* Malthus argued that food production would not be able to keep up with the rapidly growing populations of Europe and America. He concluded that mass starvation would soon hit these continents. His prediction did not come true, primarily because advances in technology *did* allow food production to keep pace with population growth.

The power of population is indefinitely greater than the power in the Earth to produce subsistence for man.

—Thomas Malthus

Think About It Some people argue that while Malthus's immediate predictions didn't come true, his overall point about a limit to population is still valid. Others cite Malthus as a classic example of underestimating the ingenuity of our species. What do *you* think? Defend your opinion.

CASE STUDY
The Population of Egypt

Over long periods of time, real population growth patterns tend to be quite complex. Sometimes the growth may look exponential, while at other times it may appear logistic or seem to exhibit overshoot and collapse. It may even appear to be some combination of these possibilities all at once.

One of the few places for which long-term population data are available is Egypt. Figure 8.6 shows these data, along with a few historical events that affected the population. (The graph uses an exponential vertical scale, with each tick mark representing a number twice as big as the previous one.) Note the complexity of the pattern. Even with the best mathematical models, it is hard to imagine that scientists in ancient Egypt could have predicted the future population of the region over a period of even a hundred years, let alone several thousand years. Note also the unprecedented increase in Egypt's population during the past two centuries, illustrating that modern population growth has no counterpart in the rest of human history.

This example offers an important lesson about mathematical models. They are useful for gaining insight into the processes being modeled. However, mathematical models can be used to *predict* future changes only when the processes are relatively simple. For example, it is easy to use mathematical modeling to predict the path of a spaceship because the law of gravity is relatively simple. But the growth of human population is such a complex phenomenon that we have little hope of ever being able to predict it reliably.

The Population of Egypt, 750 B.C.E. to Present

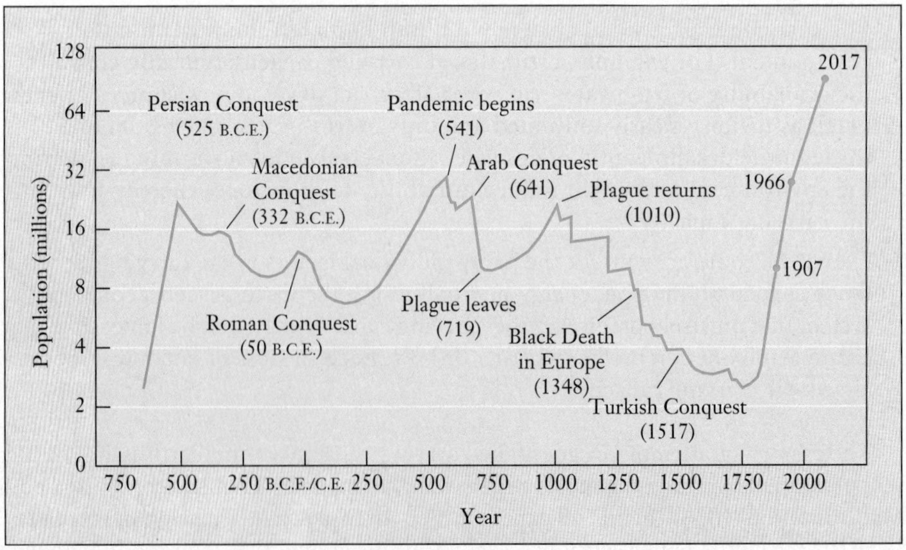

FIGURE 8.6 The historical population of Egypt. Notice that each tick mark on the vertical axis represents a doubling of the population. *Source:* T. H. Hollingsworth, *Historical Demography* (Ithaca, NY: Cornell University Press, 1969), with data added through 2017.

IN YOUR WORLD

Choosing Our Fate

As the parable of the bacteria in a bottle (Unit 8A) showed, exponential growth cannot continue indefinitely. The exponential growth of human population will stop. The only questions are when and how.

First, consider the question of when. The highest estimates put the Earth's carrying capacity around 15 billion to 20 billion, which we would reach within this century at today's 1.1% annual growth rate. Most other estimates of the carrying capacity are considerably lower, suggesting that we are rapidly approaching it. A few estimates suggest we have already exceeded the carrying capacity. Whatever estimate you believe, the major conclusion is still the same: The rapid rise of human population that has occurred during the past couple of centuries will come to a stop very soon on the scale of human history, likely within no more than a few decades.

As for how, there are only two basic ways to slow the growth of a population:

- a decrease in the birth rate or
- an increase in the death rate.

As individuals, most people are already choosing the first option, which is why birth rates today are at historic lows. Indeed, population actually is *decreasing* in about 20 nations, most of them in Europe. Nevertheless, worldwide birth rates are still much higher than worldwide death rates, so exponential growth continues.

If a decrease in the birth rate doesn't slow the growth, an increase in the death rate will. If population significantly overshoots the carrying capacity by the time this process begins, the increase in the death rate will be significant—probably on a scale never before seen. This forecast is not a threat, a warning, or a prophecy of doom. It is simply a law of nature: *Exponential growth always stops.*

As human beings, we can choose to slow our population growth through intelligent and careful decisions. Or, we can choose to do nothing, leaving ourselves to the mercy of natural forces over which we have no more control than we do over hurricanes, tornadoes, earthquakes, or the explosions of distant stars. Either way, it's a choice that each and every one of us must make—and upon which our entire future depends.

Quick Quiz 8C

Choose the best answer to each of the following questions. Explain your reasoning with one or more complete sentences.

1. World population is currently rising by about 80 million people per year. About how many people are added to the population each minute?

   a. 15 b. 50 c. 150

2. The most recent doubling of world population took about

   a. 20 years. b. 50 years. c. 90 years.

3. The primary reason for the rapid growth of human population over the past century has been

   a. an increasing birth rate.

   b. a decreasing death rate.

   c. a combination of an increasing birth rate and a decreasing death rate.

4. The *carrying capacity* of the Earth is defined as

   a. the maximum number of people who could fit elbow to elbow on the planet.

   b. the maximum population that could be sustained for a long period of time.

   c. the peak population that would be reached just before a collapse in the population size.

5. Which of the following would cause estimates of Earth's carrying capacity to increase?

   a. the discovery of a way to make people live longer

   b. the spread of a disease that killed off many crops

   c. the development of a new, inexpensive, and nonpolluting energy source

6. Recall the bacteria in a bottle parable from Unit 8A, in which the number of bacteria in a bottle doubles each minute until the bottle is full and the bacteria all die. The full history of this population of bacteria, including their death, is an example of

   a. overshoot and collapse.

   b. unending exponential growth.

   c. logistic growth.

7. If researchers project that human population will level out at 11 billion later in this century, what type of growth model are they assuming?

   a. overshoot and collapse

   b. exponential

   c. logistic

8. Consider a projection in which population levels out at 11 billion people later this century. This projection assumes that the worldwide birth rate will fall from its current level. Suppose instead that the birth rate returns to what it was in 1950. If the death rate remains steady, then

   a. population will grow to far more than 11 billion.

   b. population will level off before reaching 11 billion.

   c. population will still level off at 11 billion, but a little sooner than otherwise expected.

9. Suppose that the world population continues to grow at the 2017 rate of 1.1% per year. Given the 2017 population of 7.5 billion, when will the population double to about 15 billion?

   a. around 2080

   b. around 2220

   c. around 2450

10. Consider a projection that world population will level out later in this century at 11 billion people. Which of the following is *not* a requirement for this to happen?

    a. We must significantly increase food production.

    b. The average woman must give birth to fewer children than she does at present.

    c. We must find a way to increase life expectancies.

Exercises 8C

REVIEW QUESTIONS

1. Based on Figure 8.3, contrast the changes in human population for the 10,000 years preceding C.E. 1 and for the approximately 2000 years since. What has happened over the past few centuries?
2. Briefly describe how the overall population growth rate is related to birth and death rates.
3. How do today's birth and death rates compare to those in the past? Why is human population growing?
4. What do we mean by *carrying capacity*? Why is it so difficult to determine the carrying capacity of Earth?
5. What is *logistic growth*? Why would it be good if human population growth followed a logistic growth pattern in the future?
6. What is *overshoot and collapse*? Under what conditions does it occur? Why would it be a bad thing for the human race?

DOES IT MAKE SENSE?

Decide whether each of the following statements makes sense (or is clearly true) or does not make sense (or is clearly false). Explain your reasoning.

7. Within the next 10 years, world population will grow by more than twice the current population of the United States.
8. If birth rates fall more than death rates, the growth rate of world population will fall.
9. The carrying capacity of our planet depends only on our planet's size.
10. Thanks to rapid increases in computing technology, we should be able to pin down the carrying capacity of the Earth to a precise number within just a few years.
11. In the wild, we expect the population of any animal species to follow a logistic growth pattern.
12. Past civilizations that no longer exist must have followed logistic population growth patterns.

BASIC SKILLS & CONCEPTS

13–16: Varying Growth Rates. Starting from a 2017 world population of 7.5 billion, use the given growth rate to find the approximate doubling time (use the rule of 70) and to predict world population in 2050.

13. Use the average annual growth rate between 1850 and 1950, which was about 0.9%.
14. Use the average annual growth rate between 1950 and 2000, which was about 1.8%.
15. Use the average annual growth rate between 1970 and 2000, which was about 1.6%.
16. Use the current annual growth rate of the United States, which is about 0.7%.

17–20: Birth and Death Rates. The following table gives the birth and death rates for four countries in three different years.

	Birth rate (per 1000)			Death rate (per 1000)		
Country	1980	1995	2016	1980	1995	2016
Afghanistan	51.8	52.6	38.3	24.1	20.1	13.7
China	21.5	18.7	12.4	7.1	7.0	7.7
Russia	16.0	10.9	11.3	11.3	14.6	13.6
United States	15.5	15.1	12.5	8.7	8.6	8.2

For the country named in each exercise, do the following:

a. Find the country's net population growth rate due to births and deaths (i.e., neglect immigration) in 1980, 1995, and 2016.

b. Describe in words the general trend in the country's population growth rate. Based on this trend, predict how the country's population will change over the next 20 years. Do you think your prediction is reliable? Explain.

17. Afghanistan
18. China
19. Russia
20. United States

21. **Logistic Growth.** Consider a population that begins growing exponentially at a base rate of 4.0% per year and then follows a logistic growth pattern. If the carrying capacity is 60 million, find the actual fractional growth rate when the population is 10 million, 30 million, and 50 million.

22. **Logistic Growth.** Consider a population that begins growing exponentially at a base rate of 6.0% per year and then follows a logistic growth pattern. If the carrying capacity is 80 million, find the actual fractional growth rate when the population is 10 million, 50 million, and 70 million.

FURTHER APPLICATIONS

23–26: U.S. Population. Starting from an estimated U.S. population of 325 million in 2017, use the given growth rate to estimate U.S. population in 2050 and 2100. Use the approximate doubling time formula.

23. Use the current U.S. annual growth rate of 0.7% (includes immigration as well as births and deaths).

24. Use a growth rate of 0.5%.

25. Use a growth rate of 1.0%.

26. Use a growth rate of 0.4%.

27. **Population Growth in Your Lifetime.** Starting from the 7.5 billion world population in 2017, assume that the population maintains an annual growth rate of 1.1%. What will be the world population when you are 50 years old? 80 years old? 100 years old? Use the approximate doubling time formula.

28. **Slower Growth.** Repeat Exercise 27, but use a growth rate of 0.9%.

29–32: World Carrying Capacity. For the given carrying capacities, use the 1960 annual growth rate of 2.1% and population of 3 billion to predict the base growth rate and current growth rate with a logistic model. Assume a current world population of 7.5 billion. How does the predicted growth rate compare to the actual growth rate of about 1.1% per year?

29. Assume the carrying capacity is 9 billion.

30. Assume the carrying capacity is 10 billion.

31. Assume the carrying capacity is 15 billion.

32. Assume the carrying capacity is 20 billion.

33. **Growth Control Mediation.** A city with a 2017 population of 100,000 has a growth control policy that limits the increase in residents to 2% per year. Naturally, this policy causes a great deal of dispute. On one side, some people argue that *growth* costs the city its small-town charm and clean environment. On the other side, some people argue that *growth control* costs the city jobs and drives up housing prices. Finding their work limited by the policy, developers suggest a compromise of raising the allowed growth rate to 5% per year. Contrast the populations of this city in 2027, 2037, and 2077 for 2% annual growth and 5% annual growth. Use the approximate doubling time formula. If you were asked to mediate the dispute between growth control advocates and opponents, explain the strategy you would use.

IN YOUR WORLD

34. **Population in the News.** Find a recent news story that concerns population growth. Does the story consider the long-term effects of the growth? If so, do you agree with the claims? If not, discuss a few possible effects of future growth.

35. **Population Predictions.** Find population predictions from an organization that studies population, such as the United Nations or the U.S. Census Bureau. Read about how the predictions are made. Write a short summary of the methods used to predict future population. Be sure to discuss the uncertainties in the predictions.

36. **Global Variations.** The website for the United Nations Department of Economic and Social Affairs has data and future projections for various factors affecting population growth for every nation and various regions of the world. Choose one nation or region, and investigate its demographic trends. Write a short report on your findings and how you think they will affect the future of the nation you are studying.

37. **Carrying Capacity.** Find several different opinions concerning the carrying capacity for Earth's human population. Based on your research, draw some conclusions about whether overpopulation presents an immediate threat. Write a short essay detailing the results of your research and clearly explaining your conclusions.

38. **U.S. Population Growth.** Research population growth in the United States to determine the relative proportions of the growth resulting from birth rates and from immigration. Then research both the problems and the benefits of the growing U.S. population. Form your own opinions about whether the United States has a population problem. Write an essay covering the results of your research and stating and defending your opinions.

39. **Thomas Malthus.** Find more information about Thomas Malthus and his famous predictions about population. Write a short paper on either his life or his work.

40. **Extinction.** Choose an endangered species, and research why it is in decline. Is the decline a case of overshoot and collapse? Is human activity changing the carrying capacity for the species? Write a short summary of your findings.

UNIT 8D Logarithmic Scales: Earthquakes, Sounds, and Acids

You probably know that we measure the strength of earthquakes in *magnitudes*, the loudness of sounds in *decibels*, and the acidity of household cleansers by *pH*. In all three cases, the measurement scale involves exponential growth, because successive numbers on the scale increase by the same relative amount. For example, a liquid with pH 5 is ten times more acidic than one with pH 6. In this unit, we will explore these three important scales. They are commonly called **logarithmic scales**, which makes sense if you remember that logarithms are powers (see the Brief Review on p. 506).

The Magnitude Scale for Earthquakes

Earthquakes are a fact of life for much of the world's population (Figure 8.7). In the United States, California and Alaska are most prone to earthquakes, although earthquakes can strike almost anywhere. Most earthquakes are so minor that they can hardly be felt, but severe earthquakes can kill tens of thousands of people. Table 8.4 lists the frequencies of earthquakes of various strengths according to standard categories defined by geologists.

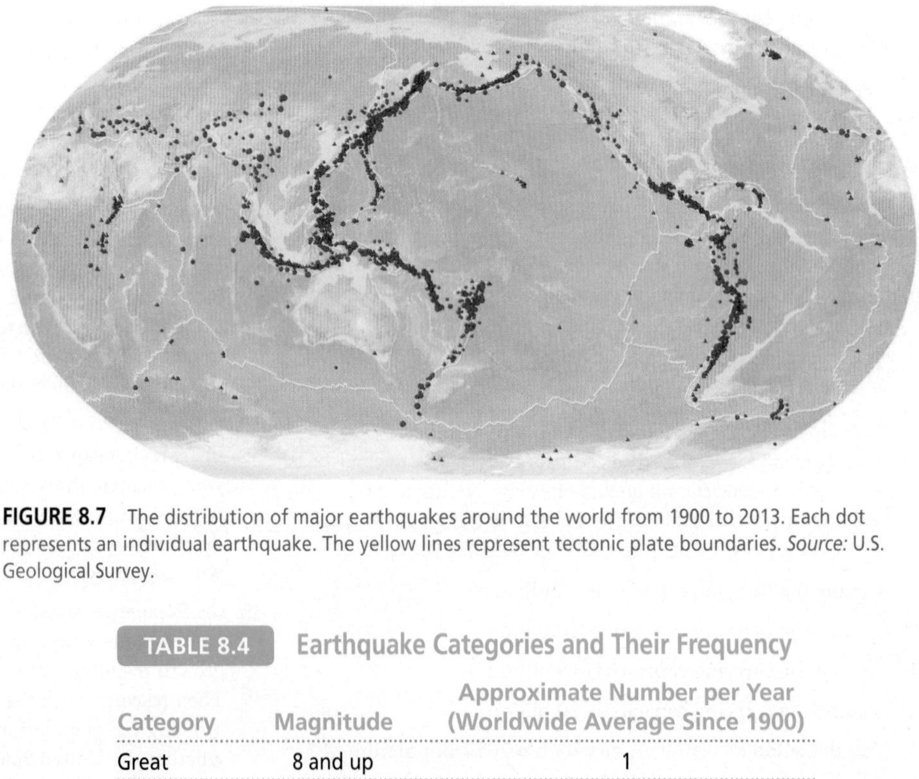

FIGURE 8.7 The distribution of major earthquakes around the world from 1900 to 2013. Each dot represents an individual earthquake. The yellow lines represent tectonic plate boundaries. *Source:* U.S. Geological Survey.

TABLE 8.4 Earthquake Categories and Their Frequency

Category	Magnitude	Approximate Number per Year (Worldwide Average Since 1900)
Great	8 and up	1
Major	7–8	18
Strong	6–7	120
Moderate	5–6	800
Light	4–5	6000
Minor	3–4	50,000
Very minor	less than 3	magnitude 2–3: 1000 per day magnitude 1–2: 8000 per day

Scientists measure earthquake strength with the earthquake **magnitude scale**. The magnitude is related to the energy released by the earthquake. But each magnitude represents about 32 times as much energy as the prior magnitude. For example, a magnitude 8 earthquake releases 32 times as much energy as a magnitude 7 earthquake. More technically, the magnitude scale is defined as follows.

HISTORICAL NOTE

The original magnitude scale was created by Charles Richter in 1935. This Richter scale measured the up and down motion of the ground during a quake. Magnitude 0 was defined as the smallest detectable quake, and each increase of 1 magnitude corresponded to a factor of 10 increase in ground motion. Most earthquakes have nearly the same magnitude on the Richter scale and on the modern magnitude scale, but only the latter quantifies the actual energy released by the earthquake.

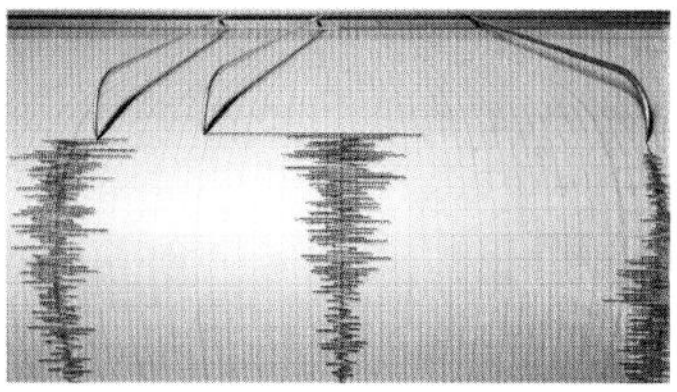

The Magnitude Scale for Earthquakes

The **magnitude scale** for earthquakes is defined so that each magnitude represents about 32 times as much energy as the prior magnitude. More technically, the magnitude, M, is related to the released energy, E, by the following equivalent formulas:

$$\log_{10} E = 4.4 + 1.5M \quad \text{or} \quad E = (2.5 \times 10^4) \times 10^{1.5M}$$

The energy is measured in joules (see Unit 2B); magnitudes have no units.

Earthquakes of the same magnitude may cause vastly different amounts of damage depending on *how* their energy is released. Every earthquake releases some of its energy into the Earth's interior, where it is fairly harmless, and some along the Earth's surface, where it shakes the ground up and down. A moderate earthquake that releases most of its energy along the surface can do more damage than a strong earthquake that releases most of its energy into the interior.

Deaths from earthquakes generally arise indirectly. Ground shaking can cause buildings to collapse, which is why the worst earthquake disasters tend to occur in regions where people cannot afford the high cost of earthquake-resistant construction. Other earthquake-related disasters occur when the shaking triggers landslides or tsunamis.

Think About It Surface waves from earthquakes make the ground roll up and down like ripples moving outward on a pond. Given this motion, suggest a few ways that buildings can be designed to withstand earthquakes. Do you think it is possible to make a building that could withstand *any* earthquake? Why or why not?

EXAMPLE 1 The Meaning of a Magnitude Increase of 1

Using the formula for earthquake magnitudes, calculate precisely how much more energy is released for each increase of 1 magnitude on the earthquake scale. Also find the energy change for a 0.5 increase in magnitude.

Solution We look at the formula that gives the energy:

$$E = (2.5 \times 10^4) \times 10^{1.5M}$$

The first term, 2.5×10^4, is a constant number that is the same no matter what value we use for M. The magnitude appears only in the second term, $10^{1.5M}$. Each time we raise the magnitude by 1, such as from 5 to 6 or from 7 to 8, the total energy E increases by a factor of $10^{1.5}$. Therefore, each successive magnitude represents $10^{1.5} \approx 31.623$ times as much energy as the prior magnitude. That is, each increase of 1 magnitude corresponds to approximately 32 times as much energy. Similarly, an increase in magnitude of 0.5 corresponds to an increase in energy by a factor of $10^{1.5 \times 0.5} = 10^{0.75} \approx 5.6$.

▶ Now try Exercises 9–10.

EXAMPLE 2 Comparing Disasters

The 1989 San Francisco earthquake, in which 90 people were killed, had a magnitude of 7.1. Calculate the energy released, in joules. Compare the energy of this earthquake to that of the 2010 earthquake in Haiti, which had a magnitude of 7.0 and killed an estimated 316,000 people.

Solution The energy released by the San Francisco earthquake was

$$E = (2.5 \times 10^4) \times 10^{1.5M} = (2.5 \times 10^4) \times 10^{1.5 \times 7.1} \approx 1.1 \times 10^{15} \text{ joules}$$

The San Francisco earthquake was 0.1 magnitude greater than the Haiti earthquake. It therefore released about $10^{1.5 \times 0.1} = 10^{0.15} \approx 1.4$ times as much energy. Nevertheless, the earthquake in Haiti killed many more people, primarily because of lower-quality building construction and lack of resources in the immediate aftermath of the quake. ▶ Now try Exercises 11–14.

Measuring Sounds in Decibels

The **decibel scale** is used to compare the loudness of sounds. It is defined so that a sound of 0 decibels, abbreviated 0 dB, represents the softest sound audible to the human ear. Table 8.5 lists the approximate loudness of some common sounds.

Technical Note

In absolute terms, the intensity of the softest audible sound is about 10^{-12} watt per square meter.

TABLE 8.5 Typical Sounds in Decibels

Decibels	Times as Loud as Softest Audible Sound	Example
140	10^{14}	jet at 30 meters
120	10^{12}	strong risk of damage to human ear
100	10^{10}	siren at 30 meters
90	10^{9}	threshold of pain for human ear
80	10^{8}	busy street traffic
60	10^{6}	ordinary conversation
40	10^{4}	background noise in average home
20	10^{2}	whisper
10	10^{1}	rustle of leaves
0	1	softest audible sound
−10	0.1	inaudible sound

HISTORICAL NOTE

A decibel is $\frac{1}{10}$ of a *bel*, a unit named for Alexander Graham Bell (1847–1922). Bell's mother was deaf, and his father was a pioneer in teaching speech to the deaf. Bell became a professor of vocal physiology at Boston University and married a deaf pupil. He patented his most famous invention, the telephone, in 1876.

The Decibel Scale for Sound

The loudness of a sound in decibels is defined by the following equivalent formulas:

$$\text{loudness in dB} = 10 \log_{10}\left(\frac{\text{intensity of the sound}}{\text{intensity of softest audible sound}}\right)$$

or

$$\left(\frac{\text{intensity of the sound}}{\text{intensity of softest audible sound}}\right) = 10^{(\text{loudness in db})/10}$$

EXAMPLE 3 Computing Decibels

Suppose a sound is 100 times as intense as the softest audible sound. What is its loudness, in decibels?

Solution We are looking for the loudness in decibels, so we use the first form of the decibel scale formula:

$$\text{loudness in dB} = 10 \log_{10}\left(\frac{\text{intensity of the sound}}{\text{intensity of softest audible sound}}\right)$$

The ratio in parentheses is 100, because we are given that the sound is 100 times as intense as the softest audible sound. We find

$$\text{loudness in dB} = 10 \log_{10} 100 = 10 \times 2 = 20 \text{ dB}$$

A sound that is 100 times as intense as the softest possible sound has a loudness of 20 dB, which is equivalent to a whisper (see Table 8.5). ▸ Now try Exercises 15–18.

EXAMPLE 4 Sound Comparison

How does the intensity of a 57-dB sound compare to that of a 23-dB sound?

Solution We can compare the loudness of two sounds by working with the second form of the decibel scale formula. You should confirm for yourself that by dividing the intensity of Sound 1 by the intensity of Sound 2, we find

$$\frac{\text{intensity of Sound 1}}{\text{intensity of Sound 2}} = 10^{[(\text{loudness of Sound 1 in dB}) - (\text{loudness of Sound 2 in dB})]/10}$$

Substituting 57 dB for Sound 1 and 23 dB for Sound 2, we have

$$\frac{\text{intensity of Sound 1}}{\text{intensity of Sound 2}} = 10^{[(\text{loudness of Sound 1 in dB}) - (\text{loudness of Sound 2 in dB})]/10}$$

$$= 10^{(57-23)/10} = 10^{3.4} \approx 2512$$

A sound of 57 dB is about 2500 times as intense as a sound of 23 dB.
▸ Now try Exercises 19–20.

The Inverse Square Law for Sound

You've probably noticed that sounds get weaker with distance. If you sit right in front of the speakers at an outdoor concert, the sound may be almost deafening, while people a mile away may not hear the music at all. It's easy to understand why.

Figure 8.8 shows the idea. The sound from the speaker gets spread over a larger and larger *area* as it moves farther from the speaker. This area increases with the *square* of the distance from the speaker. For example, at 2 meters the sound is spread over an area $2^2 = 4$ times as large as the area at 1 meter, at 3 meters the sound is spread over an area $3^2 = 9$ times as large as the area at 1 meter, and so on. That is, the intensity of the sound weakens with the *square* of the distance.

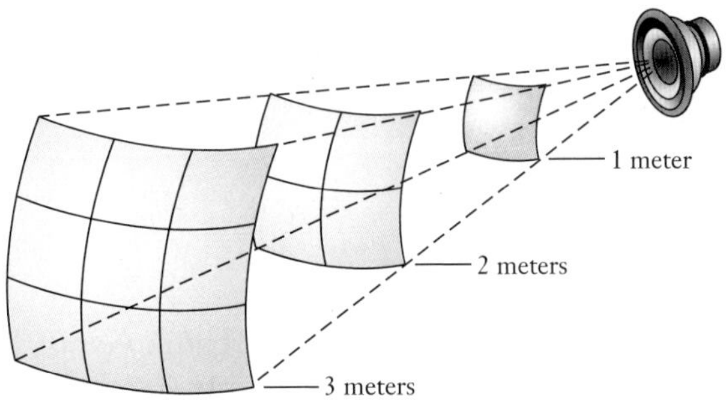

FIGURE 8.8 This figure shows why the intensity of sound decreases with the *square* of the distance from the source. Notice that the area at 2 meters is $2^2 = 4$ times as large as the area at 1 meter, and the area at 3 meters is $3^2 = 9$ times as large.

Because the intensity of sound decreases with the square of the distance, we say that sound follows an **inverse square law**. In other words, the intensity of a sound at a distance d from its source is proportional to $1/d^2$. Many other quantities also follow inverse square laws with distance, including the brightness of a light and the strength of gravity.

The Inverse Square Law for Sound

The intensity of sound decreases with the square of the distance from the source, meaning that the intensity is proportional to $1/d^2$. We therefore say that sound intensity follows an **inverse square law** with distance.

BY THE WAY

What we perceive as sound is actually tiny pressure changes in the air. Sound travels through air as a wave of pressure changes, called a *sound wave*. When a sound wave strikes the eardrum, its energy causes the eardrum to move in response. The brain analyzes the motions of the eardrum, perceiving sound.

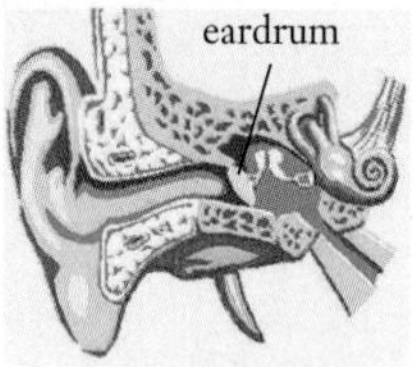

EXAMPLE 5 Sound Advice

How far should you be from a jet to avoid a strong risk of damage to your ear?

Solution Table 8.5 shows that the sound from a jet at a distance of 30 meters is 140 dB, and 120 dB is the level of sound that poses a strong risk of ear damage. The ratio of the intensity of these two sounds is

$$\frac{\text{intensity of 140-dB sound}}{\text{intensity of 120-dB sound}} = \frac{10^{140/10}}{10^{120/10}} = \frac{10^{14}}{10^{12}} = 10^2 = 100$$

The sound of the jet at 30 meters is 100 times as intense as a sound that presents a strong risk of ear damage. To prevent ear damage, you must therefore be far enough from the jet to weaken this sound intensity by at least a factor of 100. Because sound intensity follows an inverse square law with distance, moving 10 times as far away weakens the intensity by a factor of $10^2 = 100$. You should therefore be more than $10 \times 30\text{ m} = 300$ meters from the jet to be safe. ▶ Now try Exercises 21–24.

The pH Scale for Acidity

If you check the labels of many household products, including cleansers, drain openers, and shampoo, you will see that they state a quantity called the **pH**. The pH is used by chemists to classify substances as *neutral*, *acidic*, or *basic* (also called *alkaline*). By definition

- Pure water is **neutral** and has a pH of 7.
- An **acid** has a pH lower than 7.
- A **base** has a pH higher than 7.

Table 8.6 gives a few typical pH values.

TABLE 8.6 Typical pH Values

Solution	pH	Solution	pH
lemon juice	2	pure water	7
stomach acid	2–3	baking soda	8.4
vinegar	3	household ammonia	10
drinking water	6.5	drain opener	10–12

Think About It Check around your house, apartment, or dorm for labels that state a pH. Are the substances acids or bases?

Chemically, acidity is related to the concentration of positively charged hydrogen ions, which are hydrogen atoms without their electron. A hydrogen ion is symbolized by H^+, for hydrogen with a positive charge. The concentration of hydrogen ions is denoted as $[H^+]$ and is usually measured in units of *moles per liter*. A **mole** is simply a special number of particles: 1 mole $\approx 6 \times 10^{23}$ particles. (The numerical value of a mole, which is about 6×10^{23}, is known as *Avogadro's number*.)

BY THE WAY
The word *acid* comes from the Latin *acidus*, meaning "sour." Some fruits taste sour because they are acidic. Bases are substances that can neutralize the action of acids. Common antacid tablets (for upset stomachs) are bases and work by neutralizing stomach acid.

The pH Scale

The **pH scale** is defined by the following equivalent formulas:

$$\text{pH} = -\log_{10}[H^+] \quad \text{or} \quad [H^+] = 10^{-\text{pH}}$$

where $[H^+]$ is the hydrogen ion concentration in moles per liter. Pure water is neutral; its pH is 7. Acids have a pH lower than 7 and bases have a pH higher than 7.

EXAMPLE 6 Finding pH

What is the pH of a solution with a hydrogen ion concentration of 10^{-12} mole per liter? Is it an acid or a base?

Solution Using the first version of the pH formula with a hydrogen ion concentration of $[H^+] = 10^{-12}$ mole per liter, we find

$$\text{pH} = -\log_{10}[H^+] = -\log_{10} 10^{-12} = -(-12) = 12$$

(Recall that $\log_{10} 10^x = x$.) A solution with a hydrogen ion concentration of 10^{-12} mole per liter has a pH of 12. Because this pH is well above 7, the solution is a strong base.

▶ Now try Exercises 25–30.

Acid Rain

Normal raindrops are mildly acidic, with a pH slightly under 6. However, the burning of fossil fuels releases sulfur or nitrogen that can form sulfuric or nitric acids in the air, and these acids can make raindrops far more acidic than normal, creating the problem known as **acid rain**. Acid rain is a problem around the world, sometimes causing raindrops (or the related *acid fog*) to have a pH as low as 2—the same acidity as pure lemon juice!

Acid rain can cause great ecological damage. Forests in many areas of the world, including the northeastern United States and southeastern Canada, have been severely damaged by acid rain. Acid rain can also "kill" lakes by making the water so acidic that nothing can survive. Thousands of lakes in the northeastern United States and southeastern Canada have suffered this fate. Surprisingly, you can often recognize a dead lake by its exceptionally clear water—clear because it lacks the living organisms that usually make the water murky.

BY THE WAY
Acid rain is caused primarily by emissions from coal-burning power plants and industries. The problem is particularly bad when coal with a high sulfur content is burned. Today, China has the worst acid rain problems in the world, due to heavy use of coal.

EXAMPLE 7 Acid Rain vs. Normal Rain

In terms of hydrogen ion concentration, compare acid rain with a pH of 2 to ordinary rain with a pH of 6.

Solution For acid rain with pH 2, the hydrogen ion concentration is

$$[H^+] = 10^{-\text{pH}} = 10^{-2} \text{ mole per liter}$$

For ordinary rain with pH 6, the hydrogen ion concentration is

$$[H^+] = 10^{-\text{pH}} = 10^{-6} \text{ mole per liter}$$

Therefore, the hydrogen ion concentration in the acid rain is greater than that in ordinary rain by a factor of

$$\frac{10^{-2}}{10^{-6}} = 10^{-2-(-6)} = 10^4$$

Acid rain is 10,000 times as acidic as ordinary rain. ▶ Now try Exercises 31–32.

IN YOUR WORLD
Ocean Acidification

Most people are aware that pollutants from fossil fuels can cause acid rain and that carbon dioxide released by the burning of fossil fuels causes global warming. Less well known is a related effect called *ocean acidification*, which has been implicated in a global decline in fish stocks and in the health of coral reefs.

As its name suggests, ocean acidification refers to an increase in the acidity of ocean water. Its cause is well understood by scientists: Roughly a third of the carbon dioxide released by the burning of fossil fuels ends up dissolved in the oceans, where it undergoes chemical reactions that increase the seawater's acidity. The accompanying graph shows measurements confirming this process. The red curve shows the rising concentration of carbon dioxide in the atmosphere, and the green curve shows the corresponding rise in the concentration of carbon dioxide dissolved in the oceans. The blue curve shows the measured ocean pH over the period for which the most reliable records are available; remember that a *decrease in pH* means an *increase in acidity*. The downward trend in the pH is clear. Looking back further, scientists estimate that the ocean pH was about 8.25 before the beginning of the industrial era, so the overall decline in pH has been about 0.2, representing about a 60% increase in the hydrogen ion concentration.

Scientists are only beginning to understand the consequences of ocean acidification, but many potentially harmful consequences are already known. For example, greater acidity makes it more difficult for shellfish to produce their shells and for corals to build their skeletons; the latter problem explains why ocean acidification has been tied to the global decline of coral reefs, which are also under pressure from rising water temperatures. Greater acidity has also been shown to cause changes in the metabolism and immune responses of many marine organisms, and these effects can reverberate throughout the ocean food chain.

Because more than a billion people around the world depend on the oceans as a primary food source, and billions more get a significant fraction of their protein from fish, ocean acidification poses a great threat to our ability to feed the world's growing population. While it may get less media attention than acid rain or rising temperatures, keep in mind that ocean acidification is another consequence of fossil fuel use that poses a threat to our future.

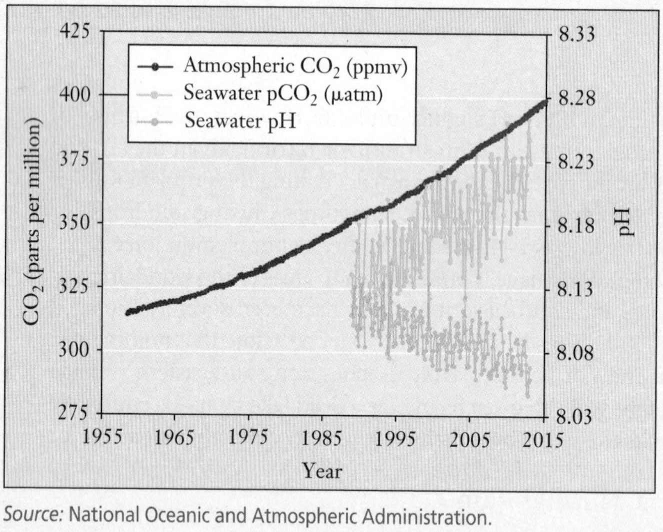

Source: National Oceanic and Atmospheric Administration.

Quick Quiz 8D

Choose the best answer to each of the following questions. Explain your reasoning with one or more complete sentences.

1. The energy released by a magnitude 8 earthquake is about how many times as great as that from a magnitude 6 earthquake?

   a. 32×32 times as great b. $10^{(8-6)}$ times as great c. $8/6$ times as great

2. Why do individual earthquakes in less developed countries tend to cause more deaths than those in developed countries?

   a. Less developed countries have earthquakes with higher magnitudes.

   b. Buildings are more likely to collapse in less developed countries.

   c. The population density is higher in less developed countries.

3. What is a 0-decibel sound?

   a. the softest sound that a human ear can hear

   b. a sound with zero intensity

   c. a sound that cannot be stated on the decibel scale

4. A sound of 85 decibels is defined to be
   a. 85 times as loud as the softest audible sound.
   b. $10^{8.5}$ times as loud as the softest audible sound.
   c. 10^{85} times as loud as the softest audible sound.
5. How does the intensity of a 20-decibel sound compare to that of a 0-decibel sound?
   a. It is 20 times as great. b. It is 100 times as great.
   c. It is 10^{20} times as great.
6. Like sound intensity, gravity strength follows an inverse square law with distance. This means that if you quadruple the distance between two objects, the strength of gravity between them
   a. increases by a factor of 4. b. decreases by a factor of 4.
   c. decreases by a factor of 16.
7. Which of the following indicates the strongest acid?
   a. a pH of 17 b. a pH of 5 c. a pH of 1
8. A pH of 1 means that the hydrogen ion concentration is
   a. 1 mole per liter. b. 10 moles per liter.
   c. the same as that in pure water.
9. A hydrogen ion concentration of 10^{-5} mole per liter means a pH of
   a. −5. b. $\log_{10} 5$. c. 5.
10. Suppose you wish to revive a lake that has been damaged by acid rain. You should add to the lake something that
   a. increases its pH. b. decreases its pH.
   c. changes its pH from positive to negative.

Exercises 8D

REVIEW QUESTIONS

1. What is the *magnitude scale* for earthquakes? What increase in energy is represented by an increase of 1 magnitude on the earthquake scale?
2. What is the *decibel scale*? Describe how it is defined.
3. What is *pH*? What pH values define an *acid*, a *base*, and a *neutral substance*?
4. What is *acid rain*? Why is it a serious environmental problem?

DOES IT MAKE SENSE?

Decide whether each of the following statements makes sense (or is clearly true) or does not make sense (or is clearly false). Explain your reasoning.

5. An earthquake of magnitude 6 will do twice as much damage as an earthquake of magnitude 3.
6. A 120-dB sound is 20% louder than a 100-dB sound.
7. If I double the amount of water in the cup, I'll also double the pH of the water in the cup.
8. The lake water was crystal clear, so it could not possibly have been affected by acid rain.

BASIC SKILLS & CONCEPTS

9–14: Earthquake Magnitudes. Use the earthquake magnitude scale to answer the questions.

9. How much energy, in joules, is released by an earthquake of magnitude 7?
10. How many times as much energy is released by an earthquake of magnitude 6 as by one of magnitude 4?
11. How much energy, in joules, was released by the December 2016 earthquake in the Solomon Islands that had a magnitude of 7.8?
12. How much energy, in joules, was released by the 2008 earthquake in Sichuan province, China (magnitude 7.9), which killed at least 68,000 people, including many schoolchildren?
13. Compare the energy of a magnitude 6 earthquake to that released by a 1-megaton nuclear bomb (5×10^{15} joules).
14. What would be the magnitude of an earthquake that released the same energy as a 1-megaton nuclear bomb (5×10^{15} joules)? Which would be more destructive, the bomb or the earthquake? Why?

15–20: The Decibel Scale. Use the decibel scale to answer the questions.

15. How many times as loud as the softest audible sound is the sound of busy street traffic?
16. How many times as loud as the softest audible sound is the pain threshold for the human ear?
17. What is the loudness, in decibels, of a sound 1 million times as loud as the softest audible sound?
18. What is the loudness, in decibels, of a sound 10 billion times as loud as the softest audible sound?
19. How much louder (more intense) is a 55-decibel sound than a 10-decibel sound?
20. Suppose that a sound is 100 times as loud as (more intense than) a whisper. What is its loudness in decibels?

21–24: Inverse Square Law. Use the inverse square law for sound intensity to answer the questions.

21. How many times greater is the intensity of sound from a concert speaker at a distance of 1 meter than the intensity at a distance of 4 meters?

22. How many times greater is the intensity of sound from a concert speaker at a distance of 10 meters than the intensity at a distance of 100 meters?

23. How many times greater is the intensity of sound from a concert speaker at a distance of 10 meters than the intensity at a distance of 80 meters?

24. How many times greater is the intensity of sound from a concert speaker at a distance of 20 meters than the intensity at a distance of 200 meters?

25–32: The pH Scale. Use the pH scale to answer the questions.

25. If the pH of a solution increases by 5 (e.g., from 4 to 9), how much does the hydrogen ion concentration change? Does the change make the solution more acidic or more basic?

26. If the pH of a solution decreases by 2.5 (e.g., from 6.5 to 4), how much does the hydrogen ion concentration change? Does the change make the solution more acidic or more basic?

27. What is the hydrogen ion concentration of a solution with pH 9.5?

28. What is the hydrogen ion concentration of a solution with pH 2.5?

29. What is the pH of a solution with a hydrogen ion concentration of 0.0001 mole per liter? Is this solution an acid or a base?

30. What is the pH of a solution with a hydrogen ion concentration of 10^{-13} mole per liter? Is this solution an acid or a base?

31. How many times more acidic is acid rain with a pH of 3 than ordinary rain with a pH of 6?

32. How many times more acidic is acid rain with a pH of 3.5 than ordinary rain with a pH of 6?

FURTHER APPLICATIONS

33–38: Logarithmic Thinking. Briefly describe, in words, the effects you would expect in the situations given.

33. An earthquake of magnitude 2.8 strikes the Los Angeles area.

34. You have your ear against a new speaker when it emits a sound with an intensity of 160 decibels.

35. A young child (too young to know better) finds and drinks from an open bottle of drain opener with pH 12.

36. An earthquake of magnitude 8.5 strikes the Tokyo area.

37. Your friend is calling to you from across the street in New York City, with a shout that registers 90 decibels. Traffic is heavy, and several emergency vehicles are passing by with sirens.

38. A forest situated a few hundred miles from a coal-burning industrial area is subjected regularly to acid rain, with pH 4, for many years.

39. **Sound and Distance.**

    a. The decibel level for busy street traffic in Table 8.5 is based on the assumption that you stand very close to the noise source—say, 1 meter from the street. If your house is 100 meters from a busy street, how loud will the street noise be, in decibels?

    b. At a distance of 10 meters from the speakers at a concert, the sound level is 135 decibels. How far away should you sit to reduce the level to 120 decibels?

    c. Imagine that you are a spy in a restaurant. The conversation you want to hear is taking place in a booth across the room, about 8 meters away. The people are speaking softly, so they hear each other's voice at about 20 decibels (they are sitting about 1 meter apart). How loud is the sound of their voices when it reaches your table? If you have a miniature amplifier in your ear and want to hear their voices at 60 decibels, by what factor must you amplify their voices?

40. **Variation in Sound with Distance.** Suppose that a siren is placed 0.1 meter from your ear.

    a. How many times as loud as the sound from a siren at 30 meters will the sound you hear be?

    b. How loud will the siren next to your ear sound, in decibels?

    c. How likely is it that this siren will cause damage to your eardrum? Explain.

41. **Toxic Dumping in Acidified Lakes.** Consider a situation in which acid rain has heavily polluted a lake to a level of pH 4. An unscrupulous chemical company dumps some acid into the lake illegally. Assume that the lake contains 100 million gallons of water and that the company dumps 100,000 gallons of acid with pH 2.

    a. What is the hydrogen ion concentration, $[H^+]$, of the lake polluted by acid rain alone?

    b. Suppose that the unpolluted lake, without acid rain, would have pH 7. If the lake were then polluted by company acid alone (no acid rain), what hydrogen ion concentration, $[H^+]$, and pH would it have?

    c. What is the hydrogen ion concentration, $[H^+]$, after the company dumps the acid into the acid rain–polluted lake (pH 4)? What is the new pH of the lake?

    d. If the U.S. Environmental Protection Agency can test for changes in pH of only 0.1 or greater, could the company's pollution be detected?

42. **Ocean pH.** Scientists estimate that the ocean pH was about 8.25 before the beginning of the industrial era, around 1750. Today it is about 8.05, and if current trends continue, it may fall to 7.9 by 2050. Use these data to calculate the percent change in the hydrogen ion concentration $([H^+])$ of the oceans from 1750 to today, and from 1750 to 2050. Briefly comment on your results.

IN YOUR WORLD

43. **Earthquakes in the News.** Find a recent news story dealing with some aspect of earthquakes, such as their destructive power, attempts to predict them, or ways of building to withstand them. How does the magnitude of an earthquake affect the issue? Explain.

44. **Earthquake Disasters.** Find the death tolls for some of the worst earthquake disasters in history. How strongly do the death tolls correlate with the earthquake magnitudes? Discuss the factors that determine the devastation caused by an earthquake.

45. **Sound Effects.** Research the effects of loud sounds on hearing; for example, you might look into evidence tying loud music to hearing loss, or evidence of hearing damage in people who work in loud environments. Write a brief report on your findings.

46. **Acid Rain.** Investigate the problem of acid rain in a region where it has been a particular problem, such as the northeastern United States, southeastern Canada, the Black Forest in Germany, eastern Europe, or China. Write a report on your findings. The report should include a description of the acidity of the rain, the source of the acidity, the damage being caused by the acid rain, and the status of efforts to alleviate this damage.

47. **Ocean Acidification.** Research ocean acidification and its impacts on both marine ecosystems and human society. Write a short report that summarizes the issue and that discusses your personal recommendations about how we should work to address it.

Chapter 8 Summary

UNIT	KEY TERMS	KEY IDEAS AND SKILLS
8A	linear growth (p. 492) exponential growth (p. 492) doublings (p. 493)	In **linear growth,** a quantity grows by the same *absolute* amount in each unit of time. (p. 492) In **exponential growth,** a quantity grows by the same *relative* amount in each unit of time. (p. 492) **Understand** the impact of doublings and why exponential growth cannot continue indefinitely. (p. 493)
8B	doubling time (p. 499) rule of 70 (p. 501) half-life (p. 502) fractional growth rate (p. 504) logarithm (p. 506)	**Doubling time calculations:** After time t, for a quantity growing exponentially with doubling time T_{double}: (p. 500) $\text{new value} = \text{initial value} \times 2^{t/T_{double}}$ The **approximate doubling time formula** for a quantity growing exponentially at a rate of $P\%$ per time period: (p. 501) $T_{double} \approx \frac{70}{P}$ **Half-life calculations:** After time t, for a quantity decaying exponentially with half-life T_{half}, (p. 503) $\text{new value} = \text{initial value} \times \left(\frac{1}{2}\right)^{t/T_{half}}$ **The approximate half-life formula** for a quantity decaying exponentially at a rate of $P\%$ per time period: (p. 504) $T_{half} \approx \frac{70}{P}$ The **exact doubling time formula** for a quantity growing exponentially at a fractional rate r: (p. 504) $T_{double} = \frac{\log_{10} 2}{\log_{10}(1 + r)}$ **The exact half-life formula** for a quantity decaying exponentially at a fractional rate (r must be negative): (p. 505) $T_{half} = -\frac{\log_{10} 2}{\log_{10}(1 + r)}$
8C	overall growth rate (p. 512) birth rate (p. 512) death rate (p. 512) carrying capacity (p. 512) logistic growth (p. 513) overshoot and collapse (p. 514)	**Logistic growth:** (p. 513) $\text{logistic growth rate} = r \times \left(1 - \frac{\text{population}}{\text{carrying capacity}}\right)$ **Contrast** exponential growth, logistic growth, and overshoot and collapse. (p. 514) **Understand** factors affecting the carrying capacity. (p. 514)

UNIT	KEY TERMS	KEY IDEAS AND SKILLS
8D	logarithmic scale (p. 520) magnitude scale (p. 521) decibel scale (p. 522) inverse square law (p. 523) pH (p. 524) neutral (p. 524) acid (p. 524) base (p. 524) mole (p. 525) acid rain (p. 525)	**Earthquake magnitude scale:** (p. 521) $\log_{10} E = 4.4 + 1.5M \quad \text{or} \quad E = (2.5 \times 10^4) \times 10^{1.5M}$ **Decibel scale:** (p. 522) $\text{loudness in dB} = 10 \log_{10}\left(\frac{\text{intensity of the sound}}{\text{intensity of softest audible sound}}\right)$ or $\left(\frac{\text{intensity of the sound}}{\text{intensity of softest audible sound}}\right) = 10^{(\text{loudness in dB})/10}$ **Inverse square law for sound:** The intensity of a sound *decreases* with the *square* of the distance from the source. (p. 524) **pH scale:** (p. 525) $\text{pH} = -\log_{10}[\text{H}^+] \quad \text{or} \quad [\text{H}^+] = 10^{-\text{pH}}$

9 MODELING OUR WORLD

It may not be possible to predict the future precisely, but we needn't go forward blindly. Mathematics provides tools for analyzing relationships between variables and building models that can help us make educated guesses about the future. Although many mathematical models are quite complex, the basic principles are easy to understand. In this chapter, we discuss the principles involved in using mathematics to model our world and explore examples of both linear and exponential models.

Q Mathematical modeling relies on identifying relationships that describe how one variable changes with respect to another. These relationships can be described in words as well as with equations. Which one of the following statements does *not* describe a relationship in which one variable changes with respect to another?

- **A** The total distance a car has traveled after various amounts of time during a road trip
- **B** The way the price of a new smart phone affects the number of people who buy it
- **C** The list on your grocery store receipt giving prices of items you purchased
- **D** The effect of income tax rates on government revenue
- **E** The way in which the population of endangered elephants depends on the number of park rangers enforcing anti-poaching laws

Nothing endures but change.

—Heraclitus, c. 500 B.C.E.

A This text began with a promise to focus on topics that will be important to you in other college courses, in your potential career, and in your daily life. You might wonder how the featured question relates to these goals, as it's not the kind of question you often hear outside of a mathematical context. However, mathematical models may well have a greater effect on your daily life than almost anything else we've covered in this text, because they play such an enormous role in our society today. Indeed, virtually every major decision that we make as citizens today—decisions about economics, tax policies, environmental laws, military policies, and much more—is informed by mathematical models.

The models we discuss in this chapter will be relatively simple ones, but they will help you understand the principles used in the many models that have greater impacts on our lives. To help you get started, choose your answer to the above question, and see if you can explain why it is correct. You'll find the solution in Example 1 in Unit 9A.

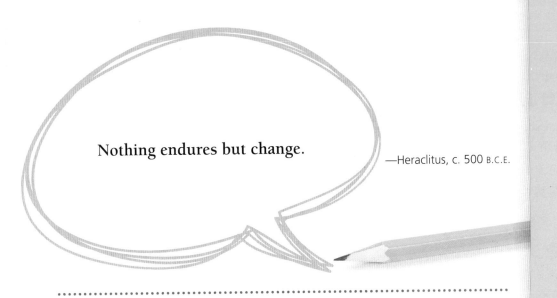

UNIT 9A

Functions: The Building Blocks of Mathematical Models: Understand the concept of a function, which expresses a mathematical relationship between two or more variables.

UNIT 9B

Linear Modeling: Learn to write and use linear functions, which describe straight-line graphs.

UNIT 9C

Exponential Modeling: Explore functions that describe exponential growth and decay, which have applications that include models of population growth and techniques for radiometric dating.

Climate Modeling

Use this activity to gain a sense of the kinds of problems this chapter will enable you to analyze. Additional activities are available online in MyLab Math.

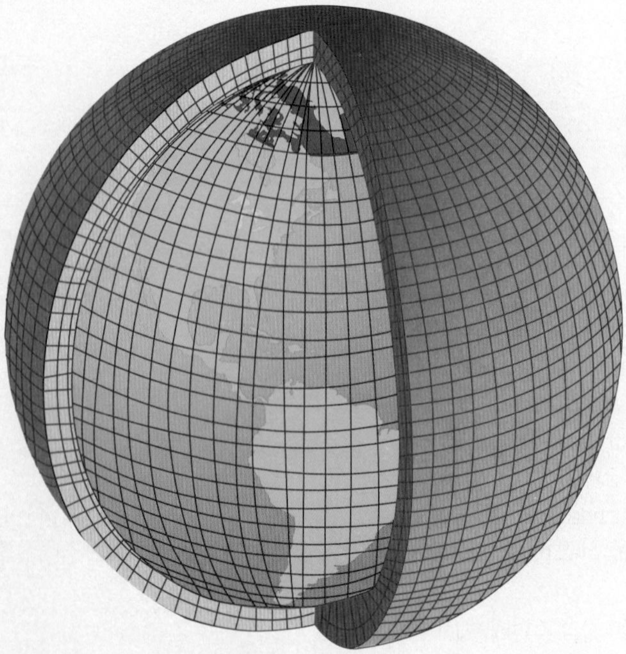

FIGURE 9.A

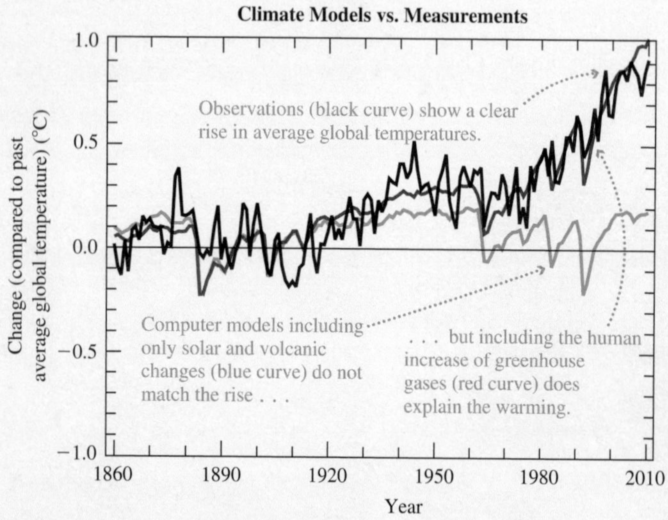

FIGURE 9.B This graph compares observed temperature changes (black curve) with the predictions of climate models that include only natural factors such as changes in the brightness of the Sun and effects of volcanoes (blue curve) and models that also include the human contribution to increasing greenhouse gas concentration (red curve). *Source:* Intergovernmental Panel on Climate Change.

The fact that increasing the atmospheric carbon dioxide concentration causes global warming was first discovered more than 150 years ago through laboratory measurements of the heat-trapping effects of carbon dioxide. This fact is confirmed by the fact that temperatures for all the planets in our solar system can be explained only when we take into account the warming effects of carbon dioxide. But while there's no doubt that continuing to add carbon dioxide will warm the Earth, we'd like to know exactly *how much* the planet will warm and how this warming will affect local climates in different parts of the world. The best way to address these questions is through mathematical modeling.

The principle behind a mathematical model of the climate is relatively simple. The model uses a grid of cubes like those shown in Figure 9.A. The "initial conditions" for the model consist of a description of the climate within each cube at one moment in time; for example, the initial climate in each cube can be described with data giving the temperature, air pressure, wind speed and direction, and humidity at the time the model begins. The model then uses equations that govern the climate (for example, equations that describe how heat and air flow from one cube to neighboring cubes) to predict how the conditions in each cube will change in a small time interval, such as the next hour. The model repeats the process to predict the conditions after another hour, and so on. In this way, the model can simulate climate changes over any period of time.

The practical difficulty with such models comes from the complexity of the climate. Modern climate models run on supercomputers and use millions of little cubes whose changes are governed by thousands of equations. Despite the complexity of the challenge, today's climate models have proven remarkably accurate at "predicting" the past climate, giving scientists confidence that they can also predict the future with reasonable accuracy. Explore climate models further by discussing the following questions in small groups.

1. Suppose you want to improve a climate model. Would you increase or decrease the number of cubes in the grid?
2. Suggest a way to test a model by looking at past climate data. How would you decide if your model was working well? What would you do if it were *not* working well?

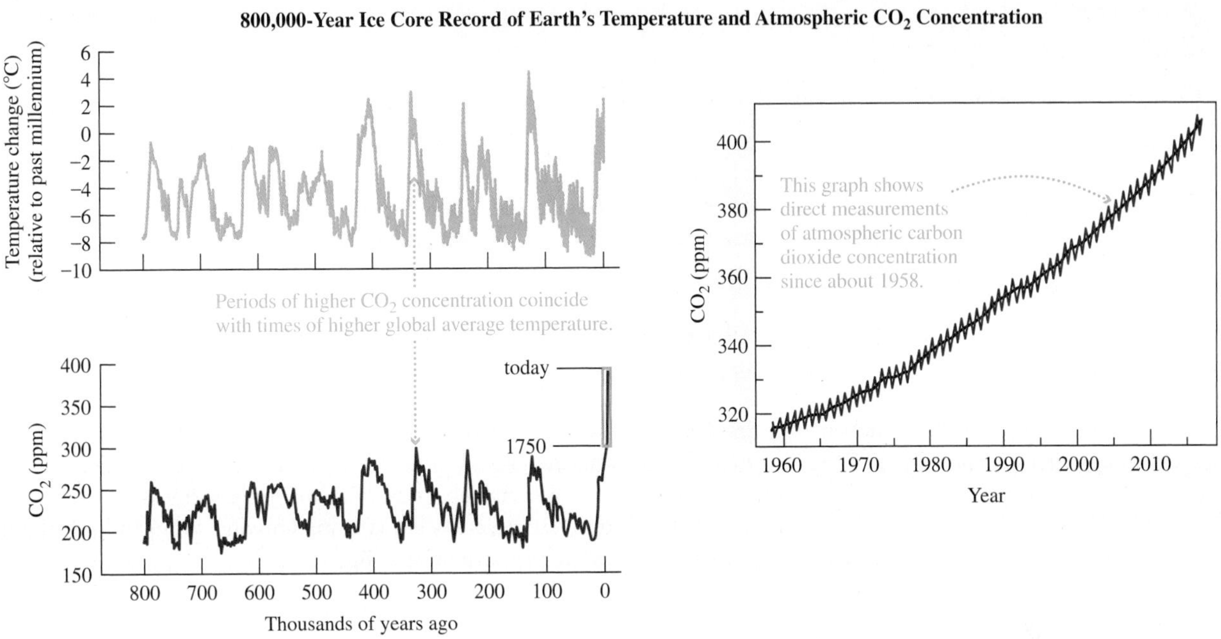

FIGURE 9.C The graphs show variations in global temperature (top) and the atmospheric carbon dioxide concentration (bottom) over the past 800,000 years, based on measurements made on air bubbles trapped in Antarctic ice cores; "ppm" stands for "parts per million." The inset at right shows direct measurements of the carbon dioxide concentration since the late 1950s. *Sources:* European Project for Ice Coring in Antarctica; National Oceanic and Atmospheric Administration.

3. Study the three curves shown in Figure 9.B. Explain why the close agreement between the black and red curves gives scientists confidence in their climate models. What can you conclude from the mismatch between the black and blue curves?

4. Figure 9.C shows the atmospheric concentration of carbon dioxide and the global average temperature over the past 800,000 years. How does the figure support the idea that changes in the carbon dioxide concentration lead to changes in Earth's temperature?

5. Based on Figure 9.C, how much has the carbon dioxide concentration varied naturally over the past 800,000 years? How does today's carbon dioxide concentration compare to the highest concentrations that have occurred naturally in the period shown?

6. Use the carbon dioxide data for recent decades, shown in the inset of Figure 9.C, to predict the carbon dioxide concentration in the years 2050 and 2100. Explain how you made your prediction and what uncertainties your predicted levels have.

7. Suppose that a particular climate model predicts that a carbon dioxide concentration of 560 parts per million (twice the pre-industrial value of 280 parts per million) will cause Earth to warm by 2°C. Based on the temperature and carbon dioxide changes shown in Figure 9.C, does the 2°C warming seem reasonable, too high, or too low? Explain your reasoning, and discuss how it influences your view of the validity of the model.

8. Do a Web search on predicted climate change in your local region (for example, if you live in California, you could search with the phrase "climate change California"). What changes do models predict for your region over the next 50 years? Based on your understanding of climate models, how much confidence do you have in these predictions? Explain.

UNIT 9A

Functions: The Building Blocks of Mathematical Models

She had not understood mathematics until he had explained to her that it was the symbolic language of relationships. "And relationships," he had told her, "contained the essential meaning of life."
—Pearl Buck, *The Goddess Abides*, Part I

A real office complex may not look exactly like the scale model used to design it, but the model helps the architects create the design. A road map doesn't look at all like a real landscape, but it serves as a model of a road system and can be extremely useful when you travel. The purpose of a **mathematical model** is similar to that of an architectural model or a road map: It represents something real and helps us understand that real thing.

More specifically, mathematical models are based on relationships between quantities that can change, such as the relationship between wind speed and stress on a bridge, or the relationship between worker productivity and unemployment. These relationships are described by mathematical tools called **functions**. In essence, functions are the building blocks of mathematical models. Some mathematical models consist of only a single function, which we can represent with a simple formula or graph. Other models, such as those used to study Earth's climate, may involve thousands of functions and require supercomputers for their analysis. But the basic idea of a function is the same in all cases.

Language and Notation of Functions

Technical Note

Not all relationships are functions. Functions have the important property that for each value of the independent variable in the domain, there corresponds exactly one value of the dependent variable.

We have already used functions several times in this book, without calling them by name. For example, in Chapter 4, we saw exactly how the *balance* in a savings plan is related to the *interest rate* and the *monthly deposits*. In Chapter 8, we saw how *population* is related to *growth rate* and *time*. These relationships are functions because they tell us specifically how one quantity varies with respect to another quantity. We are now ready to explore the language and notation of functions.

Dependent and Independent Variables

Suppose we want to model the variation in temperature over the course of a day, based on the data in Table 9.1. The first step is to recognize that two quantities are involved in this model: *time* and *temperature*. Our goal is to express the relationship between time and temperature in the form of a function.

The quantities related by a function are called **variables** because they change, or *vary*. In this case, *temperature* is the **dependent variable** because it *depends on* the time of day. *Time* is the **independent variable** because time varies *independently* of the temperature. We say that the temperature varies *with respect to* time. Notice that there is exactly one value of the temperature for each time of day.

TABLE 9.1 Temperature Data for One Day

Time	Temperature	Time	Temperature
6:00 a.m.	50°F	1:00 p.m.	73°F
7:00 a.m.	52°F	2:00 p.m.	73°F
8:00 a.m.	55°F	3:00 p.m.	70°F
9:00 a.m.	58°F	4:00 p.m.	68°F
10:00 a.m.	61°F	5:00 p.m.	65°F
11:00 a.m.	65°F	6:00 p.m.	61°F
12:00 noon	70°F		

EXAMPLE 1 Identifying Variables

Look back at the chapter-opening question (p. 532). Which statement does *not* describe a relationship in which one variable changes with respect to another? For the remaining statements, identify the dependent and independent variables and briefly explain the expected relationship.

Solution The answer to the chapter-opening question is C, because there is no obvious or ordered relationship between the items on the grocery store receipt and their prices. For the remaining statements:

- Statement: "The total distance a car has traveled after various amounts of time during a road trip." The dependent variable is *distance* and the independent variable is *time*, because the distance the car has traveled depends on how much time has elapsed in the road trip. We expect the distance to increase as more time passes.
- Statement: "The way the price of a new smart phone affects the number of people who buy it." The dependent variable is *number of people* (who buy the smart phone) and the independent variable is *price* (of the smart phone), because the number of people buying the smart phone depends on its price. We expect the number of people making the purchase to decrease as the price increases.
- Statement: "The effect of income tax rates on government revenue." The dependent variable is *government revenue* and the independent variable is *income tax rates*, because government revenue depends on the income tax rates. We expect government revenue to increase as income tax rates increase, at least up to a point; at some point, the economic effects of higher tax rates might hurt the economy enough to reduce government revenues.
- Statement: "The way in which the population of endangered elephants depends on the number of park rangers enforcing anti-poaching laws." The dependent variable is *population* (of elephants) and the independent variable is *number of rangers*, because we expect the population to depend on the number of rangers working to prevent poaching. The population should increase with the number of rangers, at least up to the point at which all poaching has been stopped.

▶ Now try Exercises 11–14.

Writing Functions

We often write related variables in the form of an ordered pair, with the independent variable first and the dependent variable second:

(*time*, *temperature*)

We use special notation to represent functions. For example, we might use x to represent the independent variable and y to represent the dependent variable. Then we write $y = f(x)$, read "y is a function of x," which means that y is related to x by the function f. In the case of our (*time*, *temperature*) example from Table 9.1, we let t represent time and T represent temperature. We write $T = f(t)$ to indicate that temperature varies with respect to time, or that temperature is a function of time.

It may be helpful to think of a function as a box with two slots, one for input and one for output (Figure 9.1). A value of the independent variable can be put into the box through the input slot. The function inside the box "operates" on the input and produces one value of the dependent variable, which appears as output from the box.

HISTORICAL NOTE

Mathematicians worked with functions for centuries before developing a standard notation. In about 1670, the German philosopher Gottfried Leibniz (1646–1716) used the notation $f(x)$ to denote a function. But it wasn't until about 1734, when the Swiss mathematician Leonhard Euler (1707–1783) adopted the same notation, that $f(x)$ became widely accepted for expressing functions.

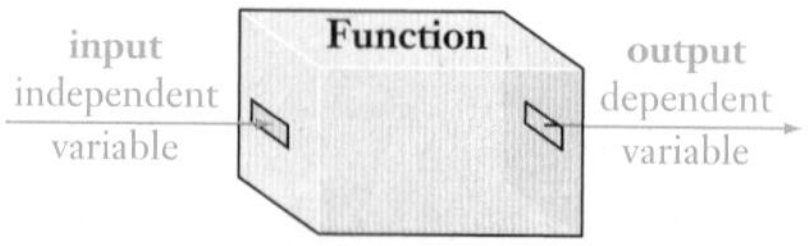

FIGURE 9.1 A pictorial representation of a function.

Functions

A function describes how a **dependent variable** changes *with respect to* one or more **independent variables.** When there are only two variables, we may denote their relationship as an ordered pair with the independent variable first:

(*independent variable, dependent variable*)

We say that the dependent variable is *a function of* the independent variable. If x is the independent variable and y is the dependent variable, we write the function as

$$y = f(x)$$

Many functions describe changes with respect to time. The weight of a child and the Consumer Price Index both vary with respect to time. But not all functions involve time. Mortgage payments are a function of the interest rate for the mortgage, because the payment amount depends on the interest rate. Similarly, the gas mileage of a car is a function of the car's speed, because the mileage is different at different speeds.

Think About It From your everyday experiences, identify several pairs of variables that appear to be related and might be described by a function. Include at least one pair that does not involve time.

EXAMPLE 2 Writing Functions

For each situation, express the given function in words. Write the two variables as an ordered pair and write the function using the notation $y = f(x)$.

a. You are riding in a hot-air balloon. As the balloon rises, the surrounding atmospheric pressure decreases (causing your ears to pop).

b. You're on a barge headed south down the Mississippi River. You notice that the width of the river changes as you travel southward with the current.

BY THE WAY

The Mississippi River runs about 2300 miles (3700 kilometers) from Lake Itasca, Minnesota, to the Gulf of Mexico. The Mississippi River system, which includes the Red Rock River in Montana and the Missouri River, has a length of about 3700 miles (6000 kilometers).

Solution

a. The pressure *depends on* your altitude, so we say that the pressure changes *with respect to* altitude. *Pressure* is the dependent variable and *altitude* is the independent variable, so the ordered pair of variables is (*altitude, pressure*). If we let A stand for altitude and P stand for pressure, then we write the function as

$$P = f(A)$$

b. The river width *depends on* your distance from the river's source, so we say that the river width changes *with respect to* the distance from the source. *River width* is the dependent variable and *distance from the source* is the independent variable, so the ordered pair of variables is (*distance from source, river width*). Letting d represent distance and w represent river width, we have the function

$$w = f(d)$$

▶ Now try Exercises 15–22

Think About It Does dependence imply causality? That is, do changes in the independent variable *cause* changes in the dependent variable? Give a few examples to make your case.

Representing Functions

There are three basic ways to represent a function.

1. We can represent a function with a **data table**, such as Table 9.1. A table provides detailed information, but can become unwieldy when we have large quantities of data.
2. We can draw a *picture*, or **graph**, of a function. A graph is easy to interpret and consolidates a great deal of information.
3. We can write a compact mathematical representation of a function in the form of an equation (or formula).

In the remainder of this unit, we explore the use of graphs to represent functions. We'll discuss equations in Unit 9B.

Brief Review The Coordinate Plane

The most common way to draw a graph of a function is to use a **coordinate plane**, which is made by drawing two perpendicular number lines. Each number line is called an **axis** (plural: **axes**). Normally, numbers increase to the right on the horizontal axis and upward on the vertical axis. The intersection point of the two axes, where both number lines show the number zero, is called the **origin**. If we are working with general functions, the horizontal axis is called the **x-axis** and the vertical axis is called the **y-axis**.

Points in the coordinate plane are described by two **coordinates** (called an *ordered pair*). The **x-coordinate** gives the point's horizontal position relative to the origin. Points to the right of the origin have positive *x*-coordinates and points to the left of the origin have negative *x*-coordinates. The **y-coordinate** gives the point's vertical position relative to the origin. Points above the origin have positive *y*-coordinates and points below the origin have negative *y*-coordinates. We express the location of a particular point by writing its *x*- and *y*-coordinates in parentheses in the form (x, y). When working with functions, we always use the *x*-axis for the independent variable and the *y*-axis for the dependent variable.

Figure 9.2a shows a coordinate plane with several points identified by their coordinates. Note that the origin is the point (0, 0). Figure 9.2b shows that the axes divide the coordinate plane into four **quadrants**, numbered counterclockwise starting from the upper right.

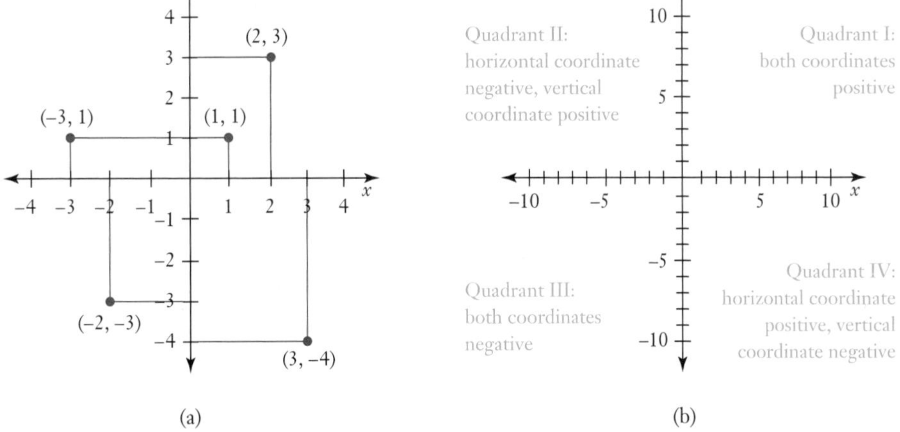

FIGURE 9.2

▸ Now try Exercises 9–10.

Domain and Range

Before we draw a graph of any function, we must determine the variables that we should show on each axis. Mathematically, each axis on a graph extends to infinity in both directions. However, most functions are meaningful only over a small region of the coordinate plane. In the case of the (*time, temperature*) function, based on the data in Table 9.1, negative values of time do not make sense. In fact, the only times of interest for this function are those over which data were collected—from 6 a.m. to 6 p.m. The times that make sense and are of interest make up the **domain** of the function.

Similarly, the only temperatures of interest for this function are those that occurred between 6:00 a.m. and 6:00 p.m. The lowest temperature recorded in this period was 50°F and the highest was 73°F. We therefore say that temperatures between 50°F and 73°F make up the **range** of the function. More generally, we can make the following definitions, which are illustrated in Figure 9.3.

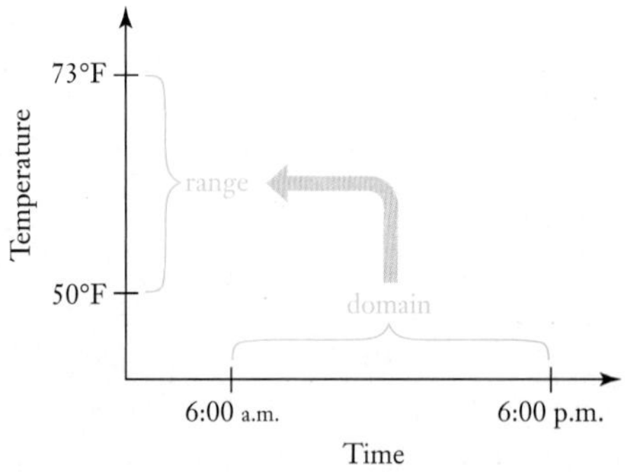

FIGURE 9.3 Each value in the domain gives one value in the range.

Domain and Range

The **domain** of a function is the set of values of the independent variable that both make sense and are of interest for the function.

The **range** of a function consists of the values of the dependent variable that correspond to the values in the domain.

Now that we've identified the domain and range, we can draw the graph. We use the horizontal axis for the time, t, and label it "Time (hours after 6:00 a.m.)." Therefore, $t = 0$ corresponds to the first measurement at 6:00 a.m., and $t = 12$ corresponds to the last measurement at 6:00 p.m. We use the vertical axis for the temperature, T, and label this axis: "Temperature (°F)." Figure 9.4a shows the result, with each point plotted individually. To show details more clearly, Figure 9.4b zooms in on the region covered by the domain and the range.

HISTORICAL NOTE

The coordinate system, called *Cartesian coordinates*, for locating points on a graph is attributed to the French thinker René Descartes (1596–1650). Descartes laid the foundation for analytical geometry. He is also known for his philosophical meditations on the theory of mind/body dualism, which asserts that the mind and the body are distinct but interacting entities. His philosophy is often summarized by his phrase "cogito ergo sum," or "I think, therefore I am."

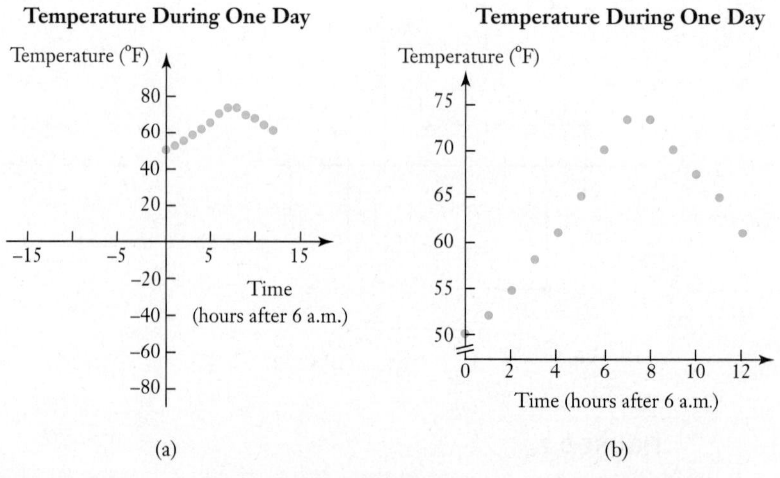

FIGURE 9.4 (a) Graph of data from Table 9.1. (b) Zooming in on the region of interest.

Completing the Model

So far, we have plotted only the 13 data points from Table 9.1. However, every instant of time has a temperature, and the temperature changes continuously throughout the day. If we want the graph to give us a realistic *model*, we should fill in the gaps between the data points. Because we don't expect any sudden spikes or dips in temperature during the day, it is reasonable to connect the data points with a smooth curve, as shown in Figure 9.5.

The graph is now a model that we can use to predict the temperature at *any* time of day. For example, the model predicts that the temperature was about 67°F at 11:30 a.m. (5.5 hours after 6 a.m.). Keep in mind that this prediction may not be exact. We cannot even check it, because Table 9.1 does not provide data for 11:30 a.m. Nevertheless, the prediction seems reasonable, given our assumptions in drawing the smooth curve. This example illustrates an important lesson that applies to all mathematical models: *A model's predictions can be only as good as the data and the assumptions from which the model is built.*

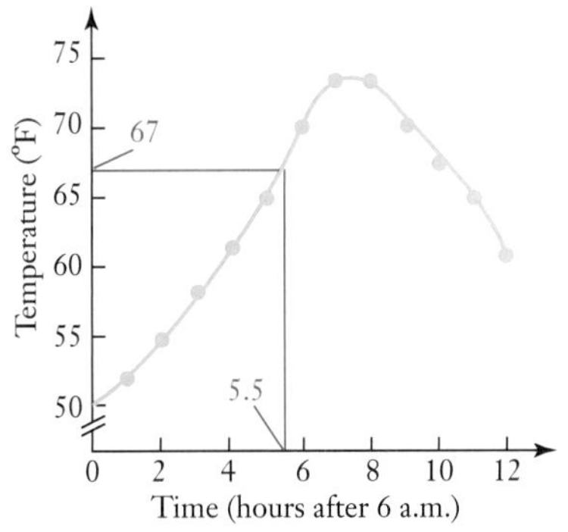

FIGURE 9.5 The graph from Figure 9.4 with a smooth curve connecting the data points.

Summary

Creating and Using Graphs of Functions

Step 1. Identify the independent and dependent variables of the function.

Step 2. Identify the domain (values of the independent variable) and the range (values of the dependent variable) of the function. Use this information to choose the scale and labels on the axes. Zoom in on the region of interest to make the graph easier to read.

Step 3. Make a graph using the given data to plot points. If appropriate, draw a line or curve to connect the data points.

Step 4. Before accepting any predictions of the model, be sure to evaluate the data and the assumptions from which the model was built.

EXAMPLE 3 Pressure-Altitude Function

Imagine measuring the atmospheric pressure as you rise upward in a hot-air balloon. Table 9.2 shows typical values you might find for the pressure at different altitudes, with the pressure given in units of *inches of mercury*. (This pressure unit is used with barometers that measure pressure by the height of a column of mercury in a tube.) Use these data to graph a function showing how atmospheric pressure depends on altitude. Use the graph to predict the atmospheric pressure at an altitude of 15,000 feet, and discuss the validity of your prediction.

TABLE 9.2 Typical Values of Pressure at Different Altitudes

Altitude (feet)	Pressure (inches of mercury)
0	30
5,000	25
10,000	22
20,000	16
30,000	10

BY THE WAY

The atmospheric pressure at the summit of Mt. Everest (elevation 29,029 feet) is about one-third of the pressure at sea level. Therefore, climbers inhale only about one-third as much oxygen (in each breath) as they would at sea level. A rule of thumb is that the pressure drops by approximately one-half with a 20,000-foot increase in altitude.

Solution In part (a) of Example 2, we identified *altitude*, A, as the independent variable and *pressure*, P, as the dependent variable. The *domain* is the set of relevant values of

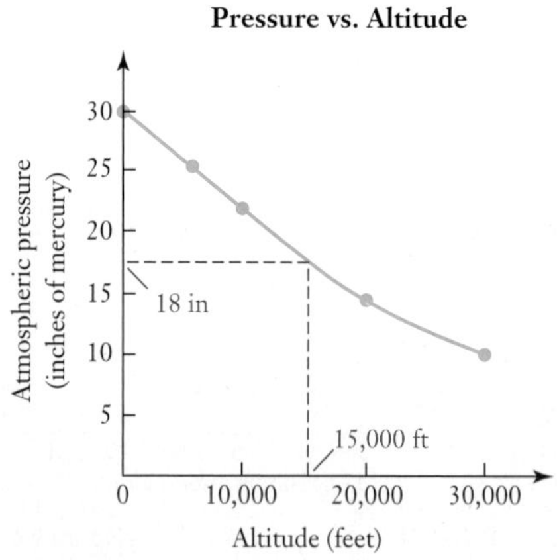

FIGURE 9.6 Pressure-altitude function.

the independent variable, A. For the altitude values in Table 9.2, the domain extends from 0 feet (sea level) to 30,000 feet. The *range* is the set of values of the dependent variable, P, that correspond to the domain. The range therefore extends from 10 to 30 inches of mercury.

We plot the five data points, as shown in Figure 9.6. Between any two data points, we can reasonably assume that pressure decreases smoothly with increasing altitude. Furthermore, because Earth's atmosphere doesn't end abruptly, the pressure must decrease more gradually at higher altitudes. We therefore complete the model by adding a smooth curve to the graph.

Using this graph, we predict that the atmospheric pressure at 15,000 feet is about 18 inches of mercury. Because we've sketched the function only roughly, we should expect this prediction to be approximately, but not exactly, correct. ▶ **Now try Exercises 23–24.**

EXAMPLE 4 Hours of Daylight

The number of hours of daylight varies with the seasons. Use the following data for 40°N latitude (the latitude of San Francisco, Denver, Philadelphia, and Rome) to model the change in the number of daylight hours with time.

- The number of hours of daylight is greatest on the *summer solstice* (about June 21), when it is about 14 hours.
- The number of hours of daylight is smallest on the *winter solstice* (about December 21), when it is about 10 hours.
- On the spring and fall *equinoxes* (about March 21 and September 21, respectively), there are about 12 hours of daylight.

According to the model, at what times of year does the number of daylight hours change most gradually? Most quickly? Discuss the validity of the model.

Solution We expect the number of hours of daylight to be a function of the time of year. *Time* is the independent variable, because time marches on regardless of other events; we denote it by t. *Hours of daylight* is the dependent variable, because it depends on the time of year; we denote it by h.

The times of interest are all the days in a year. The *domain* is all times of interest, which is the entire year. The *range* extends from 10 hours to 14 hours of daylight. We know from experience that the number of hours of daylight changes smoothly with the seasons, so we can connect the four given data points for each year with a smooth curve. Because the same pattern repeats from one year to the next, we can extend the graph for additional years (Figure 9.7); we chose three years because that is long enough for the

BY THE WAY

The seasons arise because of the tilt of the Earth's axis. The northern and southern hemispheres alternately get more and less direct sunlight as the Earth orbits the Sun, so the seasons are opposite in the two hemispheres.

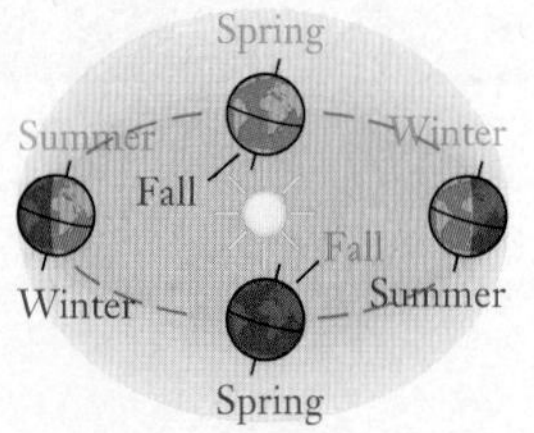

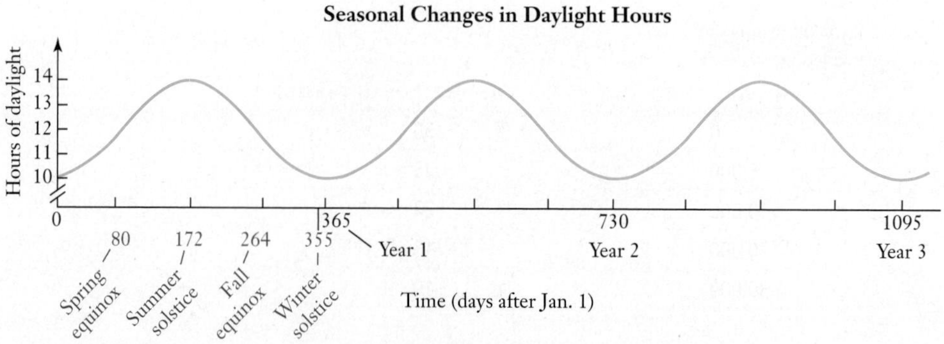

FIGURE 9.7 Hours of daylight over a 3-year period, for latitude 40° N.

pattern to be clear. This type of function, in which a particular pattern repeats over and over, is called a **periodic function**. Because this function is based on simple seasonal patterns, we can expect it to be quite accurate.

Note that the curve "tops out" at each summer solstice and "bottoms out" at each winter solstice. Because the curve is relatively flat around the two solstices, the number of hours of daylight varies most gradually near the solstices. That is, we have a couple of months with long daylight hours around the summer solstice and a couple of months with short daylight hours around the winter solstice. In contrast, the number of hours of daylight increases rapidly around the time of each spring equinox and decreases rapidly around the time of each fall equinox. In other words, the length of day changes most quickly near the equinoxes. You can observe these facts easily if you observe the changing number of daylight hours over the course of a year. ▶ **Now try Exercises 25–30.**

Think About It What is the current date? Based on the function shown in Figure 9.7, should the number of hours of daylight be changing rapidly or slowly at this time of year? Try to notice the change from one day to the next, and confirm that it matches the prediction of our model.

Quick Quiz 9A

Choose the best answer to each of the following questions. Explain your reasoning with one or more complete sentences.

1. In mathematics, a *function* tells us
   a. how one variable depends on another.
   b. how an operation, such as multiplication or division, works.
   c. how inputs and outputs are affected by machines.
2. The notation $r = f(s)$ implies that values of the variable r depend on
   a. values of the variable s.
   b. values of the variable f.
   c. values of the variables s and f.
3. The fact that the value of the Dow Jones Industrial Average (DJIA) changes from day to day tells us that
   a. the DJIA and time are both functions.
   b. time is a function of the DJIA.
   c. the DJIA is a function of time.
4. When you make a graph of a function, values of the *independent* variable are plotted
   a. in the range.
   b. along the horizontal axis.
   c. along the vertical axis.
5. When you make a graph of the function $z = f(w)$, which variable is plotted on the vertical axis?
   a. w b. z c. f
6. The values taken on by the *dependent* variable in a function belong to the function's
   a. domain. b. range. c. limits.
7. Consider a function that describes how a particular car's gas mileage depends on its speed. An appropriate *domain* for this function would be
   a. 0 to 100 miles per hour.
   b. 0 to 50 miles per gallon.
   c. times from 0 to 10 minutes.
8. A function *cannot* accept as inputs values of its independent variable that
   a. are very large.
   b. are outside its domain.
   c. are outside its range.
9. All of the following are functions of time. Which one would you expect to be closest to being a *periodic* function of time?
   a. the price of gasoline
   b. the population of the United States
   c. the volume of traffic on an urban freeway
10. Suppose that two groups of scientists have created mathematical models that they are using to predict future global warming. In general, which model would you expect to be more trustworthy?
    a. the one with more functions
    b. the one that predicts past temperature values that are closer to actual past temperature values
    c. the one that extends further into the future with its predictions

Exercises 9A

REVIEW QUESTIONS

1. What is a *mathematical model*? Explain this statement: *A model's predictions can be only as good as the data and the assumptions from which the model is built.*
2. What is a *function*? How do you decide which variable is the independent variable and which is the dependent variable?
3. What are the three basic ways to represent a function?
4. Define *domain* and *range*, and explain how to determine them for a particular function.

DOES IT MAKE SENSE?

Decide whether each of the following statements makes sense (or is clearly true) or does not make sense (or is clearly false). Explain your reasoning.

5. Scientists at the National Center for Atmospheric Research use mathematical models to learn about the Earth's climate.
6. The demand for concert tickets is a function of their price.
7. I graphed a function showing how my heart rate depends on my running speed. The domain was heart rates from 60 to 180 beats per minute.
8. My mathematical model fits the data perfectly, so I can be confident it will work equally well in any new situations we encounter.

BASIC SKILLS & CONCEPTS

9–10: Coordinate Plane Review. Use the skills covered in the Brief Review on p. 539.

9. Draw a set of axes in the coordinate plane. Plot and label the following points: $(0, 1)$, $(-2, 0)$, $(1, 5)$, $(-3, 4)$, $(5, -2)$, $(-6, -3)$.
10. Draw a set of axes in the coordinate plane. Plot and label the following points: $(0, -1)$, $(2, -1)$, $(6, 5)$, $(3, -4)$, $(-5, -2)$, $(-6, 2)$.

11–14: Identifying Functions. In each of the following situations, state whether two variables are related in a way that might be described by a function. If so, identify the independent and dependent variables.

11. You jump from an airplane (with a parachute) and want to know how far you have traveled at various times during your descent.
12. Each day for several weeks, you record the number of calories you consumed during the day and the number of hours you slept that night.
13. You are a bakery owner and want to know how the demand for muffins (the number you can sell) depends on the price you charge for them.
14. You walk through a used car lot and list the price and model year of each car you see.

15–22: Related Quantities. Write a short statement that expresses a possible relationship between the variables.

Example: (age, shoe size)

Solution: As a child ages, shoe size increases. Once the child is full-grown, shoe size remains constant.

15. (*volume of gas tank, cost to fill the tank*)
16. (*time, price of a Ford sedan*), where time represents years from 1975 to 2017
17. (*latitude, ocean temperature on a given day*)
18. (*area of a parking lot, number of cars that can be parked in the lot*)
19. (*travel time from Denver to Chicago, average speed of a car*)
20. (*rate of pedaling, speed of bicycle*)
21. (*gas mileage of car, cost of driving 500 miles*)
22. (*annual percentage rate (APR), balance in savings account after 10 years*)

23–24. Pressure Function. Use Figure 9.6 to answer the following questions.

23. a. Use the graph to estimate the pressure at altitudes of 6000 feet, 18,000 feet, and 29,000 feet.

    b. Use the graph to estimate the altitudes at which the pressure is 23, 19, and 13 inches of mercury.

    c. Extending the graph, at what altitude do you think the atmospheric pressure reaches 5 inches of mercury? Is there an altitude at which the pressure is exactly zero? Explain your reasoning.

24. Let z represent altitude in feet, with $z = 0$ being sea level, and let $p(z)$ be the atmospheric pressure at the altitude z feet. Consider the formula $p = 30 \times 10^{-z/64{,}000}$.

    a. Evaluate the pressure formula with $z = 0, 10{,}000, 20{,}000$, and $30{,}000$. How well does the formula agree with the pressures given in Table 9.2?

    b. Use the pressure formula to estimate the pressure at $z = 12{,}500$ feet and $z = 27{,}300$ feet.

25. **Daylight Function.** Study Figure 9.7, which applies to 40°N latitude.

    a. Use the graph to estimate the number of hours of daylight on April 1 (the 91st day of the year) and October 31 (the 304th day of the year).

    b. Use the graph to estimate the dates on which there are 13 hours of daylight.

    c. Use the graph to estimate the dates on which there are 10.5 hours of daylight.

    d. The graph in Figure 9.7 is valid at 40°N latitude. How do you think the graph would be different at 20°N latitude, 60°N latitude, and 40°S latitude? Why?

26–27: Functions from Graphs. Consider the graphs of the following functions.

a. Identify the independent and dependent variables, and describe the domain and range.

b. Describe the function in words.

26.

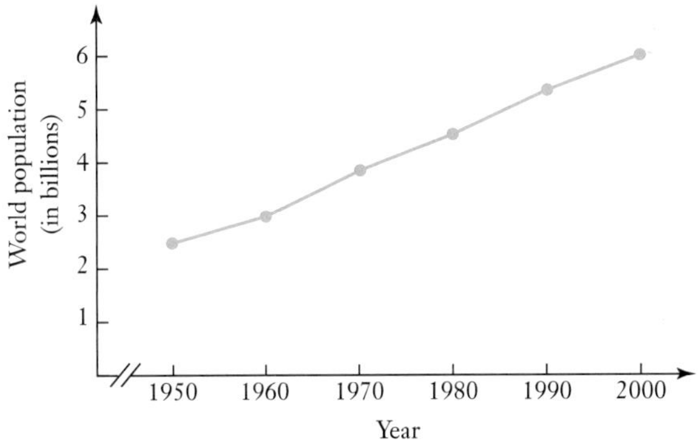

27.

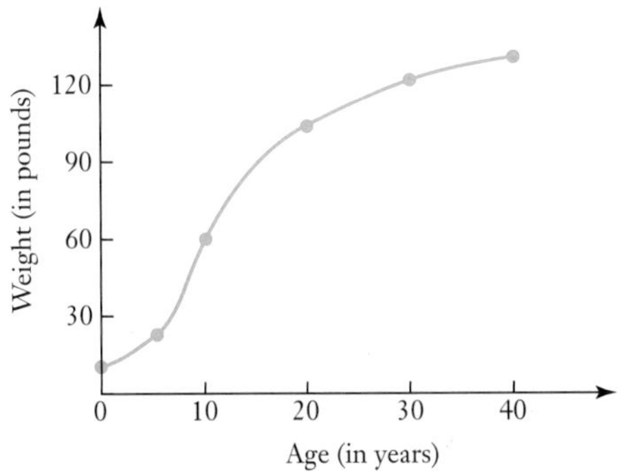

28–30: Functions from Data Tables. Each of the following data tables represents a function.

a. Identify the independent and dependent variables, and describe the domain and range.

b. Make a clear graph of the function. Explain how you decide on the shape of the curve used to fill in the gaps between the data points.

c. Describe the function in words.

28.

Date	Average high temperature
Jan. 1	42°F
Feb. 1	38°F
Mar. 1	48°F
Apr. 1	58°F
May 1	69°F
June 1	76°F
July 1	85°F
Aug. 1	83°F
Sep. 1	80°F
Oct. 1	69°F
Nov. 1	55°F
Dec. 1	48°F
Dec. 31	44°F

29.

Altitude (feet)	Boiling point of water (°F)
0	212.0
1000	210.2
2000	208.4
3000	206.6
4000	204.8
5000	203.0
6000	201.0
7000	199.3
8000	195.5
9000	193.6

30.

Speed (mi/hr)	Stopping distance (in feet)
10	13
20	39
30	75
40	117
50	169
60	234
70	312

FURTHER APPLICATIONS

31–42: Rough Sketches of Functions. For each function involving the variables listed, use your intuition or additional research, if necessary, to do the following.

a. Describe an appropriate domain and range for the function.

b. Make a rough sketch of a graph of the function. Explain the assumptions that go into your sketch.

c. Briefly discuss the validity of your graph as a model of the true function.

31. (*altitude, temperature*) when climbing a mountain
32. (*day of year, average high temperature*) over a 2-year period for the town in which you are living
33. (*blood alcohol content, reflex time*) for a single person
34. (*number of pages in a book, time to read the book*) for a single person
35. (*time of day, traffic flow*) at a busy intersection over a full day
36. (*price of gasoline, number of tourists in Yellowstone*)
37. (*number of people in a room, total number of different handshakes between two people*)
38. (*minutes after lighting, length of candle*)
39. (*time, population of China*), where time is measured in years after 1900
40. (*time of day, height of tide*) at a particular seaport over 2 days
41. (*angle of cannon, horizontal distance traveled by cannonball*)
42. (*weight of car, average gas mileage*)

IN YOUR WORLD

43. **Everyday Models.** Describe three different models (mathematical or other) that you use or encounter frequently in everyday life. What is the underlying "reality" that those models represent? What simplifications are made in constructing those models?

44. **Functions and Variables in the News.** Identify three different variables in recent news stories. For each variable, specify another related variable and then write a paragraph that describes a function relating the two variables. At least one of your three functions should *not* use time as the independent variable.

45. **Daylight Hours.** Investigate websites that give the length of day (hours of daylight) on various days throughout the year for different latitudes (a table of sunrise and sunset times would also work). Make graphs similar to Figure 9.7 showing the variation of hours of daylight over a year for several different latitudes.

46. **Variable Tables.** Find data on the Web for two variables that are clearly related in some way. Make a table (between 10 and 20 entries) of data values. Graph the data and describe in words the function that relates the variables. *Hint:* Some possible variable pairs are (*time, population of a city*), (*team batting average, average team salary*) for Major League baseball teams, and (*blood alcohol content, reaction time*) for a study of effects of alcohol.

UNIT 9B Linear Modeling

In Unit 9A, we represented functions with tables and graphs. We now turn our attention to a more common and versatile way of representing functions: with equations. Although equations are more abstract than pictures, they are easier to manipulate mathematically and give us greater power when creating and analyzing mathematical models. We can understand the basic principles of mathematical modeling by focusing on the simplest models: *linear models*, which can be represented by *linear functions*, meaning functions that have straight-line graphs.

Linear Functions

Imagine that we measure the depth of rain accumulating in a rain gauge as a steady rain falls (Figure 9.8a). The rain stops after 6 hours, and we want to describe how the rain depth varied with time during the storm. In this situation, *time* is the independent variable and *rain depth* is the dependent variable. Suppose that, based on our

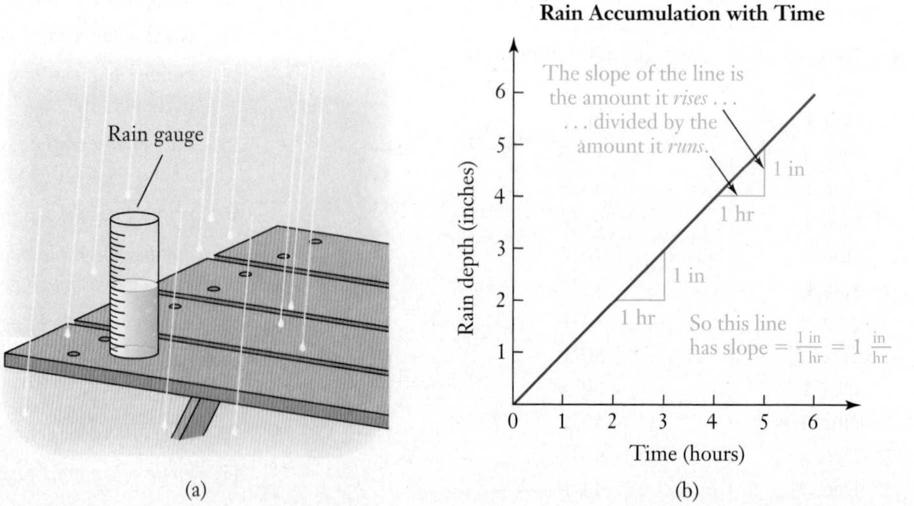

FIGURE 9.8 (a) A rain gauge. (b) Graph of a function showing how rain depth varies with time during a storm.

measurements with the rain gauge, we find the rain depth function shown in Figure 9.8b. Because the graph is a straight line, we are dealing with a **linear function**. If we use this linear function as a model to predict rain depth at different times, then we are using it as a **linear model**.

Rate of Change

The graph shows that, during the storm, the rain depth increased by 1 inch each hour. We say that the **rate of change** of the rain depth with respect to time was 1 inch per hour, or 1 in/hr. This rate of change was constant throughout the storm: No matter which 1-hour interval we choose, the rain depth increased by 1 inch. This illustrates a key fact about linear functions: *A linear function has a constant rate of change and a straight-line graph.*

Figure 9.9 shows graphs for three other steady rainstorms. The constant rate of change is 0.5 in/hr in Figure 9.9a, 1.5 in/hr in Figure 9.9b, and 2 in/hr in Figure 9.9c. Comparing the three graphs in Figure 9.9 leads to another crucial observation: *The greater the rate of change, the steeper the graph.*

The small triangles on the graphs show the **slope** of each line, defined as the amount that the graph *rises* vertically for a given distance that it *runs* horizontally. That is, *the slope is the rise over the run*. More importantly, we see that the slope is equal to the rate of change.

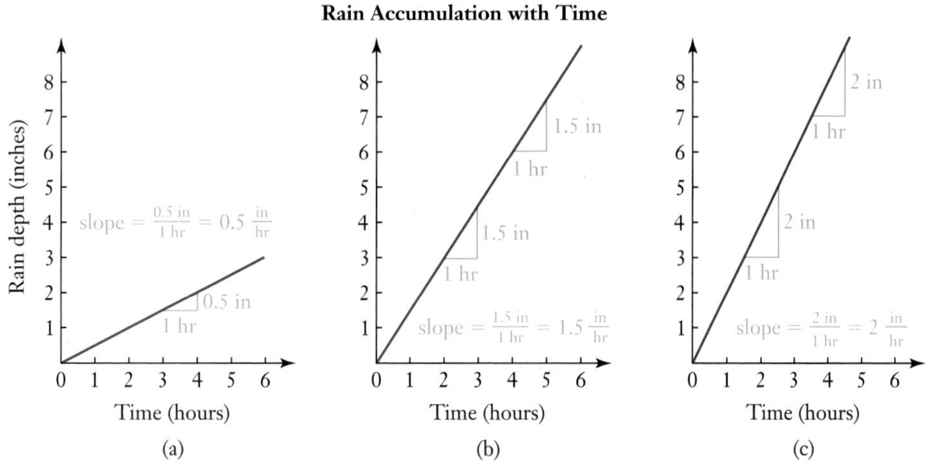

FIGURE 9.9 Three more rain depth functions, with slopes increasing from (a) to (c).

Linear Functions

A **linear function** has a constant rate of change and a straight-line graph. For all linear functions:

- The rate of change is equal to the slope of the graph.
- The greater the rate of change, the steeper the slope.
- We can calculate the rate of change by finding the slope between any two points on the graph (Figure 9.10):

$$\text{rate of change} = \text{slope} = \frac{\text{change in } \textit{dependent variable}}{\text{change in } \textit{independent variable}}$$

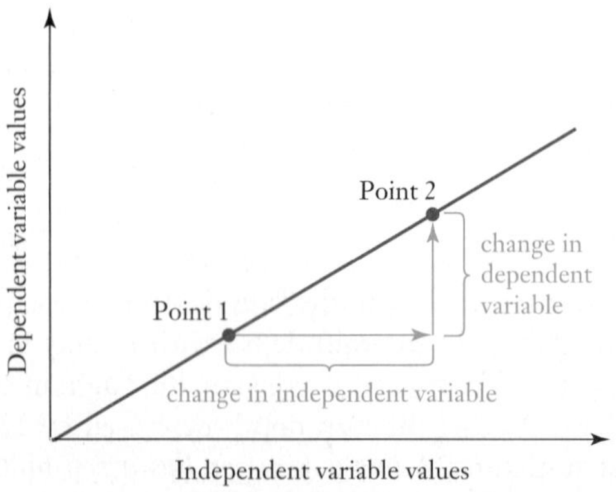

FIGURE 9.10 To find the slope of a straight line, we can look at any two points and divide the change in the dependent variable by the change in the independent variable.

EXAMPLE 1 Drawing a Linear Model

You hike a 3-mile trail, starting at an elevation of 8000 feet. Along the way, the trail gains elevation at a constant rate of 650 feet per mile. The elevation along the trail (in feet) can be viewed as a function of distance walked (in miles). What is the domain of the elevation function? From the given data, draw a graph of a linear function that gives your elevation as you hike along the trail. Does this model seem realistic?

Solution Because your elevation depends on the distance you've walked, *distance* is the independent variable and *elevation* is the dependent variable. The domain is 0 to 3 miles, which represents the length of the trail. We are given one data point: (0 mi, 8000 ft) represents the 8000-foot elevation at the start of the trail. We are also given that the rate of change of elevation with respect to distance is 650 feet per mile. Therefore, a second point on the graph is (1 mi, 8650 ft). We draw the graph by connecting these two points with a straight line and extending the line over the domain from 0 to 3 miles (Figure 9.11). As we expect, the rate of change is the slope of the graph.

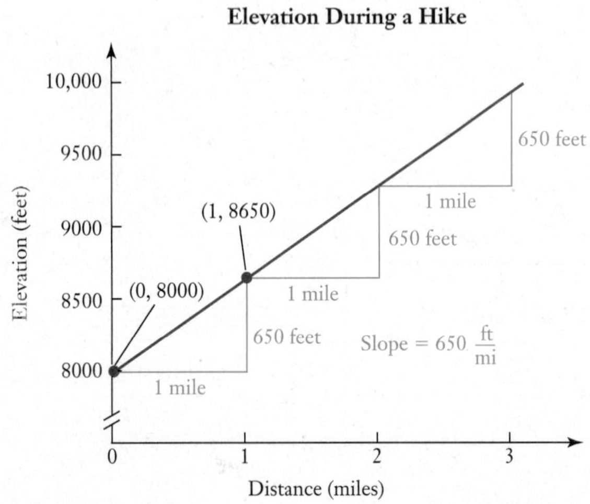

FIGURE 9.11 Linear function for Example 1.

▲ CAUTION! Because the two axes in Figure 9.11 have different units (miles and feet), the slope of the real trail (650 ft/mi) would not look as steep as the slope of the line on the graph. ▲

This model assumes that elevation increases at a constant rate along the entire 3-mile trail. While an elevation change of 650 feet per mile seems reasonable as an average, the actual rate of change probably varies from point to point along the trail. The model's predictions are likely to be reasonable *estimates*, rather than exact values, of your elevation at different points along the trail.

▶ **Now try Exercises 11–12.**

EXAMPLE 2 A Price-Demand Function

A small store sells fresh pineapples. Based on data for pineapple prices between \$2 and \$5, the storeowners created a model in which a linear function is used to describe how the demand (number of pineapples sold per day) varies with the price (Figure 9.12). For example, the point (\$2, 80 pineapples) means that, at a price of \$2 per pineapple, 80 pineapples can be sold per day. What is the rate of change for this function? Discuss the validity of this model.

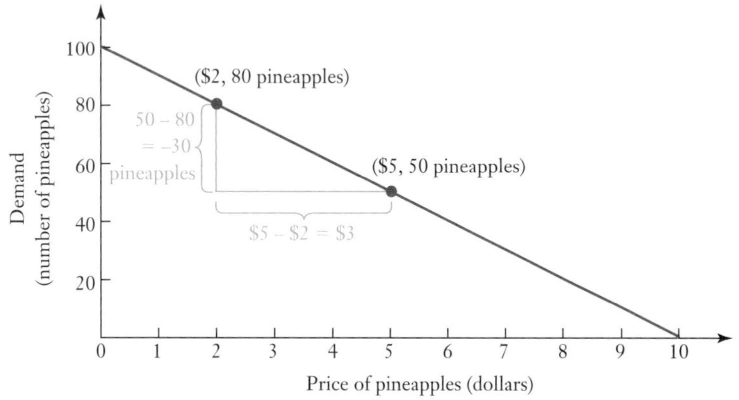

FIGURE 9.12 Linear functions for Example 2.

Solution The rate of change of the demand function is the slope of its graph. We identify *price* as the independent variable and *demand* as the dependent variable. We can calculate the slope using any two points on the graph. Let's choose Point 1 as ($2, 80 pineapples) and Point 2 as ($5, 50 pineapples). The change in *price* between the two points is $\$5 - \$2 = \$3$. The change in *demand* between the two points is

$$50 \text{ pineapples} - 80 \text{ pineapples} = -30 \text{ pineapples}$$

The change in demand is negative because demand *decreases* from Point 1 to Point 2. The rate of change is

$$\text{rate of change} = \frac{\text{change in demand}}{\text{change in price}} = \frac{-30 \text{ pineapples}}{\$3} = \frac{-10 \text{ pineapples}}{\$1}$$

The rate of change of the demand function is -10 pineapples per dollar: For every dollar that the price *increases,* the number of pineapples sold *decreases* by 10.

This model seems reasonable within the domain for which the storeowners gathered data: between prices of $2 and $5. Outside this domain, the model's predictions probably are not valid. For example, the model predicts that the store could sell one pineapple per day at a price of $9.90, but could *never* sell a pineapple at a price of $10. On the other extreme, the model predicts that the store could "sell" only 100 pineapples if they were free! Like many models, this price-demand model is useful only in a limited domain. ▶ **Now try Exercises 13–16.**

The Change in the Dependent Variable

Consider again the rain depth function in Figure 9.8. Suppose we want to know how much the rain depth changes in a 4-hour period. Because the rate of change for this function is 1 in/hr, the total change after 4 hours is

$$\text{change in rain depth} = \underbrace{1\frac{\text{in}}{\cancel{\text{hr}}}}_{\substack{\text{rate of}\\ \text{change}}} \times \underbrace{4\ \cancel{\text{hr}}}_{\substack{\text{elapsed}\\ \text{time}}} = 4 \text{ in}$$

Notice how the units work out. Note also that the *elapsed time* is the change in the independent variable and the change in rain depth is the change in the dependent variable. We can generalize this idea to other functions.

She knew only that if she did or said thus-and-so, men would unerringly respond with the complimentary thus-and-so. It was like a mathematical formula and no more difficult, for mathematics was the one subject that had come easy to Scarlett in her schooldays.
—Margaret Mitchell, *Gone with the Wind*

The Rate of Change Rule

The rate of change rule allows us to calculate the change in the dependent variable from the change in the independent variable:

$$\text{change in } \textit{dependent} \text{ variable} = \begin{pmatrix}\text{rate of} \\ \text{change}\end{pmatrix} \times \begin{pmatrix}\text{change in} \\ \textit{independent} \text{ variable}\end{pmatrix}$$

EXAMPLE 3 Change in Demand

Using the linear demand function in Figure 9.12, predict the change in demand for pineapples if the price increases by \$3.

Solution The independent variable is the *price* of the pineapples, and the dependent variable is the *demand* for pineapples. In Example 2, we found that the rate of change of demand with respect to price is −10 pineapples per dollar. The change in demand for a price increase of \$3 is

$$\begin{aligned}\text{change in } \textit{demand} &= \text{rate of change} \times \text{change in } \textit{price} \\ &= -10 \frac{\text{pineapples}}{\$} \times \$3 \\ &= -30 \text{ pineapples}\end{aligned}$$

This model predicts that a \$3 price increase will lead to 30 *fewer* pineapples being sold per day. ▶ Now try Exercises 17–22.

General Equation for a Linear Function

Suppose your job is to oversee an automated assembly line that manufactures computer chips. You arrive at work one day to find a stock of 25 chips that were produced during the night. If chips are produced at a constant rate of 4 chips per hour, how large is the stock of chips at any particular time during your shift?

Answering this question requires finding a function that describes how the number of chips depends on the time of day. We identify *time*, which we'll denote by t, as the independent variable. *Number of chips*, which we'll denote by N, is the dependent variable. At the start of your shift, $t = 0$ and your initial stock is $N = 25$ chips. Because the stock grows by 4 chips every hour, the *rate of change* of this function is 4 chips per hour. We construct a graph by starting at the initial point (0 hr, 25 chips) and drawing a straight line with a slope of 4 chips per hour (Figure 9.13).

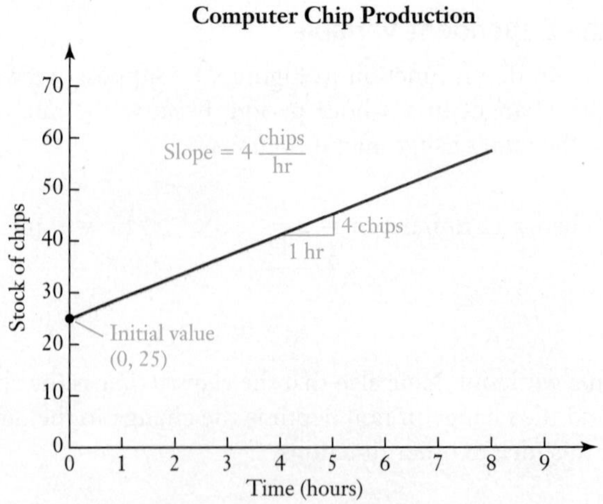

FIGURE 9.13 Linear function with initial value of 25 chips and slope of 4 chips/hour.

The goal is to write an equation for the function. First, let's describe the stock of chips at any particular time with a word equation:

$$\textit{number of chips} = \text{initial number of chips} + \text{change in number of chips}$$

For the last term above, the change in the number of chips, we have

$$\underbrace{\text{change in } \textit{number of chips}}_{\text{change in dependent variable}} = \underbrace{4\,\frac{\text{chips}}{\text{hr}}}_{\text{rate of change}} \times \underbrace{\textit{elapsed time}}_{\text{change in independent variable}}$$

Putting the two equations together and using 25 for the initial number of chips, we find:

$$\textit{number of chips} = 25 \text{ chips} + \left(4\,\frac{\text{chips}}{\text{hr}} \times \textit{elapsed time}\right)$$

We can write the equation more compactly by replacing the *elapsed time* by t and the *number of chips* by N:

$$N = 25 + 4t$$

Note that, because we no longer show the units explicitly, we must remember that 25 represents a number of chips and 4 represents a rate of change in units of chips/hr. We can use this equation to find the number of chips at any time. For example, after $t = 3.5$ hours, the number of chips is

$$N = 25 + (4 \times 3.5) = 39$$

Think About It Use the equation for chip production to find the number of chips produced after 4 hours. Does the result agree with the answer you read from the graph in Figure 9.13?

To generalize from this example to any linear function, note that

- The number of chips, N, is the **dependent variable**.
- The time, t, is the **independent variable**.
- The initial stock of 25 chips represents the **initial value** of the dependent variable when $t = 0$.
- The term 4 chips/hr is the **rate of change** of N with respect to t, so chips/hr $\times\ t$ is the *change* in N.

General Formula for a Linear Function

$$\textit{dependent variable} = \text{initial value} + (\text{rate of change} \times \textit{independent variable})$$

The Equation of a Line

If you have taken a course in algebra, you may be familiar with the equation for a linear function in a slightly different form. In algebra, x is commonly used for the independent variable and y for the dependent variable. For a straight line, the slope is usually denoted by m and the initial value, or **y-intercept**, is denoted by b. With these symbols, the equation for a linear function becomes

$$y = mx + b$$

which has the same form as the general equation of a linear function given above. For example, the equation $y = 4x - 4$ represents a straight line with a slope of 4 and a y-intercept of -4. As shown in Figure 9.14a, the y-intercept tells us where the line crosses the y-axis.

BY THE WAY

There are pros and cons to using either full names or symbols for variables. Mathematicians like the simplicity of symbols. Computer programmers often use full names. It's a matter of taste.

USING TECHNOLOGY

Graphing Functions

A graphing calculator makes it easy to graph almost any function, but there are other ways to accomplish this task. A search on "graphing calculator" will turn up numerous websites offering applets that mimic graphing calculators. You can graph functions in Excel by making a table of (x, y) data points for the function. Create the table with x values in one column and corresponding y values in a second column; then you can use Excel's chart type "Scatter" to make the graph (see Using Technology on p. 359).

Technical Note

The equation of a line can be written in other forms. For example, any equation of the form $Ax + By + C = 0$ (where A, B, and C are constants) describes a straight line. We can see why by solving the equation for y (assuming $B \neq 0$), which gives $y = -(A/B)x - (C/B)$. In this form, we identify the slope as $-(A/B)$ and the y-intercept as $-(C/B)$.

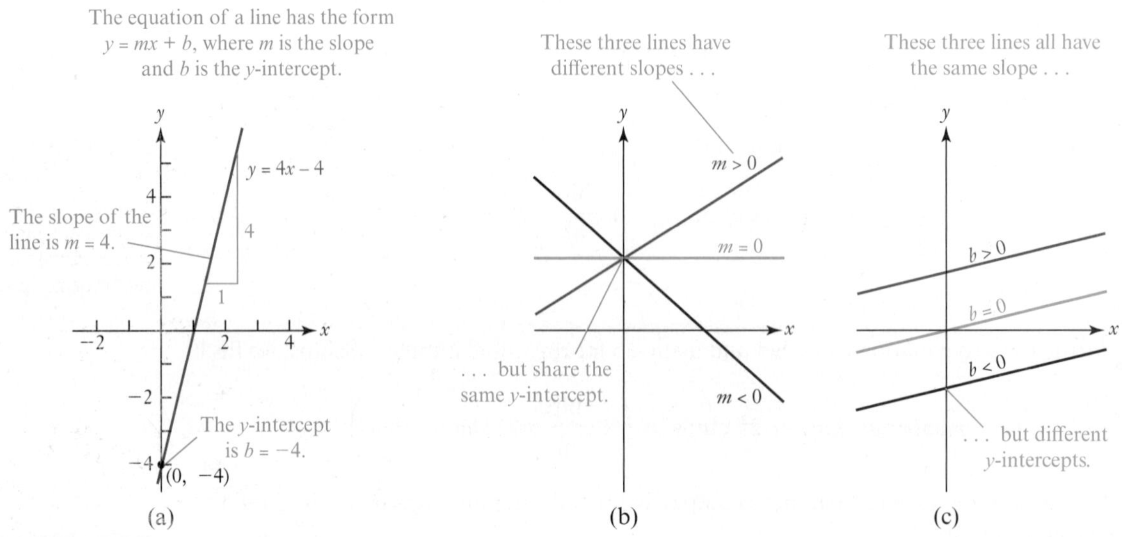

FIGURE 9.14 (a) Graph of $y = 4x - 4$. (b) Lines with same y-intercept but different slopes. (c) Lines with the same slope but different y-intercepts.

Figure 9.14b shows the effects of keeping the same y-intercept but changing the slope. A positive slope ($m > 0$) means the line rises to the right. A negative slope ($m < 0$) means the line falls to the right. A zero slope ($m = 0$) means a horizontal line.

Figure 9.14c shows the effects of changing the y-intercept for a set of lines that have the same slope. All the lines rise at the same rate but cross the y-axis at different points.

EXAMPLE 4 Rain Depth Equation

Using the function shown in Figure 9.8, write an equation that describes the rain depth at any time after the storm began. Use the equation to find the rain depth 3 hours after the storm began.

Solution For the rain depth function in Figure 9.8, the rate of change is 1 in/hr and the initial value of the rain depth when the storm begins is 0 inches. The general equation for this function is

$$\underbrace{\textit{rain depth}}_{\substack{\text{dependent}\\ \text{variable}}} = \underbrace{0 \text{ in}}_{\substack{\text{initial}\\ \text{value}}} + \underbrace{1\frac{\text{in}}{\text{hr}}}_{\substack{\text{rate of}\\ \text{change}}} \times \underbrace{\textit{time}}_{\substack{\text{independent}\\ \text{variable}}}$$

We can write this equation more compactly by letting r represent *rain depth* (in inches) and t represent *time* (in hours):

$$r = 0 + (1 \times t), \quad \text{or} \quad r = t$$

Substituting $t = 3$ hours in this equation, we find that the rain depth 3 hours after the storm began is $r = 3$ inches. ▶ **Now try Exercises 23–24.**

EXAMPLE 5 Alcohol Metabolism

The purpose of models is not to fit the data but to sharpen the questions.
—Samuel Karlin, mathematician

Alcohol is metabolized by the body (using enzymes in the liver) in such a way that the blood alcohol content (see Unit 2B) decreases linearly. A study by the National Institute on Alcohol Abuse and Alcoholism showed that, for a group of fasting males who consumed four drinks rapidly, the blood alcohol content rose to a maximum of 0.08 g/100 mL about an hour after the drinks were consumed. Three hours later, the blood alcohol content had decreased to 0.04 g/100 mL. Find a linear model that describes the elimination of alcohol from the blood after the peak blood alcohol content is reached. According to the model, what is the blood alcohol content 5 hours after the peak is reached?

Solution We seek a linear function that relates the independent variable *time* to the dependent variable *blood alcohol content* (BAC, for short). To keep the notation simpler, we'll write the BAC values without the units. If we let $t = 0$ represent the time at which the BAC reached a peak, the study gives us two points: (0 hr, 0.08) and (3 hr, 0.04). We can use these two points to find the rate of change, or slope, of the function:

$$\text{slope} = \frac{\text{change in BAC}}{\text{change in } time} = \frac{(0.04 - 0.08)}{(3 - 0)\text{ hr}} = \frac{-0.04}{3\text{ hr}} \approx \frac{-0.0133}{\text{hr}}$$

That is, for each hour, the BAC *decreases* by about 0.0133. The initial value for the BAC is 0.08. Therefore, the general linear equation for this function is

$$\underbrace{\text{BAC}}_{\text{dependent variable}} = \underbrace{0.08}_{\text{initial value}} + \left(\underbrace{-0.0133\frac{1}{\text{hr}}}_{\text{rate of change}} \times \underbrace{time}_{\text{independent variable}} \right)$$

We can write the equation more compactly as

$$\text{BAC} = 0.08 - (0.0133 \times t)$$

Figure 9.15 shows the graph of this function. The slope is -0.0133, and the vertical intercept is 0.08. To determine the BAC 5 hours after the peak BAC is reached, we set $t = 5$ and evaluate BAC:

$$\text{BAC} = 0.08 - (0.0133 \times 5) = 0.0135$$

Five hours after the peak BAC is reached (6 hours after the drinks were consumed), the BAC is still significant. In fact, it's roughly 15% of the legal limit in most states.

▶ Now try Exercises 25–26.

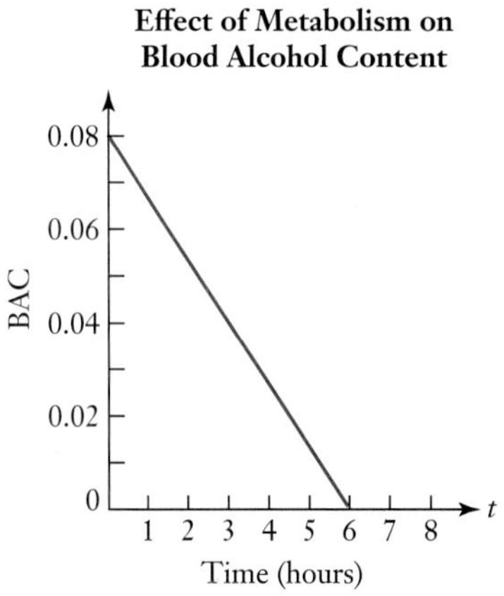

FIGURE 9.15 Linear function for Example 5.

EXAMPLE 6 Price from Demand

Write an equation for the linear demand function in Figure 9.12. Then determine the price that should result in a demand of 75 pineapples per day.

Solution Earlier, we found that the rate of change, or slope, for this function is -10 pineapples per dollar. The initial value, or y-intercept, is 100 pineapples, because that is the demand predicted for a price of \$0. Let's use p for the independent variable *price* and d for the dependent variable *demand*. With the slope of -10 and y-intercept of 100, the equation for this function is

$$d = 100 - 10p$$

Because we are asked to find the price for a given demand, we must solve this equation for the price, p. We do so by first subtracting 100 from both sides, then dividing both sides by -10 and finally interchanging the left and right sides. You should confirm that this gives the following result:

$$p = \frac{d - 100}{-10}$$

Substituting a demand of $d = 75$ pineapples, we find

$$p = \frac{75 - 100}{-10} = \frac{-25}{-10} = 2.5$$

IN YOUR WORLD

Algebra's Baghdad Connection

After the fall of ancient Rome in the 5th century, European civilization entered the period known as the Dark Ages. However, it was not a dark time in the Middle East, where a new center of intellectual achievement arose in the city of Baghdad (in modern-day Iraq). Jews, Christians, and Muslims in Baghdad worked together in scholarly pursuits during this period.

One of the greatest scholars was a Muslim named Muhammad ibn Musa al-Khwarizmi (780–850 C.E.). Al-Khwarizmi wrote several books on astronomy and mathematics, including one entitled *Hisab al-jabr wal-muqabala*, which translates roughly as "the science of equations." This book preserved and extended the work of the Greek mathematician Diophantus (210–290 C.E.) and thereby laid the foundations of algebra. In fact, the word *algebra* comes directly from the Arabic words *al-jabr* in the book's title. The introduction to the book states its purpose to teach

what is easiest and most useful in arithmetic, such as [people] constantly require in cases of inheritance, legacies, partition, lawsuits, and trade, and in all their dealings with one another, or where the measuring of lands, the digging of canals, geometrical computations, and other objects of various sorts and kinds are concerned.

Notice the practical intent of this first algebra book, which stands in stark contrast to the abstract nature of many of today's algebra books.

In another of his works, al-Khwarizmi described the numeral system developed by Hindu mathematicians, thereby popularizing the decimal system and the use of the numeral zero. Although he did not claim credit for the Hindu work, later writers often attributed it to him. That is why modern numerals are known as *Hindu-Arabic* rather than solely Hindu. Some later authors even attributed the numerals to al-Khwarizmi personally. In a sloppy writing of his name, the use of Hindu numerals became known as *algorismi*, which later became the English words *algorism* (arithmetic) and *algorithm*. Historians consider al-Khwarizmi one of the most important mathematicians of all time, and his work lies at the foundation of modern mathematics.

Based on this model, the price should be set at \$2.50 if the storeowners want to sell 75 pineapples each day. Note that we must find the units (dollars) for the answer by looking back to the original data. ▶ **Now try Exercises 27–28.**

Linear Functions from Two Data Points

Suppose we have two data points and want to find a linear function that fits them. We can find the equation for this linear function by using the two data points to determine the rate of change (slope) and the initial value for the function. The following three steps summarize the procedure, which is described for variables (x, y); of course, you can use any symbols you wish.

Creating a Linear Function from Two Data Points

Step 1. Let x be the independent variable and y be the dependent variable. Find the change in each variable between the two given points, and use these changes to calculate the slope, or rate of change:

$$\text{slope} = \frac{\text{change in } y}{\text{change in } x}$$

Step 2. Substitute this slope and the *numerical values* of y and x from either data point into the equation $y = mx + b$. You can then solve for the y-intercept, b, because it will be the only unknown in the equation.

Step 3. Now use the slope and y-intercept to write the equation of the linear function in the form $y = mx + b$.

EXAMPLE 7 100-Meter Progression

In 1912, the world record in the men's 100-meter run was 10.6 seconds. The world record dropped to more than a full second below that in 2009, when Usain Bolt ran 100 meters in 9.58 seconds. Use these two data points to create a linear model that describes the change in the men's 100-meter world record with time. Interpret the model and discuss its validity.

Solution We seek a function that describes how the *100-meter world record* varies with *time*. We identify *time* as the independent variable, which we'll denote by t and measure in *years since 1912*, when data begin. *World record* is the dependent variable, so we'll call it y and measure it in seconds.

We are given two data points. The first is (0, 10.6), representing the world record of 10.6 seconds in 1912. The second is (97, 9.58), representing the world record of 9.58 seconds in 2009, which was 97 years after 1912. We now follow the procedure from the above box to find a linear function fitting the two points, except that we use t rather x for the independent variable.

Find the slope, m:	$m = \text{slope} = \dfrac{\text{change in } y}{\text{change in } t} = \dfrac{9.58 - 10.6}{97 - 0}$ $= -\dfrac{1.02}{97} \approx -0.01052 \text{ (s/yr)}$
Write the equation in the form $y = mt + b$:	$y = -0.01052t + b$
Substitute the values from the first data point, (0, 10.6), to find b:	$10.6 = -0.01052 \times 0 + b \rightarrow b = 10.6$
Use the values of m and b to write the general linear equation:	$y = -0.01052t + 10.6$

We can check that this equation is correct by making sure it gives an accurate value for the second data point. Substituting $t = 97$ for the year 2009, we find $y = -0.01052 \times 97 + 10.6 \approx 9.58$, which is the correct world record for that year.

▲ **CAUTION!** Notice that we kept 5 decimal places when rounding the value for the slope, because with further rounding the equation would no longer produce the correct value for the world record in 2009. In general, you should keep plenty of decimal places during intermediate steps of a problem, and round only at the end. ▲

The straight line in Figure 9.16 (on the next page) shows the graph of this linear function. The slope is negative because the world record has fallen with time, and the slope's value of approximately −0.01 tells us that the world record has dropped an average of about 0.01 second per year.

To evaluate the model's validity, we can compare its predictions to actual data for the world record, which are represented by the dots in Figure 9.16. The model is clearly not perfect, but seems to do a decent job overall. One trend that jumps out is a rather steep drop in the world record from 2000 to 2009, which might make us wonder whether this period is an aberration. The model also has some obvious limitations. For example, if the world record continued to drop at a rate of about 0.01 second per year, then the record would fall below zero in about 1000 years, which is obviously impossible.

BY THE WAY

Jamaican sprinter Usain Bolt won both the 100-meter and 200-meter races in three consecutive Olympic games (2008, 2012, 2016). His 2009 time of 9.58 seconds for 100 meters broke his own world record from the previous year of 9.69 seconds, and represented the largest single drop in the 100-meter world record since electronic timing was instituted in 1975.

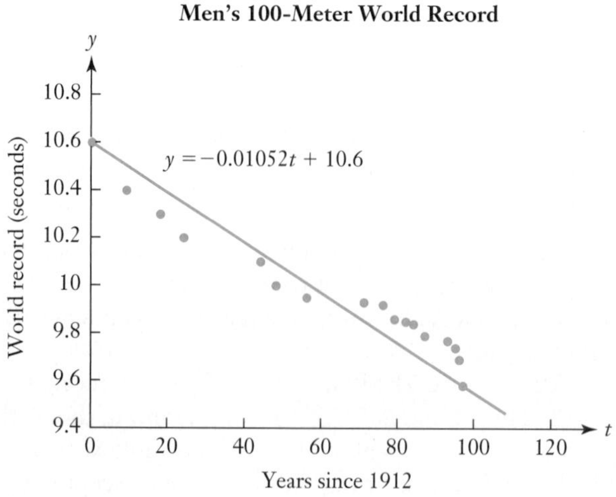

FIGURE 9.16 The line shows a linear model based on the data points for 1912 and 2009. The dots show the actual progression of world record times for the 100-meter race since 1912.

▶ Now try Exercises 29–34.

Think About It What does the model in Example 7 predict the men's 100-meter world record should be today? What is the actual world record today?

Quick Quiz 9B

Choose the best answer to each of the following questions. Explain your reasoning with one or more complete sentences.

1. A *linear function* is characterized by
   a. an increasing slope. b. a decreasing slope.
   c. a constant slope.
2. You have a graph of a linear function. To determine the function's rate of change, you should
   a. identify the domain.
   b. measure the slope.
   c. compare the function to another, closely related linear function.
3. The graph of a linear function is sloping downward (from left to right). This tells us that
   a. its domain is decreasing.
   b. its range is decreasing.
   c. it has a negative rate of change.
4. Suppose that Figure 9.11 is an accurate representation of elevation changes for the first 3 miles of a 6-mile trail and the trail has the same slope for the next 3 miles. What do you predict for the elevation at mile 5 of the trail?
   a. The elevation is $(8000 + 5 \times 650)$ feet.
   b. You should not make a prediction, because the elevation at 5 miles must be higher than the 10,000-foot maximum shown on the graph.
   c. You should not make a prediction, because mile 5 of the trail is not within the domain of the function shown.
5. Which town would have the steepest slope on a graph showing its population as a function of time?
   a. a town growing at a constant rate of 50 people per year
   b. a town growing at a constant rate of 75 people per year
   c. a town growing at a constant rate of 100 people per year
6. Consider the function $price = \$100 - (\$3/\text{yr}) \times time$. The initial value of this function is
   a. \$100. b. \$3. c. \$0.
7. Consider the demand function given in Example 6, which is $d = 100 - 10p$. A graph of this function would
   a. slope upward, starting from a price of \$100.
   b. slope upward, starting from a price of \$0.
   c. slope downward, starting from a price of \$100.
8. A line intersects the y-axis at a value of $y = 7$ and has a slope of -2. The equation of this line is
   a. $y = -2x + 7$.
   b. $y = 7x - 2$.
   c. $y = 2x - 7$.

9. Consider a line with the equation $y = 12x - 3$. Which of the following lines has the same slope but a different y-intercept?

   a. $y = \frac{12}{3}x - \frac{3}{3}$ b. $y = 12x + 3$
   c. $y = -12x - 3$

10. Charlie picks apples in an orchard at a constant rate. By 9:00 a.m. he has picked 150 apples, and by 11:00 a.m. he has picked 550 apples. If we use A for the number of apples and t for time measured in hours since 9:00 a.m., which of the following functions describes his apple picking?

   a. $A = 150t + 2$ b. $A = 550t + 150$ c. $A = 200t + 150$

Exercises 9B

REVIEW QUESTIONS

1. What does it mean to say that a function is *linear*?
2. Define *rate of change*, and describe how a rate of change is stated in words (that is, using "with respect to").
3. How is the rate of change of a linear function related to the slope of its graph?
4. How do you find the change in the dependent variable, given a change in the independent variable? Give an example.
5. Describe the general equation for a linear function. How is it related to the standard algebraic form $y = mx + b$?
6. Describe the procedure for creating the equation of a linear function from two data points. How are such models useful?

DOES IT MAKE SENSE?

Decide whether each of the following statements makes sense (or is clearly true) or does not make sense (or is clearly false). Explain your reasoning.

7. When I graphed the linear function, it turned out to be a wavy curve.
8. I graphed two linear functions, and the one with the greater rate of change had the greater slope.
9. My freeway speed is the rate of change in my distance traveled with respect to time.
10. It's possible to make a linear model from any two data points, but there's no guarantee that the model will fit other data points.

BASIC SKILLS & CONCEPTS

11–16: Linear Functions. Consider the following graphs.

a. In words, describe the function shown on the graph.

b. Find the slope of the line in the graph and express it as a rate of change (be sure to include units).

c. Briefly discuss the conditions under which a linear function is a realistic model for the given situation.

11.
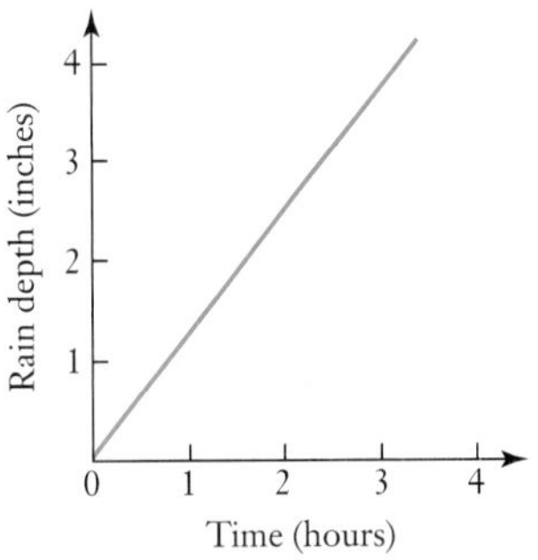

12.
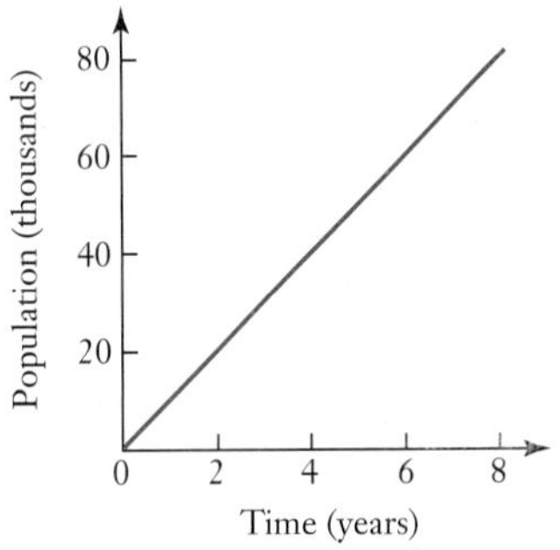

13.
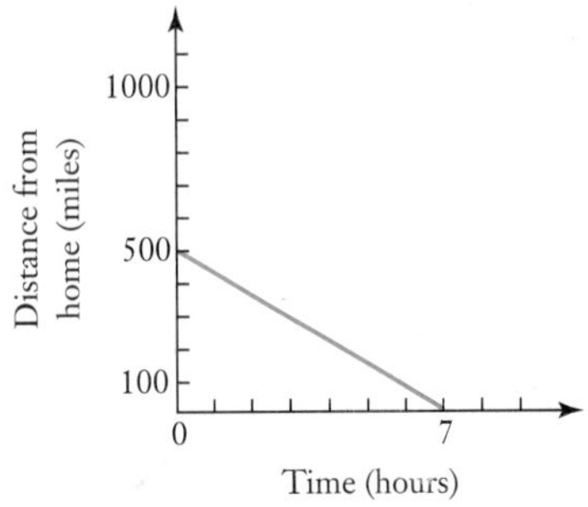

14.
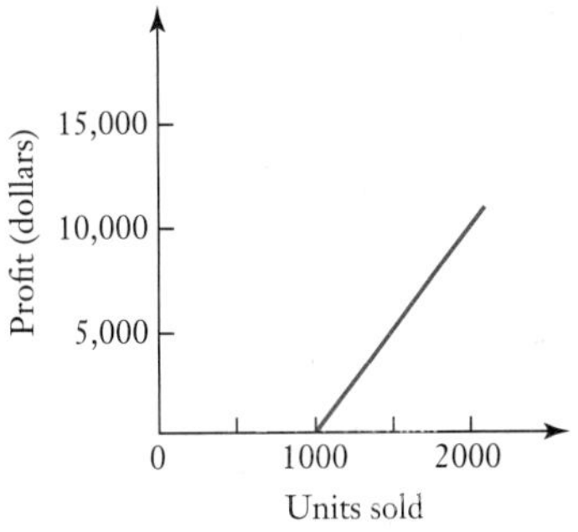

15.

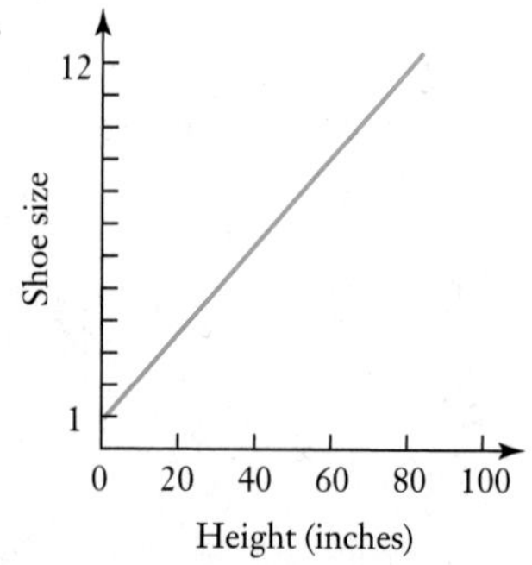

16.

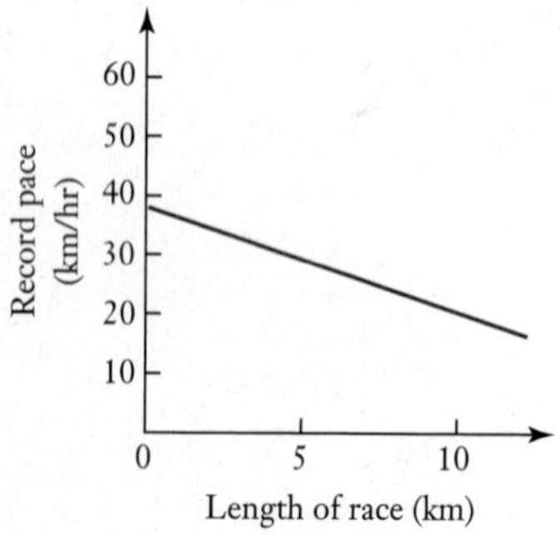

17–22: Rate of Change Rule. The following situations involve a rate of change that is constant. Write a statement that describes how one variable varies with respect to the other, give the rate of change numerically (with units), and use the rate of change rule to answer the questions.

Example: Every week your fingernails grow 5 millimeters (mm). How much will your fingernails grow in 2.5 weeks?

Solution: The *length of your fingernails* varies with respect to *time*, with a rate of change of 5 mm/wk. In 2.5 weeks, your fingernails will grow 5 mm/wk × 2.5 wk = 12.5 mm.

17. The water depth in a lake decreases at a rate of 2 inches per day because of evaporation. How much does the water depth change in 8 days? In 15 days?

18. Your driverless car maintains a constant speed of 63 miles per hour along a highway. How far do you travel in 1.5 hours? In 3.8 hours?

19. A 1-degree change (increase or decrease) on the Celsius temperature scale is equivalent to a 9/5-degree change on the Fahrenheit temperature scale. How much does the Fahrenheit temperature increase if the Celsius temperature increases 5 degrees? How much does the Fahrenheit temperature decrease if the Celsius temperature decreases 25 degrees?

20. A gas station owner finds that for every penny increase in the price of gasoline, she sells 80 fewer gallons of gas per week. How much more or less gas will she sell if she raises the price by 8 cents per gallon? If she decreases the price by 6 cents per gallon?

21. During a 6-week weight-loss regimen, Tony lost weight at a rate of 0.6 pounds per day. How much did he lose after 15 days? After 4 weeks?

22. According to one formula, your maximum heart rate (in beats per minute) is 220 minus your age (in years). How much does your maximum heart rate change from age 25 to age 40? What is your maximum heart rate at age 70?

23–28: Linear Equations. The following situations can be modeled by linear functions. In each case, write an equation for the linear function and use it to answer the given question. Be sure you clearly identify the independent and dependent variables. Then briefly discuss whether a linear model is reasonable for the situation described.

23. The price of a particular model of car is $18,000 today and rises with time at a constant rate of $900 per year. How much will a new car of this model cost in 3.5 years?

24. In 2016, Katy Ledecky set the women's world record in the 800-meter freestyle (swimming) with a time of 8:04.79, or 8 minutes 4.79 seconds. Assume that the record time falls at a constant rate of 1.15 seconds per year (which was the average rate from 2008 to 2016). What does the model predict for the record time in 2024? (*Hint*: Convert the 2016 record time of 8:04.79 into seconds before writing your equation.)

25. A snowplow has a maximum speed of 40 miles per hour on a dry highway. Its maximum speed decreases by 1.1 miles per hour for every inch of snow on the highway. According to this model, at what snow depth will the plow be unable to move?

26. The cost of leasing a car is $1000 for the down payment and processing fee plus $360 per month. For how many months can you lease the car with $3680?

27. You can rent time on computers at the local copy center for a $10 setup charge and an additional $2 for every 5 minutes. How much time can you rent for $25?

28. In 2010, the population of Boom Town was 1650 people and began increasing at a rate of 250 people per year. What is your projection for the town's population in 2030?

29–34: Equations from Two Data Points. Create the required linear function, and use it to answer each question.

29. Suppose your pet dog weighed 5.5 pounds at birth and weighed 20 pounds a year later. Based on these two data points, find a linear function that describes how weight varies with age. Use this function to predict your dog's weight at 5 and 10 years of age. Comment on the validity of the model.

30. You can purchase a motorcycle for $8300 or lease it for a down payment of $400 and $250 per month. Find a function that describes how the cost of the lease depends on time. Assuming that you make monthly payments, how long can you lease the motorcycle before you've paid more than its purchase price?

31. A Campus Republicans fundraiser offers raffle tickets for $10 each. The prize for the raffle is a $350 television set, which must be purchased with proceeds from the ticket sales. Find a function that gives the profit (or loss) for the raffle as it varies with the number of tickets sold. How many tickets must be sold for the raffle sales to equal the cost of the prize?

32. The Campus Democrats plan to pay a visitor $100 to speak at a fundraiser. Tickets will be sold for $4 apiece. Find a function that gives the profit (or loss) for the event as it varies with the number of tickets sold. How many people must attend the event for the club to break even?

33. A $1500 washing machine in a laundromat is depreciated for tax purposes at a rate of $125 per year. Find a function for the depreciated value of the washing machine as it varies with time. When does the depreciated value reach $0?

34. A mining company can extract 2000 tons of gold ore per day with a purity of 3 ounces of gold per ton. The cost of extraction is $1000 per ton. If p is the price of gold in dollars per ounce, find a function that gives the daily profit/loss of the mine as it varies with the price of gold. What is the minimum price of gold that makes the mine profitable?

FURTHER APPLICATIONS

35–42: Algebraic Linear Equations. For the following functions, find the slope of the graph and the intercept. Then sketch the graph for values of x between -10 and 10.

35. $y = 2x + 6$
36. $y = -3x + 3$
37. $y = -5x - 5$
38. $y = 4x + 1$
39. $y = 3x - 6$
40. $y = -2x + 5$
41. $y = -x + 4$
42. $y = 2x + 4$

43–48: Linear Graphs. The following situations can be modeled by linear functions. In each case, draw a graph of the function and use the graph to answer the given question. Be sure you clearly identify the independent and dependent variables. Then briefly discuss whether a linear model is reasonable for the situation described.

43. A group of climbers begin climbing at an elevation of 6500 feet and ascend at a steady rate of 600 vertical feet per hour. What is their elevation after 3.5 hours?

44. The diameter of a tree increases by 0.2 inch with each passing year. When you started observing the tree, its diameter was 4 inches. Estimate the time at which the tree started growing.

45. The cost of publishing a poster is $2000 for setting up the printing equipment, plus $3 per poster printed. What is the total cost to produce 2000 posters?

46. The amount of sugar in a fermenting batch of beer decreases with time at a rate of 0.1 gram per day, starting from an initial amount of 5 grams. When is the sugar gone?

47. The cost of a particular private school includes a one-time initiation fee of $2000, plus annual tuition of $10,000. How much will it cost to attend this school for 6 years?

48. The maximum speed of a semitrailer truck up a steep hill varies with the weight of its cargo. With no cargo, it can maintain a maximum speed of 50 miles per hour. With 20 tons of cargo, its maximum speed drops to 40 miles per hour. At what load does a linear model predict a maximum speed of 0 miles per hour?

49. **Wildlife Management.** A common technique for estimating animal populations is to tag and release individual animals in two different outings. This procedure is called *catch and release*. If the wildlife remain in the sampling area and are randomly caught, a fraction of the animals tagged during the first outing are likely to be caught again during the second outing. Based on the number tagged and the fraction caught twice, the total number of animals in the area can be estimated.

a. Consider a case in which 200 fish are tagged and released during the first outing. During a second outing in the same area, 200 fish are again caught and released, of which one-half are already tagged. Estimate N, the *total* number of fish in the entire sampling area. Explain your reasoning.

b. Consider a case in which 200 fish are tagged and released during the first outing. During a second outing in the same area, 200 fish are again caught and released, of which one-fourth are already tagged. Estimate N the *total* number of fish in the entire sampling area. Explain your reasoning.

c. Generalize your results from parts (a) and (b) by letting p be the fraction of tagged fish that are caught during the second outing. Find a formula for the function $N = f(p)$ that relates the total number of fish, N, to the fraction tagged during the second outing, p.

d. Graph the function obtained in part (c). What is the domain? Explain.

e. Suppose that 15% of the fish in the second sample are tagged. Use the formula from part (c) to estimate the total number of fish in the sampling area. Confirm your result on your graph.

f. Locate a real study in which the catch and release method was used. Report on the specific details of the study and describe how closely it followed the theory outlined in this problem.

IN YOUR WORLD

50. **Linear Models.** Describe at least two situations from the news or your own life in which predictions must be made and a linear model seems appropriate. Briefly discuss why the linear model works well.

51. **Nonlinear Models.** Describe at least one situation from the news or your own life in which predictions must be made but a linear function is *not* a good model. Briefly discuss the shape (on a graph) that you would expect the function to take.

52. **Alcohol Metabolism.** Most drugs are eliminated from the blood through an exponential decay process with a constant half-life (see Unit 9C). Alcohol is an exception in that it is metabolized through a linear decay process. Find data showing how the blood alcohol content (BAC) decreases over time, and use the data to develop a linear model. Discuss the validity of the model. What assumptions (for example, gender, weight, number of drinks) were used in creating the model?

53. **Property Depreciation.** Go to the IRS website, and examine the rules for depreciation of some type of property, such as a rental property or a piece of business equipment. Create a linear model that describes the depreciation function.

UNIT 9C Exponential Modeling

In Unit 9B, we investigated the use of linear functions as mathematical models. The hallmark of linear functions is that they describe quantities that have a constant *absolute* growth rate. As we discussed in Chapter 8, another common and important growth pattern is exponential growth, in which the *relative* growth rate is constant. In this unit, we investigate exponential functions and some of their many applications in mathematical models.

Exponential Functions

Our first task is to find a general form for an exponential function. Consider a town that begins with a population of 10,000 and grows at a rate of 20% per year. As we saw in Unit 8A, this population grows exponentially because it increases the same relative amount, 20%, each year (see Figure 8.1). In the terminology of Chapter 8, the *percentage growth rate* is $P\% = 20\%$ per year, and the **fractional growth rate** is $r = P/100 = 0.2$ per year.

The initial population is 10,000. With a 20% growth rate, the population at the end of the first year is 20% *more than* the initial value of the population; that is, it is 120% of, or 1.2 *times*, the initial population (see the "*of* versus *more than*" rule in Unit 3A):

$$\text{population after 1 year} = 10{,}000 \times 1.2 = 12{,}000$$

During the second year, the population again grows by 20%, so we can find the population at the end of the second year by multiplying again by 1.2:

$$\begin{aligned}\text{population after 2 years} &= \text{population after 1 year} \times 1.2\\ &= (10{,}000 \times 1.2) \times 1.2\\ &= 10{,}000 \times 1.2^2 = 14{,}400\end{aligned}$$

With another 20% increase in the third year, we find the population at the end of the third year by multiplying once more by 1.2:

$$\begin{aligned}\text{population after 3 years} &= \text{population after 2 years} \times 1.2\\ &= (10{,}000 \times 1.2^2) \times 1.2\\ &= 10{,}000 \times 1.2^3 = 17{,}280\end{aligned}$$

This growth is illustrated in Figure 9.17. The pattern may now be clear. If we let t represent the time in years, we find that

$$\text{population after } t \text{ years} = \text{initial population} \times 1.2^t$$

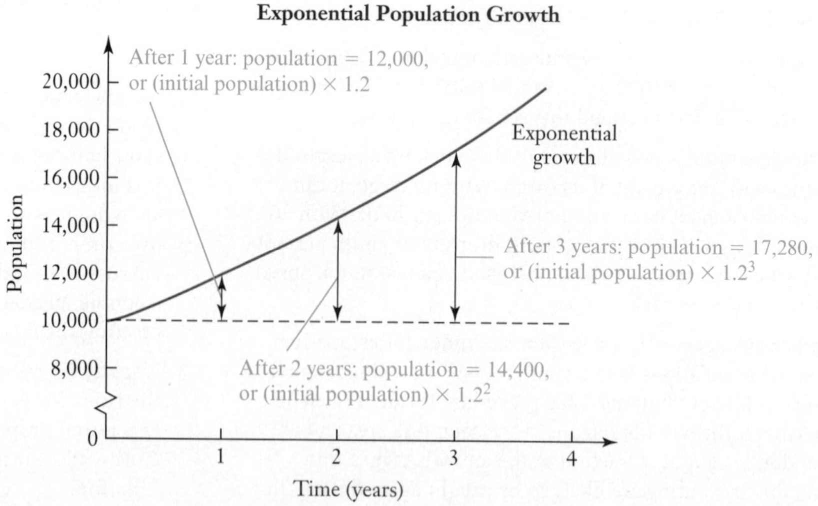

FIGURE 9.17 Exponential growth for 3 years with a growth rate of 20% per year. Each year, the population increases by a factor of 1.2.

For example, after $t = 25$ years, the population is $10{,}000 \times 1.2^{25} = 953{,}962$.

We can extend this idea to *any* exponentially growing quantity Q, and write a general exponential function.

Exponential Functions

An **exponential function** grows (or decays) by the same *relative* amount per unit time, at a rate described by the **fractional growth rate** r (the "growth" rate is negative, $r < 0$, for decay). The exponential function has the general form

$$Q = Q_0 \times (1 + r)^t$$

where

Q = value of the exponentially growing (or decaying) quantity at time t

Q_0 = **initial value** of the quantity (at $t = 0$)

r = fractional growth rate for the quantity (or fractional decay rate if $r < 0$)

t = time

Key notes:

- *The units of time used for t and r must be the same.* For example, if the fractional growth rate is 0.05 per *month*, then t must also be measured in months.
- Remember that while an exponentially growing quantity has a constant *relative* growth rate, its *absolute* growth rate increases. For example, a population may increase at a constant 20% per year, but the absolute change in the population each year is always increasing.

You may notice that the exponential function is identical to the *compound interest formula* (Unit 4B) if we identify Q as the accumulated balance (called A in the compound interest formula), Q_0 as the starting principal, r as the interest rate, and t as the number of times interest is paid. In other words, compound interest is a form of exponential growth.

EXAMPLE 1 U.S. Population Growth

The 2010 census found a U.S. population of about 309 million, with an estimated growth rate of 0.9% per year. Write an equation for the U.S. population that assumes exponential growth at this rate. Use the equation to predict the U.S. population in 2100.

Solution The quantity Q is the U.S. population. The initial value is the 2010 population, $Q_0 = 309$ million. The percentage growth rate is $P\% = 0.9\%$ per year, so the fractional growth rate is $r = P/100 = 0.009$ per year. The equation for the exponential function takes the form

$$Q = Q_0 \times (1 + r)^t = 309 \text{ million} \times (1 + 0.009)^t$$
$$= 309 \text{ million} \times (1.009)^t$$

Note that, because the units of r are *per year*, t must be measured in years. The year 2100 is $t = 90$ years after 2010. Our exponential function therefore predicts a 2100 population of

$$Q = 309 \text{ million} \times (1.009)^{90} \approx 692 \text{ million}$$

At a growth rate of 0.9% per year, the U.S. population will swell to almost 700 million by 2100.

▶ Now try Exercises 27–34, part (a).

Think About It Note that the projected U.S. population in Example 1 is more than double the current population. Do you think this projection is realistic? If so, how would you expect it to affect the economy, the environment, and other quality-of-life issues? If not, why not?

BY THE WAY

Between 1950 and 2017, China's population growth *rate* decreased by nearly a factor of 5, from about 2% per year to 0.4% per year. However, China's actual population more than doubled (it was about 550 million in 1950) during this period, and the positive growth rate means the population is still growing.

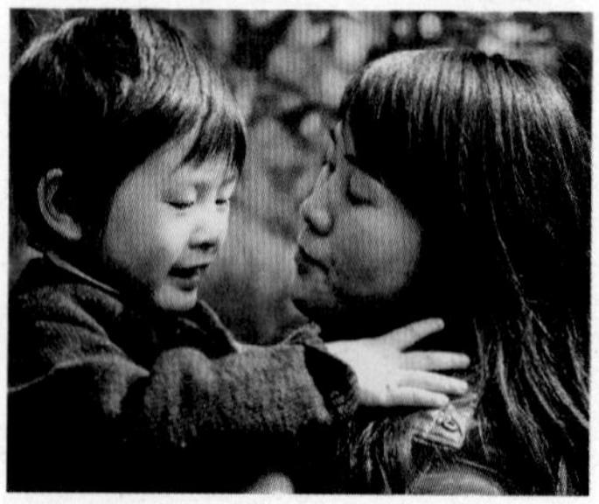

EXAMPLE 2 Declining Population

China's one-child policy (see Example 7 in Unit 2C), which was in effect from 1979 to 2015, was implemented with the goal of reducing China's population to 700 million by 2050. China's 2017 population was estimated to be about 1.4 billion. Suppose China's population were to decline at a rate of 0.5% per year. Write an equation for the exponential decay of the population. Would this rate of decline be sufficient to meet the original goal?

Solution The quantity Q is China's population. The population is *decreasing* at a rate of $P\% = 0.5\%$ per year, so the fractional growth rate r should be negative; it is $r = -P/100 = -0.005$ per year. The initial value is the 2017 population, which is given as $Q_0 = 1.4$ billion. The equation takes the form

$$Q = Q_0 \times (1 + r)^t = 1.4 \text{ billion} \times (1 - 0.005)^t = 1.4 \text{ billion} \times (0.995)^t$$

Again, because the units of r are *per year*, t must be measured in years. When we substitute $t = 33$ years (from 2017 to 2050), this exponential function predicts a 2050 population of

$$Q = 1.4 \text{ billion} \times (0.995)^{33} \approx 1.2 \text{ billion}$$

A rate of decrease of 0.5% per year would reduce China's population to about 1.2 billion by 2050, well in excess of the 700 million goal. Moreover, China's population is still growing, not shrinking. ▶ Now try Exercises 27–34, part (b).

Graphing Exponential Functions

The easiest way to graph an exponential function is to use points corresponding to several doubling times (or half-lives, in the case of decay). We start at the point $(0, Q_0)$ that represents the initial value at $t = 0$. For an exponentially growing quantity, we know that the value of Q is $2Q_0$ (double its initial value) after one doubling time (T_{double}), $4Q_0$ after two doubling times ($2T_{double}$), $8Q_0$ after three doubling times ($3T_{double}$), and so on. We simply fit a steeply rising curve between these points, as shown in Figure 9.18a.

HISTORICAL NOTE

The study of functions with nonconstant slopes—such as the exponential function—is the subject of calculus, which provides a remarkable example of apparently independent discovery. Sir Isaac Newton developed calculus in 1666 in England, but did not publish his work until 1693. Meanwhile, Gottfried Wilhelm Leibniz developed calculus in 1675 in Germany and published it in 1684. Newton accused Leibniz of stealing his work, but most historians believe that Leibniz was unaware of Newton's work and developed calculus independently.

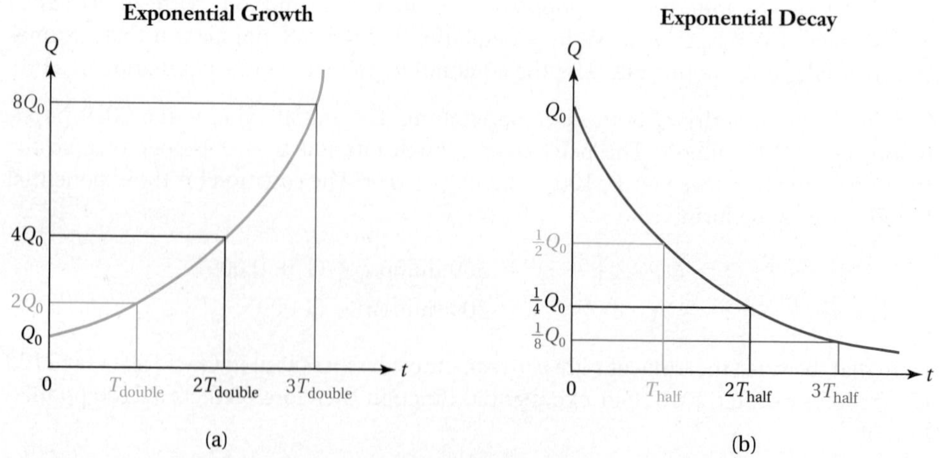

FIGURE 9.18 (a) We can graph exponential growth by plotting points for repeated doublings. (b) We can graph exponential decay by plotting points for repeated halvings.

For an exponentially decaying quantity, we know that the value of Q decreases to $Q_0/2$ (half its initial value) after one half-life (T_{half}), $Q_0/4$ after two half-lives ($2T_{half}$), $Q_0/8$ after three half-lives ($3T_{half}$), and so on. Fitting a falling curve to these points gives the graph shown in Figure 9.18b. Note that this curve gets closer and closer to the horizontal axis, but never reaches it.

EXAMPLE 3 Sensitivity to Growth Rate

The growth rate of the U.S. population has varied substantially during the past century. It depends on the immigration rate, as well as the birth and death rates. Starting from the 2010 census population of 309 million, project the population in 2100 using growth rates that are 0.2 percentage point lower and higher than the 0.9% used in Example 1. Make a graph showing the population through 2100 for each growth rate.

Solution A growth rate 0.2 percentage point lower than 0.9% is 0.7%, or $r = 0.007$. At this growth rate, the 2100 population would be

$$Q = 309 \text{ million} \times (1.007)^{90} \approx 579 \text{ million}$$

A rate 0.2 percentage point higher than 0.9% is 1.1%, or $r = 0.011$, which gives a 2100 population of

$$Q = 309 \text{ million} \times (1.011)^{90} \approx 827 \text{ million}$$

Within a range of less than half a percentage point in the growth rate, from 0.7% to 1.1%, the projected population for 2100 varies by nearly 250 million people

Brief Review Algebra with Logarithms

The basic properties of logarithms (see the Brief Review on p. 506) lead to two very useful algebraic techniques. These techniques apply to logarithms with any base, but we will focus on common (base 10) logarithms:

1. We can solve for a variable that appears in an exponent by taking the logarithm of both sides of the equation (as long as both sides are positive) and applying the rule that $\log_{10} a^x = x \log_{10} a$.
2. We can solve for a variable that appears within a logarithm by making both sides of the equation into powers of 10 and applying the rule that $10^{\log_{10} x} = x$ (as long as $x > 0$).

Example: Solve for x in the equation $2^x = 50$.

Solution: Taking the logarithm of both sides, we have

$$\log_{10} 2^x = \log_{10} 50$$

By the rule that $\log_{10} a^x = x \log_{10} a$, this equation becomes

$$x \log_{10} 2 = \log_{10} 50$$

Now we divide both sides by $\log_{10} 2$ to get the solution:

$$x = \frac{\log_{10} 50}{\log_{10} 2} = \frac{1.69897}{0.30103} \approx 5.644$$

Example: Solve for x in the equation $2 \log_{10} x = 15$.

Solution: We isolate the term containing x by dividing both sides by 2:

$$\log_{10} x = \frac{15}{2} = 7.5$$

Because x is within a logarithmic expression, we make both sides of the equation into powers of 10:

$$10^{\log_{10} x} = 10^{7.5}$$

Applying the rule $10^{\log_{10} x} = x$ (as long as $x > 0$), we find

$$x = 10^{7.5} \approx 3.1623 \times 10^7$$

▶ Now try Exercises 11–26.

(from 579 million to 827 million). Clearly, population projections are very sensitive to changes in the growth rate.

To make the graphs, note that we already have two points for each growth rate: the initial population in 2010, $Q_0 = 309$ million, and the population calculated for 2100. We find a third point in each case from the doubling time, which is the time when the population reaches $2Q_0 = 618$ million. You should use the doubling time formula (see Unit 8B) to confirm that the doubling times are about 77 years for $r = 0.009$, 99 years for $r = 0.007$, and 63 years for $r = 0.011$. Figure 9.19 shows the curves for all three growth rates.

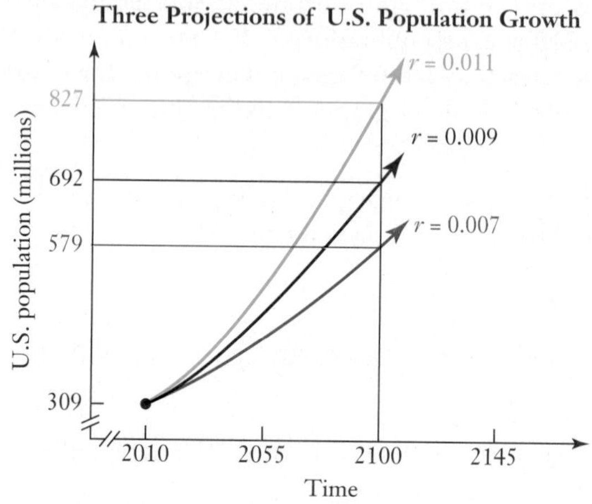

FIGURE 9.19 Future U.S. population growth with three growth rates.

▶ **Now try Exercises 27–34, part (c).**

Think About It Some people advocate increasing the birth rate in the United States to increase the population growth rate so that there will be more workers in the future. Others advocate decreasing the birth rate to keep the future population lower. What is *your* opinion? Defend it clearly.

Alternative Forms of the Exponential Function

Our general equation for the exponential function, $Q = Q_0 \times (1 + r)^t$, contains the growth rate r but not the doubling time or half-life. Because we are often given the doubling time or half-life, it's useful to rewrite the equation in forms that use T_{double} or T_{half} rather than r.

To find the first alternative form, we recognize that after a time $t = T_{double}$ the quantity Q has grown to twice its initial value, or $Q = 2Q_0$. The following algebraic steps then lead to an alternative form:

Start with the general equation:	$Q = Q_0 \times (1 + r)^t$
Substitute $Q = 2Q_0$ and $t = T_{double}$:	$2Q_0 = Q_0 \times (1 + r)^{T_{double}}$
Interchange the left and right sides and divide both sides by Q_0:	$(1 + r)^{T_{double}} = 2$
Solve for $(1 + r)$ by raising both sides to the power $1/T_{double}$:	$(1 + r) = 2^{1/T_{double}}$
Substitute this expression for $(1 + r)$ in the general equation:	$Q = Q_0 \times \left(2^{1/T_{double}}\right)^t = Q_0 \times 2^{t/T_{double}}$

Notice that the alternative form we just found matches the exponential growth equation we used in Unit 8B. Similar steps can give us the general exponential equation in terms of half-life. The following box summarizes three common forms of the general exponential equation.

Forms of the Exponential Function

- If given the growth or decay rate, r, use the exponential function in the form
$$Q = Q_0 \times (1 + r)^t$$
Remember that r is positive for growth and negative for decay.
- If given the doubling time, T_{double}, use the exponential function in the form
$$Q = Q_0 \times 2^{t/T_{double}}$$
- If given the half-life, T_{half}, use the exponential function in the form
$$Q = Q_0 \times \left(\frac{1}{2}\right)^{t/T_{half}}$$

Selected Applications

We have so far used exponential functions to model population growth and decline in this chapter and for compound interest calculations in Chapter 4. But they have much broader applications, a few of which we will now explore.

MATHEMATICAL INSIGHT

Doubling Time and Half-Life Formulas

In Unit 8B, we used both approximate and exact formulas for the doubling time and half-life. Now that we have an equation for the exponential function, we can see where these formulas came from.

For exponential growth, the doubling time is the time required for a quantity to double in size. That is, after a time $t = T_{double}$, the quantity has grown to twice its initial value, or $Q = 2Q_0$. Substituting these values into the exponential equation gives

$$2Q_0 = Q_0 \times (1 + r)^{T_{double}}$$

Dividing both sides by Q_0 and interchanging the two sides yields

$$(1 + r)^{T_{double}} = 2$$

To solve this equation for the doubling time, we must "bring down" the exponent by taking the logarithm of both sides (see the Brief Review on p. 563):

$$\log_{10}\left[(1 + r)^{T_{double}}\right] = \log_{10} 2$$

By the rule that $\log_{10} a^x = x \log_{10} a$, this equation becomes

$$T_{double} \log_{10}(1 + r) = \log_{10} 2$$

Dividing both sides by $\log_{10}(1 + r)$, we find the exact formula for the doubling time, as given in Unit 8B:

$$T_{double} = \frac{\log_{10} 2}{\log_{10}(1 + r)}$$

We find the half-life for exponential decay similarly. For decay, the quantity decreases to half its original value, or $Q = \left(\frac{1}{2}\right)Q_0$, in a time $t = T_{half}$. The algebra is the same as for the growth case, except we have T_{half} instead of T_{double} and $\log_{10}\left(\frac{1}{2}\right)$ instead of $\log_{10} 2$. The result is

$$T_{half} = \frac{\log_{10}\left(\frac{1}{2}\right)}{\log_{10}(1 + r)}$$

We can simplify this result by recognizing that $\left(\frac{1}{2}\right) = 2^{-1}$. Using the rule that $\log_{10} a^x = x \log_{10} a$, we find that $\log_{10} 2^{-1} = -\log_{10} 2$. With this substitution, the formula takes the form given in Unit 8B:

$$T_{half} = -\frac{\log_{10} 2}{\log_{10}(1 + r)}$$

Remember that r is negative for exponential decay, so $1 + r$ is less than 1 in this case. (For example, if $r = -0.05$, then $1 + r = 0.95$.) Because the logarithm of a number between 0 and 1 is negative, the formula gives a positive value for the half-life.

Inflation

Because prices tend to change with time, price comparisons from one time to another are meaningful only if the prices are adjusted for the effects of inflation (see Unit 3D). We can model the effects of inflation with an exponential function in which r represents the rate of inflation.

EXAMPLE 4 Monthly and Annual Inflation Rates

Suppose that, for a particular month in a particular country, the *monthly* rate of inflation is 0.8%. What is the *annual* rate of inflation? Is the annual rate 12 times the monthly rate? Explain.

Solution We write an exponential function for the inflation by letting Q_0 represent a price at time $t = 0$ and Q represent a price t months later. Using the monthly inflation rate $r = 0.8\% = 0.008$, we have

$$Q = Q_0 \times (1 + r)^t = Q_0 \times (1 + 0.008)^t = Q_0 \times 1.008^t$$

Note that, because r is given *per month*, t must be measured in months in this equation. Therefore, to find the *annual* inflation rate, we substitute $t = 12$ into the equation. We find that the price, Q, after 12 months is

$$Q = Q_0 \times \underbrace{1.008^{12}}_{\approx 1.100} = Q_0 \times 1.100 = 1.100Q_0$$

The price after a year is about 1.1 times the initial price, which means an annual inflation rate of 10%. Note that this annual rate is more than 12 times the monthly rate, which would be $12 \times 0.8\% = 9.6\%$. The reason is that inflation compounds with each passing month, much like compound interest. In this case, the 9.6% found by multiplying the monthly rate by 12 is analogous to the APR quoted for an interest-bearing account with monthly compounding (see Unit 4B), while the annual inflation rate is analogous to the annual yield, APY. ▶ Now try Exercises 35–38.

Environment and Resources

Among the most important applications of exponential models are those involving the environment and resource depletion. Global concentrations of many pollutants in the water and atmosphere have increased exponentially, as has consumption of nonrenewable resources such as oil and natural gas. Two basic factors can be responsible for such exponential growth.

1. The *per capita* demand for a resource often increases exponentially. For example, per capita energy consumption in the United States increased exponentially during most of the 20th century (though it has since slowed).
2. An exponentially increasing population will mean an exponentially increasing demand for a resource even if per capita demand remains constant.

In most cases, the growth rate in demand is determined by a combination of both factors.

EXAMPLE 5 China's Coal Consumption

China's rapid economic development has led to an exponentially growing demand for energy, and China generates the majority of its energy by burning coal. During the period 2000 to 2012, China's coal consumption increased at an average rate of 8% per year, and the 2012 consumption was about 3.8 billion tons of coal.

a. Suppose the growth rate in coal consumption had continued at 8% per year. What would China's coal consumption have reached by 2020?

b. Use the 8% growth rate to make a graph projecting how China's coal consumption would have changed through 2050.

Solution

a. We are modeling the growth of China's coal consumption with an exponential function for a growth rate of 8% per year, which means we set $r = 0.08$. We let $t = 0$ represent 2012, and the initial value is $Q_0 = 3.8$ (billion tons of coal). The resulting exponential function describing China's coal consumption (in billions of tons) is

$$\begin{aligned} Q &= Q_0 \times (1 + r)^t \\ &= 3.8 \times (1 + 0.08)^t = 3.8 \times 1.08^t \end{aligned}$$

To predict China's coal consumption in 2020, we set $t = 8$ because 2020 is 8 years after 2012. The predicted consumption is

$$Q = 3.8 \times 1.08^t = 3.8 \times 1.08^8 \approx 7.0 \text{ (billion tons of coal)}$$

b. There are several ways to make the graph, but let's do it by finding the doubling time. Using the exact doubling time formula (see p. 504), we find

$$T_{\text{double}} = \frac{\log_{10} 2}{\log_{10}(1 + r)} = \frac{\log_{10} 2}{\log_{10}(1.08)} \approx 9.0 \text{ yr}$$

We can now make the graph exactly as we made Figure 9.18a, starting from 2012 as $t = 0$. Figure 9.20 shows the result. This model predicts that China's coal consumption in 2050 would be more than 16 times its 2012 coal consumption. Fortunately, China has implemented policies that have decreased its coal consumption.

BY THE WAY

To reduce pollution and emissions that contribute to global warming, China is investing heavily in technologies designed to replace coal. As a result, China's coal consumption has actually declined since 2013 (at least through 2017).

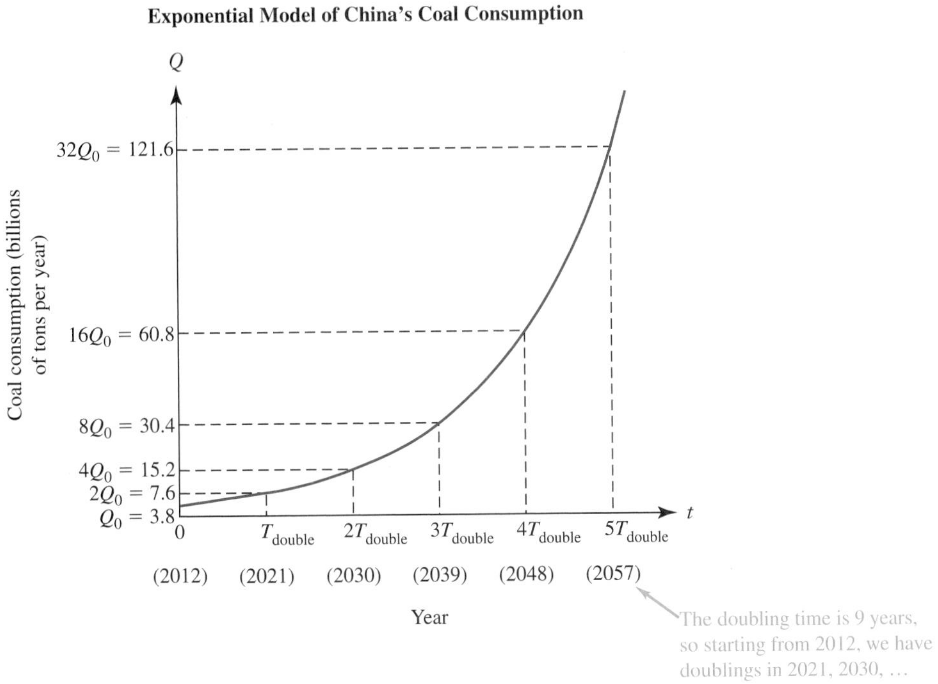

FIGURE 9.20 A model of the growth in China's coal consumption, based on a growth rate of 8% per year and starting from the 2012 value.

▶ Now try Exercises 39–40.

Physiological Processes

Many physiological processes are exponential. For example, a cancer tumor grows exponentially, at least in its early stages, and the concentration of many drugs in the bloodstream decays exponentially. (Alcohol is a notable exception, as its concentration decays linearly. See Example 5 in Unit 9B.)

EXAMPLE 6 Drug Metabolism

Consider an antibiotic that has a half-life in the bloodstream of 12 hours. A 10-milligram injection of the antibiotic is given at 1:00 p.m. How much antibiotic remains in the blood at 9:00 p.m.? Draw a graph that shows the amount of antibiotic remaining as the drug is eliminated by the body.

Solution Because we are given a half-life rather than a growth rate *r*, we use the form of the exponential function that contains the half-life:

$$Q = Q_0 \times \left(\frac{1}{2}\right)^{t/T_{\text{half}}}$$

In this case, $Q_0 = 10$ milligrams is the initial dose given at $t = 0$ and Q is the amount of antibiotic in the blood t hours later. The half-life is given as $T_{\text{half}} = 12$ hours, so the equation is

$$Q = 10 \times \left(\frac{1}{2}\right)^{t/12}$$

We must remember that Q is measured in milligrams and t in hours. At 9:00 p.m., which is $t = 8$ hours after the injection, the amount of antibiotic remaining is

$$Q = 10 \times \left(\frac{1}{2}\right)^{8/12}$$

$$= 10 \times \left(\frac{1}{2}\right)^{2/3}$$

$$\approx 6.3 \text{ mg}$$

Eight hours after the injection, 6.3 milligrams of the antibiotic remain in the bloodstream. Graphing this exponential decay function up to $t = 100$ hours, we see that the amount of antibiotic decreases steadily toward zero (Figure 9.21).

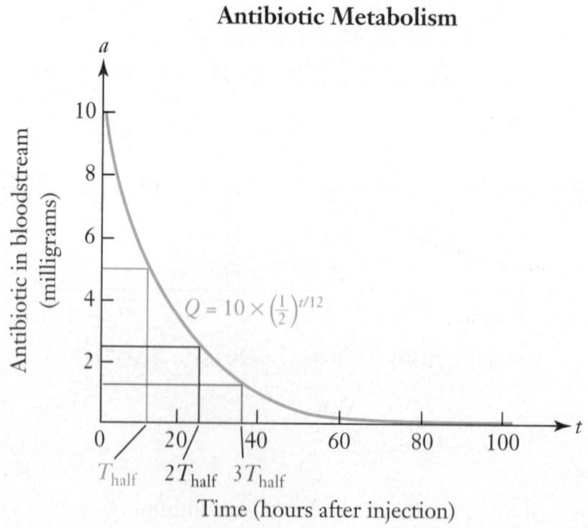

FIGURE 9.21 Exponential decay of an antibiotic with a half-life of 12 hours in the bloodstream.

▶ **Now try Exercises 41–42.**

HISTORICAL NOTE

American scientist Willard Libby won the Nobel Prize in Chemistry in 1960 for inventing the method of radiometric dating.

Radiometric Dating

As we discussed in Unit 8B, we can take advantage of the exponential nature of radioactive decay to measure the ages of rocks, bones, pottery, or other solid objects that contain radioactive elements. The process is called **radiometric dating**.

The basic idea of radiometric dating is simple: If we know both current and original amounts of the radioactive substance in the object, then we can use an exponential function to calculate the time since the object formed. Of course, we must know the half-life of the substance, but half-lives for nearly all radioactive substances have been carefully measured in laboratories. The major difficulty in radiometric dating is determining how much of the radioactive substance was originally in the object. Fortunately, because radioactive materials decay in very specific ways, we can often determine the original amount by studying the overall chemical composition of the object today.

EXAMPLE 7 The Allende Meteorite

The famous Allende meteorite lit up the skies of Mexico as it fell to Earth on February 8, 1969. Scientists melted and chemically analyzed small pieces of the meteorite and found traces of both radioactive potassium-40 and argon-40. Laboratory studies have shown that potassium-40 decays into argon-40 with a half-life of about 1.25 billion (1.25×10^9) years and that all the argon-40 in the meteorite must be a result of such decay. By comparing the amounts of the two substances in the meteorite samples, scientists determined that only 8.5% of the potassium-40 originally present in the rock remains today (the rest has decayed into argon-40). How old is the rock that makes up the Allende meteorite?

BY THE WAY
Most atomic nuclei are *stable*, meaning they do not spontaneously decay. By definition, radioactive substances have unstable nuclei that decay spontaneously, producing other nuclei as products.

Solution We can model radioactive decay with an exponential function in which Q is the current amount of the radioactive substance and Q_0 is the original amount. Because we are given the half-life, we use the function in the form

$$Q = Q_0 \times \left(\frac{1}{2}\right)^{t/T_{\text{half}}}$$

For radiometric dating, we compare the current amount Q to the original amount Q_0, so the equation is more useful if we divide both sides by Q_0:

$$\frac{Q}{Q_0} = \left(\frac{1}{2}\right)^{t/T_{\text{half}}}$$

In this case, our goal is to find t, which is the age of the rock. We are given that the half-life of potassium-40 is $T_{\text{half}} = 1.25 \times 10^9$ years and that 8.5% of the original potassium-40 remains, which means $Q/Q_0 = 0.085$.

Start with the radiometric decay equation:	$\frac{Q}{Q_0} = \left(\frac{1}{2}\right)^{t/T_{half}}$
Interchange left and right sides and begin solving for t by taking the log of both sides:	$\log_{10}(1/2)^{t/T_{\text{half}}} = \log_{10}(Q/Q_0)$
Simplify the left side by applying the rule that $\log_{10} a^x = x \log_{10} a$:	$\frac{t}{T_{\text{half}}} \times \log_{10}(1/2) = \log_{10}(Q/Q_0)$
Finish solving for t by multiplying both sides by T_{half} and dividing both sides by $\log_{10}(1/2)$:	$t = T_{\text{half}} \times \frac{\log_{10}(Q/Q_0)}{\log_{10}(1/2)}$
Substitute the given values for T_{half} and Q/Q_0:	$t = (1.25 \times 10^9 \text{ yr}) \times \frac{\log_{10} 0.085}{\log_{10}(1/2)} \approx 4.45 \times 10^9 \text{ yr}$

We conclude that the Allende meteorite is about 4.45 billion years old.

▶ Now try Exercises 43–44.

BY THE WAY
More precise measurements show that many meteorites date to about 4.55 billion years ago, but none has ever been found that is older. Scientists therefore conclude that the solar system itself must have formed 4.55 billion years ago. (The photo shows the Ahnighito meteorite at the American Museum of Natural History in New York City.)

IN YOUR WORLD

Changing Rates of Change

In this chapter, we have studied two very different families of functions. Linear functions have straight-line graphs and constant rates of change. Exponential functions have graphs that rise or fall steeply, and their (absolute) rates of change are not constant. This latter fact is illustrated in Figure 9.22a, which is the graph of an exponential growth function. Superimposed on the graph are several *tangent* lines. These lines touch the graph at just a single point, so they provide a good measure of the steepness of the graph. Note that the tangent lines at $t = 0$, $t = 1$, $t = 2$, and $t = 3$ get progressively steeper (have larger slopes). In other words, the rate of change of the function increases as *t* increases. This is a universal property of exponential growth functions: As the independent variable increases, the rate of change of the function also increases. We say that exponential growth functions "increase at an increasing rate."

Figure 9.22b shows the changing slope of an exponential decay function. In this case, as *t* increases, the tangent lines become less steep (have smaller slopes), which causes the graph to flatten out. We say that exponential decay functions "decrease at a decreasing rate."

We now see that linear functions are special: They are the *only* functions with constant rates of change. All other functions are more like exponential functions in that they have *variable* rates of change. The rate of change of a function is a very important property, particularly if the function models a real quantity, such as a population or a balance in an investment account. The rate of change tells us not only whether the function value is increasing or decreasing, but also how quickly it is increasing or decreasing, which is crucial information for modeling and prediction.

Given the importance of rates of change, it's not surprising that they are the subject of an entire branch of mathematics, called *calculus*. It's fair to say that calculus is the study of rates of change. And because the world around us is constantly changing, calculus has a lot to say about the world around us.

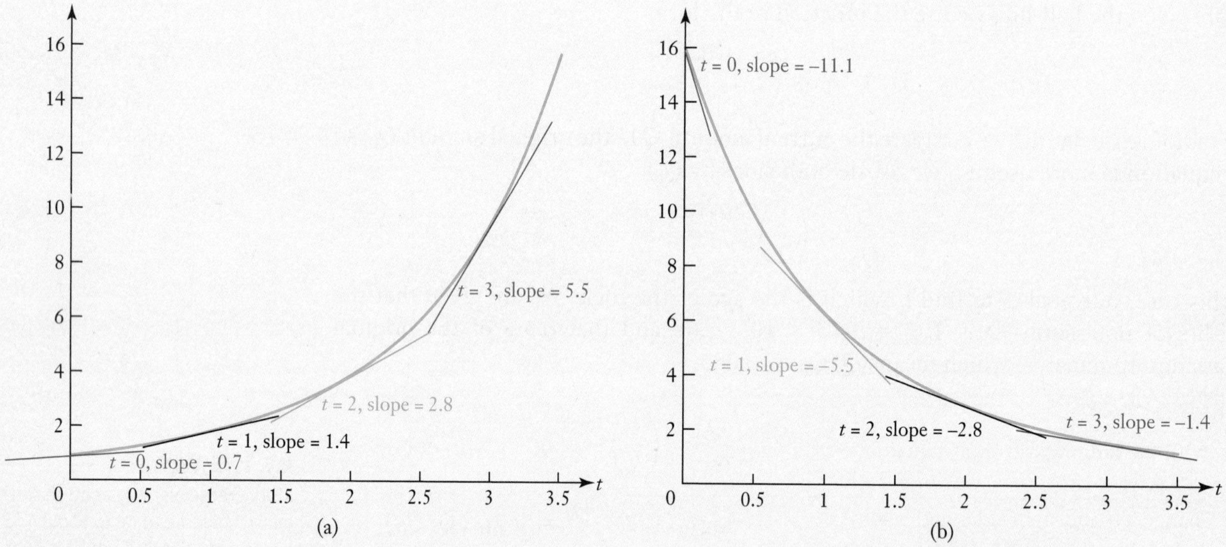

FIGURE 9.22 (a) An exponential growth function with a rate of change that increases (slopes are positive because the function increases). (b) An exponential decay function with a rate of change that decreases (slopes are negative because the function decreases).

Quick Quiz 9C

Choose the best answer to each of the following questions. Explain your reasoning with one or more complete sentences.

1. Which statement is true about exponential growth?
   a. The absolute growth rate is constant.
   b. The relative growth rate is constant.
   c. The relative growth rate is increasing.

2. A city's population starts at 100,000 people and grows 3% per year for 7 years. For this situation, in the general exponential equation $Q = Q_0 \times (1 + r)^t$, what is Q_0?
   a. 100,000 b. 3 c. 7

3. A city's population starts at 100,000 people and grows 3% per year for 7 years. For this situation, in the general exponential equation $Q = Q_0 \times (1 + r)^t$, what is r?

a. 3 b. 0.03 c. 7

4. India's 2017 population was estimated to be 1.34 billion, with a growth rate of 1.2% per year. If the growth rate remains steady, its 2027 population will be

a. 1.34 billion $\times$ 1.012^{10}

b. 1.34 billion $\times$ $10^{1.2}$

c. 2027 $\times$ $(1.34 \text{ billion})^{0.012}$

5. Suppose that inflation causes the value of a dollar to decrease at a rate of 3.5% per year. To use a general exponential model to find the value of the dollar at some future time compared to its present value, you would set r to

a. 3.5. b. 0.035. c. −0.035.

6. Figure 9.18b shows the graph of an exponentially decaying quantity. In theory, how many half-lives would it take for the value of Q to reach zero?

a. 6 b. 12

c. The value of Q never reaches zero.

7. Polly received a large dose of an antibiotic and wants to know how much antibiotic remains in her body after 3 days. Which two pieces of information are sufficient to calculate the answer?

a. her body weight and the rate at which the antibiotic is metabolized

b. the amount of the initial dose and the half-life of the antibiotic in the bloodstream

c. the rate at which the antibiotic is metabolized and the half-life of the antibiotic in the bloodstream

8. The half-life of carbon-14 is 5700 years, and carbon-14 is incorporated into the bones of a living organism only while it is alive. Suppose you have found a human bone at an archaeological site and you want to use carbon-14 to determine how long ago the person died. Which of the following additional pieces of information would allow you to do the calculation?

a. only the amount of carbon-14 in the bone today

b. both the amount of carbon-14 in the bone today and the rate at which carbon-14 decays

c. both the amount of carbon-14 in the bone today and the amount it contained at the time the person died

9. Radioactive uranium-235 has a half-life of about 700 million years. Suppose you find a rock and chemical analysis tells you that only 1/8 of the rock's original uranium-235 remains. How old is the rock?

a. 1.4 billion years old

b. 5.6 billion years old

c. 2.1 billion years old

10. Compare the first two forms of the exponential function in the Mathematical Insight box on p. 565. Given that these two forms are equivalent, what can you conclude?

a. $(1 + r)^t = 2^{t/T_{\text{double}}}$

b. $r = 1 - 2^{t/T_{\text{double}}}$

c. $Q = Q_0$ whenever $t = T_{\text{double}}$

Exercises 9C

REVIEW QUESTIONS

1. Describe the meanings of all the variables in the exponential function $Q = Q_0(1 + r)^t$. Explain how the function is used for exponential growth and decay.

2. Briefly explain how to find the doubling time and half-life from the exponential equation.

3. Describe how you can graph an exponential function with the help of the doubling time or half-life. What is the general shape of an exponential growth function? What is the general shape of an exponential decay function?

4. Describe the meaning of each of the three forms of the exponential function given in this unit. Under what circumstance is each form useful?

5. Briefly describe how exponential functions are useful for modeling inflation, environmental and resource issues, physiological processes, and radioactive decay.

6. Briefly describe the process of radiometric dating. What makes it difficult? How can the difficulty be alleviated?

DOES IT MAKE SENSE?

Decide whether each of the following statements makes sense (or is clearly true) or does not make sense (or is clearly false). Explain your reasoning.

7. After 100 years, a population growing at a rate of 2% per year will have grown by twice as many people as a population growing at a rate of 1% per year.

8. When I used the exponential function to model the decay of the medicine in my bloodstream, the growth rate r was negative.

9. We can use the fact that radioactive materials decay exponentially to determine the ages of ancient bones from archaeological sites.

10. I used the exponential function to figure how much money I'd have in a bank account that earns compound interest.

BASIC SKILLS & CONCEPTS

11–26: Review of logarithms. Use the skills covered in the Brief Review on p. 563 to solve the following equations for the unknown quantity x.

11. $2^x = 128$
12. $10^x = 23$
13. $3^x = 99$
14. $5^{2x} = 240$
15. $7^{3x} = 623$
16. $3 \times 4^x = 180$
17. $9^x = 1748$
18. $3^{x/4} = 444$
19. $\log_{10} x = 4$
20. $\log_{10} x = -3$
21. $\log_{10} x = 3.5$
22. $\log_{10} x = -2.2$
23. $3 \log_{10} x = 4.2$
24. $\log_{10} (3x) = 5.1$
25. $\log_{10} (4 + x) = 1.1$
26. $4 \log_{10} (4x) = 4$

27–34. Exponential growth and decay laws. Consider the following cases of exponential growth and decay.

a. Create an exponential function of the form $Q = Q_0 \times (1 + r)^t$ (where $r > 0$ for growth and $r < 0$ for decay) to model the situation described. Be sure to clearly identify both variables in your function.

b. Create a table showing the value of the quantity Q for the first 10 units of time (either years, months, or hours) of growth or decay.

c. Make a graph of the exponential function.

27. The population of a town with an initial population of 60,000 grows at a rate of 2.5% per year.

28. The number of restaurants in a city that had 800 restaurants in 2015 increases at a rate of 3% per year.

29. A privately owned forest that had 1 million acres of old growth is being clear cut at a rate of 7% per year.

30. A town with a population of 10,000 loses residents at a rate of 0.3% per month because of a poor economy.

31. The average price of a home in a town was \$175,000 in 2013, but home prices are rising by 5% per year.

32. A certain drug breaks down in the human body at a rate of 15% per hour. The initial amount of the drug in the bloodstream is 8 milligrams.

33. Your starting salary at a new job is \$2000 per month. At the end of each year, your monthly salary increases 5%.

34. You hid 100,000 rubles in a mattress at the end of 1991, when they had a value of \$10,000. However, the value of the ruble against the dollar then fell 50% per year.

35–36: Annual vs. Monthly Inflation. Answer the following questions about monthly and annual inflation rates.

35. If prices increase at a monthly rate of 2.5%, by what percentage do they increase in a year?

36. If the price of gold decreases at a monthly rate of 1.5%, by what percentage does it decrease in a year?

37. **Hyperinflation in Germany.** In 1923, Germany underwent one of the worst periods in history of *hyperinflation*—extraordinarily large inflation in prices. At the peak of the hyperinflation, prices rose 30,000% per month. At this rate, by what percentage would prices have risen in 1 year? In 1 day?

38. **Hyperinflation in North Korea.** Experts watching rice prices suspect that North Korea underwent hyperinflation in 2010 and 2011. Suppose that the rate of inflation was 90% per month (which is unconfirmed). At this rate, by what percentage would prices rise in 1 year? In 1 day?

39. **Extinction by Poaching.** Suppose that poaching reduces the population of an endangered animal by 6% per year. Further suppose that when the population of this animal falls below 50, its extinction is inevitable (owing to the lack of reproductive options without severe in-breeding). If the current population of the animal is 1500, when will it face extinction? Comment on the validity of the exponential model.

40. **World Oil Production.** Annual world oil production was about 520 million tons in 1950. Production increased at a rate of 7% per year between 1950 and 1972, but the rate of growth then slowed. World oil production reached approximately 4.8 billion tons per year in 2016.

    a. Using the growth rate of 7% for the period 1950 to 1972, approximately what was world oil production in 1972? Round your answer to two significant digits.

    b. Starting from the result of part (a), approximately how much oil would have been produced in 2016 if growth in production had continued at a rate of 7% between 1972 and 2016. Compare this result to the actual 2016 figure given above.

    c. Starting from the result of part (a), approximately how much oil would have been produced in 2016 if growth in production had proceeded at a rate of 3% between 1972 and 2016. Compare this result to the actual 2016 figure given above.

d. By trial and error, estimate the annual growth rate in world oil production between 1972 and 2016 with an exponential function.

41. **Oxycodone Metabolism.** The pain-killer oxycodone is eliminated from the bloodstream exponentially with a half-life of 3.5 hours. Suppose that a patient receives an initial dose of 10 milligrams of oxycodone at noon.

a. How much oxycodone is in the patient's blood at 6:00 p.m. the same day?

b. Estimate when the oxycodone concentration will reach 10% of its initial level.

42. **Aspirin Metabolism.** Assume that for the average individual, aspirin has a half-life of 8 hours in the bloodstream. At 12:00 noon, you take a 300-milligram dose of aspirin.

a. How much aspirin will be in your blood at 6:00 p.m. the same day? At midnight? At 12:00 noon the next day?

b. Estimate when the amount of aspirin will decay to 5% of its original amount.

43–44: Radiometric Dating. Use the radiometric dating formula to answer the following questions.

43. Uranium-238 has a half-life of 4.5 billion years.

a. You find a rock containing a mixture of uranium-238 and lead. You determine that 60% of the original uranium-238 remains; the other 40% decayed into lead. How old is the rock?

b. Analysis of another rock shows that it contains 55% of its original uranium-238; the other 45% decayed into lead. How old is the rock?

44. The half-life of carbon-14 is about 5700 years.

a. You find a piece of cloth painted with organic dyes. By analyzing the dye in the cloth, you find that only 63% of the carbon-14 originally in the dye remains. When was the cloth painted?

b. A well-preserved piece of wood found at an archaeological site has 12.3% of the carbon-14 that it must have had when it was alive. Estimate when the wood was cut.

c. Is carbon-14 useful for establishing the age of the Earth? Why or why not?

FURTHER APPLICATIONS

45. **Radioactive Waste.** A toxic radioactive substance with a density of 3 milligrams per square centimeter is detected in the ventilating ducts of a nuclear processing building that was used 55 years ago. If the half-life of the substance is 20 years, what was the density of the substance when it was deposited 55 years ago?

46. **Metropolitan Population Growth.** A small city had a population of 110,000 in 2010. Concerned about rapid growth, the residents passed a growth control ordinance limiting population growth to 2% each year. If the population grows at this 2% annual rate, what will the population be in 2020? What is the maximum growth rate cap that will prevent the population from reaching 150,000 in 2025?

47. **Rising Home Prices.** In 2000, the median home price in New York City was about $300,000, and from 2000 to 2006, home prices in New York City rose at an average rate of about 11% per year. If prices had continued to rise at that rate, what would the median home price have been in 2017? Compare to the actual median price of about $400,000 in 2017.

48. **Periodic Drug Doses.** It is common to take a drug (such as aspirin or an antibiotic) repeatedly at fixed time intervals. Suppose that an antibiotic has a half-life of 8 hours and a 100-milligram dose is taken every 8 hours.

a. Write an exponential function that represents the decay of the antibiotic from the moment of the first dose to just *prior* to the next dose (i.e., 8 hours after the first dose). How much antibiotic is in the bloodstream just *prior* to this next dose? How much antibiotic is in the bloodstream just *after* this next dose?

b. Following a procedure similar to that in part (a), calculate the amounts of antibiotic in the bloodstream just prior to and just after the doses at 16 hours, 24 hours, and 32 hours.

c. Make a graph of the amount of antibiotic in the bloodstream for the first 32 hours after the first dose of the drug. What do you predict will happen to the amount of drug if the doses every 8 hours continue for several days or weeks? Explain.

d. Consult a pharmacist (or read the fine print on the information sheet enclosed with many medicines) to find the half-lives of some common drugs. Create a model for the metabolism of one drug using the above procedure.

49. **Increasing Atmospheric Carbon Dioxide.** Direct measurements of the atmospheric carbon dioxide (CO_2) concentration have been made since about 1959, when the measured concentration was about 316 parts per million. In 2017, the concentration was about 407 parts per million. Assume that this growth can be modeled with an exponential function of the form $Q = Q_0 \times (1 + r)^t$.

a. By experimenting with various values of the fractional growth rate r, find an exponential function that fits the given data for 1959 and 2017.

b. Use this exponential model to predict when the CO_2 concentration will be 560 parts per million (twice its preindustrial level).

c. Research recent trends in the carbon dioxide concentration. Does your model seem to fit, overestimate, or underestimate its recent rise? Explain.

50. **Radioiodine Treatment.** Roughly 12,000 Americans are diagnosed with thyroid cancer every year, which accounts for about 1% of all cancer cases. It occurs three times as

frequently in women as in men. Fortunately, thyroid cancer can be treated successfully in many cases with radioactive iodine, or I-131. This unstable form of iodine has a half-life of 8 days and is given in small doses measured in millicuries.

a. Suppose a patient is given an initial dose of 100 millicuries. Find the exponential function that gives the amount of I-131 in the body t days after the initial dose.

b. How long does it take for the initial dose to reach a level of 10 millicuries?

c. Finding the initial dose to give a particular patient is a critical calculation. How does the time to reach 10 millicuries in part (b) change if the initial dose is increased by 10% (to 110 millicuries)?

IN YOUR WORLD

51. **Inflation Rate in the News.** Find a news report that states both monthly and annual inflation rates. Using the methods in this unit, check whether the rates agree. Explain.

52. **Exponential Process in the News.** Find a news account of a process that illustrates either exponential growth or decay. Is the term *exponential* used in the account? How do you know that there is an exponential process at work? Describe how you could use an exponential function to model the process.

53. **Radiometric Dating in the News.** Find a news report that gives the age of an archaeological site, a fossil, or a rock based on radiometric dating. Briefly describe how the dating process worked and how accurate we can expect it to be.

54. **Resource Consumption.** Choose a particular natural resource (such as natural gas or oil), and find data on the consumption of that resource either nationally or worldwide. Use an exponential growth model together with the data to make the case that consumption of the resource has or has not been increasing exponentially.

55. **Renewable Energy.** Find data on the power production from some source of renewable energy, such as wind or solar. Do you see evidence that production is increasing exponentially? Based on current trends, what can you predict about the future importance of this energy source?

Chapter 9 Summary

UNIT	KEY TERMS	KEY IDEAS AND SKILLS
9A	mathematical model (p. 536) function (p. 536) dependent variable (p. 536) independent variable (p. 536) domain (p. 540) range (p. 540) periodic function (p. 543)	**Represent** the ordered pair of variables in a function as (*independent variable, dependent variable*). (p. 537) **Understand** the notation $y = f(x)$. (p. 537) **Represent** functions with a data table, graph, or equation. (p. 539) **Create and use** graphs of functions. (p. 541) **Understand** how functions are models. (p. 541)
9B	linear model (p. 546) linear function (p. 546) rate of change (p. 547) slope (p. 547) initial value (p. 551) *y*-intercept (p. 551)	**Rate of change** for a linear function: (p. 547) $\text{rate of change} = \text{slope} = \dfrac{\text{change in } \textit{dependent} \text{ variable}}{\text{change in } \textit{independent} \text{ variable}}$ **Change in the dependent variable:** (p. 549) change in *dependent* variable = (rate of change) × (change in *independent* variable) **General equation for a linear function:** (p. 550) *dependent* variable = initial value + (rate of change × *independent* variable) **Algebraic equation of a line:** (p. 551) $y = mx + b$ **Slope** from two data points: (p. 554) $\text{slope} = \dfrac{\text{change in } y}{\text{change in } x}$
9C	fractional growth rate, *r* (p. 560) exponential function (p. 560) initial value (p. 561) radiometric dating (p. 568)	**Equation for an exponential function:** (p. 561) $Q = Q_0 \times (1 + r)^t$ **Equation for an exponential function** given the doubling time: (p. 565) $Q = Q_0 \times 2^{t/T_{\text{double}}}$ **Equation for an exponential function** given the half-life: (p. 565) $Q = Q_0 \times \left(\frac{1}{2}\right)^{t/T_{\text{half}}}$

10 MODELING WITH GEOMETRY

We live in a three-dimensional world, and much of our understanding of that world is rooted in geometry. Surveying land for a new park, analyzing a satellite photograph of the Earth, navigating a ship at sea, and creating a medical image all rely on ideas that originated with the ancient Greeks. This chapter reveals how both ancient and modern geometry help us understand the world around us.

Q

You can find the perimeter of a geometric shape like a circle or square with a simple formula. But suppose you want to find the perimeter of a natural object with jagged edges such as a fern leaf. You'll need some kind of ruler that you can lay along the edges of the leaf, but different rulers have different resolutions (that is, the minimum length that you can measure). For example, most rulers allow you to measure to about the nearest millimeter, but with a microscope and a ruler with finer markings you could measure much smaller distances. Which of the following statements correctly describes the results you will find when you measure the perimeter of a fern leaf?

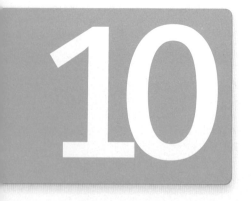

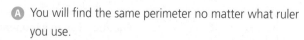

- **A** You will find the same perimeter no matter what ruler you use.
- **B** If you measure with one ruler and then re-measure with another that measures smaller lengths, the re-measurement will yield a smaller perimeter.
- **C** If you measure with one ruler and then re-measure with another that measures smaller lengths, the re-measurement will yield a larger perimeter.
- **D** Using a ruler that measures smaller lengths will affect your measurements only up to a point. For a fern leaf, your measurements won't change once you use a ruler that measures about a tenth of a millimeter.
- **E** Measuring is too difficult for a fern leaf, so you'll get a better estimate of the perimeter by tracing a simple geometric shape around it and calculating the perimeter of that shape.

> A work of morality, politics, criticism, perhaps even eloquence will be more elegant, other things being equal, if it is shaped by the hand of geometry.
>
> —Bernard Le Bovier de Fontenelle, 1729

A

For more than 2000 years after the ancient Greeks developed the basic ideas of geometry, it was generally assumed that the correct answer was D. That is, as long as you used a ruler with sufficiently fine markings, you could measure the exact perimeter of the leaf. However, during the 20th century mathematicians began to realize that many natural objects, like fern leaves, display ever finer structure as you look on smaller and smaller scales. As a result, the correct answer is C, because the ability to measure smaller lengths means you'll measure more features and thereby find a larger perimeter.

This insight gave rise to an entirely new form of geometry, called *fractal geometry*, which can give us a deeper understanding of many natural objects than we can gain with classical (or Euclidean) geometry. The discovery of fractal geometry has not made classical geometry any less important, which is why we'll devote the first two units of this chapter to it. But fractal geometry has opened a whole new world of geometric possibilities, including the strange idea that there are fractional dimensions lying between the one, two, and three dimensions of classical geometry. Unit 10C gives you the opportunity to learn more about this fascinating new type of geometry.

UNIT 10A

Fundamentals of Geometry: Study fundamental ideas of geometry, including formulas for finding the perimeter, area, and volume of common objects.

UNIT 10B

Problem Solving with Geometry: Investigate examples that use geometry to solve problems that arise in everyday life.

UNIT 10C

Fractal Geometry: Explore the ideas that underlie *fractal geometry,* a type of geometry that has become important both in the arts and in our understanding of nature.

ACTIVITY

Eyes in the Sky

Use this activity to gain a sense of the kinds of problems this chapter will enable you to analyze. Try working through it before you begin the chapter, then return to it after you've learned the chapter material.

It's easy to find satellite images of almost any location on Earth with Google Earth and similar services. These images come from telescopes on satellites that orbit Earth; unlike astronomical telescopes (such as the Hubble and the James Webb Space Telescopes), these Earth-observing telescopes look downward. We can investigate the capabilities of downward-looking telescopes with some simple geometry—the topic of this chapter.

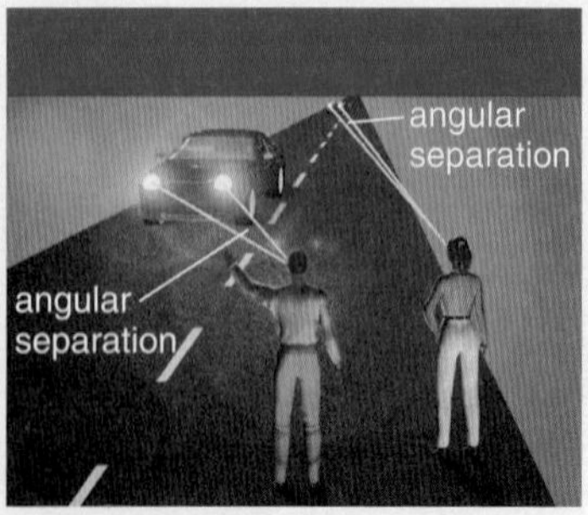

1. The *angular separation* of two points depends on their actual separation and their distance. For example, the figure on the left shows that the angular separation of the two headlights on a car is smaller when the car is farther away. Your eyes can tell that two points are distinct if their angular separation is greater than about $\frac{1}{60}°$, so we say that the *angular resolution* of your eyes is about $\frac{1}{60}°$, or 1 arcminute. Similarly, we define the angular resolution of a telescope to be the smallest angular separation that the telescope can detect. To maximize the observable detail, is it better to have a telescope with large or small angular resolution? Why?

2. Assuming that a telescope is well made and the viewing conditions are ideal, the angular resolution of a telescope depends only on its size and the wavelength of light it is observing. For visible light (average wavelength of 500 nanometers), the angular resolution of a telescope is given by

$$\text{angular resolution in degrees} \approx \frac{3.5 \times 10^{-5}}{\text{telescope diameter in meters}}$$

Use this formula to find (a) the angular resolution of the Hubble Space Telescope, which is 2.4 meters in diameter; (b) the angular resolution of a 6.5-meter-diameter telescope for visible-light observations (the James Webb Space Telescope has this size but is designed to observe infrared light).

3. Suppose a telescope the size of the Hubble Space Telescope is looking down at Earth from an altitude of 300 kilometers. What minimum distance must separate two points on the ground for the telescope to be able to distinguish them? How does this compare to the minimum sizes of objects that can be distinguished in satellite images on the Web? (*Hint:* You will need the *small-angle formula* given in Unit 10B.)

4. The military is presumed to have spy satellites with much better angular resolution than that available to the public. Suppose the military wanted a telescope that could read a newspaper from an altitude of 300 kilometers. How big would the telescope have to be? Discuss any assumptions you must make. (*Hint:* Think of each letter as being composed of a series of individual points that together form the letter's shape.)

5. Lower altitudes obviously make it easier for spy satellites to see details on the ground, but satellites in low orbits circle Earth about every 90 minutes, with each orbit taking them over different places on Earth. In order to watch an area continuously, a satellite has to be in geostationary orbit—where it matches Earth's rotation speed and therefore stays fixed over a single location on Earth's equator—which means an altitude of about 35,600 kilometers. How large would a telescope have to be to read a newspaper from geostationary altitude? Do you think it is practical for the military to use satellites to keep a continuous watch on, say, suspected terrorist locations?

UNIT 10A Fundamentals of Geometry

The word *geometry* literally means "earth measure." Many ancient cultures developed geometrical methods to survey flood basins around agricultural fields and to establish patterns of planetary and star motion. However, geometry was always more than just a practical science, as we can see from the use of geometric shapes and patterns in ancient art.

The Greek mathematician Euclid (c. 325–270 B.C.E.) summarized Greek knowledge of geometry in a 13-volume textbook called *Elements*. The geometry described in Euclid's work, now called **Euclidean geometry**, is the familiar geometry of lines, angles, and planes. In this unit, we'll explore the foundations of Euclidean geometry.

HISTORICAL NOTE

Euclid's *Elements* was the primary geometry textbook used throughout the Western world for almost 2000 years. Until recently, it was the second-most reproduced book of all time (after the Bible), and by almost any measure it was the most successful textbook in history.

There is no royal road to geometry.
—Euclid, to King Ptolemy I

Think About It Euclid worked at a university called the Museum, so named because it honored the Muses—the patron goddesses of the sciences and the fine arts, which went hand in hand for the Greeks. Do the sciences and the arts still seem so clearly linked today? Why or why not?

Points, Lines, and Planes

Geometric objects, such as points, lines, and planes, represent *idealizations* that do not exist in the real world (Figure 10.1). A geometric **point** is imagined to have zero size. No real object has zero size, but many real phenomena approximate geometric points. Stars, for example, appear to our eyes as points of light in the night sky.

A geometric **line** is formed by connecting two points along the shortest possible path. It has infinite length and no thickness. Because no physical object is infinite in length, we usually work with **line segments**, or pieces of a line. A long taut wire is a good approximation to a line segment.

A geometric **plane** is a perfectly flat surface that has infinite length and width, but no thickness. A smooth tabletop is a good approximation to a portion of a plane.

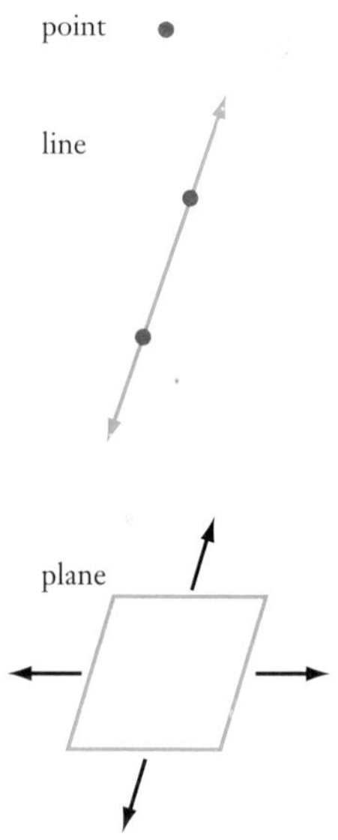

FIGURE 10.1 Representations of a point, a line, and a plane.

Think About It Describe at least three additional everyday realizations of points, line segments, and portions of planes. How does each real object compare to its geometric idealization?

Dimension

The **dimension** of an object can be thought of as the number of independent directions in which you could move if you were on the object. If you were a prisoner confined to a point, you would have no place to go, so a point has zero dimensions. A line is one-dimensional because, if you walk on a line, you can move in only one direction. (Forward and backward count as the same direction—one positive and the other negative.) In a plane, you can move in two independent directions, such as north/south and east/west; all other directions are a combination of those two independent directions. Therefore, a plane is two-dimensional. In a three-dimensional **space**, such as the world around us, you can move in three independent directions: north/south, east/west, and up/down.

We can also think about dimension in terms of the number of **coordinates** required to locate a point (Figure 10.2). A line is one-dimensional because it requires only one coordinate, such as x, to locate a point on a line. A plane is two-dimensional because it requires two coordinates, such as x and y, to locate a point on a plane. Three coordinates, such as x, y, and z, are needed to locate a point in three-dimensional space.

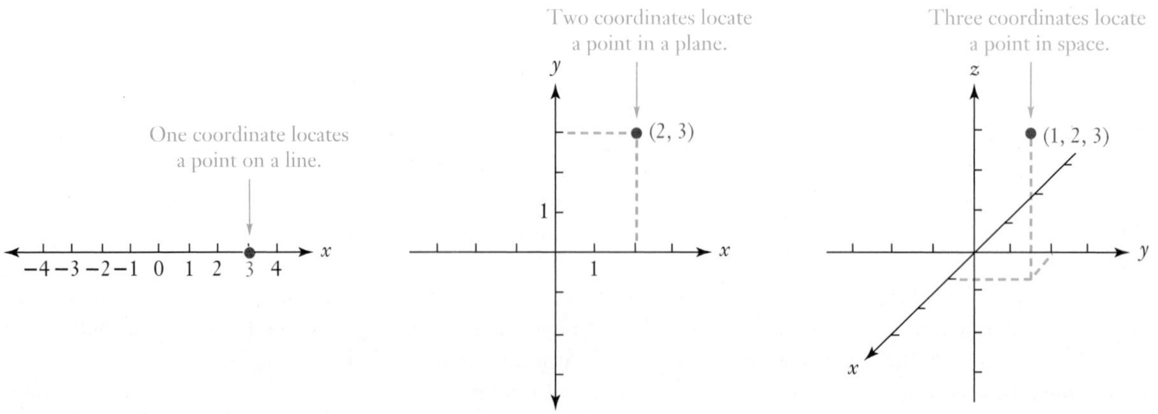

FIGURE 10.2

Angles

The intersection of two lines or line segments forms an **angle**. The point of intersection is called the **vertex**. Figure 10.3a shows an arbitrary angle with its vertex at point *A*, so we call it *angle A*, denoted as $\angle A$. The most common way to measure angles is in *degrees* (°) derived from the ancient base-60 numeral system of the Babylonians. By definition, a full circle encompasses an angle of 360°, so an angle of 1° represents $\frac{1}{360}$ of a circle (Figure 10.3b). To measure an angle, we imagine its vertex as the center of a circle. Figure 10.3c shows that $\angle A$ subtends $\frac{1}{12}$ of a circle, which means that $\angle A$ measures $\frac{1}{12} \times 360°$, or 30°.

Technical Note

Angles can also be measured in radians. The angle that subtends a full circle (360 degrees) is defined to be 2π radians. Therefore,

$$1° = \frac{2\pi}{360} \text{ radians}$$

and

$$1 \text{ radian} = \frac{360°}{2\pi} \approx 57.3°$$

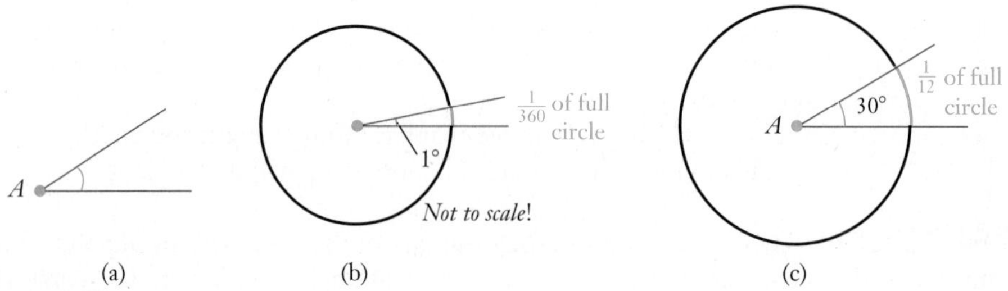

FIGURE 10.3 Measuring angles.

Some angles have special names, as shown by the examples in Figure 10.4.

- A **right angle** measures 90°.
- A **straight angle** is formed by a straight line and measures 180°.
- An **acute angle** is any angle whose measure is less than 90°.
- An **obtuse angle** is any angle whose measure is between 90° and 180°.

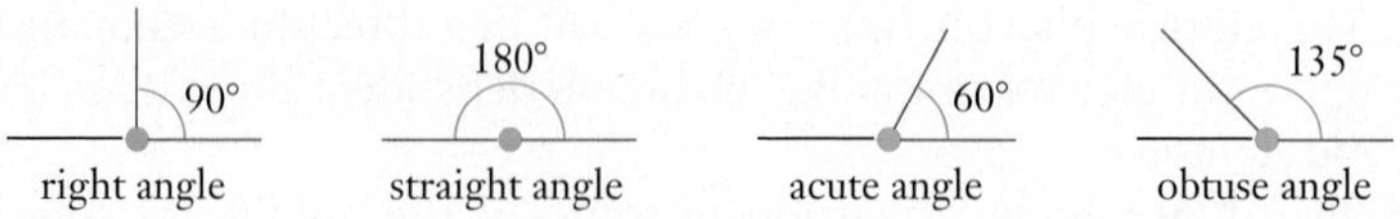

FIGURE 10.4

Think About It Draw an acute angle and an obtuse angle different from the ones shown in Figure 10.4. How is the meaning of the term *acute* in *acute illness* related to its meaning in *acute angle*? If we say that someone is *being obtuse*, does the meaning of *obtuse* bear any relation to its meaning in *obtuse angle*? Explain.

Two lines or line segments that meet in a right (90°) angle are said to be **perpendicular** (Figure 10.5a). Two lines or line segments in a plane that are the same distance apart everywhere along their lengths are said to be **parallel** (Figure 10.5b). Parallel lines in a plane never meet.

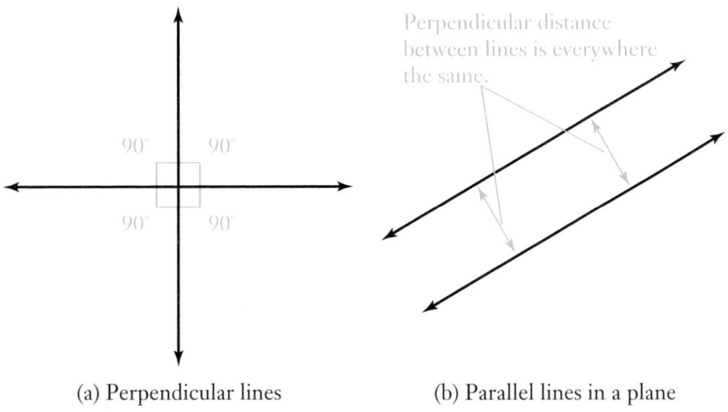

FIGURE 10.5

HISTORICAL NOTE

Formal geometry usually is traced back to the Greek philosopher Thales (624–546 B.C.E.). He is regarded as the first person to introduce ideas of abstraction into geometry, envisioning lines of zero thickness and perfect straightness.

EXAMPLE 1 Angles

Find the degree measure of the angle that subtends (spans):

a. a semicircle (half a circle)
b. a quarter circle
c. an eighth of a circle
d. a hundredth of a circle

Solution

a. The angle that subtends a semicircle measures $\frac{1}{2} \times 360° = 180°$.
b. The angle that subtends a quarter circle measures $\frac{1}{4} \times 360° = 90°$.
c. The angle that subtends an eighth of a circle measures $\frac{1}{8} \times 360° = 45°$.
d. The angle that subtends a hundredth of a circle measures $\frac{1}{100} \times 360° = 3.6°$.

▶ Now try Exercises 17–30.

Plane Geometry

Plane geometry is the geometry of two-dimensional objects. Here we examine features of circles and polygons, which are the most common two-dimensional objects.

All points on a circle are located at the same distance—the **radius**—from the circle's **center** (Figure 10.6). The **diameter** of a circle is twice its radius, which means the diameter is the distance across the circle on a line passing through its center.

A **polygon** is any closed shape in a plane made from straight line segments (Figure 10.7). The root *poly* comes from a Greek word for "many," so a *polygon* is a many-sided figure. A **regular polygon** is a polygon in which all the sides have the same length and all interior angles are equal. Table 10.1 shows several common regular polygons and gives their names.

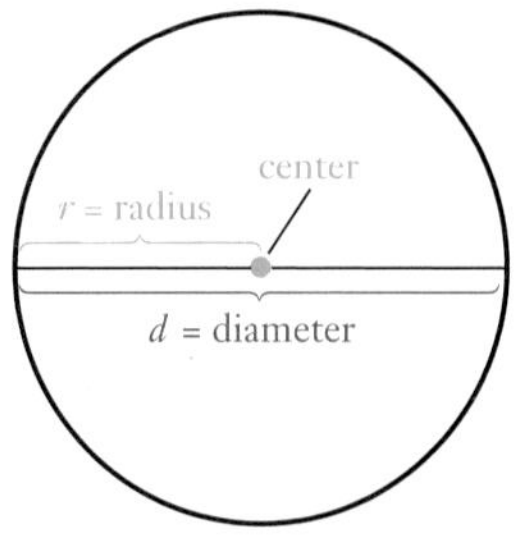

FIGURE 10.6 Definitions for a circle.

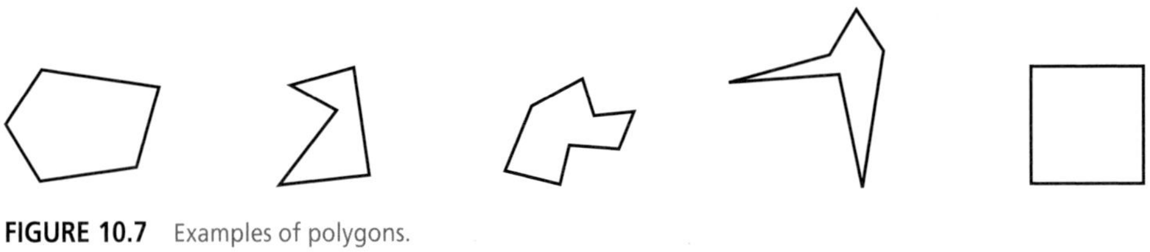

FIGURE 10.7 Examples of polygons.

TABLE 10.1 A Few Regular Polygons

Sides	Name	Picture	Sides	Name	Picture
3	Equilateral triangle		6	Regular hexagon	
4	Square		8	Regular octagon	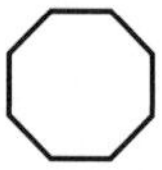
5	Regular pentagon	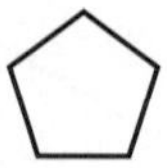	10	Regular decagon	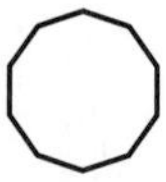

HISTORICAL NOTE

The ancient Greeks considered geometry to be the ultimate human endeavor. Above the doorway to the Academy founded in Athens by Plato in 387 B.C.E., which was in effect the world's first university, an inscription read: "Let no one ignorant of mathematics enter here." Plato's Academy remained a center of learning for more than 900 years, until it was ordered closed by the Eastern Roman Emperor Justinian in 529 C.E. The painting shows the Academy as imagined by the Renaissance painter Raphael; the central figures on the stairs are Plato and Aristotle.

Think About It Give several other English words that use the Greek root *poly*. What does *poly* mean in each case?

Triangles are among the most important of all polygons, and they take many different forms. All three sides of an **equilateral triangle** have equal length (Figure 10.8a), making it a regular polygon. An **isosceles triangle** (Figure 10.8b) has exactly two sides of equal length. A **right triangle** (Figure 10.8c) contains one right (90°) angle. We will return to right triangles and their many uses in Unit 10B. By sketching some triangles—particularly triangles with one very large angle and two very small angles (Figure 10.8d)—you can probably convince yourself that all triangles have the following property: *The sum of the three angle measures is always* 180°.

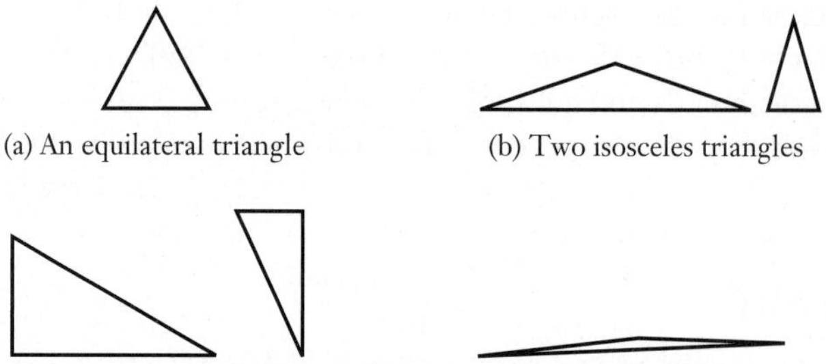

(a) An equilateral triangle

(b) Two isosceles triangles

(c) Two right triangles

(d) A triangle that has a large obtuse angle

FIGURE 10.8 Different types of triangles. In all cases, the sum of the three angles is 180°.

Perimeter

The **perimeter** of a plane object is simply the length of its boundary (Table 10.2). We can find the perimeter of a polygon by adding the lengths of the individual edges. The perimeter of a circle, called the **circumference**, is related to the circle's diameter or radius by the universal constant π (pronounced "pie"), which has a value of approximately 3.14:

$$\begin{aligned}\text{circumference of circle} &= \pi \times \text{diameter} = \pi \times d \\ &= 2 \times \pi \times \text{radius} = 2 \times \pi \times r\end{aligned}$$

▶ Now try Exercises 31–46.

TABLE 10.2 Perimeter and Area Formulas for Familiar Two-Dimensional Objects

Object	Picture	Perimeter	Area
Circle	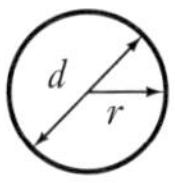	$2\pi r = \pi d$	πr^2
Square	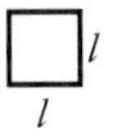	$4l$	l^2
Rectangle	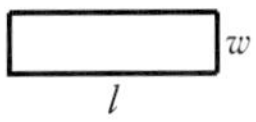	$2l + 2w$	lw
Parallelogram	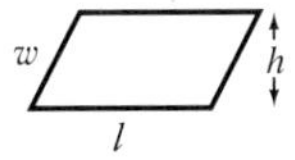	$2l + 2w$	lh
Triangle	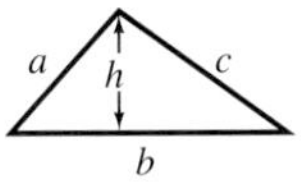	$a + b + c$	$\frac{1}{2}bh$

MATHEMATICAL INSIGHT

Archimedes and Pi

Ancient people recognized that the circumference of any circle is proportional to its radius. The first known attempt to find an exact formula for the circumference was made by the Greek scientist Archimedes (287–212 B.C.E.). His strategy began with two squares: one *inscribed* in a circle and the other *circumscribed* around the circle (Figure 10.9). He reasoned that the circumference of the circle must be greater than the perimeter of the inscribed square and less than the perimeter of the circumscribed square. Next, he doubled the number of sides of his figures, so that each square became an octagon. He repeated the process, producing inscribed and circumscribed 16-sided polygons (or hexadecagons), 32-sided polygons, and so on.

The following table shows the perimeters (in inches) for the inscribed and circumscribed polygons around a circle with a diameter of 1 inch. Notice that both sets of perimeters converge to π—one set from above and one from below. Therefore, if we could apply Archimedes' strategy to polygons with an infinite number of sides, we would find an exact value for π.

Number of sides of polygon	Perimeter of inscribed polygon	Perimeter of circumscribed polygon
4	2.8284	4.0000
8	3.0615	3.3137
16	3.1214	3.1826
32	3.1365	3.1517
64	3.1403	3.1441

Archimedes' approximation to π was about 3.14. Because π is irrational, it cannot be written exactly. The first few digits are 3.141592653589793. . . . As of 2013, the most precise approximation to π has more than 10 trillion decimal digits.

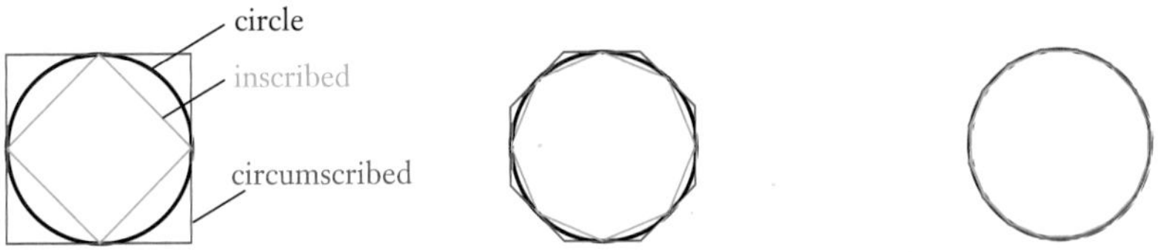

FIGURE 10.9 Circles with inscribed (blue) and circumscribed (red) polygons. Left to right, the polygons are squares, octagons, and hexadecagons, with their perimeters making successively better approximations to the circumference of a circle.

FIGURE 10.10

EXAMPLE 2 Interior Design

A window consists of a 4-foot-by-6-foot rectangle capped by a semicircle (Figure 10.10). How much trim is needed to go around the window?

Solution The trim must line the 4-foot base of the window, two 6-foot sides, and the semicircular cap. The three straight edges have a total length of 4 ft + 6 ft + 6 ft = 16 ft. The perimeter of the semicircular cap is half of the circumference of a full circle with a diameter of 4 feet, or

$$\frac{1}{2} \times \pi \times 4 \text{ ft} \approx \frac{1}{2} \times 3.14 \times 4 \text{ ft} \approx 6.3 \text{ ft}$$

The total length of trim needed for the window is about

$$16 \text{ ft} + 6.3 \text{ ft} = 22.3 \text{ ft}$$

▶ Now try Exercises 47–48.

Areas

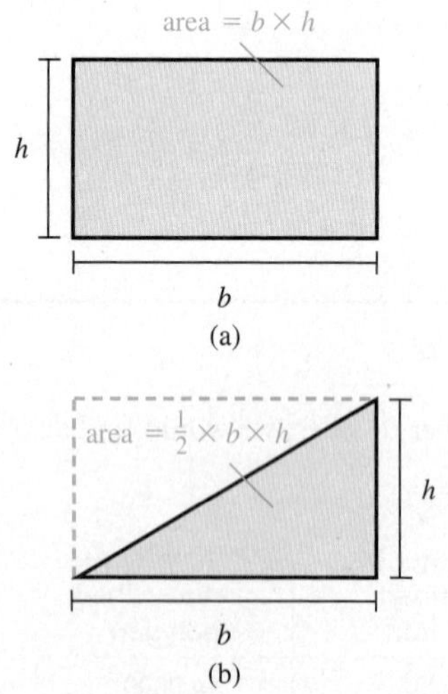

FIGURE 10.11

We can find the areas of many geometrical objects with fairly simple formulas (see Table 10.2). For example, the area of any circle is related to its radius by the following formula:

$$\text{area of circle} = \pi \times \text{radius}^2 = \pi \times r^2$$

Another familiar formula gives the area of a rectangle (Figure 10.11a):

$$\text{area of rectangle} = \text{base} \times \text{height} = b \times h$$

Cutting a rectangle along its diagonal produces two identical right triangles, as shown in Figure 10.11b. As the figure makes clear, the area of each triangle is

$$\text{area of triangle} = \frac{1}{2} \times \text{base} \times \text{height} = \frac{1}{2} \times b \times h$$

This area formula holds for all triangles.

Similarly, we can find the area formula for a **parallelogram**—a four-sided polygon in which the opposite sides are parallel. Figure 10.12 shows how a parallelogram is transformed into a rectangle with the same area. We can then see that the area formula for a parallelogram is the same as that for a rectangle:

$$\text{area of parallelogram} = \text{base} \times \text{height} = b \times h$$

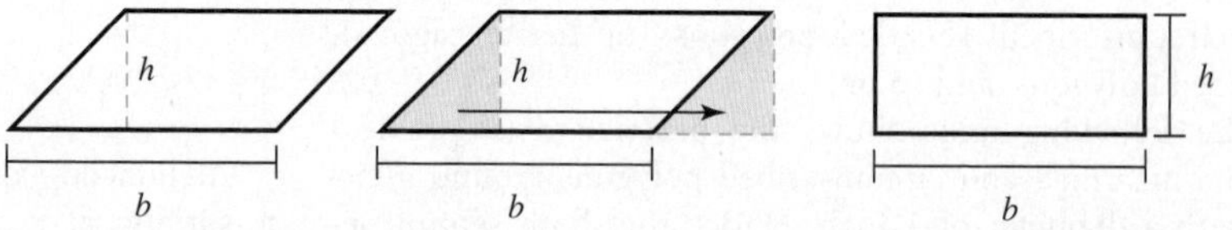

FIGURE 10.12 We can transform a parallelogram into a rectangle by shifting one triangular segment from one side to the other.

Many other areas can be computed from these basic formulas. For example, we can find the area of any **quadrilateral** (a four-sided polygon) by dividing it into two triangles and adding the areas of the two triangles (Figure 10.13).

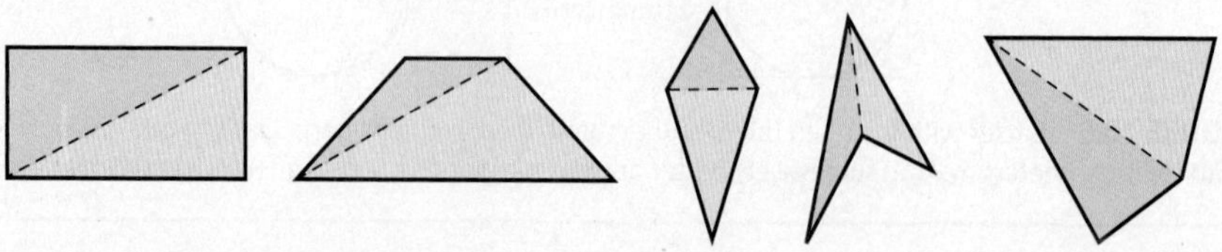

FIGURE 10.13 The area of any quadrilateral can be computed as the sum of the areas of two triangles.

EXAMPLE 3 Building Stairs

You have built a stairway in a new house and want to cover the space beneath the stairs with plywood. Figure 10.14 shows the region to be covered. What is the area of this region?

Solution The region to be covered is triangular, with a base of 12 feet and height of 9 feet. The area of this triangle is

$$\text{area} = \frac{1}{2} \times b \times h = \frac{1}{2} \times 12 \text{ ft} \times 9 \text{ ft} = 54 \text{ ft}^2$$

The area of the region to be covered with plywood is 54 square feet.

▶ Now try Exercises 49–50.

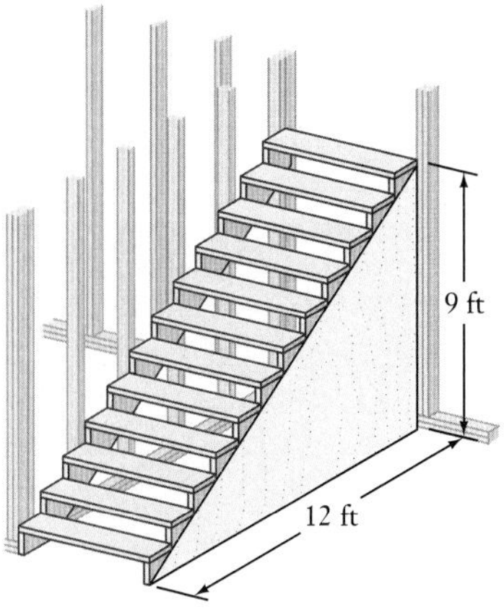

FIGURE 10.14

EXAMPLE 4 City Park

A one-block city park is bounded by two sets of parallel streets (Figure 10.15). The streets along the block are each 55 yards long and the perpendicular distance between the streets is 39 yards. How much sod should be purchased to cover the entire park in grass?

Solution The city park is a parallelogram with a base of 55 yards and height of 39 yards. The area of the parallelogram is

$$\text{area} = b \times h = 55 \text{ yd} \times 39 \text{ yd} = 2145 \text{ yd}^2$$

The city will need to purchase 2145 square yards of sod for the park.

▶ Now try Exercises 51–52.

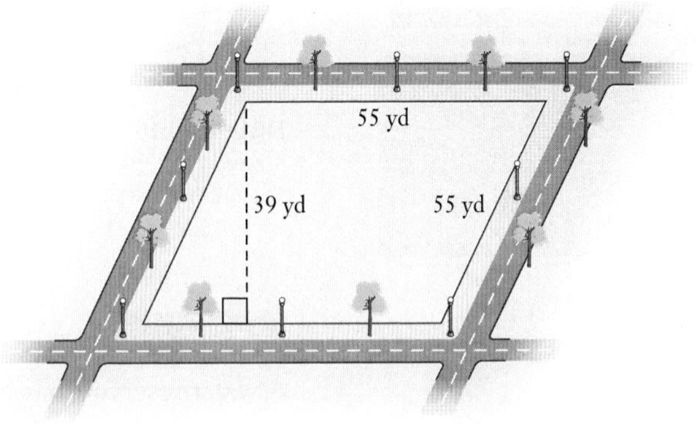

FIGURE 10.15

Three-Dimensional Geometry

Two of the most important properties of a three-dimensional object, such as a box or a sphere, are its *volume* and its *surface area*. Table 10.3 gives the names of several familiar three-dimensional objects, along with their volume and surface area formulas.

Some of the formulas in Table 10.3 are easily understood. For example, the volume formula for a box or a cube is just the familiar *length* × *width* × *height* formula. The surface area formula for a box or a cube is the result of adding the areas of the six (rectangular) faces of the object.

BY THE WAY

A solid with plane (flat) faces is called a *polyhedron*, from the Greek for "many faces."

TABLE 10.3 Three-Dimensional Objects

Object	Picture	Surface Area	Volume
Sphere	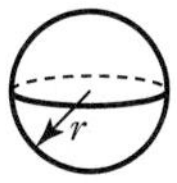	$4\pi r^2$	$\frac{4}{3}\pi r^3$
Cube	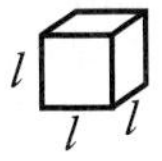	$6l^2$	l^3
Rectangular prism (box)	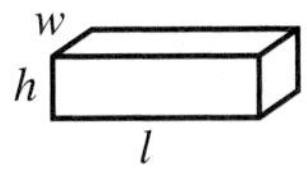	$2(lw + lh + wh)$	lwh
Right circular cylinder	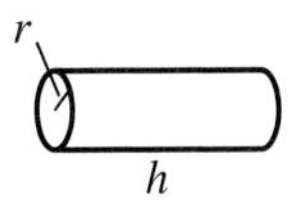	$2\pi r^2 + 2\pi rh$	$\pi r^2 h$

BY THE WAY

Many famous buildings use simple geometrical shapes, including the cylindrical leaning Tower of Pisa and the Pentagon.

The volume formula for a cylinder should make sense because it is just the area of the circular base (πr^2) multiplied by the height, h. The surface area formula for a cylinder has two parts. The area of each circular end of the cylinder is πr^2. The area of the curved surface of the cylinder is found by using a clever trick. As shown in Figure 10.16, when the cylinder is cut and unfolded, the curved surface becomes a rectangle. The length of the rectangle is the circumference of the original cylinder $(2\pi r)$, and the height of the rectangle is the height, h, of the cylinder. So the area of the curved surface of the cylinder is $2\pi rh$.

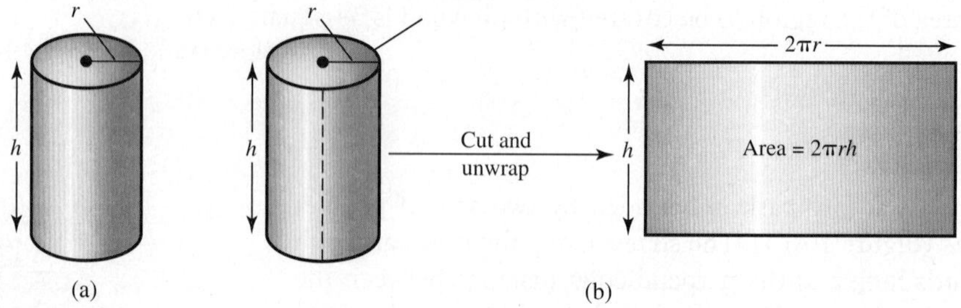

FIGURE 10.16 (a) A circular cylinder has a height h and a circular base with a radius r. (b) To find the area of the curved surface, imagine cutting the cylinder lengthwise and unfolding it to form a rectangle. The area of the rectangle equals the area of the curved surface.

▶ Now try Exercises 53–57.

EXAMPLE 5 Water Reservoir

A water reservoir has a rectangular base that measures 30 meters by 40 meters and vertical walls that are 15 meters high. At the beginning of the summer, the reservoir was filled to capacity. At the end of the summer, the water depth was 4 meters. How much water was used?

Solution The reservoir has the shape of a rectangular prism, so the volume of water in the reservoir is its length times its width times its depth. When the reservoir was filled at the beginning of the summer, the volume of water was

$$30 \text{ m} \times 40 \text{ m} \times 15 \text{ m} = 18{,}000 \text{ m}^3$$

At the end of the summer, the amount of water remaining was

$$30 \text{ m} \times 40 \text{ m} \times 4 \text{ m} = 4800 \text{ m}^3$$

Therefore, the amount of water used was $18{,}000 \text{ m}^3 - 4800 \text{ m}^3 = 13{,}200 \text{ m}^3$.

▶ Now try Exercises 58–59.

EXAMPLE 6 Comparing Volumes

FIGURE 10.17

Which holds more soup (Figure 10.17): a can with a diameter of 3 inches and a height of 4 inches or a can with a diameter of 4 inches and a height of 3 inches?

Solution Recall that radius $= \frac{1}{2}$ diameter. Soup cans have the shape of right circular cylinders, so the volumes of the two cans are

$$\textit{Can 1:} \quad V = \pi \times r^2 \times h = \pi \times (1.5 \text{ in})^2 \times 4 \text{ in} \approx 28.27 \text{ in}^3$$

$$\textit{Can 2:} \quad V = \pi \times r^2 \times h = \pi \times (2 \text{ in})^2 \times 3 \text{ in} \approx 37.70 \text{ in}^3$$

The second can, with the larger radius but smaller height, has the larger volume.

▶ Now try Exercises 60–61.

IN YOUR WORLD

Plato, Geometry, and Atlantis

Plato (427–347 B.C.E.) was one of the many ancient Greeks who sought to find geometric patterns in nature. He believed that the heavens must exhibit perfect geometric form and therefore argued that the Sun, Moon, planets, and stars must move in perfect circles—an idea that held sway until it was proven false by Johannes Kepler in the early 17th century.

Another of Plato's ideas about geometry and the universe involved the five *perfect solids*. Each of these perfect solids has the special property that all of its faces are the same regular polygon (Figure 10.18). Plato believed that four of the perfect solids represented the four elements that the Greeks thought made up the universe: earth, water, fire, and air. The dodecahedron, he believed, represented the universe as a whole.

Plato presented his ideas in several of his dialogues. In the dialogue *Timaeus*, in which he discussed the role of the perfect solids in the universe, he also invented a moralistic tale about a fictitious land called Atlantis. Interestingly, while Plato's ideas about the universe were abandoned long ago, millions of people today believe that Atlantis really existed. Plato's fiction, in the end, has more adherents than the ideas in which he firmly believed. Commenting on this irony, the popular author Isaac Asimov (1920–1992) wrote:

> *If there is a Valhalla for philosophers, Plato must be sitting there in endless chagrin, thinking of how many foolish thousands, in all the centuries since his time... who have never read his dialogues or absorbed a sentence of his serious teachings... believed with all their hearts in the reality of Atlantis.*

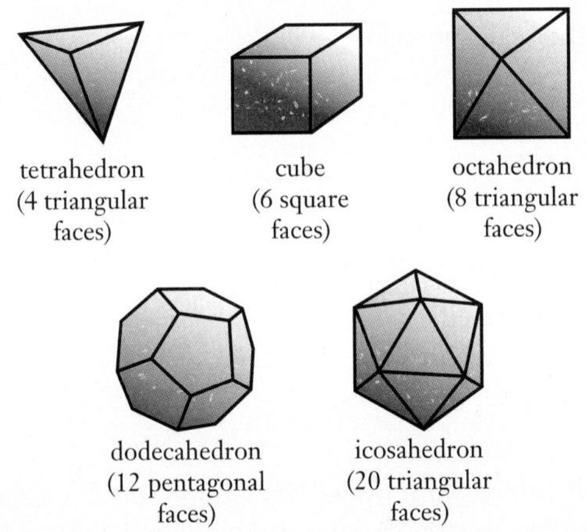

FIGURE 10.18 The five perfect solids.

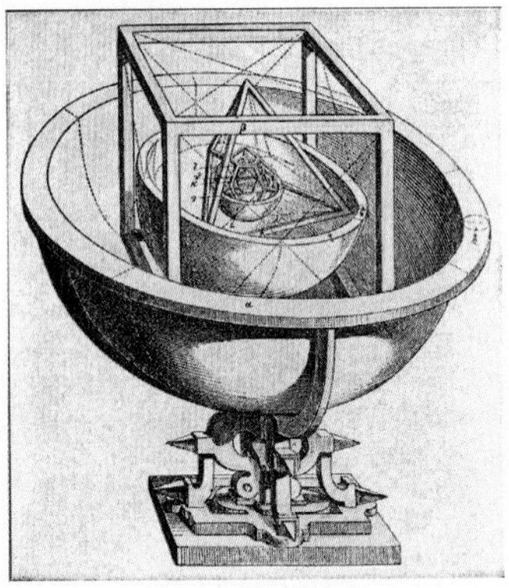

Scaling Laws

As we saw in Unit 3B, scaling is a process by which a real object is modeled by a similar object whose dimensions are proportionally larger or smaller. For example, an architect might make a scale model of an auditorium in which all lengths are smaller by a factor of 100. Or a biologist might make a scale model of a cell in which all lengths are larger by a factor of 10,000.

Suppose that we make an engineering model of a car with a *scale factor* of 10 (Figure 10.19). The actual car is 10 times as long, 10 times as wide, and 10 times as tall as the model car. How are the surface area and volume of the actual car related to the surface area and volume of the model?

Consider the area of the car roof, which is its length times its width. Because the length and width of the actual car roof are each 10 times their size in the model, the area of the actual car roof is

$$\begin{aligned}\text{actual roof area} &= \text{actual roof length} \times \text{actual roof width}\\ &= (10 \times \text{model roof length}) \times (10 \times \text{model roof width})\\ &= 10^2 \times \text{model roof length} \times \text{model roof width}\\ &= 10^2 \times \text{model roof area}\end{aligned}$$

That is, the actual roof area is greater than the model roof area by the *square* of the scale factor ($10^2 = 100$, in this case).

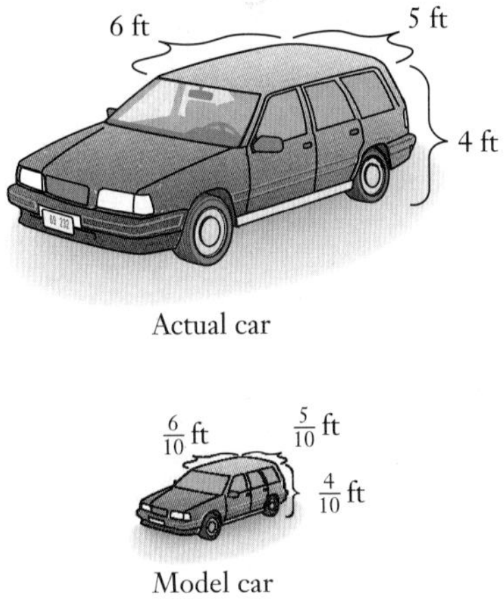

FIGURE 10.19 All dimensions of the model car are smaller than those of the actual car by a *scale factor* of 10. The *area* of any model car surface is therefore smaller than the actual car area by a factor of 10^2, and the model car *volume* is smaller than the actual car volume by a factor of 10^3.

We can do a similar calculation for the volume by considering, for example, the box-shaped (rectangular prism) passenger compartment.

$$\begin{aligned}\text{actual volume} &= \text{actual length} \times \text{actual width} \times \text{actual height}\\ &= (10 \times \text{model length}) \times (10 \times \text{model width}) \times (10 \times \text{model height})\\ &= 10^3 \times \text{model length} \times \text{model width} \times \text{model height}\\ &= 10^3 \times \text{model volume}\end{aligned}$$

The actual car volume is greater than the model car volume by the *cube* of the scale factor ($10^3 = 1000$, in this case).

Scaling Laws

- *Lengths* scale with the scale factor.
- *Areas* scale with the *square* of the scale factor.
- *Volumes* scale with the *cube* of the scale factor.

BY THE WAY

Scaling laws explain the appearance of many animals. Pressure, which determines how much weight an animal's body can support, is defined as the weight divided by the area bearing the weight. Because weight scales as the cube and area scales as the square (of length), making an animal larger increases the pressure on its joints. Nature therefore gives larger animals proportionally larger joints to bear the pressure. That is why, for example, elephants have relatively thicker legs than deer.

EXAMPLE 7 Doubling Your Size

Suppose that, magically, your size suddenly doubled—that is, your height, width, and depth doubled. For example, if you were 5 feet tall before, you now are 10 feet tall.

a. By what factor has your waist size increased?
b. How much more material will be required for your clothes?
c. By what factor has your weight changed?

Solution

a. Waist size is like a perimeter. Therefore, your waist size simply doubles, just like your other *linear* dimensions of height, width, and depth. If you had a 30-inch waist before, it is now a 60-inch waist.
b. Clothing covers surface *area* and therefore scales with the *square* of the scale factor. The scale factor by which you have grown is 2 (doubling), so your surface area grows by a factor of $2^2 = 4$. If your shirt used 2 square yards of material before, it now uses $4 \times 2 = 8$ square yards of material.
c. Your weight depends on your *volume*, which scales with the *cube* of the scale factor. Your new volume and new weight are therefore $2^3 = 8$ times their old values. If your old weight was 100 pounds, your new weight is $8 \times 100 = 800$ pounds.

▶ Now try Exercises 62–74.

The Surface-Area-to-Volume Ratio

Another important concept in scaling is the *relative* scaling of areas and volumes. We define the **surface-area-to-volume ratio** for any object as its surface area divided by its volume:

$$\text{surface-area-to-volume ratio} = \frac{\text{surface area}}{\text{volume}}$$

Because surface area scales with the square of the scale factor and volume scales with the cube of the scale factor, the surface-area-to-volume ratio must scale with the reciprocal of the scale factor:

$$\text{scaling of surface-area-to-volume ratio} = \frac{(\text{scale factor})^2}{(\text{scale factor})^3} = \frac{1}{\text{scale factor}}$$

Therefore, when an object is "scaled up," its surface-area-to-volume ratio *decreases*. When an object is "scaled down," its surface-area-to-volume ratio *increases*.

Surface-Area-to-Volume Ratio

- Larger objects have *smaller* surface-area-to-volume ratios than similarly proportioned small objects.
- Smaller objects have *larger* surface-area-to-volume ratios than similarly proportioned large objects.

EXAMPLE 8 Chilled Drink

Suppose you have a few ice cubes and you want to cool your drink quickly. Should you crush the ice before you put it into your drink? Why or why not?

Solution A drink is cooled by contact between the liquid and the ice surface. Thus, the greater the surface area of the ice, the more rapidly the drink will cool. Because smaller objects have *larger* surface-area-to-volume ratios, the crushed ice will have more total surface area than the same volume of ice cubes. The crushed ice therefore will cool the drink more quickly.

▶ Now try Exercises 75–80.

(The) knowledge at which geometry aims is of the eternal, and not of the perishing and transient.

—Plato

Quick Quiz 10A

Choose the best answer to each of the following questions. Explain your reasoning with one or more complete sentences.

1. Any two points can be used to define
   a. a line. b. a plane. c. an angle.
2. An example of a two-dimensional object is
   a. a straight line. b. the surface of a wall. c. a chair.
3. An obtuse angle is
   a. less than 90°. b. exactly 90°. c. more than 90°.
4. A *regular* polygon always has
   a. four sides. b. at least four sides. c. all sides the same length.
5. A *right* triangle always has
   a. three equal-length sides.
   b. one 90° angle.
   c. two 90° angles.
6. The *circumference* of a circle of radius r is
   a. $2\pi r$. b. πr^2. c. $2\pi r^2$.
7. The *volume* of a sphere of radius r is
   a. πr^2. b. $4\pi r^2$. c. $\frac{4}{3}\pi r^3$.
8. If you double the lengths of all the sides of a square, the area of the square
   a. doubles.
   b. triples.
   c. quadruples.
9. If you triple the radius of a sphere, the volume of the sphere goes up by a factor of
   a. 3. b. 3^2. c. 3^3.
10. Suppose you cut a large stone block into four equal-sized pieces. The four pieces combined are *not* different from the original block in
    a. total volume of stone.
    b. total surface area of stone.
    c. surface-area-to-volume ratio.

Exercises 10A

REVIEW QUESTIONS

1. What do we mean by *Euclidean geometry*?
2. Give a geometric definition for each of the following: *point*, *line*, *line segment*, *plane*, and *space*. Give an example of an everyday object that approximates each of these geometrical objects.
3. What do we mean by *dimension*? How is dimension related to the number of coordinates needed to locate a point?

4. Define an *angle* in geometric terms. What is the *vertex*? What do we mean when we say an angle subtends some portion of a circle? Distinguish among *right angles*, *straight angles*, *acute angles*, and *obtuse angles*.

5. What is *plane geometry*? What does it mean for lines to be *perpendicular* or *parallel* in a plane?

6. What is a *polygon*? How do we measure the *perimeter* of a polygon? Describe how we calculate the areas of a few simple polygons.

7. What are the formulas for the circumference and area of a circle?

8. Describe how we calculate the volumes and surface areas of a few simple three-dimensional objects.

9. What are the scaling laws for area and volume? Explain.

10. What is the *surface-area-to-volume ratio*? How does this ratio change if we make an object bigger? Smaller?

DOES IT MAKE SENSE?

Decide whether each of the following statements makes sense (or is clearly true) or does not make sense (or is clearly false). Explain your reasoning.

11. The two highways look like straight lines and they intersect twice.

12. The city park is triangular in shape and has two right angles.

13. My bedroom is a rectangular prism that measures 12 feet by 10 feet by 8 feet.

14. Kara walked around the circular pond to a point on the opposite side, and Jamie swam at the same speed directly across the pond to the same point. Jamie must have arrived before Kara.

15. This basketball is shaped like a right circular cylinder.

16. By building a fence across my rectangular backyard, I can create two triangular yards.

BASIC SKILLS & CONCEPTS

17–22: Angles and Circles. Find the degree measure of the angle that subtends each of the following parts of a circle.

17. $\frac{1}{2}$ circle

18. $\frac{1}{6}$ circle

19. $\frac{1}{5}$ circle

20. $\frac{2}{3}$ circle

21. $\frac{2}{9}$ circle

22. $\frac{5}{6}$ circle

23–30: Fractions of Circles. Find the fraction of a circle subtended by each angle.

23. 45°

24. 6°

25. 120°

26. 90°

27. 36°

28. 60°

29. 300°

30. 225°

31–36: Circle Practice. Find the circumference and area of each circle. Round answers to the nearest tenth.

31. A circle with a radius of 6 meters

32. A circle with a radius of 4 kilometers

33. A circle with a diameter of 23 feet

34. A circle with a radius of 8.2 meters

35. A circle with a diameter of 40 millimeters

36. A circle with a diameter of 1.5 kilometers

37–42: Perimeters and Areas. Use Table 10.2 to find the perimeter and area of each figure. Round answers to the nearest tenth.

37. A square state park with sides of length 8 miles

38. A rectangular envelope with a length of 4 inches and a width of 8 inches

39. A parallelogram with sides of length 8 feet and 30 feet and a distance of 4 feet between the 30-foot sides

40. A square with sides of length 5.3 centimeters

41. A rectangular postage stamp with a length of 2.2 centimeters and a width of 2.0 centimeters

42. A parallelogram with sides of length 4.5 feet and 12.2 feet and a distance of 3.6 feet between the 12.2-foot sides

43–46: Triangle Geometry. Find the perimeter and area of each triangle.

43.

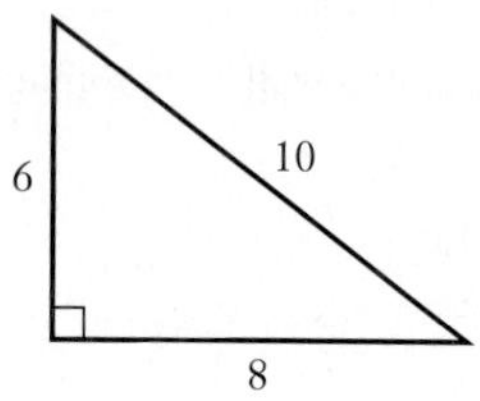

44.

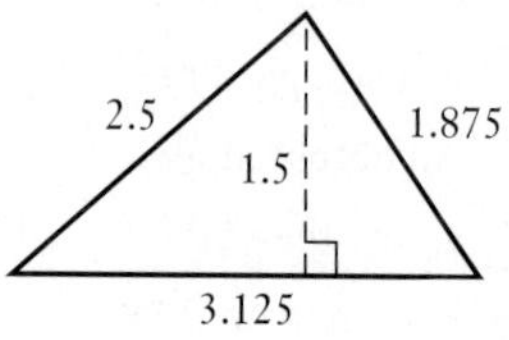

45.

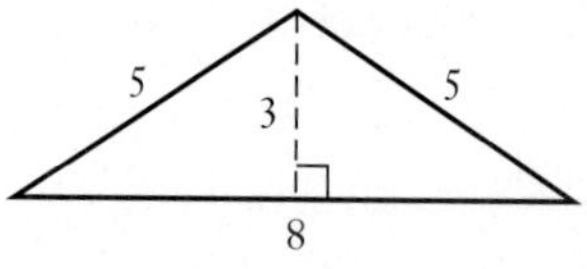

46.

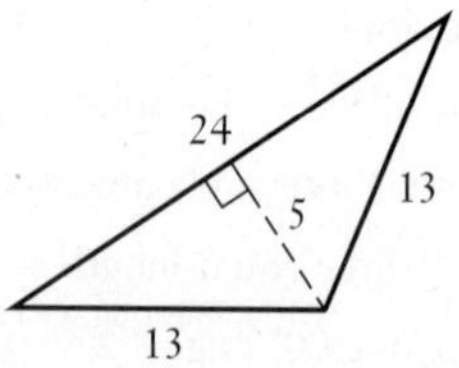

47. **Window Space.** A picture window has a length of 8 feet and a height of 6 feet, with a semicircular cap on each end (see Figure 10.20). How much metal trim is needed for the perimeter of the entire window, and how much glass is needed for the opening of the window?

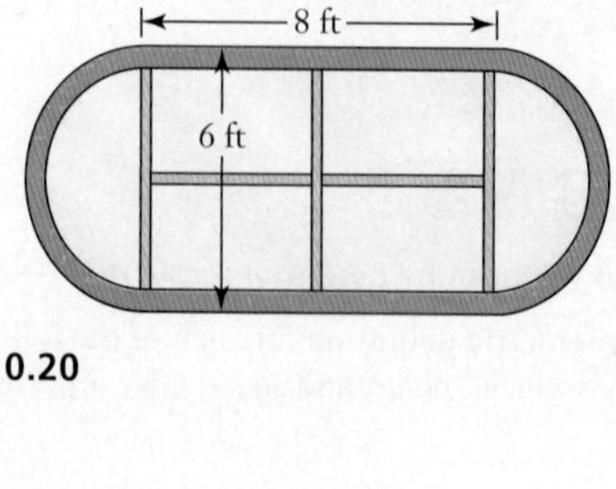

FIGURE 10.20

48. **A Running Track.** A running track has straight legs of length 100 yards that are 60 yards apart (see Figure 10.21). What is the length of the (inner lane of the) track, including the semicircular legs, and what is the area of the infield of the track?

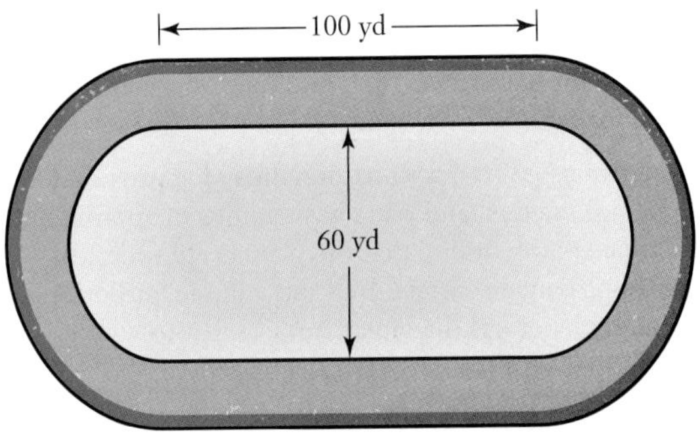

FIGURE 10.21

49. **Building Stairs.** Refer to Figure 10.14, showing the region to be covered with plywood under a set of stairs. Suppose that the stairs rise at a steeper angle and are 11 feet tall. What is the area of the region to be covered in that case?

50. **No Calculation Required.** The end views of two different barns are shown in Figure 10.22. Without calculating, decide which end has the greater area. Explain how you know.

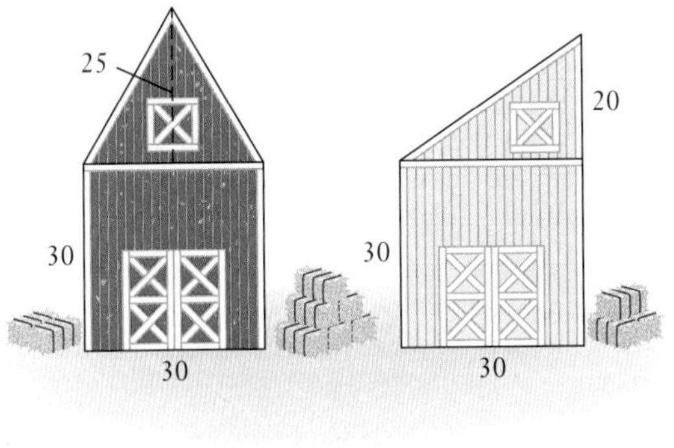

FIGURE 10.22

51. **Parking Lot.** A parking lot is shaped like a parallelogram and bounded on four sides by streets, as shown in Figure 10.23. How much asphalt (in square yards) is needed to pave the parking lot?

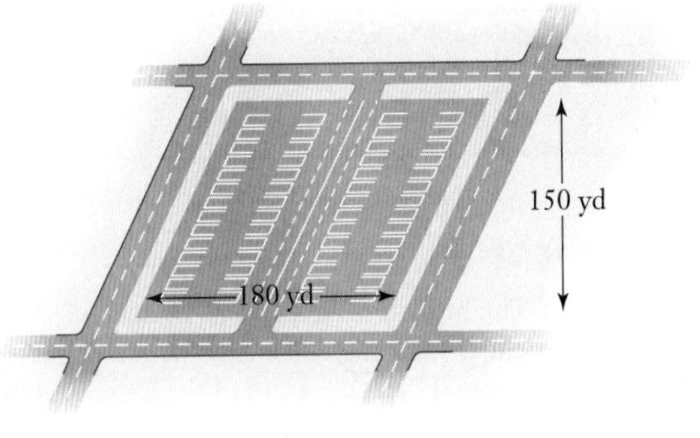

FIGURE 10.23

52. **City Park.** Figure 10.24 shows a city park in the shape of a parallelogram with a rectangular playground in the center. If all but the playground is covered with grass, what area is covered with grass?

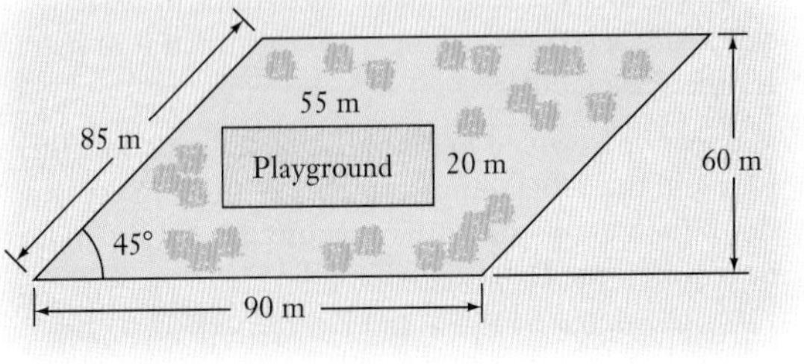

FIGURE 10.24

53–57: Three-Dimensional Objects. Use the formulas in Table 10.3 to answer the following questions.

53. A competition swimming pool is 50 meters long, 30 meters wide, and 2.5 meters deep. How much water does the pool hold?

54. A warehouse has a floor that measures 30 meters by 60 meters, with a ceiling 9 meters high. How much air does it hold, in cubic meters? In liters? (*Hint*: 1 cubic meter = 1000 liters.)

55. An air duct in a stadium has a circular cross section with a radius of 18 inches and a length of 40 feet. What is the volume of the duct, and how much paint (in square feet) is needed to paint the exterior of the duct?

56. A grain storage building is a hemispherical shell with a radius of 25 meters. What is the volume of the building? How much paint is needed to cover the exterior of the building?

57. Three tennis balls fit perfectly when stacked in a cylindrical can. Which is greater: the circumference of the can or the height of the can? Explain your reasoning.

58. **Water Canal.** A water canal has a rectangular cross section 3 meters wide and 2 meters deep. How much water is contained in a 30-meter length of the canal? How much water does this segment of the canal hold after 60% of the water has evaporated?

59. **Water Reservoir.** The water reservoir for a city is shaped like a rectangular prism 250 meters long, 60 meters wide, and 12 meters deep. At the end of the day, the reservoir is 70% full. How much water must be added overnight to fill the reservoir?

60. **Oil Drums.** Which holds more: an oil drum with a radius of 2 feet and a height of 3 feet or an oil drum with a radius of 3 feet and a height of 2 feet?

61. **Tree Volumes.** Is there more wood in a 40-foot-high tree trunk with a radius of 2.5 feet or in a 50-foot-high tree trunk with a radius of 2.1 feet? Assume that the tree trunks can be regarded as right circular cylinders.

62–64: Architectural Model. Suppose you build an architectural model of a new concert hall using a scale factor of 30.

62. How will the height of the actual concert hall compare to the height of the scale model?

63. How will the surface area of the actual concert hall compare to the surface area of the scale model?

64. How will the volume of the actual concert hall compare to the volume of the scale model?

65–67: Architectural Model: Suppose you build an architectural model of a new office complex using a scale factor of 80.

65. How will the height of the actual office complex compare to the height of the scale model?

66. How will the amount of paint needed for the exterior of the actual office complex compare to the amount of paint needed for the scale model?

67. Suppose you wanted to fill both the scale model office complex and the actual office complex with marbles. How many times the number of marbles required to fill the model would be required to fill the actual building?

68–71: Quadrupling Your Size. Suppose you magically quadrupled in size—that is, your height, width, and depth increased by a factor of 4.

68. By what factor has your arm length increased?

69. By what factor has your waist size increased?

70. By what factor has the amount of material required for your clothes increased?

71. By what factor has your weight increased?

72–74: Comparing People. Consider a person named Sam, who is 20% taller than you but proportioned in exactly the same way. (That is, Sam looks like a larger version of you.)

72. How tall are you? How tall is Sam?

73. What size is your waist? What size is Sam's waist?

74. How much do you weigh? How much does Sam weigh?

75–76: Squirrels or People? Squirrels and humans are both mammals that stay warm through metabolism that takes place in the body *volume*. Mammals must constantly generate internal heat to replace the heat that they lose through the *surface area* of their skin.

75. In general terms, how does the surface-area-to-volume ratio of a squirrel compare to that of a human being?

76. Which animal must maintain a higher rate of metabolism to replace the heat lost through the skin: squirrels or humans? Based on your answer, which animal would you expect to eat more food in proportion to its body weight each day? Explain.

77–78: Earth and Moon. Both the Moon and Earth are thought to have had similar internal temperatures at the time they formed about 4.5 billion years ago. Both worlds gradually lose this internal heat to space as the heat passes out through their surfaces.

77. The diameter of Earth is about four times the diameter of the Moon (see Figure 10.25). How does the surface-area-to-volume ratio of Earth compare to that of the Moon?

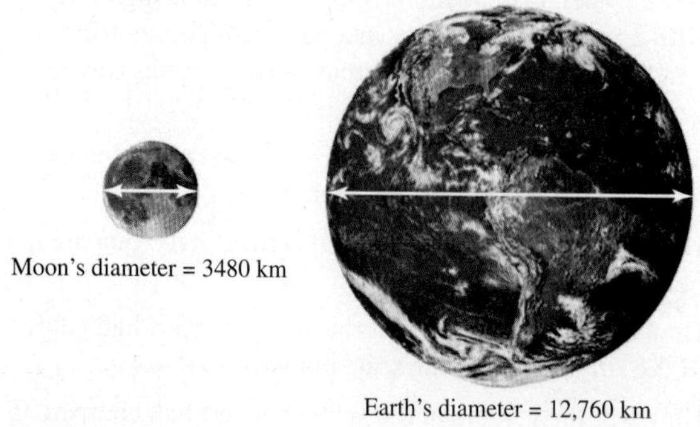

FIGURE 10.25

78. Based on your answer to Exercise 77, which would you expect to have a hotter interior today: Earth or the Moon? Why? Use your answer to explain why Earth remains volcanically active today, while the Moon has no active volcanoes.

79. **Comparing Balls.** Consider a softball with a radius of approximately 2 inches and a bowling ball with a radius of approximately 6 inches. Compute the approximate surface area and volume for both balls. Then find the surface-area-to-volume ratio for both balls. Which ball has the larger ratio?

80. **Comparing Planets.** Earth has a radius of approximately 6400 kilometers, and Mars has a radius of approximately 3400 kilometers (assuming that the planets are spherical). Compute the approximate surface area and volume for both planets. Which planet has the larger surface-area-to-volume ratio?

FURTHER APPLICATIONS

81. **Dimension.** Examine a closed book.

a. How many dimensions are needed to describe the book? Explain.

b. How many dimensions describe the surface (cover) of the book? Explain.

c. How many dimensions describe an edge of the book? Explain.

d. Describe some aspect of the book that represents zero dimensions.

82. **Perpendicular and Parallel.** Suppose you mark a single point on a line that lies in a fixed plane. Can you draw any other lines in that plane that pass through the point and are *perpendicular* to the original line? Can you draw any other lines that pass through the point and are *parallel* to the original line? Explain.

83. **Perpendicular and Parallel.** Suppose you draw two parallel lines in a plane. If a third line is perpendicular to one of the two parallel lines, is it necessarily perpendicular to the other line? Explain.

84. **Backyard.** Figure 10.26 shows the layout of a backyard that is to be seeded with grass except for the patio and flower garden. What is the area of the region that is to be seeded with grass?

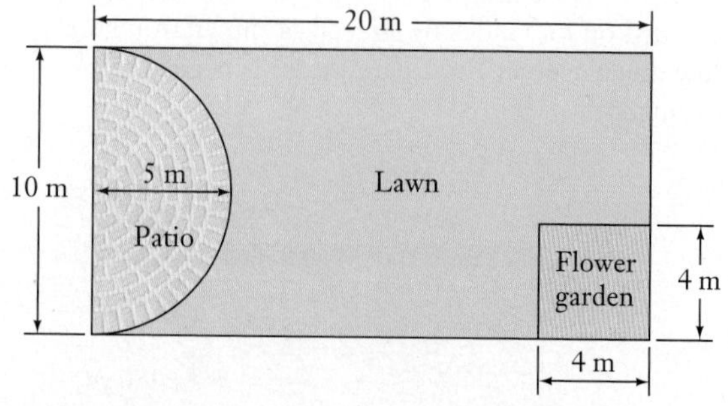

FIGURE 10.26

85. **Human Lung.** The human lung has approximately 300 million nearly spherical air sacs (alveoli), each with a diameter of about $\frac{1}{3}$ millimeter. The key feature of the air sacs is their surface area, because on their surfaces gas is exchanged between the bloodstream and the air.

a. What is the total surface area of the air sacs? What is the total volume of the air sacs?

b. Suppose a single sphere were made that had the same volume as the total volume of the air sacs. What would be the radius and surface area of such a sphere? How would this surface area compare to that of the air sacs?

c. If a single sphere had the same surface area as the total surface area of the air sacs, what would be its radius? Based on your results, comment on the design of the human lung.

86. **Automobile Engine Capacity.** The size of a car engine is often stated as the total volume of its cylinders.

a. American car manufacturers often state engine sizes in cubic inches. Suppose that a six-cylinder car has cylinders with a radius of 2.22 inches and a height of 3.25 inches. What is the engine size?

b. Foreign car manufacturers often state engine sizes in liters. Compare the engine size in part (a) to that of a foreign car with a 2.2-liter engine.

c. Look up the number of cylinders and the engine size for your car (if you don't own a car, choose a car that you'd like to own). Estimate the dimensions (radius and height) of the cylinders in your car. Explain your work.

87. **The Chunnel.** The English Channel Tunnel, or "Chunnel," runs a distance of 50.45 kilometers from Dover, England, to Calais, France. The Chunnel actually consists of three individual, adjacent tunnels, each shaped like a half-cylinder with a radius of 4 meters (the height of the tunnel). How much earth (volume) was removed to build the entire Chunnel?

IN YOUR WORLD

88. **Geometry in the News.** Find a recent news report describing a project that requires geometric ideas or techniques. Explain how geometry is used in the project.

89. **Circles and Polygons.** Describe at least three instances of circles or polygons that play an important role in your daily life.

90. **Three-Dimensional Objects.** Describe at least three instances of three-dimensional objects that play an important role in your daily life.

91. **The Geometry of Ancient Cultures.** Research the use of geometry in an ancient culture of your choice. Some possible areas of focus: (1) study the use of geometry in ancient Chinese art and architecture; (2) investigate the geometry and purpose of Stonehenge; (3) compare and contrast the geometry of the Egyptian pyramids and those of Central America; (4) study the geometry and possible astronomical orientations of Anasazi buildings and communities; or (5) research the use of geometry in the ancient African empire of Aksum (in modern-day Ethiopia).

92. **Surveying and GIS.** Surveying is one of the oldest and most practical applications of geometry. A modern technology for surveying is called Geographical Information Systems, or GIS. Use the Web to learn about GIS. In two pages or less, describe how GIS works, along with a few of its applications.

93. **Platonic Solids.** Why are there five and only five perfect, or Platonic, solids? What special significance did these objects have to the Greeks? Research these questions, and write short answers to them.

UNIT 10B Problem Solving with Geometry

In Unit 10A, we surveyed the geometrical properties of two- and three-dimensional objects. In this unit, we'll investigate additional geometrical ideas that can be used to solve many practical problems.

Uses of Angles

Many geometrical problems require working with angles. For example, we use angles in navigation and architectural design, for measuring the steepness of mountains and the orientation of roads, and for characterizing triangles. In Unit 10A, we worked with angle measurement in degrees, in which a full circle represents 360°. We need greater precision for many geometrical problems, so let's briefly review how we measure angles to a fraction of a degree.

We can write fractions of a degree with decimals (for example, 45.23°), but it is more common to subdivide each degree into 60 *minutes of arc* and each minute into 60 *seconds of arc* (see Figure 10.27). We use the symbols ′ and ″ for minutes and seconds of arc, respectively. For example, we read 30°33′31″ as "30 degrees, 33 minutes, and 31 seconds."

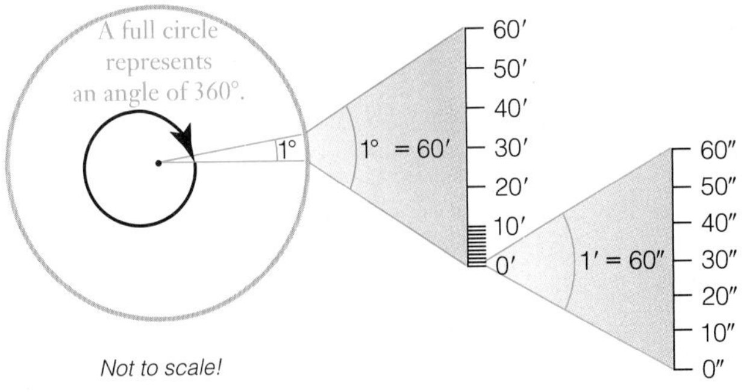

FIGURE 10.27 Each degree is subdivided into 60 minutes of arc, and each minute is subdivided into 60 seconds of arc.

EXAMPLE 1 Fractional Degrees

a. Convert 3.6° into degrees, minutes, and seconds of arc.
b. Convert 30°33′31″ into decimal form.

Solution

a. Because $3.6° = 3° + 0.6°$, we first convert 0.6° into minutes of arc:

$$0.6° \times \frac{60'}{1°} = 36'$$

We have found that $3.6° = 3° + 36' = 3°36'$.

b. Remember that 1 minute of arc is $\left(\frac{1}{60}\right)°$, and 1 second of arc is $\left(\frac{1}{60}\right)$ of a minute of arc, or $\left(\frac{1}{60 \times 60}\right)°$. Therefore,

$$30°33'31'' = 30° + 33' + 31'' = 30° + \left(\frac{33}{60}\right)° + \left(\frac{31}{60 \times 60}\right)°$$
$$\approx 30° + 0.55° + 0.00861° = 30.55861°$$

▶ Now try Exercises 15–28.

Latitude and Longitude

One common use of angle measures is to locate positions on Earth (Figure 10.28). **Latitude** measures positions north or south of the equator, which is defined to have latitude 0°. Locations in the Northern Hemisphere have latitudes denoted N (for north) and locations in the Southern Hemisphere have latitudes denoted S (for south). For example, the North Pole has latitude 90°N and the South Pole has latitude 90°S. Note that lines of latitude (also called *parallels* of latitude) are actually circles running parallel to the equator.

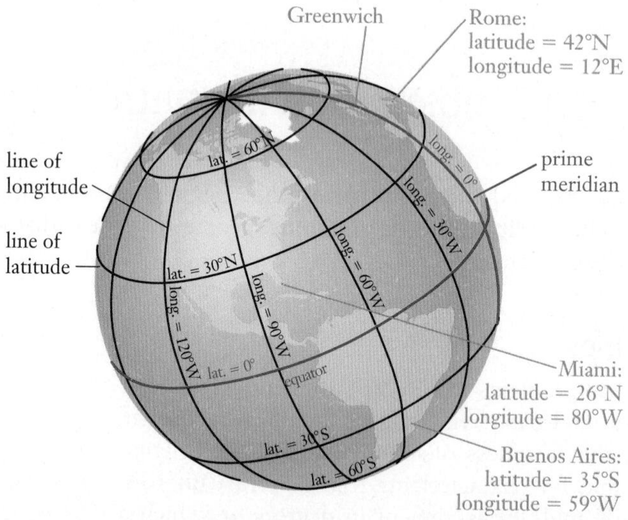

FIGURE 10.28 We can locate any place on the Earth's surface by its latitude and longitude.

BY THE WAY

The line of longitude for Greenwich is defined more specifically as a line emerging from the door of the Old Royal Observatory (see the photo). This line was adopted as the prime meridian at an international conference in 1884.

Longitude measures east-west position, so lines of longitude (also called *meridians* of longitude) are semicircles extending from the North Pole to the South Pole. The line of longitude passing through Greenwich, England, is defined to be longitude 0° and is called the **prime meridian**. Longitudes are usually given with an angle less than (or equal to) 180°, so a location less than halfway around the world to the east of the prime meridian has a longitude designation of E (for east) and a location to the west has a longitude designation of W (for west).

Stating a latitude and a longitude pinpoints a location on Earth. Figure 10.28 shows latitudes and longitudes for Miami, Rome, and Buenos Aires.

EXAMPLE 2 Latitude and Longitude

Answer each of the following questions, and explain your answers clearly.

a. Suppose you could drill from Miami straight through the center of Earth and continue in a straight line to the other side of Earth. At what latitude and longitude would you emerge?

b. Perhaps you've heard that if you dug a straight hole from the United States through the center of Earth, you'd come out in China. Is this true?

c. Suppose you travel through 1° of latitude either north or south. How far have you traveled? (*Hint:* The circumference of Earth is about 25,000 miles.)

d. Suppose you travel through 1° of longitude either east or west. How far have you traveled?

Solution

a. The point on Earth directly opposite Miami must have the opposite latitude (26° S rather than 26° N) and must be 180° away in longitude. Let's consider going 180° eastward from Miami's longitude of 80° W. The first 80° eastward would take us to the prime meridian; then we would continue to longitude 100° E to be a full 180° away from Miami. Therefore, the position of the point on Earth opposite Miami is latitude 26° S and longitude 100° E (which is in the Indian Ocean).

b. It is *not* true. The United States is in the Northern Hemisphere, so the points on the globe opposite the United States must be in the Southern Hemisphere. China is also in the Northern Hemisphere, so it cannot be opposite the United States.

c. If you study Figure 10.28 (or a globe), you'll see that lines of latitude are all parallel to each other. Therefore, every degree of latitude represents the same north or south distance. If you were to traverse the Earth's 25,000-mile circumference, you would travel through 360° of latitude. Therefore, each degree of latitude represents

$$\frac{25{,}000 \text{ mi}}{360^\circ} \approx 69.4 \text{ mi per degree}$$

Traveling through 1° of latitude north or south means traveling a distance of approximately 70 miles.

d. If you study Figure 10.28 (or a globe), you'll see that lines of longitude are *not* parallel to each other. Instead, they are much closer together near the poles than near the equator. Therefore, we cannot answer the question without also knowing the latitude. If you are at the equator, 1° of longitude represents about 70 miles [by the same calculation used in part (c)]; however, 1° of longitude represents shorter distances as you head north or south. ▶ Now try Exercises 29–36.

Angular Size and Distance

If you hold a coin in front of one eye, it can block your entire field of view. But as you move it farther away, it appears to get smaller and it blocks less of your view. The true size of the coin does not change, of course. Instead, what changes is its **angular size**—the angle that it covers as seen from your eye. Figure 10.29 shows this effect.

To learn how angular size and physical size are related, we can use a little trick. Imagine that, as shown in Figure 10.30, there is a vertical circle all the way around your eye, with a radius equal to the distance from your eye to the coin. Notice that, because the coin's angular size is small, its physical size (diameter) is approximately equal to the arc length of the small piece of the circle that it subtends. In other words, the ratio of

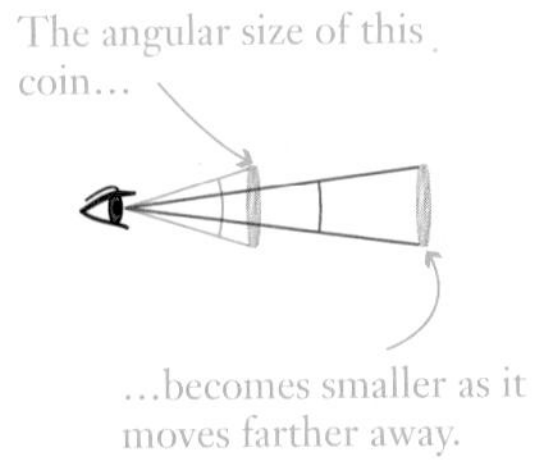
The angular size of this coin...
...becomes smaller as it moves farther away.

FIGURE 10.29 The angular size of an object depends on both its physical size and its distance from the eye.

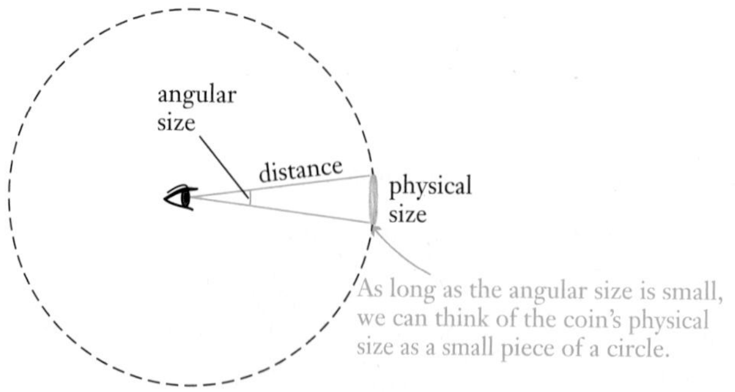
angular
size
distance
physical
size
As long as the angular size is small,
we can think of the coin's physical
size as a small piece of a circle.

FIGURE 10.30 Approximating an object's size as a piece of a circle allows us to find a formula relating angular size, physical size, and distance.

the coin's angular size to the full 360° circle approximately equals the ratio of its physical size to the circle's circumference—which is $2\pi \times (\text{distance})$. In equation form, we write this equivalence as

$$\frac{\text{angular size}}{360^\circ} = \frac{\text{physical size}}{2\pi \times \text{distance}}$$

Multiplying both sides by 360° and rearranging, we have a formula that allows us to determine angular size when we know physical size and distance:

$$\text{angular size} = \text{physical size} \times \frac{360^\circ}{2\pi \times \text{distance}}$$

This formula is sometimes called the *small-angle formula*, because it is valid only when the angular size is small (less than a few degrees).

The Small-Angle Formula for Angular Size, Physical Size, and Distance

The farther away an object is located from you, the smaller it will appear to you: Its angular size depends on its distance from you. As long as an object's angular size is less than a few degrees, the following formula relates its angular size, physical size, and distance:

$$\text{angular size} = \text{physical size} \times \frac{360^\circ}{2\pi \times \text{distance}}$$

EXAMPLE 3 Angular Size and Distance

a. A quarter is about 1 inch in diameter. Approximately how big will its angular size be if you hold it 1 yard (36 inches) from your eye?

b. The angular diameter of the Moon as seen from Earth is approximately 0.5°, and the Moon is approximately 380,000 kilometers from Earth. What is the real diameter of the Moon?

Solution

a. We use the small-angle formula with 1 inch as the quarter's physical size (diameter) and 36 inches as its distance:

$$\text{angular size} = (1 \text{ in}) \times \frac{360^\circ}{2\pi \times (36 \text{ in})} \approx 1.6^\circ$$

The angular size (diameter) of a quarter at a distance of 1 yard is about 1.6°.

b. In this case, we are asked to find the Moon's physical size (diameter) from its angular size and distance, so we must solve the small-angle formula for physical size. You should confirm that the formula becomes

$$\text{physical size} = \text{angular size} \times \frac{2\pi \times \text{distance}}{360°}$$

Now we substitute 380,000 kilometers for the distance and 0.5° for the angular size:

$$\text{physical size} = 0.5° \times \frac{2\pi \times 380{,}000 \text{ km}}{360°} \approx 3300 \text{ km}$$

The Moon's real diameter is about 3300 kilometers. ▶ Now try Exercises 37–40.

BY THE WAY

The Sun is about 400 times as large as the Moon in diameter, but it is also about 400 times as far away from Earth. As a result, the Moon and Sun both have about the same angular size, which is what makes total solar eclipses possible. The last total solar eclipse in the United States occurred on August 21, 2017, and the next will occur on April 8, 2024. You can find a map of the eclipse path with the free app called Totality (created by one of the authors of this text).

Pitch, Grade, and Slope

Consider a road that rises uniformly 2 feet in the vertical direction for every 20 feet in the horizontal direction (Figure 10.31). The road makes an angle with the horizontal that we could measure in degrees. However, it is more common to say that the road has a **pitch** of 2 in 20, or 1 in 10. It is also common to describe the rise of the road in terms of its **slope** (see Chapter 9). In this case, the road's slope, or *rise over run*, is $2/20 = 0.1$. If we express the slope as a percentage, it's called a **grade**. The grade of the road in this case is $0.1 = 10\%$.

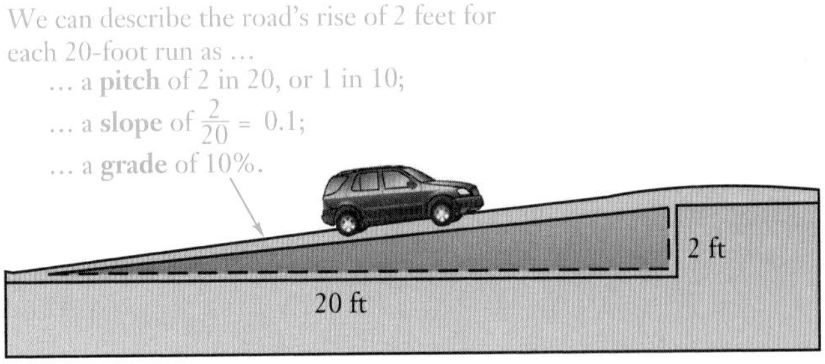
We can describe the road's rise of 2 feet for each 20-foot run as ...
... a pitch of 2 in 20, or 1 in 10;
... a slope of 2/20 = 0.1;
... a grade of 10%.
2 ft
20 ft

FIGURE 10.31

BY THE WAY

On Earth's round surface, the shortest distance between two places is not a straight line on a map but rather a piece of a *great circle*—a circle whose center is at the center of the Earth. That is why airplanes usually try to fly great circle routes. For example, Philadelphia and Beijing, China, are both at 40°N latitude, but the shortest path between them is *not* along the 40° line of latitude. Instead it is a path going almost directly over the North Pole.

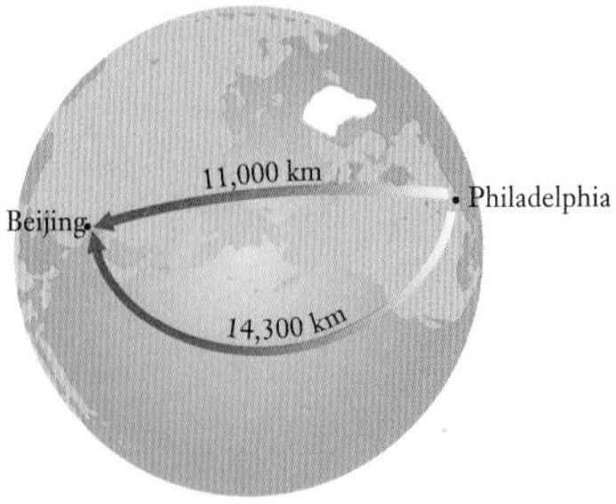
11,000 km
Philadelphia
Beijing
14,300 km

EXAMPLE 4 How Steep?

a. Suppose a road has a 100% grade. What is its slope? What is its pitch? What angle does it make with the horizontal?
b. Which is steeper: a road with an 8% grade or a road with a pitch of 1 in 9?
c. Which is steeper: a roof with a pitch of 2 in 12 or a roof with a pitch of 3 in 15?

Solution

a. A 100% grade means a slope of 1 and a pitch of 1 in 1. If you look at a line with a slope of 1 on a graph, you'll see that it makes an angle of 45° with the horizontal.
b. A road with a pitch of 1 in 9 has a slope of $1/9 = 0.11$, or 11%. Therefore, this road is steeper than a road with an 8% grade.
c. The 2 in 12 roof has a slope of $2/12 = 0.167$, while the 3 in 15 roof has a slope of $3/15 = 0.20$. The second roof is steeper. ▶ Now try Exercises 41–50.

Further Uses of Triangles

In Unit 10A, we used the area formula for a triangle to solve practical problems. Here we investigate two other ways in which we can use triangles in modeling.

Using the Pythagorean Theorem

The Pythagorean theorem is one of the most famous theorems in mathematics. We already saw a proof of the theorem in Unit 1D and put it to use in Unit 2C (see Examples 5 and 6). Because it is so important—and because it is rooted in geometry—we will now explore a few other uses of this theorem.

The Pythagorean Theorem

The Pythagorean theorem applies only to right triangles (those with one 90° angle). For a right triangle with side lengths *a*, *b*, and *c*, in which *c* is the longest side (or *hypotenuse*), the Pythagorean theorem states that

$$a^2 + b^2 = c^2$$

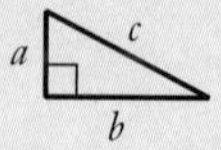

EXAMPLE 5 Distance Measurements

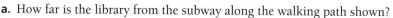

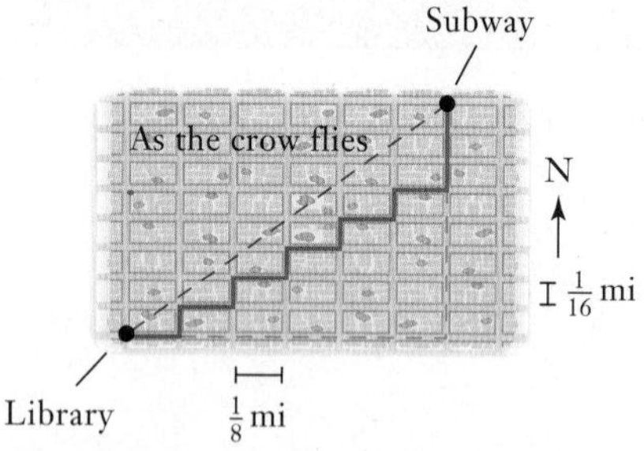

FIGURE 10.32 A map of several city blocks, showing the locations of the library and the subway.

Consider the map in Figure 10.32, showing several city streets in a rectangular grid. The lengths of the individual city blocks are $\frac{1}{8}$ of a mile in the east-west direction and $\frac{1}{16}$ of a mile in the north-south direction.

a. How far is the library from the subway along the walking path shown?

b. How far is the library from the subway "as the crow flies" (that is, along a straight diagonal path)?

Solution

a. If you follow the walking path, you'll see that it goes 6 blocks east and 8 blocks north. The total distance along this path is

$$\underbrace{\left(6 \times \frac{1}{8}\text{ mi}\right)}_{\substack{\text{6 east-west blocks}\\ \text{each 1/8 mi long}}} + \underbrace{\left(8 \times \frac{1}{16}\text{ mi}\right)}_{\substack{\text{north-south blocks}\\ \text{each 1/16 mi long}}} = \frac{3}{4}\text{ mi} + \frac{1}{2}\text{ mi} = 1\frac{1}{4}\text{ mi}$$

b. As shown in the figure, the distance "as the crow flies" is the hypotenuse, or side *c*, of a right triangle. The horizontal and vertical sides of this triangle have lengths $\frac{3}{4} = 0.75$ mile and $\frac{1}{2} = 0.5$ mile, so we can set these as *a* and *b* in the Pythagorean formula, respectively. We can then find the hypotenuse length as follows:

Start with the Pythagorean theorem, with c^2 on the left side:	$c^2 = a^2 + b^2$
Solve for *c* by taking the square root of both sides:	$c = \sqrt{a^2 + b^2}$
Substitute the given values:	$c = \sqrt{(0.75\text{ mi})^2 + (0.5\text{ mi})^2} \approx 0.90\text{ mi}$

We've found that the straight-line distance between the library and the subway is about 0.9 mile, which is considerably shorter than the 1.25-mile distance along the walking path found in part (a).

▶ **Now try Exercises 51–56.**

Think About It Are there other routes along the streets that have the same length (1.25 miles) as the route shown with the heavy red line in Figure 10.32? If so, how many? Are there any *shorter* routes that follow the streets? Explain.

EXAMPLE 6 Lot Size

Find the area, in acres, of the triangular mountain lot shown in Figure 10.33.

Solution We can find the area of the lot using the formula for the area of a triangle: $A = \frac{1}{2} \times \text{base} \times \text{height}$ (see Unit 10A). The 250-foot frontage along the stream is the base of the triangle. The height is the unlabeled side in Figure 10.33. We can find this height with the Pythagorean theorem as follows.

Start with the Pythagorean theorem:	$\text{base}^2 + \text{height}^2 = \text{hypotenuse}^2$
Solve for the height by subtracting base² from both sides, then taking the square root of both sides:	$\text{height} = \sqrt{\text{hypotenuse}^2 - \text{base}^2}$
Substitute the given values:	$\text{height} = \sqrt{(1200\text{ ft})^2 - (250\text{ ft})^2} \approx 1174\text{ ft}$
Use this height to find the area of the triangle:	$\text{area} = \frac{1}{2} \times \text{base} \times \text{height} = \frac{1}{2} \times 250\text{ ft} \times 1174\text{ ft} = 146{,}750\text{ ft}^2$
Convert to acres (1 acre = 43,560 ft²):	$146{,}750\text{ ft}^2 \times \frac{1\text{ acre}}{43{,}560\text{ ft}^2} \approx 3.4\text{ acres}$

The lot has an area of about 3.4 acres. ▶ Now try Exercises 57–60.

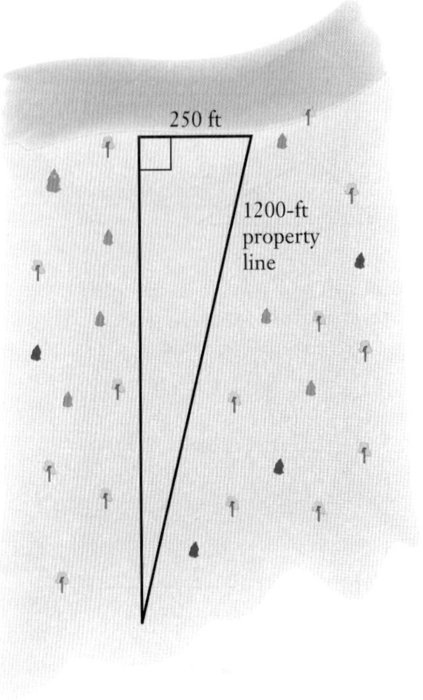

FIGURE 10.33 A mountain lot shaped like a right triangle, with frontage on a stream.

Similar Triangles

Two triangles are **similar** if they have the same shape, but not necessarily the same size. Having the same shape means that one triangle is a scaled-up or scaled-down version of the other triangle. Similar triangles provide a powerful tool for geometrical modeling.

Figure 10.34 shows two similar triangles. The angles are labeled with capital letters (A, B, C and A', B', C') and the sides with lowercase letters (a, b, c and a', b', c'). By convention, angle A is opposite side a, angle B' is opposite side b', and so forth. Notice that the labels were chosen so that it is easy to see the similarity. For example, angle A and angle A' look the same. If you measure the angles and sides in the two similar triangles, you'll find they have the following key properties.

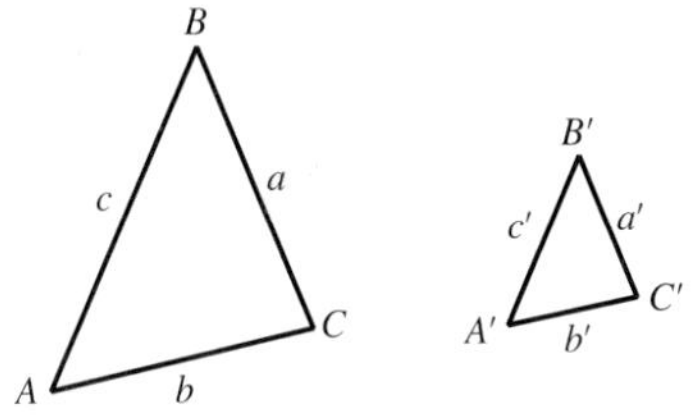

FIGURE 10.34 These two triangles are similar because they have the same shape. That is, their angles are the same and the ratios of their side lengths are the same.

Similar Triangles

Two triangles are similar if they have the same shape (but not necessarily the same size), meaning that one is a scaled-up or scaled-down version of the other.

- For two similar triangles, corresponding pairs of angles in the triangles are equal. That is, angle A = angle A', angle B = angle B', and angle C = angle C'.
- The ratios of the corresponding side lengths of two similar triangles are all equal:

$$\frac{a}{a'} = \frac{b}{b'} = \frac{c}{c'}$$

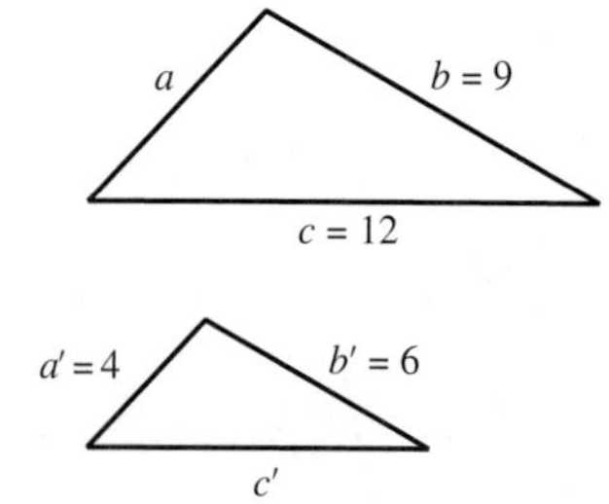

FIGURE 10.35 Two similar triangles with two sides of unknown length.

EXAMPLE 7 Practice with Similar Triangles

Figure 10.35 shows two similar triangles. Find the lengths of the sides labeled a and c'.

Solution We can find the side lengths from the properties of similar triangles.

Start with the property relating side lengths for similar triangles:	$\frac{a}{a'} = \frac{b}{b'} = \frac{c}{c'}$
Insert the known side lengths from Figure 10.35:	$\frac{a}{4} = \frac{9}{6} = \frac{12}{c'}$
Separate the first two terms from the other term and solve for a by multiplying both sides by 4:	$\frac{a}{4} = \frac{9}{6} \Rightarrow a = \frac{9 \times 4}{6} = 6$
Solve for c' using the last two terms (in Step 2):	$\frac{9}{6} = \frac{12}{c'} \Rightarrow c' = \frac{12 \times 6}{9} = 8$

The "unknown" side lengths are $a = 6$ and $c' = 8$. ▸ Now try Exercises 61–68.

EXAMPLE 8 Solar Access

BY THE WAY

Shadows are longest on the shortest day of the year, which in the Northern Hemisphere is the winter solstice (about December 22). The shadow lengths given in Example 8 correspond to latitude 33°N, approximately the latitude of Santa Barbara, California; Albuquerque, New Mexico; and Charlotte, North Carolina.

Some cities have policies that prevent property owners from constructing new houses and additions that cast shadows on neighboring houses. The intent of these policies is to allow everyone access to the Sun for powering solar energy devices. Consider the following solar access policy:

> *On the shortest day of the year, a house cannot cast a noontime shadow that reaches farther than the shadow that would be cast by a 12-foot fence on the property line.*

Suppose your house is set back 30 feet from the north property line and that, on the shortest day of the year, a 12-foot fence on that property line would cast a noontime shadow 20 feet in length. If you build an addition onto your house, how high can the north side of the remodeled house be under this policy?

Solution The key to solving this problem is drawing a good picture and identifying similar triangles. Figure 10.36 shows the geometry. Notice these features of the figure:

- The smaller triangle (at left) shows the 12-foot fence and the 20-foot length of its shadow along the ground.
- The 30-foot setback lies between the fence and the north side of the house. We are trying to determine h, the maximum allowed height of the north side of the house.
- The policy defines the maximum allowed height h of the house such that the shadow cast by the house reaches the same point as the shadow of the 12-foot fence. The figure illustrates this property by showing one line that passes through the tip of the fence's shadow, the top of the fence, the top of the house, and the Sun. We see that the house has a 50-foot shadow along the ground: 30 feet for the setback plus 20 feet for the fence shadow.

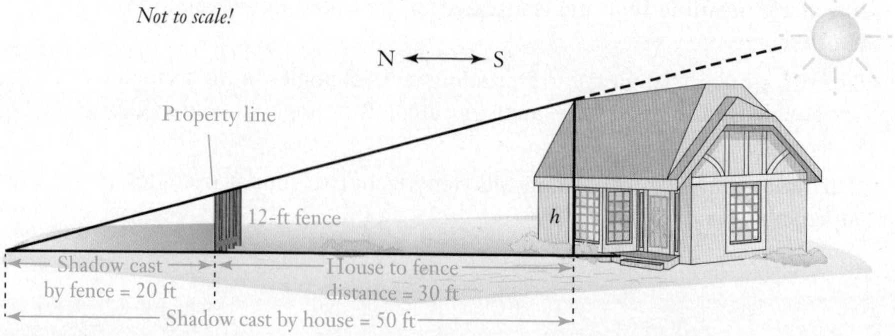

FIGURE 10.36 The figure shows a 12-foot fence on the property line and the 20-foot shadow it casts on the shortest day of the year. Set back 30 feet from the property line, the house is drawn at the maximum height at which its shadow extends to the same point as the fence shadow.

- The end result is a figure with two similar triangles. The smaller one is formed by the 12-foot fence and its 20-foot shadow, and the larger one is formed by the house with unknown height h and its 50-foot shadow.

We can now apply the ratio property for the side lengths of similar triangles. In this case, the ratio of the house height to the fence height must equal the ratio of the house shadow length to the fence shadow length:

$$\frac{\text{house height}}{\text{fence height}} = \frac{\text{house shadow length}}{\text{fence shadow length}}$$

Substituting the known values, we find

$$\frac{h}{12\text{ ft}} = \frac{50\text{ ft}}{20\text{ ft}} \xrightarrow{\text{multiply both sides by 12 ft}} h = \frac{50\text{ ft}}{20\text{ ft}} \times 12\text{ ft} = 30\text{ ft}$$

The maximum height of the north side of the house after the remodeling is 30 feet.

▶ Now try Exercises 69–72.

Optimization Problems

Optimization problems seek a "best possible" solution. For example, they may seek to find the largest volume or lowest cost that can be obtained under particular conditions. These problems have many applications, so let's look at two examples.

EXAMPLE 9 Optimizing Area

You have 132 meters of fence that you plan to use to enclose a corral on a ranch. What shape should you choose if you want the corral to have the greatest possible area? What is the area of this "optimized" corral?

Solution A good way to start this problem is to experiment with a loop of string, making it into various shapes. Notice that a long narrow shape doesn't enclose as much area as a simpler shape like a circle or a rectangle (Figure 10.37). After some experimentation, you can probably convince yourself that the largest area occurs with either a square or a circle. We can decide between these two possibilities by calculating the areas.

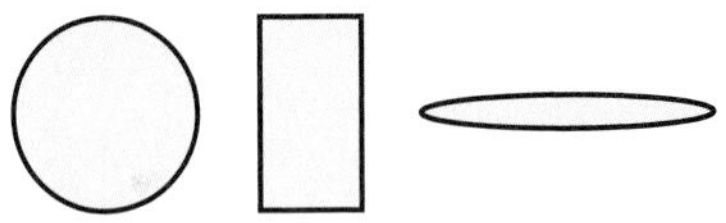

FIGURE 10.37 All three of these shapes have the same perimeter, but different areas. The circle has the greatest area. In fact, a circle has the greatest area of all two-dimensional figures with a fixed perimeter.

If you use your 132 meters of fence to make a square, each side will have length $132/4 = 33$ meters. The enclosed area of the square will be

$$A = (33\text{ m})^2 = 1089\text{ m}^2$$

If you make a circle, the 132 meters of fence will be the circumference of the circle. Because circumference $= 2\pi r$, the radius of the circle will be $r = 132/2\pi$ meters (or approximately 21.008 meters). The area of the circle will therefore be

$$A = \pi r^2 = \pi\left(\frac{132}{2\pi}\text{ m}\right)^2 \approx 1387\text{ m}^2$$

In fact, a circle is the shape with the greatest possible area for a fixed perimeter. In this case, the best possible choice is to use the 132 meters of fence to make a circular corral with an area of about 1387 square meters.

▶ Now try Exercises 73–76.

EXAMPLE 10 Optimal Container Design

You are designing a wooden crate (rectangular prism) that must have a volume of 2 cubic meters. The cost of the wood is \$12 per square meter. What dimensions give the least expensive design? With that optimal design, how much will the material for each crate cost?

Solution The wood makes the surfaces of the crate, so to minimize cost we are looking for the minimum *surface area* for a crate with a volume of 2 cubic meters. As in

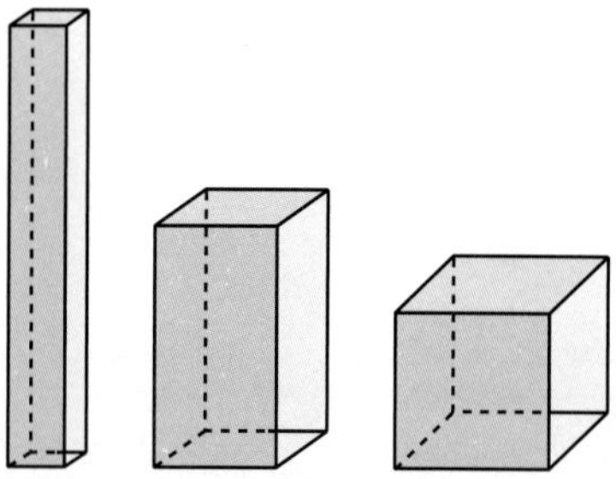

FIGURE 10.38 All three of these boxes have the same volume, but the cube has the *smallest* total *surface area*.

Example 9, you might begin by experimentation, making either little paper boxes or good drawings of boxes (Figure 10.38). With experimentation or calculation, you will find that for a given volume, a cube has the minimum surface area. The crates should therefore be cubes, and we find the dimensions and cost as follows.

Start with the formula for the volume of a cube:	$V = (\text{side length})^3$
Solve for side length and substitute $V = 2$ cubic meters:	$\text{side length} = \sqrt[3]{V} = \sqrt[3]{2\text{ m}^3} \approx 1.26\text{ m}$
Cost is based on the total surface area of the cube, which has 6 sides of equal side length:	$\text{surface area} = 6 \times (\text{side length})^2 = 6 \times (1.26\text{ m})^2 \approx 9.53\text{ m}^2$
Find the total cost from the surface area and the price of \$12 per square meter:	$\text{total cost} \approx 9.53\text{ m}^2 \times \frac{\$12}{\text{m}^2} \approx \$114$

The optimal design for the crates is a cube of side length 1.26 meters, in which case each crate costs about \$114.

▶ Now try Exercises 77–80.

Quick Quiz 10B

Choose the best answer to each of the following questions. Explain your reasoning with one or more complete sentences.

1. The number of minutes of arc in a full circle is

   a. 60. b. 360. c. 60×360.

2. The number of seconds of arc in 1° is

   a. 60.

   b. 60×60.

   c. 60×360.

3. If you travel due east, you are traveling

   a. along a line of constant latitude.

   b. along a line of constant longitude.

   c. along the prime meridian.

4. If you are located at latitude 30°S and longitude 120°W, you are in

   a. North America.

   b. the south Pacific Ocean.

   c. the south Atlantic Ocean.

5. What would be different about the Sun if you viewed it from Mars (which is farther from the Sun than Earth is) instead of from Earth?

   a. its radius

   b. its angular size

   c. its volume

6. The Sun is about 400 times as far away from Earth as the Moon is, but the Sun and the Moon have the same angular size in our sky. This means that the Sun's diameter is larger than the Moon's diameter by a factor of

   a. $\sqrt{400}$. b. 400. c. 400^2.

7. If you are bicycling eastward up a hill with a 10% grade, you know that

   a. for every 100 yards you ride eastward, you gain 10 yards in altitude.

   b. the hill slopes upward at a 10° angle.

   c. the effort required is 10% greater than the effort required on a flat road.

8. The length of side x in the right triangle below is

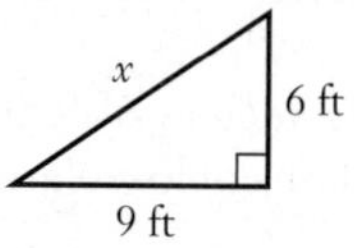

   a. $\sqrt{6^2 + 9^2}$. b. $6^2 + 9^2$. c. $\sqrt{6^2 \times 9^2}$.

9. For the two similar triangles below, which statement is true?

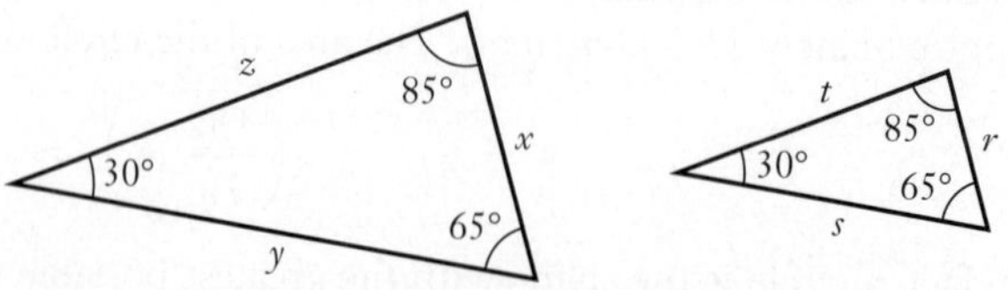

   a. $\frac{x}{y} = \frac{r}{s}$ b. $\frac{x}{r} = \frac{y}{t}$ c. $z = t$

10. You have a piece of string and you want to lay it out on a flat surface so that it encloses the greatest possible area. What shape should you make with the string?

    a. a square b. a circle

    c. It doesn't matter, because all shapes you make with it will have the same area.

Exercises 10B

REVIEW QUESTIONS

1. How do we describe fractions of a degree of angle?
2. Explain how a position on Earth can be specified using latitude and longitude.
3. How is *angular size* related to physical size?
4. Give four different ways to describe the angle that an inclined surface (such as a road) makes with the horizontal.
5. Give at least two examples of ways in which the Pythagorean theorem can be useful in solving a practical problem.
6. Make a sketch of two *similar triangles*, and explain the properties that make the triangles similar.
7. Give an example of a practical problem that can be solved with similar triangles.
8. What is an optimization problem? Give an example.

DOES IT MAKE SENSE?

Decide whether each of the following statements makes sense (or is clearly true) or does not make sense (or is clearly false). Explain your reasoning.

9. In December, it is winter at 70°W and 44°S.
10. When I looked at the Moon in the sky, it seemed to be a mile wide.
11. I should have no trouble riding my bike up a road with a grade of 7%.
12. The angles in triangle *A* have a sum of 180°, as do the angles in triangle *B*. Therefore, the two triangles are similar.
13. The sides of triangle *A* are half as long as the corresponding sides of triangle *B*. Therefore, the two triangles are similar.
14. There are many different rectangular boxes that have the same volume, but only one combination of length, width, and depth gives the smallest possible surface area.

BASIC SKILLS & CONCEPTS

15–20: Angle Conversions I. Convert the given degree measure into degrees, minutes, and seconds of arc.

15. 32.5°
16. 280.1°
17. 12.33°
18. 0.08°
19. 149.83°
20. 47.6723°

21–26: Angle Conversions II. Convert the given angle measure into degrees and decimal fractions of a degree. For example, $30°30' = 30.5°$.

21. 30°10′
22. 60°30′30″
23. 123°10′36″
24. 2°2′2″
25. 8°59′10″
26. 150°14′28″
27. **Minutes in a Circle.** Calculate the number of minutes of arc in a full circle.
28. **Seconds in a Circle.** Calculate the number of seconds of arc in a full circle.

29–36: Latitude and Longitude. Consult an atlas, globe, or website to answer the following questions.

29. Find the latitude and longitude of Madrid, Spain.
30. Find the latitude and longitude of Sydney, Australia.
31. Find the latitude and longitude of the location on Earth precisely opposite Toronto, Canada (latitude 44°N, longitude 79°W).
32. Find the latitude and longitude of the location on Earth precisely opposite Portland, Oregon (latitude 46°N, longitude 123°W).
33. Which is farther from the North Pole: Buenos Aires, Argentina, or Cape Town, South Africa? Explain.
34. Which is farther from the South Pole: Guatemala City, Guatemala, or Lusaka, Zambia? Explain.
35. Buffalo, New York, is at nearly the same longitude as Miami, Florida, but Buffalo's latitude is 43°N while Miami's latitude is 26°N. About how far away is Buffalo from Miami? Explain.
36. Washington, DC, is at about latitude 38°N and longitude 77°W. Lima, Peru, is at about latitude 12°S and longitude 77°W. About how far apart are the two cities? Explain.

37–40: Angular Size. Use the formula relating angular size, physical size, and distance.

37. What is the angular size of a quarter viewed from a distance of 3 yards?
38. What is the angular size of a quarter viewed from a distance of 20 yards?
39. The Sun has an angular diameter of about 0.5° viewed from Earth and a distance from Earth of about 150 million kilometers. What is its true diameter?
40. You are looking at a tree on a hillside that is 0.5 mile away. You measure the tree's angular size (apparent height) to be about $\frac{1}{2}$°. How tall is the tree?

41–44: Slope, Pitch, Grade. Determine which of the following pairs of surfaces is steeper.

41. A roof with a pitch of 1 in 4 or a roof with a slope of $\frac{2}{10}$
42. A road with a 12% grade or a road with a pitch of 1 in 8
43. A railroad track with a 3% grade or a railroad track with a slope of $\frac{1}{25}$
44. A sidewalk with a pitch of 1 in 6 or a sidewalk with a 15% grade
45. **Slope of a Roof.** What is the slope of a roof with an 8 in 12 pitch? How much does the roof rise in 15 horizontal feet?
46. **Grade of a Road.** How much does a road with a 5% grade rise for each horizontal foot? If you drive along this road for 6 miles, how much elevation will you gain?

47. **Pitch of a Roof.** What is the angle (relative to the horizontal) of a roof with a 6 in 6 pitch? Is it possible for a roof to have a 7 in 6 pitch? Explain.

48. **Grade of a Path.** What is the approximate grade (expressed as a percentage) of a path that rises 1500 feet every mile?

49. **Grade of a Road.** What is the grade of a road that rises 20 feet for every 150 horizontal feet?

50. **Grade of a Trail.** How much does a trail with a 22% grade rise for each 200 horizontal yards?

51–55: Map Distances. Refer to the map in Figure 10.39. Assume that the length of each east-west block is $\frac{1}{8}$ mile and the length of each north-south block is $\frac{1}{5}$ mile.

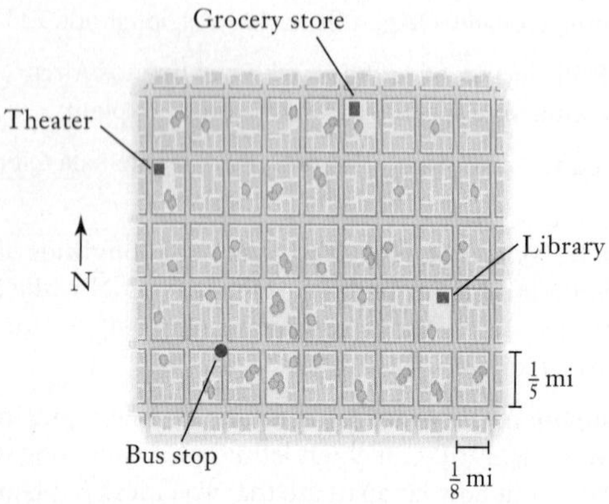

FIGURE 10.39

For the locations given in each exercise, do the following:

a. Find the shortest possible walking distance (following the streets) between the two locations.

b. Find the straight-line distance ("as the crow flies") between the two locations.

51. The bus stop and the library

52. The bus stop and the grocery store

53. The bus stop and the theater

54. The theater and the library

55. The grocery store and the library

56. **Different Shortest Paths.** How many different paths between the bus stop and the library in Figure 10.39 have the shortest possible walking distance?

57–60: Acreage Problems. Refer to Figure 10.33, but use the lengths given in each exercise. Find the area in acres of the property for each set of lengths.

57. The stream frontage is 200 feet in length and the property line is 800 feet in length.

58. The stream frontage is 300 feet in length and the property line is 1800 feet in length.

59. The stream frontage is 600 feet in length and the property line is 3800 feet in length.

60. The stream frontage is 0.45 mile in length and the property line is 1.2 miles in length.

61–64: Determining Similarity. Determine which pairs of triangles are similar, and explain how you know.

61.

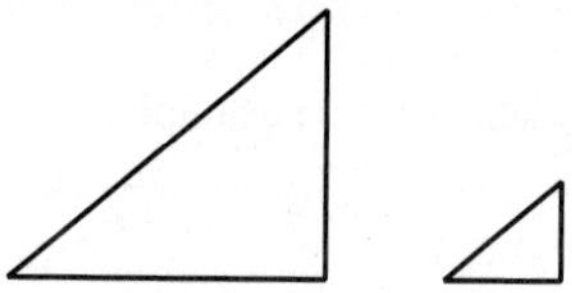

62.

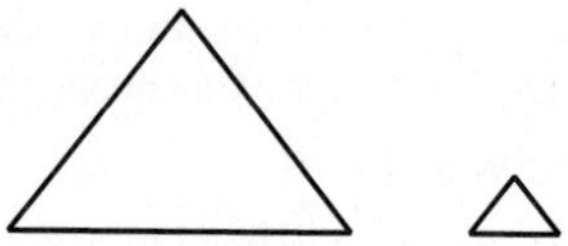

63.

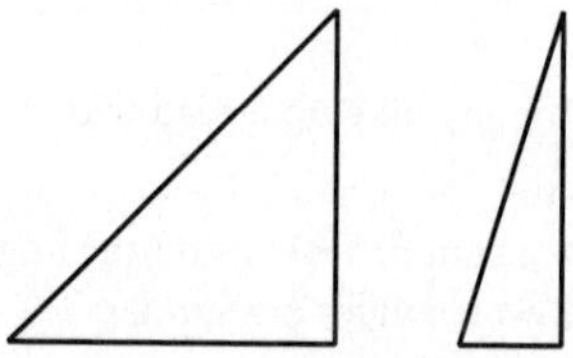

64.

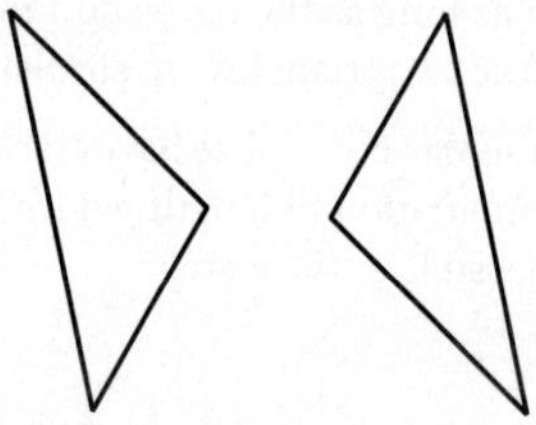

65–68: Analyzing Similar Triangles. Determine the lengths of the unknown sides in the following pairs of similar triangles.

65.

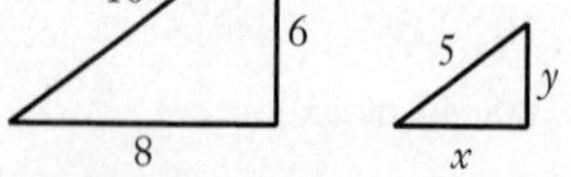

66.

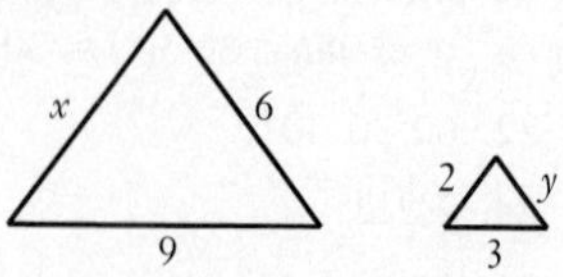

67.

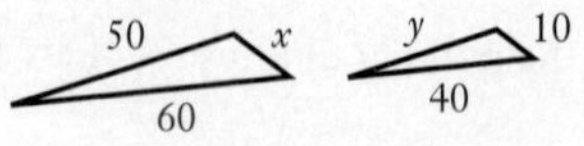

68.

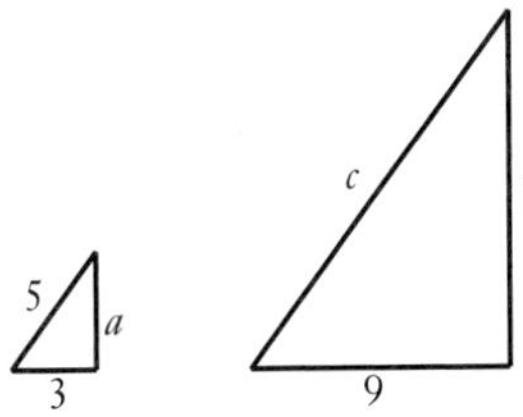

69–72: Solar Access. Assume that the solar access policy given in Example 8 is in force, and find the maximum allowed height for each house.

69. A 12-foot fence on the property line casts a 25-foot shadow on the shortest day of the year, and the north side of the house is set back 60 feet from the property line.

70. A 12-foot fence on the property line casts a 20-foot shadow on the shortest day of the year, and the north side of the house is set back 30 feet from the property line.

71. A 12-foot fence on the property line casts a 30-foot shadow on the shortest day of the year, and the north side of the house is set back 50 feet from the property line.

72. A 12-foot fence on the property line casts a 10-foot shadow on the shortest day of the year, and the north side of the house is set back 30 feet from the property line.

73–76: Optimizing Area. Determine the area of a circular enclosure and a square enclosure made with the given amount of fence. Compare the areas and comment.

73. 50 meters of fence

74. 800 feet of fence

75. 150 meters of fence

76. 0.27 mile of fence

77. **Designing Cans.** Suppose you work for a company that manufactures cylindrical cans. Which will cost more to manufacture: Can 1, with a radius of 4 inches and a height of 5 inches, or Can 2, with a radius of 5 inches and a height of 4 inches? Assume that the cost of material for the tops and bottoms is \$1.00 per square inch and the cost of material for the curved surfaces is \$0.50 per square inch.

78. **Designing Plastic Buckets.** A company manufactures plastic buckets that are shaped like cylinders without a lid. Which will cost more to manufacture: Bucket 1, with a radius of 6 inches and a height of 18 inches, or Bucket 2, with a radius of 9 inches and a height of 15 inches? Assume that the plastic material costs \$0.50 per square foot, but the bottom of each bucket must have double thickness.

79. **Designing Cardboard Boxes.** Suppose you are designing a cardboard box that must have a volume of 8 cubic feet. The cost of the cardboard is \$0.15 per square foot. What is the most economical design for the box (the one that minimizes the cost), and how much will the material for each box cost?

80. **Designing Steel Safes.** A large steel safe with a volume of 4 cubic feet is to be designed in the shape of a rectangular prism. The cost of the steel is \$6.50 per square foot. What is the most economical design for the safe, and how much will the material for each such safe cost?

FURTHER APPLICATIONS

81. **Blu-ray Geometry.** The capacity of a single-sided, dual-layer Blu-ray Disc is approximately 50 billion bytes (50 GB). The inner and outer radii that define the storage region of a Blu-ray Disc are $r = 2.5$ cm and $R = 5.9$ cm, respectively.

a. What is the area of the storage region in square centimeters (cm^2)?

b. What is the density of the data on a Blu-ray Disc in millions of bytes/cm^2?

c. A Blu-ray Disc consists of a single long track, or "groove," that spirals outward from the inner edge to the outer edge of the storage region. The width of each turn of the spiral (essentially the thickness of the groove) is $d = 0.3$ micrometer (1 micrometer $= 10^{-6}$ meter). It can be shown that the length of the entire groove is approximately $L = \pi(R^2 - r^2)/d$. What is the length of the track on a Blu-ray Disc in centimeters? In miles?

82. **Angular Size of a Star.** Imagine a star that is the same size as the Sun (diameter about 1.4 million km) but is located 10 light-years away from Earth. What is the star's angular size as seen from Earth? Given that the most powerful telescopes available today can see details no smaller than about 0.01 second of arc, is it possible to see any details on the surface of this star? Explain. (*Hint:* 1 light-year $\approx 10^{13}$ km.)

83. **Baseball Geometry.** The distance between bases on a baseball diamond (which is really a square) is 90 feet. How far does a baseball travel when the catcher throws it from home plate to second base (along the diagonal of the diamond)?

84. **Saving Costs.** A phone line runs east along a field for 1.5 miles and then north along the edge of the same field for 2.75 miles. If the phone line cost \$3500 per mile to install, how much could have been saved if the phone line had been installed diagonally across the field? Draw a good picture!

85. **Travel Times.** To get to a cabin, you have a choice of either riding a bicycle west from a parking lot along the edge of a rectangular reservoir for 1.2 miles and then south along the edge of the reservoir for 0.9 mile or rowing a boat directly from the parking lot. If you can ride 1.5 times as fast as you can row, which is the faster route?

86. **Keep Off the Grass.** An old principle of public landscaping says "Build sidewalks last." In other words, let the people find their chosen paths and then build the sidewalks on the paths. Figure 10.40 shows a campus quadrangle that measures 40 meters by 30 meters. The locations of the doors of the library, chemistry building, and humanities building are shown. What is the combined length of the new sidewalks (gold lines) that connect the three buildings?

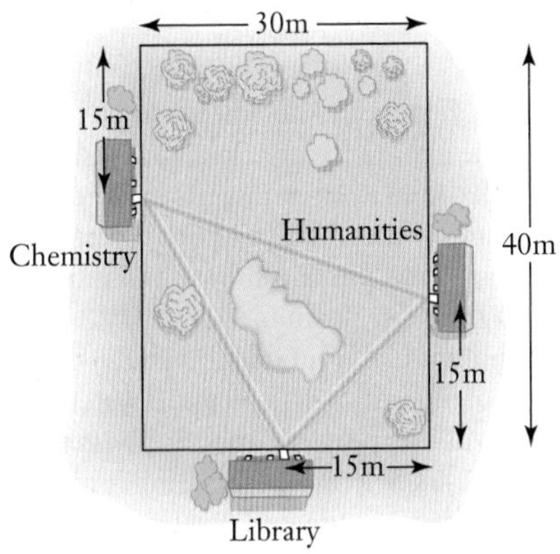

FIGURE 10.40

87. **Water Bed Leak.** Suppose you have, in a second-floor bedroom, a water bed whose mattress measures 8 ft × 7 ft × 0.75 ft. One day the mattress leaks and all the water drains into the room below the bedroom.

a. If the lower room measures 10 ft × 8 ft, how deep is the water in that room (assuming all of the water accumulates in the room)?

b. What is the weight of the water that was in the water bed? (Water has a density of 62.4 pounds per cubic foot.)

88. **Filling a Pool.** A spherical water tank has a radius of 25 feet. Can it hold enough water to fill a swimming pool 50 meters long and 25 meters wide to a constant depth of 2 meters?

89. **Optimal Fencing.** A rancher must design a rectangular corral with an area of 400 square meters. She decides to make a corral that measures 10 meters by 40 meters (which has the correct total area). How much fencing is needed for this corral? Has the rancher found the most economical solution? Can you find another design for the corral that requires less fencing but still provides 400 square meters of corral?

90. **Optimal Boxes.** You are making boxes and begin with a rectangular piece of cardboard that measures 1.75 meters by 1.25 meters. From each corner of that rectangular piece you cut out a square piece that is 0.25 meter on a side, as shown in Figure 10.41. You then fold up the "flaps" to form a box without a lid. What is the outside surface area of the box? What is its volume? If the corner cuts were squares 0.3 meter on a side, would the resulting box have a larger or smaller volume than the first box? What is your estimate of the size of the square corner cuts that will give a box with the largest volume?

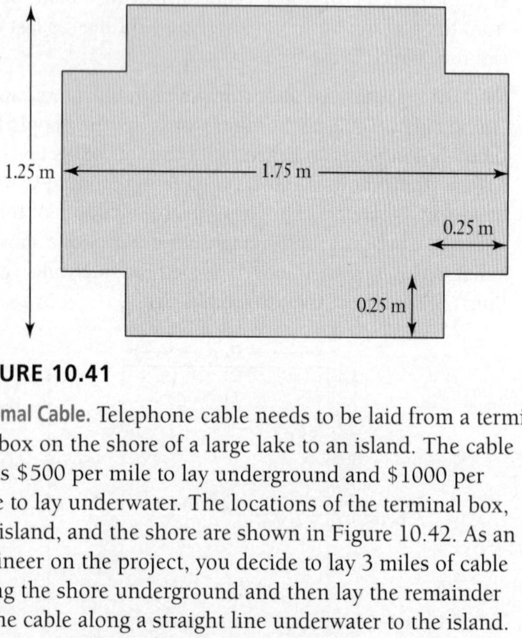

FIGURE 10.41

91. **Optimal Cable.** Telephone cable needs to be laid from a terminal box on the shore of a large lake to an island. The cable costs \$500 per mile to lay underground and \$1000 per mile to lay underwater. The locations of the terminal box, the island, and the shore are shown in Figure 10.42. As an engineer on the project, you decide to lay 3 miles of cable along the shore underground and then lay the remainder of the cable along a straight line underwater to the island. How much will this project cost? Your boss examines your proposal and asks whether laying 4 miles of cable underground before starting the underwater cable would be more economical. How much would your boss's proposal cost? Will you still have a job?

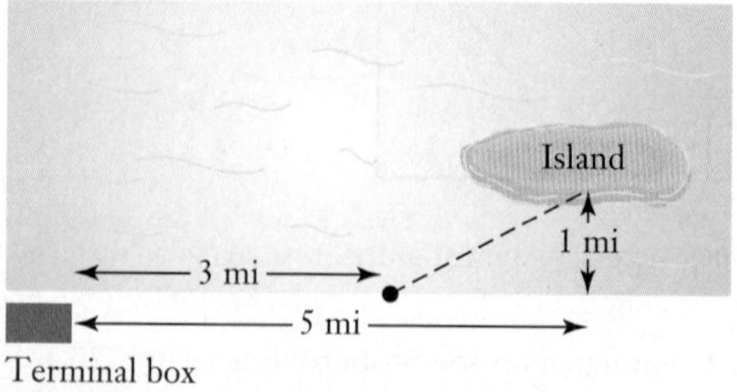

FIGURE 10.42

92. **Estimating Heights.** In trying to estimate the height of a nearby building, you make the following observations. If you stand 15 feet away from a lamp post, you can line up the top of the building with the top of the lamp post. Furthermore, you can easily determine that the top of the lamp post is 10 feet above your head and that the building is 50 feet from your sighting position. What is the height of the building? Draw a good picture!

93. **Soda Can Design.** Standard soda cans hold 12 ounces, or 355 milliliters, of soda. Thus, their volume is 355 cm^3. Assume that soda cans must be right circular cylinders.

a. The cost of materials for a can depends on its surface area. Through trial and error with cans of different sizes, find the dimensions of a 12-ounce can that has the lowest cost for materials. Explain how you arrive at your answer.

b. Compare the dimensions of the can from part (a) to the dimensions of a real soda can from a vending machine or store. Suggest some reasons why real soda cans might not have the dimensions that minimize their use of material.

94. **Melting Ice.** A glacier's surface is approximately rectangular, with a length of about 100 meters and a width of about 20 meters. The ice in the glacier averages about 3 meters in depth. Suppose that the glacier melts into a lake that is roughly circular with a radius of 1 kilometer. Assuming that the area of the lake does not expand significantly, about how much would the water level rise if the entire glacier melted into the lake?

95. **Sand Cones.** Think back to your days in the sandbox, and imagine pouring sand through a funnel. The stream of sand forms a cone on the ground that becomes higher and wider as sand is added. The proportions of the sand cone remain the same as the cone grows, and the height of the cone is roughly one-third the radius of the circular base (see Figure 10.43). The formula for the volume of a cone with height h and a base with radius r is $V = \frac{1}{3}\pi r^2 h$.

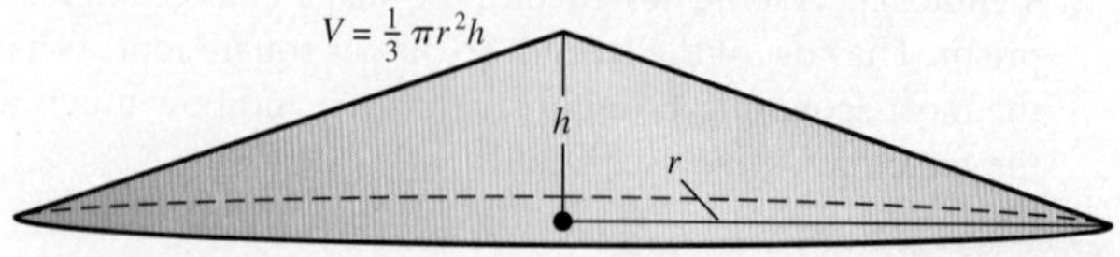

FIGURE 10.43

a. If you build a sand cone 2 feet high, how many cubic feet of sand does it hold?

b. Suppose that you excavate a basement and extract 1000 cubic feet of dirt. If the dirt is stored in a conical pile, how high will the cone be? Assume that a dirt cone has the same proportions as a sand cone.

c. Estimate the total number of grains of sand in the cone in part (a). Explain your assumptions and uncertainties.

96. **The Trucker's Dilemma.** It is a dark and stormy night. Through the beat of the wipers on your truck, you see a rickety country bridge marked "Load Limit 40 Tons." You know your truck, *White Lightning*, like the back of your hand; it weighs 16.3 tons with you and your gear. Trouble is, you are carrying a cylindrical steel water tank that is full of water. The empty weight of the tank is printed on its side: 1750 pounds. But what about the water? Fortunately, the Massey-Fergusson trucker's almanac in your glove box tells you that every cubic inch of water weighs 0.03613 pound. So you dash out into the rain with a tape measure to find the dimensions of the tank: length = 22 ft, diameter = 6 ft 6 in. Back in the truck, dripping and calculating, do you risk crossing? Explain.

97. **The Great Pyramids of Egypt.** Egypt's Old Kingdom began in about 2700 B.C.E. and lasted for 550 years. During that time, at least six pyramids were built as monuments to the life and afterlife of various pharaohs. These pyramids remain among the largest and most impressive structures constructed by any civilization. The building of the pyramids required a mastery of art, architecture, engineering, and social organization at a level unknown before that time. The collective effort required to complete the pyramids transformed Egypt into the first nation-state in the world. Of the six pyramids, the best known are those on the Giza plateau outside Cairo, and the largest of those is the Great Pyramid built by Pharaoh Khufu (or Cheops to the Greeks) in about 2550 B.C.E. With a square base of 756 feet on a side and a height of 481 feet, the pyramid is laced with tunnels, shafts, corridors, and chambers, all leading to and from the deeply concealed pharaoh's burial chamber (Figure 10.44). The stones used to build the pyramids were transported, often hundreds of miles, with sand sledges and river barges, by a labor force of 100,000, as estimated by the Greek historian Herodotus. Historical records suggest that the Great Pyramid was completed in approximately 25 years.

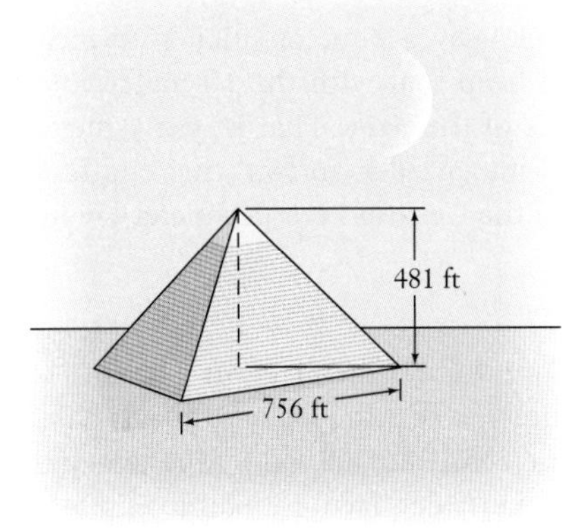

FIGURE 10.44

a. To appreciate the size of the Great Pyramid of Khufu, compare its height to the length of a football field (which is 100 yards).

b. The volume of a pyramid is given by the formula $V = \frac{1}{3} \times$ area of base $\times$ height. Use this formula to estimate the volume of the Great Pyramid. State your answer in both cubic feet and cubic yards.

c. The average size of a limestone block in the Great Pyramid is 1.5 cubic yards. How many blocks were used to construct this pyramid?

d. A modern research team, led by Mark Lehner of the University of Chicago, estimated that the use of winding ramps to lift the stones, desert clay, and water would allow placing one stone every 2.5 minutes. If the pyramid workers labored 12 hours per day, 365 days per year, how long would it have taken to build the Great Pyramid? How does this estimate compare with historical records? Why did Lehner's team conclude that the Great Pyramid could have been completed with only 10,000 laborers, rather than the 100,000 laborers estimated by Herodotus?

e. Constructed in 1889 for the Paris Exposition, the Eiffel Tower is a 980-foot iron lattice structure supported on four arching legs. The legs of the tower stand at the corners of a square with sides of length 120 feet. If the Eiffel Tower were a solid pyramid, how would its volume compare to the volume of the Great Pyramid?

IN YOUR WORLD

98. **Great Circles.** A great circle route is the shortest path between two points on the surface of the Earth. It is the optimal route for an airliner. Pick several pairs of cities, and find the length of the great circle route between the cities in each pair. Estimate the length of other possible routes between the cities to verify that the great circle route is the shortest (this calculation will be easiest if you choose two cities with the same longitude). Explain in words how to find the great circle route between two points on the Earth.

99. **Sphere Packing.** Investigate the sphere packing problem, which has a history dating from the 17th century to the present day. Summarize the problem, early conjectures about its solution, and recent developments. You might also want to explore packing problems with nonspherical objects.

100. **Eclipse!** A total solar eclipse is a spectacular sight, made possible by the coincidence of the Moon and Sun both having very nearly the same angular size when viewed from Earth. However, not all eclipses are total, because there are times when the Moon's angular size is slightly smaller than the Sun's. Moreover, the Moon is gradually moving farther from Earth, so there will come a day in the distant future when total solar eclipses will no longer be possible. Find out why the Moon's angular size is sometimes smaller than the Sun's even today, and approximately when the Moon will have moved far enough from Earth so that eclipses can no longer occur. Summarize your findings in 1 or 2 paragraphs, using diagrams as needed.

UNIT 10C Fractal Geometry

The geometry discussed in the previous two units is classical geometry, developed largely by the ancient Greeks. While this geometry still serves us well in many applications, it is less applicable to forms that arise in nature. In recent decades, a new type of geometry called **fractal geometry** has emerged. It has proved to be so effective at describing nature that it is now used in art and film to create realistic imagery. Figure 10.45 shows an imaginary landscape generated entirely on a computer using fractal geometry.

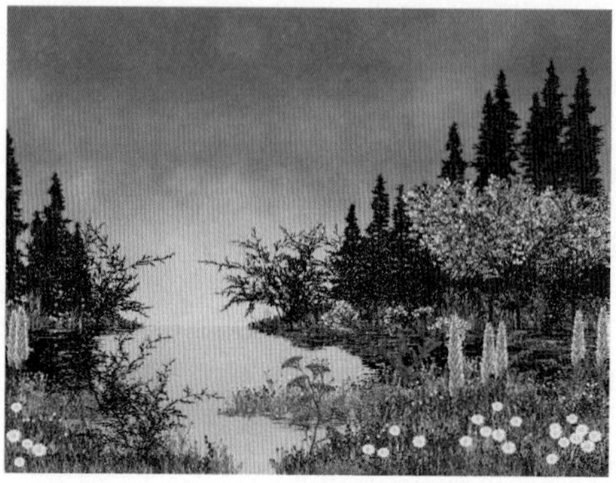

FIGURE 10.45 A fractal landscape. *Source*: Created by Anne Burns.

What Are Fractals?

We can investigate fractals by envisioning measurements made with "rulers" of different lengths, such as 10 meters, 1 meter, 1 millimeter, and so on. Each length laid out by a ruler is called an **element** (of length). Suppose we use a ruler of length 10 meters and find that an object is 50 elements long. Then the total length of the object is $50 \times 10 \text{ m} = 500 \text{ m}$. More generally, the length of any object is

$$\text{total length} \approx \text{number of elements} \times \text{length of each element}$$

Measuring the Perimeter of Central Park

FIGURE 10.46 The rectangular layout of Central Park.

Imagine that you are asked to measure the perimeter of Central Park in New York City, which was designed in the shape of a rectangle (Figure 10.46). Suppose you start with a 10-meter ruler. Multiplying the number of elements by their 10-meter length gives you a measurement of the park perimeter.

Now, suppose you use a shorter ruler—say, a 1-meter ruler. Your measurement with the 1-meter ruler will not differ much from that with the 10-meter ruler, because you are simply measuring the straight sides of the park. That is, ten 1-meter rulers will fit along the same straight line as one 10-meter ruler. In fact, the length of the ruler will not significantly affect measurement of the Central Park perimeter (Figure 10.47).

Measuring an Island Coastline

Next, imagine that you want to measure the perimeter of an island in a large lake. To make the task easier, you wait until winter, when the water around the island is frozen. Then you can measure the length of the coastline defined by the ice-land boundary.

Suppose you begin by using a 100-meter ruler, laying it end to end around the island. The 100-meter ruler (the single blue line segment in Figure 10.48) will adequately measure large-scale features such as bays and estuaries, but will miss features such as promontories

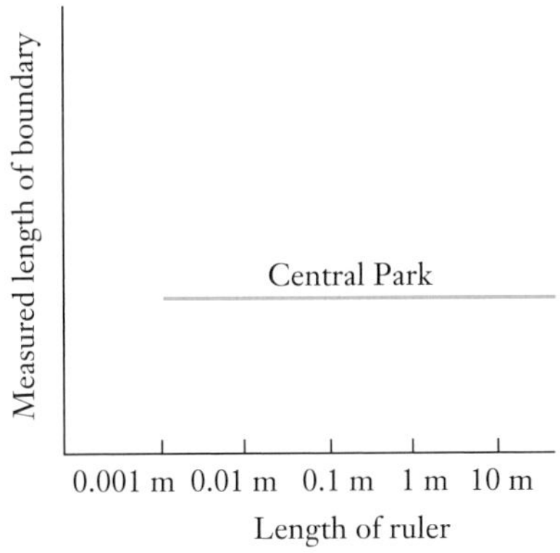

FIGURE 10.47 Ruler length does not affect measurement of the Central Park perimeter.

and inlets that are less than 100 meters across. Switching to a 10-meter ruler (the short red line segments in Figure 10.48) will allow you to follow many features that were missed by the 100-meter ruler. As a result, you'll measure a *longer* perimeter.

The 10-meter ruler is still too long to measure all the features along the coastline, so you'll find an even greater perimeter if you switch to a 1-meter ruler. In fact, as you use shorter and shorter rulers to measure the coastline, you'll get longer and longer estimates of the perimeter because shorter rulers measure new levels of detail. Figure 10.49 shows how the measured length of the coastline increases when measured with shorter rulers. We cannot agree on the "true" length of the coastline because, unlike the clearly defined perimeter of Central Park, it depends on the length of the ruler used.

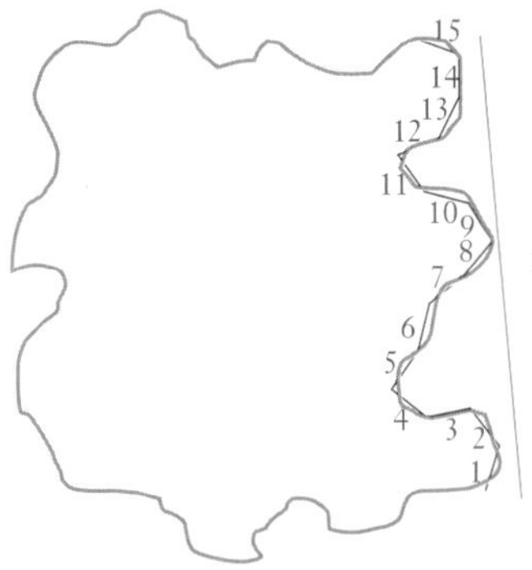

FIGURE 10.48 A single long ruler (blue) cannot measure details that can be measured with shorter rulers (red).

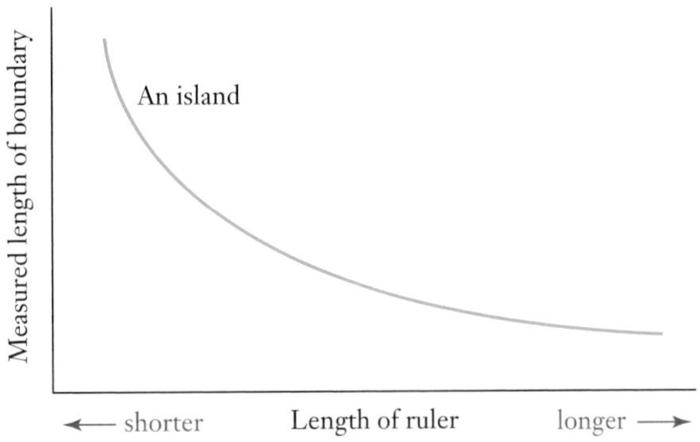

FIGURE 10.49 The measured perimeter of an island gets longer (vertical axis) as the ruler length gets shorter (horizontal axis).

Rectangles and Coastlines Under Magnification

Imagine viewing a piece of the rectangular perimeter of Central Park under a magnifying glass. No new details will appear—it is still a straight-line segment (Figure 10.50a). In contrast, if you view a piece of the coastline under a magnifying glass, you will see details that were not visible without magnification (Figure 10.50b).

Objects like the coastline that continually reveal new features at smaller scales are called **fractals**. Many natural objects have fractal structure. For example, both coral (Figure 10.51a) and mountain ranges (Figure 10.51b) reveal more and more features when viewed under greater magnification and hence are fractals.

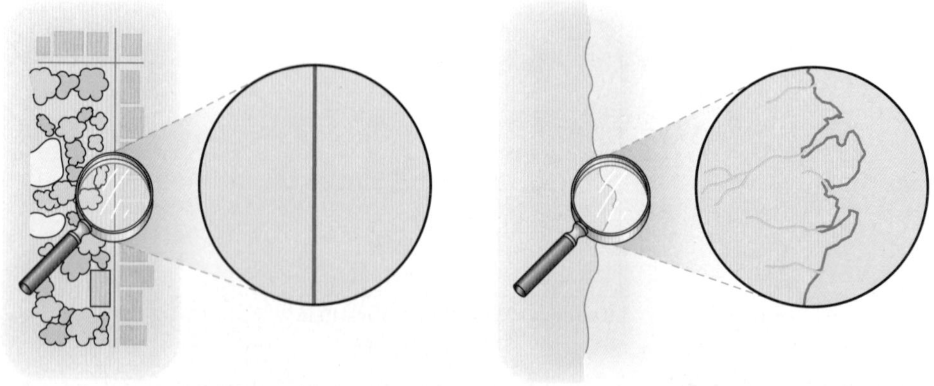

(a) (b)

FIGURE 10.50 (a) A segment of a rectangle looks the same under magnification. (b) A segment of coastline reveals new details under magnification.

(a) (b)

FIGURE 10.51 Coral (a) and mountains (b) show fractal structure.

Fractal Dimension

The boundary of Central Park is one-dimensional because a single number locates any point. For example, if you tell people to meet 375 meters from the park's northwest corner, going clockwise, they'll know exactly where to go.

In contrast, if you tell people to meet 375 meters along the coastline from a particular point on the island, different people will end up in different places depending on the length of the ruler they use. We conclude that the coastline is *not* an ordinary one-dimensional object with a clearly defined length. But it clearly isn't two-dimensional, either, because the land/water boundary is not an area. We say that the coastline has a **fractal dimension** that lies *between* one and two, indicating that the coastline has some properties that are like length (one dimension) and others that are more like area (two dimensions).

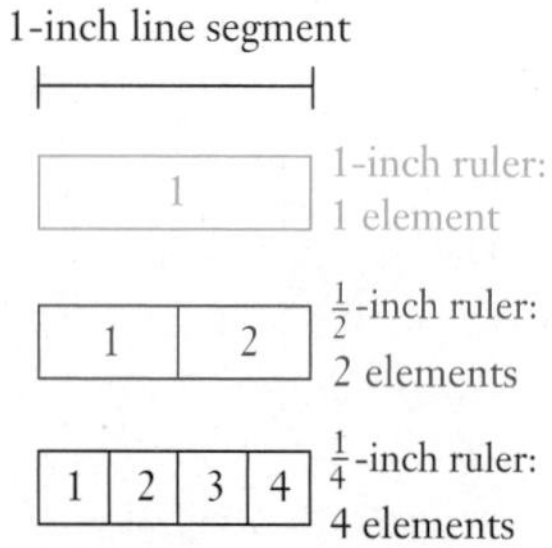

FIGURE 10.52 For a line segment, reducing ruler length by a factor R increases the number of elements by a factor $N = R$.

A New Definition of Dimension

To consider fractal dimensions that are between the ordinary dimensions of 0, 1, 2, and 3, we need to define *dimension* in a way that goes beyond the basic definition introduced in Unit 10A. To find this new definition, we begin by considering what happens when we measure straight lines with rulers of varying length. If we measure a 1-inch line segment with a 1-inch ruler, we find *one* 1-inch element (Figure 10.52). Using a ruler that is smaller by a factor of 2, or $\frac{1}{2}$ inch in length, we find *two* elements along the 1-inch line segment. If we choose a ruler that is smaller by a factor of 4, or $\frac{1}{4}$ inch in length, we find *four* elements of length along the 1-inch line segment.

Let's use R to represent the *reduction factor* in the length of the ruler and N to represent the factor by which the *number of elements* increases. We can now restate our results for the line segment:

- Reducing the ruler length by a *reduction factor* $R = 2$ (to $\frac{1}{2}$ inch) leads to an increase in the *number of elements* by a factor $N = 2$.
- Reducing the ruler length by a *reduction factor* $R = 10$ (to $\frac{1}{10}$ inch) leads to an increase in the *number of elements* by a factor $N = 10$.

Generalizing for the line segment, we find that decreasing the ruler length by a *reduction factor* R leads to an increase in the *number of elements* by a factor $N = R$.

We can use a similar process to determine the area of a square by counting the number of *area elements* that fit within it (Figure 10.53). Using a ruler that has the same 1-inch length as a side of the square, we can measure a single area element that fits in the square. If we reduce the ruler length by a factor $R = 2$, to $\frac{1}{2}$ inch, we find that we can fit $N = 4$ times as many elements in the square. Reducing the ruler length by $R = 4$, to $\frac{1}{4}$ inch, allows us to fit $N = 16$ times as many elements in the square. Generalizing for the square, we find that reducing the ruler length by a reduction factor R increases the number of area elements by a factor $N = R^2$.

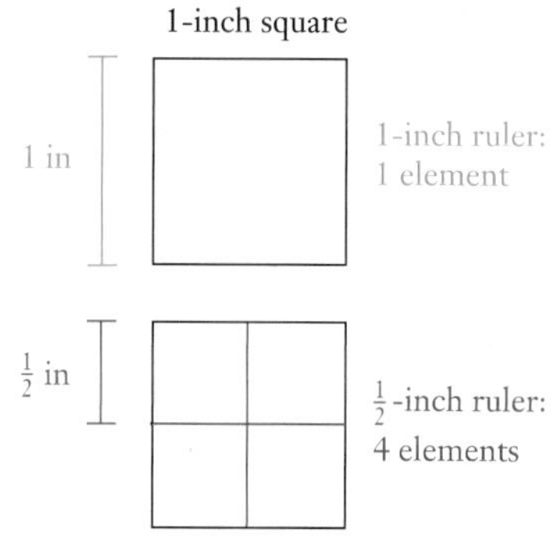

FIGURE 10.53 For a square, reducing ruler length by a factor R increases the number of elements by a factor $N = R^2$.

For a cube, we can count the number of *volume elements* that fit within it (Figure 10.54). A ruler with the same 1-inch length as a side of the cube measures a single volume element that fits in the cube. This time, reducing the ruler length by a factor $R = 2$, to $\frac{1}{2}$ inch, allows us to fit $N = 8$ times as many volume elements in the cube. Reducing the ruler length by $R = 4$, to $\frac{1}{4}$ inch, we find we can fit $N = 64$ times as many elements in the cube. Generalizing for the cube, we find that reducing the ruler length by a factor R increases the number of volume elements by a factor $N = R^3$.

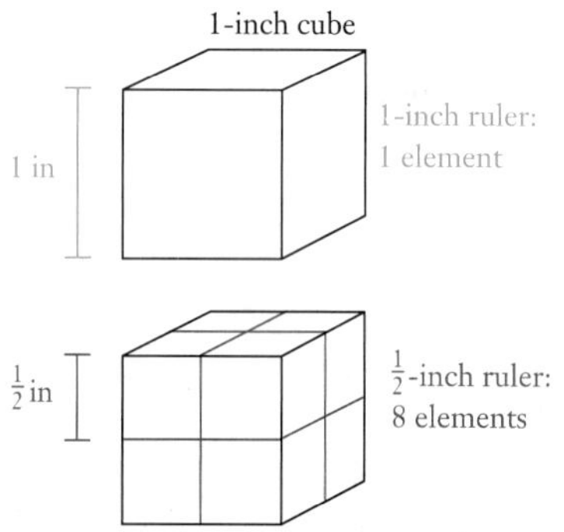

FIGURE 10.54 For a cube, reducing ruler length by a factor R increases the number of elements by a factor $N = R^3$.

Let's summarize our results:

- For a one-dimensional object (such as a line segment), we found $N = R^1$.
- For a two-dimensional object (such as a square), we found $N = R^2$.
- For a three-dimensional object (such as a cube), we found $N = R^3$.

In each case, the dimension of the object shows up as a power of R. We use this idea to define an object's *fractal dimension*.

Definition

The **fractal dimension** of an object is defined as a number D such that

$$N = R^D$$

where N is the factor by which the number of elements increases when we shorten the ruler by a reduction factor R.

EXAMPLE 1 Finding a Fractal Dimension

In measuring an object, every time you reduce the length of your ruler by a factor of 3, the number of elements increases by a factor of 4. What is the fractal dimension of this object?

Solution Reducing the ruler length by a factor $R = 3$ increases the number of elements by a factor $N = 4$. We are therefore looking for a fractal dimension D such that

$$4 = 3^D$$

We now solve for D as follows.

Start with the given equation:	$4 = 3^D$
Take the logarithm of both sides:	$\log_{10} 4 = \log_{10} 3^D$
On the right side, apply the rule that $\log_{10} a^x = x \log_{10} a$:	$\log_{10} 4 = D \log_{10} 3$
Interchange the left and right sides and divide both sides by $\log_{10} 3$:	$D = \dfrac{\log_{10} 4}{\log_{10} 3} \approx 1.2619$

The *fractal dimension* of this object is about 1.2619. ▶ Now try Exercises 15–26.

The Snowflake Curve

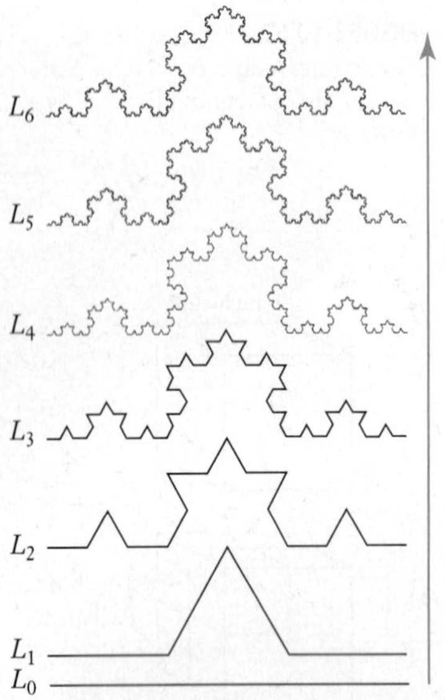

FIGURE 10.55 The generating process for the snowflake curve, from L_0 (bottom) to L_6 (top).

What kind of object could have a fractal dimension like the one we calculated in Example 1? Let's consider a special object called a **snowflake curve**, generated by a process that begins with a straight-line segment. As shown at the bottom of Figure 10.55, we designate the starting line segment L_0, which has a length of 1. We then generate L_1 with the following three steps:

1. Divide the line segment L_0 into three equal pieces.
2. Remove the middle piece.
3. Replace the middle piece with two segments of the same length arranged as two sides of an equilateral triangle.

Note that L_1 consists of four line segments and that each has a length of $\frac{1}{3}$ (because L_0 was divided into three equal pieces).

Next, we repeat the three steps on *each* of the four segments of L_1. The result is L_2 which has 16 line segments, each of length $\frac{1}{9}$.

Think About It Count the segments shown in Figure 10.55 to confirm that L_2 has 16 line segments. Measure to confirm that each is $\frac{1}{9}$ the length of L_0. Why are the segments $\frac{1}{9}$ the length of L_0? How long are the segments of L_3?

BY THE WAY The snowflake curve is sometimes called a *Koch curve* after Helge von Koch, who first described it in 1906.

Repeating the three-step process on each segment of L_2 generates L_3, repeating it again on each segment of L_3 produces L_4, and so forth. If we could repeat this process an *infinite* number of times, the snowflake curve, denoted L_∞, would be the ultimate result. Any figure that we actually draw, whether L_6 or $L_{1,000,000}$, can only be an approximation to the true snowflake curve.

Now imagine measuring the length of the complete snowflake curve, L_∞. A ruler with a length of 1 would simply lie across the base of the snowflake curve, missing all fine detail and measuring only the straight-line distance between the endpoints. This ruler would yield only one element along the snowflake curve. Reducing the ruler length by a factor $R = 3$, to $\frac{1}{3}$, would make it the same length as each of the four segments of L_1. This ruler would find four elements along the snowflake curve, or $N = 4$ times as many elements as the first ruler. Reducing the ruler length by another factor $R = 3$, to $\frac{1}{9}$, would make it the length of each of the 16 segments of L_2. This new ruler would find 16 elements along the snowflake curve, or $N = 4$ times as many elements as the previous ruler.

In general, every time we reduce the ruler length by another factor $R = 3$, we find $N = 4$ times as many elements. This is exactly the situation described in Example 1, which means that the snowflake curve has the fractal dimension $D \approx 1.2619$.

What does it mean to have a fractal dimension of 1.2619? The fact that the fractal dimension is greater than 1 means that the snowflake curve has more "substance" than an ordinary one-dimensional object. In a sense, the snowflake curve begins to fill the part of the plane in which it lies. The closer the fractal dimension of an object is to 1, the more closely it resembles a collection of line segments. The closer the fractal dimension is to 2, the closer it comes to filling a part of a plane.

EXAMPLE 2 How Long Is a Snowflake Curve?

How much longer is L_1 than L_0? How much longer is L_2 than L_0? Generalize your results, and discuss the length of a complete snowflake curve.

Solution Recall that L_0 has a length of 1. Figure 10.55 shows that L_1 consists of four line segments, each of length $\frac{1}{3}$. Its length is therefore $\frac{4}{3}$ times the length of L_0 or $\frac{4}{3}$:

$$\text{length of } L_1 = \tfrac{4}{3}$$

L_2 has 16 line segments, each of length $\frac{1}{9}$. Its length is

$$\text{length of } L_2 = \tfrac{16}{9} = \left(\tfrac{4}{3}\right)^2$$

Generalizing, we find that

$$\text{length of } L_n = \left(\tfrac{4}{3}\right)^n$$

Because $\left(\frac{4}{3}\right)^n$ grows without bound as n gets larger, we conclude that the complete snowflake curve, L_∞, must be infinitely long. ▶ Now try Exercise 27.

The Snowflake Island

The **snowflake island** is a *region* (an island) bounded by three snowflake curves. The process of drawing the snowflake island begins with an equilateral triangle (Figure 10.56). Then we convert *each* of the three sides of the triangle into a snowflake curve, L_∞. We cannot draw the complete snowflake island because it would require an infinite number of steps, but Figure 10.56 shows the results where the sides are L_0, L_1, L_2, and L_6.

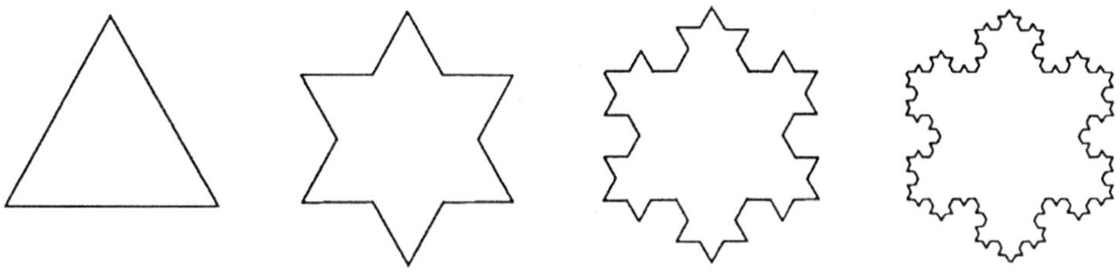

FIGURE 10.56 Left to right, successive approximations to a snowflake island, with sides made from L_0, L_1, L_2, and L_6, respectively.

The snowflake island illustrates some truly extraordinary properties. In Example 2, we found that a single snowflake curve is infinitely long. Therefore, the coastline of the snowflake island also must be infinitely long, because it is made of three snowflake curves. However, Figure 10.56 shows that the island's area is clearly contained within the bounds of the page. We have the intriguing result that a snowflake island is an object with a *finite area* and an *infinitely long boundary.*

Real Coastlines and Borders

Note that each of the four pieces of L_2 in Figure 10.55 looks exactly like L_1, except smaller. Similarly, L_3 consists of four pieces that each look like L_2. In fact, if we magnify *any* piece of the snowflake curve, L_∞, it will look exactly like one of the earlier curves, $L_0, L_1, L_2, \ldots$, used in its generation. Because the snowflake curve looks similar to itself when examined at different scales, we say that it is a **self-similar fractal**. The snowflake curve's self-similarity is due to repeated application of a simple set of rules.

BY THE WAY

Lewis Fry Richardson was an eccentric English scientist who proposed computer methods for predicting the weather in 1920—before computers were invented!

Natural objects, such as real coastlines, also reveal new details under higher magnification. Unlike a self-similar fractal, a natural object isn't likely to have segments that look *exactly* the same when magnified. Nevertheless, if we have data about how measured lengths change with different "ruler" sizes, we can still assign a fractal dimension to a natural object.

The first significant data concerning the fractal dimensions of natural objects were collected by Lewis Fry Richardson in about 1960. Richardson's data represented measurements and estimates of the lengths of various coastlines and international borders measured by "rulers" of varying sizes. His data suggest that most coastlines have a fractal dimension of about $D = 1.25$, which is very close to the fractal dimension of a snowflake curve.

EXAMPLE 3 The Fractal Border of Spain and Portugal

Portugal claims that its international border with Spain is 987 kilometers in length. Spain claims that the border is 1214 kilometers in length. However, the two countries agree on the location of the border. How is this possible?

Solution The border follows a variety of natural objects including rivers and mountain ranges. It is therefore a fractal, with more and more detail revealed on closer and closer examination. Like the snowflake curve, the border will be *longer* if it is measured with a shorter "ruler." It is therefore possible for Spain and Portugal to agree on the border's location but disagree on its length if they measured the border with "rulers" of different length. Spain claims a longer border length, so it must have used a shorter ruler for the measurement. ▶ **Now try Exercise 28.**

The Fascinating Variety of Fractals

The process of repeatedly applying a rule over and over to generate a self-similar fractal is called **iteration**. Different sets of rules can produce a fascinating variety of self-similar fractals. Let's look at just a few.

FIGURE 10.57 Several steps (top to bottom) in generating the Cantor set.

Consider a fractal generated from a line segment to which the following rule is applied repeatedly: *Delete the middle third of each line segment of the current figure* (Figure 10.57). With each iteration, the line segments become shorter until eventually the line turns to dust. The limit (after infinitely many iterations) is a fractal called the **Cantor set**. Because this ephemeral structure results from diminishing a one-dimensional line segment, its fractal dimension is less than 1.

Another interesting fractal, called the **Sierpinski triangle**, is produced by starting with a solid black equilateral triangle and iterating with the following rule: *For each black triangle in the current figure, connect the midpoints of the sides and remove the resulting inner triangle* (Figure 10.58). In the complete Sierpinski triangle, which would require *infinitely* many iterations to produce, every remaining black triangle would be infinitesimally small.

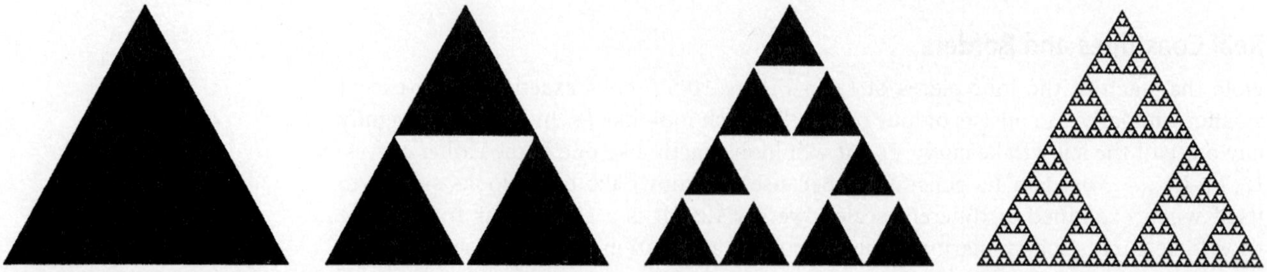

FIGURE 10.58 Several steps (left to right) in generating the Sierpinski triangle.

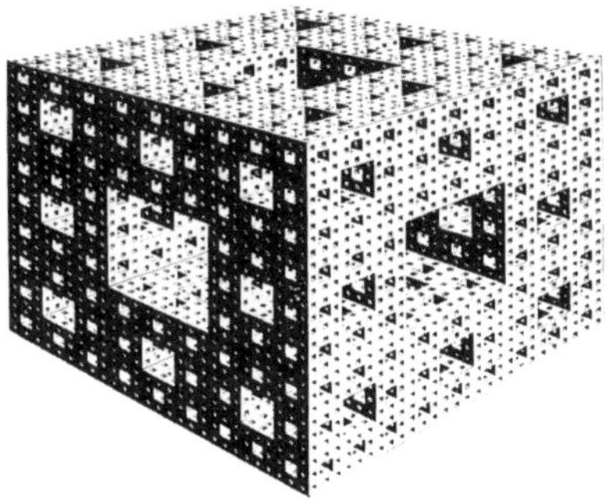

FIGURE 10.59 An approximation to a Sierpinski sponge.

Therefore, the total area of the black regions in the complete Sierpinski triangle is zero. The fractal dimension of the Sierpinski triangle is between 1 and 2. It is less than 2 because "material" has been removed from the initial two-dimensional triangle.

A closely related object is the **Sierpinski sponge** (Figure 10.59). It is generated by starting with a solid cube and iterating with this rule: *Divide each cube of the current object into 27 identical subcubes and remove the central subcube and the center cube of each face*. The resulting object has a fractal dimension between 2 and 3. It has less than a full three dimensions because material has been removed from the space occupied by the sponge.

An infinite variety of fractals can be generated by iteration. Some are remarkably beautiful, such as the famous **Mandelbrot set** (Figure 10.60).

All the self-similar fractals we have considered so far are created by applying the exact same set of rules in each iteration. An alternative approach, called **random iteration**, introduces slight random variation in every iteration. The resulting fractals therefore are *not* precisely self-similar, but they are close. Such fractals often appear remarkably realistic. For example, Barnsley's fern (Figure 10.61) is a fractal produced by random iteration, and it looks very much like a real fern.

The fact that fractals so successfully replicate natural forms suggests an intriguing possibility: Perhaps nature produces the many diverse forms that we see around us through simple rules that are applied repeatedly and with a hint of randomness. Because of this observation and because modern computers can test iterative processes, fractal geometry surely will remain an active field of research for decades to come.

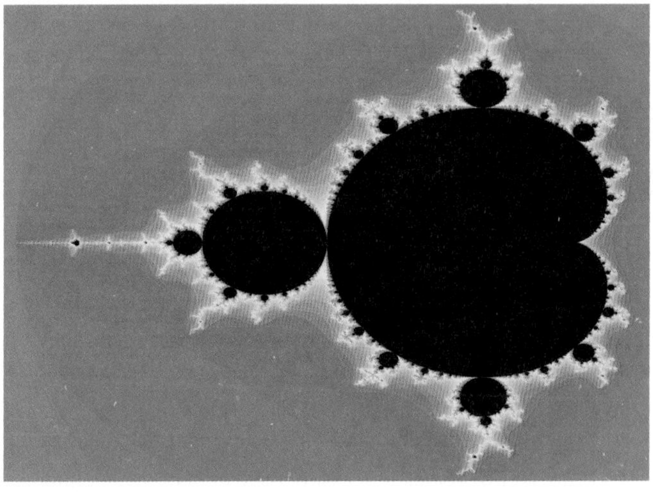

FIGURE 10.60 The Mandelbrot set.

FIGURE 10.61 Barnsley's fern.

Quick Quiz 10C

Choose the best answer to each of the following questions. Explain your reasoning with one or more complete sentences.

1. Fractal geometry is useful because
   a. it is the only type of geometry in which fractional answers (rather than only integer answers) are possible.
   b. it can be used to create shapes that look like naturally existing shapes.
   c. it is the well-tested geometry invented by the ancient Greeks.
2. Suppose you measure the length of a coastline with a ruler that is precise only to the nearest meter, then measure it again with a ruler that is precise to the nearest millimeter. The second measurement will be
   a. larger than the first. b. smaller than the first.
   c. the same as the first.
3. Which of the following has a shape that makes it a fractal?
   a. a perfect square b. a leaf
   c. a table top
4. Which of the following is a general characteristic of a fractal?
   a. It has an infinite area.
   b. You need a ruler to measure its length.
   c. When you look at it with greater magnification, you see new details.
5. How do fractal dimensions differ from dimensions in Euclidean geometry?
   a. They can be greater than 3.
   b. They can be negative.
   c. They can have fractional values.
6. An island coastline has a fractal dimension
   a. less than 0.
   b. between 0 and 1.
   c. between 1 and 2.
7. Which statement about the fractal object known as a snowflake curve is *not* true?
   a. It is the curve labeled L_6 in Figure 10.55.
   b. It is infinite in length.
   c. It can be laid out within a flat plane.
8. According to fractal geometry, which of the following is *not* possible?
   a. a finite area bounded by an infinitely long curve
   b. a finite area bounded by a finite curve
   c. an infinite area bounded by a finite curve
9. What characterizes a self-similar fractal?
   a. The same pattern is seen under greater magnifications.
   b. It can be built entirely from similar triangles.
   c. Its fractal dimension is the same as its ordinary (Euclidean) dimension.
10. Which object represented in this unit has an infinite surface area but a finite volume?
    a. the snowflake island
    b. the Sierpinski sponge
    c. Barnsley's fern

Exercises 10C

REVIEW QUESTIONS

1. What is a *fractal*? Explain why measuring the length or perimeter of a fractal with a shorter ruler leads to a longer measurement.
2. Why do fractal dimensions fall in between the ordinary dimensions of 0, 1, 2, and 3?
3. Explain the meaning of the factors R and N used in calculating fractal dimensions.
4. What is the *snowflake curve*? Explain why we cannot actually draw it, but can draw only partial representations of it.
5. What is the *snowflake island*? Explain how it can have an infinitely long coastline, yet have a finite area.
6. What do we mean by a *self-similar fractal*? How is a self-similar fractal, like the snowflake curve, similar to a real coastline? How is it different?
7. Briefly describe what we mean by the process of *iteration* in generating fractals. Describe the generation of the Cantor set, the Sierpinski triangle, and the Sierpinski sponge. Describe the fractal dimension of each.
8. What is *random iteration*? Why do objects generated by random iteration make scientists think that fractals are important in understanding nature?

DOES IT MAKE SENSE?

Decide whether each of the following statements makes sense (or is clearly true) or does not make sense (or is clearly false). Explain your reasoning.

9. I can use a yardstick to find the area of my rectangular patio.
10. I can use a yardstick to measure the length of the mountain skyline accurately.

11. The area of the snowflake island is given by its length times its width.

12. The measured length of a fractal increases as the ruler length increases.

13. The edge of this leaf has a fractal dimension of 1.34.

14. This entire leaf, riddled with holes, has a fractal dimension of 1.87.

BASIC SKILLS & CONCEPTS

15–26: Ordinary and Fractal Dimensions. Find the dimension of each object, and state whether or not it is a fractal.

15. In measuring the length of the object, when you reduce the length of your ruler by a factor of 2, the number of length elements increases by a factor of 2.

16. In measuring the area of the object, when you reduce the length of your ruler by a factor of 2, the number of area elements increases by a factor of 4.

17. In measuring the volume of the object, when you reduce the length of your ruler by a factor of 2, the number of volume elements increases by a factor of 8.

18. In measuring the length of the object, when you reduce the length of your ruler by a factor of 2, the number of length elements increases by a factor of 3.

19. In measuring the area of the object, when you reduce the length of your ruler by a factor of 2, the number of area elements increases by a factor of 6.

20. In measuring the volume of the object, when you reduce the length of your ruler by a factor of 2, the number of volume elements increases by a factor of 12.

21. In measuring the length of the object, when you reduce the length of your ruler by a factor of 5, the number of length elements increases by a factor of 5.

22. In measuring the area of the object, when you reduce the length of your ruler by a factor of 5, the number of area elements increases by a factor of 25.

23. In measuring the volume of the object, when you reduce the length of your ruler by a factor of 5, the number of volume elements increases by a factor of 125.

24. In measuring the length of the object, when you reduce the length of your ruler by a factor of 5, the number of length elements increases by a factor of 7.

25. In measuring the area of the object, when you reduce the length of your ruler by a factor of 5, the number of area elements increases by a factor of 30.

26. In measuring the volume of the object, when you reduce the length of your ruler by a factor of 5, the number of volume elements increases by a factor of 150.

27. **The Quadric Koch Curve and Quadric Koch Island.** To draw the *quadric Koch curve* (one of many variations of the snowflake curve), one begins with a horizontal line segment and applies the following rule: *Divide each line segment into four equal pieces; replace the second piece with three line segments of equal length that make the shape of a square above the original piece; and replace the third piece with three line segments making a square below the original piece.* The quadric Koch curve would result from infinite applications of the rule on each resulting segment (applying the rule to both horizontal and vertical segments); the first three stages of the construction are shown in Figure 10.62.

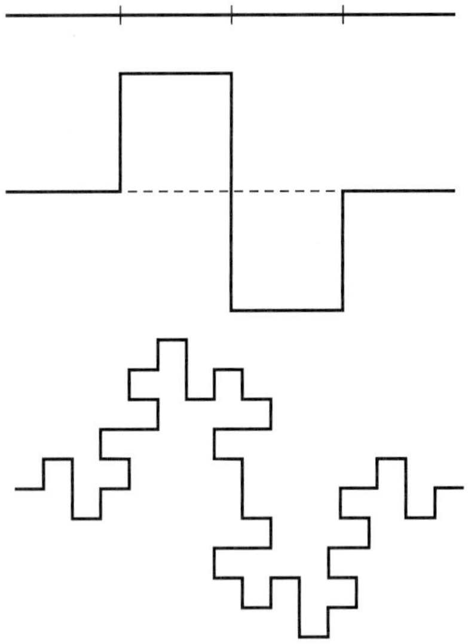

FIGURE 10.62

a. Determine the relation between N and R for the quadric Koch curve.

b. What is the fractal dimension of the quadric Koch curve? Can you draw any conclusions about the length of the quadric Koch curve? Explain.

c. The *quadric Koch island* is constructed by beginning with a square and then replacing each of the four sides of the square with a quadric Koch curve. Explain why the total area of the quadric Koch island is the same as the area of the original square. How long is the coastline of the quadric Koch island?

28. **Fractal Dimension from Measurements.** An ambitious and patient crew of surveyors has used various rulers to measure the length of the coastline of Dragon Island. The following table gives the measured length of the coastline, L, and the length of the ruler used, r.

r meters	100	10	1	0.1	0.01	0.001
L meters	315	1256	5000	19,905	79,244	315,479

a. Extend the table by adding rows of entries for the values of $\log_{10} r$ and $\log_{10} L$.

b. Graph these data $(\log_{10} r, \log_{10} L)$ on a set of axes. Connect the data points.

c. If the graph of the data is close to a straight line, it is an indication that the coastline is a self-similar fractal. Does the coastline of Dragon Island appear to be a self-similar fractal?

d. What is the approximate slope s of the line on your graph? The fractal dimension of the coastline is $D = 1 - s$. What is the fractal dimension of the coastline of Dragon Island?

FURTHER APPLICATIONS

29. **The Cantor Set.** Recall that the Cantor set is formed by starting with a line segment and removing the middle one-third of it, then repeating this rule on each successive figure (Figure 10.57). If a ruler the length of the original line segment is used, it detects one element in the Cantor set because it can't "see" details smaller than itself. If the ruler is reduced in size by a factor of $R = 3$, it finds two elements (only solid pieces of line, not holes, are measured). If the ruler is reduced in size by a factor of $R = 9$, how many elements does it find? Based on these results, what is the fractal dimension of the Cantor set? Explain why this number is less than 1.

30. **Ordinary Dimensions for Ordinary Objects.**

    a. Suppose you want to measure the length of the sidewalk in front of your house. Describe a thought process by which you can conclude that $N = R$ for the sidewalk and that its fractal dimension is therefore the same as its ordinary dimension of 1.

    b. Suppose you want to measure the area of your living room floor, which is square. Describe a thought process by which you can conclude that $N = R^2$ for the living room and that its fractal dimension is therefore the same as its ordinary dimension of 2.

    c. Suppose you want to measure the volume of a cubical swimming pool. Describe a thought process by which you can conclude that $N = R^3$ for the pool and that its fractal dimension is therefore the same as its ordinary dimension of 3.

31. **Fractal Dimensions for Fractal Objects.**

    a. Suppose you are measuring the length of the stream frontage along a piece of mountain property. You begin with a 15-meter ruler and find just one element along the length of the stream frontage. When you switch to a 1.5-meter ruler, you are able to trace finer details of the stream edge and you find 20 elements along its length. Switching to a 15-centimeter ruler, you find 400 elements along the stream frontage. Based on these measurements, what is the fractal dimension of the stream frontage?

    b. Suppose you are measuring the area of a very unusual square leaf with many holes, perhaps from hungry insects, in a fractal pattern (e.g., similar to the Sierpinski triangle, Figure 10.58). You begin with a 10-centimeter ruler and find that it lies over the entire square, making just one element. When you switch to a 5-centimeter ruler, you are better able to cover areas of leaf while skipping areas of holes and you find 3 area elements. You switch to a 2.5-centimeter ruler and find 9 area elements. Based on these measurements, what is the fractal dimension of the leaf? Explain *why* the fractal dimension is less than 2.

    c. Suppose you are measuring the volume of a cube cut from a large rock that contains many cavities forming a fractal pattern. Beginning with a 10-meter ruler, you find just one volume element. Smaller rulers allow you to ignore cavities, gauging only the volume of rock material. With a 5-meter ruler, you find 6 volume elements. With a 2.5-meter ruler, you find 36 volume elements. Based on these measurements, what is the fractal dimension of the rock? Explain why a fractal dimension between 2 and 3 is reasonable. (Ignore the practical difficulties caused by the fact that you cannot see through a rock to find all its holes!)

32. **Fractal Patterns in Nature.** Identify at least five natural objects that exhibit fractal patterns. In each case, explain the structure that makes the pattern a fractal, and estimate its fractal dimension.

33. **Natural Fractals Through Branching.** One way natural objects reveal fractal patterns is by branching. For example, the intricate structure of the human lung, the web of capillaries in muscle tissue, the branches or roots of a tree, and the successive divisions of streams in a river delta all involve branching at different spatial scales. Explain why structures formed by branching resemble self-similar fractals. Further, explain why fractal geometry rather than ordinary geometry leads to a greater understanding of such structures.

IN YOUR WORLD

34. **Fractal Research.** Locate at least two websites devoted to fractals, and use them to write a two- to three-page paper on either a specific use of fractals or a technique for generating fractals.

35. **Fractal Art.** Visit a website that features fractal art. Choose a specific piece of art, explain how it was generated, and discuss its visual impact.

Chapter 10 Summary

UNIT	KEY TERMS	KEY IDEAS AND SKILLS

10A	Euclidean geometry (p. 579)
	point, line, plane (p. 579)
	dimension (p. 579)
	angle (p. 580)
	  right angle (p. 580)
	  straight angle (p. 580)
	  acute angle (p. 580)
	  obtuse angle (p. 580)
	circle (p. 581)
	  radius (p. 581)
	  diameter (p. 581)
	  circumference (p. 582)
	polygon (p. 581)
	  triangle (p. 582)
	  square (p. 582)
	  parallelogram (p. 584)
	perimeter (p. 582)
	area (p. 584)
	surface area (p. 585)
	volume (p. 585)
	sphere (p. 585)
	cube (p. 585)
	rectangular prism (p. 585)
	cylinder (p. 585)
	scaling laws (p. 588)
	surface-area-to-volume ratio (p. 588)
		Understand basic concepts of Euclidean geometry. (p. 579)
		Know and use area and perimeter formulas for two-dimensional figures. (p. 582)
		Know and use surface area and volume formulas for three-dimensional figures. (p. 585)
		Understand the use of scaling laws and scale factors: (p. 587)
		  Lengths scale with the scale factor.
		  Areas scale with the square of the scale factor.
		  Volumes scale with the cube of the scale factor.
		Understand the implications of the surface-area-to-volume ratio. (p. 588)

10B	degrees, minutes, seconds (p. 593) latitude, longitude (p. 594) prime meridian, equator (p. 594) angular size (p. 595) pitch (p. 597) slope (p. 597) grade (p. 597) similar triangles (p. 599) optimization (p. 601)	**Understand** various ways to describe angles. (p. 593) **Know how** to locate points and find distances on the Earth given values for longitude and latitude. (p. 594) **Understand** angular size. (p. 595) **Know how** to use the Pythagorean theorem. (p. 598) **Understand** various ways to measure distances. (p. 598) **Understand** and use properties of similar triangles to solve problems. (p. 599) **Understand** the goal of optimization problems. (p. 601)
10C	fractal geometry (p. 608) fractal dimension (p. 610) snowflake curve (p. 612) snowflake island (p. 613) self-similar fractal (p. 613) iteration (p. 614)	**Explain** why different rulers can give different length measurements of natural objects. (p. 608) **Fractal dimension:** $N = R^D$, where N is the factor by which the number of elements increases and R is the factor by which the ruler length is reduced. (p. 610) **Understand** the uses of fractal geometry. (p. 614)

11 MATHEMATICS AND THE ARTS

The connections between art and mathematics go deep into history. The ties are most evident in architecture: the Great Pyramids in Egypt, the Eiffel Tower in France, and modern-day skyscrapers all required mathematics in their design. But mathematics has contributed in equally profound ways to music, painting, and sculpture. In this chapter, we explore a few of the many connections between mathematics and the arts.

Q

Which of the following statements best describes the role mathematics played for great Renaissance painters such as Leonardo da Vinci and Raphael?

- **A** We can analyze some aspects of Renaissance paintings with the tools of mathematics, but the painters themselves were unfamiliar with the mathematics we use.
- **B** Almost all Renaissance painters worked strictly "by eye" and did not employ any mathematical techniques.
- **C** Renaissance painters employed a few techniques of ancient Greek geometry.
- **D** Renaissance painters learned and in some cases invented mathematical techniques that they used in their paintings.
- **E** Renaissance painters calculated precise positions for virtually every drop of paint they applied to their canvases.

(Mathematics) seems to stand for all that is practical, poetry for all that is visionary, but in the kingdom of the imagination you will find them close akin, and they should go together as a precious heritage to every youth.

—Florence Milner, *School Review,* 1898

UNIT 11A

Mathematics and Music: Explore the connections between mathematics and music, particularly as they apply to the ideas of musical tones and scales.

UNIT 11B

Perspective and Symmetry: Investigate the mathematics of perspective and symmetry both in classical art and in modern tilings.

UNIT 11C

Proportion and the Golden Ratio: Understand the concept and use of proportion in art, and explore the famous golden ratio.

A Today, many people think of the arts as being very distinct from mathematics and science. In fact, mathematics and the arts have always gone hand in hand, and many of the great artists in history have also made important contributions to mathematics and science. Leonardo da Vinci is one of the best known of these *polymaths*, a word that means someone who has learned about many subjects (reminding us that the Greek word *mathematikos* means "inclined to learn"). The correct answer to the above question is therefore D, which means you cannot truly appreciate art or art history without an understanding of the underlying mathematics. You'll find a more detailed analysis of some of the mathematical techniques used in both Renaissance and modern art in Unit 11B.

Digital Music Files

Use this activity to gain a sense of the kinds of problems this chapter will enable you to analyze. Additional activities are available online in MyLab Math.

Have you wondered why you have so many choices for digital music, such as AAC versus MP3 or 128 kbps versus 256 kbps? You can understand the answer by exploring how music is encoded, and in the process you'll see how the long historical connection between mathematics and music has deepened in the digital age.

Music consists of sound waves, and sound waves are characterized by two variables: *frequency* (how fast the wave vibrates) and amplitude (which determines the volume of the sound). We can graph a sound wave by showing volume as a function of time (see Figure 11.A); that is, the height of the graph at each time represents volume. (Simple as it looks, the wave in the figure actually has several different frequencies.)

Recording and storing music digitally means converting an actual sound wave into a list of numbers that can be stored in a computer. Doing so requires *sampling* the wave at specific time intervals and measuring the volume of each sample. The dots in the figure show the volume of the sound wave at sample intervals of $0.2 = 1/5$ second, which is a *sampling rate* of 5 samples per second, or 5 hertz.

1. Connect the dots in the figure with straight line segments. Is the result a good representation of the original wave? Now use a sampling rate of 10 samples per second by adding five more dots along the wave (so there is a dot for every 0.1 second). Connect the ten dots with line segments. Does the representation improve? How high do you think the sampling rate should be to represent the wave faithfully?

2. A mathematical theorem (the Nyquist-Shannon sampling theorem) states that a faithful representation requires a sampling rate *at least* twice the maximum frequency of the wave. Real music has frequencies up to about 20,000 samples per second. What is the minimum sampling rate required to encode real music digitally?

3. A computer must store a value for the volume of every individual sample. Computers store numbers in binary format as *bits*: 1 bit represents $2^1 = 2$ possible values (usually denoted 0 and 1), 2 bits represents $2^2 = 4$ possible values (00, 10, 01, and 11), and so on. The horizontal green lines in the figure show the 4 possible values $(\text{volume} = 1,\ 2,\ 3,\ \text{or } 4)$ with a *bit depth* of 2 bits. Show where each of the five sample dots in the figure would go if you had to move it to the nearest green line. Is a bit depth of 2 bits enough to represent a sound wave faithfully? You can represent a bit depth of 3 bits ($2^3 = 8$ possible values) by adding another line between each pair of green lines (making 8 lines total). How does increasing the bit depth improve the representation of the wave?

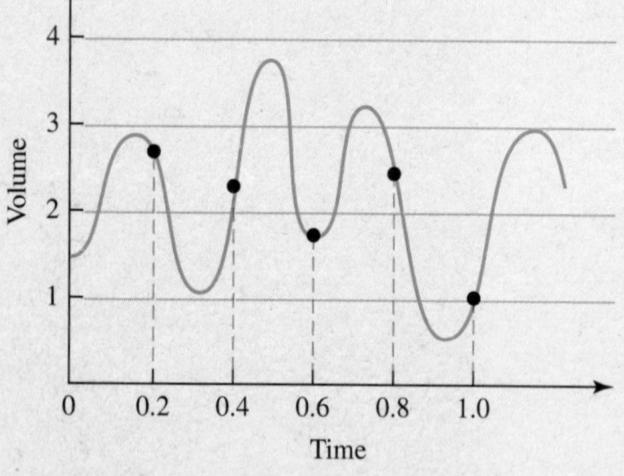

FIGURE 11.A A simple sound wave. The dots represent samples of the sound wave taken 5 times during 1 second, which means a sampling rate of 5 samples per second (5 hertz).

4. You should now see that the quality of a digital music file depends on both the sampling rate and the bit depth. Standard CDs encode data with a sampling rate of 44,100 samples per second (44,100 hertz), and each sample is stored with a bit depth of 16 bits (which allows $2^{16} = 65{,}536$ possible values). Stereo CDs have two stereo channels, so the total number of bits required to store music at "CD quality" is $44{,}100 \times 16 \times 2 = 1{,}411{,}200$ bits per second of data. Most music downloads have some of these bits removed so that the file sizes are smaller. A 128 kilobit per second (kbps) MP3 file contains about 128,000 bits per second of music. What fraction of the data needed for "CD quality" is retained in this MP3 file? How does this fact affect the number of songs you can fit onto a particular music player?

5. Suppose you are given the option of recording your MP3 files at 128, 256, or 512 kbps. How would you decide which to choose? If possible, record a favorite song at each of those three bit rates. Do you notice a difference in the sound quality?
6. As you can see, compressed music files are missing a lot of the original data from a CD recording. But there is more than one way to compress the data, and the difference among AAC, MP3, and other formats lies in the program used to do the compression. For example, a 128-kbps AAC file requires the same storage space as a 128-kbps MP3 file, but it stores a different set of bits and therefore the two files will not sound exactly alike. Try listening to the same song recorded in different formats; do you notice any difference? Do you think there is a "best choice" for a compression format? Do you think that it is worth the extra storage space required to use a *lossless format* that retains the original CD quality?

UNIT 11A Mathematics and Music

Music is the universal language of mankind.
—Henry Wadsworth Longfellow

The roots of mathematics and music are entwined in antiquity. Pythagoras (c. 500 B.C.E.) claimed that "all nature consists of harmony arising out of number." He imagined that the planets circled Earth on invisible heavenly spheres, obeying specific numeric laws and emitting the ethereal sounds known as the "music of the spheres." He therefore saw a direct connection between geometry and music. Connections between mathematics and music have been explored ever since, and the modern era of digital music has made the connections deeper than ever.

Sound and Music

Any vibrating object produces sound. The vibrations produce a **wave** (much like a water wave) that propagates through the surrounding air in all directions. When such a wave impinges on the ear, we perceive it as sound. Of course, some sounds, such as speech and screeching tires, do not qualify as music. Most musical sounds are made by vibrating strings (violins, cellos, guitars, and pianos), vibrating reeds (clarinets, oboes, and saxophones), or vibrating columns of air (pipe organs, horns, and flutes).

HISTORICAL NOTE

Evidence of music is found in nearly all ancient cultures. Indeed, based on evidence dating back 30,000 years, some archaeologists suspect that music may predate speech. String and wind instruments were designed at least 5000 years ago in Mesopotamia, and Egyptians played in small ensembles of lyres and flutes by 2700 B.C.E.

One of the most basic qualities of sound is **pitch**. For example, a tuba has a "lower" pitch than a flute, and a violin has a "higher" pitch than a bass guitar. To understand pitch, find a taut string (a guitar string works best, but a stretched rubber band will do). When you pluck the string, it produces a sound with a certain pitch. Next, use your finger to hold the midpoint of the string in place, and pluck either half of the string. A higher-pitched sound is produced, demonstrating an ancient musical principle discovered by the Greeks: *The shorter the string, the higher the pitch.*

It was many centuries before anyone understood why a shorter string creates a sound with a higher pitch, but we now know that it is due to a relationship between pitch and **frequency**. The frequency of a vibrating string is the rate at which it moves up and down. For example, a string that vibrates up and down 100 times each second (reaching the high and low points 100 times) has a frequency of 100 **cycles per second (cps)**; cycles or samples per second are also called *hertz* (Hz), which is the official metric (SI) unit of frequency. The sound waves produced by the string vibrate with the same frequency as the string, and the relationship between pitch and frequency of the sound wave is quite simple: *The higher the frequency, the higher the pitch.*

The lowest possible frequency for a particular string, called its **fundamental frequency**, occurs when it vibrates up and down along its full length (Figure 11.1a).

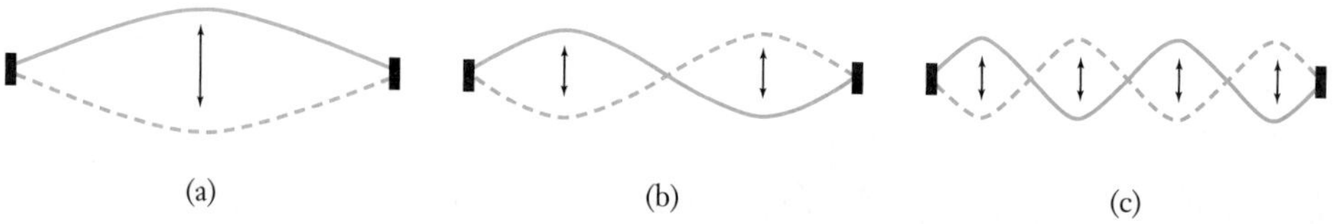

FIGURE 11.1 A string vibrating up and down at (a) its fundamental frequency, (b) twice its fundamental frequency (waves 1/2 as long as those at the fundamental frequency), and (c) four times its fundamental frequency (waves 1/4 as long as those at the fundamental frequency).

Every string has its own fundamental frequency, which depends on the length, density, and tension of the string. If you generate a wave that vibrates up and down along each half of the length of the string, that wave will have a frequency that is twice the fundamental frequency (Figure 11.1b). For example, if a string has a fundamental frequency of 100 cps, a wave that is half as long as the original wave has a frequency of 200 cps. Similarly, a wave one-quarter the length of the original wave has a frequency four times the fundamental frequency (Figure 11.1c). Waves like those shown in Figure 11.1b and Figure 11.1c are called **harmonics** of the fundamental frequency for that string; note that harmonics have frequencies that are integer multiples of the fundamental frequency. The same ideas apply to other types of musical instruments, with the pitch related to the frequency of vibration of a reed (for example, in a clarinet) or air column (for example, in an organ).

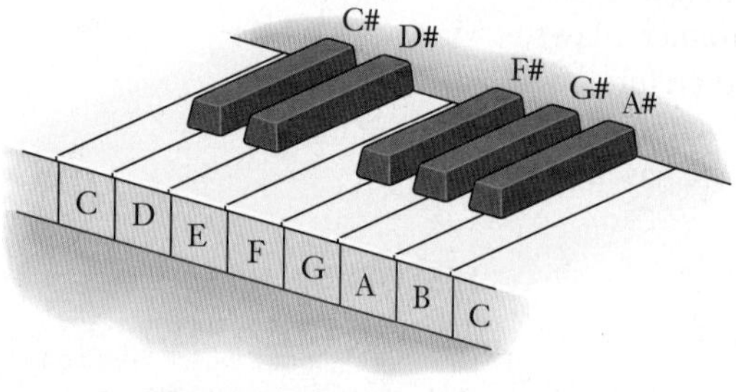

FIGURE 11.2 Piano keys.

The relationship between pitch and frequency helps explain another discovery of the ancient Greeks: Pairs of notes sound particularly pleasing and natural together when one note is an **octave** higher than the other note. We now know that raising the pitch by an octave corresponds to a doubling of frequency. The piano keyboard (Figure 11.2) is helpful here. An octave is the interval between, say, middle C and the next higher C. For example, middle C has a frequency of 260 cps, the C above middle C has a frequency of 2×260 cps $= 520$ cps, and the next higher C has a frequency of 2×520 cps $= 1040$ cps. Similarly, the C below middle C has a frequency of about $\frac{1}{2} \times 260$ cps $= 130$ cps.

Think About It The note middle A (above middle C) has a frequency of about 440 cps. What are the frequencies of the A notes an octave higher and an octave lower?

Scales

The musical tones that span an octave comprise a **scale**. The Greeks invented the 7-note (or diatonic) scale that corresponds to the white keys on the piano. In the 17th century, Johann Sebastian Bach adopted a 12-tone scale, which corresponds to both the white and the black keys on a modern piano. With Bach's music, the 12-tone scale spread throughout Europe, becoming a foundation of Western music. Many other scales are possible. For example, 3-tone scales are common in African music, scales with more than 12 tones occur in Asian music, and 19-tone scales are sometimes used in contemporary music.

On the 12-tone scale, two consecutive notes on the piano keyboard are separated by a **half-step**. For example, E and F are separated by a half-step, as are F and F# (read "F sharp"). For each half-step, the frequency increases by some *multiplicative* factor; let's call it f. The frequency of C# is the frequency of C times the factor f, the frequency of D is the frequency of C# times the factor f, and so on. The frequencies of the notes across the entire scale are related as follows:

$$C \xrightarrow{f} C\# \xrightarrow{f} D \xrightarrow{f} D\# \xrightarrow{f} E \xrightarrow{f} F \xrightarrow{f} F\# \xrightarrow{f} G \xrightarrow{f} G\# \xrightarrow{f} A \xrightarrow{f} A\# \xrightarrow{f} B \xrightarrow{f} C$$

BY THE WAY

Human speech consists of sounds with frequencies of 200 to 400 cycles per second. The range of a piano extends from about 27 to 4200 cycles per second. The maximum frequency audible to the human ear declines gradually with age, from about 20,000 cycles per second in children and young teenagers to about 12,000 cycles per second at age 50.

Because an octave corresponds to an increase in frequency by a factor of 2, the factor f must have the property

$$\underbrace{f \times f \times f \times f \times f \times f \times f \times f \times f \times f \times f \times f}_{12 \text{ times}} = f^{12} = 2$$

Because $f^{12} = 2$, we conclude that f is the *twelfth root* of 2, written $f = \sqrt[12]{2} \approx 1.05946$.

We can now calculate the frequency of every note of a 12-tone scale. Starting from middle C, with its frequency of 260 cps, we multiply by $f = \sqrt[12]{2}$. to find that the frequency of C# is 260 cps $\times f \approx$ 275 cps. Multiplying again by f gives the frequency of D as 275 cps $\times f \approx$ 292 cps. Continuing in this way generates Table 11.1.

The fourth column of Table 11.1 shows that a few tones have simple ratios with respect to middle C. For example, the frequency of G is approximately $\frac{3}{2}$ times the frequency of middle C (musicians call this interval a *fifth*), and the frequency of F is approximately $\frac{4}{3}$ times the frequency of middle C (musicians call this interval a *fourth*). Many musicians find the most pleasing combinations of notes, called **consonant tones**, to be those whose frequencies have a simple ratio. Referring to consonant tones, the Chinese philosopher Confucius observed that small numbers are the source of perfection in music.

TABLE 11.1 Frequencies of Notes in the Octave Above Middle C

Note	Frequency (cps)	Ratio to Frequency of Preceding Note	Ratio to Frequency of Middle C
C	260	$\sqrt[12]{2} \approx 1.05946$	1.00000 = 1
C#	275	$\sqrt[12]{2} \approx 1.05946$	1.05946
D (second)	292	$\sqrt[12]{2} \approx 1.05946$	1.12246
D#	309	$\sqrt[12]{2} \approx 1.05946$	1.18921
E (third)	328	$\sqrt[12]{2} \approx 1.05946$	$1.25992 \approx \frac{5}{4}$
F (fourth)	347	$\sqrt[12]{2} \approx 1.05946$	$1.33484 \approx \frac{4}{3}$
F#	368	$\sqrt[12]{2} \approx 1.05946$	1.41421
G (fifth)	390	$\sqrt[12]{2} \approx 1.05946$	$1.49831 \approx \frac{3}{2}$
G#	413	$\sqrt[12]{2} \approx 1.05946$	1.58740
A (sixth)	437	$\sqrt[12]{2} \approx 1.05946$	$1.68179 \approx \frac{5}{3}$
A#	463	$\sqrt[12]{2} \approx 1.05946$	1.78180
B (seventh)	491	$\sqrt[12]{2} \approx 1.05946$	1.88775
C (octave)	520	$\sqrt[12]{2} \approx 1.05946$	2.00000 = 2

BY THE WAY

All the entries in the third column are the same because the same factor, $f \approx \sqrt[12]{2}$ relates every pair of notes. The parenthetical terms in the first column are names used by musicians to describe intervals between the note shown and middle C.

EXAMPLE 1 The Dilemma of Temperament

Because the whole-number ratios in Table 11.1 are not exact, tuners of musical instruments have the problem of *temperament,* which can be demonstrated as follows. Start at middle C with a frequency of 260 cps. Using the whole-number ratios, find the frequency if you raise C by a sixth to A, raise A by a fourth to D, lower D by a fifth to G, and lower G by a fifth to C. Having returned to the same note, have you also returned to the same frequency?

Solution According to Table 11.1, raising a note by a sixth increases its frequency by a factor of approximately $\frac{5}{3}$. For example, the frequency of A above middle C is

$\frac{5}{3} \times 260$ cps ≈ 433.33 cps. Raising this note by a fourth increases its frequency by $\frac{4}{3}$, producing D with a frequency of $\frac{4}{3} \times 433.33$ cps ≈ 577.77 cps. Lowering D by a fifth (a factor of $\frac{2}{3}$) to G gives a frequency of $\frac{2}{3} \times 577.77 \approx 385.18$ cps. Finally, lowering G by another fifth (a factor of $\frac{2}{3}$) puts us back to middle C, but with a frequency of $\frac{2}{3} \times 385.18$ cps ≈ 256.79. Note that, by using whole-number ratios, we have not quite returned to the proper frequency of 260 cps for middle C. The problem is that the whole-number ratios are not exact. That is, $\frac{5}{3} \times \frac{4}{3} \times \frac{2}{3} \times \frac{2}{3} = \frac{80}{81}$ is close to, but not *exactly*, 1.

▶ **Now try Exercises 11–14.**

Exponential Growth and Musical Scales

Human speech is like a cracked kettle on which we tap crude rhythms for bears to dance to, while we long to make music that will melt the stars.

—Gustave Flaubert

The increase in frequencies in a scale is an example of exponential growth. Each successive frequency is $f \approx 1.05946$ times, or approximately 5.9% more than, the previous frequency. In other words, the frequencies increase at a fixed relative growth rate. We can therefore use an exponential function (see Unit 9C) to find any frequency on the scale. Suppose we start at a frequency Q_0. Then the frequency Q of the note n half-steps higher is given by

$$Q = Q_0 \times f^n \approx Q_0 \times 1.05946^n$$

EXAMPLE 2 **Exponential Growth in Musical Scales**

Use the exponential growth law to find the frequency of the note a fifth above middle C, the note one octave and a fifth above middle C, and the note two octaves and a fifth above middle C.

Solution We let the frequency of middle C be the initial value for the scale; that is, we set $Q_0 = 260$ cps. Table 11.1 shows that the note a fifth above middle C is G, which is seven half-steps above middle C. Therefore, we let $n = 7$ in the exponential law and find that the frequency of G is

$$Q \approx Q_0 \times 1.05946^7 \approx 390 \text{ cps}$$

The note one octave and a fifth above middle C is $12 + 7 = 19$ half-steps above middle C. Letting $n = 19$, we find that the frequency of this note is

$$Q \approx Q_0 \times 1.05946^{19} \approx 779 \text{ cps}$$

The note two octaves and a fifth above middle C is $(2 \times 12) + 7 = 31$ half-steps above middle C. Letting $n = 31$, we find that the frequency of this note is

$$Q \approx Q_0 \times 1.05946^{31} \approx 1558 \text{ cps}$$

▶ **Now try Exercises 15–16.**

From Tones to Music

BY THE WAY

For more than a thousand years, the standard secondary school curriculum consisted of what was called the *quadrivium* (Latin for "crossroads"), made up of the four subject areas arithmetic, geometry, music, and astronomy. The choice of these four studies was first outlined by Plato in *The Republic*, and they became the standard for the equivalent of a master's degree at medieval universities. They continued to be the core subjects taught in high school until the late 19th century.

Although the simple frequencies of "pure" tones are the building blocks of music, the sounds of music are far richer and more complex. For example, a plucked violin string does much more than produce a single frequency. The vibration of the string is transferred through the bridge of the violin to its top, and the ribs transfer those vibrations to the back of the instrument. With the top and back of the violin in oscillation, the entire instrument acts as a resonating chamber, which excites and amplifies many harmonics of the original tone.

Similar principles generate rich and complex sounds in all instruments. The wave on the left side of Figure 11.3 represents a typical sound wave that might be produced by an instrument. It isn't a simple wave like those pictured in Figure 11.1. Instead, it consists of a combination of simple waves that are the harmonics of the fundamental. In this case, the wave on the left is the *sum* of the three simple waves shown on the right in Figure 11.3. The fact that a musical sound can be expressed as a sum of

IN YOUR WORLD

Music Just for You

In this unit, we focus primarily on the mathematics that is actually *in* music through harmonics and scales, but today music intersects mathematics in a second and very different way: through your music listening choices. If you use any type of music streaming service, you are probably offered the option to let the service choose songs that you might like. Creating these "personal playlists" involves some very sophisticated mathematics and data analysis.

To get a sense of the challenge, first consider some numbers. Music services typically offer tens of millions of songs; let's say 30 million to use a round number. If we assume that the average song is $2\frac{1}{2}$ minutes long, then it would take about 75 million minutes—which is a little more than 140 years—to listen to all of a service's songs once, and that's assuming you could listen continuously, including while you're sleeping. Clearly, it's not possible for anyone to listen to even a small fraction of all the music that's out there.

So how does a music service guess at what you might like from the millions of songs that you've never heard? The answer lies with data and mathematical algorithms that sift the data in search of songs for you. In the case of music playlists, you can think of the data as consisting of two major parts: you and everyone else. The data for "you" consists of the music you already listen to, how often you listen to various songs or types of music, and any other musical preferences you may have made known (for example, by assigning ratings that indicate how much you like various songs). Mathematical algorithms then compare your data to those of all other people using the same music service to find songs that other people with similar taste are listening to, but which you might have missed.

Your personalized recommendations can be further refined as time goes on, based on how you respond to selections that have previously been offered. For example, if the service recommends a new song that you then listen to over and over, it can recognize that it made a good choice. Conversely, if you listen to a new song once and then never again, it can assume you did not like it. In this way, the service can refine your personal profile over time, enabling it to make better recommendations for you in the future.

Overall, the software a music service uses is really a form of artificial intelligence, because it learns how to make better selections for you as time goes on. Artificial intelligence has become increasingly important not only in music selection, but in many other areas of our lives. In the end, though, it is built from a combination of data, computer programming, and mathematics. As a result, just as you'll find that the music selected for you is not always perfect, you'll want to understand both the benefits and limitations of artificial intelligence in the many other areas in which it will affect your life—and that will always mean having at least some understanding of data, statistics, and mathematics.

simple harmonics is surely the deepest connection between mathematics and music. The French mathematician Jean Baptiste Joseph Fourier first enunciated this principle in about 1810. It was one of the most profound discoveries in mathematics.

Although mathematics helps in understanding music, many mysteries remain. For example, in about 1700, an Italian craftsman known as Stradivarius made what are still considered to be the finest violins and cellos ever produced. Despite years of study by mathematicians and scientists, no one has succeeded in reproducing the unique sounds of a Stradivarius instrument.

> *Stradivarius was essentially a craftsman of science, one with considerable, demonstrable knowledge of mathematics and acoustical physics.*
>
> —Thomas Levenson, *Measure for Measure: A Musical History of Science*, pp. 207–208

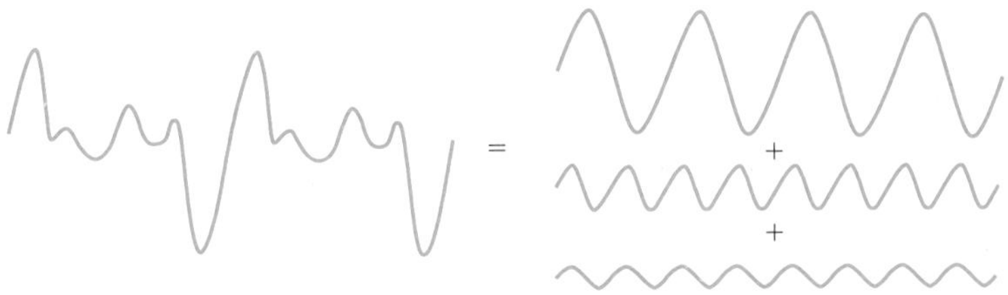

FIGURE 11.3 The complex sound wave on the left is the sum of the three simple waves on the right.

The Digital Age

When we talk about sound waves and imagine music to consist of waves, we are working with the **analog** picture of music. Until the early 1980s, nearly all musical recordings (phonograph cylinders, records, and tape recordings) were based on the analog picture of music. Storing music in the analog mode requires storing analogs of sound waves. For example, on vinyl records, the grooves in the vinyl surface are etched with the shape of the original musical sound wave. If you have listened to analog recordings, you know that this shape can easily become distorted or damaged.

Today, most of us listen to **digital** recordings of music. When a recording is made, music (played by performers) passes through an electronic device that converts the sound waves into an analog electrical signal (with varying voltage). This analog signal is then *digitized* by a computer. As discussed in the chapter-opening activity (p. 622), converting an analog wave into a digital representation requires two basic steps:

1. The wave must be *sampled* at regular time intervals so that a number can be used to represent the volume at each sampled point. The *sampling rate* describes the number of samples taken each second; it is usually given in units of hertz, which in this case means "samples per second." For example, a sampling rate of 10 hertz means 10 samples per second, so samples are taken at intervals of 1/10 second. Standard CDs are recorded at a sampling rate of 44,100 hertz, which means 44,100 samples per second, or a sample every 1/44,100 second.
2. Each individual sample is represented by a single number, which essentially tells us the volume (height) of the wave at that point. CDs are recorded in a 16-bit format, which allows $2^{16} = 65{,}536$ possible values for each volume.

This process of digitization converts the analog wave into a list of numbers. The list can then be stored on a CD or as a music file in computer memory. Inside a playback device such as a CD player or smart phone, a computer uses the reverse process to convert the numbers back into an analog electrical signal, which speakers then convert back into sound waves.

In addition to making it possible to store music on a CD or a computer, digitizing also makes it easy to "process" music. For example, *digital signal processing* allows extraneous sounds (such as background noise) to be detected and removed. It also makes it possible to correct errors made by musicians, combine different pieces of music (or tracks recorded at different times), and even add musical sounds without actually playing a musical instrument. In the digital age, the dividing line between mathematics and music has all but vanished.

Quick Quiz 11A

Choose the best answer to each of the following questions. Explain your reasoning with one or more complete sentences.

1. Musical sounds are generally produced by
   a. one object hitting another.
   b. objects that vibrate.
   c. objects that are being stretched to greater length.
2. If a string is vibrating up and down with a frequency of 100 cycles per second, the middle of the string is at its high point
   a. 100 times per second.
   b. 50 times per second.
   c. once every 100 seconds.
3. To make a sound with a higher pitch, you need to make a string vibrate with a
   a. higher frequency.
   b. lower frequency.
   c. higher maximum height.
4. The frequency of the lowest pitch you can hear from a particular string is the string's
   a. cycles per second.
   b. octave.
   c. fundamental frequency.

5. When you raise the pitch of a sound by an octave, the frequency of the sound
   a. doubles.
   b. goes up by a factor of 4.
   c. goes up by a factor of 8.
6. On a 12-tone scale, the frequency of each note is higher than that of the previous note by a factor of
   a. 2. b. 12. c. $\sqrt[12]{2}$.
7. The last entry in Table 11.1 shows that a frequency of 520 cps represents a note of C. What note has a frequency closest to $520 \times 1.5 = 780$ cps?
   a. D# b. E c. G
8. Suppose you made a graph by plotting notes in half-steps along the horizontal axis and the frequencies of these notes on the vertical axis. The general shape of this graph would be the same as that of a graph showing
   a. a linearly growing population.
   b. an exponentially growing population.
   c. a logistically growing population.
9. All musical sounds have a wave pattern that can be represented as
   a. a very simple constant-frequency wave.
   b. the sum of one or more individual constant-frequency waves.
   c. a series of half-step waves.
10. If you could look at the underlying computer code, you would find that the music stored on a CD or iPod was represented by
    a. lists of numbers.
    b. pictures of different-shaped waves.
    c. notes of a 12-step scale.

Exercises 11A

REVIEW QUESTIONS

1. What is *pitch*? How is it related to the frequency of a musical note?
2. Define *fundamental frequency, harmonics,* and *octave*. Why are these concepts important in music?
3. What is a *12-tone scale*? How are the frequencies of the notes on a 12-tone scale related to one another?
4. Explain how the notes of the scale are generated by exponential growth.
5. How do the wave forms of real musical sounds differ from the wave forms of simple tones? How are they related?
6. What is the difference between analog and digital recordings of music? What are the advantages of digital recording?

DOES IT MAKE SENSE?

Decide whether each of the following statements makes sense (or is clearly true) or does not make sense (or is clearly false). Explain your reasoning.

7. If I pluck this string more often, then it will have a higher pitch.
8. Jack made the length of the string one-fourth of its original length, and the pitch went up two octaves.
9. Exponential growth is found even in musical scales.
10. A piano has 88 keys, so it must have a range of about 7 octaves.

BASIC SKILLS & CONCEPTS

11. **Octaves.** Starting with a tone having a frequency of 220 cycles per second, find the frequencies of the tones that are one, two, three, and four octaves higher.
12. **Octaves.** Starting with a tone having a frequency of 1760 cycles per second, find the frequencies of the tones that are one, two, three, and four octaves lower.
13. **Notes of a Scale.** Find the frequencies of the 12 notes of the scale that starts at the F above middle C; this F has a frequency of 347 cycles per second.
14. **Notes of a Scale.** Find the frequencies of the 12 notes of the scale that starts at the G above middle C; this G has a frequency of 390 cycles per second.

15. **Exponential Growth and Scales.** Starting at middle C, with a frequency of 260 cps, find the frequency of the following notes:
    a. seven half-steps above middle C
    b. a sixth (nine half-steps) above middle C
    c. an octave and a fifth (seven half-steps) above middle C
    d. 25 half-steps above middle C
    e. three octaves and three half-steps above middle C

16. **Exponential Growth and Scales.** Starting at middle G, with a frequency of 390 cps, find the frequency of the following notes:

    a. six half-steps above middle G

    b. a third (four half-steps) above middle G

    c. an octave and a fourth (five half-steps) above middle G

    d. 25 half-steps above middle G

    e. two octaves and two half-steps above middle G

FURTHER APPLICATIONS

17. **Exponential Decay and Scales.** What is the frequency of the note seven half-steps *below* middle A (which has a frequency of 437 cps)? Of the note ten half-steps *below* middle A?

18. **The Dilemma of Temperament.** Start at middle A, with a frequency of 437 cps. Using the whole-number ratios in Table 11.1, find the frequency if you raise A by a fifth to E. What is the frequency if you raise E by a fifth to B? What is the frequency if you lower B by a sixth to D? What is the frequency if you lower D by a fourth to A? Having returned to the same note, have you also returned to the same frequency? Explain.

19. **Circle of Fifths.** A *circle of fifths* is generated by starting at a particular musical note and stepping upward by intervals of a fifth (seven half-steps). For example, starting at middle C, a circle of fifths includes the notes $C \rightarrow G \rightarrow D' \rightarrow A' \rightarrow E'' \rightarrow B'' \rightarrow \ldots$, where each $'$ symbol denotes a higher octave. Eventually the circle comes back to C several octaves higher.

    a. Show that the frequency of a tone increases by a factor of $2^{7/12} = 1.498$ if it is raised by a fifth. (*Hint:* Recall that each half-step corresponds to an increase in frequency by a factor of $f = \sqrt[12]{2}$.)

    b. By what factor does the frequency of a tone increase if it is raised by two fifths?

    c. Starting with middle C, at a frequency of 260 cycles per second, find the frequencies of the other notes in the circle of fifths.

    d. How many fifths are required for the circle of fifths to return to a C? How many octaves are covered by a complete circle of fifths?

    e. How does the frequency of the C at the end of the circle compare to that of the C at the beginning of the circle?

20. **Circle of Fourths.** A *circle of fourths* is generated by starting at any note and stepping upward by intervals of a fourth (five half-steps). By what factor is the frequency of a tone increased if it is raised by a fourth? How many fourths are required to complete the entire circle of fourths? How many octaves are covered in a complete circle of fourths?

21. **Rhythm and Mathematics.** In this unit, we focused on musical sounds, but rhythm and mathematics are also closely related. For example, in *4/4 time*, there are four *quarter notes* in a measure. If two quarter notes have the duration of a *half note*, how many half notes are in one measure? If two *eighth notes* have the duration of a quarter note, how many eighth notes are in one measure? If two *sixteenth notes* have the duration of an eighth note, how many sixteenth notes are in one measure?

IN YOUR WORLD

22. **Mathematics and Music.** Visit a website devoted to connections between music and mathematics. Write a one- to two-page essay that describes at least one connection between mathematics and music.

23. **Mathematics and Composers.** Many musical composers, both classical and modern, have used mathematics in their compositions. Research the life of one such composer. Write a one- to two-page essay discussing the role of mathematics in the composer's life and music.

24. **Digital Music.** The ease with which digital music can be copied has increased the importance of the copyright issue, especially with respect to music available on the Web. Find a recent article on issues of copying digital music. Discuss the article and its conclusions.

25. **Your Music Format.** In what format (e.g, AAC or MP3) is most of the music you listen to stored? What is its bit rate? Are you satisfied with its sound? How could you get improved sound quality?

26. **Digital Processing.** A variety of apps and software programs allow digital processing of music, photos, and movies. Find an app and experiment with it. Discuss uses you might find for digital processing, now or in the future.

27. **How Old Is Your Hearing?** Find an online "hearing age test," which will provide tones of various frequencies. What is the highest frequency that you can hear? Does your hearing age match your actual age? Have a few friends or family members of different ages take the test. Make a table in which you list each person's age and maximum hearing frequency, and write a short summary of your conclusions.

UNIT 11B Perspective and Symmetry

We now turn our attention to the connections between mathematics and the visual arts, such as painting, sculpture, and architecture. As we saw in the chapter-opening activity (p. 622), mathematics has played a crucial role in art history. The deepest connections are found in three particular aspects of the visual arts: *perspective*, *symmetry*, and *proportion*. In this unit, we will explore how Renaissance mathematicians and artists discovered techniques for painting perspective, and we'll also explore the idea of symmetry. We'll save *proportion* for Unit 11C.

Perspective

People of all cultures have used geometrical ideas and patterns in their artwork. The ancient Greeks developed strong ties between the arts and mathematics because both endeavors were central to their view of the world. Much of the Greek outlook was lost during the Middle Ages, but the Renaissance brought at least two new developments that made mathematics an essential tool of artists. First, there was a renewed interest in natural scenes, which led to a need to paint with realism. Second, many of the artists of the day also worked as engineers and architects.

The desire to paint landscapes with three-dimensional realism brought Renaissance painters face to face with the matter of perspective. In their attempts to capture depth and volume on a two-dimensional canvas, these artists made a science of painting. The painters Brunelleschi (1377–1446) and Alberti (1404–1472) are generally credited with developing, in about 1430, a system of perspective that involved geometrical thinking. Alberti's principle that a painting "is a section of a projection" lies at the heart of drawing with perspective.

Suppose you want to paint a simple view looking down a hallway with a checkerboard tile floor. Figure 11.4 shows a *side view* of the artist's eye, the canvas, and the hallway. Note the four lines labeled L_1, L_2, L_3 and L_4. The two side walls of the hallway intersect the floor and the ceiling along these four lines. These lines are important because they are parallel to each other in the scene and perpendicular to the canvas (or the plane containing the canvas).

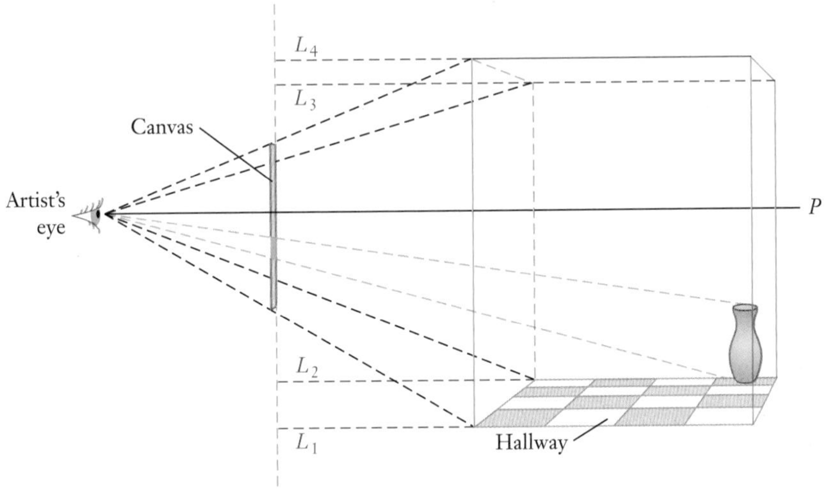

FIGURE 11.4 Side view of a hallway, showing perspectives.

Let's now look at the scene as the artist sees and paints it. The artist looks down the hallway with the point of view shown in Figure 11.5.

The lines L_1, L_2, L_3 and L_4 are parallel in the actual scene (in reality). However, they do not appear parallel to the artist and they should not be parallel in the painting. In fact, they all meet at a single point, labeled P, which is called the **principal vanishing point**. This fact leads us to the first principle of perspective discovered by the Renaissance painters:

All lines that are parallel in the actual scene and perpendicular to the canvas must intersect at the principal vanishing point of the painting.

Other lines that are parallel to lines L_1, L_2, L_3 and L_4, such as the lines going straight down the hallway along the floor tiles, also meet at the principal vanishing point. For example, the line connecting the points B and b intersects P, as do the line connecting C and c and the line connecting D and d.

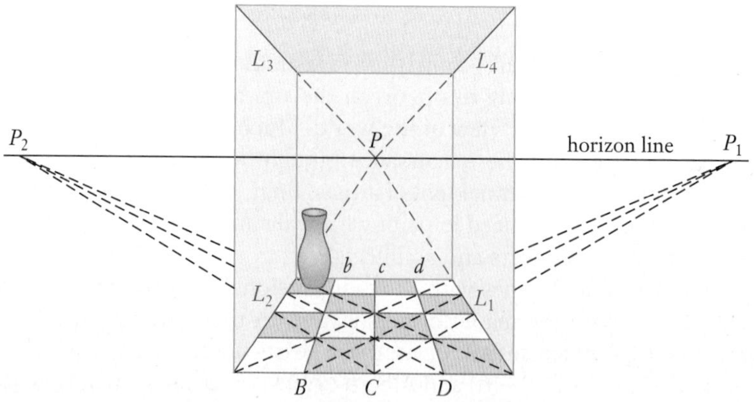

FIGURE 11.5 An artist's view of the hallway from Figure 11.4, showing perspectives. *Source:* Adapted from figure 23 in Morris Kline, *Mathematics in Western Culture* (Oxford University Press, 1953).

What happens to lines that are parallel in the actual scene but *not* perpendicular to the canvas, such as the dashed diagonal lines along the floor tiles? Figure 11.5 shows that these lines intersect at their own vanishing points, which are all on the horizontal line passing through the principal vanishing point. This line is called the **horizon line**. For example, the right-slanting diagonals of the floor tiles are parallel in the actual scene but meet in the painting at the vanishing point labeled P_1 on the horizon line. Similarly, the left-slanting diagonals meet at the vanishing point P_2 on the horizon line. In fact, all sets of lines that are mutually parallel in the real scene (except those parallel to the horizon line) must meet at their own vanishing point on the horizon line.

Let no one who is not a mathematician read my works.
—Leonardo da Vinci

Leonardo da Vinci (1452–1519) contributed greatly to the science of perspective. We can see da Vinci's mastery of perspective in many of his paintings. If you study *The Last Supper* (Figure 11.6), you will notice several parallel lines in the actual scene intersecting at the principal vanishing point of the painting, which is directly behind the central figure of Christ.

Think About It Imagine looking along a set of long parallel lines that stretches far into the distance, such as a set of train tracks or a set of telephone lines. The lines will appear to your eyes to get closer to each other as you look into the distance. If

HISTORICAL NOTE

Leonardo da Vinci was not only an artist but also a great scientist and engineer. His notebooks contain many ideas that were far ahead of his time. A century before Copernicus, he wrote that Earth was not the center of the universe. He correctly recognized fossils as remains from extinct species and suggested the possibility of long-term geological change on Earth. Unfortunately, he wrote his notebooks in code, so his contemporaries knew little of his ideas.

FIGURE 11.6 *The Last Supper*, by Leonardo da Vinci, shown with several lines that are parallel in the real scene and therefore converge at the principal vanishing point behind Christ.

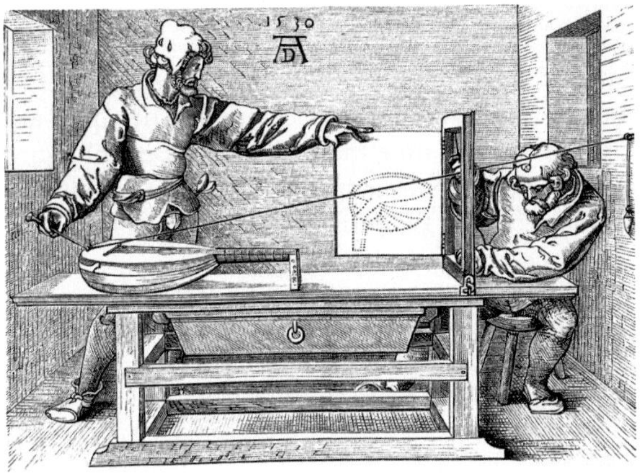

FIGURE 11.7 Woodcut by Albrecht Dürer.

FIGURE 11.8 Sketch by Hans Vredeman de Vries. Note the lines showing vanishing points for sets of parallel lines in the real scene.

you were painting a picture of the scene, where would you put the principal vanishing point? Why?

The German artist Albrecht Dürer (1471–1528) further developed the science of perspective. Near the end of his life, he wrote a popular book that stressed the use of geometry and encouraged artists to paint according to mathematical principles. Figure 11.7 is one of Dürer's woodcuts, showing an artist using Dürer's principles of perspective. A string from a point on the lute is attached to the wall at the point corresponding to the artist's eye. At the point where the string passes through the frame, a point is placed on the canvas. As the string is moved to different points on the lute, a drawing of the lute is created in perfect perspective on the canvas.

The Dutch artist Hans Vredeman de Vries (1527–1607) summarized much of the science of perspective in a book he published in 1604. Figure 11.8 shows a sketch from his book that illustrates how thoroughly perspective can be analyzed.

Think About It Identify some of the vanishing points in the sketch in Figure 11.8. What parallel lines from the real scene converge at each vanishing point?

Perspective drawing is sometimes abused deliberately. The engraving *False Perspective* (Figure 11.9) by English artist William Hogarth (1697–1764) reminds us that perspective is essential in art. Note where the fishing line of the man in the foreground lands and how the woman in the window appears to be lighting the pipe of a man on a distant hill. ▶ **Now try Exercises 15–20.**

FIGURE 11.9 *False Perspective*, by William Hogarth.

Symmetry

The term *symmetry* has many meanings. Sometimes it refers to a kind of balance. For example, *The Last Supper* (see Figure 11.6) is symmetrical because the disciples are grouped in four groups of three, with two groups on either side of the central figure of Christ. A human body is symmetrical because a vertical line drawn through the head and navel divides the body into two (nearly) identical parts (Figure 11.10).

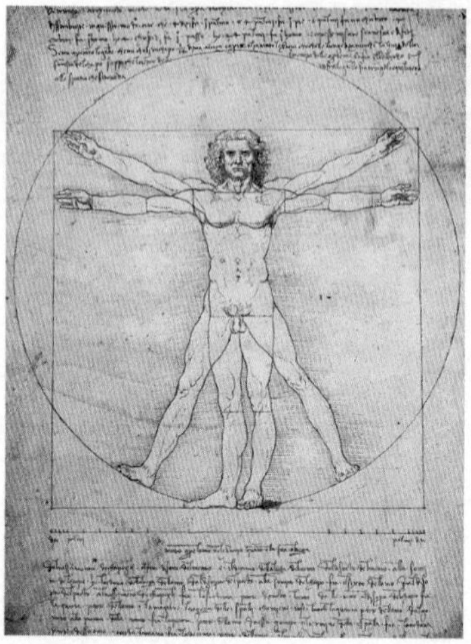

FIGURE 11.10 Leonardo da Vinci's sketch showing the symmetry of the human body.

Symmetry can also refer to repetition of patterns. Native American pottery is often decorated with simple borders that use repeating patterns. Similar symmetries are found in African, Islamic, and Moorish art, such as the tiling shown in Figure 11.11.

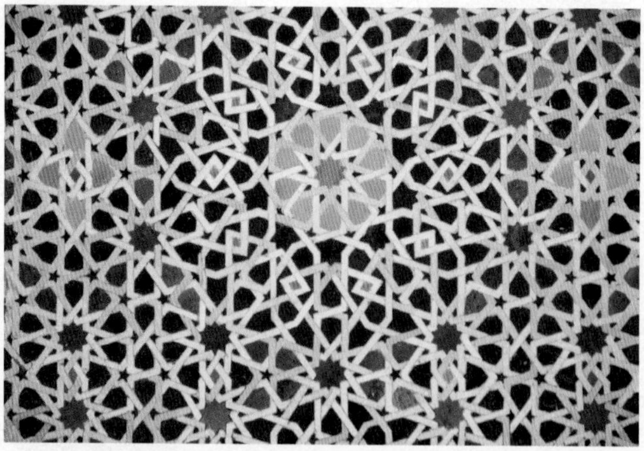

FIGURE 11.11 This Islamic tiling shows many elaborate symmetries. It is located in the Medersa Attarine, a 14th-century school of Muslim theology in Fez, Morocco.

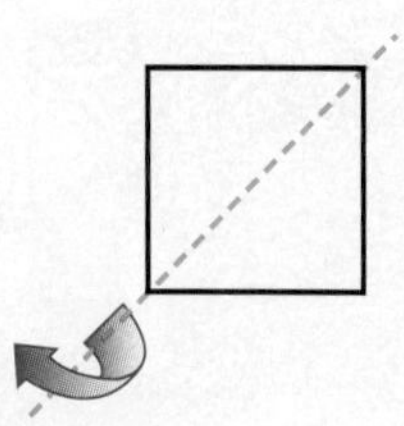

FIGURE 11.12

In mathematics, *symmetry* is a property of an object that remains unchanged under certain operations. For example, a circle still looks the same if it is rotated about its center. A square still looks the same if it is flipped across one of its diagonals (Figure 11.12). Many mathematical symmetries are quite subtle. However, three symmetries are easy to identify:

- **Reflection symmetry:** An object has reflection symmetry if it remains unchanged when reflected across a straight line. For example, the letter *A* has reflection symmetry about a vertical line, while the letter *H* has reflection symmetry about a vertical and a horizontal line (Figure 11.13).

FIGURE 11.13 Reflection symmetry.

- **Rotation symmetry:** An object has rotation symmetry if it remains unchanged when rotated through some angle about a point. For example, the letters *O* and *S* have rotation symmetry because they are unchanged when rotated 180° (Figure 11.14).

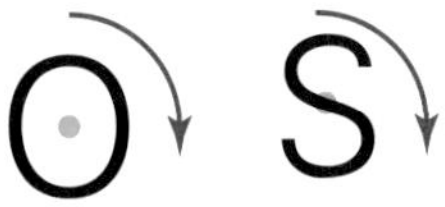

FIGURE 11.14 Rotation symmetry.

- **Translation symmetry:** A pattern shows translation symmetry if it remains the same when shifted, say, to the right or left. The pattern...XXX...(with the Xs continuing in both directions) has translation symmetry because it still looks the same if we shift it to the left or to the right (Figure 11.15). (In mathematics and physics, *translating* an object means moving it in a straight line, without rotating it.)

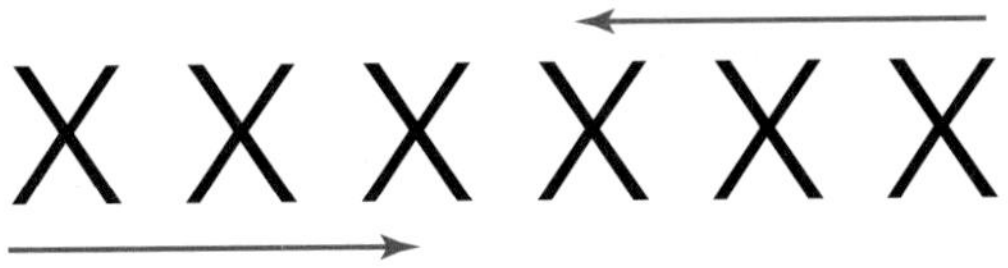

FIGURE 11.15 Translation symmetry.

EXAMPLE 1 Finding Symmetries

Identify the types of symmetry in each star in Figure 11.16.

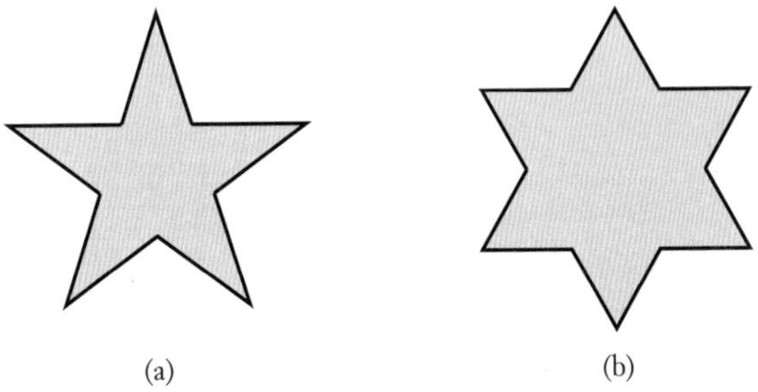

FIGURE 11.16

Solution

a. The five-pointed star has five lines about which it can be flipped (reflected) without changing its appearance, so it has five reflection symmetries (Figure 11.17a). Because it has five vertices that all look the same, it can be rotated by $\frac{1}{5}$ of a full circle, or $360°/5 = 72°$, and it still looks the same. Similarly, its appearance remains unchanged if it is rotated by $2 \times 72° = 144°$, $3 \times 72° = 216°$, or $4 \times 72° = 288°$. Therefore, this star has four rotation symmetries.

b. The six-pointed star has six reflection lines about which it can be flipped (reflected) without changing its appearance, so it has six reflection symmetries

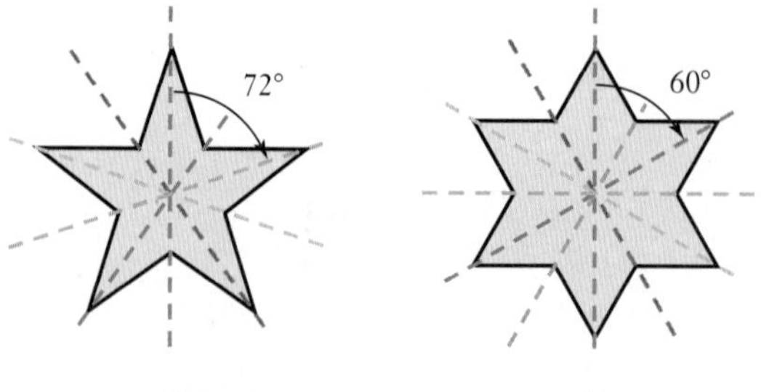

FIGURE 11.17 Each dashed line represents a reflection symmetry for the star. Rotation symmetries require rotation by multiples of the indicated angles.

(Figure 11.17b). Because of its six vertices, it has rotation symmetry when rotated by $\frac{1}{6}$ of a full circle, or $360°/6 = 60°$. It also has symmetry if rotated by $2 \times 60° = 120°$, $3 \times 60° = 180°$, $4 \times 60° = 240°$, or $5 \times 60° = 300°$. Therefore, this star has five rotation symmetries. ▶ Now try Exercises 21–27.

Symmetry in Art

Gustave Doré's (1832–1883) engraving *The Vision of the Empyrean* offers a dramatic illustration of rotation symmetry (Figure 11.18). This grand image of the cosmos can be rotated by many different angles and, at least on a large scale, appears much the same.

Sometimes, it is the *departures* from symmetry that make art effective. The 20th-century work *Supernovae* (Figure 11.19), by the Hungarian painter Victor Vasarely (1908–1997), might have started as a symmetric arrangement of circles and squares, but the gradual deviations from that pattern make a powerful visual effect.

FIGURE 11.18 *The Vision of the Empyrean*, by Gustave Doré.

FIGURE 11.19 *Supernovae* (1959–61), Victor Vasarely. *Source:* Tate Gallery, London/© 2013 ARS, NY.

Given the strong ties between mathematics and art, you may not be surprised that mathematical algorithms (recipes) can generate art on computers. Figure 11.20 shows an intricate Persian rug design generated on a computer by Anne Burns. The algorithm can be varied to give an endless array of patterns and symmetries.

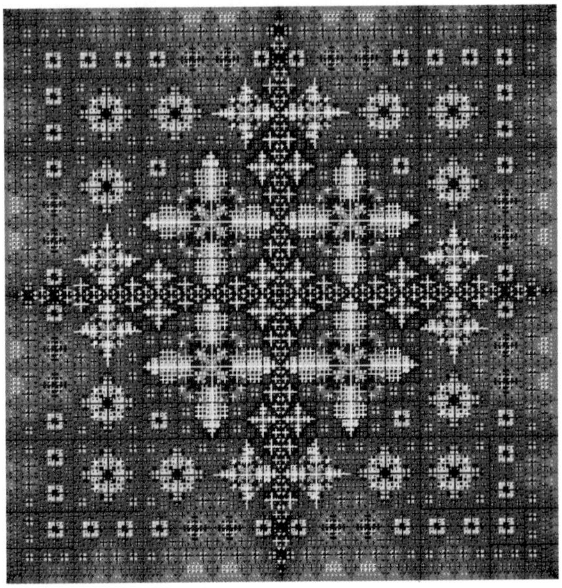

FIGURE 11.20 A "Persian rug," generated on a computer. *Source:* Created by Anne Burns.)

Tilings

A form of art called *tilings* (or *tessellations*) involves covering a flat area, such as a floor, with geometrical shapes. Tilings usually have regular or symmetric patterns. Tilings are found in ancient Roman mosaics, stained glass windows, and the elaborate courtyards of Arab mosques—as well as in many modern kitchens and bathrooms.

More precisely, a **tiling** is an arrangement of *polygons* (see Unit 10A) that interlock perfectly with no overlapping. The simplest tilings use just one type of regular polygon. Figure 11.21 shows three such tilings made with equilateral triangles, squares, and regular hexagons, respectively. Note that there are no gaps or overlaps between the polygons in any of the three cases. In each case, the tiling is made by translating (shifting) the same basic polygon in various directions. That is, these tilings have translation symmetry.

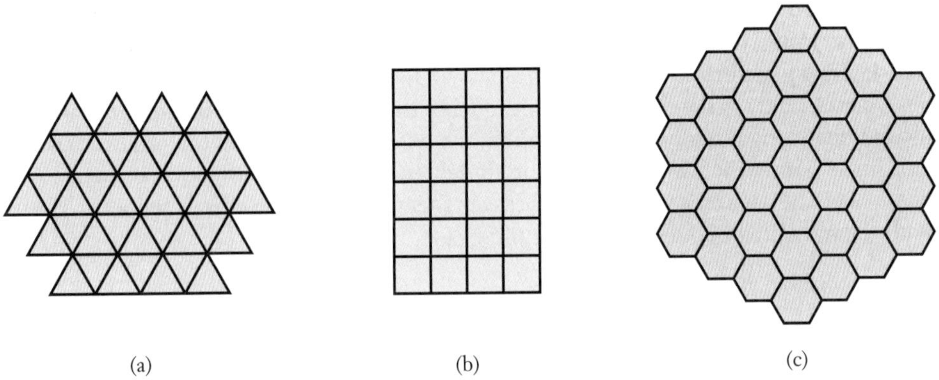

FIGURE 11.21 The three possible tilings made from regular polygons: (a) triangles, (b) squares, and (c) hexagons.

BY THE WAY

The light-sensitive photoreceptors in the human eye "tile" the retina in a hexagonal array, much like that in Figure 11.21c.

What happens if you try to make a tiling with, say, regular pentagons? If you try, you'll find that it simply does not work. The interior angles of a regular pentagon measure 108°. As Figure 11.22 shows, the angle that remains when three regular pentagons are placed next to each other is too small to fit another regular pentagon. In fact, a mathematical theorem states that tilings with a single regular polygon are possible only with equilateral triangles, squares, and hexagons (as shown in Figure 11.21).

More tilings are possible if we remove the restriction of using only a single type of regular polygon. For example, if we allow different regular polygons, but still require that the arrangement of polygons look the same around each vertex (intersection point), it is possible to make the eight tiling patterns in Figure 11.23.

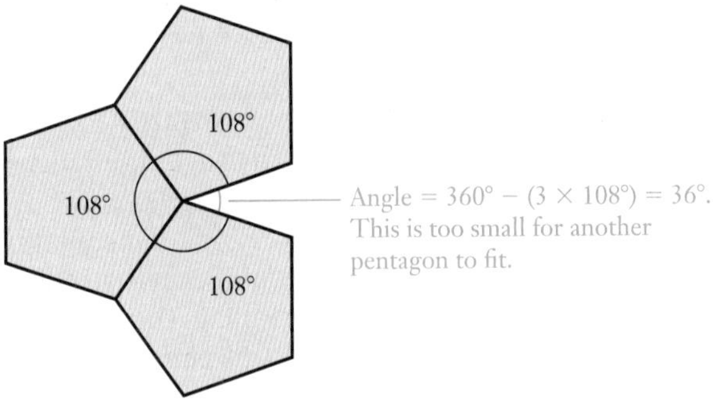

FIGURE 11.22 Regular pentagons cannot be used to make a tiling.

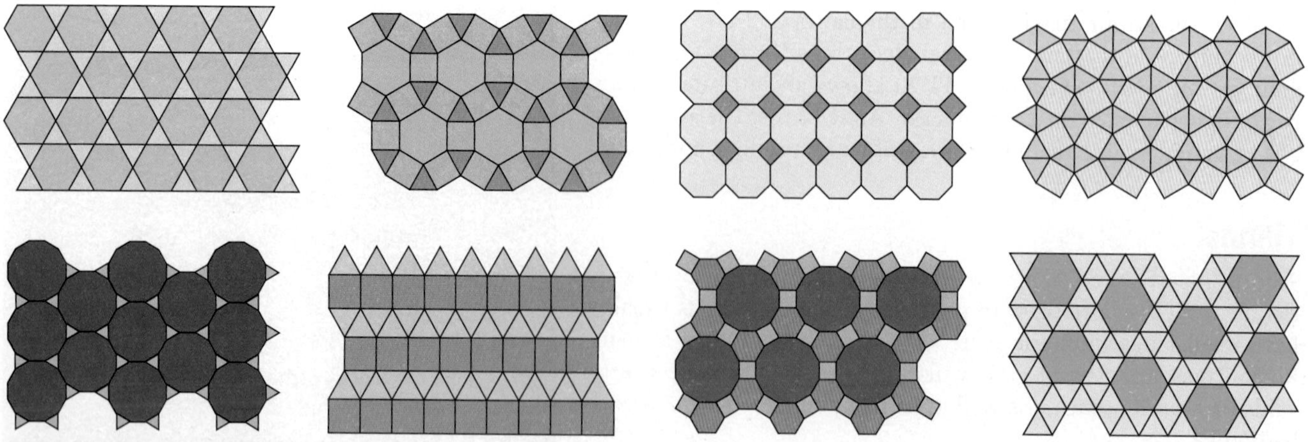

FIGURE 11.23 Eight tilings, each made by combining two or more types of regular polygons. These are sometimes called *Archimedean tilings*, or *semi-regular tilings.*

Think About It Verify that each of the tilings in Figure 11.23 uses only regular polygons. How many different regular polygons are used in each of these tilings? Verify that the same arrangement of polygons appears around each vertex. (Look carefully at the polygons; there are no circles in this figure.)

Tilings that use *irregular* polygons (those with sides of different lengths) are endless in number. As an example, suppose we start with an arbitrary triangle that has no special properties (other than three sides). The easiest way to tile a region with this triangle is by translating it parallel to two of its sides, as shown in Figure 11.24. We shift the original triangle to the right so that the new triangle touches the original triangle at a single point. We also shift the original triangle down so that the new triangle touches the original triangle at a single point. Then we repeat these right/left and up/down translations as many times as we like. The gaps created in this process are themselves triangles that interlock perfectly with the translated triangles to create a tiling.

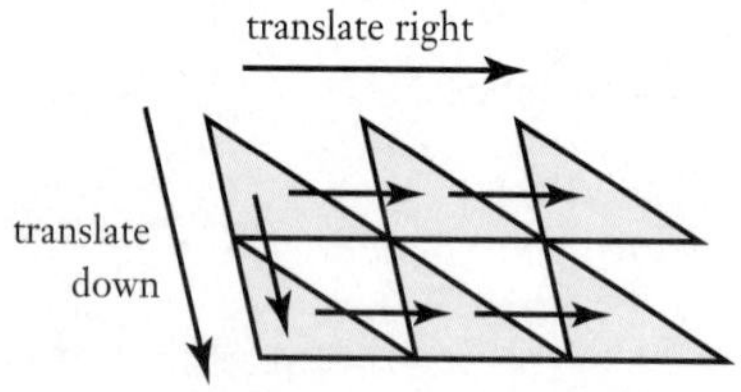

FIGURE 11.24 A tiling made by translating a triangle in two directions.

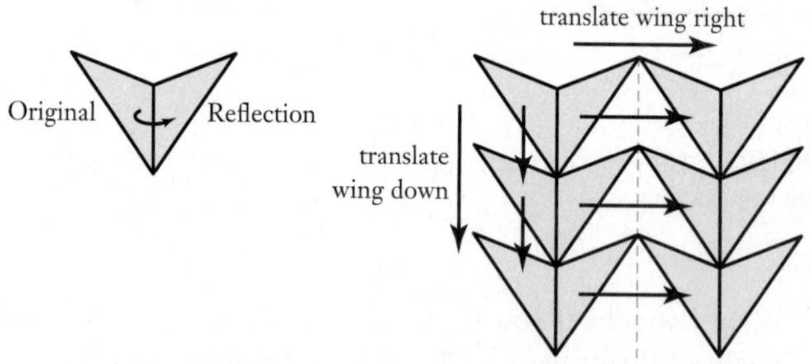

FIGURE 11.25 A tiling made by first reflecting a triangle to make a "wing," then translating the wing in two directions.

Figure 11.25 shows another example. This time, we begin by reflecting an arbitrary triangle to produce a wing-shaped object, and then we translate this object up/down and right/left.

All of the tilings discussed so far are called *periodic tilings* because they have a pattern that is repeated throughout the tiling. In recent decades, mathematicians have explored tilings that are *aperiodic*, meaning that they do not have a pattern that repeats throughout the entire tiling. Figure 11.26 shows an aperiodic tiling created by British mathematician Roger Penrose. If you look at the center of the figure, there appears to be a fivefold symmetry (a rotational symmetry that is found in a pentagon). However, if the figure were extended indefinitely in all directions, the same pattern would never be repeated.

FIGURE 11.26 An aperiodic tiling by Roger Penrose.

Tilings can be beautiful and practical for such things as floors and ceilings. However, recent research also shows that tilings may be very important in nature. Many molecules and crystals apparently have patterns and symmetries that can be understood with the same mathematics used to study tilings in art.

EXAMPLE 2 Quadrilateral Tiling

Create a tiling by translating the quadrilateral shown in Figure 11.27. As you translate the quadrilateral, make sure that the gaps left behind have the same quadrilateral shape.

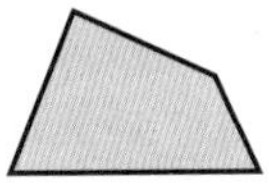

FIGURE 11.27

Solution We can find the solution by trial and error, translating the quadrilateral in different directions until we have correctly shaped gaps. Figure 11.28 shows the solution. Note that the translations are along the directions of the two diagonals of the quadrilateral. The gaps between the translated quadrilaterals are themselves quadrilaterals that interlock perfectly to complete the tiling.

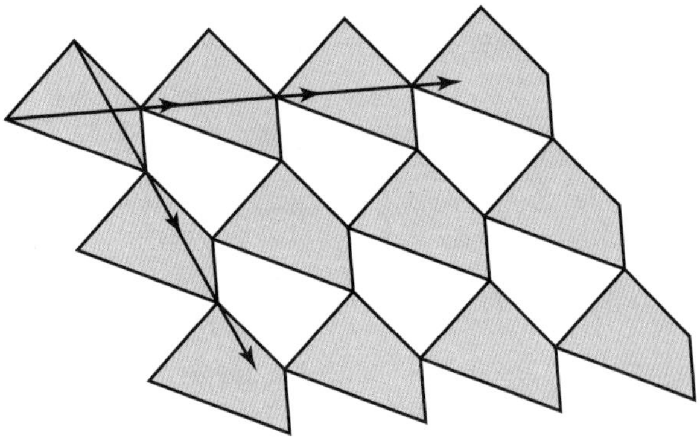

FIGURE 11.28 A quadrilateral tiling.

▸ Now try Exercises 28–33.

Quick Quiz 11B

Choose the best answer to each of the following questions. Explain your reasoning with one or more complete sentences.

1. In a painting of train tracks that run perpendicular to the canvas, the principal vanishing point is

   a. the place in the real scene where the train tracks disappear from view.

   b. the place in the painting where the two rails meet.

   c. the place in the painting where the rails first appear to touch the sky.

2. *All* lines that are parallel in a real scene converge in a painting

   a. at the principal vanishing point.

   b. somewhere along the horizon line.

   c. somewhere along the top edge of the painting.

3. Study *The Last Supper* in Figure 11.6. Which of the following converge at the principal vanishing point?

   a. parallel beams in the ceiling

   b. lines connecting the heads of the disciples

   c. the vertical sides of the doors and windows

4. The symmetry in da Vinci's sketch of the human body in Figure 11.10 arises because

   a. he shows two sets of arms and legs.

   b. he has enclosed the bodies in a circle.

   c. the left and right sides of the sketch are nearly mirror images of each other.

5. The letter W has reflection symmetry along a line

   a. going diagonally through its middle.

   b. going horizontally through its middle.

   c. going vertically through its middle.

6. The letter Z has

   a. reflection symmetry.

   b. rotation symmetry.

   c. translation symmetry.

7. A perfect circle has

   a. both reflection and rotation symmetry.

   b. reflection symmetry only.

   c. rotation symmetry only.

8. Which of the following regular polygons *cannot* be used to make a tiling?

   a. equilateral triangles

   b. regular hexagons

   c. regular octagons

9. Suppose you arrange regular pentagons so that they are as close together as possible on a flat floor. The pattern will not be a tiling because

   a. it will have gaps between some of the pentagons.

   b. regular pentagons lack any type of symmetry.

   c. tilings are possible only with three- and four-sided polygons.

10. A *periodic* tiling is one in which

   a. only regular polygons are used.

   b. only triangles are used.

   c. the same pattern repeats over and over.

Exercises 11B

REVIEW QUESTIONS

1. Describe the ideas of perspective and symmetry.
2. How is the *principal vanishing point* in a picture determined?
3. What is the *horizon line*? How is it important in a painting showing perspective?
4. Briefly describe and distinguish among reflection symmetry, rotation symmetry, and translation symmetry. Draw a simple picture that shows each type of symmetry.
5. What is a *tiling*? Draw a simple example.
6. Briefly explain why there are only three possible tiling patterns that consist of a single regular polygon. What are the three patterns?
7. Briefly explain why more tilings are possible if we remove the restriction of using regular polygons.
8. What is the difference between periodic and aperiodic tilings?

DOES IT MAKE SENSE?

Decide whether each of the following statements makes sense (or is clearly true) or does not make sense (or is clearly false). Explain your reasoning.

9. The principal vanishing point of the painting is so far away that it cannot be seen in the painting.
10. Sid does not need to use perspective in her ground-level painting of a flat desert because the desert is two-dimensional.

11. Jane wants near objects to look nearby and far objects to look far away in her painting, so she should use perspective.
12. Kenny likes symmetry, so he prefers the letter *R* to the letter *O*.
13. Susan found a sale on octagonal (eight-sided) floor tiles, so she bought them to tile her kitchen. (Assume she uses no other tiles on her floor.)
14. Frank always liked the symmetry of the Washington Monument (in Washington, DC).

BASIC SKILLS & CONCEPTS

15. **Vanishing Points.** Consider the simple drawing of a road and a telephone pole in Figure 11.29.

FIGURE 11.29

a. Locate a vanishing point for the drawing. Is it the principal vanishing point?

b. With proper perspective, draw three more telephone poles receding into the distance.

16. **Correct Perspective.** Consider the two boxes shown in Figure 11.30. Which one is drawn with proper perspective relative to a single vanishing point? Explain.

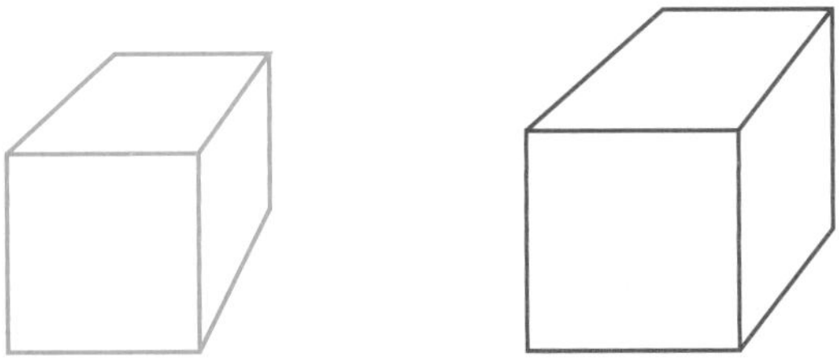

FIGURE 11.30

17. **Drawing with Perspective.** Make the square, circle, and triangle in Figure 11.31 into three-dimensional solid objects: a box, a cylinder, and a triangular prism, respectively. The given objects should be used as the front faces of the three-dimensional objects, and all figures should be drawn with correct perspective relative to the given vanishing point, *P*.

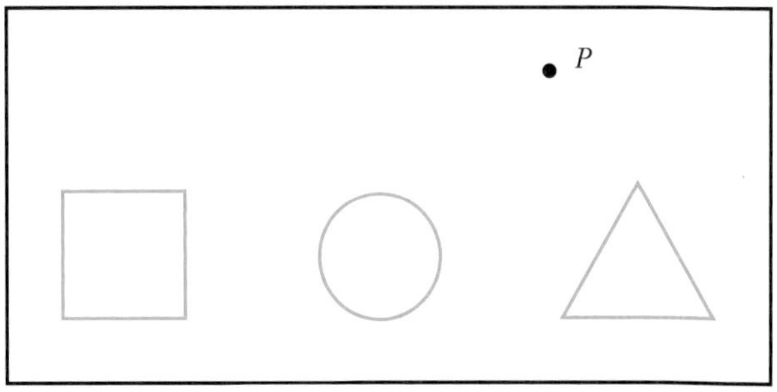

FIGURE 11.31

18. **Drawing MATH with Perspective.** Make the letters *M*, *A*, *T*, and *H* in Figure 11.32 into three-dimensional solid letters. The given letters should be used as the front faces of three-dimensional letters as deep as the *T* is wide, and all letters should be drawn with correct perspective relative to the given vanishing point, *P*.

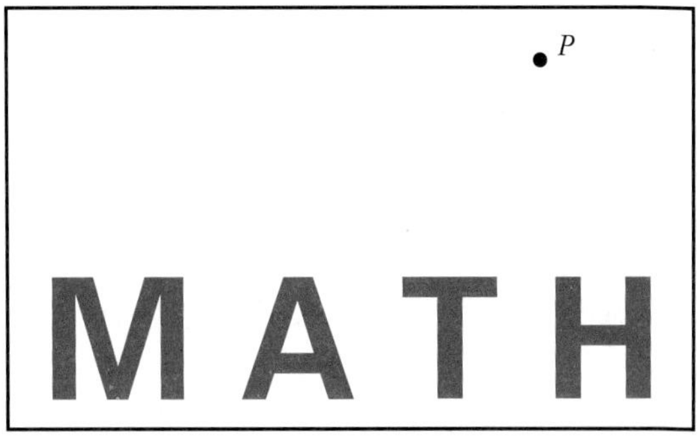

FIGURE 11.32

19. **Proportion and Perspective.** The drawing in Figure 11.33 shows two poles drawn with correct perspective relative to a single vanishing point. As you can check, the first pole is 2 centimeters tall in the drawing and the second pole is 2 centimeters away (measured base to base) with a height of 1.5 centimeters.

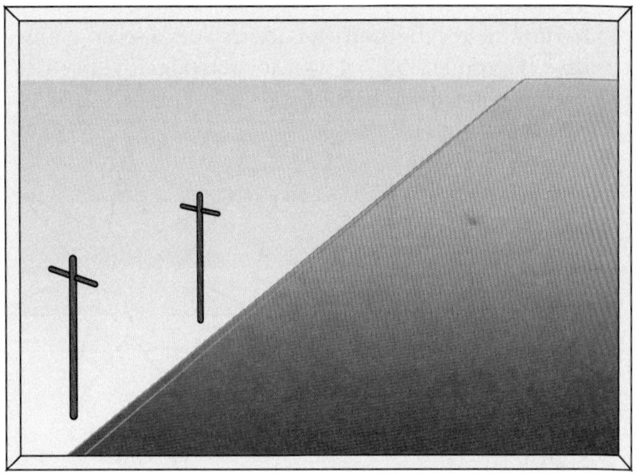

FIGURE 11.33

a. Draw two more vertical poles with correct perspective that are equally spaced along the baseline in the drawing. Assume the poles are of equal height in the real scene.

b. Estimate the heights of the two new poles in your drawing.

c. In the actual scene, would these four poles be equally spaced? Explain.

20. **Two Vanishing Points.** Figure 11.34 shows a road receding into the distance. In the direction of the arrow, draw a second road that intersects the first road. Be sure that the vanishing points of the two roads lie on the horizon line.

FIGURE 11.34

21. **Symmetry in Letters.** Find all of the capital letters of the alphabet that have

a. right/left reflection symmetry (such as *A*).

b. top/bottom reflection symmetry (such as *H*).

c. both right/left and top/bottom reflection symmetry.

d. a rotational symmetry.

22. **Star Symmetries.**

a. How many reflection symmetries does a four-pointed star have (Figure 11.35a)? How many rotational symmetries does a four-pointed star have?

b. How many reflection symmetries does a seven-pointed star have (Figure 11.35b)? How many rotational symmetries does a seven-pointed star have?

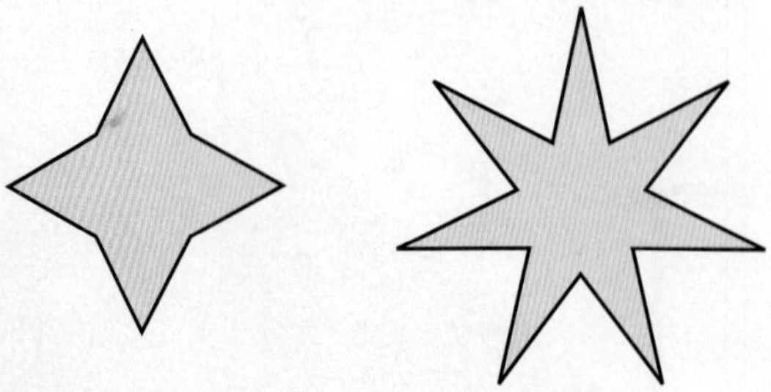

(a) (b)

FIGURE 11.35

23. **Symmetries of Geometric Figures.**

a. Draw an equilateral triangle (all three sides have equal length). How many degrees can the triangle be rotated about its center and remain unchanged in appearance? (There is more than one correct answer.)

b. Draw a square (all four sides have equal length). How many degrees can the square be rotated about its center and remain unchanged in appearance? (There are several correct answers.)

c. Draw a regular pentagon (all five sides have equal length). How many degrees can the pentagon be rotated about its center and remain unchanged in appearance? (There are several correct answers.)

d. Can you see a pattern in parts (a), (b), and (c)? How many degrees can a regular *n*-gon (a regular polygon with *n* sides) be rotated about its center so that it remains unchanged in appearance? How many different angles answer this question for an *n*-gon?

24–27: Identifying Symmetries. Identify all of the symmetries in the following figures.

24.

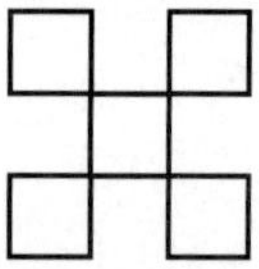

25.

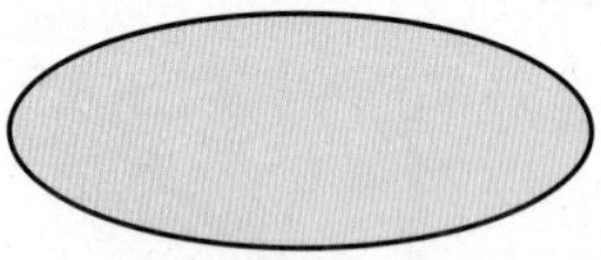

26.

Pattern continues in both directions.

27.

28–29: Tilings from Translating Triangles. Make a tiling from each triangle using translations only, as in Figure 11.24.

28.

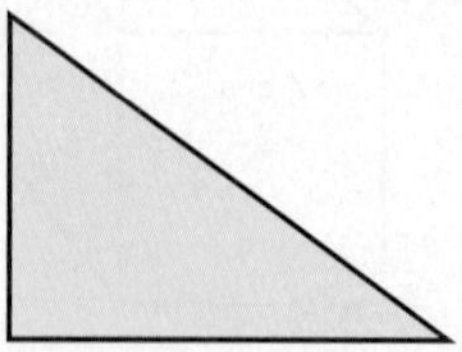

29.

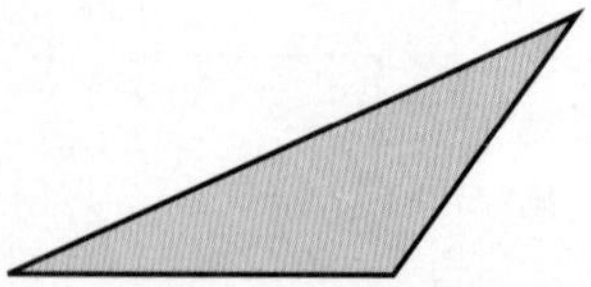

30–31: Tilings from Translating and Reflecting Triangles. Make a tiling from the given triangle using translations and reflections, as in Figure 11.25.

30. The triangle in Exercise 28

31. The triangle in Exercise 29

32–33: Tilings from Quadrilaterals. Make a tiling from each quadrilateral using translations, as in Figure 11.28.

32.

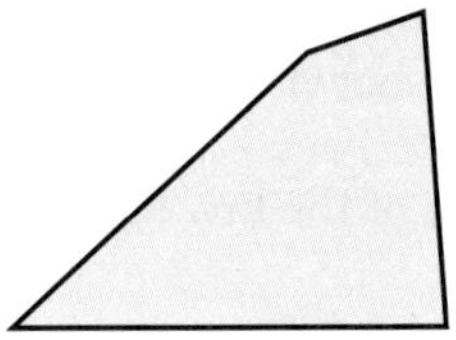

33.

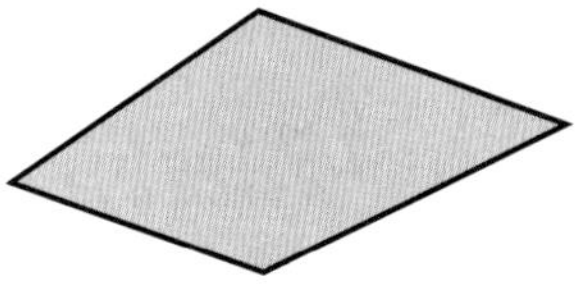

FURTHER APPLICATIONS

34. **Desargues' Theorem.** An early theorem of projective geometry, proved by French architect and engineer Girard Desargues (1593–1662), says that if two triangles (*ABC* and *abc* in Figure 11.36) are drawn so that the straight lines joining corresponding vertices (*Aa*, *Bb*, and *Cc*) all meet in a point *P* (corresponding to a vanishing point), then the corresponding sides (*AC* and *ac*, *AB* and *ab*, *BC* and *bc*), if extended, will meet in three points that all lie on the same line *L*. Draw two triangles of your own in such a way that the conditions of Desargues' theorem are satisfied. Verify that the conclusions of the theorem are true.

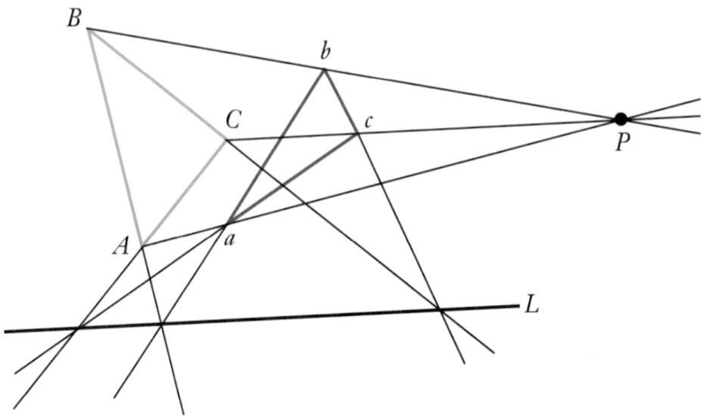

FIGURE 11.36

35. **Why Quadrilateral Tilings Work.** Consider the tiling with quadrilaterals shown in Figure 11.28. Look at any of the points at which four quadrilaterals meet in the tiling. Call such a point *P*. Given that the sum of the inside angles of any quadrilateral is 360°, show that the sum of the angles around the point *P* is also 360°, thus proving that the quadrilaterals interlock perfectly.

36. **Tiling with a Rhombus.** A *rhombus* is a quadrilateral in which all four sides have the same length and opposite sides are parallel. Show how a tiling can be made from a rhombus using translations only (as in Figure 11.24) and an initial reflection and then translations (as in Figure 11.25).

IN YOUR WORLD

37. **Perspective in Life.** Describe at least three ways in which the use of visual perspective affects your life.

38. **Symmetry in Life.** Find at least three objects from your daily life that exhibit some type of mathematical symmetry. Describe the symmetry in each case.

39. **Art and Mathematics.** Visit a website devoted to connections between art and mathematics. Write a one- to two-page essay that describes one or more connections between mathematics and art.

40. **Art Museums.** Choose an art museum, and study its online collection. Describe a few pieces in which ideas of perspective or symmetry are important.

41. **M. C. Escher.** The work of Maurits C. Escher (1898–1972) is designed to confound us with its use and abuse of perspective. Choose one work by Escher, and write a short essay about the use of perspective in the piece.

42. **Penrose Tilings.** Learn more about the nature and uses of Penrose (aperiodic) tilings. Write a short essay describing your findings.

43. **Symmetry and Proportion in Art.** Find one piece of pre-20th-century art and one piece of 20th-century or 21st-century art that you like. Use as many ideas from this unit as possible (involving perspective and symmetry) to write a two- to three-page analysis and comparison of these two pieces of art.

UNIT 11C Proportion and the Golden Ratio

The senses delight in things duly proportioned.
—St. Thomas Aquinas (1225–1274)

In Unit 11B, we studied how mathematics affects art through the ideas of symmetry and perspective. In this unit, we turn our attention to the third major mathematical idea involved with art: proportion.

The importance of proportion was expressed well by the astronomer Johannes Kepler (1571–1630):

> *Geometry has two great treasures: one is the theorem of Pythagoras; the other the division of a line into extreme and mean ratio. The first we may compare to gold; the second we may name a precious jewel.*

Kepler's statement about *the division of a line into extreme and mean ratio* describes one of the oldest principles of proportion. This principle dates back to the time of Pythagoras

(c. 500 B.C.E.), when scholars asked the following question: How can a line segment be divided into two pieces that have the most appeal and balance?

FIGURE 11.37

Although this was a question of beauty, there seemed to be general agreement on the answer. Suppose a line segment is divided into two pieces, as shown in Figure 11.37. We call the length of the long piece L and the length of the short piece 1.

The Greeks claimed that the most visually pleasing division of the line had the following property for the ratios of its lengths:

ratio of long piece to short piece = ratio of entire line segment to long piece

That is, $$\frac{L}{1} = \frac{L+1}{L}$$

This statement of proportion can be solved (see Exercise 19) to find that L has a special value, denoted by the Greek letter ϕ (phi, pronounced "fie" or "fee"), which is

$$\phi = \frac{1+\sqrt{5}}{2} = 1.61803\ldots$$

FIGURE 11.38

The number ϕ is commonly called the **golden ratio**; alternative names include the *golden mean*, the *golden section*, and the *divine proportion*. It is an irrational number often approximated as 1.6, or $\frac{8}{5}$. Figure 11.38 shows that, for any line segment divided in two pieces according to the golden ratio, the ratio of the long piece to the short piece is

$$\frac{x}{y} = \frac{1.61803\ldots}{1} = \phi \approx \frac{8}{5}$$

EXAMPLE 1 Calculating with the Golden Ratio

Suppose the line segment in Figure 11.39 is divided according to the golden ratio. If the length of the longer piece labeled x is 5 centimeters, how long is the entire line segment?

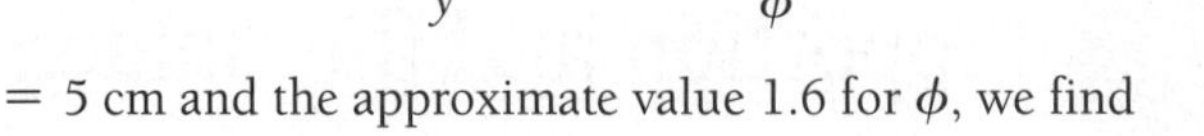

FIGURE 11.39

Solution Because the line segment is divided in the golden ratio, we know that $x/y = \phi$. We solve for y by multiplying both sides of this equation by y and dividing both sides by ϕ:

$$\frac{x}{y} = \phi \rightarrow y = \frac{x}{\phi}$$

Substituting $x = 5$ cm and the approximate value 1.6 for ϕ, we find

$$y = \frac{5\text{ cm}}{\phi} \approx \frac{5\text{ cm}}{1.6} \approx 3.1\text{ cm}$$

The entire segment has a length of $x + y$, so its total length is approximately 5 cm + 3.1 cm = 8.1 cm. ▶ **Now try Exercises 11–12.**

BY THE WAY
The Greek letter ϕ is the first letter in the Greek spelling of Phydias, the name of a Greek sculptor who may have used the golden ratio in his work.

The Golden Ratio in Art History

Although the ancient Greeks struggled with the notion of irrational numbers, they embraced the golden ratio. Consider the *pentagram* (Figure 11.40), which is a five-pointed star inscribed in a circle, producing a pentagon at its center. The pentagram was the seal of the mystical Pythagorean Brotherhood. The golden ratio occurs in at least ten different ways in the pentagram. For example, if the length of each side of the pentagon is 1, then each arm of the star has length ϕ.

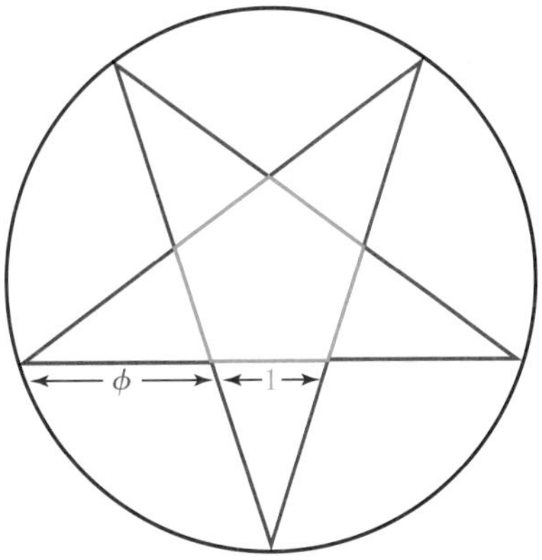

FIGURE 11.40 A pentagram.

BY THE WAY

The pentagram, also called a *pentacle*, has appeared in numerous works of literature throughout history and gained recent fame as one of the key symbols in both the novel and movie adaptation of *The Da Vinci Code*, written by Dan Brown.

Think About It Using a ruler, find at least one other place in the pentagram of Figure 11.40 where the ratio of the lengths of two line segments is the golden ratio.

From the golden ratio it is a short step to another famous Greek expression of proportion, the **golden rectangle**—a rectangle whose long side is ϕ times as long as its short side. A golden rectangle can be of any size, but its sides must have a ratio of $\phi \approx \frac{8}{5}$. Figure 11.41 shows a golden rectangle.

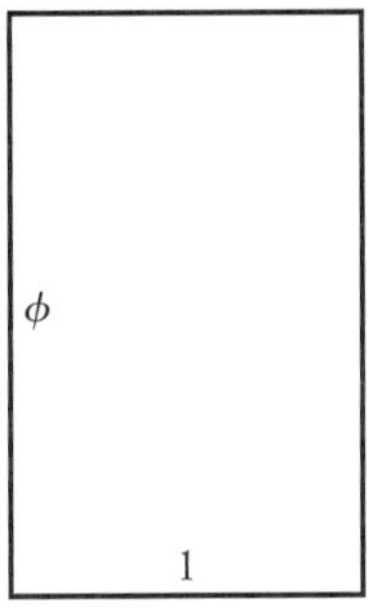

FIGURE 11.41 A golden rectangle—the side lengths have a ratio $\phi \approx \frac{8}{5}$.

The golden rectangle had both practical and mystical importance to the Greeks. It became a cornerstone of their philosophy of *aesthetics*—the study of beauty. There is considerable speculation about the uses of the golden rectangle in art and architecture in ancient times. For example, it is widely claimed that many of the great monuments of antiquity, such as the Pyramids in Egypt, were designed in accordance with the golden rectangle. And, whether by design or by chance, the proportions of the Parthenon (in Athens, Greece) closely match those of the golden rectangle (Figure 11.42).

BY THE WAY

The Parthenon was completed in about 430 B.C.E. as a temple to Athena Parthenos, the Warrior Maiden. It stands on the Acropolis (which means "the uppermost city"), about 500 feet above Athens.

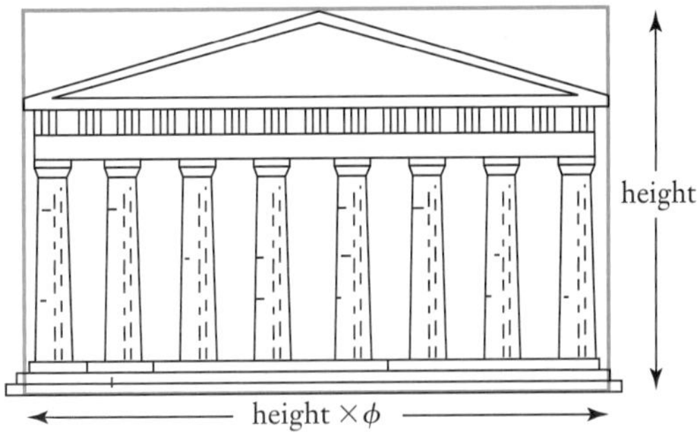

FIGURE 11.42 The proportions of the Parthenon have been claimed to closely match those of a golden rectangle as shown, though the claim has been disputed.

FIGURE 11.43 *St. Jerome*, by Leonardo da Vinci.

FIGURE 11.44 *Circus Sideshow*, by Georges Seurat, shown with an overlay that calls attention to the use of golden rectangles, which make a logarithmic spiral.

The golden rectangle also appears in many other works of art and architecture. The book *De Divina Proportione*, illustrated by Leonardo da Vinci in 1509, is filled with references to and uses of ϕ. Da Vinci's unfinished painting *St. Jerome* (Figure 11.43) appears to place the central figure inside an imaginary golden rectangle. More recently, the French impressionist painter Georges Seurat is said to have used the golden ratio on every canvas (Figure 11.44). The abstract geometric paintings of the 20th-century Dutch painter Piet Mondrian are also filled with golden rectangles.

Today, the golden rectangle appears in many everyday items. Photographs, note cards, cereal boxes, posters, and windows often have proportions close to those of the golden rectangle. But the question remains as to whether the golden rectangle is really more pleasing than other rectangles. In the late 19th century, the German psychologist Gustav Fechner (1801–1887) studied the question statistically. He showed several rectangles with various length-to-width ratios to hundreds of people and recorded their choices for the most and least visually pleasing rectangles. The results, given in Table 11.2, show that almost 75% of the participants chose one of the three rectangles with proportions closest to those of the golden rectangle.

BY THE WAY

Despite evidence of the appeal of the golden rectangle and golden ratio, there are theories to the contrary. A 1992 study by George Markowsky discredits claims that the golden ratio was used in art and architecture, attributing them to coincidence and bad science. He also claims that statistical studies of people's preferences, such as Fechner's research, are not conclusive (see Exercise 26).

TABLE 11.2 Fechner's Data

Length-to-Width Ratio	Most Pleasing Rectangle (Percentage Response)	Least Pleasing Rectangle (Percentage Response)
1.00	3.0%	27.8%
1.20	0.2%	19.7%
1.25	2.0%	9.4%
1.33	2.5%	2.5%
1.45	7.7%	1.2%
1.50	20.6%	0.4%
$\phi \approx 1.62$	**35.0%**	**0.0%**
1.75	20.0%	0.8%
2.00	7.5%	2.5%
2.50	1.5%	35.7%

Think About It Do you think that the golden rectangle is visually more pleasing than other rectangles? Explain.

EXAMPLE 2 Household Golden Ratios

Consider the following household items with the given dimensions. Which item comes closest to having the proportions of the golden ratio?

- Standard sheet of paper: 8.5 in $\times$ 11 in
- 8 $\times$ 10 picture frame: 8 in $\times$ 10 in
- HDTV (high-definition television) screen, which comes in many sizes but always with a 16:9 ratio of width to height

Solution The ratio of the sides of a standard sheet of paper is $11/8.5 \approx 1.29$, which is 20% less than the golden ratio. The ratio of the sides of a standard picture frame is $10/8 = 1.25$, which is 23% less than the golden ratio. The width-to-height ratio of an HDTV screen is $16/9 \approx 1.78$ which is about 10% more than the golden ratio. Of the three objects, the HDTV screen is closest to being a golden rectangle.

▶ Now try Exercises 13–18.

The Golden Ratio in Nature

The golden ratio appears to be common in the "artwork" of nature. One striking example is spirals created from golden rectangles. We begin by dividing a golden rectangle so that there is a square on its left side, as shown in Figure 11.45a. If you measure the sides of the remaining smaller rectangle to the right of the square, you'll see that it is a smaller golden rectangle.

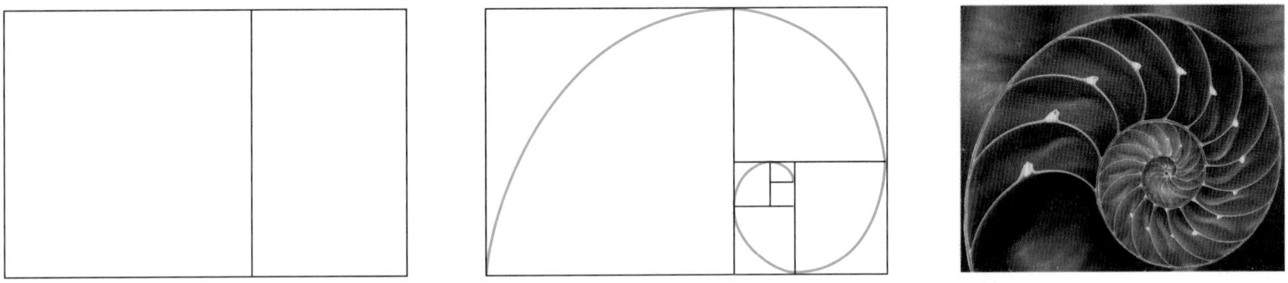

(a) (b) (c)

FIGURE 11.45 (a) A logarithmic spiral begins with a golden rectangle that is divided to make a square on the left. (b) The process is repeated in each successive golden rectangle, and a spiral is created by connecting the corners of all the squares. (c) A chambered nautilus shell looks much like a logarithmic spiral.

We now repeat this splitting process on the second golden rectangle, this time making the square on the top (instead of the left) of the golden rectangle. This split makes a third, even smaller, golden rectangle. Continuing to split each new golden rectangle in this manner generates the result shown in Figure 11.45b. We then connect opposite corners of all the squares with a smooth curve. The result is a continuous curve called a **logarithmic spiral** (or *equiangular spiral*). This spiral very closely matches the spiral shape of the beautiful chambered nautilus shell (Figure 11.45c).

Another intriguing connection between the golden ratio and nature comes from a problem in population biology, first posed by a mathematician known as Fibonacci in 1202. Fibonacci's problem essentially asked the following question about the reproduction of rabbits.

Suppose that a pair of baby rabbits takes one month to mature into adults, then produces a new pair of baby rabbits the following month and each subsequent month. Further suppose that each newly born pair of rabbits matures and gives birth to additional pairs with the same reproductive pattern. If no rabbits die, how many pairs of rabbits are in the population at the beginning of each month?

BY THE WAY
Fibonacci, also known as Leonardo of Pisa (b. 1170), is credited with popularizing the use of Hindu-Arabic numerals in Europe. His book *Liber Abaci* (*Book of the Abacus*), published in 1202, explained their use and the importance of the number zero.

Figure 11.46 shows the solution to this problem for the first six months. The number of pairs of rabbits at the beginning of each month forms a sequence that begins 1, 1, 2, 3, 5, 8. Fibonacci found that the numbers in this sequence continue to grow with the following pattern:

$$1, 1, 2, 3, 5, 8, 13, 21, 34, 55, \ldots$$

This sequence of numbers is known as the **Fibonacci sequence**.

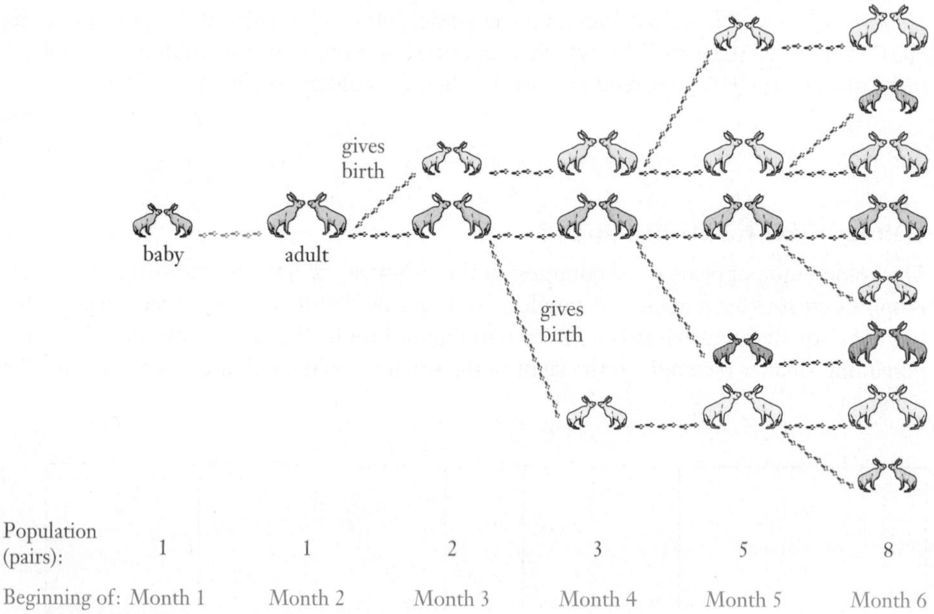

FIGURE 11.46

If we let F_n denote the nth Fibonacci number, then we have $F_1 = 1, F_2 = 1, F_3 = 2, F_4 = 3$, and so forth. The most basic property of the Fibonacci sequence is that the next number in the sequence is the sum of the previous two numbers. For example, note that

$$F_3 = F_2 + F_1 = 1 + 1 = 2$$

and

$$F_4 = F_3 + F_2 = 2 + 1 = 3$$

We can express this rule in general as $F_{n+1} = F_n + F_{n-1}$.

Think About It Confirm that the above rule works for Fibonacci numbers F_3 through F_{10}. Use the rule to determine the eleventh Fibonacci number (F_{11}).

The connection between the Fibonacci numbers and the golden ratio becomes clear when we compute the ratios of successive Fibonacci numbers, as shown in Table 11.3. Note that, as we go further along in the sequence, the ratios of successive Fibonacci numbers get closer and closer to the golden ratio $\phi = 1.61803. \ldots$

TABLE 11.3 Ratios of Successive Fibonacci Numbers

$F_3/F_2 = 2/1 = 2.0$	$F_{11}/F_{10} = 89/55 \approx 1.618182$
$F_4/F_3 = 3/2 = 1.5$	$F_{12}/F_{11} = 144/89 \approx 1.617978$
$F_5/F_4 = 5/3 \approx 1.667$	$F_{13}/F_{12} = 233/144 \approx 1.618056$
$F_6/F_5 = 8/5 = 1.600$	$F_{14}/F_{13} = 377/233 \approx 1.618026$
$F_7/F_6 = 13/8 = 1.625$	$F_{15}/F_{14} = 610/377 \approx 1.618037$
$F_8/F_7 = 21/13 \approx 1.6154$	$F_{16}/F_{15} = 987/610 \approx 1.618033$
$F_9/F_8 = 34/21 \approx 1.61905$	$F_{17}/F_{16} = 1597/987 \approx 1.618034$
$F_{10}/F_9 = 55/34 \approx 1.61765$	$F_{18}/F_{17} = 2584/1597 \approx 1.618034$

There are many examples of the Fibonacci sequence in nature. The heads of sunflowers and daisies consist of a clockwise spiral superimposed on a counterclockwise spiral (both of which are logarithmic spirals), as shown in Figure 11.47. The number of individual florets in each of these intertwined spirals is a Fibonacci number—for example, 21 and 34 or 34 and 55. Biologists have also observed that the number of petals on many common flowers is a Fibonacci number (for example, irises have 3 petals, primroses have 5 petals, ragworts have 13 petals, and daisies have 34 petals). The arrangement of leaves on the stems of many plants also exhibits the Fibonacci sequence. And spiraling Fibonacci numbers can be identified on pine cones and pineapples.

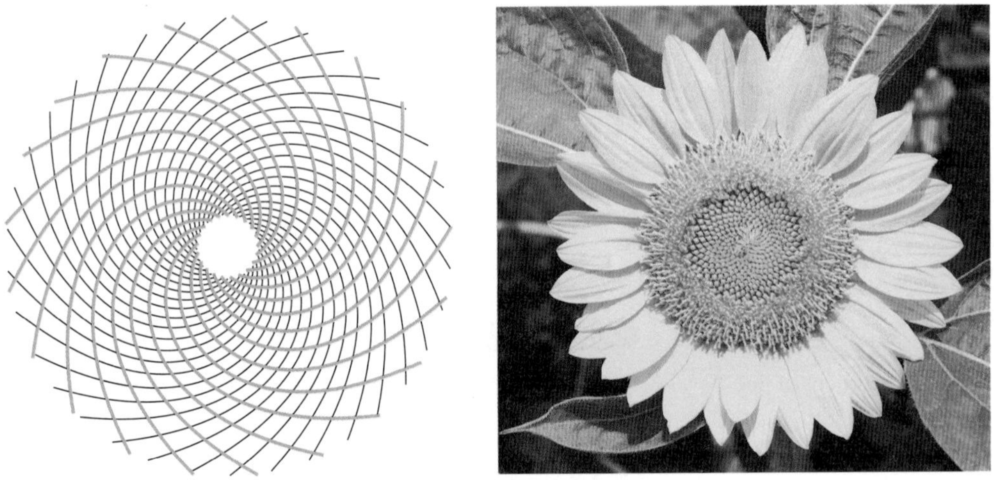

(a) (b)

FIGURE 11.47

Quick Quiz 11C

Choose the best answer to each of the following questions. Explain your reasoning with one or more complete sentences.

1. The golden ratio is
   a. exactly 1.6.
   b. a perfect number discovered by Pythagoras.
   c. $\frac{1 + \sqrt{5}}{2}$.
2. Which of the following is *not* a characteristic of the golden ratio?
   a. It is an irrational number.
   b. It is between 1 and 2.
   c. It is the fourth number in the Fibonacci sequence.
3. If a 1-foot line segment is divided according to the golden ratio, the two pieces
   a. have equal length.
   b. have lengths of roughly 2 inches and 10 inches.
   c. have lengths of roughly 0.4 foot and 0.6 foot.
4. To make a golden rectangle, you should
   a. inscribe a rectangle in a circle.
   b. draw a rectangle so that the ratio of the long side to the short side is the golden ratio.
   c. draw a rectangle so that the ratio of the diagonal to the short side is the golden ratio.
5. Suppose you want a bay window to have the proportions of a golden rectangle. If the window is to be 10 feet high, approximately how wide should it be?
   a. 5 feet
   b. $6\frac{1}{4}$ feet
   c. 12 feet
6. Why did the Greeks tend to build rectangular buildings with the proportions of golden rectangles?
   a. They thought these buildings were the most visually pleasing.
   b. They thought these buildings were structurally stronger.
   c. They thought these buildings could be constructed at the lowest possible cost.
7. Suppose you start with a golden rectangle and cut the largest possible square from one end of the rectangle. The remaining piece of the golden rectangle will be
   a. another golden rectangle.
   b. another square.
   c. a logarithmic spiral.
8. The rabbit problem of Fibonacci describes an example of
   a. a linearly growing population.
   b. an exponentially growing population.
   c. a logistically growing population.
9. The 18th, 19th, and 20th numbers in the Fibonacci sequence are, respectively, 2584, 4181, and 6765. The 21st number is
   a. 2584×4181. b. $4181 + 6765$. c. 6765×1.6.
10. In what way does the golden ratio appear in the Fibonacci sequence?
    a. It is the ratio between all pairs of successive Fibonacci numbers.
    b. The ratio of successive Fibonacci numbers gets ever closer to the golden ratio as the numbers get higher.
    c. It is the ratio of the last Fibonacci number to the first one.

Exercises 11C

REVIEW QUESTIONS

1. Explain the golden ratio in terms of proportions of line segments.
2. How is a golden rectangle formed?
3. What evidence suggests that the golden ratio and golden rectangle hold particular beauty?
4. What is a *logarithmic spiral*? How is it formed from a golden rectangle?
5. What is the *Fibonacci sequence*?
6. What is the connection between the Fibonacci sequence and the golden ratio? Give some examples of the Fibonacci sequence in nature.

DOES IT MAKE SENSE?

Decide whether each of the following statements makes sense (or is clearly true) or does not make sense (or is clearly false). Explain your reasoning.

7. Maria cut her 4-foot walking stick into two 2-foot sticks in keeping with the golden ratio.
8. Dan attributes his love of his special set of playing cards to the fact that their shape is a golden rectangle.
9. The circular pattern in the floor is attractive because it exhibits the golden ratio.
10. Each year, Juliet's age is another Fibonacci number.

BASIC SKILLS & CONCEPTS

11. **Golden Ratio.** Draw a line segment 6 inches long. Now subdivide it according to the golden ratio. Verify your work by computing the ratio of the whole segment length to the long segment length and the ratio of the long segment length to the short segment length.

12. **Golden Ratio.** A line is subdivided according to the golden ratio, with the smaller piece having a length of 5 meters. What is the length of the entire line?

13. **Golden Rectangles.** Measure the sides of each rectangle in Figure 11.48, and compute the ratio of the long side to the short side for each rectangle. Which ones are golden rectangles?

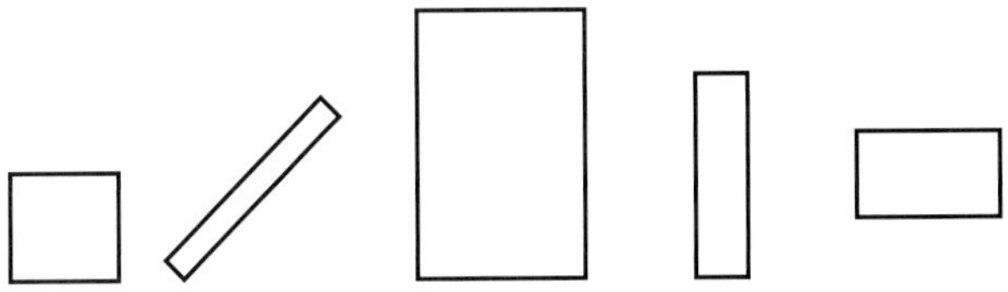

FIGURE 11.48

14–17: Dimensions of Golden Rectangles. Consider the following lengths of one side of a golden rectangle. Find the length of the other side. Notice that the other side could be either longer or shorter than the given side. Use the approximation $\phi \approx 1.62$ for your work.

14. 2.7 inches

15. 5.8 meters

16. 12.6 kilometers

17. 0.66 centimeter

18. **Everyday Golden Rectangles.** Find at least three everyday objects with rectangular shapes (for example, cereal boxes, windows). In each case, measure the side lengths and calculate the ratio. Are any of these objects golden rectangles? Explain.

FURTHER APPLICATIONS

19. **Finding ϕ.** The property that defines the golden ratio is

$$\frac{L}{1} = \frac{L+1}{L}$$

a. Show that, if we multiply both sides by L and rearrange, this equation becomes

$$L^2 - L - 1 = 0$$

Confirm that substituting the value of ϕ for L satisfies this equation.

b. The quadratic formula states that, for any equation of the form $ax^2 + bx + c = 0$, the solutions are given by

$$x = \frac{-b + \sqrt{b^2 - 4ac}}{2a}$$

and

$$x = \frac{-b - \sqrt{b^2 - 4ac}}{2a}$$

Use the quadratic formula to solve for L in the formula for the golden ratio. Show that one of the roots is ϕ.

20. **Properties of ϕ.**

a. Enter $\phi = (1 + \sqrt{5})/2$ into your calculator. Show that $1/\phi = \phi - 1$.

b. Now compute ϕ^2. How is this number related to ϕ?

21. **Logarithmic Spirals.** Draw a rectangle that is 10 centimeters on each short side. Follow the procedure described in the text for subdividing the rectangle until you can draw a logarithmic spiral. Do all work carefully, and show your measurements with your work.

22. **The Lucas Sequence.** A sequence called the *Lucas sequence* is closely related to the Fibonacci sequence. The Lucas sequence begins with the numbers $L_1 = 1$ and $L_2 = 3$ and then uses the same relation $L_{n+1} = L_n + L_{n-1}$ to generate $L_3, L_4, \ldots$.

a. Generate the first ten Lucas numbers.

b. Compute the ratio of successive Lucas numbers $L_2/L_1, L_3/L_2, L_4/L_3$, and so on. Can you determine if these ratios approach a single number? What number is it?

23. **Graphing Fechner's Data.** Consider Gustav Fechner's data shown in Table 11.2. Make a histogram that displays the responses for both the most pleasing and the least pleasing rectangle proportions.

24. **The Golden Navel.** An old theory claims that, on average, the ratio of the height of a person to the height of his or her navel is the golden ratio. Collect "navel ratio data" from as many people as possible. Graph the ratios in a histogram, find the average ratio over your entire sample, and discuss the outcome. Do your data support the theory?

25. **Mozart and the Golden Ratio.** Each movement of Mozart's 19 piano sonatas is clearly divided into two parts (the exposition and the development and recapitulation). In a paper called "The Golden Section and the Piano Sonatas of Mozart" (*Mathematics Magazine*, Vol. 68, No. 4, 1995), John Putz gives the lengths (in measures) of the first part (a) and the second part (b) of each movement. Some of the data are given below.

a = length of first part	b = length of second part
38	62
28	46
56	102
56	88
24	36
77	113
40	69
46	60
15	18
39	63
53	67

a. The ratio of the length of the whole movement to the length of the longer segment is $(a + b)/b$. Add a third column to the table and compute this ratio for the given data.

b. Briefly comment on how well the ratios that you computed in part (a) are approximated by ϕ.

c. Read the article by John Putz. Do you believe that Mozart composed with the golden ratio in mind?

26. **Debunking the Golden Ratio.** Find the article "Misconceptions about the Golden Ratio" by George Markowsky (*College Mathematics Journal*, Vol. 23, No. 1, 1992). Choose at least one of the misconceptions that Markowsky discusses, summarize it, and then explain whether you find his argument convincing. Discuss your opinion of whether the golden ratio has been consciously used by artists and architects in their work.

IN YOUR WORLD

27. **Proportion in Life.** Describe at least three ways in which the use of visual proportion affects your life.

28. **The Golden Ratio.** Find a recent picture of a new building or architectural design. Study the picture and decide whether the golden ratio is involved in any way.

29. **Golden Controversies.** Many websites are devoted to the controversy concerning the role of the golden ratio in art. Find a specific argument on one side of this controversy, and summarize it in a one- to two-page essay.

30. **Fibonacci Numbers.** Learn more about Fibonacci numbers and possible occurrences of the Fibonacci sequence in nature. Write a short essay about one aspect of Fibonacci numbers that you find interesting.

Chapter 11 Summary

UNIT	KEY TERMS	KEY IDEAS AND SKILLS
11A	sound wave (p. 623) pitch (p. 623) frequency (p. 623) harmonics (p. 624) octave (p. 624) musical scale (p. 624) digital recording (p. 628)	**Understand** how a plucked string produces sound. (p. 623) **Measure** frequency in cycles per second, and find harmonics of a frequency. (p. 623) **Understand** the musical scale and the ratios of frequencies among musical notes. (p. 624) **Understand** how the frequencies of a scale exhibit exponential growth. (p. 626) **Explain** the difference between analog and digital representations of music. (p. 628)
11B	perspective (p. 631) vanishing point (p. 631) horizon line (p. 632) symmetry (p. 634) reflection (p. 634) rotation (p. 635) translation (p. 635)	**Understand** the use of perspective in painting. (p. 631) **Find** symmetries in paintings and tilings. (p. 634) **Create** tilings with regular or irregular polygons. (p. 637)
11C	proportion (p. 643) golden ratio (p. 644) golden rectangle (p. 645) Fibonacci sequence (p. 648)	**The golden ratio:** (p. 644) $\phi = \frac{1 + \sqrt{5}}{2} = 1.61803\ldots$ **Understand** claimed uses of the golden ratio in both art and nature. (p. 644) **The Fibonacci sequence:** (p. 648) 1, 1, 2, 3, 5, 8, 13, 21, 34, 55, . . .

12 MATHEMATICS AND POLITICS

Mathematics and politics have been part of human culture for millennia, so it's not surprising that the two have been interwoven throughout history. We have already studied several instances in which mathematics plays an important role in political decision making, including the case of the federal budget (Unit 4F). But the connections between mathematics and politics go deep into the heart of the democratic process. In this chapter, we'll discuss the prominent role of mathematics in understanding systems of voting and apportionment.

Q For all 435 Congressional House districts combined, the overall results of the 2016 elections showed that Republican candidates for House seats received 49.1% of the national vote and Democratic candidates received 48.0% of the national vote. (The remaining 2.9% went to candidates from other parties.) What would you guess was the average (mean) margin of victory among the 435 individual House races?

- (A) less than 2 percentage points
- (B) 2 to 5 percentage points
- (C) 6 to 10 percentage points
- (D) 11 to 20 percentage points
- (E) more than 20 percentage points

I must study politics and war that my sons may have liberty to study mathematics and philosophy.

—John Adams, letter to Abigail Adams, May 12, 1780

A This question asks for a guess because there's no way for you to know the answer without looking at the election data. Nevertheless, given the fact that the national vote for the two major parties differed by just over 1 percentage point (49.1% to 48.0%), most people guess that the answer would be A. However, the correct answer is E, and the actual mean margin of victory was about 37 percentage points. In other words, the mean outcome of the vote was more than 68% to the winner and less than 32% to the loser, for a more than 2-to-1 margin of victory. In fact, many of the winners had such a clear path to victory that no serious candidate even bothered to challenge them.

How could individual races be decided by such an enormous average margin when the national vote was so closely split? The answer lies in the fact that elections are not nearly as simple as we often assume them to be, and as a result it is often possible for partisan politicians to game the system to their advantage. We'll explore some of the ways this is done in the activity on the next page and in Unit 12D, but the basic message should already be clear: Unless you understand the mathematics of voting, you may be very surprised by how modern elections are decided.

UNIT 12A

Voting: Does the Majority Always Rule? Investigate methods for choosing a winner in elections with more than two candidates and learn why different methods can lead to different winners.

UNIT 12B

Theory of Voting: Explore issues of fairness in voting, leading to the surprising conclusion that no single system is absolutely fair in all cases.

UNIT 12C

Apportionment: The House of Representatives and Beyond: Study several acceptable ways to apportion seats in the House of Representatives, and again see that no single method is always fair.

UNIT 12D

Dividing the Political Pie: Investigate some of the mathematical issues that surround redistricting—the way in which Congressional district maps are drawn—and why this process has become one of the most important and contentious issues in U.S. politics.

ACTIVITY

Partisan Redistricting

Use this activity to gain a sense of the kinds of problems this chapter will enable you to analyze. Additional activities are available online in MyLab Math.

By law, Congressional district maps must be redrawn every 10 years, following the national census. However, states can decide exactly how they draw district boundaries, and in most states the process is controlled by politicians. Figure 12.A shows one result of this process. Notice the bizarre shape of this district, which was carefully drawn in order to maximize the likelihood that it would be won by a Democrat. We'll study the process of drawing district boundaries in some detail in Unit 12D, but let's begin by exploring how the process can be used to partisan advantage.

Figure 12.B shows a simple "state" that has only 64 voters, each represented by a red or blue house. Assume that each blue house represents a voter who will vote Democratic and each red house represents a voter who will vote Republican. Note that there are equal numbers (32 each) of Democrats and Republicans. Work in small groups to answer the following questions.

1. Suppose the state must be divided into 8 congressional districts, and every district must have the same number of voters. Based on the overall numbers of Democratic and Republican voters, how many congressional seats would you expect to go to each party?
2. Now look at the set of 8 districts drawn in Figure 12.B. Based on these district boundaries, how many of the districts would be represented by a Republican? How many would be represented by a Democrat? Does this representation match the "expected" representation from Question 1?
3. Study the 8 districts in Figure 12.B carefully. How many would you say are "strongly Republican," how many are "strongly Democratic," and how many are "swing" districts in which you might expect voting results to be close? Based on your answers, what is the key factor that allows one party to gain more congressional seats than you would expect from the overall proportions of voters in the state?
4. Suppose you are in charge of redistricting (drawing new district boundaries) for the state in the figure. Experiment with various possible boundaries (remembering that each district must have 8 voters). Can you come up with a set of boundaries that reflects the overall proportions of voters in the state? Can you come up with another set of boundaries (different from those in the figure) that strongly favors one of the parties?

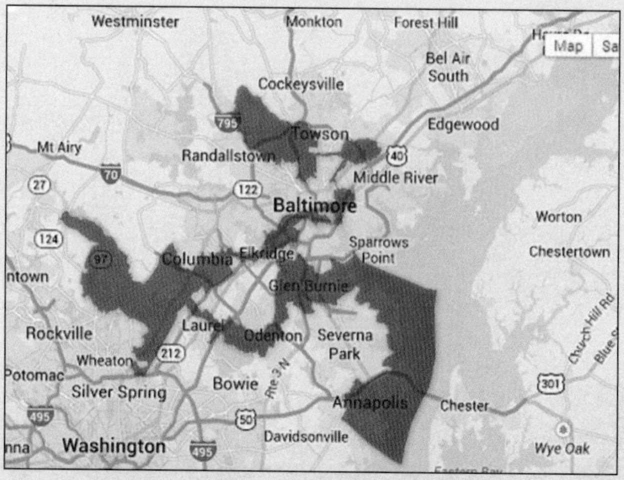

FIGURE 12.A Maryland's Third Congressional District.

FIGURE 12.B A simple "state" with 64 voters. Blue houses represent the homes of Democrats, and red houses represent the homes of Republicans.

5. As you have seen, one of the consequences of political redistricting is that a large number of districts are drawn so that they tend to lean strongly to one party or the other, leaving relatively few "swing" districts with competitive elections. How does this fact explain the answer to the chapter-opening question (p. 654)?
6. In the 2016 elections, Republicans won 241 seats in the House of Representatives to the Democrats 194 seats. What percentage of the seats did each party win? How does this distribution of House seats compare to the overall national vote for the House (all districts combined), which was 49.1% for Republicans and 48.0% for Democrats?
7. In the 2016 elections, the average (mean) margin of victory for Democrats who won was about 41.5 percentage points, compared to 33.5 percentage points for Republican winners. Does this difference help account for the discrepancy you found in Question 6? Explain.
8. Find the district map for your state. Who drew the district boundaries in your state, and what factors did they consider in drawing them? Do you think the boundaries are drawn fairly? Defend your opinion.

UNIT 12A Voting: Does the Majority Always Rule?

We usually assume that the winner of an election is the person with the most votes, but elections are not always so simple. For example, Donald Trump won the 2016 presidential election, despite receiving nearly 3 million fewer votes than Hillary Clinton. To understand how this and other surprising outcomes can arise, let's begin by investigating general principles of voting systems.

Majority Rule

The simplest type of voting involves only two candidates or two choices, so voters can choose the winner by **majority rule**: The choice receiving more than 50% of the votes wins. Mathematically, majority rule has three important properties:

- Every vote has the same *weight*. That is, no person's vote counts more or less than any other person's vote.

BY THE WAY

In 1952, mathematician Kenneth May proved that majority rule is the *only* voting system for two candidates that satisfies all three properties.

- There is *symmetry* between the candidates: If all the votes were reversed, the loser would become the winner.
- If a vote for the loser were changed to a vote for the winner, the outcome of the election would not be changed.

We can illustrate these three properties with a simple example. Suppose that Olivia and Rafael are running for captain of the debate team. Olivia wins with 10 votes to Rafael's 9 votes. The first property, that all votes have the same weight, means we care only about the vote totals—not about the names of the people who cast the individual votes. The second property tells us that if we reverse all the votes, so that Olivia receives 9 votes and Rafael receives 10, then Rafael becomes the winner. The third property tells us that if a vote for Rafael (the loser) is changed to a vote for Olivia (the winner), Olivia still wins: This change raises Olivia's vote total to 11 and reduces Rafael's to 8.

Majority Rule

Majority rule says that the choice receiving more than 50% of the votes is the winner.

Think About It Suppose that a vote for the *winner* is changed to a vote for the loser. Can the outcome of the election change in that case? Explain.

U.S. Presidential Elections

U.S. presidential elections are settled by majority rule, but with a twist. The **popular vote** in a presidential election reflects the total number of votes received by each candidate. However, Presidents are elected by **electoral votes**, which are cast by people known as *electors*. When you cast a ballot for President, you are actually voting for your state's electors; the electors meet to cast their votes a few weeks after the general election. It is a majority rule system, because the winner of the presidential election must receive a majority of the electoral votes. If no candidate receives a majority of the electoral votes, the House of Representatives chooses the President. This has happened twice in U.S. history: The House elected Thomas Jefferson following the undecided election of 1800 and John Quincy Adams following the undecided election of 1824.

Article II of the U.S. Constitution mandates that each state gets as many electoral votes as the state has members of Congress (senators plus representatives), and the 23rd Amendment to the Constitution provides for electors to represent the District of Columbia. In most states (and the District of Columbia), the electoral votes are chosen by a winner-take-all system: *All* the electoral votes of that state go to the candidate with the most popular votes in that state. (The winner-take-all system is *not* required by the Constitution. However, as of 2017, there were only two exceptions to the winner-take-all system: Nebraska and Maine.) Today, there are a total of 538 electors, so 270 electoral votes are required to make a majority.

The electoral system explains why it is possible for the President to lose the popular vote. However, prior to the year 2000, this possibility occurred only twice in U.S. history: In 1876, Rutherford B. Hayes won the presidency by an electoral vote of 185 to 184 while receiving about 264,000 fewer popular votes than Samuel J. Tilden, and in 1888, Benjamin Harrison won the electoral vote by a margin of 233 to 168 while losing the popular vote by about 90,000 votes to Grover Cleveland.

The difference between the electoral and popular votes has become more consequential in recent years. The popular vote winner lost the election in both 2000 (when George W. Bush won the electoral vote while receiving about a half million fewer popular votes than Al Gore) and 2016. It came close to happening a third time: In 2004,

HISTORICAL NOTE

The 1800 election pitted the Republican Party ticket of Thomas Jefferson and Aaron Burr against the Federalist Party ticket of John Adams and Charles Pinckney. At that time, the ballot did not distinguish between the presidential and vice-presidential candidates; electors could vote for two, with the first-place winner becoming President and second-place becoming Vice President, regardless of their party. Those favoring the Jefferson-Burr ticket intended that one elector would pick Jefferson only, making him the winner, but instead Jefferson and Burr ended up tied for first place. It took 7 days and 36 ballots before the House of Representatives finally decided the election in favor of Jefferson. The 12th Amendment to the U.S. Constitution, passed in 1804, instituted separate balloting for President and Vice President.

George W. Bush won the popular vote by more than 3 million votes over John Kerry, but if just 60,000 voters in Ohio (out of some 5.6 million voters in that state) had chosen Kerry rather than Bush, Kerry would have won an electoral majority. (These facts have led numerous people from both parties to advocate electing the president by popular vote; see In Your World on page 663.)

EXAMPLE 1 2016 Presidential Election

Table 12.1 shows the official results of the 2016 presidential election. Determine the margins of victory in both absolute and percentage terms for the popular and electoral votes.

TABLE 12.1 2016 Presidential Election

Candidate	Popular Vote	Electoral Vote
Donald Trump	62,979,984	304
Hillary Clinton	65,844,969	227
Gary Johnson	4,492,919	0
Jill Stein	1,449,370	0
Other	1,684,908	7*
Total	136,452,150	538

*These 7 electors represented 2 who had originally committed to vote for Trump and 5 for Clinton, but then voted for other candidates when the Electoral College met.

Solution The absolute difference in popular vote totals between Clinton and Trump was

$$65{,}844{,}969 - 62{,}979{,}984 = 2{,}864{,}985$$

That is, Clinton received almost 3 million more votes than Trump. We find Clinton and Trump's percentages of the total popular vote by dividing their individual totals by the combined total for all candidates:

$$\textit{Clinton:}\quad \frac{65{,}844{,}969}{136{,}452{,}150} \approx 0.4825 = 48.25\%$$

$$\textit{Trump:}\quad \frac{62{,}979{,}984}{136{,}452{,}150} \approx 0.4616 = 46.16\%$$

Notice that, as a result of the votes that went to other candidates, neither of the two major candidates received a majority of the popular vote.

In the electoral vote, Trump won by a margin of $304 - 227 = 77$ electoral votes. As percentages of the total 538 electoral votes available, the candidates received:

$$\textit{Trump:}\quad \frac{304}{538} \approx 0.5651 = 56.51\%$$

$$\textit{Clinton:}\quad \frac{227}{538} \approx 0.4219 = 42.19\%$$

Because he won a majority of the electoral vote, Donald Trump became the President of the United States. ▶ **Now try Exercises 15–24.**

Think About It Notice in Table 12.1 that some electors did not vote for the candidates they had originally committed to vote for. Do you think this should be allowed? Discuss this question, along with your overall opinion of the electoral system.

BY THE WAY

Senate rules do not limit what a senator can say. During some filibusters, senators have read novels aloud in order to keep speaking. In 1964, the Civil Rights Act was passed by Congress only after 75 days of filibuster by southern senators.

Variations on Majority Rule

Many voting systems use slight variations on majority rule. For example, the United States Senate passes bills by majority rule, but sometimes a majority is not enough. Senators are generally allowed to speak about a legislative bill for as long as they wish before it is brought to a vote. If a senator opposes a particular bill, but fears that it will gain a majority vote, he or she may choose to speak continuously (during the hours that the issue is under consideration)—and thereby prevent the vote from taking place. This technique, called a **filibuster**, can be ended only by a vote of 3/5, or 60%, of the senators, which means that when a filibuster is allowed, just 41% of senators can prevent the majority from acting. (For most of U.S. history, senators actually spoke during filibusters; more recently, the mere threat of a filibuster generally leads to the tabling of a bill unless 60% of senators vote to end the filibuster, a vote called *cloture*.)

In other cases, a candidate or issue must receive *more* than a majority of the vote to win—such as 60% of the vote, 75% of the vote, or a unanimous vote. We then say that a **super majority** is required. Criminal trials offer an example: All states require a super majority vote of the jury to reach a verdict, and many states require that the vote be unanimous. A jury that cannot reach this super majority or unanimous agreement is called a *hung jury*. When a jury is hung, the judge generally declares a mistrial, and the case must be tried again or dropped.

The U.S. Constitution requires super majority votes for many specific issues. For example, an international treaty can be ratified only by a 2/3 super majority vote of the Senate. Amending the Constitution first requires a 2/3 super majority vote for the amendment in both the House and the Senate, and then approval of the amendment by 3/4 of the states (generally by their state legislatures, often after a popular vote has taken place).

Another variation on majority rule occurs with **veto** power. For example, a bill proposed in the U.S. Congress can be signed into law by the President if it receives a majority vote in both the House and the Senate. However, if the President vetoes the bill, it can become law only if it then receives a 2/3 super majority vote in both the House and the Senate to override the veto. The courts can also effectively veto a popular vote. For example, even if a proposition in a state election receives a huge majority of the popular vote, it will not become law if the courts declare that the proposition violates the U.S. Constitution.

BY THE WAY

The U.S. Constitution also specifies that it may be amended if 2/3 of the states call for "a convention for proposing amendments." However, such a call for a convention has never occurred.

EXAMPLE 2 Majority Rule?

Evaluate the outcome in each of the following cases.

a. Of the 100 senators in the U.S. Senate, 59 favor a new bill on campaign finance reform. The other 41 senators are adamantly opposed and start a filibuster. Will the bill pass?

b. A criminal conviction in a particular state requires a vote by 3/4 of the jury members. On a nine-member jury, seven jurors vote to convict. Is the defendant convicted?

c. A proposed amendment to the U.S. Constitution has passed both the House and the Senate with more than the required 2/3 super majority. Each state holds a vote on the amendment. The amendment garners a majority vote in 36 of the 50 states. Is the Constitution amended?

d. A bill limiting the powers of the President has the support of 73 out of 100 senators and 270 out of 435 members of the House of Representatives. But the President promises to veto the bill if it is passed. Will it become law?

Solution

a. The filibuster can be ended only by a vote of 3/5 of the Senate, or 60 out of the 100 senators. The 59 senators in favor of the bill cannot stop the filibuster. The bill will not become law.

b. Seven out of nine jurors represents a super majority of $7/9 = 77.8\%$. This percentage is more than the required 3/4 (75%), so the defendant is convicted.

c. The amendment must be approved by 3/4, or 75%, of the states. But 36 out of 50 states represent only $36/50 = 72\%$ of the states, so the amendment does not become part of the Constitution.

d. The bill has the support of $73/100 = 73\%$ of the Senate and $270/435 = 62\%$ of the House. But overriding a presidential veto requires a super majority vote of $2/3 = 66.7\%$ in *both* the House and the Senate. The 62% support in the House is not enough to override the veto, so the bill will not become law. (Note: The law requires a vote of 2/3 of those physically present in Congress; this example assumes that all members are present.)

▶ Now try Exercises 25–26.

BY THE WAY
The proposed Equal Rights Amendment (ERA) to the U.S. Constitution, first introduced to Congress in 1923, was designed to guarantee equal rights for women. It passed both the House and Senate with the required super majorities in 1972, but was then ratified by only 35 states, leaving it 3 states short of the 38 states required to represent a 3/4 super majority of the 50 states.

Voting with Three or More Choices

Table 12.2 shows results for a hypothetical election for governor in which there are three candidates:

TABLE 12.2 Three-Way Governor Race

Candidate	Percentage of Vote
Smith	32%
Kim	33%
Garcia	35%

No candidate has a majority, so who should become governor?

The most common method for deciding such an election is to award the governorship to the person who received the most votes, called a **plurality** of the vote. Garcia has the highest percentage and becomes governor if we decide this election by plurality. However, there are alternative methods for deciding this election. For example, in many political elections, any race in which no candidate receives a majority is followed by a **runoff** between the top two vote getters.

In the case of Table 12.2, a runoff would mean a follow-up election between Kim and Garcia. Because Smith's voters can no longer vote for Smith in the runoff, the outcome depends on whether these voters prefer Kim or Garcia as their second choice. For example, suppose that all of Smith's supporters prefer Kim to Garcia. The 32% of voters who chose Smith would then give their votes to Kim in the runoff, so that Kim would win the runoff easily with 65% of the vote (her own 33% plus Smith's 32%).

Note that if all of Smith's voters prefer Kim to Garcia, the runoff method seems to give a better result than the plurality method, because the plurality method leads to a victor (Garcia) who is the *last* choice for 65% of the electorate. For this reason, many people believe that a runoff system is better than a plurality system. However, as we'll see in Unit 12B, runoffs do not always lead to a fairer result, and a runoff generally requires holding a second election, which means added expense and a longer campaign season.

EXAMPLE 3 Potential Impact of Runoffs

Table 12.3 shows the results of the 2016 U.S. presidential election for Pennsylvania, Michigan, and Wisconsin, which accounted for a combined total of 46 electoral votes. Suppose that these states had each held a runoff between the top two candidates before allotting their electoral votes. Is it possible that this would have changed the national election outcome? Explain.

Solution Donald Trump won a plurality but not a majority in each of these three states. Therefore, in a runoff, we would assume that the voters who initially voted for other candidates would have either switched their votes to Trump or Clinton, or not voted at all. We cannot predict exactly what these voters would have done in a runoff, but the very small margins between Clinton and Trump in these three states mean that it is at least possible that runoffs would have changed the outcomes (see Exercise 42). Because these states accounted

BY THE WAY
In the United States, presidential elections are decided by plurality (within each state). However, many state and local elections require a runoff if no candidate receives a majority on the first ballot. Runoffs are also widely used internationally.

TABLE 12.3 2016 Presidential Election Results for Pennsylvania, Michigan, and Wisconsin

Candidate	Pennsylvania (20 Electoral Votes)	Michigan (16 Electoral Votes)	Wisconsin (10 Electoral Votes)
Donald Trump	2,970,733 (48.18%)	2,279,543 (47.50%)	1,405,284 (47.22%)
Hillary Clinton	2,926,441 (47.46%)	2,268,839 (47.27%)	1,382,536 (46.45%)
Gary Johnson	146,715 (2.38%)	172,136 (3.59%)	106,674 (3.58%)
Jill Stein	49,941 (0.81%)	51,463 (1.07%)	31,072 (1.04%)
Other	71,648 (1.16%)	27,303 (0.57%)	50,584 (1.70%)
Total	6,165,478	4,799,284	2,976,150

for a combined 46 electoral votes, a swing of those electoral votes from Trump to Clinton would have reduced Trump's electoral vote total to $304 - 46 = 258$ and increased Clinton's electoral vote total to $227 + 46 = 273$, in which case Hillary Clinton would have won the presidency.

▶ **Now try Exercises 27–28.**

Think About It Do you think that presidential elections should have runoffs? Why or why not?

Preference Schedules

As we've seen, the outcome of a runoff can depend on voters' second-choice preferences in an election. With more than three candidates, it can also depend on their third choices, fourth choices, and so on. We could therefore create ballots that allow voters to record all their preferences among the candidates. Figure 12.1 shows what such a ballot might look like for the three-way governor race between Smith, Kim, and Garcia.

Tabulating results from ballots like those in Figure 12.1 requires a special type of table, called a **preference schedule**, that tells us how many voters chose each particular ranking order among the candidates. The following example illustrates how we construct a preference schedule.

FIGURE 12.1 Sample ballot for showing preferences in a three-way governor race.

EXAMPLE 4 Making a Preference Schedule

Table 12.2 showed first-place percentages in the three-way governor race between Smith, Kim, and Garcia. Suppose that there were a total of 1000 voters, and their full preferences were as follows:

- For 320 voters, Smith is their first choice and Kim is their second choice, leaving Garcia as their third (last) choice.
- For 330 voters, Kim is their first choice and Smith is their second choice, leaving Garcia as their third choice.
- For 175 voters, Garcia is their first choice, Smith is their second choice, and Kim is their third choice.
- For 175 voters, Garcia is their first choice, Kim is their second choice, and Smith is their third choice.

Make a preference schedule for this election.

Technical Note

For Example 4, there are two other possible orders that no voters chose: Smith-Garcia-Kim and Kim-Garcia-Smith. More generally, the total number of possible rank orders for an election with n candidates is $n!$ (n factorial). To keep the work manageable, the examples in this text use only some of the possible rank orders.

Solution Because we are given four different arrangements of voter preferences, our preference table will need four columns to show them. We start the table with rows for each choice (first, second, third), followed by a row for the number of voters choosing each arrangement. We then put each arrangement and its total number of voters in one

IN YOUR WORLD

Counting Votes—Not as Easy as It Sounds

Many of the examples and exercises in this chapter give vote totals from real elections, but counting votes is not as easy as it might sound. The classic example in recent history occurred in the 2000 presidential election, in which the two major candidates were George W. Bush and Al Gore. Gore won the popular vote by a fairly wide margin (about a half million votes), but the national electoral vote was close enough that the final outcome hinged on disputed results in the state of Florida.

The dispute revolved primarily around the question of whether votes were accurately counted. Part of the question concerned the accuracy of the original vote count; because humans are involved in the counting, mistakes are always possible (for example, accidentally putting a vote for one candidate into a different candidate's tally), which is why close elections often lead to a recount as a way of double checking the original count. A more difficult question arose from the fact that Florida used punch-style voting machines, in which a vote is cast by pressing a lever that punches out a tiny piece of paper called a "chad." This can lead to ambiguity when the chad is left partially attached (a "hanging chad") or even fully attached but with the markings of the punch around it (indicating that the voter at least started to press the lever). For such chads, it's difficult to know whether the voter intended to vote for the candidate but didn't push the lever hard enough, in which case the vote should count, or stopped pressing the lever on purpose, in which case the vote should not count. These and other ballot counting issues led to legal challenges that delayed the final 2000 election results until nearly 6 weeks after election day. The legal challenges ended when, in a 5-4 decision (of the case called *Bush v. Gore*), the United States Supreme Court halted a statewide recount that had been ordered by the Florida Supreme Court. At that point, Florida certified Bush as the winner by 537 votes out of more than 6 million votes cast, thereby giving him the national electoral victory and the presidency. Subsequent studies of the Florida ballots showed that even if a recount had been conducted, there was no unambiguous way to decide who had really received the most votes. In other words, the nature of the ballots used for this election made it impossible to get an exact count of the vote totals.

Beyond those direct counting issues, there were other reasons to wonder whether the Florida outcome accurately represented the choices of voters. In particular, the election was almost certainly affected by the "butterfly" ballot used in Palm Beach County, in which candidates were listed on both sides of the punch holes running down the middle. As you can see in the reproduction of the ballot shown below, Al Gore was listed in the *second* position on the left side of the ballot, but voting for him required punching the *third* hole from the top (labeled "5"); if a voter accidently punched the second hole, it was a vote for Pat Buchanan. Statistical evidence (see Exercise 28) and follow-up interviews with voters suggested that Buchanan ended up with at least about 1000 votes that were intended for Gore—which would have been enough to change the outcome, given the statewide margin of only 537 votes. Additional issues concerned whether some eligible voters were improperly denied an opportunity to vote and whether some votes were cast by people who were ineligible to vote.

Are there better ways to count votes? After the 2000 election, there was a great deal of discussion about instituting national voting standards, so that all states would use the same types of ballots and voting machines, but this effort stalled before any legislation was enacted. If the effort is revived, the most likely outcome would be to use electronic (computer) voting machines that also produce a "paper trail"; the electronic machines can count votes much more accurately than human counters can, while the paper trail can serve as a backup in the event of hacking or other concerns about the integrity of the electronic vote. Of course, even if the counting issues are resolved, there's still the matter of eligibility to vote, which evokes strong emotions in both those who believe voters are being disenfranchised and those who believe that voter fraud is a problem. One way or other, the problem of accurately collecting and counting votes is likely to remain with us for a long time.

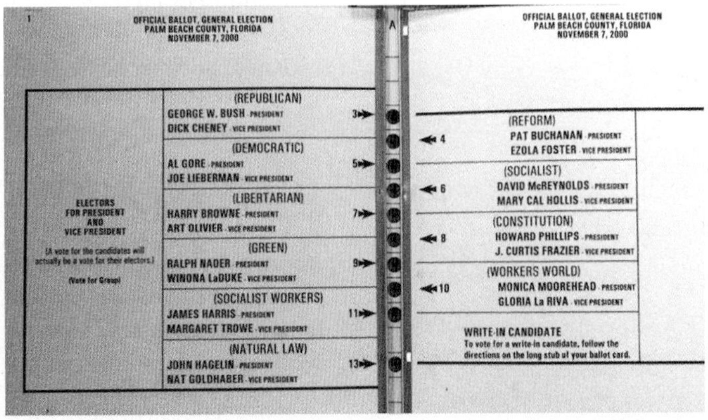

column. The first arrangement we are given is for the 320 voters whose choices in order are Smith, Kim, and Garcia; therefore, the first column in the preference schedule lists the candidates in this order and shows at the bottom that 320 voters ranked the

TABLE 12.4 Preference Schedule for Three-Way Governor Race

First Choice	Smith	Kim	Garcia	Garcia
Second Choice	Kim	Smith	Smith	Kim
Third Choice	Garcia	Garcia	Kim	Smith
Number of Voters	**320**	**330**	**175**	**175**

Each column shows one preference order that some voters chose.

The last row shows the number of voters who chose each particular preference order.

candidates in this order. Similarly, the second column shows the preference order for the 330 voters who put Kim as first choice, and the last two columns show the preference orders for the two groups of voters that had Garcia as first choice. Table 12.4 shows the full preference table.

▶ **Now try Exercises 29–30.**

Five Voting Methods

So far we have noted that a race with three or more candidates could be decided by plurality or by a single runoff between the top two candidates. In fact, there are many other possible ways to decide the outcome of a multi-candidate race. We will study three additional methods in this chapter:

- We could choose a winner through a **sequential runoff**. In this method, the candidate with the fewest first-place votes is eliminated after the first round, and voters' other choices move up. If there is still no candidate with a majority of the first-place votes, then the process is repeated until someone claims a majority. This method is often used in clubs and corporate elections and is also used to select nominees for the Academy Awards.
- We could choose a winner with a **point system**, often called a **Borda count**, in which we assign points to different rankings. For example, we might assign 3 points for a first-place vote, 2 points for a second-place vote, and 1 point for a third-place vote. Point systems are particularly common in sports. For example, swim meets and track meets assign different numbers of points to various placings in each individual event, and the winner of the meet is the team with the most total points. Similarly, voting for "most valuable player" awards and "top 25" rankings are generally based on a point system.
- We could choose a winner through **pairwise comparisons**, also known as the **Condorcet method**. In this method, each candidate is pitted one on one against each other candidate. For example, in the three-way governor race between Smith, Kim, and Garcia, we would have three one-on-one contests: Smith versus Kim, Smith versus Garcia, and Kim versus Garcia. The winner is the candidate who wins the most one-on-one contests.

Technical Note

A true Borda count is a point system that assigns points for every ranking—that is, points are assigned to third place for a three-way race, to fourth place for a four-way race, and so on. A standard Borda count assigns 1 point for last place and an additional point for each higher place.

The following box summarizes these five methods of selecting a winner, and Examples 5 through 7 present simple applications of them.

Voting Methods with Three or More Candidates

Plurality method: The candidate with the most first-place votes wins.

Single (top-two) runoff method: The two candidates with the most first-place votes have a runoff. The winner of this runoff is the winner of the election.

Sequential runoff method: A series of runoffs is held, eliminating the candidate with the *fewest* first-place votes at each stage. Runoffs continue until one candidate has a *majority* of the first-place votes and is declared the winner.

Point system (Borda count): Points are awarded according to the rank of each candidate on each ballot (first, second, third, ...). The candidate with the most points wins.

Pairwise comparisons (Condorcet method): The candidate who wins the most pairwise (one-on-one) contests is the winner of the election.

EXAMPLE 5 Applying the Five Methods

Consider the preference schedule in Table 12.4 for the three-way governor race between Smith, Kim, and Garcia. Determine the winner by each of the five methods we have discussed.

Solution Here are the results for the five methods, with the winner in each case highlighted in blue:

Plurality: Table 12.4 shows Garcia in first place in columns 3 and 4, with a total of 175 + 175 = 350 votes. Garcia therefore has the most first-place votes and is the plurality winner.

Single runoff: A runoff would pit first-place Garcia against second-place Kim. Because Smith is not in the runoff, voters who chose Smith as their first-choice will now select their second choice. Table 12.4 shows that *all* of Smith's voters have Kim as their second choice, so Kim receives 330 + 320 = 650 votes in the runoff, easily beating Garcia's 350 votes.

Sequential runoff: This method tells us to eliminate the last-place finisher, Smith, and then redo the election with only the remaining candidates. Because the only remaining candidates are Kim and Garcia, the situation is exactly the same as with a single runoff, with Kim the winner. In general, a sequential runoff is the same as a single runoff for elections with three candidates. If there were more than three candidates, the first runoff would be followed by additional runoffs, each eliminating the last-place finisher from the prior round.

Point system: With three candidates, we assign 3 points for first-place votes, 2 points for second-place votes, and 1 point for third-place votes. Based on Table 12.4, we find:

- Smith received 320 first-place votes, 330 + 175 = 505 second-place votes, and 175 third-place votes. Therefore, Smith's point total is

$$(320 \times 3) + (505 \times 2) + (175 \times 1) = 2145$$

- Kim received 330 first-place votes, 320 + 175 = 495 second-place votes, and 175 third-place votes. Therefore, Kim's point total is

$$(330 \times 3) + (495 \times 2) + (175 \times 1) = 2155$$

- Garcia received 175 + 175 = 350 first-place votes, 0 second-place votes, and 320 + 330 = 650 third-place votes. Therefore, Garcia's point total is

$$(350 \times 3) + (0 \times 2) + (650 \times 1) = 1700$$

Kim has the most points and therefore is the winner by the point system.

Pairwise comparisons: We evaluate the three pairwise comparisons:

- Smith versus Kim: Smith is ranked above Kim in columns 1 and 3, which total 320 + 175 = 495 votes. Kim is ranked above Smith in columns 2 and 4, which total 330 + 175 = 505 votes. Kim wins the comparison.
- Smith versus Garcia: Smith is ranked above Garcia in columns 1 and 2, which total 320 + 330 = 650 votes. Garcia is ranked above Smith in columns 3 and 4, which total 175 + 175 = 350 votes. Smith wins the comparison.
- Kim versus Garcia: Kim is ranked above Garcia in columns 1 and 2, which total 320 + 330 = 650 votes. Garcia is ranked above Kim in columns 3 and 4, which total 175 + 175 = 350 votes. Kim wins the comparison.

Kim wins two out of the three pairwise comparisons and therefore is the winner by this method. ▶ Now try Exercises 31–32, parts (a) to (c).

EXAMPLE 6 Majority Loser

The seven sportswriters of Seldom County rank the county's three women's volleyball teams—which we'll call teams A, B, and C—according to the following preference schedule. Determine the winner by plurality and by a Borda count. Discuss the results.

First	A	C	B
Second	B	B	C
Third	C	A	A
Number of voters	4	1	2

HISTORICAL NOTE
The Borda count is named for French mathematician and astronomer Jean-Charles de Borda (1733–1799). Borda also has a crater on the Moon named for him and his name is one of 72 inscribed on the Eiffel Tower.

Solution By plurality, Team A is the winner, having received 4 out of the 7 first-place votes—which is not just a plurality, but also a majority. For the Borda count, we assign 3 points for a first-place vote, 2 points for a second-place vote, and 1 point for a third-place vote and find the total number of points.

$$\text{Team A gets } (4 \times 3) + (1 \times 1) + (2 \times 1) = 15 \text{ points}$$
$$\text{Team B gets } (4 \times 2) + (1 \times 2) + (2 \times 3) = 16 \text{ points}$$
$$\text{Team C gets } (4 \times 1) + (1 \times 3) + (2 \times 2) = 11 \text{ points}$$

Although the majority of the sportswriters chose Team A as the best team, Team B gets ranked first by the Borda count. The fact that the Borda count winner can be different from the majority winner is a well-known shortcoming of the method.

▶ Now try Exercises 31–32, part (d).

Think About It Which team from Example 6 would *you* say is the "top-ranked" team? Defend your opinion.

EXAMPLE 7 Condorcet Paradox

Consider the following three-candidate preference schedule for candidates A, B, and C. Can you find a winner through pairwise comparisons? Explain.

First	A	C	B
Second	B	A	C
Third	C	B	A
Number of voters	14	12	10

Solution Three pairwise comparisons are possible: A versus B, B versus C, and A versus C. Columns 1 and 2 show A ranked ahead of B, but B is ahead of A in column 3. Therefore, A beats B in a one-on-one contest by 26 to 10. In the B versus C contest, B is ranked ahead of C in columns 1 and 3, so B beats C by 24 to 12. Similarly, A is ranked ahead of C only in column 1, so C beats A by 22 to 14. Summarizing, we have the following results for the pairwise comparisons:

- A beats B
- B beats C
- C beats A

HISTORICAL NOTE
Pairwise comparisons are called the Condorcet method after Marie Jean Antoine Nicholas de Caritat, Marquis de Condorcet (1743–1794), who did pioneering work in probability and calculus before becoming a leader of the French Revolution. He argued forcefully for equal rights for women, for universal free education, and against capital punishment. In 1794, as extremists took control of the revolution, he was arrested because of his aristocratic background. He died in prison the next day, in a death labeled a suicide by his captors.

Because A beats B and B beats C, the first two results suggest that A should also beat C. However, the third result shows that A actually loses to C. The results illustrate what is often called the *Condorcet paradox*, in which pairwise comparisons do not produce a clear winner.

▶ Now try Exercises 31–32, part (e).

Think About It Does Example 7 have a plurality winner? Does it have a winner by a Borda count? Explain.

Different Methods and Different Winners

We have already seen examples in which not all voting methods lead to the same winner. In extreme cases, it is even possible to produce a different winner with each of the five methods.

Imagine a club of 55 people that holds an election among five candidates for president; we'll call the candidates A, B, C, D, and E. Each ballot asks the voter to rank these candidates in order of preference (Figure 12.2). Suppose the results come out as shown by the preference schedule in Table 12.5, which was set up carefully to show you how different methods can lead to different winners.

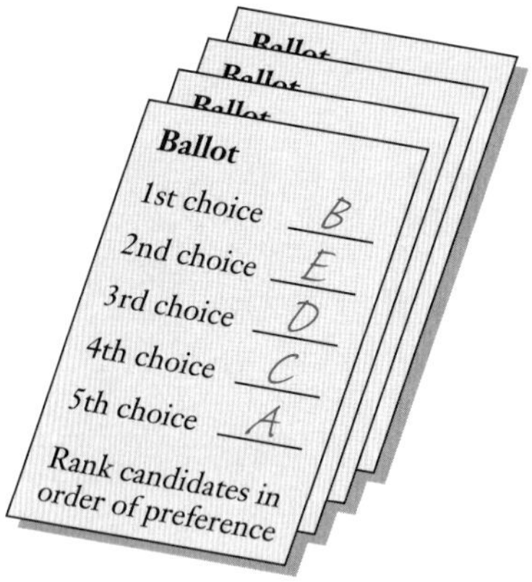

FIGURE 12.2 Sample ballot for the club election.

TABLE 12.5 Preference Schedule for the Club Election

First	A	B	C	D	E	E
Second	D	E	B	C	B	C
Third	E	D	E	E	D	D
Fourth	C	C	D	B	C	B
Fifth	B	A	A	A	A	A
Number of Voters	**18**	**12**	**10**	**9**	**4**	**2**

EXAMPLE 8 Reading the Preference Schedule

Answer the following questions to be sure you understand the preference schedule in Table 12.5.

a. How many voters ranked candidates in the order E, B, D, C, A?
b. How many voters had Candidate E as their first choice?
c. How many voters preferred Candidate C over Candidate A?

Solution

a. The order E, B, D, C, A appears in the second-to-last column of the table, and the last entry of that column shows that 4 voters preferred this order.
b. Candidate E appears as first choice in the last two columns, which represent a total of $4 + 2 = 6$ voters.
c. Notice that the first column shows Candidate A as first choice and Candidate C as fourth choice, which means the 18 voters who chose this order prefer A to C. However, all the remaining columns show Candidate C ranked higher than Candidate A; for example, the second column shows 12 voters who put C in fourth place and A in fifth place. Therefore, the total number of voters who prefer C over A is $12 + 10 + 9 + 4 + 2 = 37$. ▸ Now try Exercises 33–36.

Plurality Winner

Let's now find the winner from the preference schedule in Table 12.5 by each of our five methods. We begin with plurality, which simply requires counting first-place votes:

- A received 18 first-place votes (column 1).
- B received 12 first-place votes (column 2).
- C received 10 first-place votes (column 3).
- D received 9 first-place votes (column 4).
- E received $4 + 2 = 6$ first-place votes (columns 5 and 6).

The supporters of Candidate A can argue that A received a plurality and should be declared the winner. ▸ Now try Exercises 37–41, parts (a) and (b).

Single Runoff

"Not so fast!" yell the supporters of Candidate B. They suggest a runoff between A and B because they were the top two candidates in terms of first-place votes. We can use Table 12.5 to see what would happen in the runoff:

- Candidate A will still receive the 18 votes of the people who had A ranked first.
- Candidate B will still receive the 12 votes of the people who had B ranked first.
- In the runoff, anyone who ranked C, D, or E in first place must choose between A and B. Notice that all 25 people who originally chose C, D, or E have B ranked above A in Table 12.5. (In fact, they all have A ranked in last place.)
- Therefore, all 25 of the votes that originally went to C, D, or E will go to B in the runoff. Adding these votes to B's 12 original votes will give a total of 37 votes to B in the runoff.

We conclude that, in the runoff, Candidate B will win by a vote of 37 to 18. Supporters of Candidate B can now proclaim victory for their candidate.

▸ **Now try Exercises 37–41, part (c).**

BY THE WAY

A slight variation on a sequential runoff, called an *instant runoff,* allows voters to rank as many candidates as they wish on a multi-candidate ballot. The election is then decided as with a sequential runoff, by eliminating low-ranking candidates until one candidate is left with a majority. This system has been used in numerous countries around the world, including Australia, India, and Ireland.

Sequential Runoffs

Now the supporters of Candidate C chime in. They claim that a single runoff is unfair because it ignores rankings below the top two and instead suggest a sequential runoff. Because Candidate E had the fewest first-place votes (E's first-place votes are on the $4 + 2 = 6$ ballots represented by the last two columns), we eliminate E for the first runoff. To see what happens, we repeat the original preference schedule (Table 12.5), but with votes for E highlighted.

TABLE 12.5 (repeated) Preference Schedule with Votes for Candidate E Highlighted

First	A	B	C	D	E	E
Second	D	E	B	C	B	C
Third	E	D	E	E	D	D
Fourth	C	C	D	B	C	B
Fifth	B	A	A	A	A	A
Number of Voters	**18**	**12**	**10**	**9**	**4**	**2**

When we eliminate E, all the votes below E move up, giving us the new rankings shown in Table 12.6. Notice, for example, that the first column originally had five choices in the order A, D, E, C, B; eliminating E leaves four choices in the order A, D, C, B.

TABLE 12.6 Rankings After the First Runoff (Candidate E Eliminated)

First	A	B	C	D	B	C
Second	D	D	B	C	D	D
Third	C	C	D	B	C	B
Fourth	B	A	A	A	A	A
Number of Voters	**18**	**12**	**10**	**9**	**4**	**2**

Notice that D (highlighted) now has the fewest first-place votes (with 9) and is therefore eliminated for the second runoff.

With this new ranking, Candidate D has 9 first-place votes, which is fewer first-place votes than A (18), B ($12 + 4 = 16$), or C ($10 + 2 = 12$). We therefore eliminate D for the second runoff, which leaves only three choices (A, B, C) to be ranked. Removing the highlighted entries for D from Table 12.6, we end up with the new rankings shown in Table 12.7. Note that the 9 people who previously had D in first place now have C in first place.

TABLE 12.7 Rankings After the Second Runoff (Candidate D Eliminated)

First	A	B	C	C	B	C
Second	C	C	B	B	C	B
Third	B	A	A	A	A	A
Number of Voters	18	12	10	9	4	2

B (highlighted) now has the fewest first-place votes and is eliminated for the third runoff.

Candidate C is now the leader in first-place votes, with 10 + 9 + 2 = 21 (columns 3, 4, and 6), followed by Candidate A with 18 first-place votes (column 1). Candidate B has the fewest first-place votes (12 + 4 = 16) and is therefore eliminated for the final runoff. Table 12.8 shows the results after eliminating Candidate B.

TABLE 12.8 Rankings After the Third Runoff (Candidate B Eliminated)

First	A	C	C	C	C	C
Second	C	A	A	A	A	A
Number of Voters	18	12	10	9	4	2

There are now 18 first-place votes for A (column 1 only) and 12 + 10 + 9 + 4 + 2 = 37 first-place votes for C.

Notice that Candidate C now has a majority of first-place votes, with 37, and is therefore the winner by the sequential runoff method. ▸ **Now try Exercises 37–41, part (d).**

Point System

Next we turn to Candidate D's supporters, who suggest using a point system. Because there are five candidates, first-place votes are worth 5 points, second-place votes are worth 4 points, and so on, down to 1 point for fifth-place votes. We can calculate the total number of points won by each candidate by multiplying the number of votes in each column of Table 12.5 by the assigned point value and adding these products together:

A gets $(18 \times 5) + (12 \times 1) + (10 \times 1) + (9 \times 1) + (4 \times 1) + (2 \times 1) = 127$ points

B gets $(18 \times 1) + (12 \times 5) + (10 \times 4) + (9 \times 2) + (4 \times 4) + (2 \times 2) = 156$ points

C gets $(18 \times 2) + (12 \times 2) + (10 \times 5) + (9 \times 4) + (4 \times 2) + (2 \times 4) = 162$ points

D gets $(18 \times 4) + (12 \times 3) + (10 \times 2) + (9 \times 5) + (4 \times 3) + (2 \times 3) = 191$ points

E gets $(18 \times 3) + (12 \times 4) + (10 \times 3) + (9 \times 3) + (4 \times 5) + (2 \times 5) = 189$ points

Candidate D is the winner, by virtue of having the largest number of points.
▸ **Now try Exercises 37–41, part (e).**

Pairwise Comparisons

Now it is time for Candidate E's supporters, who point out an important fact about the election rankings. Suppose, they say, that the vote had been only between E and A, without the other candidates. Column 1 of Table 12.5 has A ranked higher than E, so these 18 voters would choose A over E. However, E is ranked higher than A in all the other columns, so E would get the remaining 37 votes (out of the 55 total). That is, E beats A by 37 to 18 in a one-on-one contest.

Now suppose that the vote had been only between E and B. Candidate E is ranked higher than B in columns 1, 4, 5, and 6 (of Table 12.5), so E gets 18 + 9 + 4 + 2 = 33 votes. The remaining 22 votes go to B. That is, E beats B by 33 to 22. A similar analysis of the race between only E and C shows that E is the winner by 36 to 19, and E comes out ahead in a one-on-one contest with D by 28 to 27.

Because Candidate E beats every other candidate in one-on-one contests, E's supporters declare victory by the method of pairwise comparisons.

▸ Now try Exercises 37–41, part (f).

Summary: Choosing a Winner Is Not So Easy

We have analyzed the preference schedule in Table 12.5 by five different methods, and we found a different winner with each method. Therefore, for this case, all five candidates could reasonably claim to be the winner. Other election results may not be so ambiguous, but the point should be clear: Different people can reasonably disagree about who should be declared the winner in an election with more than two candidates. In fact, as we will see in Unit 12B, a mathematical theorem *proves* that there is *no absolutely fair* way of deciding elections among more than two candidates.

▸ Now try Exercises 37–41, part (g).

Quick Quiz 12A

Choose the best answer to each of the following questions. Explain your reasoning with one or more complete sentences.

1. Is it possible to decide all elections by majority rule?

a. Yes.

b. No; a majority rule winner is guaranteed only in elections with only two candidates.

c. No; a majority rule winner is possible only when you use a runoff method.

2. According to Table 12.1, in the 2016 presidential election Hillary Clinton received

a. a majority of the popular vote.

b. a majority of the electoral vote.

c. a plurality of the popular vote.

3. In the U.S. Senate, a *filibuster* allows

a. a minority of the senators to pass legislation over the objection of the majority.

b. a minority of the senators to prevent the majority from passing a bill.

c. a majority of the senators to disregard the will of the minority.

4. Consider the results of the 2016 presidential election in Table 12.3. If the election had been decided by a runoff, who could *not* have won?

a. Trump b. Clinton c. Johnson

5. What is the basic purpose of a *preference schedule?*

a. to allow voters to indicate their first choice among many candidates

b. to allow voters to rank every candidate in an election in order from their favorite to their least favorite

c. to allow voters to rate each candidate in an election on a scale of 1 to 5

6. Suppose there are four candidates in an election. If you were voting with a preference schedule, you would need to indicate

a. your first choice only.

b. your first and second choices only.

c. your first, second, third, and last choices.

7. Study Table 12.5. How many *second*-place votes did Candidate A receive?

a. 0 b. 12 c. 18

8. Study Table 12.5. How many *third*-place votes did Candidate D receive?

a. 9 b. 12 c. 18

9. Study Table 12.5. Which candidate received the *fewest* first-place votes?

a. A b. E

c. A, B, C, and D all tied for the fewest first-place votes.

10. What is the primary lesson of the preference schedule in Table 12.5?

a. If you plan well, you can always come up with a clear way to decide an election.

b. The winner of an election can depend on the method you use for deciding the vote.

c. A point system is the fairest way of deciding an election.

Exercises 12A

REVIEW QUESTIONS

1. What is *majority rule*? When can it definitively decide an election?
2. Contrast the *popular vote* and the *electoral vote* in a U.S. presidential election.
3. What is a *filibuster*? What percentage of the vote is required to end one?
4. What is a *super majority*? Give several examples in which a super majority is required to decide a vote.
5. What is a *veto*? How does a veto affect the idea of majority rule?
6. Describe how a three-way election can be decided either by plurality or by runoff. Will both methods necessarily give the same results? Explain.
7. What is a *preference schedule*? Give an example of how to make one.
8. Using the preference schedule in Table 12.5, explain how the winner can be determined in five different ways.

DOES IT MAKE SENSE?

Decide whether each of the following statements makes sense (or is clearly true) or does not make sense (or is clearly false). Explain your reasoning.

9. In an election with only two candidates, both candidates received more than 50% of the votes.
10. Susan received only 43% of the vote, but she won a plurality.
11. Herman won a plurality of the vote, but Hanna won the election in a sequential runoff.
12. Fred beat Fran using the point system (Borda count), but Fran won by the method of pairwise comparisons (Condorcet method).
13. Candidate Reagan won the popular vote for the U.S. presidency and also won the electoral vote.
14. The defendant was found not guilty, even though four of the nine jurors voted against conviction.

BASIC SKILLS & CONCEPTS

15–24: Presidential Elections. The following tables give the popular and electoral votes for the two major candidates for various presidential elections. The total popular vote count, including votes that went to other candidates, is also given. All electoral votes are shown.

a. Compute each candidate's percentage of the total popular vote. Did either candidate receive a popular majority?

b. Compute each candidate's percentage of the electoral vote. Was the electoral winner also the winner of the popular vote?

15.

Year	Candidate	Electoral votes	Popular votes
1876	Rutherford B. Hayes	185	4,034,142
	Samuel J. Tilden	184	4,286,808
	Total popular votes		8,418,659

16.

Year	Candidate	Electoral votes	Popular votes
1880	James Garfield	214	4,453,337
	Winfield Hancock	155	4,444,267
	Total popular votes		9,217,410

17.

Year	Candidate	Electoral votes	Popular votes
1888	Benjamin Harrison	233	5,443,633
	Grover Cleveland	168	5,538,163
	Total popular votes		11,388,846

18.

Year	Candidate	Electoral votes	Popular votes
1916	Woodrow Wilson	277	9,126,868
	Charles Hughes	254	8,548,728
	Total popular votes		18,536,585

19.

Year	Candidate	Electoral votes	Popular votes
1992	Bill Clinton	370	44,909,806
	George H. W. Bush	168	39,104,550
	Total popular votes		104,423,923

20.

Year	Candidate	Electoral votes	Popular votes
1996	Bill Clinton	379	47,400,125
	Robert Dole	159	39,198,755
	Total popular votes		96,275,401

21.

Year	Candidate	Electoral votes	Popular votes
2000	George W. Bush	271	50,456,002
	Al Gore	266	50,999,897
	Total popular votes		105,405,100

22.

Year	Candidate	Electoral votes	Popular votes
2004	George W. Bush	286	62,040,610
	John Kerry	251	59,028,439
	Total popular votes		122,293,548

23.

Year	Candidate	Electoral votes	Popular votes
2008	Barack Obama	365	69,498,516
	John McCain	173	59,948,323
	Total popular votes		131,313,820

24.

Year	Candidate	Electoral votes	Popular votes
2012	Barack Obama	332	65,907,213
	Mitt Romney	206	60,931,767
	Total popular votes		129,064,662

25. **Super Majorities.**

a. Of the 100 senators in the U.S. Senate, 62 favor a new bill on health care reform. The opposing senators start a filibuster. Is the bill likely to pass?

b. A criminal conviction in a particular state requires that 2/3 of the jury members vote to convict. On an 11-member jury, 7 jurors vote to convict. Will the defendant be convicted?

c. A proposed amendment to the U.S. Constitution has passed both the House and the Senate with more than the required 2/3 super majority. Each state holds a vote on the amendment, and it receives a majority vote in all but 14 of the 50 states. Is the Constitution amended?

d. A tax increase bill has the support of 68 out of 100 senators and 270 out of 435 members of the House of Representatives. The President promises to veto the bill if it is passed. Is it likely to become law?

26. **Super Majorities.**

a. According to the by-laws of a corporation, a 2/3 vote of the shareholders is needed to approve a merger. Of the 10,100 shareholders voting on a certain merger, 6650 approve of the merger. Will the merger happen?

b. A criminal conviction in a particular state requires that 3/4 of the jury members vote to convict. On a 12-member jury, 8 jurors vote to convict. Will the defendant be convicted?

c. A proposed amendment to the U.S. Constitution has passed both the House and the Senate with more than the required 2/3 super majority. Each state holds a vote on the amendment, and it receives a majority vote in 35 of the 50 states. Is the Constitution amended?

d. A tax increase bill has the support of 68 out of 100 senators and 292 out of 435 members of the House of Representatives. The President promises to veto the bill if it is passed. Is it likely to become law?

27. **1992 Presidential Election.** The 1992 U.S. presidential election featured three major candidates, with the vote split as follows. For this exercise, assume that all votes were cast for one of these three candidates.

Candidate	Popular votes	Electoral votes
Bill Clinton	44,909,889	370
George H. W. Bush	39,104,545	168
Ross Perot	19,742,267	0

a. Calculate each candidate's percentage of the popular vote. Who won a plurality? Did any candidate win a majority?

b. Calculate each candidate's percentage of the electoral vote. Who won a plurality? Did any candidate win a majority?

c. Suppose that Perot had dropped out of the election. Is it possible that Bush would have won the popular vote? Could Bush have become President in that case? Explain.

d. Suppose that Bush had dropped out of the election. Is it possible that Perot would have won the popular vote? Could Perot have become President in that case? Explain.

28. **Florida 2000.** The following table gives the results of the 2000 presidential election in Florida, which ultimately determined the overall winner.

Candidate	Votes
George W. Bush	2,912,790
Al Gore	2,912,253
Ralph Nader	97,421
Pat Buchanan	17,484
Other	23,102

a. Calculate the percentages of the total vote that went to Bush and Gore. What was Bush's percentage margin of victory?

b. Polls suggested that most Nader voters preferred Gore to Bush, while most Buchanan voters preferred Bush to Gore. Suppose that Nader and Buchanan had both dropped out of the election, and Nader voters split their votes 60% to Gore and 40% to Bush while Buchanan voters split 60% to Bush and 40% to Gore. What would the vote totals and outcome have been?

c. Repeat part (b), but with the Nader voters splitting 51% to 49% for Gore and the Buchanan voters splitting 51% to 49% for Bush.

d. The confusing butterfly ballot (see In Your World, p. 663) was used in Palm Beach County, where Buchanan received 0.8% of the overall vote, or 3407 votes total. How did Buchanan's vote percentage in Palm Beach County compare to his statewide vote percentage?

e. The butterfly ballot design made it easy for voters to accidentally vote for Buchanan when they intended to vote for Gore. What fraction of Buchanan's vote in Palm Beach County would Gore have needed to get to win the presidency? Buchanan's own party (the Reform Party) estimated his support in Palm Beach County at no more than about 0.3%. If the rest of Buchanan's votes in Palm Beach County had been intended for Gore, would Gore have won with a better-designed ballot?

29. **Three-Way Race Preference Schedule.** Consider a three-way race between candidates called A, B, and C. Make a preference schedule for the following results:

- 22 voters rank the candidates in the order (first choice to last choice) A, B, C.
- 20 voters rank the candidates in the order C, B, A.

- 16 voters rank the candidates in the order B, C, A.
- 8 voters rank the candidates in the order C, A, B.

30. **Four-Way Race Preference Schedule.** Consider a four-way race between candidates called A, B, C, and D. Make a preference schedule for the following results:
    - 39 voters rank the candidates in the order (first choice to last choice) D, C, A, B.
    - 32 voters rank the candidates in the order B, C, D, A.
    - 27 voters rank the candidates in the order C, D, A, B.
    - 21 voters rank the candidates in the order A, C, B, D.
    - 12 voters rank the candidates in the order D, A, B, C.

31–32. Finding Winners. In Exercises 31–32, find the winners by each of the five methods:

a. plurality

b. single runoff

c. sequential runoff

d. point system (Borda count)

e. pairwise comparisons (Condorcet method)

31. Find the winners using the preference schedule from Exercise 29.
32. Find the winners using the preference schedule from Exercise 30.

33–36. Interpreting Preference Schedules. Answer the following questions about the preference schedule in Table 12.5.

33. How many voters preferred Candidate B to Candidate E?
34. How many voters preferred Candidate D to Candidate C?
35. If Candidate E withdrew from the election (and votes for the other candidates were moved up in the table), how many votes would the other four candidates receive?
36. If Candidate C withdrew from the election (and votes for the other candidates were moved up in the table), how many votes would the other four candidates receive?

37–41: Preference Schedules. Consider the following preference schedules.

a. How many votes were cast?

b. Find the plurality winner. Did the plurality winner also receive a majority? Explain.

c. Find the winner of a runoff between the top two candidates.

d. Find the winner of a sequential runoff.

e. Find the winner by a Borda count.

f. Find the winner, if any, by the method of pairwise comparisons.

g. Summarize the results of the various methods of determining a winner. Based on these results, is there a clear winner? If so, why? If not, which candidate should be selected as the winner, and why?

37.

First	B	D	C	A	D	C
Second	D	A	D	D	A	A
Third	C	C	A	C	B	B
Fourth	A	B	B	B	C	D
Number of voters	20	15	10	8	7	6

38.

First	B	D	D	C	E
Second	A	B	B	A	A
Third	C	A	E	B	D
Fourth	D	C	C	D	B
Fifth	E	E	A	E	C
Number of voters	9	7	6	4	3

39.

First	A	A	B	B	C	C
Second	B	C	A	C	A	B
Third	C	B	C	A	B	A
Number of voters	30	5	20	5	10	30

40.

First	A	B	D
Second	B	A	C
Third	C	D	B
Fourth	D	C	A
Number of voters	10	10	10

41.

First	E	B	D
Second	D	C	A
Third	A	E	B
Fourth	B	A	C
Fifth	C	D	E
Number of voters	40	30	20

FURTHER APPLICATIONS

42. **How Close Was 2016?** Consider Table 12.3, which shows the results of the 2016 presidential election in Pennsylvania, Michigan, and Wisconsin.

    a. If Clinton had won Pennsylvania, Michigan, and Wisconsin, what would the final electoral vote result have been?

    b. What was the total popular vote difference in these three states for the top two candidates?

    c. Suppose there had been runoff elections between Trump and Clinton in those three states, and that all of the voters who originally selected other candidates then voted for either Trump or Clinton. For each of the three states, calculate the percentage of the votes for other candidates that Clinton would have needed to have won in order to have won the statewide vote.

43. **Three-Candidate Elections.** Consider an election in which the votes were cast as follows.

Candidate	Percentage of votes
Able	35%
Best	42%
Crown	23%

a. Who won a plurality? Does any candidate have a majority? Explain.

b. What percentage of Crown's votes would Able need to win a runoff election?

44. **Three-Candidate Elections.** Consider an election in which the votes were cast as follows.

Candidate	Percentage of votes
Davis	26%
Earnest	27%
Fillipo	47%

a. Who won a plurality? Does any candidate have a majority? Explain.

b. What percentage of Davis's votes would Earnest need to win a runoff election?

45. **Three-Candidate Elections.** Consider an election in which the votes were cast as follows.

Candidate	Number of votes
Giordano	120
Heyduke	160
Irving	205

a. Who won a plurality? Does any candidate have a majority? Explain.

b. How many of Giordano's votes would Heyduke need to win a runoff election?

46. **Three-Candidate Elections.** Consider an election in which the votes were cast as follows.

Candidate	Number of votes
Joker	255
King	382
Lord	306

a. Who won a plurality? Does any candidate have a majority? Explain.

b. How many of Joker's votes would Lord need to win a runoff election?

47. **Condorcet Winner.** If a candidate wins *all* head-to-head (two-way) races with other candidates, that candidate is called the *Condorcet winner.* A Condorcet winner automatically wins by the method of pairwise comparisons (Condorcet method). Consider the following preference schedule for a four-candidate election. Is there a Condorcet winner? Explain.

First	B	B	A	A
Second	A	A	C	D
Third	C	D	D	C
Fourth	D	C	B	B
Number of voters	30	30	30	20

48. **Condorcet Paradox.** Consider the following preference schedule. Can you find a winner through pairwise comparisons? Explain.

First	C	B	A
Second	A	C	B
Third	B	A	C
Number of voters	8	9	10

49. **Pairwise Comparisons.**

a. How many pairs of candidates must be examined to carry out the method of pairwise comparisons with four candidates?

b. How many pairs of candidates must be examined to carry out the method of pairwise comparisons with five candidates?

c. How many pairs of candidates must be examined to carry out the method of pairwise comparisons with six candidates?

50. **Borda Question.** In a preference schedule with six candidates and 30 voters, what is the total number of points awarded to the candidates using the usual Borda count weights?

51. **Borda Question.** Suppose 30 voters rank four candidates: A, B, C, and D. With the usual Borda count weights, A receives 100 points, B receives 80 points, and C receives 75 points. How many points does D have? Who wins the election? Explain.

IN YOUR WORLD

52. **U.S. Voting Equipment.** Investigate at least four types of voting equipment used in different places throughout the United States (a good starting source is VerifiedVoting.org). What are the pros and cons of each? Form an opinion as to what type of equipment you would recommend if the United States were to institute uniform voting procedures nationwide. Defend your choice.

53. **Election Controversy.** Pick one election from recent history that had a controversial outcome, and examine the evidence on both sides of the debate. Write a short report discussing the controversy and giving your opinion as to whether or not the outcome reflected the will of the voters.

54. **Voting Around the World.** Choose a country that is considered to have free and fair elections, and find out how it conducts its elections. In what ways are the voting procedures similar to those in the United States? In what ways are they different? Write a one-page summary of lessons that you believe the United States could learn from the country you have selected.

55. **Academy Awards.** The election process for the Academy Awards (for films) involves several stages and several different voting methods. Use the website for the Academy Awards to investigate the full election procedure for the Academy Awards. Describe the procedure and comment on its fairness.

56. **Sports Polls.** Most men's and women's major college sports have regular polls while the sport is in season. Choose a particular sport and investigate how teams are ranked. Describe the methods used to rank teams, give some typical results, and discuss the fairness of the method.

UNIT 12B Theory of Voting

In Unit 12A, we saw that different methods of counting votes can lead to different results. Mathematicians, economists, and political scientists have discovered many other surprises about voting. In this unit, we investigate just a few of these surprises.

Which Method Is Fairest?

When there are only two candidates in an election, the clear winner is the candidate who gets a majority of the votes. As we saw in Unit 12A, where we studied five different methods for deciding an election (see the box on p. 663), choosing a winner can be much more difficult in elections with three or more candidates.

Sometimes, all five methods give the same winner. Other times, different methods give different winners. In extreme cases, such as the election results in Table 12.5, the five methods can produce five different winners. The key question becomes: Which method is the fairest?

Criteria of Fairness

Judgments of fairness are necessarily subjective. Nevertheless, mathematicians and political scientists have come up with four basic **fairness criteria** that should be met by a fair voting method, all listed in the following box.

Technical Note
Voting theorists refer to these four criteria as, respectively, the *majority criterion*, the *Condorcet criterion*, the *monotonicity criterion*, and the *independence of irrelevant alternatives criterion*.

Fairness Criteria

Criterion 1: If a candidate receives a majority of the first-place votes, that candidate should be the winner.

Criterion 2: If a candidate is favored over every other candidate in pairwise races, that candidate should be declared the winner.

Criterion 3: Suppose that Candidate X is declared the winner of an election, and then a second election is held. If some voters rank X *higher* in the second election than in the first election (without changing the order of other candidates), then X should also win the second election.

Criterion 4: Suppose that Candidate X is declared the winner of an election, and then a second election is held. If voters do not change their preferences but one (or more) of the losing candidates drops out, then X should also win the second election.

Because the election results in Table 12.5 (Unit 12A) gave five different winners with the five methods, you may already have guessed that none of these methods satisfies all four fairness criteria in all elections. However, to understand the these criteria better, let's examine a few more examples in which we test each criterion.

EXAMPLE 1 An Unfair Plurality

Consider the preference schedule below. Suppose the winner is chosen by the plurality method. Does this method satisfy the four fairness criteria?

First	A	B	C
Second	B	C	B
Third	C	A	A
Number of voters	5	4	2

Solution Candidate A has the most first-place votes and is therefore the plurality winner. We now apply the four fairness criteria to this result.

Criterion 1: No candidate received a majority of first-place votes in this election, so Criterion 1 does not apply.

Criterion 2: To check whether this criterion is satisfied, we need to examine pairwise races between the candidates. There are three pairs to consider: A versus B, A versus C, and B versus C. Here are the results:

- A is ranked ahead of B in column 1 (5 votes), but behind B in columns 2 and 3 ($4 + 2 = 6$ votes). Therefore, B beats A by 6 to 5.
- A is ranked ahead of C in column 1 (5 votes), but behind C in columns 2 and 3 ($4 + 2 = 6$ votes). Therefore, C beats A by 6 to 5.
- B is ranked ahead of C in columns 1 and 2 ($5 + 4 = 9$ votes), but behind C in column 3 (2 votes). Therefore, B beats C by 9 to 2.

By Criterion 2, B should be the winner because B beat both A and C in pairwise contests. But B is *not* the plurality winner, so the plurality method is unfair in this case.

Criterion 3: We evaluate the plurality method on this criterion by imagining a second election in which some voters move A higher in their rankings, but do not change the order in which they rank B and C. Column 1 cannot change because A already is ranked first. Moving A higher in column 2 or 3 may give A even more first-place votes, but cannot reduce A's number of first-place votes. Therefore, A would still win a plurality of the votes, and Criterion 3 is satisfied.

Criterion 4: This time we imagine a second election in which one of the losers drops out. Suppose that C were to drop out. Then B would move up to first place in column 3, picking up the 2 first-place votes in this column. Because B already has 4 first-place votes in column 2, B would then have 6 first-place votes and would win the election by 6 to 5 over A. That is, the plurality method is unfair according to Criterion 4 because A would lose to B if C were to drop out.

In summary, the plurality method violates two of the four fairness criteria in this election. The basic problem is clear if we look back at the criteria that are violated: Most voters prefer Candidate B over Candidate A, but A is the winner by plurality.

▶ Now try Exercises 9–14.

EXAMPLE 2 An Unfair Runoff

Consider the preference schedule below. Who wins by a single runoff? Does the result violate any of the four fairness criteria?

First	C	A	C	B
Second	B	C	A	A
Third	A	B	B	C
Number of voters	9	13	5	11

Solution The first-place totals are 13 for A (column 2), 11 for B (column 4), and $9 + 5 = 14$ for C (columns 1 and 3). Therefore, the runoff is between A and C. The 11 voters who chose B (column 4) ranked A second, so A picks up these first-place votes and wins the runoff over C by 24 to 14. Now let's test this result with the fairness criteria.

Criterion 1 does not apply because no candidate received a majority in the original election. Criterion 2 does not apply, either, because no candidate wins all the pairwise comparisons. (You should confirm that, in pairwise comparisons, B beats A, A beats C, and C beats B.)

Criterion 3 says that A should still be the winner if A picks up additional first-place votes. But suppose that the five voters in column 3 move A up to first place. Then A is in first place in columns 2 and 3 and has a total of 13 + 5 = 18 votes, while C is in first place only in column 1, which decreases C's first-place votes to 9. B's first-place total remains at 11. Because C now has the fewest first-place votes, the runoff is between A and B. Notice that B ranks ahead of A in columns 1 and 4, so B gets the 9 + 11 = 20 votes from these columns in the runoff; this is a majority of the 38 total votes, making B the winner. We see that Criterion 3 is violated because a gain in first-place votes ends up costing A the election.

Criterion 4 says that A should still be the winner if one (or more) of the losing candidates drops out. Candidate B was eliminated in the original runoff election, so the results are not affected if B drops out. However, if C drops out, B picks up the 9 first-place votes from column 1 and A picks up the 5 first-place votes from column 3. This change increases A's first-place total to 13 + 5 = 18 and B's first-place total to 9 + 11 = 20, making B the winner instead of A. Therefore, Criterion 4 is also violated.

▶ Now try Exercises 15–21.

EXAMPLE 3 A Fair Election

Consider the preference schedule below. Does the plurality method satisfy the four fairness criteria?

First	A	B	C
Second	B	C	B
Third	C	A	A
Number of voters	10	4	2

Solution The winner by the plurality method is A, with 10 votes. Because this plurality is also a majority of the 16 votes cast, Criterion 1 is satisfied. Examining pairwise matchups shows that A beats B by 10 to 6 and A beats C by 10 to 6. Therefore, A is the pairwise winner and Criterion 2 is satisfied. Criterion 3 is also satisfied because, if we imagine a second election in which A picks up additional votes, that will only increase A's majority. Finally, Criterion 4 asks what happens in a second election in which one (or more) of the losers drops out. Eliminating either B or C cannot reduce A's majority, so A still wins. In summary, the plurality method satisfies all four fairness criteria in this particular election.

▶ Now try Exercises 22–33.

Think About It Example 1 presents an election decided by the plurality method that *does not* meet all four fairness criteria. Example 3 presents an election decided by the plurality method that *does* meet the criteria. Note that the plurality is also a *majority* in Example 3. Is it generally true that an election decided by a majority will be fair? Explain.

Arrow's Impossibility Theorem

Suppose we continue to test the fairness criteria on many different elections decided by any of the five methods we have discussed. Sometimes, as in Example 3, we will find that all four criteria are satisfied and we can declare the election to be fair. Other times, as in Examples 1 and 2, the results will violate one or more of the fairness criteria.

Despite the fact that some elections are fair and others are not, we can identify some general rules. For example, elections decided by plurality always satisfy Criteria 1 and 3, but sometimes violate Criteria 2 and 4. A similar analysis of the other voting methods is presented in Table 12.9.

BY THE WAY
Kenneth Arrow received the 1972 Nobel Prize in Economics for his mathematical analysis of voting systems that led him to discover the impossibility theorem.

TABLE 12.9 Fairness Criteria and Voting Methods

	Plurality	Top-Two Runoff	Sequential Runoff	Borda Count	Pairwise Comparisons
Criterion 1	Y	Y	Y	N	Y
Criterion 2	N	N	N	N	Y
Criterion 3	Y	N	N	Y	Y
Criterion 4	N	N	N	N	N

Note: Y = criterion always satisfied; N = criterion may be violated.

No one pretends that democracy is perfect or all-wise. Indeed it has been said that democracy is the worst form of government except for all those that have been tried from time to time.

—Winston Churchill

Table 12.9 conveys the disconcerting message that *none* of the five voting methods always gives a fair outcome. During the two centuries following the American and French revolutions, political theorists attempted to devise a better voting system—one that would always give fair results. Unfortunately, their quest was futile, because a perfect voting system will never be found. In 1952, economist Kenneth Arrow proved mathematically that it is *impossible* to find a voting system that always satisfies all four fairness criteria. This result is one of the landmark applications of mathematics to social theory and is now known as **Arrow's impossibility theorem**.

Arrow's Impossibility Theorem

No voting system can satisfy all four fairness criteria in all cases.

Think About It Arrow's impossibility theorem tells us that no voting system is perfect. However, some systems may still be superior to others. Among the voting methods we have discussed, which would *you* choose for electing a U.S. President? Why?

Approval Voting

Democratic voting systems have traditionally been based on the principle of *one person, one vote*. However, in light of the fact that no voting system can be perfect, some political theorists have proposed alternative methods of voting. Arrow's impossibility theorem assures us that none of these methods can be perfect, but a new method *might* give fair results more often than traditional methods.

One such alternative voting system, called **approval voting**, asks voters to specify whether they approve or disapprove of each candidate. Voters may approve as many candidates as they like, and the candidate with the most approval votes wins.

As an example, consider a race for governor among Candidates A, B, and C. With approval voting, the ballot would look something like Figure 12.3.

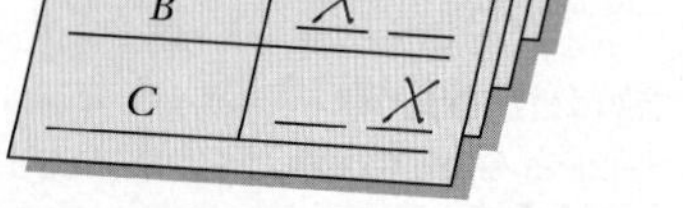

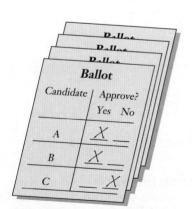

FIGURE 12.3 A sample ballot for approval voting.

Suppose that voters' opinions about the candidates are as follows (perhaps because Candidates A and B are closely aligned politically, while Candidate C is on the opposite end of the political spectrum):

- 32% want A as their first choice, but would also approve of B.
- 32% want B as their first choice, but would also approve of A.
- 1% want A as their first choice and approve of neither B nor C.
- 35% want C as their first choice and approve of neither A nor B.

Note that the total is 100%, as it must be. In terms of approval votes, we find that

A is approved by 32% + 32% + 1% = 65%.
B is approved by 32% + 32% = 64%.
C is approved by 35%.

By approval voting, A becomes the new governor and C is clearly in last place. However, if this election had been decided by plurality, C would have been elected by virtue of the most first-place votes, with 35%.

Think About It Do you think that approval voting is a better way to decide this particular election than plurality? Do you think it is a better method in general? Defend your opinion.

EXAMPLE 4 A Drawback to Approval Voting

The opinions of voters in a particular three-way race for governor are as follows:

- 26% want A as their first choice, but would also approve of B.
- 25% want A as their first choice and approve of neither B nor C.
- 15% want B as their first choice and approve of neither A nor C.
- 18% want C as their first choice, but would also approve of B.
- 16% want C as their first choice and approve of neither A nor B.

Note that the total is 100%. Contrast the results if the election is decided by approval voting or by the plurality method.

Solution By approval voting, we find the following results:

A is approved by 26% + 25% = 51%.
B is approved by 26% + 15% + 18% = 59%.
C is approved by 18% + 16% = 34%.

Therefore, B is the winner by approval voting. However, if we count only first choices, we find the following outcome by the plurality method:

A is the first choice for 26% + 25% = 51%.
B is the first choice for 15%.
C is the first choice for 18% + 16% = 34%.

A *majority* of voters want A as their first choice for governor. However, because a larger majority finds B "acceptable," B wins in the approval voting. In essence, in this election approval voting violates Criterion 1, because A would have received a majority of first-place votes but did not win the election. ▸ Now try Exercises 34–35.

Think About It Example 4 shows that while approval voting ensures that the winning candidate is acceptable to the largest number of people, another candidate might be the first choice of the majority. Proponents of approval voting consider this drawback less serious than the drawbacks of other voting systems. What's your opinion?

Voting Power

So far, we have taken for granted that every voter has the same amount of power to influence an election as every other voter. However, this is not necessarily the case. For example, shareholders in a corporation generally are given votes in proportion to the number of shares they own. A shareholder with 10 shares gets 10 votes, and a shareholder with 1000 shares gets 1000 votes. Not every voter at a shareholder meeting is equally empowered to influence the outcome of an election.

A similar situation arises in politics when voters form groups, or **coalitions**, that agree to vote the same way on a particular issue. Coalition building can drastically affect the power of individual voters to influence the outcome of a vote.

Several techniques have been devised for measuring the effective power wielded by different voters when not all voters have the same power. Although we will not discuss the details of these techniques, we can take a brief look at the type of situation that can arise.

IN YOUR WORLD

The Electoral College and the Presidency

The Electoral College system, in which the U.S. President is chosen by the electoral vote rather than by the popular vote, has long been controversial. The discussion remained largely academic during the 20th century, because every presidential election prior to 2000 had the same winner by both the popular and electoral votes. However, the issue has taken on new significance since the popular vote winner lost the electoral vote in two of the past five presidential elections (2000 and 2016) and, as discussed in Unit 12A (on p. 663), that *almost* occurred a third time (2004). Note that the issue affects both major political parties: Although the 2000 and 2016 cases both ended up benefiting a Republican candidate, the "almost" case in 2004 would have given the presidency to a Democrat (John Kerry) who lost the popular vote by a wide margin (more than 3 million votes).

Those who favor eliminating the Electoral College generally cite three major reasons:

1. As discussed in Example 6, the Electoral College gives different amounts of voting power to citizens in different states, which opponents of this system argue to be unfair.
2. The Electoral College system leads candidates to focus their campaign messages on a small number of *swing states*—states in which the vote is expected to be close—at the expense of the larger number of states that tend to be either solidly Republican or solidly Democratic. Opponents of the Electoral College claim that this causes candidates to focus on narrow, locally important issues while ignoring issues that may be more important to the majority of Americans.
3. Statistical data suggest that the Electoral College system reduces voter turnout in non-swing states, because voters in those states know that their presidential vote is unlikely to change the outcome. Opponents of the Electoral College argue that this effectively disenfranchises voters and may have negative consequences for non-presidential races that appear on the same ballot.

The counterarguments from those who support keeping the Electoral College generally follow similar lines but with different interpretations of the consequences. For example, supporters argue that the differences in voting power are beneficial to smaller states and hence are in line with the founding principles of the United States, and that the focus on swing states ensures that candidates pay attention to concerns of rural voters and minorities, who would be ignored in favor of major population centers if the Electoral College were eliminated.

Overall, polls show large majorities of people in both parties favoring elimination of the Electoral College, a change that has generally been assumed to require a constitutional amendment (because the Electoral College was established in Article II of the Constitution).

Suppose, for example, that the 100 members of the U.S. Senate happened to be divided as follows: 49 Democrats, 49 Republicans, and 2 Independents. Moreover, suppose that all 49 Democrats favor a particular bill, while all 49 Republicans are opposed. The 2 Independents may not care which way the vote goes, but if they vote together their two votes will decide the outcome of the election. Both the Democrats and the Republicans are likely to work hard to woo the votes of the Independents, perhaps by agreeing to support other bills that the Independents favor. The power of the 2 Independents to influence decision making is therefore much greater than their 2 votes out of 100 would seem to imply.

EXAMPLE 5 Missing the Big Vote

A small corporation has four shareholders. The 10,000 shares in this corporation are divided among the shareholders as follows:

Shareholder A owns 2650 shares (26.5% of the company).
Shareholder B owns 2550 shares (25.5% of the company).
Shareholder C owns 2500 shares (25% of the company).
Shareholder D owns 2300 shares (23% of the company).

The corporation's board of directors has scheduled a key vote about buying out another company. Each shareholder's vote is counted in proportion to the number of shares the person owns. Suppose that Shareholder D misses the vote. Does it matter?

Electoral Votes by State (2012–2020)

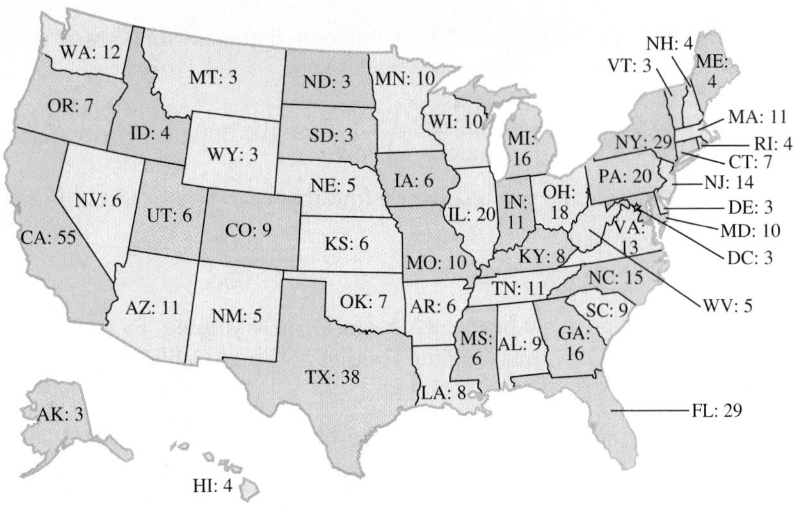

However, a proposal called the National Popular Vote Interstate Compact suggests an alternative way of ensuring that the winner of a presidential election is the winner of the popular vote without amending the Constitution. This proposal asks individual states to pass legislation that would do the following: Once enough states have signed on to represent the 270 electoral votes required to win the electoral vote count, the states that have agreed to the Compact would assign *all* of their electors to the winner of the national popular vote. This would ensure that the winner of the national popular vote would automatically also win the electoral vote.

As of 2017, the National Popular Vote Interstate Compact had been adopted by 10 states and the District of Columbia, representing a total of 165 electoral votes. If enough additional states sign on to reach the 270 electoral vote threshold, its constitutionality is likely to be challenged, so the final decision might well rest with the Supreme Court. Of course, even if it is invalidated, the option of a constitutional amendment remains. One way or another, in the coming years, you are likely to be hearing a lot about whether the Electoral College should be kept or abolished.

Solution At first, it would seem disastrous for a major shareholder to miss the chance to influence the future of the company. However, note that any two of the three largest shareholders can form a majority by voting together:

A and B: 26.5% + 25.5% = 52%
A and C: 26.5% + 25% = 51.5%
B and C: 25.5% + 25% = 50.5%

In contrast, Shareholder D cannot be part of a majority unless at least two other shareholders vote the same way. But in that case, the other two shareholders already determine the outcome by themselves. In this case, Shareholder D has *no power* to affect the election outcome.

▶ Now try Exercise 36.

EXAMPLE 6 Electoral Power

In the 2016 presidential election, California had 55 electoral votes and Wyoming had 3 electoral votes. The populations of the two states were approximately 39,800,000 and 590,000, respectively. In terms of electoral votes per resident, contrast the voting power of California and Wyoming.

Solution In California, where 39,800,000 people had 55 electoral votes, the number of people represented by each electoral vote was

$$\frac{39{,}800{,}000 \text{ people}}{55 \text{ electoral votes}} \approx 720{,}000 \text{ people per electoral vote}$$

BY THE WAY

Another measure of voting power was invented in the 1950s by Georgetown University law professor John Banzhaf. Called the *Banzhaf power index*, it analyzes voting power in terms of the number of ways an individual or group can cast a critical vote that changes an election outcome. By this measure, California's voting power is quite high despite its large number of voters per elector. The reason is the winner-take-all system, in which California's 55 electoral votes have a much better chance of swaying the election than the fewer electoral votes of small states.

In Wyoming, where 590,000 people had 3 electoral votes, the number of people represented by each electoral vote was

$$\frac{590{,}000 \text{ people}}{3 \text{ electoral votes}} \approx 197{,}000 \text{ people per electoral vote}$$

Dividing the two results, we find that each elector in California represented

$$\frac{720{,}000}{197{,}000} \approx 3.7$$

times as many people as an elector in Wyoming. By this analysis, Wyoming voters had almost four times as much voting power as California voters.

▶ Now try Exercises 37–41.

Quick Quiz 12B

Choose the best answer to each of the following questions. Explain your reasoning with one or more complete sentences.

1. How many of the four fairness criteria (see p. 675) must be satisfied for an election to be considered fair?

   a. at least one b. at least two c. all of them

Questions 2–7 refer to the following preference schedule for a three-candidate election in which 100 people cast ballots:

First	Berman	Freedman	Goldsmith
Second	Freedman	Goldsmith	Freedman
Third	Goldsmith	Berman	Berman
Number of voters	43	39	18

2. If this election is decided by the plurality method, who wins?

   a. Berman b. Freedman c. Goldsmith

3. If this election is decided by the single runoff method, who wins?

   a. Berman b. Freedman c. Goldsmith

4. For this election, Criterion 1 is

   a. satisfied. b. violated. c. not applicable.

5. Suppose that Berman is declared the winner of this election. Is Criterion 2 satisfied?

   a. yes

   b. no, because a majority of the voters ranked Freedman higher than Berman and Goldsmith

   c. no, because a majority of the voters did not give Berman first-place votes

6. Suppose that Freedman is declared the winner of this election. Is Criterion 2 satisfied?

   a. yes

   b. no, because Berman received more first-place votes than Freedman

   c. no, because a majority of the voters did not give Freedman first-place votes

7. Notice that if Goldsmith dropped out, Freedman would then have a majority of the first-place votes. This means that Criterion 4 would be

   a. satisfied only if Freedman had already been declared the winner.

   b. satisfied only if Berman had already been declared the winner.

   c. satisfied in all circumstances.

8. Which of the following does *not* follow from Arrow's impossibility theorem?

   a. No election system can be fair under all circumstances.

   b. It is impossible to have a fair election.

   c. Government by democracy is mathematically imperfect.

9. Which of the following is *not* an advantage of approval voting?

   a. It ensures that the winner is the candidate acceptable to the most voters.

   b. It satisfies all the fairness criteria.

   c. It tends to prevent a candidate who is opposed by a majority from ending up as the winner.

10. All 50 states of the United States have two senators. In general, this means that the voters who have the most voting power in Senate elections are

    a. those who vote for winning candidates.

    b. those who live in the most populous states.

    c. those who live in the least populous states.

Exercises 12B

REVIEW QUESTIONS

1. Briefly summarize each of the four fairness criteria. For each one, give an example that describes why an election would be unfair if the criterion were violated.
2. What is *Arrow's impossibility theorem*? Summarize its meaning and its importance.
3. What is *approval voting*? How is it different from the traditional idea of one person, one vote?
4. Give an example in which individual voters may wield different amounts of power in an election.

DOES IT MAKE SENSE?

Decide whether each of the following statements makes sense (or is clearly true) or does not make sense (or is clearly false). Explain your reasoning.

5. Karen won a one-on-one election against each of the other candidates. She believes that she should win even though she lost in a runoff election.
6. Kai won a majority of the votes and yet lost the election by the point system (Borda count). He believes that he should be declared the winner of the election.
7. Wendy demanded a second election because she lost to Walt by the plurality method in a close vote. In the second election, some of Wendy's supporters voted for Walt, but Wendy managed to win. (Assume that other votes stayed the same.)
8. Approval voting is the best method for conducting elections because it satisfies all four fairness criteria.

BASIC SKILLS & CONCEPTS

9. **Plurality and Criterion 1.** Explain in words why the plurality method always satisfies Fairness Criterion 1.
10. **Plurality and Criterion 2.** Consider the preference schedule shown below. Which candidate is the winner by the plurality method? Does this method satisfy Fairness Criterion 2? Explain.

First	A	B	C
Second	B	C	B
Third	C	A	A
Number of voters	3	2	2

11. **Plurality and Criterion 2.** Devise a preference schedule with three candidates (A, B, and C) and 11 voters in which C is the plurality winner and yet A beats C and A beats B in one-on-one races. Explain your work.
12. **Plurality and Criterion 3.** Explain in words why the plurality method always satisfies Fairness Criterion 3.
13. **Plurality and Criterion 4.** Suppose the plurality method is used with the following preference schedule. Is Fairness Criterion 4 satisfied? Explain.

First	A	B	C
Second	B	C	B
Third	C	A	A
Number of voters	6	2	5

14. **Plurality and Criterion 4.** Devise a preference schedule with three candidates and nine votes for which the plurality method violates Fairness Criterion 4. Explain your work.
15. **Runoff Methods and Criterion 1.** Explain in words why both the single runoff and the sequential runoff methods always satisfy Fairness Criterion 1.
16. **Sequential Runoff and Criterion 2.** Suppose the sequential runoff method is used with the following preference schedule. Is Fairness Criterion 2 satisfied? Explain.

First	A	B	C
Second	B	C	B
Third	C	A	A
Number of voters	10	7	2

17. **Sequential Runoff and Criterion 2.** Suppose the sequential runoff method is used with the following preference schedule. Is Fairness Criterion 2 satisfied? Explain.

First	A	B	C
Second	B	C	B
Third	C	A	A
Number of voters	10	7	8

18. **Sequential Runoff and Criterion 2.** Devise a preference schedule with three candidates and three rankings for which the sequential runoff method violates Fairness Criterion 2. Explain your work.
19. **Sequential Runoff and Criterion 3.** Suppose the sequential runoff method is used with the following preference schedule. Is Fairness Criterion 3 satisfied? Explain.

First	A	B	A	C
Second	B	C	C	A
Third	C	A	B	B
Number of voters	7	8	4	10

20. **Sequential Runoff and Criterion 4.** Suppose the sequential runoff method is used with the following preference schedule. Is Fairness Criterion 4 satisfied? Explain.

First	A	B	C
Second	C	C	B
Third	B	A	A
Number of voters	8	6	3

21. **Sequential Runoff and Criterion 4.** Devise a preference schedule with three candidates for which the sequential runoff method violates Fairness Criterion 4. Explain your work.

22. **Point System and Criterion 1.** Consider the following preference schedule for three candidates. Which candidate wins by the point system (Borda count)? Is Fairness Criterion 1 satisfied? Explain.

First	A	B
Second	B	C
Third	C	A
Number of voters	3	2

23. **Point System and Criterion 1.** Devise a preference schedule with three candidates, three rankings, and seven voters for which the point system (Borda count) violates Fairness Criterion 1. Explain your work.

24. **Point System and Criterion 2.** Suppose the point system (Borda count) is used on the following preference schedule. Is Fairness Criterion 2 violated? Explain.

First	A	B	C
Second	B	C	B
Third	C	A	A
Number of voters	5	2	2

25. **Point System and Criterion 2.** Devise a preference schedule with four candidates for which the point system (Borda count) violates Fairness Criterion 2.

26. **Point System and Criterion 3.** Explain in words why the point system (Borda count) always satisfies Fairness Criterion 3.

27. **Point System and Criterion 4.** Suppose the point system (Borda count) method is used with the following preference schedule. Is Fairness Criterion 4 satisfied? Explain.

First	C	B	A
Second	A	C	B
Third	B	A	C
Number of voters	5	4	3

28. **Point System and Criterion 4.** Devise a preference schedule with three candidates for which the point system (Borda count) violates Fairness Criterion 4. Explain your work.

29. **Pairwise Comparisons and Criterion 1.** Explain in words why the method of pairwise comparisons always satisfies Fairness Criterion 1.

30. **Pairwise Comparisons and Criterion 2.** Explain in words why the method of pairwise comparisons always satisfies Fairness Criterion 2.

31. **Pairwise Comparisons and Criterion 3.** Explain in words why the method of pairwise comparisons always satisfies Fairness Criterion 3.

32. **Pairwise Comparisons and Criterion 4.** Suppose the method of pairwise comparisons is used with the following preference schedule. Is Fairness Criterion 4 satisfied? Explain.

First	A	A	E	C	D
Second	E	C	B	B	B
Third	C	D	A	A	A
Fourth	D	E	C	D	E
Fifth	B	B	D	E	C
Number of voters	1	1	1	1	1

33. **Pairwise Comparisons and Criterion 4.** Devise a preference schedule with five candidates for which the method of pairwise comparisons violates Fairness Criterion 4.

34. **Approval Voting.** Suppose that Candidates A and B have moderate political positions, while Candidate C is relatively conservative. Voter opinions about the candidates are as follows:

- 30% want A as their first choice, but would also approve of B.
- 29% want B as their first choice, but would also approve of A.
- 1% want A as their first choice and approve of neither B nor C.
- 40% want C as their first choice and approve of neither A nor B.

a. If all voters could vote only for their first choice, which candidate would win by the plurality method?

b. Which candidate wins by approval voting?

35. **Approval Voting.** Suppose that Candidates A and B have moderate political positions, while Candidate C is quite liberal. Voter opinions about the candidates are as follows:

- 28% want A as their first choice, but would also approve of B.
- 29% want B as their first choice, but would also approve of A.
- 1% want B as their first choice, and approve of neither A nor C.
- 42% want C as their first choice, and approve of neither A nor B.

a. If all voters could vote only for their first choice, which candidate would win by the plurality method?

b. Which candidate wins by approval voting?

36. **Power Voting.** Imagine that a small company has four shareholders who hold 26%, 26%, 25%, and 23% of the company's stock. Assume that votes are assigned in proportion to shares held (e.g., if there are a total of 100 votes, the four shareholders get 26, 26, 25, and 23 votes, respectively). Also assume that decisions are made by strict majority vote. Explain why, although each individual holds roughly one-fourth of the company's stock, the individual with 23% holds *no* effective power in voting.

37–41: Electoral Power. Use the table below to answer the following questions.

State	Population (in 2016)	Electoral votes
Alaska	742,000	3
Illinois	12,802,000	20
New York	19,745,000	29
Rhode Island	1,056,000	4

37. Which state has more voting power per person: Alaska or Illinois?

38. Which state has more voting power per person: Alaska or New York?

39. Which state has more voting power per person: Rhode Island or Illinois?

40. Which state has more voting power per person: Illinois or New York?

41. Rank the four states in order of voting power per person, from most voting power to least voting power.

FURTHER APPLICATIONS

42–46: Fairness Criteria. Consider the following preference schedule for four candidates.

First	A	B	C	D
Second	D	A	B	C
Third	B	D	A	A
Fourth	C	C	D	B
Number of voters	16	10	8	7

42. Suppose the winner is decided by the plurality method. Analyze whether this choice satisfies the four fairness criteria.

43. Suppose the winner is decided by a single runoff. Analyze whether this choice satisfies the four fairness criteria.

44. Suppose the winner is decided by a sequential runoff. Analyze whether this choice satisfies the four fairness criteria.

45. Suppose the winner is decided by a Borda count (point system). Analyze whether this choice satisfies the four fairness criteria.

46. Suppose the winner is decided by pairwise comparisons. Analyze whether this choice satisfies the four fairness criteria.

47–51: Fairness Criteria. Consider the following preference schedule for five candidates (Table 12.5 in Unit 12A).

First	A	B	C	D	E	E
Second	D	E	B	C	B	C
Third	E	D	E	E	D	D
Fourth	C	C	D	B	C	B
Fifth	B	A	A	A	A	A
Number of voters	18	12	10	9	4	2

47. Suppose the winner is decided by the plurality method. Analyze whether this choice satisfies the four fairness criteria.

48. Suppose the winner is decided by a single runoff. Analyze whether this choice satisfies the four fairness criteria.

49. Suppose the winner is decided by a sequential runoff. Analyze whether this choice satisfies the four fairness criteria.

50. Suppose the winner is decided by a Borda count (point system). Analyze whether this choice satisfies the four fairness criteria.

51. Suppose the winner is decided by pairwise comparisons. Analyze whether this choice satisfies the four fairness criteria.

52. **Swing Votes.** Suppose that the Senate has the following breakdown in party representation: 49 Democrats, 49 Republicans, and 2 Independents. Further suppose that all Democrats and Republicans vote along party lines. Assuming that a majority is required to pass a bill, explain why the 2 Independents, despite holding only 2% of the Senate seats, effectively hold power equal to that of either of the larger parties.

IN YOUR WORLD

53. **Elections Gone Bad.** Find a news report about an election in which the results were disputed because of corruption or because the election system failed. Discuss the incident in light of the fairness criteria presented in this unit.

54. **Other Fairness Criteria.** The fairness criteria discussed in this unit are not the only ones. Investigate other criteria (such as the favorite-betrayal criterion, the strong adverse results criterion, and the weak defensive strategy criterion). Discuss the merits of these criteria and their shortcomings.

55. **Approval Voting.** Many political systems use approval voting, and it is strongly advocated by many political organizations. Investigate the details of approval voting, cite countries in which it is used, and discuss its strengths and weaknesses.

56. **Power Voting and Coalitions.** Use the Web to investigate the political coalitions at the national level in a particular country (for example, Israel). Describe the coalitions, their sizes, and their effective voting power.

57. **General Voting Power.** Find a news report about any election in which not all participants had the same voting power. What factors affected voting power?

58. **The National Popular Vote Interstate Compact.** Find the current status of the National Popular Vote Interstate Compact (discussed in In Your World on p. 681). Has it passed or is it under consideration in *your* state? What are its prospects nationally? Do you think it is a good idea? Defend your opinion.

UNIT 12C Apportionment: The House of Representatives and Beyond

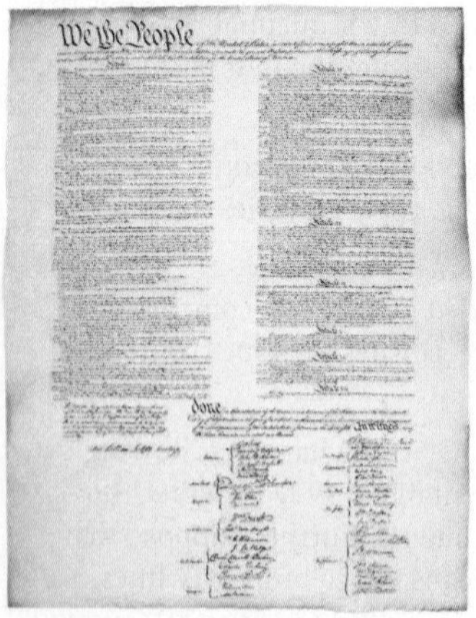

Each state in the United States has two senators, but the number of representatives varies from state to state. Who decides how many of the 435 seats in the House of Representatives each state gets? This question is one of **apportionment**, because the seats in the House of Representatives must be apportioned (divided) among the states in a fair way. In this unit, we'll discuss the mathematics that underlies apportionment.

> **Definition**
>
> For the U.S. House of Representatives, **apportionment** is a process used to divide the available seats among the states. More generally, apportionment is a process used to divide a set of people or objects among various groups or individuals.

All legislative Powers herein granted shall be vested in a Congress of the United States, which shall consist of a Senate and House of Representatives.
—Article 1, Section 1, of the United States Constitution

The Constitutional Context

Much of the mathematics of apportionment arose from historical attempts to meet the letter and spirit of the United States Constitution. According to the Constitution, the legislative (law-making) branch of the United States government consists of two bodies: the **Senate** and the **House of Representatives** (called "the House" for short).

The Constitution (Article 1, Section 3) specifies a simple method of apportionment for senators: Each state gets two senators. According to the original Constitution, senators were chosen by state legislatures. However, the 17th amendment to the Constitution, adopted in 1913, changed the method of selecting senators to direct election by the people. Senators are elected for six-year terms.

The requirements for the House of Representatives are much more complicated. The Constitution mandates that seats in the House be apportioned to the states according to their populations, subject to a minimum of one seat for every state. It allows Congress to set the total number of representatives, so long as the total number does not exceed one for each 30,000 people.

The Constitution directs Congress to reapportion seats in the House every 10 years, following a census. However, the Constitution does not specify a particular procedure for apportionment. As we will soon see, it is not easy to choose an apportionment procedure, and no single procedure is always fair.

HISTORICAL NOTE

The first Congress, which met on March 4, 1789, in the Federal Hall in New York City, had 59 members in the House of Representatives. The number gradually rose, reaching 435 in 1912. It has stayed at 435 since 1912, except when two representatives were temporarily added when Hawaii and Alaska became states in 1959. The number returned to 435 at the next apportionment.

EXAMPLE 1 United States House of Representatives

The House has 435 representatives and is currently apportioned based on the 2010 census, which reported a U.S. population of 309 million. On average, using the 2010 population, how many people are represented by each representative? Suppose that the total number of representatives were the constitutional limit of one for every 30,000 people. How many representatives would there be in that case?

Solution Dividing the population of 309 million by 435 representatives, we find that the average number of people served by each representative in 2010 was

$$\frac{\text{population}}{\text{number of representatives}} = \frac{309{,}000{,}000}{435} \approx 710{,}000$$

On average, each representative served approximately 710,000 people in 2010. If there were one representative for every 30,000 people, the total number of representatives in 2010 would have been

$$\frac{\text{population}}{30{,}000} = \frac{309{,}000{,}000}{30{,}000} = 10{,}300$$

That is, the constitutional limit allows more than 10,000 representatives, in contrast to the actual number of 435. ▶ Now try Exercises 13–14.

The Apportionment Problem

Example 1 shows that, using the 2010 census data, each House member represents an average of 710,000 people. Apportionment would be easy if every state's population were a simple multiple of 710,000. For example, a state with a population of 1,420,000, which is $2 \times 710{,}000$, would get 2 representatives. Similarly, a state with a population of 7,100,000, which is $10 \times 710{,}000$, would get 10 representatives.

Of course, states do not have such convenient populations. For example, Rhode Island had a 2010 census population of about 1,050,000, which is about $1.5 \times 710{,}000$. We might therefore say that Rhode Island is "entitled" to 1.5 representatives, but representatives are people and cannot be divided fractionally. Rhode Island could have one representative or two representatives, but not 1.5 representatives.

The people of Rhode Island would prefer to have two representatives, because that strengthens their voice in the House. However, people of other states might prefer that Rhode Island have only one representative, thereby leaving one more representative for another state.

In essence, the apportionment problem deals with finding a systematic way of deciding whether Rhode Island gets one or two representatives. The system must be as fair to all states as possible, while also ensuring that precisely 435 House seats are awarded in total.

BY THE WAY

The 2010 apportionment changed the House representation of 18 states. Texas gained four seats, Florida gained two, and Arizona, Georgia, Nevada, South Carolina, Utah, and Washington each gained one seat. On the losing side, New York and Ohio each lost two seats, while Illinois, Iowa, Louisiana, Massachusetts, Michigan, Missouri, New Jersey, and Pennsylvania each lost one seat.

The Standard Divisor and Quota

Look again at how we determined that Rhode Island was "entitled" to 1.5 representatives. First, we divided the total U.S. population by the total number of representatives to find that each representative serves an average of 710,000 people. We then divided Rhode Island's population by 710,000 to find the 1.5 representatives.

In the terminology of apportionment, the 710,000 average is called the **standard divisor** for this problem. The number 1.5 is called the **standard quota** of representatives for Rhode Island. It is the number (quota) that Rhode Island would get if it were possible to have fractional representatives. Note that there is only *one* standard divisor. In contrast, each state has its own standard quota.

Standard Divisor and Quota

The **standard divisor** is the average number of people per seat (in the House of Representatives) for the entire population of the United States:

$$\text{standard divisor} = \frac{\text{total U.S. population}}{\text{number of seats}}$$

The **standard quota** for a state is the number of seats it would be entitled to *if* fractional seats were allowed:

$$\text{standard quota} = \frac{\text{state population}}{\text{standard divisor}}$$

BY THE WAY

Besides the 100 senators and 435 representatives, the U.S. Congress also includes a resident commissioner from Puerto Rico and a delegate from American Samoa, the District of Columbia, Guam, the Northern Mariana Islands, and the Virgin Islands. These six individuals may take part in the floor discussions and vote in committees, but as of 2017 have no vote in the full House.

The standard divisor and standard quota also apply to apportionment problems besides that of the House of Representatives. Simply replace the number of seats by the number of items to be apportioned and the state and total populations by the relevant populations in the problem.

EXAMPLE 2 Finding Standard Quotas

The 2010 census found populations of 989,000 for Montana and 564,000 for Wyoming. Using these census populations, find the standard quota for each state. The 2010 apportionment left each of these states with the constitutional minimum of one representative. Compare their standard quotas to their actual representations.

Solution The standard divisor is the same for all states. As we found earlier, it is 710,000 people per representative. For Montana, the standard quota based on its 2010 population is

$$\text{standard quota} = \frac{\text{state population}}{\text{standard divisor}} = \frac{989{,}000}{710{,}000} \approx 1.4$$

For Wyoming, the standard quota is

$$\text{standard quota} = \frac{\text{state population}}{\text{standard divisor}} = \frac{564{,}000}{710{,}000} \approx 0.79$$

Montana's standard quota is 1.4, but it has only one representative. Montana's people can rightfully feel underrepresented in Congress. Wyoming also has the constitutional minimum of one representative, even though its standard quota is only 0.79. Wyoming's people are relatively overrepresented in Congress, at least according to population.

▶ Now try Exercises 15–18.

EXAMPLE 3 School Teacher Apportionment

A small school district is reapportioning its 14 elementary teachers among its three elementary schools, which have the following enrollments: Washington Elementary, 197 students; Lincoln Elementary, 106 students; and Roosevelt Elementary, 145 students. Find the standard quota of teachers for each school.

Solution This problem is just like the House apportionment problem, except we are apportioning teachers rather than representatives. The standard divisor is the average number of students per teacher in the entire district. We find it by dividing the total student population by the number of teachers:

$$\text{standard divisor} = \frac{\text{total number of students}}{\text{number of teachers}} = \frac{197 + 106 + 145}{14} = 32$$

We find the standard quota for each school by dividing the school's enrollment by the standard divisor. Table 12.10 shows the calculations and results. Note that the total (sum) of the standard quotas is the number of teachers to be apportioned.

TABLE 12.10 Finding Standard Quotas for a School Teacher Apportionment

School	Washington	Lincoln	Roosevelt	Total
Enrollment	197	106	145	448
Standard Quota	$\frac{197}{32} = 6.15625$	$\frac{106}{32} = 3.3125$	$\frac{145}{32} = 4.53125$	14

Enrollments divided by standard divisor of 32 → Standard Quota

▶ Now try Exercises 19–20.

Think About It Notice that the three standard quotas in Table 12.10 are all fractions, but each school must get a whole number of teachers. Based on the standard quotas, how would *you* apportion the 14 teachers among the three schools? With your apportionment, what average class size would each school have?

The Challenge of Apportionment

The 435 representatives and 50 states make apportionment in the United States House of Representatives complex. For the purposes of understanding apportionment, it's easier to work with a smaller set of states. Suppose there were only four states, A, B, C, and D, with populations as shown in Table 12.11 and a total population of 10,000.

Suppose these four states decide to create a legislature with 100 seats. Then the standard divisor is

$$\text{standard divisor} = \frac{\text{total population}}{\text{number of seats}} = \frac{10{,}000}{100} = 100$$

The second row of Table 12.11 shows the standard quotas calculated with this standard divisor. As it must, the total of the standard quotas equals the 100 available seats.

TABLE 12.11 Standard Quotas for Apportioning 100 Seats Among Four States

State	A	B	C	D	Total
Population	936	2726	2603	3735	10,000
Standard Quota	9.36	27.26	26.03	37.35	100

State populations divided by standard divisor of 100 → Standard Quota

Note: The total population is 10,000, and the standard divisor is 100.

We cannot use the fractional standard quotas for the actual apportionment. Instead, we must find a way to make integers from the standard quotas. The most obvious solution would be to round all the standard quotas according to standard rounding rules. In that case, all four standard quotas would round down, because all four have fractional parts less than 0.5. However, rounding in this way leads to a total of $9 + 27 + 26 + 37 = 99$ seats, one short of the 100 seats that are supposed to be apportioned. This failure to attain 100 seats using standard rounding rules means that we must find a different way to make integers from the standard quotas. Herein lies the challenge of apportionment: There are many reasonable ways to make integers from the standard quotas, but they do not all lead to the same apportionment results.

Hamilton's Method

The United States conducted its first census in 1790, presenting Congress with its first task of apportionment almost immediately after ratification of the Constitution. Alexander Hamilton, then Secretary of the Treasury, championed a simple apportionment procedure that worked as follows.

Hamilton's Method of Apportionment

After finding the standard quota for each state:

- First, give each state the number of seats found by rounding its standard quota *down*. (For example, 3.99 would round down to 3.) This number is the state's **minimum quota** (or *lower quota*).
- If there are any extra seats after each state has been given its minimum quota, look at each state's **fractional remainder**—the fraction of the standard quota that remains after subtracting the minimum quota. Give the first extra seat to the state with the highest fractional remainder, the next to the state with the second highest fractional remainder, and so on until all the seats are gone.

BY THE WAY

Alexander Hamilton (1757–1804) recently gained renewed fame as a result of the award-winning show *Hamilton: An American Musical*, written by and originally starring Lin-Manuel Miranda. Miranda's inspiration for the show came from reading the biography *Alexander Hamilton* by Ron Chernow. The show opened on Broadway in August 2015, and began its first national tour in 2017.

Let's apply Hamilton's method to the four states in Table 12.11. The easiest way is by extending the table, as shown in Table 12.12. The first two rows (population and standard quota) are unchanged. The third row shows the minimum quotas found by rounding down the standard quotas. Note that the total of the minimum quotas is 99, leaving 1 extra seat out of the 100 to be allotted. The fourth row shows the fractional remainders after rounding down. The extra seat goes to the state with the largest fractional remainder, which is State A. The last row shows the final apportionment. Note that State A gets one seat more than its minimum quota, while the other states get precisely their minimum quotas.

TABLE 12.12 Applying Hamilton's Method to the Four-State Data from Table 12.11

State	A	B	C	D	Total
Population	936	2726	2603	3735	10,000
Standard Quota	9.36	27.26	26.03	37.35	100
Minimum Quota	9	27	26	37	99
Fractional Remainder	0.36 (largest)	0.26	0.03	0.35	1
Final Apportionment	10	27	26	37	100

State populations divided by standard divisor of 100 → Standard Quota

Standard quotas rounded down → Minimum Quota

Remainders after rounding → Fractional Remainder

The extra seat goes to the state with the largest fractional remainder (State A). → Final Apportionment

EXAMPLE 4 Applying Hamilton's Method

Apply Hamilton's method to determine the apportionment of teachers among the schools in Example 3.

Solution Table 12.13 repeats the enrollments and standard quotas found in Example 3. Hamilton's method tells us to first round down all the standard quotas; we get the minimum quotas in the third row. The total of the minimum quotas is 13, one less than the 14 teachers to be apportioned. The school with the largest fractional remainder (fourth row), which is Roosevelt, gets the extra teacher. In the final apportionment, Washington and Lincoln get their minimum quotas of 6 and 3 teachers, respectively; Roosevelt gets 5 teachers, rather than its minimum quota of 4.

TABLE 12.13 Applying Hamilton's Method to the School Teacher Apportionment from Table 12.10

School	Washington	Lincoln	Roosevelt	Total
Enrollment	197	106	145	448
Standard Quota	$\frac{197}{32} = 6.15625$	$\frac{106}{32} = 3.3125$	$\frac{145}{32} = 4.53125$	14
Minimum Quota	6	3	4	13
Fractional Remainder	0.15625	0.3125	0.53125 (largest)	1
Final Apportionment	6	3	5	14

Enrollments divided by standard divisor of 32 → Standard Quota

Standard quotas rounded down → Minimum Quota

Remainders after rounding → Fractional Remainder

The extra teacher goes to the school with the largest fractional remainder (Roosevelt). → Final Apportionment

▶ Now try Exercises 21–24.

The First Presidential Veto

Hamilton's method is simple and appears reasonably fair. Perhaps as a result, both the Senate and the House voted in 1791 to adopt this method for apportionment. However, recall that for a bill to become law, it must not only be passed by both the Senate and the House, but must also be signed by the President. If the President vetoes a bill, super majority votes of 2/3 of both House and Senate members are needed to override the veto (see Unit 12A). In 1792, the bill authorizing Hamilton's method received the first presidential veto in the history of the United States. In his veto message of April 5, 1792, President George Washington wrote:

> *I have maturely considered the act passed by the two Houses entitled "An act for an apportionment of representatives among the several States according to the first enumeration," and I return it to your House, wherein it originated, with the following objections:*
>
> *First. The Constitution has prescribed that representatives shall be apportioned among the several States according to their respective numbers, and there is no one proportion or divisor which, applied to the respective numbers of the States, will yield the number and allotment of representatives proposed by the bill.*
>
> *Second. The Constitution has also provided that the number of representatives shall not exceed 1 for every 30,000, which restriction is by the context and by fair and obvious construction to be applied to the separate and respective numbers of the States; and the bill has allotted to eight of the States more than 1 for every 30,000.*

Congress was unable to override the veto and instead passed a bill authorizing a different apportionment method proposed by Thomas Jefferson, then Secretary of State. We'll discuss Jefferson's method shortly.

HISTORICAL NOTE

President Washington's veto of Hamilton's method and the subsequent adoption of Jefferson's method affected the representation of only two states. Delaware would have had 2 seats by Hamilton's method, but ended up with only 1 by Jefferson's method. Virginia would have had 18 seats by Hamilton's method, but ended up with 19 by Jefferson's. Perhaps not coincidentally, Virginia was Jefferson's home state.

Think About It The Constitution says, "The number of representatives shall not exceed one for every thirty thousand, but each state shall have at least one representative...." Read Washington's veto statement carefully. Why did he have to include the words "which restriction is by the context and by fair and obvious construction ..." in order to justify his veto?

Fairness Issues with Hamilton's Method

Although Hamilton's method did not become law at the time it was proposed, it was reintroduced and adopted as the apportionment procedure in 1850. It remained in use until 1900. During this time, several problems with Hamilton's method emerged; the most famous is called the **Alabama paradox**.

After the 1880 census, the Chief Clerk of the Census Office, C. W. Seaton, used Hamilton's method to compute apportionments for various possible House sizes. (Unlike today, in those days the House size often changed with each apportionment.) He discovered that Alabama would get 8 seats in a House of 299 representatives, but only 7 seats in a House of 300 representatives. This curious result seems unfair, because Alabama loses a seat when the total number of seats increases.

HISTORICAL NOTE

Hamilton's method was reintroduced to Congress by Representative Samuel Vinton of Ohio. However, Vinton was apparently unaware of Hamilton's proposal more than 50 years earlier. As a result, the method was temporarily called *Vinton's method*.

The Alabama Paradox

In a fair apportionment system, adding extra seats must not result in fewer seats for any state. The **Alabama paradox** occurs when the total number of available seats increases, yet one state (or more) loses seats as a result. It can occur with Hamilton's method.

Alabama did not actually lose a seat in the 1882 apportionment, because Congress chose a larger house size (325) for which the Alabama paradox did not present itself. But knowing that the paradox was possible reduced support for Hamilton's method, and it was abandoned by Congress in 1900.

Hamilton's method has not been used for House apportionment since 1900, but further study of it led to the discovery of two other paradoxes. The **population paradox** was discovered around 1900, when it was found that Hamilton's method would award a seat to Maine at Virginia's expense, even though Virginia was growing much faster than Maine.

HISTORICAL NOTE

In the 1876 election, Democrat Samuel J. Tilden won the popular vote but lost the presidency to Republican Rutherford B. Hayes by one electoral vote. The electoral vote was vigorously disputed, with some states appointing rival slates of electors. Tilden eventually stepped aside in the "Compromise of 1877," which led to Hayes's inauguration on March 4, 1877. Interestingly, if electors had been apportioned by Hamilton's method, as the law required, Tilden would have won in the first place. Instead, electors had been apportioned according to a compromise from a few years earlier, even though the compromise violated the legal requirement.

The Population Paradox

When apportionment changes because of population growth, we expect faster-growing states to gain seats at the expense of slow-growing states. When the opposite occurs—a slow-growing state gains a seat at the expense of a faster-growing state—we have the **population paradox.**

The **new states paradox** was discovered in 1907, when Oklahoma became the 46th state in the United States. Because it was not yet time for reapportionment, Congress decided to add *new* seats for Oklahoma. Based on its population, Oklahoma was clearly entitled to 5 seats, so Congress increased the number of seats in the House from 386 to 391. Surprisingly, calculations with Hamilton's method showed that the addition of Oklahoma's 5 seats would have caused New York to lose a seat while Maine gained one.

The New States Paradox

When additional seats are added to accommodate a new state, we do not expect this addition to change the apportionment for existing states. If it does, we have the **new states paradox.**

EXAMPLE 5 Four-State Alabama Paradox

Using Hamilton's method, recompute the apportionment in Table 12.12 if there are 101 seats instead of 100. Does the Alabama paradox occur? Explain.

Solution The total population is 10,000, so with 101 seats the standard divisor is

$$\text{standard divisor} = \frac{\text{total population}}{\text{number of seats}} = \frac{10{,}000}{101} \approx 99.0099$$

We find the standard quota for each state by dividing its population by this standard divisor. You should confirm that the standard quotas are shown in Table 12.14. Note that they are slightly different from the standard quotas shown in Table 12.12 for 100 seats. The third row of Table 12.14 shows the minimum quotas, which are unchanged from Table 12.12. However, the fractional remainders (fourth row) have changed, which changes the final apportionment.

TABLE 12.14 Recomputed Apportionment for Table 12.12 with 101 Seats

State	A	B	C	D	Total
Population	936	2726	2603	3735	10,000
Standard Quota	9.4536	27.5326	26.2903	37.7235	101
Minimum Quota	9	27	26	37	99
Fractional Remainder	0.4536	0.5326 (2nd largest)	0.2903	0.7235 (largest)	2
Final Apportionment	9	28	26	38	101

State populations divided by standard divisor of 99.0099 → Standard Quota

Standard quotas rounded down → Minimum Quota

Remainders after rounding → Fractional Remainder

The extra seats go to the two states with the largest fractional remainders (D and B). → Final Apportionment

This time, there are two extra seats after the minimum quotas have been filled. By Hamilton's method, these two seats go to States D and B, because they have the two largest fractional remainders. Meanwhile, State A ends up with its minimum quota of 9 seats, even though it had 10 seats in Table 12.12. In other words, increasing the total number of seats from 100 to 101 has cost State A a seat—an illustration of the Alabama paradox.

▶ Now try Exercises 25–28.

Jefferson's Method

Thomas Jefferson's rival apportionment method became law after President Washington vetoed Hamilton's method. Jefferson's method essentially tries to avoid decisions about fractional remainders by seeking minimum quotas that use up all available seats. As we've seen, the minimum quotas used in Hamilton's method—which are simply the standard quotas rounded down—often leave extra seats. Jefferson realized that he could achieve his goal by changing the standard quotas to new values, called **modified quotas**, so that the new set of minimum quotas would use up all the seats.

Finding modified quotas means dividing the state populations by a **modified divisor** that is different (lower) than the standard divisor. The trick in Jefferson's method is choosing a modified divisor that leads to the desired result, which is a set of minimum quotas that uses all the seats. Choosing a modified divisor generally requires trial and error: If your first choice doesn't work, try again until you find one that does.

Statue of Thomas Jefferson in the Jefferson Memorial, Washington, DC.

Jefferson's Method of Apportionment

Begin by finding the standard divisor, standard quotas, and minimum quotas (as in Hamilton's method). If there are no extra seats after each state has been given its minimum quota, then the apportionment is complete. If there *are* extra seats, *start the computations over* as follows.

- Choose a **modified divisor** that is *lower* than the standard divisor. Use it to compute **modified quotas** by dividing the state populations by the modified divisor. Round these modified quotas down to find a new set of minimum quotas.
- If there are just enough seats to fill all the minimum quotas, the apportionment is complete. Otherwise, do one of the following:
  1. If there are still extra seats with the new minimum quotas, start again with a lower modified divisor.
  2. If there are not enough total seats to fill all the minimum quotas, start again with a higher modified divisor (but still smaller than the standard divisor).

Let's apply Jefferson's method to the same four states to which we applied Hamilton's method. The first three rows of Table 12.15 are unchanged from Table 12.12, showing the minimum quotas (third row) that result from the standard divisor of 100. Because the minimum quotas leave an extra seat, Jefferson's method tells us to try a new (lower) divisor. Study Table 12.15 carefully to see how we implement Jefferson's method.

- The standard divisor is 100, so we try a lower *modified divisor* of 99. We divide the state populations by 99 to find the modified quotas in the fourth row. The fifth

TABLE 12.15 Applying Jefferson's Method to the Four-State Data from Table 12.12

	State	A	B	C	D	Total
	Population	936	2726	2603	3735	10,000
State populations divided by standard divisor of 100 →	Standard Quota	9.36	27.26	26.03	37.35	100
Standard quotas rounded down →	Minimum Quota	9	27	26	37	99
State populations divided by modified divisor of 99 →	Modified Quota (with divisor 99)	9.45	27.54	26.29	37.73	101.01
Modified quotas rounded down →	Minimum Quota (with divisor 99)	9	27	26	37	99
State populations divided by modified divisor of 98 →	Modified Quota (with divisor 98)	9.55	27.82	26.56	38.11	102.04
Modified quotas rounded down; all 100 seats are used, so this is final apportionment. →	Minimum Quota (with divisor 98)	9	27	26	38	100

row shows the new set of minimum quotas, found by rounding down the modified quotas.

- Because the new set of modified quotas still leaves an extra seat, we try an even lower modified divisor of 98. The last two rows show the modified quotas and minimum quotas found with this divisor. This time they add up to 100, so the apportionment is complete.

Note that State A gets 9 seats by Jefferson's method, as opposed to 10 seats by Hamilton's method (see Table 12.12). Meanwhile, State D gets 38 seats by Jefferson's method, one more than it gets by Hamilton's method.

Is Jefferson's Method Fair?

We've seen that Hamilton's method is susceptible to several paradoxes that make it appear unfair in some cases. Is Jefferson's method better?

Mathematicians have studied these methods carefully. Jefferson's method avoids all three of the paradoxes found with Hamilton's method (the Alabama, population, and new states paradoxes). However, Jefferson's method sometimes leads to a different problem.

BY THE WAY

Hamilton's method always satisfies the quota criterion because it starts with the standard quota rounded down, then at most adds one—which gives the standard quota rounded up. Hamilton's method and other apportionment methods that always satisfy the quota criterion are generically called *quota methods.* In contrast, methods that change the divisor, like Jefferson's method, are called *divisor methods* and sometimes violate the quota criterion.

Remember that standard quotas represent the number of seats that would be apportioned *if* fractional apportionment were allowed. For example, we earlier found a standard quota of 1.5 for Rhode Island, suggesting that it is "entitled" to 1.5 seats. We might therefore expect that any "fair" apportionment method would either round this standard quota up to give Rhode Island two seats or round it down to give Rhode Island one seat. If instead Rhode Island ended up with zero seats, or with three or more seats, we would probably conclude that the apportionment was unfair. This idea, which holds that the actual apportionment for any state should be its standard quota rounded either up or down, is called the **quota criterion**. An apportionment that fails the quota criterion is generally considered unfair. As the following example shows, Jefferson's method sometimes fails the quota criterion.

The Quota Criterion

For a fair apportionment, the number of seats assigned to each state should be its standard quota rounded either up or down to the nearest integer.

EXAMPLE 6 Jefferson's Method and the Quota Criterion

Consider a four-state legislature with 100 seats, in which the states have the following populations: State A, 680; State B, 1626; State C, 1095; and State D, 6599. Use Jefferson's method to apportion the 100 seats. Is the quota criterion satisfied?

Solution We begin by finding the standard divisor, using the fact that the total population is 10,000 and the number of seats is 100.

$$\text{standard divisor} = \frac{\text{total population}}{\text{number of seats}} = \frac{10{,}000}{100} = 100$$

Table 12.16 shows the computations by Jefferson's method. The second row gives the standard quotas and the third row gives the resulting minimum quotas. There are three extra seats, so Jefferson's method tells us to try a lower modified divisor. The fourth row shows the modified quotas with a divisor of 98, and the last row shows the new minimum quotas. Because this new set of minimum quotas uses all 100 seats, it represents the final apportionment.

TABLE 12.16 A Case Where Jefferson's Method Violates the Quota Criterion

	State	A	B	C	D	Total
	Population	680	1626	1095	6599	10,000
State populations divided by standard divisor of 100 →	**Standard Quota**	6.80	16.26	10.95	65.99	100
Standard quotas rounded down →	**Minimum Quota**	6	16	10	65	97
State populations divided by *modified divisor* of 98 →	**Modified Quota (with divisor 98)**	6.94	16.59	11.17	67.34	102.04
Modified quotas rounded down →	**Minimum Quota (with divisor 98)**	6	16	11	67	100

Note that State D has a standard quota of 65.99. By the quota criterion, State D should get either 65 or 66 seats. However, State D ends up with 67 seats, so this apportionment violates the quota criterion. ▶ **Now try Exercises 29–32.**

Usage of Jefferson's Method

After being signed into law by President Washington in 1792, Jefferson's method was used through the 1830s. However, violations of the quota criterion occurred in the apportionments following both the 1820 and the 1830 census. Moreover, Congress learned that the violations tended to give extra seats to larger states at the expense of smaller states. As a result, Congress abandoned Jefferson's method for the apportionment following the 1840 census.

Other Apportionment Methods

As we've discussed, Jefferson's method was abandoned by 1840 and Hamilton's method, adopted in 1850, was abandoned in 1900. Many other apportionment methods have been proposed, but only two others have been used for the United States House of Representatives.

Webster's Method

In 1832, the famous senator and orator Daniel Webster, of Massachusetts, proposed a variation on Jefferson's method. Webster's method was adopted for the apportionment following the 1840 census. It was then abandoned in favor of Hamilton's method in 1850, but resurrected in 1900. Webster's method remained in use until 1940.

What a man does for others, not what they do for him, gives him immortality.
—Daniel Webster (1782–1852)

Webster's method is similar to Jefferson's method with one exception: Instead of looking for a set of modified quotas that can all be rounded *down* to give the correct total number of seats, Webster's method seeks modified quotas that give the correct total number of seats by using standard rounding rules. That is, we round *up* when the fractional part of a modified quota is 0.5 or more and round *down* when the fractional part is less than 0.5. Note that, while the modified divisor in Jefferson's method is always less than the standard divisor, the modified divisor in Webster's method may be either greater or less than the standard divisor.

EXAMPLE 7 Applying Webster's Method

Consider a four-state legislature with 100 seats, in which the states have the following populations: State A, 948; State B, 749; State C, 649; and State D, 7654. Use Webster's method to apportion the 100 seats.

Solution The total population is 10,000 and the number of seats is 100, so the standard divisor is 100 (as in Example 6). Table 12.17 shows the computations by Webster's method. The second row gives the standard quotas, and the third row gives the resulting minimum quotas; the result is a total of 98 seats, which means we still need to fill two extra seats to make 100 total. Webster's method tells us to try a modified divisor and then round the modified quotas to the nearest integer. The fourth row shows the modified quotas with a divisor of 99.85. Finding a divisor that works generally requires some trial and error (not shown here). The last row shows the rounded quotas, which represent the final apportionment because it uses all 100 seats.

TABLE 12.17 A Four-State Apportionment with Webster's Method

State	A	B	C	D	Total
Population	948	749	649	7654	10,000
Standard Quota	9.48	7.49	6.49	76.54	100
Minimum Quota	9	7	6	76	98
Modified Quota (with divisor 99.85)	9.4942	7.5013	6.4997	76.6550	100.1502
Rounded Quota	9	8	6	77	100

State populations divided by standard divisor of 100 → Standard Quota

Standard quotas rounded down → Minimum Quota

State populations divided by *modified* divisor of 99.85 → Modified Quota

Modified quotas rounded to nearest integer → Rounded Quota

▶ **Now try Exercises 33–34.**

Think About It Is the quota criterion satisfied by the final apportionment in Example 7? Explain.

The Hill-Huntington Method

In 1911, Joseph Hill, who served as Chief Statistician of the Census Bureau, proposed an alternative apportionment method that was further developed by Harvard mathematician Edward Huntington. Their method of apportionment, called the Hill-Huntington method, replaced Webster's method in 1941 and remains in use today. As is often the case in politics, adoption of the Hill-Huntington method was based at least as much on political calculations as on fairness issues. When working on the 1941 apportionment (based on the 1940 census), Congress found that Webster's method would have given one extra seat to Michigan, which tended to vote Republican, while the Hill-Huntington method gave the seat to Arkansas, which tended to vote Democratic. Because Democrats had a majority in Congress, they voted to use the Hill-Huntington method, thereby giving the extra seat to Democratic-leaning Arkansas. President Roosevelt (also a Democrat) signed the bill into law.

The Hill-Huntington method is almost identical to Webster's method, except it uses a different rule to decide whether modified quotas should be rounded up or down. In Webster's method, rounding follows the usual rule in which we round up if the fractional part is 0.5 or more and round down if it is less than 0.5. In the Hill-Huntington method, rounding is based instead on the **geometric mean** of the integers on either side of the modified quota.

Definitions

The **geometric mean** of any two numbers x and y is $\sqrt{x \times y}$. The more familiar mean, $(x + y)/2$, is called the **arithmetic mean.** In general, we assume a "mean" is an arithmetic mean unless told otherwise. (The mean we have used elsewhere in this text is the arithmetic mean.)

If the modified quota is less than the geometric mean of the two nearest integers, it gets rounded down. If it is more than the geometric mean, it gets rounded up. As an example, suppose a state has a modified quota of 2.47, which we could potentially round to either 2 or 3. Under Webster's method, we would round it down to 2, because the fractional part of 2.47 is less than 0.5. Under the Hill-Huntington method, we first find the geometric mean of 2 and 3, which is

$$\sqrt{2 \times 3} = \sqrt{6} \approx 2.45$$

Because the modified quota of 2.47 is *greater* than this geometric mean, it gets rounded up to 3 in the Hill-Huntington method.

Note that the geometric mean of two consecutive integers is always less than their arithmetic mean. For example, the geometric mean of 2 and 3 is 2.45, which is less than their arithmetic mean of 2.5 by 0.05. However, the two means tend to be closer for larger consecutive integers. For example, the geometric mean of 10 and 11 is $\sqrt{10 \times 11} = \sqrt{110} \approx 10.488$, which is only 0.012 less than their arithmetic mean of 10.5. In practical terms, use of the geometric mean in the Hill-Huntington method therefore increases the chance that extra seats will go to smaller states rather than larger states.

EXAMPLE 8 Applying the Hill-Huntington Method

Apply the Hill-Huntington method to the case described in Example 7. Is the resulting apportionment the same as that found with Webster's method?

Solution Table 12.18 shows the calculations. The first three rows are the same as in Table 12.17, because the standard quotas and minimum quotas are computed the same way by all methods. The fourth row shows modified quotas with a modified divisor of 100.06. Again, finding a modified divisor that works generally requires some trial and error, which is not shown here. To determine the rounded quotas, we first find the geometric means (fifth row) of the modified quotas. The Hill-Huntington method then tells us to round down if the modified quota is less than the geometric mean and to round up otherwise. The last row shows the rounded quotas; because they use all 100 seats, the apportionment is complete. Note that States C and D end up with different numbers of seats with this apportionment than with Webster's apportionment (Example 7).

TABLE 12.18 Applying the Hill-Huntington Method to the Four-State Data from Table 12.17

	State	A	B	C	D	Total
	Population	948	749	649	7654	10,000
State populations divided by standard divisor of 100 →	**Standard Quota**	9.48	7.49	6.49	76.54	100
Standard quotas rounded down →	**Minimum Quota**	9	7	6	76	98
State populations divided by *modified divisor* of 100.06 →	**Modified Quota (with divisor 100.06)**	9.47	7.49	6.49	76.49	99.94
Geometric mean of integers on either side of modified quota →	**Geometric Mean**	$\sqrt{9 \times 10} \approx 9.487$	$\sqrt{7 \times 8} \approx 7.483$	$\sqrt{6 \times 7} \approx 6.481$	$\sqrt{76 \times 77} \approx 76.498$	
Modified quotas rounded down if less than geometric mean, up otherwise →	**Rounded Quota**	9	8	7	76	100

▸ Now try Exercises 35–36.

Think About It Consider the results in Examples 7 and 8. Which apportionment method would State C prefer? Which would State D prefer? Explain.

Is There a Best Method for Apportionment?

Hamilton's method proved to be unfair because of the three paradoxes (Alabama, population, and new states). Jefferson's method proved to be unfair because it can violate the quota criterion. How do Webster's method and the Hill-Huntington method compare?

Mathematicians have studied the methods by running simulations of many possible apportionments. Results show that, like Jefferson's method, Webster's method and the Hill-Huntington method can also violate the quota criterion. However, the violations occur much less frequently with Webster's or the Hill-Huntington method than with Jefferson's method, making them arguably fairer.

For example, if Jefferson's method had remained in use, nearly every apportionment since 1850 would have violated the quota criterion. In contrast, no violations have occurred with Webster's method or the Hill-Huntington method. Simulations show that, by chance alone, Webster's and Hill-Huntington are expected to violate the quota criterion only once in several hundred apportionments. Interestingly, Webster's appears slightly less prone to violations of the quota criterion than Hill-Huntington.

Is there any apportionment system that is unquestionably fair in all circumstances? Such an apportionment system would have to satisfy the quota criterion in all cases, while also being immune to the three paradoxes that affect Hamilton's method. Unfortunately, a theorem proved by mathematicians M. L. Balinsky and H. P. Young states that such a system is impossible.

In essence, the **Balinsky and Young theorem** for apportionment is analogous to Arrow's impossibility theorem for voting (see Unit 12B). It tells us that, in the end, we cannot choose between apportionment procedures on the basis of fairness alone. As a result, apportionment will always involve political decisions, and we can expect it to be a subject of continued debate.

Quick Quiz 12C

Choose the best answer to each of the following questions. Explain your reasoning with one or more complete sentences.

1. Which of the following is *not* mandated by the U.S. Constitution?
   a. Every state has two senators.
   b. Every state has at least one representative in the House.
   c. The House has a total of 435 members.

2. What do we mean by *apportionment* for the House of Representatives?
   a. deciding the total number of House seats
   b. choosing how to divide the total number of House seats among the states
   c. setting the boundaries for each House seat's district within a state

3. Suppose that the 2020 census finds a U.S. population of 335 million. If the House were reapportioned based on this population, the *standard divisor* would be
   a. 335 million ÷ 435. b. 435 ÷ 335 million.
   c. still the 2010 value of 710,000.

4. Suppose that in 2030 the census shows that the average House member represents 1 million people. The standard quota for a state with a population of 1.5 million would be
   a. 1. b. 1.5. c. 2.

5. Consider a school district with 50 schools, 1000 teachers, and 25,000 students. If the goal is to apportion teachers among the schools so that the teacher-student ratio is the same in every school, the standard divisor should be
   a. 20. b. 25. c. 50.

6. Consider the school district described in Question 5. If a school had 220 students, its standard quota of teachers would be
   a. 11. b. 8.8. c. 4.4.

7. Consider three schools with the following standard quotas for teachers: Douglass Elementary, 7.2 teachers; King Elementary, 7.3 teachers; Parks Elementary, 7.4 teachers. Suppose that 22 teachers are available for the three schools together. Under Hamilton's method, which school gets eight teachers?
   a. Douglass b. King c. Parks

8. In this unit, four different apportionment methods were discussed. What is special about these four?
   a. They are the only four known methods of apportionment.
   b. They are the four fairest methods of apportionment.
   c. They are the four methods of apportionment that have actually been used to apportion seats in the U.S. House of Representatives.

9. Based on current law, what method of apportionment will be used to reapportion seats in the U.S. House of Representatives after the 2020 census?
   a. Jefferson's method b. Hamilton's method
   c. the Hill-Huntington method

10. A method of apportionment that always satisfies the quota criterion and is not susceptible to the Alabama, population, and new state paradoxes is
    a. not possible. b. possible, but not yet in use.
    c. Webster's method.

Exercises 12C

REVIEW QUESTIONS

1. What is *apportionment*? What does it mean for the U.S. House of Representatives?
2. Briefly describe the nature of the apportionment problem. How does this problem arise from the requirements of the Constitution?
3. Explain how *Hamilton's method* apportions House seats. Briefly describe the history of Hamilton's method.
4. What is the *Alabama paradox*? What other paradoxes arise with Hamilton's method? Why do these paradoxes make the method seem unfair in the cases where they arise?
5. Explain how *Jefferson's method* apportions House seats. Briefly describe the history of Jefferson's method.
6. What is the *quota criterion*? Why are violations of this criterion considered unfair?
7. Briefly describe how *Webster's method* and the *Hill-Huntington method* differ from Jefferson's method.
8. Explain why Webster's method and the Hill-Huntington method are considered fairer than Jefferson's method, but still not fair in all cases. What is the significance of the *Balinsky and Young theorem*?

DOES IT MAKE SENSE?

Decide whether each of the following statements makes sense (or is clearly true) or does not make sense (or is clearly false). Explain your reasoning.

9. Mike is the president of a large company with 12 divisions. He plans to use an apportionment method to decide how many staff support persons to assign to each division.
10. Charlene is the head judge in a figure skating competition. She plans to use an apportionment method to decide how the judges' points should be allocated to the skaters.
11. The Hill-Huntington method is superior to other apportionment methods because it requires more advanced mathematics.
12. Ten new teachers were hired in the Meadowlark School District this year. Horizon School has the same proportion of the district's students as it did last year, but it now has three fewer teachers. The apportionment method used to allocate teachers was not fair.

BASIC SKILLS & CONCEPTS

13. **Representation in Congress.** If the population of the United States increased to 350 million (the projected population in 2030) and the number of representatives remained 435, how many Americans, on average, would each representative serve? With this population, if the number of representatives were set at the constitutional limit of one representative for every 30,000 people, how many representatives would there be in Congress?
14. **Representation in Congress.** If the population of the United States increased to 400 million (the projected population in 2050) and the number of representatives increased to 500, how many Americans, on average, would each representative serve? With this population, if the number of representatives were set at the constitutional limit of one representative for every 30,000 people, how many representatives would there be in Congress?

15–18: State Representation. The following table shows the 2010 populations of four states and their numbers of seats in the House of Representatives. Find the standard quota for each state, and compare it to the actual number of seats for the state. Then explain whether the state is relatively under- or overrepresented in the House. Assume a total U.S. population of 309 million in 2010 and 435 House seats.

State	Population	House seats
Connecticut	3,574,097	5
Georgia	9,687,653	14
Florida	18,801,310	27
Ohio	11,353,140	16

15. Connecticut
16. Georgia
17. Florida
18. Ohio
19. **Standard Quotas in Business.** A large company has four divisions with 250, 320, 380, and 400 employees, respectively. A total of 35 computer technicians must be allocated to the four divisions according to their size. Find the standard quota for each division.
20. **Standard Quotas in Education.** Capital University has five colleges with 560, 1230, 1490, 1760, and 2340 students, respectively. A total of 18 academic advisors must be allocated to the five colleges according to their size. Find the standard quota for each college.

21–22. Practice with Hamilton's Method. Fill out the following tables with the calculations required for an apportionment by Hamilton's method. In each case, assume that 100 representatives need to be apportioned.

21.

State	A	B	C	D	Total
Population	914	1186	2192	708	5000
Standard quota					100
Minimum quota					
Fractional remainder					—
Final apportionment					100

22.

State	A	B	C	D	Total
Population	1342	2408	4772	1478	10,000
Standard quota					100
Minimum quota					
Fractional remainder					—
Final apportionment					100

23. **Hamilton's Method.** Use Hamilton's method to determine the allocation of computer technicians in Exercise 19.

24. **Hamilton's Method.** Use Hamilton's method to determine the allocation of academic advisors in Exercise 20.

25–28: Alabama Paradox. Assume that 100 representatives must be apportioned to the following sets of states with the given populations. Determine the number of representatives for each state using Hamilton's method. Then assume that the number of representatives is increased to 101. Determine the new number of representatives for each state using Hamilton's method. State whether the change in total number of representatives results in the Alabama paradox.

25. A: 950; B: 670; C: 246

26. A: 2540; B: 1140; C: 6330

27. A: 770; B: 155; C: 70; D: 673

28. A: 562; B: 88; C: 108; D: 242

29–32: Jefferson's Method. Apply Jefferson's method to the following sets of states with the given populations. Assume that 100 seats are to be apportioned. In each case, state whether the quota criterion is satisfied.

29. A: 98; B: 689; C: 212 (modified divisor 9.83)

30. A: 1280; B: 631; C: 2320 (modified divisor 42.00)

31. A: 69; B: 680; C: 155; D: 75 (modified divisor 9.60)

32. A: 1220; B: 5030; C: 2460; D: 690 (modified divisor 92.00)

33. **Webster's Method.** Use Webster's method to determine the allocation of computer technicians in Exercise 19.

34. **Webster's Method.** Use Webster's method to determine the allocation of academic advisors in Exercise 20.

35. **Hill-Huntington Method.** Use the Hill-Huntington method to determine the allocation of computer technicians in Exercise 19.

36. **Hill-Huntington Method.** Use the Hill-Huntington method to determine the allocation of academic advisors in Exercise 20.

FURTHER APPLICATIONS

37–38: New States Paradox. The populations of three states are given. Determine how 100 seats should be apportioned among these states using Hamilton's method. Then suppose that a new state with a population of 500 is added to the system, along with 5 new seats. Determine the apportionment for the four states (assuming the populations of the first three states remain the same). Does the addition of the new state decrease the representation of any of the original states?

37. A: 1140; B: 6320; C: 250

38. A: 5310; B: 1330; C: 3308

39–42: Comparing Methods. Assume 100 seats are to be apportioned among the following sets of states with the given populations.

a. Find the apportionment by Hamilton's method.
b. Find the apportionment by Jefferson's method.
c. Find the apportionment by Webster's method.
d. Find the apportionment by the Hill-Huntington method.
e. Compare the results of the various methods. Which methods give the same results? Do any of the results violate the quota criterion? Overall, which method do you think is best for this apportionment? Why?

39. A: 535; B: 344; C: 120

40. A: 144; B: 443; C: 389

41. A: 836; B: 2703; C: 2626; D: 3835

42. A: 1234; B: 3498; C: 2267; D: 5558

43–46: Non-House Apportionments. The following exercises describe apportionment problems for situations besides the House of Representatives. In each case, do the following:

a. Find the apportionment by Hamilton's method.
b. Find the apportionment by Jefferson's method.
c. Find the apportionment by Webster's method.
d. Find the apportionment by the Hill-Huntington method.
e. Compare the results of the various methods.

43. A high school is creating a student committee to allocate use of classrooms after hours. The committee is to consist of 10 student members, chosen from three interest groups: social groups, which have 48 members; political groups, which have 97 members; and athletic groups, which have 245 members.

44. A city plans to purchase 16 new emergency vehicles. They are to be apportioned among 485 members of the police department, 213 members of the fire department, and 306 members of the paramedic squad.

45. A chain of hardware stores is reapportioning its 25 managers among stores in four locations according to the stores' monthly gross sales. The sales at the four stores are as follows: Boulder, \$2.5 million; Denver, \$7.6 million; Broomfield, \$3.9 million; Ft. Collins, \$5.5 million.

46. The city parks commission plans to build nine new parks, to be apportioned among three neighborhoods by population. The neighborhoods are Greenwood, population 4300; Willowbrook, population 3040; and Cherryville, population 2950.

IN YOUR WORLD

47. **Census Apportionment.** How did your home state fare in the most recent House apportionment? How have changes in population affected your state's standard quota since that apportionment? Overall, do you think the apportionment was fair to your state?

48. **Local Apportionment.** Find a recent news story concerning an apportionment problem that applies locally or at the state level. Discuss the apportionment procedure used. Does the apportionment seem fair?

49. **Your State's Representatives.** How many seats does your state have in the House of Representatives? How does this number compare to your state's standard quota? Based on your findings, is your state overrepresented or underrepresented in the House? Explain.

UNIT 12D Dividing the Political Pie

Apportionment, which we discussed in Unit 12C, is the process used to determine the number of representatives that each state gets when the House of Representatives is reapportioned every 10 years. Politically, however, apportionment is only the beginning of the practical issues associated with electing representatives.

In states with more than one seat in the House, the seats are divided among congressional *districts* within the state, with the goal of having an equal number of people in each district within a state. For example, a state with six seats has six districts, with the people in each district electing one representative. Every 10 years, after a census, district boundaries must be redrawn to reflect shifts in population. The process of redrawing district boundaries, called **redistricting**, is one of the most contentious political issues of our times. Like apportionment, it is a process that relies heavily on mathematics.

The Contemporary Problem

If the opposite of pro is con, what's the opposite of progress?
—Joke circulating after public approval of Congress reached record lows

Redistricting has always been politically charged, but most political observers agree that it has become even more contentious in recent years. The two major parties, Republican and Democratic, now routinely engage in what amounts to open political warfare in each state as the district boundaries are redrawn at the beginning of each decade.

Of course, partisan squabbling is nothing new, and some people argue that it can even be good for the country as a whole. So the question to ask is this: Is there any evidence that redistricting is creating problems for our democracy? The answer seems to be yes.

Table 12.19 compares election statistics for President and for the House of Representatives in recent presidential election years. Notice that popular votes for the presidency tend to be quite close; in fact, no President has won the popular vote by more than 10 percentage points since 1984. In contrast, and as we discussed in the chapter-opening question (p. 654), the average (mean) margin of victory in House races has been enormous, and relatively few House seats have changed party even in years when the presidency has changed party (2000, 2008, 2016). These facts indicate that while the nation as a whole is fairly closely split in political opinion, House districts have been drawn in such a way as to group like-minded voters together. As a result, many political observers believe that representatives tend to have more highly partisan views than the population as a whole.

TABLE 12.19 Comparison of Election Statistics for the Presidency and the House

Year	President's Popular-Vote Margin of Victory (percentage points)	House of Representatives Mean Margin of Victory (percentage points)	Percentage of House Seats Changing Party
2000	−0.5*	39.9	3.9
2004	2.4	40.5	3.0
2008	7.3	37.1	7.1
2012	3.9	31.9	10.3
2016	−2.1%*	36.7	4.4

*The margin is negative because the electoral vote winner lost the popular vote.

Sources: U.S. Clerk's Office; Fairvote.com.

Think About It Find the margin of victory in the most recent election for the representative from *your* congressional district. Does your district appear to be competitive, or does it lean strongly to one side? What influence do you think this fact has on whether your representative is more moderate or more partisan?

BY THE WAY

Different states use different methods for redistricting. Seven states have only one representative, so the district map is automatically the entire state; 37 states use maps chosen primarily by the state legislature, though sometimes with input from advisory commissions and sometimes subject to approval by the governor. As of 2017, only six states use commissions considered to be fully independent or bipartisan: Arizona, California, Hawaii, Idaho, New Jersey, and Washington.

EXAMPLE 1 Partisan Advantage in North Carolina

North Carolina has 13 seats in the U.S. House of Representatives. Its redistricting process is controlled by the state legislature. Following the 2000 census, the legislature was controlled by Democrats, who drew the district maps for elections from 2002–2010 (though these maps were modified following court challenges). Following the 2010 census, the legislature was controlled by Republicans, who drew the redistricting maps for elections beginning in 2012 (Figure 12.4). Table 12.20 shows the overall North Carolina results for the elections immediately before and after the maps based on 2010 census took effect. Analyze the data to decide how partisan advantage in the state legislature affected results of elections for seats in the House of Representatives.

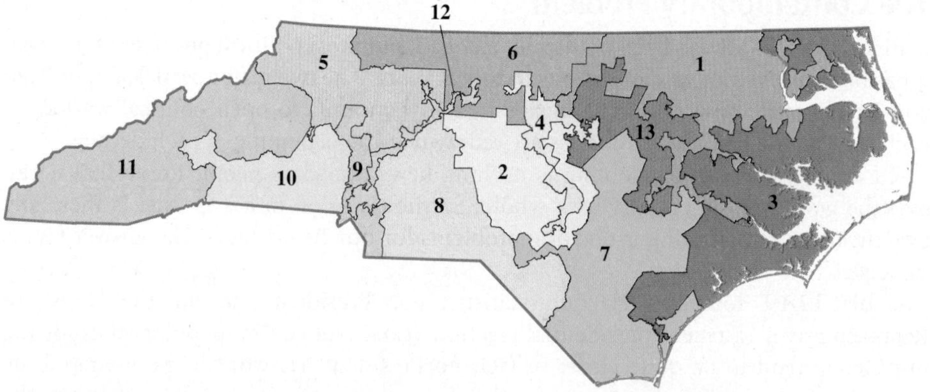

FIGURE 12.4 The district boundaries for North Carolina's 13 seats in the U.S. House of Representatives, as drawn after the 2010 census. Notice the strange shapes of some of the districts, chosen to maximize partisan advantage. The four seats won by Democrats in 2012 were Districts 1, 4, 7, and 12. (Ordinarily, these boundaries would have remained in place until after the 2020 census, but some of them were ruled unconstitutional by the U.S. Supreme Court in 2017, thereby requiring them to be redrawn.)

TABLE 12.20 North Carolina Election Results

	Statewide Vote for Democratic House Candidates (percentage)	Statewide Vote for Republican House Candidates (percentage)	Democratic House Members Elected	Republican House Members Elected
2010 (Democratic maps)	45%	54%	7	6
2012 (Republican maps)	51%	49%	4	9

Solution There are many ways to analyze the data, but here a few things to notice:

- In 2010, Democrats received only 45% of the statewide vote but won $7/13 \approx 54\%$ of the North Carolina seats in the House of Representatives. In other words, Democrats were *over*represented in the House delegation by 9 percentage points compared to their proportion of the vote.
- In 2012, Democrats received 51% of the statewide vote but their congressional representation decreased to $4/13 \approx 31\%$ of North Carolina's House seats. They were therefore *under*represented in the 2012 delegation by about 20 percentage points compared to their proportion of the vote.

- Overall, Democrats' share of the congressional vote was 6 percentage points higher in 2012 than in 2010, yet their representation in North Carolina's House delegation decreased by about 23 percentage points.

Mathematically, the lesson is clear: Simply by changing the map of district boundaries, it is possible to cause a great change in a state's House delegation. As this case shows, it is even possible for a party to gain seats while losing voters.

▶ Now try Exercises 13–17.

Gerrymandering

The practice of drawing district boundaries for political advantage is so common that it has its own name—**gerrymandering**. The term originated in 1812, when Massachusetts Governor Elbridge Gerry created a district that critics ridiculed as having the shape of a salamander. A famous political cartoon attached the label "Gerry-mander" (Figure 12.5), which has been used ever since.

FIGURE 12.5 This 1812 political cartoon (by Elkanah Tinsdale) depicts the salamander-like shape of a Massachusetts district created under Governor Elbridge Gerry, with dragon characteristics presumably indicating what the cartoonist thought of it.

> **Definition**
>
> **Gerrymandering** is the drawing of district boundaries so as to serve the political interests of the politicians in charge of the drawing process.

To understand how gerrymandering can affect election outcomes, consider a state that has equal numbers of Democratic and Republican voters, but the Democrats have a majority in the state legislature and are in charge of drawing district boundaries. In that case, the Democrats can draw boundaries that concentrate the Republican voters in one or a few districts. That way, the Republicans will win huge victories in those few districts, leaving the Democrats in the majority everywhere else. The following example illustrates the idea.

EXAMPLE 2 The Principle of Gerrymandering

Consider a state that has six House seats and elects its six representatives in six districts (District 1 through District 6) with 1 million people each. Assume that half the voters are Democrats and half are Republicans, and they always vote on party lines.

a. If district boundaries were drawn randomly, what would be the most likely distribution of the six House seats?

b. Suppose that, somehow, the legislature could draw the boundaries of District 1 so that *all* of its 1 million people were Democrats. If the remaining population were randomly distributed among the other five districts, what would be the most likely distribution of the six House seats?

Solution

a. The most likely outcome with random districts is that the House representation will reflect the voter representation: The six House seats will be divided evenly, with three Democrats and three Republicans.

b. Since District 1 consists of 1 million Democrats (and no Republicans), this district will elect a Democrat. The remaining five districts then have an overall population of 3 million Republicans and 2 million Democrats. If the boundaries of these districts are drawn so that their populations reflect this general distribution, Republicans will outnumber Democrats by a 3-to-2 margin in every one of these five districts—so the Republicans will win all five of these districts. The end result: Despite having equal numbers of Democratic and Republican voters, the state ends up with five Republican representatives and only one Democratic representative.

▶ Now try Exercises 18–23.

BY THE WAY

In the 2016 vote for the House of Representatives, Republican candidates received 49.13% of the total vote nationally to 48.03% for Democrats, a margin of just over 1 percentage point. However, Republicans won the House by a margin of 241 to 194, or 55.4% to 44.6%, which is a margin of more than 10 percentage points.

Sample Cases of Boundary Drawing

The case described in part (b) of Example 2 is extreme and could not happen in reality because there are rules that must be obeyed in drawing district boundaries. Indeed, the Supreme Court has ruled that districts may *not* be drawn for such overtly partisan purposes. In practice, however, the Court has placed few restrictions on gerrymandering, in part because it's very difficult to prove that boundaries are drawn for partisan purposes rather than other purposes that might be more reasonable. For example, suburbs tend to have a greater concentration of Republicans than do urban areas. So if a district wraps around an urban area in a complex pattern, does that mean it was drawn in order to concentrate Republican voters or to give people with common interests—those living in suburbs—a stronger voice?

Given the difficulty of answering such questions, the courts have generally allowed very convoluted district boundaries to stand, as long as they do not violate any specific laws and meet the following two criteria:

1. All districts within a particular state must have very nearly equal populations. This requirement is based on the principle of "one person, one vote," which was enunciated in a series of Supreme Court decisions starting in the 1960s.
2. Each district must be *contiguous*, meaning that every part of the district must be connected to every other part. You cannot, for example, have a district that consists of two separate pieces in two different parts of a state.

The best way to see how gerrymandering can be accomplished within these constraints is to consider sample districts that are much simpler than real congressional districts. Figure 12.6 shows a "state" consisting of only 64 voters, half Democratic (blue houses) and half Republican (red houses), in which there are to be eight districts with eight voters each. Notice that there is some geographical concentration of voters by party, just as is usually the case in the real world. (This is the same "state" used in the chapter-opening activity on p. 656.)

Think About It Before you read on, draw a simple set of district boundaries on Figure 12.6, making sure each district has eight voters. With your boundaries, how many districts would be won by Democrats? How many would be won by Republicans? How many would be ties?

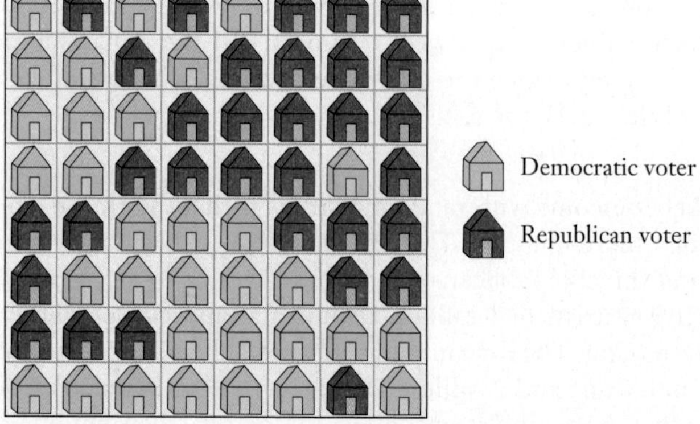

FIGURE 12.6 This simple state has just 64 voters (each represented by his or her own house), half of whom are Democrats and the other half Republicans. The state must be divided into eight districts with eight voters each. *Source:* Adapted from "On Partisan Fairness," by Brian Gaines, in *Redistricting Illinois*, from the Institute of Government and Public Affairs, University of Illinois.

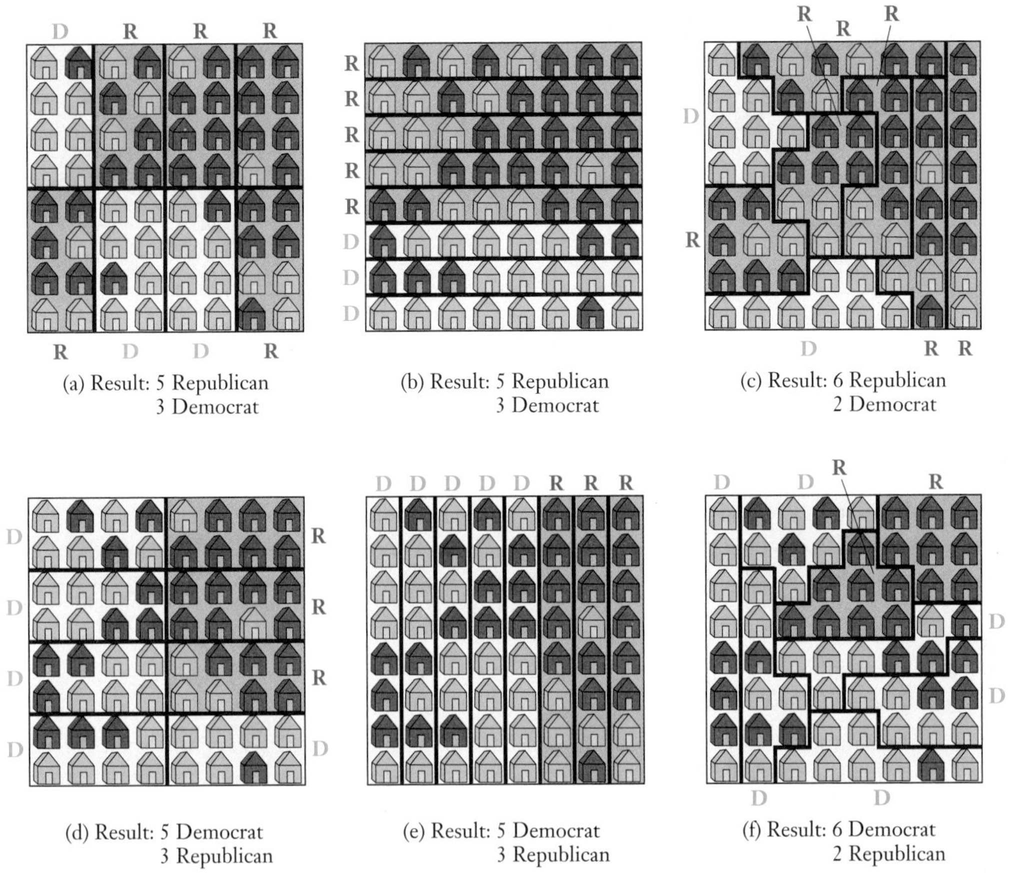

(a) Result: 5 Republican
3 Democrat

(b) Result: 5 Republican
3 Democrat

(c) Result: 6 Republican
2 Democrat

(d) Result: 5 Democrat
3 Republican

(e) Result: 5 Democrat
3 Republican

(f) Result: 6 Democrat
2 Republican

FIGURE 12.7 Six different sets of possible district boundaries for the state shown in Figure 12.6. *Source:* Adapted from "On Partisan Fairness," by Brian Gaines, in *Redistricting Illinois*, from the Institute of Government and Public Affairs, University of Illinois.

If election results reflected the overall party affiliation of this state's population, the state would end up with equal numbers of Democrats and Republicans being elected. But the actual results can be quite different, depending on how the boundaries are drawn. Figure 12.7 shows six different sets of possible district boundaries that result in one or the other party winning a majority, despite the fact that the population is equally divided in its political preferences. Indeed, two of the cases (Figure 12.7c and Figure 12.7f) allow one party to get a 6-to-2 advantage in representation.

EXAMPLE 3 Strangely Shaped Districts

The districts in Figure 12.7 all have boundaries that make relatively simple and compact polygons, but gerrymandering can be much more clever if you allow a greater range of shapes. Figure 12.8 shows a state with 16 voters, half Democrats and half Republicans. Suppose the state has four districts with four voters each. Can you find a way to draw boundaries so that one of the districts is *all* Republican, leaving Democrats with either a majority or a tie in the other three districts? Remember that a district must be contiguous, but assume no other restrictions.

Solution Figure 12.9 shows one of many possible ways to draw the district boundaries and accomplish the goal of giving the Democrats an edge in representation. Notice that District 1 is now entirely Republican, while Democrats outnumber Republicans in Districts 2 and 4. District 3 is a tie. Depending on how the tie is broken, Democrats will

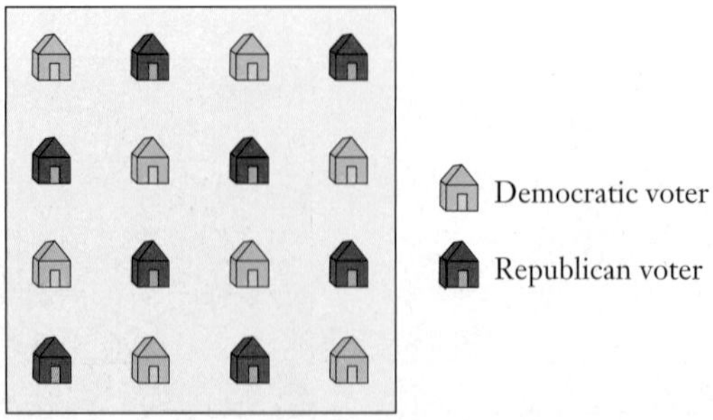

FIGURE 12.8 This state has 16 voters, half of whom are Democrats and the other half Republicans. The state must be divided into four districts with four voters each. District boundaries may go wherever you like, as long as each district is contiguous.

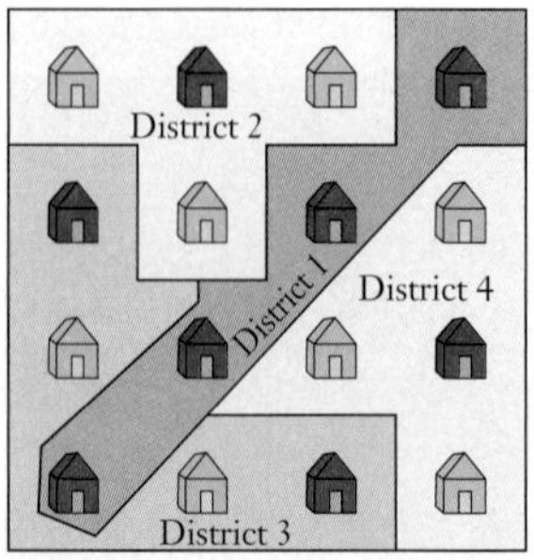

FIGURE 12.9 District boundaries for the state in Figure 12.8 that give Democrats a majority in two districts and a tie in a third, while Republicans have only one majority district.

at worst win two of the four districts, and they might win three of the four. In contrast, Republicans will win two at best, and they may win only one. This is an example of gerrymandering, because Democrats have an edge in the overall outcome even though the preferences of voters are evenly split between the two parties.

▶ Now try Exercises 24–32.

Ideas for Reform

Voters no longer choose members of the House, the people who draw the lines do.

—Samuel Issacharoff, Columbia Law School

The sample cases we have considered have small numbers of voters, but the principles apply to actual redistricting. When members of a partisan group are in charge of producing new district maps, they collect data that include precinct-by-precinct results from past elections, party registrations of individual voters, and detailed geographical maps of population based on census data (which include information about income level, ethnic group, and more). These data allow them to estimate the numbers of voters likely to vote Democratic and Republican on very small geographical scales—often down to individual blocks. With sophisticated computer programs, the group uses these data to draw tens, hundreds, or thousands of different sets of district maps and then chooses the one expected to maximize its party's representation in Congress.

As we have seen, this type of computer-aided redistricting is remarkably effective, and it helps explain why so few congressional districts are competitive today. That alone might be cause for concern, but many political observers believe this fact also lies at the root of the increasingly wide partisan divide in the United States Congress today. In an election for a competitive seat, the parties have an incentive to produce candidates who will appeal to the large political middle. But in an election for a seat that, say, a Democrat is almost guaranteed to win, the real contest occurs in the primary election rather than the general election. Because primaries tend to draw smaller numbers of voters, many of whom have more clearly partisan interests, the result is that noncompetitive districts tend to elect representatives with more extreme partisan views, rather than representatives who appeal to the broad political middle.

BY THE WAY

Starting in 2010, California enacted two reforms in hopes of reducing partisanship: (1) A bipartisan commission now handles redistricting; (2) primaries are nonpartisan, with the top two candidates advancing to the general election regardless of party affiliation. Political analysts are watching closely to see whether these reforms are having their intended effects.

Is there any way to reform the system to prevent such overtly partisan redistricting? Two approaches have been suggested. First, redistricting can be turned over to an independent, nonpartisan panel, such as a panel of judges. Many other countries use such panels to handle redistricting—including Great Britain, Australia, and Canada—and a handful of states do the same. However, some people argue that no one is truly nonpartisan and that this system still leaves too much room for political manipulation.

The second reform approach is to come up with a mathematical algorithm that would draw the boundaries independent of any human input, thereby guaranteeing that no partisan advantage could be taken. Unfortunately, this turns out to be a very difficult mathematical problem. It isn't possible to use a particular set of simple district shapes (such as triangles or rectangles), because they won't always fit together properly (see the discussion of tiling in Unit 11B) or have equal numbers of voters. Once we move away from simple geometrical shapes, the mathematics becomes far more complex. For example, one requirement that has been suggested for district shapes is that they should be "compact" rather than spread out in strange ways, but there is no known mathematical definition of "compactness" that matches our intuition.

The bottom line is that redistricting is an intensely mathematical issue that, for the present at least, offers mathematically sophisticated politicians the opportunity to gain huge partisan advantages. Perhaps in the future mathematics will also offer a way out of this dilemma, when someone comes up with an acceptable mathematical algorithm that can take redistricting out of partisan hands.

Quick Quiz 12D

Choose the best answer to each of the following questions. Explain your reasoning with one or more complete sentences.

1. What do we mean by *redistricting* for the U.S. House of Representatives?

   a. deciding the total number of House seats

   b. choosing how to divide the total number of House seats among the 50 states

   c. setting the boundaries for each House seat's district within a state

2. Which of the following best summarizes why redistricting is an important political issue?

   a. Proposals for redistricting must always be voted on by voters in the general election.

   b. Redistricting can be done in a way that gives one party more House seats than would be expected based on its overall representation among a state's voters.

   c. Redistricting can prevent some people from being able to vote at all.

3. If we compare results in presidential elections to those in elections for the House of Representatives, we find that

   a. Presidential and House elections are on average decided by very similar margins of victory.

   b. House elections are on average decided by much larger margins of victory.

   c. House elections are on average decided by much smaller margins of victory.

4. In 2010, Republicans in North Carolina received 54% of the statewide vote and won 6 of North Carolina's 13 House seats. This implies that

   a. the district boundaries fairly represented the preferences of North Carolina voters.

   b. the district boundaries were drawn to favor Republicans.

   c. the district boundaries were drawn to favor Democrats.

5. In 2012, Democrats in North Carolina received 51% of the statewide vote and won 4 of North Carolina's 13 House seats. This implies that

   a. the district boundaries fairly represented the preferences of North Carolina voters.

   b. the district boundaries were drawn to favor Republicans.

   c. the district boundaries were drawn to favor Democrats.

6. What is *gerrymandering?*

   a. another name for redistricting

   b. drawing district boundaries that result in unusual shapes

   c. the drawing of district boundaries for partisan advantage

7. Suppose you are in charge of redistricting for a state that has equal numbers of Republican and Democratic voters and 25 House seats. If you wish to draw boundaries that will maximize the number of Democrats winning House seats, you should

   a. draw boundaries that concentrate very large majorities of Republicans in a few districts.

   b. draw boundaries that concentrate very large majorities of Democrats in a few districts.

   c. draw boundaries that make all districts have equal numbers of Democrats and Republicans.

8. Which of the following is *not* a general requirement for district boundaries drawn within a particular state?

   a. All districts should have nearly equal populations.

   b. Districts should have simple geometrical shapes, such as rectangles or pentagons.

   c. Every point within each district should be connected to every other point in that district.

9. Consider a state with equal numbers of Democratic and Republican voters and 30 House seats. Which outcome is *not* possible, no matter how you draw the districts, assuming everyone votes along party lines?

   a. 30 Democrats and 0 Republicans

   b. 15 Democrats and 15 Republicans

   c. 12 Democrats and 18 Republicans

10. A likely effect of districts in which one party has such a large majority that its candidate is almost guaranteed to be elected is

    a. the election of candidates who represent the large political middle.

    b. the election of candidates who represent more extreme partisan views.

    c. the election of third-party candidates.

Exercises 12D

REVIEW QUESTIONS

1. What is *redistricting*, and when must it be done?
2. How has the competitiveness of elections for the House changed over the past few decades? How might this be bad for our democracy?
3. What is *gerrymandering*? Where does this term come from?
4. Briefly describe how the drawing of boundaries can be used to give one party an advantage, even when voters are split evenly between the two parties.
5. What requirements must be met in drawing district boundaries?
6. Briefly describe two ideas for reforming the redistricting process and some potential pros and cons of each.

DOES IT MAKE SENSE?

Decide whether each of the following statements makes sense (or is clearly true) or does not make sense (or is clearly false). Explain your reasoning.

7. In the last election in my home state, 48% of the people voted for a Democrat, and Democrats won 65% of our state's seats in the House of Representatives.
8. In my home state, 46% of the voters are registered Republicans, but I live in a district in which 72% of the voters are Republican.
9. Polls show that half the voters in our state plan to vote for Democrats in the House elections. Therefore, I can definitely expect half of our representatives to be Democrats.
10. My state has a population of 8 million and 8 House seats, and I live in a district with a population of 200,000 people.
11. My district occupies a rural area in the northwest corner of our state and another rural area in the southeast corner, but none of the middle sections of the state.
12. If we could stop the practice of gerrymandering, we'd have fewer members of the House of Representatives holding extreme views.

BASIC SKILLS & CONCEPTS

13–17: Redistricting and House Elections. The 2010 census led to the loss of two House seats in New York and Ohio, the loss of one House seat in Pennsylvania, the gain of two House seats in Florida, and the gain of four House seats in Texas. Consider the following approximate vote counts for all House districts in these states in 2010 (before redistricting) and 2012 (after redistricting). All figures are in thousands of votes. Votes for other parties are neglected.

a. Find the percentages of votes cast for Republican and Democratic House candidates in 2010 and 2012.

b. Find the percentages of House seats that were won by Republican and Democratic candidates in 2010 and 2012.

c. Explain whether the distributions of votes reflect the distributions of House seats.

d. Discuss whether redistricting based on the 2010 census affected the distributions of votes and representatives.

13. Ohio

	Votes for Republican candidates (thousands)	Votes for Democratic candidates (thousands)	Republican seats	Democratic seats
2010	2053	1611	13	5
2012	2620	2412	12	4

14. Florida

	Votes for Republican candidates (thousands)	Votes for Democratic candidates (thousands)	Republican seats	Democratic seats
2010	2234	1529	19	6
2012	4157	3679	17	10

15. Texas

	Votes for Republican candidates (thousands)	Votes for Democratic candidates (thousands)	Republican seats	Democratic seats
2010	3058	1450	23	9
2012	4429	2950	24	12

16. New York

	Votes for Republican candidates (thousands)	Votes for Democratic candidates (thousands)	Republican seats	Democratic seats
2010	1854	2601	8	21
2012	2252	4127	6	21

17. Pennsylvania

	Votes for Republican candidates (thousands)	Votes for Democratic candidates (thousands)	Republican seats	Democratic seats
2010	2034	1882	12	7
2012	2710	2794	13	5

18–23: Average and Extreme Districts. Consider the following demographic data for hypothetical states. In each case, answer the following questions. Assume that everyone votes along party lines.

a. If districts were drawn randomly, what would be the most likely distribution of House seats?

b. If the districts could be drawn without restriction (unlimited gerrymandering), what would be the *maximum* and *minimum* number of Republican representatives who could be sent to Congress? Explain how each result could be achieved.

18. The state has 10 representatives and a population of 6 million; party affiliations are 50% Republican and 50% Democrat.

19. The state has 16 representatives and a population of 10 million; party affiliations are 50% Republican and 50% Democrat.

20. The state has 10 representatives and a population of 10 million; party affiliations are 50% Democrat and 50% Republican.

21. The state has 12 representatives and a population of 8 million; party affiliations are 50% Republican and 50% Democrat.

22. The state has 10 representatives and a population of 5 million; party affiliations are 70% Republican and 30% Democrat.

23. The state has 15 representatives and a population of 7.5 million; party affiliations are 20% Democrat and 80% Republican.

24–25: Drawing Districts Set I. Refer to Figure 12.10, which shows the geographical distribution of voters in a state with 64 voters and 8 districts. The voters are half Democrat and half Republican. *Assume that district boundaries must follow the grid lines shown in the figure and that each district must be contiguous.*

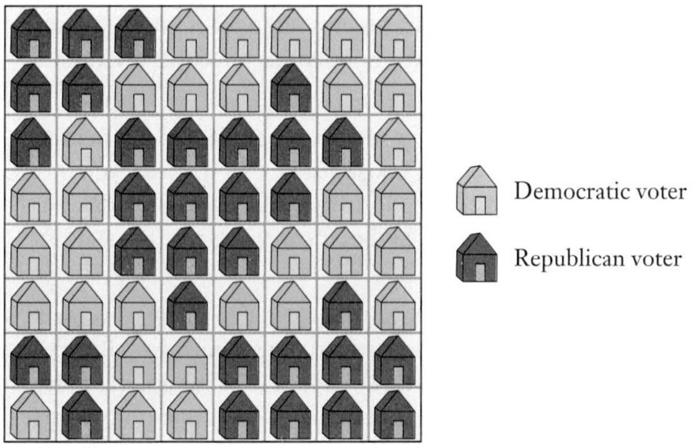

FIGURE 12.10

24. Draw boundaries expected to result in the election of five Republican and three Democratic representatives.

25. Draw boundaries expected to result in the election of four Republican and four Democratic representatives.

26–27: Drawing Districts Set II. Refer to Figure 12.11, which shows the geographical distribution of voters in a state with 64 voters and 8 districts. The voters are half Democrat and half Republican. *Assume that district boundaries must follow the grid lines shown in the figure and that each district must be contiguous.*

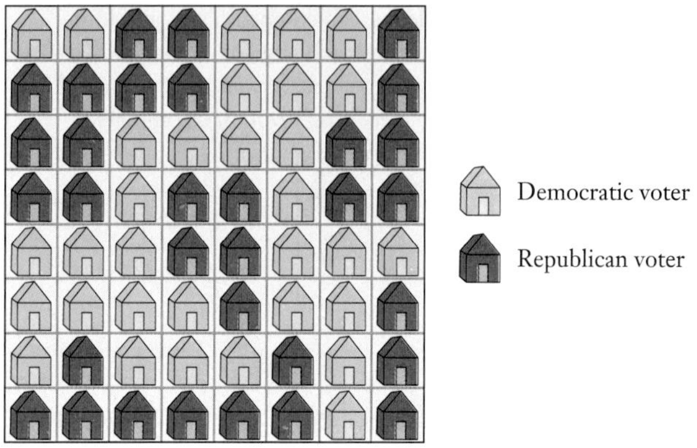

FIGURE 12.11

26. Draw boundaries expected to result in the election of four Republican and four Democratic representatives.

27. Draw boundaries expected to result in the election of five Republican and three Democratic representatives.

28–29: Drawing Districts Set III. Refer to Figure 12.8 (used in Example 3 of this unit). For these exercises, you may draw districts with any shape you wish, as long as each district is contiguous. Draw district boundaries that accomplish the following distributions of representatives; if you think that the distribution is not possible, explain why not.

28. Two Republicans and two Democrats

29. One Republican and three Democrats

30–32: Drawing Districts Set IV. Refer to Figure 12.12, which shows the distribution of voters in a state with 15 Democratic voters and 10 Republican voters and five voters per district. For these exercises, you may draw districts with any shape you wish, as long as the district is contiguous.

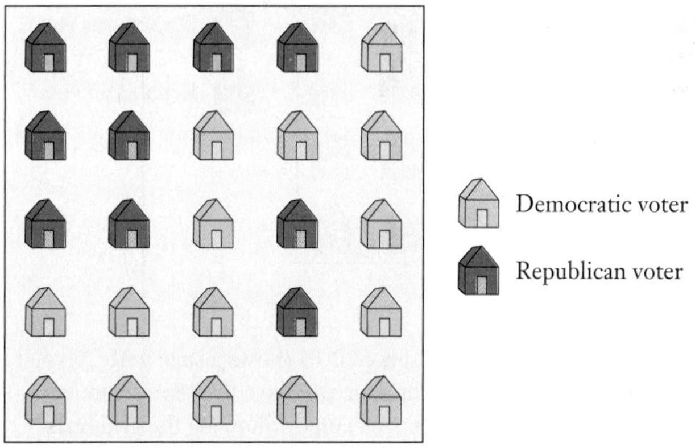

FIGURE 12.12

30. Draw district boundaries so that four Democrats and one Republican are elected.

31. Draw district boundaries so that three Democrats and two Republicans are elected.

32. Is it possible to draw district boundaries so that two Democrats and three Republicans are elected? Explain.

FURTHER APPLICATIONS

33. **District Possibilities on a Grid.** For the voter distribution in Figure 12.13 (36 voters, 4 districts, 50% Republican), determine whether straight-line boundaries can be drawn so that one

Republican, two Republicans, three Republicans, and four Republicans are elected. Show each case, or explain why it is not possible. *Assume that district boundaries must follow the grid lines shown in the figure and that each district must be contiguous.*

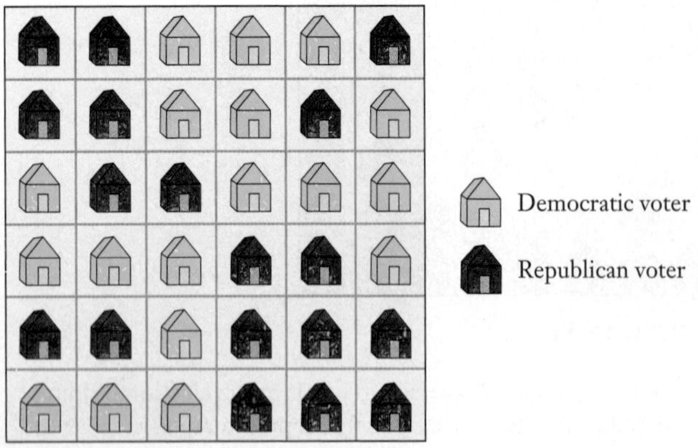

FIGURE 12.13

34. **District Boundaries with Any Shape.** For the voter distribution in Figure 12.14 (36 voters, 4 districts, 50% Republican), determine whether boundaries can be drawn so that one Republican, two Republicans, three Republicans, and four Republicans are elected. You may draw districts with any shape you wish, as long as the district is contiguous. Show each case, or explain why it is not possible.

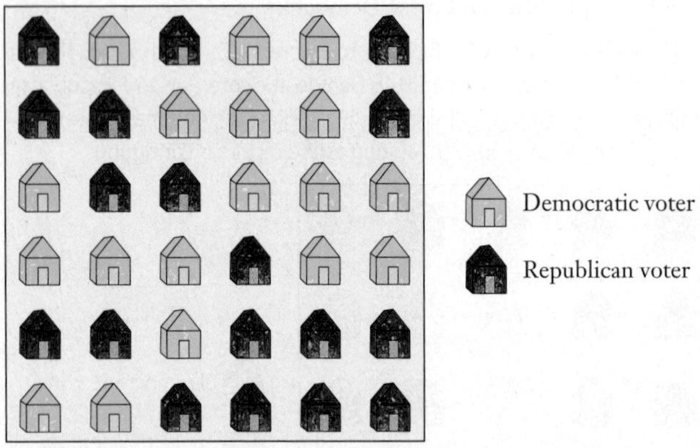

FIGURE 12.14

35. **Assign Party Affiliations.** Figure 12.15 shows a state with 25 voter locations. Assume that the state will have five representatives, and each district must be a rectangle following the grid lines shown in the figure. Given that there are 15 Republicans and 10 Democrats, write party affiliations (D or R) into the 25 locations in such a way that no Democrats can be elected, regardless of how the rectangular district boundaries are drawn.

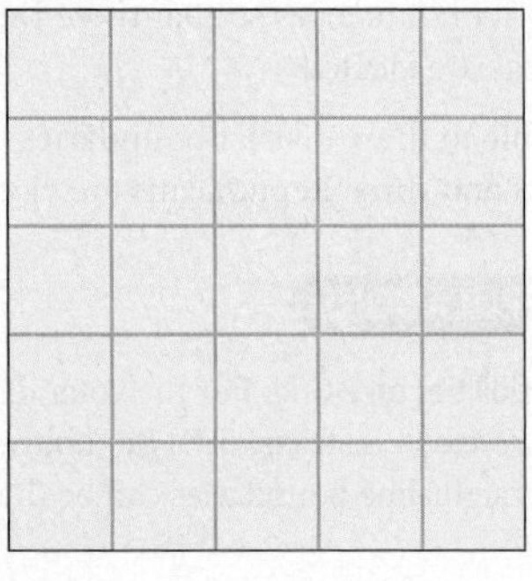

FIGURE 12.15

36. **Draw Your Own State.** Create your own hypothetical state with 36 voters (half Democrat and half Republican) and 6 districts. You may arrange the voters geographically any way you wish. Show at least four possible sets of districts, and explain the expected results in each case.

37. **Unanimous Delegation.** Suppose a state with two or more districts has exactly half Republican and half Democratic voters, but with an odd number of voters in each district (for example, 20 total voters and 4 districts, so there are 5 voters in each district). Is it possible for one party to win every House seat? Why or why not?

38. **Project: Lost Seat.** Consider the state we worked with in Figures 12.6 and 12.7. Suppose that, after the next census, the state in the figure lost one voter and as a result lost its 8th congressional district (so it ends up with 63 voters and 7 districts). Experiment with taking away one voter in the figure and drawing boundaries for 7 districts. Discuss the possible effects of the loss of a district.

IN YOUR WORLD

39. **Redistricting Controversy.** Find a news report about a redistricting controversy in a particular state. Summarize the issues involved in light of the discussion in this unit.

40. **Voter and Population Data.** The Office of the Clerk of the U.S. House of Representatives keeps final results of all elections for congressional representatives from all states (1920–present). Find a state that interests you and analyze the House results for a pre-2010 election and a post-2010 election. (Or, if you are using this text after 2022, do this for results pre- and post-2020 redistricting.) Determine the percentages of voters and representatives from each party. Discuss whether redistricting had an effect on the number of representatives.

41. **District Maps.** Maps of the current congressional districts can be found on the Web. Choose a state that interests you, print a map of the districts, collect voter and population data for the state, and discuss whether the districts for that state are fairly drawn.

42. **Redistricting Procedures.** Choose a state that interests you and find out about its current redistricting procedures. For example, who draws district boundaries? What criteria must be met in creating the boundaries? Write a one- to two-page summary of your findings.

43. **Reform Efforts.** Investigate the current status of efforts to reform redistricting procedures. Have any significant changes occurred in the past few years? Are any major changes being considered for the near future? Write a short essay summarizing your findings.

44. **Mathematical Algorithms for Reform.** Search for proposals for using mathematical algorithms in redistricting. Investigate one such proposal, and write a short report explaining its goals, limitations, and prospects for coming into actual use.

45. ***Gill v. Whitford.*** Find the Supreme Court ruling in the 2017 gerrymandering case *Gill v. Whitford*. What impact has the ruling had? Does the ruling affect your opinion about gerrymandering and what might be done to address it?

Chapter 12 Summary

UNIT	KEY TERMS	KEY IDEAS AND SKILLS
12A	majority rule (p. 657) popular vote (p. 658) electoral vote (p. 658) filibuster (p. 660) super majority (p. 660) veto (p. 660) plurality (p. 661) preference schedule (p. 662)	**Understand** U.S. presidential elections. (p. 658) **Understand** variations on majority rule. (p. 660) **Apply** five methods for deciding an election with three or more candidates: (p. 664) plurality method single (top-two) runoff sequential runoff point system (Borda count) pairwise comparisons (Condorcet method)
12B	fairness criteria (p. 675) Arrow's impossibility theorem (p. 677) approval voting (p. 678)	**Apply** the four fairness criteria. (p. 675) **Know** that no voting method can satisfy all fairness criteria in all cases (Arrow's impossibility theorem). (p. 677) **Understand** approval voting as an alternative voting system. (p. 678) **Understand** variations in voting power when not all voters have equal weight. (p. 679)
12C	apportionment (p. 686) standard divisor (p. 687) standard quota (p. 687) minimum quota (p. 689) Alabama paradox (p. 691) population paradox (p. 692) new states paradox (p. 692) modified divisor (p. 693) modified quota (p. 693) quota criterion (p. 694)	**Know** the history of apportionment mathematics. (p. 686) **Apply** four apportionment methods: (p. 689) Hamilton's method Jefferson's method Webster's method Hill-Huntington method **Know** the potential flaws of each apportionment method. (p. 697) **Understand** the significance of the Balinsky and Young theorem. (p. 698)
12D	districts (congressional) (p. 701) redistricting (p. 701) gerrymandering (p. 703)	**Understand** why redistricting is both political and mathematical. (p. 701) **Know** how redistricting can affect the competitiveness of House elections. (p. 701) **Understand** the principles behind gerrymandering. (p. 703) **Apply** the legal requirements for redistricting, including equal-size populations and contiguous districts. (p. 704) **Be aware** of consequences of current redistricting practices and opportunities for reform. (p. 706)

CREDITS

About the Authors

p. ix: (Jeffrey Bennett) Jeffrey Bennett; (William Briggs) Buzz Burrell.

Prologue

p. P-1: LuciaP/Fotolia.

Chapter-Opening Quotes

pp. 3, 69, 121, 191, 293, 371, 425, 489, 533, 577, 621, 655: (pencil) Ruslan Ivantsov/Shutterstock; (speech bubble) Pearson Education, Inc.

In Your World Boxes

pp. P-6, 11, 60, 84, 90, 99, 173, 220, 237, 251, 255, 314, 414, 465, 517, 526, 554, 570, 587, 627: (binoculars) Windu/Fotolia.

Chapter 1

p. 2: Sadovnikova Olga/Shutterstock. **p. 3:** (students in library) VM/E+/Getty Images. **p. 4:** Jeffrey Bennett. **p. 6:** Amanda Hall/Robertharding/Alamy Stock Photo. **p. 11:** (Facts check mark) Alexmillos/Shutterstock. **p. 13:** Jeffrey Bennett. **p. 14:** Paris Pierce/Alamy Stock Photo. **p. 15:** Panos Karas/Shutterstock. **p. 16:** Mdgn/Shutterstock. **p. 17:** Roman Sakhno/Shutterstock. **p. 25:** PJF Military Collection/Alamy Stock Photo. **p. 34:** Francois Etienne du Plessis/Shutterstock. **p. 41:** Thuy Mai/NASA. **p. 52:** Mark Skalny/Shutterstock. **p. 60:** Echo/Juice Images/Getty Images. **p. 65:** Tabbycat/Fotolia.

Chapter 2

p. 68: Pearson Education, Inc. **p. 69:** Monkey Business Images/Shutterstock. **p. 70:** NASA. **p. 83:** Gena96/Shutterstock. **p. 87:** NASA. **p. 90:** Sebastian Duda/Shutterstock. **p. 91:** (scales) Mch67/Fotolia; (Jefferson) WDC Photos/Alamy Stock Photo. **p. 93:** Novastock/Stock Connection Blue/Alamy Stock Photo. **p. 94:** Archive Image/Alamy Stock Photo. **p. 99:** Kenjii/Fotolia. **p. 101:** RosaIrene-Betancourt 6/Alamy Stock Photo. **p. 103:** Mike Lyvers/Moment/Getty Images. **p. 106:** Majeczka/Shutterstock. **p. 114:** Zulufoto/Shutterstock. **p. 118:** Picturepartners/Shutterstock.

Chapter 3

p. 120: Piai/Fotolia. **p. 121:** Holbox/Shutterstock. **p. 122:** Clarinda Maclow/Courtesy of Jeffrey Bennett. **p. 138:** Hero Images/Getty Images. **p. 141:** C. Ehlinger/Abacausa.com/Newsom. **p. 145:** Yuriy Kulik/Shutterstock. **p. 147:** Jeffrey Bennett. **p. 148:** ESA/M. Livio/Hubble 20th Anniversary Team/NASA. **p. 159:** Egd/Shutterstock. **p. 161:** Hubble Heritage Team/NASA. **p. 162:** Fotos593/Shutterstock. **p. 165:** Pinyo Bonmark/Shutterstock. **p. 167:** Jose Gil/Shutterstock. **p. 172:** AP Images. **p. 180:** Antoniodiaz/Shutterstock. **p. 181:** Collection Christophel/Alamy Stock Photo.

Chapter 4

p. 190: Pearson Education, Inc. **p. 191:** Creativa Images/Shutterstock. **p. 192:** WavebreakMediaMicro/Fotolia. **p. 196:** Iofoto/Fotolia. **p. 206:** PjrTravel/Alamy Stock Photo. **p. 219:** Ruth Jenkinson/Dorling Kindersley, Ltd. **p. 235:** Andersphoto/Shutterstock. **p. 237:** Jupiterimages/Stockbyte/Getty Images. **p. 239:** Don Farrell/Photodisc/Getty Images. **p. 251:** Aleksandra Gigowska/Shutterstock. **p. 253:** Gino Santa Maria/Fotolia. **p. 255:** (couple applying for loan) Goodluz/Shutterstock; (mortgage ad) Robert F. Bukaty/AP Images. **p. 258:** Hongqi Zhang/123RF. **p. 266:** Tiffany Bryant/Shutterstock. **p. 268:** Zimmytws/Fotolia.

Chapter 5

p. 292: Pearson Education, Inc. **p. 293:** Robert Daly/Caiaimage/Getty Images. **p. 294:** Boogich/iStock/Getty Images. **p. 295:**Spotmatikphoto/Fotolia. **p. 299:** Jose Angel Astor/123RF. **p. 306:** Andrey Popov/Shutterstock. **p. 310:** Danuta Mayer/Dorling Kindersley Ltd. **p. 313:** Doug Martin/Science Source. **p. 314:** B Brown/Shutterstock. **p. 317:** Deyan Georgiev/Shutterstock. **p. 323:** Lou Linwei/Alamy Stock Photo. **p. 361:** Trinity Mirror/Mirrorpix/Alamy Stock Photo. **p. 362:** Dave & Les Jacobs/Blend Images/Getty Images. **p. 363:** AP Images.

Chapter 6

p. 370: Pearson Education, Inc. **p. 371:** Rido/Shutterstock. **p. 376:** Timothy A. Clary/Getty Images. **p. 393:** Jonathan Wiess/Shutterstock. **p. 395:** Eduardo Mariano Rivero/Alamy Stock Photo. **p. 397:** Beth Anderson/Pearson Education, Inc. **p. 402:** Asia Images Group Pte Ltd/Alamy Stock Photo. **p. 406:** Anna Tamila/Shutterstock. **p. 411:** Sayyid Azim/AP Images. **p. 414:** Patrimonio Designs Ltd/Shutterstock. **p. 415:** Jeffrey Bennett. **p. 416:** Zuzule/Fotolia. **p. 420:** (baby and dog) Jeffrey Bennett; (rainbow flag) BeeZee-Photo/Shutterstock.

Chapter 7

p. 424: Pearson Education, Inc. **p. 425:** Mishoo/123RF. **p. 426:** Jaschin/Fotolia. **p. 428:** Story Time From Space/Courtesy of Jeffrey Bennett. **p. 430:** Purestock/Getty Images. **p. 432:** Caitlin Mirra/Shutterstock. **p. 437:** Xaoc/Dreamstime. **p. 442:** Fuse/Corbis/Getty Images. **p. 449:** Syantse/Dreamstime. **p. 457:** Album/Art Resource, NY. **p. 458:** IS692/Image Source Plus/Alamy Stock Photo. **p. 459:** Cezary Wojtkowski/Shutterstock. **p. 462:** Michael Warwick/Shutterstock. **p. 464:** Brt Photo/Alamy Stock Photo. **p. 465:** Lan Dagnall/Alamy Stock Photo. **p. 467:** Dreamnikon/Fotolia. **p. 480:** Bcubic/Dreamstime. **p. 484:** Jeffrey Bennett.

Chapter 8

p. 488: Pearson Education, Inc. **p. 489:** Frans Lemmens/Alamy Stock Photo. **p. 490:** Mark Higgins/Shutterstock. **p. 494:** Courtesy of Al Bartlett. **p. 503:** National Archives and Records Administration (NARA). **p. 509:** David Stelle/Fotolia. **p. 511:** (crowd in New Delhi) Brianindia/Alamy Stock Photo; (new citizens being sworn in) Jim West/Alamy Stock Photo. **p. 512:** NASA. **p. 517:** Kali9/E+/Getty Images. **p. 519:** Chad McDermott/Shutterstock. **p. 521:** Albund/123RF. **p. 522:** (Haiti earthquake damage) Arindam Banerjee/123RF; (Alexander Graham Bell) Photos.com/Getty Images. **p. 525:** (citrus fruits) Torsten Märtke/Fotolia; (chemical factory) Nickolay Khoroshkov/Fotolia. **p. 527:** Ariel Skelley/Bled Images/Getty Images. **p. 528:** Olga Khoroshunova/Fotolia.

Chapter 9

p. 532: (elephant) Freshidea/Fotolia; (red car) Zentilia/Fotolia; (receipt) Doomu/Fotolia. **p. 533:** Alphaspirit/Shutterstock. **p. 538:** Rudi1976/Fotolia. **p. 541:** MyWorld/Fotolia. **p. 554:** J. Marshall/Tribaleye Images/Alamy Stock Photo. **p. 555:** Aflo Co. Ltd./Nippon News/Alamy Stock Photo. **p. 562:** Ji Zhou/Fotolia. **p. 569:** Jonathan Blair/Corbis Documentary/Getty Images. **p. 572:** Petr Kovalenkov/Shutterstock.

Chapter 10

p. 576: Raskolnikova/Fotolia. **p. 577:** Judie Long/AGE Fotostock/Alamy Stock Photo. **p. 582:** Interfoto/Fine Arts/Alamy Stock Photo. **p. 586:** (Leaning Tower of Pisa) David Buffington/Stockbyte/Getty Images; (Pentagon) Hisham Ibrahim/Photodisc/Getty Images. **p. 587:** North Wind Picture Archives/Alamy Stock Photo. **p. 588:** Guy Sagi/123RF. **p. 592:** (Earth) Reto Stockli with the help of Alan Nelson, under the leadership of Fritz Hasler/NASA. **p. 594:** Geogphotos/Alamy Stock Photo. **p. 597:** Sebastien Decoret/123RF. **p. 600:** Southern Stock/Photodisc/Getty Images. **p. 604:** Africa Studio/Shutterstock. **p. 608:** Anne Burns. **p. 610:** (coral) Yang Yu/Fotolia; (mountains) William Briggs. **p. 615:** (Mandelbrot set) Ian Evans/Alamy Stock Photo; (Barnsley's fern) Claudio Divizia/Shutterstock.

Chapter 11

p. 620: Photogolfer/Shutterstock. **p. 621:** Hill Street Studios/Blend Images/Getty Images. **p. 623:** James Steidl/123RF. **p. 624:** Jason Stitt/Fotolia. **p. 627:** Mtsaride/123RF. **p. 629:** Friday/Fotolia. **p. 632:** Scala/Art Resource, NY. **p. 633:** (wood engraving by Dürer) Universal Images Group/SuperStock; (perspective sketch by de Vries) Pearson Education, Inc; (engraving by Hogarth) Historical Picture Archive/Corbis/Getty Images. **p. 634:** (drawing by da Vinci) Oronoz/Album/SuperStock; (Islamic tiling) John R. Jones/Corbis Documentary/Getty Images. **p. 636:** (engraving by Doré) Mestral/Prisma/Album/SuperStock; (painting by Vasarely) Tate, London/Art Resource, NY. **p. 637:** Anne Burns. **p. 639:** Paul J. Steinhardt/Princeton University/Jeffrey Bennett. **p. 645:** Michele Burgess/Corbis. **p. 646:** (painting by da Vinci) Scala/Art Resource, NY; (painting by Seurat) The Metropolitan Museum of Art/Art Resource, NY. **p. 647:** Kaz Chiba/Stockbyte/Getty Images. **p. 649:** Kathathep/Shutterstock.

Chapter 12

p. 654: Hafakot/123RF. **p. 655:** Blend Images/Alamy Stock Photo. **p. 659:** Jeremy Woodhouse/Photodisc/Getty Images. **p. 660:** Bettmann/Getty Images. **p. 661:** MPVHistory/Alamy Stock Photo. **p. 663:** Marc Serota/Reuters/Alamy Stock Photo. **p. 675:** Hill Street Studios/Blend Images/Getty Images. **p. 686:** (Constitution) Onur Ersin/Shutterstock; (first Congress) MPI/Archive Photos/Getty Images. **p. 687:** Lawrence Jackson/The White House Photo Office. **p. 689:** Ed Rooney/Alamy Stock Photo. **p. 693:** Hisham Ibrahim/Photodisc/Getty Images. **p. 703:** Bettmann/Getty Images.

Answers to Quick Quizzes and Odd-Numbered Exercises

CHAPTER 1

Unit 1A Quick Quiz

1. a 2. c 3. b 4. a 5. b 6. b 7. b 8. c 9. b 10. a

Unit 1A Exercises

5. Does not make sense
7. Makes sense
9. Does not make sense
11. a. *Premise:* Apple's iPhones outsell all other smart phones. *Conclusion:* They must be the best smart phone on the market.
    b. The fact that many people buy iPhones does not necessarily mean they are the best smart phones.
13. a. *Premise:* Decades of searching have not revealed life on other planets. *Conclusion:* Life in the universe must be confined to Earth.
    b. Failure to find life on other planets does not imply that life does not exist elsewhere in the universe.
15. a. *Premise:* He refused to testify by invoking his Fifth Amendment rights. *Conclusion:* He must be guilty.
    b. The conclusion is stated as if it were the only possible conclusion.
17. a. *Premise:* Senator Smith has accepted contributions from companies that sell genetically modified crop seeds. *Conclusion:* Senator Smith's bill on agricultural policy is a sham.
    b. A claim about Senator Smith's personal behavior is used to criticize his bill.
19. a. *Premise:* Good grades are needed to get into college, and a college diploma is necessary for a successful career. *Conclusion:* Attendance should count in high school grades.
    b. The premise (which is often true) directs attention away from the conclusion.
21. False
23. False
25. *Premise:* A nightmare followed eating oysters for dinner. *Conclusion:* Oysters cause nightmares. False cause
27. *Premise:* All the nurses in a particular hospital are women. *Conclusion:* Women are better qualified for medical jobs. Hasty generalization
29. *Premise:* My uncle never drank alcohol and lived to be 93. *Conclusion:* Avoiding alcohol leads to greater longevity. False cause
31. *Premise:* Five hundred million copies of *Don Quixote* have been sold. *Conclusion: Don Quixote* is popular. Appeal to popularity
33. *Premise:* An audit of the last charity I gave to showed that most of the money was used to pay administrators in the front office. *Conclusion:* Charities cannot be trusted. Hasty generalization
35. *Premise:* The senator is a member of the National Rifle Association. *Conclusion:* I'm sure she opposes a ban on large-capacity magazines. Straw man
37. *Premise:* Some Democrats support doubling the federal minimum wage. *Conclusion:* Democrats think that everyone should have the same income. Straw man
39. *Premise:* My little boy loves dolls and my little girl loves trucks. *Conclusion:* It's not true that little boys are more interested in mechanical toys and girls prefer maternal toys. Appeal to ignorance or hasty generalization
41. The example shows the fallacy of division because the fact that Jake is an American does not mean that he acts the same as all other Americans.
43. The example shows the fallacy of slippery slope because it assumes that the fact that troops have been sent to three countries means it's inevitable that they'll be sent to more.

Unit 1B Quick Quiz

1. c 2. a 3. c 4. c 5. c 6. a 7. b 8. c 9. b 10. a

Unit 1B Exercises

7. Does not make sense
9. Makes sense
11. Does not make sense
13. Proposition
15. Not a proposition
17. Not a proposition
19. The negation is *Asia is not in the northern hemisphere*. The original proposition is true; the negation is false.
21. The negation is *The Beatles were a German band*. The original proposition is true; the negation is false.
23. Sarah did go to dinner.
25. Taxes will not be lowered.
27. Sue wants new trees planted in the park.
29.

q	*r*	*q and r*
T	T	T
T	F	F
F	T	F
F	F	F

31. p = *dogs are animals*. q = *oak trees are plants*. Both propositions are true, so the conjunction is true.
33. p = *Venus is a planet*. q = *the Sun is a star*. Both propositions are true, so the conjunction is true.
35. p = *all birds can fly*. q = *some fish live in trees*. Both propositions are false, so the conjunction is false.
37.

q	*r*	*s*	*q and r and s*
T	T	T	T
T	T	F	F
T	F	T	F
T	F	F	F
F	T	T	F
F	T	F	F
F	F	T	F
F	F	F	F

39. Exclusive
41. Exclusive
43. Inclusive

45.

r	s	r or s
T	T	T
T	F	T
F	T	T
F	F	F

47.

p	not p	p and (not p)
T	F	F
F	T	F

49.

p	q	r	p or q or r
T	T	T	T
T	T	F	T
T	F	T	T
T	F	F	T
F	T	T	T
F	T	F	T
F	F	T	T
F	F	F	F

51. p = *elephants are animals*. q = *elephants are plants*. Proposition p is true and q is false, so the disjunction (p *or* q) is true.
53. $p = 3 \times 5 = 15$. $q = 3 + 5 = 8$. Both propositions are true, so the disjunction p *or* q is true.
55. p = *cars swim*. q = *dolphins fly*. Neither proposition is true, so the disjunction p *or* q is false.
57.

p	r	if p, then r
T	T	T
T	F	F
F	T	T
F	F	T

59. *Hypothesis:* Trout can swim. *Conclusion:* Trout are fish. Both propositions are true, and the conditional proposition (implication) is true.
61. *Hypothesis:* Paris is in France. *Conclusion:* New York is in China. The hypothesis is true, the conclusion is false, and the conditional proposition is false.
63. *Hypothesis:* Trees can walk. *Conclusion:* Birds wear wigs. The hypothesis is false, the conclusion is false, and the conditional proposition is true.
65. *Hypothesis:* Dogs can swim. *Conclusion:* Dogs are fish. The hypothesis is true, the conclusion is false, and the conditional proposition is false.
67. If it snows, then I get cold.
69. If you are breathing, then you are alive.
71. If you are pregnant, then you are a woman.
73. *Converse:* If Tara owns a car, then she owns a Cadillac. *Inverse:* If Tara does not own a Cadillac, then she does not own a car. *Contrapositive:* If Tara does not own a car, then she does not own a Cadillac. The original proposition and the contrapositive are equivalent. The converse and inverse are equivalent.
75. *Converse:* If Helen is a U.S. citizen, then she is the U.S. President. *Inverse:* If Helen is not the U.S. President, then she is not a U.S. citizen. *Contrapositive:* If Helen is not a U.S. citizen, then she is not the U.S. President. The original proposition and the contrapositive are equivalent. The converse and inverse are equivalent.
77. *Converse:* If there is gas in the tank, then the engine is running. *Inverse:* If the engine is not running, then there is no gas in the tank. *Contrapositive:* If there is no gas in the tank, then the engine is not running. The original proposition and the contrapositive are equivalent. The converse and inverse are equivalent.
79. If you don't have passion, then you don't have energy. If you don't have energy, then you have nothing.
81. If you are excellent at flipping fries at McDonald's, then everyone will want to be in your line.
83. If Sue lives in Cleveland, then she lives in Ohio. (When, in fact, Sue lives in Cincinnati.)
85. If Ramon lives in Albuquerque, then he lives in New Mexico.
87. If it is a fruit, then it is an apple.
89. (a) Believing is sufficient for achieving.
(b) Achieving is necessary for believing.
91. (a) Having six children is sufficient for being committed.
(b) Being committed is necessary for having six children.

93.

p	q	p and q	not (p and q)	(not p) or (not q)
T	T	T	F	F
T	F	F	T	T
F	T	F	T	T
F	F	F	T	T

The statements are equivalent.

95.

p	q	p and q	not (p and q)	(not p) and (not q)
T	T	T	F	F
T	F	F	T	F
F	T	F	T	F
F	F	F	T	T

The statements are not equivalent.

97.

p	q	r	p and q	(p and q) or r	p or r	p or q	(p or r) and (p or q)
T	T	T	T	T	T	T	T
T	T	F	T	T	T	T	T
T	F	T	F	T	T	T	T
T	F	F	F	F	T	T	T
F	T	T	F	T	T	T	T
F	T	F	F	F	F	T	F
F	F	T	F	T	T	F	F
F	F	F	F	F	F	F	F

The statements are not equivalent.

99. Given the conditional proposition *if p, then q*, the contrapositive is *if not q, then not p*. The converse is *if q, then p*, and the inverse of the converse is *if not q, then not p*, which is also the contrapositive. Similarly, the contrapositive is the converse of the inverse.

Unit 1C Quick Quiz

1. b 2. c 3. a 4. b 5. a 6. c 7. a 8. c 9. a 10. c

Unit 1C Exercises

7. Does not make sense
9. Does not make sense
11. Does not make sense
13. *natural numbers*
15. *rational numbers*
17. *rational numbers*
19. *real numbers*
21. *rational numbers*
23. *real numbers*
25. *rational numbers*
27. *real numbers*
29. {1, 2, 3, ..., 30, 31}
31. {Louisiana, Arkansas, Tennessee, Alabama}
33. {16, 25, 36}
35. {6, 12, 18, 24, 30}
37.

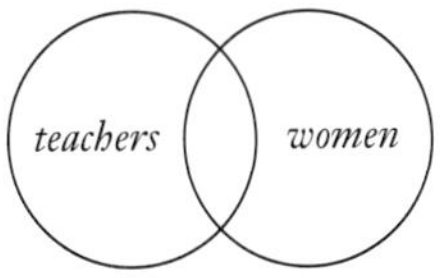

39.

41.

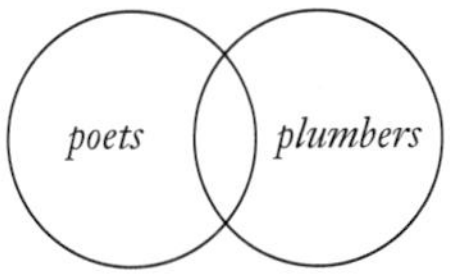

43.

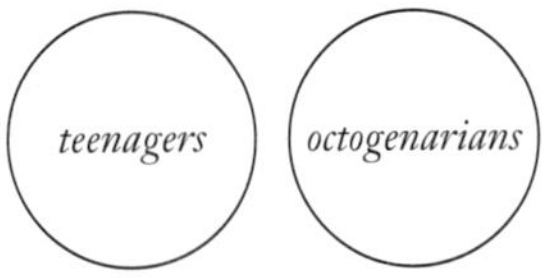

45. a.
b. *Subject: S* = kings. *Predicate: P* = men.
c.

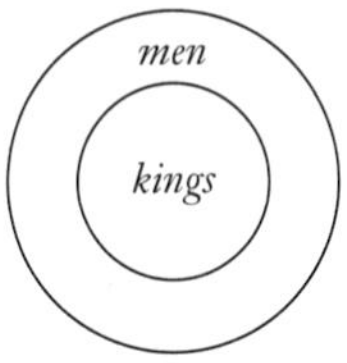

d. No

47. a.
b. *Subject: S* = surgeons. *Predicate: P* = fishermen.
c.

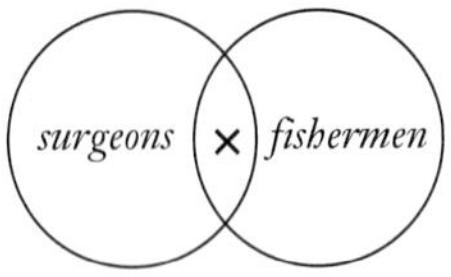

d. No

49. a. No monks are swearers.
b. *Subject: S* = monks. *Predicate: P* = swearers.
c.

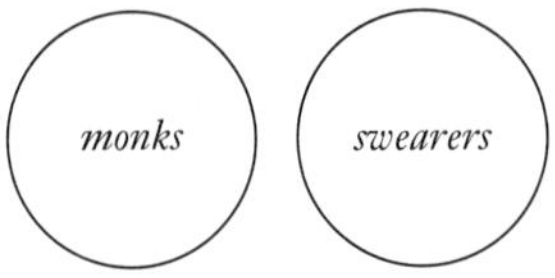

d. No

51. a.
b. *Subject: S* = sharpshooters. *Predicate: P* = men.
c.

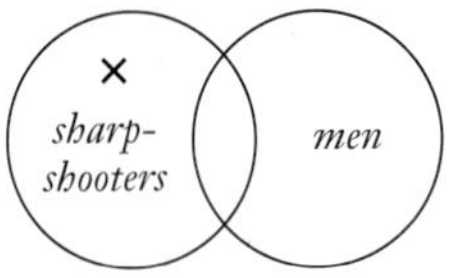

d. No

53.

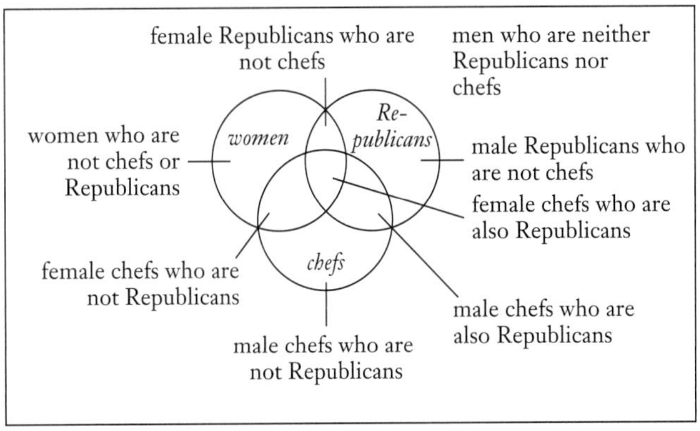

55.

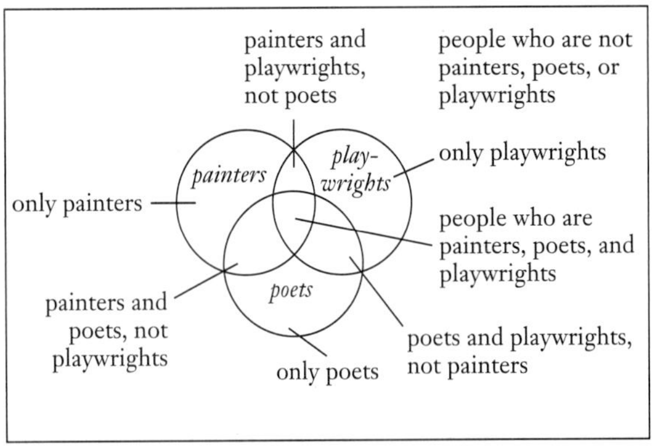

57.

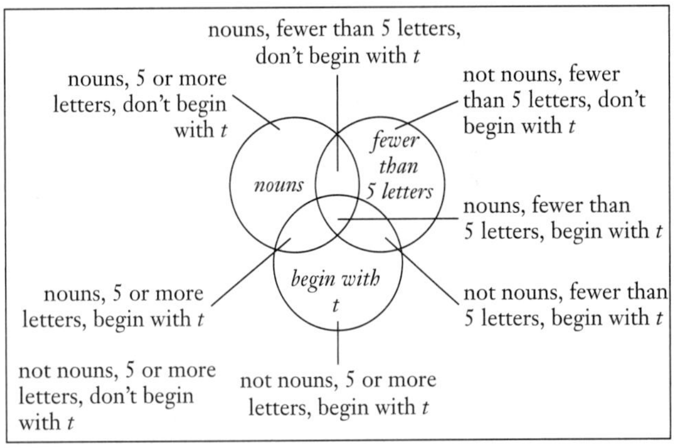

59.

	Women	Men
Right-handed	182	132
Left-handed	18	18

61. a. 31 b. 28 c. 37 d. 81
63. a. 16 b. 24 c. 8 d. 43
65. a.

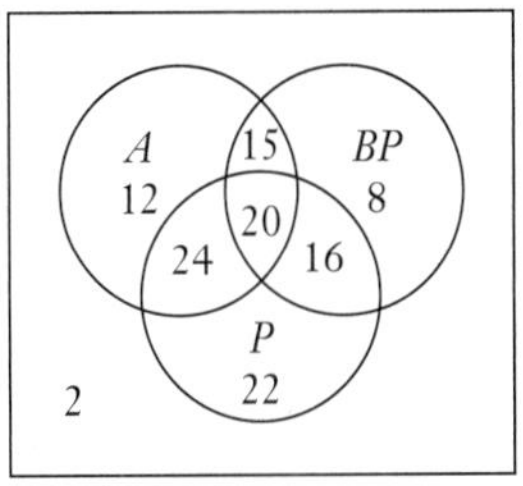

b. 105 c. 24 d. 59
e. 22 f. 55

67. a.

	Favorable	Unfavorable
Documentaries	8	4
Feature films	18	6

b. 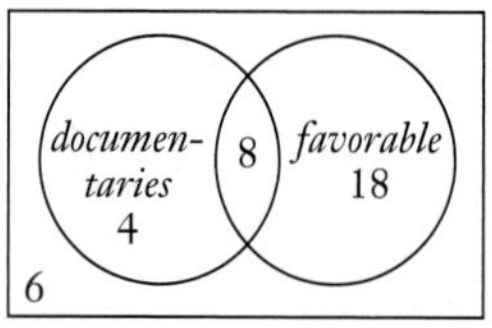

c. 4 d. 18

69. a.

	Blues	Country
Nashville	16	35
San Francisco	30	19

b. 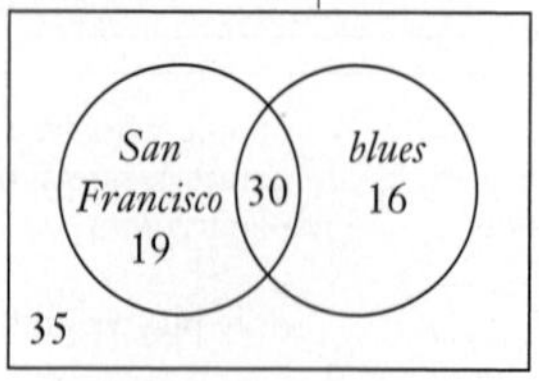

c. 16 d. 46

71. 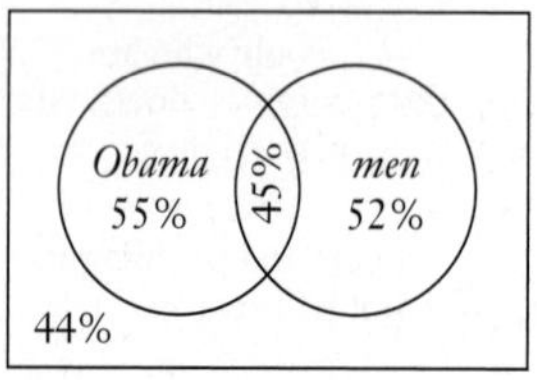

73. a.

	Subject did not lie.	Subject lied.	Total
Polygraph: lie	15	42	57
Polygraph: no lie	32	9	41
Total	47	51	98

b. 75.5% c. 24.5%

75. 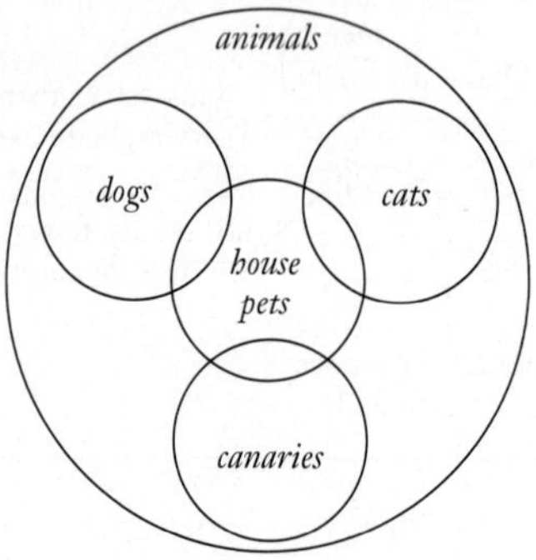

77. 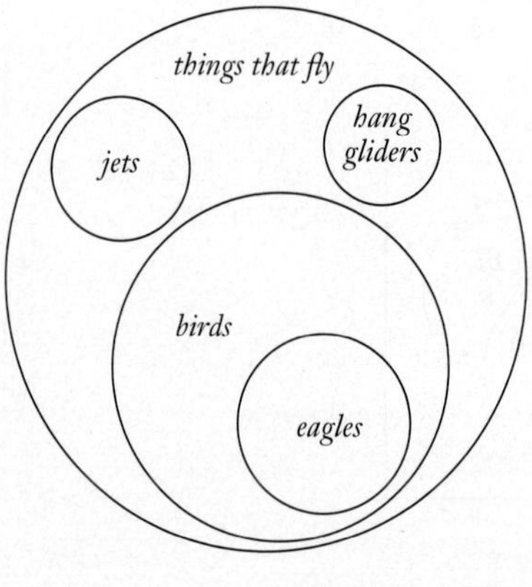

Unit 1D Quick Quiz

1. b 2. c 3. c 4. a 5. c 6. b 7. c 8. c 9. b 10. b

Unit 1D Exercises

9. Does not make sense
11. Makes sense
13. Does not make sense
15. Inductive
17. Inductive
19. Deductive
21. Deductive
23. Premise is true; argument is weak; conclusion is false.
25. Premises are true; argument *seems* moderately strong; conclusion is false.
27. Premises are true; argument is moderately strong; conclusion is true.
29. a. Valid

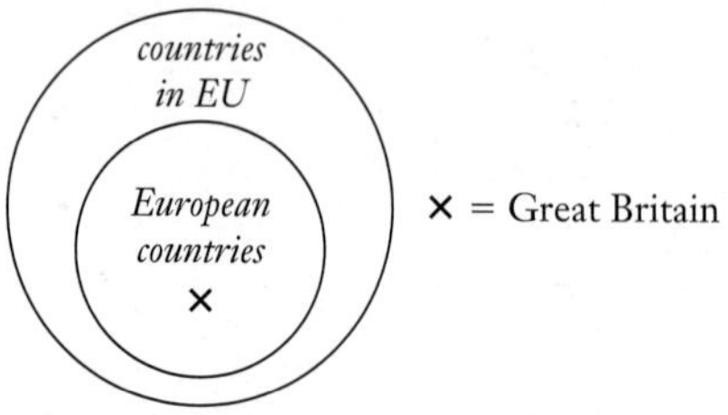

b. The first premise is false, so the argument is not sound.

31. a. Not valid

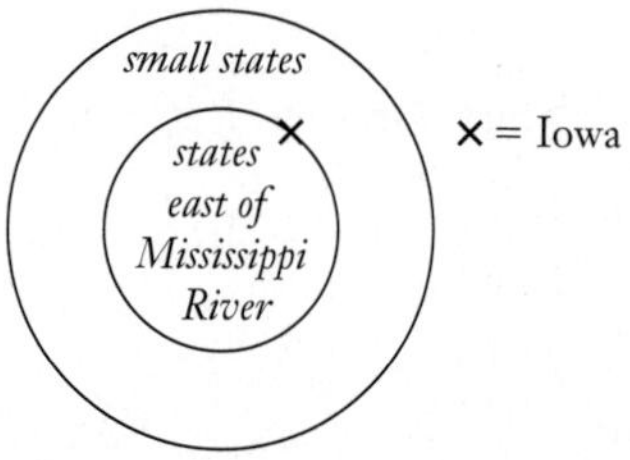

b. The premises are questionable, but the argument is not valid so it is not sound.

33. a. Valid

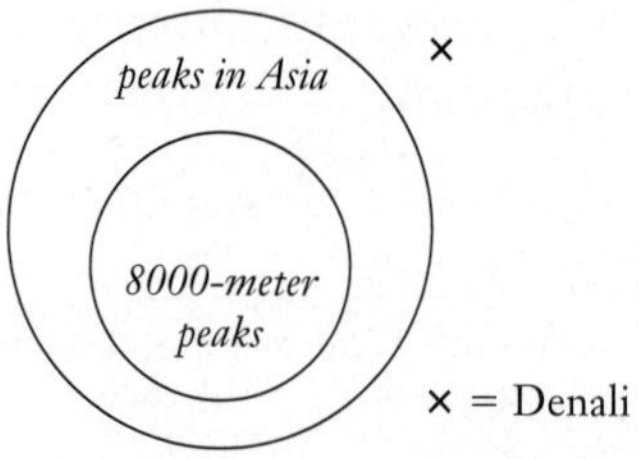

b. The premises are true, and the argument is sound.

35. a. Not valid

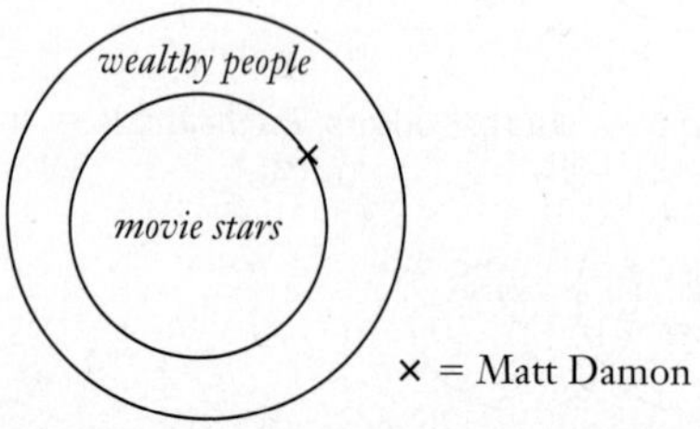

b. Even if the first premise is true, the argument is not valid and not sound.

37. a. Affirming the hypothesis; valid

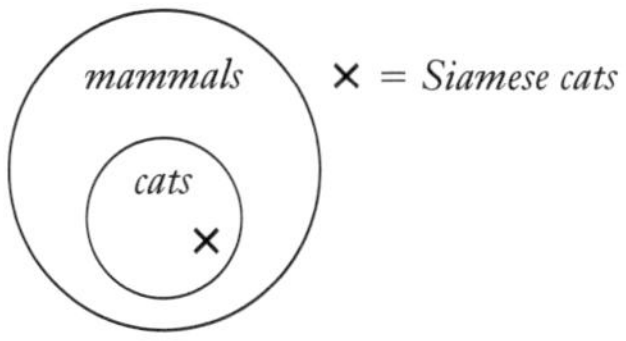

b. The premises are true, and the argument is sound.

39. a. Affirming the conclusion; not valid

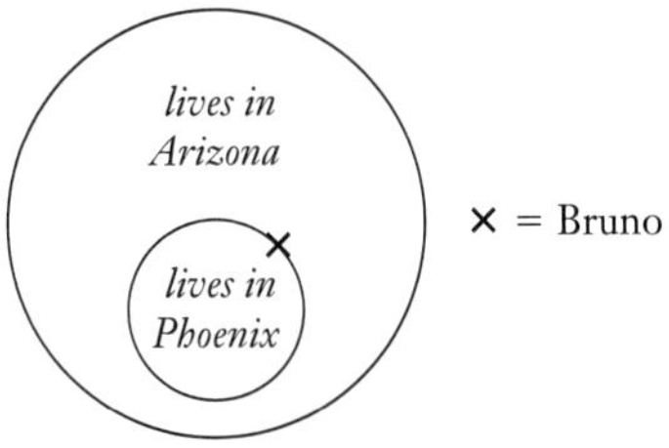

b. The argument is not sound.

41. a. Denying the hypothesis; not valid

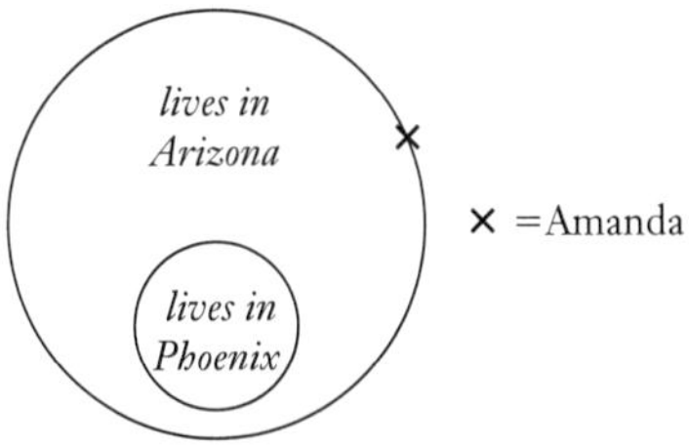

b. The argument is not sound.

43. a. Denying the conclusion; valid

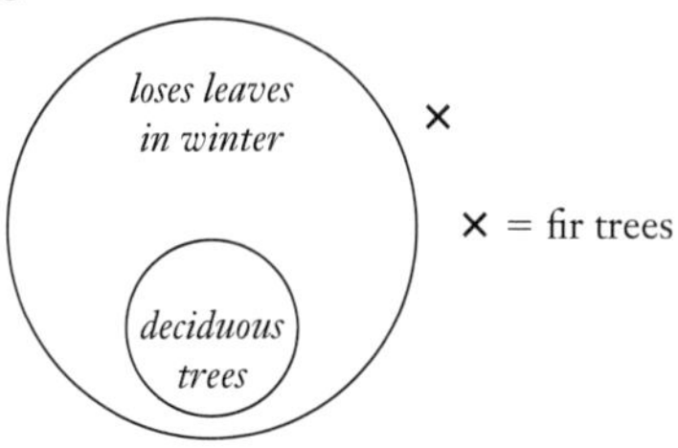

b. The premises are true, and the argument is sound.

45. p = a natural number is divisible by 18, q = a natural number is divisible by 9, r = a natural number is divisible by 3. Valid

47. Not valid

49. True

51. Not true

53. Example of valid and sound argument:
Premise: All living mammals breathe.
Premise: All monkeys are mammals.
Conclusion: All living monkeys breathe.

55. Example of valid argument that is not sound:
Premise: All mammals fly. (false)
Premise: All monkeys are mammals. (true)
Conclusion: All monkeys fly. (false)

57. Example of argument that is not valid with true premises and conclusion:
Premise: All mammals breathe. (true)
Premise: All mammals have hair. (true)
Conclusion: All hairy animals breathe. (true)

59. Example of affirming the hypothesis (valid):
Premise: If I am in Phoenix, then I am in Arizona.
Premise: I am in Phoenix.
Conclusion: I am in Arizona.

61. Example of denying the hypothesis (not valid):
Premise: If I am in Phoenix, then I am in Arizona.
Premise: I am not in Phoenix.
Conclusion: I am not in Arizona.

63. a. We will sacrifice both free enterprise and security.
b. We did not insist too adamantly on protecting our privacy.
c. No conclusion

65. a. Apes act as if they don't understand the problem at hand.
b. The usual procedures were followed.
c. No conclusion

Unit 1E Quick Quiz

1. b 2. b 3. c 4. a 5. c 6. b 7. c 8. a 9. b 10. c

Unit 1E Exercises

5. Makes sense
7. Does not make sense
9. Makes sense
11. A "yes" vote is a vote *against* gay rights, while a "no" vote is a vote *for* gay rights.
13. a. Yes, if it has not expired and has a picture.
b. No, it doesn't have a picture.
c. Apply for a voter identification card or sign a sworn statement and cast a provisional ballot.
d. The initiative does not explain how to get a voter identification card.
15. *Hidden assumptions:* Buying a house will continue to be a good investment. You will spend less out-of-pocket on your mortgage payments than you would on rent.
17. *Hidden assumptions:* The governor will keep his promise on tax cuts. You consider tax cuts to be more important than other issues.
19. The speaker may have a fundamental ideological opposition to paying taxes.
21. Option A costs \$2200 if you go and \$0 if you cancel. Option B costs \$1200 if you go and \$300 if you cancel. Option A costs \$1000 more if you go. Option B costs \$300 more if you cancel. If the likelihood of cancellation is low, then Option B is better, but if the likelihood of cancellation is high, then Option A is better.
23. A legitimate sweepstakes would not ask you to pay a processing fee in order to claim your prize. Note also that the notice never says that your vacation will be fully paid for. There are many "red flags" that should cause you to delete this message as spam.
25. Pyramidologists will use real things, such as the way ancient structures often follow astronomical alignments, to support their belief that ancient people had deep knowledge, then simply assume that much more must be hidden, even though there is no evidence of it.
27. Popes have never taken a position in American politics. Of the two major candidates in 2016, one might question why the pope would support Trump's positions on a variety of issues.
29. 4
31. Roosters don't lay eggs.
33. 22 35. 3

37. 1, 2, 3, 4
39. No
41. The current policy will cost you $5035 over nine months. The upgrade will cost $2475 over nine months.
43. For 24 flights, Airline A ($7000) costs less than Airline B ($7800).
45. a. Must file b. Must file c. Needn't file d. Must file
47. a. $45
b. Yes, you must pay within 25 days of the statement closing date.
c. Yes
49. a. Accepting a campaign contribution from someone you have never met conforms with the law.
b. Accepting a contribution from a government campaign fund would conform with the law. Accepting money from a CEO who will benefit from a bill you are sponsoring would violate the law.
55. Answers will vary.

CHAPTER 2

Unit 2A Quick Quiz

1. b 2. a 3. b 4. a 5. b 6. c 7. b 8. c 9. a 10. b

Unit 2A Exercises

7. Does not make sense
9. Makes sense
11. Does not make sense
13. a. 3/8 b. 2/5 c. 2 d. 5/6 e. 1/6 f. 5/8 g. 3/8 h. 1
15. a. 7/2 b. 3/10 c. 1/20 d. 41/10 e. 43/20 f. 7/20 g. 49/50 h. 401/100
17. a. 0.25 b. 0.375 c. 0.667 d. 0.6 e. 6.5 f. 3.833 g. 2.06 h. 1.615
19. 1797.6 lb
21. $3.13
23. $18.26/hr
25. a. Area = 240 ft^2; volume = 1920 ft^3
b. Surface area = 500 yd^2; volume = 1000 yd^3
c. Area = 180 ft^2; volume = 216 ft^3
27. mi/hr; miles per hour
29. in^3/sec; cubic inches per second
31. $; dollars
33. $28.80
35. $3000
37. 186 deaths/100,000 people
39. $2.39
41. Wrong; the student's division gives an answer with units of $pounds^2$ per dollar, but the answer should be in dollars (or cents). Instead, multiply: $(0.11 \text{ lb}) \times \frac{\$7.70}{1 \text{ lb}} = \$0.85$
43. Wrong; the student's division gives an answer with units of pounds per dollar, but the answer should be in dollars (or cents) per pound. Instead, divide dollars by pounds: $\frac{\$11}{50 \text{ lb}} = \$0.22/\text{lb}$, or 22 cents/pound, which is less than the 39 cents/pound for the smaller bag.
45. 384 in
47. 2100 s
49. 15,120 s
51. 35,040 hr
53. $1 \text{ ft}^2 = 144 \text{ in}^2; \frac{144 \text{ in}^2}{1 \text{ ft}^2}; \frac{1 \text{ ft}^2}{144 \text{ in}^2}$
55. 152,460 ft^2
57. $1 \text{ m}^3 = 10^6 \text{ cm}^3; \frac{10^6 \text{ cm}^3}{1 \text{ m}^3}; \frac{1 \text{ m}^3}{10^6 \text{ cm}^3}$
59. 13.0 yd^3
61. $100.12
63. $338.56
65. $6.01/gal
67. 79.4 mi/hr
69. 7.6 births/min; 12.5 births/1000 people
71. Approximately 2.7 billion heart beats
73. a. Car A: $30.60 per tank; car B: $51.00 per tank
b. Car A: 6.25 tanks; car B: 5 tanks
c. Car A: $191; car B: $255
75. a. 25 hr; 20 hr b. $119.53; $153.00
77. 4320 L
79. Answers will vary.

Unit 2B Quick Quiz

1. a 2. a 3. b 4. b 5. c 6. a 7. c 8. b 9. c 10. b

Unit 2B Exercises

7. Makes sense
9. Does not make sense
11. Does not make sense
13. a. 10^{11} b. 10^2 c. 10^4 d. 10^{-9} e. 10^{16} f. $10^0 = 1$ g. 10,100 h. 10^{20}
15. a. 1.25 mi b. 2200 yd
17. 373.9 lb; 0.19 ton
19. 34.5 mi/hr; 92,517 metric tonnes
21. 6-ounce bottle for $3.99
23. 15-gallon tank for $35.25
25. 10^3
27. 10^6
29. 10^6
31. 13.7 qt
33. 15.4 kg
35. 1.2 mi^2
37. 21.0 m/s
39. 0.8 lb/in^3
41. 0.007 km = 7 m
43. a. 7.2°C b. 68.0°F c. 5.0°F d. −22.0°F e. 21.1°C
45. a. −223.15°C b. −33.15°C c. 283.15 K
47. 1743 watt
49. $0.16; $42.71 saved
51. 0.6 gm/cm^3; floats
53. County A
55. New Jersey: 1213 people/mi^2; Alaska: 1.3 people/mi^2
57. a. 56 tablets b. 280 mL
59. a. 1.0 g/100 mL; would result in a lethal concentration
b. 0.25 g/100 mL; not safe to drive
61. a. 93.2%
b. 16.13 mi/hr (mile); 16.28 mi/hr (1500 m)
c. 14.25 mi/hr (mile); 14.56 mi/hr (1500 m)
d. Yes for both
63. $10.82
65. The apartment in Santa Fe is less expensive.
67. 9.1 g; 0.32 oz
69. 1.47 oz
71. 621,200 mg; 1.37 lb
73. a. Shower 2.33 ft^3; bath 22.5 ft^3
b. 96 min
c. Put the plug in the drain when you take a shower.
75. a. 13,889 cfs; about 46% of the flow rate of the Colorado River
b. Approximately 7.2 ft
77. a. 7069 fbm b. 42 c. Approximately 619 fbm
79. Case A: $9250; Case B: $9500
81. a. 3,240,000,000 joules
b. 1210 watts
c. 270 L ≈ 71.33 gal
83. a. 121 watts
b. 3.8 billion joules per year; about 1% of total energy used
85. About 857 million kilowatt-hours per month; 53.6 kg of uranium; 857,000 homes
87. 4.32 million joules per day; 50 watts

89. 21,900,000 kilowatt-hr in a year; 2190 households
91. a. 37.5 g; 1000 mL b. 10.8 mg; 1667 mL
93. a. 125 mL/hr; 1.125 mg/hr b. 2500 gtt/hr c. 4.5 mg
95. a. 1 capsule every 6 hours b. Approximately 50 gtt/hr
97. a. 10,000 gtt b. 60,000 gtt c. 15,000 gtt, 50 gtt/min

Unit 2C Quick Quiz

1. c 2. a 3. b 4. b 5. b 6. a 7. c 8. c 9. b 10. b

Unit 2C Exercises

3. Does not make sense
5. Does not make sense
7. (cars, buses): (16, 0), (13, 2), (10, 4), (7, 6), (4, 8), (1, 10)
9. a. Jordan b. Jordan c. Amari d. 10.53 m
11. Answers will vary.
13. Approximately 7.1 m
15. Yes; imagine that at the same time the monk leaves the monastery to walk up the mountain, his twin brother leaves the temple and walks down the mountain. Clearly, the two must pass each other somewhere along the path.
17. a. Yes; 21 m^2 b. Yes; 16 m^2 c. 5 m by 5 m
19. 20.6 ft
21. 75 s
23. 4 apples
25. Select a fruit from the box labeled *Apples and Oranges*.
27. 7 crossings
29. 3 socks
31. 11.25 s

CHAPTER 3

Unit 3A Quick Quiz

1. b 2. b 3. c 4. c 5. c 6. c 7. a 8. c 9. c 10. b

Unit 3A Exercises

7. Makes sense
9. Does not make sense
11. Makes sense
13. Makes sense
15. $2/5 = 0.4 = 40\%$
17. $0.20 = 1/5 = 20\%$
19. $150\% = 1.5 = 3/2$
21. $4/9 = 0.444\ldots = 44.44\ldots\%$
23. $5/8 = 0.625 = 62.5\%$
25. $69\% = 0.69 = 69/100$
27. $7/5 = 1.4 = 140\%$
29. $4/3 = 1.333\ldots = 133.33\ldots\%$
31. 4.24; 0.24; *A* is 424% of *B*.
33. 0.76; 1.31; *A* is 76% of *B*.
35. 1.10; 0.91; *A* is 110% of *B*.
37. 56.1%
39. 127%
41. 115%
43. Helen's salary increased more in absolute terms ($12,000 vs. $10,000). Both salaries increased by 40%.
45. Absolute change = 7.4 million; relative change = 85.1%
47. Absolute change = −12.4; relative change = −18.3%
49. 6.3
51. 7.2
53. 163
55. 48
57. 0.7
59. 1.2
61. 14.4 percentage points; 35.2%
63. −14.7 percentage points; −29.4%
65. 33%; 40%
67. $1187.96
69. 37.4%
71. False; the rate decreased by 5.9%.
73. False; sales increased by about 7.7%.
75. Not possible
77. Possible
79. Possible
81. No, you cannot average averages.
83. True
85. False; some cars may have both Bluetooth and GPS.
87. 5353
89. 192,000
91. 0.69
93. 19.7%
95. a. −6.98% (loss of 6.98%) b. −508 (loss of 508 points)

Unit 3B Quick Quiz

1. a 2. a 3. b 4. b 5. c 6. a 7. c 8. b 9. c 10. b

Unit 3B Exercises

9. Makes sense
11. Makes sense
13. Does not make sense
15. a. 60,000 = sixty thousand
 b. 300,000 = three hundred thousand
 c. 340,000 = three hundred forty thousand
 d. 0.002 = two thousandths
 e. 0.067 = sixty-seven thousandths
 f. 0.000003 = three millionths
17. a. 4.68×10^2 b. 1.26547×10^5 c. 4×10^{-2} d. 9.73623×10^3 e. 1.256×10^1 f. 8.642×10^{-1}
19. a. 6×10^5 b. 8×10^{10} c. 3.2×10^3 d. 2×10^5
21. a. 10^{24} is 10^6, or 1 million, times larger than 10^{18}.
 b. 10^{27} is 10^{10}, or 10 billion, times as large as 10^{17}.
 c. 1 trillion is 10^6, or 1 million, times as large as 1 million.
23. 1.2×10^{12} bytes
25. 5×10^{-10} meters
27. a. 5×10^7; exact value is 5.2×10^7.
 b. 10^{10}; exact value is 9.69×10^9.
 c. 200; a more precise value is 193.5.
29. Answers will vary. One cup of coffee per day at $3/cup (plus tip) amounts to about $90 per month. One tank of gasoline per week at $35/tank (e.g., 12 gallons at $3/gallon) amounts to about $150 per month.
31. Yes; 800 quarters (0.2 oz each) weigh 10 lb.
33. $18 billion (100 million pizza eaters, 1 pizza/person/month, $15/pizza)
35. 200 gallons per year (recommended amount is 2 liters per day)
37. 800 gallons (20,000 miles/yr and 25 mi/gal)
39. 32 candy bars
41. Fission of 1 kg of uranium-235 releases 35,000 times as much energy as burning 1 kg of coal.
43. About 0.007 L
45. About 4900 kg
47. 2,000,000 to 1
49. 50,000,000 to 1
51.

Planet	Model diameter	Model distance from Sun
Mercury	0.5 mm	6 m
Venus	1.2 mm	11 m
Earth	1.3 mm	15 m
Mars	0.7 mm	23 m
Jupiter	14.3 mm	78 m
Saturn	12.0 mm	143 m
Uranus	5.2 mm	287 m
Neptune	4.8 mm	450 m

53. a. 11.1 m b. 0.22 mm
55. Approximately 5.1 deaths per minute
57. Approximately 11,590 passengers per hour
59. Approximately $62,300 per person
61. Approximately 31 cars per minute
63. a. About 10^{10} cells per cm^3 b. About 10^{13} cells per liter
 c. About 7×10^{14} cells per person

65. About 9000 gallons per month (assuming 100 million households)
67. a. About 1821 kg/cm^3 b. 7300 kg, or the mass of a tank c. 7×10^{11} kg/cm^3, more than the total mass of Mt. Everest
69. a. \$38,300/person/yr b. \$105/person/day c. 67% d. 17% e. Total spending increased 81%; health care spending increased 129%.
71. 1.5 mm; 1.8 mm; 1.3 mm; 1.7 mm; in quarters
73. Answers will vary depending on assumptions.
81. a. 5.87×10^{12} miles b. 2,598,960 hands c. 4.9 metric tons d. 5.5 g/cm^3 e. 4.4×10^{17} sec

Unit 3C Quick Quiz

1. b 2. a 3. b 4. a 5. b 6. c 7. b 8. a 9. a 10. a

Unit 3C Exercises

7. Does not make sense
9. Does not make sense
11. Makes sense
13. a. 6 b. 98 c. 0 d. 357 e. 12,784 f. 3 g. 7387 h. −16 i. −14
15. Four significant digits; precise to the nearest 1 (whole number)
17. One significant digit; precise to the nearest hundred
19. Four significant digits; precise to the nearest thousandth of a mile
21. Two significant digits; precise to the nearest thousand seconds
23. Seven significant digits; precise to the nearest ten-thousandth of a pound
25. Five significant digits; precise to the nearest tenth of a km/s
27. 285
29. 96
31. 1.18×10^2
33. Random errors could occur due to not counting some birds and double counting other birds.
35. Random errors could occur when taxpayers make honest mistakes or when the income amounts are recorded incorrectly. Systematic errors could occur when dishonest taxpayers report income amounts that are lower than their true income amounts.
37. Random errors could occur with an inaccurate radar gun or with honest mistakes made when the officer records the speeds. Systematic errors could occur if a radar gun is incorrectly calibrated and consistently reads too high or too low.
39. (1) is a random error; (2) is a systematic error (most likely due to underreporting).
41. All altitude readings will be about 500 feet too high; this is a systematic error.
43. Absolute error = 0.03 m; relative error = 1.7%
45. Absolute error = 2 mi/hr; relative error = 8.3%
47. Absolute error = −3.5 ft; relative error = −10.4%
49. Absolute error = −0.5°F; relative error = −0.5%
51. Laser device is more precise; tape measure is more accurate.
53. Digital scale is more precise and more accurate.
55. 137 lb
57. 71 mi/hr
59. 147 oz
61. \$23 per person
63. Random or systematic errors could be present; believable with the given precision.
65. Random or systematic errors could be present; not believable with the given precision.
67. Random or systematic errors could be present; also Asia is not well-defined; not believable with the given precision.
69. Random or systematic errors could be present; the precision is a bit too high for the claim to be believable. For example, are lost and stolen books included in the count?
71. a. Between 43 in and 53 in
    b. Between 43.42 in and 53.03 in

Unit 3D Quick Quiz

1. b 2. c 3. b 4. c 5. c 6. b 7. a 8. a 9. c 10. c

Unit 3D Exercises

5. Does not make sense
7. Makes sense
9. Does not make sense
11. 212.5
13. \$25.20
15. 0.23 of the same tank
17. \$54,745
19. 28.1%
21. \$0.90
23. \$3.43
25. \$3.60
27. \$528,000
29. \$426,000
31. Health care spending increased by about 3665%; the rate of inflation measured by the CPI was about 291%.
33. Private college cost increased by 298.7%, while the rate of inflation was 83.6%.
35. \$11.12
37. In 1996, actual dollars are 1996 dollars.
39. Yes, it is consistent (\$6.27 either way).
41. The 1968 minimum wage of \$7.21 (1996 dollars) would be \$11.03 in 2016, which is more than the current minimum wage in 2016 and less than a minimum wage of \$15.
43. a. An index is usually the ratio of two quantities with the same units (such as prices), which means it has no units.
    b.

Team	FCI
Boston	164.3
NY (Yankees)	153.6
Chicago (Cubs)	142.3
Colorado	88.4
San Diego	83.3
Arizona	60.2
Major League average	100.0

45. Answers will vary depending on when the index was last recalculated.
51. \$260
53. \$321
55. $A > B$

Unit 3E Quick Quiz

1. a 2. a 3. b 4. a 5. c 6. c 7. b 8. c 9. c 10. b

Unit 3E Exercises

5. Makes sense
7. Does not make sense
9. Does not make sense
11. a. Josh b. Josh c. Jude
13. a. New Jersey; Nebraska
    b. The percentage of nonwhites is significantly lower in Nebraska than in New Jersey.

15. a. Whites: 0.18%; nonwhites: 0.54%; total: 0.19%
b. Whites: 0.16%; nonwhites: 0.34%; total: 0.23%
c. The rate for both whites and nonwhites was higher in New York than in Richmond, yet the overall rate was higher in Richmond than in New York. The percentage of nonwhites was significantly lower in New York than in Richmond.
17. a. Your work should clearly verify that the numbers are correct.
b. 8.3%, or about 8 in 100
c. 90%, or 9 in 10, which is the accuracy of the test
d. 0.11%, or about 11 in 10,000
19. a. Your work should clearly verify the table entries
b. 232 athletes; 36 athletes were users and 196 were non-users; 84.5% were falsely accused.
c. 1768 athletes; 4 athletes were users; 0.2% were wrongly cleared.
21. The housing program will get a 1% increase in actual dollars. However, when adjusted for inflation, the program gets a cut.
23. a. Spelman had a better record for home games (34.5% vs. 32.1%) and away games (75.0% vs. 73.7%) individually.
b. Morehouse has the better overall average (62.5% vs. 48.9%).
c. Teams are generally rated on their overall record.
25. a. Show the work you did to verify the incidence and detection rates.
b. 95% of those at risk for HIV test positive; 67.9% of those at risk who test positive have HIV.
c. The chance of the patient having HIV is 67.9%, which is greater than the overall "at risk" incidence rate of 10%.
d. 95% of those in the general population with HIV test positive; 5.4% of those in the general population who test positive have HIV.
e. The chance of the patient having HIV is 5.4%, compared to the overall incidence rate of 0.3%.
27. a. Excelsior Airline had the higher on-time percentage in each of the five cities taken individually.
b. Averaged over all five cities, the on-time percentage was 86.7% for Excelsior and 89.1% for Paradise.
c. Excelsior had the higher on-time percentage in the five cities individually, while Paradise had the higher overall on-time percentage.

CHAPTER 4

Unit 4A Quick Quiz

1. a 2. a 3. b 4. c 5. b 6. c 7. a 8. a 9. a 10. c

Unit 4A Exercises

7. Does not make sense
9. Makes sense
11. Does not make sense
13. $3100; 72%
15. $1220; 510%
17. $1104; 130%
19. $2204; 176%
21. $11.25
23. $72
25. $1183.33
27. $316.67
29. $95.83
31. −$270
33. −$105
35. Below average
37. At average
39. Above average
41. Old car: $20,333; new car: $25,056
43. Buy: $12,000; lease: $10,000
45. In-state: $12,400; out-of-state: $11,900
47. $2,868,480 more
49. The man earns approximately 29% more per year than the woman, or $1,635,840 more over 40 years.
51. $3000 ($1500 cost for the credit hours plus $1500 you could have earned with the job)
53. a. $2300 b. $2780
55. Plan A: $5800; Plan B: $4184
57. About 62 months
59. a. With policy: $2050; without policy: $1375
b. With policy: $1700; without policy: $1650
c. With policy: $1800; without policy: $1700
61. Plan A: $20,600; Plan B: $19,000; Plan C: $16,400
63–66. Answers will vary.

Unit 4B Quick Quiz

1. b 2. a 3. c 4. a 5. c 6. b 7. a 8. a 9. a 10. c

Unit 4B Exercises

9. Does not make sense
11. Does not make sense
13. Makes sense
15. 9
17. 32
19. 5
21. 1/4
23. 9
25. 25
27. $x = 20$
29. $z = 16$
31. $y = 4$
33. $z = 4$
35. $x = 10$
37. $a = -2$
39. $q = 40$
41. $t = 80$
43. $x = \pm 7$
45. $x = 10$ or $x = -2$
47. $t = \pm 12$
49. $u = 2$
51. $1000
53. $3760
55.

End of year	Suzanne's annual interest	Suzanne's balance	Derek's annual interest	Derek's balance
1	$75	$3075	$75	$3075
2	$75	$3150	$77	$3152
3	$75	$3225	$79	$3231
4	$75	$3300	$81	$3312
5	$75	$3375	$83	$3395

57. $7401.22
59. $32,967.32
61. $15,464.83
63. $6103.97
65. $29,045.68
67. $13,240.82
69. $75,472.89
71. 4.18%
73. 1.24%
75. 1 year: $5230.14; 5 years: $6261.61; 20 years: $12,298.02; APY = 4.60%
77. 1 year: $7322.20; 5 years: $8766.26; 20 years: $17,217.22; APY = 4.60%
79. 1 year: $3185.51; 5 years: $4049.58; 20 years: $9960.35; APY = 6.18%
81. $15,685.31
83. $15,488.10
85. $53,751.97
87. $78,375.20
89. After 10 years, Chang has $705.30; after 30 years, he has $1403.40. After 10 years, Kio has $722.52; after 30 years, she has $1508.74. Kio has $17.22, or 2.4%, more than Chang after 10 years. Kio has $105.34, or 7.5%, more than Chang after 30 years.
91. Quarterly: 5.41%; monthly: 5.43%; daily: 5.44%

93.

	Account 1		Account 2	
Year	Interest	Balance	Interest	Balance
0	—	$1000	—	$1000
1	$55	$1055	$57	$1057
2	$58	$1113	$59	$1116
3	$61	$1174	$63	$1179
4	$65	$1239	$67	$1246
5	$68	$1307	$70	$1316
6	$72	$1379	$74	$1390
7	$76	$1455	$79	$1469
8	$80	$1535	$83	$1552
9	$84	$1619	$88	$1640
10	$89	$1708	$93	$1733

After 10 years, Account 1 has increased in value by $708, or 70.8%; Account 2 has increased in value by $733, or 73.3%.

95. a. Rosa: $3649.96, $6573.37; Julian: $3190.70, $6633.24
b. Rosa: 18%, 54%; Julian 22%, 62%

97. Plan A: $27,765.29; Plan B: $28,431.33

99. 18.9 years

101. 68.1 years

103. a. $92,516,718
b. Yes, the interest alone would be $76,742,847.
c. About $5.8 million per week

109. a. $161.05 b. $8,940,049.24

111. a. $5808.08 b. $3078.16 c. $6520.19

113. a. 24.5325 b. 1.0672 c. 4.0811%

Unit 4C Quick Quiz

1. b 2. b 3. c 4. b 5. a 6. a 7. b 8. a 9. b 10. c

Unit 4C Exercises

9. Does not make sense
11. Does not make sense
13. Does not make sense
15. $912.48
17. $7636.31
19. $228,903.02; $72,000
21. $90,091.51; $64,800
23. $636.02
25. $768.38
27. $1659.18
29. 56.7%; 9.4%
31. 73.8%, 2.8%
33. −42.9%; −17.0%
35. 68.0%; 5.3%
37. Stocks: $400,267; bonds: $2984; cash: $756
39. a. INTC b. $35.24 c. About $368 million d. 0.22% e. $108.92 f. $2.30 g. $10.85 billion
41. $8.23 per share; slightly overpriced
43. $12.16 per share; priced about right
45. Answers will vary.
47. 2.11%
49. 5.00%
51. $40.95
53. $70.87
55. a. 454.96 shares b. $5320.41 c. $6369.27
57. Yolanda: balance = $31,056.46; deposits = $24,000
Zach: balance = $30,186.94; deposits = $24,000
59. Juan: balance = $65,551.74; deposits = $48,000
Maria: balance = $67,472.11; deposits = $50,000
61. Balance of $15,848.11 will not be enough.
63. Balance of $29,081.87 will not be enough.
65. 36.4% per share
67. Total return = 439,900%; annual return = 18.3%; much higher than the average annual return for stocks
69. a. $88,548 b. $66,439
c. Mitch deposited $10,000; Bill deposited $30,000.
77. a. $51,412.95
b. $95,102.64; less than double
c. $190,935.64; more than double
79. a. 1.293569 b. 4.932424 c. 14.27%

Unit 4D Quick Quiz

1. a 2. a 3. b 4. b 5. c 6. c 7. c 8. b 9. c 10. a

Unit 4D Exercises

7. Makes sense
9. Does not make sense
11. Makes sense
13. a. Amount borrowed = $120,000; APR = 6%; number of payments per year = 12; loan term = 15 years; payment amount = $1013
b. Number of payments = 180; amount paid = $182,340
c. Principal = 65.8%; interest = 34.2%
15. a. $716.43 b. $171,943.44
c. Principal = 58.2%, interest = 41.8%
17. a. $1796.18 b. $646,624.80
c. Principal = 61.9%; interest = 38.1%
19. a. $690.58 b. $124,304.40
c. Principal = 80.4%; interest = 19.6%
21. a. $313.36 b. $11,281
c. Principal = 88.6%; interest = 11.4%
23. a. $1186.19 b. $213,514
c. Principal = 70.3%; interest = 29.7%
25. Monthly payment = $716.12

End of ...	Interest	Payment toward principal	New principal
Month 1	$500.00	$216.12	$149,783.88
Month 2	$499.28	$216.84	$149,567.04
Month 3	$498.56	$217.56	$149,349.48

27. Monthly payments: $463.16, $362.68, $304.15. You can afford only option 3.
29. a. $916.80 b. $11,001.60 c. 9.1%
31. a. $376.76 b. $13,563.36 c. 26.3%
33.

Month	Payment	Expenses	Interest	New balance
0				$1200.00
1	$200	$75	$18.00	$1093.00
2	$200	$75	$16.40	$984.40
3	$200	$75	$14.77	$874.17
4	$200	$75	$13.11	$762.28
5	$200	$75	$11.43	$648.71
6	$200	$75	$9.73	$533.44
7	$200	$75	$8.00	$416.44
8	$200	$75	$6.25	$297.69
9	$200	$75	$4.47	$177.16
10	$200	$75	$2.66	$54.82

A partial 11th payment will pay off the loan.

35.

Month	Payment	Expenses	Interest	Balance
0				$300.00
1	$300	$175	$4.50	$179.50
2	$150	$150	$2.69	$182.19
3	$400	$350	$2.73	$134.92
4	$500	$450	$2.02	$86.94
5	0	$100	$1.30	$188.24
6	$100	$100	$2.82	$191.06
7	$200	$150	$2.87	$143.93
8	$100	$80	$2.16	$126.09

37. Option 1: Monthly payment = $2935.06; total payment = $1,056,621.60. Option 2: monthly payment = $3708.05; total payment = $667,449.00.
39. Option 1: monthly payment = $405.24; total payment = $145,886.40. Option 2: monthly payment = $530.95; total payment = $95,571.
41. Choice 1: monthly payment = $572.90; closing costs = $1200; Choice 2: monthly payments = $538.85, closing costs = $3600
43. Choice 1: monthly payment = $608.02; closing costs = $2400. Choice 2: monthly payment = $590.33; closing costs = $4800.
45. a. $269.92 b. $380.03
    c. Total payment for 20-year term = $64,780.80; total payment for 10-year term = $45,603.60
47. $187.50, $250.00 per month
49. $107,964, $134,956
51. a. $87.11, $116.30, $148.38 b. $61,397
    c. $326.67, $58,801
53. a. $100 b. 20% c. 520%
55. a. $8718.46 every year, $716.43 every month, $330.43 every two weeks, $165.17 every week
    b. Yearly: $174,369.20; monthly: $171,943.20; biweekly: $171,823.60; weekly: $171,776.80
65. a. For the first 10 months of the loan:

Month	Interest	Principal	Balance
0			7500.00
1	56.25	38.76	7461.24
2	55.96	39.05	7422.19
3	55.67	39.34	7382.85
4	55.37	39.64	7343.21
5	55.07	39.94	7303.27
6	54.77	40.24	7263.04
7	54.47	40.54	7222.50
8	54.17	40.84	7181.66
9	53.86	41.15	7140.51
10	53.55	41.46	7099.05

    b. Interest = $56.25; principal = $38.76
    c. Interest = $0.70; principal = $94.31

Unit 4E Quick Quiz

1. a 2. c 3. a 4. b 5. b 6. a 7. c 8. a 9. b 10. c

Unit 4E Exercises

11. Does not make sense
13. Makes sense
15. Makes sense
17. Makes sense
19. Gross income = $49,600; adjusted gross income = $46,100; taxable income = $35,700
21. Gross income = $93,650; adjusted gross income = $87,450; taxable income = $74,350
23. Do not itemize; your itemized deductions total $11,945, which is less than your standard deduction.
25. Gross income = $33,550; adjusted gross income = $33,050; taxable income = $22,650 (standard deduction)
27. Gross income and adjusted gross income = $33,900; taxable income = $15,400 (standard deduction)
29. $4844
31. $15,086
33. $15,578
35. $15,853
37. $500
39. $0
41. $280
43. Cheaper to own (home cost including amount saved through mortgage interest deduction is $1406, which is less than the rent of $1600)
45. Maria's true cost = $6700; Steven's true cost = $8500
47. FICA tax = $2142; income tax = $2174; total tax = $4316; tax rate = 15.4%
49. FICA tax = $3427; income tax = $4881; total tax = $8308; tax rate = 18.0%
51. FICA tax = $3687; income tax = $5204; total tax = $8891; tax rate = 18.4%
53. Pierre: FICA tax = $9180; income tax = $23,670; total tax = $32,850; tax rate = 27.4%. Katarina: FICA tax = $0; income tax = $10,748; total tax = $10,748; tax rate = 9.0%.
55. Your take-home pay decreases by $340, rather than by $400.
57. Your take-home pay decreases by $600, rather than by $800.
59. Total tax at single rate = $31,028; total tax at married rate = $31,181; marriage penalty
61. Total tax at single rate = $83,740; total tax at married rate = $90,453; marriage penalty
63. a. Deirdre: $6885; Robert: $6885; Jessica and Frank: $13,770
    b. Deirdre: $22,524; Robert: $18,263; Jessica and Frank: $45,221
    c. Deirdre: 25.0%; Robert: 20.3%; Jessica and Frank: 25.1%
    d. In order of increasing tax rates: Serena, Robert, Deirdre, Jessica and Frank
    e. Serena receives by far the greatest tax break because her income is from investments.
65. a. 62.3%; 5.6% b. 2.7%; 51.6%

Unit 4F Quick Quiz

1. b 2. c 3. a 4. b 5. a 6. a 7. b 8. b 9. b 10. c

Unit 4F Exercises

9. Makes sense
11. Does not make sense
13. Does not make sense
15. a. Surplus b. Reduces your surplus c. Yes

17. a. \$63,000
b. Outlays = \$1,113,000; deficit = \$63,000; debt = \$836,000
c. \$69,000
d. Outlays = \$869,000; surplus = \$231,000; debt = \$605,000
19. \$125,000 per worker
21. 2000: surplus, 2.3%; debt, 55.4%. 2010: deficit, −8.7%; debt, 91.9%
23. 2010: 8.7%; 2020: 2.5%; −71.2%
25. \$383 billion; interest payment increases to \$495 billion, or 29.4%.
27. \$1.62 trillion; \$1.09 trillion
29. Revenue increases by \$66.2 billion; spending increases by \$62.2 billion; deficit decreases
31. a. \$1.48 trilllion b. \$1.21 trillion c. \$77.8 billion; 6.4%
33. \$140 billion deficit
35. Cut government spending, borrow money, raise taxes
37. About 634,000 years
39. \$22.1 trillion; \$32.9 trillion
41. \$1.70 trillion/yr
43. About 2360 years

CHAPTER 5

Unit 5A Quick Quiz

1. a 2. c 3. a 4. c 5. b 6. c 7. a 8. c 9. b 10. b

Unit 5A Exercises

9. Does not make sense
11. Does not make sense
13. Does not make sense
15. *Population*: all Americans; *sample*: 1001 Americans surveyed by telephone; *population parameters*: opinions of all Americans on the plan; *sample statistics*: opinions on the plan among those in the sample
17. *Population*: all stars in the galaxy; *sample*: the few stars selected for measurements; *population parameter*: mean of distances between all stars in the galaxy and Earth; *sample statistics*: mean of distances between the stars in the sample and Earth
19. *Population*: all senior executives; *sample*: 150 senior executives; *population parameters*: most common job interview mistake according to all senior executives; *sample statistics*: most common job interview mistake according to the 150 executives in the sample (little or no knowledge of the company)
21. *Step 1*: Population is all high school seniors; goal is to determine the percentage of seniors who use a cell phone while driving. *Step 2*: Choose a representative sample of seniors. *Step 3*: Determine the percentage of those in the sample who claim to use a cell phone regularly while driving. *Step 4*: Infer the percentage of all seniors who use a cell phone regularly while driving. *Step 5*: Assess results and formulate conclusion.
23. *Step 1*: Population is American college students whose campuses have a basketball program; goal is to determine the percentage of those students who attend home basketball games. *Step 2*: Choose a representative sample of students. *Step 3*: Determine the percentage of students in the sample who attend home basketball games. *Step 4*: Infer the percentage of all students at such campuses who attend home basketball games. *Step 5*: Assess results and formulate conclusion.
25. *Step 1*: Population is all golden retrievers; goal is to determine the average lifetime of all golden retrievers. *Step 2*: Choose a representative sample of retrievers, possibly from records of randomly selected vets. *Step 3*: Determine the average lifetime of golden retrievers in the sample. *Step 4*: Infer the average lifetime for all golden retrievers. *Step 5*: Assess results and formulate conclusion.
27. The third sample is most representative, with respect to attendance. The other samples are biased toward certain subgroups within the population.
29. Stratified
31. Stratified
33. Simple random
35. Observational, retrospective; cases are volunteers with a tendency to lie; controls are volunteers without such a tendency.
37. Observational, retrospective; cases are runners with a specific type of injury; controls are runners without that injury.
39. Observational, not retrospective
41. Experiment. There are two treatment groups; it would also be useful to include a control group given no treatment. No blinding is necessary.
43. Observational; retrospective. Cases are teams with high-altitude home courts; controls are the other teams.
45. Experiment. The treatment group gets multivitamins; the control group gets no multivitamins. Blinding is not necessary, since it should be clear whether a person suffers a stroke.
47. 50.5% to 55.5%; yes
49. 43% to 49%; yes
51. a. The treatment group received the new drug; the control group received the placebo.
b. Significantly more patients improved with the treatment compared to the placebo. However, fewer than half of the patients in the treatment group improved.
c. Possibly; nearly 20% of the patients showed improvement with the placebo.
d. Answers will vary, but might refer to side effects, alternative treatments, and/or cost.
53. a. Population is all Americans; population parameters are percentages of men and women who regift.
b. Sample is 900 registered voters; sample statistics are percentages of men and women who regift.
c. Observational
55. a. Population is registered American voters; population parameter is percentage of registered American voters who are concerned about fake news.
b. Sample is the 1006 registered voters surveyed; sample statistic is percentage of voters in the sample who are concerned about fake news.
c. Observational
57. a. Population is all adult Americans; population parameter is percentage of Americans who approved of the ruling.
b. Sample is the adults who were surveyed; sample statistic is the percentage of respondents in the sample who approved of the ruling (30.8% in 1974 and 39.1% in 2014).
c. Observational
67. a, b. Answers will vary. c. The mean will be near 0.5.

Unit 5B Quick Quiz

1. c 2. b 3. a 4. b 5. b 6. a 7. c 8. c 9. b 10. b

Unit 5B Exercises

5. Does not make sense
7. Does not make sense
9. The method is appropriate, so the results should be valid.
11. You should question the results of the study because of possible bias.
13. You should question the results of this study because call-in polls tend to be subject to participation bias.
15. You should question the results of the study unless it explains how optimism is defined and how it is measured.
17. You should question the results of this study because other factors should also be considered.
19. You should question the results of the study; self-reporting is often not accurate.
21. This is a reasonable claim.
23. Depending on the reliability of the president's evidence, this could be a reasonable claim. You should look for more research on the subject.
25. The Chamber of Commerce would have no reason to distort its data, so the claim is believable.
27. Political bias may have been present. Other estimates of attendance ranged from 250,000 to 900,000.
29. Selection and participation bias may be present.
31. While false self-reporting may be a problem, the results should be free of bias.
33. While false self-reporting may be a problem, the results should be free of bias.
35. Answers will vary.
37. How were respondents chosen? How was the question asked?
39. Who responded to the survey? How were the respondents selected? How was quality of restaurants measured?
41. How were respondents chosen? How was the question asked? Were respondents given the potato as choice, or did they suggest the potato without prompting?
43. 98% of all movies is different from 98% of the top movie rentals.
45. The questions have a different population. The population for the first question is all people who have dated on the Internet. The population for the second question is all married people.
47. a. 14% b. 38 people
    c. The 0.1% error rate assumed in part (b) would account for 38 of the 48 people who said they were noncitizens who voted, so that in fact there were only 10 noncitizen voters in the survey. If the error rate were 0.13%, it would account for all those who claimed to be noncitizen voters.
    d. The fact that there were zero voters among the 85 people who gave consistent answers about being noncitizens suggests that no noncitizens voted.

Unit 5C Quick Quiz

1. b 2. c 3. b 4. c 5. a 6. c 7. a 8. c 9. c 10. b

Unit 5C Exercises

7. Does not make sense
9. Does not make sense
11. Makes sense

13.

Grade	Freq.	Rel. freq.	Cum. freq.
A	6	0.20	6
B	6	0.20	12
C	10	0.33	22
D	5	0.17	27
F	3	0.10	30
Totals	**30**	**1.00**	**30**

15. Qualitative
17. Quantitative
19. Qualitative
21. Quantitative

23.

Bin	Freq.	Rel. freq.	Cum. freq.
95–99	3	0.15	3
90–94	2	0.10	5
85–89	3	0.15	8
80–84	2	0.10	10
75–79	4	0.20	14
70–74	1	0.05	15
65–69	3	0.15	18
60–64	2	0.10	20
Totals	**20**	**1.00**	**20**

25.

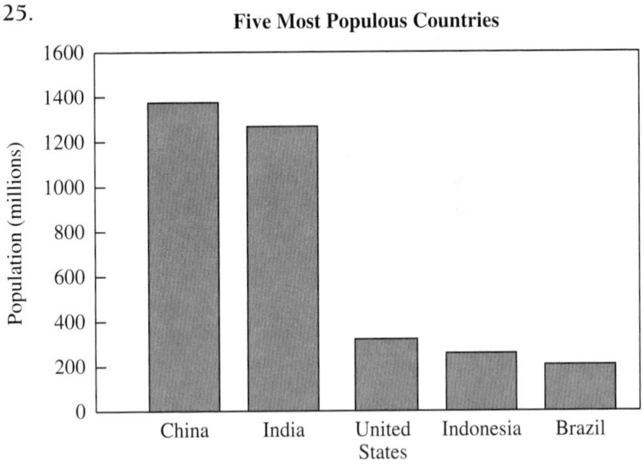

27. **Number of Children for Mothers Aged 40–44**

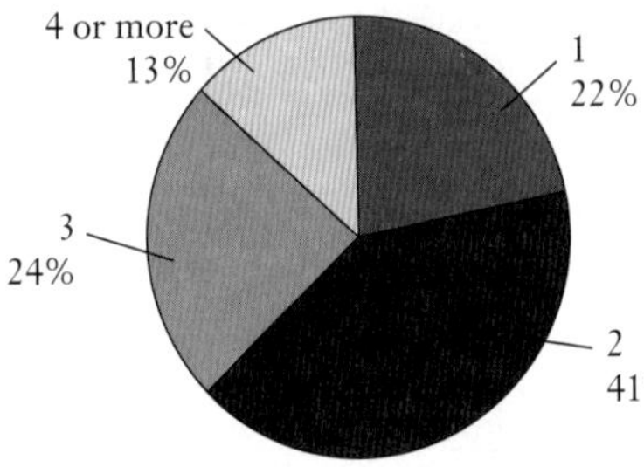

29.

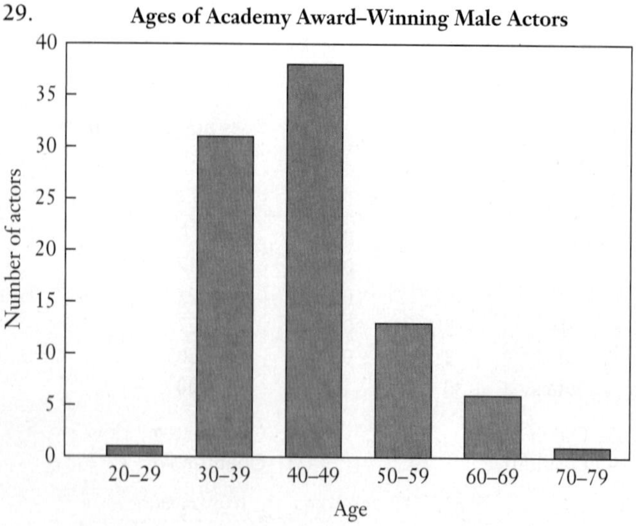

31. The growth is linear.

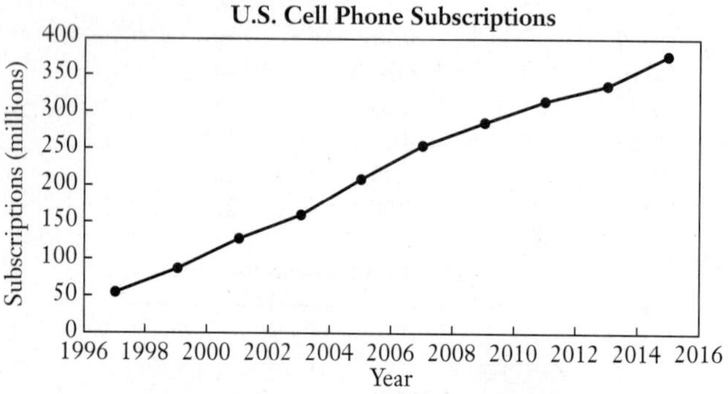

33.

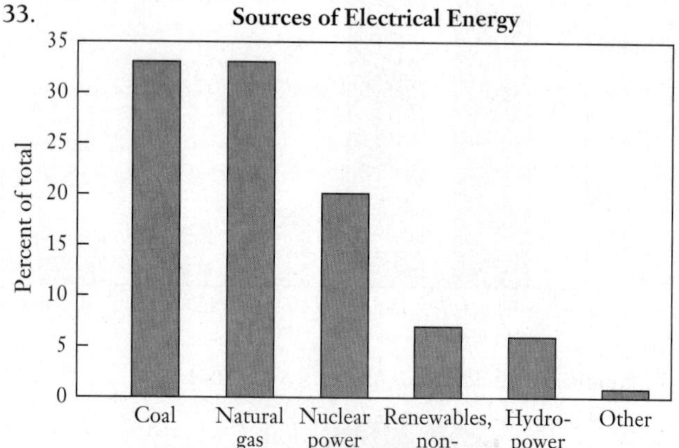

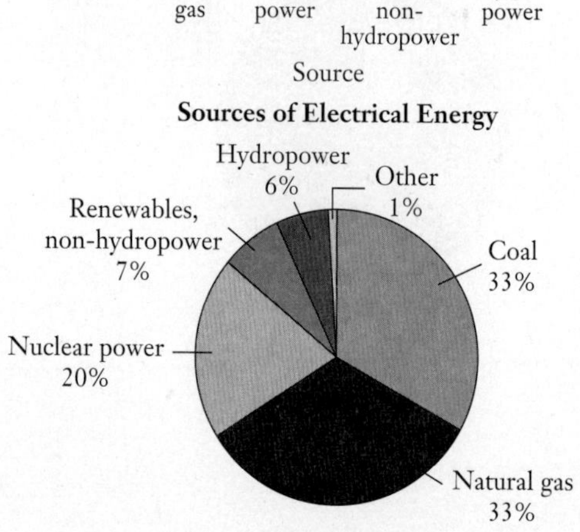

35.

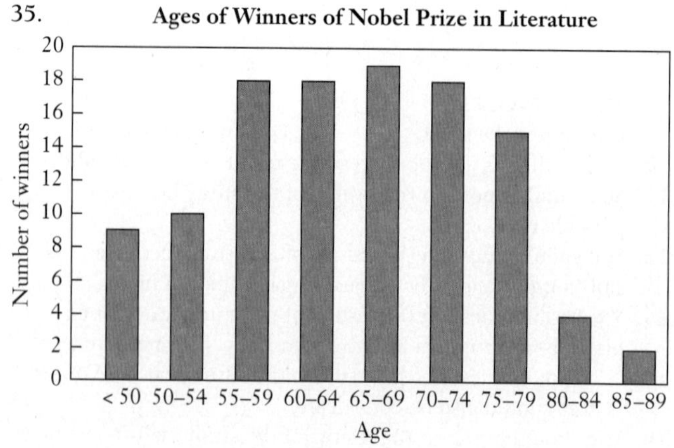

37.

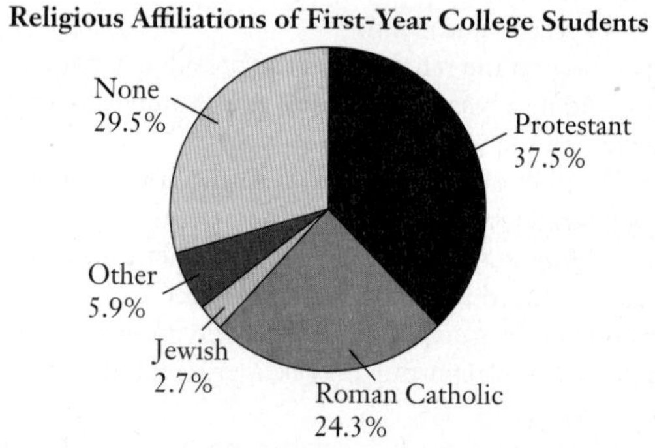

39.

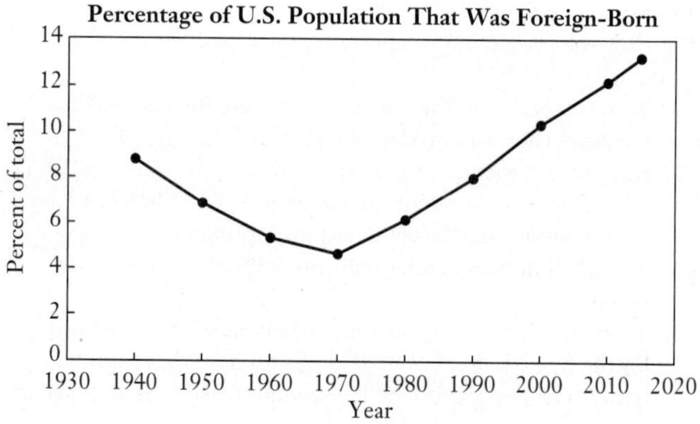

47. a.

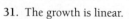

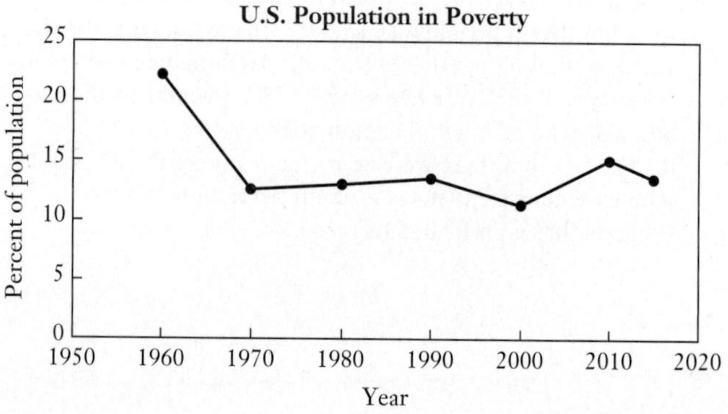

b.

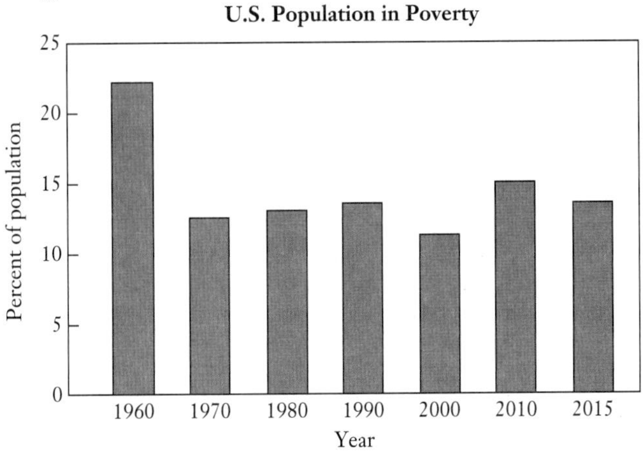

49. a.

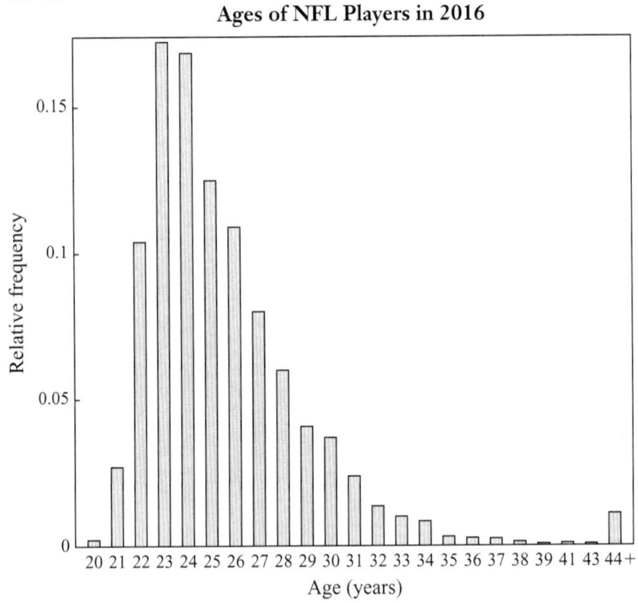

b.

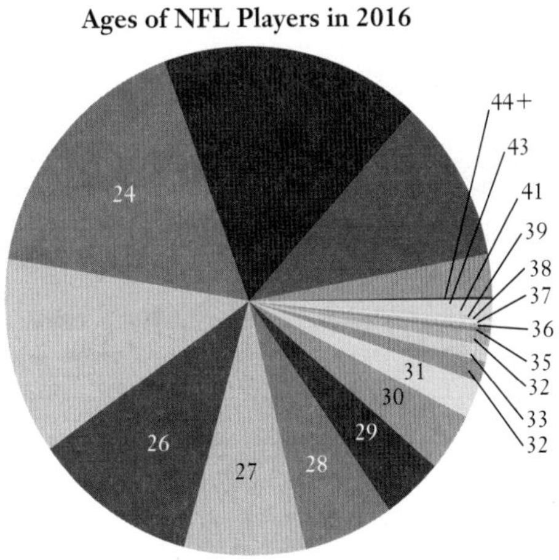

c. 23 pie charts are generated—one for each position listed.

Unit 5D Quick Quiz

1. c 2. c 3. a 4. a 5. c 6. c 7. a 8. b 9. c 10. b

Unit 5D Exercises

9. Does not make sense
11. Makes sense
13. a. 38.1% higher for men b. 59.1% higher for men
15. a. \$1,341,720 b. \$1,385,640
17. a. Boys in Canada (523) scored lower than girls in South Korea (544).
    b. Boys in Finland (517) scored lower than girls in Japan (527).
19. a. Books and supplies vary the least, probably because students at all types of colleges need the same number of books and other supplies on average, and these don't vary much in cost.
    b. Transportation, because commuters must spend more to get to and from school, while students living on campus probably spend more time on campus.
21. a. Cancer, increased; tuberculosis, decreased; cardiovascular, increased; pneumonia, decreased
    b. 510 in 1950 c. 193
    d. If the lines of the graph are extended, cancer will overtake cardiovascular as the leading cause of death.
23. a. 1976, 5%; 2016, 16% b. 1967–1969
    c. Increased from about 10% to about 25%
    d. No, because dollar figures are not given.
25. With a few isolated exceptions, the highest rates occur in the southern part of the country.
27. a. Downhill b. B to D
29. a. Approximately 12% b. Approximately 20%
    c. It decreases as a percentage of the population.
31. a. 290 b. Yes, the two groups must overlap.
33. No. The height of the barrels represents oil consumption, but the eye is drawn to the volume of the barrels.
35. a. 0.32, 0.09
    c.

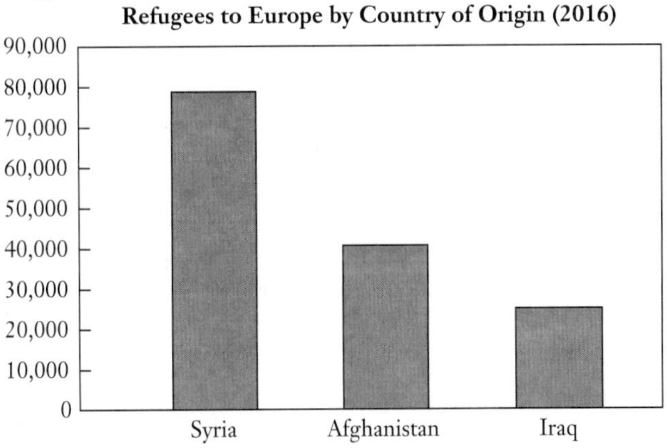

37. a.

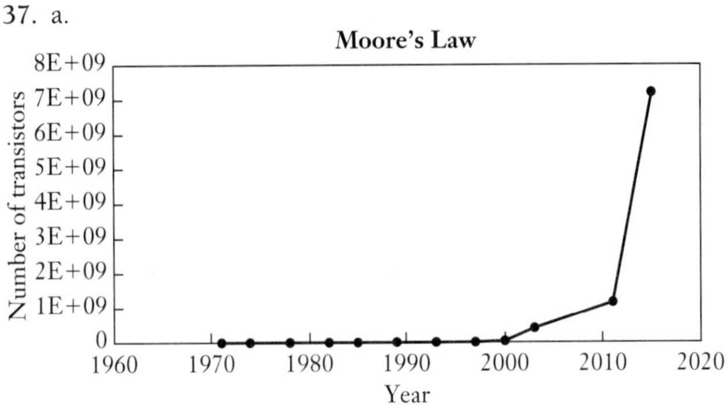

b.

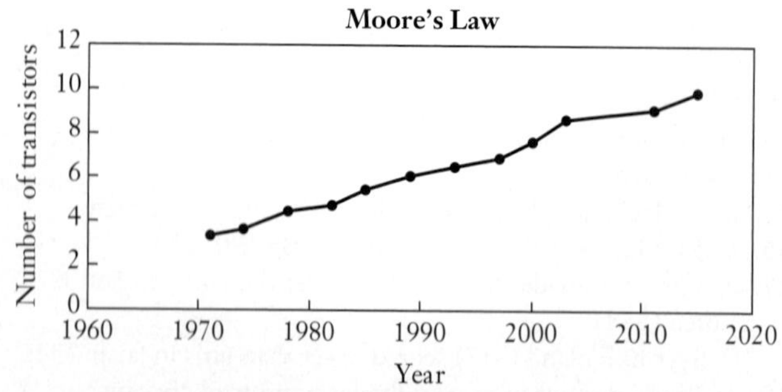

39. a. 2003–2004; 13% b. 5%
c. Private college tuition rose more in terms of dollars.

41. Most of the growth occurred after 1950.

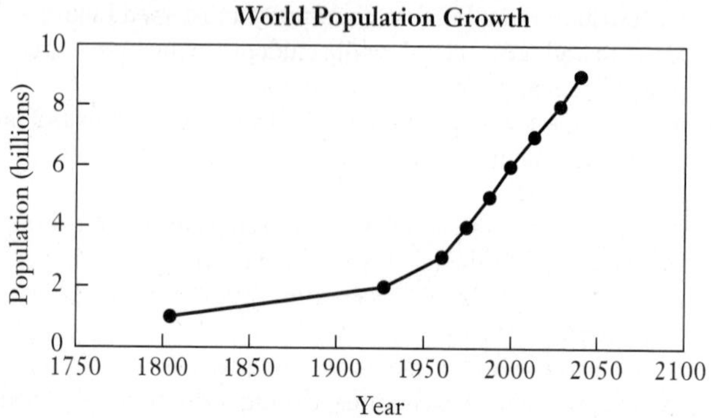

43. a. China, India, Russia b. China, India, Russia

45. a. 1957 b. 1990

47.

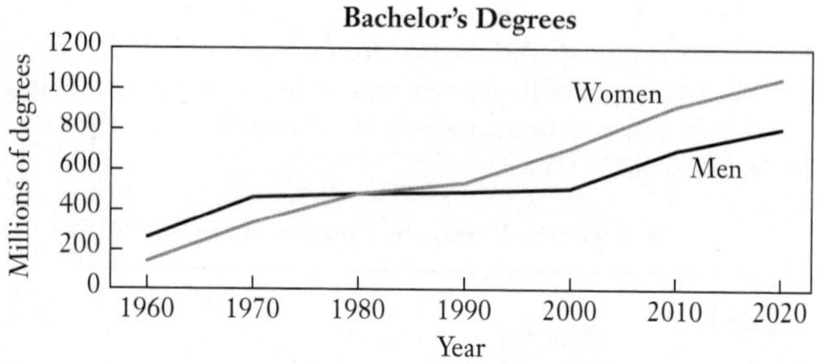

49.

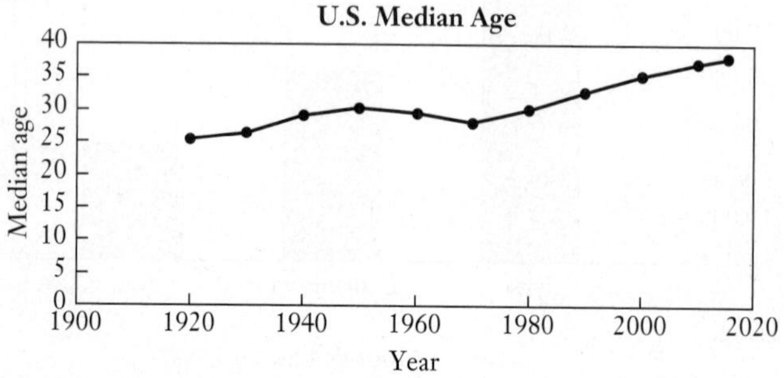

51.

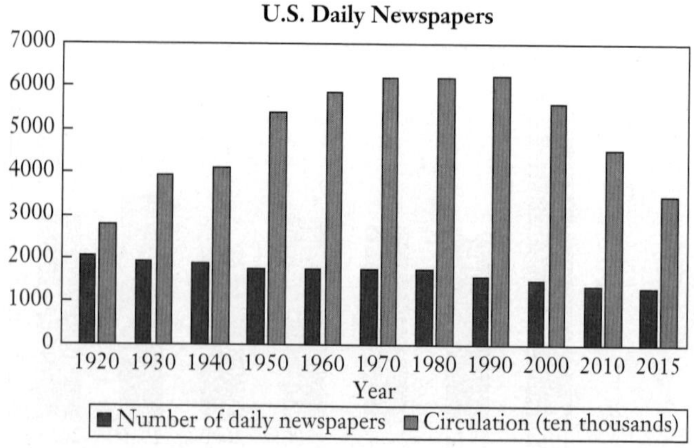

53. a.

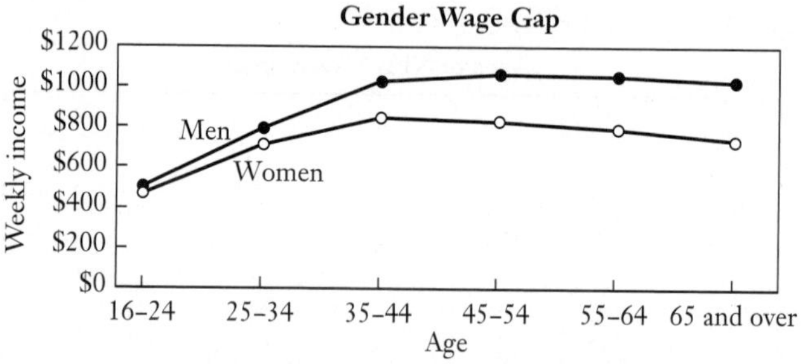

b. Wage ratios: ages 16–24, 0.93; ages 25–34, 0.90; ages 35–44, 0.83; ages 45–54, 0.78; ages 55–64, 0.75; age 65+, 0.71

61.

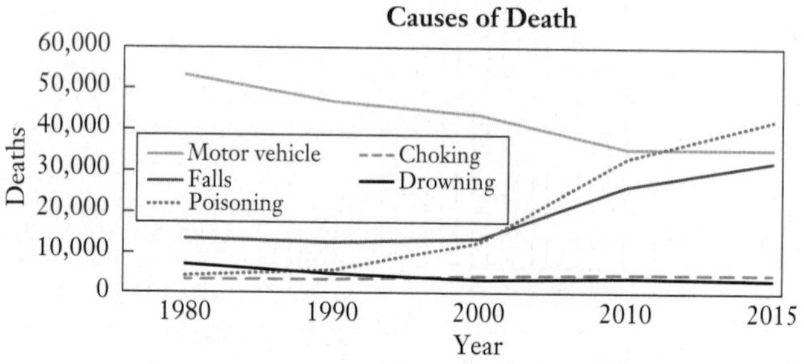

63.

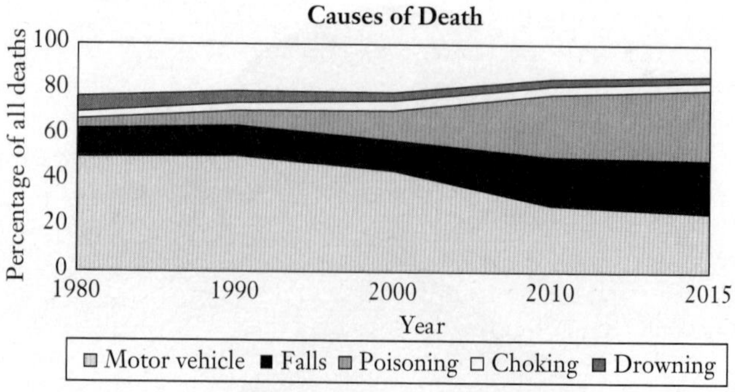

65.

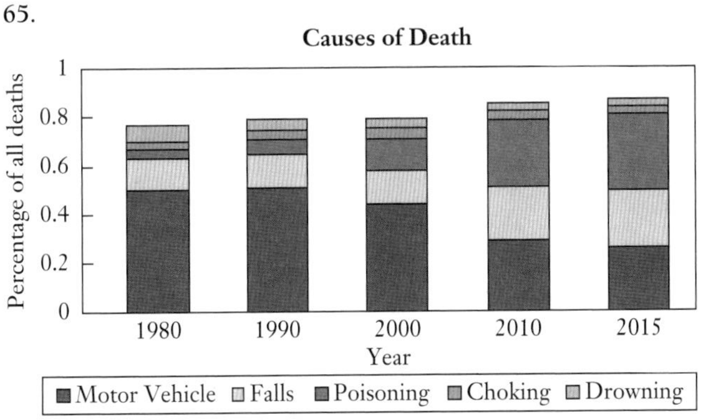

67.

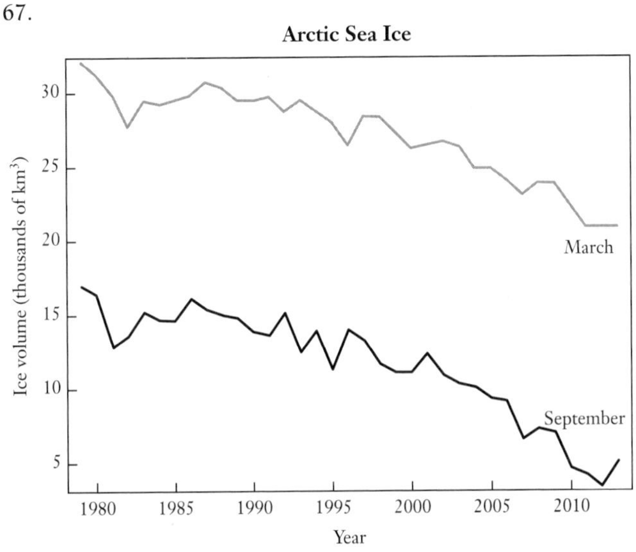

Unit 5E Quick Quiz

1. c 2. b 3. a 4. b 5. a 6. c 7. a 8. a 9. c 10. c

Unit 5E Exercises

7. Makes sense
9. Does not make sense
11. Makes sense
13. a. Strong negative correlation
    b. Heavier cars get lower city gas mileage.
15. a. Possible weak positive correlation
    b. Higher AGI may imply slightly higher charitable giving as a percentage of AGI.
17. Degrees of latitude, degrees Fahrenheit (or Celsius); strong negative correlation
19. Years, hours; possible negative correlation over some ages
21. Year, square miles; positive correlation
23. Children per woman, years; strong negative correlation
25. a.

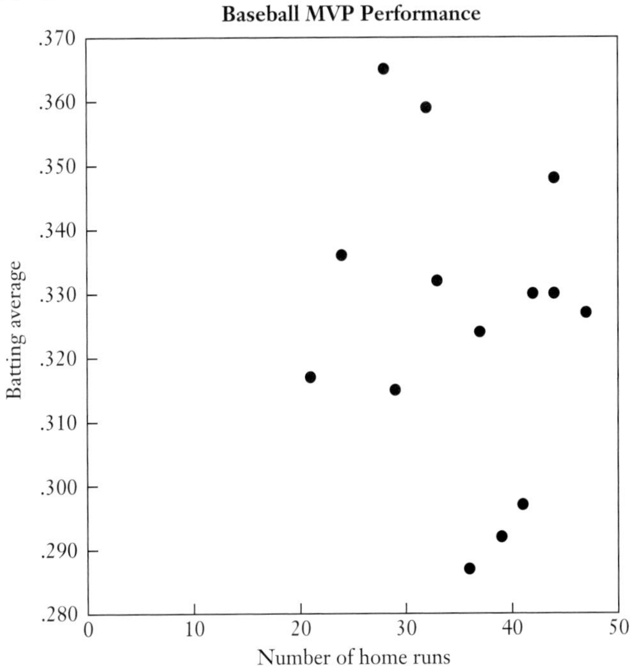

    b. No correlation
27. a.

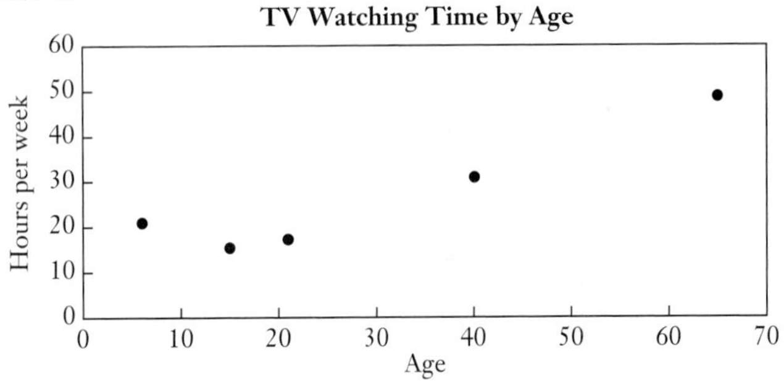

    b. Strong positive correlation
29. a.

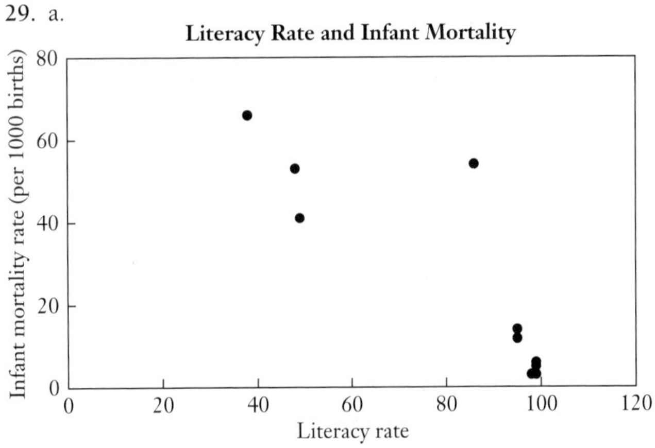

    b. Strong negative correlation

31. a. There is no correlation between the data, so there is little evidence for the claim.
b. Neither the horizontal nor vertical scale includes the zero point, so the spread in the data is exaggerated for both variables.
33. There is a positive correlation between the miles of freeway and the amount of traffic congestion; common underlying cause.
35. There is a negative correlation between carbon dioxide levels and pirates; coincidence.
37. There is a positive correlation between the number of ministers and bartenders; common underlying cause.
39. The causes of cancer are often random. A cancer cell is produced when the growth control mechanisms of a normal cell are altered by a random mutation. Smoking may increase the chance of such a mutation occurring, but the mutation will not occur in all individuals. Therefore, by chance, some smokers escape cancer.
41. Cause cannot be established until a mechanism is confirmed. There could be an underlying cause.
49. a.

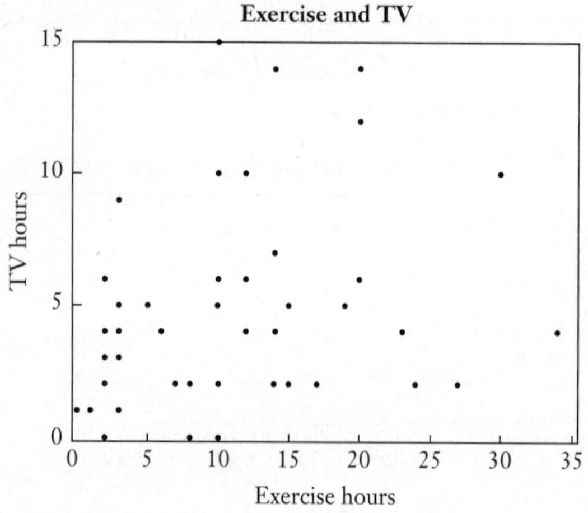

b. Correlation coefficient = 0.282

CHAPTER 6

Unit 6A Quick Quiz

1. a 2. b 3. c 4. a 5. c 6. c 7. a 8. a 9. b 10. b

Unit 6A Exercises

7. Makes sense
9. Makes sense
11. Does not make sense
13. Mean = 354.6; median = 355.0; mode = 345
15. Mean = 0.188; median = 0.165; mode = 0.16
17. Mean = 58.3 sec; median = 55.5 sec; mode = 49 sec
19. Mean = 0.81237 pound; median = 0.8161 pound; 0.7901 is an outlier; excluding the outlier, mean = 0.81608 pound and median = 0.8163 pound
21. Median
23. Median
25. Mean
27. a. One peak b.

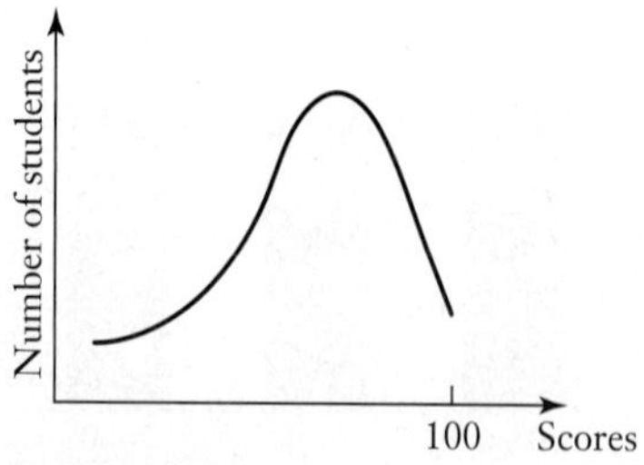

c. Left-skewed d. Large variation
29. a. One peak b.

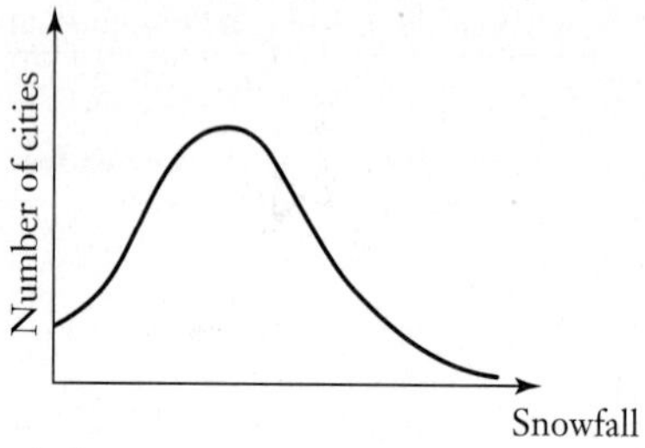

c. Right-skewed d. Large variation
31. a. One peak b.

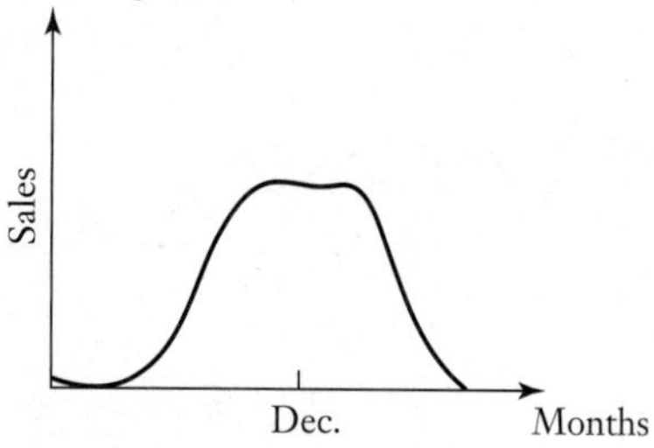

c. Symmetric d. Moderate variation
33. a. One peak b.

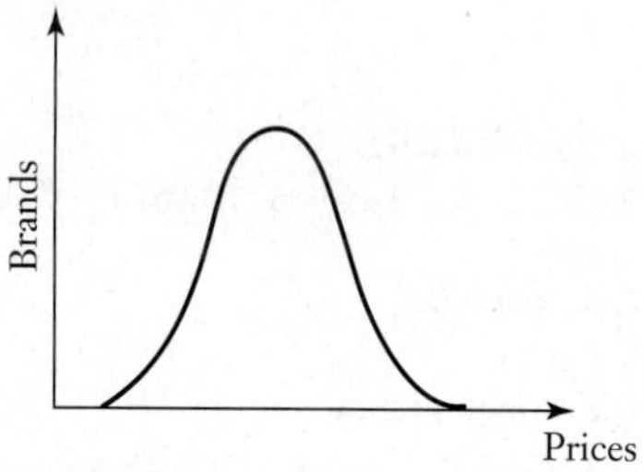

c. Symmetric d. Moderate variation
35. Symmetric, moderate variation
37. The distribution has two peaks (bimodal), is not symmetric, and has large variation.
39. The distribution has one peak, is symmetric, and has moderate variation.
41. a. Because the mean is greater than the median, the distribution is right-skewed.

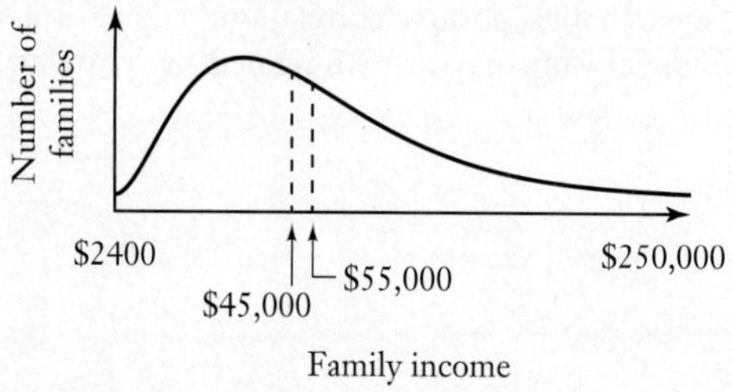

b. About 50% of the families earned less than $45,000.
c. No

43. a. 3.17 b. At least an A–

51. a. Team A: mean 134, median 125, mode 115; Team B: mean 131, median 135, no mode. Coach A can claim to have the team with the greater mean weight. Coach B can claim to have the team with the greater median weight.
b. Yes, in this case, because both data sets have the same number of values.
c. Yes, in this case, but not in general.

53. Answers will vary.

Unit 6B Quick Quiz

1. a 2. c 3. a 4. b 5. b 6. b 7. c 8. a 9. b 10. c

Unit 6B Exercises

7. Does not make sense
9. Makes sense
11. Makes sense
13. Mean = 7.2; median = 7.2
15. a. High-cost cities: mean = 166.0; median = 149; range = 83.
Low-cost cities: mean = 80.7; median = 81; range = 7.
b. High-cost: (145, 148, 149, 177, 228); low-cost: (76, 80, 81, 83, 83)
c. Stand. dev.: High-cost, 32.6; low-cost, 2.7

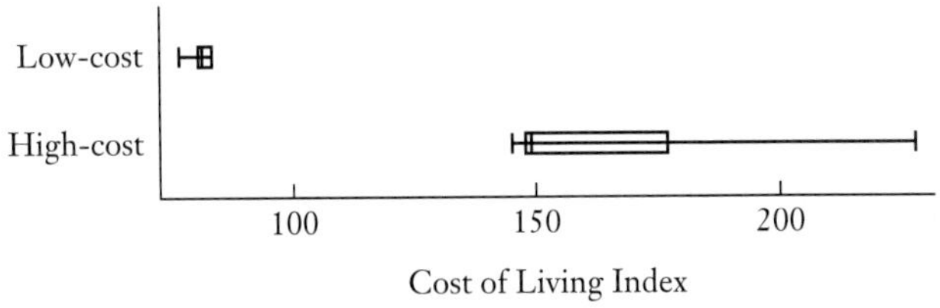

d. Range rule of thumb: High-cost, 20.8; low-cost, 1.8

17. a. National League (NL): mean = 0.5217; median = 0.5105; range = 0.220.
American League (AL): mean = 0.5005; median = 0.5245; range = 0.222.
b. NL: (0.420, 0.463, 0.5105, 0.586, 0.640);
AL: (0.364, 0.420, 0.5245, 0.584, 0.586)

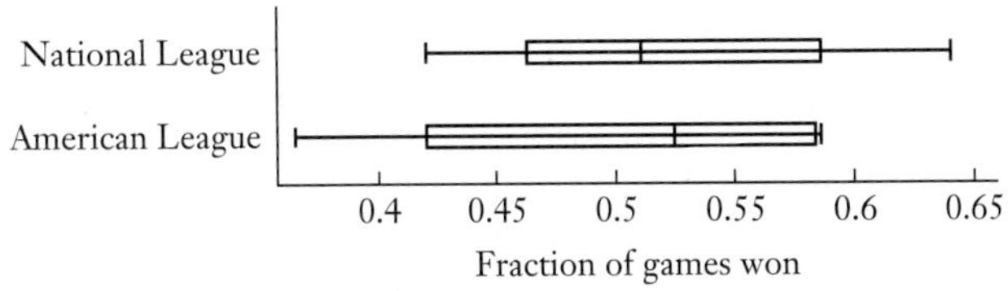

c. Stand. dev.: NL, 0.0819; AL, 0.0914
d. Range rule of thumb: NL, 0.0550; AL, 0.0555

19. a.

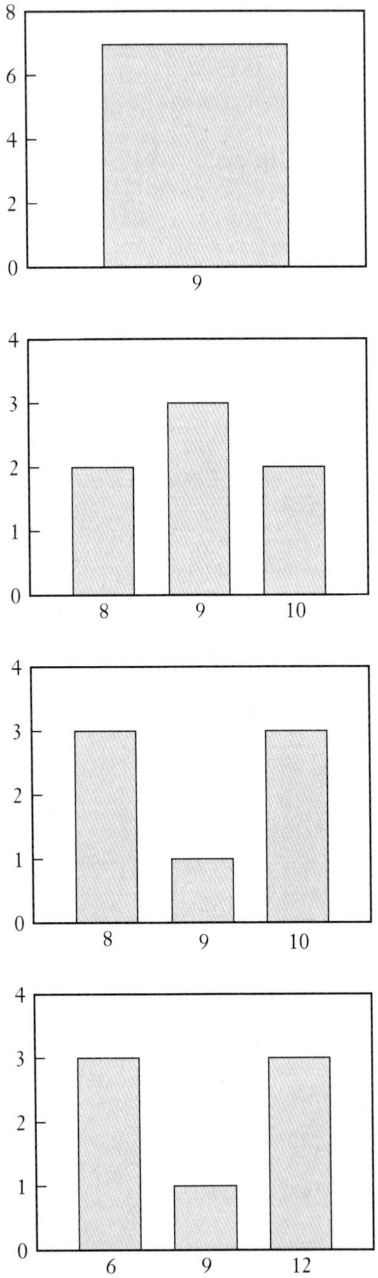

b.

	Set 1	Set 2	Set 3	Set 4
Lowest value	9	8	8	6
Lower quartile	9	8	8	6
Median	9	9	9	9
Upper quartile	9	10	10	12
Highest value	9	10	10	12

(Boxplot not shown.)

c. Stand. dev. = 0.000, 0.816, 1.000, 3.000

21. The means are nearly equal, but the variation is significantly greater for the second shop than for the first. If you want a reliable delivery time, choose the first shop.
23. A lower standard deviation means more certainty in the return and less risk.
25. The means are nearly equal. However, because Jerry's standard deviation is larger, he likely serves more small portions.
27. a. 2000: mean = 10.053, standard deviation = 0.100; 2008: mean = 9.923, standard deviation = 0.105; 2016: mean = 9.943, standard deviation = 0.080
29. a. Supplier A: mean = 16.32, standard deviation = 0.057; Supplier B: mean = 16.33, standard deviation = 0.098
b.

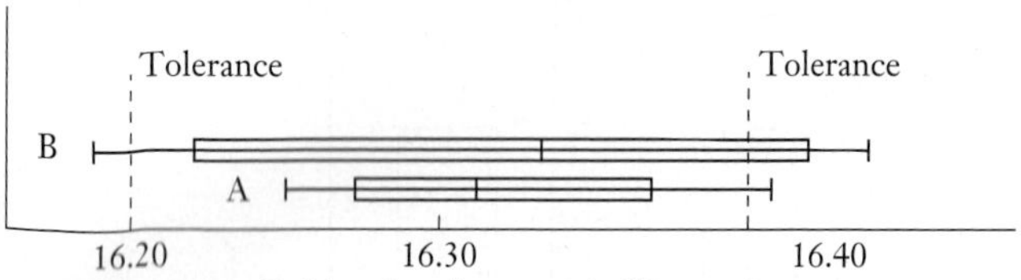

c. A: 86%; B: 57%
35. a. Range = 34; standard deviation = 1.14
b. Range = 15; standard deviation = 0.54
c.

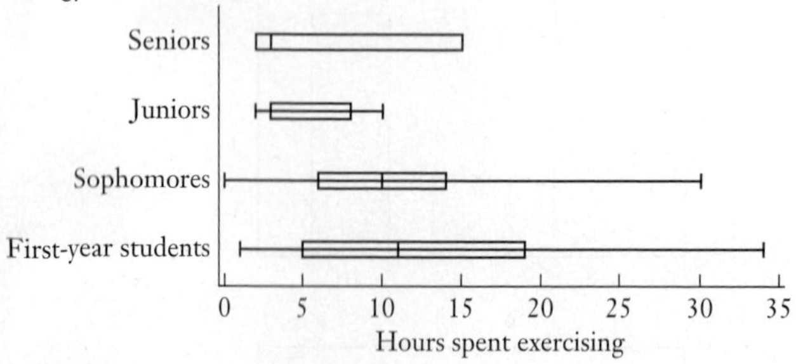

d.

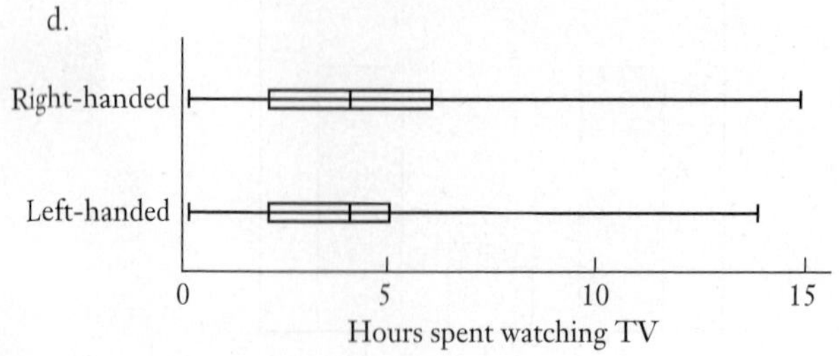

Unit 6C Quick Quiz

1. b 2. a 3. a 4. c 5. a 6. b 7. a 8. c 9. c 10. c

Unit 6C Exercises

5. Makes sense 7. Does not make sense
9. Makes sense
11. (a) and (c) are normal; (c) has the larger standard deviation.
13. Not normal 15. Normal
17. Not normal
19. a. 50% b. 16% c. 16% d. 97.5% e. 2.5% f. 84% g. 84% h. 68%
21. 16% 23. 51; 2.5%
25. 0 27. 1.5
29. a. $z = -1.0$; 15.87 percentile
b. $z = -0.5$; 30.85 percentile
c. $z = 1.5$; 93.32 percentile
31. a. $z = -0.52$; 0.52 standard deviations below the mean
b. $z = 0.52$; 0.52 standard deviations above the mean
c. $z = 0.12$; 0.12 standard deviations above the mean
33. 68% 35. 16%
37. (These answers are from software; answers using Table 6.3 will be approximate.)
a. $z = 0.16$; 56.4 percentile
b. $z = -1.04$; 14.9 percentile
c. $z = -1.52$; 6.4 percentile
d. $z = 0.48$; 68.4 percentile
39. Not likely 41. Verbal 161, quantitative 164
43. 18.4%
45. a. 41.1 percentile b. 79.5 percentile
47. 0.9 percentile; 99.1 percentile
51. a. Answers will vary because the data are random.

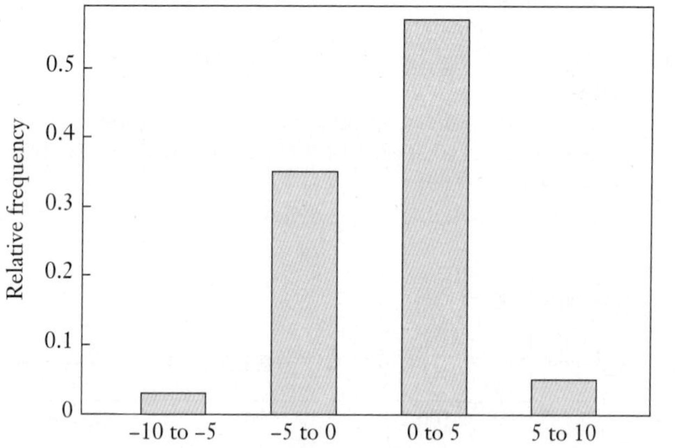

b.

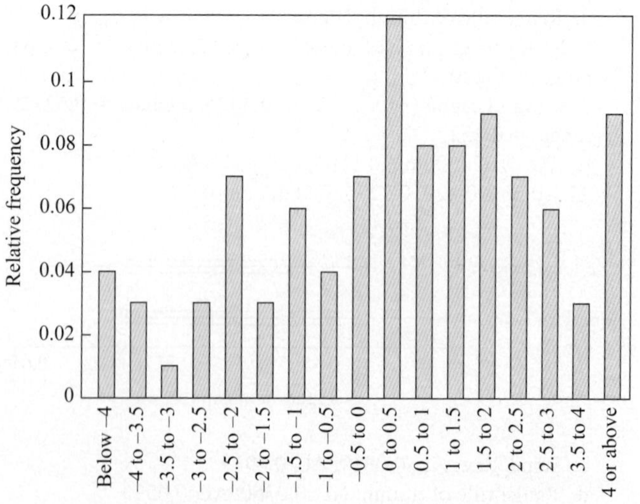

c. In general, the distribution should look more normal with more data values and less normal with fewer data values.

Unit 6D Quick Quiz

1. c 2. b 3. a 4. c 5. b 6. c 7. c 8. a 9. c 10. b

Unit 6D Exercises

9. Makes sense
11. Does not make sense
13. Makes sense
15. Significant
17. Not significant
19. Significant
21. Significant at the 0.05 and 0.01 levels
23. Not significant at the 0.05 or 0.01 level
25. Margin of error = 0.031, or 3.1 percentage points; confidence interval is 28.9% to 35.1%.
27. Margin of error = 0.022, or 2.2 percentage points; confidence interval is 80.8% to 85.2%.
29. Margin of error = 0.032, or 3.2 percentage points; confidence interval is 49.8% to 56.2%.
31. Margin of error = 0.025, or 2.5 percentage points; confidence interval is 79.5% to 84.5%.
33. a. Null hypothesis: graduation rate = 60%; alternative hypothesis: graduation rate > 60%
b. Rejecting the null hypothesis means there is evidence that the graduation rate exceeds 60%. Failing to reject the null hypothesis means there is insufficient evidence to conclude that the graduation rate exceeds 60%.
35. a. Null hypothesis: median salary = $57,900; alternative hypothesis: median salary > $57,900
b. Rejecting the null hypothesis means there is evidence that the median teacher salary in the state exceeds $57,900. Failing to reject the null hypothesis means there is insufficient evidence to conclude that the median teacher salary exceeds $57,900.
37. a. Null hypothesis: percentage of delayed flights = 15%; alternative hypothesis: percentage of delayed flights > 15%
b. Rejecting the null hypothesis means there is evidence that the percentage of delayed flights exceeds 15%. Failing to reject the null hypothesis means there is insufficient evidence to conclude that the percentage of delayed flights exceeds 15%.
39. Null hypothesis: mean annual mileage of cars in the fleet = 11,725 miles; alternative hypothesis: mean annual mileage of cars in the fleet is greater than 11,725 miles. The result is significant at the 0.01 level and provides good evidence for rejecting the null hypothesis.
41. Null hypothesis: mean stay = 2.1 days; alternative hypothesis: mean stay > 2.1 days. The result is not significant at the 0.05 level, and there are no grounds for rejecting the null hypothesis.
43. Null hypothesis: mean income = $50,000; alternative hypothesis: mean income > $50,000. The result is significant at the 0.01 level and provides good evidence for rejecting the null hypothesis.
45. Margin of error = 0.014, or 1.4 percentage points; confidence interval is 11.4% to 14.2%.
47. The sample size must be 4 times as great.
49. The margin of error, 3 percentage points, is consistent with the sample size.
55. Answers will vary.

CHAPTER 7

Unit 7A Quick Quiz

1. b 2. b 3. a 4. c 5. c 6. a 7. c 8. b 9. a 10. a

Unit 7A Exercises

7. Makes sense
9. Makes sense
11. Does not make sense
13. 48
15. 96
17. WW, WL, LW, LL; 0, 1, 2
19. 3/4
21. 1/13
23. 1/4
25. 30/365
27. 1/31
29. 5/9
31. A relative frequency of 15/100 = 0.15 is not near the expected frequency probability of 0.5. The coin is likely not fair.
33. Theoretical
35. 5/6
37. 0.31
39.

Result	Probability
0 boys	1/8
1 boy	3/8
2 boys	3/8
3 boys	1/8

41.

Sum	Probability
2	1/16
3	2/16
4	3/16
5	4/16
6	3/16
7	2/16
8	1/16

43. Odds for: 1 to 5; odds against: 5 to 1
45. Odds for: 1 to 35; odds against: 35 to 1
47. Gain of $12
49. 3/26
51. 5/13
53. 1/12
55. 1/5
57. 1/8
59. 1/10
61. 1/4
63. 1/15
65. 762/889, or about 6/7
67. a. WW, WB, WR, BW, BB, BR, RW, RB, RR
b.

Result	Probability
0 B	4/9
1 B	4/9
2 B	1/9

69. 112 combinations
71. 693 systems
73. a. 0.48 b. 0.41 c. 0.9 d. 0.28 e. 0.79
75. a. 0.51 b. 0.46
c. No, unless you also know the total numbers of men and women in the population.
77. Answers will vary.
85. Answers will vary.

Unit 7B Quick Quiz

1. c 2. c 3. b 4. c 5. a 6. c 7. a 8. b 9. c 10. b

Unit 7B Exercises

5. Makes sense
7. Does not make sense
9. Does not make sense
11. Experiment results will vary, but you will *not* always get at least one head in your two tosses.

13. Independent; 1/32
15. Independent; 1/216
17. Independent; 1/343
19. Dependent, 6/203
21. Dependent, 21/1292
23. Non-overlapping, 5/18
25. Non-overlapping, 1/13
27. Overlapping, 3/4
29. 7/8
31. Approximately 0.657
33. Approximately 0.140
35. 1/4
37. 31/32
39. 11/4165
41. 11/26
43. 0.922
45. Approximately 0.381
47. 1/32,768
49. Approximately 0.606
51. 13/16
53. 1/4
55. Approximately 0.0776
57. a. 0.986 b. 0.619
59. a. Dependent b. $10/33 \approx 0.303$
c. $25/81 \approx 0.309$ d. Nearly equal
61. Approximately 0.506
63. a. 1/20 b. $1/400 = 0.0025$ c. Approximately 0.401
65. a. 0.3 b. 0.21
c. 0.49. Making both is the most likely outcome.

Unit 7C Quick Quiz

1. b 2. c 3. c 4. c 5. b 6. a 7. c 8. c 9. c 10. a

Unit 7C Exercises

7. Makes sense
9. Does not make sense
11. Does not make sense
13. You shouldn't expect to get exactly 5000 heads. The proportion of heads should approach 0.5 as the number of tosses increases.
15. Expected value = −\$0.25; outcome of 1 game cannot be predicted; over 100 games, you can expect to lose.
17. Expected value = \$0; outcome of 1 game cannot be predicted; over 100 games; you can expect to break even.
19. Expected value = \$0; expected profit = \$0
21. 10 minutes
23. a. 46% heads; net loss of \$8
b. Yes; net loss of \$18
c. You need 59 heads, which is possible but not likely.
d. The probability of a head on any toss is always 0.5.
25. a. If you toss a head, the difference becomes 15; if you toss a tail, the difference becomes 17.
b. On each toss, the difference in heads and tails is equally likely to increase or decrease. After 1000 tosses, the difference is equally likely to be greater than 16 or less than 16.
c. Once you have fewer heads than tails, the deficit of heads is likely to remain.
27. Equally likely
29. The chance of hitting five free throws in a row is about 0.17, so by chance Maria could have such a game in a 25-game season.
31. a. −\$0.014; house edge is \$0.014.
b. Lose \$1.40
c. Lose \$7
d. \$14,000
33. Expected value = −\$0.58; loss over a year = \$300
35. Expected value = −\$0.77; loss = \$77
37. a. 0.9848, 0.9848
b. 0.9391, 0.9391
c. 0.4791, 0.9582
d. Before the change, one-point conversions were best. After the change, two-point conversions are best.
39. 2.6 people

Unit 7D Quick Quiz

1. b 2. c 3. b 4. c 5. c 6. a 7. c 8. c 9. b 10. a

Unit 7D Exercises

5. Does not make sense
7. Does not make sense
9. About once every 13 minutes; about once every 7 seconds
11. About \$1320 per person
13. a. 1.7 deaths per 100 million vehicle-miles in 1995; 1.1 deaths per 100 million vehicle-miles in 2015
b. 15.9 deaths per 100,000 people in 1995; 10.9 deaths per 100,000 people in 2015
c. 23.6 deaths per 100,000 licensed drivers in 1995; 16.1 deaths per 100,000 licensed drivers in 2015
d. They are generally consistent, as all three measures declined by similar amounts from 1995 to 2015. (The decline per vehicle-mile was slightly greater than the declines per person or per driver.)
15. a. 2000: 2.9 accidents per 1 million hours flown; 2015: 1.6 accidents per 1 million hours flown
b. 2000: 6.9 accidents per 1 billion miles flown; 2015: 3.6 accidents per 1 billion miles flown
17. 0.00024; 1.6 times greater
19. 205 people
21. Assuming a death rate of 11 per 1000, approximately 156,200 deaths
23. Loss of \$50 million
25. 133 years
27. a. 140 people b. 34 people
c. 1705 births per 100,000 people
d. 946 births per 100,000 people
29. a. 4.01 million births
b. 2.70 million deaths
c. By 1.31 million people
d. 1.70 million people; 57%
31. a.

Year	1950	1960	1970	1980	1990	2000	2010	2015
Percentage of Americans over age 65	8.4	9.6	10.3	11.5	12.8	12.4	13.1	14.9

b. 113%
c. 276%
d. Assuming half of Americans aged 25–65 worked, 84.5 million people contributed in 2015, which was less than double the over-65 population in that year.

Unit 7E Quick Quiz

1. c 2. c 3. c 4. b 5. b 6. c 7. c 8. b 9. c 10. c

Unit 7E Exercises

5. Makes sense
7. Makes sense
9. Makes sense
11. 720
13. 20
15. 3,326,400
17. 165
19. 56
21. 10,080
23. $10^7 = 10{,}000{,}000$
25. 7,893,600
27. 1680
29. 151,200
31. 924
33. 210
35. 175,760,000
37. 64
39. 64

41. 0.0297; 0.167
43. a. 160 sundaes b. 8000 cones c. 6840 cones d. 1140 cones
45. 8 toppings; 9 toppings
47. 1/906,192
49. 1/649,740
51. 1/495
53. 0.000240
55. a. 0.025; 50 people b. 0.00065; 1.3 people
57. 1 in 258,890,850, or about 0.000000004
63. a. 1/1,086,008 b. 1/3,838,380
65. Answers will vary.

CHAPTER 8

Unit 8A Quick Quiz

1. b 2. c 3. b 4. b 5. c 6. c 7. b 8. b 9. a 10. b

Unit 8A Exercises

5. Makes sense
7. Makes sense
9. Linear; 3600
11. Exponential; 3842 bolívars
13. Exponential; \$30.71
15. Linear; \$110,000
17. 32,768 grains; 65,535 grains; about 9.36 pounds
19. 2.64×10^{15} pounds
21. \$20,971.52
23. About 37 days
25. 2^{50} bacteria; 1/1024 full
27. 2.5 meters; more than knee-deep
29. a. (Only 100-year intervals are shown here.)

Year	Population
2200	9.6×10^{10}
2300	3.8×10^{11}
2400	1.5×10^{12}
2500	6.1×10^{12}
2600	2.5×10^{13}
2700	9.8×10^{13}
2800	3.9×10^{14}
2900	1.6×10^{15}
3000	6.3×10^{15}

b. Between 2800 and 2850 c. About 2150 d. No

31. a, b

	Month			
	Dec. 2013	Dec. 2014	Dec. 2015	Dec. 2016
Monthly active users (millions)	1228	1393	1591	1860
Absolute change over previous year (millions)	—	165	198	269
Percent change over previous year	—	13.4%	14.2%	16.9%

c. Both the absolute change and the percent change are increasing. So the growth is closer to exponential.

Unit 8B Quick Quiz

1. a 2. b 3. a 4. b 5. c 6. b 7. a 8. c 9. c 10. a

Unit 8B Exercises

9. Does not make sense
11. Does not make sense
13. True
15. False
17. False
19. False
21. True
23. a. 1.204 b. 4.301 c. −1.301 d. 2.107 e. −1.699 f. −1.505
25. 8; 16,384
27. 40 years
29. About 34,461 people; about 79,170 people
31. 10,321 cells; 16,777,216 cells
33. 8.9 billion; 17.8 billion; 42.4 billion
35.

Month	Population	Month	Population
0	100	8	172
1	107	9	184
2	114	10	197
3	123	11	210
4	131	12	225
5	140	13	241
6	150	14	258
7	161	15	276

Doubling time is just over 10 months.

37. 21.9 years; 1.10
39. 58.3 months; 1.15; 3.13
41. 1/4; 1/8
43. About 0.35; 0.044
45. About 420,000 animals; about 132,000 animals
47. 0.41 kg; 0.17 kg
49. 11.7 years; 0.05
51. 8.75 years; about 24,000 elephants
53. Approximate: 11.7 years; exact: 11.90 years; price = \$757
55. Approximate: 20 years; exact: 20.15 years; population = 281 million
57. The age of the universe is approximately 191,667 Pu-239 half-lives, so the amount of Pu-239 remaining today is negligible.
59. Higher by a factor of 3.95
61. About 23 years
63. (Approximate formula)
a. 906 mb b. 416 mb c. About 10%
69. a. 2.698970 b. 1.397940 c. −2.096910 d. −0.698970 e. −2.301030 f. 2.795880

Unit 8C Quick Quiz

1. c 2. b 3. b 4. b 5. c 6. a 7. c 8. a 9. a 10. c

Unit 8C Exercises

7. Makes sense
9. Does not make sense
11. Does not make sense
13. Doubling time ≈ 78 years; 2050 population ≈ 10.1 billion
15. Doubling time ≈ 44 years; 2050 population ≈ 12.6 billion
17. a. 1980 growth rate = 27.7 per 1000; 1995 growth rate = 32.5 per 1000; 2016 growth rate = 24.6 per 1000
19. a. 1980 growth rate = 4.7 per 1000; 1995 growth rate = −3.7 per 1000; 2016 growth rate = −2.3 per 1000
21. 3.3%; 2.0%; 0.67%

23. 2050: 409 million; 2100: 578 million
25. 2050: 451 million; 2100: 739 million
27. Answers will vary with student's age. For a student who was born in 2000: population at age 50 = 10.7 billion; population at age 80 = 14.9 billion; population at age 100 = 18.6 billion.
29. Base growth rate = 3.15%; predicted current growth rate = 0.53%; predicted rate is much lower than actual growth rate.
31. Base growth rate = 2.63%; predicted current growth rate = 1.32%; predicted rate is higher than actual growth rate.
33. 2% growth rate: 122,000; 149,000; 328,000; 5% growth rate: 164,000; 269,000; 1,950,000

Unit 8D Quick Quiz

1. a 2. b 3. a 4. b 5. b 6. c 7. c 8. b 9. c 10. a

Unit 8D Exercises

5. Does not make sense
7. Does not make sense
9. 7.9×10^{14} joules
11. 1.3×10^{16} joules
13. The bomb releases 200 times as much energy.
15. 10^8 times as loud
17. 60 dB
19. 31,623 times as loud
21. 16 times as great at 1 meter as at 4 meters
23. 64 times as great at 10 meters as at 80 meters
25. Decreases by a factor of 100,000; more basic
27. 3.16×10^{-10} mole per liter
29. pH = 4; acid
31. 1000 times
33. Earthquake's category: very minor; few effects
35. Serious effects
37. Shout could not be heard.
39. a. 40 dB b. 56 m c. 1.94 dB; factor of 640,000
41. a. 10^{-4} mole per liter
    b. 10^{-5} mole per liter; pH = 5
    c. 1.1×10^{-4} mole per liter; pH = 3.96
    d. The change in part (b) could be detected; the change in part (c) could not be detected.

CHAPTER 9

Unit 9A Quick Quiz

1. a 2. a 3. c 4. b 5. b 6. b 7. a 8. b 9. c 10. b

Unit 9A Exercises

5. Makes sense
7. Does not make sense
9.

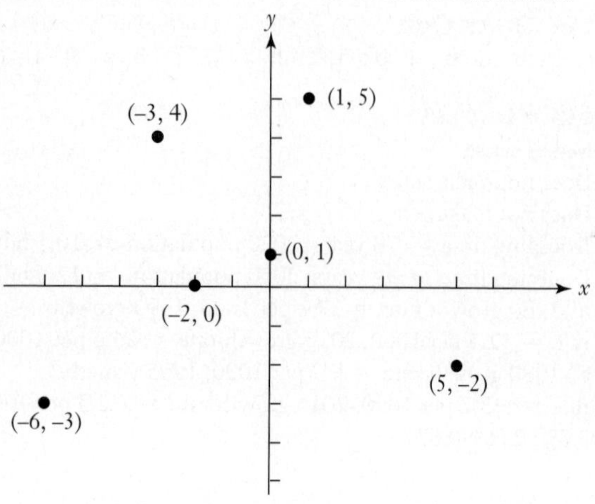

11. Independent variable, *time*; dependent variable, *distance fallen*
13. Independent variable, *price*; dependent variable, *demand*
15. As the volume of the tank increases, the cost of filling it also increases.
17. The ocean temperature decreases with increasing latitude.
19. As the average speed of the car increases, the travel time decreases.
21. As the gas mileage increases, the cost of driving a fixed distance decreases.
23. a. 24 in, 17 in, and 11 in
    b. 8000 ft, 14,000 ft, and 25,000 ft
    c. It appears that the pressure reaches 5 inches of mercury at an altitude of about 50,000 feet. The pressure approaches zero, but theoretically never reaches zero.
25. a. About 12.4 hours and 10.7 hours
    b. About April 19 and August 20 (Your answers may vary by a few days.)
    c. About January 31 and November 8 (Your answers may vary by a few days.)
    d. At 20°N latitude (closer to the equator), the graph would be flatter. At 60°N latitude (closer to the North Pole), the graph would be steeper. At 40°S latitude, the graph would look the same except that it would be shifted by half a year.
27. a. The independent variable is *age*, measured in years, and the dependent variable is *weight*, measured in pounds. We can take the domain to be ages between 0 and 40 years. The range is all weights between 8 pounds and about 130 pounds.
    b. The function shows that weight increases with age, increasing rapidly in the early years and then leveling off in the later years.
29. a. The independent variable is *altitude*; the dependent variable is *boiling temperature*. The domain is all altitudes between 0 and 9000 feet. The range is temperatures between about 190°F and 212°F.
    b.

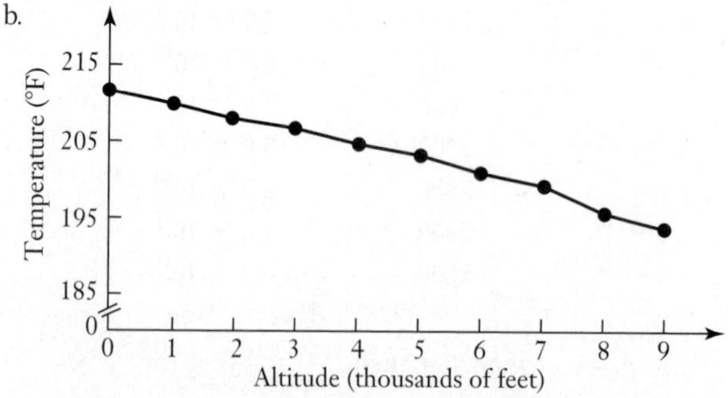

    c. The boiling point of water decreases with altitude.
31. a. The domain of the function relating *altitude* and *temperature* is all altitudes of interest—say, 0 to 15,000 feet (or 0 to 4000 meters). The range is all temperatures associated with the altitudes in the domain; the interval 30°F to 90°F (or about 0°C to 30°C) would cover all temperatures of interest.
    b.

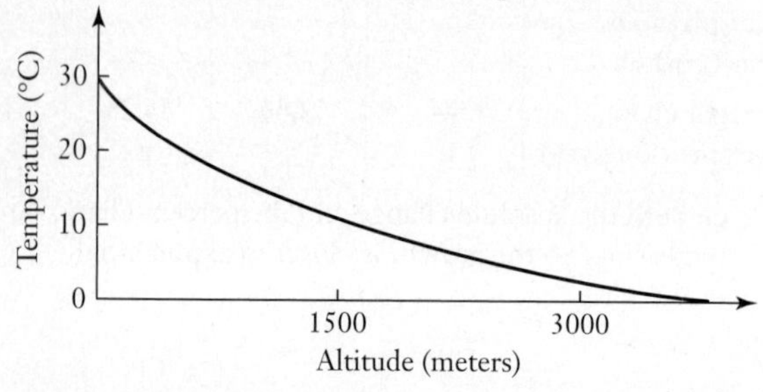

c. If based on reliable data, this graph is a good model of how the temperature varies with altitude.

33. a. The domain of the function relating *blood alcohol content* and *reflex time* consists of all reasonable BACs (in g/100 mL); for example, numbers between 0 and 0.25 g/mL would be appropriate. The range consists of the reflex times (in seconds) associated with those BACs.

b.

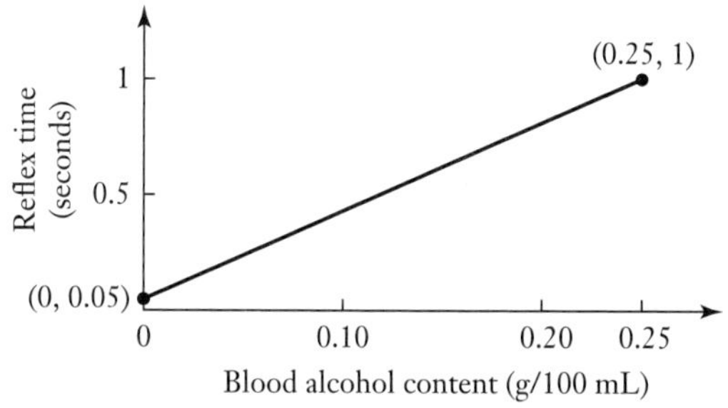

c. The validity of this graph as a model of alcohol impairment will depend on how accurately reflex times can be measured.

35. a. The domain of the function relating *time of day* and *traffic flow* consists of all times over a full day. It could be all the hours between 0 and 24 hours or all times between, say, 6:00 a.m. on Monday and 6:00 a.m on Tuesday. The range consists of all traffic flows (in units of number of cars per minute) observed at the times during the day.

b.

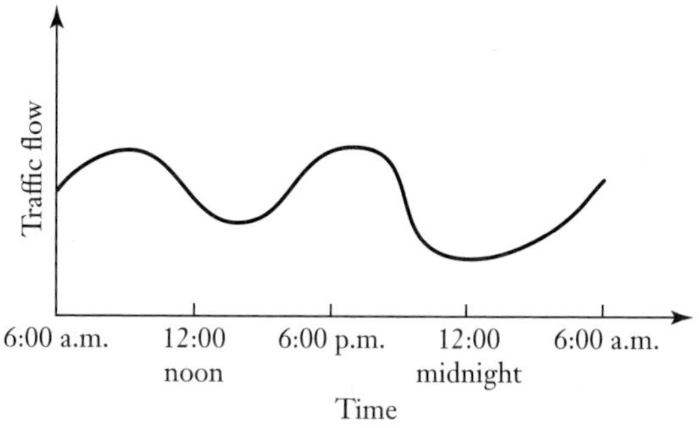

c. We expect light traffic flow at night, medium traffic flow during the midday hours, and heavy traffic flow during the two rush hours. The graph of this function would be a good model only if based on reliable data.

37. a. The domain of the function relating *number of people in a room* and *number of different handshakes* consists of the number of people of interest, say, $n = 2$ through $n = 10$. The domain consists of positive integers. The range consists of the number of different possible handshakes. Note that $f(2) = 1, f(3) = 1 + 2$, and in general,

$$f(n) = 1 + 2 + \cdots + (n - 1) = \frac{n(n-1)}{2}$$

b.

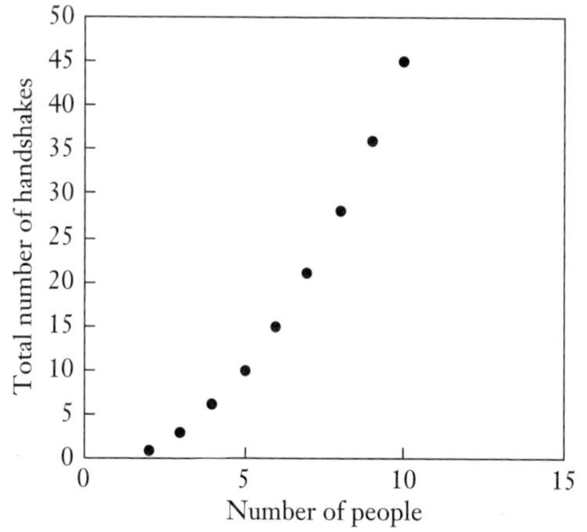

c. The model is exact.

39. a. The domain of the function relating *time* and *population of China* is all years from 1900 to, say, 2010. The range consists of the populations of China during those years—roughly from 400 million to 1.3 billion.

b.

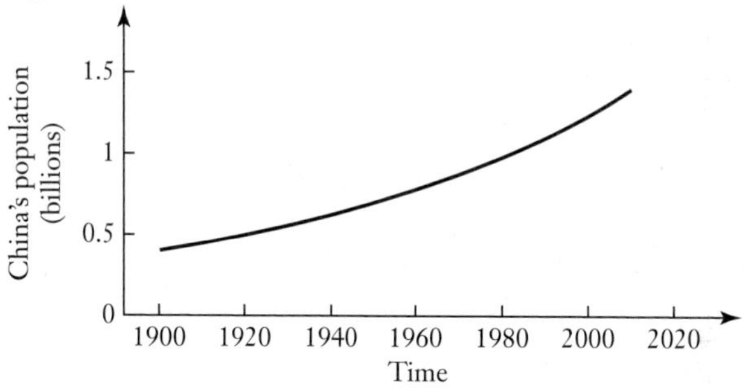

c. With accurate yearly data, this graph would be a good model of population growth in China.

41. a. The domain of the function relating *angle of cannon* and *horizontal distance traveled by cannonball* is all cannon angles between 0° and 90°. The range would consist (literally) of all ranges of the cannon (the horizontal distance traveled by the cannonball) for the various angles in the domain.

b.

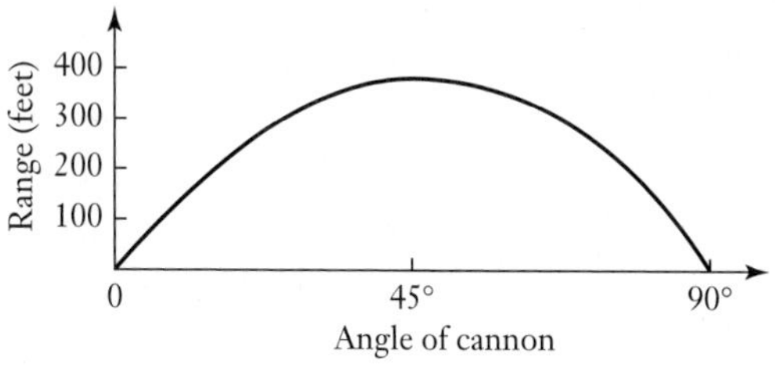

c. It is well known that a projectile has maximum range when the angle of firing is about 45°, so the graph above shows a peak at about 45°. This is a good qualitative model.

Unit 9B Quick Quiz

1. c 2. b 3. c 4. a 5. c 6. a 7. c 8. a 9. b 10. c

Unit 9B Exercises

7. Does not make sense
9. Makes sense
11. a. Rain depth increases linearly with time.
    b. $\frac{4}{3}$ in/hr
    c. Good model if rainfall rate is constant for 4 hours
13. a. On a trip back home, distance from home decreases linearly with time.
    b. −71.4 mi/hr
    c. Good model if speed is constant for 7 hours
15. a. Shoe size increases linearly with the height of the individual.
    b. 0.1375 size/in
    c. The model is a rough approximation at best.
17. The water depth decreases with respect to time at a rate of 2 inches per day. The rate of change is −2 in/day. In 8 days, the water depth decreases by 16 inches. In 15 days, the water depth decreases by 30 inches.
19. The Fahrenheit temperature changes with respect to the Celsius temperature at a rate of 9/5 degrees Fahrenheit per degree Celsius. The rate of change is 9/5 degrees Fahrenheit/degree Celsius. An increase of 5 degrees Celsius results in an increase of 9 degrees Fahrenheit. A decrease of 25 degrees Celsius results in a decrease of 45 degrees Fahrenheit.
21. Tony's weight decreases by 0.6 pounds per day. The rate of change is −0.6 pounds per day. After 15 days, he lost 9 pounds. After 4 weeks (28 days), he lost 16.8 pounds.
23. The independent variable is *time* (or t) measured in years. We let $t = 0$ represent today. The dependent variable is *price* (or p). The equation of the price function is $p = 18{,}000 + 900t$. The price of the car in 3.5 years will be $21,150. This function doesn't give a good model of car prices.
25. The variables in this situation are (*snow depth, maximum speed*), or (d, s), where snow depth is measured in inches and maximum speed is measured in mi/hr. The equation for the function is $s = 40 - 1.1d$. The plow has zero maximum speed when the snow depth reaches 36 inches, or 3 feet. The plowing rate most likely is not a constant, so this model is an approximation.
27. The variables in this problem are (*time, rental cost*), or (t, r), where t is measured in minutes. The cost per minute is $2/5 min = $0.40/min. The equation for the function is $r = 10 + 0.40t$. The number of minutes you can rent for $25 is almost 37.5 minutes. This function gives a good model of rental costs, *if* you are allowed to rent for any amount of time at the rate of $0.40/minute. However, many rentals require that you pay the full $2 for each 5-minute period, even if you do not use the full 5 minutes.
29. $w = 5.5 + 14.5t$; 78 pounds; 150.5 pounds. Model is accurate for small ages only.
31. $P = -350 + 10n$; 35 tickets
33. $V = 1500 - 125t$; 12 yr
35. The equation $y = 2x + 6$ describes a straight line with y-intercept (0, 6) and slope 2.

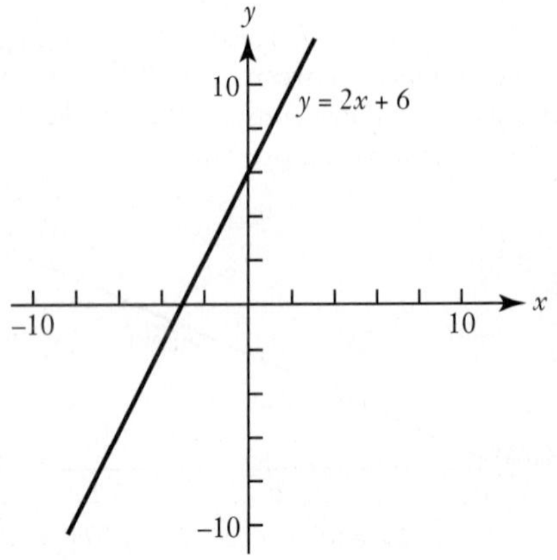

37. The equation $y = -5x - 5$ describes a straight line with y-intercept (0, −5) and slope −5.

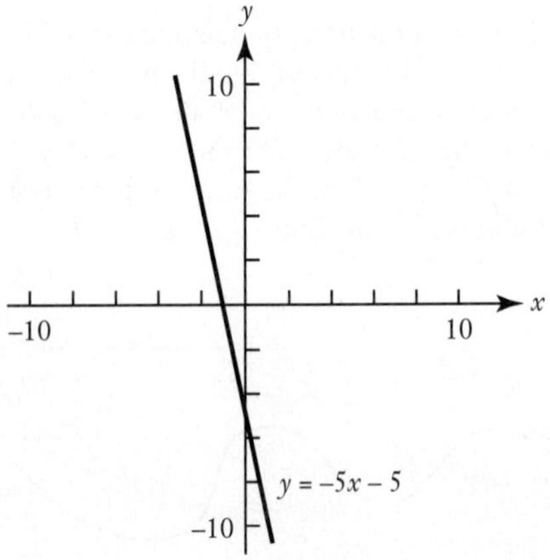

39. The equation $y = 3x - 6$ describes a straight line with y-intercept (0, −6) and slope 3.
41. The equation $y = -x + 4$ describes a straight line with y-intercept (0, 4) and slope −1.

Sketches for Exercises 39 and 41:

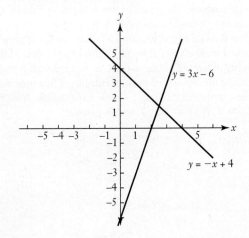

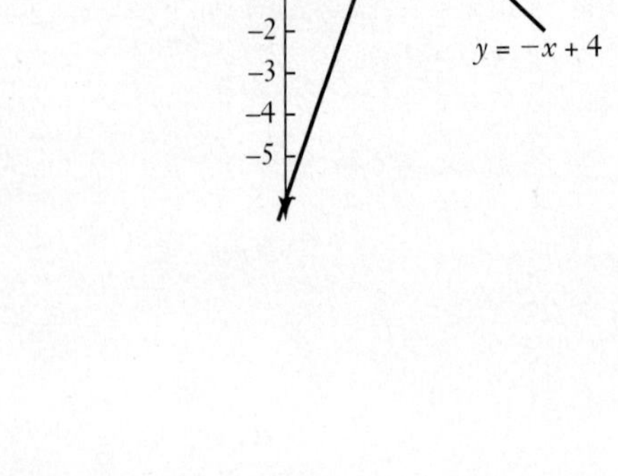

43. The variables are (*time, elevation*).

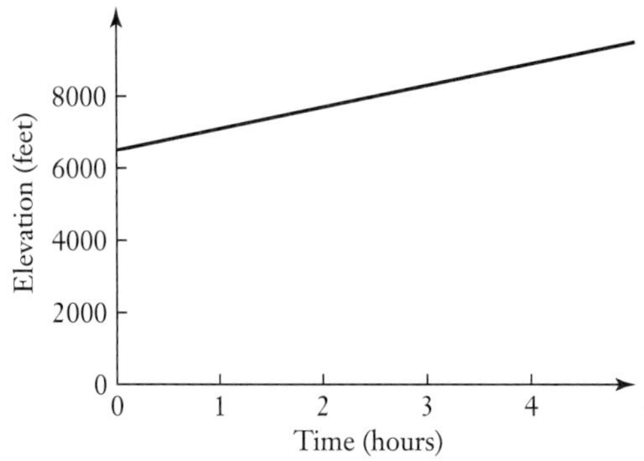

After 3.5 hours, their elevation is 8600 feet. Provided the rate of ascent is really a constant, this linear equation gives a good model of the climb.

45. The variables are (*number of posters, cost*).

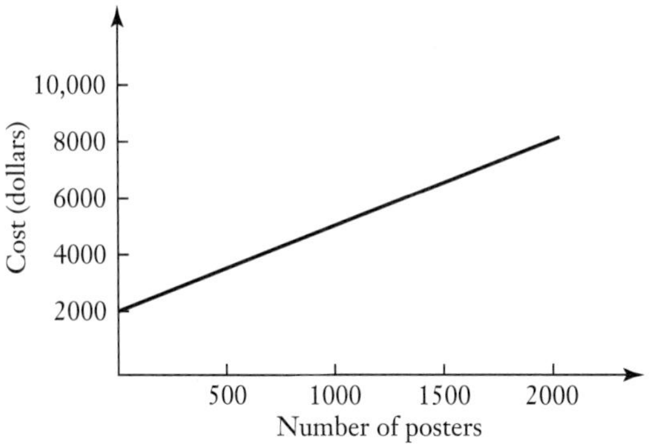

The cost of printing 2000 posters is \$8000. This function probably gives a fairly realistic estimate of printing costs.

47. The variables are (*time, cost*).

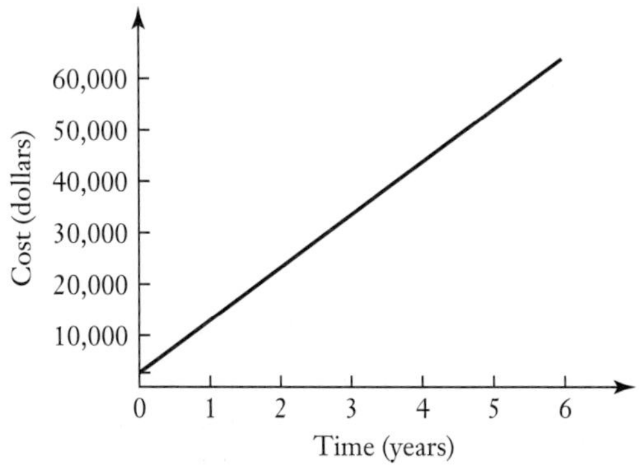

The cost of attending the school for 6 years is \$62,000. Provided costs do not change during the 6-year period, this function is an accurate model of the cost.

49. a. $N \approx 400$ b. $N \approx 800$ c. $N \approx 200/p$
d. The domain is all numbers between 0 and 1 (excluding 0 and including 1).

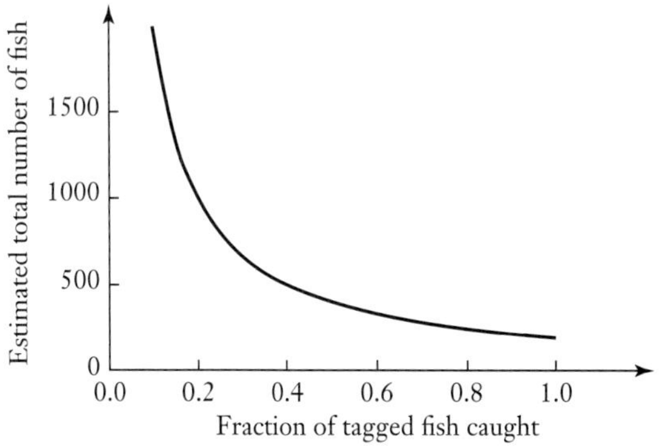

e. $N \approx 1333$ f. Answers will vary.

Unit 9C Quick Quiz

1. b 2. a 3. b 4. a 5. c 6. c 7. b 8. c 9. c 10. a

Unit 9C Exercises

7. Does not make sense
9. Makes sense
11. $x = 7$
13. $x = 4.18$
15. $x = 1.10$
17. $x = 3.40$
19. $x = 10{,}000$
21. $x = 3162.28$
23. $x = 25.12$
25. $x = 8.59$
27. a. $Q = 60{,}000 \times (1.025)^t$

b.

Year	Population
0	60,000
1	61,500
2	63,038
3	64,613
4	66,229
5	67,884
6	69,582
7	71,321
8	73,104
9	74,932
10	76,805

c.

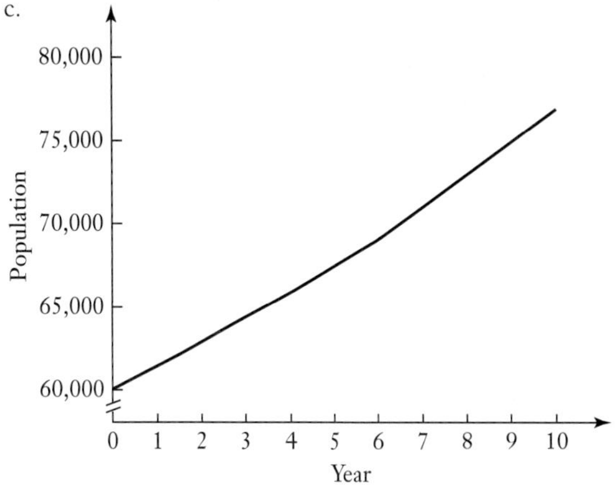

29. a. $Q = 10^6 \times (0.93)^t$

b.

Year	Millions of acres
0	1
1	0.93
2	0.86
3	0.80
4	0.75
5	0.70
6	0.65
7	0.60
8	0.56
9	0.52
10	0.48

c.

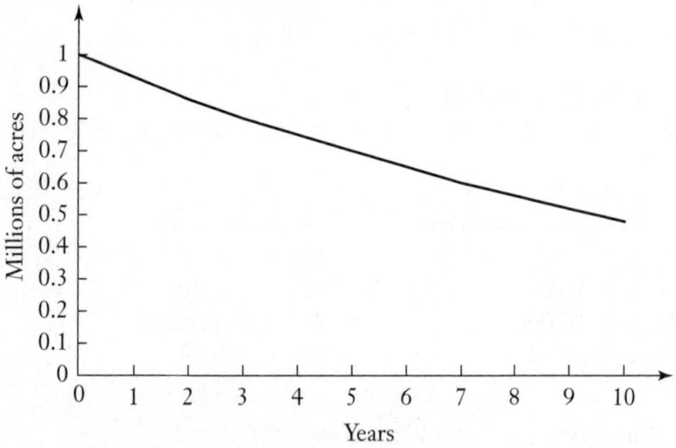

31. a. $Q = 175{,}000 \times (1.05)^t$

b.

Year	Average price
0	$175,000
1	$183,750
2	$192,938
3	$202,584
4	$212,714
5	$223,349
6	$234,517
7	$246,243
8	$258,555
9	$271,482
10	$285,057

c.

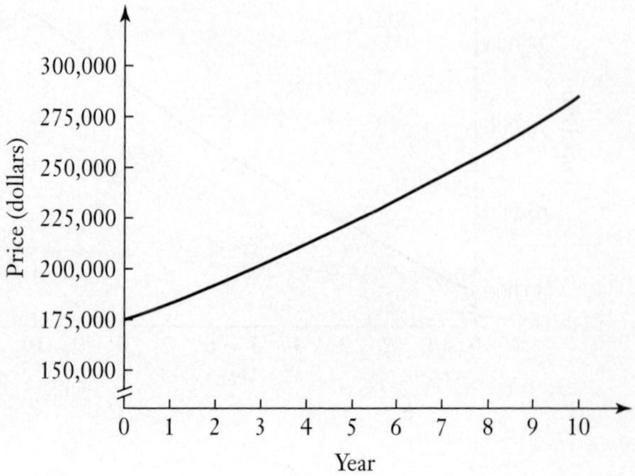

33. a. $Q = 2000 \times (1.05)^t$

b.

Year	Monthly salary
0	$2000.00
1	$2100.00
2	$2205.00
3	$2315.25
4	$2431.01
5	$2552.56
6	$2680.19
7	$2814.20
8	$2954.91
9	$3102.66
10	$3257.79

c.

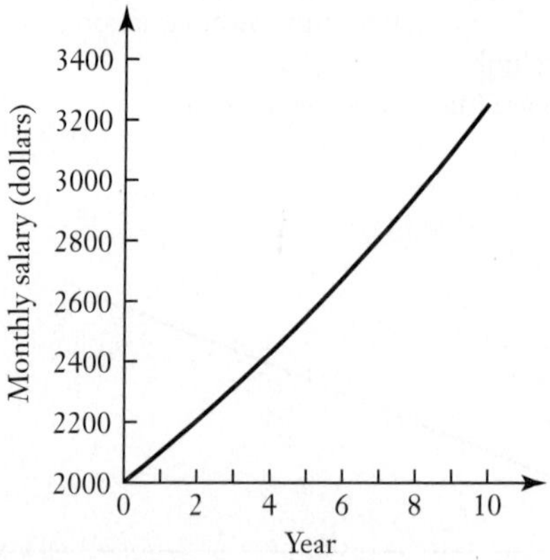

35. 34.5%
37. Approximately 5×10^{31}% per year; approximately 21% per day
39. In 54.97 years
41. a. 3.05 mg b. After 11.6 hours
43. a. 3.3 billion years old b. 3.9 billion years old
45. 20.2 mg/cm^2
47. $1.77 million
49. a. $Q = 316 \times (1.0044)^t$
b. In the year 2090

CHAPTER 10

Unit 10A Quick Quiz

1. a 2. b 3. c 4. c 5. b 6. a 7. c 8. c 9. c 10. a

Unit 10A Exercises

11. Does not make sense
13. Makes sense
15. Does not make sense
17. 180°
19. 72°
21. 80°
23. $\frac{1}{8}$
25. $\frac{1}{3}$
27. $\frac{1}{10}$
29. $\frac{5}{6}$
31. 37.7 m; 113.1 m^2
33. 72.3 ft; 415.5 ft^2
35. 125.7 mm; 1256.6 mm^2
37. 32 mi; 64 mi^2
39. 76 ft; 120 ft^2
41. 8.4 cm; 4.4 cm^2
43. 24 units; 24 units^2
45. 18 units; 12 units^2
47. 34.85 ft; 76.27 ft^2
49. 66 ft^2
51. 27,000 yd^2
53. 3750 m^3
55. 282.7 ft^3; 377.0 ft^2
57. Circumference is greater.
59. 54,000 m^3

61. The first tree (volumes are 785.4 ft^3 vs. 692.7 ft^3)
63. 900 times as great
65. 80 times as great
67. 512,000 times as many
69. 4
71. 64
73. Answers will vary. Example: 32 in; 38.4 in
75. Squirrels have a higher surface-area-to-volume ratio.
77. The Moon's surface-area-to-volume ratio is four times as great as Earth's.
79. Softball: 50.3 in^2, 33.5 in^3, ratio $\approx$ 1.5; bowling ball: 452.4 in^2, 904.8 in^3, ratio $\approx$ 0.5
81. a. 3 b. 2 c. 1 d. A corner of a page
83. The third line must be perpendicular to both lines.
85. a. 1.05×10^8 mm^2 = 105 m^2; 5.8×10^6 mm^3
b. 112 mm; 1.56×10^5 mm^2; air sacs have about 670 times as much surface area.
c. 2.9 m; for their relatively small volume, the air sacs have a very large surface area.
87. 0.0038 km^3 = 3.8×10^6 m^3

Unit 10B Quick Quiz

1. c 2. b 3. a 4. b 5. b 6. b 7. a 8. a 9. a 10. b

Unit 10B Exercises

9. Does not make sense
11. Makes sense
13. Makes sense
15. 32°30′0″
17. 12°19′48″
19. 149°49′48″
21. 30.17°
23. 123.18°
25. 8.99°
27. 21,600′
29. 40°N, 4°W
31. 44°S, 101°E
33. Buenos Aires
35. 1200 mi
37. 0.53°
39. 1.31×10^6 km
41. Roof with a 1 in 4 pitch
43. Railroad track with a 1/25 slope
45. 8/12 $\approx$ 0.67; approximately 10 ft
47. 45°; yes, it is possible.
49. 13.3%
51. a. 0.95 mi b. 0.78 mi
53. a. 0.85 mi b. 0.65 mi
55. a. 0.98 mi b. 0.71 mi
57. 1.78 acres
59. 25.84 acres
61. Similar
63. Not similar
65. $x = 4, y = 3$
67. $x = 15, y = 33.3$
69. 40.8 ft
71. 32 ft
73. Circular: 199 m^2; square: 156 m^2
75. Circular: 1790 m^2; square: 1406 m^2
77. Can 1 costs \$163.36; Can 2 costs \$219.91.
79. Cube; \$3.60
81. a. 89.7 cm^2 b. 557 million bytes/cm^2
c. 2,990,796 cm, 18.6 mi
83. 127.3 ft
85. Riding the bicycle is faster.
87. a. 0.525 ft deep b. 2621 lb
89. 100 m; no; a square 20 m × 20 m corral would require 80 m of fence.
91. \$3736; \$3414
93. a. Radius = 3.8 cm, height = 7.7 cm
95. a. 75.4 ft^3 b. 4.7 ft high
c. Assuming 10,000 grains per cubic inch, there are about 1.3 billion grains.
97. a. About 1.6 times the length of a football field
b. 91,636,272 ft^3, 3,393,936 yd^3
c. About 2,263,000 blocks
d. 21.5 years
e. About 5% of the volume of the Great Pyramid

Unit 10C Quick Quiz

1. b 2. a 3. b 4. c 5. c 6. c 7. a 8. c 9. a 10. b

Unit 10C Exercises

9. Makes sense
11. Does not make sense
13. Makes sense
15. The dimension is 1 and the object is ordinary (non-fractal).
17. The dimension is 3 and the object is ordinary (non-fractal).
19. The dimension is 2.585 and the object is a fractal.
21. The dimension is 1 and the object is ordinary (non-fractal).
23. The dimension is 3 and the object is ordinary (non-fractal).
25. The dimension is 2.113 and the object is a fractal.
27. a. When the ruler size is reduced by a factor of $R = 4$, there are $N = 8$ times as many elements.
b. 1.5
c. The objects that lead to the quadric Koch island all have the same area, because each time a piece of the boundary juts out (adding area) another piece juts in (removing the same amount of area). The length of the coastline of the island is infinite.
29. When the ruler's length is reduced by a factor of $R = 9$, the ruler will find 4 elements; fractal dimension = 0.631.
31. a. 1.301 b. 1.585 c. 2.585
33. The branching in many natural objects has the same pattern repeated on many different scales. This is the process by which self-similar fractals are generated. Euclidean geometry is not equipped to describe the repetition of patterns on many scales.

CHAPTER 11

Unit 11A Quick Quiz

1. b 2. a 3. a 4. c 5. a 6. c 7. c 8. b 9. b 10. a

Unit 11A Exercises

7. Does not make sense
9. Makes sense
11. 440 cps; 880 cps; 1760 cps; 3520 cps
13.

Note	Frequency (cps)
F	347
F#	368
G	389
G#	413
A	437
A#	463
B	491
C	520
C#	551
D	584
D#	618
E	655

15. a. 390 cps b. 437 cps c. 779 cps
d. 1102 cps e. 2474 cps

17. 292 cps; 245 cps

19. b. 2.245

c.

Note	Frequency (cps)
C	260
G	390
D	584
A	875
E	1310
B	1963
F#	2942
C#	4407
G#	6604
D#	9894
A#	14,825
F	22,212
C	33,280

d. 12 fifths; 7 octaves e. 128 times the frequency

21. In 4/4 time, there are 4 quarter notes in a measure. This means that there are 2 half notes in a measure, 8 eighth notes in a measure, and 16 sixteenth notes in a measure.

Unit 11B Quick Quiz

1. b **2.** b **3.** a **4.** c **5.** c **6.** b **7.** a **8.** c **9.** a **10.** c

Unit 11B Exercises

9. Does not make sense

11. Makes sense **13.** Does not make sense

15. a. A vanishing point for the drawing is that point at which the edges of the road appear to meet. According to the definition given in the text, this vanishing point is not the principal vanishing point. The principal vanishing point is the apparent intersection point in the picture of all the lines that are parallel in the real scene and perpendicular to the canvas. The road in this scene is not perpendicular to the canvas.

b.

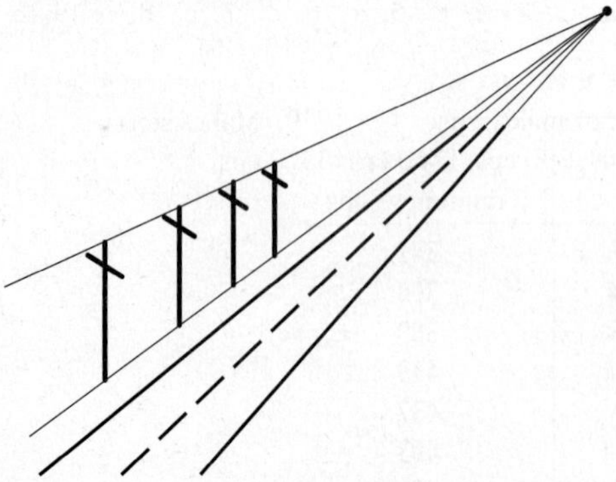

17.

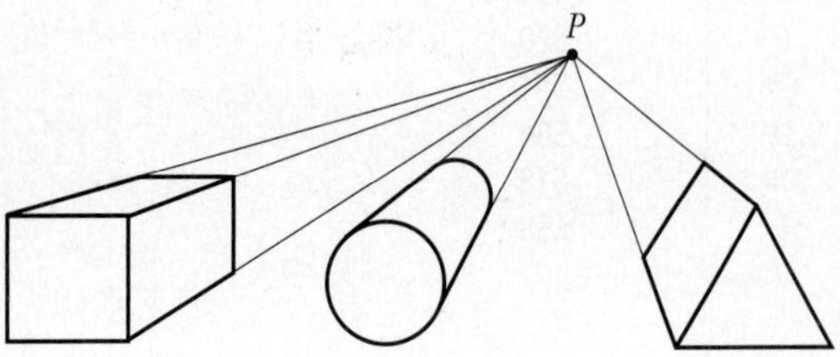

19. a. Scale has been reduced:

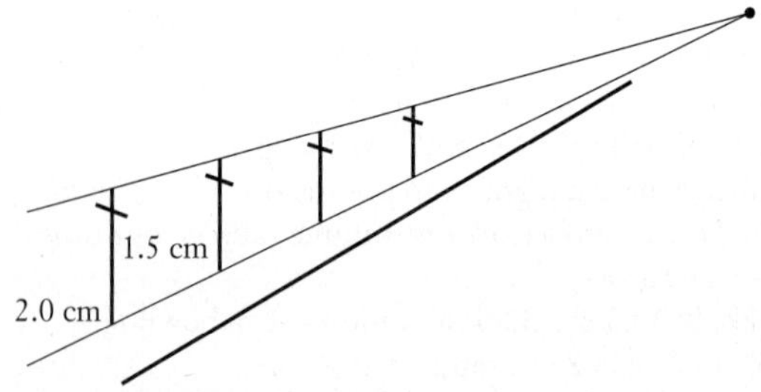

b. Approximately 1.28 cm and 1.02 cm
c. No

21. a. *A, H, I, M, O, T, U, V, W, X, Y* b. *B, C, D, E, H, I, K, O, X*
c. *H, I, O, X* d. *H, I, N, O, S, X, Z*

23. a. 120°; 240° b. 90°; 180°; 270° c. 72°; 144°; 216°; 288°
d. A regular polygon with n sides can be rotated through $360°/n$ and multiples of this angle, and its appearance will remain the same; $n - 1$ different angles for an n-gon.

25. It has reflection symmetries; it can be reflected across a vertical line through its center or a horizontal line through its center, and its appearance remains the same. It has rotation symmetry; it can be rotated through 180°, and its appearance remains the same.

27. It has rotation symmetry with angles of $360° \div 6 = 60°$ and multiples of 60°. It can also be reflected across six lines through its center and retain its appearance.

29.

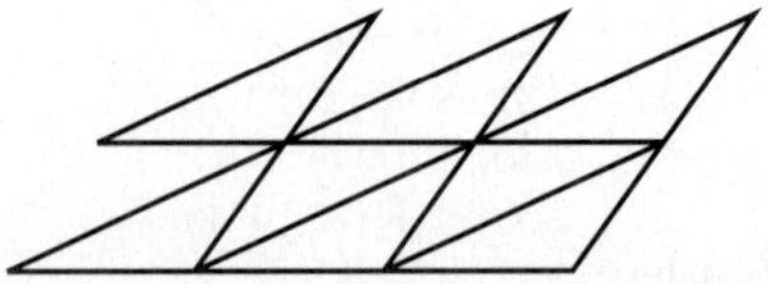

31.

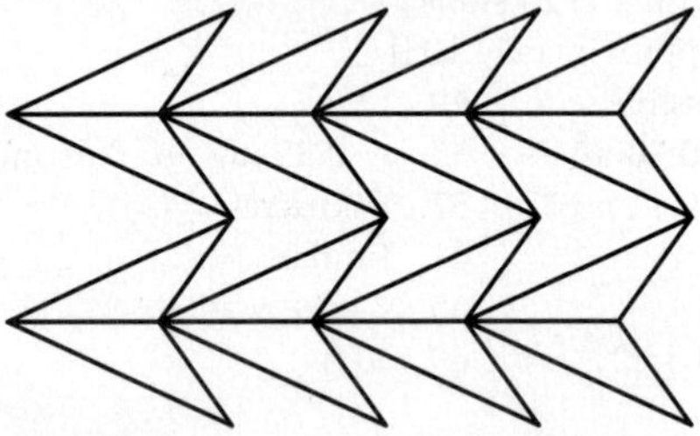

33.

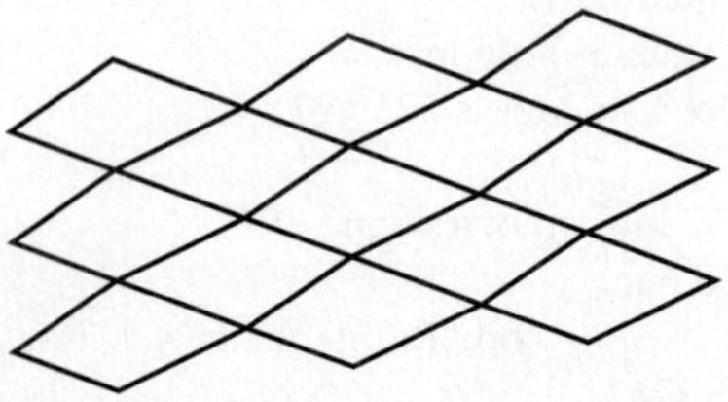

35. The angles around a point *P* are precisely the angles that appear inside of a single quadrilateral. Thus, the angles around *P* have a sum of 360°, and the quadrilaterals around *P* fit together perfectly.

Unit 11C Quick Quiz

1. c **2.** c **3.** c **4.** b **5.** b **6.** a **7.** a **8.** b **9.** b **10.** b

Unit 11C Exercises

7. Does not make sense
9. Does not make sense
11. The longer segment should have a length of 3.71 inches, and the shorter segment should have a length of 2.29 inches.
13. Third and fifth rectangles
15. 9.40 m or 3.58 m
17. 1.07 cm or 0.41 cm
19. Answers will vary but should show work clearly.
21. Drawing should look like Figure 11.45b, with measurements shown clearly.
23.

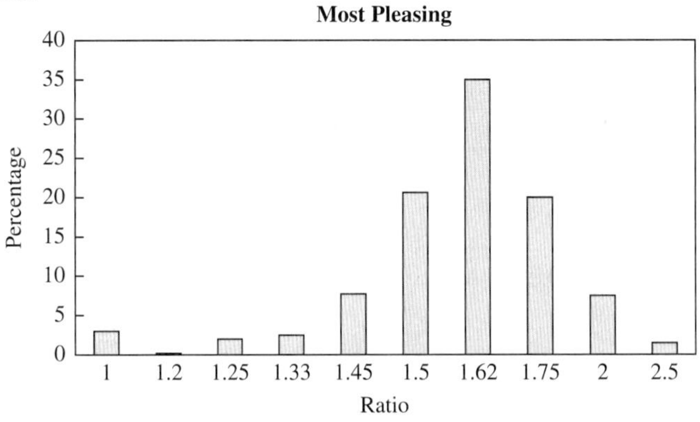

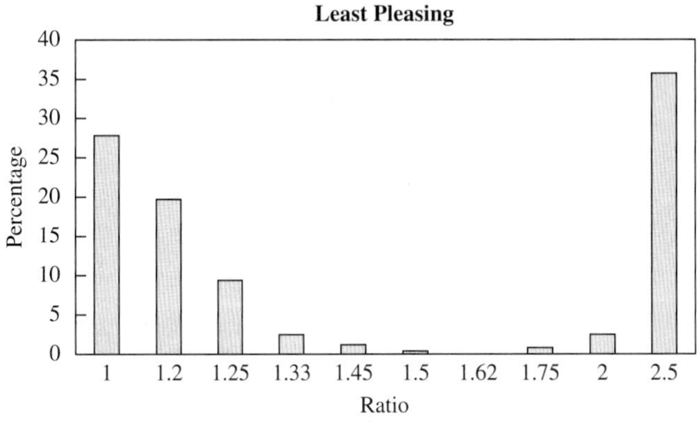

25. a.

a	b	$\frac{a+b}{b}$
38	62	1.6129
28	46	1.6087
56	102	1.549
56	88	1.6364
24	36	1.6667
77	113	1.6814
40	69	1.5797
46	60	1.7667
15	18	1.8333
39	63	1.619
53	67	1.791

b. While the values of $(a + b)/b$ are clustered around ϕ, their average is 1.668 and there are more values greater than ϕ. Therefore, the values of $(a + b)/b$ are not well predicted by ϕ.

CHAPTER 12

Unit 12A Quick Quiz

1. b 2. c 3. b 4. c 5. b 6. c 7. a 8. c 9. b 10. b

Unit 12A Exercises

9. Does not make sense
11. Makes sense
13. Makes sense
15. a. Hayes 47.9%, Tilden 50.9%; Tilden had a popular majority.
b. Hayes 50.1%, Tilden 49.9%; electoral winner was not the popular winner.
17. a. Harrison 47.8%, Cleveland 48.6%; no majority
b. Harrison 58.1%, Cleveland 41.9%; no
19. a. Clinton 43.0%, Bush 37.4%; no majority
b. Clinton 68.8%, Bush 31.2%; yes
21. a. Bush 47.9%, Gore 48.4%; no majority
b. Bush 50.5%, Gore 49.5%; electoral winner was not the popular winner.
23. a. Obama 52.9%, McCain 45.7%; Obama had a popular majority.
b. Obama 67.8%, McCain 32.1%; yes
25. a. $62/100 = 62\%$ of the senators would vote to end the filibuster, so the filibuster would likely end and the bill could come to a vote.
b. To get 2/3 of an 11-member jury requires at least 8 votes. Therefore, there will be no conviction.
c. Only 72% of the states support the amendment, so it fails to pass.
d. The override will get $68/100 = 68\%$ of the votes in the Senate and $270/435 \approx 62\%$ of the votes in the House, so the veto will not be overturned.
27. a. Clinton 43.3%, Bush 37.7%, Perot 19.0%; Clinton won the popular vote by a plurality but not by a majority.
b. Clinton 68.8%, Bush 31.2%, Perot 0%; Clinton won the electoral vote by a plurality and also by a majority.
c. If Perot had dropped out of the election and Bush had won most of Perot's popular votes, then Bush could have won the popular vote. For Bush to win, his additional popular votes would need to have been distributed among the states in a way that put him ahead of Clinton in many of the states that Clinton had won.
d. If Bush had dropped out of the election and Perot had won most of Bush's popular votes, then Perot could have won the popular vote. However, his additional popular votes would need to have been distributed among states in a way that allowed him to win the electoral vote.
29.

First	A	C	B	C
Second	B	B	C	A
Third	C	A	A	B
No. of voters	22	20	16	8

31. a. Plurality: C
b. Single runoff: C
c. Sequential runoff: C
d. Point system: B
e. Pairwise comparisons: B

33. 22
35. A 18, B 16, C 12, D 9
37. a. A total of 66 votes were cast.
 b. D is the plurality winner (but not by a majority).
 c. D is the winner of the top-two runoff.
 d. The winner of a sequential runoff is D.
 e. D is the winner by the Borda count.
 f. D wins by the method of pairwise comparisons.
 g. As the winner by all five methods, Candidate D is clearly the winner of the election.
39. a. A total of 100 votes were cast.
 b. C is the plurality winner (but not by a majority).
 c. A is the winner of the top-two runoff.
 d. The winner of a sequential runoff is A.
 e. B is the winner by the Borda count.
 f. B wins by the method of pairwise comparisons.
 g. There is no clear winner.
41. a. A total of 90 votes were cast.
 b. E is the plurality winner (but not by a majority).
 c. B is the winner of the top-two runoff.
 d. Because only three candidates received first-place votes, the sequential runoff and the top-two runoff give the same result: B wins.
 e. E is the winner by the Borda count.
 f. D is the winner by the method of pairwise comparisons.
 g. Candidates B and E each win by two of the five methods, so the outcome is debatable.
43. a. Best won a plurality but not a majority.
 b. Imagine that there were 100 votes cast in the entire election. Then Able would need 15 votes to win 50% of the votes. Fifteen votes represents $15/23 \approx 65.2\%$ of Crown's votes.
45. a. Irving has a plurality but not a majority.
 b. Heyduke would need 83 additional votes.
47. B is the Condorcet winner.
49. a. 6 pairs b. 10 pairs c. 15 pairs
51. D has 45 points, and A wins the election.

Unit 12B Quick Quiz

1. c 2. a 3. b 4. c 5. b 6. a 7. a 8. b 9. b 10. c

Unit 12B Exercises

5. Makes sense
7. Does not make sense
9. Assume that a candidate receives a majority of the votes. Then she or he is the only candidate to receive a plurality and wins by the plurality method. Thus, Criterion 1 is satisfied.
11. The following preference schedule is just one possible example.

First	B	A	C	C
Second	A	B	A	B
Third	C	C	B	A
No. of voters	2	4	2	3

13. Criterion 4 is violated when C drops out.
15. Assume that a candidate receives a majority of the votes. With either runoff method, votes are redistributed as candidates are eliminated. But it is impossible for another candidate to accumulate enough votes to overtake a candidate who already has a majority.
17. Criterion 2 is violated.
19. Criterion 3 is violated (if the four voters in the third column of the table move C above A).
21. The following preference schedule is just one possible example.

First	A	B	C
Second	B	A	B
Third	C	C	A
No. of voters	4	3	5

23. The following preference schedule is just one possible example.

First	A	B	C
Second	B	C	B
Third	C	A	A
No. of voters	4	2	1

25. The following preference schedule is just one possible example.

First	A	D	C
Second	B	B	B
Third	C	A	D
Fourth	D	C	A
No. of voters	4	8	3

27. Criterion 4 is violated.
29. Assume Candidate A wins a majority of the first-place votes. Then in every head-to-head race with another candidate, A must win (by a majority). Thus, A wins every head-to-head race and is the winner by pairwise comparisons, so Criterion 1 is satisfied.
31. Suppose Candidate A wins by the method of pairwise comparisons and in a second election moves up above Candidate B in at least one ballot. A's position relative to B remains the same or improves and A's position and B's position relative to the other candidates remain the same, so A must win the second election.
33. The following preference schedule is just one possible example.

First	A	A	B	E	E
Second	B	C	A	B	D
Third	C	D	C	A	B
Fourth	D	E	D	C	A
Fifth	E	B	E	D	C
No. of voters	1	3	2	1	2

35. a. Candidate C wins by the plurality method.
 b. Candidate B wins by approval voting.
37. Alaska
39. Rhode Island
41. Alaska, Rhode Island, Illinois, New York
43. Criterion 1 does not apply; Criteria 2 and 3 are satisfied; in this case (but not in general), Criterion 4 is satisfied.
45. Criterion 1 does not apply; Criterion 2 is satisfied; the point system always satisfies Criterion 3; Criterion 4 is satisfied.

47. Criterion 1 does not apply; Criterion 2 is violated; the plurality method always satisfies Criterion 3; Criterion 4 is violated.
49. Criterion 1 does not apply; Criterion 2 is violated; Criterion 3 is satisfied; Criterion 4 is violated.
51. Criterion 1 does not apply; the method of pairwise comparisons always satisfies Criteria 2 and 3; Criterion 4 is satisfied.

Unit 12C Quick Quiz

1. c 2. b 3. a 4. b 5. b 6. b 7. c 8. c 9. c 10. a

Unit 12C Exercises

9. Makes sense 11. Does not make sense
13. 804,598 people per representative; 11,667 representatives
15. 5.03; represented fairly (very slightly underrepresented)
17. 26.47; overrepresented 19. 6.48, 8.30, 9.85, 10.37
21.

State	A	B	C	D	Total
Population	914	1186	2192	708	5000
Standard quota	18.28	23.72	43.84	14.16	100
Minimum quota	18	23	43	14	98
Fractional remainder	0.28	0.72	0.84	0.16	—
Final apportionment	18	24	44	14	100

23. 7, 8, 10, 10
25. 51, 36, 13; 52, 36, 13; paradox does not occur.
27. 46, 9, 4, 41; 47, 9, 4, 41; paradox does not occur.
29. 9, 70, 21; quota criterion is violated because State B's standard quota is 68.97.
31. 7, 70, 16, 7; quota criterion is satisfied.
33. 7, 8, 10, 10; modified divisor 38.4
35. 7, 8, 10, 10; modified divisor 38.4
37. 15, 82, 3; 15, 81, 3, 6; State B's representation decreases.
39. a. 54, 34, 12
 b. 54, 34, 12; modified divisor 9.90
 c. 54, 34, 12; modified divisor 9.99
 d. 54, 34, 12; modified divisor 9.99
41. a. 9, 27, 26, 38
 b. 8, 27, 26, 39; modified divisor 98.3
 c. 8, 27, 26, 39; modified divisor 99.5
 d. 8, 27, 26, 39; modified divisor 99.5
43. a. 1, 3, 6
 b. 1, 2, 7; modified divisor 35.0
 c. 1, 3, 6; modified divisor 38.0
 d. 1, 3, 6; modified divisor 38.0
45. a. 3, 10, 5, 7
 b. 3, 10, 5, 7; modified divisor 0.76
 c. 3, 10, 5, 7; modified divisor 0.77
 d. 3, 10, 5, 7; modified divisor 0.78

Unit 12D Quick Quiz

1. c 2. b 3. b 4. c 5. b 6. c 7. a 8. b 9. a 10. b

Unit 12D Exercises

7. Makes sense 9. Does not make sense
11. Does not make sense
13. a. 2010: Republican 56%, Democrat 44%; 2012: Republican 52%, Democrat 48%
 b. 2010: Republican 72%, Democrat 28%; 2012: Republican 75%, Democrat 25%
15. a. 2010: Republican 68%, Democrat 32%; 2012: Republican 60%, Democrat 40%
 b. 2010: Republican 72%, Democrat 28%; 2012: Republican 67%, Democrat 33%
17. a. 2010: Republican 52%, Democrat 48%; 2012: Republican 49%, Democrat 51%
 b. 2010: Republican 63%, Democrat 37%; 2012: Republican 72%, Democrat 28%
19. a. 8 Republicans, 8 Democrats
 b. Maximum: 15 Republicans; minimum: 1 Republican
21. a. 6 Republicans, 6 Democrats
 b. Maximum: 11 Republicans; minimum: 1 Republican
23. a. 12 Republicans, 3 Democrats
 b. Maximum: 15 Republicans; minimum: 10 Republicans
25. The boundaries should divide the state into eight districts; each district is 4 blocks wide (in the horizontal direction) and 2 blocks high (in the vertical direction).
27. The boundaries should divide the state into eight districts; each district is 4 blocks wide (in the horizontal direction) and 2 blocks high (in the vertical direction).
29. It is not possible. 31. Answers will vary.
33. Answers will vary. All cases are possible except for 4 Republicans.
35. Answers will vary. 37. It is not possible.

INDEX

INDEX OF APPLICATIONS

Note: A = Activity, CS = Case Study, E = Example, IE = In-text Example, MI = Mathematical Insight, P = Problem, TE = Technology Exercise, UT = Using Technology, YW = In Your World

Arts and Entertainment

Biological and Health Sciences

Business and Economics

Education

Environmental Science

Miscellaneous

Personal Finance, Investments, and Taxes